T3-BNX-614

HANDBOOK OF COMBINATORICS

Volume 1

HANDBOOK OF COMBINATORICS

Volume 1

edited by

R.L. GRAHAM
AT&T Bell Laboratories, Murray Hill, NJ, USA

M. GRÖTSCHEL
Konrad-Zuse-Zentrum für Informationstechnik, Berlin-Wilmersdorf, Germany

L. LOVÁSZ
Yale University, New Haven, CT, USA

1995

ELSEVIER
AMSTERDAM - LAUSANNE - NEW YORK
OXFORD - SHANNON - TOKYO

1995

THE MIT PRESS
CAMBRIDGE, MASSACHUSETTS

ELSEVIER SCIENCE B.V.
Sara Burgerhartstraat 25
P.O. Box 211, 1000 AE Amsterdam, The Netherlands

Co-publishers for the United States and Canada:

The MIT Press
55 Hayward Street
Cambridge, MA 02142, U.S.A.

Elsevier Science B.V.
ISBN: 0-444-82346-8 (Volume 1)
ISBN: 0-444-88002-X (Set of vols 1 and 2)

The MIT Press
ISBN: 0-262-07170-3 (Volume 1)
ISBN: 0-262-07169-X (Set of vols 1 and 2)

This book is printed on acid-free paper.

Printed in The Netherlands

Preface

Combinatorics belongs to those areas of mathematics having experienced a most impressive growth in recent years. This growth has been fuelled in large part by the increasing importance of computers, the needs of computer science and demands from applications where discrete models play more and more important roles. But also more classical branches of mathematics have come to recognize that combinatorial structures are essential components of many mathematical theories.

Despite the dynamic state of this development, we feel that the time is ripe for summarizing the current status of the field and for surveying those major results that in our opinion will be of long-term importance. We approached leading experts in all areas of combinatorics to write chapters for this Handbook. The response was overwhelmingly enthusiastic and the result is what you see here.

The intention of the Handbook is to provide the working mathematician or computer scientist with a good overview of basic methods and paradigms, as well as important results and current issues and trends across the broad spectrum of combinatorics. However, our hope is that even specialists can benefit from reading this Handbook, by learning a leading expert's coherent and individual view of the topic.

As the reader will notice by looking at the table of contents, we have structured the Handbook into five sections: Structures, Aspects, Methods, Applications, and Horizons. We feel that viewing the whole field from different perspectives and taking different cross-sections will help to understand the underlying framework of the subject and to see the interrelationships more clearly. As a consequence of this approach, a number of the fundamental results occur in more than one chapter. We believe that this is an asset rather than a shortcoming, since it illustrates different viewpoints and interpretations of the results.

We thank the authors not only for writing the chapters but also for many helpful suggestions on the organization of the book and the presentation of the material. Many colleagues have contributed to the Handbook by reading the initial versions of the chapters and by making proposals with respect to the inclusion of topics and results as well as the structuring of the chapters. We are grateful for the significant help we received.

Even though this Handbook is quite voluminous, it was inevitable that some areas of combinatorics had to be left out or were not covered in the depth they deserved. Nevertheless, we believe that the Handbook of Combinatorics presents a comprehensive and accessible view of the present state of the field and that it will prove to be of lasting value.

Ronald Graham
Martin Grötschel
Lászlό Lovász

Contents

Volume I

Symmetric Structures

Combinatorial Structures in Geometry and Number Theory

Volume II

Part II: Aspects 1019

List of Contributors

Alon, N., *Tel Aviv University, Tel Aviv* (Ch. 32).
Babai, L., *Eötvös University, Budapest and University of Chicago, Chicago, IL* (Ch. 27).
Beck, J., *Rutgers University, New Brunswick, NJ* (Ch. 26).
Biggs, N.L., *University of London, London* (Ch. 44).
Bixby, R.E., *Rice University, Houston, TX* (Ch. 11).
Björner, A., *Royal Institute of Technology, Stockholm* (Ch. 34).
Bollobás, B., *University of Cambridge, Cambridge and Louisiana State University, Baton Rouge, LA* (Ch. 23).
Bondy, J.A., *Université Claude Bernard Lyon 1, Villeurbanne and University of Waterloo, Waterloo, Ont.* (Ch. 1).
Brouwer, A.E., *Eindhoven University of Technology, Eindhoven* (Chs. 14, 15).
Cameron, P.J., *Queen Mary and Westfield College, London* (Chs. 12, 13).
Cunningham, W.H., *University of Waterloo, Waterloo, Ont.* (Ch. 11).
Duchet, P., *IMAG, Laboratoire de Structures Discrètes et de Didactique, Grenoble Cedex* (Ch. 7).
Erdős, P., *Hungarian Academy of Sciences, Budapest* (Ch. 17).
Frank, A., *Eőtvős University, Budapest* (Ch. 2).
Frankl, P., *CNRS, Paris* (Ch. 24).
Gessel, I.M., *Brandeis University, Waltham, MA* (Ch. 21).
Godsil, C.D., *University of Waterloo, Waterloo, Ont.* (Chs. 31, 37).
Grötschel, M., *Konrad-Zuse-Zentrum für Informationstechnik, Berlin-Wilmersdorf* (Chs. 28, 37).
Guy, R.K., *University of Calgary, Calgary, Alta.* (Ch. 43).
Haemers, W.H., *Tilburg University, Tilburg* (Ch. 15).
Hajnal, A., *Hungarian Academy of Sciences, Budapest* (Ch. 42).
Karoński, M., *Adam Mickiewicz University, Poznań and Emory University, Atlanta, GA* (Ch. 6).
Klee, V., *University of Washington, Seattle, WA* (Ch. 18).
Kleinschmidt, P., *Universität Passau, Passau* (Ch. 18).
Kolen, A.W.J., *University of Limburg, Maastricht* (Ch. 35).
Lagarias, J.C., *AT&T Bell Laboratories, Murray Hill, NJ* (Ch. 19).
Lenstra, J.K., *Eindhoven University of Technology, Eindhoven and Centre for Mathematics and Computer Science, Amsterdam* (Ch. 35).
Lloyd, E.K., *University of Southampton, Southampton* (Ch. 44).
Lovász, L., *Yale University, New Haven, CT* (Chs. 28, 31, 40, 41).
Nešetřil, J., *Charles University, Prague* (Ch. 25).
Odlyzko, A.M., *AT&T Bell Laboratories, Murray Hill, NJ* (Ch. 22).
Pomerance, C., *University of Georgia, Athens, GA* (Ch. 20).

Pulleyblank, W.R., *IBM, Thomas J. Watson Research Center, Yorktown Heights, NY* (Ch. 3).
Purdy, G., *University of Cincinnati, Cincinnati, OH* (Ch. 17).
Pyber, L., *Hungarian Academy of Sciences, Budapest* (Ch. 41).
Recski, A., *Technical University of Budapest, Budapest* (Ch. 36).
Rouvray, D.H., *The University of Georgia, Athens, GA* (Ch. 38).
Sárközy, A., *Hungarian Academy of Sciences, Budapest* (Ch. 20).
Schrijver, A., *Centrum voor Wiskunde en Informatica, Amsterdam* (Ch. 30).
Seymour, P.D., *Bell Communications Research, Morristown, NJ* (Appendix to Ch. 4, Ch. 10).
Shmoys, D.B., *Cornell University, Ithaca, NY* (Chs. 29, 40).
Sós, V.T., *Hungarian Academy of Sciences, Budapest* (Ch. 26).
Spencer, J., *New York University, New York, NY* (Ch. 33).
Stanley, R.P., *Massachusetts Institute of Technology, Cambridge, MA* (Ch. 21).
Tardos, É., *Cornell University, Ithaca, NY* (Chs. 29, 40).
Thomassen, C., *The Technical University of Denmark, Lyngby* (Ch. 5).
Toft, B., *Odense University, Odense* (Ch. 4).
Trotter, W.T., *Arizona State University, Tempe, AZ* (Ch. 8).
Van Lint, J.H., *Eindhoven University of Technology, Eindhoven* (Ch. 16).
Waterman, M.S., *University of Southern California, Los Angeles, CA* (Ch. 39).
Welsh, D.J.A., *University of Oxford, Oxford* (Chs. 9, 37, 41).
Wilson, R.J., *The Open University, Milton Keynes* (Ch. 44).
Ziegler, G.M., *Konrad-Zuse-Zentrum für Informationstechnik, Berlin-Wilmersdorf* (Ch. 41).

Part I
Structures

CHAPTER 1

Basic Graph Theory: Paths and Circuits

J.A. BONDY

Institut de Mathématiques et Informatique, Université Claude Bernard Lyon 1, 69622 VILLEURBANNE Cedex, France

and

Faculty of Mathematics, University of Waterloo, Waterloo, Ont. N2L 3G1, Canada

Contents

HANDBOOK OF COMBINATORICS
Edited by R. Graham, M. Grötschel and L. Lovász

Our purpose in this chapter is twofold: to introduce the basic concepts and techniques of graph theory, and to develop that part of the theory concerned with paths and circuits. We have attempted to weave these two strands together, proceeding from fundamental notions and elementary observations to more sophisticated concepts and methods, much as one does when engaged in research. One of the attractive features of the subject is the wealth of simply stated but challenging open problems, and we have included a large number of these. It must be emphasized that the literature on paths and circuits is extensive. Thus what is presented here is necessarily a selection. Nevertheless, our hope is that, after studying this chapter, the reader will be well equipped to embark on the chapters dealing with particular aspects of graph theory. Circuits in graphs are treated in depth in the books of Walther and Voss (1974) and Voss (1991). There are also many survey articles on particular aspects of the topic. We recommend, in particular, the articles of Carsten Thomassen. Besides their substantial contributions to the theory of graphs, each includes an informed and stimulating discussion of the question under consideration and its links to related matters.

1. Basic concepts

1.1. Graphs

A *graph* G consists of a nonempty set $V(G)$ of elements, called *vertices* or *nodes*, and a set $E(G)$ of elements called *edges*, together with a relation of *incidence* that associates with each edge two vertices, called its *ends*. Two graphs G and H are *isomorphic* if there are bijections $\theta : V(G) \to V(H)$ and $\phi : E(G) \to E(H)$ such that vertex v and edge e are incident in G if and only if vertex $\theta(v)$ and edge $\phi(e)$ are incident in H. The pair of mappings (θ, ϕ) is an *isomorphism* from G to H.

An edge with identical ends is a *loop* and one with distinct ends is a *link*. Two or more edges with the same pair of ends are *multiple* edges. The two ends of an edge are said to be *joined* by the edge and to be *adjacent* to one another; adjacent vertices are also referred to as *neighbours*. An edge with ends x and y is denoted by the unordered pair xy. The number of vertices of a graph G is called its *order*, the number of edges its *size*; these parameters are denoted by $v(G)$ and $e(G)$, respectively.

A *multigraph* is one with no loops, a *simple* graph one with neither loops nor multiple edges. A graph is *finite* if both its vertex set and its edge set are finite, and *infinite* otherwise. This chapter is devoted entirely to finite graphs. Quite different techniques are needed for dealing with infinite graphs. These are discussed in the survey by Thomassen (1983e) and also in chapter 42 by Hajnal.

Graphs are so named because they can be represented pictorially, a vertex being indicated by a point and an edge by a line joining the points representing its ends; fig. 1 depicts the five *Platonic graphs*, the graphs of the Platonic solids. The graph of the tetrahedron in fig. 1 is an example of a *complete* graph: any two of its vertices

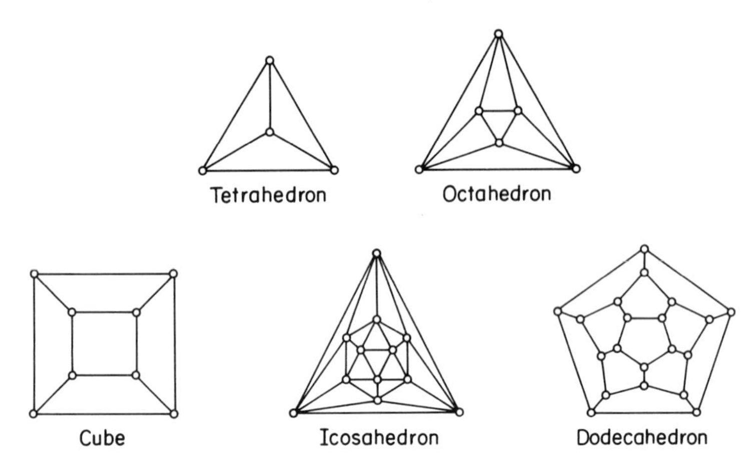

Figure 1. The Platonic graphs.

are joined by an edge. An *empty* graph, by contrast, is one with no edges. The simple complete graph on n vertices is denoted by K_n.

A graph is *bipartite* if its vertex set can be partitioned into subsets X and Y so that each edge has one end in X and one end in Y; such a partition (X, Y) is called a *bipartition* of the graph. A bipartite graph G with bipartition (X, Y) is denoted by $G(X, Y)$. Such a graph is *balanced* if $|X| = |Y|$ and *complete* if each vertex of X is joined to each vertex of Y; the graph of the cube in fig. 1 is a balanced bipartite graph. The simple complete bipartite graph $G(X, Y)$ in which $|X| = m$ and $|Y| = n$ is denoted by $K_{m,n}$; if $m = 1$, such a graph is called a *star*. The graphs K_5 and $K_{3,3}$ are known as the *Kuratowski graphs* because they feature in a fundamental theorem of K. Kuratowski (Theorem 5.4); $K_{3,3}$ is also sometimes referred to as the *Thomsen graph*.

It is sometimes convenient to represent the relations of adjacency and incidence in a graph by matrices. The *adjacency matrix* $A(G)$ is the matrix (a_{uv}) whose rows and columns are indexed by the vertices of G, a_{uv} being the number of edges of G joining vertices u and v. The *incidence matrix* $B(G)$ is the matrix (b_{ve}) whose rows are indexed by the vertices of G and columns by the edges of G, b_{ve} being the number of times vertex v and edge e are incident; so $b_{ve} := 2$ if e is a loop at v, $b_{ve} := 1$ if e is a link at v, and $b_{ve} := 0$ otherwise.

As a general rule, the symbol specifying the graph (typically G) is dropped from notation if the context makes clear which graph is being discussed; vertex and edge sets are then denoted simply by V and E, and adjacency and incidence matrices by $\boldsymbol{A}$ and $\boldsymbol{B}$. By the same token, the notation $G(V, E)$ specifies a graph G with vertex set V and edge set E. An exception to this rule is the frequent use of n for the order, and occasionally m for the size, the letters v and e always denoting vertices and edges, respectively.

The set of neighbours of a vertex v in a graph G is denoted by $N_G(v)$. In a simple graph, the number of neighbours of a vertex is called its *degree*. More generally, the *degree* of a vertex v in a graph G is the number of edges of G incident with v, a loop counting as two edges. A graph is *d-regular* if each vertex has degree d, and *regular* if it is d-regular for some d. A graph is *even* if each vertex is of even degree. The degree of a vertex v in a graph G is denoted by $d_G(v)$, the minimum, maximum, and average degrees of the vertices of a graph G by $\delta(G)$, $\Delta(G)$, and $d(G)$, respectively. An *isolated vertex* is a vertex of degree zero, a *leaf* one of degree one.

The number of edges in a graph and the degrees of its vertices are linked by a simple identity due to Euler (1736).

Proposition 1.1. *For any graph G,*

$$\sum_{v \in V} d(v) = 2e(G).$$

Proof (by *counting in two ways*). We consider the incidence matrix $B := (b_{ve})$ of G and compute the sum of its entries in two ways:

$$\sum_{v \in V} d(v) = \sum_{v \in V} \sum_{e \in E} b_{ve} = \sum_{e \in E} \sum_{v \in V} b_{ve} = \sum_{e \in E} 2 = 2e(G). \qquad \square$$

We have drawn attention to the proof technique used here, *counting in two ways*, because it is one that is frequently employed in combinatorial arguments. Further techniques will be highlighted as they are encountered; some of these are of general applicability, while others are designed specifically to study paths and circuits in graphs.

Corollary 1.2. *In any graph, the number of vertices of odd degree is even.*

1.2. Subgraphs and minors

Let G and H be graphs. If $V(H) \subseteq V(G)$ and $E(H) \subseteq E(G)$, and if every edge of H has the same pair of ends in H as it has in G, the graph H is called a *subgraph* of G and the graph G a *supergraph* of H; these relationships are denoted by $H \subseteq G$ and $G \supseteq H$, respectively. We also say that H is *contained* in G and that G *contains* H. A subgraph or supergraph H of G *spans* G if $V(H) = V(G)$.

The *union* of two subgraphs G_1 and G_2 of G is the subgraph $G_1 \cup G_2$ with vertex set $V(G_1) \cup V(G_2)$ and edge set $E(G_1) \cup E(G_2)$. Likewise the *intersection* of G_1 and G_2 is the subgraph $G_1 \cap G_2$ with vertex set $V(G_1) \cap V(G_2)$ and edge set $E(G_1) \cap E(G_2)$.

If S is a set of edges of a graph $G(V,E)$, the graph whose vertex set is V and whose edge set is $E \setminus S$ is a subgraph of G. Likewise, if S is a set of vertices of G, the subgraph of G whose vertex set is $V \setminus S$ and whose edge set consists of those edges of G with both ends in $V \setminus S$ is a subgraph of G. In both instances, this subgraph is denoted by $G \setminus S$, and is said to be obtained by *deleting* S from G. If H is a subgraph of G, we denote the subgraph $G \setminus V(H)$ simply by $G \setminus H$.

If S is a set of unordered pairs of vertices of $G(V,E)$, the graph whose vertex set is V and whose edge set is $E \cup S$ is a spanning supergraph of G. This graph is denoted by $G+S$, and is said to be obtained by *adding* S to G; addition and deletion of edges are inverse operations. One may also add a set S of vertices to a graph. In this case, the resulting graph depends on which edges incident to the vertices of S are also added. If all pairs uv with $u \in S$ and $v \in V$ are added, and no others, the resulting graph is denoted by $G+S$.

If S is a set of edges of G, the graph whose vertex set consists of all ends of edges in S and whose edge set is S is a subgraph of G. Likewise, if S is a set of vertices of G, the graph whose vertex set is S and whose edge set consists of those edges of G with both ends in S is a subgraph of G. In both instances, this subgraph is denoted by $G[S]$ and is called the subgraph of G *induced* by S.

Any subgraph of a graph may be obtained by deleting appropriate vertices and edges. Similarly, any spanning subgraph may be obtained by deleting edges, any vertex-induced subgraph by deleting vertices, and any edge-induced subgraph by deleting edges and isolated vertices.

The subgraphs of a graph can be represented algebraically by their incidence matrices, but it turns out to be more fruitful to identify a subgraph with the characteristic function of its edge set. Let E be a set. The *characteristic function* of a subset S of E is the function $\chi_S : E \to \mathrm{GF}(2)$ defined by

$$\chi_S(e) := \begin{cases} 1 & \text{if } e \in S, \\ 0 & \text{if } e \notin S. \end{cases}$$

The functions χ_S form a vector space over GF(2). When E is the edge set of a graph G, this vector space is called the *edge space* of G, and may be regarded as the set of edge-induced subgraphs of G, the sum of two such subgraphs $G[S]$ and $G[T]$ being $G[S \Delta T]$, where Δ denotes symmetric difference. Families of subgraphs that are closed under this operation are thus of particular interest. The even subgraphs of G constitute such a family, and form a vector space $\mathscr{C}(G)$ over GF(2) called the *cycle space* of G. (Originally, even graphs were called *cycles*. To avoid confusion, we have chosen not to adopt this name, because it is also commonly used for what we shall call a circuit.)

As noted earlier, deletion and addition of edges are inverse operations. Another operation closely related to deletion, but in a more subtle way, is that of contraction. If S is a set of edges of a graph G, the graph derived from G by deleting S and identifying each pair of ends of the edges of S is said to be obtained by *contracting* S, and is denoted by G/S. If H is a subgraph of G, we denote the contraction $G/E(H)$ simply by G/H. A *minor* of a graph G is any graph obtainable from G by means of a sequence of vertex deletions, edge deletions and edge contractions. Minors play a very important role in the structural theory of graphs. In particular, the absence of a particular graph as a minor has a profound effect on the properties of a graph. The survey by Robertson and Seymour (1985) gives a good overview of this topic. Minors of graphs are discussed also in chapter 5 by Thomassen.

1.3. Walks

A *walk* in a graph G is a sequence $W := v_0e_1v_1e_2\cdots e_\ell v_\ell$, where $v_0, v_1, \dots, v_\ell$ are vertices of G, $e_1, e_2, \dots, e_\ell$ are edges of G, and v_{i-1} and v_i are the ends of e_i, $1 \leqslant i \leqslant \ell$. In a simple graph, a walk is completely specified by its sequence of vertices, $v_0v_1\cdots v_\ell$. The vertex v_0 is the *tail* of W, the vertices $v_1, \dots, v_{\ell-1}$ its *internal vertices*, and the vertex v_ℓ its *head*; the tail and head are often referred to as the *ends* of W. The walk W is described as a v_0*-walk* or a v_0v_ℓ*-walk* if, as is often the case, one wishes to specify explicitly one or both of its ends; one also says that W *connects* v_0 and v_ℓ. The walk W is *open* if $v_0 \neq v_\ell$ and *closed* if $v_0 = v_\ell$. The *length* of W is its number of edges, namely ℓ; the *parity* of W is the parity of its length.

We say that vertex v_i *precedes* vertex v_j on W if $i < j$, and indicate this relationship by the notation $v_i \prec v_j$. If v_i precedes v_j, the walk $v_ie_{i+1}v_{i+1}\cdots v_{j-1}e_jv_j$ is referred to as the v_iv_j*-segment* of W, and may be denoted by $W[v_i, v_j]$, $W(v_{i-1}, v_j]$, $W[v_i, v_{j+1})$ or $W(v_{i-1}, v_{j+1})$, as convenient; or by $\overleftarrow{W}[v_j, v_i]$, to indicate traversal of the segment in the opposite sense.

The walk W is a *trail* if its edges $e_1, e_2, \dots, e_\ell$ are distinct, and a *path* if its vertices $v_0, v_1, \dots, v_\ell$ are distinct. Note that a v_0v_ℓ-walk of minimum length is necessarily a v_0v_ℓ-path. A closed trail of positive length whose vertices (apart from its ends) are distinct is called a *circuit*. Dirac (1952) found a link between the length of a longest path or circuit in a graph and the minimum degree.

Theorem 1.3. *Let G be a simple graph of minimum degree δ. Then*
(i) *G contains a path P of length at least δ;*
(ii) *if $\delta \geqslant 2$, G contains a circuit C of length at least $\delta + 1$.*

Proof (by *choosing a maximal path*). Let $P := v_0v_1\cdots v_\ell$ be a maximal path of G (one that cannot be extended from either end). Then all the neighbours of v_0 lie on P. Setting $k := \max\{i\colon v_0v_i \in E\}$, we have $k \geqslant d(v_0) \geqslant \delta$. Thus P is of length at least δ and, if $\delta \geqslant 2$, $C := v_0v_1\cdots v_kv_0$ is a circuit of length at least $\delta + 1$. □

A *circuit decomposition* of a graph is a partition of its edge set into circuits. For a graph to have a circuit decomposition, each vertex must clearly be of even degree. Veblen (1912) proved the converse statement.

Theorem 1.4. *A graph is even if and only if it admits a circuit decomposition.*

Proof. A graph with a circuit decomposition is even because each vertex is incident with either zero or two edges of any circuit. On the other hand, a nonempty even graph has at least one circuit, by Theorem 1.3(ii), and the subgraph obtained on deleting the edges of this circuit is even. Sufficiency now follows by induction on $e(G)$. □

1.4. Connection

A graph is *connected* if any two of its vertices are connected by a path. A *component* of a graph G is a maximal connected subgraph of G (that is, a connected subgraph contained in no other connected subgraph). The number of components of a graph G is denoted by $c(G)$. The vertices and edges of an xy-path form a connected subgraph in which the ends x and y have degree one and the internal vertices have degree two (unless the path is of length zero, in which case $x = y$ has degree zero); similarly, the vertices and edges of a circuit form a connected 2-regular subgraph. The terms *path* and *circuit* will frequently be used to refer to such subgraphs. Circuits are very often called *cycles*; they are also sometimes referred to as *polygons*, short ones being given suggestive geometrical names: *digon, triangle, quadrilateral*, and so on.

The *distance* $d(x,y)$ between two vertices x and y in a graph is the length of a shortest xy-path in G, if there is one; otherwise their distance is defined to be infinite. The *diameter* of a graph is the maximum distance between its vertices. The *girth* of a graph is the length of a shortest circuit, if there is one; otherwise the girth is defined to be zero.

A connected graph G is *separable* if it contains subgraphs G_1 and G_2, and a vertex v, such that (a) $G_1 \cup G_2 = G$, (b) $G_1 \cap G_2 = \{v\}$, and (c) $G_i \neq \{v\}$, $i = 1,2$. Otherwise, G is *nonseparable*. A *block* of a connected graph is a maximal nonseparable subgraph; the blocks of a disconnected graph are the blocks of its components.

A *cut vertex* of a connected graph G is any vertex v satisfying conditions (b) and (c) above for some pair G_1, G_2 of subgraphs of G satisfying (a). The cut vertices of a disconnected graph are defined to be the cut vertices of its components. Thus v is a cut vertex of G if and only if $G \setminus v$ has more components than G, unless v is incident to a loop and at least one other edge. Analogously, we define a *cut edge* of a graph G to be an edge e such that $G \setminus e$ has more components than G; equivalently, a cut edge is one that belongs to no circuit.

The circuits of a graph lie entirely within its blocks. Thus, where circuits are concerned, nonseparability is a natural condition to impose. Hartman (1983) strengthened Theorem 1.3 (ii) for nonseparable graphs by showing that the circuits of length at least $\delta + 1$ in a simple nonseparable graph G of minimum degree δ generate the cycle space of G, unless δ is odd and $G = K_{\delta+1}$. If such a graph has an odd circuit, it therefore has an odd circuit of length at least $\delta + 1$, unless δ is odd and the graph is $K_{\delta+1}$.

A graph is *acyclic*, or a *forest*, if it contains no circuit or, equivalently, if each of its edges is a cut edge. The following basic property of forests was first noted by Listing (1862).

Proposition 1.5. *Let G be a forest. Then*

$$e(G) - v(G) + c(G) = 0.$$

Proof (by *induction*). The assertion holds if $e(G) = 0$, since in this case each component is a single vertex. Assume that $e(G) > 0$ and that the assertion holds for all forests on fewer than $e(G)$ edges. Let G' be a subgraph of G obtained by deleting any one edge. Then G' is a forest with $e(G) - 1$ edges, $v(G)$ vertices and $c(G) + 1$

components. By the induction hypothesis, applied to G',

$$\begin{aligned} e(G) - v(G) + c(G) &= e(G) - 1 - v(G) + c(G) + 1 \\ &= e(G') - v(G') + c(G') = 0. \end{aligned}$$

Thus the assertion holds for all forests. □

A *tree* is a connected acyclic graph. It follows from Propositions 1.1 and 1.5 that every tree on two or more vertices has at least two vertices of degree one; such vertices are called *leaves*. Any connected graph, when viewed as an assemblage of its blocks, has the structure of a tree. More precisely, the *block cut-vertex tree* of a connected graph G is the tree H whose vertices are the blocks and the cut vertices of G, a block and cut vertex being adjacent in H if the block includes the cut vertex in G: H is connected because G is, and acyclic because a block is a maximal nonseparable subgraph. The blocks of G corresponding to the leaves of H are the *endblocks* of G.

Every connected graph contains a *spanning tree*, a spanning subgraph which is a tree. In fact, the spanning trees are precisely the spanning subgraphs that are connected and are minimal with respect to the property. They are also the spanning subgraphs that are acyclic and are maximal with respect to the property, and thus can be "grown" by starting with one vertex and adding edges and vertices one by one. This latter viewpoint is of fundamental importance in both theoretical and applied contexts, as we shall see shortly in the section on search trees.

Each spanning tree T of a connected graph G determines a basis $\mathscr{C}_T$ of its cycle space $\mathscr{C}$ in a natural way. For each edge e of G not in T, the subgraph $T+e$ contains a unique circuit C_e. The set $\mathscr{C}_T$ of all such circuits is a basis for $\mathscr{C}$: it is independent because C_e is the only circuit in $\mathscr{C}_T$ that includes e, and it generates $\mathscr{C}$ because any even subgraph H of G can be expressed in the form

$$H = \sum_{e \in E(H) \setminus E(T)} C_e.$$

To see this, it suffices to observe that $H + \sum_{e \in E(H) \setminus E(T)} C_e$ is an even subgraph contained in T, and hence is empty by Theorem 1.4.

Because the cycle space of a disconnected graph is the direct sum of the cycle spaces of its components, Proposition 1.5 shows that the dimension of the cycle space of a graph G is given by the formula

$$\dim \mathscr{C}(G) = e(G) - v(G) + c(G).$$

1.5. Extremal graphs

Proposition 1.5 has the following immediate consequence.

Corollary 1.6. *Let G be a graph on n vertices and m edges, where $m \geqslant n$. Then G contains a circuit.*

Corollary 1.6 is a very simple instance of a result in *extremal graph theory.* It asserts that every graph on n vertices and sufficiently many edges (at least n) contains a circuit. Furthermore, the bound on the number of edges is best possible because there exist graphs on n vertices and $n-1$ edges that do not contain circuits (namely trees); these graphs (trees) are referred to as *extremal graphs* for the property in question (containing a circuit).

While a graph on n vertices and at least n edges is guaranteed to contain a circuit, the length of the circuit may take on any value between 1 and n. The following theorem, due to Mantel (1907), specifies how many edges are needed to guarantee the existence of a triangle in a simple graph.

Theorem 1.7. *Let G be a simple graph on n vertices and m edges, where $m > n^2/4$. Then G contains a triangle.*

Proof (by *contradiction*). Suppose that G contains no triangle. Then adjacent vertices have disjoint sets of neighbours, so

$$d(u) + d(v) \leqslant n, \quad \text{for each edge } uv \in E.$$

Summing over all edges of G, we have

$$\sum_{v \in V} d(v)^2 \leqslant mn.$$

On the other hand, by Chebyshev's inequality and Proposition 1.1,

$$\sum_{v \in V} d(v)^2 \geqslant \frac{(\sum_{v \in V} d(v))^2}{n} = \frac{4m^2}{n}.$$

These two inequalities imply that $m \leqslant n^2/4$, contradicting the hypothesis. □

Equality holds in the above inequalities only if G is regular of degree $n/2$. It follows readily that the extremal graphs in this case are the complete bipartite graphs $K_{n/2,n/2}$. Theorem 1.7 is a special case of Turán's Theorem, the cornerstone of extremal graph theory. This topic is the subject of chapter 23 by Bollobás.

1.6. Search trees

Theorem 1.3, Corollary 1.6 and Theorem 1.7 assert only that certain types of paths and circuits *may* be found in graphs with appropriate properties. They do not address the question, of paramount importance in practical applications, of *how* to find such subgraphs. Of course, an exhaustive search of a finite graph can always, in theory, be completed in finite time. In practice, however, this is not good enough – time is of the essence; to be effective, an algorithm must be efficient. The running time of an algorithm is measured by the total number of basic operations (arithmetic operations, comparisons, and so on) required to implement it, and depends on the size and nature of the input. The *complexity* of a graph-theoretical algorithm is its running time, in the worst case, on graphs of a given size (meaning, usually, their number of vertices and/or edges). A *polynomial-time* algorithm is one whose complexity is bounded above by a polynomial in the size of the input graph, an *exponential-time* algorithm one whose complexity is bounded below by an exponential function in the size of the input graph. Polynomial-time algorithms are regarded as efficient because their running time grows relatively slowly with the size of the input; this means that fairly large graphs (perhaps with several hundred vertices) can be handled effectively. Exponential-time algorithms, on the other hand, may take inordinately long to process graphs of very modest size (say, twenty vertices).

Spanning trees provide an efficient way to search graphs, and are the key to many polynomial-time algorithms. A *search tree* in a connected graph is a spanning tree whose vertex set $\{v_0, v_1, \ldots, v_{n-1}\}$ and edge set $\{e_1, e_2, \ldots, e_{n-1}\}$ are ordered in such a way that, for each j, $1 \leqslant j \leqslant n-1$, the subgraph with vertex set $\{v_0, v_1, \ldots, v_j\}$ and edge set $\{e_1, e_2, \ldots, e_j\}$ is a tree; the vertex v_0 is the *root* of the search tree. In other words, a search tree is the *growth process* of a spanning tree, not just the tree itself. Starting at the root v_0, the tree is grown by adding a vertex v_j and edge $e_j = v_i v_j$ at each stage of the process; we call v_j a *child* of v_i, and v_i the *parent* of v_j, in the tree. In general, there is some freedom as to the choice of v_j and e_j, but if they are chosen appropriately, important structural information about the graph is encoded in the resulting search tree. The most basic type of tree search, breadth-first search, was developed by Dijkstra (1959).

A *breadth-first search tree* is a search tree with vertex set $\{v_0, v_1, \ldots, v_{n-1}\}$ and edge set $\{e_1, e_2, \ldots, e_{n-1}\}$ where, for each j, $1 \leqslant j \leqslant n-1$, v_j and e_j are chosen in such a way that the parent of v_j is the vertex v_i with smallest possible index i. Such a tree can be grown in linear time. The key property of breadth-first search is that the distance in a breadth-first search tree between its root and any other vertex is the same as their distance in the graph itself. Thus, by growing a breadth-first search tree from each vertex, shortest paths between all pairs of vertices in a graph can be found. In particular, the diameter can be determined.

Breadth-first search also enables one to decide easily whether or not a graph is bipartite. Let T be a breadth-first search tree with root x in a graph G, which we assume to be connected. (If it is not, the procedure may be applied to each component in turn.) Let X be the set of vertices whose distance from x is even, and Y the set of vertices whose distance from x is odd. If no two vertices of X and no two vertices of Y are adjacent, (X, Y) is a bipartition of G and so G is bipartite. If, on the other hand, two of these vertices, u and v, are adjacent, consider the ux-path P and vx-path Q in T. Let w denote the first vertex common to P and Q. Then $P[u, w]$ and $Q[v, w]$ are paths of the same parity and so, together with the edge uv, form an odd circuit of G. We conclude that G is not bipartite in this case, since a circuit in a bipartite graph alternates between the sets of its bipartition and thus is necessarily even.

The above algorithm outputs a bipartition if the input graph is bipartite and an odd circuit if it is not. Thus it supplies, as a byproduct, the following characterization of bipartite graphs, due to König (1916).

Theorem 1.8. *A graph is bipartite if and only if it contains no odd circuit.*

If a graph is not bipartite, a shortest odd circuit can be found by growing a breadth-first search tree with root x from each vertex x. It is less easy to find a shortest even circuit, if there is one; an algorithm for doing so is described in section 4 (at the end of the discussion of stability number).

The above examples illustrate the power of breadth-first search. Another highly effective tree search procedure, depth-first search, was developed by Tarjan (1972). A *depth-first search tree* is a search tree with vertex set $\{v_0, v_1, \ldots, v_{n-1}\}$ and edge set $\{e_1, e_2, \ldots, e_{n-1}\}$ where, for each j, $1 \leqslant j \leqslant n-1$, v_j and e_j are chosen in such a way that the parent of v_j is the vertex v_i with largest possible index i. Depth-first search also takes linear time. It can be used to find the blocks and cut vertices of a graph, and has many other important applications. These are described in chapter 28 by Grötschel, and in the book by Aho et al. (1974).

In applications, the edges of a graph often have associated weights, or costs. An *edge-weighted graph* is a graph G together with a *weight function* w that assigns to each edge of G a *weight* $w(e)$. Usually, the range of w is the set of nonnegative real numbers; in some instances, however, a different choice is appropriate: the set of all real numbers, the set of integers, or the set of integers modulo m, for example. The *weight* of an edge-weighted graph is the sum of the weights on its edges. An unweighted graph can be regarded as an edge-weighted graph in which each edge has weight one; its weight is then simply its size. In this way, results about graphs can sometimes be generalized to weighted graphs. Dijkstra (1959) showed, for instance, not only how to find shortest paths in graphs but, more generally, how to find minimum-weight paths in edge-weighted graphs with nonnegative weights.

Most of the theorems to be mentioned in this chapter have proofs that can be adapted fairly easily to yield polynomial-time algorithms. However, there are many basic problems for which polynomial-time algorithms have yet to be found, and indeed might well not exist. In fact, determining which problems are solvable in polynomial time is one of the main preoccupations of theoretical computer scientists. An important class of problems in this regard is the class $\mathcal{NP}$ (standing for *nondeterministic polynomial-time*). We give here an informal definition; a precise treatment can be found in chapter 29, or in the book by Garey and Johnson (1979).

A *decision problem* is a question whose answer is either "yes" or "no"; any object to which the question is addressed is an *instance* of the problem. A decision problem belongs to the class $\mathcal{P}$ if there is a polynomial-time algorithm that solves any instance of the problem in polynomial time. A decision problem belongs to the class $\mathcal{NP}$ if, given any instance of the problem whose answer is 'yes', there is a certificate validating this fact which can be checked in polynomial time, and one, moreover, whose length is bounded above by a polynomial in the size of the input; such a certificate is said to be *succinct*. It follows from these definitions that $\mathcal{P} \subseteq \mathcal{NP}$, since a polynomial-time algorithm is, in itself, a succinct certificate.

The problem of determining whether a graph is bipartite belongs to $\mathcal{P}$ and hence also to $\mathcal{NP}$; likewise, the problem of determining whether a graph is nonbipartite belongs to both $\mathcal{P}$ and $\mathcal{NP}$. On the other hand, even though no polynomial-time algorithm is known for determining whether a graph contains a spanning circuit, this problem belongs to $\mathcal{NP}$, an appropriate certificate being the spanning circuit itself. Verifying that a graph contains no spanning circuit, by contrast, appears to be much harder; indeed, its status as regards membership in $\mathcal{NP}$ is unresolved.

As noted above, $\mathcal{P} \subseteq \mathcal{NP}$. Whether $\mathcal{P} = \mathcal{NP}$ or not is the most fundamental open question in theoretical computer science. The so-called *$\mathcal{NP}$-complete* problems, which form a subclass of $\mathcal{NP}$, are of much interest in this regard. These problems, which include the problem of determining whether a graph contains a spanning circuit, are algorithmically the most difficult in $\mathcal{NP}$, in the sense that if any one of them can be solved in polynomial time, then so can every problem in $\mathcal{NP}$ (that is, $\mathcal{P} = \mathcal{NP}$). Several $\mathcal{NP}$-complete problems will be encountered in this chapter. The $\mathcal{NP}$-complete problems belong to a larger class of seemingly intractable problems, the class of *$\mathcal{NP}$-hard* problems. The concept of $\mathcal{NP}$-completeness was set on a sound theoretical footing by Cook (1971) and further developed by Karp (1972).

1.7. Directed graphs

A *directed graph*, or *digraph*, D is a graph G in which each edge is assigned a direction, one end being designated its *tail* and the other its *head*. We call D an *orientation* of G, and write $D := \overrightarrow{G}$. An edge with tail x and head y is denoted by the ordered pair (x, y); we say that x *dominates* y and write $x \to y$. In diagrams, the direction of an edge is indicated by an arrow pointing towards the head of the edge. Two or more edges with the same tail and head are *multiple* edges. A *strict* digraph is one with neither loops nor multiple edges.

The *associated digraph* $D(G)$ of a graph G is the digraph obtained from G by replacing each edge by two oppositely directed edges with the same ends. An example is the *strict complete digraph*, the associated digraph $D(K_n)$ of the simple complete graph K_n. The *underlying graph* $G(D)$ of a digraph D is the graph obtained from D by ignoring the orientations of its edges. An *oriented graph* is a digraph whose underlying graph is simple. A *tournament* is a digraph whose underlying graph is simple and complete.

The *adjacency matrix* $A(D)$ of a digraph D is the matrix (a_{uv}) whose rows and columns are indexed by the vertices of D, a_{uv} being the number of edges of D with tail u and head v.

Let v be a vertex of a digraph D. The *outdegree* $d^+_D(v)$ of v in D is the number of edges of D whose tail is v, the *indegree* $d^-_D(v)$ the number of edges of D whose head is v. A *sink* is a vertex of outdegree zero, a *source* a vertex of indegree zero. The analogue for digraphs of Proposition 1.1 asserts that, for any digraph D,

$$\sum_{v \in V} d^-(v) = e(D) = \sum_{v \in V} d^+(v).$$

A *directed walk* in a digraph D is a sequence $W := (v_0, e_1, v_1, \ldots, e_\ell, v_\ell)$, where $v_0, v_1, \ldots, v_\ell$ are vertices of D, $e_1, \ldots, e_\ell$ are edges of D, and v_{i-1} and v_i are the tail and head, respectively, of e_i, $1 \leqslant i \leqslant \ell$. The vertices v_0 and v_ℓ are the *tail* and *head* of W; when one wishes to refer to them explicitly, W is described as a (v_0, v_ℓ)-*walk*. The notions of *directed trail*, *directed path*, and *directed circuit* are defined in an analogous manner, as are *directed girth* and *directed diameter*. A digraph is *strong* (or *strongly connected*) if any two vertices x and y are connected by an (x,y)-path. A *strong component* of a digraph D is a maximal strong subdigraph of D.

Graphs can be regarded as a subclass of digraphs, a graph being identified with its associated digraph. This leads to a basic question, one that we shall examine with respect to paths and circuits: *which properties of graphs extend to digraphs?* Another fundamental question has to do with the relationship between a graph and its orientations: *how are the properties of a graph reflected in the properties of its orientations?* This latter question was studied by Chvátal and Thomassen (1978) with regard to girth and diameter. They established the existence of an integer-valued function f and an orientation function $\longrightarrow$ such that, for any graph G and any edge e of G that belongs to a circuit of length at most k, the image $\overrightarrow{e}$ of e belongs to a directed circuit of length at most $f(k)$ in $\overrightarrow{G}$. This implies, in particular, that a connected graph with no cut edge has a strongly connected orientation, a result due to Robbins (1939). Chvátal and Thomassen (1978) obtained a similar but sharper result for the diameter.

Theorem 1.9. *Let G be a graph of diameter d with no cut edge. Then G admits an orientation of directed diameter at most $2d^2+2d$.*

A digraph is *acyclic* if it contains no directed circuit. An acyclic tournament is called a *transitive tournament*, the one of order n being denoted TT_n. The digraph D^* obtained from D by contracting each strong component to a single vertex is an acyclic digraph, the *condensation* of D; if D is a tournament, D^* is a transitive tournament. A strong component of D is called *maximal* if it has indegree zero in D^* and *minimal* if it has outdegree zero in D^*.

An *arborescence* is an orientation of a tree in which one vertex, called the *root*, has indegree zero and every other vertex has indegree one. The analogue for digraphs of a breadth-first search tree is a *breadth-first search arborescence*. Shortest directed paths between pairs of vertices in a digraph can be found efficiently by growing such arborescences. In a strong digraph, a directed odd circuit can also be found in this way, if there is one. Consider a breadth-first search arborescence with root x. Let X be the set of vertices whose distance from x is even, and Y the set of vertices whose distance from x is odd. If some edge e joins two vertices in the same set, x is connected to the head y of e by directed paths of both parities. The concatenation of one of these paths with a (y,x)-path is a directed closed walk of odd length, which necessarily contains a directed odd circuit. On the other hand, if each edge has one end in X and the other end in Y, the digraph is bipartite. Thus a digraph has a directed odd circuit if and only if some strong component is not bipartite.

This characterization leads directly to a theorem of Richardson (1953) on the existence of kernels in digraphs. A *kernel* in a digraph D is a set S of pairwise nonadjacent vertices that together dominate every vertex of $V \setminus S$. Many digraphs fail to have kernels, the simplest being the directed odd circuits. What Richardson proved is that a digraph without a kernel necessarily contains a directed odd circuit.

Theorem 1.10. *Let D be a digraph with no directed odd circuit. Then D has a kernel.*

Proof. By induction on the order of D. If D is strong, then D is bipartite and each class of the bipartition is a kernel of D. If D is not strong, let D_1 be a maximal strong component of D, with vertex set V_1. By the induction hypothesis, D_1 has a kernel, S_1. Let V_2 be the set of vertices of D dominated by vertices of S_1, and define $D_2 := D \setminus (V_1 \cup V_2)$. Again by induction, D_2 has a kernel, S_2. The set $S_1 \cup S_2$ is a kernel of D. □

Kernels originated in the analysis of games on graphs, a kernel representing a set of winning positions. But they have other applications. We refer the reader to Berge (1985, chapter 14) for further information.

A shortest directed odd circuit is a shortest directed closed odd walk, and such a walk, if one exists, can be found in polynomial time by taking odd powers of the adjacency matrix until a nonzero diagonal entry is encountered. Finding a directed even circuit appears to be a much harder proposition. Friedland (1989) gave a simple necessary and sufficient condition in terms of the adjacency matrix for a digraph to contain such a circuit.

Theorem 1.11. *Let D be a digraph, with adjacency matrix A. Then D has a directed even circuit if and only if*

$$\operatorname{per}(I + A) \neq \det(I + A),$$

where per *denotes the permanent and* det *the determinant.*

However, no polynomial-time algorithm has been developed to determine when this condition is satisfied. On the other hand, the problem of finding a directed even circuit in a digraph (posed by D.H. Younger – see Bermond and Thomassen 1981) has not been proved $\mathcal{NP}$-complete.

Problem 1.12. Is there a polynomial-time algorithm for finding a directed even circuit in a digraph?

Thomassen (1993) has described such an algorithm for planar digraphs. Besides being a fundamental question, Problem 1.12 has an intriguing knife edge quality: the slightly harder question of deciding whether there is a directed odd (or even) circuit through a given edge in a digraph is already $\mathcal{NP}$-complete (Thomassen 1985a).

Problem 1.12 turns out to be equivalent to a question about a special class of digraphs considered by Seymour and Thomassen (1987). To *subdivide* an edge $e := xy$ in a graph is to replace e by a path $P := xvy$, where v is a new vertex. A *subdivision* of a graph is any graph derivable from it by recursively subdividing

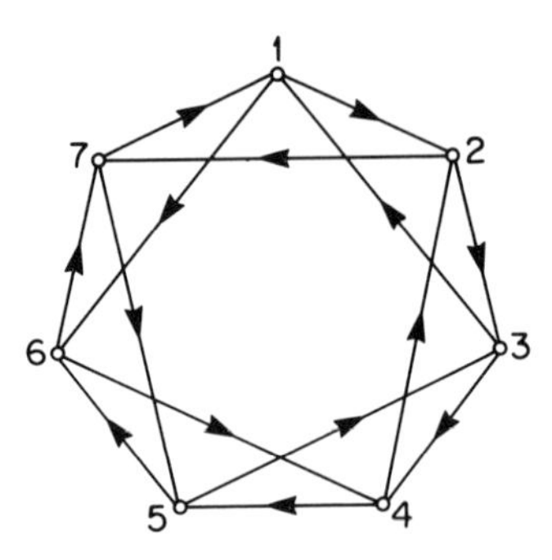

Figure 2. The Koh digraph.

edges. Analogous definitions apply to digraphs, the path P being a directed path in this case. Seymour and Thomassen (1987) characterized the digraphs with the property that every subdivision contains a directed even circuit. This characterization does not, however, lead to a polynomial-time recognition algorithm. Indeed, Seymour and Thomassen showed that the existence of such an algorithm is equivalent to the existence of a polynomial-time algorithm for finding a directed even circuit in a digraph.

Thomassen (1992) found a sufficient condition in terms of the minimum indegree and outdegree for the existence of a directed even circuit in a strong digraph.

Theorem 1.13. *Let D be a strict strong digraph of minimum indegree δ^- and minimum outdegree δ^+, where $\delta^- \geqslant 3$ and $\delta^+ \geqslant 3$. Then D contains a directed even circuit.*

The bound on the degrees in Theorem 1.13 is sharp. Figure 2 shows a strong 2-diregular digraph on seven vertices without directed even circuits, found by Koh (1976).

1.8. Hypergraphs

A *hypergraph* H consists of a set $V(H)$ of elements, called *vertices*, and a family $E(H)$ of subsets of $V(H)$, called *hyperedges* (or simply *edges*, when there is no ambiguity). An example is the *Fano hypergraph*, shown in fig. 3, which has as vertices and edges the seven points and seven lines of the *Fano plane*, the finite projective plane PG(2,2) (see, for example, Biggs and White 1979).

Note that the seven 3-subsets defined by the vertices of the Koh digraph (fig. 2) and their outneighbours constitute the edges of the Fano hypergraph; this is an instance of a connection established by Seymour (1974) between strong digraphs without directed even circuits and a certain class of hypergraphs of which the Fano hypergraph is prototypical.

A vertex v and edge e of a hypergraph are *incident* if $v \in e$. The *incidence graph* of a hypergraph H is the bipartite graph with bipartition $(V(H), E(H))$ in which

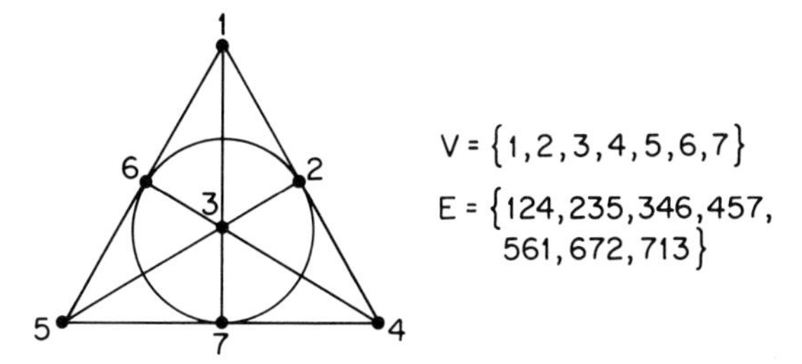

Figure 3. The Fano hypergraph.

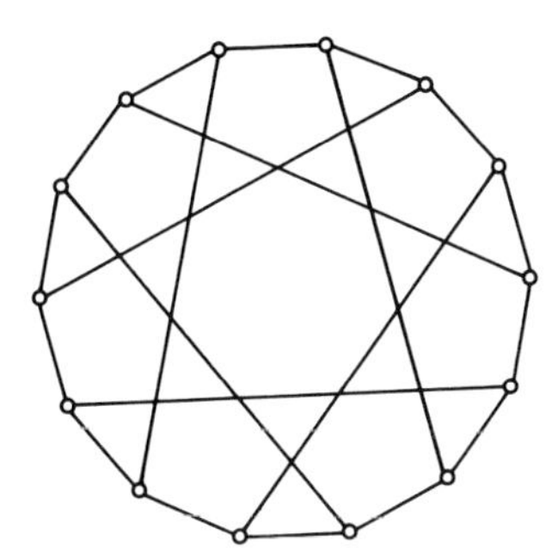

Figure 4. The Heawood graph.

$v \in V(H)$ and $e \in E(H)$ are adjacent if and only if they are incident in H. The incidence graph of the Fano hypergraph is a 3-regular graph on fourteen vertices known as the *Heawood graph*, shown in fig. 4.

With each hypergraph, we may thus associate a bipartite graph. Conversely, to each bipartite graph $G(X, Y)$, there corresponds a hypergraph H whose vertex set is X and whose hyperedges are the sets of neighbours of the vertices of Y. In this way, we can recover a hypergraph from its incidence graph, so the two structures are, in a sense, one and the same. Many other hypergraphs of interest can be associated with a graph G. We have already encountered one, the cycle space $\mathscr{C}(G)$, which may be regarded as the hypergraph whose hyperedges are the edge sets of the even subgraphs of G. Retaining only the minimal non-null elements of $\mathscr{C}(G)$, we obtain the hypergraph whose hyperedges are the edge sets of the circuits of G. This hypergraph has two special properties: (a) no proper subset of a hyperedge is a hyperedge, and (b) if C_1 and C_2 are hyperedges and e is an element of $C_1 \cap C_2$, then $(C_1 \cup C_2) \setminus e$ contains a hyperedge. A hypergraph with property (a) is termed a *clutter*, one with properties (a) and (b) a *matroid*. The matroid defined by the circuits of a graph G is called the *cycle matroid* of G. Although they are a substantial abstraction of graphs, matroids capture perfectly certain graph-theoretical concepts and properties; indeed, they are the appropriate framework for the study of graph minors and for the treatment of a number of significant questions about circuits, forests and spanning trees. They are also of fundamental importance in the related field of combinatorial optimization. Their properties are discussed in chapters 9–11 and, more fully, in the book by Welsh (1976). Hypergraphs appear in various guises throughout combinatorics. They are treated in several chapters of this handbook, in particular in chapter 7, as well as in the book by Berge (1989).

2. Hamilton paths and circuits in graphs

2.1. Dirac's theorem

For a simple graph G, the minimum degree δ is a parameter of some importance. The larger it is, as a function of n, the more likely it is that G contains a particular subgraph. For instance, Theorem 1.3(ii) implies that if $\delta \geqslant 2$ then G contains a circuit, and it follows from Proposition 1.1 and Theorem 1.7 that if $\delta > n/2$ then G contains a triangle; yet another illustration is provided by Theorem 1.13. The theorem that initiated the study of long circuits (Theorem 2.1 below) is of this type.

A path or circuit that includes every vertex of the graph is called a *Hamilton path* or *circuit*. A *hamiltonian* graph is one that contains a Hamilton circuit, a *traceable* graph one that contains a Hamilton path. As we noted in section 1, the problem of determining whether a graph contains a Hamilton circuit is $\mathcal{NP}$-complete. The more general problem of finding a minimum-weight Hamilton circuit in an edge-weighted complete graph is also $\mathcal{NP}$-complete. Known as the *Travelling Salesman Problem*, this latter question has been the subject of much study. Its many aspects are surveyed in chapters 28 and 35, and in the book edited by Lawler et al. (1985). Whether it is $\mathcal{NP}$-hard to determine the parity of the number of Hamilton circuits, a question raised by K.A. Berman (personal communication 1986b), is not known.

Dirac (1952) determined how large the minimum degree must be to guarantee the existence of a Hamilton circuit.

Theorem 2.1 (Dirac's Theorem). *Let G be a simple graph of minimum degree δ on n vertices, where $\delta \geqslant n/2$ and $n \geqslant 3$. Then G contains a Hamilton circuit.*

We present three proofs of this fundamental result. All are based on an observation that we shall refer to as the Lollipop Lemma (2.2). An (x,y)-*lollipop* is a graph $C \cup P$, where C is a circuit and P an xy-path such that $C \cap P = \{y\}$; lollipops have also been called *loops* and *lassos*.

Let C be a circuit in a graph G, with a specified orientation, and let x be a vertex of C. The predecessor of x on C is denoted by x^-, the successor of x on C by x^+. This notation is extended to a set S of vertices of C in the natural way:

$$S^- := \{x^- | x \in S\} \quad \text{and} \quad S^+ := \{x^+ | x \in S\}.$$

Lemma 2.2 (The Lollipop Lemma). *Let $C \cup P$ be an (x,y)-lollipop in a graph G, where C is a circuit of length ℓ and P a path of length m. Then G contains the xy^--*

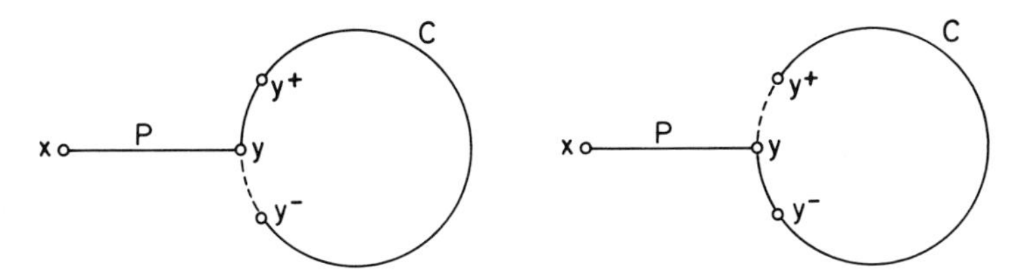

Figure 5. Two paths in a lollipop.

path $PC[y, y^-]$ and the xy^+-path $P\overleftarrow{C}[y, y^+]$, both of length $\ell + m - 1$. In particular, if C is a longest circuit of G and P is nontrivial, x is nonadjacent to both y^- and y^+.

The Lollipop Lemma (2.2) is illustrated in fig. 5. Simple though it is, this lemma is the basis of many theorems on paths and circuits. We shall appeal to it frequently, often without explicit reference.

The first proof that we give of Dirac's Theorem (2.1) is essentially the one given by Dirac himself.

Proof 1 (by *choosing a longest circuit).* Let C be a longest circuit in G, of length ℓ. By Theorem 1.3(ii), $\ell \geqslant \delta + 1$. If C is a Hamilton circuit, there is nothing to prove, so suppose that $G \setminus C \neq \emptyset$. Let $C \cup P$ be an (x, y)-lollipop, where P is as long as possible. Let m be the length of P. Note that since

$$\delta \geqslant n - \delta > n - \ell = v(G \setminus C),$$

each vertex of $G \setminus C$ is adjacent to some vertex of C, and so $m \geqslant 1$. Each vertex of P is a potential neighbour of x. On the other hand, no vertex of $C \setminus y$ whose distance from y along C is m or less can be adjacent to x; and of the remaining vertices of C, no two consecutive ones can be adjacent to x. We deduce that

$$d(x) \leqslant m + \frac{\ell - 2m}{2} = \frac{\ell}{2} < \frac{n}{2} \leqslant \delta,$$

a contradiction. □

The second proof of Dirac's Theorem (2.1) is due to Newman (1958).

Proof 2 (by *choosing a longest path).* Let P be a longest path in G, of length ℓ. We first show that G has a circuit of length $\ell + 1$. If the ends, x and y, of P are adjacent, $C := P + xy$ is such a circuit. So we may assume that x and y are not adjacent. By the choice of P, all the neighbours of x and y lie on P. Set

$$P := v_0 v_1 \cdots v_\ell,$$

where $v_0 := x$ and $v_\ell := y$, and define

$$X := \{v_i \colon v_0 v_{i+1} \in E\}, \qquad Y := \{v_i \colon v_i v_\ell \in E\}.$$

We shall prove that $X \cap Y \neq \emptyset$. Since $\delta \geqslant n/2$,

$$|X| + |Y| = d(v_0) + d(v_\ell) \geqslant 2\delta \geqslant n,$$

and, since $y \notin X \cup Y$,

$$|X \cup Y| \leqslant \ell \leqslant n - 1.$$

Therefore

$$|X \cap Y| = |X| + |Y| - |X \cup Y| \geqslant 1.$$

Let $v_k \in X \cap Y$. Then

$$C := P[v_0, v_k]v_k v_\ell \overleftarrow{P}[v_\ell, v_{k+1}]v_{k+1}v_0$$

is a circuit of length $\ell + 1$.

We now show that C is a Hamilton circuit of G. If not, let $C \cup P$ be a lollipop, where P has length m and, as in Proof 1, $m \geqslant 1$. By the Lollipop Lemma (2.2), G contains a path of length $\ell + m$, contradicting the choice of P. Therefore C is indeed a Hamilton circuit of G. □

The third proof that we give of Dirac's Theorem, due to Bondy (1981), incorporates elements of each of the other two.

Proof 3. Let C be a longest circuit in G. If C is a Hamilton circuit, there is nothing to prove, so suppose that $G \setminus C \neq \emptyset$ and let $C \cup P$ be an (x, y)-lollipop. Set

$$V_1 := V(C), \qquad V_2 := V(P) \setminus y, \qquad V_3 := V(G) \setminus V(C \cup P),$$

and let $d_i(x)$ and $d_i(y^+)$ denote the numbers of neighbours of x and y^+, respectively, in V_i, $i = 1, 2, 3$. Since C is a longest circuit, we deduce that

(i) x and y^+ cannot be adjacent to consecutive vertices v and v^+, respectively, on C;

(ii) y^+ cannot be adjacent to any vertex of V_2;

(iii) x and y^+ cannot be adjacent to a common vertex of V_3.

By (i),

$$d_1(x) + d_1(y^+) \leqslant |V_1|,$$

by (ii),

$$d_2(y^+) = 0,$$

and, by (iii),

$$d_3(x) + d_3(y^+) \leqslant |V_3|.$$

Since

$$d_2(x) \leqslant |V_2| - 1,$$

we have

$$2\delta \leqslant d(x) + d(y^+) = \sum_{i=1}^{3}(d_i(x) + d_i(y^+)) \leqslant |V_1| + |V_2| + |V_3| - 1 = n - 1,$$

contradicting the hypothesis. Thus C is indeed a Hamilton circuit of G. □

No inductive proof of Dirac's Theorem (2.1) is known. A false one was concocted by Woodall (1975) in order to illustrate the potential pitfalls of this fundamental technique.

By the simple idea of transforming one problem into another, further results can be derived very easily. We give one small illustration; others can be found in chapter 29 by Shmoys and Tardos.

Corollary 2.3. *Let G be a simple graph of minimum degree δ on n vertices, where $\delta \geqslant (n-1)/2$. Then G contains a Hamilton path.*

Proof. For $n = 1$, the result is trivial. For $n \geqslant 2$, we form the graph $H := G + v$. This graph H has $n+1$ vertices and minimum degree at least $(n+1)/2$. By Dirac's Theorem (2.1), H contains a Hamilton circuit C. The subgraph $C \setminus v$ is a Hamilton path of G. □

2.2. Generalizations of Dirac's theorem

Dirac's Theorem (2.1) is important because it is amenable to generalization in a great variety of ways. To start with, the hypothesis can be weakened. Proof 2 does not use the full strength of the hypothesis $\delta \geqslant n/2$, but just the weaker condition that the ends x and y of a longest path satisfy the inequality $d(x) + d(y) \geqslant n$. Moreover, this latter condition is applied only to nonadjacent vertices x and y. Thus the very same proof yields the following stronger theorem, due to Ore (1960).

Theorem 2.4 (Ore's Theorem). *Let G be a simple graph on n vertices, where $n \geqslant 3$, in which the degree sum of any two nonadjacent vertices is at least n. Then G contains a Hamilton circuit.*

In fact, if one examines the proof more carefully still, one sees that it is based on the following observation.

Proposition 2.5. *Let G be a simple graph on n vertices, where $n \geqslant 3$, and let x and y be nonadjacent vertices of G with degree sum at least n. If $G + xy$ is hamiltonian, then so is G.*

This observation leads naturally to the definition of a closure operation on graphs. The *closure* of a graph G on n vertices is the graph derived from G by recursively joining pairs of nonadjacent vertices having degree sum at least n. Repeated application of Proposition 2.5 yields the following result of Bondy and Chvátal (1976).

Lemma 2.6 (The Closure Lemma). *A simple graph G is hamiltonian if and only if its closure is hamiltonian.*

In particular, a graph is hamiltonian if its closure is complete: the graphs satisfying the hypothesis of Dirac's Theorem (2.1) are hamiltonian for this reason. Chvátal (1972) found a wider class of graphs defined in terms of their degree sequences (the widest, in a certain sense) whose closures are complete and which are consequently hamiltonian.

Theorem 2.7. *Let G be a simple graph on n vertices with degree sequence $(d_1, d_2, \ldots, d_n)$, where $n \geqslant 3$ and $d_1 \leqslant d_2 \leqslant \cdots \leqslant d_n$. If there is no value of $k \leqslant (n-1)/2$ such that $d_k \leqslant k$ and $d_{n-k} < n-k$, then G contains a Hamilton circuit.*

Proof. We shall show that the closure of G is complete; the conclusion will then follow directly from the Closure Lemma (2.6). Suppose, to the contrary, that this is not so. Let H be the closure of G. Then the degree sum of any two nonadjacent vertices of H is at most $n-1$. Let x and y be two such vertices of H whose degree sum is maximum. Without loss of generality, we may assume that $d_H(x) \leqslant d_H(y)$. Set $d_H(x) := k$. Then $k \leqslant (n-1)/2$ and $d_H(y) \leqslant n-k-1$. Denote by X the set of vertices nonadjacent to y in H. Then $|X| = n-1-d_H(y) \geqslant k$. Also, by the choice of x and y, $d_G(v) \leqslant d_H(v) \leqslant k$ for all $v \in X$. Similarly, if Y denotes the set of vertices that are nonadjacent to x in H, we have $|Y| = n-k-1$ and $d_G(v) \leqslant d_H(v) \leqslant n-k-1$ for all $v \in Y$. Thus G has at least k vertices of degree at most k (namely those in the set X) and at least $n-k$ vertices of degree at most $n-k-1$ (namely those in the set $Y \cup \{x\}$). This contradicts the hypothesis on G. We deduce that the closure of G is indeed complete, as claimed. □

Tian and Shi (1986) observed that the Closure Lemma (2.6) can also be used to give an easy proof of a generalization of Ore's Theorem (2.4) due to Fan (1984): *if G is a simple graph on n vertices, where $n \geqslant 3$, in which no two vertices of degree less than $n/2$ have a common neighbour, then G contains a Hamilton circuit.*

Shi (1992) and Bollobás and Brightwell (1993) gave the following elegant generalization of Dirac's Theorem (2.1).

Theorem 2.8. *Let G be a simple graph on n vertices, and let S denote the set of vertices of degree at least $n/2$ in G. If $|S| \geqslant 3$, G has a circuit that includes every vertex of S.*

Proof. Let C be a circuit that includes as many vertices of S as possible. Suppose that $S \setminus V(C) \neq \emptyset$, and let $C \cup P$ be an (x,y)-lollipop, where $x \in S$. Note that C includes at least two vertices of S because any two nonadjacent vertices of S lie in a 4-circuit of G. Let z be the first vertex of S following y on C. Then, as in Proof 3 of Dirac's Theorem (2.1), the choice of C implies that $d(x) + d(z) \leqslant n-1$, a contradiction. □

Another way in which Dirac's Theorem (2.1) can be strengthened is by refining its conclusion. Corollary 2.10 below is such a result. A graph on n vertices is *pancyclic* if it contains circuits of every length ℓ, $3 \leqslant \ell \leqslant n$. In particular, a pancyclic graph is hamiltonian. The following theorem of Bondy (1971a) gives a condition under which the converse holds. The proof presented here is due to C. Thomassen (see Bollobás 1978).

Theorem 2.9. *Let G be a simple hamiltonian graph on n vertices and at least $n^2/4$ edges. Then either G is pancyclic or n is even and $G = K_{n/2,n/2}$.*

Proof (by *induction*). There is nothing to prove when $n = 3$, so assume that $n \geqslant 4$. Suppose, first, that G has a circuit of length $n-1$,

$$C := v_1 v_2 \cdots v_{n-1} v_1.$$

Let $v := V(G) \setminus V(C)$. If $d(v) \leqslant (n-1)/2$, then

$$e(G \setminus v) = e(G) - d(v) \geqslant \frac{n^2}{4} - \frac{n-1}{2} > \frac{(n-1)^2}{4}.$$

By the induction hypothesis, $G \setminus v$ is pancyclic. Since G is hamiltonian, by hypothesis, G is pancyclic too. If $d(v) > (n-1)/2$, there exists for each ℓ, $3 \leqslant \ell \leqslant n$, an index i such that both vv_i and $vv_{i+\ell-2}$ are edges of G, and G contains the circuit $vv_iC[v_i, v_{i+\ell-2}]v_{i+\ell-2}v$, of length ℓ. Thus G is pancyclic in this case too.

Suppose, now, that G has no circuit of length $n-1$. Let $v_1 v_2 \cdots v_n v_1$ be a Hamilton circuit of G. Then, for all i, j such that $1 \leqslant i, j \leqslant n$ and $j \neq i-1, i$, at most one of the pairs $v_i v_j$ and $v_{i+1} v_{j+2}$ can be an edge of G. Thus

$$d(v_i) + d(v_{i+1}) \leqslant n.$$

Summing over all i, $1 \leqslant i \leqslant n$, we have

$$4e(G) = 2\sum_{i=1}^{n} d(v_i) \leqslant n^2,$$

whence

$$e(G) \leqslant \frac{n^2}{4}.$$

Equality holds only if n is even and, for all i, j such that $1 \leqslant i, j \leqslant n$ and $j \neq i-1, i$, exactly one of the pairs $v_i v_j$ and $v_{i+1} v_{j+2}$ is an edge of G. It follows easily that $G = K_{n/2,n/2}$ in this case. □

Corollary 2.10. *Let G be a simple graph of minimum degree δ on n vertices, where $\delta \geqslant n/2$. Then either G is pancyclic or n is even and $G = K_{n/2,n/2}$.*

Corollary 2.10 was refined substantially by Amar et al. (1991), who proved that a simple hamiltonian graph of minimum degree δ on n vertices, where $\delta > 2n/5$ and n is sufficiently large, is either pancyclic or bipartite. Nash-Williams (1971a) strengthened the conclusion of Dirac's Theorem (2.1) in a very different manner.

Theorem 2.11. *Let G be a simple graph of minimum degree δ on n vertices, where $\delta \geqslant n/2$ and $n \geqslant 3$. Then G contains at least $\lfloor cn \rfloor$ edge-disjoint Hamilton circuits, where $c = 5/224 \approx 0.0223$.*

Weakening its hypothesis and strengthening its conclusion are two ways to generalize a theorem. A more common way is simply to extend the domain of the hypothesis, including the theorem as a special case. There are many such generalizations of Dirac's Theorem (2.1). We begin with one due to Pósa (1963). A *path system* is a graph each of whose components is a path.

Theorem 2.12. *Let s be a nonnegative integer, G a simple graph of minimum degree δ on n vertices, where $\delta \geqslant (n+s)/2$ and $n \geqslant 3$, and S a set of s edges of G that induce a path system. Then G contains a Hamilton circuit that includes every edge of S.*

Proof (by *considering a maximal counterexample*). Suppose the theorem is false for some value of n, and let G be a maximal graph on n vertices satisfying the hypotheses but not the conclusion (for some set S of edges). Since a complete graph clearly satisfies the conclusion (for any set S), G is not complete. Let x and y be two nonadjacent vertices of G. Then $G + xy$ also satisfies the hypotheses of the theorem (for the set S). By the maximality of G, there is a Hamilton circuit C in $G + xy$ that includes S. Because G does not contain such a circuit, $xy \in E(C)$. Let $P := C \setminus xy$. Then P is a Hamilton path in G that connects x and y and includes S. Let $P := v_1 v_2 \cdots v_n$, where $v_1 := x$ and $v_n := y$. A counting argument similar to the one used in Proof 2 of Dirac's Theorem (2.1) shows that, for some k, $1 \leqslant k \leqslant n-1$, $xv_{k+1} \in E$, $v_k y \in E$ and $v_k v_{k+1} \notin S$. But now $P[v_1, v_k] v_k v_n \overleftarrow{P}[v_n, v_{k+1}] v_{k+1} v_1$ is a Hamilton circuit in G that includes S, contradicting our initial supposition and establishing the theorem. □

Note that the above proof, restricted to the case $s = 0$, closely resembles Proof 2 of Dirac's Theorem (2.1) but is more succinct. In the case $s = 1$, a slightly stronger conclusion can be deduced. A graph is *Hamilton-connected* if any two of its vertices are connected by a Hamilton path. Implicit in the work of Erdős and Gallai (1959, Lemma 1.6) is the following theorem.

Corollary 2.13. *Let G be a simple graph of minimum degree δ on n vertices, where $\delta \geqslant (n+1)/2$ and $n \geqslant 3$. Then G is Hamilton-connected.*

Proof. That any two adjacent vertices in G are connected by a Hamilton path follows directly from Theorem 2.12. For a nonadjacent pair, it suffices to add an edge e joining them and apply Theorem 2.12 to the graph $G + e$ and the set $S := \{e\}$. □

A Hamilton circuit is one that is as long as possible, passing through each vertex of the graph. Dirac's Theorem (2.1) has been generalized to results asserting the existence of long (but not necessarily Hamilton) circuits. We mention two instances, the first due to Egawa and Miyamoto (1989) and Bollobás and Brightwell (1993), the second to Dirac himself (1952).

Theorem 2.14. *Let k be a positive integer and G a simple graph of minimum degree δ on n vertices, where $\delta \geqslant n/(k+1)$ and $n \geqslant 3$. Then G contains a circuit of length at least n/k.*

Theorem 2.15. *Let G be a simple nonseparable graph of minimum degree δ on n vertices, where $n \geqslant 3$. Then G contains either a circuit of length at least 2δ or a Hamilton circuit.*

Strictly speaking, Theorem 2.15 is not a generalization of Dirac's Theorem (2.1) because of the additional requirement that G be nonseparable. However, the latter property is an easy consequence of the hypothesis of Dirac's Theorem (2.1). Thus we may indeed regard the theorem as a generalization. A similar comment applies to the following result of Bondy and Nash-Williams (see Nash-Williams 1971a, p. 159), where the degree condition of Dirac's Theorem (2.1) is relaxed considerably. This result is a useful adjunct to the Hopping Lemma (6.7), a powerful proof technique developed by Woodall (1973), and will be employed in section 6.

Theorem 2.16. *Let G be a simple nonseparable graph of minimum degree δ on n vertices, where $\delta \geqslant (n+2)/3$ and $n \geqslant 3$, and let C be a longest circuit of G. Then no two vertices of $G \setminus C$ are adjacent.*

Theorems 2.15 and 2.16 can both be derived by the longest-circuit technique, the strategy being to examine the structure of the components of $G \setminus C$, where C is a longest circuit. A proof of Theorem 2.15 will be given in section 4 (see Theorem 4.11).

A Hamilton circuit can also be viewed as a spanning walk that is as short as possible, passing through each vertex just once. Jackson and Wormald (1990a) and Pruesse (1990) generalized Dirac's Theorem (2.1) to spanning walks that pass through each vertex no more than k times; such walks are called *k-walks*.

Theorem 2.17. *Let k be a positive integer and G a simple connected graph of minimum degree δ on n vertices, where $\delta \geqslant n/(k+1)$ and $n \geqslant 3$. Then G has a k-walk.*

A closely related concept is that of a *k-tree*, a spanning tree in which each vertex is of degree at most k; 2-trees are thus Hamilton paths. It is easy to see that a k-tree admits a k-walk in which each edge of the tree is traversed twice. Jackson and Wormald (1990a) showed, conversely, that a graph which admits a k-walk necessarily contains a $(k+1)$-tree. Theorem 2.17 thus implies that, for $k \geqslant 2$, a simple connected graph of minimum degree δ on n vertices, where $\delta \geqslant n/k$ and $n \geqslant 3$, has a k-tree. Win (1975) obtained a slightly sharper bound on the minimum degree.

Theorem 2.18. *Let k be an integer, $k \geqslant 2$, and G a simple connected graph of minimum degree δ on n vertices, where $\delta \geqslant (n-1)/k$. Then G contains a k-tree.*

Finally, we turn to digraphs and another theorem of Nash-Williams (1969).

Theorem 2.19. *Let D be a strict digraph on n vertices with minimum indegree δ^- and minimum outdegree δ^+, where* $\min\{\delta^-, \delta^+\} \geqslant n/2$. *Then D contains a directed Hamilton circuit.*

This theorem, too, implies Dirac's Theorem (2.1), since if a graph G satisfies the hypothesis of Dirac's Theorem (2.1), its associated digraph $D(G)$ satisfies the hypothesis of Theorem 2.19. We defer a proof to section 3 (see Meyniel's Theorem, 3.11), observing only that the proof techniques we have described thus far fail for digraphs.

The above examples serve to illustrate not only the central role of Dirac's Theorem (2.1) in the study of circuits in graphs but also, more importantly, the process of generalization in mathematics. They can be thought of as basic elements in an ever-growing structure of generalizations of Dirac's Theorem (2.1). And, just as they generalize Dirac's Theorem (2.1), so they too are susceptible to generalization; indeed, Corollary 2.3 (same proof), Corollary 2.10 (similar proof), Theorem 2.12 (same proof), Corollary 2.13 (see Ore 1963), Theorem 2.14 (see Egawa and Miyamoto 1989, or Bollobás and Brightwell, 1993), Theorem 2.15 (Theorem 4.11), Theorem 2.16 (Theorem 4.23, case $k = 2$), Theorem 2.17 (Theorem 4.26), Theorem 2.18 (Theorem 4.27) and Theorem 2.19 (Meyniel's Theorem, 3.11) all admit generalizations of Ore type.

Many other generalizations of Dirac's Theorem (2.1) can be found in the literature; we shall encounter some in later sections. Surveys on hamiltonian graphs appear periodically. Those by Nash-Williams (1975), Lesniak-Foster (1977), Bermond (1978b), Chvátal (1985) and Gould (1991) represent a variety of viewpoints on the topic.

3. Hamilton paths and circuits in digraphs

3.1. Rédei's theorem

In this section, we consider Hamilton paths and circuits in digraphs. The central result here, playing the analogous role to Dirac's Theorem (2.1), is a theorem on tournaments due to Rédei (1934).

Theorem 3.1 (Rédei's Theorem). *Every tournament contains a directed Hamilton path.*

This theorem is an immediate consequence of the following lemma, the basic idea of which is due to D. König and P. Veress (see Rédei 1934).

Lemma 3.2. *Let D be a strict digraph, P an (x,y)-path in D of length ℓ, and $v \in V(D) \setminus V(P)$. Then*

(i) $d_P(v) \leqslant \ell + 2$ *if there is no (x,y)-path P' in D such that $V(P') = V(P) \cup \{v\}$;*

(ii) $d_P(v) \leqslant \ell$ *if P is a longest directed path in D.*

Proof. (i) Let $P := (v_0, v_1, \ldots, v_\ell)$, where $v_0 := x$ and $v_\ell := y$, and set

$$X := \{v_i \in V(P) \setminus v_0\colon \ (v, v_i) \in E\},$$
$$Y := \{v_i \in V(P) \setminus v_0\colon \ (v_{i-1}, v) \in E\}.$$

Then

$$X \cup Y \subseteq \{v_1, v_2, \ldots, v_\ell\} \quad \text{and} \quad X \cap Y = \emptyset,$$

for if $v_i \in X \cap Y$, the (x,y)-path $P' := (v_0, v_1, \ldots, v_{i-1}, v, v_i, \ldots, v_\ell)$ contradicts the hypothesis on P. Thus

$$d_P(v) \leqslant |X| + |Y| + 2 = |X \cup Y| + |X \cap Y| + 2 \leqslant \ell + 2.$$

(ii) If P is a longest directed path in D, neither (v, v_0) nor (v_ℓ, v) can be an edge of D, and the above argument yields

$$d_P(v) \leqslant |X| + |Y| \leqslant \ell. \qquad \square$$

Proof of Theorem 3.1. Let T be a tournament and P a longest directed path in T, of length ℓ. If P is not a Hamilton path of T, let $v \in V(T) \setminus V(P)$. Since T is a tournament, $d_P(v) = \ell + 1$. On the other hand, by Lemma 3.2, $d_P(v) \leqslant \ell$, a contradiction. Thus P is, indeed, a Hamilton path of T. $\square$

3.2. Generalizations of Rédei's theorem

Rédei's Theorem (3.1), like Dirac's Theorem (2.1), admits a number of interesting generalizations. Thomassen (1982), for instance, refined it by imposing a rather natural degree condition on the ends of the Hamilton path.

Theorem 3.3. *Every tournament contains a directed Hamilton path whose tail is a vertex of maximum outdegree and whose head is a vertex of maximum indegree.*

Another generalization is a theorem of Rédei (1934) on the parity of the number of directed Hamilton paths; it was from this result that Rédei's Theorem (3.1) was originally deduced.

Theorem 3.4. *Every tournament contains an odd number of directed Hamilton paths.*

Theorem 3.4 is established by means of a proof technique known as *inclusion–exclusion*, an inversion formula with applications throughout mathematics. The following version of the formula suffices for our purpose; more general versions are described in chapter 21 by Gessel and Stanley. We omit the proof.

Formula 3.5 (Inclusion–Exclusion). *Let Z be a finite set and $f: 2^Z \to \mathbb{R}$ a real-valued function defined on the subsets of Z. Define the function $g: 2^Z \to \mathbb{R}$ by*

$$g(X) := \sum_{X \subseteq Y \subseteq Z} f(Y).$$

Then

$$f(X) = \sum_{X \subseteq Y \subseteq Z} (-1)^{|Y|-|X|} g(Y).$$

If D is a strict digraph with vertex set $V := \{1, 2, \ldots, n\}$ and edge set E, and if π is a permutation of V, the subset E_π of E defined by

$$E_\pi := E \cap \{(\pi(i), \pi(i+1)): \ i = 1, 2, \ldots, n-1\}$$

induces a digraph each of whose components is a directed path. Such a digraph we call a *directed-path system.*

Proposition 3.6. *Let D be a strict digraph with vertex set $V := \{1, 2, \ldots, n\}$ and edge set E. Denote by Π the set of permutations of V and by $\mathcal{H}$ the set of directed Hamilton paths of D. For each subset X of E, define*

$$f(X) := |\{\pi \in \Pi: \ X = E_\pi\}| \quad \text{and} \quad h(X) := |\{H \in \mathcal{H}: \ X \subseteq E(H)\}|.$$

Then

$$f(X) \equiv h(X) \pmod 2.$$

Proof. For each subset X of E, define $g(X) := |\{\pi \in \Pi: \ X \subseteq E_\pi\}|$. Then

$$g(X) = \sum_{X \subseteq Y \subseteq E} f(Y),$$

and so, by the Inclusion–Exclusion Formula (3.5),

$$f(X) = \sum_{X \subseteq Y \subseteq E} (-1)^{|Y|-|X|} g(Y).$$

Observe that $g(Y) = r!$ if and only if the spanning subdigraph of D with edge set Y is a directed-path system with r components. Thus $g(Y)$ is odd if and only if Y induces a directed Hamilton path H of D, and

$$f(X) \equiv \sum_{\{H \in \mathcal{H}: \ X \subseteq E(H)\}} (-1)^{n-1-|X|} \equiv h(X) \pmod 2. \qquad \square$$

Proof of Theorem 3.4. The theorem is valid for transitive tournaments, each of which has precisely one directed Hamilton path. Since any tournament T on n vertices can be derived from the transitive tournament TT_n by reorienting appropriate edges, it suffices to prove that the parity of $h(T)$, the number of directed Hamilton paths in T, is unaltered by the reorientation of any one edge e. Taking $X := \{e\}$ in Proposition 3.6, we have

$$f(e) \equiv h(e) \pmod 2.$$

Thus, if T' denotes the tournament derived from T by reorienting e,

$$h(T') = h(T) + f(e) - h(e) \equiv h(T) \pmod 2. \qquad \square$$

A third generalization of Rédei's Theorem (3.1) by Thomason (1986), asserting the existence in all sufficiently large tournaments of oriented Hamilton paths of every type (not just directed Hamilton paths). The following definition makes this notion precise.

Let D be a digraph, and let $P := v_0e_1v_1 \cdots e_\ell v_\ell$ be a (not necessarily directed) path in D. The *type* of P is the sequence $(\varepsilon_1, \varepsilon_2, \ldots, \varepsilon_\ell)$, where

$$\varepsilon_i := \begin{cases} - & \text{if } e_i \text{ has head } v_{i-1}, \\ + & \text{if } e_i \text{ has head } v_i. \end{cases}$$

The *type* of a circuit $v_1e_1 \cdots v_\ell e_\ell v_1$ is defined similarly. A directed path or circuit is thus one of type (+,+, ...,+).

Theorem 3.7. *Let T be a tournament on n vertices, where $n \geqslant 2^{128}$. Then T contains a Hamilton circuit of each type except, possibly, the directed Hamilton circuit. In particular, T contains a Hamilton path of each type.*

An analogous generalization of Theorem 3.4 was obtained by Forcade (1973), who discovered the remarkable fact that the parity of the number of Hamilton paths of any type in a tournament is independent of the tournament, depending only on the path type. Perhaps more surprising still is the following beautiful corollary of Forcade's theorem.

Theorem 3.8. *Let T be a tournament of order 2^n. Then T contains an odd number of Hamilton paths of each type.*

Excluding the directed Hamilton circuit in Theorem 3.7 is essential; the transitive tournament, for instance, contains no directed circuits whatsoever. Indeed, an obvious necessary condition for the existence of a directed Hamilton circuit in a digraph is that it be strong. Camion (1959) showed that for tournaments this condition is also sufficient.

Theorem 3.9 (Camion's Theorem). *Every strong tournament contains a directed Hamilton circuit.*

Camion's Theorem (3.9) is yet another generalization of Rédei's Theorem (3.1). To see this, one can use a construction similar to that employed to deduce Corollary 2.3 from Dirac's Theorem (2.1). Alternatively, one can apply Camion's Theorem (3.9) to the strong components of the tournament, thereby obtaining a directed Hamilton path in each, and then piece these paths together to form a directed Hamilton path in the entire tournament.

Camion's Theorem (3.9) is proved by means of a longest-circuit technique. The following simple lemma is the basis of this and other theorems on directed circuits in strong digraphs. Let D be a digraph, C a subgraph of D, and X and Y sets of vertices of D. A *C-bypass* is a directed path of length at least two whose head and tail lie in C and whose internal vertices lie in $D \setminus C$. An *(X,Y)-path* is a directed path whose tail lies in X and whose head lies in Y.

Lemma 3.10 (The Bypass Lemma). *Let D be a strong nonseparable digraph, and let C be a nontrivial proper subgraph of D. Then D contains a C-bypass.*

Proof. Since D is nonseparable, there is a path in D of length at least two whose tail x and head y belong to C and whose internal vertices belong to $D \setminus C$. Among all such paths, we choose one in which the number of edges directed the wrong way (that is, towards x) is as small as possible. We shall show that this path P is a C-bypass. Assume the contrary, and let (u,v) be an edge of P directed towards x. Since D is strong, there exist in D a (C,u)-path P_u and a (v,C)-path P_v. It cannot be the case that the tail of P_u is y and the head of P_v is x, since then the (y,x)-walk $P_u(u,v)P_v$ would contain a C-bypass, contradicting the choice of P. Without loss of generality, therefore, we may assume that the tail of P_u is z, where $z \neq y$. But now the zy-walk $P_uP[u,y]$ contains a zy-path that contradicts the choice of P. Thus P is, indeed, a C-bypass. □

Proof of Theorem 3.9. Let T be a strong tournament and C a longest directed circuit in T. If C is not a Hamilton circuit, T contains a C-bypass by the Bypass Lemma (3.10). We choose such a path P that the distance along C from its tail x to its head y is as small as possible. By the choice of C, this distance, the length of $C[x,y]$, is no less than the length of P. Let u and v be internal vertices of $C[x,y]$ and P, respectively. Since T is a tournament, one of these vertices dominates the other. If u dominates v, the (u,y)-path $(u,v)P[v,y]$ contradicts the choice of P. If, on the other hand, v dominates u, the (x,u)-path $P[x,v](v,u)$ contradicts the choice of P. We conclude that C is, indeed, a directed Hamilton circuit of T. □

Moon (1968, p. 6) strengthened the conclusion of Camion's Theorem (3.9) by showing that every strong tournament is *vertex pancyclic*: each vertex lies in a directed circuit of every length. This can be proved easily using the Bypass Lemma (3.10) (see Bondy 1977). The Bypass Lemma (3.10) is also effective in establishing a far-reaching simultaneous generalization of Ore's Theorem (2.4), Theorem 2.19 and Camion's Theorem (3.9), due to Meyniel (1973). The proof we present is by Bondy and Thomassen (1977).

Theorem 3.11 (Meyniel's Theorem). *Let D be a strict strong digraph on n vertices, where $n \geqslant 2$, in which the degree sum of any two nonadjacent vertices is at least $2n-1$. Then D contains a directed Hamilton circuit.*

Proof. The degree condition implies that D is nonseparable. Let $C := (v_1, v_2, \ldots, v_\ell, v_1)$ be a longest directed circuit in D. If C is not a Hamilton circuit, there is a C-bypass in D by the Bypass Lemma (3.10). As in the proof of Camion's Theorem (3.9), we choose such a path P that the distance along C from its tail x to its head y is as small as possible. Let

$$V_1 := V(P) \setminus \{x,y\}, \quad V_2 := V(C(x,y)), \quad V_3 := V(C[y,x]), \tag{3.1}$$

$$V_4 := V(D) \setminus (V_1 \cup V_2 \cup V_3). \tag{3.2}$$

By the definition of a C-bypass, $V_1 \neq \emptyset$, and by the choice of C, $V_2 \neq \emptyset$. Let $u \in V_1$ and $v \in V_2$, and denote by $d_i(u)$ and $d_i(v)$ the numbers of neighbours of u and v, respectively, in V_i, $i = 1,2,3,4$. Then

$$d_1(u) \leqslant 2|V_1| - 2.$$

By the choice of P, u and v are nonadjacent; moreover,

$$d_1(v) = 0 \quad \text{and} \quad d_2(u) = 0.$$

Also by the choice of P, there is no directed path (u,w,v) or (v,w,u) with $w \in V_4$, and so

$$d_4(u) + d_4(v) \leqslant 2|V_4|.$$

Finally, by Lemma 3.2 and the choice of C,

$$d_3(u) \leqslant |V_3| + 1.$$

Denote by P' a longest (y,x)-path in D such that $V_3 \subseteq V(P') \subseteq V_2 \cup V_3$. Let

$$V_2' := V_2 \setminus V(P') \quad \text{and} \quad V_3' := V(P'),$$

and denote by $d_i'(u)$ and $d_i'(v)$ the numbers of neighbours of u and v, respectively, in V_i', $i = 2,3$. By the choice of C, $V_2' \neq \emptyset$. For $v \in V_2'$,

$$d_2'(v) \leqslant 2|V_2'| - 2 \quad \text{and} \quad d_3'(v) \leqslant |V_3'| + 1,$$

the latter inequality following from Lemma 3.2 and the choice of P'. Summing the above inequalities yields

$$\begin{aligned} d(u) + d(v) &= d_1(u) + d_1(v) + d_2(u) + d_3(u) + d_2'(v) + d_3'(v) + d_4(u) + d_4(v) \\ &\leqslant 2|V_1| - 2 + |V_3| + 2|V_2'| + |V_3'| + 2|V_4| \\ &\leqslant 2|V_1| + 2|V_2| + 2|V_3| + 2|V_4| - 2 = 2n - 2. \end{aligned}$$

This contradicts the hypothesis and establishes the theorem. □

4. Fundamental parameters

In this section, we introduce three important measures of the complexity of a graph: the connectivity, the stability number, and the chromatic number. These parameters are discussed in detail in chapters 2 and 4. Here we highlight their relationship to paths and circuits. We also introduce the edge analogues of the above concepts: the edge connectivity, the matching number, and the edge chromatic number, as well as one parameter of particular relevance to paths and circuits, the toughness.

4.1. Connectivity and edge connectivity

A connected graph is, by definition, one in which any two vertices are linked by a path. Alternatively, a connected graph can be regarded as one whose vertex set admits no partition into two proper subsets, where vertices in different subsets are nonadjacent. These two viewpoints lead to two different ways of measuring the degree of connectedness of a graph.

We denote a graph G with two specified vertices x and y by $G(x,y)$. The members of a collection of xy-paths in a graph $G(x,y)$ are *edge-disjoint* if no two have an edge in common, and *internally-disjoint* if no two have an internal vertex in common. A graph G on at least two vertices is *k-edge-connected* if any two vertices are connected by at least k edge-disjoint paths, and *k-connected* if any two vertices are connected by at least k internally-disjoint paths; a graph on one vertex is defined to be both k-edge-connected and k-connected for $k = 0, 1$, but not for $k \geqslant 2$. Thus every graph is 0-connected, a graph is 1-connected if and only if it is connected, and a loopless graph with at least two edges is 2-connected if and only if it is nonseparable. Also, since internally-disjoint paths are edge-disjoint, k-connected graphs are k-edge-connected.

Analogous definitions apply to digraphs. A digraph D is *strongly k-edge-connected* if any two vertices u and v are connected by at least k edge-disjoint (u,v)-paths, and *strongly k-connected* or *k-strong* if any two vertices u and v are connected by at least k internally-disjoint (u,v)-paths.

The *edge connectivity* of a graph G is the greatest integer k for which G is k-edge-connected, the *connectivity* the greatest integer k for which G is k-connected. Likewise, in a digraph D, the *strong connectivity* is the greatest integer k for which D is k-strong, and the *strong edge connectivity* is the greatest integer k for which D is strongly k-edge-connected. The edge connectivity of G is denoted by $\kappa'(G)$, the connectivity of G by $\kappa(G)$, the strong edge connectivity of D by $\kappa'(D)$ and the strong connectivity of D by $\kappa(D)$. These parameters can be computed in polynomial time (see chapter 2).

If X and Y are sets of vertices in a graph G, the set of edges of G with one end in X and the other end in Y is denoted by $[X, Y]$. When $X \cup Y$ is a partition of $V(G)$, the set $[X, Y]$ is called an *edge cut* of G; if $X, Y \neq \emptyset$, the edge cut $[X, Y]$ is *nontrivial*; if $x \in X$ and $y \in Y$, it *separates* x and y. The edge cut $[X, Y]$ is also called the *coboundary* of X (or of Y). Every nonempty edge cut is a disjoint union of minimal nonempty edge cuts, or *bonds*. In a connected graph G, a set S of edges is a bond if and only if $G \setminus S$ has exactly two components. The edge cuts of a graph G form a vector space $\mathscr{B}(G)$ over GF(2) called the *bond space* of G. The bond space and cycle space are orthogonal complements in the edge space. Because of this relationship, edge cuts are sometimes called *cocycles* and bonds *cocircuits*. Edge cuts are also referred to as *cuts* or *cutsets*.

Analogously, if $X \cup S \cup Y$ is a partition of V, where $X, Y \neq \emptyset$ and no edge of G has one end in X and the other end in Y, the set S is called a *vertex cut* of G; thus a set S of vertices in a graph G is a vertex cut if and only if $G \setminus S$ is disconnected. If $x \in X$ and $y \in Y$, we say that the vertex cut S *separates* x and y;

note that only nonadjacent vertices are separated by vertex cuts. Vertex cuts are also called *articulation sets.*

An important and extremely useful theorem due to Menger (1927) links these concepts.

Theorem 4.1 (Menger's Theorem).
– Edge version: *Let $G(x,y)$ be a graph. Then the maximum number of edge-disjoint xy-paths in G is equal to the minimum number of edges in an edge cut of G separating x and y.*
– Vertex version: *Let $G(x,y)$ be a graph, where x and y are nonadjacent vertices of G. Then the maximum number of internally-disjoint xy-paths in G is equal to the minimum number of vertices in a vertex cut of G separating x and y.*

There is a corresponding version of Menger's Theorem (4.1) for directed paths in digraphs. Proofs of both versions are given in chapter 2: Connectivity and Network Flows. Menger's Theorem (4.1) is the standard tool for dealing with questions involving connectivity. It is frequently used in the following form.

Lemma 4.2 (The Fan Lemma). *Let G be a k-connected graph, where $k \geqslant 1$, let x be a vertex of G, and let Y be a set of at least k vertices of G, where $x \notin Y$. Then there exist distinct vertices $y_1, y_2, \ldots, y_k$ in Y and internally-disjoint paths $P_1, P_2, \ldots, P_k$, such that*
(i) *P_i is an xy_i-path, $1 \leqslant i \leqslant k$, and*
(ii) *$V(P_i) \cap Y = \{y_i\}$, $1 \leqslant i \leqslant k$.*

Remark 4.3. Menger's Theorem (4.1) likewise guarantees that, given any two k-sets of vertices $X := \{x_1, x_2, \ldots, x_k\}$ and $Y := \{y_1, y_2, \ldots, y_k\}$ in a k-connected graph G, there are k disjoint paths P_1, $P_2, \ldots,$ P_k connecting X and Y in G. It does not, however, guarantee that path P_i can be so chosen as to connect vertices x_i and y_i, $1 \leqslant i \leqslant k$. Such a system is called a *k-linkage*, and a graph is *k-linked* if it has a k-linkage for arbitrary k-sets X and Y. The 2-linked graphs were characterized by Seymour (1980) and Thomassen (1980c). The corresponding concept for edge-disjoint paths is called a *weak k-linkage.* Thomassen (1980c) conjectured that a k-edge-connected graph is weakly k-linked if k is odd. This would imply that a k-edge-connected graph is weakly $(k-1)$-linked if k is even, a result established by Huck (1991). Further information on linkages can be found in chapter 2.

The Fan Lemma (4.2) is a very versatile tool. We give here one illustration of its use, by Dirac (1960).

Theorem 4.4. *Let G be a k-connected graph, where $k \geqslant 2$, and let S be a set of k vertices in G. Then G contains a circuit C that includes every vertex of S.*

Proof. The theorem is true for $k = 2$, since the union of two internally-disjoint xy-paths is a circuit that includes both x and y. We proceed by induction on k.

Suppose that the theorem holds for $k < n$, where $n \geqslant 3$. Let G be an n-connected graph, let S be a set of n vertices of G, and let $x \in S$. Put $S' := S \setminus x$. Then S' is a set of $n-1$ vertices in the n-connected graph G. By the induction hypothesis, there is a circuit C' in G that includes every vertex of S'. If $x \in V(C')$, we may take $C := C'$, and there is nothing more to prove. Suppose, then, that $x \notin V(C')$. Note that the vertices of S' partition C' into $n-1$ segments. There are two cases. If C' is of length at least n, we apply the Fan Lemma (4.2) with $k := n$ and $Y := V(C')$. By the pigeonhole principle, two of the paths whose existence is guaranteed by the lemma, say P_i and P_j, must terminate at vertices y_i and y_j that lie in the same segment of C'. The desired circuit C can now be obtained from C' by replacing the $y_i y_j$-segment of C' by the $y_i y_j$-path $P_i \overleftarrow{P}_j$. If C' is of length $n-1$, we apply the Fan Lemma (4.2) with $k := n-1$ and $Y := V(C')$, and use the above argument. □

Remark 4.5. Watkins and Mesner (1967) characterized the extremal graphs for Theorem 4.4, that is, those k-connected graphs in which some set of $k+1$ vertices is included in no circuit. For $k \geqslant 3$, they are the k-connected graphs having a set of k vertices whose deletion leaves a graph with more than k components; there are two additional families of extremal graphs when $k = 2$.

Bermond and Lovász (1975) conjectured a digraph analogue of Theorem 4.4 for the case $k = 2$, namely that in a digraph of sufficiently high strong connectivity, any two vertices lie on a common directed circuit. This was disproved by Thomassen (1991). Bondy and Lovász (1981) generalized Theorem 4.4 by showing that, if $k \geqslant 2$, the circuits through a set S of k vertices in a k-connected graph G generate the cycle space of G (except in certain specified cases); any circuit of G can, moreover, be expressed as the sum of an odd number of circuits through S. This implies that there is an even circuit through S and, if G is not bipartite, also an odd circuit through S.

Egawa et al. (1991) obtained a strong common generalization of Theorem 4.4 and Theorem 2.15.

Theorem 4.6. *Let G be a simple k-connected graph of minimum degree δ, where $k \geqslant 2$, and let S be a set of k vertices of G. Then G contains either a circuit of length at least 2δ that includes every vertex of S or a Hamilton circuit.*

Establishing the existence of circuits through specified edges is far more challenging. Let $e_1 := x_1y_1$ and $e_2 := x_2y_2$ be any two edges in a 2-connected graph G. Form a new graph G' by subdividing e_i by v_i, $i = 1,2$. Then G' is 2-connected and, by Theorem 4.4, there is a circuit in G' that includes both v_1 and v_2. Thus, in G, there is a circuit that includes both e_1 and e_2. This observation cannot be extended to 3-connected graphs; three edges that are incident with the same vertex, or that form an edge cut, do not lie on a common circuit. However, these are the only exceptions, as Adám (1963) proved. Lovász (1974) was led to formulate the following general conjecture.

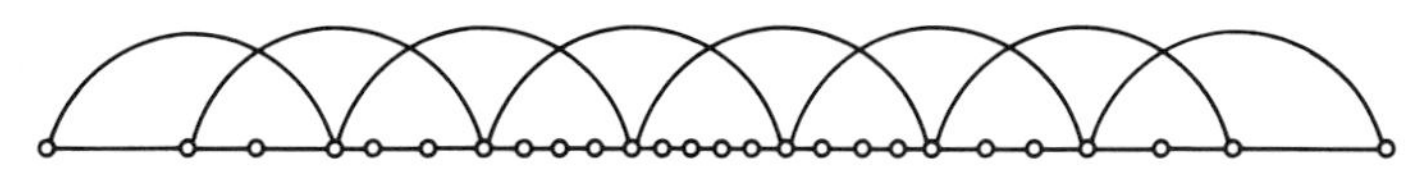

Figure 6. A vine on a path.

Conjecture 4.7. Let G be a k-connected graph, where $k \geqslant 2$, and let S be a set of k edges in G that induce a path system. Then there is a circuit of G that includes every edge of S, unless k is odd and S is an edge cut of G.

A slightly weaker version of this conjecture, due to Woodall (1977), was verified by Häggkvist and Thomassen (1982). Their proof uses a variant of the Hopping Lemma (6.7), to be discussed in section 6.

Theorem 4.8. *Let G be a k-connected graph, where $k \geqslant 2$, and let S be a set of $k-1$ edges in G that induce a path system. Then there is a circuit of G that includes every edge of S.*

A related conjecture, for 3-regular graphs, was formulated by Holton and Thomassen (1986). The connectivity hypothesis is replaced here by the less stringent one of cyclic edge connectivity. A connected graph G that contains two disjoint circuits is *cyclically k-edge-connected* if it has no bond S of fewer than k edges such that both components of $G \setminus S$ contain circuits.

Conjecture 4.9. Let G be a 3-regular cyclically k-edge-connected graph and let S be a set of $k-1$ edges in G that induce a path system. Then there is a circuit of G that includes every edge of S.

In Theorem 4.4, the Fan Lemma (4.2) was used to establish the existence of certain circuits in k-connected graphs. Another very useful tool in this context, developed by Dirac (1952), is that of a vine (see fig. 6). A *vine* on a path $P := P[x,y]$ is a set $\mathscr{P} := \{P_i[x_i,y_i]\colon 1 \leqslant i \leqslant m\}$ of internally-disjoint paths such that

(a) $P_i \cap P = \{x_i, y_i\}$, $1 \leqslant i \leqslant m$;

(b) $x = x_1 \prec x_2 \prec y_1 \preccurlyeq x_3 \prec y_2 \preccurlyeq x_4 \prec \cdots \preccurlyeq x_m \prec y_{m-1} \prec y_m = y$ on P.

Dirac (1952) showed that there is a vine on any path in a 2-connected graph. A generalization of this result was given by Bondy and Locke (1981). Vines $\mathscr{P}_1$ and $\mathscr{P}_2$ on a path $P[x,y]$ are *disjoint* if

(a) the paths of $\mathscr{P}_1$ and $\mathscr{P}_2$ are internally-disjoint;

(b) the only vertex that is the tail of a path of $\mathscr{P}_1$ and the tail of a path of $\mathscr{P}_2$ is x;

(c) the only vertex that is the head of a path of $\mathscr{P}_1$ and the head of a path of $\mathscr{P}_2$ is y.

For sets of vertices X and Y in a graph, a path whose tail lies in X and whose head lies in Y is called an *$[X,Y]$-path*.

Lemma 4.10 (The Vine Lemma). *Let P be a path in a k-connected graph G. Then there are $k-1$ pairwise-disjoint vines $\mathscr{P}_1, \mathscr{P}_2, \ldots, \mathscr{P}_{k-1}$ on P.*

Proof. Let $P := v_0 v_1 \cdots v_\ell$. Since G is k-connected, there are k internally-disjoint $v_0 v_\ell$-paths $Q_1, Q_2, \ldots, Q_k$ in G. From each path Q_i, we construct a family $\mathcal{Q}_i$ of paths as follows. If $V(P) \cap V(Q_i) := \{w_1, w_2, \ldots, w_m\}$, where $w_j \prec w_{j+1}$ on Q_i, $1 \leqslant j \leqslant m-1$, we set

$$\mathcal{Q}_i := \{Q_i[w_j, w_{j+1}]:\ w_j \prec w_{j+1} \text{ on } P\}.$$

We now define

$$\mathcal{Q} := \cup_{i=1}^{k} \mathcal{Q}_i.$$

It follows from these definitions that if

$$X_j := \{v_0, v_1, \ldots, v_{j-1}\} \text{ and } Y_j := \{v_{j+1}, v_{j+2}, \ldots, v_\ell\}, \quad \text{where } 0 < j < \ell,$$

then $\mathcal{Q}_i$ contains an $[X_j, Y_j]$-path avoiding v_j, for any j such that $v_j \notin V(Q_i)$. Because the paths Q_i are internally-disjoint, at most one of them includes any particular v_j, and so $\mathcal{Q}$ contains at least $k-1$ $[X_j, Y_j]$-paths avoiding v_j, for each j, $0 < j < \ell$.

We now construct the desired vines inductively. Initially, we select from $\mathcal{Q}$ $k-1$ $[X_1, Y_1]$-paths $P_1, P_2, \ldots, P_{k-1}$ avoiding v_1 and set $\mathcal{P}_i := \{P_i\}$, $1 \leqslant i \leqslant k-1$. At each subsequent stage, we modify the families $\mathcal{P}_i$ as follows. Let p_i denote the vertex of $\mathcal{P}_i$ that is furthest from v_0 on P. If $p_1 = p_2 = \cdots = p_{k-1} = v_\ell$, the families $\mathcal{P}_i$ are vines on P satisfying the conditions of the lemma. Otherwise, we define $v_j := p_1$, where we assume (relabelling if necessary) that $p_1 \preccurlyeq p_2 \preccurlyeq \cdots \preccurlyeq p_{k-1}$. Each $\mathcal{P}_i$, $2 \leqslant i \leqslant k-1$, contains at most one $[X_j, Y_j]$-path avoiding v_j. Therefore, there is at least one more such path Q in $\mathcal{Q} \setminus \cup_{i=1}^{k-1} \mathcal{P}_i$. Let u_j be the tail of Q. We modify the family $\mathcal{P}_1$ by adding Q and deleting each path whose head lies in $P(u_j, v_j]$ except the one whose tail is closest to v_0; the families $\mathcal{P}_i$, $2 \leqslant i \leqslant k-1$, are not modified. This procedure terminates with $k-1$ pairwise-disjoint vines $\mathcal{P}_1, \mathcal{P}_2, \ldots, \mathcal{P}_{k-1}$ on P. □

Dirac (1952) used the Vine Lemma (4.10) to prove that, if a 2-connected graph has a path of length ℓ, it necessarily has a circuit of length at least $2\ell^{1/2}$. (The graph of fig. 6 is an extremal graph for this theorem, with $\ell = 25$.) The lemma was also applied by Bondy and Locke (1981) and Locke (1982) to establish analogous results for graphs of higher connectivity; for example, if a 3-connected graph has a path of length ℓ, it necessarily has a circuit of length at least $2\ell/5$. Of particular note here is the marked difference in the behaviour of 2-connected and 3-connected graphs; we shall see further instances of this phenomenon shortly. A similar change may well occur in the passage from 3-connected to 4-connected graphs; several intriguing conjectures support this view, although there is little hard evidence.

Dirac (1952) also used the Vine Lemma (4.10) to prove Theorem 2.15, a generalization of his Theorem 2.1. The same proof technique (see Pósa 1963, proof of Theorem 3) yields the corresponding generalization of Ore's Theorem (2.4) first stated explicitly by Bermond (1976) and Linial (1976).

Theorem 4.11. *Let G be a simple 2-connected graph in which the degree sum of any two nonadjacent vertices is at least d. Then G contains either a circuit of length at least d or a Hamilton circuit.*

Proof. Let $P[x,y]$ be a longest path in G and $\mathcal{P} := \{P_i[x_i, y_i]: 1 \leqslant i \leqslant m\}$ a vine on P; note that P_1 and P_m are of length one, by the choice of P. Consider the circuit

$$C := \sum_{i=1}^{m} C_i,$$

where $C_i := P[x_i, y_i]\overleftarrow{P}_i$, $1 \leqslant i \leqslant m$, and assume that $\mathcal{P}$ is chosen so that
(i) $|\mathcal{P}| = m$ is as small as possible;
(ii) subject to (i), $|V(C) \cap V(P)|$ is as large as possible. Set $P := v_0v_1v_2\cdots v_\ell$, where $v_0 := x$, $v_k := y_1$ and $v_\ell := y$, and define

$$X := \{v_i\colon\ xv_{i+1} \in E\} \quad \text{and} \quad Y := \{v_i\colon\ v_iy \in E\}.$$

If $X \cap Y \neq \emptyset$, then (as in Proof 2 of Dirac's Theorem, 2.1) C is a Hamilton circuit of G, since P is a longest path. If $X \cap Y = \emptyset$, then x and y are nonadjacent. Moreover, by the choice of $\mathcal{P}$, C contains every vertex of $(X \cup Y \cup \{y\}) \setminus \{v_{k-1}\}$ and thus is of length at least $d(x) + d(y)$. By hypothesis, this is at least d. □

One consequence of Theorem 4.11 is a frequently used theorem of Erdős and Gallai (1959).

Corollary 4.12. *Let $G(x,y)$ be a simple 2-connected graph such that $d(v) \geqslant d$, for all $v \neq x,y$. Then there is an xy-path of length at least d in G.*

Proof. Denote by H the graph obtained from G by adding a set S of $d-1$ vertices and joining them to one another and to both x and y. Then H is a simple nonseparable graph on n vertices, where $n \geqslant 2d \geqslant 4$, in which the degree sum of any two nonadjacent vertices is at least $2d$. By Theorem 4.11, H contains a circuit C of length at least $2d$. If C includes vertices of S, $C \setminus S$ is an xy-path in G of length at least d. Otherwise $C \subseteq G$ and, because G is 2-connected, there are disjoint paths P_1 and P_2 in G connecting $\{x,y\}$ and C. In this case, the subgraph $C \cup P_1 \cup P_2$ of G contains two xy-paths, at least one of which is of length at least d. □

Theorem 4.11 is an Ore-type generalization of Theorem 2.15. Voss and Zuluaga (1977) generalized the latter theorem in another direction.

Theorem 4.13. *Let G be a simple nonseparable nonbipartite graph of minimum degree δ on n vertices, where $\delta \leqslant n/2$. Then G contains both an odd circuit of length at least $2\delta - 1$ and an even circuit of length at least 2δ.*

This clearly implies Theorem 2.15 in the case of nonbipartite graphs. For bipartite graphs, it suffices to add an edge between two vertices in the same part of the bipartition and then apply Theorem 4.13. We propose the following conjecture, a strengthening of one mentioned in Locke (1985). If true, it would imply Theorem 4.13 in the case of 3-connected graphs.

Conjecture 4.14. Let G be a simple 3-connected graph of minimum degree δ on n vertices, where $\delta \leqslant n/2$. Then the circuits of length at least $2\delta - 1$ generate the cycle space of G; moreover, every circuit of G can be expressed as the sum of an odd number of circuits of length at least $2\delta - 1$.

We close this discussion of connectivity with an attractive conjecture of S. Smith (see Grötschel 1984).

Conjecture 4.15. In a k-connected graph, where $k \geqslant 2$, any two longest circuits have at least k vertices in common.

4.2. Stability number

A set S of vertices of a graph G is *stable* if no two elements of S are adjacent in G. The maximum cardinality of a stable set of G is called the *stability number* of G and denoted by $\alpha(G)$. Stable sets are also called *independent* sets and the stability number the *independence number*. Determining whether a graph has stability number at most k is an $\mathcal{NP}$-complete problem.

A simple and surprising relationship linking the stability number, the connectivity, and the existence of Hamilton circuits was discovered by Chvátal and Erdős (1972).

Theorem 4.16 (The Chvátal–Erdős Theorem). *Let G be a graph on at least three vertices with stability number α and connectivity κ, where $\alpha \leqslant \kappa$. Then G is hamiltonian.*

The proof of the Chvátal–Erdős Theorem (4.16) is based on the fundamental concept of a bridge, illustrated in fig. 7. Let G be a graph and H a subgraph of G. Consider the equivalence relation $\sim$ on $E(G) \setminus E(H)$ defined by $e \sim f$ if and only if there is a walk internally-disjoint from H whose first edge is e and whose last edge is f. The subgraphs of G induced by the equivalence classes under this relation are the *bridges* of H in G; isolated vertices of G in $G \setminus H$ are also regarded as bridges of H. A bridge having at most one edge is *degenerate*; all other bridges are *proper*. A *chord* of H is a degenerate bridge linking two distinct vertices of H. If B is a bridge of H in G, the elements of $V(B) \setminus V(H)$ are the *internal vertices* of B and the elements of $V(B) \cap V(H)$ the *vertices of attachment* of B to H. The edges of B incident to its vertices of attachment are its *edges of attachment* to H.

For the proof of the Chvátal–Erdős Theorem (4.16), we shall also need the undirected version of the concept of a bypass, defined for digraphs in section 3. Let G be a graph, and let C be a subgraph of G. A *C-bypass* in G is a path of length at least two whose ends lie in C and whose internal vertices lie in $G \setminus C$.

Proof of Theorem 4.16. Let C be a longest circuit in G. Suppose that C is not a Hamilton circuit. Let B be a proper bridge of C in G and denote by S its set of vertices of attachment to C. Any two vertices of S are connected by a C-bypass contained in B. Because C is a longest circuit, we deduce from the Lollipop Lemma

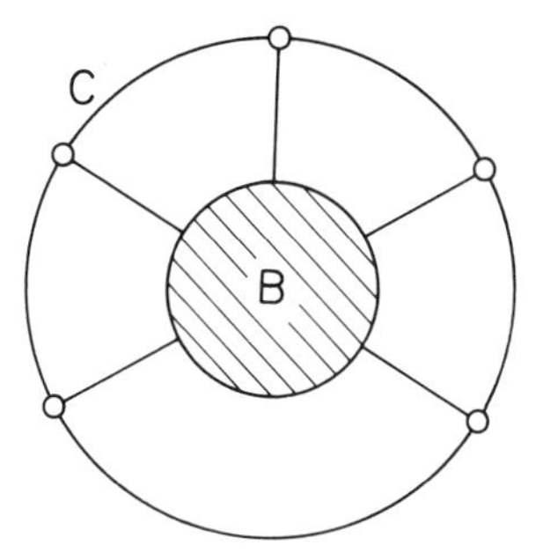

Figure 7. A bridge of a circuit.

(2.2) that S^+ is a stable set of G and that no internal vertex x of B is adjacent to any vertex of S^+. Thus $S^+ \cup \{x\}$ is a stable set and

$$|S^+| < \alpha.$$

Also, S is a vertex cut of G ($B \setminus S$ being one component of $G \setminus S$) and so

$$|S| \geqslant \kappa.$$

But now

$$\kappa \leqslant |S| = |S^+| < \alpha,$$

contradicting the hypothesis. Therefore, C is indeed a Hamilton circuit of G. □

For future reference, we note the following consequence of the Chvátal–Erdős Theorem (4.16), derived by the same technique as was used to establish Corollary 2.3 from Dirac's Theorem (2.1).

Corollary 4.17. *Let G be a graph with stability number α and connectivity κ, where $\alpha \leqslant \kappa + 1$. Then G has a Hamilton path.*

Bondy (1978a) observed that the Chvátal–Erdős Theorem (4.16) implies Ore's Theorem (2.4) and also the following result of Nash-Williams (1971b).

Corollary 4.18. *Let G be a simple d-regular graph on $2d+1$ vertices, where $d \geqslant 2$. Then G is hamiltonian.*

The Chvátal–Erdős Theorem (4.16) applies only to graphs G which satisfy $\alpha \leqslant \kappa$. Kouider (1994) extended it to all 2-connected graphs.

Theorem 4.19. *Let G be a graph with connectivity κ, where $\kappa \geqslant 2$, and let H be a subgraph of G. Then G has a circuit C such that either $V(H) \subseteq V(C)$ or $\alpha(H \setminus C) \leqslant \alpha(H) - \kappa(G)$.*

A conjecture of Amar, Fournier, Germa, Häggkvist and Thomassen (see Fournier 1982) on the number of circuits needed to cover the vertex set of a 2-connected graph follows easily from the above theorem. It, too, is an extension of the Chvátal–Erdős Theorem (4.16). A *vertex covering* of a graph or digraph is a family of subgraphs whose union spans the graph or digraph.

Corollary 4.20. *Let G be a graph with stability number α and connectivity κ, where $\kappa \geqslant 2$. Then G admits a vertex covering by at most $\lceil \alpha/\kappa \rceil$ circuits.*

The 5-circuit is one example of a graph which satisfies the inequality $\alpha \leqslant \kappa$ of the Chvátal–Erdős Theorem (4.16) but is not pancyclic. There are many others; indeed, if G has this property, so does the composition $G[\bar{K}_r]$. Jackson and Ordaz (1990) conjecture, however, that in all such examples $\alpha = \kappa$.

Conjecture 4.21. Let G be a graph with stability number α and connectivity κ, where $\alpha \leqslant \kappa - 1$. Then G is pancyclic.

The idea of considering the minimum degree sum of two nonadjacent vertices was introduced by Ore (1960). More generally, the minimum degree sum of a stable set of k vertices, denoted σ_k, turns out to be a parameter of relevance to circuits. Bondy (1981) proposed a common generalization of Ore's Theorem (2.4) and the Chvátal–Erdős Theorem (4.16) involving this parameter.

Conjecture 4.22. Let G be a simple k-connected graph on n vertices such that $\sigma_{k+1} \geqslant n + k(k-1)$, and let C be a longest circuit in G. Then $G \setminus C$ contains no path of length $k-1$.

Conjecture 4.22 has been verified for $k = 1$ (Ore's Theorem 2.4) and $k = 2$ (a generalization of Theorem 2.16); these are special cases of Theorem 4.23 below, which establishes a weaker form of the conjecture. The case $k = 3$ of Conjecture 4.22 is pertinent to Conjecture 5.20 and can be proved by slightly modifying the proof of Theorem 4.23 (Bondy 1981).

Theorem 4.23. *Let G be a simple k-connected graph on n vertices such that $\sigma_{k+1} \geqslant n + k(k-1)$, and let C be a longest circuit in G. Then $G \setminus C$ contains no complete subgraph of order k.*

Bauer et al. (1989) applied the case $k = 2$ of Theorem 4.23 to derive a sufficient condition for a graph to be hamiltonian. This is yet another generalization (of a generalization) of Dirac's Theorem (2.1).

Theorem 4.24. *Let G be a simple graph of connectivity κ on n vertices such that $\sigma_3 \geqslant n + \kappa$, where $\kappa \geqslant 2$. Then G contains a Hamilton circuit.*

Another variant of Conjecture 4.22 was established by Fraisse (1986). It concerns k-dominating circuits rather than longest circuits, and also generalizes Ore's Theorem (2.4). A *dominating circuit* of a graph G is a circuit C such that $V(G) \setminus V(C)$ is a stable set of G. More generally, if k is a positive integer, a *k-dominating circuit* of a graph G is a circuit C such that each component of $G \setminus C$ has fewer than k vertices.

Theorem 4.25. *Let G be a simple k-connected graph on n vertices, where $n \geqslant 3$, such that $\sigma_{k+1} \geqslant n + k(k-1)$. Then G has a k-dominating circuit.*

The concepts of k-dominating circuit and k-walk are intimately related via a graph product called composition. The *composition* (or *lexicographic product*) of simple graphs G and H is the graph $G[H]$ whose vertex set is $V(G) \times V(H)$, vertices (u_1, v_1) and (u_2, v_2) being adjacent if and only if either $u_1 u_2 \in E(G)$ or both $u_1 = u_2$ and $v_1 v_2 \in E(H)$. Jackson and Wormald (1990a) observed that a graph G has a k-walk if and only if the composition $G[K_k]$ has a k-dominating circuit. Applying Theorem 4.25, they deduced the following common generalization of Theorem 2.17 and Theorem 2.4.

Corollary 4.26. *Let G be a simple connected graph on n vertices such that $\sigma_{k+1} \geqslant n$. Then G has a k-walk.*

Closely related is a common generalization of Theorem 2.18 and the Ore-type analogue of Corollary 2.3, due to Win (1975).

Theorem 4.27. *Let G be a simple connected graph on n vertices such that $\sigma_k \geqslant n-1$, where $k \geqslant 2$. Then G contains a k-tree.*

Fournier and Fraisse (1985) proved a common generalization of Theorem 4.11 and the Chvátal–Erdős Theorem (4.16) involving degree sums of stable sets of vertices.

Theorem 4.28. *Let G be a simple k-connected graph, where $k \geqslant 2$, such that $\sigma_{k+1} \geqslant d$. Then G contains either a circuit of length at least $2d/(k+1)$ or a Hamilton circuit.*

The stability number also plays an important role in digraphs. We define a *stable set* of a digraph to be a stable set of its underlying graph. Gallai and Milgram (1960) found a beautiful link between the stability number of a digraph and partitions of its vertex set into directed paths. A *partition* of a graph G is a family of disjoint subgraphs whose vertices partition $V(G)$. A *path partition* is a partition into paths, and the minimum number of paths in a path partition of a graph G is denoted by $\pi(G)$. Analogous definitions apply to digraphs, a path partition of a digraph being a partition into directed paths.

For a path partition $\mathcal{Q}$ in a digraph D, we denote by $\alpha(\mathcal{Q})$ the cardinality of a maximum stable set whose elements belong to different paths of $\mathcal{Q}$, by $h(\mathcal{Q})$ the set of heads of the paths in $\mathcal{Q}$ and by $t(\mathcal{Q})$ the set of tails of the paths in $\mathcal{Q}$. The following lemma (Bondy 1995) is a slight refinement of one due to Gallai and Milgram (1960).

Lemma 4.29. *Let D be a digraph and $\mathcal{Q}$ a path partition of D such that $|\mathcal{Q}| > \alpha(\mathcal{Q})$. Then there is a path partition $\mathcal{P}$ of D such that $|\mathcal{P}| = |\mathcal{Q}| - 1$, $h(\mathcal{P}) \subset h(\mathcal{Q})$ and $t(\mathcal{P}) \subset t(\mathcal{Q})$.*

Proof (by *induction on $v(D)$, the case $v(D)=1$ holding vacuously*). Let $\mathcal{Q}$ be a path partition of D such that $|\mathcal{Q}| > \alpha(\mathcal{Q})$. Then $h(\mathcal{Q})$ is not a stable set, so there exist $y, z \in h(\mathcal{Q})$ such that $(y, z) \in E$. If (z) is a path of $\mathcal{Q}$, we define $\mathcal{P}$ to be the path partition of D obtained from $\mathcal{Q}$ by deleting (z) and extending the path of $\mathcal{Q}$ with head y by the edge (y, z). Thus we may assume that (z) is not a path of $\mathcal{Q}$. Let x be the predecessor of z on the path of $\mathcal{Q}$ with head z, set $D' := D \setminus z$, and let $\mathcal{Q}'$ be the restriction of $\mathcal{Q}$ to D'. Note that

$$|\mathcal{Q}'| = |\mathcal{Q}| > \alpha(\mathcal{Q}) \geqslant \alpha(\mathcal{Q}'),$$
$$h(\mathcal{Q}') = (h(\mathcal{Q}) \setminus \{z\}) \cup \{x\} \quad \text{and} \quad t(\mathcal{Q}') = t(\mathcal{Q}).$$

By the induction hypothesis, there is a path partition $\mathcal{P}'$ of D' such that $|\mathcal{P}'| = |\mathcal{Q}'| - 1$, $h(\mathcal{P}') \subset h(\mathcal{Q}')$ and $t(\mathcal{P}') \subset t(\mathcal{Q}')$. If $x \in h(\mathcal{P}')$, we define $\mathcal{P}$ to be the path partition of D obtained from $\mathcal{P}'$ by extending the path of $\mathcal{P}'$ with head x by the edge (x, z). If $x \notin h(\mathcal{P}')$, then $y \in h(\mathcal{P}')$ and we define $\mathcal{P}$ to be the path partition of D obtained from $\mathcal{P}'$ by extending the path of $\mathcal{P}'$ with head y by the edge (y, z). In both cases, $|\mathcal{P}| = |\mathcal{Q}| - 1$, $h(\mathcal{P}) \subset h(\mathcal{Q})$ and $t(\mathcal{P}) \subset t(\mathcal{Q})$. □

An immediate consequence of this lemma is the theorem of Gallai and Milgram (1960) alluded to above.

Theorem 4.30 (The Gallai–Milgram Theorem). *Let D be a digraph with stability number α. Then D admits a partition into at most α directed paths. That is,*

$$\pi \leqslant \alpha.$$

Remark 4.31. The Gallai–Milgram Theorem (4.30) can be viewed as a formula for the stability number of a graph in terms of the path partition numbers of its orientations:

$$\alpha(G) = \max\, \{\pi(\overrightarrow{G}): \ \overrightarrow{G} \text{ an orientation of } G\}.$$

To see this, observe that $\alpha(G) \geqslant \pi(\overrightarrow{G})$ for every orientation $\overrightarrow{G}$ of G, by the Gallai–Milgram Theorem (4.30), and that the reverse inequality holds when $\overrightarrow{G}$ is an orientation of G in which every vertex of some maximum stable set is a source.

The Gallai–Milgram Theorem (4.30) includes a well-known theorem of Dilworth (1950) as a special case.

Theorem 4.32 (Dilworth's Theorem). *Let $P := (X, \prec)$ be a partially ordered set. Then the minimum number of chains of P into which X can be partitioned is equal to the maximum number of elements in an antichain of P.*

Proof. Denote by $D(P)$ the digraph whose vertex set is X and whose edges are the ordered pairs (u, v) such that $u \prec v$ in P. Chains and antichains in P then correspond to directed paths and stable sets, respectively, in $D(P)$, and the Gallai–Milgram Theorem (4.30) implies the existence of a chain partition of cardinality no greater than the maximum number of elements in an antichain. The reverse inequality follows from the observation that no two elements of an antichain can belong to a common chain. □

An analogue of the Gallai–Milgram Theorem (4.30) for strong digraphs was proposed by Gallai (1964).

Conjecture 4.33. Let D be a strong digraph with stability number α. Then D admits a vertex covering by at most α directed circuits.

Two special cases of Conjecture 4.33 are easily verified. When D is the associated digraph of a 2-edge-connected graph G, this follows from a result of Pósa (1963) which asserts that any graph admits a partition into at most α circuits, edges and vertices; alternatively, one may apply Corollary 4.20 to an endblock of G and use induction. And when $\alpha = 1$, it is just Camion's Theorem (3.9). Chen and Manalastas (1983) (see also Bondy 1995) verified the case $\alpha = 2$.

Theorem 4.34. *Let D be a strong digraph with stability number two. Then D admits a vertex covering by two directed circuits $\{C_1, C_2\}$, where $C_1 \cap C_2$ is either empty or a directed path of D.*

One interesting consequence of Theorem 4.34 is the following.

Corollary 4.35. *Let D be a strong digraph with stability number two. Then D contains a directed Hamilton path.*

This corollary leads to two conjectures. One, a simultaneous generalization of Corollary 4.35 and the Gallai–Milgram Theorem (4.30), was proposed (in a stronger form) by M. Las Vergnas (see Berge 1983).

Conjecture 4.36. Let D be a strong digraph with stability number α, where $\alpha \geqslant 2$. Then D admits a partition into at most $\alpha - 1$ directed paths.

The second conjecture, put forward by Jackson and Ordaz (1990), involves a less stringent definition of stability in digraphs. A stable set in a digraph D may equally well be defined as a set S of vertices such that $D[S]$ contains no directed circuit, or such that $D[S]$ contains no directed 2-circuit. The corresponding stability numbers are called the *acyclic stability number* and the *oriented stability number*, respectively, and are denoted by α_1 and α_2. Clearly, $\alpha \leqslant \alpha_1 \leqslant \alpha_2$. The following conjecture would thus generalize Corollary 4.17.

Conjecture 4.37. Let D be a digraph with oriented stability number α_2 and strong connectivity κ, where $\alpha_2 \leqslant \kappa + 1$. Then D contains a directed Hamilton path.

Remark 4.38. It is not true in general that a k-strong digraph with stability number $k+1$ contains a directed Hamilton path; while this is so for $k = 0$, by Rédei's Theorem (3.1), and also for $k = 1$, by Corollary 4.35, it is false for $k = 2$, a counterexample being the digraph derived from the composition of $K_{1,3}$ with K_2 by orienting each copy of $K_{2,2}$ that corresponds to an edge of $K_{1,3}$ as a directed 4-circuit.

Jackson (1987) also made use of the concept of oriented stability number to establish a directed analogue of the Chvátal–Erdős Theorem (4.16).

Theorem 4.39. *Let D be a digraph with strong connectivity κ, where $\kappa \geqslant 2^{\alpha_2}(\alpha_2+2)!$ Then D contains a directed Hamilton circuit.*

Heydemann (1985) likewise proved the directed analogue of Conjecture 4.33 for the oriented stability number.

4.3. Matching number

In this section, we consider the edge analogue of a stable set. A set M of edges of a graph G is a *matching* if no two elements of M are adjacent in G. The matching M is perfect if each vertex of G is incident to an edge of M. The maximum cardinality of a matching of G is called the *matching number* of G and denoted by $\nu(G)$.

The Ore-type generalization of Corollary 2.3 referred to at the end of section 2 asserts that a simple graph on n vertices in which the degree sum of any two nonadjacent vertices is at least $n-1$ has a Hamilton path. Häggkvist (1979) showed that when n is even, the very same hypothesis implies something much stronger.

Theorem 4.40. *Let G be a simple graph on $2m$ vertices in which the degree sum of any two nonadjacent vertices is at least $2m-1$. Then each perfect matching of G is contained in a Hamilton path of G.*

Proof. Let $M := \{x_1y_1, x_2y_2, \ldots, x_my_m\}$ be a perfect matching of G, where the sets $X := \{x_1, x_2, \ldots, x_m\}$ and $Y := \{y_1, y_2, \ldots, y_m\}$ are chosen so as to maximize the number of edges in the bipartite subgraph of $G \setminus M$ induced by the bipartition (X, Y). Denote this subgraph by B. By the choice of X and Y,

$$d_B(x_i) + d_B(y_i) \geqslant \tfrac{1}{2}(d_{G\setminus M}(x_i) + d_{G\setminus M}(y_i)) = \tfrac{1}{2}(d_G(x_i) + d_G(y_i)) - 1, \quad 1 \leqslant i \leqslant m.$$

Let D be the digraph obtained from B by first orienting each edge from X to Y and then identifying x_i and y_i, denoting the resulting vertex by v_i, $1 \leqslant i \leqslant m$. For any two nonadjacent vertices v_i and v_j of D,

$$\begin{aligned} d_D(v_i) + d_D(v_j) &= d_B(x_i) + d_B(y_i) + d_B(x_j) + d_B(y_j) \\ &\geqslant \tfrac{1}{2}(d_G(x_i) + d_G(y_j)) + \tfrac{1}{2}(d_G(x_j) + d_G(y_i)) - 2 \geqslant 2m - 3. \end{aligned}$$

It now follows from Meyniel's Theorem (3.11), by applying the digraph analogue of the trick used to establish Corollary 2.3 from Dirac's Theorem, 2.1, that D has a directed Hamilton path. Without loss of generality, let this path be $(v_1, v_2, \ldots, v_m)$. Then $x_1y_1x_2y_2\cdots x_my_m$ is a Hamilton path of G through each edge of M. □

An analogous theorem on Hamilton circuits containing perfect matchings also due to Häggkvist (1979), was extended by Berman (1983) as follows.

Theorem 4.41. *Let G be a simple graph on n vertices, in which the degree sum of any two nonadjacent vertices is at least $n+1$, where $n \geqslant 3$. Then each matching of G is contained in a circuit of G.*

Theorem 4.41 implies Theorem 4.40. Let G satisfy the hypotheses of Theorem 4.40. Form G' from G by adding two new vertices, u and v, and joining each to every vertex of G and also to one another. Define $M' := M \cup \{e\}$, where $e := uv$. Then M' is a perfect matching of G'. Also, the degree sum of any two nonadjacent vertices of G' is at least $2m+3 = (2m+2)+1$. By Theorem 4.41, G' has a Hamilton circuit through each edge of M', so G has a Hamilton path through each edge of M.

Jackson and Wormald (1990b) characterized the extremal graphs for Theorem 4.41 and derived a strengthening of Dirac's Theorem (2.1) for graphs whose number of vertices is divisible by four.

Theorem 4.42. *Let G be a simple graph of minimum degree δ on n vertices, where $\delta \geqslant n/2$ and $n \equiv 0 \pmod 4$. Then each perfect matching of G is contained in a Hamilton circuit of G.*

Matchings enjoy a rich and extensive theory. This is discussed in detail in chapter 3 and in the treatise by Lovász and Plummer (1986). The algorithmic aspects of matchings are also well developed. In particular, a maximum-weight matching in an edge-weighted graph can be found in polynomial time. This leads to an efficient method for finding a shortest path of specified parity between two vertices of a graph, if there is one. The idea, due to J. Edmonds (see Grötschel and Pulleyblank 1981), is most easily explained with the aid of the following definition. The *cartesian product* of simple graphs G and H is the graph $G \square H$ whose vertex set is $V(G) \times V(H)$, vertices (u_1, v_1) and (u_2, v_2) being adjacent if and only if either $u_1 = u_2$ and $v_1 v_2 \in E(H)$ or $v_1 = v_2$ and $u_1 u_2 \in E(G)$. The cartesian product $G \square K_2$ is referred to as the *prism* over G; it can be viewed as consisting of two disjoint copies of G, G_1 and G_2, joined by a perfect matching $M := \{v_1 v_2 \mid v \in V(G)\}$, where v_i is the vertex of copy G_i corresponding to vertex v of G, $i = 1, 2$.

There is a natural bijection between the odd xy-paths of G and the perfect matchings of the graph H derived from the prism over G by deleting the vertices x_2 and y_2 of G_2 that correspond to x and y. Moreover, an odd xy-path of length ℓ in G corresponds to a perfect matching of H that includes exactly $n-1-\ell$ edges of M. Therefore, if each edge of G_i in H, $i = 1, 2$, is assigned weight $n-1$, and each edge of M in H is assigned weight n, where $n := v(G)$, a matching of weight $n(n-1) - \ell$ in H is necessarily a perfect matching of H and corresponds to an odd xy-path of length ℓ in G. Thus a maximum-weight matching in H corresponds to a shortest odd xy-path in G. A shortest even xy-path in G can be found by simply adding a vertex y' and edge yy' to G and finding a shortest odd xy'-path in the resulting graph G'.

4.4. Chromatic number and edge chromatic number

A *k-colouring* of a graph G is a function $c: V(G) \to S$, where S is a k-set; commonly, S is the set $\{1, 2, \ldots, k\}$. The element $c(v)$ is called the *colour* of vertex v, and the sets $C_i := \{v\colon c(v) = i\}$, $i \in S$, the *colour classes* of G with respect to c. A k-colouring of a graph is thus a partition of its vertex set into colour classes. A colouring c is *proper* if each colour class of G with respect to c is a stable set of G; only loopless graphs admit proper colourings, and we implicitly assume here that all graphs are loopless. A graph is *k-colourable* if it admits a proper k-colouring. A 2-colourable graph is one which is bipartite, and thus can be recognized as such in polynomial time (see section 1.6, Search Trees). On the other hand determining whether a graph is k-colourable when $k \geqslant 3$ is an $\mathcal{NP}$-complete problem.

The *chromatic number* of a graph G is the minimum value of k for which G is k-colourable, and is denoted by $\chi(G)$. A graph G is *k-chromatic* if $\chi(G) = k$, and *χ-critical* if $\chi(G \setminus v) < \chi(G)$ for all vertices v of G. Two basic properties of χ-critical k-chromatic graphs are easily established: they are nonseparable and have minimum degree at least $k - 1$.

A graph with chromatic number at least three is nonbipartite and so contains an odd circuit. Erdős and Hajnal (1966) extended this observation by showing that a k-chromatic graph, where $k \geqslant 3$, contains an odd circuit of length at least $k - 1$. A word of caution is in order here. Despite its appearance, the theorem of Erdős and Hajnal is linked only superficially to the chromatic number; indeed, the theorem is an immediate consequence of Theorem 4.13: a k-chromatic graph contains a χ-critical k-chromatic subgraph and thus has minimum degree at least $k - 1$.

A more essential use of the chromatic number was made by Gallai (1968a) and Roy (1967), who discovered a simple relationship between the chromatic number of a digraph and the length of a longest directed path in the digraph, where the *chromatic number* $\chi(D)$ of a digraph D is defined to be the chromatic number of its underlying graph $G(D)$. We denote the number of vertices in a longest directed path of a digraph D by $\lambda(D)$.

Theorem 4.43 (The Gallai–Roy Theorem). *Let D be a digraph with chromatic number χ. Then D contains a directed path of length at least $\chi - 1$. That is,*

$$\lambda \geqslant \chi.$$

Proof. Let D' be a maximal acyclic subdigraph of D, λ' the number of vertices in a longest directed path of D', and $\lambda'(v)$ the number of vertices in a longest directed v-path of D'. Noting that $V(D') = V(D)$, define a λ'-colouring c of D by setting $c(v) := \lambda'(v)$, $v \in V(D')$. We claim that c is a proper colouring of D.

Observe, first, that if there is an (x, y)-path P in D', then $c(x) > c(y)$. For if Q is a y-path in D', PQ is an x-path in D', because D' is acyclic. In particular, if $(x, y) \in E(D')$, $c(x) \neq c(y)$. On the other hand, if $(x, y) \in E(D) \setminus E(D')$, then there is an (y, x)-path in D', by the choice of D', and again $c(x) \neq c(y)$. □

Remark 4.44. The Gallai–Roy Theorem (4.43) can be regarded as a "dual" of the Gallai–Milgram Theorem (4.30), the roles of directed paths and stable sets being interchanged; this duality is well understood in the case of acyclic digraphs (Frank 1980; see also chapter 2), but how far it extends to all digraphs is not clear. The Gallai–Roy Theorem (4.43) can also be viewed as a formula for the chromatic number of a graph in terms of the lengths of longest directed paths in orientations of the graph:

$$\chi(G) = \min \{\lambda(\overrightarrow{G}) : \overrightarrow{G} \text{ an orientation of } G\}.$$

To see this, observe that $\chi(G) \leqslant \lambda(\overrightarrow{G})$ for every orientation $\overrightarrow{G}$ of G, by the Gallai–Roy Theorem (4.43), and that the reverse inequality holds when $\overrightarrow{G}$ is the orientation of G in which u dominates v whenever $c(u) < c(v)$, where $c: V \rightarrow \{1, 2, \ldots, \chi\}$ is any χ-colouring of G (cf. Remark 4.31).

Laborde et al. (1983) proposed the following rather strong refinement of the Gallai–Roy Theorem (4.43).

Conjecture 4.45. Let D be a nontrivial digraph and $\lambda_1 + \lambda_2$ a partition of $\lambda(D)$ into two positive integers. Then there is a partition (V_1, V_2) of $V(D)$ such that $\lambda(D[V_i]) = \lambda_i$, $i = 1, 2$.

The case $\lambda_2 = 1$ of Conjecture 4.45 asserts that in any digraph there is a stable set meeting every longest directed path; applied recursively, this would yield the Gallai–Roy Theorem (4.43). Interestingly, the dual statement is conjectured by Hahn and Jackson (1990) to be false.

Conjecture 4.46. For any positive integer k, there exists a digraph D with stability number k such that $\alpha(D \setminus \cup_{i=1}^{k-1} P_i) = \alpha(D)$ for any family $\{P_1, P_2, \ldots, P_{k-1}\}$ of $k-1$ disjoint directed paths of D.

A generalization of the Gallai–Roy Theorem (4.43) to strong digraphs was obtained, with the aid of the Bypass Lemma (3.10), by Bondy (1976).

Theorem 4.47. *Let D be a strong digraph with chromatic number χ. Then D contains a directed circuit of length at least χ.*

Linial (1981) proposed a common generalization of the Gallai–Roy Theorem (4.43) and the Gallai–Milgram Theorem (4.30). In order to state it, we need two definitions. Let D be a digraph and k a positive integer. A path partition $\mathcal{P}$ of D that minimizes the function

$$\sum_{P \in \mathcal{P}} \min\{v(P), k\}$$

is called *k-optimal*, and the minimum value of this function is denoted by $\pi_k(D)$. In particular, $\pi_1 = \pi$, the minimum number of directed paths in a path partition of D, and $\pi_k = v(D)$ if $k \geqslant \lambda(D)$. A *partial k-colouring* of D is a family $\mathcal{C}$ of

k pairwise-disjoint stable sets of D. A partial k-colouring $\mathscr{C}$ that maximizes the function

$$|\cup_{C\in\mathscr{C}} C|$$

is called *optimal*, and the maximum value of this function is denoted by $\alpha_k(D)$. In particular, $\alpha_1 = \alpha$, the stability number of D, and $\alpha_k = v(D)$ if and only if $k \geqslant \chi(D)$.

Conjecture 4.48. For any digraph D and any positive integer k,

$$\alpha_k \geqslant \pi_k.$$

Conjecture 4.48 was first posed, as a question, by Linial (1981), and was proved by him for acyclic digraphs; it achieved the status of a conjecture with Berge (1982). When $k = 1$, it is the Gallai–Milgram Theorem (4.30); when $k \geqslant \lambda(D)$, it is the Gallai–Roy Theorem (4.43). It is unproved for all other values of k. A still stronger conjecture was proposed by Berge (1982). A path partition $\mathscr{P}$ and a partial k-colouring $\mathscr{C}$ are *orthogonal* if every directed path $P \in \mathscr{P}$ meets $\min\{v(P), k\}$ different colour classes of $\mathscr{C}$.

Conjecture 4.49. Let D be a digraph, k a positive integer and $\mathscr{P}$ a k-optimal path partition of D. Then there is a partial k-colouring of D orthogonal to $\mathscr{P}$.

Like its weaker variant, this conjecture is valid for $k = 1$ (Linial 1978) and for $k \geqslant \lambda(D)$ (Berge 1982). It has also been verified for acyclic digraphs by several authors, some using network flow techniques (see, for example, Aharoni et al. 1985).

The duality between directed paths and stable sets referred to in Remark 4.44 suggests potential duals of Conjectures 4.48 and 4.49. The stronger of these dual conjectures turns out to be false (Aharoni et al. 1985), but the dual of Conjecture 4.48, posed as a question by Linial (1981), is still open. Let D be a digraph and k a positive integer. A *partial k-path partition* of D is a family $\mathscr{P}$ of k pairwise-disjoint directed paths of D. A partial k-path partition $\mathscr{P}$ that maximizes the function

$$|\cup_{P\in\mathscr{P}} V(P)|$$

is called *optimal*, and the maximum value of this function is denoted by $\lambda_k(D)$. In particular, $\lambda_1 = \lambda$, the number of vertices in a longest directed path of D, and $\lambda_k = v(D)$ if and only if $k \geqslant \pi(D)$. A proper vertex colouring $\mathscr{C}$ of D that minimizes the function

$$\sum_{C\in\mathscr{C}} \min\{|C|, k\}$$

is called *k-optimal*, and the minimum value of this function is denoted by $\chi_k(D)$. In particular, $\chi_1 = \chi$, the chromatic number of D, and $\chi_k = v(D)$ if $k \geqslant \alpha(D)$.

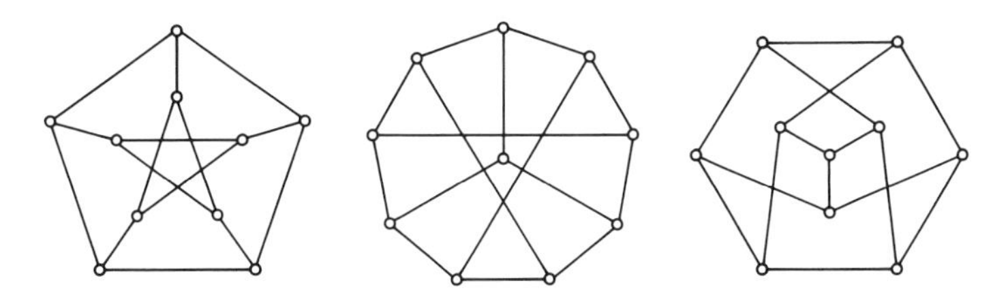

Figure 8. The Petersen graph: three drawings.

Conjecture 4.50. For any digraph D and any positive integer k,

$$\lambda_k \geqslant \chi_k.$$

The restriction of Conjecture 4.50 to acyclic digraphs was verified by Hoffman (1983). When $k = 1$, the conjecture is the Gallai–Roy Theorem (4.43); when $k \geqslant \alpha(D)$, it is the Gallai–Milgram Theorem (4.30). The conjecture is unproved for all other values of k.

We conclude by defining the edge analogue of vertex colouring. A *k-edge colouring* of a graph G is a function $c : E(G) \to S$, where S is a k-set. An edge colouring c is *proper* if each colour class of G with respect to c is a matching of G; only loopless graphs admit proper edge colourings. A graph is *k-edge-colourable* if it admits a proper k-edge colouring. The *edge chromatic number* of a graph G is the minimum value of k for which G is k-edge-colourable, and is denoted by $\chi'(G)$. The edge chromatic number of a bipartite graph is equal to its maximum degree, and thus can be determined in polynomial time. For graphs in general, however, it is an $\mathcal{NP}$-complete problem to determine whether a graph is k-edge-colourable when $k \geqslant 3$. The topic of circuits in edge-coloured graphs is touched upon in Remark 8.39. Edge and vertex colourings are discussed in detail in chapter 4: Colouring, Stable Sets and Perfect Graphs.

4.5. Toughness

We have concentrated up to now on sufficient conditions for the existence of Hamilton circuits. Necessary conditions are unfortunately much harder to come by. A simple one can be derived as follows. Let G be a hamiltonian graph, H a Hamilton circuit in G and S a vertex cut of G. Clearly $H \setminus S$ has at most $|S|$ components. Because $H \setminus S$ is a spanning subgraph of $G \setminus S$, the same is true of $G \setminus S$. Thus $c(G \setminus S) \leqslant |S|$ for every vertex cut S of G. A graph G with this property is called *tough*. Hamiltonian graphs are tough, as we have just observed, but not every tough graph is hamiltonian, the *Petersen graph* of fig. 8 being a counterexample.

The Petersen graph was, in fact, first constructed by Kempe (1886); it was rediscovered twelve years later by Petersen (1898). It is a graph with many remarkable properties. It is not 3-edge-colourable, and is the smallest simple 3-regular graph with this property. This implies that it is nonhamiltonian, since the edges of a

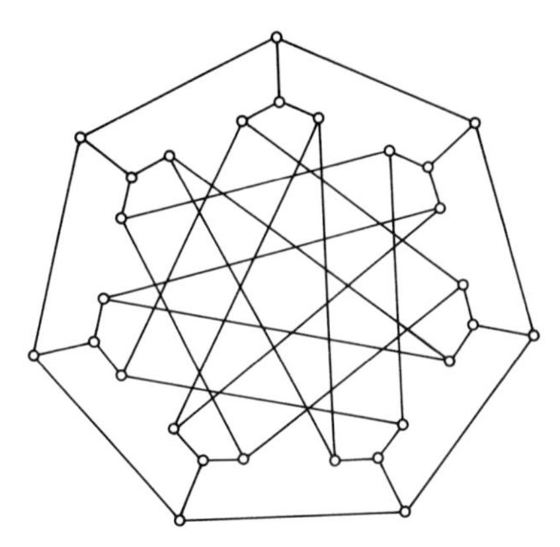

Figure 9. The Coxeter graph.

Hamilton circuit could be properly 2-coloured and the remaining edges assigned a third colour, yielding a proper 3-edge colouring. It is, moreover, maximally nonhamiltonian: the addition of any one edge yields a hamiltonian graph. Another very interesting 3-regular graph is the *Coxeter graph*, discovered by H.S.M. Coxeter (see Tutte 1960) and shown in fig. 9; the beautiful drawing is due to M. Randíc (personal communication 1981). This graph, like the Petersen graph, is not only nonhamiltonian (Tutte 1960) but also maximally nonhamiltonian (Clark and Entringer 1983). In addition, it is tough.

Faced with such examples, Chvátal (1973) was prompted to consider a refinement of the concept of toughness. Let t be a positive real number. A graph G is *t-tough* if

$$c(G \setminus S) \leqslant \frac{|S|}{t}$$

for every vertex cut S of G. In particular, a 1-tough graph is one that is tough. The *toughness* of a graph G is the largest value of t for which G is t-tough. Chvátal (1973) conjectured the existence of a constant t such that every t-tough graph is hamiltonian. Examples constructed by Enomoto et al. (1985) show that, if such a constant t does indeed exist, $t \geqslant 2$. A natural conjecture, then, is the following one.

Conjecture 4.51. Every 2-tough graph is hamiltonian.

This conjecture, in a slightly weaker form (restricted to locally-connected graphs), was formulated by Chvátal (1973), who noted that it would imply a theorem of Fleischner (1974) on squares of graphs.

Let G be a simple graph and m a positive integer. The graph with vertex set $V(G)$ in which two vertices are adjacent if and only if their distance in G is at most m is called the *mth power* of G and denoted by G^m. The *square* of G is the graph G^2, the *cube* of G the graph G^3. A fairly easy induction argument shows that the cube of a connected graph on at least three vertices is hamiltonian (Sekanina 1960). Fleischner proved something much stronger, namely that the square of a 2-connected graph is hamiltonian. Since squares of k-connected graphs are k-tough, Fleischner's theorem is implied by Conjecture 4.51. A simpler proof of Fleischner's theorem was found by Řiha (1991). Underground (1978) noted that the problem of determining whether the square of a graph is hamiltonian is $\mathcal{NP}$-complete.

Win (1989) established a link between toughness and k-trees.

Theorem 4.52. *Let k be a positive integer. Then every $(1/k)$-tough graph has a $(k+2)$-tree.*

A refinement, in terms of k-walks, was conjectured by Jackson and Wormald (1990a).

Conjecture 4.53. Let k be a positive integer. Then every $(1/k)$-tough graph has a $(k+1)$-walk.

Attractive though Conjectures 4.51 and 4.53 are, the insight they would provide is questionable in light of the fact proved by Bauer et al. (1990) that, for any positive constant t, it is $\mathcal{NP}$-hard to determine whether a graph is t-tough; each of these conjectures thus relates two properties whose verification is $\mathcal{NP}$-hard. What can be hoped for, nevertheless, is a polynomial-time algorithm that either finds a Hamilton circuit (respectively, a $(k+1)$-walk) or else exhibits a set S of vertices showing that the input graph is not 2-tough (respectively, not $(1/k)$-tough).

Many sufficient conditions for hamiltonicity can be weakened marginally by imposing a toughness condition (see, for instance, the survey by Bauer et al. 1991). Improvement is possible, of course, only if the extremal graphs for the condition in question are themselves not tough; imposing toughness then serves to exclude them. Häggkvist (1992) achieved more remarkable results by imposing a stronger necessary condition, which one might call path-toughness. A graph G is *path-tough* if

$$\pi(G \setminus S) \leqslant |S|$$

for every nonempty proper subset S of $V(G)$, or equivalently, if each vertex-deleted subgraph is traceable. Hamiltonian graphs are clearly path-tough. Häggkvist's insight was that Dirac's minimum-degree condition can be relaxed substantially for path-tough graphs. His lower bound of $8(n-1)/17$ on the minimum degree was further reduced to $(2n-3)/5$ by Dankelmann et al. (preprint). Schiermeyer (1992) made the following conjecture as to the best possible bound.

Conjecture 4.54. Let G be a simple path-tough graph of minimum degree δ on n vertices, where $\delta \geqslant (n-2)/3$ and $n \geqslant 3$. Then G is either hamiltonian or isomorphic to the Petersen graph.

Relatively little is known about nonhamiltonian path-tough graphs. Two special classes have been studied, however. A graph G is *hypohamiltonian* if every vertex-deleted subgraph $G \setminus v$ is hamiltonian but G is not, and *hypotraceable* if every vertex-deleted subgraph $G \setminus v$ is traceable but G is not. R. Sousselier (see Berge 1963) observed that the Petersen graph (fig. 8) is hypohamiltonian; Bondy (1972) showed likewise that the Coxeter graph (fig. 9) has this property. But there are many other examples. The definitive sources on this topic are Thomassen (1978, 1981a) where a variety of constructions are described, including one for planar hypohamiltonian graphs and another for surprisingly dense hypohamiltonian graphs. Still, a number of basic questions remain. For instance, while it is easy to see that the minimum degree in a hypohamiltonian or hypotraceable graph must be at least three, no example is known in which each vertex is of degree at least four.

Problem 4.55. Is there a hypohamiltonian graph of minimum degree at least four?

This problem was posed by Thomassen (1978), who proved, with the aid of the Bridge Lemma (5.12), that no such graph can be planar. Also worth noting is a question of Grötschel (personal communication 1978) on hypotraceable graphs.

Problem 4.56. Is there a bipartite hypotraceable graph?

The Petersen graph (fig. 8) has girth five, and the Coxeter graph (fig. 9) girth seven. Horák and Širáň (1986) showed how infinite families of maximally nonhamiltonian graphs of girths five and seven can be formed from these graphs, and asked the following question.

Problem 4.57. Is there a maximally nonhamiltonian graph of girth g for all $g \geqslant 3$?

5. Fundamental classes of graphs and digraphs

We now discuss several classes of graphs and digraphs of particular interest: bipartite graphs, planar graphs, regular graphs, vertex-transitive graphs, line graphs, claw-free graphs, oriented graphs and tournaments. As one would expect, stronger results than are valid for graphs in general often hold for members of these classes. In some instances, the improvements are significant.

5.1. Bipartite graphs

By Theorem 1.8, a graph is bipartite if and only if it contains no odd circuits. It is not surprising, therefore, that in problems relating the existence of circuits to vertex degrees, a good rule of thumb is that the conditions can be weakened by half when the graph in question is bipartite. We illustrate this general principle with two typical examples. The first is a bipartite analogue of Theorem 2.15 and a special case of a result of Jackson (1985).

Theorem 5.1. *Let G be a simple nonseparable balanced bipartite graph of minimum degree δ. Then G contains either a circuit of length at least $4\delta - 2$ or a Hamilton circuit.*

Recall that the closure of a graph G on n vertices is the graph derived from G by recursively joining pairs of nonadjacent vertices having degree sum at least n. There is an analogous concept for bipartite graphs. The *bipartite closure* of a balanced bipartite graph $G(X, Y)$ on $2n$ vertices is the graph derived from G by recursively joining pairs of vertices in different parts whose degree sum is at least $n+1$. In section 2, it was proved that a simple graph on at least three vertices is hamiltonian if and only if its closure is hamiltonian (Closure Lemma 2.6). The corresponding result for bipartite graphs can be established by a similar argument (Bondy and Chvátal 1976).

Theorem 5.2. *A simple balanced bipartite graph G is hamiltonian if and only if its bipartite closure is hamiltonian.*

5.2. Planar graphs

A *planar embedding* of a graph G is a function ϕ that assigns to each vertex a point in the plane and to each edge a simple curve in the plane, in such a way that the curves $\phi(e_1)$ and $\phi(e_2)$ meet at a point p if and only if $p = \phi(v)$ for some vertex v which is incident in G to the edges e_1 and e_2. A graph is *planar* if it admits a planar embedding; the Platonic graphs of fig. 1 are the prototypical examples. Note that a graph is planar if and only if it has an embedding in the sphere. A *plane* graph is a pair (G, ϕ) where G is planar and ϕ is a planar embedding of G. The (topological) closures of the connected components of the complement of $\phi(G)$ are the *faces* of G. An edge and face are *incident* if the edge is contained in the face. The edges incident to a face f form a closed walk in which cut edges are each traversed twice. The length of this walk is the *degree* $d(f)$ of f. If the walk is a circuit, it is called a *facial circuit*. Two faces are *adjacent* if they are incident with a common edge; a face which is incident to a cut edge is adjacent to itself. The *boundary* of a face consists of those edges incident with the face and with some other face. The boundary of a face is always an even subgraph.

If G is a plane graph, the *planar dual* of G is the planar graph G^* whose vertices are the faces of G, two vertices being joined by an edge e^* if and only if the corresponding faces of G are incident to a common edge e. The bijection between $E(G)$ and $E(G^*)$ which maps e to e^* has the striking property that it takes circuits to bonds and bonds to circuits. The cycle space of G and the bond space of G^* are therefore isomorphic vector spaces.

In a 2-connected plane graph, the boundary of each face is a circuit, and these facial circuits generate the cycle space; moreover, omitting any one of them results in a cycle basis in which each edge occurs at most twice (because each edge is incident to two faces). MacLane (1937) showed that this property characterizes planar graphs.

Theorem 5.3. *A graph is planar if and only if there is a basis for its cycle space in which each edge occurs at most twice.*

Kuratowski (1930) gave a quite different, and more significant, characterization of planar graphs.

Theorem 5.4 (Kuratowski's Theorem). *A graph is planar if and only if it contains no subdivision of $K_{3,3}$ or K_5.*

Despite these characterizations, and the existence of linear-time algorithms to test for planarity (see chapter 5), planar graphs remain something of an enigma, on account of the notorious Four-Colour Problem. A *k-face colouring* of a plane graph G with face set $F(G)$ is a function $c: F(G) \to S$, where S is a k-set. The face colouring c is *proper* if adjacent faces receive different colours; thus a graph has a proper face colouring if and only if it has no cut edge. A plane graph is *k-face-colourable* if it admits a proper k-face colouring. In 1852, F. Guthrie (see Biggs et al. 1976) asked whether every 2-edge-connected plane graph is 4-face-colourable. This innocent question was the prime motivation for much of the research undertaken in graph theory during the hundred years which followed. Known as the *Four-Colour Problem*, and later as the *Four-Colour Conjecture*, it was given an affirmative answer by Appel and Haken (1976). While the basic ideas of their proof are well understood, it is fair to say that the details are still a matter of some dispute, on account of their complexity.

Theorem 5.5 (The Four-Colour Theorem). *Every 2-edge-connected plane graph is 4-face-colourable.*

By duality, an equivalent statement is that every connected loopless planar graph is 4-vertex-colourable. Tait (1880) derived yet another equivalent formulation of the Four-Colour Conjecture: *every 3-connected 3-regular planar graph is 3-edge-colourable.* This settled the question, he firmly believed, it being widely supposed at the time that every 3-connected 3-regular planar graph was hamiltonian (and hence 3-edge-colourable). However, Tutte (1946) showed the latter assertion to be false. Later, Grinberg (1968) gave a simple necessary condition for a plane graph to be hamiltonian and used it to produce a variety of nonhamiltonian 3-connected 3-regular planar graphs, including the cyclically 5-edge-connected one shown in fig. 10. Ironically, Grinberg's approach was familiar already to Kirkman (1881), but Kirkman, sharing Tait's belief that every 3-connected 3-regular planar graph had at least one Hamilton circuit, employed it merely as an aid in searching for these elusive objects.

Theorem 5.6 (Grinberg's Condition). *Let G be a plane graph on n vertices with a Hamilton circuit C. Denote by f_i' and f_i'' the numbers of faces of G of degree i contained in the interior and exterior of C, respectively. Then*

$$\sum_{i=1}^{n}(i-2)(f_i' - f_i'') = 0.$$

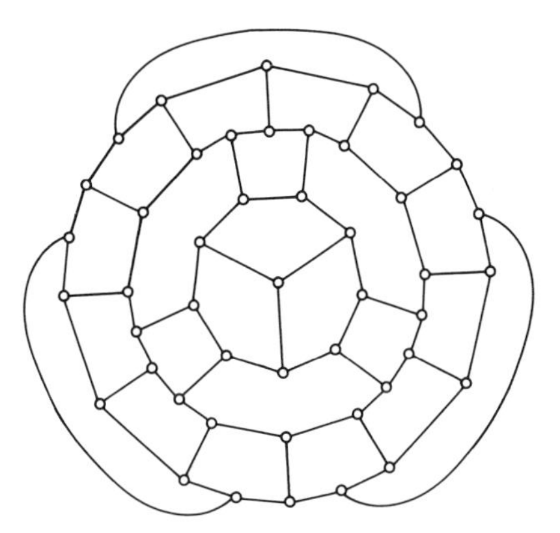

Figure 10. The Grinberg graph.

Proof. Denote by E' the subset of $E(G) \setminus E(C)$ contained in the interior of C. There are $|E'|+1$ faces of G in the interior of C, and so

$$\sum_{i=1}^{n} f_i' = |E'| + 1.$$

Each edge in E' lies on the boundary of two faces in the interior of C, and each edge of C lies on the boundary of one such face. Therefore

$$\sum_{i=1}^{n} i f_i' = 2|E'| + n.$$

Eliminating $|E'|$, we have

$$\sum_{i=1}^{n} (i-2) f_i' = n - 2.$$

Similarly,

$$\sum_{i=1}^{n} (i-2) f_i'' = n - 2.$$

These two equations yield the stated identity. □

To see that the graph of fig. 10 is nonhamiltonian, it suffices to consider the face degrees modulo three and observe that Grinberg's condition cannot hold. It seems hard to construct bipartite counterexamples in this way, and Barnette (1969) has conjectured that there are none. (That the assumption of planarity is necessary here was shown by J.D. Horton; see Bondy and Murty, 1976, p. 240.)

Conjecture 5.7. Every 3-connected 3-regular bipartite planar graph is hamiltonian.

Increasing the connectivity by one, we arrive at a major theorem of Tutte (1956).

Theorem 5.8 (Tutte's Theorem). *Every 4-connected planar graph is hamiltonian.*

Theorem 5.8 is an easy consequence of a lemma about bridges of circuits in planar graphs (the Bridge Lemma, 5.12). Before stating this lemma, we briefly review the important role of bridges in planar graphs. (Bridges were introduced in section 4, before Theorem 4.16.)

A characterization of planar graphs in terms of their bridge structure was given by Tutte (1959). This characterization is the basis of most planarity-testing algorithms. Let G be a graph and C a circuit of G. Two bridges B_1 and B_2 of C in G *overlap* if C cannot be divided into two segments C_1 and C_2 with the property that the vertices of attachment of B_i to C lie on C_i, $i = 1, 2$. The *overlap graph* of G with respect to C has as vertices the bridges of C in G, two bridges being adjacent if they overlap.

Theorem 5.9. *A graph G is planar if and only if, for every circuit C of G, the overlap graph of G with respect to C is bipartite.*

A circuit C in a graph G is *nonseparating* if it has at most one proper bridge, and *peripheral* if it is nonseparating and has no chords. Tutte (1963) showed that in a 3-connected plane graph a circuit is peripheral if and only if it bounds a face. It follows that a 3-connected planar graph has a unique plane embedding (in the sense that there is a unique planar dual) and that its peripheral circuits generate the cycle space. In fact, the latter assertion is true even for nonplanar graphs, as Tutte (1963) proved.

Theorem 5.10. *In a 3-connected graph the peripheral circuits generate the cycle space.*

An insightful exposition of the theory of bridges, leading to a refinement of his original proof of Theorem 5.8, is given by Tutte (1977). Bridges clearly play a very important role in the study of paths and circuits, and it can be argued that their role is central. This is the point of view adopted by Voss (1991) in his treatise on circuits. Of relevance to the study of long circuits is the concept of the span of a bridge, introduced by Voss (1991). Let G be a graph, H a subgraph of G and B a bridge of H in G. The *span* of B (Voss uses the term *length*) is the maximum number of edges in a tree of B whose end vertices are the vertices of attachment of B to H. Voss (1991) made the following conjecture, and proved it for $m \leqslant 3$.

Conjecture 5.11. Let G be a 2-connected graph, C a longest circuit of G and B_i, $1 \leqslant i \leqslant m$, disjoint bridges of C which induce a tree in the overlap graph of G with respect to C. Then

$$\sum_{i=1}^{m} s(B_i) \leqslant \ell(C)/2,$$

where $s(B_i)$ denotes the span of B_i and $\ell(C)$ the length of C.

We now return to the statement of the Bridge Lemma.

Lemma 5.12 (The Bridge Lemma). *Let G be a 2-connected plane graph, e an edge of G, C' and C'' the facial circuits of G which include e, and e' any edge of C'. Then there is a circuit C in G including both e and e' such that*

(i) *each bridge of C in G has either two or three vertices of attachment;*

(ii) *each bridge of C in G that includes an edge of C' or C'' has exactly two vertices of attachment.*

Let G be a simple 4-connected plane graph, and define e, C', C'' and e' as in the Bridge Lemma. Then G has a circuit C which includes e and e' and satisfies conditions (i) and (ii). This circuit is necessarily a Hamilton circuit of G. Suppose, to the contrary, that C has a proper bridge. By (i), each bridge of C has at most three vertices of attachment. Because G is 4-connected, C must be a triangle and have exactly one proper bridge B, with three vertices of attachment. Moreover, because G is simple, C can have no degenerate bridge. It follows that $C = C'$ and that B includes an edge of C''. But this contradicts (ii). Thus every 4-connected planar graph is hamiltonian (Theorem 5.8). Thomassen (1983b) obtained an analogue (and strengthening) of the Bridge Lemma for bridges of paths in plane graphs, and deduced that every 4-connected planar graph is Hamilton-connected. A related class of 4-connected graphs are the graphs of simple 4-polytopes. D.W. Barnette (see Grünbaum 1970, p. 1145) has conjectured that these too are hamiltonian.

While they need not be hamiltonian, 3-connected planar graphs always have long circuits. Jackson and Wormald (1992a) employed the Bridge Lemma (5.12) to prove that every 3-connected planar graph on n vertices has a circuit whose length is of order at least n^c, where $c \approx 0.2$. Examples due to Moon and Moser (1963), constructed from K_4 by recursively inserting a vertex of degree three in each face, show on the other hand that such a graph need not have any circuit whose length is of order more than n^c, where $c := \log 2/\log 3 \approx 0.63$.

Another property enjoyed by 3-connected planar graphs was discovered by Barnette (1966), who proved that every 3-connected planar graph has a 3-tree. A stronger conjecture, proposed by Jackson and Wormald (1990a), was confirmed by Gao and Richter (1994).

Theorem 5.13. *Every 3-connected planar graph has a 2-walk.*

Properties of graphs embeddable in the plane and other surfaces are discussed in chapter 5: Embeddings and Minors.

5.3. Regular graphs and vertex-transitive graphs

Regular graphs are a particularly fascinating class. The classical examples are the graphs of the Platonic solids (fig. 1). Regular graphs of degree three are often referred to as *cubic* graphs. They appear to capture much of the complexity of graphs in general, and interesting examples abound, the Heawood graph (fig. 4), the Petersen graph (fig. 8), and the Coxeter graph (fig. 9) being three notable ones.

Nonhamiltonian d-regular d-connected graphs exist for all $d \geqslant 3$. A construction, based on the Petersen graph, was first given by Meredith (1973); moreover, an infinite family of such graphs in which the longest circuit is of length no more than n^c, where n is the order and c (< 1) depends only on d, was described by Jackson and Parsons (1982). Nevertheless, as we have remarked, the imposition of regularity often has surprising consequences. A convincing illustration is provided by the following theorem of C.A.B. Smith (see Tutte 1946) on the parity of the number of Hamilton circuits in a 3-regular graph.

Theorem 5.14 (Smith's Theorem). *Let G be a 3-regular graph and e an edge of G. Then the number of Hamilton circuits of G containing e is even.*

Smith's Theorem (5.14) implies that a 3-regular graph with one Hamilton circuit has at least three such circuits. But, as we shall see, the proof is not constructive, and Chrobak and Poljak (1988) pose the intriguing problem of efficiently finding a second Hamilton circuit in such a graph. The proof we give makes use of the concept of an even 2-factor. A *d-factor* of a graph is a d-regular spanning subgraph. A 2-factor is *even* if each of its components is an even circuit.

Proof. Consider the bipartite graph $H(X, Y)$, where X is the set of even 2-factors of G containing e, Y is the set of 3-edge colourings of G, and an even 2-factor F containing e is joined to a 3-edge colouring (E_1, E_2, E_3) if and only if $F = E_i \cup E_j$ for some $\{i, j\} \subset \{1,2,3\}$. A vertex F of X with c components has degree 2^{c-1} in H, and each vertex of Y has degree two. Thus the vertices of H of odd degree are precisely the Hamilton circuits of G containing e. By Corollary 1.2, their number is even. □

The hamiltonian problem remains $\mathcal{NP}$-complete even when restricted to 3-connected 3-regular graphs. Nevertheless, almost all such graphs are indeed hamiltonian, by a theorem of Robinson and Wormald (1992), who also established the following more general result (Robinson and Wormald 1991).

Theorem 5.15. *For every integer $d \geqslant 3$, almost all d-regular graphs are hamiltonian.*

Surprisingly, planarity changes the picture completely, as was shown by Richmond et al. (1985).

Theorem 5.16. *Almost all 3-connected 3-regular planar graphs are not hamiltonian.*

Theorems 5.15 and 5.16 are probabilistic statements. For their precise meaning, we refer the reader to chapter 6: Random Graphs, or to the book of Bollobás (1985).

Jackson (1980a) obtained an analogue of Dirac's Theorem (2.1) for regular graphs. The effect of regularity here is quite striking.

Theorem 5.17 (Jackson's Theorem). *Let G be a 2-connected d-regular graph on n vertices, where $d \geqslant n/3$. Then G is hamiltonian.*

The bound on d, already close to best possible, was reduced to $(n/3) - 1$ for all $d \geqslant 3$ by Hilbig (1986); there are two exceptions, the Petersen graph (fig. 8) and the graph obtained from it by *expanding* one vertex *to a triangle* (that is, deleting the vertex, adding a triangle, and joining the three former neighbours of the vertex to the three vertices of the triangle by a matching).

Jackson's Theorem (5.17) indicates that regular graphs can be expected to contain considerably longer paths and circuits than graphs of the same minimum degree, and one may ask how long a path or circuit there must be in a 2-connected d-regular graph on n vertices. The following conjecture, due to Bondy (1978b), includes Jackson's Theorem (5.17) as a special case.

Conjecture 5.18. Let G be a 2-connected d-regular graph on n vertices, where $d \geqslant n/r$, $r \geqslant 3$, and n is sufficiently large. Then G contains a circuit of length at least $2n/(r-1)$.

This conjecture is of interest only when r is small compared to n; when r is of order n, regularity plays essentially no role and can be replaced by a condition on the maximum degree. Voss (1991) showed that a 2-connected graph of maximum degree Δ on n vertices, where $3 \leqslant \Delta \leqslant n-2$, has a circuit of length at least $4 \log_{\Delta-1} n - c$, where c is an appropriate constant, while Lang and Walther (1968) constructed regular examples containing no circuits of length more than $4 \log_{\Delta-1} n + 4$. For bipartite graphs, we have the following theorem of Jackson and Li (1994); Häggkvist (1978a) conjectures that the bound on d can be reduced to $n/6$.

Theorem 5.19. *Let G be a 2-connected d-regular bipartite graph on n vertices, where $d \geqslant (n/6) + 8$. Then G is hamiltonian.*

As noted earlier, the passage from 2-connected to 3-connected graphs often has a substantial effect on circuit structure. Regular graphs are no exception. Fraisse (1986) conjectures that the lower bound on the degree in Jackson's Theorem (5.17) can be reduced significantly for 3-connected regular graphs. The following slightly sharper formulation of Fraisse's conjecture is due to B. Jackson (personal communication 1989).

Conjecture 5.20. Let G be a 3-connected d-regular graph on n vertices, where $d \geqslant n/4$. Then G is hamiltonian unless G is either the Petersen graph (fig. 8) or the graph obtained from it by expanding one vertex to a triangle.

Broersma et al. (preprint) showed that, if $d \geqslant (n+3)/4$, such a graph has a dominating circuit. Using this fact, they proved that every 3-connected d-regular graph on n vertices, where $d \geqslant (2n/7) + 2$, is hamiltonian. Surprisingly, raising the connectivity further has little effect. For $d \equiv 0 \pmod 4$, B. Jackson (see Jackson et al. 1991) and H.A. Jung (see Min Aung 1989) independently found an example of a nonhamiltonian $(d/2)$-connected d-regular graph on $4d+1$ vertices.

Of relevance to Conjecture 5.20 is the following theorem of Fan (1985) and Jung (1984).

Theorem 5.21. *Let G be a 3-connected d-regular graph. Then G contains either a circuit of length at least 3d or a Hamilton circuit.*

When d is small, this theorem is weak. Jackson (1986) proved that a 3-connected 3-regular graph on n vertices contains a circuit whose length is of order at least n^c, where $c := \log_2(1+\sqrt{5}) - 1 \approx 0.69$. Examples based on the Petersen graph (fig. 8) show, on the other hand, that such a graph need not have any circuit whose length is of order more than n^c, where $c := \log 8/\log 9 \approx 0.96$. However, these constructions work because the resulting graphs are not cyclically 4-edge-connected; they have many 3-edge cuts, which serve to restrict the possible routes taken by a circuit. The following conjecture is due to the author (see Jackson 1986), and is implied by successively stronger conjectures of Thomassen (1986), Matthews and Sumner (1984) and Jackson (1992) (see section 5.4, below).

Conjecture 5.22. Let G be a 3-regular cyclically 4-edge-connected graph on n vertices. Then G contains a circuit of length at least cn, where c is a positive absolute constant.

The question of circuits through specified vertices in regular graphs was considered by Holton et al. (1982) and by Kelmans and Lomonosov (1982a). Both groups of authors proved the following theorem on 3-regular graphs.

Theorem 5.23. *Let G be a 3-connected 3-regular graph, and let S be a set of nine vertices of G. Then there is a circuit of G that includes every vertex of S.*

The Petersen graph (fig. 8) is 3-connected and 3-regular but has no circuit including all ten vertices. Ellingham et al. (1984) showed that any 3-connected 3-regular graph having a set S of ten vertices not lying on a common circuit necessarily contains a subdivision of the Petersen graph in which S is the set of vertices of degree three (see, also, Kelmans and Lomonosov 1982b). One may ask the following general question.

Problem 5.24. Does there exist a constant $c > 1$ such that, for every d-regular d-connected graph G and for every set S of $\lfloor cd \rfloor$ vertices of G, where $d \geqslant 2$, there is a circuit of G that includes every vertex of S?

Regular digraphs, also, have been studied with regard to circuit structure. For instance, Thomassen (1981b) generalized Corollary 4.18 to digraphs.

Theorem 5.25. *Let D be a strict d-diregular digraph on $2d+1$ vertices, where $d \geqslant 1$. Then either D has a directed Hamilton circuit or D is isomorphic to one of two exceptional digraphs of orders five and seven.*

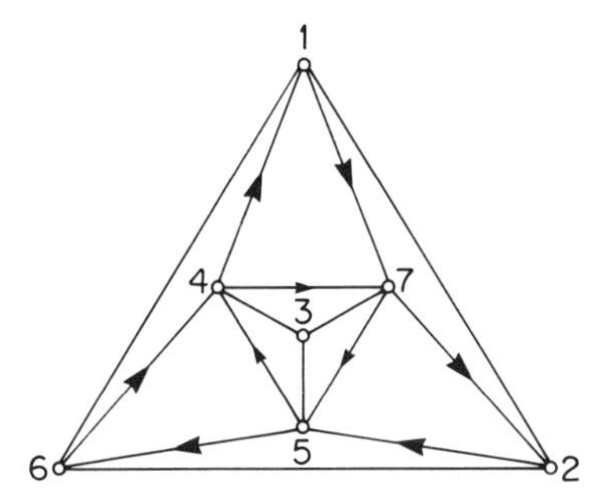

Figure 11. A digraph analogue of the Petersen graph.

The larger of the two exceptional digraphs in Theorem 5.25 is shown in fig. 11. Found by Bondy (1978b), it has several properties reminiscent of the Petersen graph (fig. 8), being 3-diregular, maximally nonhamiltonian, and also hypohamiltonian. Moreover, as R. Häggkvist (personal communication 1978b) observed, the sets of outneighbours of its seven vertices constitute the edges of the Fano hypergraph (fig. 3).

In this chapter, we have chosen to concentrate on longest circuits in graphs. It is appropriate here, however, to say a word about the girth of regular graphs. Regular graphs of diameter k and girth $2k+1$ are called *Moore graphs*, regular graphs of given girth with the minimum number of vertices *cages*. The Petersen graph (fig. 8) is a Moore graph and also a cage; the Heawood graph (fig. 4) is a cage, too. In general, members of both these classes tend to have much symmetry, and are mainly studied by algebraic methods. We refer the reader to chapter 31 or to the survey by Wong (1982) for further details.

The directed analogue of the cage appears to have a far simpler structure. For $d \geqslant 1$ and $k \geqslant 2$, consider the digraph whose vertices are the integers modulo $d(k-1)+1$, (i,j) being an edge if and only if $1 \leqslant j-i \leqslant d$. This digraph is d-diregular and has girth k. Behzad et al. (1970) conjectured that no digraph with these properties has fewer vertices.

Conjecture 5.26. Let D be a strict d-diregular digraph on n vertices. Then D has girth at most $\lceil n/d \rceil$.

We conclude this section by noting that several of the results mentioned here do not depend on strict regularity for their conclusions but rather on bounds on the minimum and/or maximum degrees. For instance, Jackson (1980a) gave an extension of Theorem 5.17 to almost-regular graphs, and Jackson and Wormald (1993) generalized the theorem of Jackson (1986) on longest circuits in 3-connected 3-regular graphs to graphs of arbitrary maximum degree as follows.

Theorem 5.27. *Let G be a 3-connected graph of maximum degree Δ on n vertices. Then G has a circuit of length at least $n^c/2$, where $c := (1+\log_2 3 + 2\log_2 \Delta)^{-1}$.*

Caccetta and Häggkvist (1978) proposed a generalization of Conjecture 5.26.

Conjecture 5.28. Let D be a strict digraph of minimum outdegree δ^+ on n vertices. Then D has girth at most $\lceil n/\delta^+ \rceil$.

This conjecture has been verified for $\delta^+ \leqslant 5$ (see Hoáng and Reed 1987) and a near-optimal upper bound, of $(n/\delta^+) + 2500$, was established by Chvátal and Szemerédi (1983); the constant has since been reduced, but not eliminated. Surprisingly, the case when $\delta^+ \geqslant n/3$ is still unresolved.

An important subclass of regular graphs, with correspondingly stronger properties, are those that are vertex-transitive. An *automorphism* of a graph is an isomorphism from the graph to itself. A graph is *vertex-transitive* if its automorphism group is transitive on its vertices. Vertex-transitive graphs can be constructed in a variety of ways. Two particularly interesting families are the Kneser graphs and the Levi graphs.

Let m be a positive integer and S an n-set, where $n > 2m$. The simple graph whose vertices are the m-subsets of S, two being adjacent if they are disjoint, is called a *Kneser graph* and denoted by $\mathrm{KG}(m,n)$. Such graphs, first defined by Kneser (1955), are clearly vertex-transitive. The Petersen graph (fig. 8) is $\mathrm{KG}(2,5)$; the Coxeter graph (fig. 9) can be derived from $\mathrm{KG}(3,7)$ by deleting the seven edges of a Fano hypergraph (fig. 3).

Let $\mathrm{PG}(2,n)$ be a finite projective plane of order n, with point set P and line set L. The incidence graph of the hypergraph with vertex set P and edge set L is called the *Levi graph* of $\mathrm{PG}(2,n)$ and denoted by $\mathrm{LG}(2,n)$; the Heawood graph (fig. 4) is $\mathrm{LG}(2,2)$. If $n = q$, a prime power, the Levi graph $\mathrm{LG}(2,q)$ is vertex-transitive (see, for example, Biggs and White 1979).

Lovász (1970) considered the hamiltonicity of vertex-transitive graphs, and asked the following question.

Problem 5.29. Does every connected vertex-transitive graph have a Hamilton path?

This is so for graphs on p, $2p$, $3p$, $4p$, $5p$, p^2, $2p^2$ and p^3 vertices, where p is prime (see Alspach 1981). In fact, besides K_1 and K_2, only four simple connected vertex-transitive graphs without Hamilton circuits are known: the Petersen graph (fig. 8), the Coxeter graph (fig. 9), and the 3-regular graphs derived from these by expanding each vertex to a triangle. C. Thomassen (see Bermond 1978b) conjectures that there are only finitely many such graphs. Babai (1979) has shown that every connected vertex-transitive graph on n vertices has a circuit whose length is of order at least $n^{1/2}$.

Groups give rise to vertex-transitive graphs in a natural way. Let Γ be a group and X a set of elements of Γ. The digraph $D(\Gamma,X)$ with vertex set Γ and edge set $\{(g,gx)\colon g \in \Gamma,\ x \in X\}$ is the *Cayley digraph* of Γ with respect to X. If $X^{-1} = X$, $D(\Gamma,X)$ is the associated digraph of a simple graph $G(\Gamma,X)$, called the *Cayley graph* of Γ with respect to X. The graph $G(\Gamma,X)$ and digraph $D(\Gamma,X)$ are loopless if and only if $1 \notin X$ and connected if and only if X generates Γ. When Γ is a cyclic group, the resulting Cayley graph or digraph is called a *circulant* or *directed circulant*, respectively. The Koh digraph (fig. 2) is a directed circulant on the group $\Gamma := \mathscr{Z}_7$ with generating set $X := \{1,5\}$. T.D. Parsons (see Witte and Gallian 1984) proposed the following conjecture.

Conjecture 5.30. Every Cayley graph has a Hamilton circuit.

A review of progress on this conjecture is given in the survey by Witte and Gallian (1984). Vertex-transitive graphs and other highly symmetric graphs are discussed in chapter 27.

5.4. Line graphs and claw-free graphs

The *line graph* of a graph G is the graph with vertex set $E(G)$, two vertices being adjacent if and only if they are adjacent (as edges) in G. (Line graphs should more properly be called *edge graphs*, but the name is by now too well entrenched to be changed.) The line graph of G is denoted by $L(G)$.

The question of hamiltonicity of line graphs is intimately related to the notion of an *Euler tour*, a closed trail that includes every edge of the graph. A graph is *eulerian* if it admits an Euler tour. Such graphs were characterized, not surprisingly, by Euler (1736).

Theorem 5.31. *A graph admits an Euler tour if and only if it is connected and even.*

A proof of Theorem 5.31 was given only much later, by Hierholzer (1873). Hierholzer's proof yields a linear-time algorithm for finding an Euler tour in an eulerian graph. Harary and Nash-Williams (1965) made the key observation linking Hamilton circuits in line graphs to Euler tours.

Proposition 5.32. *Let G be a graph on at least three edges. Then $L(G)$ is hamiltonian if and only if G has a dominating eulerian subgraph.*

Jaeger (1979) established a sufficient condition for a graph to contain a spanning (and hence dominating) eulerian subgraph.

Theorem 5.33. *Every 4-edge-connected graph has a spanning eulerian subgraph.*

As Thomassen (1986) observed, it follows from Theorems 5.32 and 5.33 that line graphs of 4-edge-connected graphs are hamiltonian. By similar methods, Zhan (1991) showed that 7-connected line graphs are Hamilton-connected. These results lend support to the following conjecture of Thomassen (1986).

Conjecture 5.34. Every 4-connected line graph is hamiltonian.

A basic property of line graphs is that they contain no induced subgraph isomorphic to $K_{1,3}$. Such a subgraph is referred to as a *claw*, and a graph which contains no claw is called *claw-free*. The original motivation for studying claw-free graphs was that they include line graphs as a proper subclass; Minty (1980) and Sbihi (1980) described polynomial-time algorithms for finding a maximum stable set in a claw-free graph, thereby extending Edmonds's maximum matching algorithm (1965), and a third such algorithm can be found in Lovász and Plummer (1986, pp. 471–480). The evidence now, however, is that claw-free graphs are of interest in their own right.

In a claw-free graph, the subgraph induced by the neighbours of a vertex has stability number one or two. If this subgraph is disconnected, it necessarily consists of two disjoint cliques; if it is connected, it contains a Hamilton path by Corollary 4.17. Claw-free graphs therefore include two quite distinct subclasses: those in which each neighbourhood induces a disconnected subgraph (for instance, line graphs of triangle-free simple graphs of minimum degree at least two), and those in which each neighbourhood induces a connected subgraph (the octahedron and icosahedron of fig. 1 being two examples). Graphs with this latter property are called *locally-connected*.

Graphs in the former class do not necessarily have Hamilton circuits; indeed, if G is an arbitrary simple 3-regular graph, the graph H obtained from G by expanding each vertex to a triangle is hamiltonian if and only if G is hamiltonian; moreover, G and H have the same connectivity. Since it is $\mathcal{NP}$-hard to determine whether an arbitrary 3-connected 3-regular graph is hamiltonian, we deduce that it is $\mathcal{NP}$-hard to determine whether an arbitrary 3-connected 3-regular claw-free graph is hamiltonian. On the other hand, Oberly and Sumner (1979) proved that every connected, locally-connected claw-free graph is hamiltonian.

One may ask how long a circuit a k-connected claw-free graph on n vertices must contain. For $k = 2$, an upper bound can be derived from the work of Lang and Walther (1968), who constructed 2-connected 3-regular graphs on n vertices with no circuit of length more than $4 \log_2 n + 4$ (see the discussion following Conjecture 5.18). By expanding each vertex to a triangle, one obtains 2-connected claw-free graphs on $3n$ vertices with no circuit of length more than $12 \log_2 n + 12$. Jackson and Wormald (personal communication 1992b) proved that this lower bound is of the right order of magnitude. They also obtained an analogous result for 3-connected claw-free graphs.

Theorem 5.35. *Let G be a 3-connected claw-free graph on n vertices. Then G contains a circuit of length at least n^c, where c is a positive constant.*

A similar construction shows that this theorem is sharp, too, apart from the value of the constant c. The case of 4-connected graphs was addressed by Matthews and Sumner (1984), who proposed the following conjecture, a strengthening of Conjecture 5.34.

Conjecture 5.36. Every 4-connected claw-free graph is hamiltonian.

A still stronger conjecture was put forward by Jackson (1992). A *Tutte circuit* is a circuit of length at least four each of whose bridges has at most three vertices of attachment. The Bridge Lemma (5.12) guarantees the existence of such circuits in 2-connected planar graphs on at least four vertices. Jackson conjectures their existence in 2-connected claw-free graphs as well.

Conjecture 5.37. Every 2-connected claw-free graph on at least four vertices contains a Tutte circuit.

A similar strengthening of Theorem 5.33 was also proposed by Jackson (1992).

Conjecture 5.38. Every simple 2-edge-connected graph on at least four edges contains a connected even subgraph on at least four edges each of whose bridges has at most three edges of attachment.

Conjecture 5.38 would imply that every 2-connected 3-regular graph on at least four vertices has a Tutte circuit, and this would imply the truth of Conjecture 5.22 (with $c = 3/4$).

Li (1988) proved that the bound on the degree of regularity in Jackson's Theorem (5.17) can be reduced even further when the graphs are assumed also to be claw-free.

Theorem 5.39. *Let G be a 2-connected d-regular claw-free graph on n vertices, where $d \geqslant n/5$ and $d \geqslant 10$. Then G is hamiltonian.*

5.5. Oriented graphs and tournaments

Whereas two edges (one in each direction) may link a pair of vertices in a strict digraph, at most one edge is permitted in an oriented graph. For this reason, and by analogy with the situation that obtains in bipartite graphs, one might expect conditions for the existence of directed paths and circuits in oriented graphs to be roughly half as demanding as their digraph counterparts. This is sometimes the case, but there are exceptions. Moreover, best possible results are much harder to come by. Jackson (1980b) showed that an oriented graph with minimum indegree and outdegree k has a directed path of length at least $2k$. However, the corresponding question for directed circuits, also considered by Jackson (1980b), remains unresolved.

Conjecture 5.40. Let D be a strong oriented graph with minimum indegree and outdegree at least k. Then D has a directed circuit of length at least $2k+1$.

This conjecture is sharp when k is small compared to the order of D, but not when k is large. Häggkvist (1993) proved that an oriented graph on n vertices with minimum indegree and outdegree at least cn, where $c := (1/2) - 2^{-18} \approx 0.499996$, has a directed Hamilton circuit, and conjectured that a much smaller constant will do.

Conjecture 5.41. Let D be an oriented graph of minimum outdegree δ^+ on n vertices, where $\delta^+ \geqslant (3n-2)/8$. Then D contains a directed Hamilton circuit.

An analogous conjecture for diregular oriented graphs was proposed by Jackson (1981b).

Conjecture 5.42. Every k-diregular oriented graph on at most $4k+1$ vertices, where $k \neq 2$, contains a directed Hamilton circuit.

Evidently, questions about oriented graphs are much easier to come by than are answers, and we mention here just one more. Due to Thomassen (1987a), it concerns a property which resembles maximal nonhamiltonicity but is meaningful only for oriented graphs.

Problem 5.43. Is there a nonhamiltonian oriented graph in which the reversal of any edge results in a hamiltonian graph?

Remark 5.44. There do exist oriented graphs with the slightly weaker property that reversing any edge increases the length of a longest directed circuit; one such family, found by Thomassen (1987b), are the oriented graphs $\boldsymbol{C}_5 \Box \boldsymbol{C}_{10k+7}$, $k \geqslant 0$.

While oriented graphs in general are poorly understood, the theory of tournaments is rich and highly developed, as the beautiful theorems of Rédei (3.4), Thomason (3.7) and Forcade (3.8) on Hamilton paths and circuits in tournaments demonstrate. Two directed analogues of Hamilton-connectedness were considered by Thomassen (1980a).

A digraph is *weakly Hamilton-connected* if any two vertices are connected by a directed Hamilton path (in at least one direction) and *strongly Hamilton-connected* if any two vertices are connected by directed Hamilton paths in both directions. Thomassen (1980a) characterized the weakly Hamilton-connected tournaments and proved that 4-connected tournaments are strongly Hamilton-connected. He also showed that each edge of a 3-connected tournament lies in a directed Hamilton circuit. Thomassen's methods were adapted by Bang-Jensen et al. (1992), resulting in a polynomial-time algorithm for finding, in a tournament, a directed Hamilton path with prescribed head and tail, if there is one. In particular, the algorithm can be applied to determine whether a tournament is weakly or strongly Hamilton-connected and whether there is a directed Hamilton circuit through a prescribed edge in a tournament. Bang-Jensen et al. (1992) also made the following conjecture.

Conjecture 5.45. For each fixed k, there exists a polynomial-time algorithm for deciding if there is a directed Hamilton circuit through k prescribed edges in a tournament.

Bang-Jensen and Thomassen (1992) proved that the above problem is $\mathcal{NP}$-complete if k is not fixed, and they described a polynomial-time algorithm for finding a directed circuit in a tournament through two prescribed edges, if there is one. This latter problem is $\mathcal{NP}$-complete for digraphs in general, by a result of Fortune et al. (1980).

We close this brief discussion of oriented graphs by defining the bipartite analogue of a tournament and noting a bipartite analogue of Camion's Theorem (3.9) due to Gutin (1984) and, independently, Häggkvist and Manoussakis (1989). A *bipartite tournament* is an orientation of a simple complete bipartite graph. A *directed d-factor* of a digraph is a d-diregular spanning subgraph.

Theorem 5.46. *A strong bipartite tournament has a directed Hamilton circuit if and only if it has a directed* 1*-factor.*

The proof by Häggkvist and Manoussakis (1989) of Theorem 5.46 makes use of a neat trick. The bipartite tournament $T(V_1, V_2)$ is transformed into a 2-edge-coloured complete graph G on $V_1 \cup V_2$ by assigning colour i to each edge of $G[V_i]$ and to each edge of T whose tail lies in V_i, $i = 1, 2$. Noting now that T has a directed Hamilton circuit if and only if G has a Hamilton circuit whose edges alternate in colour, Häggkvist and Manoussakis (1989) apply a necessary and sufficient condition due to Bánkfalvi and Bánkfalvi (1968) for a 2-edge-coloured complete graph to possess such a circuit.

6. Special proof techniques for paths and circuits

Thus far, a variety of proof techniques have been introduced. Some, such as counting in two ways (Proposition 1.1), induction (Proposition 1.5), contradiction (Theorem 1.7), search trees (Theorem 1.8), maximal counterexample (Theorem 2.12), and inclusion–exclusion (Theorem 3.4) are of general applicability, while others are expressly designed to handle questions on paths and circuits. Among the latter are the use of maximal paths (Theorem 1.3), longest circuits (Theorem 2.1, Proof 1), longest paths (Theorem 2.1, Proof 2), lollipops (Theorem 2.1), the closure (Theorem 2.7), bypasses (Theorem 3.9) and vines (Theorem 4.11). In this section, we discuss three additional proof techniques specific to paths and circuits.

6.1. Thomason's lemma

Let G be a graph and x a vertex of G. Recall that an *x-path* is a path in G with tail x. If P is an x-path in G whose head y is adjacent to a vertex z of P other than its immediate predecessor on P, then $P + yz$ is an (x, z)-lollipop and thus, by the Lollipop Lemma (2.2), contains a second spanning x-path Q. The x-paths P and Q are said to be related by an *elementary exchange*. The graph whose vertices are the longest x-paths of G, two such paths being adjacent if and only if they are related by an elementary exchange, is called the *x-path graph* of G. Thomason (1978) proved a basic lemma on x-paths with the aid of this concept.

Lemma 6.1 (Thomason's Lemma). *Let G be a connected graph on at least two vertices, and let x be a vertex of G. Then the number of longest x-paths of G that terminate in a vertex of even degree is even.*

Proof. Let H be the x-path graph of G and P a longest x-path of G. If P terminates in y,

$$d_H(P) = d_G(y) - 1.$$

Thus y is of even degree in G if and only if P is of odd degree in H. Since the number of vertices of odd degree in H is even, by Corollary 1.2, the number of longest x-paths in G that terminate in a vertex of even degree is also even. □

Corollary 6.2. *Let G be a graph on at least three vertices, and let x and y be two vertices of G. If each vertex of G other than x and y is of odd degree, the number of Hamilton xy-paths in G is even.*

Proof. We may assume that G has at least one Hamilton xy-path; otherwise, the conclusion is trivial. The longest x-paths of $G \setminus y$ are then Hamilton paths of $G \setminus y$, and each Hamilton xy-path of G is an extension of such a path. Let P be a longest x-path of $G \setminus y$ terminating in z. If z is of odd degree in $G \setminus y$, the number of edges between y and z is even, since z is of odd degree in G. Thus P gives rise to an even number of Hamilton xy-paths in this case. On the other hand, if z is of even degree in $G \setminus y$, then P gives rise to an odd number of Hamilton xy-paths; by Thomason's Lemma (6.1), however, the number of such paths P is even. Hence the total number of Hamilton xy-paths in G is also even. □

Smith's Theorem (5.14) is an immediate consequence of Corollary 6.2; it suffices to choose u and v as the ends of the given edge e. Thomason (1978) also applied his Lemma (6.1) to derive a generalization of the theorem of J. Bosák and A. Kotzig (see Bosák 1967) asserting that every 3-regular bipartite graph of order at least four has an even number of Hamilton circuits.

The x-path graph has other applications. Hendry (1984, 1986) considered the class of graphs in which there is a unique Hamilton path from one vertex x to every other vertex. He showed that the x-path graph of such a graph is a forest with $d(x)/2$ components, and deduced that every such graph on n vertices has precisely $3(n-1)/2$ edges.

6.2. Pósa's lemma

The naive approach to finding a long path in a graph is to grow a maximal path P and consider the paths obtainable from P by means of elementary exchanges (applied at either end); if one of these is not maximal, it can be extended to a maximal path longer than P. The procedure may then be repeated. While this approach fails badly on certain graphs, Pósa (1976) proved that it works remarkably well on most graphs of sufficiently high edge density. His argument hinges on the following lemma.

Lemma 6.3 (Pósa's Lemma). *Let G be a graph, P a longest path in G and $\mathcal{P}$ the set of all paths of G obtainable from P by elementary exchanges. Denote by S the* *set of ends of paths in $\mathcal{P}$ and by S^- and S^+ the sets of vertices immediately preceding and following the vertices of S on P, respectively. Then*

$$N(S) \subseteq S^- \cup S \cup S^+.$$

Proof. Let $x \in S$ and $y \in V(G) \setminus (S^- \cup S \cup S^+)$, and consider an x-path Q in $\mathcal{P}$. If $y \in V(G) \setminus V(P)$, x and y cannot be adjacent because Q is a longest path in G. Suppose, then, that $y \in V(P) \setminus (S^- \cup S \cup S^+)$. Then y has the same neighbours on each path in $\mathcal{P}$, because an elementary exchange that removed one of these neighbours would, at the same time, establish either it or y as an element of S. If x and y were adjacent, an elementary exchange applied to Q would thus yield a path whose head was a neighbour of y on P, a contradiction. Thus, in both cases, x and y are nonadjacent. □

Pósa (1976) used a simple consequence of his Lemma 6.3, most easily stated using the concept of the degree of a set of vertices. Let S be a set of vertices in a graph G. The *degree* of S in G, denoted by $d(S)$, is the number of neighbours of S in $G \setminus S$, that is,

$$d(S) := |N(S) \setminus S|.$$

(A priori, it might seem more natural to define the degree of a set S of vertices as the total number of neighbours of S, $|N(S)|$, but the definition given here appears to be the right one. Note that the two definitions coincide when S is a stable set.)

Corollary 6.4. *Let G be a graph and k and ℓ integers, where $k \geqslant -1$, $\ell \geqslant 1$ and $k \leqslant \ell$. If $d(S) \geqslant 2|S| + k$ for every nonempty set S of vertices such that $|S| \leqslant \lceil(\ell - k + 1)/3\rceil$, then G contains a path of length ℓ.*

Proof. Let P be a longest path in G and define S as in Pósa's Lemma (6.3). Then $N(S) \setminus S \subseteq S^- \cup S^+$, and so $d(S) \leqslant 2|S| - 2$. By the hypothesis of Corollary 6.4, $|S| > \lceil(\ell - k + 1)/3\rceil$. Let S' be a subset of S such that $|S'| = \lceil(\ell - k + 1)/3\rceil$. Then $d(S') \geqslant 2|S'| + k$. Since $V(P) \supseteq S' \cup N(S')$,

$$v(P) \geqslant |S' \cup N(S')| = |S'| + d(S') \geqslant 3|S'| + k \geqslant \ell + 1.$$

Thus P is of length at least ℓ. □

A graph on n vertices in which $d(S) \geqslant c|S|$ for all sets S of at most $\lceil dn \rceil$ vertices is called a *(c, d)-expander*. The case $k = 0$ of Corollary 6.4 asserts that $(2, 1/3)$-expanders possess Hamilton paths. Expanders have many useful properties (see, for example, the survey article by Bien 1989). Ways of constructing them are discussed in chapter 32. Note that the case $k = \ell - 2$ of Corollary 6.4 is simply Theorem 1.3(i).

Using Corollary 6.4, Pósa proved that a random graph with n vertices and $cn \log n$ edges, where c is a sufficiently large constant, is almost surely hamiltonian. Apart from the value of the constant c, this result is best possible. Pósa's bound was sharpened fully by Komlós and Szemerédi (1983). Corollary 6.4 has been invoked to help resolve a number of other problems on paths and circuits in random graphs. We limit ourselves here to one remarkable application by Beck (1983); further applications can be found in the stimulating survey article by Frieze (1989).

Theorem 6.5. *There exists a family of graphs $\{G_\ell\colon \ell \geqslant 3\}$ such that $e(G_\ell) \leqslant 900\ell$ and every subgraph of G_ℓ with at least $e(G_\ell)/2$ edges contains a path of length ℓ.*

Remark 6.6. Theorem 6.5 implies that in any 2-edge-colouring of such a graph G, there is a monochromatic circuit of length m. Statements of this type, asserting the existence of a monochromatic configuration in an edge-coloured graph or hypergraph, are studied under the collective name of Ramsey Theory. Chapter 25 by Nešetřil is devoted to this topic, as is the book by Graham et al. (1980).

6.3. Woodall's hopping lemma

Pósa's Lemma (6.3) demonstrates the effectiveness of iterative procedures in establishing sufficient conditions for Hamilton circuits; so too does the Closure Lemma (2.6). Here, we discuss a powerful technique of this kind, developed by Woodall (1973).

Lemma 6.7 (The Hopping Lemma). *Let G be a graph and C a longest circuit of G such that $G \setminus C$ has as few components as possible. Suppose that some component of $G \setminus C$ is an isolated vertex, u. Put $X_0 := N(u)$ and define, recursively, sets X_i, $i \geqslant 1$, by*

$$X_i := N((X_{i-1}^- \cap X_{i-1}^+) \cup \{u\}),$$

where S^- and S^+, respectively, denote the sets of vertices immediately preceding and following the vertices of S on C with respect to a fixed orientation. Set

$$X := \lim_{i\to\infty} X_i.$$

Then (i) *$X \subset V(C)$, and* (ii) *$X \cap X^+ = \emptyset$.*

Corollary 6.8. *Let G, C, u and X be defined as in the Hopping Lemma* (6.7) *and set*

$$Y := X^- \cap X^+.$$

Then (i) *$N(Y \cup \{u\}) = X$,* (ii) *$Y \cup \{u\}$ is a stable set, and* (iii) *$v(C) \geqslant 3|X| - |Y|$.*

Proof. Assertion (i) follows directly from the definitions of X and Y. Assertion (ii) is implied by assertion (i), part (ii) of the Hopping Lemma (6.7) and the definition of Y. To prove (iii), consider the partition of $V(C)$ induced by the $|X|$ disjoint segments of C whose tails belong to X and whose heads belong to X^-. By part (ii) of the Hopping Lemma (6.7) each such segment has at least two vertices. Moreover, by the definition of Y, a segment has exactly two vertices if and only if it includes a vertex of Y. Therefore

$$v(C) \geqslant 3(|X| - |Y|) + 2|Y| = 3|X| - |Y|. \qquad \square$$

We give one illustration, due to Woodall (1978a), of the use of the Hopping Lemma (6.7). [The following statement is slightly stronger than the one given by Woodall, who considered only the case $\ell = n$ and imposed the condition $d(S) \geqslant (n + |S| - 1)/3$ on all nonempty sets S of vertices. The proof, however, is the same.]

Theorem 6.9. *Let G be a simple 2-connected graph of minimum degree δ on n vertices, where $\delta \geqslant (n+2)/3$, and let ℓ be a positive integer. If $d(S) \geqslant (\ell + |S| - 1)/3$ for every stable set S of vertices of G such that $2 \leqslant |S| \leqslant (\ell+1)/2$, then G contains a circuit of length at least ℓ.*

Proof. Let C be a longest circuit of G. Suppose that C is of length less than ℓ. By Theorem 2.16, $V(G) \setminus V(C)$ is a stable set. Let $u \in V(G) \setminus V(C)$, and define the sets X and Y as in the Hopping Lemma (6.7) and Corollary 6.8. By part (ii) of Corollary 6.8, $Y \cup \{u\}$ is a stable set. Since $Y \subset V(C)$, we deduce that

$$|Y \cup \{u\}| \leqslant \tfrac{1}{2}v(C) + 1 \leqslant \tfrac{1}{2}(\ell + 1).$$

By parts (i) and (iii) of Corollary 6.8,

$$d(Y \cup \{u\}) = |X| \leqslant \tfrac{1}{3}(v(C) + |Y|) \leqslant \tfrac{1}{3}(\ell + |Y| - 1),$$

contradicting the hypothesis, with $S := Y \cup \{u\}$. □

Theorem 6.9 implies that a simple 2-connected graph on n vertices in which the minimum degree is at least $(n+2)/3$ and the degree of any stable set of two vertices is at least $n/2$ is hamiltonian. It thus generalizes Dirac's Theorem (2.1) and, for graphs of minimum degree at least $(n+2)/3$, Ore's Theorem (2.4). The Petersen graph (fig. 8) is an example of a simple nonhamiltonian 2-connected graph on n vertices in which the degree of any stable set of two vertices is at least $n/2$. Broersma et al. (1993) characterized all such graphs, thereby generalizing Ore's Theorem (2.4) fully. Of related interest is a generalization of Dirac's Theorem (2.1) by Häggkvist (1985b), restated here in terms of degrees of sets of vertices.

Theorem 6.10. *Let G be a simple graph of minimum degree δ on n vertices. If there is a positive integer $k \leqslant \delta$ such that $d(S) \geqslant n - \delta$ for every set S of k vertices, then G has a Hamilton circuit.*

The Hopping Lemma (6.7) (and variants of it) has diverse applications, among them the theorems of Häggkvist and Thomassen (4.8) and Jackson (5.17) referred to earlier. Suitably generalized, it may well be the appropriate tool to handle open questions such as Conjectures 4.7, 4.9, 4.22, 4.51, 5.11, 5.18, 5.20, 5.22 and 5.36 and Problems 4.55 and 5.24, as well as others to be discussed in later sections, such as Problem 7.14. The following conjectures, due to Thomassen (personal communication 1976; see also Alspach and Godsil 1985, p. 466) and to Fleischner and Jackson (1989), respectively, should also be mentioned in this context.

Conjecture 6.11. Let C be a longest circuit in a 3-connected graph. Then C has at least one chord.

Conjecture 6.12. Let G be a 3-regular cyclically 4-edge-connected graph. Then G has a dominating circuit.

Conjecture 6.12 is implied by Conjecture 5.36 and implies Conjecture 5.22. Fleischner and Jackson (1989) have shown that it is equivalent to the conjecture of Thomassen (1986) that every 4-connected line graph is hamiltonian, Fouquet and Thuillier (1990) that it is equivalent to the assertion that every 3-regular cyclically 4-edge-connected graph has a dominating circuit containing two specified non-adjacent edges. (Fouquet and Thuillier 1990, establish a similar strengthening of Conjecture 5.7.) Conjecture 6.12 is true for planar graphs by the Bridge Lemma (5.12).

The toughness is a measure of how well-knit a graph is. Another such measure is the *binding number*, introduced by Woodall (1973). This is the largest value of t such that $|N(S)| \geqslant t|S|$ for all sets S of vertices such that $N(S) \neq V$. Unlike the toughness, the binding number can be computed in polynomial time, using network flow techniques (Cunningham 1990). Woodall (1973) applied the Hopping Lemma (6.7) to prove that a loopless graph is hamiltonian if its binding number is at least $3/2$. His conjecture that such a graph is pancyclic under the same condition was verified by Shi (1987).

Theorem 6.13. *A loopless graph with binding number at least* $3/2$ *is pancyclic.*

7. Lengths of circuits

Most of the discussion so far has centred on long circuits, and we have been content to establish lower bounds on the lengths of circuits in graphs with given properties, rather than their exact values (except, of course, in the case of Hamilton circuits). Here, we consider questions of the latter type. We discuss conditions on the minimum degree and the number of edges guaranteeing the existence of various types of circuits.

Two basic lemmas, both due to Erdős (1965a,b), prove to be very useful in attacking extremal questions such as these. Roughly speaking, the first asserts that a graph with many edges has a subgraph of large minimum degree, the second that a graph of large minimum degree has a bipartite spanning subgraph of large minimum degree.

Lemma 7.1. *Let G be a graph on n vertices and more than $(c-1)n$ edges, where c is a positive integer. Then G has a subgraph H of minimum degree at least c.*

Proof. Any minimal subgraph H with more than $(c-1)v(H)$ edges has the required property: if H had a vertex v of degree at most $c-1$, the subgraph $H \setminus v$ would contradict the choice of H. $\square$

Lemma 7.2. *Every graph G has a bipartite spanning subgraph B such that $d_B(v) \geqslant d_G(v)/2$ for all vertices v.*

Proof. Any bipartite spanning subgraph $B(X,Y)$ with the maximum number of edges has the required property: if B had a vertex v of degree less than $d_G(v)/2$, where without loss of generality $v \in X$, the bipartite spanning subgraph with bipartition $(X \setminus v, Y \cup \{v\})$ would contradict the choice of B. □

These lemmas are often invoked in proofs of extremal theorems. If a best possible result is not sought, but only one in which the number of edges is of the right order of magnitude, their use serves to restrict the class of graphs under consideration to bipartite graphs of large degree, and such graphs are easier to treat. One reason for this is that breadth-first search trees, which are very useful in the analysis of circuit structure, take on a particular form in bipartite graphs of large degree: the sets of vertices equidistant from the root are stable and initially tend to grow rapidly in size with increasing distance from the root.

7.1. Circuits of given length modulo k

The existence of odd and even circuits has already been covered in some detail (see, for example, Theorems 1.8 and 4.13 and the discussions thereon). Barefoot et al. (1991) considered the case of circuits of length 0 modulo 3. They proved that simple 3-regular graphs necessarily have such circuits, and made a stronger conjecture, proved by Chen and Saito (1994).

Theorem 7.3. *Every simple graph of minimum degree at least three contains a circuit of length* 0 *modulo* 3.

A still more general conjecture was formulated by N. Dean (see Chen and Saito 1994).

Conjecture 7.4. Every simple graph of minimum degree at least k, where $k \geqslant 3$, contains a circuit of length 0 modulo k.

The case $k = 4$ of this conjecture was verified by Dean et al. (1993). Other circuit lengths modulo 3 were considered by Dean et al. (1991), who showed that a simple 2-connected graph of minimum degree at least three contains a circuit of length 1 modulo 3 unless it is the Petersen graph (fig. 8), and a circuit of length 2 modulo 3 unless it is K_4 or $K_{3,3}$ (see also Saito 1992).

Thomassen (1983c) proved that every simple graph of minimum degree at least $2\ell(k+1)$ contains a circuit of even length ℓ modulo k. Note that a bipartite graph has no circuit of odd length modulo k if k is even, no matter how large its minimum degree. On the other hand, for odd k, Bollobás (1977) established the existence of a function f such that every simple graph with minimum degree at least $f(k)$ contains circuits of all lengths modulo k; Thomassen (1983c) likewise established the existence of a function g such that every simple 2-connected nonbipartite graph with minimum degree at least $g(k)$ contains circuits of all lengths modulo k, and speculated as to the best possible such function g.

Conjecture 7.5. Every simple 2-connected nonbipartite graph of minimum degree at least $k+2$ contains circuits of all lengths ℓ modulo k.

A degree condition for the existence of a directed circuit of length 0 modulo k in a strict digraph was given by Alon and Linial (1989). The same condition guarantees, more generally, a directed circuit of weight 0 modulo k in a graph with integral edge weights. Alon (1991) has asked whether there exists a polynomial-time algorithm for finding such a circuit; the proof of Alon and Linial, which uses probabilistic methods, does not supply one.

7.2. Circuits of given length

The complete bipartite graph $K_{n/2,n/2}$ has $n^2/4$ edges and no odd circuits. Theorem 1.7 tells us that any simple graph on n vertices and more than this number of edges contains a triangle. Bondy (1971b), with the aid of Theorem 2.9, showed that any such graph contains circuits of all lengths ℓ, $3 \leqslant \ell \leqslant \lfloor (n+3)/2 \rfloor$. This number of edges is extremal for odd ℓ but not for even ℓ. In the range $\lfloor (n+3)/2 \rfloor \leqslant \ell \leqslant n$, Woodall (1972) determined the exact bound, namely $\binom{\ell-1}{2} = \binom{n-\ell+2}{2}$; the connected graph with blocks $K_{\ell-1}$ and $K_{n-\ell+2}$ has this number of edges but no circuit of length ℓ.

The situation as regards short even circuits is less clear. Reiman (1958) observed that only $n^{3/2}/2 + n/4$ edges are needed to guarantee the existence of a circuit of length four; a simple graph G with this number of edges satisfies the inequality

$$\sum_{v \in V} \binom{d(v)}{2} > \binom{n}{2},$$

and thus has two vertices with a pair of neighbours in common, that is, a 4-circuit. By exploiting the structure of breadth-first search trees, Bondy and Simonovits (1974) showed, more generally, that a simple graph on n vertices and more than $c\, n^{1+1/\ell}$ edges, where $c := c(\ell)$ is a function only of ℓ, contains a 2ℓ-circuit. Apart from the value of c, this number of edges is probably extremal; it is known to be so for $\ell = 2$, 3 and 5. For instance, the Levi graph $\mathrm{LG}(2,q)$ of a finite projective plane of order q has $2(q^2+q+1)$ vertices, $(q+1)(q^2+q+1)$ edges and no 4-circuit. Brown (1966) and Erdős et al. (1966b) constructed denser examples in the case $\ell = 2$, also based on finite projective planes, and Füredi (1983, preprint) proved that these examples are precisely the extremal graphs when $n = q^2+q+1$ and $q \notin \{1,7,9,11,13\}$. Benson (1966) settled the cases $\ell = 3$ and $\ell = 5$, also by constructing graphs from finite geometries. Conceptually simpler examples, with fewer edges but still of the right order of magnitude, were constructed by Wenger (1991) for the same three values of ℓ.

Problem 7.6. For each integer $\ell \geqslant 2$ and for infinitely many values of n, construct a simple graph on n vertices and $c\, n^{1+1/\ell}$ edges, where c is a constant depending only on ℓ, containing no 2ℓ-circuit.

By Lemma 7.2, the extremal problem is no easier for bipartite graphs. For digraphs, on the other hand, the extremal question was completely settled by Häggkvist and Thomassen (1976). They determined the number of edges needed in a digraph on n vertices to guarantee the existence of a directed circuit of length ℓ, and also derived bounds that are close to best possible for the restriction of this problem to strong digraphs. These and many other extremal questions on circuits in digraphs are discussed in the survey by Bermond and Thomassen (1981) and in the book by Bollobás (1978).

7.3. *Circuits of many lengths*

The proof of Theorem 1.3 shows that a simple graph of minimum degree k has at least $k-1$ different circuit lengths; this is best possible, in view of the complete graph K_{k+1}. A more sensitive measure of the richness of circuit lengths in a graph – the sum of their reciprocals – was proposed by Erdős and Hajnal (1966) (see also Erdős 1975). Defining

$$\mathscr{L}(G) := \sum\{1/\ell\colon\ G \text{ has a circuit of length } \ell\},$$

and noting that

$$\mathscr{L}(K_{k+1}) \approx \log\ k + \gamma - \tfrac{3}{2} \qquad \text{and} \qquad \mathscr{L}(K_{k,k}) \approx \tfrac{1}{2}(\log\ k + \gamma - \tfrac{3}{2}),$$

where γ is the Mascheroni constant ($\gamma \approx 0.5772$), they asked if there exists a constant c such that the sum of the reciprocals of the circuit lengths in a simple graph of minimum degree k is at least $c \log k$, regardless of the order of the graph. An affirmative answer was provided by Gyarfás et al. (1984). Note that, by virtue of Lemma 7.1, the same remarkable property holds (with a different constant) in a graph of average degree k. Sparse graphs, with average degree $1+(1/k)$, where k is sufficiently large, exhibit a similar behaviour. Gyarfás et al. (1985) proved that the sum of the reciprocals of the circuit lengths in such a graph is at least $1/(300k\ \log\ k)$.

7.4. *Circuits of all lengths*

Pancyclic graphs, those that have circuits of all possible lengths, were introduced in section 2. Sufficient conditions for hamiltonicity frequently imply pancyclicity. Corollary 2.10 is a typical instance of this phenomenon, which provides convincing evidence that it is hard to capture the essence of hamiltonicity. This is not surprising, of course, hamiltonicity being an $\mathscr{N}\mathscr{P}$-complete property.

A digraph analogue of Theorem 2.9 was established by Häggkvist and Thomassen (1976), and Corollary 2.10 was generalized to digraphs by Thomassen (1977). A bipartite analogue of Theorem 2.9 was considered by Entringer and Schmeichel (1988). Using similar methods, they proved that a hamiltonian bipartite graph on $2n$ vertices and more than $n^2/2$ edges is *bipancyclic*, that is, has circuits of every even length 2ℓ, $2 \leqslant \ell \leqslant n$. This bound on the number of edges is not thought to be best possible, however, and could conceivably be reduced by a factor of two. Specifically, one has the following conjecture of Mitchem and Schmeichel (1985).

Conjecture 7.7. Let G be a simple hamiltonian bipartite graph on $2n$ vertices and at least $n^2/4+n+1$ edges. Then G is bipancyclic.

Mitchem and Schmeichel (1985) also proposed a related conjecture, which can be regarded as the bipartite analogue of Corollary 2.10.

Conjecture 7.8. Let G be a simple hamiltonian bipartite graph of minimum degree δ on $2n$ vertices, where $\delta^2 - \delta \geqslant n$. Then G is bipancyclic.

The Levi graphs of finite projective planes (see sections 5.3 and 7.2) show that the lower bound on δ here is necessary.

We turn now to pancyclicity in tournaments, recalling that every strong tournament is vertex pancyclic. A digraph on n vertices is *edge-ℓ-cyclic* if each edge lies in a directed circuit of length ℓ, and *edge pancyclic* if each edge lies in a directed circuit of each length ℓ, $3 \leqslant \ell \leqslant n$. Wu et al. (1981) proved that every edge-3-cyclic tournament on n vertices is edge-ℓ-cyclic for all ℓ, $3 \leqslant \ell \leqslant \lfloor (n+5)/2 \rfloor$. This somewhat surprising result was extended by Tian et al. (1982).

Theorem 7.9. *An edge-3-cyclic tournament is edge pancyclic unless it belongs to one of two specified families.*

One corollary of Theorem 7.9 is the theorem of Alspach (1967) that regular tournaments are edge pancyclic. A number of other properties stronger than pancyclicity have been studied. Faudree and Schelp (1975) define a graph on n vertices to be *panconnected* if any two vertices are connected by paths of every length ℓ, $d \leqslant \ell \leqslant n-1$, where d is their distance. Hendry (1990) calls a circuit C *extendable* if there is a circuit C' such that $v(C') = v(C) + 1$ and $V(C') \supset V(C)$, and a graph G *circuit extendable* if G contains at least one circuit and every nonspanning circuit of G is extendable. He also introduces the concept of a *pancyclic ordering*, a linear ordering $v_1, v_2, \ldots, v_n$ of the vertices of a graph such that the subgraph induced by $\{v_1, v_2, \ldots, v_i\}$ is hamiltonian for all i, $3 \leqslant i \leqslant n$.

7.5. The number of circuits

Results on the existence of circuits lead naturally to questions about their number, and to some intriguing conjectures. It is lower bounds on the number of circuits that are of principal interest. We consider, first, two problems on the total number of circuits.

The number of circuits in 3-connected 3-regular graphs was investigated by Barefoot et al. (1986), who observed that the total number of circuits in a 3-regular graph on n vertices which has no circuit of length more than n^c, where $c < 1$, is necessarily subexponential in n. Such graphs exist (see the discussion following Theorem 5.21). Barefoot et al. (1986) conjectured, however, that the number of circuits is always moderately large.

Conjecture 7.10. The number of circuits in a 3-connected 3-regular graph on n vertices is superpolynomial in n.

A long-standing conjecture on the number of directed circuits in a digraph is the following one, due to Adám (1964).

Conjecture 7.11. Let D be a strict digraph with at least one directed circuit. Then D has an edge whose reversal reduces the total number of directed circuits.

Although not explicit in Adám's formulation, the requirement that D be strict is essential. Examples of digraphs with multiple edges in which the reversal of any edge increases the number of circuits have been described by Thomassen (1987b) and by Grinberg (1987). Thomassen's examples are based on the existence of digraphs such as those described in Remark 5.44, in which the reversal of any edge increases the length of a longest directed circuit. Multiplying each edge suitably many times yields a digraph with the desired property.

Theorems asserting the existence of Hamilton circuits in graphs are numerous, prominent among them being those of Dirac (2.1), Ore (2.4), Chvátal and Erdős (4.16), Jackson (5.17), and Tutte (5.8). In each case, one may ask how many Hamilton circuits there must be. Surprisingly, of these theorems, only Tutte's has been investigated from this aspect (although the theorems of Dirac and Ore have been studied with regard to the number of edge-disjoint Hamilton circuits; see the discussion following Theorem 8.8). Hakimi et al. (1979) showed that a 4-connected planar graph on n vertices has at least $n/\log_2 n$ Hamilton circuits and conjectured that this bound can be improved to $2(n-2)(n-4)$.

A tournament has at least one directed Hamilton path, by Rédei's Theorem (3.1), but need not have more. Moon (1972) showed, however, that a strong tournament on n vertices has at least c^{n-1} directed Hamilton paths, where $c := 6^{1/4} \approx 1.565$. Thomassen (1985b) examined the same question for directed Hamilton circuits. By Camion's Theorem (3.9), a strong tournament has at least one directed Hamilton circuit but, again, need have no more. Thomassen nevertheless found an exponential lower bound in terms of the minimum outdegree: a strong tournament with minimum outdegree at least $3k+3$ has at least $4^k k!$ directed Hamilton circuits. Thomassen (1980b) also obtained an analogue of Moon's theorem, proving that a 2-strong tournament on n vertices has at least $c^{n/32-1} \approx 1.014^n$ directed Hamilton circuits. Thomassen (personal communication 1990) remarks that no construction is known of a tournament with more than $(n-1)!/2^n$ directed Hamilton circuits, although an elementary counting argument establishes the existence of such a tournament. A related question posed by Thomassen (personal communication 1990) is the following one.

Problem 7.12. Is there a diregular tournament on n vertices with at least $(n-1)!/2^n$ directed Hamilton circuits?

As already noted, Smith's Theorem (5.14) implies that a 3-regular hamiltonian graph has at least three Hamilton circuits. One may ask which have precisely three. The smallest example is the graph on two vertices joined by three edges, $K_2^{(3)}$. Bosák (1967) observed that any graph obtainable from this graph by recursively expanding vertices to triangles has the same property and conjectured, incorrectly, that there are no others; Ninčák (1974) found a nonplanar counterexample, a generalization of the Petersen graph, and asked whether Bosák's conjecture is true when restricted to planar graphs.

Problem 7.13. Let G be a 3-regular planar hamiltonian graph with precisely three Hamilton circuits. Can G be obtained from $K_2^{(3)}$ by recursively expanding vertices to triangles? Equivalently, if $G \neq K_2^{(3)}$, must G contain a triangle?

If G is a 3-regular graph with precisely three Hamilton circuits H_1, H_2, H_3, the sets $E_i := E(G) \setminus E(H_i)$, $i = 1, 2, 3$, constitute a proper edge colouring in which the edges of any two colours induce a Hamilton circuit. The graphs possessing such an edge colouring were introduced by Kotzig (1962) under the name of *Hamiltonian* graphs; they are now referred to as *strongly hamiltonian*, *perfectly* 1-*factorable* or *Hamilton* 1-*factorable*. A selection of open problems about them is given in Kotzig and Labelle (1979).

It is not known whether a simple 4-regular hamiltonian graph must have at least two Hamilton circuits. Sheehan (1975) conjectures that this is so. Fleischner (1994) has constructed a 4-regular uniquely hamiltonian multigraph; thus the condition that the graph be simple cannot be dropped. Also, Entringer and Swart (1980) have constructed uniquely hamiltonian simple graphs of minimum degree three. No example with minimum degree four is known.

Problem 7.14. Is there a uniquely hamiltonian simple graph of minimum degree four?

8. Packings and coverings by paths and circuits

This is an important and extensive topic, rich in challenging open questions. The concepts of packing and covering are best explained in the context of hypergraphs.

8.1. Packings and coverings of hypergraphs

Recall that a hypergraph H consists of a set $V(H)$ of elements, called vertices, and a family $E(H)$ of subsets of $V(H)$, called edges. A *packing* of H is a set of disjoint edges, a *covering* of H a set of edges whose union is $V(H)$. A *partition* of H is a set of edges that constitutes at the same time a packing and a covering.

Given a hypergraph H, several questions naturally arise. How many edges can there be in a packing of H? How few edges can there be in a covering of H? Does H admit a partition? If so, how many or how few edges can there be in a partition of H? Consideration of the first question leads to the concept of a *transversal* of a hypergraph, a set of vertices that meets every edge in at least one vertex. Packings and transversals are linked in a simple way.

Proposition 8.1. *Let H be a hypergraph, $\mathcal{P}$ a packing of H, and T a transversal of H. Then*

$$|\mathcal{P}| \leqslant |T|.$$

Proof. Because T is a transversal, each edge of $\mathcal{P}$ contains at least one vertex of T. On the other hand, because $\mathcal{P}$ is a packing, no two edges in $\mathcal{P}$ contain the same vertex of T. The inequality follows. □

Proposition 8.1 implies the following inequality for any hypergraph:

$$\max\{|\mathcal{P}|: \mathcal{P} \text{ a packing of } H\} \leqslant \min\{|T|: T \text{ a transversal of } H\}.$$

Hypergraphs H for which equality holds in this inequality are said to have the *minmax property*. Such hypergraphs are of particular import in the field of combinatorial optimization. In order to find a packing of maximum cardinality or a transversal of minimum cardinality in a hypergraph with the minmax property, it suffices to find a packing $\mathcal{P}$ and a transversal T of the same cardinality: Proposition 8.1 guarantees that $\mathcal{P}$ is then maximum and T minimum. Finding such optimal packings and transversals efficiently is, of course, another matter. However, it is a striking fact that almost all packing problems for hypergraphs satisfying the minmax property do indeed admit polynomial-time algorithms. This phenomenon has, as yet, no theoretical foundation.

We now define what we mean by a packing or covering of a graph. Associated with a graph G and a family $\mathcal{F}$ of subgraphs of G are two hypergraphs, one having vertex set $V(G)$ and edge set $\{V(F): F \in \mathcal{F}\}$, the second having vertex set $E(G)$ and edge set $\{E(F): F \in \mathcal{F}\}$. Packings and coverings of these hypergraphs correspond in a natural way to packings and coverings of G by members of $\mathcal{F}$, and lead to the following definitions.

A family $\mathcal{F}$ of subgraphs of a graph G is a *vertex packing* of G if the members of $\mathcal{F}$ are vertex-disjoint, an *edge packing* if they are edge-disjoint, a *vertex covering* if $\cup_{F\in\mathcal{F}} V(F) = V(G)$ and an *edge covering* if $\cup_{F\in\mathcal{F}} E(F) = E(G)$. A *partition* (or *vertex partition*) is a vertex packing which is also a vertex covering, a *decomposition* (or *edge partition*) an edge packing which is also an edge covering. We shall be interested in the case where the family $\mathcal{F}$ consists of paths and/or circuits.

8.2. Packings by paths and circuits

For paths, the vertex-packing problem is trivial unless restrictions are placed on the paths. If they are required to be of length one, we have the maximum matching problem referred to in section 4 (at the end of the discussion of stability number). In this case, the minmax property is valid for bipartite graphs, but not in general – a subtler minmax property holds for all graphs (see chapter 3). If the paths are required to be of any other fixed length, the problem is $\mathcal{NP}$-complete, as is the vertex-packing problem for circuits of any fixed length (Kirkpatrick and Hell 1983).

Vertex packings by circuits do not satisfy the minmax property: only one circuit can be packed into K_5, for instance, whereas a transversal requires at least three vertices. Nevertheless, a relationship does exist between the cardinalities of a maximum packing and a minimum transversal.

Theorem 8.2. *Let G be a graph and k a positive integer. Either G has k disjoint circuits or there is a set of at most $ck \log k$ vertices in G meeting all circuits, where c is an absolute constant.*

This theorem, due to Erdős and Pósa (1962), leads to a straightforward polynomial-time algorithm for determining if a graph has k disjoint circuits, for fixed k. Let $s := \lceil ck \log k \rceil$. One first checks each subgraph $G \setminus S$, where $S \subset V$ and $|S| = s$; there are no more than n^s of these, where n is the order of G. If no such subgraph is a forest, Theorem 8.2 guarantees that G has k disjoint circuits. If, on the other hand, some $G \setminus S$ is a forest, every circuit of G meets S. The circuits of G can then be enumerated directly. If C is a circuit meeting S in a subset R, there are at most n^2 choices for the two edges of C incident to any vertex v of R, and so at most n^{2r} choices for all the edges of C incident to R, where $r := |R|$. Moreover, each such choice gives rise to at most one circuit of G; if there were two, their symmetric difference would be a nonempty even subgraph of the forest $G \setminus S$. Thus the number of circuits of G is polynomial in n. Each set of k circuits can now be checked individually.

Whether there is also a function f such that every digraph has either k disjoint directed circuits or a set of $f(k)$ vertices meeting all directed circuits is not known. This question was first asked by Gallai (1968b). Thomassen (1983d) noted that the existence of such a function would imply a polynomial-time algorithm for determining whether a digraph has k disjoint directed circuits. Younger (1973) conjectured that $f(2) = 3$. This was confirmed by McCuaig (1993).

Theorem 8.3. *Let D be a digraph. Either D has two disjoint directed circuits or there is a set of at most three vertices in D meeting all directed circuits.*

Corrádi and Hajnal (1963) gave a sharp bound on the minimum degree for a graph to have k disjoint circuits.

Theorem 8.4. *Let k be a positive integer and G a simple graph of minimum degree δ on n vertices, where $\delta \geqslant 2k$ and $n \geqslant 3k$. Then G has k disjoint circuits.*

Thomassen (1983a) conjectured that if n is a sufficiently large function of k, the hypothesis of Theorem 8.4 guarantees the existence of k disjoint circuits of the same length. This was proved by Egawa (preprint). Häggkvist (1985c) showed that regular graphs have many disjoint circuits of the same length; there are at least $n/64(\log_2 n)^2$, for instance, in a 3-regular graph on n vertices. For digraphs, Thomassen (1983d) proved the existence of a function f such that any strict digraph of minimum outdegree at least $f(k)$ has k disjoint directed circuits, and further conjectured the existence of a function g such that any strict digraph of minimum outdegree at least g has k disjoint directed circuits of the same length. He also ventured a conjecture as to the smallest value of the function f (see, also, Bermond and Thomassen 1981).

Conjecture 8.5. Let k be a positive integer and D a strict digraph of minimum outdegree at least $2k - 1$. Then D has k disjoint directed circuits.

Edge packings by xy-paths in graphs, and by (x, y)-paths in digraphs, satisfy the minmax property, by Menger's Theorem (4.1). Lucchesi and Younger (1978) established that edge packings by circuits also have the minmax property, in planar digraphs.

Theorem 8.6. *Let D be a planar digraph. Then the maximum number of edge-disjoint directed circuits in D is equal to the minimum number of edges of D whose deletion destroys all directed circuits.*

A related minmax equality for planar digraphs was conjectured by Woodall (1978b).

Conjecture 8.7. Let D be a planar digraph. Then the minimum number of edges in a directed circuit of D (the girth of D) is equal to the maximum number of disjoint sets of edges of D the deletion of each of which destroys all directed circuits.

Theorem 8.6 and Conjecture 8.7 are special cases of the Lucchesi–Younger Theorem and Woodall's conjecture, respectively. Stated in dual form, in terms of directed bonds rather than directed circuits, the latter apply to all digraphs, not just planar ones. A related result, on edge transversals of maximal directed edge cuts, was obtained by Maamoun (1985).

Theorem 8.8. *In any digraph, there is a directed path or directed circuit whose edges meet every maximal directed edge cut.*

McCuaig (1993) obtained an edge analogue of Theorem 8.3.

Theorem 8.9. *Let D be a digraph. Either D has two edge-disjoint directed circuits or there is a set of at most three edges in D meeting all directed circuits.*

Theorem 2.11 refines Dirac's Theorem (2.1) by giving a lower bound on the number of edge-disjoint Hamilton circuits in a graph satisfying the hypotheses of Dirac's Theorem (2.1). A similar refinement of Ore's Theorem (2.4), but valid only for graphs of relatively small minimum degree, was given by Li (1989). Corollary 4.18 shows that a simple $2d$-regular graph on at most $4d+1$ vertices has at least one Hamilton circuit. Jackson (1979) proved that every simple d-regular graph on n vertices, where $d \geqslant (n-1)/2$ and $n \geqslant 14$, contains at least $\lfloor (3d-n+1)/6 \rfloor$ edge-disjoint Hamilton circuits, deducing this from his Theorem 5.17. Thomassen (1982) conjectured the existence of a function f such that every $f(k)$-connected tournament has k edge-disjoint directed Hamilton circuits. The case of triangles was examined by Tuza (1984), who made the following conjecture.

Conjecture 8.10. Let G be a simple graph and k a positive integer. Either G has $k+1$ edge-disjoint triangles or there is a set of at most $2k$ edges in G meeting all triangles.

Tuza (1990) made a similar conjecture for directed triangles in tournaments and established a minmax equality for directed triangles in planar oriented graphs (Tuza 1995).

We have omitted mention here of conditions on the number of edges guaranteeing the existence of k disjoint or edge-disjoint circuits. For these, the reader may consult either chapter 23 or the book of Bollobás (1978).

8.3. Coverings by circuits

We have already seen a variety of ways in which theorems on the existence of long circuits can be generalized. Yet another way is to determine how few circuits it takes to cover the vertices or edges of the graph in question. For instance, Theorem 4.47 implies that a strong digraph on n vertices with stability number α has a directed circuit of length at least n/α, while Conjecture 4.33 makes the stronger assertion that α directed circuits suffice to cover the vertex set of the digraph. In the same vein, Kouider and Lonc (preprint) established the following generalization of Theorem 2.14.

Theorem 8.11. *Let k be a positive integer and G a simple graph on n vertices such that $\sigma_{k+1} \geqslant n$. Then V can be covered by k circuits, edges or vertices.*

When $\alpha(G) \leqslant k$, this is a consequence of Corollary 4.20. The following conjecture on edge coverings by circuits was proposed by Bondy (1990). If true, it would imply the theorem of Pyber (1985) that the edges of a simple graph on n vertices can be covered by $n-1$ circuits and edges.

Conjecture 8.12. Let G be a simple 2-connected graph on n vertices. Then $E(G)$ can be covered by at most $(2n-1)/3$ circuits.

Thomassen (1985b) proved that the edges of a diregular tournament on n vertices can be covered by $12n$ directed Hamilton circuits. This result is of interest with regard to a question of P.J. Kelly on decompositions of diregular tournaments into directed Hamilton circuits (see Conjecture 8.25 and the subsequent discussion).

The cardinality $|\mathscr{C}|$ of a circuit covering $\mathscr{C}$ is but one measure of the efficiency of the covering; another is the *length* $\|\mathscr{C}\|$ of the covering, defined by

$$\|\mathscr{C}\| := \sum_{C \in \mathscr{C}} e(C).$$

We denote the minimum length of a circuit covering of a 2-edge-connected graph G by $\mathrm{cc}(G)$.

Thomassen (preprint) conjectured that the problem of determining whether $\mathrm{cc}(G) \leqslant k$ is $\mathcal{NP}$-complete. Nevertheless, a good lower bound for $\mathrm{cc}(G)$ can be found in polynomial time by determining a shortest postman tour of G. A *postman tour* of a connected graph G is a closed walk that includes every edge of G. We denote the length of a shortest postman tour of G by $\mathrm{pt}(G)$. A polynomial-time algorithm for finding a minimum-weight postman tour in an edge-weighted graph was developed by Edmonds and Johnson (1973). It is based on the shortest-path algorithm of Dijkstra (1959) and the weighted matching algorithm of Edmonds (1965). The idea is to find shortest paths between all pairs of vertices of odd degree, and then determine a pairing of these vertices for which the set of connecting shortest paths is of minimum total weight. A postman tour that traverses the edges of the selected paths twice and each other edge once is one of minimum weight. If $\mathscr{C}$ is a circuit covering of a 2-edge-connected graph G, the edge-disjoint union of the circuits in $\mathscr{C}$ is an eulerian spanning supergraph of G and thus admits an Euler tour of length $\|\mathscr{C}\|$. The corresponding tour in G is a postman tour of the same length. It follows that

$$\mathrm{cc}(G) \geqslant \mathrm{pt}(G).$$

Clearly, $\mathrm{cc}(G) = \mathrm{pt}(G) = e(G)$ if G is eulerian, and $\mathrm{cc}(G) \geqslant \mathrm{pt}(G) > e(G)$ otherwise. An upper bound for $\mathrm{cc}(G)$ in terms of $e(G)$ was conjectured by J.C. Bermond, B. Jackson and F. Jaeger (F. Jaeger, personal communication, 1989).

Conjecture 8.13. Every 2-edge-connected graph on e edges admits a circuit covering of length at most $7e/5$.

The bound of $7e/5$ is attained by the Petersen graph (fig. 8). Bermond et al. (1983) proved that a 2-edge-connected graph on e edges admits a circuit covering of length at most $5e/3$; this was generalized to edge-weighted graphs by Fan (1990).

8.4. Partitions into paths and circuits

The question of partitioning a graph or digraph into as few paths or circuits as possible is strongly related to – indeed includes – the hamiltonian problem. We have already studied many aspects of this question. Moreover, path partitions of digraphs and their connections to stable sets and colourings were discussed in section 4.

Path partitions of 3-regular graphs were examined by Reed (1989), who proved that any simple 2-connected 3-regular graph on n vertices can be partitioned into at most $n/9$ paths and conjectured a slightly stronger conclusion.

Conjecture 8.14. Let G be a simple 2-connected 3-regular graph on n vertices. Then G admits a partition into at most $n/10$ paths.

One may consider, more specifically, partitions of a graph or digraph into paths or circuits of prescribed length. El-Zahar (1984) proposed an appealing generalization of Dirac's Theorem (2.1) in this vein.

Conjecture 8.15. Let n be a positive integer, $n \geqslant 3$, and let $n_1 + n_2 + \cdots + n_k$, where $n_i \geqslant 3$, be a partition of n into k parts, s of which are odd. If G is a simple graph of minimum degree δ on n vertices, where $\delta \geqslant (n+s)/2$, then G admits a partition into k circuits, of lengths $n_1, n_2, \ldots, n_k$.

The case $k=2$ was verified by El-Zahar. The case $n=3k$ is true by virtue of Theorem 8.4, but would also be a consequence of the following conjecture of L. Pósa (see Erdős 1964).

Conjecture 8.16. Let G be a simple graph of minimum degree δ on n vertices, where $\delta \geqslant 2n/3$. Then G contains the square of a Hamilton circuit.

Fan and Kierstead (1995) established an asymptotic version of this conjecture by showing that G contains the square of a Hamilton circuit if, for some $\varepsilon > 0$, $\delta \geqslant (2+\varepsilon)n/3$ and n is sufficiently large (as a function of ε). They also (Fan and Kierstead, preprint) proved the analogue (and corollary) of Conjecture 8.16 for Hamilton paths.

Theorem 8.17. *Let G be a simple graph of minimum degree δ on n vertices, where $\delta \geqslant (2n-1)/3$. Then G contains the square of a Hamilton path.*

Aigner and Brandt (1993) verified a weaker form of Conjecture 8.16 in which the bound $(n+s)/2$ is replaced by $(2n-1)/3$. This result is implied by Theorem 8.17.

8.5. Decompositions into paths and circuits

A *path decomposition* or *circuit decomposition* is a decomposition into paths or circuits, respectively, a *Hamilton decomposition* a decomposition into Hamilton circuits. Similar definitions apply to digraphs, a path or circuit decomposition connoting a decomposition into directed paths or circuits as the case may be. Circuit decompositions were touched on briefly in section 1. Just as Theorem 3.4 refines Rédei's Theorem (3.1), so Theorem 1.4 was refined by Toida (1973).

Theorem 8.18. *An even graph admits an odd number of circuit decompositions.*

Theorem 8.18 can be proved using Thomason's Lemma (6.1) (Bondy and Halberstam 1986); for a different approach, see Woodall (1990). Another refinement of Theorem 1.4, on decompositions of planar graphs into even circuits, was found by Seymour (1981).

Theorem 8.19. *Let G be a 2-connected planar even graph on an even number of edges. Then G admits a decomposition into even circuits.*

The number of circuits in a decomposition is the subject of a longstanding conjecture of G. Hajós (see Lovász 1968, and Dean 1986).

Conjecture 8.20. Every simple even graph on n vertices admits a decomposition into at most $(n-1)/2$ circuits.

Hajós's conjecture was verified for planar graphs by Tao (1984). T. Gallai (see Lovász 1968) proposed an analogous conjecture on path decompositions.

Conjecture 8.21. Every simple connected graph on n vertices admits a decomposition into at most $(n+1)/2$ paths.

Lovász (1968) proved that every simple connected graph on n vertices admits a decomposition into at most $n/2$ paths and circuits. If the graph is odd, such a decomposition must necessarily consist of $n/2$ paths; thus Lovász's theorem implies

Conjecture 8.21 in this case. Donald (1980) refined Lovász's argument, establishing that every simple connected graph on n vertices admits a decomposition into at most $(3n_0/4)+(n_1/2)$ paths, where n_0 and n_1 denote the numbers of vertices of even and odd degree, respectively.

A related question, on decompositions into circuits and edges, was considered by P. Erdős and T. Gallai (P. Erdős, personal communication, 1990; see also Erdős et al. 1966a, where it is discussed without attribution).

Conjecture 8.22. Every simple graph on n vertices admits a decomposition into at most cn circuits and edges, where c is an absolute constant.

Dean (1986) reviews analogues of Conjecture 8.20 for digraphs and poses several conjectures, of which the following is typical.

Conjecture 8.23. Let D be an eulerian oriented graph on n vertices. Then D admits a decomposition into at most $2n/3$ directed circuits.

Theorem 8.8, applied inductively, shows that a strict digraph on n vertices admits a decomposition into at most k directed paths or circuits, where k is the number of edges in a largest directed edge cut, and is thus at most $n^2/4$. This can be viewed as a digraph analogue of the theorem of Lovász noted above. O'Brien (1977) obtained a sharper result, a digraph analogue of Conjecture 8.21.

Theorem 8.24. *Let D be a strict digraph on n vertices. Then D admits a decomposition into at most $n^2/4$ directed paths.*

Alspach and Pullman (1974) defined the *excess* of a digraph D by the formula

$$\mathrm{ex}(D) := \sum_{v \in V} \max\,\{d^+(v) - d^-(v), 0\},$$

noting that any decomposition of D into directed paths must have at least this cardinality because at least $\max\,\{d^+(v) - d^-(v), 0\}$ of the paths in the decomposition have v as their tail. Alspach et al. (1976) conjectured that this bound is tight if D is a tournament of even order. N.J. Pullman (see Reid and Wayland 1987) extended the conjecture to all regular oriented graphs of odd degree.

Conjecture 8.25. Let D be a regular oriented graph of odd degree. Then D admits a decomposition into $\mathrm{ex}(D)$ directed paths.

Häggkvist (1985a) found very general results on circuit decompositions of almost-complete graphs and complete bipartite graphs. The basis of his approach is the following observation.

Proposition 8.26. *If G is a path or circuit of length n, the composition $G[2K_1]$ is decomposable into two circuits of length $2n$.*

Define a list $\{F_1, F_2, \ldots, F_m\}$ of graphs to be *even* if each entry occurs an even number of times, and to be *proper*, for a graph H, if

(1) $e(H) = \sum_{i=1}^{m} e(F_i)$, and

(2) H has a subgraph isomorphic to F_i, $1 \leqslant i \leqslant m$.

An immediate consequence of Proposition 8.26 is that, if G has a Hamilton decomposition, $G[2K_1]$ has a decomposition into any proper even list of even 2-factors; in particular, $K_{4n+2} - M$, where M is a perfect matching, and $K_{4n,4n}$ both have this latter property, the first being isomorphic to $K_{2n+1}[2K_1]$, the second to $K_{2n,2n}[2K_1]$. Another (less immediate) consequence is that $K_{4n+2} - M$ admits a decomposition into any proper even list of even circuits not including circuits of length $4n$, and that $K_{4n,4n}$ admits a decomposition into any proper even list of even circuits not including circuits of lengths $8n-2$, $8n-4$ or $8n-6$.

To conclude, we note a conjecture on triangle decompositions due to Nash-Williams (1970).

Conjecture 8.27. Let G be a simple even graph of minimum degree δ on n vertices and m edges, where $\delta \geqslant 3n/4$. Then G admits a decomposition into triangles provided that $m \equiv 0 \pmod 3$ and n is sufficiently large.

This appears to be a very hard problem. The result of Gustavsson (1991) that the assertion holds if $\delta \geqslant (1 - 10^{-24})n$ is therefore significant.

8.6. Decompositions into Hamilton circuits

The decomposition of graphs and digraphs is a topic with a wealth of problems and results; indeed, a chapter could easily be devoted to decompositions of the complete graph alone. In a sense, such questions belong to the realm of combinatorial design (see chapter 14) where algebraic techniques are employed as a rule. However, graph-theoretical methods are often highly effective, as Proposition 8.26 well illustrates. The theory of edge colourings was applied by Hilton (1984) to establish the following elegant result on Hamilton decompositions of complete graphs (subsequently generalized by Nash-Williams (1987) using network flow techniques).

Theorem 8.28. *Let r and n be positive integers, $r \leqslant 2n$, let G and H be complete graphs of orders r and $2n+1$, respectively, and let $\{E_1, E_2, \ldots, E_n\}$ be an n-edge colouring of G. Then there is a Hamilton decomposition $\{H_1, H_2, \ldots, H_n\}$ of H such that $E_i \subset H_i$, $1 \leqslant i \leqslant n$, if and only if the spanning subgraph of G with edge set E_i is a path system with at most $2n+r-1$ components, $1 \leqslant i \leqslant n$.*

That complete graphs of odd order admit Hamilton decompositions is a classical result; one such construction, due to a Monsieur Walecki, is described in the book by Lucas (1891, pp. 161–164). Hilton (1984) explains how Theorem 8.28 can be used to find all such decompositions. An extension of Walecki's theorem to regular graphs of high degree (and at the same time a strengthening of Corollary 4.18 for even d) was proposed by Nash-Williams (1971b).

Conjecture 8.29. Let G be a simple $2d$-regular graph on at most $4d+1$ vertices, where $d \geqslant 1$. Then G admits a Hamilton decomposition.

Hamilton decompositions of cartesian products of graphs are the subject of a conjecture of Bermond (1978a), several special cases of which have been settled by Stong (1991), and a problem of Alspach and Rosenfeld (1986).

Conjecture 8.30. If G and H admit Hamilton decompositions, so does $G \Box H$.

Problem 8.31. If G is a simple hamiltonian 3-regular graph, does the prism over G, $G \Box K_2$, admit a Hamilton decomposition?

In fact, no simple 2-connected 3-regular graph is known whose prism fails to admit a Hamilton decomposition. Hamilton decompositions of 4-regular graphs were considered also by Thomason (1978), who discovered an analogue of Smith's Theorem (5.14) for such graphs.

Theorem 8.32. *Let G be a 4-regular graph, and let e and f be two edges of G. Then the number of Hamilton decompositions of G in which e and f belong to different Hamilton circuits is even.*

Berman (1986a) applied methods similar to those of Thomason (1978) to obtain yet another parity theorem: *the number of decompositions of a graph into a Hamilton circuit and a spanning tree containing a specified edge is even* (see also Chrobak and Poljak 1988). A corollary of Theorem 8.32 is that a $2d$-regular graph with one Hamilton decomposition has at least $3^{d-1}(d-1)!$ such decompositions (Thomason 1978). The question as to which graphs admit Hamilton decompositions has received much study. A graph that can be decomposed into d Hamilton circuits must be regular of degree $2d$ and have edge connectivity $2d$. Petersen (1891) proved that a regular graph of even degree necessarily admits a decomposition into 2-factors. The examples of Meredith (1973) show, however, that a decomposition into Hamilton circuits is not guaranteed. Kotzig (1957) found an elegant necessary and sufficient condition for certain 4-regular graphs to admit Hamilton decompositions.

Theorem 8.33. *The line graph of a simple 3-regular graph admits a Hamilton decomposition if and only if the graph is 3-edge-colourable.*

Many 4-regular graphs without Hamilton decompositions can be produced using this theorem, the smallest being the line graph of the Petersen graph (fig. 8). But no 2-connected planar example can be so obtained because, by virtue of the Four-Colour Theorem (5.5), every simple 2-connected 3-regular planar graph is 3-edge-colourable. Such examples can, however, be found with the aid of a necessary condition akin to Grinberg's condition (Theorem 5.6), for a 4-regular planar graph to admit a Hamilton decomposition (Bondy and Häggkvist 1981). A partial analogue of Theorem 8.33 for 4-regular graphs was found by Jaeger (1983).

Theorem 8.34. *The line graph of a simple 4-regular graph admits a Hamilton decomposition if the graph itself admits a Hamilton decomposition.*

The above theorem is not a true analogue of Theorem 8.33 because the condition is not both necessary and sufficient. Indeed, the line graph of any simple 3-connected 4-regular graph has a Hamilton decomposition, by a result of Jackson (1991), whereas some 3-connected 4-regular graphs are not even hamiltonian. Jackson (1991) put forward the following general conjecture.

Conjecture 8.35. The line graph of a simple $2d$-regular graph admits a Hamilton decomposition if and only if it has edge connectivity $4d-2$.

Conjecture 8.25 includes a longstanding question of P.J. Kelly (see Moon 1968, p. 7, Exercise 9) as to whether every diregular tournament admits a decomposition into directed Hamilton circuits: deleting a vertex from any such tournament T on $2m+1$ vertices, one is left with a tournament T' on $2m$ vertices and excess m; a decomposition of T' into m directed paths is necessarily a decomposition into Hamilton paths, and these can be extended to yield a decomposition of T into directed Hamilton circuits. The corresponding question for bipartite tournaments was put forward by Jackson (1981a).

Conjecture 8.36. Every diregular bipartite tournament admits a decomposition into directed Hamilton circuits.

Hamilton decompositions of graphs and digraphs are surveyed by Alspach et al. (1990).

8.7. Compatible circuit decompositions

We discuss here a question on circuit decompositions of eulerian graphs which has important links to coverings. A *transition* in a graph G is a pair of adjacent edges. If v is a vertex of even degree in G, the edges incident to v can be partitioned into transitions; if v is a vertex of odd degree, all but one of these edges can be so partitioned. Specifying such a partition for each vertex v determines a *transition system* for G. Two sets of transitions are *compatible* if they are disjoint. Transition systems were first studied by Kotzig (1968).

An Euler tour induces a transition system whose elements are the pairs of consecutive edges of the tour. Likewise, a circuit decomposition induces a transition system whose elements are the pairs of consecutive edges of circuits of the decomposition. The tour and decomposition are *compatible* if the systems of transitions they induce are compatible (except at vertices of degree two). Kotzig (1968) proved that to each circuit decomposition of an eulerian graph there corresponds a compatible Euler tour. The converse was conjectured by G. Sabidussi (see Fleischner 1980).

Conjecture 8.37. Let G be an eulerian graph and T an Euler tour of G. Then G admits a circuit decomposition compatible with T.

Fleischner (1980) observed that the conjecture is true if $d(v) \equiv 0 \pmod 4$ for each vertex v. Colouring the edges of T alternately red and blue induces a decomposition of G into two even subgraphs, one red and one blue. Each of these subgraphs can be decomposed into circuits, resulting in a circuit decomposition of G compatible with T. Fleischner (1980) also proved a stronger theorem in the case of planar graphs. A transition system $\mathcal{T}$ is *nonseparating* if $G \setminus \{e_1, e_2\}$ is connected for each transition $\{e_1, e_2\} \in \mathcal{T}$.

Theorem 8.38. *Let G be a planar eulerian graph, and let $\mathcal{T}$ be a nonseparating transition system of G. Then G admits a circuit decomposition compatible with $\mathcal{T}$.*

This readily implies Theorem 8.19 in the case of 4-regular graphs, as Fleischner and Frank (1990) remark. As before, one colours the edges of an Euler tour alternately red and blue. Since the number of edges is even, each vertex is then incident with two edges of each colour. The pairs of edges of the same colour define a nonseparating transition system $\mathcal{T}$ of G. Any circuit decomposition compatible with $\mathcal{T}$ consists of circuits whose edges alternate in colour, and which are therefore even.

Remark 8.39. More generally, any edge colouring of a graph induces a set of transitions, namely the pairs of adjacent edges of the same colour; a circuit is then compatible with the edge colouring if no two consecutive edges of the circuit have the same colour. Such compatibility questions have been considered by a number of authors (see, for example, Grossman and Häggkvist 1983). In particular, Fleischner and Frank (1990) generalized Theorem 8.38 to sets of transitions induced by edge colourings in this way. An alternative set of transitions determined by an edge colouring is the set of all pairs of adjacent edges of different colour; in this case, a compatible circuit is one that is monochromatic. Compatibility questions of this latter type are touched upon in Remark 6.6.

A generalization of Dirac's Theorem (2.1) involving the notion of compatibility was proposed by R. Häggkvist (see Hellgren 1988, p. 15).

Conjecture 8.40. Let G be a simple graph of minimum degree δ on n vertices, where $\delta \geqslant n/2$ and n is sufficiently large, and let $\mathcal{T}$ be a transition system for G. Then G contains a Hamilton circuit compatible with $\mathcal{T}$.

The theory of eulerian graphs is extensive, and we refer the reader to the treatise by Fleischner (1990) for further details.

8.8. Double covers

By Theorem 1.4, every even graph admits a circuit decomposition, but no graph with a vertex of odd degree admits such a decomposition, however. In order to formulate an analogue of Theorem 1.4 valid for all graphs, one is led to the concept of a *circuit double cover*, that is, a collection $\mathcal{C}$ of circuits such that each edge of the graph belongs to exactly two members of $\mathcal{C}$.

A necessary condition for a connected graph to admit a circuit double cover is that it be 2-edge-connected, because each edge must lie in some circuit. Szekeres (1973) and Seymour (1979), motivated by quite different considerations, conjectured that this condition is also sufficient.

Conjecture 8.41 (The Circuit Double Cover Conjecture). Every 2-edge-connected graph admits a circuit double cover.

According to Tutte (personal communication 1992), the Circuit Double Cover Conjecture was already known in the 1950s. It has been studied extensively. A recommended reference is the survey article by Jaeger (1985); additional material can be found in Goddyn (1988), including a proof that the girth of a minimal counterexample is at least ten. The conjecture has close links to several other basic questions. It is implied by the combination of Conjectures 8.37 and 6.12 (Fleischner 1984). Perhaps more surprisingly, it is also implied by Conjecture 8.13 (Jamshy and Tarsi 1992). Moreover, to each 2-connected loopless graph can be associated an eulerian graph that has an even circuit decomposition if and only if the original graph has a circuit double cover (Jaeger 1985).

Several refinements of the Circuit Double Cover Conjecture have been formulated, of which we mention four. First, a conjecture on embeddings, due to Haggard (1977). An *embedding* of a graph G in a surface Σ is a function ϕ that assigns to each vertex a point of Σ and to each edge a simple curve in Σ, in such a way that curves $\phi(e_1)$ and $\phi(e_2)$ meet at point $\phi(v)$ of Σ if and only if edges e_1 and e_2 are incident with vertex v in G. The connected components of $\Sigma \setminus \phi(G)$ are the *faces* of G.

Conjecture 8.42. Every 2-edge-connected graph can be embedded in some surface in such a way that each face is bounded by a circuit.

Conjecture 8.42 implies the Circuit Double Cover Conjecture because the circuits bounding the faces in such an embedding constitute a circuit double cover. A still stronger conjecture is that the surface can be chosen to be orientable and the embedded graph 5-face-colourable (see Jaeger 1985).

Next, a conjecture of Seymour (1979) on edge-weighted graphs. A collection $\mathscr{C}$ of circuits in a graph G induces a weighting w of G, the weight $w(e)$ being the number of circuits of $\mathscr{C}$ that include edge e. Seymour asked which edge-weighted graphs arise in this way, that is, which edge-weighted graphs are *sums of circuits*. One obvious necessary condition is that:

(i) for each edge e, $w(e)$ is a nonnegative integer.

There are also two simple necessary conditions involving edge cuts, both consequences of the fact that a circuit meets an edge cut in an even number of edges:

(ii) for each edge cut S, $\sum_{e \in S} w(e)$ is even;

(iii) for each edge cut S and each edge $f \in S$, $w(f) \leqslant \sum_{e \in S \setminus f} w(e)$.

Seymour (1979) proved that conditions (i)–(iii) are sufficient for an edge-weighted planar graph to be a sum of circuits. They are not sufficient in general: a Petersen graph (fig. 8) in which the edges of a perfect matching have weight two and the remaining edges have weight one is not a sum of circuits. Alspach et al. (1994) extended Seymour's result by proving that this configuration is present in any edge-weighted graph that is not a sum of circuits.

Theorem 8.43. *Let G be an edge-weighted graph whose weight function w satisfies* *conditions* (i)–(iii) *above. Then either G is a weighted sum of circuits or there is a weighting w' of G such that*

(a) $w'(e) \leqslant w(e)$ *for each edge e of G,*

(b) $w'(e) \in \{0, 1, 2\}$ *for each edge e of G, and*

(c) *the support of w' can be obtained from the edge-weighted Petersen graph described above by recursively applying the following two operations: subdividing an edge of weight one or two, and replacing an edge of weight two by two multiple edges of weight one.*

Seymour (1979) conjectured that conditions (i)–(iii) are sufficient, however, if all weights are even (in which case, condition (ii) is automatically satisfied).

Conjecture 8.44. An edge-weighted graph with nonnegative even weights is a sum of circuits if and only if, for each edge cut S and each edge $f \in S$, $w(f) \leqslant \sum_{e \in S \setminus f} w(e)$.

Thirdly, an analogue of Conjecture 8.20 for double covers, proposed by Bondy (1990).

Conjecture 8.45. Every simple 2-edge-connected graph on n vertices admits a circuit double cover $\mathscr{C}$ such that $|\mathscr{C}| \leqslant n - 1$.

Two implications of this conjecture have been established. One, concerning path double covers, was obtained by Li (1990). A *path double cover* of a graph G is a collection $\mathscr{P}$ of paths such that each edge of G belongs to exactly two members of $\mathscr{P}$.

Theorem 8.46. *Every simple graph G admits a path double cover $\mathscr{P}$ such that each vertex of G occurs exactly twice as an end of a path of $\mathscr{P}$.*

The other is an implication for edge-weighted graphs, proved by Bondy and Fan (1991).

Theorem 8.47. *Let G be a simple 2-edge-connected edge-weighted graph on n vertices. Then G contains a circuit of weight at least $2w(G)/(n - 1)$.*

Fourthly, a conjecture due to Celmins (1984) and Preissmann (1981) on the minimum number of even subgraphs, rather than circuits, in a double cover.

Conjecture 8.48. Every 2-edge-connected graph admits a double cover by five even subgraphs.

Jaeger (1976) and Kilpatrick (1975) obtained a theorem on coverings by even subgraphs that is pertinent to Conjecture 8.48 and has other interesting implications.

Theorem 8.49. (i) *Every 4-edge-connected graph is the union of two even subgraphs.*

(ii) *Every 2-edge-connected graph is the union of three even subgraphs.*

Theorem 8.49(i) implies that every 4-edge-connected graph admits a double cover by three even subgraphs: such a graph is the union of two even subgraphs, and these subgraphs together with their symmetric difference constitute a double cover of the graph. Bermond et al. (1983) noted that Theorem 8.49(ii) likewise implies that every 2-edge-connected graph admits a quadruple cover by seven even subgraphs. This is not thought to be best possible, however. Fulkerson (1971) conjectured that every 2-edge-connected 3-regular graph admits a double cover by six perfect matchings, and Jaeger (1985) observed that Fulkerson's conjecture can be reformulated in terms of even subgraphs as follows.

Conjecture 8.50. Every 2-edge-connected graph admits a quadruple cover by six even subgraphs.

Fan (1992) proved that every 2-edge-connected graph has a sextuple cover by ten even subgraphs; for 3-regular non-3-edge-colourable graphs, this number is least possible. Many results on circuit covers, such as this one, are proved by the methods of integer flows. This topic is discussed in the appendix to chapter 4 by Seymour; another excellent source is the survey by Jaeger (1988).

We close by noting an attractive result of Chen (1971) and T. Gallai (see Lovász 1979, p. 254) akin to the two assertions of Theorem 8.49. An inductive proof due to L. Pósa is described in Lovász (1979); for a very short proof, see Woodall (1990).

Theorem 8.51. *Every graph can be decomposed into an even subgraph and an edge cut.*

References

Ádám, A.

[1963] The quasi-series decomposition of two-terminal graphs, *Publ. Math. Debrecen* **10**, 96–107. MR 30#574(E); correction MR 37#1270.

[1964] Problem 2, in: *Theory of Graphs and its Applications,* ed. M. Fiedler (Czech. Acad. Sci. Publ., Prague) p. 157.

Aharoni, R., I.B.A. Hartman and A.J. Hoffman

[1985] Path partitions and packs of acyclic digraphs, *Pacific J. Math.* **118**, 249–259. MR 86i:05067.

Aho, A.V., J.E. Hopcroft and J.D. Ullman

[1974] *The Design and Analysis of Computer Algorithms* (Addison-Wesley, Reading, MA) MR 54#1706.

Aigner, M., and S. Brandt

[1993] Embedding arbitrary graphs of maximum degree two, *J. London Math. Soc.* **48**, 39–51.

Alon, N.

[1991] Non-constructive proofs in combinatorics, in: *Proc. Int. Congr. Math.,* Kyoto, 1990 (Springer, Tokyo) pp. 1421–1429.

Alon, N., and N. Linial

[1989] Cycles of length 0 modulo k in directed graphs, *J. Combin. Theory B* **47**, 114–119. MR 90g:05089.

Alspach, B.

[1967] Cycles of each length in regular tournaments, *Canad. Math. Bull.* **10**, 283–286. MR 35#4125.

[1981] The search for long paths and cycles in vertex-transitive graphs and digraphs, in: *Combinatorial Mathematics VIII, Lecture Notes in Mathematics,* Vol. 884, ed. K.L. McAveney (Springer, Berlin) pp. 14–22. MR 83b:05080.

Alspach, B., and C.D. Godsil

[1985] *Cycles in Graphs, Ann. Discrete Math.* **27**.

Alspach, B., and N.J. Pullman

[1974] Path decomposition of digraphs, *Bull. Austral. Math. Soc.* **10**, 421–427. MR 50#169.

Alspach, B., and M. Rosenfeld

[1986] On Hamilton decompositions of prisms over simple 3-polytopes, *Graphs and Combinatorics* **2**, 1–8.

Alspach, B., D.W. Mason and N.J. Pullman

[1976] Path number of tournaments, *J. Combin. Theory B* **20**, 222–228. MR 54#10065.

Alspach, B., J.-C. Bermond and D. Sotteau

[1990] Decomposition into cycles I: Hamilton decompositions, in: *Cycles and Rays*, eds. G. Hahn, G. Sabidussi and R. Woodrow, NATO ASI Ser. C (Kluwer Academic Publishers, Dordrecht) pp. 9–18.

Alspach, B., L.A. Goddyn and C.Q. Zhang

[1994] Graphs with the circuit cover property, *Trans. Amer. Math. Soc.* **344**, 131–154.

Amar, D., E. Flandrin, I. Fournier and A. Germa

[1991] Pancyclism in hamiltonian graphs, *Discrete Math.* **89**, 111–131.

Appel, K., and W. Haken

[1976] Every planar map is four colorable, *Bull. Amer. Math. Soc.* **82**, 711–712. MR 54#12561.

Babai, L.

[1979] Long cycles in vertex-transitive graphs, *J. Graph Theory*, **3**, 301304. MR 80m:05059.

Bang-Jensen, J., and C. Thomassen

[1992] A polynomial algorithm for the 2-path problem for semicomplete digraphs, *SIAM J. Discrete Math.* **5**, 366–376.

Bang-Jensen, J., Y. Manoussakis and C. Thomassen

[1992] A polynomial algorithm for hamiltonian-connectedness in semicomplete digraphs, *J. Algorithms*, **13**, 114–127.

Bánkfalvi, M., and Zs. Bánkfalvi

[1968] Alternating hamiltonian circuits in two-coloured complete graphs, in: *Theory of Graphs*, eds. P. Erdős and G. Katona (Academic Press, New York) pp. 11–18. MR 38#2052.

Barefoot, C.A., L. Clark and R.C. Entringer

[1986] Cubic graphs with the minimum number of cycles, in: *Congress. Numerantium*, Vol. 53, eds. F. Hoffman, R.C. Mullin, R.G. Stanton and K.B. Reid (Utilitas Math., Winnipeg) pp. 49–62. MR 88k:05102.

Barefoot, C.A., L.H. Clark, J. Douthett, R.C. Entringer and M.R. Fellows

[1991] Cycles of length 0 modulo 3 in graphs, in: *Graph Theory, Combinatorics and Applications*, eds. Y. Alavi, G. Chartrand, O. Oellerman and A.J. Schwenk (Wiley, New York) pp. 87–101.

Barnette, D.

[1966] Trees in polyhedral graphs, *Canad. J. Math.* **18**, 731–736. MR 33#3951.

[1969] Conjecture 5, in: *Recent Progress in Combinatorics*, ed W.T. Tutte (Academic Press, New York) p. 343.

Bauer, D., H.J. Broersma, R. Li and H.J. Veldman

[1989] A generalization of a result of Häggkvist and Nicoghossian, *J. Combin. Theory B* **47**, 237–243. MR 91a:05066.

Bauer, D., S.L. Hakimi and E.F. Schmeichel

[1990] Recognizing tough graphs is NP-hard, *Discrete Appl. Math.* **28**, 191–195.

Bauer, D., E.F. Schmeichel and H.J. Veldman

[1991] Some recent results on long cycles in tough graphs, in: *Graph Theory, Combinatorics and Applications*, eds. Y. Alavi, G. Chartrand, O. Oellerman and A.J. Schwenk (Wiley, New York) pp. 113–123.

Beck, J.

[1983] On size Ramsey number of paths, trees, and circuits, I, *J. Graph Theory* **7**, 115–129.

[1990] On size Ramsey number of paths, trees, and circuits. II, in: *Mathematics of Ramsey Theory*, eds. J. Nešetřil and V. Rödl (Springer, Berlin) pp. 34–45.

Behzad, M., G. Chartrand and C.E. Wall
[1970] On minimal regular digraphs with given girth, *Fund. Math.* **69**, 27–231. MR 44#2666.

Benson, C.T.
[1966] Minimal regular graphs of girths eight and twelve, *Canad. J. Math.* **18**, 1091–1094. MR 33#5507.

Berge, C.
[1963] Problèmes plaisans et délectables, *Rev. Française Rech. Opér.* **29**(1), 405–408.
[1982] k-optimal partitions of a directed graph, *European J. Combin.* **3**, 97–101. MR 84d:05086.
[1983] Path partitions in directed graphs, in: *Combinatorial Mathematics,* eds. C. Berge, D. Bresson, P. Camion and F. Sterboul, *Ann. Discrete Math.* **17**, 59–63. MR 87f:05096.
[1985] *Graphs* (North-Holland, Amsterdam). MR 87e:05050.
[1989] *Hypergraphs* (North-Holland, Amsterdam). MR 90h:05090.

Berman, K.A.
[1983] Proof of a conjecture of Häggkvist on cycles and independent edges, *Discrete Math.* **46**, 9–13. MR 85a:05051.
[1986a] Parity results on connected f-factors, *Discrete Math.* **59**, 1–8. MR 87f:05127.
[1986b] oral communication.

Bermond, J.-C.
[1976] On Hamiltonian walks, in: *Proc. Fifth British Combinatorial Conference,* eds. C.St.J.A. Nash-Williams and J. Sheehan (Utilitas Math., Winnipeg) pp. 41–51. MR 53#2742.
[1978a] Hamilton decomposition of graphs, directed graphs and hypergraphs, in: *Advances in Graph Theory,* ed. B. Bollobás, *Ann. Discrete Math.* **3**, 21–28. MR 58#21803.
[1978b] Hamiltonian graphs, in: *Selected Topics in Graph Theory,* eds. L.W. Beineke and R.J. Wilson (Academic Press, New York) pp. 127–167. (See MR 81e:05059.).

Bermond, J.-C., and L. Lovász
[1975] Problem 3, in: *Recent Advances in Graph Theory,* ed. M. Fiedler (Academia, Prague) p. 541.

Bermond, J.-C., and C. Thomassen
[1981] Cycles in digraphs – a survey, *J. Graph Theory* **5**, 1–43. MR 82k:05053.

Bermond, J.-C., B. Jackson and F. Jaeger
[1983] Shortest coverings of graphs with cycles, *J. Combin. Theory B* **35**, 297–308. MR 86a:05078.

Bien, F.
[1989] Construction of telephone networks by group representations, *Notices Amer. Math. Soc.* **36**, 5–22. MR 90a:90052.

Biggs, N.L., and A.T. White
[1979] *Permutation Groups and Combinatorial Structures, London Math. Soc. Lecture Note Series,* Vol. 33 (Cambridge University Press, Cambridge). MR 80k:20005.

Biggs, N.L., E.K. Lloyd and R.J. Wilson
[1976] *Graph Theory 1736–1936* (Clarendon Press, Oxford). MR 56#2771.

Bollobás, B.
[1977] Cycles modulo k, *Bull. London Math. Soc.* **9**, 97–98. MR 56#181.
[1978] *Extremal Graph Theory* (Academic Press, New York). MR 80a:05120.
[1985] *Random Graphs* (Academic Press, London). MR 87f:05152.

Bollobás, B., and C. Brightwell
[1993] Cycles through specified vertices, *Combinatorica* **13**, 147–155.

Bondy, J.A.
[1971a] Pancyclic graphs, *J. Combin. Theory B* **11**, 80–84. MR 44#2642.
[1971b] Large cycles in graphs, *Discrete Math.* **1**, 121–132. MR 44#3903.
[1972] Variations on the hamiltonian theme, *Canad. Math. Bull.* **15**, 57–62. MR 47#3241.
[1976] Diconnected orientations and a conjecture of Las Vergnas, *J. London Math. Soc.(2),* **14**, 277–282. MR 56#8414.
[1977] Cycles in digraphs, in: *Congress. Numerantium,* Vol. 19, eds. F. Hoffman, L. Lesniak-Foster, D. McCarthy, R.C. Mullin, K.B. Reid and R.G. Stanton (Utilitas Math., Winnipeg) pp. 91–98. MR 58#21805.

[1978a] A remark on two sufficient conditions for Hamilton cycles, *Discrete Math.* **22**, 191–193. MR 80j:05084.

[1978b] Hamilton cycles in graphs and digraphs, in: *Congress. Numerantium,* Vol. 21, eds. F. Hoffman, D. McCarthy, R.C. Mullin and R.G. Stanton (Utilitas Math., Winnipeg) pp. 3–28. MR 80k:05074.

[1981] Integrity in graph theory, in: *The Theory and Applications of Graphs,* eds G. Chartrand, Y. Alavi, D.L. Goldsmith, L. Lesniak-Foster and D.R. Lick (Wiley, New York) pp. 117–125. MR 83e:05070.

[1990] Small cycle double covers of graphs, in: *Cycles and Rays,* eds. G. Hahn, G. Sabidussi and R. Woodrow, NATO ASI Ser. C (Kluwer Academic Publishers, Dordrecht) pp. 21–40.

[1995] A short proof of the Chen–Manalastas theorem, *Discrete Math.,* to appear.

Bondy, J.A., and V. Chvátal

[1976] A method in graph theory, *Discrete Math.* **15**, 111–135. MR 54#2531.

Bondy, J.A., and G. Fan

[1991] Cycles in weighted graphs, *Combinatorica* **11**, 191–205.

Bondy, J.A., and R. Häggkvist

[1981] Edge-disjoint Hamilton cycles in 4-regular planar graphs, *Aequationes Math.* **22**, 42–45. MR 83a:05090.

Bondy, J.A., and F.Y. Halberstam

[1986] Parity theorems for paths and cycles in graphs, *J. Graph Theory* **10**, 107–115. MR 87f:05097.

Bondy, J.A., and S.C. Locke

[1981] Relative lengths of paths and cycles in 3-connected graphs, *Discrete Math.* **33**, 111–122. MR 82b:05088.

Bondy, J.A., and L. Lovász

[1981] Cycles through specified vertices of a graph, *Combinatorica* **1**, 117–140. MR 82k:05073.

Bondy, J.A., and U.S.R. Murty

[1976] *Graph Theory with Applications* (Macmillan, London). MR 54#117.

Bondy, J.A., and M. Simonovits

[1974] Cycles of even length in graphs, *J. Combin. Theory B* **16**, 97–105. MR 49#4851.

Bondy, J.A., and C. Thomassen

[1977] A short proof of Meyniel's theorem, *Discrete Math.* **19**, 195–197. MR 57#5811.

Bosák, J.

[1967] Hamiltonian lines in cubic graphs, in: *Theory of Graphs,* ed. P. Rosenstiehl (Gordon and Breach, New York) pp. 35–46. MR 36#5022.

Broersma, H.J., J. van den Heuvel and H.J. Veldman

[1993] A generalization of Ore's theorem involving neighbourhood unions, *Discrete Math.* **122**, 37–49.

Broersma, H.J., J. van den Heuvel, B. Jackson and H.J. Veldman

[preprint] *Hamiltonicity of Regular 2-Connected Graphs.*

Brown, W.G.

[1966] On graphs that do not contain a Thomsen graph, *Canad. Math. Bull.* **9**, 281–285. MR 34#81.

Caccetta, L., and R. Häggkvist

[1978] On minimal digraphs with given girth, in: *Congress. Numerantium,* Vol. 21, eds. F. Hoffman, D. McCarthy, R.C. Mullin and R.G. Stanton (Utilitas Math., Winnipeg) pp. 181–187. MR83h:05055.

Camion, P.

[1959] Chemins et circuits hamiltoniens des graphes complets, *C.R. Acad. Sci. Paris* **249**, 2151–2152. MR 23#A75.

Celmins, U.

[1984] *On cubic graphs that do not have an edge 3-colouring,* Ph.D. Thesis (University of Waterloo).

Chen, C.C., and P. Manalastas Jr

[1983] Every finite strongly connected digraph of stability 2 has a Hamiltonian path, *Discrete Math.* **44**, 243–250. MR 84g:05065.

Chen, G.T., and A. Saito

[1994] Graphs with a cycle of length divisible by three, *J. Combin. Theory B* **60**, 277–292.

Chen, W.K.
[1971] On vector spaces associated with a graph, *SIAM J. Appl. Math.* **20**, 26–529. MR 44#3905.

Chrobak, M., and S. Poljak
[1988] On common edges in optimal solutions to traveling salesman and other optimization problems, *Discrete Appl. Math.* **20**, 101–111. MR 89h:90083.

Chvátal, V.
[1972] On Hamilton's ideals, *J. Combin. Theory B* **12**, 163–168. MR 45#3228.
[1973] Tough graphs and Hamiltonian circuits, *Discrete Math.* **5**, 215–228. MR 47#4849.
[1985] Hamiltonian cycles, in: *The Traveling Salesman Problem: A Guided Tour through Combinatorial Optimization,* eds. E.L. Lawler, J.K. Lenstra, A.H.G. Rinnooy Kan and D.B. Shmoys (Wiley, New York) 403–430.

Chvátal, V., and P. Erdős
[1972] A note on Hamiltonian circuits, *Discrete Math.* **2**, 111–113. MR 45#6654.

Chvátal, V., and E. Szemerédi
[1983] Short cycles in directed graphs, *J. Combin. Theory B* **35**, 323–327. MR 85f:05065.

Chvátal, V., and C. Thomassen
[1978] Distances in orientations of graphs, *J. Combin. Theory B* **24**, 61–75. MR 58#10589.

Clark, L.H., and R.C. Entringer
[1983] Smallest maximally nonhamiltonian graphs, *Period. Math. Hungar.* **14**, 57–68. MR 84i:05065.

Cook, S.A.
[1971] The complexity of theorem-proving procedures, in: *Proc. 3rd Annu. ACM Symp. on Theory of Computing,* eds. P.M. Lewis, M.J. Fisher, J.E. Hopcroft, A.L. Rosenberg, J.W. Thatcher and P.R. Young (ACM, New York) pp. 151–158.

Corrádi, K., and A. Hajnal
[1963] On the maximal number of independent circuits in a graph, *Acta Math. Acad. Sci. Hungar.* **14**, 423–439. MR 34#84.

Cunningham, W.H.
[1990] Computing the binding number of a graph, *Discrete Appl. Math.* **27**, 283–285. MR 91m:05175.

Dankelmann, P., T. Niessen and I. Schiermeyer
[preprint] *Properties of Path-Tough Graphs.*

Dean, N.
[1986] What is the smallest number of dicycles in a dicycle decomposition of an eulerian digraph? *J. Graph Theory* **10**, 299–308. MR 87i:05135.

Dean, N., A. Kaneko, K. Ota and B. Toft
[1991] *Cycles modulo 3,* DIMACS Technical Report 91-32, May.

Dean, N., L. Lesniak and A. Saito
[1993] Cycles of length 0 modulo 4 in graphs, *Discrete Math.* **121**, 37–49.

Dijkstra, E.W.
[1959] A note on two problems in connection with graphs, *Numer. Math.* **1**, 269–271. MR 21#6334.

Dilworth, R.P.
[1950] A decomposition theorem for partially ordered sets, *Ann. of Math.* **51**, 161–166. MR 11–309.

Dirac, G.A.
[1952] Some theorems on abstract graphs, *Proc. London Math. Soc.* **2**, 69–81. MR 13–856.
[1960] In abstrakten Graphen vorhandene vollständige 4-Graphen und ihre Unterteilungen, *Math. Nachr.* **22**, 61–85. MR 22#12053.

Donald, A.
[1980] An upper bound for the path number of a graph, *J. Graph Theory* **4**, 189–201. MR 81f:05102.

Edmonds, J., and E.L. Johnson
[1973] Matching, Euler tours and the Chinese postman, *Math. Programming* **5**, 88–124. MR 48#168.

Edmonds, J.R.
[1965] Paths, trees and flowers, *Canad. J. Math.* **17**, 449–467. MR 31#2165.

Egawa, Y.
[preprint] *Vertex-disjoint Cycles of the same Length.*
Egawa, Y., and T. Miyamoto
[1989] The longest cycles in a graph G with minimum degree at least $|G|/k$, *J. Combin. Theory B* **46**, 356–362. MR 90d:05139.
Egawa, Y., R. Glas and S.C. Locke
[1991] Cycles and paths through specified vertices in k-connected graphs, *J. Combin. Theory B* **52**, 20–29.
El-Zahar, M.H.
[1984] On circuits in graphs, *Discrete Math.* **50**, 227–230. MR 85g:05092.
Ellingham, M.N., D.A. Holton and C.H.C. Little
[1984] Cycles through ten vertices in 3-connected cubic graphs, *Combinatorica* **4**, 265–273. MR 86j:05087.
Enomoto, H., B. Jackson, P. Katerinis and A. Saito
[1985] Toughness and the existence of k-factors, *J. Graph Theory* **9**, 87–95. MR 86f:05075.
Entringer, R.C., and E.F. Schmeichel
[1988] Edge conditions and cycle structure in bipartite graphs, *Ars Combin.* **26**, 229–232. MR 89m:05069.
Entringer, R.C., and H. Swart
[1980] Spanning cycles of nearly cubic graphs, *J. Combin. Theory B* **29**, 303–309. MR 82e:05093.
Erdős, P.
[1964] Problem 9, in: *Theory of Graphs and its Applications,* ed. M. Fiedler (Czech. Acad. Sci. Publ., Prague) p. 159.
[1965a] On an extremal problem in graph theory, *Colloq. Math.* **13**, 251254. MR 31#3353.
[1965b] On some extremal problems in graph theory, *Israel J. Math.* **3**, 113–116. MR 32#7443.
[1975] Some recent progress on extremal problems in graph theory, in: *Congress. Numerantium,* Vol. 14, eds. F. Hoffman, R.C. Mullin, R.B. Levow, D. Roselle, R.G. Stanton and R.S.D. Thomas (Utilitas Math., Winnipeg) pp. 3–14. MR 52#13488.
[1990] Letter, June 4.
Erdős, P., and T. Gallai
[1959] On maximal paths and circuits of graphs, *Acta Math. Acad. Sci. Hungar.* **10**, 337–356. MR 22#5591.
Erdős, P., and A. Hajnal
[1966] On chromatic numbers of graphs and set-systems, *Acta Math. Acad. Sci. Hungar.* **17**, 61–99. MR 33#1247.
Erdős, P., and L. Pósa
[1962] On the maximal number of disjoint circuits of a graph, *Publ. Math. Debrecen* **9**, 3–12. MR 27#743.
Erdős, P., A. Goodman and L. Pósa
[1966a] The representation of a graph by set intersections, *Canad. J. Math.* **18**, 106–112. MR 32#4034.
Erdős, P., A. Rényi and V.T. Sós
[1966b] On a problem of graph theory, *Studia Sci. Math. Hungar.* **1**, 51–57. MR 36#6310.
Euler, L.
[1736] Solutio problematis ad geometriam situs pertinentis, *Comm. Acad. Sci. Imp. Petropol.* **8**, 128–140.
Fan, G.
[1984] New sufficient conditions for cycles in graphs, *J. Combin. Theory B* **37**, 221–227. MR 86c:05083.
[1985] Longest cycles in regular graphs, *J. Combin. Theory B* **39**, 325–345. MR 87b:05078.
[1990] Covering weighted graphs by even subgraphs, *J. Combin. Theory B* **49**, 137–141. MR 91e:05062.
[1992] Integer flows and cycle covers, *J. Combin. Theory B* **54**, 113–122.
Fan, G., and H. Kierstead
[preprint] *Hamiltonian Square-Paths.*
Fan, G., and H.A. Kierstead
[1995] The square of paths and cycles, *J. Combin. Theory B,* to appear.
Faudree, R.J., and R.H. Schelp
[1975] The entire graph of a bridgeless connected plane graph is panconnected, *J. London Math. Soc.* **12**, 59–66. MR 52#5460.

Fleischner, H.

[1974] The square of every two-connected graph is Hamiltonian, *J. Combin. Theory B* **16**, 29–34. MR 48#10899b.

[1980] Eulersche Linien und Kreisüberdeckungen die vorgegebene Durchgänge in den Kanten vermeiden, *J. Combin. Theory B* **29**, 145–167. MR 82e:05094.

[1984] Cycle decompositions, 2-coverings, removable cycles and the four-color-disease, in: *Progress in Graph Theory*, eds. J.A. Bondy and U.S.R. Murty (Academic Press, Toronto) pp. 233–246. MR 86d:05072.

[1990] *Eulerian Graphs and Related Topics, Ann. Discrete Math.* **45**. MR 91g:05086.

[1994] Uniqueness of maximal dominating cycles in 3-regular graphs and of Hamiltonian cycles in 4-regular graphs, *J. Graph Theory* **18**, 449–459.

Fleischner, H., and A. Frank

[1990] On circuit decomposition of planar Eulerian graphs, *J. Combin. Theory B* **50**, 245–253. MR 91i:05076.

Fleischner, H., and B. Jackson

[1989] A note concerning some conjectures on cyclically 4-edge-connected 3-regular graphs, in: *Graph Theory in Memory of G.A. Dirac*, eds. L.D. Andersen, C. Thomassen and B. Toft, *Ann. Discrete Math.* **41**, 171–178. MR 90c:05135.

Forcade, R.

[1973] Parity of paths and circuits in tournaments, *Discrete Math.* **6**, 115–118. MR 47#8352.

Fortune, S., J. Hopcroft and J. Wyllie

[1980] The directed subgraph homeomorphism problem, *Theor. Comput. Sci.* **10**, 111–121. MR 81e:68079.

Fouquet, J.L., and H. Thuillier

[1990] On some conjectures on cubic 3-connected graphs, *Discrete Math.* **80**, 41–57. MR 91h:05077.

Fournier, I.

[1982] *Cycles et Numérotations de Graphes*, Thèse de Doctorat de 3ème cycle (Université de Paris-Sud).

Fournier, I., and P. Fraisse

[1985] On a conjecture of Bondy, *J. Combin. Theory B* **39**, 17–26. MR 87a:05094.

Fraisse, P.

[1986] *D_λ-Cycles and their Applications for Hamiltonian Graphs*, Thèse de Doctorat d'état (Université de Paris-Sud).

Frank, A.

[1980] On chain and antichain families of a partially ordered set, *J. Combin. Theory B* **29**, 176–184. MR 81m:06004.

Friedland, S.

[1989] Every 7-regular digraph contains an even cycle, *J. Combin. Theory B* **46**, 249–252. MR 90c:05126.

Frieze, A.M.

[1989] On matchings and Hamilton cycles in random graphs, in: *Surveys in Combinatorics, 1989, London Math. Soc. Lecture Note Series*, Vol. 141 (Cambridge University Press, Cambridge) pp. 84–114. MR 90k:05125.

Fulkerson, D.R.

[1971] Blocking and antiblocking pairs of polyhedra, *Math. Programming* **1**, 168–194. MR 45#3222.

Füredi, Z.

[1983] Graphs without quadrilaterals, *J. Combin. Theory B* **34**, 187–190. MR 85b:05104.

[preprint] *On the Number of Edges of Quadrilateral-free Graphs.*

Gallai, T.

[1964] Problem 15, in: *Theory of Graphs and its Applications*, ed. M. Fiedler (Czech. Acad. Sci. Publ., Prague) p. 161.

[1968a] On directed paths and circuits, in: *Theory of Graphs*, eds. P. Erdős and G. Katona (Academic Press, New York) pp. 115–118. MR 38#2054.

[1968b] Problem 6, in: *Theory of Graphs*, eds. P. Erdős and G. Katona (Academic Press, New York) p. 362.

Gallai, T., and A.N. Milgram
[1960] Verallgemeinerung eines graphentheoretischen Satzes von Rédei, *Acta Sci. Math. Szeged* **21**, 181–186. MR 25#3862.

Gao, Z.C., and R.B. Richter
[1994] 2-walks in circuit graphs, *J. Combin. Theory B* **62**, 259–267.

Garey, M.R., and D.S. Johnson
[1979] *Computers and Intractability: A Guide to the Theory of NP-Completeness* (Freeman, San Francisco). MR 80g:68056.

Goddyn, L.A.
[1988] *Cycle covers of graphs,* Ph.D. Thesis (University of Waterloo).

Gould, R.J.
[1991] Updating the hamiltonian problem – a survey, *J. Graph Theory* **15**, 121–157.

Graham, R.L., B.L. Rothschild and J.H. Spencer
[1980] *Ramsey Theory* (Wiley, New York). MR 82b:05001.

Grinberg, E.Ja.
[1968] Plane homogeneous graphs of degree three without Hamiltonian circuits, *Latvian Math. Yearbook* **4**, 51–58 (in Russian). MR 39#96.
[1987] Examples of non-Ádám multigraphs, *Latvian Math. Yearbook* **31**, 28–138 (in Russian). MR 89d:05087.

Grossman, J.W., and R. Häggkvist
[1983] Alternating cycles in edge-partitioned graphs, *J. Combin. Theory B* **34**, 77–81. MR 84h:05044.

Grötschel, M.
[1978] oral communication.
[1984] On intersections of longest cycles, in: *Graph Theory and Combinatorics,* ed. B. Bollobás (Academic Press, London) pp. 171–189. MR 86d:05073.

Grötschel, M., and W.R. Pulleyblank
[1981] Weakly bipartite graphs and the max-cut problem, *Oper. Res. Lett.* **1**, 23–27. MR 83e:05048.

Grünbaum, B.
[1970] Polytopes, graphs and complexes, *Bull. Amer. Math. Soc.* **76**, 1131–1201. MR 42#959.

Gustavsson, T.
[1991] *Decompositions of large graphs and digraphs with high minimum degree,* Doctoral Dissertation (Stockholm University). ISBN 91–7146–883–8.

Gutin, G.
[1984] A criterion for complete bipartite digraphs to be Hamiltonian, *Vestsi Acad. Navuk BSSR Ser. Fiz.-Mat. Navuk* **1**, 99–100 (in Russian).

Gyárfás, A., J. Komlós and E. Szemerédi
[1984] On the distribution of cycle lengths in graphs, *J. Graph Theory* **4**, 441–462. MR 85k:05068.

Gyárfás, A., H.J. Prömel, E. Szemerédi and B. Voigt
[1985] On the sum of the reciprocals of cycle lengths in sparse graphs, *Combinatorica* **6**, 41–52. MR 86m:05059.

Haggard, G.
[1977] Edmonds' characterization of disc embeddings, in: *Congress. Numerantium,* Vol. 19, eds. F. Hoffman, L. Lesniak-Foster, D. McCarthy, R.C. Mullin, K.B. Reid and R.G. Stanton (Utilitas Math., Winnipeg) pp. 291–302. MR 58#10466.

Häggkvist, R.
[1978a] Conjecture 4, in: *Combinatorics,* eds. A. Hajnal and V.T. Sós, *Colloq. Math. Soc. János Bolyai* **18**, 1204.
[1978b] oral communication.
[1979] On F-Hamiltonian graphs, in: *Graph Theory and Related Topics,* eds. J.A. Bondy and U.S.R. Murty (Academic Press, New York) pp. 219–231. MR 82c:05066.
[1985a] A lemma on cycle decompositions, in: *Cycles in Graphs,* eds. B. Alspach and C.D. Godsil, *Ann. Discrete Math.* **27**, 227–232. MR 87b:05079.

[1985b] A note on Hamilton cycles, in: *Cycles in Graphs,* eds. B. Alspach and C.D. Godsil, *Ann. Discrete Math.* **27**, 233–234. MR 87a:05095.

[1985c] Equicardinal disjoint cycles in sparse graphs, in: *Cycles in Graphs,* eds B. Alspach and C.D. Godsil, *Ann. Discrete Math.* **27**, 269–274. MR 87b:05080.

[1992] On the structure of non-hamiltonian graphs, 1, *Combinatorics, Probability and Computing* **1**, 27–34.

[1993] Hamilton cycles in oriented graphs, *Combinatorics, Probability and Computing* **2**, 25–32.

Häggkvist, R., and Y. Manoussakis

[1989] Cycles and paths in bipartite tournaments with spanning configurations, *Combinatorica* **9**, 33–38. MR 9Of:05065.

Häggkvist, R., and C. Thomassen

[1976] On pancyclic digraphs, *J. Combin. Theory B* **20**, 20–40. MR 52#10481.

[1982] Circuits through specified edges, *Discrete Math.* **41**, 29–34. MR 84g:05095.

Hahn, G., and B. Jackson

[1990] A note concerning paths and independence number in digraphs, *Discrete Math.* **82**, 327–329. MR 91f:05061.

Hakimi, S.L., E.F. Schmeichel and C. Thomassen

[1979] On the number of Hamiltonian cycles in a maximal planar graph, *J. Graph Theory* **3**, 365–370. MR 80k:05075.

Harary, F., and C.St.J.A. Nash-Williams

[1965] On Eulerian and Hamiltonian graphs and line graphs, *Canad. Math. Bull.* **8**, 701–710. MR 33#66.

Hartman, I.B.A.

[1983] Long cycles generate the cycle space of a graph, *European J. Combin.* **4**, 237–246. MR 85b:05113.

Hellgren, T.B.

[1988] *Hypergraph Colourings, Compatible Hamiltonian Cycles and Restricted 2-factors in Dense Graphs,* Doctoral Dissertation (University of Stockholm).

Hendry, G.R.T.

[1984] Graphs uniquely hamiltonian-connected from a vertex, *Discrete Math.* **49**, 61–74. MR 85e:05118.

[1986] The size of graphs uniquely hamiltonian-connected from a vertex, *Discrete Math.* **61**, 57–60. MR 87j:05108.

[1990] Extending cycles in graphs, *Discrete Math.* **85**, 59–72. MR 91h:05074.

Heydemann, M.C.

[1985] Minimum number of circuits covering the vertices of a strong digraph, in: *Cycles in Graphs,* eds. B. Alspach and C.D. Godsil, *Ann. Discrete Math.* **27**, 287–296.

Hierholzer, C.

[1873] Über die Möglichkeit, einen Linienzug ohne Wiederholung und ohne Unterbrechnung zu umfahren, *Math. Ann.* **6**, 30–32.

Hilbig, F.

[1986] *Kantenstrukturen in nichthamiltonischen Graphen,* Dissertation (Technical University, Berlin).

Hilton, A.J.W.

[1984] Hamiltonian decompositions of complete graphs, *J. Combin. Theory B* **36**, 125–134. MR 85e:05119.

Hoáng, C.T., and B. Reed

[1987] A note on short cycles in digraphs, *Discrete Math.* **66**, 103–107. MR 89a:05072.

Hoffman, A.J.

[1983] Extending Greene's theorem to directed graphs, *J. Combin. Theory A* **34**, 102–107. MR 85b:05090.

Holton, D.A., and C. Thomassen

[1986] Problem 81, *Discrete Math.* **62**, 111–112.

Holton, D.A., B.D. McKay, M.D. Plummer and C. Thomassen

[1982] A nine point theorem for 3-connected graphs, *Combinatorica* **2**, 53–62. MR 84:05069.

Horák, P., and J. Širáň

[1986] On a construction of Thomassen, *Graphs and Combinatorics* **2**, 347–350. MR 89j:05045.

Huck, A.

[1991] A sufficient condition for graphs to be weakly k-linked, *Graphs and Combinatorics* **7**, 323–351.

Itai, A., R.J. Lipton, C.H. Papadimitriou and M. Rodeh
[1981] Covering graphs by simple circuits, *SIAM J. Comput.* **10**, 746–750. MR 83d:68062.

Jackson, B.
[1979] Edge-disjoint Hamilton cycles in regular graphs of large degree, *J. London Math. Soc.* **19**, 13–16. MR 80k:05078.
[1980a] Hamilton cycles in regular 2-connected graphs, *J. Combin. Theory B* **29**, 27–46. MR 82e:05096a.
[1980b] Paths and cycles in oriented graphs, in: *Combinatorics 79,* eds. M. Deza and I.G. Rosenberg, *Ann. Discrete Math.* **8**, 275–277. MR 82a:05051.
[1981a] Cycles in bipartite graphs, *J. Combin. Theory B* **30**, 332–342. MR 83h:05057.
[1981b] Long paths and cycles in oriented graphs, *J. Graph Theory* **5**, 145–157. MR 82i:05050.
[1985] Long cycles in bipartite graphs, *J. Combin. Theory B* **38**, 118–131. MR 86g:05056.
[1986] Longest cycles in 3-connected cubic graphs, *J. Combin. Theory B* **41**, 17–26. MR 87k:05106.
[1987] A Chvátal–Erdős condition for Hamilton cycles in digraphs, *J. Combin. Theory B* **43**, 245–252. MR 89a:05073.
[1989] oral communication.
[1991] A characterization of graphs having three pairwise compatible Euler tours, *J. Combin. Theory B* **53**, 80–92.
[1992] Concerning the circumference of certain families of graphs, in: *Contributions to the Twente Workshop on Hamiltonian Graph Theory,* eds. H.J. Broersma, J. van den Heuvel and H.J. Veldman, Memorandum No. 1043 (Twente University, Enschede) pp. 75–82.

Jackson, B., and H. Li
[1994] Hamilton cycles in 2-connected regular bipartite graphs, *J. Combin. Theory B* **62**, 236–258.

Jackson, B., and O. Ordaz
[1990] Chvátal–Erdős conditions for paths and cycles in graphs and digraphs. A survey, *Discrete Math.* **84**, 241–254.

Jackson, B., and T.D. Parsons
[1982] A shortness exponent for r-regular, r-connected graphs, *J. Graph Theory* **6**, 169–176. MR 83g:05044.

Jackson, B., and N.C. Wormald
[1990a] k-walks in graphs, *Australasian J. Combinatorics* **2**, 135–146.
[1990b] Cycles containing matchings and pairwise compatible Euler tours, *J. Graph Theory* **14**, 127–138. MR 91g:05087.
[1992a] Longest cycles in 3-connected planar graphs, *J. Combin. Theory B* **54**, 291–321.
[1992b] electronic mail.
[1993] Longest cycles in 3-connected graphs of bounded maximum degree, in: *Graphs, Matrices and Designs, Lecture Notes in Pure and Applied Mathematics,* Vol. 139, ed. R.S. Rees (Marcel Dekker, New York) pp. 237–254.

Jackson, B., H. Li and Y.J. Zhu
[1991] Dominating cycles in regular 3-connected graphs, *Discrete Math.* **102**, 163–176.

Jaeger, F.
[1976] On nowhere-zero flows in multigraphs, in: *Proc. Fifth British Combinatorial Conference,* eds. C.St.J.A. NashWilliams and J. Sheehan (Utilitas Math., Winnipeg) pp. 373–378. MR 52#16570.
[1979] A note on sub-Eulerian graphs, *J. Graph Theory* **3**, 91–93. MR 80k:05036.
[1983] The 1-factorization of some line graphs, *Discrete Math.* **46**, 89–92. MR 84i:05087.
[1985] A survey of the cycle double cover conjecture, in: *Cycles in Graphs,* eds. B.R. Alspach and C.D. Godsil, *Ann. Discrete Math.* **27**, 1–12. MR 87b:05082.
[1988] Nowhere-zero flow problems, in: *Selected Topics in Graph Theory 3,* eds. L.W. Beineke and R.J. Wilson (Academic Press, London) pp. 71–95.
[1989] oral communication.

Jamshy, U., and M. Tarsi
[1992] Short cycle covers and the cycle double cover conjecture, *J. Combin. Theory B* **56**, 197–204.

Jung, H.A.
[1984] *Degree and longest circuits in 3-connected graphs,* Unpublished Manuscript.

Karp, R.M.
[1972] Reducibility among combinatorial problems, in: *Complexity of Computer Computations*, eds. R.E. Miller and J.W. Thatcher (Plenum Press, New York) pp. 85–103. MR 51#14644.

Kelmans, A.K., and M.V. Lomonosov
[1982a] When m vertices in a k-connected graph cannot be walked round along a simple cycle, *Discrete Math.* **38**, 317–322. MR 83m:05083.
[1982b] A cubic 3-connected graph having no cycle through given 10 vertices has the "Petersen form", *J. Graph Theory* **6**, 495–496.

Kempe, A.B.
[1886] A memoir on the theory of mathematical form, *Philos. Trans. Roy. Soc. London* **177**, 1–70.

Kilpatrick, P.A.
[1975] *Tutte's first colour-cycle conjecture*, Ph.D. Thesis (Cape Town).

Kirkman, T.P.
[1881] Question 6610, solution by the proposer, *Math. Quest. Solut. Educ. Times* **35**, 112–116.

Kirkpatrick, D.G., and P. Hell
[1983] On the complexity of general graph factor problems, *SIAM J. Comput.* **12**, 601–609. MR 85c:68070.

Kneser, M.
[1955] Aufgabe 300, *Jber. Deutsch. Math.-Verein.* **58**, 27.

Koh, K.M.
[1976] Even circuits in directed graphs and Lovász' conjecture, *Bull. Malaysian Math. Soc.* **7**, 47–52. MR 56#2863.

Komlós, J., and E. Szemerédi
[1983] Limit distribution for the existence of Hamilton cycles in a random graph, *Discrete Math.* **43**, 55–63. MR 85g:05124.

König, D.
[1916] Über Graphen und ihre Anwendung auf Determinanten-theorie und Mengenlehre, *Math. Ann.* **77**, 453–465.

Kotzig, A.
[1957] From the theory of finite regular graphs of degree three and four, *Časopis Pěst. Mat.* **82**, 76–92 (in Slovak). MR 19#876.
[1962] Construction of Hamiltonian graphs of third degree, *Časopis Pěst. Mat.* **87**, 148–168 (in Russian). MR 25#2004.
[1968] Moves without forbidden transitions, *Mat.-Fyz. Časopis* **18**, 76–80. MR 39#4038.

Kotzig, A., and J. Labelle
[1979] Quelques problèmes ouverts concernant les graphes fortement hamiltoniens, *Ann. Sci. Math. Québec* **3**, 95–106. MR 80g:05047.

Kouider, M.
[1994] Cycles in graphs with prescribed stability number and connectivity, *J. Combin. Theory B* **60**, 315–318.

Kouider, M., and Z. Lonc
[preprint] *Covering Cycles and k-Term Degree Sums.*

Kuratowski, K.
[1930] Sur le problème de courbes gauches en topologie, *Fund. Math.* **15**, 271–283.

Laborde, J.M., C. Payan and N.H. Xuong
[1983] Independent sets and longest directed paths in digraphs, in: *Graphs and Other Combinatorial Topics*, ed. M. Fiedler (Teubner, Leipzig) pp. 173–177. MR 85e:05082.

Lang, R., and H. Walther
[1968] Über langste Kreise in regulären Graphen, in: *Beiträge zur Graphentheorie*, eds. H. Sachs, H.J. Voss and H. Walther (Teubner, Leipzig) pp. 91–98. MR 39#6773.

Lawler, E.L., J.K. Lenstra, A.H.G. Rinnooy Kan and D.B. Shmoys
[1985] eds., *The Traveling Salesman Problem: A Guided Tour through Combinatorial Optimization* (Wiley, New York).

Lesniak-Foster, L.
[1977] Some recent results in Hamiltonian graphs, *J. Graph Theory* **1**, 27–36. MR 56#15489.

Li, H.
[1989] Edge disjoint cycles in graphs, *J. Graph Theory* **13**, 313–322. MR 91e:05052.
[1990] Perfect path double covers in every simple graph, *J. Graph Theory* **14**, 645–650. MR 91h:05098.

Li, M.C.
[1988] *Hamiltonian Cycles in Regular 2-connected Claw-free Graphs*, Preprint.

Linial, N.
[1976] A lower bound for the circumference of a graph, *Discrete Math.* **15**, 297–300. MR 54#2524.
[1978] Covering digraphs by paths, *Discrete Math.* **23**, 257–272. MR 80f:05046.
[1981] Extending the Greene–Kleitman theorem to directed graphs, *J. Combin. Theory A* **30**, 331–334. MR 82h:05026.

Listing, J.B.
[1862] Der Census räumlicher Complexe oder Verallgemeinerung des Euler'schen Satzes von den Polyëdern, *Abh. K. Ges. Wiss. Göttingen Math. Cl.* **10**, 97–182.

Locke, S.C.
[1982] Relative lengths of paths and cycles in k-connected graphs, *J. Combin. Theory B* **32**, 206–222. MR 83f:05036.
[1985] A basis for the cycle space of a 2-connected graph, *European J. Combin.* **6**, 253–256. MR 87g:05138.

Lovász, L.
[1968] On covering of graphs, in: *Theory of Graphs*, eds. P. Erdős and G. Katona (Academic Press, New York) pp. 231–236. MR 38#2044.
[1970] Problem 11, in: *Combinatorial Structures and their Applications*, eds. R. Guy, H. Hanani, N. Sauer and J. Schönheim (Gordon and Breach, New York) p. 497.
[1974] Problem 5, *Period. Math. Hungar.* **4**, 82.
[1979] *Combinatorial Problems and Exercises* (Akadémiai Kiadó/North-Holland, Budapest/Amsterdam). MR 80m:05001.

Lovász, L., and M.D. Plummer
[1986] *Matching Theory, Ann. Discrete Math.* **29**. MR 88b:90087.

Lucas, É.
[1891] *Récréations Mathématiques*, Vol. 2 (Gauthier-Villars, Paris).

Lucchesi, C.L., and D.H. Younger
[1978] A minimax theorem for directed graphs, *J. London Math. Soc.* **17**, 369–374. MR 80e:05062.

Maamoun, M.
[1985] Decompositions of digraphs into paths and cycles, *J. Combin. Theory B* **38**, 97–101. MR 87m:05097.

MacLane, S.
[1937] A structural characterization of planar combinatorial graphs, *Duke Math. J.* **3**, 460–472.

Mantel, W.
[1907] Problem 28 (solution by H. Gouwentak, W. Mantel, J. Texeira de Mattes, F. Schuh and W.A. Wythoff), *Wiskundige Opgaven* **10**, 60–61.

Matthews, M., and D. Sumner
[1984] Hamiltonian results in $K_{1,3}$-free graphs, *J. Graph Theory* **8**, 139–146. MR 85f:05083.

McCuaig, W.
[1993] Intercyclic graphs, in: *Graph Structure Theory, Contemp. Math.* **147**.

Menger, K.
[1927] Zur allgemeinen Kurventheorie, *Fund. Math.* **10**, 96–115.

Meredith, G.H.J.
[1973] Regular n-valent, n-connected, non-Hamiltonian, non-n-edge colourable graphs, *J. Combin. Theory B* **14**, 55–60. MR 47#65.

Meyniel, H.
[1973] Une condition suffisante d'existence d'un circuit hamiltonien dans un graphe orienté, *J. Combin. Theory B* **14**, 137–147. MR 47#6546.

Min Aung
[1989] Circumference of a regular graph, *J. Graph Theory* **13**, 149–155.

Minty, G.J.
[1980] On maximal independent sets of vertices in claw-free graphs, *J. Combin. Theory B* **28**, 284–304. MR 81f:68076.

Mitchem, J., and E. Schmeichel
[1985] Pancyclic and bipancyclic graphs – a survey, in: *Graphs and Applications*, eds. F. Harary and J.S. Maybee (Wiley, New York) pp. 271–278. MR 86c:05085.

Moon, J.W.
[1968] *Topics on Tournaments* (Holt, Rinehart and Winston, New York). MR 41#1574.
[1972] The minimum number of spanning paths in a strong tournament, *Publ. Math. Debrecen* **19**, 101–104. MR 48#8294.

Moon, J.W., and L. Moser
[1963] On hamiltonian bipartite graphs, *Israel J. Math.* **1**, 163–165. MR 28#4540.

Nash-Williams, C.St.J.A.
[1969] Hamiltonian circuits in graphs and digraphs, in: *The Many Facets of Graph Theory, Lecture Notes in Mathematics*, Vol. 110, eds. G. Chartrand and S.F. Kapoor (Springer, Berlin) pp. 237–243. MR 40#5484.
[1970] Problem, in: *Combinatorial Theory and its Applications*, eds. P. Erdős, A. Rényi and V.T. Sós, *Colloq. Math. Soc. János Bolyai* **4**, 1179–1182.
[1971a] Edge-disjoint Hamiltonian circuits in graphs with vertices of large valency, in: *Studies in Pure Mathematics*, ed. L. Mirsky (Academic Press, London) pp. 157–183. MR 44#1594.
[1971b] Hamiltonian arcs and circuits, in: *Recent Trends in Graph Theory, Lecture Notes in Mathematics*, Vol. 186, eds. M. Capobianco, J.B. Frechen and M. Krolik (Springer, Berlin) pp. 197–210. MR 43#3150.
[1975] Hamiltonian circuits, in: *Studies in Graph Theory*, ed. D.R. Fulkerson (Mathematical Association of America, Washington, DC) pp. 301–360. MR 53#10649.
[1987] Amalgamations of almost regular edge-colourings of simple graphs, *J. Combin. Theory B* **43**, 322–342. MR 89c:05063.

Newman, D.J.
[1958] A problem in graph theory, *Amer. Math. Monthly* **65**, 611. MR 20#5487.

Ninčák, J.
[1974] Hamilton circuits in cubic graphs, *Comment. Math. Univ. Carolinae* **15**, 627–630. MR 50#6922.

Oberly, D.J., and D.P. Sumner
[1979] Every connected, locally connected nontrivial graph with no induced claw is Hamiltonian, *J. Graph Theory* **3**, 351–356. MR 80j:05086.

O'Brien, R.C.
[1977] An upper bound on the path number of a digraph, *J. Combin. Theory B* **22**, 168–174. MR 58#21774.

Ore, O.
[1960] Note on Hamilton circuits, *Amer. Math. Monthly* **67**, 55. MR 22#9454.
[1963] Hamilton connected graphs, *J. Math. Pures Appl.* **42**, 21–27. MR 26#4336.

Petersen, J.
[1891] Die Theorie der regulären Graphs, *Acta Math.* **15**, 193–220.
[1898] Sur le théorème de Tait, *Interméd. Math.* **5**, 225–227.

Pósa, L.
[1963] On the circuits of finite graphs (Russian summary), *Magyar Tud. Akad. Mat. Kutató Int. Közl.* **8**, 355–361. MR 33#5520.
[1976] Hamiltonian circuits in random graphs, *Discrete Math.* **14**, 359–364. MR 52#10497.

Preissmann, M.
[1981] *Sur les colorations des arêtes des graphes cubiques*, Thèse de Doctorat de 3ème cycle (Université de Grenoble).

Pruesse, G.
[1990] *A generalization of Hamiltonicity*, M.Sc. Thesis (University of Toronto).

Pyber, L.
[1985] An Erdős–Gallai conjecture, *Combinatorica* **5**, 67–79. MR 87a:05099.

Randíc, M.
[1981] Letter, January 18.

Rédei, L.
[1934] Ein kombinatorischer Satz, *Acta Litt. Szeged* **7**, 39–43.

Reed, B.
[1989] *Paths, Stars and the Number Three,* Preprint.

Reid, K.B., and K. Wayland
[1987] Minimum path decomposition of oriented cubic graphs, *J. Graph Theory* **11**, 113–118. MR 87m:05116.

Reiman, I.
[1958] Über ein Problem von K. Zarankiewicz, *Acta Math. Acad. Sci. Hungar.* **9**, 269–279. MR 21#63.

Richardson, M.
[1953] Solutions of irreflexive relations, *Ann. of Math.* **58**, 573–590. MR 17#704.

Richmond, L.B., R.W. Robinson and N.C. Wormald
[1985] On Hamilton cycles in 3-connected cubic maps, in: *Cycles in Graphs,* eds. B.R. Alspach and C.D. Godsil, *Ann. Discrete Math.* **27**, 141–149. MR 87d:05115.

Řiha, S.
[1991] A new proof of the theorem by Fleischner, *J. Combin. Theory B* **52**, 117–123.

Robbins, H.E.
[1939] A theorem on graphs, with an application to a problem of traffic control, *Amer. Math. Monthly* **46**, 281–283.

Robertson, N., and P.D. Seymour
[1985] Graph minors – a survey, in: *Surveys in Combinatorics,* ed. I. Anderson (Cambridge University Press, Cambridge) pp. 153–171. MR 87e:05130.

Robinson, R.W., and N.C. Wormald
[1991] *Almost all Regular Graphs are Hamiltonian,* Preprint.
[1992] Almost all cubic graphs are Hamiltonian, *Random Structures and Algorithms* **3**, 117–125. MR 93d:05105.

Roy, B.
[1967] Nombre chromatique et plus longs chemins d'un graphe, *Rev. Française Automat. Informat. Rech. Opér.* **1**, 127–132. MR 37#1276.

Saito, A.
[1992] Cycles of length 2 modulo 3 in graphs, *Discrete Math.* **101**, 285–289.

Sbihi, N.
[1980] Algorithme de recherche d'un stable de cardinalité maximum dans un graphe sans étoile, *Discrete Math.* **29**, 53–76. MR 81e:68087.

Schiermeyer, I.
[1992] Hamilton cycles in path tough graphs, in: *Contributions to the Twente Workshop on Hamiltonian Graph Theory,* eds. H.J. Broersma, J. van den Heuvel and H.J. Veldman, Memorandum No. 1043 (Twente University, Enschede) pp. 85–87.

Sekanina, M.
[1960] On an ordering of the set of vertices of a connected graph, *Publ. Fac. Sci. Univ. Brno* **412**, 137–142. MR 25#3518.

Seymour, P.D.
[1974] On the two-colouring of hypergraphs, *Quart. J. Math.* **25**, 303–312. MR 51#7927.
[1979] Sums of circuits, in: *Graph Theory and Related Topics,* eds. J.A. Bondy and U.S.R. Murty (Academic Press, New York) pp. 341–355. MR 81b:05068.
[1980] Disjoint paths in graphs, *Discrete Math.* **29**, 293–309. MR 82b:05091.
[1981] Even circuits in planar graphs, *J. Combin. Theory B* **31**, 327–338. MR 82m:05061.

Seymour, P.D., and C. Thomassen
[1987] Characterization of even directed graphs, *J. Combin. Theory B* **42**, 36–45. MR 88c:05089.
Sheehan, J.
[1975] The multiplicity of Hamiltonian circuits in a graph, in: *Recent Advances in Graph Theory,* ed. M. Fiedler (Academia, Prague) pp. 447–480. MR 53#2747.
Shi, R.H.
[1987] The binding number of a graph and its pancyclism, *Acta Math. Appl. Sinica* **3**, 257–269.
[1992] 2-neighbourhoods and hamiltonian conditions, *J. Graph Theory* **16**, 267–271.
Stong, R.
[1991] Hamilton decompositions of cartesian products of graphs, *Discrete Math.* **90**, 169–190.
Szekeres, G.
[1973] Polyhedral decomposition of cubic graphs, *Bull. Austral. Math. Soc.* **8**, 367–387. MR 48#3785.
Tait, P.G.
[1880] Remarks on colouring of maps, *Proc. Roy. Soc. Edinburgh Sec. A* **10**, 729.
Tao, J.
[1984] On Hajós' conjecture, *J. China Univ. Sci. Tech.* **14**, 585–592 (in Chinese). MR 87g:05178.
Tarjan, R.E.
[1972] Depth first search and linear graph algorithms, *SIAM J. Comput.* **1**, 146–160. MR 46#3313.
Thomason, A.G.
[1978] Hamiltonian cycles and uniquely edge colourable graphs, in: *Advances in Graph Theory,* ed. B. Bollobás, *Ann. Discrete Math.* **3**, 259–268. MR 80e:05077.
[1986] Paths and cycles in tournaments, *Trans. Amer. Math. Soc.* **296**, 167–180. MR 87i:05106.
Thomassen, C.
[preprint] *On the Complexity of Minimum Cycle Covers of Graphs.*
[1976] oral communication.
[1977] An Ore-type condition implying a digraph to be pancyclic, *Discrete Math.* **19**, 85–92. MR 58#21776.
[1978] Hypohamiltonian graphs and digraphs, in: *Theory and Applications of Graphs, Lecture Notes in Mathematics,* Vol. 642, eds. Y. Alavi and D.R. Lick (Springer, Berlin) pp. 557–571. MR 80e:05079.
[1980a] Hamiltonian-connected tournaments, *J. Combin. Theory B* **28**, 142–163. MR 82d:05065.
[1980b] On the number of Hamiltonian cycles in tournaments, *Discrete Math.* **31**, 315–323. MR 81h:05073.
[1980c] 2-linked graphs, *European J. Combin.* **1**, 371–378. MR 82c:05086.
[1981a] Planar cubic hypo-Hamiltonian and hypotraceable graphs, *J. Combin. Theory B* **30**, 36–44. MR 83f:05042.
[1981b] Long cycles in digraphs, *Proc. London Math. Soc.* **42**, 231–251. MR 83f:05043.
[1982] Edge-disjoint Hamiltonian paths and cycles in tournaments, *Proc. London Math. Soc.* **45**, 151–168. MR 83k:05075.
[1983a] Girth in graphs, *J. Combin. Theory B* **35**, 129–141. MR 85f:05076.
[1983b] A theorem on paths in planar graphs, *J. Graph Theory* **7**, 169–176. MR 84i:05075.
[1983c] Graph decomposition with applications to subdivisions and path systems modulo k, *J. Graph Theory* **7**, 261–271. MR 85e:05106.
[1983d] Disjoint cycles in digraphs, *Combinatorica* **3**, 393–396. MR 85e:05087.
[1983e] Infinite graphs, in: *Selected Topics in Graph Theory 2,* eds. L.W. Beineke and R.J. Wilson (Academic Press, New York) pp. 129–160. MR 87b:05045.
[1985a] Even cycles in directed graphs, *European J. Combin.* **6**, 85–89. MR 86i:05098.
[1985b] Hamilton circuits in regular tournaments, in: *Cycles in Graphs,* eds. B. Alspach and C.D. Godsil, *Ann. Discrete Math.* **27**, 159–162. MR 87c:05083.
[1986] Reflections on graph theory, *J. Graph Theory* **10**, 309–324. MR 87i:05077.
[1987a] On digraphs with no two disjoint directed cycles, *Combinatorica* **7**, 145–150. MR 88g:05067.
[1987b] Counterexamples to Ádám's conjecture on arc reversals in directed graphs, *J. Combin. Theory B* **42**, 128–130. MR 88c:05076.
[1990] Letter, October 22.
[1991] Highly connected non-2-linked digraphs, *Combinatorica* **11**, 393–395.
[1992] The even cycle problem for directed graphs, *J. Amer. Math. Soc.* **5**, 217–229.

[1993] The even cycle problem for planar digraphs, *J. Algorithms* **15**, 61–75.
Tian, F., and R.H. Shi
[1986] A new class of pancyclic graphs, *J. Systems Sci. Math. Sci.* **6**, 258–262. MR 88a:05125.
Tian, F., Z.S. Wu and C.Q. Zhang
[1982] Cycles of each length in tournaments, *J. Combin. Theory B* **33**, 245–255. MR 84c:04059.
Toida, S.
[1973] Properties of an Euler graph, *J. Franklin Inst.* **295**, 343–346. MR 48#10906.
Tutte, W.T.
[1946] On hamiltonian circuits, *J. London Math. Soc.* **21**, 98–101. MR 8397.
[1956] A theorem on planar graphs, *Trans. Amer. Math. Soc.* **82**, 99–116. MR 18–408.
[1959] Matroids and graphs, *Trans. Amer. Math. Soc.* **90**, 527–552. MR 21#337.
[1960] A non-Hamiltonian graph, *Canad. Math. Bull.* **3**, 1–5. MR 22#4646.
[1963] How to draw a graph, *Proc. London Math. Soc.* **3**, 743–768. MR 28#1610.
[1977] Bridges and hamiltonian circuits in planar graphs, *Aequationes Math.* **15**, 1–33. MR 57#5826.
[1992] oral communication.
Tuza, Zs.
[1984] Conjecture, in: *Finite and Infinite Sets*, eds. A. Hajnal, L. Lovász and V.T. Sós, *Colloq. Math. Soc. János Bolyai* **37**, 888.
[1990] A conjecture on triangles of graphs, *Graphs and Combinatorics* **6**, 373–380.
[1995] Perfect triangle families, *Bull. London Math. Soc.*, to appear.
Underground, P.
[1978] On graphs with hamiltonian squares, *Discrete Math.* **21**, 323. MR 80c:05097.
Veblen, O.
[1912] An application of modular equations in Analysis Situs, *Ann. Math.* **14**, 86–94.
Voss, H.J.
[1991] *Cycles and Bridges in Graphs* (VEB Deutscher Verlag der Wissenschaften/Kluwer Academic Publishers, Berlin/Dordrecht).
Voss, H.J., and C. Zuluaga
[1977] Maximale gerade und ungerade Kreise in Graphen I, *Wiss. Z. Tech. Hochsch. Ilmenau* **23**, 57–70. MR 58#396.
Walther, H., and H.J. Voss
[1974] *Über Kreise in Graphen* (VEB Deutscher Verlag der Wissenschaften, Berlin).
Watkins, M.E., and D.M. Mesner
[1967] Cycles and connectivity in graphs, *Canad. J. Math.* **19**, 1319–1328. MR 36#1355.
Welsh, D.J.A.
[1976] *Matroid Theory* (Academic Press, London). MR 55#148.
Wenger, R.
[1991] Extremal graphs with no C^4's, C^6's or C^{10}'s, *J. Combin. Theory B* **52**, 113–116.
Win, S.
[1975] Existenz von Gerüsten mit vorgeschriebenem Maximalgrad in Graphen, *Abh. Math. Sem. Univ. Hamburg* **43**, 263–267. MR 54#2529.
[1989] On a connection between the existence of k-trees and the toughness of a graph, *Graphs and Combinatorics* **5**, 201–205. MR 90d:05081.
Witte, D., and J.A. Gallian
[1984] A survey: hamiltonian cycles in Cayley graphs, *Discrete Math.* **51**, 293–304. MR 86a:05084.
Wong, P.K.
[1982] Cages – a survey, *J. Graph Theory* **6**, 1–22. MR 83c:05080.
Woodall, D.R.
[1972] Sufficient conditions for circuits in graphs, *Proc. London Math. Soc.* **24**, 739–755. MR 47#6549.
[1973] The binding number of a graph and its Anderson number, *J. Combin. Theory B* **15**, 225–255. MR 48#5915.
[1975] Inductio ad absurdum?, *Math. Gaz.* **59**, 64–70. MR 57#5531.

[1977] Circuits containing specified edges, *J. Combin. Theory B* **22**, 274–278. MR 55#12572.

[1978a] A sufficient condition for Hamiltonian circuits, *J. Combin. Theory B* **25**, 184–186. MR 80e:05080.

[1978b] Menger and König systems, in: *Theory and Applications of Graphs, Lecture Notes in Mathematics,* Vol. 642, eds. Y. Alavi and D.R. Lick (Springer, Berlin) pp. 620–635. MR 80a:05064.

[1990] A proof of McKee's Eulerian-bipartite characterization, *Discrete Math.* **84**, 217–220. MR 91f:05081.

Wu, Z.S., K.M. Zhang and Y. Zou

[1981] Arc-k-cycle of Tss graphs, *J. Nanjing Teachers College,* 1–3.

Younger, D.H.

[1973] Graphs with interlinked directed circuits, in: *Proc. 16th Midwest Symp. on Circuit Theory,* Vol. 2, *Systems Design* (University of Waterloo) pp. 2.1–2.7.

Zhan, S.M.

[1991] On hamiltonian line graphs and connectivity, *Discrete Math.* **89**, 89–95.

CHAPTER 2

Connectivity and Network Flows

András FRANK

Department of Computer Science, Eőtvős University, Múzeum krt. 6–8, Budapest H-1088, Hungary

Contents

HANDBOOK OF COMBINATORICS
Edited by R. Graham, M. Grötschel and L. Lovász

1. Introduction, preliminaries

Intuitively, a graph is felt to be connected if there is no way to separate it into two parts with no connection between the two parts. Or, equivalently, for any two nodes of the graph there is a path connecting them.

If one is interested in various properties of graphs, it is often useful to dismantle the graph into connected components and then investigate those components separately. For example, to decide whether a graph is k-colorable it suffices to deal with the connected components. A similar idea works for higher connectivity, as well. Namely, we separate first the graph into highly connected parts, then establish properties of these parts, and finally, using these properties we try to obtain information about the whole graph. For example, to prove Kuratowski's theorem on plane representation of graphs one can assume first that the graph is 3-connected since otherwise the graph can be decomposed along a 2-separation and the planar representations of the smaller parts can be pieced together. Second, by exploiting stronger properties of 3-connected graphs one can more easily deduce Kuratowski's theorem for 3-connected graphs. (See Thomassen 1980b.) (We remark that the same program proved extremely useful for matroids, as well.)

But what does "higher connectivity" mean? Intuitively, one may have at least two possible definitions for k-connectivity. First, the graph is not only connected but remains so after deleting any set of at most $k-1$ nodes. Second, a graph is k-connected if there are k (openly) disjoint paths between any two of its nodes. Fortunately, by a theorem of Whitney, these seemingly different concepts coincide. Whitney's (1932) theorem is an easy consequence of what can be considered the fundamental result of this whole chapter: Menger's theorem. It states that either there are k openly disjoint paths between two specified nodes or there is a set of less than k elements separating these two nodes.

Another source of the theory to be surveyed here is network flows. Its basic result, the max-flow min-cut theorem, can be considered as a capacitated (and directed) counterpart of Menger's theorem. Network flow theory is a systematic treatment of combinatorial optimization problems. It found a great number of applications in practice as well as in other branches of mathematics.

In this chapter we try to provide a rather comprehensive overview of results belonging to this area. Certain (mostly easier) parts are discussed in greater detail in order to give some hints on the general techniques used. Other parts are more difficult so we confine ourselves to give a general framework filled with various results but no proofs. In some cases, however, when it did not need too much space, proofs of some deep theorems (e.g., Tutte's wheel theorem or Nash-Williams' theorem on covering trees) have been included.

Throughout the chapter we use the following notation. Let V be a finite set and s, t elements of V. A subset X of V is called an $s\bar{t}$-set if $s \in X \subseteq V-t$. We often do not distinguish between a one-element set and its single element. Let $D=(V, A)$ be a digraph. The elements of A are called directed edges or arcs. For $S \subseteq V$ let $\Delta^+(S)$ denote the set of arcs with tail in S and head in $V-S$. We use the notation

$\Delta^-(S) := \Delta^+(V-S)$, $\delta^+(S) =: |\Delta^+(S)|$, $\delta^-(S) := \delta^+(V-S)$. For a graph or digraph G and vector $x: A \to \mathbb{R}$, let $d_x(X, Y)$ denote the sum $\sum (x(a)$: $a \in A$, one end of a is in $X-Y$, the other end is in $Y-X)$. When $x \equiv 1$ we use $d(X, Y)$ for $d_x(X, Y)$.

If it is not ambiguous we will use the notation $e = uv$ for an edge with endpoints u and v. Similarly $e = (u, v)$ stands for an arc with tail u and head v. An edge uv or an arc (u, v) is said to leave (enter) S if S is a $u\bar{v}$-set ($v\bar{u}$-set).

To complete this introductory section let us draw attention to some other survey papers and books. Concerning connectivity two books of Tutte (1966, 1984) deserve special mentioning. Mader's survey paper (1979) includes a long list of connectivity results and their relationship. Bollobás' (1978) book is also an excellent reference that includes the proof of many difficult theorems.

As far as network flows are concerned the classical book of Ford and Fulkerson (1962) is even today a refreshing reading. Recently, Ahuja et al. (1993) provided a comprehensive book on network flow techniques. Another useful survey on network flow theory, given by Goldberg et al. (1990), appeared in a book entitled "Paths, Flows, and VLSI-Layout" (B. Korte et al., eds., Springer, 1990). This book includes other important surveys concerning connectivity results. One of them, due to A. Schrijver, is concerned with the homotopic paths packing problem. The survey of N. Robertson and P.D. Seymour outlines the authors' very complex disjoint paths method. A third survey paper from the same book, written by the present author, provides an overview on packing paths, circuits, and cuts.

2. Reachability

2.1. Paths and walks

In a graph by a *walk* W we mean an alternating sequence $(v_0, e_1, v_1, e_2, \ldots, e_n, v_n)$ consisting of nodes and edges where e_i is an edge between v_{i-1} to v_i. The nodes v_0 and v_n are called the endpoints of W. Sometimes we say that W is a walk between v_0 and v_n or that W is a walk from v_0 to v_n (or from v_n to v_0). In a digraph by a (*directed*) *walk* W we mean an alternating sequence $(v_0, e_1, v_1, e_2, \ldots, e_k, v_k)$ where e_i is an arc from v_{i-1} to v_i. We say that W is a walk from v_0 to v_n or that v_n is reachable from v_0 by W.

The following definitions concern both graphs and digraphs. The number k of edges of a walk is called the *length* of the walk. The *distance* of t from s is the minimum length of a path from s to t. Obviously, if W_1 is a walk (directed or undirected) from u to v and W_2 is a walk from v to w, then the concatenation $W = W_1W_2$ is a walk from u to w.

We say that a walk is *simple* if all its defining terms are distinct. A simple walk is called a *path*. A walk is called *closed* if its endpoints coincide. If otherwise the terms of a closed walk are distinct, it is called a *circuit*.

Let $W = (v_0, e_1, \ldots, e_k, v_k)$ be a walk. Suppose that $v_i = v_j$ for some $i, j, 0 \leq i < j \leq k$, and the subsequence $C = (v_i, e_{i+1}, \ldots, e_{j-1}, v_j)$ is a circuit. *Reducing* W by circuit C means that we define a new walk $W' := (v_0, e_1, \ldots, v_i, e_{j+1}, \ldots, e_k, v_k)$. *Simplifying* W means that one reduces W as long as possible. The final walk is a path from v_0 to v_k (that may depend on the order of reductions). Thus we have the following.

Proposition 2.1. (a) *In a graph if there is a walk between two nodes u and v, there is a path between u and v.*

(b) *In a digraph if there is a directed walk from u to v, there is a directed path from u to v.*

Let us call two nodes of a graph $G = (V, E)$ equivalent if there is a path connecting them. This is an equivalence relation: from the definition of path it is symmetric and reflexive, by Proposition 2.1 it is transitive. An equivalence class is called a *component* of G. If G has exactly one equivalence class, G is called *connected*. Equivalently, G is *connected* if there is a path between any two of its nodes.

We call two nodes u and v of a digraph $D = (V, A)$ equivalent if there is a directed path from u to v and one from v to u. This is again an equivalence relation. An equivalence class is called a *strong component*. If D has exactly one equivalence class, D is called *strongly connected*. Equivalently, a digraph D is *strongly connected* if there is a directed path from every node to any other.

Proposition 2.2. *Let s and t be two specified nodes of a digraph $D = (V, A)$. There is a directed path from s to t if and only if for all $s\bar{t}$-sets S there is an edge leaving S.*

Proof. The necessity is straightforward. The sufficiency follows by observing that, if there is no path from s to t, the set S of nodes reachable from s has no leaving edges. □

Let us introduce some further notions. An undirected graph is called a *tree* if it is connected but deleting any of its edges disconnects the graph.

Proposition 2.3. *For a graph $G = (V, E)$ the following are equivalent*:

(a) *G is a tree.*

(b) *G is a connected graph containing no circuit.*

(c) *In G there is a unique path between any pair of nodes.*

(d) *G is connected and $|E| = |V| - 1$.*

(e) *G can be built up from any of its nodes by consecutively adjoining edges so that one end of the currently added edge belongs to the graph having already been constructed while the other endpoint does not.*

A graph is called a *forest* if each of its components is a tree. A digraph $D = (V, A)$ is called an *arborescence* if D arises from a tree by orienting the edges in such a way that every node but one has one entering arc. The exceptional node, called the *root*, has no entering arc. The union of node-disjoint arborescences is called a *branching*. (Equivalently, a branching is a directed forest such that the in-degree of each node is at most one.)

Proposition 2.4. *For a digraph D the following are equivalent:*

(a) *D is an arborescence.*

(b) *D contains a node r such that every node can be reached from r by a unique path.*

(c) *D contains a node r such that every node can be reached from r and deleting any edge yields a node that is not reachable from r.*

(d) *D can be built from a node r by adjoining sequentially arcs so that the tail of the currently added new arc belongs to the digraph having already been constructed while the head is a new node.*

Let $D = (V, A)$ be a digraph with a specified node s. We describe now a simple device, the labeling technique, to determine the set S of nodes reachable from s along with a sub-arborescence of D rooted at s that spans S.

We use a label called R-label for every node v showing if v has already been reached or not. If not, the label has entry "NON-REACHED". If v is reached its R-label says REACHED and contains the arc $(u, v) \in A$ along which v has been reached. The only exception is the source node s: the entry of its label is always REACHED. At the beginning every node but s has NON-REACHED in its R-label.

We also use another label called S-label for every node to indicate whether v is SCANNED or UNSCANNED.

At the beginning of the algorithm for every node the entry of every S-label is UNSCANNED. In a general step we pick up an unscanned node u that has already been reached (at the beginning only the source is such) and decide if there is a non-reached node v such that (u, v) is an arc of D. If there is none, declare u SCANNED and repeat. Otherwise declare v REACHED and put (u, v) into its R-label and repeat.

The algorithm terminates if there is no more unscanned node which is reached.

Proposition 2.5. *The set S of nodes that have REACHED in their R-label has no leaving arcs and consists precisely of nodes reachable from s. The set of arcs occurring in their R-labels forms an arborescence rooted at s with node set S.*

Note that the procedure can be applied to undirected graphs as well. In the algorithm there is much freedom in choosing a reached and unscanned node. One possible strategy is to choose each time an unscanned node u which has been reached earliest. In this case the procedure is called *breadth first search* (BFS).

An application of BFS is to compute the distance of the nodes in S from s. The only modification in the above algorithm is that we need a third variable $\text{dist}(v)$ at every node v to store the distance of v from s. At the beginning this is 0 at s and ∞ at all other nodes. When a node v is reached from u we define $\text{dist}(v)$ to be $\text{dist}(u) + 1$.

Another natural strategy is to choose each time an unscanned node that has been reached latest. In this case the procedure is called *depth first search* (DFS). Depth first search has a great number of important applications and we will mention three of them in the next section.

Finally, we mention a third kind of search, the so-called maximum cardinality search. Here each time an unscanned node is chosen that has a maximum number of already reached neighbours. This search was introduced by Tarjan and Yannakakis (1984) in order to find a simplicial ordering of the nodes of a chordal graph. Nagamochi and Ibaraki (1992) showed how maximum cardinality search can be used to find a sparse k-connected subgraph of a k-connected graph.

We remark that it is not difficult to implement these search procedures so as to run in linear-time. (For details, see Tarjan 1983.)

2.2. 2-Connectivity and strong connectivity

Given a graph or digraph $G=(V,E)$, a node $v \in V$ is called a *cut node* if E can be partitioned into two non-empty subsets E_1 and E_2 such that $V(E_1)$ and $V(E_2)$ have just the node v in common. (For $F \subseteq E$, $V(F)$ denotes the set of nodes incident to at least one element of F.) In particular, a node incident to a loop and to another edge is a cut node. If G is loopless, then v is a cut node if and only if its deletion increases the number of components: $c(G-v) > c(G)$.

A connected graph is called a *block* if it has no cut node. A graph is called *2-connected* if it is a block and has at least three nodes.

Proposition 2.6. *For a loopless graph $G=(V,E)$ with $|V| \geq 3$ the following are equivalent*:

(a) *G is 2-connected.*

(b) *For any two nodes there is a circuit containing them.*

(c) *For any two sets A, $B \subseteq V$ with $|A|$, $|B| \geq 2$ there are two disjoint paths connecting nodes of A and B.*

(d) *Any pair of edges is contained in a circuit.*

(e) *G can be built up from a circuit by sequentially adjoining edges* (*loops are not allowed*) *and subdividing edges* (*in any order*).

The following is a useful reduction property of 2-connected graphs.

Proposition 2.7. *For every edge e of a 2-connected graph with at least four nodes either the deletion or the contraction of e results in a 2-connected graph.*

(*Contracting* an edge uv means that we identify u and v into a new node z and for each edge uw or vw of G we introduce an edge zw. *Deleting* an edge $e=uv$ means that we leave out e from E.)

A strongly connected digraph with at least three nodes is called a *strong block* if it has no cut node.

Proposition 2.8. *For a digraph $D=(V, A)$ with $|V| \geq 3$ the following are equivalent*:

(a) *D is a strong block.*

(b) *D can be built up from a directed circuit by sequentially adding arcs (no loops allowed) and subdividing arcs.*

Subdividing an arc (u, v) means that we replace (u, v) by a path P from u to v where the inner nodes of P are new nodes of the graph.

(Note that it is not true that every pair of nodes of a strong block lies on a directed circuit.)

Let $G=(V, E)$ be a connected but not necessarily 2-connected graph. A *block of graph* G is a subgraph that is a block and is maximal with respect to this property. The blocks of G form a tree-like structure in the following sense. Let $B_1, B_2, \ldots, B_k$ be the blocks of G. Form a bipartite graph $T=(V, B; F)$ where the elements of $B=\{b_1, b_2, \ldots, b_k\}$ correspond to the blocks of G. In T let nodes v_i and b_j be connected by an edge if $v_i \in B_j$.

Proposition 2.9. *The blocks of a graph $G=(V, E)$ partition the set E of edges. Two edges belong to the same block if and only if there is a circuit containing both. Any two blocks have at most one node in common and the nodes belonging to more than one block are cut nodes. The graph $T=(V, B; F)$ is a tree.*

We call an edge e of a graph $G=(V, E)$ a *cut edge* or an *isthmus* if $G-e$ has more components than G. A connected graph is called 2-*edge connected* if it contains no cut edges.

Proposition 2.10. *For a connected graph $G=(V, E)$ the following are equivalent*:

(a) *G is 2-edge-connected.*

(b) *For any pair of nodes there are two edge-disjoint paths connecting them.*

(c) *Any edge is contained in a circuit.*

(d) *G can be built up from a node by sequentially adjoining edges (loops are allowed) and subdividing edges.*

Property (d) is sometimes formulated in another way. By an *ear-decomposition* of G we mean a sequence $G_0, G_1, \ldots, G_t=G$ of subgraphs of G where G_0 consists of one node and no edge, and each G_i arises from G_{i-1} by adding a path P_i for which the two (not necessarily distinct) end-nodes belong to G_{i-1} while the inner nodes of P_i do not. The paths P_i are called *ears*. (P_i may consist of a single edge). Now (d) is equivalent to saying that a graph is 2-edge-connected if and only if it has an ear-decomposition. There are several other ear-decomposition theorems. One is mentioned in the next proposition. Another asserts that a graph is 2-connected if and only if there is an ear-decomposition using only open ears. Yet another (due to L. Lovász) says that a graph is factor-critical if and only if there is an ear-decomposition using only ears of odd length. (See the chapter on matchings by W. Pulleyblank.)

Proposition 2.11. *Let $D = (V, A)$ be a digraph whose underlying graph is connected. The following are equivalent:*

(a) *D is strongly connected.*

(b) *There is at least one arc leaving each set $X \subset V(D)$, $X \neq \emptyset$, that is, there is no directed cut.*

(c) *Every arc is in a directed circuit.*

(d) *D can be built up from a node by sequentially adding arcs (loops are allowed) and subdividing arcs.*

Let $D = (V, A)$ be a digraph whose underlying graph is connected. Let $C_1, C_2, \ldots, C_k$ be maximal strongly connected subgraphs of D. These subgraphs are called the *strong components* of G. The name is justified by the following proposition.

Proposition 2.12. *The node sets $V(C_i)$ $(i = 1, \ldots, k)$ form a partition of V. By contracting each C_i into a node one obtains an acyclic digraph.*

Using depth first search, both the blocks of an undirected graph and the strong components of a directed graph can be found in linear time (Tarjan 1972).

Propositions 2.10 and 2.11 indicate that the analogous concepts for graphs and digraphs are 2-edge-connectivity and strong connectivity. Actually, parts (d) of these propositions immediately imply a theorem of Robbins (1939).

Corollary 2.13. *A graph G has a strongly connected orientation if and only G is 2-edge-connected.*

Here we provide another proof of this result that gives rise to a linear time algorithm (Tarjan 1972). Let s be an arbitrary node of G. Let T be a spanning tree determined by depth first search. Define an arborescence F by orienting the edges of T away from s.

Claim. *The unique path in T connecting the endpoints of any edge $e = uv \in E - T$ is a directed path P_e in F.*

The orientation of G obtained by orienting each edge $e \in E - T$ so as to form a directed circuit with P_e is strongly connected.

The following slight extension of Corollary 2.13 also holds (Boesch and Tindell 1980).

Theorem 2.13a. *The undirected edges of a mixed graph G (i.e., a graph having directed and undirected edges) can be oriented in such a way that the resulting digraph is strongly connected if and only if there is no cut edge in G and there is no directed cut.*

Corollary 2.13 naturally gives rise to the following question. Given a digraph

$D = (V, A)$, what is the minimum number of arcs the reversal of which makes D strongly connected? We must require that the underlying graph of D is 2-edge-connected. Obviously, the required subset of arcs meets all the directed cuts. Conversely, from Theorem 2.13a one can derive the following.

Proposition 2.14. *If F is a minimal set of edges covering all the directed cuts, the reversal of the elements of F leaves a strongly connected digraph.*

Therefore the following deep theorem of Lucchesi and Younger (1978) answers the question above. (For a relatively simple proof, see Lovász 1976b, for a constructive proof yielding a polynomial-time algorithm, see Frank 1981.)

Theorem 2.15 (Lucchesi and Younger 1978). *Let $D = (V, A)$ be an arbitrary digraph. The minimum number of arcs covering all the directed cuts is equal to the maximum number of pairwise disjoint directed cuts.*

3. Directed walks and paths of minimum cost

3.1. Walks and paths

Throughout this section we use the terms walk, path, circuit to mean diwalk, dipath, dicircuit, respectively. The *length* of a path is the number of edges of the path. Let $D = (V, A)$ be a loop-free digraph on n nodes and s and t two specified nodes called *source* and *sink*, respectively. Given a cost function $c: A \rightarrow \mathbb{R}$, find a path from s to t of minimum cost. Here the *cost* $c(P)$ of a path P is the sum of the cost of its arcs. A path is called a *min-cost path* if it has minimum cost among the paths from its origin to its terminus.

In this section we survey this problem and its variants. (For more detailed analysis, see Lawler 1976 and Tarjan 1983.) It turns out that computing one min-cost path from s to t is not simpler than computing a min-cost path from s to every other node reachable from s. So we focus mainly on that problem. We are going to describe several algorithms but the emphasis will be put on ideas and we do not seek for finding the most efficient procedures. The complexity of an algorithm heavily depends on the representation of the problem and the data structure used.

Let $w_k(v)$ denote the minimum cost of a walk of length at most k from s to v. If there is no such a walk, let $w_k(v) := \infty$. We will assume that each node of D is reachable from s. The following recursion is straightforward.

Proposition 3.1. $w_{k+1}(v) = \min(w_k(v), \min(w_k(u) + c(u, v)\colon (u, v) \in A))$ *for* $v \in V$.

(The minimum taken over the empty set is defined to be ∞.)

Relying on this proposition one can easily design an $O(k|A|)$ algorithm to compute $w_k(v)$ for $v \in V$ as well as a walk $W_k(v)$ from s to v of length at most k with cost $w_k(v)$.

To outline this let us assume for convenience that for every node $v \in V-s$ there is an arc (s, v). If this is not the case, add a new arc $e=(s, v)$ with $c(s, v)=\infty$. At the beginning let $w_1(v):=c(s, v)$ for $v \in V-s$ and $w_1(s)=0$. Furthermore, let $W_1(v)=\{s,(s, v), v)\}$ for $v \in V-s$ and $W_1(s):=\{s\}$. In the $(k+1)$th phase of the algorithm $w_{k+1}(v)$ and $W_{k+1}(v)$ is computed with the help of the formula in the proposition.

What about minimum cost paths? If the cost function c is arbitrary, there is no hope for a good algorithm since the problem of finding a longest path from s to t is NP-complete and that problem can be formulated as a minimum cost path problem by choosing $c(e)=-1$ for every arc e.

Therefore it would be natural to suppose that $c \geqslant 0$. However, not everything is lost if we do not require that much. The minimum cost path problem is tractable under the weaker assumption that there are no circuits of negative total cost (or, shortly, negative circuit). In particular, in acyclic digraphs the min-cost paths can be computed in polynomial time for arbitrary cost functions.

We call a cost-function c *conservative* if there is no directed circuit with negative total weight. A function $\pi: V \rightarrow \mathbb{R}$ is called a *feasible potential* (subject to c) if $\pi(v)-\pi(u) \leqslant c(u, v)$ for every arc $(u, v) \in A$.

Theorem 3.2. *Given a digraph $D=(V, A)$ and a cost-function $c: A \rightarrow \mathbb{R}$, c is conservative if and only if there is a feasible potential. The potential can be chosen integer-valued if c is integer-valued.*

Proof. Suppose first that π is a feasible potential and $C=(v_0, E_1, v_1, e_2, \ldots, e_n, v_n=v_0)$ is an arbitrary circuit. Then we have $c(C)=\sum c(e_i) \geqslant \sum[\pi(v_i)-\pi(v_{i-1})]=0$.

Conversely, suppose that there is no negative circuit. We can assume that there is a node s from which every other node is reachable. For otherwise adjoin a new node s to D and an arc (s, v) for each $v \in V$. Define the cost of the new arcs to be 0. Since there is no circuit containing the new node, the extended digraph contains no negative circuit.

We claim that $\pi(v):=w_n(v)$ (where $n=|V|$) is a feasible potential. Indeed, if there is no negative circuit, then $w_n(v)$ can be realized by a path P_v which has at most $n-1$ arcs. Then $\pi(u)+c(u, v)=c(P_u)+c(u, v) \geqslant \pi(v)$. □

Let $p_k(v)$ denote the minimum cost of a path of length at most k from s to v. Obviously, $p_n(v)$ is the minimum cost of a path from s to v. The preceding proof also shows the following.

Proposition 3.3. *If none of the walks $W_n(v)$ $(v \in V)$ induces a negative circuit, then there is no negative circuit in D, i.e., c is conservative.*

Since for a node $v \in V$ simplifying $W_n(v)$ can easily be carried out (in linear time) we have obtained an algorithm, due to Bellman (1958) and Ford (1956), that either finds a negative circuit or computes a min-cost path from s to every other node v. The complexity of the algorithm is $O(|A||V|)$.

In section 5 we will make use of the following problem, also interesting for its own sake. Suppose that there are negative circuits in a digraph $D = (V, A)$ and we want to eliminate all of the negative circuits by increasing the cost of every edge by the same value ε so that ε is as small as possible. It is an easy exercise to see that ε has the following interpretation: $-\varepsilon$ is the minimum circuit mean. (The *mean* of a circuit C is $c(C)/|C|$.)

To compute ε revise the algorithm mentioned after Proposition 3.1 as follows. Whenever a negative circuit C is detected by the algorithm, compute the current mean cost ε' of C and update the cost function by increasing the cost of each arc by $|\varepsilon'|$. Obviously C becomes a circuit of zero cost. The algorithm halts at the nth stage when there is no more circuit of negative (current) cost. One can see that the minimal ε is the sum of increments of costs and that a circuit C that became of zero cost last has the minimum mean cost. (See Karp 1978.)

Next we list some basic properties of minimum cost paths. For convenience, let us assume that every node of D can be reached from a specified node s.

Proposition 3.4. *Suppose that c is conservative. If $P = (s = v_0, e_1, \ldots, e_i, v_i, \ldots, e_j, v_j, \ldots, t)$ is a min-cost path from s to t, then a subpath $R = (v_i, \ldots, e_j, v_j)$ is a min-cost path from v_i to v_j. (Note that the corresponding statement for undirected graphs does not hold.)*

Proof. Let R' be a path from v_i to v_j for which $c(R') < c(R)$. Construct a walk W by replacing the segment R in P by R' and let P' be a path obtained by simplifying W. Since there is no negative circuit we have $c(P') \leqslant c(W) < c(P)$. □

Theorem 3.5. *If c is conservative, the minimum cost of a path from s to t is equal to $\max(\pi(t) - \pi(s)$: π a feasible potential).*

Proof. Let $P = (s = v_0, \ldots, v_k = t)$ be a path from s to t and π a feasible potential. Then $(*)$ $c(P) = \sum (c(v_{i-1}, v_i): i = 1, \ldots, k) \geqslant \sum (\pi(v_i) - \pi(v_{i-1}): i = 1, \ldots, k) = \pi(t) - \pi(s)$ and so $\max \leqslant \min$.

To see the other direction let $p_n(v)$ denote the minimum cost of a path from s to v. We have seen that p_n is a feasible potential. Since $p_n(t) - p_n(s) = p_n(t) = c(P)$ we have equality in $(*)$. □

For a feasible potential π we say an arc (u, v) to be *tight* if $\pi(v) - \pi(u) = c(u, v)$. By Theorem 3.5 a path P is a min-cost path if and only if there is a feasible potential π such that P consists of arcs which are tight with respect to π.

A *min-cost-path s-arborescence F* is an arborescence of D rooted at s for which the unique path in F from s to any other node v is a minimum cost path in D.

Proposition 3.6. *If there is no negative circuit, there is a spanning min-cost-path s-arborescence.*

Proof. Let L be the union of arcs belonging to any min-cost path starting from s. The arcs in L are tight with respect to p_n and L contains a spanning s-arborescence. □

Proposition 3.7. *A spanning s-arborescence F is a min-cost-path s-arborescence if and only if* $c_F(v) - c_F(u) \leq c(u, v)$ *for every* $(u, v) \in A$ ($c_F(v)$ *denotes the cost of the unique path in F from s to v*).

Above we outlined a method of complexity $O(|A||V|)$ to decide if a cost function c is conservative and, if so, to compute a min-cost-path s-arborescence. It is not known whether there is a method of complexity $O(|V|^2)$. There are, however, two special cases when such an algorithm exists.

3.2. Non-negative costs and acyclic digraphs

Assume the cost function c is non-negative. We briefly summarize Dijkstra's (1959) method. The basic observation is the following.

Lemma 3.8. *If T is a min-cost-path s-arborescence (not necessarily spanning) and* $m_T := \min(p_n(u) + c(u, v)\colon (u, v) \in \Delta^+(V(T))$ *is attained at an arc* $a = (u_a, v_a)$, *then* $T + a$ *is a min-cost-path s-arborescence.*

Proof. Let P_1 denote the path obtained from the path in T from s to u_a by adding a. Let P be any path from s to v_a and $e = u_e v_e$ the first arc on P that leaves $V(T)$. Since $c \geq 0$, $c(P') \leq c(P)$ where P' is the subpath of P from s to v_e. By the choice of a, $c(P_1) \leq c(P')$ so P_1 is a min-cost path. □

Dijkstra's method 3.9 consists of $n - 1$ phases. Starting at s we build up, arc by arc, a min-cost-path s-arborescence T. In order to compute m_T we maintain a label $l(v) = \min(p_n(u) + c(u, v)\colon (u, v) \in \Delta^+(V_T))$ for $v \in V - V(T)$. This label tells us which arc $a = (u_a, v_a)$ has to be added to the current T. When an arc $a = (u_a, v_a)$ has been added to T, label $l(v)$ is updated by $l(v) := \min(l(v), l(v_a) + c(v_a, v))$. Thus updating all $l(v)$ needs $O(n)$ time and the overall complexity of the algorithm is $O(n^2)$.

As an application, let us return for a moment to the general case when c may not be non-negative but there is no negative circuit. Dijkstra's algorithm can be used to show that there is an $O(n^2)$ algorithm for computing a min-cost-path s-arborescence provided that a feasible potential π is available. Indeed, define $c'(u, v) := c(u, v) - \pi(v) + \pi(u)$.

Claim 3.10. *A path P from s to t is a min-cost path with respect to c if and only if P is a min-cost path with respect to c'.*

Proof. For any path R from s to t the difference $c(R) - c'(R) = \pi(t) - \pi(s)$ does not depend on R whence the claim follows. □

Therefore one can apply Dijkstra's algorithm to c'.

Another important special case when an $O(n^2)$ algorithm is available is the one of acyclic digraphs. We can suppose again that every node is reachable from s. The following slight modification of Dijkstra's algorithm works.

Lemma 3.11. *Let T be a min-cost-path s-arborescence and $v \in V - V(T)$ such that $\Delta^-(v) \subseteq \Delta^+(V(T))$. If $m_T(v) := \min(p_n(u) + c(u, v)$: $(u, v) \in \Delta^+(V(T))$ is attained at an arc e, then $T + e$ is a min-cost-path s-arborescence.*

The proof is straightforward. To implement the algorithm one has to find an ordering $v_1 = s, v_2, \ldots, v_n$ of the nodes such that (v_i, v_j) can be an arc only if $i < j$ and then build the s-arborescence along this ordering. (Using depth first search such an ordering can easily be found in $O(|E|)$ time.)

This algorithm enables us to find a maximum path in an acyclic digraph, in particular, a maximum weight chain in a weighted poset.

3.3. Shortest paths in undirected graphs

Finally, we are interested in finding a min-cost path between two nodes s and t of an undirected graph $G = (V, E)$. If the cost function c is non-negative, the min-cost path problem can easily be reduced to the directed case by replacing each edge by a pair of oppositely directed arcs. This reduction, however, does not work in the general case, even if there is no negative circuit, since then we would introduce a negative (2-element) circuit. To overcome this difficulty one has to invoke matching theory, in particular, the theory of T-joins.

Let $E^- := \{e \in E: w(e) < 0\}$. Let $T_w := \{v \in V$: an odd number of edges from E^- is incident to $v\}$ and $T := \{s, t\} \oplus T_w$ where $\oplus$ denotes the symmetric difference. Define $w'(e) := |w(e)|$ for each $e \in E$. If there is no circuit of negative w-cost, then, for any T-join F of minimum w'-cost, $F \oplus E^-$ consists of an st-path of minimum w-cost and some disjoint circuit of zero w-cost. Therefore in order to compute a min-cost st-path it suffices to compute F. This, in turn, can be done with the help of a weighted matching algorithm. (While solving the min-cost path problem for directed graphs was not too difficult, one may be wondering if it is indeed necessary to invoke such a sophisticated tool, the matching algorithm, for solving the minimum cost path problem in undirected graphs. However this is not surprising anymore once one observes that an algorithm solving the latter problem can easily be used to compute a minimum weight perfect matching.)

For the structure of distances, see Sebő (1993).

4. Circulations and flows

4.1. Feasible circulations and maximum flows

In the theory of network flows there are several models which are, on one hand, equivalent via elementary constructions. On the other hand, for different type of applications it is convenient to have various models. We are going to survey the two basic ones: flows from a source to a sink and circulations. Although historically flows came earlier here we start with circulations. For more detailed network flow theory, see Ford and Fulkerson (1962), Lawler (1976), Phillips and Garcia-Diaz (1981), Lovász and Plummer (1986), Ahuja et al. (1993), Goldberg et al. (1990).

Throughout this section we work with a digraph $D=(V,A)$. Let $f:A\to\mathbb{R}\cup\{-\infty\}$ be a lower capacity, $g:A\to\mathbb{R}\cup\{+\infty\}$ an upper capacity such that $f\leq g$. For a vector $x:A\to\mathbb{R}$ and a subset $S\subseteq V$ let $\delta_x^-(S):=\sum(x(u,v)\colon (u,v)\in A,$ (u,v) enters $S)$ and let $\delta_x^+(S):=\delta_x^-(V-S)$. Vector x is called a *circulation* if the *conservation rule* $\delta_x^-(v)=\delta_x^+(v)$ holds at every node v. It is easily seen that, given a circulation x, $\delta_x^-(X)=\delta_x^+(X)$ for all $X\subseteq V$. A circulation x is *feasible* if $f\leq x\leq g$.

Theorem 4.1 (Hoffman 1960). *There exists a feasible circulation if and only if*

$$\delta_f^-(X)\leq\delta_g^+(X)\quad \textit{for every } X\subseteq V\,. \tag{4.1}$$

If f and g are integer-valued and (4.1) *holds, there is an integer-valued feasible circulation.*

Proof. *Necessity*. If x is a feasible circulation, then $\delta_g^+(X)-\delta_f^-(X)\geq\delta_x^+(X)-\delta_x^-(X)=0$ and (4.1) follows.

Sufficiency. Let $\gamma(X):=\delta_g^+(X)-\delta_f^-(X)$. Then (4.1) is equivalent to $\gamma(X)\geq 0$.

Lemma 4.2. $\gamma(X)+\gamma(Y)=\gamma(X\cap Y)+\gamma(X\cup Y)+d_{g-f}(X,Y)$.

Proof. The contribution of any arc to the two sides is the same. □

Choose a counter-example for which the number q of arcs with $f(a)<g(a)$ is minimum. There is such an arc $a=(s,t)$ since otherwise $x:=f(=g)$ is a feasible circulation (by (4.1)). Modify f by increasing $f(a)$ as much as possible without violating (4.1). By the minimal choice of q the modified $f(a)$ is still smaller than $g(a)$. Furthermore, there is a $t\bar{s}$-set T for which $\delta_f^-(T)=\delta_g^+(T)$, that is, $\gamma(T)=0$. Similarly, reduce $g(a)$ as much as possible without violating (4.1). Again, the modified $g(a)$ is bigger than (the modified) $f(a)$ and there is an $s\bar{t}$-set S for which $\gamma(S)=0$.

Because of arc a the value $d_{f-g}(S,T)$ is strictly positive. Thus, by Lemma 4.2 and by (4.1) we have $0+0=\gamma(S)+\gamma(T)>\gamma(A\cap T)+\gamma(S\cup T)\geq 0+0$, a contradiction. The same proof shows that if f and g are integer-valued, then there is an integer-valued feasible circulation. □

Let $D=(V,A)$ be a digraph with a specified source s and sink t. We assume, without restricting generality, that no arcs enter s and no arcs leave t. Let $g:A\to\mathbb{R}_+$ be a capacity function that is positive everywhere. A vector $x:A\to\mathbb{R}_+$

is called a *flow* from s to t, or an *st-flow*, if $\delta_x^-(v) = \delta_x^+(v)$ holds for every $v \in V - \{s, t\}$. A flow x is called *feasible* if $0 \leq x \leq g$.

It can easily be seen that for a flow x and for any $s\bar{t}$-set S the *netflow* leaving S defined by $\delta_x^+(S) - \delta_x^-(S)$ does not depend on the choice of S. This common value $\delta_x^+(s)$ is called the *flow value* of x and denoted by val(x). The value $x(u, v)$, $(u, v) \in A$, is called the *arc-flow*. A flow x is called a *path-flow* if x is positive only along a path from s to t.

The fundamental theorem of network flows, called the max-flow min-cut (MFMC) theorem, is due to Ford and Fulkerson (1956) and to Elias et al. (1956).

Theorem 4.3. *The maximum value of a feasible st-flow is the minimum of* $\delta_g^+(S)$ *over all* $s\bar{t}$*-sets* S. *If* g *is integer-valued, there is an integer-valued maximum flow.*

Proof. Let x be a feasible *st*-flow and S an $s\bar{t}$-set. Then we have $\mathrm{val}(x) = \delta_x^+(S) - \delta_x^-(S) \leq \delta_g^+(S)$ from which max $\leq$ min follows.

To see the other direction let m denote the minimum in question. Adjoin a new arc $e = (t, s)$ to D and define $f(e) := g(e) := m$. Let f be zero on all old arcs. It is easy to see that (4.1) holds for that choice of f and g so by Theorem (4.1) there is a feasible circulation (integral if f and g are integral). This circulation, without the new arc e, is an *st*-flow of value m. □

From the first part of the proof we see that an x is a maximum flow and $\delta_g^+(S)$ is of minimum if and only if

(a) $x(a) = g(a)$ for every arc a leaving S and

(b) $x(a) = 0$ for every arc a entering S.

These optimality criteria are crucial for the next algorithm due to Ford and Fulkerson. This provides an algorithmic proof of Theorem 4.3 for the case when the capacity function g is integer-valued (and thus, when g is rational). A refinement of that method, due to J. Edmonds and R. Karp and to E.A. Dinits, will provide a strongly polynomial algorithm and an algorithmic proof for arbitrary capacities.

4.2. Augmenting paths method

The algorithm of Ford and Fulkerson starts with an arbitrary feasible *st*-flow x (for example $x \equiv 0$) and iteratively improves it. To describe one iteration let x be a feasible *st*-flow. Construct an auxiliary digraph $D_x = (V, A_x)$ as follows. An arc (u, v) belongs to A_x if either (i) $(u, v) \in A$ and $x(u, v) < g(u, v)$ and then this arc of D_x is called a *forward* arc, or (ii) $(v, u) \in A$ and $x(v, u) > 0$ and then (u, v) is a *backward* arc.

Let S denote the set of nodes reachable from s in D_x.

Case 1: $t \notin S$. Since no arc of D_x leaves S, optimality criteria (a) and (b) hold and the algorithm terminates.

Case 2: $t \in S$. Let P be any path in D_x from s to t.

Let $\Delta_1 := \min(g(u, v) - x(u, v)$: (u, v) is a forward arc of $P)$ and $\Delta_2 =$

$\min(x(v,u)\colon (u,v)$ is a backward arc of $P)$. Let $\Delta = \min(\Delta_1, \Delta_2)$. Then Δ is positive. Call an arc of P *critical* if Δ is attained at that arc.

Update x as follows. If (u,v) is a forward arc of P, increase $x(u,v)$ by Δ. If (u,v) is a backward arc of P, decrease $x(v,u)$ by Δ. An easy consideration shows that the revised x' is a feasible st-flow with $\mathrm{val}(x') = \mathrm{val}(x) + \Delta$. Consequently, if g is integer-valued case 2 can occur only finitely many times.

In case of rational g we can easily reduce the problem by multiplying through the components of g by the least common denominator. When g is irrational, the algorithm above may not terminate. An example to show this pathological situation is given in Ford and Fulkerson (1962) (a simpler one occurs in Lovász and Plummer 1986). Another drawback of the algorithm is that even for integer capacities the number of iterations may be proportional to the largest occurring capacity M, as the following example shows, see fig. 4.1.

If among the possible augmenting paths we always choose the one of three arcs, then the flow augmentation Δ is just one in every step. Therefore the complexity of the algorithm is exponential in $\log M$, the size of the input.

4.3. The method of Edmonds, Karp and Dinits

To overcome these difficulties Edmonds and Karp (1972) and Dinits (1970) proposed to choose a shortest augmenting path at each iteration. This simple modification makes it possible to bound the number of iterations in the Ford–Fulkerson algorithm by a polynomial of $|V|$ and $|A|$, irrespective of the capacities.

Let $\sigma_x(v)$ denote the distance of v from s in D_x. (If there is no path from s to v, then $\sigma_x(v) = \infty$). Let P be a shortest path in D_x from s to t. Then for any arc (u,v) of P, $\sigma_x(v) = \sigma_x(u) + 1$.

Lemma 4.4. *Performing an augmentation along P does not decrease $\sigma_x(v)$.*

Proof. Let us consider how an augmentation affects D_x. Since the flow has been changed only at the arcs of D corresponding to the arcs of P, D_x may be changed at the arcs of P. Namely, the (possibly) new arcs of D'_x are the arcs of P in reverse orientation (while the critical arcs of P disappear). The distance of a node v from s could decrease only if we adjoin an arc (u,w) for which $\sigma_x(w) > \sigma_x(u) + 1$ and the lemma follows. □

The sequence of augmentations can be divided into phases. In one phase $\sigma_x(t)$ remains the same. By Lemma 4.4 there may be at most $|V| - 1$ phases.

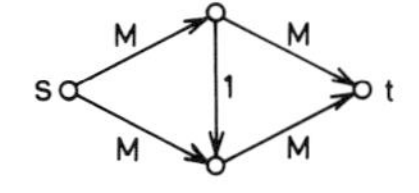

Figure 4.1.

Lemma 4.5. *Within one phase at most $|A|$ augmentations may occur.*

Proof. Let $\sigma_i(v)$ denote the distance of v from s in the auxiliary digraph at the beginning of a given phase i. Call an arc (u, v) i-tight if $\sigma_i(v) = \sigma_i(u) + 1$. Within phase i only i-tight arcs may be used. By Lemma 4.4 an augmentation eliminates at least one i-tight arc from the current auxiliary digraph and no new i-tight arc arises. Since an auxiliary digraph may have at most $|A|$ arcs, the lemma follows. □

By Lemma 4.5 the algorithm needs at most $|V||A|$ augmentations. One augmentation can be performed in $O(|V|)$ time so the overall complexity is $O(|V|^2 |A|)$.

Remark. The original augmenting paths method leaves freedom in choosing augmenting paths. The modification above imposes certain restrictions but still the algorithm may have different runs and, as a result, it may end up with different maximum flows. The final minimum cut $\delta^+(S)$, however, provided by the augmenting paths method is independent of the run of the algorithm. It is easy to show that if both X and Y minimize $\delta_g^+(Z)$ over $s\bar{t}$-sets, then both $X \cap Y$ and $X \cup Y$ are minimizing $s\bar{t}$-sets and therefore there is a unique minimal minimizing set S. The augmenting path method ends up with this S.

Since the algorithm of Edmonds, Karp and Dinits quite a few improvements on the complexity of max-flow min-cut algorithms have been devised. See Cherkasskij (1977a), Malhotra et al. (1978), Galil (1980), Shiloach (1978), Sleator (1980).

Max-flow algorithms using augmenting paths have the characteristic feature that the current flow is always changed in a "big piece": along all the edges of an augmenting path. An important conceptional development was the introduction of preflows by Karzanov (1974). A *preflow* is a non-negative function x on the edges of a digraph so that $\delta_x^-(v) \geq \delta_x^+(v)$ holds for every node v distinct from the source. A *preflow-push algorithm* changes the current preflow each time along just one edge. This is the basis of the greater flexibility and efficiency of preflow-push algorithms. See Shiloach and Vishkin (1982), Goldberg and Tarjan (1986), Cheriyan and Maheshwari (1989). Alon (1990) described a deterministic version of a randomized algorithm of Cheriyan and Hagerup (1990) whose complexity is $O(nm \log n)$

The interested reader may find a much more detailed comparison of the max-flow algorithms in the survey paper of Goldberg et al. (1990) and in the textbook of Ahuja et al. (1993).

4.4. Finding feasible circulations

We have seen how the max-flow min-cut theorem could be derived from Hoffman's circulation theorem. We show now the reverse direction. This way we will have a tool by which a feasible circulation can be found (or a set violating

(4.1)) with the help of one max-flow min-cut computation (on a slightly larger digraph).

For simplicity we restrict ourselves to finite f and g (the general case is left to the reader). For $v \in V$ denote $\gamma(v) = \delta_f^-(v) - \delta_f^+(v)$. If γ is zero everywhere, then f is a feasible circulation. Otherwise, the sets $S = (v\colon \gamma(v) > 0)$ and $T = (v\colon \gamma(v) < 0)$ are non-empty. Let $D' = (V', A')$ where $V' = V \cup \{s, t\}$ and $A' = A \cup (s, v)\colon v \in S \cup ((v, t)\colon v \in T)$. Define a capacity function g' as follows. $g'(s, v) := \gamma(v)$ if $v \in S$, $g'(v, t) = -\gamma(v)$ if $v \in T$ and $g'(a) = g(a) - f(a)$ if $a \in A$. Let $M = \sum (\gamma(v)\colon v \in S)$.

Lemma 4.6. (a) *x is an st-flow of value M in D' (with respect to g') if and only if $f + x$ (restricted circulation).*

(b) *A set $X \subseteq V$ violates* (4.1) *if and only if $\delta_{g'}^+(X + s) < M$.*

4.5. Other models and applications

Via elementary constructions flows and circulations are equivalent. There are other, more sophisticated variants. For example, one can impose lower bound on the arc-flows in the maximum flow problem. Or, lower and upper bounds can be requested on the in-flows $(\delta_x^-(v))$ at nodes. Moreover, instead of one source and one sink multiple sources and sinks can be specified. This generalization is sometimes called a *transshipment* problem. The *transportation* problem consists of finding a minimum cost degree-constrained subgraph of a bipartite graph. With relatively simple elementary constructions all these models go back to flows or circulations. We refer to the classical book of Ford and Fulkerson (1962).

Next, we are going to survey some of the combinatorial consequences of the flow theory.

Theorem 4.7 (Menger's theorem arc-version). *Let $D = (V, A)$ be a digraph with two specified nodes s and t. The maximum number of arc-disjoint paths from s to t is the minimum of $\delta^+(S)$ over all $s\bar{t}$-sets S.*

Proof. Apply Theorem 4.3 with $g \equiv 1$ and notice that every flow from s to t is the sum of path-flows from s to t and a non-negative circulation. □

In section 7 some other versions of Menger's theorem will be discussed.

Theorem 4.8 (König 1915). *The maximum number of disjoint edges of a bipartite graph $G = (V_1, V_2; E)$ is the minimum number of nodes covering all the edges.*

Proof. Orient the edges of G from V_1 to V_2. Then extend G by two new nodes s and t and new arcs (s, v) $(v \in V_1)$ and $(v, t)(v \in V_2)$. Let the capacities of all the new arcs be 1 and the other capacities M, a big number. A maximum st-flow of value k corresponds to k independent edges of G. By Theorem 4.7 there is an

$s\bar{t}$-set S for which $\delta_g^+(S) = k$. No arc of capacity M can leave S therefore $(V_1 - S) \cup (V_2 \cap S)$ is a covering of E with cardinality k. □

4.6. Gomory–Hu trees

In many applications the ability to compute the minimum cut separating two nodes is more important than finding a maximum flow. In undirected graphs the minimum cuts separating every pair of nodes has an especially attractive structure.

Let $G = (V, E)$ be a connected undirected graph and $g: E \rightarrow \mathbb{R}_+$ a non-negative capacity function. Let $\lambda(u, v)$ denote the maximum flow value between u and v. We say that a set Z separating u and v is *uv-minimal* if $d_g(Z)$ is minimal over all subsets separating u and v. Equivalently, $d_g(Z) = \lambda(u, v)$.

A set X is called *critical* if X is uv-minimal for some u and v. In order to have some insight into the structure of critical sets, one may be interested in a list of critical sets that contains a uv-minimal set for each pair u, v. How short can this list be? By choosing a separate uv-minimal set for each pair there is a list of $n(n-1)/2$ sets. But one can do much better.

Let $G_T = (V, F)$ be any tree on node set V (not necessarily a subgraph of G). For every edge e in F let $m(e) := d_g(X_e)$ where X_e and $V - X_e$ are the two components of $G_T - e$. G_T is called a *Gomory–Hu tree* of G (with respect to the given capacity function g) if (a) for every pair $\{s, t\}$ of nodes $\lambda(s, t)$ is the minimum of m-values over the edges of the unique path in G_T connecting s and t and (b) if e is an edge where the minimum is attained, then X_e is st-minimal.

For example, if $G = K_{3,3}$ and $g \equiv 1$, then a star of five edges forms a Gomory–Hu tree (and there is no other one showing that a Gomory–Hu tree cannot be chosen, in general, as a subgraph of G).

Theorem 4.9 (Gomory and Hu 1961). *Every graph possesses a Gomory–Hu tree.*

We are going to consider only the case $g \equiv 1$. For general g the proof goes along the same line. We need the following terminology. A family $\mathcal{F}$ of subsets of nodes is called *laminar* if for any two non-disjoint members of $\mathcal{F}$ one of them includes the other. We say that $\mathcal{F}$ *separates* nodes u and v if at least one member of $\mathcal{F}$ separates u and v.

Let $\mathcal{F}$ be a laminar family and $\{u, v\}$ a pair of nodes not separated by $\mathcal{F}$. We say that a subset X of nodes separating u and v is *uv-minimal with respect to* $\mathcal{F}$ if $\mathcal{F} \cup \{X\}$ is laminar and $d(X)$ is as small as possible. Note that such an X can be computed by one MFMC computation in a graph obtained from G by contracting the complement of the smallest member X of $\mathcal{F}$ containing u and v and contracting the maximal members of $\mathcal{F}$ included in X.

Proof. Let us construct a laminar family $\mathcal{F}$ of $n-1$ sets as follows. Let $\mathcal{F}_0$ be empty. Suppose we have constructed a laminar family $\mathcal{F}_{k-1} = \{A_1, \ldots, A_{k-1}\}$ for some $k = 1, \ldots, n-1$. Let $\{u_k, v_k\}$ be any pair of nodes not separated by

$\mathcal{F}_{k-1}$. Determine a set A_k that is $u_k v_k$-minimal with respect to $\mathcal{F}_{k-1}$ and let $\mathcal{F}_k := \mathcal{F}_{k-1} \cup \{A_k\}$.

Let $\mathcal{F} := \mathcal{F}_{n-1}$. Let $E_1 := \{u_i v_i : i = 1, \ldots, n-1\}$. From the construction we see that $T_1 := (V, E_1)$ is a tree. (This is just an auxiliary tree for the proof.)

Claim 1. *Let X be an xx'-minimal set and Y a critical set. If Y does not contain x and x', then either $X - Y$ or $X \cup Y$ is xx'-minimal. If Y contains x and x', then either $X \cap Y$ or $Y - X$ is xx'-minimal.*

Proof. To prove the first statement suppose that Y is yy'-minimal. Then either $Y - X$ or $Y \cap X$ separates $\{y, y'\}$. In the first case one has $\lambda(x, x') + \lambda(y, y') = d(X) + d(Y) \geq d(X - Y) + d(Y - X) \geq \lambda(x, x') + \lambda(y, y')$. Hence equality follows everywhere showing that $X - Y$ is xx'-minimal.

In the second case $\lambda(x, x') + \lambda(y, y') = d(X) + d(Y) \geq d(X \cup Y) + d(X \cap Y) \geq \lambda(x, x') + \lambda(y, y')$. Hence equality follows everywhere showing that $X \cup Y$ is xx'-minimal, as required for the first statement.

The second statement follows from the first one if we replace Y by its complement. □

Claim 2. *$A_i \in \mathcal{F}$ is $u_i v_i$-minimal for each $i = 1, \ldots, n-1$.*

Proof. The claim is clear for $i = 1$. Suppose we have already proved it for $1, 2, \ldots, i-1$ and let $x := u_i$, $x' := v_i$. By induction on $j = 0, 1, \ldots, i-1$ we are going to show that $(*)$ there is an xx'-minimal set X for which $\mathcal{F}_j \cup \{X\}$ is laminar. From this the claim will clearly follow.

$(*)$ obviously holds for $j = 0$. Suppose we have already shown $(*)$ for some $0 \leq j < i-1$. Let X' be the xx'-minimal set assured by Claim 1 when applied to X and $Y := A_j$. Then $\mathcal{F}_j \cup \{X'\}$ is laminar, as required. □

Claim 3. *For every pair $\{s, t\}$ of nodes there is an st-minimal member of $\mathcal{F}$.*

Proof. Let P be the unique path in T_1 connecting s and t and let $M := \min(\lambda(u_i, v_i) : u_i v_i$ an edge of $P)$. By the MFMC theorem $\lambda(s, t) \geq M$.

Let j be the smallest subscript for which $u_j v_j$ is an edge of P and $\lambda(u_j, v_j) = M$. We claim that A_j does not separate any other edge of P. Indeed, if A_j separates an edge $u_i v_i$ of P, then $i < j$ by the construction of $\mathcal{F}$. Furthermore, $M \leq \lambda(u_i, v_i) \leq d(A_j) = M$ and hence $\lambda(u_i, v_i) = M$, contradicting the minimal choice of j.

Therefore A_j must separate $\{s, t\}$ and hence we have $M \geq \lambda(u_j, v_j) = d(A_j) \geq \lambda(s, t) \geq M$. We can conclude that A_j is an st-minimal set. □

Let $A_0 := V$ and $\mathcal{F}' := \mathcal{F} \cup \{V\}$. For each $A_i \in \mathcal{F}'$ the union of maximal sets of $\mathcal{F}$ included in A_i is precisely one element smaller than A_i. Let t_i denote this element and for $i \geq 1$ let $s_i := t_j$ where A_j is the unique minimal element of $\mathcal{F}'$ including A_i. Let $F' := \{(s_i, t_i) : i = 1, \ldots, n-1\}$. Then $G'_T = (V, F')$ is an arborescence such that each arc of it enters one member of $\mathcal{F}$. Let G_T denote the underlying (undirected) tree. By this construction and by Claim 3, G_T is a Gomory–Hu tree. □

The following corollary, due to Padberg and Rao (1982), found a nice application in a linear programming approach to matching problems. It basically asserts that a Gomory–Hu tree encodes not only a minimum cut separating any given pair of nodes but also a minimum T-cut for any even subset T of nodes. (For the definition of T-cut and T-join, see the chapter on matchings.)

Corollary 4.10 (Padberg and Rao 1982). *Let G_T be a Gomory–Hu tree and T an even subset of nodes. Then a minimum T-cut can be obtained by choosing an edge e of G_T for which the cut determined by the two components of $G_T - e$ is a T-cut and $m(e)$ is as small as possible.*

Proof. Let C be a minimum T-cut and let J be the set of edges e of G_T for which the cut C_e determined by $G_T - e$ is a T-cut. Clearly J is a T-join and therefore there is an edge $e = uv \in C \cap J$. We have $|C| \geq \lambda(u, v) = m(u, v) = |C_e| \geq |C|$ showing that C_e is also a minimum T-cut. □

Another corollary states that if the degree of each node of a graph is at least k, then there are two distinct nodes which are connected by at least k edge-disjoint paths. Indeed, if u is a node having degree one in the Gomory–Hu tree and v is its neighbour in the tree, then u and v will do.

5. Minimum cost circulations and flows

5.1. *Min-cost circulations*

In the previous section we got to know how to find feasible circulations. It is not a less striking problem to find a possible circulation that minimizes a specified linear cost function. To attack this problem let us consider the set Q of all feasible circulations. Q forms a polyhedron in $\mathbb{R}^E$ called a *circulation polyhedron* and denoted by $C(D; f, g)$. (That is, $C(D; f, g) = \{x \in \mathbb{R}^E : \delta_x^-(v) = \delta_x^+(v)$ for $v \in V$ and $f(u, v) \leq x(u, v) \leq g(u, v)$ for $(u, v) \in A\}$. Q is said to be *integral* if f and g are integer-valued.

Theorem 4.1 implies that a (non-empty) integral circulation polyhedron contains an integer point. Since the face of an (integral) circulation polyhedron is obviously an (integral) circulation polyhedron, we have proved the following.

Theorem 5.1. *Every face of an integral circulation polyhedron Q contains an integer point.*

Let $Q = C(D; f, g)$ be a non-empty circulation polyhedron and $c: A \to \mathbb{R}$ a cost function. The cost cx of a circulation x is defined by $\sum (x(a)c(a): a \in A)$. What is the minimum cost of a feasible circulation and when does this minimum exist?

Define a digraph $D' = (V, A')$ and a cost function c' on A' as follows. An arc (u, v) belongs to A' if either (i) $(v, u) \in A$ and $f(v, u) = -\infty$ or (ii) $(u, v) \in A$ and $g(u, v) = \infty$. In case (i) let $c'(u, v) = -c(v, u)$, in case (ii) let $c'(u, v) = c(u, v)$.

Theorem 5.2. *The following are equivalent:*

(1a) *There is a feasible circulation of minimum cost.*

(1b) *There is no negative circuit in D' with respect to c'.*

(1c) *There is a potential $\pi: V \to \mathbb{R}$ such that*

$$\pi(v) - \pi(u) \leq c(u,v) \quad \textit{whenever } (u,v) \in A,\ g(u,v) = \infty \tag{5.1i}$$

and

$$\pi(v) - \pi(u) \geq c(u,v) \quad \textit{whenever } (u,v) \in A,\ f(u,v) = -\infty\,. \tag{5.1ii}$$

Note that a potential satisfying (1c) can be found (if one exists) by a min-cost path computation.

How can we characterize optimal feasible circulations? For $x \in Q$ define a digraph $D_x = (V, A_x)$ and a cost function c_x as follows. Let an arc (u,v) belong to A_x if either (i) $(u,v) \in A, x(u,v) < g(u,v)$ and then let $c_x(u,v) = c(u,v)$ (forward arc of A_x) or (ii) $(v,u) \in A, x(v,u) > f(v,u)$ and then let $c_x(u,v) = -c(v,u)$ (backward arc of A_x).

Theorem 5.3. *For a feasible circulation x the following are equivalent:*

(2a) *x is of minimum cost.*

(2b) *There is no negative circuit in D_x with respect to c_x.*

(2c) *There is a potential $\pi: V \to \mathbb{R}$ such that*

$$\pi(v) - \pi(u) \leq c(u,v) \quad \textit{if } x(u,v) < g(u,v)((u,v) \in A) \tag{5.2i}$$

and

$$\pi(v) - \pi(u) \geq c(u,v) \quad \textit{if } x(u,v) > f(u,v)((u,v) \in A)\,. \tag{5.2ii}$$

We call (5.2i) and (5.2ii) *optimality criteria.*

5.2. Min-cost circulation algorithm

The algorithm (Ford and Fulkerson 1962) starts with a feasible circulation x and a potential π satisfying (1c). If x and π satisfy (2c) as well, we are done. Otherwise, let $(t,s) \in A$ violate, say, (5.2i), that is $x(t,s) < g(t,s)$ and $\pi(s) - \pi(t) > c(t,s)$. (The case when an arc violates (5.2ii) is analogous.)

Define a digraph $D_x = (V, A_x)$ and a capacity function $g_x: A_x \to \mathbb{R}_+$ as follows. An arc (u,v) belongs to A_x if either

$$(u,v) \in A,\ \pi(v) - \pi(u) \geq c(u,v) \quad \text{and} \quad x(u,v) < g(u,v) \tag{3i}$$

or

$$(v,u) \in A,\ \pi(u) - \pi(v) \leq c(v,u) \quad \text{and} \quad x(v,u) > f(v,u)\,. \tag{3ii}$$

Let $g_x(u,v) = g(u,v) - x(u,v)$ in case (i) and $g_x(u,v) = x(v,u) - f(v,u)$ in case (ii). Let M denote the maximum value of an st-flow and let $\Delta = \min(M, g(t,s) - x(t,s))$. With the help of a max-flow min-cut computation determine an st-flow z

of value Δ. Revise x as follows.

$$x'(u,v) = x(u,v) + \Delta \quad \text{if } (u,v) \text{ is a forward arc of } D_x\,,$$

$$x'(u,v) = x(u,v) - \Delta \quad \text{if } (v,u) \text{ is a backward arc of } D_x\,.$$

Claim. *x' is a feasible circulation. An arc $(u,v) \in A$ satisfies* (5.2i) *and* (5.2ii) *with respect to x' if it satisfies with respect to x.*

There may be two cases.

Case 1: $\text{val}(z) = g(t,s) - x(t,s)$. In this case arc (t,s) no longer violates the optimality criteria.

Case 2: $M(=\text{val}(z)) < g(t,s) - x(t,s)$. Let S be the $s\bar{t}$-set determined by the max-flow min-cut computation. Let $\varepsilon_1 = \min(c(u,v) - \pi(v) + \pi(u)\colon (u,v)$ leaves S, $x(u,v) < g(u,v))$. Let $\varepsilon_2 = \min(\pi(u) - \pi(v) + c(u,v)\colon (u,v)$ enters S, $x(u,v) > f(u,v))$. Define $\varepsilon = \min(\varepsilon_1, \varepsilon_2)$ and revise π as follows: $\pi'(v) = \pi(v) + \varepsilon$ if $v \notin S$ and $= \pi(v)$ if $v \in S$.

Repeat the procedure with x' and π'.

It can be shown that this algorithm is finite for any cost and capacity functions. It is not necessarily of polynomial time, however, even if c, f, g are integer-valued. We describe a machinery, the scaling technique, to make the above algorithm of polynomial time. The method is due to Edmonds and Karp (1972).

5.3. Scaling technique

Suppose that the cost function c is integer-valued and that a feasible circulation x and an integer-valued potential π satisfying the optimality criteria are available. Let c' be another cost function differing from c on one arc by 1. The basic observation is that the algorithm above finds in polynomial time a feasible circulation x' and an integer-valued potential π' satisfying the optimality criteria with respect to f, g, c' provided that we start with the available x and π. Indeed, if case 2 occurs, then $\varepsilon = 1$ and arc (t,s) violates no longer the optimality criteria. In other words, with one max-flow min-cut computation (whether case 1 or case 2 occurs) the required x' and π' can be obtained.

Consequently, if c'' is an integer-valued cost function differing from c on every arc by at most one, then, starting with x and π, at most $|A|$ max-flow min-cut computations yield an x'' and π'' satisfying the optimality criteria with respect to c''.

Another, trivial, observation is that if x and π satisfy the optimality criteria with respect to c, then so do x and 2π with respect to $2c$.

Assume now for convenience that c is non-negative and let c be given in binary base. Let the maximum of $c(a)$ have K digits. Then there are K 0–1 vectors $c_0, \ldots, c_{K-1}$ in $\mathbb{Z}_+^A$ such that $c = \sum (2^i c_i\colon i = 0, \ldots, K-1)$.

First solve the min-cost circulation problem for c_{K-1}. This needs at most $|A|$ MFMC computations. Let the solution be x_{K-1} and π_{K-1}. Starting with x_{K-1}, $2\pi_{K-1}$ solve the min-cost circulation problem for $2c_{K-1} + c_{K-2}$. This also needs at most $|A|$ MFMC computations. Continuing this way, after at most $K|A|$ applications of the MFMC algorithm, we obtain a feasible circulation x and a potential π satisfying the optimality criteria with respect to c.

Remark. Let us draw attention to a small technical difficulty which can, however, be easily overcome: it may happen that there is a min-cost circulation with respect to c but there is none with respect to an intermediate cost function.

5.4. *Strongly polynomial algorithm*

Comparing the complexity of the scaling technique and the maximum flow algorithm of Edmonds, Karp and Dinits there is a significant difference. Namely, the complexity of the latter algorithm does not depend on the magnitude of the numbers (if we assume that adding and comparing two numbers is one step) and in this sense this algorithm is "strongly" polynomial while the complexity of the scaling technique is proportional to the number of digits.

Tardos (1985) was the first who constructed a strongly polynomial algorithm for finding a min-cost circulation. Since her work many other strongly polynomial algorithms have been developed. The fastest one is due to Orlin (1988). Here we briefly outline the algorithm of Goldberg and Tarjan (1989) that seems to be conceptionally the most attractive.

Optimality criterion (2b) suggests the following procedure. Start with a feasible circulation x. If (2b) holds, x is optimal. Otherwise, choose a circuit C in D_x violating (2b) and cancel along C. *Canceling along* C means that we increase $x(u, v)$ by Δ if (u, v) is a forward arc of C and decrease $x(u, v)$ by Δ if (v, u) is a backward arc of C. Here Δ is the smaller value of $\min(g(u, v) - x(u, v)$: uv a forward arc of C) and $\min(x(u, v) - f(u, v)$: (v, u) a backward arc of C).

Clearly, the modified x' is a feasible circulation and its cost is smaller than that of x. The algorithm consists of repeating this canceling procedure as long as (2b) is violated. This procedure is not necessarily of polynomial time. However, Goldberg and Tarjan proved that the following selection rule makes the algorithm strongly polynomial: each time choose a circuit C in G_x to be one of minimum mean cost. In section 3 we indicated how to compute such a circuit.

A beautiful feature of the algorithm of Goldberg and Tarjan is that it can be considered as a straight generalization of the Edmonds–Karp–Dinits algorithm for computing a maximum flow. Indeed, in the proof of Theorem 4.3 it was shown how a max-flow problem can be formulated as a minimum cost circulation problem. The Goldberg–Tarjan algorithm, when applied to this special min-cost circulation problem, yields precisely the Edmonds–Karp–Dinits algorithm.

5.5. *Minimum cost flows*

Let us be given again a digraph $D = (V, A)$ with a source s and a sink t. A non-negative capacity function g and a non-negative cost function c are given on A. We assume that both g and c are integer-valued. We have seen how to compute the maximum value M of an st-flow. This time we are interested in finding a minimum cost st-flow of value m for all possible integers m, $0 \leq m \leq M$. The cost of a flow z is defined by $cz = \sum (c(e)z(e): e \in A)$. We say that an st-flow z is a *min-cost flow* if z has the minimum cost among the feasible st-flows of value $\text{val}(z)$.

We have seen the equivalence between the feasible circulation problem and the maximum flow problem. Using the same elementary construction the min-cost st-flow problem could be solved in strongly polynomial time with the help of a strongly-polynomial min-cost circulation algorithm.

Here we briefly survey a direct algorithm due to Ford and Fulkerson (1962). This algorithm is strongly polynomial only if the capacities are small integers. The reason why we include this algorithm is that it has a nice combinatorial application.

A flow z is of minimum cost if and only if there is a function $\pi: V \to Z_+$, called a *potential*, ($\pi(s) = 0 \leq \pi(v) \leq \pi(t)$ for $v \in V$) for which the following two optimality criteria hold:

$$\pi(v) - \pi(u) < c(u, v) \Rightarrow z(u, v) = 0\,, \tag{5.3i}$$

$$\pi(v) - \pi(u) > c(u, v) \Rightarrow z(u, v) = g(u, v)\,. \tag{5.3ii}$$

We use the notation $\bar{c}(u, v) = c(u, v) - \pi(v) + \pi(u)$ for $(u, v) \in A$. The method (Ford and Fulkerson 1962) can be considered as a refinement of the max-flow min-cut algorithm of Ford and Fulkerson. It constructs a min-cost flow for all possible (integer) flow values m.

The algorithm starts with the identically zero flow and the identically zero potential. Then the flow value is increased one by one and the potential is appropriately increased so that the optimality criteria are throughout maintained. The algorithm terminates when a maximum flow (and a minimum cut) is found.

Iterative step. At the general step we are given a flow z and a potential π satisfying (i) and (ii). Construct an auxiliary digraph $D' = (V, A')$ as follows. D' has two types of arcs: forward and backward. An arc (u, v) is a forward arc if $uv \in A$, $\bar{c}(u, v) = 0$ and $z(u, v) < g(u, v)$. An arc (u, v) is a backward arc if $(v, u) \in A$, $\bar{c}(v, u) = 0$ and $z(v, u) > 0$. Let S be the set of nodes reachable from s in D'. There are two cases.

Case 1: $t \notin S$. Define $\varepsilon_1 = \min(\bar{c}(u, v): (u, v) \in \delta^+(S), z(u, v) < g(u, v))$ and $\varepsilon_2 = \min(-\bar{c}(u, v): uv \in \delta^+(V - S), z(u, v) > 0)$ where the minimum is defined to be ∞ if it is taken over the empty set. Let $\varepsilon = \min(\varepsilon_1, \varepsilon_2)$. The optimality criteria and the construction of S imply that ε is positive.

If $\varepsilon = \infty$, the algorithm terminates since we have $\delta_g^+(S) = \text{val}(z)$ and thus the current flow z is maximum and $\delta^+(S)$ is a minimum cut.

If $\varepsilon < \infty$, revise π by increasing $\pi(v)$ for every $v \in V - S$ by ε.

Claim. *The revised potential and the unchanged flow satisfy the optimality criteria.*

Repeat the procedure. Observe that in the new auxiliary digraph the set of reachable nodes from s is strictly larger than S. Therefore, after at most $|V|-1$ occurrences of case 1 either $\varepsilon=\infty$ or case 2 occurs.

Case 2: $t \in S$. Let P be a path in D' from s to t. Modify z as follows. Let $z'(u,v)=z(u,v)+1$ if (u,v) is a forward arc of P and let $z'(u,v)=z(u,v)-1$ if (v,u) is a backward arc of P.

Claim. *The revised flow and the unchanged potential satisfy the optimality criteria.*

What can we say about the complexity of the algorithm? We need roughly M flow augmentations. So the algorithm is polynomial if all the M minimum cost flows (of value $1,2,\ldots,M$) are required. The algorithm is not necessarily polynomial if one wants to compute only a min-cost flow of value M since the complexity is proportional to M.

If the maximum capacity is not too big (namely, its value can be bounded by a polynomial of $|V|$, then the algorithm is (strongly) polynomial, irrespective of the cost function.

5.6. An application to partially ordered sets

Let $P=\{p_1,p_2,\ldots,p_n\}$ be a partially ordered set. Dilworth's (1950) famous theorem asserts that the maximum cardinality a of an antichain is equal to the minimum number of covering chains. Another result of this type asserts that the maximum cardinality c of a chain is equal to the minimum number of covering antichains. In this section we discuss common generalizations of these results.

For a family $\mathscr{B}=\{B_1,B_2,\ldots,B_k\}$ denote $\bigcup\mathscr{B}=\bigcup(B_i: i=1,\ldots,k)$. By a *chain family* $\mathscr{C}_\gamma=\{C_1,C_2,\ldots,C_\gamma\}$ we mean a set of γ disjoint non-empty chains. Let $\boldsymbol{C}_\gamma$ denote the set of chain families of γ chains and $\boldsymbol{C}$ the set of all chain families. Let $c_\gamma=\min(|\bigcup C_\gamma|: \mathscr{C}\in\boldsymbol{C}_\gamma)$.

By an *antichain family* $\mathscr{A}_\alpha=\{A_1,A_2,\ldots,A_\alpha\}$ we mean a set of α disjoint non-empty antichains. Let $\boldsymbol{A}_\alpha$ denote the set of antichain families of α antichains and $\boldsymbol{A}$ the set of all antichains. Let $a_\alpha=\min(|\bigcup\mathscr{A}_\alpha|: \mathscr{A}_\alpha\in\boldsymbol{A}_\alpha)$.

By Dilworth's theorem $c_a=n$, by its polar $a_c=n$. What can be said about c_γ $(1\leq\gamma\leq a)$ and about a_α $(1\leq\alpha\leq c)$?

Theorem 5.4a (Greene and Kleitman 1976). $a_\alpha=\min(q\alpha+|P-\bigcup\mathscr{C}_q|: \mathscr{C}_q\in\boldsymbol{C})$.

Theorem 5.5a (Greene 1976). $c_\gamma=\min(q\gamma+|P-\bigcup\mathscr{A}_q|:\mathscr{A}_q\in\boldsymbol{A})$.

Since a chain and an antichain can share at most one element, a_α and c_γ do not exceed the minimum in question.

Definition. An antichain family $\mathscr{A}_\alpha=\{A_1,A_2,\ldots,A_\alpha)$ and a chain family $\mathscr{C}_\gamma=\{C_1,C_2,\ldots,C_\gamma\}$ are said to be *orthogonal* if

$$P=(\bigcup\mathscr{A}_\alpha)\cup(\bigcup\mathscr{C}_\gamma) \tag{a}$$

and

$$A_i \cap C_j \neq \emptyset \quad \text{for } 1 \leq i \leq \alpha\ ,\ 1 \leq j \leq \gamma\ . \tag{b}$$

The non-trivial parts of Theorems 5.4a and 5.5a can be reformulated as follows.

Theorem 5.4b. *For every α, $1 \leq \alpha \leq c$, there are $\mathscr{A}_\alpha \in \mathbf{A}_\alpha$ and $\mathscr{C}_\gamma \in \mathbf{C}$, for some γ, which are orthogonal.*
Theorem 5.5b. *For every γ, $1 \leq \gamma \leq a$, there are $\mathscr{C}_\gamma \in \mathbf{C}$ and $\mathscr{A}_\alpha \in \mathbf{A}$, for some α, which are orthogonal.*

A common generalization is due to Frank (1980).

Theorem 5.6. *There exists a sequence $\mathscr{C}_a | \mathscr{A}_1, \mathscr{A}_2, \ldots, \mathscr{A}_{i_1} | \mathscr{C}_{a-1}, \mathscr{C}_{a-2}, \ldots, \mathscr{C}_{a-j_1} | \mathscr{A}_{i_1+1}, \ldots$ which arises as a combination of two sequences $\mathscr{C}_a, \mathscr{C}_{a-1}, \ldots, \mathscr{C}_1$ and $\mathscr{A}_1, \mathscr{A}_2, \ldots, \mathscr{A}_c$, where $\mathscr{C}_j \in \mathbf{C}$ and $\mathscr{A}_i \in \mathbf{A}$, with the property that any member of the sequence (whether $\mathscr{C}_j$ or $\mathscr{A}_i$) is orthogonal to the last member of other type preceding it. (That is, $\mathscr{A}_1, \mathscr{A}_2, \ldots, \mathscr{A}_{i_1}$ are orthogonal to $\mathscr{C}_a$ and $\mathscr{C}_{a-1}, \mathscr{C}_{a-2}, \ldots, \mathscr{C}_{a-j_1}$ are orthogonal to $\mathscr{A}_{i_1}$, and so on.)*

Proof. Associate a digraph $D = (V, A)$ with P where $V := \{s, t, x_1, x_2, \ldots, x_n, y_1, y_2, \ldots y_n\}$, and $A := \{(s, x_i): i = 1, 2, \ldots, n\} \cup \{(y_i, t): i = 1, 2, \ldots, n\} \cup \{(x_i, y_j): \text{if } p_i \geq p_j\}$. Define all arc capacities $g(a)$ to be 1, while the costs are $c(e) = 1$ if $e = (x_i, y_i)$ and 0 otherwise.

Apply the min-cost flow algorithm to this network and let z and π be a flow and a potential at an intermediate stage of the algorithm. By analyzing the effect of a flow augmentation and a potential change and using the optimality criteria the following lemma can easily be proved.

Lemma 5.7. (a) *Either $\pi(y_i) = \pi(x_i)$ or $\pi(y_i) = \pi(x_i) + 1$.*
(b) *If $p_i > p_j$ and $z(x_i, y_j) = 1$, then $\pi(x_i) = \pi(y_j)$.*

The arcs (x_i, y_j) $(i < j)$ for which $z(x_i, y_j) = 1$ correspond to a chain family $\mathscr{C}_\gamma$ where $\gamma = n - \text{val}(z)$. For $\alpha = \pi(t)$ define a family $\mathscr{A}_\alpha = \{A_1, A_2, \ldots, A_\alpha\}$ where $A_i = \{p_j: \pi(x_j) + 1 = \pi(y_j) = i\}$.

Lemma 5.8. *$\mathscr{A}_\alpha$ is an antichain family and is orthogonal to $\mathscr{C}_\gamma$.*

The proof easily follows from the optimality criteria and from Lemma 5.7.

Now the Ford–Fulkerson algorithm and Lemma 5.8 immediately imply Theorem 5.6. □

6. Trees and arborescences

6.1. *Minimum cost trees and arborescences*

Given a connected graph $G = (V, E)$ and a cost function $c: E \rightarrow \mathbb{R}$, find a minimum cost spanning tree. This problem is one of the earliest combinatorial optimization problem that has been solved. For an excellent historical survey, see Graham and Hell (1985). For algorithmic details, see Tarjan (1983).

The following property of trees is crucial.

Lemma 6.1. *Let T_1 and T_2 be the edge sets of two trees on the same node set. Then for any edge $e \in T_1$ there is an edge $f \in T_2$ such that both $T_1 - e + f$ and $T_2 - f + e$ are trees.*

Proof. If $e \in T_2$, then $f := e$ will do. Suppose that $e = st \not\in T_2$. $T_1 - e$ has two components C_1 and C_2. T_2 contains a path P connecting s and t. Let f be an edge of P that connects C_1 and C_2. This f satisfies the requirement of the lemma. □

Let T be a spanning tree of a graph $G = (V, E)$. A *fundamental cut* belonging to an element e of F is a cut of G determined by the two components of $T - e$. A *fundamental circuit* belonging to an edge $f = uv \in E - T$ is a circuit consisting of f and the unique path in T connecting u and v. Clearly, an edge $e \in T$ is in the fundamental cut of an edge $f \in E - T$ if and only if e is in the fundamental circuit of f.

Lemma 6.1 immediately implies the following.

Theorem 6.2. *For a spanning tree T of G the following are equivalent:*

(a) *T is of minimum cost.*

(b) *$c(e) \leq c(f)$ for any edge $e \in T$ and edge f of the fundamental cut of e.*

(c) *$c(e) \geq c(f)$ for any edge $e \not\in T$ and edge e of the fundamental circuit of f.*

One of the simplest (and earliest) algorithms in combinatorial optimization is the greedy algorithm to construct a minimum-cost spanning tree of a connected graph $G = (V, E)$.

Greedy algorithm (Boruvka 1926, Kruskal 1956). The procedure consists of building a spanning forest by adding edges one by one. It starts with a forest of node-set V that has no edges and stops when the current forest is a spanning tree. The general step consists of adding an edge of minimum cost that connects two distinct components of the current forest.

There is another version of the greedy algorithm.

Dijkstra–Prim algorithm (Dijkstra 1959, Prim 1957). Choose an arbitrary node

x_0. Starting at x_0 build a tree edge by edge. At a general step choose a least cost edge to be added that has exactly one end in the current tree.

The following algorithm does not build a tree or forest directly but gets rid of edges of big cost and the remaining graph is the desired tree.

Reverse greedy algorithm. The procedure consists of discarding edges one by one so that the remaining graph is connected. At a general step choose an edge of maximum cost that is not a cut-edge of the current graph and delete it. The algorithm stops when the remaining graph is a spanning tree.

All of these algorithms can be formulated in a general framework.

General algorithm. The algorithm consists of applications of the following two operations in arbitrary order. The first operation builds a spanning forest F by adding edges one by one while the second operations deletes edges one by one. More precisely, let F denote the forest already constructed (at the beginning F is the forest of no edges.)

Step 1. If F is a spanning tree, halt. Otherwise, choose an arbitrary cut B disjoint from F and let $e \in B$ be a cheapest edge of B. Add e to F.

Step 2. Choose an arbitrary circuit C (if there is none, step 2 does not apply any longer) and let $e \in C$ be the most expensive edge of $C - F$. Delete e from G.

Theorem 6.3. *The final tree of the algorithm is of minimum cost.*

Proof. Any stage of the algorithm can be specified by a pair (F, D) of disjoint subsets of E where F denotes the forest constructed so far and D denotes the set of edges deleted so far.

We prove by induction that at each stage (F, D) of the algorithm there is a minimum cost tree T of G for which $F \subseteq T \subseteq E - D$. Any min-cost tree will do when $F = D = \emptyset$. Suppose we have already proved the statement for (F, D), that is, there is min-cost tree T with $F \subseteq T \subseteq E - D$.

Assume first that step 1 is applied and let $e \in B$ be the newly added edge and $F' := F + e$. If $e \in T$, we are done. Otherwise, let C_e be the fundamental circuit of e with respect to T. Edge e is in cut B and in circuit C_e therefore there must be another edge f in $B \cap C_e$. Since $e, f \in B$, by the rule in step 1 we have $c(e) \leqslant c(f)$. Since e, $f \in C_e$ and T is of minimum cost we have $c(f) \geqslant c(e)$. Hence $c(e) = c(f)$ and $T' := T - f + e$ is another min-cost tree for which $F' \subseteq T' \subseteq E - D$.

Second, assume that step 2 is applied and let $e \in C$ be the newly deleted edge. If T does not contain e, we are done. Otherwise let B_e be the fundamental cut of e belonging to T. There is an edge $f \neq e$ with $f \in C \cap B_e$. Since e, $f \in C$, by the rule in step 2 we have $c(e) \geqslant c(f)$. Since e, $f \in B_e$ and T is of minimum cost we have $c(e) \leqslant c(f)$. Hence $c(e) = c(f)$ and $T' := T - f + e$ is another min-cost tree for which $F \subseteq T' \subseteq E - (D + e)$. $\square$

Remark. The above algorithm can be extended to matroids. That is, there is a greedy algorithm for finding a minimum cost basis of a matroid. (See chapter 11.)

Let us turn to a directed counter-part of the minimum cost spanning tree problem. Let us be given a digraph $D=(V, A)$ and a cost function $c: A \to \mathbb{R}_+$. Assume that every node can be reached from a specified node s, that is, D includes a spanning s-arborescence. Our next problem is to find a minimum cost s-arborescence.

This problem has been solved by Fulkerson. Note that the min-cost tree problem can be reduced to a min-cost arborescence problem: replace each edge of G by a pair of oppositely directed arcs.

Call a set-function $z: 2^{V-s} \to \mathbb{R}_+$ *c-feasible* if

$$c(a) \geqslant \sum (z(X)\colon a \text{ enters } X) \quad \text{for every } a \in A\,. \tag{6.1}$$

Theorem 6.4 (Fulkerson 1974). *The minimum cost of a spanning s-arborescence is* $\max(\sum (z(x)\colon X \subseteq V-s)\colon z$ *is c-feasible). Furthermore, if c is integer-valued, the optimal z can be chosen integer-valued.*

Proof. Let F be a spanning s-arborescence and z a c-feasible vector. We have

$$\begin{aligned} c(F) &= \sum (c(a)\colon a \in F) \\ &\geqslant \sum \left(\sum (z(x)\colon a \text{ enters } X)\colon a \in F\right) \\ &\geqslant \sum (z(X)\colon X \subseteq V-s) \end{aligned} \tag{6.2}$$

from which max $\leqslant$ min follows. In (6.2) we have equality if the following *optimality criteria* hold.

$$c(a) = \sum (z(X)\colon a \text{ enters } X) \text{ for every } a \in F\,, \tag{6.3a}$$

$$z(X) > 0 \text{ implies } \delta_F^-(X) = 1\,. \tag{6.3b}$$

The algorithm below finds a spanning s-arborescence F and a feasible z for which (6.3a–b) holds. It consists of two parts. The first part constructs z while the second constructs F. In the course of the first part we revise the cost function. The current cost function is denoted by c'. We call an arc a a 0-arc if $c'(a)=0$.

Part 1. Iterate the following step. Choose a minimal set $X \subseteq V-s$ with no entering 0-arc. Define $z(X) := \min(c'(a)\colon a$ enters $X)$ and revise c' as follows. $c'(a) := c'(a) - z(X)$ if a enters X. The new c' is non-negative and its value is zero on at least one more arc.

Part 1 terminates if every set $X \subseteq V-s$ has an entering 0-arc, equivalently, there is a spanning s-arborescence consisting of 0-arcs.

Part 2. Starting at s and using only 0-arcs build up a spanning s-arborescence F. If, during the building process, there is more than one 0-arc leaving the sub-arborescence already constructed, choose that one which became a 0-arc earliest during the first part.

Obviously the constructed z is c-feasible and (6.3a) holds.

Lemma. *z and F satisfy* (6.3b).

Proof. Let X be a set with $z(X) > 0$. If, indirectly, $\delta_F^-(X) > 1$, there is a moment during part 2 when a sub-arborescence F' is at hand for which $\delta_{F'}^-(X) = 1$ and the arc e currently added to F' enters X. Consider the moment of part 1 when $z(X)$ became positive. Then no 0-arc entered X and every proper non-empty subset of X had an entering 0-arc. In particular, there is a 0-arc f entering $X - V(F')$ which does not enter X. Therefore, when $z(X)$ became positive, f was a 0-arc while e was not and we are in a contradiction with the rule of part 2: arc f should have been chosen instead of e. □

A related problem is as follows. Given a digraph $D = (V, A)$ and a cost function d on A, find a maximum cost branching. We call a pair (p, y) a *covering*, where $p: V \to R_+$ is a non-negative function on V and $y: 2^V \to \mathbb{R}_+$ a non-negative function on the subsets of V if $d(u, v) \leq p(v) + \sum (y(B): u, v \in B \subseteq V)$ for every $(u, v) \in A$. The *value* of a covering is $\sum (p(v): v \in V) + \sum (y(B)(|B| - 1): B \subseteq V)$.

Theorem 6.5 (Edmonds 1967, Chu and Liu 1965). *The maximum cost of a branching of D is the minimum value of a covering. If d is integer-valued, the optimal covering can be chosen integer-valued.*

Proof. Obviously max $\leq$ min. To see the other direction, extend D by a new node s and new arcs (s, v) $(v \in V)$. Define a cost function c on the arcs of the extended digraph D', as follows. Let the cost of the new arcs of D' be $M := \max(d(a)$: $a \in A)$ and $c(a) = M - d(a)$ $(a \in A)$. By Theorem 6.4 there is a c-feasible vector $z: 2^V \to \mathbb{R}_+$ and a spanning s-arborescence F of D' satisfying 6.3. Define $p(v) := M - \sum (z(B): v \in B)$ and for $|B| > 1$ define $y(B) := z(B)$ $(B \subseteq V)$. It is straightforward that (p, y) is a covering and its value is equal to the d-cost of the branching $F \cap A$ of D. □

6.2. Packing and covering

A basic result on this field is due to Edmonds (1973).

Theorem 6.6. *Given a digraph $D = (V, A)$ with a specified node s, there are k pairwise arc-disjoint spanning arborescences rooted at s if and only if $\delta^-(X) \geq k$ for every $X \subseteq V - s$.*

Proof (Lovász 1976a, *sketch*). Starting at s we build up an arborescence F such that

$$\delta_{A-F}^-(X) \geq k - 1 \quad \text{for every } X \subseteq V - s . \tag{$*$}$$

By induction on k, this will prove the sufficiency of the condition. Suppose that F is sub-arborescence satisfying ($*$) which is not spanning (that is, $V(F) \neq V$). Call a set $X \subseteq V-s$ *critical* (with respect to F) if it satisfies ($*$) with equality. By submodularity, if X and Y are critical and $X \cap Y \neq \emptyset$, then $X \cap Y$ and $X \cup Y$ are critical. By the hypothesis there is no critical subset of $V - V(F)$. Let M be a minimal critical subset for which $M - V(F) \neq \emptyset$. Then there is an arc $(u, v) \in A$ such that $u \in M \cap V(F)$ and $v \in M - V(F)$. This arc does not enter any critical set therefore the sub-arborescence $F' = F + (u, v)$ continues to satisfy ($*$). (If no such an M exists, any arc (u, v) leaving $V(F)$ will do.) □

Note that Theorem 6.6 of Edmonds immediately implies the directed arc-disjoint version of Menger's theorem. Indeed, adjoin k parallel arcs from t to v for every $v \in V - \{s, t\}$ and apply Edmonds' theorem.

The following conjecture is a kind of node-disjoint counterpart of Theorem 6.6.

Conjecture 6.7. (a) Suppose that, given a digraph $D = (V, A)$ and a specified node $s \in V$, there are k openly disjoint paths from s to any other node of D. Then there are k arc-disjoint s-arborescences of D such that for any node $v \in V - s$ the k paths from s to v uniquely determined by the arborescences are openly disjoint.

(b) Suppose that, given a graph $G = (V, E)$ and a specified node $s \in V$, there are k openly disjoint paths from s to any other node of G. Then there are k spanning trees of G such that for any node $v \in V - s$ the k paths from s to v uniquely determined by the k trees are openly disjoint.

(c) The same as (b) except replace "openly disjoint" by "edge-disjoint".

Conjecture 6.7 (a) easily implies Conjecture 6.7 (b). Whitty (1986) proved Conjecture 6.7 (a) for $k = 2$ while Conjecture 6.7 (b) has been proved for $k = 3$ by Cheriyan and Maheshwari (1988) and by Zehavi and Itai (1989). Conjecture 6.7 (c) is proved for $k = 2$ (using the ear-decomposition of 2-edge-connected graphs).

Recently, A. Huck proved (a) for acyclic digraphs and disproved it for general digraphs when $k \geq 3$.

One may be interested in finding k arc-disjoint spanning arborescences which need not be rooted at the same node.

Theorem 6.8. *There are k arc-disjoint spanning arborescences if and only if*

$$\sum_{i=1}^{t} \delta^{-}(X_i) \geq k(t-1)$$

for every family of disjoint non-empty sets $X_1, X_2, \ldots, X_t$.

More generally, the arborescence packing problem can be solved when lower and upper bounds are imposed at every node for the number of arborescences rooted at that node.

Another extension of Edmonds' theorem is due to Schrijver (1982). Let

$D=(V, A)$ be a digraph and let V_1, V_2 be a bipartition of V. Call a subset B of arcs a *bi-branching* if $\delta_B^-(X) \geq 1$ for every $X \subseteq V_1$ and $X \supseteq V_2$.

Theorem 6.9. (Schrijver 1982). *There are k arc-disjoint bi-branchings if and only if $\delta^-(X) \geq k$ for every $X \subseteq V_1$ and for every $X \supseteq V_2$.*

When $|V_2|=1$ we are back at Theorem 6.6. When both V_1 and V_2 are independent König's edge coloring theorem is obtained as a special case. (See chapter 3.)

Schrijver used this result to prove the following conjecture of Woodall for acyclic digraphs in the special case where each sink can be reached from each source.

Conjecture 6.10. In an acyclic digraph if every directed cut contains at least k arcs, there are k disjoint sets of arcs each of which covers all directed cuts.

A counterpart of Theorem 6.6 is due to Vidyasankar (1978).

Theorem 6.11. *Let s be a specified node of a digraph $D=(V, A)$ with no entering arc. The arcs of D can be covered by k spanning s-arborescences if and only if* (i) $\delta^-(v) \leq k$ *for* $v \in V-s$ *and* (ii) $k-\delta^-(X) \leq \sum(k-\delta^-(v): v \in T(X))$ *for every* $X \subseteq V-s$ *where* $T(X):=\{v \in X$: *there is an arc* $(u, v) \in A$ *with* $u \in V-X\}$.

Proof. By elementary construction. For every $v \in V-s$ adjoin to D a copy of v, denoted by v', and k parallel arcs from v to v' and $k-\delta^-(v)$ parallel arcs from v' to v. Furthermore, for $(u, v) \in A$ adjoin k parallel arcs from u to v'. Apply Theorem 6.6 to the extended digraph D' and observe that k arc-disjoint spanning arborescences in D' correspond to k covering s-arborescences in D. Moreover, if $\delta'^-(X')<k$ for some $X' \subseteq V'-s$, then $X=\{v \in X': v' \notin X'\}$ violates (ii). □

Another interesting consequence of Edmonds' theorem is the following.

Theorem 6.12. *The arc-set of a digraph $D=(V, A)$ can be covered by k branchings if and only if* (i) $\delta^-(v) \leq k$ *for every* $v \in V$ *and* (ii) $|A(X)| \leq k(|X|-1)$ *for every* $X \subseteq V$ *(where $A(X)$ denotes the set of arcs induced by X).*

Proof. By elementary construction. Adjoin a new node s to V and for $v \in V$ adjoin $k-\delta^-(v)$ parallel arcs from s to v. In the new digraph D' we have

$$\begin{aligned}\delta'^-(X) &= \delta^-(X)+\sum(k-\delta^-(v): v \in X) \\ &= \delta^-(X)-\delta^-(X)-|A(X)|+k|X| \geq k\end{aligned}$$

for every $X' \subseteq V$. By Theorem 6.6 there are k arc-disjoint spanning s-arborescences in D'. These determine k covering branchings of D. □

For undirected graphs we have the following (historically earlier) theorem by Nash-Williams (1964).

Theorem 6.13. *The edge set of an undirected graph $G=(V,E)$ can be covered by k forests if and only if $|E(X)| \leq k(|X|-1)$ for every $X \subseteq V$.*

Proof. Theorem 6.12 and the following easy lemma imply the result. □

Lemma. *The edges of a graph $G=(V,E)$ have an orientation for which $\delta^-(v) \leq k$ for every $v \in V$ if and only if*

$$|E(X)| \leq k|X| \quad \text{for every } X \subseteq V . \tag{$*$}$$

Proof. (*Sufficiency*) In an orientation of G call a node v *bad* if $\delta^-(v) > k$. Choose an orientation where the "badness" $\sum (\delta^-(v) - k\colon v \text{ bad})$ is minimal. If there is no bad node, we are done. Otherwise, let t be a bad node and let X be the set of nodes from which t is reachable in the current orientation. Then X contains a node s with $\delta^-(s) < k$ since otherwise $|E(X)| = \sum (\delta^-(v)\colon v \in X) > k|X|$, contradicting $(*)$. Reorienting the arcs of a dipath from s to t results in an orientation with smaller badness. □

(Note that this lemma easily follows from König's theorem, as well). For connected graphs the problems of covering the edge set by k forests or by k spanning trees are clearly equivalent. The packing problem of spanning trees was solved by Tutte (1961a).

Theorem 6.14. *A connected graph $G=(V,E)$ contains k pairwise edge-disjoint spanning trees if and only if $e_t \geq k(t-1)$ holds for every partition $\{V_1, V_2, \ldots, V_t\}$ of V $(V_i \neq \emptyset)$ where e_t denotes the number of edges connecting different V_i.*

Remark. Edmonds extended Tutte's theorem to matroids by providing a good characterization of the existence of k disjoint bases of a matroid. See chapter 11.

7. Higher connectivity

7.1 Connectivity between two nodes

We start this section with a result which is undoubtedly the central theorem of this whole chapter, the Menger (1927) theorem. In what follows s and t are two specified nodes of the graph or digraph in question. A set of paths is called *openly disjoint* if the paths are pairwise disjoint except, possibly, for their end nodes.

Theorem 7.1. (a) *In a digraph (graph) the maximum number of arc-disjoint (edge-disjoint) st-paths is equal to the minimum number of arcs (edges) covering all* *st-paths. (Moreover, the minimum is attained on a set of type $\Delta^+(S)$ where $S \subseteq V$ is an $s\bar{t}$-set.)*

(b) *In a digraph (graph) if there is no arc (edge) from s to t, the maximum number of openly disjoint st-paths is equal to the minimum number of nodes distinct from s and t covering all st-paths.*

Actually this is four theorems according to whether we consider directed or undirected and edge-(arc-)disjoint or openly disjoint *st*-paths. Menger originally proved the undirected, openly disjoint version.

Proof. We have already seen two proofs for the arc-disjoint case (as a consequence of the max-flow min-cut theorem and a consequence of Edmonds' disjoint arborescences theorem). From this the other three cases follow by elementary construction. Namely, in case (a) replace each edge by a pair of oppositely directed arcs and observe that if there is a set of k arc-disjoint paths in the resulting digraph, then there is one that does not use both arcs assigned to an original edge. The same construction yields the undirected openly-disjoint version from the directed one.

To see the directed openly-disjoint version construct a new digraph D' from D, as follows. Replace each node v ($v \neq s, t$) of D by a pair of new nodes v' and v''. Let (v', v'') be an arc of D' and for an arc (u, v) of D let (u'', v') be an arc of D'. Arc-disjoint *st*-paths in D' correspond to openly disjoint paths in D. Moreover, if there are k arcs in D' covering all *st*-paths, then these arcs can be assumed to be of type (v', v'') and this set of arcs corresponds to a set of k nodes of D covering all *st*-paths. □

There exist other versions of Menger's theorem. For example, given a graph and two disjoint subsets S, T of its node set, there are k disjoint paths between S and T if and only if there are no $k-1$ nodes covering all such paths. By elementary construction this result easily follows from the original Menger theorem. Yet another version, sometimes called the *fan lemma*, is as follows. Let s be a node of a graph and T a subset of nodes not containing s. There are k paths connecting s and some elements of T so that they are disjoint except at s if and only if there are no $k-1$ nodes in $V-s$ covering all such paths.

Hoffman found the following unifying approach to the different versions of Menger's theorem. Let S be a finite set and let $\mathcal{P}$ be a set of ordered subsets of S. We call the members of $\mathcal{P}$ paths. Suppose that for any two paths $P = \{p_1, p_2, \ldots, p_k\}$ and $T = \{t_1, t_2, \ldots, t_l\}$ sharing an element $p_i = t_j$ the sequence $\{p_1, \ldots, p_i, t_{j+1}, \ldots, t_l\}$ includes a path.

Theorem 7.2 (Hoffman 1974). *The maximum number of disjoint paths is equal to the minimum number of elements covering all the paths.*

As a consequence, a Menger-type theorem can be formulated for disjoint shortest paths.

Corollary 7.3. *The maximum number of openly disjoint shortest paths from s to t is equal to the minimum number of nodes covering all shortest paths from s to t.*

This type of min–max results fails to be true for paths of bounded length in general, however we have the following.

Theorem 7.4 (Lovász et al. 1978). *Let $G=(V,E)$ be an undirected graph with two specified non-adjacent nodes s and t. The maximum number of openly disjoint st-paths of length at most k ($k \geq 2$) is at least $2/k$ times the minimum number of nodes (distinct from s, t) covering all st-paths of length at most k.*

A possible generalization of Menger's theorem (undirected, openly disjoint) is the following. In an undirected graph given a subset of nodes T, what is the maximum number of disjoint paths connecting nodes of T. This problem was answered by Gallai (see Corollary 8.25). Mader found a min–max formula for the maximum number of openly disjoint paths with end nodes in T. (See Theorem 8.24.)

7.2. Global connectivity

Let k be a positive integer. A graph $G=(V,E)$ is called *k-connected* (sometimes k-node-connected) if $|V|>k$ and for any subset $X \subseteq V$ with less than k elements $G(V-X)$ is connected. G is called *k-edge-connected* if deleting any subset of edges of less than k elements leaves a connected graph. This is equivalent to requiring $d^+(X) \geq k$ for any $\emptyset \neq X \subset V$. A digraph $D=(V,A)$ is called *k-arc-connected* (often the term strongly k-arc-connected is used) if deleting any subset of arcs of less than k elements leaves a strongly connected digraph. This is equivalent to saying that $\delta^+(X) \geq k$ for any $\emptyset \neq X \subset V$. We will say that a digraph D with a specified node s is k-arc-connected from s if $\delta^-(X) \geq k$ for any subset $X \subseteq V-s$. By Menger's theorem this is equivalent to the property that there are k arc-disjoint paths from s to any other node. This is, in turn, equivalent to that there are k spanning arborescences rooted at s (Theorem 6.6). Finally, let us call a digraph strongly k-connected if deleting any subset of nodes of less than k elements leaves a strongly connected digraph.

These definitions reflect one of our intuitive expectations for a graph to be "pretty much connected": it is not possible to destroy the connectivity by taking away a small part of the graph. Another natural definition for high connectivity is that there are many disjoint paths between any pair of nodes. The following result, due to Whitney, says that these two approaches coincide. (One may have other intuitions for high connectivity: for example, if the graph contains k edge-disjoint spanning trees. Let us call such a graph *k-tree-connected*. For a characterization of k-tree-connected graphs, see Theorems 6.13 and 7.11).

Theorem 7.5 (Whitney 1932). *A graph on more than k nodes is k-connected if and only if there are k openly disjoint paths between any two nodes. A graph (digraph)* *is k-edge-(arc-)connected if and only if there are k-edge-(arc-)disjoint paths from any node to any other.*

Proof. The second part immediately follows from Theorem 7.1. To see the non-trivial direction of the first part let s and t be two nodes. If they are not adjacent, we are done by Menger. Otherwise let e be an edge connecting s and t. If there are no $k-1$ openly disjoint st-paths in $G-e$, there is (by Menger) a subset of nodes of at most $k-2$ elements not containing s and t for which s and t belong to different components of $G-e-X$. Since G has more than k nodes either $X+s$ or $X+t$ is a disconnecting set of $k-1$ elements. □

Robbins' (1939) theorem (Corollary 2.13) described a relation between 2-edge-connected graphs and strongly connected digraphs. The following generalization is due to Nash-Williams (1960).

Theorem 7.6. *An undirected graph is 2k-edge-connected if and only if it has an orientation which is k-arc-connected.*

Actually Nash-Williams proved a much stronger result:

Theorem 7.7. *An undirected graph has an orientation such that for every ordered pair (x, y) of nodes there are $\lfloor \lambda(x, y)/2 \rfloor$ arc-disjoint xy-paths where $\lambda(x, y)$ denotes the maximum number of edge-disjoint xy-paths.*

One may be interested in the existence of a (strongly) k-connected orientation. The ("sufficiency" part of the) following conjecture is open even for $k=2$.

Conjecture 7.8. A graph $G=(V, E)$ has a k-connected orientation if and only if deleting any subset X of j nodes $(0 \leq j \leq k-1)$ results in a $2(k-j)$-edge-connected graph.

In section 2 we saw an ear-decomposition theorem for strongly-connected digraphs. This can be interpreted so that every strongly connected digraph can be obtained from a node by consecutively adding and subdividing arcs. Subdividing an arc (u, v) with a new node z means that we replace arc (u, v) by arcs (u, z) and (z, v) where z is a new node. The following generalization is due to Mader.

Theorem 7.9 (Mader 1982). *A digraph D is k-arc-connected if and only if D can be obtained from a node by adding arcs (connecting old nodes) and applying operation O_k:*

Operation O_k: Pick up k arbitrary arcs, subdivide them by nodes $z_1, \ldots, z_k$ and identify $z_1, \ldots, z_k$ to a new node z.

Corollary 7.10. *A digraph D (with $\delta^-(s)=0$) is k-arc-connected from a node s if and only if D can be built up from s by repeated applications of the following* *operation: For some* j, $0 \leq j \leq k$, *first apply* O_j *and add then* $k-j$ *new arcs entering* z.

Using this characterization one can easily derive Edmonds' Theorem 6.6 on arc-disjoint arborescences (see Mader 1983). Also we have the following.

Corollary 7.11. *A graph* $G=(V,E)$ *is* k*-tree-connected if and only if* G *can be obtained from a node by sequentially adding edges (connecting old nodes) and applying the following operation: choose* j $(0 \leq j \leq k)$ *distinct edges and* $k-j$ *(not necessarily distinct) nodes, subdivide the* j *edges by* j *nodes, identify these* j *nodes to a new node* z *and connect* z *and the* $k-j$ *old nodes by* $k-j$ *edges.*

Theorems 7.6 and 7.9 imply a result of Lovász.

Theorem 7.12 (Lovász 1979, Problem 6.53). *An undirected graph* G *is* $2k$*-edge-connected if and only if* G *can be obtained from a node by adding edges and applying Operation* Q_k: *pick up* k *arbitrary edges, subdivide them by nodes* $z_1, \ldots, z_k$ *and identify* $z_1, \ldots, z_k$ *into a new node* z.

Notice on the other hand that Theorem 7.12 implies Theorem 7.6. What about $(2k+1)$-edge-connected graphs? We need two other operations.

Operation Q_k^+: Proceed as in Q_k, then choose a node x of the graph and add a new edge joining x and the new z.

Operation Q_k^2: Proceed as in Q_k thereby constructing G', choose k distinct edges $e_1', \ldots, e_k'$ of G' not all incident to z, subdivide each e_i' by a node z_i', identify the z_i''s into a new node z' and add a new edge joining z and z'.

Theorem 7.13 (Mader 1978a). *A graph* G *is* $(2k+1)$*-edge-connected if and only if* G *can be obtained from a node by successive addition of edges and repeated applications of* Q_k^+ *and* Q_k^2.

Let us be given a k-edge-connected undirected graph $G=(V,E)$. It is not difficult to prove that for k odd the k-element cuts are pairwise non-crossing. (Two cuts $\Delta(X), \Delta(Y)$ are called *crossing* if none of $X-Y$, $Y-X$, $X \cap Y$, $V-(X \cup Y)$ is empty). Dinits et al. (1976) showed that the structure of minimum cuts can also be described when k is even. Let $k=2l$.

Let us call a 2-edge-connected (loop-free) graph $T=(U,F)$ a *circuit-tree* if each block of T is a (possibly 2-element) circuit. Intuitively, T consists of edge-disjoint circuits which are joined to each other in a tree-like manner.

Any minimum cut of T consists of two edges belonging to the same circuit of T. If we replace each edge of T by l parallel edges, we obtain a $2l$-edge-connected graph T'. Clearly, the minimum cuts of T' correspond to the minimum cuts of T.

The content of the next theorem is that the structure of minimum cuts of every $2l$-edge-connected graph can be described with the help of a circuit-tree.

Theorem 7.14 (Dinits et al. 1976). *Let $G=(V,E)$ be a 2l-edge-connected graph. There exists a circuit-tree $T=(U,F)$ and a mapping $\varphi: V \rightarrow U$ so that for every minimum cut of T determined by a partition $[X, U-X]$ of U the cut of G determined by the partition $[\varphi^{-1}(X), \varphi^{-1}(U-X)]$ is a minimum (i.e., of $2l$ elements) cut of G and every minimum cut of G arises this way.*

We close this subsection by mentioning two results on constructing k-edge connected graphs and k-arc-connected digraphs.

Theorem 7.15. (a) (Watanabe and Nakamura 1987) *An undirected graph $G=(V,E)$ can be made k-edge-connected ($k \geqslant 2$) by adding at most γ new edges if and only if*

$$\sum (k - d(X_i): i=1,\ldots,t) \leqslant 2\gamma$$

holds for every family $\{X_i\}$ of disjoint non-empty subsets of V.

(b) (Frank 1992a) *A directed graph $D=(V,A)$ can be made k-edge-connected ($k \geqslant 1$) by adding at most γ new edges if and only if*

$$\sum (k - \delta^-(X_i): i=1,\ldots,t) \leqslant \gamma \quad \text{and} \quad \sum (k - \delta^+(X_i): i=1,\ldots,t) \leqslant \gamma$$

holds for every family $\{X_i\}$ of disjoint non-empty subsets of V.

In Frank (1992a) the first part has been generalized to the case when the prescribed edge-connectivity between each pair of nodes is arbitrary (and not necessarily the same number k). Also augmentations with minimum node-costs are tractable. For a survey see Frank (1994).

7.3. 3-connected graphs

In the preceding subsection we saw how to construct all the k-edge-connected graphs. As far as node-connectivity is concerned there is an ear-decomposition result for 2-connected graphs (Proposition 2.6). Tutte (1966) developed a theory for decomposing a 2-connected graph into 3-connected "components". (The reader is refered to the original work since even formulating the result needs too much space.)

Unfortunately there are no known analogous constructions for k-connected graphs, in general. For 3-connectivity, however, the situation is much better. In order to be able to work with 3-connected graphs we must have "reductions" that preserve 3-connectivity. Two simple reductions are deleting and contracting an edge e. We use the notation $G-e$ and G/e for the graphs arising from G by deleting and contracting e, respectively. For 2-connectivity we saw (Proposition 2.7) that any edge of a 2-connected graph can be either deleted or contracted without destroying 2-connectivity. For 3-connectivity one has the following.

Theorem 7.16 (Tutte 1966, Theorem 12.65). *If e is any edge of a 3-connected* *graph on at least four nodes, then either G/e is 3-connected or $G-e$ is a subdivision of a 3-connected graph.*

Proof (Thomassen 1984). Suppose G/e is not 3-connected for an edge $e=xy$. Then there is a node z such that $G' = G - \{x, y, z\}$ is not connected. What we have to show is that there are three openly disjoint paths between x' and y' not using e for any two nodes x', y' distinct from x and y. In G there are three openly disjoint paths connecting x' and y'. If one of these contains e, then x' and y' belong to the same component of G' or one of x', y' equals z. Then there is a component C of G' not containing x' and y' and there is a path P in $C + \{x, y\} - e$ connecting x and y. But now replacing e by P we obtain three openly disjoint paths between x' and y' not using e. □

We say that an edge e of a 3-connected graph is *contractible* if G/e is 3-connected. The next result, due to Tutte (1961b) shows that there always exists a contractible edge.

Theorem 7.17. *A 3-connected graph $G=(V, E)$ with at least five nodes has a contractible edge.*

Proof (*sketch*, Thomassen 1980b). If an edge xy is not good, there is a node z such that $\{x, y, z\}$ is a disconnecting set of G. Choose xy in such a way that the largest component C of $G - \{x, y, z\}$ is as big as possible. Let C' be another component of $G - \{x, y, z\}$ and $u \in V(C')$ such that $uz \in E$. The contraction of uz leaves a 3-connected graph. □

One way to generate new 3-connected graphs is applying the following splitting operation (that may be considered as a converse to contracting an edge). Note that the graph may have parallel edges.

Operation S: Choose a node v of degree at least four. Partition the edges incident to v into two parts E_1 and E_2 so that $|\{u: uv \in E_i\}| \geq 2$ for $i=1,2$. Replace v by two nodes v_1 and v_2, replace each edge $vu \in E_i$ by an edge $v_i u$ $(i=1,2)$ and join v_1 and v_2 by an edge.

Theorem 7.17 provides a kind of converse.

Corollary 7.18. *A (not necessarily simple) graph G is 3-connected if and only if G can be obtained from K_4 by repeatedly adding edges (connecting old nodes) and applying operation S.*

A slight drawback of this theorem is that, though parallel edges do not play any role in 3-connectivity, it may happen that even if the 3-connected graph G to be constructed is simple the graphs occurring in the intermediate steps are not. This is the case, for example, if G is a wheel (bigger than K_4). (A *wheel* is a circuit plus an extra node connected to all nodes of the circuit.) In a sense wheels are the only essential examples of this type since Tutte (1961b) proved the following.

Theorem 7.19. *A simple graph is 3-connected if and only if G can be obtained from a wheel by repeatedly adding edges connecting non-adjacent old nodes and applying operation S.*

This result is a reformulation of the following one.

Theorem 7.20 (Tutte 1961b). *A 3-connected graph G is either*
(a) *a wheel or*
(b) *contains an edge e for which $G-e$ is 3-connected or*
(c) *contains a contractible edge which is not in a triangle.*

Proof (Thomassen 1984). Suppose that neither (b) nor (c) occurs. By Theorem 7.17 there is a triangle T. Let $V(T)=\{x, y, z\}$. We claim that at least two nodes of T have degree three. Suppose, indirectly, x has neighbours x_1, x_2 not in T and y has neighbours y_1, y_2 not in T. Since $G-xy$ is not 3-connected, there is a node z' such that $G'=G-\{z,z'\}-xy$ is not connected. Let G_x and G_y denote the components of G' containing x and y, respectively. Since $G-y$ is 2-connected, there are two disjoint paths P_1, P_2 from $\{y_1, y_2\}$ to $\{z', z\}$. One of x_1 and x_2, say x_1, is distinct from z'. In $G-x$ there are two openly disjoint paths Q_1, Q_2 from x_1 to $\{z', z\}$. Since P_1, P_2 are in $G_x+\{z', z\}$ and Q_1, Q_2 are in $G_y+\{z', z\}$, $P_1'=(P_1+Q_1+Q_2)$ is a path from y to z. Now P_1', P_2 and $P_3=(yx, xz)$ are three openly disjoint paths between y and z, therefore edge yz satisfies (b), a contradiction.

So T has at least two nodes of degree 3, say x and y. If x' ($\neq y, z$) is a neighbour of x, then it is easy to see that G/xx' is 3-connected. Hence xx' must be in a triangle. The third node of this triangle must be z (unless $G=K_4$) and we conclude that x' has degree 3. We then consider the neighbour x'' of x' distinct from x and z and continuing this way we see that G is a wheel with center z. □

Obviously Tutte's theorem, in turn, implies Theorem 7.17. Theorem 7.17 is a highly powerful device in proving results apparently not related to connectivity. For example, with the help of it, Thomassen found an easy proof of Kuratowski's theorem on planarity of graphs as well as Tutte's theorem stating that every 3-connected planar graph has a convex representation in the plane. Note that Tutte originally used a different approach. He relied on the following result.

Theorem 7.21 (Tutte 1963). *Every edge uv of a* 3-connected *graph is contained in two peripherical circuits C_1 and C_2 (that is a chordless circuit the deletion of which results in a connected graph) for which $V(C_1)\cap V(C_2)=\{u, v\}$ A 3-connected graph is planar if and only if every edge is contained in exactly two peripherical circuits.*

Note that the first part of the theorem is straightforward for planar graphs as the two circuits determined by the faces incident to uv satisfy the requirements. Here are three variations of Theorem 7.17.

Theorem 7.22 (Thomassen and Toft 1981). *Every simple 3-connected graph with no triangles contains a circuit C such that every edge of C is contractible.*

Theorem 7.23 (Halin 1969a). *If v is a node of degree 3 in a 3-connected graph, then there is an edge uv for which G/uv is 3-connected.*

Theorem 7.24 (Ando et al. 1987). *Every 3-connected graph $G=(V,E)$ with $|V|>4$ has at least $|V|/2$ contractible edges.*

There is a counterpart of Theorem 7.17.

Theorem 7.25 (Barnette and Grünbaum 1969, Titov 1975). *Every 3-connected graph with at least five nodes has an edge e such that $G-e$ is a subdivision of a 3-connected graph.*

Proof. If there is no such an edge, then by Theorem 7.16 every edge is contractible without destroying 3-connectivity. Let G' be a graph obtained by contracting an edge st. By induction there is an edge e of G' such that $G'-e$ is a subdivision of a 3-connected graph. If $G-e$ is not a subdivision of a 3-connected graph, then there are nodes x, y such that $G-\{x,y\}-e$ has a component consisting of two nodes and e is incident to one of them, denoted by z. Since G is 3-connected there is an edge f from z to $\{x,y\}$. But $\{z,x,y\}$ is a separating set of G so the contraction of f destroys 3-connectivity. □

Corollary 7.26. *A graph $G=(V,E)$ is 3-connected if and only if G can be obtained from K_4 by sequentially adding edges and applying the following operations*:

(a) *Pick up two non-parallel edges, subdivide them by nodes u, v and join u and v by an edge.*

(b) *Pick up an edge xy and a node v ($\neq x, y$), subdivide xy by a node u and join u and v by an edge.*

Barnette and Grünbaum (1969) used Theorem 7.25 to provide a short proof of a theorem of Steinitz stating that the 1-skeletons of the 3-dimensional polytopes are precisely the 3-connected planar graphs.

Finally, here is a theorem consisting deletion of nodes rather than edges.

Theorem 7.27 (Chartrand et al. 1972). *Every 3-connected graph of minimum degree at least 4 has a node v such that $G-v$ is 3-connected.*

7.4. Preserving connectivity

In the preceding subsection on 3-connected graphs we have encountered theorems saying that a 3-connected graph remains 3-connected under certain operations (Theorems 7.16, 7.17, 7.27). In this part we survey further operations that preserve connectivity properties of graphs. As far as node-connectivity is concerned, only a few general reduction results are known. Here is one.

Theorem 7.28 (Thomassen 1981). *A k-connected graph with no triangle contains an edge whose contraction results in a k-connected graph.*

The situation is much better for edge connectivity. Let G be a graph (or digraph) and let $e = uz$ and $f = zv$ be two edges (or arcs) of G incident to a common node z. We say that a pair $\{e, f\}$ is split off (at z) if we replace e and f by a new edge (arc) uv. The resulting graph (digraph) is denoted by G^{ef}.

The following fundamental result is due to Mader (1978a).

Theorem 7.29. *Let $G = (V, E)$ be a (not necessarily simple but loopless) graph and z a node of degree at least 4 so that there is no cut-edge incident to z. There exist edges e, f incident to z such that $\lambda(x, y; G) = \lambda(x, y; G^{ef})$ for every pair of distinct nodes x, y different from z. (Here $\lambda(x, y; G)$ denotes the maximum number of edge-disjoint paths in G connecting x and y).*

A relatively simple proof can be found in Frank (1992b). Mader used it to derive Nash-Williams' orientation result (Theorem 7.7). Theorem 7.13 was also obtained from this result. Theorem 7.6 follows already from a weaker form of Theorem 7.29 due to Lovász (1979, Problem 6.53): If z is a node of even degree and $\lambda(x, y\colon G) \geq k$ whenever $x, y \in V - z$, then for any edges e incident to z there is an edge f incident to z such that $\lambda(x, y\colon G^{ef}) \geq k$ for every x, y ($\neq z$). While the directed analogue of Theorem 7.29 is not true in general the counterpart of Lovász' version holds, as follows.

Theorem 7.30 (Mader 1983). *Let $D = (V, A)$ be a digraph and z a node for which $\delta^-(z) = \delta^+(z)$. Suppose that $\lambda(x, y\colon D) \geq k$ for every $x, y \in V - z$. For any arc entering z there is an arc f leaving z such that $\lambda(x, y\colon D^{ef}) \geq k$ for every $x, y \in V - z$.*

This result is one ingredient to Theorem 7.9. The other one will be mentioned later (Theorem 7.41).

If we restrict ourselves to Eulerian digraphs (that is, $\delta^-(z) = \delta^+(z)$ for every node z) then the counterpart of Theorem 7.28 does hold (although it is much easier).

Theorem 7.31 (Frank 1989, Jackson 1988). *Let $D = (V, A)$ be an Eulerian digraph and z a node. For any arc e entering z there is an arc f leaving z such that $\lambda(x, y\colon D) = \lambda(x, y\colon D^{ef})$ for every $x, y \in V - z$.*

Sometimes we can maintain connectivity properties under "bigger" reductions.

Theorem 7.32 (Thomassen and Toft 1981). *Every 3-connected graph with mini-*

mum degree at least 4 *contains a circuit whose contraction results in a* 3-*connected graph*.

Theorem 7.33 (Jackson 1980). *A simple* 2-*connected graph of minimum degree at least* 4 *contains a circuit* C *such that the removal of its edges leaves the graph* 2-*connected and, in addition* (Thomassen and Toft 1981), $G - V(C)$ *is connected. The first part holds for not necessarily simple planar graphs* (Fleischner and Jackson 1985).

Theorem 7.34 (Thomassen and Toft 1981). *Every simple* 2-*connected graph* G *of minimum degree at least* 3 *contains an induced circuit* C (*that is a circuit without chords*) *such that* $G - V(C)$ *is connected.*

The same conclusion was proved by Tutte for 3-connected graphs (see Theorem 7.21).

Theorem 7.35 (Thomassen and Toft 1981). *If* G *is a* 2-*connected graph with minimum degree at least* 5, *then* G *contains an induced circuit such that* $G - V(C)$ *is* 2-*connected. If* G *is* 3-*connected with minimum degree at least* 4, *then* G *has a circuit such that* $G - V(C)$ *is a block.*

For higher node connectivity we have the following results.

Theorem 7.36 (Thomassen 1981). *Every* $(k + 3)$-*connected graph* G *contains an induced circuit* C *such that* $G - V(C)$ *is* k-*connected.*

Theorem 7.37 (Egawa 1987). *Every* $(k + 2)$-*connected triangle-free graph* G *contains an induced cycle* C *such that* $G - V(C)$ *is* k-*connected.*

Theorem 7.38 (Mader 1974a). *Every* k-*connected graph* G *with minimum degree at least* $k + 2$ *contains a circuit* C *such that* $G - E(C)$ *is* k-*connected.*

For edge-connectivity Mader proved the following.

Theorem 7.39 (Mader 1985a). *For every pair of nodes* s, t *of a connected graph there is a path* P *connecting* s *and* t *such that deleting the edges of* P *the local connectivity* $\lambda(x, y)$, *for any pair* x, y *of nodes, can decrease by at most two.*

A directed counterpart of this result is also due to Mader.

Theorem 7.40 (Mader 1981). *For every pair of nodes* s, t *of a* k-*arc-connected digraph there is a path* P *from* s *to* t *such that deleting the arcs of* P *leaves the digraph* $(k - 1)$-*arc-connected.*

7.5. Minimal and critical graphs

In graph theory it is a typical way to prove things by starting with a graph critical (or minimal) with respect to a certain property. For example, if this property is "having no perfect matching" we arrive at the concept of factor-critical graphs (see chapter 3). Therefore, it is a general program to investigate "critical" graphs. Typically, we use the adjective "minimal" (resp. critical) if deleting any edge (resp. node) destroys the property considered.

Call a graph (digraph) G minimally k-edge-connected (k-arc-connected) if G is k-edge-connected (k-arc-connected) but $G-e$ is not for each edge (arc) e. Similarly, a graph G is minimally k-connected if G is k-connected but $G-e$ is not for each edge e. Strongly minimally k-connected (or, briefly, minimally k-connected digraphs) are defined analogously.

A k-connected (k-edge-connected) graph is called critically k-connected (critically k-edge-connected) if deleting any node destroys k-connectivity (k-edge-connectivity). The corresponding notions for digraphs are defined analogously.

Actually there are eight classes to be investigated corresponding to the possible choices: directed or undirected graph, edge- (arc-)connectivity or node-connectivity, critical or minimal. There are interesting results concerning each but one of these classes (critically k-arc-connected digraphs have not yet been investigated). Here we list only the most important theorems. The starting point is a theorem due to Halin.

Theorem 7.41. (a) (Halin 1969a) *Every minimally k-connected graph has at least one node of degree k.*

(b) *Every minimally k-edge-connected graph G (with at least two nodes) contains a node of degree k.*

Mader extended these results.

Theorem 7.42 (Mader 1972). *Every minimally k-connected graph contains at least $k+1$ nodes of degree k. Furthermore, every circuit of G contains a node of degree k* (Mader 1971b). *Every minimally k-edge-connected simple graph contains at least $k+1$ nodes of degree k.*

Another interesting generalization of Halin's result is also due to Mader (1973).

Theorem 7.43. *In a simple graph if every degree is at least $k+1$, there are two adjacent nodes s and t which are connected by $k+1$ openly disjoint paths.*

Let us see critical graphs.

Theorem 7.44 (Mader 1986). *Every critically k-edge-connected simple graph G contains a node of degree k. Furthermore, G contains a node x such that $G-x$ is $(k-1)$-edge-connected.*

To formulate results on critically k-connected graphs we need the following concepts. In a k-connected graph G a set X is a separating set if $G-X$ is not connected. A subset C of nodes called an *end* of G if C is one of the components in $G-X$ for a k-element separating set X. An *atom* is an end with smallest cardinality. The name "atom" is justified by the following.

Theorem 7.45 (Mader 1971a). *If C is an atom and K is an end, then either $K \cap C \neq \emptyset$ or $C \subseteq K$.*

It is not true, in general, that a critically k-connected graph G has a node of degree k (or, equivalently, the atoms of G are of cardinality one). But one has the following.

Theorem 7.46 (Mader 1985c). *Every critically k-connected (not complete) graph G contains two disjoint ends with cardinality at most $k/2$. Furthermore, G contains four disjoint ends.*

Corollary 7.47 (Chartrand et al. 1972). *Every critically k-connected simple (not complete) graph G contains a node of degree at most $\lfloor 3k/2 \rfloor$ and this bound is best possible. Actually, G contains at least two such nodes* (Hamidoune 1980 and Veldman 1983).

What about directed graphs?

Theorem 7.48 (Mader 1985b). *Every minimally k-connected digraph contains at least k nodes of in-degree k and at least k nodes of out-degree k.*

Conjecture 7.49 (Mader 1979). Every minimally k-connected digraph contains a node of in-degree and out-degree k.

The k-arc-connected version of this statement is true and plays a central role in constructing all k-arc-connected digraphs (Theorem 7.9).

Theorem 7.50 (Mader 1974b). *Every minimally k-arc-connected digraph contains at least two nodes having both in-degree and out-degree k.*

For critical strongly connected digraphs it is true again that there is a node of in degree 1. More specifically, the following holds.

Theorem 7.51 (Mader 1989). *Every critical strongly connected digraph with at least four nodes contains four distinct nodes x_1, x_2, y_1, y_2 for which $\delta^-(x_i) = \delta^+(y_i) = 1$ $(i = 1, 2)$.*

Theorem 7.52 (Mader 1991). *Every critically k-connected digraph G contains a node s for which $\delta^-(s) \leq 2k-1$ or $\delta^+(s) \leq 2k-1$. If G is antisymmetric (that is, if*

(x, y) is an arc, then (y, x) is not), then G contains a node s for which $\delta^-(s) \leq \lfloor 3k - \frac{1}{2} \rfloor$ or $\delta^+(s) \leq \lfloor 3k - \frac{1}{2} \rfloor$. These bounds are best possible.

Actually this theorem is a consequence of a result of Mader that can be considered as a directed counterpart of Theorem 7.46. It is not true that the minimum in-degree in a critically k-connected digraph ($k \geq 2$) is at most $2k - 1$.

7.6. Connected subgraphs

By Whitney's Theorem 7.5 we know that a k-connected graph contains k openly disjoint paths connecting two specified nodes. It is a natural feeling that a highly connected graphs must contain some other type of subgraphs. In this subsection we briefly summarize such results. The starting point is a theorem by Dirac.

Theorem 7.53 (Dirac 1960). *In a k-connected graph G every subset of k nodes is included in a circuit. If, in addition, G is non-bipartite, every subset of $k - 1$ nodes is included in an odd circuit* (Bondy and Lovász 1981).

Theorem 7.54 (Mesner and Watkins 1967). *In a k-connected graph ($k \geq 3$) a subset H of $k + 1$ nodes is included in a circuit if and only if there is no set $X \subseteq V - H$ with $|X| = k$ such that each node in H belongs to a different component of $G - X$.*

Suppose we want much more: find $k(k - 1)/2$ openly disjoint paths between k specified nodes $s_1, \ldots, s_k$ (one path for one pair). Such a configuration can be considered as a subdivision of K_k with principal nodes $s_1, \ldots, s_k$. With sufficiently high connectivity this property can also be guaranteed.

Theorem 7.55 (Jung 1970, Larman and Mani 1970). *If G is $2^{3k(k-1)/2}$ connected, then for distinct nodes $s_1, \ldots, s_k$ there is a subdivision of K_k in G having $s_1, \ldots, s_k$ as principal nodes.*

Theorem 7.56 (Häggkvist and Thomassen 1982). *In a k-connected graph every subset of $k - 1$ independent edges is included in a circuit.*

Conjecture 7.57 (Lovász). In a k-connected graph every subset of k independent edges is included in a circuit unless k is odd and the k edges disconnect the graph.

For $k = 3$ this was shown by Lovász. The following result, due to Lovász (1977) and Györi (1978), is about partitions of graphs or digraphs into connected parts of given size.

Theorem 7.58. *In a digraph (graph) let $S = \{s_1, s_2, \ldots, s_k\}$ be a set of k nodes and $n_1, n_2, \ldots, n_k$ positive integers such that $\sum n_i = |V|$. Suppose that for any $v \in V - S$ there are k paths from S to v pairwise disjoint except at v. There is a* *partition* $\{V_1, V_2, \ldots, V_k\}$ *of* V *into* k *parts such that* $V_i \cap S = \{s_i\}$, $|V_i| = n_i$ *and the digraph (graph) induced by* V_i *contains an arborescence rooted at* s_i *(is connected).*

7.7. *Extremal results*

In this last subsection we briefly mention some extremal-type results concerning connectivity. More detailed accounts are found in Mader (1979) and in Bollobás, (1978). Let us start with a result on digraphs.

Theorem 7.59 (Dalmazzo 1977). *A minimal k-edge-connected digraph* $D = (V, A)$ *on* n *nodes has at most* $2k(n-1)$ *arcs.*

Proof. Let s be an arbitrary node of D. By Edmonds' Theorem 6.6 there are k edge-disjoint spanning arborescences of root s. Let G_1 denote the union of these arborescences. Similarly, there are k edge-disjoint spanning co-arborescences of root s (a co-arborescence of root s is a directed tree such that re-orienting all of its edges results in an arborescence of root s). Let G_2 denote the union of these co-arborescences. Clearly, both G_1 and G_2 have $k(n-1)$ arcs and their union is k-edge-connected from which the result follows. □

Note that the bound in the theorem is sharp as is shown by a digraph obtained from any tree by replacing each edge uv by $2k$ parallel arcs among which k are in one direction and the other k are in the other direction.

Theorem 7.59 immediately implies that a minimal k-edge-connected graph on n nodes has at most $k(n-1)$ edges. Indeed, if we replace each edge by two oppositely directed arcs, we obtain a minimal k-edge-connected digraph and then Theorem 7.59 applies.

Mader proved that for simple graphs a better bound exists.

Theorem 7.60 (Mader 1974b). *A minimal k-edge-connected simple graph on n nodes has at most* $kn - k(k+1)/2$ *edges.*

Since a minimal k-edge-connected graph must not have a $(k+1)$-edge-connected subgraph, Theorem 7.60 is an immediate consequence of the following.

Theorem 7.61 (Mader 1974b). *Every simple graph on n nodes with more than* $kn - k(k+1)/2$ *edges has a* $(k+1)$*-edge-connected subgraph.*

Again, the bound is sharp as is shown by a graph constructed from a complete bipartite graph $K_{k,n-k}$ by adding all the possible edges in the k-element part.

8. Multicommodity flows and disjoint paths

8.1. *Problem formulation*

In this section we address the following problem, called *the disjoint paths problem.* Given a graph or a digraph and k pairs of nodes (s_1, t_1),

$(s_2, t_2), \ldots, (s_k, t_k)$, find k pairwise openly disjoint paths connecting the corresponding pairs (s_i, t_i). If we are interested in finding edge-disjoint paths we speak about the *edge-disjoint paths problem.* In the book "Paths, Flows, and VLSI-Layout" (B. Korte et al., eds., Springer 1990), several survey papers are included related to the material of this section (Frank 1990, Schrijver 1990, Robertson and Seymour 1990].

A capacitated version of the edge-disjoint paths problem is the following. For every edge of the graph a non-negative capacity is specified and, similarly, for every pair to be connected a non-negative demand is given. The *integer multicommodity* flow problem is that of finding as many paths between the corresponding terminals as their demands are so that every edge occurs in at most as many paths as its capacity. If we allow fractional paths as well, we speak about the *multicommodity flow problem* or, in short, *multiflow problem.* That is, a multicommodity flow is defined by paths $P_1, P_2, \ldots, P_k$ and non-negative numbers $\lambda_1, \ldots \lambda_k$ such that each path P_i is a path from s_i to t_i and for each edge e the sum of coefficients assigned to paths using e is at most the capacity of e.

Actually this kind of problem can be considered as a *feasibility* problem. The *maximization problem* is that when no demands are specified and one is interested in finding a maximum number of paths connecting the corresponding terminal pairs.

Sometimes it is convenient to mark the terminal pairs to be connected by an edge. The graph $H = (U, F)$ formed by the marking edges is called a *demand graph* while the original graph $G = (V, E)$ is the *supply graph.* Let us call a circuit of $G + H$ a *good* circuit if it contains precisely one demand edge. In this terminology the edge-disjoint paths problem is equivalent to seeking for $|F|$ edge-disjoint good circuits.

The multiflow problem can be formulated as a linear program. One way to do so is as follows. Let A be a 0–1 matrix the rows of which correspond to the edges of G the columns correspond to the good circuits. An entry (i, j) is 1 if the edge corresponding to i is in the circuit corresponding to j and 0 otherwise. Similarly let B be a 0–(-1) matrix the rows of which correspond to the edges of H, the columns correspond to the good circuits. An entry (i, j) is -1 if the edge corresponding to i is in the circuit corresponding to j and 0 otherwise. (The structure of B is simple: every column has exactly one non-zero entry.) The multiflow problem is equivalent to the following linear inequality system. $Ax \leq \mathbf{1}$, $Bx = -\mathbf{1}$, $x \geq 0$, where $\mathbf{1}$ and $-\mathbf{1}$ are appropriately sized vectors of 1s and -1s, respectively.

By Farkas's lemma this system has no solution if and only if there is a vector w in $\mathbb{R}^E_+$ and a vector z in $\mathbb{R}^F$ such that $\sum (w(e)\colon e \in E) - \sum (z(f)\colon f \in F) < 0$ and such that $\sum (w(e)\colon e \in C - f) - z(f) \geq 0$ holds for every demand edge f and every circuit C for which $C \cap F = \{f\}$. Obviously, if there is such a w and z, then z can be chosen so as to satisfy $z(f) = \mathrm{dist}_w(u, v)$ where $f = uv$ and $\mathrm{dist}_w(u, v)$ is the minimum w-weight of a path in G connecting the end nodes of demand edge f. We obtain the following.

Theorem 8.1 (Iri 1970, Onaga and Kakusho 1971). *The multiflow problem has a* *solution if and only if the*

$$\text{Distance criterion: } \sum(\text{dist}_w(u, v)\colon uv \in F) \leqslant \sum(w(e)\colon e \in E) \tag{8.1}$$

holds for every $w \in \mathbb{R}^E_+$.

As general linear programs can be solved in polynomial time, so is the multiflow problem. Since the constraint matrix above has entries 0, ± 1 there is a strongly polynomial algorithm as well (Tardos 1986). This is why we concentrate only on integer multicommodity flows or disjoint paths.

First we survey results concerning undirected graphs.

Theorem 8.2 (Karp 1975). *The undirected (edge-) disjoint path problem (when k is a part of the input) is* NP-*complete.*

Even et al. (1976) proved that the problem is NP-complete even in the special case when the demand graph consists of two sets of parallel edges. In other words, the integer 2-commodity flow problem is NP-complete. On the other hand we have the following very difficult result.

Theorem 8.3 (Robertson and Seymour 1986b). *For fixed k the undirected (edge-disjoint) disjoint paths problem can be solved in polynomial time.*

8.2. Characterizations for edge-disjoint paths

First, let us concentrate on edge-disjoint paths. A natural necessary condition is the cut-criterion:

Cut-criterion: $d_G(X) \geqslant d_H(X)$ for every $X \subseteq V$.

Note that the cut-criterion is a special case of the distance-criterion. We call the difference $d_G(X) - d_H(X)$ the *surplus* of cut $\Delta(X)$ and denote it by $s(X)$. A cut $\Delta(X)$ is called *tight* if $s(X) = 0$.

The cut criterion is not sufficient, in general, as the simple example in fig. 8.1 shows. It is sufficient, however, if the demand graph is a star (that is, the demand edges share a common endpoint). (This immediately follows from the undirected edge version of Menger's theorem.)

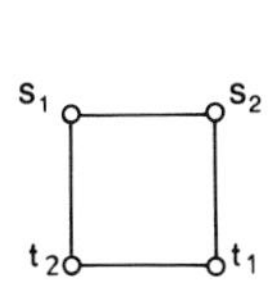

Figure 8.1

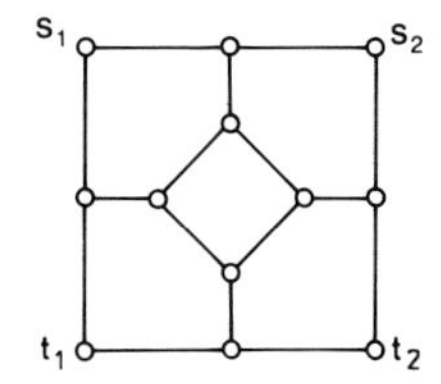

Figure 8.2.

The next two simplest demand graphs are $2K_2$ (a graph on four nodes with two disjoint edges) and C_3 (a triangle). The following characterization for $2K_2$ is due to Seymour (1980c) and Thomassen (1980a).

Theorem 8.4. *Let G be a graph such that no cut edge separates both of the two terminal pairs (s_1, t_1) and (s_2, t_2). There are no two edge-disjoint paths between the corresponding terminals if and only if some edges of G can be contracted so that the resulting graph G' is planar, the four terminals have degree two while the other nodes are of degree 3 and the terminals are positioned on the outer face in this order: s_1, s_2, t_1, t_2.*

Figure 8.2 shows a typical example where the two edge-disjoint paths do not exist.

If we want k_i paths between s_i and t_i $(i = 1, 2)$ the problem becomes NP-complete. The situation is much better for the other special H mentioned above.

Theorem 8.5 (Seymour 1980c). *If the demand graph H consists of three sets of parallel edges between three nodes v_1, v_2 and v_3, the edge-disjoint paths problem has a solution if and only if the cut criterion holds and*

$$q(V_1 \cup V_2 \cup V_3) \leqslant s(V_1) + s(V_2) + s(V_3)$$

for every choice of disjoint sets V_i with $v_i \in V_i$ $(i = 1, 2, 3)$ where $s(X)$ denotes the surplus and $q(X)$ denotes the number of components C in $G - X$ for which $d_G(X) + d_H(X)$ is odd.

This result is a rather easy consequence of a theorem of Mader (Theorem 8.23 below) on edge-disjoint T-paths (when $|T| = 3$).

Let us call a set X, given G and H, an *odd set* and the cut $\Delta(X)$ an *odd cut* (with respect to $G + H$) if $d_G(X) + d_H(X)$ is odd (or equivalently, the surplus $s(X)$ is odd). A basic feature of odd cuts is that in any solution to the edge-disjoint paths problem an odd number of edges of an odd cut, in particular, at least one edge, will not be used.

What if there are no odd cuts at all, that is, $G + H$ is Eulerian? The cut criterion is still not sufficient as is shown in fig. 8.3. Even worse, Middendorf and Pfeiffer (1990) proved that the edge-disjoint paths problem is NP-complete even if $G + H$ is Eulerian.

However, in the special cases listed below the cut criterion proves to be

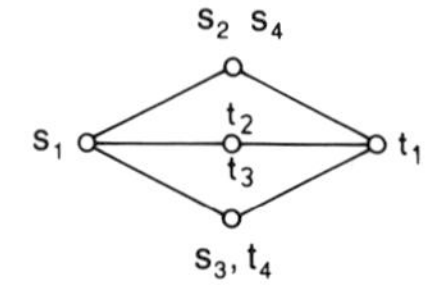

Figure 8.3.

sufficient. Given a demand graph $H = (V, A)$, H' will denote the graph arisen from H by replacing each set of parallel edges by one edge.

Theorem 8.6. *Suppose that $G + H$ is Eulerian. In the following cases the cut criterion is necessary and sufficient for the solvability of the edge-disjoint paths problem.*

(a) *H' is $2K_1$* (Rothschild and Whinston 1966b).

(b) *H' consists of two stars* (Papernov 1976, Seymour 1980a, Lomonosov 1985).

(c) *H' is K_4* (Papernov 1976, Seymour 1980a, Lomonosov 1985).

(d) *H' is C_5* (Lomonosov 1985).

(e) *G is planar and each terminal is on one face* (Okamura and Seymour 1981).

(f) *G is planar and there are two faces F_1, F_2 such that each demand edge connects two nodes of either F_1 or F_2* (Okamura 1983).

(g) *$G + H$ is planar* (Seymour 1981).

(h) *G is planar and there are two specified inner faces C_1 and C_2 of G. The demand edges $s_1t_1, \ldots, s_kt_k$ are positioned in such a way that each s_i is on C_1, each t_i is on C_2 and their cyclic order is the same* (Schrijver 1989).

Note that part (g) of this theorem immediately follows (by planar dualization) from a theorem by Seymour asserting that a ± 1 weighted bipartite graph (planar or not) has no circuit of negative total length if and only if the edge set can be partitioned into cuts such that each cut contains at most one negative edge. This is an equivalent formulation of Seymour's theorem on the maximum number of disjoint T-cuts.

As we mentioned earlier the cut criterion is not sufficient, in general, even if $G + H$ is Eulerian. Sometimes the stronger distance criterion (8.1) helps.

Theorem 8.7 (Karzanov 1987). *Suppose that $G + H$ is Eulerian and the demand edges form a graph arising from K_5 by adding parallel edges. Then the distance criterion is necessary and sufficient for the solvability of the edge-disjoint paths problem. (In other words, if there is a fractional solution, there is an integral one.)*

Theorem 8.8 (Karzanov 1994). *Suppose that G is planar, $G + H$ is Eulerian and each demand edge connects two nodes of one of three specified faces of G. Then the distance criterion is necessary and sufficient for the solvability of the edge-disjoint paths problem.*

Theorem 8.5 provided an example where parity played a basic role in a good characterization. Here are two more cases.

Theorem 8.9 (Frank 1990). *Suppose that $G + H$ is planar and the demand edges are on two faces of G. The edge-disjoint paths problem has a solution if and only if the cut criterion holds and $d_{G+H}(X \cap Y)$ is even for every pair of tight sets X, Y.*

This theorem is a generalization of an earlier theorem of Seymour (1981) where H consisted of two sets of parallel edges. Sebő (1993) proved that if $G+H$ is planar and the number of demand edges nodes is bounded by a constant, then there is a polynomial time algorithm to solve the integer multiflow problem. Another result, due to Schrijver (1990) asserts, that if $G+H$ is planar and the number of faces covering all the terminal nodes is bounded by a constant, then the edge-disjoint paths problem is polynomially solvable. On the other hand, the problem is NP-complete if there is no such a bound (Middendorf and Pfeiffer 1990).

Theorem 8.10 (Frank 1985). *Suppose that G is planar, the terminals are on the outer face and the degree of every node not on the outer face is even. The edge disjoint paths problem has a solution if and only if $\sum s(C_i) \geq q/2$ for every family $(C_1, C_2, \ldots, C_k)$ of cuts ($k \leq |V|$) where q denotes the number of odd components in $G - C_1 - C_2 - \cdots - C_k$ and $s(C)$ is the surplus of C.*

To close this subsection we mention a theorem by van Hoesel and Schrijver (1986) where topology plays a role.

Theorem 8.11. *Let G be a planar graph embedded in $\mathbb{R}^2$. Let O denote the interior of the unbounded face and I the interior of a specified bounded face. Let C_1, $C_2, \ldots, C_k$ be curves in $\mathbb{R}^2 - (I \cup O)$ each of which connects a node on $I \cup O$ with a node on $I \cup O$ so that for each node v of G the degree of v has the same parity as the number of curves ending at v. Then there exist pairwise edge-disjoint paths P_1, $P_2, \ldots, P_k$ in G so that P_i is homotopic to C_i in $\mathbb{R}^2 - (O \cup I)$ ($i = 1, 2, \ldots, k$) if and only if for each dual path Q from $I \cup O$ to $I \cup O$ the number of edges in Q is not smaller than the number of times Q necessarily intersects the curves C_i.*

Note that this theorem generalizes part (e) of Theorem 8.6. It is an open problem to find a common generalization of Theorems 8.6(f) and 8.11. This last theorem is a prototype of theorems belonging to the area one may call homotopic paths packing. An excellent survey of this topic occurs in Schrijver (1990).

8.3. Sufficient conditions for edge-disjoint paths

We call a graph *k-linked on the edges* if for any choice of k pairs of terminals there are k edge-disjoint paths connecting the corresponding terminal pairs.

Theorem 8.4 implies that a 3-edge-connected graph is 2-linked on the edges. Actually such a graph is 3-linked as the following even stronger result shows.

Theorem 8.12 (Okamura 1984). *In a graph three terminal pairs (s_i, t_i) ($i = 1, 2, 3$) are specified such that for each i there are three edge-disjoint paths connecting s_i and t_i. Then there are s_it_i-paths ($i = 1, 2, 3$) pairwise edge-disjoint.*

Okamura's theorem is an answer to the following conjecture of Thomassen, when $k = 3$.

Conjecture 8.13 A k-edge-connected graph is k-linked on the edges if k is odd and $(k-1)$-linked on the edges if k is even.

Note that by Tutte's Theorem 6.14 a $(2k)$-edge-connected graph always has k edge-disjoint spanning trees and therefore it is k-linked on the edges.

The following theorem gets very close to the conjecture.

Theorem 8.14 (Huck 1991). *A k-edge-connected graph is $(k-1)$-linked on the edges if k is odd and $(k-2)$-linked on the edges if k is even.*

We note that Thomassen's conjecture is open for $k = 5$.

In certain cases the cut condition is not strong enough to ensure the existence of the required paths but the demands can almost be met.

Theorem 8.15 (Korach and Penn 1992). *Suppose that $G + H$ is planar, for each terminal pair (s_i, t_i) an integer demand d_i is given and the cut condition holds with respect to d. Then there are $d_i - 1$ paths connecting s_i and t_i for each terminal pair such that all these paths are pairwise disjoint.*

Theorem 8.16 (Itai and Zehavi 1984). *Assume that in a graph G (s_i, t_i) are terminal pairs $(i = 1, 2)$ such that there are k edge-disjoint paths connecting s_i and t_i $(i = 1, 2)$. Then for each m, $0 \leq m < k$ there are k edge-disjoint paths P, S_1, $S_2, \ldots, S_m$, Q_1, $Q_2, \ldots, Q_{k-m-1}$ such that each S_i connects s_1 and t_1, each Q_j connects s_2 and t_2 and P connects either s_1 and t_1 or s_2 and t_2.*

8.4. Node-disjoint paths

We call a graph *k-linked* if for any choice of k pairs of terminals there are k openly disjoint paths connecting the corresponding terminal pairs. A counter-part of the cut-condition is:

Node-cut condition: No subset S of nodes can separate more than $|S|$ terminal pairs.

This condition is sufficient if the terminal pairs share a common node (a version of the node-Menger theorem) but not in general.

Theorem 8.17 (Thomassen 1980a, Seymour 1980c). *Let G be a graph such that no cut node separates s_1 from t_1 and s_2 from t_2. There are no disjoint paths between s_1 and t_1 and between s_2 and t_2 if and only if G arises from a planar graph G', where the four terminals are on the outer face in this order s_1, s_2, t_1, t_2, by placing an arbitrary graph into some faces of G' bounded by two or three edges.*

The problem was solved algorithmically by Shiloach (1980).

Corollary 8.18 (Jung 1970). *A 4-connected non-planar graph is 2-linked. A 6-connected graph is 2-linked.*

Here the second statement follows from the first one since a planar graph always has a node of degree at most 5. Note that there is a 5-connected planar graph that is not 2-linked. For higher connectivity we have the following.

Theorem 8.19 (Jung 1970, Larman and Mani 1970). *A 2^{3k} connected graph is k-linked.*

It is not known if 2^{3k} can be replaced by a linear bound. The natural $2k + 2$ is not enough as can be seen from a K_{3k-1} with edges $x_1y_1, \ldots, x_ky_k$ removed (an example due to Strange and Toft 1983).

The following pretty result is not difficult to prove.

Theorem 8.20 (Robertson and Seymour 1986a). *Suppose that G is planar and the terminals are on the outer face. The disjoint paths problem has a solution if and only if the node-cut condition holds and there are no two "crossing" terminal pairs (that is, any two pairs (s_1, t_1) and (s_2, t_2) are in this order on the outer face; s_1, t_1, s_2, t_2).*

Robertson and Seymour also found a characterization for the disjoint paths problem when G is planar and the terminals are positioned on two specified faces.

8.5. *Maximization*

So far we have studied multicommodity flow problems of feasibility type. One can also be interested in the maximization form: Given a graph $G = (V, E)$ with non-negative integer capacity function c on the edges and a set of terminal pair $(s_1, t_1), \ldots, (s_k, t_k)$, find flows between s_i and t_i $(i = 1, 2, \ldots, k)$ that maximize the sum M of flow values under the condition that for each edge e the sum of edge-values of flows on this edge is at most $c(e)$. Let us denote by M_I the maximum sum of flow values if we restrict ourselves to integer flows.

We will use the notation $d_c(v)$ for the sum of capacities of edges incident to v. Let $V_1, V_2, \ldots, V_t$ be a family $\mathcal{P}$ of disjoint subsets of V such that each demand edge connects different V_i. By a *multicut* defined by $\mathcal{P}$ we mean the set of edges uv of G such that $u \in V_i$, $v \notin V_i$ for some i. The capacity of a multicut is defined to be $\sum d(V_i)/2$. Let m denote the minimum capacity of a multicut. Let m_1 denote the minimum capacity of a cut separating each terminal pair (if there is any). Obviously, $m_1 \geqslant m \geqslant M \geqslant M_I$. If $k = 1$, then $m_1 = M_I$ by Menger's theorem.

Theorem 8.21 (Hu 1963). *If $k = 2$, $m_1 = M$. If $k = 2$ and $d_c(v)$ is even for each non-terminal node, then $m_1 = M_I$* (Rothschild and Whinston 1966a).

Theorem 8.22 (Lovász 1976b, Cherkasskij 1977b). If the demand edges form a complete graph induced by T ($T \subseteq V$), then $m = M$. In addition, if $d_c(v)$ is even for $v \in V - T$, then $m = M_I$.

Generalizing this result to non-Eulerian graphs, Mader (1978b) found the following characterization for M_I.

Theorem 8.23. *Let $G = (V, E)$ be a graph and T a specified subset of nodes. The maximum number of edge-disjoint paths connecting distinct elements of T is $\min[\sum d(V_i) - q_0(\cup V_i)]/2$ where the minimum is taken over all collections of disjoint subsets $V_1, V_2, \ldots, V_{|T|}$ for which $|V_i \cap T| = 1$. (Here $d(X)$ denotes the edges leaving X and $q_0(X)$ denotes the number of components C of $G - X$ for which $d(C)$ is odd.)*

To formulate a node-disjoint version of Theorem 8.23 suppose that T is independent. For a subset X of V and a subgraph G' of G let $b(X; G') := |\{x \in X$: there is an $xy \in E(G')$ with $y \not\in X\}|$. $E(X)$ denotes the set of edges induced by X.

Theorem 8.24 (Mader 1978c). *The maximum number of openly node-disjoint paths connecting distinct members of T is equal to* $\min(|V_0| + \sum \lfloor b(V_i; G - V_0)/2 \rfloor)$ *where the minimum is taken over all collections of disjoint subsets $V_0, V_1, \ldots, V_k$ of $V - T$ ($k \geq 0$) (where only V_0 can be empty) such that $G - V_0 - \bigcup(E(V_i): i = 1, \ldots, k)$ contains no path connecting distinct nodes of T.*

This result can be regarded as a common generalization of Menger's theorem and the Berge–Tutte theorem. An immediate corollary of Theorem 8.24 is a result of Gallai (1961).

Corollary 8.25. *The maximum number of disjoint paths having end nodes in T is $\min(|K| + \sum \lfloor |C \cap T|/2 \rfloor : K \subseteq V)$ where the sum is taken over the components C of $G - K$.*

Let us turn back to edge-disjoint paths. A common generalization of Theorems 8.21 and 8.22 is as follows.

Theorem 8.26 (Karzanov and Lomonosov 1978). *Let $H = (T, F)$ denote the demand graph. If the maximal independent sets of H can be partitioned into two classes such that both classes consist of disjoint sets (which is equivalent to saying that the complement of H is the line graph of a bipartite graph), then $m = M$. In addition, if $d_c(v)$ is even for $v \in V - T$, then $m = M_I$* (Karzanov 1985).

A proof relying on the polymatroid intersection theorem of Edmonds can be found in Frank et al. (1992). Let us continue our survey with two results where no parity restrictions are imposed.

Theorem 8.27 (Lomonosov 1983). *Suppose that $k = 2$ and $G + H$ is planar. Then*

M_I is either $m-1$ or m. $M_I = m-1$ if and only if there are three cuts of value m (each separating both (s_i, t_i) $(i=1,2)$) the union of which includes a cut B of odd capacity so that B does not separate (s_i, t_i) $(i=1,2)$.

Note that in Theorem 8.27 $m = M$ immediately follows from Theorem 8.6(g).

Theorem 8.28 (Kleitman et al. 1970). *If every node of $v \in V - (s_1, \ldots, s_k) - (t_1, \ldots, t_k)$ is adjacent to a member of at least $k-1$ terminal pairs and the terminals form an independent set of G, then $m = M_I$.*

8.6. Directed graphs

All the results we have considered so far in this section concerned undirected graphs. Let $D = (V, A)$ be a digraph and let (s_i, t_i) $(i = 1, 2, \ldots, k)$ be (ordered) pairs of terminals. The problem is to find (arc-) disjoint paths from s_i to t_i. Let $H = (U, F)$ denote the demand graph, where $F = \{(t_i, s_i): i = 1, 2, \ldots, k\}$. We call a directed circuit of $D + H$ *good* if it contains precisely one demand arc. Then the arc-disjoint paths problem is equivalent to finding k arc-disjoint good circuits of $D + H$.

Unfortunately, much less is known about directed graphs. One negative result is as follows.

Theorem 8.29 (Fortune et al. 1980). *The (arc-) disjoint paths problem is* NP-*complete for $k = 2$.*

Notice that the corresponding undirected problem is tractable (see Theorem 8.3). In what follows we briefly list some special cases when good characterizations and/or polynomial time algorithms are available.

The following criterion is clearly necessary in the arc-disjoint case:

Directed cut criterion: $\delta_D^-(X) \geq \delta_H^+(X)$ for every $X \subseteq V$. If $s_1 = \cdots = s_k$ and $t_1 = \cdots = t_k$ then the directed cut criterion is sufficient as well (directed arc-version of Menger's theorem). A counter-part of Theorem 8.8(a) is also true.

Theorem 8.30. *If H consists of two sets of parallel arcs and $D + H$ is Eulerian (that is, the in-degree of any node is equal to the out-degree), then the directed cut criterion is necessary and sufficient for the solvability of the arc-disjoint paths problem.*

Proof. Assume that H consists of α_i arcs from t_i to s_i $(i = 1, 2)$. By Menger's theorem, it follows from the hypothesis of the theorem that there are α_1 arc-disjoint paths in D from s_1 to t_1. If we leave out these paths and the α_1 demand edges from $D + H$ we obtain an Eulerian digraph. This partitions into arc-disjoint circuits, and hence it contains α_2 edge-disjoint paths from s_2 to t_2. □

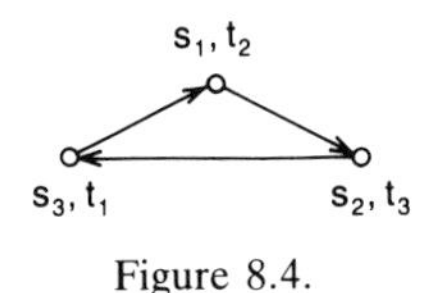

Figure 8.4.

In the example in fig. 8.4 the directed cut criterion is satisfied but there is no solution to the directed edge-disjoint paths problem.

The reason is that the following necessary condition, called *covering criterion*, is violated: the directed circuits of $D + H$ cannot be covered by less than k arcs.

Theorem 8.31. *When $D + H$ is planar and D is acyclic the directed edge-disjoint paths problem has a solution if and only if the covering criterion holds.*

Proof. By planar dualization we obtain from the Lucchesi–Younger Theorem 2.15 that in a planar digraph the maximum number of directed circuits is equal to the minimum number of arcs covering all the directed circuits. Since D is acyclic the set $A(H)$ of demand arcs covers all directed circuits of $D + H$. By the covering criterion this is a minimum covering of directed circuits and hence there are $k = |A(H)|$ arc-disjoint circuits in $D + H$ which must be good circuits. □

In the following theorem D may be non-planar but the number k of demand arcs is considered as a constant.

Theorem 8.32 (Fortune et al. 1980). *In acyclic digraphs the (arc-) disjoint paths problem can be solved in polynomial time if k is fixed.*

In the special case $k = 2$, Thomassen (1985) found a complete description of acyclic digraphs having no solution to the disjoint paths problem. The core of his result is as follows.

Theorem 8.33. *Let us be given an acyclic digraph $D = (V, A)$ and terminal pairs (s_1, t_1), (t_2, t_2) such that $|V| \geqslant 5$, $\delta^-(v)$, $\delta^+(v) \geqslant 2$ for each non-terminal node v and $\delta^-(s_1) = \delta^-(s_2) = \delta^+(t_1) = \delta^+(t_2) = 0$. If there are no disjoint paths from s_1 to t_1 and from s_2 to t_2, then D is planar and has a plane representation in such a way that s_1, t_2, t_1, s_2 are on the outer face occurring in that cyclic order.*

Ibaraki and Poljak (1991) solved the arc-disjoint paths problem when $k = 3$ and $D + H$ is Eulerian. Let D be an Eulerian digraph with three distinct specified nodes a, b, c, called *terminals*. The *three-terminal* problem consists of finding (altogether three) arc-disjoint paths from a to b, from b to c and from c to a. Clearly, this is a special case of the three arc-disjoint paths problem but Ibaraki and Poljak observed that, conversely, the three arc-disjoint paths problem can also be easily reduced to the three-terminal problem.

Suppose that D is a planar digraph with no cut-nodes that has a plane representation such that each face is bounded by a directed circuit, the terminals have degree 2, and they lie in one face where their order with respect to the orientation of the face is a, c, b. It is easy to see that in such case the three-terminal problem has no solution. Therefore we call such a representation *bad*.

Theorem 8.34 (Ibaraki and Poljak 1991). *Given an Eulerian digraph D with terminals a, b, c, the three-terminal problem has a solution if and only if D cannot be contracted to a planar digraph that has a bad plane representation.*

As far as the maximization problem is concerned for digraphs we mention the following (rather easy) counter-part of Theorem 8.22.

Theorem 8.35 (Frank 1989). *In an Eulerian digraph $D = (V, A)$ the maximum number of arc-disjoint paths connecting distinct nodes of a specified subset T of V is equal to the minimum of $\sum \delta^{-}(V_i)$ over all families of disjoint subsets $V_1, V_2, \ldots, V_{|T|}$ of V for which $|V_i \cap T| = 1$ $(i = 1, 2, \ldots, |T|)$.*

We close this section by mentioning an interesting sufficient condition by Shiloach (1979). Let us call a digraph $D = (V, A)$ *k-linked on the arcs* if for any choice of k pairs $\{s_1, t_1\}, \ldots, (s_k, t_k\}$ of (not necessarily distinct) terminals there are arc-disjoint paths P_i from s_i to t_i $(i = 1, \ldots, k)$. Obviously such a digraph is strongly k-arc connected (that is every non-empty proper subset of nodes has k entering arcs.)

Theorem 8.36. *A strongly k-arc connected digraph is k-linked on the arcs.*

Proof. Add a new node r to D and new arcs (r, s_i) $(i = 1, 2, \ldots, k)$ and apply Edmonds' disjoint arborescence Theorem 6.9. □

The theorem also easily follows from Theorem 7.30 of Mader.

References

Ahuja, R.K., T.L. Magnanti and J.B. Orlin
[1993] *Network Flows: Theory, Algorithms, and Applications* (Prentice-Hall, Englewood Cliffs, NJ).

Alon, N.
[1990] Generating pseudo-random permutations and maximum flow algorithms, *Inform. Process. Lett.* **35**, 201–204.

Ando, K., H. Enomoto and A. Saito
[1987] Contractible edges in 3-connected graphs, *J. Combin. Theory B* **42**, 87–93.

Barnette, D.W., and B. Grünbaum
[1969] On Steinitz's theorem concerning convex 3-polytopes and on some properties of planar graphs, in: *The Many Facets of Graph Theory, Lecture Notes in Mathematics,* Vol. 110, eds. G. Chartrand and S.F. Kapoor (Springer, Berlin) pp. 27–40.

Bellman, R.E.
[1958] On a routing problem, *Quart. Appl. Math.* **16**, 87–90.

Boesch, F., and R. Tindell
[1980] Robbins's theorem for mixed multigraphs, *Amer. Math. Monthly* **87**, 716–719.
Bollobás, B.
[1978] *Extremal Graph Theory* (Academic Press, New York).
Bondy, J.A., and L. Lovász
[1981] Cycles through specified vertices of a graph, *Combinatorica* **1**, 117–140.
Boruvka, O.
[1926] O jistem problemu minimalnim, *Prace Mor. Prirodoved. Spol. v Brne (Acta Societ. Scient. Natur. Moravicae)* **3**, 37–58.
Chartrand, G., A. Kaugars and D.R. Lick
[1972] Critically n-connected graphs, *Proc. Amer. Math. Soc.* **32**, 63–68.
Cheriyan, J., and T. Hagerup
[1990] A randomized maximum flow algorithm, in: *Proc. 30th IEEE Conf. on the Foundations of Computer Science*, pp. 118–123.
Cheriyan, J., and S.N. Maheshwari
[1988] Finding nonseparating induced cycles and independent spanning trees in 3-connected graphs, *J. Algorithms* **9**, 507–537.
[1989] Analysis of preflow-push algorithms for maximum network flow, *SIAM J. Comput.* **18**, 1057–1086.
Cherkasskij, B.V.
[1977a] Algorithm of construction of maximal flow in networks with complexity $O(n^2p^{1/2})$ operations, Akad. Nauk SSSR, CEMI (*Mathematical Methods for the Solution of Economical Problems*) **7**, 117–126 (in Russian).
[1977b] A solution of a problem of multicommodity flows in a network, *Ekonom. i Mat. Metody* **13**(1), 143–151 (in Russian).
Chu, Y.-J., and T.-H. Liu
[1965] On the shortest arborescence of a directed graph, *Sci. Sinica* **4**, 1396–1400.
Dalmazzo, M.
[1977] Nombre d'arcs dans les graphes k-arc-fortement connexes minimaux, *C.R. Acad. Sci. Paris A* **2853**, 341–344.
Dijkstra, E.W.
[1959] A note on two problems in connexion with graphs, *Numer. Math.* **1**, 269–271.
Dilworth, R.P.
[1950] A decomposition theorem for partially ordered sets, *Ann. of Math.* **51**, 161–166.
Dinits, E.A.
[1970] Algorithm for solution of a problem of maximum flow in a network with power estimation, *Dokl. Akad. Nauk SSSR* **194**, 754–757 (in Russian) [Soviet Math. Dokl. **111**, 277–1289].
Dinits, E.A., A.V. Karzanov and M.L. Lomonosov
[1976] On the structure of a family of minimal weighted cuts in a graph, in: *Studies in Discrete Mathematics* (in Russian), ed. A.A. Fridman (Nauka, Moscow) pp. 290–306.
Dirac, G.A.
[1960] In abstrakten Graphen vorhandene vollständige 4-Graphen und ihre Unterteilungen, *Math. Nachr.* **22**, 61–85.
Edmonds, J.
[1967] Optimum branchings, *J. Res. Nat. Bur. Standards B* **71**, 233–240.
[1973] Edge-disjoint branchings, in: *Combinatorial Algorithms*, ed. R. Rustin (Academic Press, New York) pp. 91–96.
Edmonds, J., and R.M. Karp
[1972] Theoretical improvements in algorithmic efficiency for network flow problems, *J. Assoc. Comput. Mach.* **19**, 248–264.
Egawa, Y.
[1987] Cycles in k-connected graphs whose deletion results in a $(k-2)$-connected graph, *J. Combin. Theory B* **42**, 371–377.

Elias, P., A. Feinstein and C.E. Shannon

[1956] A note on the maximum flow through a network, *IRE Trans. Information Theory* **IT-2**, 117–119.

Even, S., A. Rai and A. Shamir

[1976] On the complexity of timetable and multicommodity flow problems, *SIAM J. Comput.* **5**(4), 691–703.

Fleischner, H., and B. Jackson

[1985] Removable cycles in planar graphs, *J. London Math. Soc.* **31**(2), 193–199.

Ford, L.R., and D.R. Fulkerson

[1956] Maximum flow through a network, *Canad. J. Math.* **8**, 399–404.

[1962] *Flows in Networks* (Princeton University Press, Princeton, NJ).

Ford Jr, L.R.

[1956] *Network Flow Theory,* Paper P-923 (RAND Corporation, Santa Monica, CA).

Fortune, S., J. Hopcroft and J. Wyllie

[1980] The directed subgraph homeomorphism problem, *Theoret. Comput. Sci.* **10**, 111–121.

Frank, A.

[1980] On chain and antichain families of partially ordered sets, *J. Combin. Theory B* **29**, 176–184.

[1981] How to make a digraph strongly connected, *Combinatorica* **1**, 145–153.

[1985] Edge-disjoint paths in planar graphs, *J. Combin. Theory B* **39**(2), 164–178.

[1989] On connectivity properties of Eulerian digraphs, *Ann. Discrete Math.* **41**, 179–194.

[1990] Packing paths, circuits, and cuts – a survey, in: *Paths, Flows, and VLSI-Layout,* eds. B. Korte, L. Lovász, H.J. Prömel and A. Schrijver (Springer, Berlin) pp. 47–100.

[1992a] Augmenting graphs to meet edge-connectivity requirements, *SIAM J. Discrete Math.* **5**(1), 25–53.

[1992b] On a theorem of Mader, *Ann. Discrete Math.* **101**, 49–57.

[1994] Connectivity augmentation problems in network design, in: *Mathematical Programming: State of the Art 1994,* eds. J.R. Birge and K.G. Murty (University of Michigan) pp. 34–63.

Frank, A., A.V. Karzanov and A. Sebő

[1992] On multiflow problems, in: *Proc. Meeting on Integer Programming and Combinatorial Optimization, Pittsburgh, PA,* eds. E. Balas, G. Cornuéjols and R. Kannan, pp. 85–101.

Fulkerson, D.R.

[1974] Packing rooted directed cuts in a weighted directed graph, *Math. Programming* **6**, 1–13.

Galil, Z.

[1980] An $O(|V|^{5/3}\,|E|^{2/3})$ algorithm for the maximal flow problem, *Acta Inform.* **14**, 221–242.

Gallai, T.

[1961] Maximum–Minimum Sätze und verallgemeinerte Faktoren von Graphen, *Acta Math. Acad. Sci. Hungar.* **12**, 131–163.

Goldberg, A.V., and R.E. Tarjan

[1986] A new approach to the maximum flow problem, in: *Proc. 18th Annu. ACM Symp. on Theory of Computing,* pp. 136–146. Full paper: 1988, *J. ACM* **35**, 921–940.

[1989] Finding minimum-cost circulations by canceling negative cycles, *J. Assoc. Comput. Mach.* **36**(4), 873–886.

Goldberg, A.V., R.E. Tarjan and É. Tardos

[1990] Network flow algorithms, in: *Paths, Flows, and VLSI-Layout,* eds. B. Korte, L. Lovász, H.J. Prömel and A. Schrijver (Springer, Berlin) pp. 101–164.

Gomory, R., and T.C. Hu

[1961] Multiterminal network flows, *J. SIAM* **9**, 551–570.

Graham, R.L., and P. Hell

[1985] On the history of the minimum spanning tree problem, *Ann. Hist. Comput.* **7**(1), 43–57.

Greene, C.

[1976] Some partitions associated with a partially ordered set, *J. Combin. Theory A* **20**, 69–79.

Greene, C., and D.J. Kleitman

[1976] The structure of Sperner k-families, *J. Combin. Theory A* **20**, 41–68.

Győry, E.

[1978] On division of graphs to connected subgraphs, in: *Combinatorics 1,* eds. A. Hajnal and V.T. Sós, *Colloq. Math. Soc. János Bolyai* **18**, 485–494.

Häggkvist, R., and C. Thomassen
[1982] Circuits through specified edges, *Discrete Math.* **41**, 29–34.
Halin, R.
[1969a] A theorem on n-connected graphs, *J. Combin. Theory,* 150–154.
[1969b] Untersuchungen über minimale n-fach zusammenhängende Graphen, *Math. Ann.* **182**, 175–188.
Hamidoune, Y.O.
[1980] On critically, h-connected simple graphs, *Discrete Math.* **32**, 257–262.
Hoffman, A.
[1960] Some recent applications of the theory of linear inequalities to extremal combinatorial analysis, in: *Combinatorial Analysis,* eds. R. Bellman and M. Hall (AMS, Providence, RI) pp. 113–128.
Hoffman, A.J.
[1974] A generalization of max flow–min cut, *Math. Programming* **6**, 352–359.
Hu, T.C.
[1963] Multicommodity network flows, *Oper. Res.* **11**, 344–360.
Huck, A.
[1991] A sufficient condition for a graph to be weakly weakly k-linked, *Graphs Combin.* **7**, 323–351.
Ibaraki, T., and S. Poljak
[1991] Weak three-linking in Eulerian digraphs, *SIAM J. Discrete Math.* **4**(1), 84–89.
Iri, M.
[1970] On an extension of the max-flow min-cut theorem for multicommodity flows, *J. Oper. Res. Soc. Japan* **13**, 129–135.
Itai, A., and A. Zehavi
[1984] Bounds on path connectivity, *Discrete Math.* 25–34.
Jackson, B.
[1980] Removable cycles in 2-connected graphs of minimum degree at least four, *J. London Math. Soc.* **21**(2), 385–392.
[1988] Some remarks on arc-connectivity, vertex splitting and orientation in digraphs, *J. Graph Theory* **12**(3), 429–436.
Jung, H.A.
[1970] Eine Verallgemeinerung des n-fachen Zusammenhangs für Graphen, *Math. Ann.* **187**, 95–103.
Karp, R.M.
[1975] On the computational complexity of combinatorial problems, *Networks* **5**, 45–68.
[1978] A characterization of the minimum cycle mean in a digraph, *Discrete Math.* **23**, 309–311.
Karzanov, A.V.
[1974] The problem of finding the maximal flow in a network by the method of preflows, *Dokl. Akad. Nauk SSSR* **215**, 49–52 (in Russian) [*Soviet Math. Dokl.* **215**, 434–437].
[1985] On multicommodity flow problems with integer-valued optimal solutions, *Dokl. Akad. Nauk SSSR* **280**(4) [*Soviet Math. Dokl.* **32**, 151–154].
[1987] Half-integral five-terminus flows, *Discrete Appl. Math.* **18**, 263–278.
[1994] Paths and metrics in a planar graph with three or more holes I and II, *J. Combin. Theory B* **60**(1), 1–35.
Karzanov, A.V., and M.V. Lomonosov
[1978] Multiflows in undirected graphs, in: *Mathematical Programming, The Problems of Social and Economic Systems, Operations Research Model,* Vol. 1 (in Russian) (The Institute for System Studies, Moscow).
Kleitman, D.J., A. Martin-Löf, B. Rothschild and A. Whinston
[1970] A matching theorem for graphs, *J. Combin. Theory* **8**, 104–114.
König, D.
[1915] Line systems and determinants, *Math. Termész. Ért.* **33**, 221–229 (in Hungarian).
Korach, E., and M. Penn
[1992] Tight integral duality gap in the Chinese Postman Problem, *Math. Programming* **55**, 183–191.

Kruskal, J.B.
[1956] On the shortest spanning tree of a graph and the travelling salesman problem, *Proc. Amer. Math. Soc.* **7**, 48–50.

Larman, D.G., and P. Mani
[1970] On the existence of certain configurations within graphs and the 1-skeletons of polytopes, *Proc. London Math. Soc.* **20**, 144–160.

Lawler, E.L.
[1976] *Combinatorial Optimization: Networks and Matroids* (Holt, Rinehart & Winston, New York).

Lomonosov, M.V.
[1983] On the planar integer two-flow problem, *Combinatorica* **3**(2), 207–219.
[1985] Combinatorial approaches to multiflow problems, *Discrete Appl. Math.* **11**(1), 1–93.

Lovász, L.
[1976a] On two minimax theorems in graph theory, *J. Combin. Theory B* **21**, 96–103.
[1976b] On some connectivity properties of Eulerian graphs, *Acta Math. Acad. Sci. Hungar.* **28**, 129–138.
[1977] Homology theory for spanning trees of a graph, *Acta Math. Acad. Sci. Hungar.* **30**, 241–251.
[1979] *Combinatorial Problems and Exercises* (North-Holland, Amsterdam).

Lovász, L., and M.D. Plummer
[1986] *Matching Theory, Ann. Discrete Math.* **29**.

Lovász, L., V. Neumann-Lara and M.D. Plummer
[1978] Mengerian theorems for paths of bounded length, *Period. Math. Hungar.* **9**, 269–276.

Lucchesi, C.L., and D.H. Younger
[1978] A minimax relation for directed graphs, *J. London Math. Soc.* **17**(2), 369–374.

Mader, W.
[1971a] Eine Eigenschaft der Atome endlicher Graphen, *Arch. Math.* **22**, 257–262.
[1971b] Minimale n-fach kantenzusammenhangende Graphen, *Math. Ann.* **191**, 21–28.
[1972] Ecken vom Grad n in minimalen n-fach zusammenhangende Graphen, *Arch. Math.* **23**, 219–224.
[1973] Grad und lokaler Zusammenhang in endlichen Graphen, *Math. Ann.* **205**.
[1974a] Kreuzungsfreie (a, b)-Wege in endlichen Graphen, *Abh. Mat. Sem. Univ. Hamburg* **42**, 187–204.
[1974b] Ecken vom Innen- und Aussengrad k in minimal n-fach kantenzusammenhangenden Digraphen, *Arch. Math.* **25**, 107–112.
[1978a] A reduction method for edge-connectivity in graphs, *Ann. Discrete Math.* **3**, 145–164.
[1978b] Über die Maximalzahl kantendisjunkter A-Wege, *Arch. Math. (Basel)* **30**, 325–336.
[1978c] Über die Maximalzahl kreuzungsfreier H-Wege, *Arch. Math. (Basel)* **31**, 387–402.
[1979] Connectivity and edge-connectivity in finite graphs, in: *Surveys on Combinatorics, London Math. Soc. Lecture Notes Series,* Vol. 38, ed. B. Bollobás (London Mathematical Society, London) pp. 293–309.
[1981] On a property of n-edge connected digraphs, *Combinatorica* **1**(4), 385–386.
[1982] Konstruktion aller n-fach kantenzusammenhängenden Digraphen, *European J. Combin.* **3**, 63–67.
[1983] On n-edge-connected digraphs, *Ann. Discrete Math.* **17**, 439–441.
[1985a] Paths in graphs reducing the edge-connectivity only by two, *Graphs and Combinatorics* **1**, 81–89.
[1985b] Minimal n-fach zusammenhangende Digraphen, *J. Combin. Theory B* **38**(2), 102–117.
[1985c] Disjunkte Fragmente in kritisch n-fach zusammenhängende Graphen, *European J. Combin.* **6**, 353–359.
[1986] Kritisch n-fach kantenzusammenhängende Graphen, *J. Combin. Theory B* **40**(2), 152–158.
[1989] On critically connected digraphs, *J. Graph Theory* **13**(4), 513–522.
[1991] Ecken von kleinem Grad in kritisch n-fach zusammenhängende Digraphen, *J. Combin. Theory B* **53**(2), 260–272.

Malhotra, V.M., M.P. Kumar and S.N. Maheswari
[1978] An $O(|V|^3)$ algorithm for finding maximum flows in a network, *Inform. Process. Lett.* **7**, 277–278.

Menger, K.
[1927] Zur allgemeinen Kurventheorie, *Fund. Math.* **10**, 96–115.

Mesner, D.M., and M.E. Watkins
[1967] Cycles and connectivity in graphs, *Canad. J. Math.* **19**, 1319–1328.

Middendorf, M., and F. Pfeiffer
[1990] On the complexity of disjoint paths problem, in: *Polyhedral Combinatorics, ACM Series in Discrete Mathematics and Theoretical Computer Science,* Vol. 1, eds. W. Cook and P.D. Seymour (ACM, New York) pp. 171–178.
Nagamochi, H., and T. Ibaraki
[1992] A linear time algorithm for finding a sparse k-connnected spanning subgraph of a k-connected graph, *Algorithmica* **7**, 583–596.
Nash-Williams, C.St.J.A.
[1960] On orientations, connectivity and odd vertex pairings in finite graphs, *Canad. J. Math.* **12**, 555–567.
[1961] Edge-disjoint spanning trees of finite graphs, *J. London Math. Soc.* **36**, 445–450.
[1964] Decomposition of finite graphs into forests, *J. London Math. Soc.* **39**, 12.
Okamura, H.
[1983] Multicommodity flows in graphs, *Discrete Appl. Math.* **6**, 55–62.
[1984] Multicommodity flows in graphs II, *Japan J. Math.* **10**, 99–116.
Okamura, H., and P.D. Seymour
[1981] Multicommodity flows in planar graphs, *J. Combin. Theory B* **31**, 75–81.
Onaga, K., and O. Kakusho
[1971] On feasibility conditions of multicommodity flows in networks, *IEEE Trans. Circuit Theory* **CT-18**, 425–429.
Orlin, J.P.
[1988] A faster strongly polynomial minimum cost flow algorithm, in: *Proc. 20th ACM Symp. on Theory of Computing* (ACM, New York) pp. 377–387. Full paper: *J. Oper. Res.,* to appear.
Padberg, M.W., and M.R. Rao
[1982] Odd minimum cut-sets and b-matchings, *Math. Oper. Res.* **7**, 67–80.
Papernov, B.A.
[1976] Feasibility of multicommodity flows, in: *Studies in Discrete Optimization* (Issledovaniya po Discretnoy Optimizacii), ed. A.A. Friedman (Isdat. Nauka, Moscow) pp. 230–261 (in Russian).
Phillips, D.T., and A. Garcia-Diaz
[1981] *Fundamentals of Network Analysis* (Prentice-Hall, Englewood Cliffs, NJ).
Prim, R.C.
[1957] Shortest connection networks and some generalization, *Bell Syst. Techn. J.* **36**, 1389–1401.
Robbins, H.E.
[1939] A theorem on graphs with an application to a problem of traffic control, *Amer. Math. Monthly* **46**, 281–283.
Robertson, N., and P.D. Seymour
[1986a] Graph minors VI: Disjoint paths across a disc, *J. Combin. Theory B* **41**, 115–138.
[1986b] Graph minors XIII: The disjoint paths problem, *J. Combin. Theory B,* to appear.
[1990] An outline of a disjoint paths algorithm, in: *Paths, Flows, and VLSI-Layout,* eds. B. Korte, L. Lovász, H.J. Prömel and A. Schrijver (Springer, Berlin) pp. 267–291.
Rothschild, B., and A. Whinston
[1966a] On two-commodity network flows, *Oper. Res.* **14**, 377–387.
[1966b] Feasibility of two-commodity network flows, *Oper. Res.* **14**, 1121–1129.
Schrijver, A.
[1982] Min–max relations for directed graphs, *Ann. Discrete Math.* **16**, 261–280.
[1989] The Klein bottle and multicommodity flows, *Combinatorica* **9**(4), 375–384.
[1990] Homotopic routing methods, in: *Paths, Flows, and VLSI-Layout,* eds. B. Korte, L. Lovász, H.J. Prömel and A. Schrijver (Springer, Berlin) pp. 329–371.
Sebő, A.
[1990] Undirected distances and the postman-structure of graphs, *J. Combin. Theory B* **49**(1), 10–39.
[1993] Integer plane multicommodity flows with a fixed number of demands, *J. Combin. Theory B* **59**(2), 163–171.

Seymour, P.D.
[1980a] Four-terminus flows, *Networks* **10**, 79–86.
[1980b] Decomposition of regular matroids, *J. Combin. Theory B* **28**, 305–359.
[1980c] Disjoint paths in graphs, *Discrete Math.* **29**, 293–309.
[1981] On odd cuts and plane multicommodity flows, *Proc. London Math. Soc.* **42**, 178–192.
Shiloach, Y.
[1978] *An $O(nI(\log I)^2)$ maximum flow algorithm,* Tech. Report STAN-CS-78-802 (Stanford University Computer Science Department, Stanford, CA).
[1979] Edge-disjoint branchings in directed multigraphs, *Inform. Proc. Lett.* **8**, 24–27.
[1980] A polynomial solution to the undirected two paths problem, *J. Assoc. Comput. Mach.* **27**, 445–456.
Shiloach, Y., and U. Vishkin
[1982] An $O(n^2 \log n)$ parallel max-flow algorithm, *J. Algorithms* **3**, 128–146.
Sleator, D.D.K.
[1980] *An $O(nm\log n)$ algorithm for maximum network flow,* Tech. Report STAN-CS-80-831 (Stanford University Computer Science Department, Stanford, CA).
Strange, K.E., and B. Toft
[1983] An introduction to the subgraph homeomorphism problem, in: *Proc. Third Czech Symp. on Graph Theory* (B.G. Teubner, Stuttgart) pp. 296–301.
Tardos, É.
[1985] A strongly polynomial minimum cost circulation algorithm, *Combinatorica* **5**, 247–255.
[1986] Strongly polynomial algorithm to solve combinatorial linear programs, *Oper. Res.* **34**(2), 250–256.
Tarjan, R.E.
[1972] Depth-first search and linear graph algorithms, *SIAM J. Comput.* **1**, 146–160.
[1983] Data structures and network algorithms, *CBMS-NFS Regional Conf. Ser. in Appl. Math.,* Vol. 44 (Society for Industrial and Applied Mathematics, Philadelphia, PA).
Tarjan, R.E., and M. Yannakakis
[1984] Simple linear-time algorithms to test chordality of graphs, test acyclicity of hypergraphs, and selectively reduce acyclic hypergraphs, *SIAM J. Computing,* **13**(3).
Thomassen, C.
[1980a] 2-linked graphs, *European J. Combin.* **1**, 371–378.
[1980b] Planarity and duality of finite and infinite graphs, *J. Combin. Theory B* **29**, 244–271.
[1981] Non-separating cycles in k-connected graphs, *J. Graph Theory* **5**, 351–354.
[1984] Plane representations of graphs, in: *Progress in Graph Theory,* eds. A. Bondy and R. Murty (Academic Press) pp. 43–69.
[1985] The 2-linkage problem for acyclic digraphs, *Discrete Math.* **55**, 73–87.
Thomassen, C., and B. Toft
[1981] Non-separating induced cycles in graphs, *J. Combin. Theory B* **31**(2), 199–224.
Titov, V.K.
[1975] *A constructive description of some classes of graphs,* Doctoral Dissertation (Moscow).
Tutte, W.T.
[1961a] On the problem of decomposing a graph into n-connected factors, *J. London Math. Soc.* **36**, 221–230.
[1961b] A theory of 3-connected graphs, *Indag. Math.* **23**, 441–455.
[1963] How to draw a graph, *Proc. London Math. Soc.* **13**, 743–767.
[1966] *Connectivity in Graphs* (University of Toronto Press, Toronto).
[1984] *Graph Theory* (Addison-Wesley, Reading, MA).
Van Hoesel, C., and A. Schrijver
[1986] Edge-disjoint homotopic paths in a planar graph with one hole, *J. Combin. Theory B* **48**, 77–91.
Veldman, H.J.
[1983] *Three topics in graph theory,* Dissertation (Technical University Twente, Enschede).
Vidyasankar, K.
[1978] Covering the edge-set of a directed graph with trees, *Discrete Math.* **24**, 79–85.
Watanabe, T., and A. Nakamura
[1987] Edge-connectivity augmentation problems, *J. Computer & System Sci.* **35**(1), 96–144.

Whitney, H.
[1932] Non-separable and planar graphs, *Trans. Amer. Math. Soc.* **34**, 339–362.

Whitty, R.H.
[1986] Vertex-disjoint paths and edge-disjoint branchings in directed graphs, *J. Graph Theory,* **11**(3), 349–358.

Zehavi, A., and A. Itai
[1989] Three tree-paths, *J. Graph Theory,* **13**(2), 175–188.

CHAPTER 3

Matchings and Extensions

W.R. PULLEYBLANK

IBM, Thomas J. Watson Research Center, P.O. Box 218, Yorktown Heights, NY 10598, USA

Contents

HANDBOOK OF COMBINATORICS
Edited by R. Graham, M. Grötschel and L. Lovász

1. Introduction and preliminaries

A *matching* in a graph $G = (V, E)$ is a set of edges (links) no two of which have a common end. The most basic problems of matching theory involve establishing the existence of matchings with sufficiently many elements. The size of the largest matching is denoted by $\nu(G)$. A node v is *saturated* by a matching M if some edge of M is incident with v. A matching which saturates all nodes of G is called *perfect*.

This problem of constructing maximum matchings, or determining whether there exists a perfect matching has been one of the most extensively studied problems of graph theory. This is due in part to the wide variety of extensions and applications which exist, in part to the amount of structural information this subject provides about a graph, in part to the branches of combinatorics for which it serves as a prototype (e.g., polyhedral combinatorics) and in part to its inherent tractibility.

A *stable set* (or *node packing*) in a graph $G = (V, E)$ is a set of nodes, no two of which are adjacent. The size of a maximum stable set is denoted by $\alpha(G)$. Initially, the problems of determining $\alpha(G)$ and $\nu(G)$ sound deceptively similar. In one case we pack nodes, in the other we pack edges. However, the problem of packing a maximum number of edges can be solved in time which only grows polynomially with the size of the graph (see section 3) but the maximum stable set problem is one of the original NP-hard problems. Consequently much of the research on this latter topic has dealt with special classes of graphs for which the problem is more tractible. In this chapter we focus our attention on matchings, but also discuss stable sets, particularly with respect to their connections with matching theory. They are also treated in chapter 4.

A *node cover* in a graph is a set T of nodes such that every edge has at least one end in T. An *edge cover* is a set of edges such that every node is incident with at least one edge of the set. Then $\tau(G)$ and $\rho(G)$ denote the sizes of smallest node covers and edge covers respectively of G. Since S is a stable set if and only if $V \setminus S$ is a node cover,

$$\alpha(G) + \tau(G) = |V| = n . \tag{1.1}$$

Less obviously, the same relation holds for edges.

Proposition 1.2. *$\nu(G) + \rho(G) = |V| = n$, provided that G has no isolated nodes.*

Proof. Let M be a maximum matching, and let U be the set of nodes not saturated by M. For each $u \in U$ we can choose an incident edge $e(u)$, let S be the set of these edges. Then $M \cup S$ is an edge cover, so $|M| + |M \cup S| = 2|M| + |S| = (|V| - |U|) + |U| = |V|$, so $\nu(G) + \rho(G) \leq n$. Conversely, let C be an edge cover of cardinality $\rho(G)$. Minimality of C ensures that C is the edge set of a forest, so if k is the number of components, then $k = n - \rho(G)$. Each component contains an edge; choosing one edge from each component yields a matching of cardinality k, so $\nu(G)+\rho(G)\geq n$. Combining these inequalities gives Proposition 1.2. □

Relation (1.1) and Proposition 1.2 are called the *Gallai identities* (see Gallai 1959 and Lovász and Plummer 1986).

There is another dual relationship between these parameters. If T is a node cover and M is a matching, then no node of T can cover more than one edge of M, so

$$\nu(G)\leq\tau(G)\,. \tag{1.3}$$

Similarly, no edge of an edge cover can meet more than one node of a stable set, so

$$\rho(G)\geq\alpha(G)\,. \tag{1.4}$$

Matching theory is much easier for bipartite graphs than for nonbipartite graphs. In the next section we discuss a fundamental theorem of König which states that we have equality in (1.3) and (1.4) if G is bipartite. However, since $\nu(K_3)=1$ but $\tau(K_3)=2$ and $\rho(K_3)=2$ but $\alpha(K_3)=1$ we see that equality need not hold for nonbipartite graphs. It does hold for some nonbipartite graphs, for example, suppose we add a new node u adjacent to any node of K_3. Then equality again holds in (1.3) and (1.4). Those graphs for which equality holds are called König–Egerváry graphs, and are discussed in section 6, in the context of 2-matchings.

Suppose now we wished to modify (1.3) so that equality holds for all graphs. First we could try to obtain an equation which holds for $\nu(G)$. This was obtained by Edmonds (1965a). Let $\mathcal{K}$ be a family of pairwise disjoint odd cardinality subsets of V. We say that $\mathcal{K}$ is an *odd set cover* if every edge either is incident with a node belonging to a singleton component of $\mathcal{K}$, or else has both ends in a nonsingleton component. The weight $w(S)$ for $S\in\mathcal{K}$ is defined by $w(S)=1$ if $|S|=1$; $w(S)=(|S|-1)/2$ if $|S|\geq 3$, and $w(\mathcal{K})=\sum_{S\in\mathcal{K}} w(S)$. Let $\tilde{\tau}(G)$ denote the minimum of $w(\mathcal{K})$, over all odd set covers. Again, trivially, $\nu(G)\leq\tilde{\tau}(G)$, and moreover we have the following.

Theorem 1.5 (Edmonds 1965a). *For any graph G, $\nu(G)=\tilde{\tau}(G)$.*

We discuss this result in section 3, in connection with the matching algorithm as it was in this context that it was developed.

The analogous problem for τ would be to try and pack something more general than edges to obtain an expression for $\tau(G)$. Unless NP = coNP, no such computationally tractible condition exists for general graphs. Nevertheless, some interesting special classes do occur, the best known of which is the class of *perfect graphs*. Suppose we define the weight $w(K)$ of a clique K in G to be $|K|-1$. We let $\tilde{\nu}(G)$ be the maximum of $\{\sum_{K\in\mathcal{M}} w(K)$: $\mathcal{M}$ is a set of pairwise disjoint cliques in $G\}$. Then G is perfect if and only if $\tau(G') = \tilde{\nu}(G')$, for every node induced subgraph G'. This topic is discussed in chapter 4.

We mention one more relationship between the matching and stable set problems. Recall that the node set of the *line graph* $L(G)$ is the edge set E of G, and two nodes of $L(G)$ are adjacent if and only if the edges have a common end in G. Then $M \subseteq E$ is a matching in G if and only if M is a stable set in $L(G)$. Thus line graphs provide a class of graphs for which α can be computed in time which grows polynomially with the size of the graph. An interesting extension of this result was obtained independently by Minty (1980) and Sbihi (1980). A graph is the line graph of a simple graph if and only if it does not contain any of nine node-induced subgraphs, the simplest of which is the *claw*, $K_{1,3}$. A graph is *claw-free* if no node induced subgraph is isomorphic to $K_{1,3}$. Minty and Sbihi showed that $\alpha(G)$ could be calculated in polynomial time for claw-free graphs, see section 8. However, at present there has not been obtained the breadth of results concerning stable sets for claw-free graphs that has been obtained for matchings.

One of the fundamental ideas of matching theory is that of an augmenting path. A path or circuit P in a graph is said to be *alternating* with respect to a matching M if its edges alternately are in and not in M. (An alternating circuit will have an even number of edges.) If the ends of a path P are not saturated by M, then P is called an augmenting path, for $M' = M \triangle P$ is a larger matching then M. This notion provides one characterization of $\nu(G)$.

Theorem 1.6 (Berge 1957). *A matching M of G is maximum if and only if there exists no augmenting path with respect to M.*

Proof. We have seen that the existence of an augmenting path implies that M is not maximum. Conversely, if there exists a matching M' such that $|M'| > |M|$, then $M \triangle M'$ consists of a number of alternating paths and circuits with respect to M and at least one path must be augmenting. □

This result has an attractive analogue for maximum stable sets in claw-free graphs, which is discussed in section 8.

Note the difference between Theorem 1.5 and Theorem 1.6, both of which characterize maximum matchings. The former provides a means of showing that a matching is maximum, which can be checked in polynomial time, but does not indicate how to find such a matching. The latter indicates an approach to finding a maximum matching, but does not provide an obvious means of proving maximality. (How can one show that no augmenting path exists?) In section 3 we will present an algorithm which combines these to efficiently compute $\nu(G)$.

The best general reference for matching is the book Lovász and Plummer (1986). Most books on graph theory and integer programming contain an introduction to matching theory, especially the bipartite case. Good references are Lawler (1976), Papadimitriou and Steiglitz (1982), and Nemhauser and Wolsey (1988). Derigs (1988) discusses algorithmic and polyhedral issues and

Burkard and Derigs (1980) provide descriptions of algorithms as well as FORTRAN codes.

2. Bipartite matching

2.1. König's theorem and consequences

Let $G=(V_1 \cup V_2, E)$ be a bipartite graph with colour classes V_1 and V_2. In this case the parameters ν, ρ, τ, α can all be efficiently computed and their values are closely related. This fundamental result is due to König (1916a,b).

König's Theorem 2.1. *If $G=(V_1 \cup V_2, E)$ is bipartite, then $\nu(G)=\tau(G)$. (Or, a maximum matching and minimum node cover have the same cardinality.)*

Proof (D. de Caen, private communication). We first show that in every bipartite graph with edges, at least one end of each edge is saturated by every maximum matching. For let u and v be adjacent nodes and suppose there exist maximum matchings M^u and M^v which leave u and v, respectively unsaturated. Each of u and v is the end of a path whose edge set belongs to $M^u \triangle M^v$. Since M^u and M^v are of maximum cardinality, and hence equicardinal, these paths must be of even length. If these paths are different, then their union plus the edge uv forms an odd length path joining a node unsaturated by M^u and a node unsaturated by M^v. If we replace the edges of M^u belonging to this path with the path's odd numbered edges, we get a matching larger than M^u, a contradiction. Therefore these paths are identical and have an even number of edges, so this path, plus the edge uv, forms an odd cycle which contradicts G being bipartite.

Now the result follows by induction. If G has no edges, then $\upsilon(G)=\tau(G)=0$. Otherwise, choose a node v saturated by every maximum matching. Then $\nu(G-v)=\nu(G)-1$ and since $C\backslash\{v\}$ is a cover of G, for any cover C of $G-v$, $\tau(G)\leqslant\tau(G-v)+1$. By induction $\nu(G-v)=\tau(G-v)$, so $\tau(G)\leqslant\nu(G)$, which together with (1.3) gives the result. □

There are many other proofs of this theorem, see, e.g., Lovász and Plummer (1986). It also follows directly from the matching algorithm, which we present in the next section for nonbipartite graphs. It also follows from the characterization of the perfect matching polytope, discussed in section 5.

There is an important difference between the relationships between ν and τ established by König's Theorem 2.1, which asserts that $\nu(G)=\tau(G)$ for a bipartite graph G, and the Gallai identities (1.1) and Proposition 1.2. Let M be any matching. If $|M|$ is maximum, then by König's theorem there must exist a node cover C such that $|C|=|M|$. Moreover, in view of (1.3), the existence of such a C provides a short proof of the maximality of $|M|$. Any interested observer can readily convince himself of this fact by verifying that M is a matching, C is a node cover and $|M|=|C|$.

However, suppose we have found a matching M and an edge cover R such that

$|M| + |R| = |V|$. We have not established the maximality of $|M|$ or the minimality of $|R|$. It must be independently established for either M or R and then it will follow for the other. Similarly, the existence of a stable set S and a node cover C such that $|S| + |C| = |V|$ establishes nothing about $\alpha(G)$ or $\tau(G)$. In section 3 we describe a polynomially bounded algorithm which will compute $\nu(G)$ for an arbitrary graph G in time which only grows polynomially with the size of G. However, unless P = NP (see chapter 29), no such polynomially bounded algorithm can exist for computing $\alpha(G)$ or $\tau(G)$ for an arbitrary graph G.

One striking feature of bipartite matching theory is the number of equivalent, but different forms of this fundamental relationship. One of the best known is the so-called SDR theorem of P. Hall. Let $(Q_i\colon i \in I)$ be a finite family of subsets of a finite set S. A *system of distinct representatives* (SDR) is a function $f : I \to S$, such that for each $i \in I$, $f(i) \in Q_i$ for each i, $j \in I$, and $i \neq j$ implies $f(i) \neq f(j)$. The element $f(i)$ is called the *representative* of the set Q_i.

SDR Theorem 2.2 (Hall 1935). *The family $(Q_i\colon i \in I)$ has an SDR if and only if*

$$|\bigcup(Q_i\colon i \in H)| \geq |H|, \quad \text{for every } H \subseteq I. \tag{2.3}$$

Note that the necessity of this condition is obvious – if there is to exist an SDR, then the union of every subfamily of the sets must contain at least as many elements as exist in the subfamily. We will show that the sufficiency follows from Theorem 2.1, by first modelling the problem as a bipartite graph. Construct a graph $G = (S \cup I, E)$, where there is an edge in E joining $s \in S$ to $i \in I$ if and only if $s \in Q_i$. Then an SDR corresponds to a matching in G which saturates all nodes of I and condition (2.3), called "Hall's condition", becomes $|N(H)| \geq |H|$ for all $H \subseteq I$. Thus Theorem 2.2 is equivalent to the following.

Theorem 2.4. *A bipartite graph $G = (V_1 \cup V_2, E)$ has a matching which saturates all nodes of V_1 if and only if, for every $H \subseteq V_1$, $|N(H)| \geq |H|$.*

Proof. As we have seen, the necessity of the condition is easy. Conversely, suppose $\nu(G) < |V_1|$. Let T be a node cover of size $\nu(G)$, i.e., $|T| < |V_1|$. Let $S = V_1 \backslash T$. Then $N(S) \subseteq V_2 \cap T$, so $|S| = |V_1| - |T \cap V_1| = |V_1| - |T| + |V_2 \cap T| > |N(S)|$, so the condition is violated. □

Corollary 2.5 (The marriage theorem, Frobenius 1917). *A bipartite graph $G = (V_1 \cup V_2, E)$ has a perfect matching if and only if $|V_1| = |V_2|$ and $|H| \leq |N(H)|$ for all $H \subseteq V_1$.*

There is an interesting connection between SDRs and matroid theory. Suppose that $(Q_i\colon i \in I)$ is a finite family of subsets of a finite set S. Let $\Pi = \{J \subseteq I\colon (Q_i\colon i \in J)$ has an SDR$\}$. Edmonds and Fulkerson (1965) showed that Π is the family of independent sets of a matroid defined on I. If we consider the model of the SDR problem as a bipartite matching problem on the graph $G = (S \cup I, E)$ described previously, then we can see that this is equivalent to the assertion that the subsets of the nodes of I saturated by some matching of G form the independent sets of a matroid. By reversing the roles of S and I we can obtain one more equivalent form. A set $T \subseteq S$ is called a *partial transversal* of $(Q_i: i \in I)$ if there exists a function $f: T \rightarrow I$ such that for each $s \in T$, $f(s)$ is distinct and $s \in Q_{f(s)}$. Then we have that the family of partial transversals is the family of independent sets of a matroid of the set S. See chapter 9 of this volume.

A well-known theorem of Menger (see also chapter 2) provides an example of a theorem less obviously equivalent to Theorem 2.1. Let s and t be distinct nodes of a graph G (digraph $\vec{G}$). We say that two s–t-paths (dipaths) are internally disjoint if they have no nodes in common other than s and t. A set $X \subseteq V \setminus \{s, t\}$ is called an *s–t-separator* if s and t belong to different components of $G - X$.

Menger's Theorem 2.6 (Menger 1927). *For any graph (digraph) G and any two distinct nodes s and t such that (s, t) is not an arc of G, the maximum number of pairwise internally disjoint s–t-paths (dipaths) equals the minimum cardinality of an s–t-separator.*

Proof. We prove the directed theorem. From this the undirected result can be deduced by replacing each edge uv with arcs (u, v) and (v, u).

Construct a bipartite graph G^* from G by replacing each node v of G other than s and t with two new nodes $h(v)$ and $t(v)$. Node s is replaced with $\deg^+(s)$ new nodes and t is replaced with $\deg^-(t)$ new nodes. For each arc (s, u) of G, we construct an edge joining $t(u)$ and a distinct new node corresponding to s. For each arc (u, t) we construct an edge joining $h(u)$ and a distinct copy of t. Finally, nodes $h(u)$ and $t(v)$ of G^* are adjacent if and only if $u = v$ or G contains an arc (u, v). See fig. 2.1.

Let M be a maximum matching in G^*. Then for each node $u \in V \setminus \{s, t\}$, at least one of $h(u)$, $t(u)$ must be saturated. In fact if only one is saturated, we can replace the incident edge of M with $h(u)t(u)$, and get another maximum matching of G^* which saturates both. Then the arcs of G for which the corresponding edges of G^* are in M decompose into internally disjoint circuits and s–t-paths in G, and the number of s–t-paths in this collection is $|M| - |V \setminus \{s, t\}|$. By Theorem 2.1 there is a cover C in G^* with $|C| = |M|$ and we can assume that only nodes $h(u)$ or $t(u)$ belong to C. Let W be the set of nodes u of G for which both $h(u)$ and $t(u)$ are in C. If any s–t-dipath in G avoided all nodes of W, then some edge of G^* would not

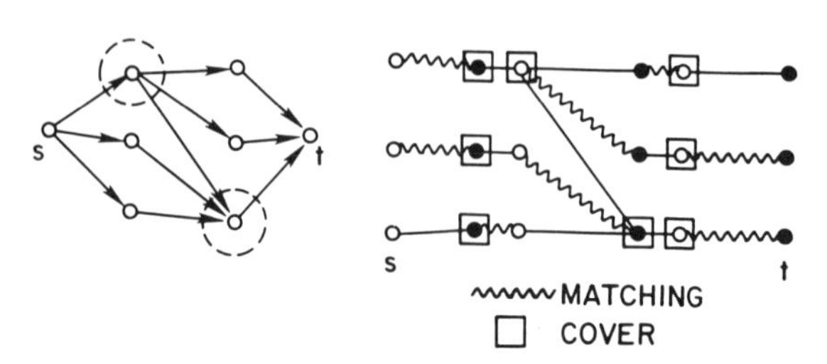

Figure 2.1.

be covered by C, a contradiction. So W is an s–t-separating set of size $|M| - |V \setminus \{s, t\}|$. □

2.2. Term rank and the Birkhoff–Von Neumann theorem

Another corollary of Theorem 2.1 can be stated in terms of matrix theory. Let A be an $m \times n$ matrix, all of whose entries have value 0 or 1. The *term rank* of A is the size of the largest subset R of the entries having value 1, such that no two members of R belong to the same row or column. The term rank of A is the same as the linear-algebra rank of the matrix obtained by replacing the 1s in A by distinct indeterminates (over an integral domain). (See Edmonds 1967.)

Theorem 2.7. *The term rank of a* (0–1)*-matrix A equals the minimum size of a set of rows and columns which contain every nonzero entry of A.*

Proof. Let I and J be the sets of row and column indices of A. Construct a bipartite graph $G = (I \cup J, E)$ where there exists an edge joining $i \in I$ and $j \in J$ if and only if $a_{ij} = 1$. Then $\nu(G)$ is the term rank and $\tau(G)$ is the size of a smallest cover of the 1s with rows and columns. □

This result does have an interesting consequence. An $n \times n$ matrix $A = (a_{ij}: i = 1, 2, \ldots, n;\ j = 1, 2, \ldots, n)$ is *doubly stochastic* if $a_{ij} \geq 0$ for all i and j and $\sum_{i=1}^{n} a_{ij} = 1$ for all j, and $\sum_{j=1}^{n} a_{ij} = 1$ for all i. We define an *assignment* to be a set of n entries of A, no two of which belong to the same row or column.

Lemma 2.8. *A doubly stochastic matrix has an assignment contained in its nonzero elements.*

Proof. Let $\tilde{A}$ be obtained by replacing each positive element of A with a 1. Then A has an assignment contained in its nonzero elements if and only if $\tilde{A}$ has term rank n. Let H and K be subsets of the rows and columns of A which contain all nonzero elements of A. Since each row sum and column sum of A is 1, the sum of all its entries is n, which must be less than or equal to the sum of the entries in each of the rows of H plus the sum of the entries in each of the columns of K. Therefore $|H| + |K| \geq n$, and so $\tilde{A}$ has term rank n. □

Using Lemma 2.8 we can obtain a basic theorem of polyhedral combinatorics, for bipartite graphs. An $n \times n$ (0–1)-matrix A is called a *permutation matrix* if the nonzeros comprise precisely an assignment. A matrix A is a convex combination of matrices $A^1, A^2, \ldots, A^k$ if there exist nonnegative reals $\lambda_1, \lambda_2, \ldots, \lambda_k$ such that $A = \sum_{i=1}^{k} \lambda_i A^i$ and $\sum_{i=1}^{k} \lambda_i = 1$.

Birkhoff–Von Neumann Theorem 2.9 (Birkhoff 1946, Von Neumann 1953). *Every doubly stochastic matrix is a convex combination of permutation matrices.*

Proof. It is sufficient to show that each $(n \times n)$ nonnegative matrix A having all row and column sums equal to some positive value can be expressed as a nonnegative linear combination of permutation matrices. We proceed by induction on the number of nonzero entries. Since all row and column sums are positive, there are at least n nonzero entries. If there are exactly n, then A is just a positive multiple of a permutation matrix. Suppose there are more than n. By Lemma 2.8 (applied to A suitably scaled), there is an assignment T contained in nonzero entries of A. Let λ_1 be the smallest entry of T, and let A^1 be the corresponding permutation matrix. Let $A' = A - \lambda_1 A^1$. Then $\tilde{A}$ has constant row and column sums, but fewer nonzero entries. By induction, there exist $\lambda_2, \lambda_3, \ldots, \lambda_k > 0$ and permutation matrices $A^2, A^3, \ldots, A^k$ such that $\tilde{A} = \sum_{i=2}^{k} \lambda_i A^i$. Therefore $A = \sum_{i=1}^{k} A^i$ as required. □

There are many different proofs of this result. It is an easy corollary of the matching polyhedron characterizations of section 5. It is also an immediate consequence of the total unimodularity of the node–edge incidence matrix of a bipartite graph. (See chapter 30.)

3. Nonbipartite matching

Matching theory for nonbipartite graphs is significantly more complicated than for bipartite graphs. As we saw, K_3 provides an example of a graph G for which $\nu(G) < \tau(G)$. Nevertheless, the matching problem for nonbipartite graphs is as well solved as it is for the simpler bipartite case.

3.1. *Theorems of Petersen and Tutte*

The following theorem is due to Petersen (1891).

Petersen's Theorem 3.1. *Let G be a cubic (regular of degree three) graph with at most three isthmuses. Then G has a perfect matching.*

We will show later how this follows from characterizations of Tutte and Edmonds.

This theorem had its roots in the attempt by Tait to prove the four-colour conjecture. Tait showed in 1880 that this conjecture was equivalent to the assertion that a cubic three connected planar graph could be edge 3-coloured, i.e., have its edge set partitioned into three perfect matchings. Moreover, he claimed that this latter fact followed from the hamiltonicity of planar cubic 3-connected graphs, which indeed it would have, were this latter "fact" true. Only much later, Tutte (1946) constructed a counterexample to Tait's claim. Note that this did not rule out the existence of the required partition into three perfect matchings, just one method of obtaining the partition.

Petersen apparently was sceptical concerning Tait's "proof", and constructed an example of a cubic isthmus-free graph whose edges could not be partitioned into three perfect matchings. This is the now-famous "Petersen graph". (Of course, this graph is nonplanar.) His Theorem 3.1 did show that planar or not, a cubic isthmus free graph must have at least one perfect matching. See Biggs et al. (1976) for a fascinating account of these events.

Tutte (1947) was able to completely characterize those graphs having perfect matchings. This result may be the most fundamental result of matching theory. For any graph $G = (V, E)$, we let $\theta(G)$ denote the number of connected components of G which have an odd number of nodes.

Tutte's Theorem 3.2. *$G = (V, E)$ has a perfect matching if and only if,*

$$\text{for every } X \subseteq V, \theta(G - X) \leq |X| . \tag{3.3}$$

Proof. The necessity of the condition is immediate. For if G has a perfect matching M, then there must be at least one edge of M joining each odd component of $G - X$ to X. If J is the set of all such edges, then $\theta(G - X) \leq |J| \leq |X|$.

We prove the sufficiency by induction on the size of G. Suppose that (3.3) holds. By taking $X = \emptyset$ we see that every component of G has an even number of nodes and $|V|$ is even. Therefore $|X|$ and $\theta(G - X)$ have the same parity, for all $X \subseteq V$. Consequently if $|X| > \theta(G - X)$ for every nonempty $X \subseteq V$, we can delete any pair u, v of adjacent nodes from G; (3.3) will still hold; by induction there exists a perfect matching M' of $G - \{u, v\}$ and $M' \cup \{uv\}$ is a perfect matching of G.

Suppose now that there exists $\emptyset \neq X \subseteq V$ such that $|X| = \theta(G - X)$. Choose X maximal with this property. Let S be the set of nodes of $V \backslash X$ belonging to even components of $G - X$.

First note that $G[S]$ must have a perfect matching, for if not, by induction, there exists $X' \subseteq S$ such that $G[S] - X'$ has more than $|X'|$ odd components. But then $X \cup X'$ violates (3.3) for G.

Second, let K be the node set of an odd component of $G - X$, and let v be any node of K. We claim that $G[K \backslash \{v\}]$ has a perfect matching. For if not, by induction, there exists $\bar{X} \subseteq K \backslash \{v\}$ such that $G[K \backslash (\bar{X} \cup \{v\})]$ has at least $|\bar{X}| + 2$ odd components, and so deleting $X \cup \bar{X} \cup \{v\}$ from G must result in at least $|X \cup \bar{X} \cup \{v\}|$ odd components, contradicting our choice of X.

Finally, let $\bar{G}$ be the bipartite graph obtained from $G - S$ by deleting all edges of $E(X)$ and contracting each odd component of $G - X$ to form a pseudonode. Let W be the set of these pseudonodes, some of which may just be single nodes of G. If $\bar{G}$ has a perfect matching we are finished, for we can combine a perfect matching of $\bar{G}$ with a perfect matching of $G[S]$ and perfect matchings of $G' - v$, for each odd component G' of $G - X$, for a suitable node v of the component. If $\bar{G}$ does not have a perfect matching, then by Hall's Theorem 2.4 there exists $W' \subseteq W$ such that $|N_{\bar{G}}(W')| < |W'|$. But then if we let $X^* = N_{\bar{G}}(W') \subseteq X$, we violate (3.3), a contradiction. □

Tutte's original proof of the sufficiency of his condition made use of the notion of the Pfaffian of a matrix. (See section 7.) The proof we gave here, which makes use of Hall's theorem, is due to Anderson (1971). Later in this section we will provide another proof by describing a polynomially bounded algorithm which either constructs a perfect matching of G or else finds a set $X \subseteq Y$ such that $\theta(G-X) > |X|$. For another proof, and references to still more proofs, see Lovász and Plummer (1986).

At first Tutte's condition (3.3) may seem of limited use, because it is clearly impractical to check all sets $X \subseteq V$ to see if it is satisfied. However, just as König's Theorem 2.1 provides us with an efficient means of establishing the value of $\nu(G)$, it does provide us with a very simple way of proving that a graph does *not* have a perfect matching – simply display $X \subseteq V$ such that $|X| < \theta(G-X)$. In the language of complexity, it proves that the class of all graphs *not* possessing perfect matchings is in NP (see chapter 29). (Trivially the class of graphs possessing perfect matchings is in NP – a perfect matching provides a succinct certificate.) Compare this with Berge's alternating path Theorem 1.6.

Petersen's Theorem 3.1 follows immediately from Theorem 3.2. All we need verify is that (3.3) holds for a cubic two-connected graph $G = (V, E)$. Let $X \subseteq V$ and let J be the set of edges joining nodes of X to nodes of odd components of $G - X$. For any $S \subseteq V$, such that $|S|$ is odd, $|\delta(S)|$ will also be odd, and if G is two-connected, then $|\delta(S)| \geqslant 3$. Therefore, $|J| \geqslant 3\theta(G-X)$. But since G is cubic, $3|X| \geqslant |J|$. Therefore $|X| \geqslant \theta(G-X)$ and so G has a perfect matching as required.

3.2. Edmonds' perfect matching algorithm

We now consider the problem of determining whether a graph has a perfect matching. This was solved for bipartite graphs by König (1916a,b) and Egerváry (1931), for both weighted and unweighted problems. Their method was termed "the Hungarian method" by Kuhn. Edmonds was able to generalize this method to nonbipartite graphs. The unweighted problem was treated in Edmonds (1965a) and the weighted in Edmonds (1965b). We discuss the unweighted version here and postpone treatment of the weighted problem until section 5.

The perfect matching algorithm starts with any matching. If all nodes are saturated, then the matching is perfect and it terminates. Otherwise, it proceeds to either find a larger matching or else show that no perfect matching exists. It chooses an unsaturated node r and proceeds to grow a tree rooted at r which will enable it to find an augmenting path with one end equal to r, if such a path exists.

An *alternating tree* T is defined relative to a matching M. It is rooted at one node r and the nodes are called *even* or *odd* depending on whether the distance to r in the tree is even or odd. It satisfies the following:

(i) r is not saturated by M, all other nodes are saturated by M;

(ii) for each odd node v, $\deg_T(v) = 2$;

(iii) for each even node u, the first edge in the path in T from u to r is in M. (See fig. 3.1).

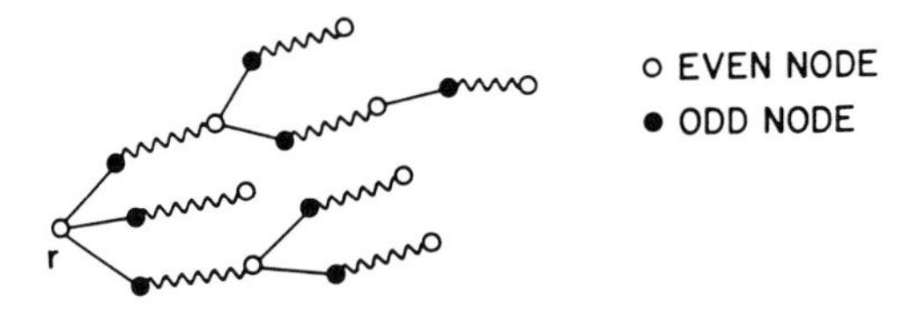

Figure 3.1. An alternating tree.

A feature of the algorithm is that it *shrinks* certain subsets of nodes. Let $G=(V,E)$ be a graph and let $S \subseteq V$. The graph $G \times S$ obtained from G by *shrinking* S is obtained by replacing all nodes in S with a single *pseudonode*, called "S". All edges of $E(S)$ are deleted, and any edge which joined $u \in V \setminus S$ and $v \in S$ in G is replaced by an edge joining u and S in $G \times S$. See fig. 3.2.

Matching Algorithm 3.4.

Input. A graph $G=(V,E)$.

Output. Either a perfect matching M, or a set $X \subseteq V$ such that $|X| < \theta(G-X)$.

Step 0. Let M be any matching, e.g., $M=\emptyset$.

Step 1. If every node is saturated by M, then M is perfect; terminate. Otherwise, choose an unsaturated node r and begin growing an alternating tree T rooted at r. Initially T consists of the single even node r.

Step 2. If every edge of G incident with an even node of T has as its other end an odd node of T, then let X be the set of odd nodes of T and terminate. Otherwise, let uv be an edge of G such that u is an even node of T and v is not an odd node of T. There are three possibilities:

Case 1. If v is not in T and is saturated, we go to Step 3 where we grow T.

Case 2. If v is in T, then it must be an even node of T. Go to Step 4 where we shrink.

Case 3. If v is not in T, and is not saturated, we go to Step 5 where we augment.

Step 3. (*Grow tree*). Let vw be the edge of M incident with v. Grow T by adding v, w, uv and vw. Note that v becomes odd and w becomes even. Go to Step 2.

Step 4. (*Shrink*). Both u and v are even nodes of T. Let p be the first common node in the paths in T from u to r and v to r. Then p will be an even node of T. (Possibly $p=u$, v or r.) The edge uv plus the paths in T from

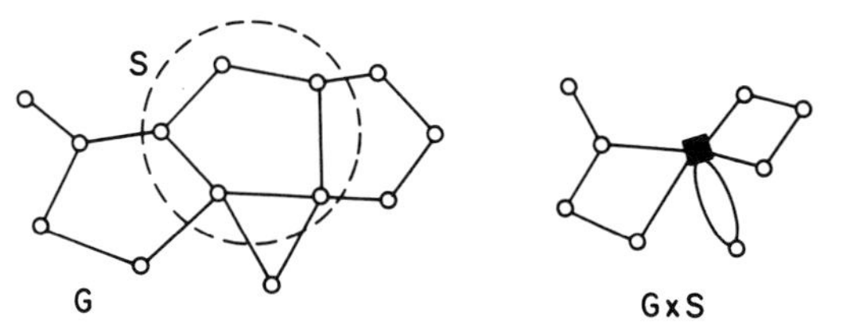

Figure 3.2. Shrinking.

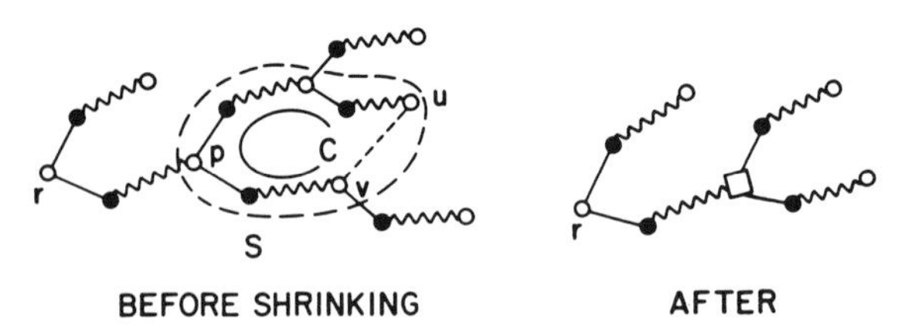

Figure 3.3.

u and v to p form an odd cycle C. Let S be the nodeset of C. Replace G by $G \times S$. Replace M by the restriction of M to the edges of $G \times S$. Replace T by $T \times S$. Go to Step 2. (See fig. 3.3.)

Step 5. (*Augment*). The edge uv plus the path in T from u to r comprises an augmenting path. Enlarge the matching using this path. (See Theorem 1.6.)

Step 5a. (*Expand pseudos*). If there exists pseudonodes (formed by shrinking in step 4), we proceed to expand them in the order reverse to that in which they were formed. Let S be the last (unexpanded) pseudonode formed and let C by the odd polygon whose nodes were shrunk to form S. There will be exactly one edge of M joining a node u of S to a node not in S. There will be exactly one node p of C not saturated by an edge of M. Let P be the even length path contained in C which joins u and p. Replace M with $M \triangle P$. The result is a matching in the graph obtained by expanding S. (See fig. 3.4.) If there remain unexpanded pseudonodes, repeat this process. If no pseudonodes remain, go to Step 1.

Now we show that the algorithm performs as claimed. It is easy to see that after each augmentation, and subsequent sequence of expansions, we are left with a matching exactly one edge larger than the one with which we began.

Suppose we terminate in Step 2 unable to grow the tree. The tree always has exactly one more even node than odd node, for it begins with a single even node and each time we grow the tree, we add one of each. Each time we shrink, we reduce the number of odd and even nodes by the same amount. Moreover, each even node of T, real or pseudo, will form a distinct odd component of $G - X$, where X is the set of odd nodes. Therefore $|X| = \theta(G - X) - 1$, and so (3.3) is violated, as required.

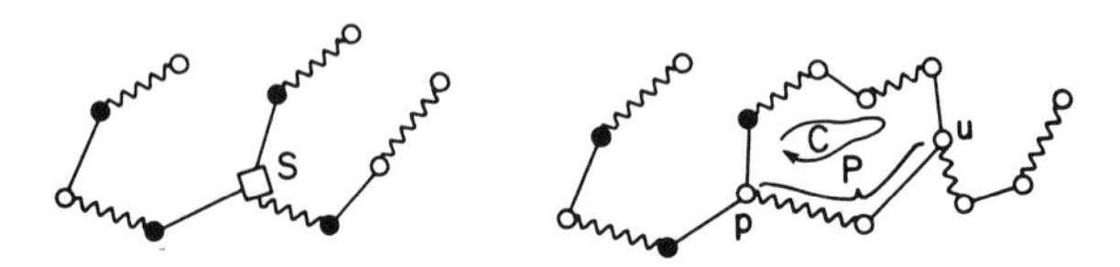

Figure 3.4.

Finally, we establish a bound on the running time of the algorithm. The algorithm can perform at most $|V|/2$ augmentations. Between augmentations, it can grow the tree at most $|V|/2$ times and shrink at most $|V|/2$ times. Using standard data structures, it is straightforward to implement tree growth and edge selection in constant time, shrinking in time $O(|V|)$ and augmentation with expansion in time $O(|V|^2)$. This gives a total running time of $O(|V|^3)$.

Hopcroft and Karp (1973) showed that the maximum matching problem for bipartite graphs could be solved in $O(|V|^{5/2})$, by systematically using all possible shortest augmenting paths in each case. This was generalized to the nonbipartite case by Even and Kariv (1975). See also Vazirani (1994). For a good survey of the issues involved in implementing both weighted and unweighted matching algorithms, plus some improvements to the above bounds, see Galil (1986). This reference also discusses some of the issues involved in parallel computation. See also Derigs (1988) for a comprehensive discussion of matching algorithms in both the weighted and unweighted cases.

Note how the algorithm simplifies in the case that G is bipartite. Case 2 of Step 2 cannot occur (because there are no odd cycles). Consequently we never need shrink or expand. This algorithm is the "Hungarian method" for constructing a maximum matching in a bipartite graph.

Moreover, note that, as claimed, the algorithm does provide a second proof of the sufficiency of Tutte's condition (3.3). However, it has the advantage that it either finds a perfect matching or else a set X which violates (3.3) in time which only grows polynomially with the size of the graph.

3.3. *Maximum matching and calculation of $\nu(G)$*

It is easy to modify the algorithm so that it will construct a maximum cardinality matching. When we reach the state in Step 2 of being unable to grow the tree, we say that the tree has become *Hungarian*. At this point we remove all nodes of the tree, or contained in pseudonodes of the tree, from the graph and resume trying to saturate another unsaturated node, if one exists. We repeat this process until all nodes not in Hungarian trees are saturated.

We will reach the following situation: We have a set of Hungarian trees called a *Hungarian forest*. The nodes of G can be partitioned into $D \cup A \cup C$, where D consists of all nodes of G which are even nodes, or contained in even pseudonodes of a Hungarian tree; A is the set of odd nodes of Hungarian trees and C is the set of nodes in no Hungarian tree. No edge joins a node of D to a node of C. An edge joins nodes of D only if they belong to the same pseudonode. The current matching induces a perfect matching on $G[C]$, and each node of D is matched to a node of A. (See fig. 3.5.) Let t be the number of Hungarian trees. Then the number of nodes unsaturated by the final matching M will be exactly t. Moreover, $G - A$ has $|A| + t$ odd components. Since the value $\theta(G - X) - |X|$ provides a lower bound on the number of nodes which must be left unsaturated by any matching, for any $X \subseteq V$, this shows that M is maximum. We thereby obtain a generalization due to Berge (1958) of Tutte's theorem.

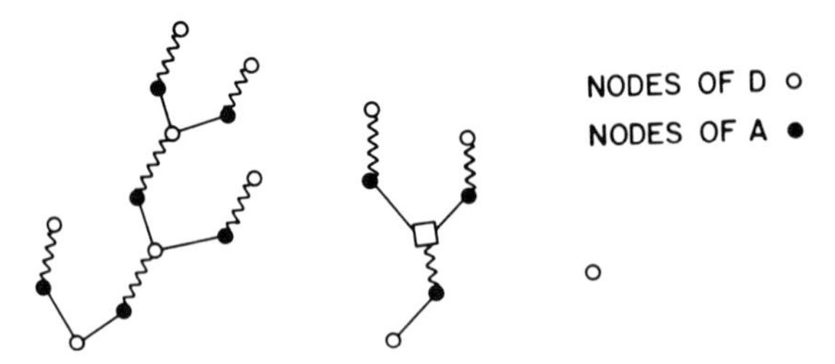

Figure 3.5. Hungarian forest.

Theorem 3.5. *For any graph G,*

$$\nu(G) = \min_{X \subseteq V} \left\{ \frac{|V| - \theta(G - X) + |X|}{2} \right\} .$$

The *deficiency* of a matching M of G is the number of nodes left unsaturated by M. We let $d(G)$ denote the *deficiency of* G, namely the deficiency of a maximum matching. Berge's Theorem 3.5 is equivalent to the following.

Corollary 3.6. *For any graph $G = (V, E)$*

$$d(G) = \max_{X \subseteq V} \{\theta(G - X) - |X|\} .$$

It is possible to deduce Theorem 3.5 and Corollary 3.6 directly, see Lovász and Plummer (1986). Also, they clearly imply Tutte's theorem. However it is also easy to deduce them directly from Tutte's theorem. Let $d = \max_{X \subseteq V}(\theta(G - X) - |X|\}$. Suppose we add d new nodes to G, each adjacent to every node in V. Then it is easy to see that Tutte's condition holds in the new graph G', so G' has a perfect matching M'. But at most d edges of M' are incident with the new nodes, so the edges of M' belonging to E form a matching having deficiency d as required.

Finally, we note how Edmonds' Theorem 1.5 follows from the algorithm for constructing a maximum matching. The final matching M produced will contain $(|V| - t/2)$ edges, where t is the number of Hungarian trees. The following is an odd set cover of E. Initially, define $\mathcal{K}$ by

$$K = \{\{v\}: v \in A\} \cup \{S: S \text{ is the set of real nodes contained in an even pseudonode of a Hungarian tree}\} .$$

If $C \neq \emptyset$, then choose some $\tilde{v} \in C$ and add to $\mathcal{K}$ the sets $\{\tilde{v}\}$ and $C \backslash \{\tilde{v}\}$. Let e and o denote the number of even and odd nodes of the Hungarian trees respectively. Then the weight of $\mathcal{K}$ is $\frac{1}{2}(|V| - e + o) = \frac{1}{2}(|V| - t)$, establishing Theorem 1.5.

In section 5 we will look at these results once again and see that they are special cases of some general weighted theorems.

4. Structure theory

A rich structure theory for graphs has been developed, based upon its matchings. Not only does this provide information concerning which edges and sets of edges belong to maximum matchings, but in addition, yields decomposition results which are useful in proving other theorems.

4.1. *Ear decompositions of bipartite and critical graphs*

One common type of theorem establishes a so-called *ear decomposition*. It asserts that all graphs in a particular class can be constructed by starting with some "simple" structure and then adding "simple" appendages. As an example, we say that a graph G is *matching covered* if every edge belongs to a perfect matching. (Note that we can test whether an edge uv belongs to a perfect matching by testing whether $G - u - v$ has a perfect matching.) Sometimes matching covered graphs are called 1-*extendable*. More generally, a graph is k-*extendable* if it has a perfect matching and every matching of cardinality k is contained in a perfect matching.

An *ear* with respect to a subgraph $\tilde{G} = (\tilde{V}, \tilde{E})$ of G consists of a path having an odd number of edges, such that the two end nodes are in $\tilde{V}$, but no internal nodes belong to $\tilde{V}$. (See fig. 4.1.)

Theorem 4.1 (Hetyei 1964, see Lovász and Plummer 1986). *Let $G = (V \cup W, E)$ be a connected bipartite graph. Then G is matching covered if and only if there exists a sequence $G_0, G_1, G_2, \ldots, G_p = G$ of graphs such that*

(i) *G_0 is K_2;*

(ii) *for $i = 1, 2, \ldots, p$, G_i is obtained from G_{i-1} by adding a single ear joining two nodes belonging to different parts of the bipartition of G_{i-1}.*

We call this a *bipartite ear decomposition*.

Proof. We first prove the necessity. Let M be a perfect matching of G and let $\bar{G} = (\bar{V}, \bar{E})$ be a maximal subgraph of G such that

(i) $\bar{G}$ has a bipartite ear decomposition;

(ii) M induces a perfect matching of $\bar{G}$.

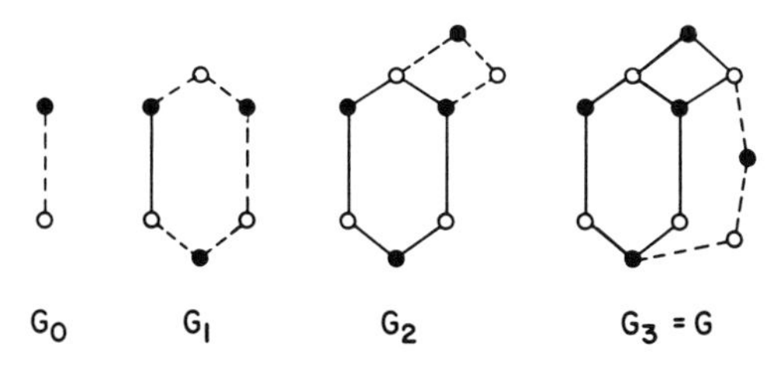

Figure 4.1. Ear decomposition.

First note that $\bar{E} = E(\bar{V})$, for if any edge of $E(\bar{V})$ has been omitted, it could be added as an ear. Second, suppose that $V \neq \bar{V}$. Let $u \in V \setminus \bar{V}$ be adjacent to $v \in \bar{V}$. By (ii), edge uv is not in M. Let $\bar{M}$ be a perfect matching containing uv. Then $\bar{M} \triangle M$ contains an even cycle containing edge uv whose edges alternately belong to M and to $\bar{M}$. Let P be the maximal portion of this cycle containing uv, but containing no edges of $\bar{E}$. Then the first and last edges of P are not in M, and P provides an ear which contradicts the maximality of $\bar{G}$.

Conversely, suppose that G has a bipartite ear decomposition. Then it is straightforward to prove that G is matching covered by induction on the number of ears, if we add to the induction hypothesis the following property: for all nodes u, v belong to different parts of the partition, $G - u - v$ has a perfect matching. □

A graph G is called (*factor*) *critical* or *hypomatchable* if $G - v$ has a perfect matching for every node v. Such graphs must have an odd number of nodes and be connected, and it is easy to see that they must also be nonbipartite. Critical graphs play important roles in matching theory. They have ear decompositions very similar to those of matching covered bipartite graphs.

Theorem 4.2 (Lovász 1972b). *Let G be a graph. Then G is critical if and only if there exists a sequence $G_0, G_1, \ldots, G_p = G$ of graphs such that*

(i) *G_0 is an odd cycle;*

(ii) *for $i = 1, 2, \ldots, p$, G_i is obtained from G_{i-1} by adding an ear joining two (not necessarily distinct) nodes of G_{i-1}.*

This theorem can be proved analogously to Theorem 4.1.

We call the sequence $G_0, G_1, \ldots, G_p$ a *critical ear decomposition* of G. Note that critical graphs may contain cutnodes. In this case, some ears become odd cycles attached to single nodes of the previous graph. However, if G is 2-connected, then these degenerate ears can be avoided. This fact will make possible a simple inductive proof of a theorem characterizing some of the essential inequalities for the matching polyhedron (Theorem 5.18).

Theorem 4.3 (Cornuéjols and Pulleyblank 1983, see also Lovász and Plummer 1986). *If G is a 2-connected critical graph, then G has a critical ear decomposition $G_0, G_1, \ldots, G_p$ such that each G_i is 2-connected.*

Proof. Suppose this is false; let $G_0, G_1, \ldots, G_p$ be an ear decomposition such that

(i) the smallest t for which G_t contains a cutnode v is as large as possible; and, subject to this,

(ii) the first s for which v is no longer a cutnode of G_s is as small as possible.

For each $i \in \{1, 2, \ldots, p\}$, let C_i be the ear added to G_{i-1} to form G_i. By our choice of s and t, each ear C_i for $t \leq i \leq s-1$ has neither end in G_{t-1}, but C_s must have one end in G_{t-1} and the other must be an internal node u of some C_i for

$t \leq i \leq s-1$. The node u splits C_i into an even length path P^0 and an odd length path P^1. We can concatenate P^0 onto C_s and obtain an ear C_i' which we can add instead of C_i. Then we can add P^1 as a separate ear. Then we can add the ears $C_{i+1}, \ldots, C_{s-1}$. We will now have constructed G_s, so we can complete the ear decomposition. But now v stopped being a cutnode after adding C_i', and $i < s$, a contradiction. □

The matching algorithm shrunk certain subgraphs of G. We say that a graph is *shrinkable* if the maximum matching algorithm, when applied to it, shrinks it to a single pseudonode. That is, G is shrinkable if either G is a single node or else G contains an odd cycle with nodeset C such that $G \times C$ is shrinkable. It follows from Theorem 4.2 that all critical graphs are shrinkable. It is easy to show by induction that for any node v of a shrinkable graph G, $G-v$ has a perfect matching, that is, G is critical. Then G is critical if and only if it is shrinkable.

4.2. Barriers and the canonical partition

It is possible to give an ear decomposition theorem for nonbipartite matching covered graphs, but if we want every intermediate graph to be matching covered, we cannot do it by adding ears one at a time. Rather we must sometimes add ears in pairs. Also, care must be exercised when adding single or double ears to ensure that each graph in the sequence is matching covered. In order to specify which pairs of nodes are suitable as attachment points for ears, we introduce the *canonical partition* of V for any matching covered graph $G=(V, E)$. Most of the material in this section is based on Hetyei (1964), Kotzig (1959a,b, 1960) and Lovász (1972a), see Lovász and Plummer (1986).

First we observed two preliminary facts.

Lemma 4.4. *Let $\bar{M}$ be a perfect matching of a connected matching covered graph $G=(V, E)$. For every pair u, v of distinct nodes, there is an alternating path from u to v, the first edge of which is in $\bar{M}$.*

Proof. Let $u \in V$ and let S be the set of nodes reachable from u by an alternating path which begins with an edge of $\bar{M}$, plus u itself. If $S \not\subseteq V$ then there is $st \in E$ such that $s \in S$ and $t \in V \backslash S$. If $st \in \bar{M}$, then t must also be reachable from u, so $t \in S$, a contradiction. Therefore $st \not\in \bar{M}$, but there exists a perfect matching $\hat{M}$ of G containing st. The symmetric difference $\bar{M} \triangle \hat{M}$ contains an alternating cycle C which uses st, and includes nodes of S. Let P be a minimal alternating path from u to a node r of C, which starts with an edge of $\bar{M}$, if u is not in C. Then P plus one of the two paths in C from r to t provides an alternating path from u to t as required, contradicting the definition of S. Therefore $S=V$. □

An alternating path with respect to a matching M is called a *decrementing* path if the first and last edges belong to M. Then $G-u-v$ has a perfect matching if and only if u and v are the ends of a decrementing path. (The sufficiency is trivial, the necessity follows by considering $M \triangle \bar{M}$, for a perfect matching $\bar{M}$ of $G-u-v$.) We define a relation $\sim$ by $u \sim v$ if and only if $G-u-v$ does *not* have a perfect matching.

Lemma 4.5. *Let $G=(V,E)$ be connected and matching covered. Then $\sim$ is an equivalence relation.*

Proof. All we need show is transitivity. Let M be any perfect matching of G. Suppose $u \sim v$ and $v \sim w$ but $u \not\sim w$. Then u and w are joined by a decrementing path P. By Lemma 4.4 there is an alternating path from v to u, which begins with an edge of M. Let P' be a minimal such alternating path from v to any node s of P. Then P', plus the portion of P from s to u or s to w is a decrementing path from v to u or v to w, contradicting either $u \sim v$ or $v \sim w$. □

The *canonical partition* of G, denoted by $\mathcal{P}(G)$, is the set of equivalence classes of this relation. If G is nonempty, bipartite and matching covered, then there are exactly two equivalence classes – the two colour classes. A graph $G=(V,E)$ is *bicritical* if $G-u-v$ has a perfect matching for each $u, v \in V$. Equivalently, G is bicritical if every class of $\mathcal{P}(G)$ is a singleton.

A *barrier* (Lovász and Plummer 1986) in a connected matching covered graph G is a set $X \subseteq V$ such that $|X| = \theta(G-X)$. For each $v \in V$, $\{v\}$ is trivially a barrier, but we are more interested in the maximal barriers.

Lemma 4.6. *Let X be a maximal barrier in a connected matching covered graph $G=(V,E)$. Then every component of $G-X$ is critical.*

Proof. Since G is matching covered, all components of $G-X$ are odd. Suppose that S is the nodeset of a component of $G-X$ which is not critical. Then there exists $v \in S$ such that $G[S]-v$ has no perfect matching. By Tutte's theorem, there exists $Y \subseteq S \backslash \{v\}$ such that $G[S]-(Y \cup \{v\})$ has more than $|Y|$ odd components. Therefore $G-(X \cup Y \cup \{v\})$ has at least $|X|+|Y|+1$ odd components, and since G is matching covered, it must have exactly this number. Therefore $X \cup Y \cup \{u\}$ is a barrier, contradicting the maximality of X. □

The following is the main result relating barriers to the canonical partition.

Theorem 4.7 (Kotzig 1959a,b, 1960 and Lovász 1972a). *For a connected matching covered graph G, the maximal barriers of G are precisely the classes of $\mathcal{P}(G)$.*

Proof. If u and v belong to a barrier X then $G-u-v$ can have no perfect matching, since deleting $X \backslash \{u, v\}$ leaves $|X|$ odd components. Therefore every maximal barrier is contained in a single class of $\mathcal{P}(G)$.

Now let X be a maximal barrier and let $v \in V \backslash X$. By Lemma 4.6, v belongs to a critical component $G[S]$ of $G \backslash X$. Let M be a perfect matching of G. Some node $w \in S$ is joined to a node $u \in X$ by an edge of M. Remove uw from M and replace the matching on $G[S]$ with a matching which leaves only v unsaturated. This yields a perfect matching of $G - \{v, u\}$. Hence $v \not\sim u$, so X must be a class of $\mathcal{P}(G)$. □

Now we can describe the ear decomposition of matching covered graphs.

Theorem 4.8 (Lovász and Plummer 1975). *Let $G = (V, E)$ be a connected graph. Then G is matching covered if and only if there exists a sequence G_0, $G_1, \ldots, G_p = G$ of graphs such that*

(i) *G_0 is a K_2,*

(ii) *for $i = 1, 2, \ldots, p$, G_i is obtained from G_{i-1} by either*

(a) *adding a single ear joining nodes belonging to different classes of $\mathcal{P}(G_{i-1})$ or*

(b) *adding two ears, such that one joins distinct nodes u_1, u_2 belonging to some class of $\mathcal{P}(G_{i-1})$ and the other joins distinct nodes v_1, v_2 belonging to a different set of $\mathcal{P}(G_{i-1})$, and $G - u_1 - u_2 - v_1 - v_2$ has a perfect matching.*

Proof. See Lovász and Plummer 1986. □

The hard part of the proof is showing that we never require more than two ears to be added at a time. An elementary proof of this was recently obtained by Little and Rendl (1989).

4.3. The Edmonds–Gallai partition

Now we describe another type of decomposition of graphs which is useful for graphs which do not have perfect matchings. The idea is to decompose the graph into simpler graphs, such that the maximum matchings of the original graph can be constructed from those of the smaller graphs. For any graph $G = (V, E)$, let $D(G)$ be the set of all nodes v such that some maximum matching of G leaves v unsaturated. Let $A(G)$ be the nodes not in $D(G)$, but adjacent to nodes of $D(G)$. Let $C(G) = V \setminus (D(C) \cup A(G))$. ("$D$" stands for "deficient", "A" stands for "adjacent" and "C" stands for "covered".) The partition $V = D(V) \cup A(G) \cup C(G)$ is called the *Edmonds–Gallai partition* of G.

Theorem 4.9 (Gallai 1963, 1964, Edmonds 1965a). *Let $G = (V, E)$ be an arbitrary graph and let $D(G)$, $A(G)$ and $C(G)$ be defined as above. Then*

$$\text{every component of } G[D(G)] \text{ is critical ;} \tag{4.10}$$

a matching M of G is maximum if and only if

$$M \text{ induces a perfect matching of } G[C(G)] \text{ ;} \tag{4.11a}$$

$$M \text{ induces a near-perfect matching of } G[K], \text{ for every component } K \text{ of } G[D(G)] \text{ ,} \tag{4.11b}$$

for each $v \in A(G)$, *there is an edge* $uv \in M$ *such that* u *belongs to* $D(G)$. (4.11c)

(Note that (4.11b) and (4.11c) imply that each node of $A(G)$ is matched to a unique component of $G[D(G)]$. Also note that if G has a perfect matching, then $V = C(G)$. If G is critical then $V = D(G)$. The interesting cases are when G neither has a perfect matching nor is critical.)

This result follows easily from linear programming duality, and we postpone the proof to the next section. However, we should note that the maximum matching algorithm of the previous section actually constructed the partition. The set $D(G)$ consists of all nodes of V which are even nodes, or contained in even pseudonodes, of the Hungarian trees. The set $A(G)$ consist of the odd nodes of the Hungarian trees and $C(G)$ consists of the nodes not belonging to Hungarian trees. Consequently, constructing the Edmonds–Gallai partition is no more difficult than finding a maximum matching.

4.4. Brick decompositions

Suppose that $G = (V, E)$ has a perfect matching. As we have already noted, in this case the Edmonds–Gallai theorem 4.9 gives no information. If we wish to decompose G based on the maximum (perfect) matchings, then edges not in any perfect matching are of no interest, so assume that G is matching covered. We now describe the so-called *brick decomposition* of G. A *brick* is a bicritical 3-connected graph. This process decomposes an arbitrary matching covered graph into a set of bricks and matching covered bipartite graphs.

Brick Decomposition Procedure 4.12. Let $G = (V, E)$ be a connected matching covered graph.

(1) If G is a brick or is bipartite then it is indecomposable.

(2) If G is nonbipartite and not bicritical, then let X be a class of the canonical partition $\mathcal{P}(G)$ for which $|X| \geq 2$, and let S be the nodeset of a component of $G - X$ such that $|S| \geq 3$. Let $G_1 = G \times S$. Let $G_2 = G[S \cup X] \times X$. See fig. 4.2.

A brick decomposition of G is the union of brick decompositions of G_1 and G_2.

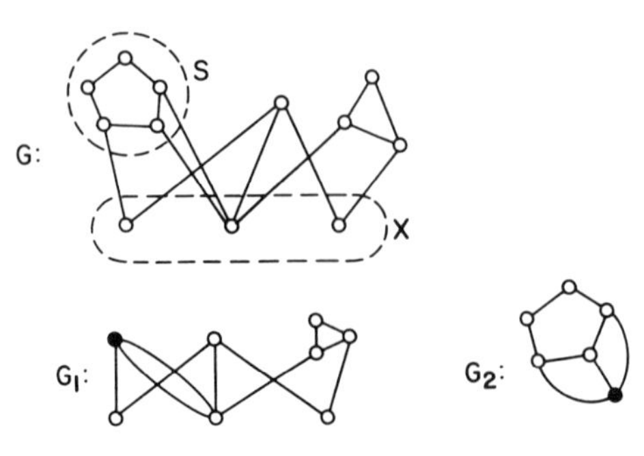

Figure 4.2.

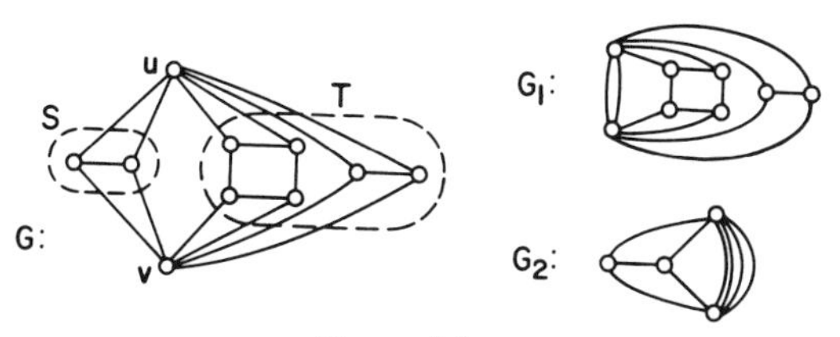

Figure 4.3.

(3) If G is bicritical, but not 3-connected, then let $\{u, v\}$ be a two node cutset, let S be the nodeset of a component of $G - u - v$ and let $T = V \backslash (S \cup \{u, v\})$. Let $G_1 = G \times (S \cup \{u\})$ and let $G_2 = G \times (T \cup \{v\})$. (See fig. 4.3.) A brick decomposition of G is the union of brick decompositions of G_1 and G_2.

Note that in both (2) and (3) above, the coboundary of the shrunk nodes is a tight cut of G, i.e., each perfect matching uses exactly one edge. Moreover, the perfect matchings of G can be obtained by composing the perfect matchings of G_1 and G_2.

When performing a brick decomposition, we have considerable freedom in our choice of X and S in Step 2 and of u, v and S in Step 3. Surprisingly the final list of bricks and bipartite graphs obtained is (almost) independent of these choices.

Theorem 4.13 (Lovász 1987). *The list of bricks and bipartite graphs obtained by a brick decomposition is independent* (*up to multiplicity of the edges*) *of the choices made in Step* 2 *and* 3).

In the next section we will see that the brick decomposition determines the rank of the incidence vectors of the perfect matchings of G, over the field of real (or rational) numbers.

5. Weighted matchings and polyhedra

5.1. Maximum weight matchings

Suppose we have a graph $G = (V, E)$ and a vector $c = (c_e\colon e \in E)$ of edge costs. We wish to find a matching, or perfect matching, for which the sum of the edge costs is maximized. Both these problems can be solved almost as easily as the maximum cardinality matching problem, by exploiting a connection with linear programming.

We first formulate these problems as integer programs. We introduce a 0–1 variable x_e for each $e \in E$ with the interpretation that $x_e = 1$ if e is in the matching and $x_e = 0$ otherwise. For any vector $x = (x_j\colon j \in J)$ and any $K \subseteq J$ we let $x(K)$ denote $\sum (x_j\colon j \in K)$. Recall that $\delta(v)$ denotes the set of edges incident with v.

The maximum weight matching problem becomes:

$$\begin{aligned} \text{maximize} \quad & cx \\ \text{subject to} \quad & x \geq 0, \text{ integer}, \\ & x(\delta(v)) \leq 1 \text{ for all } v \in V . \end{aligned}$$

Let $\mathcal{O} = \{S \subseteq V\colon |S| \geq 3, \text{ odd}\}$. For any $S \in \mathcal{O}$, if x is the incidence vector of a matching, then $x(E(S)) \leq (|S|-1)/2$. Now consider the following *linear* program:

$$\text{maximize} \quad cx$$

$$\text{subject to} \quad x \geq 0 \tag{5.1}$$

$$x(\delta(v)) \leq 1 \text{ for all } v \in V , \tag{5.2}$$

$$x(E(S)) \leq (|S|-1)/2 \text{ for all } S \in \mathcal{O} . \tag{5.3}$$

We have already noted that all incidence vectors of matchings satisfy (5.1)–(5.3). We will see that every extreme solution to (5.1)–(5.3) is the incidence vector of a matching. This we will prove by showing that for any vector c of edge weights, there exists an optimum solution to this linear program which is the incidence vector of a matching.

The dual linear program is the following:

$$\text{minimize} \quad \sum_{v \in V} y_v + \sum_{S \in \mathcal{O}} ((|S|-1)/2) z_S$$

$$\text{subject to} \quad y_v \geq 0 \text{ for all } v \in V , \tag{5.4}$$

$$z_S \geq 0 \text{ for all } S \in \mathcal{O} , \tag{5.5}$$

$$y_u + y_v + \sum (z_S; u, v \in S \in \mathcal{O}) \geq c_{uv} \text{ for all } uv \in E . \tag{5.6}$$

We will prove optimality by constructing both an incidence vector x^* of a matching, and a feasible dual solution y^*, z^* which satisfy the complementary slackness conditions for optimality. Namely, x^* satisfying (5.1)–(5.3) and y^*, z^* satisfying (5.4)–(5.6) are optimal if and only if

$$x^*_{uv} > 0 \text{ implies } y^*_u + y^*_v + \sum (z^*_S : u, v \in S \in \mathcal{O}) = c_{uv} \text{ for all } uv \in E , \tag{5.7}$$

$$z^*_S > 0 \text{ implies } x^*(E(S)) = (|S|-1)/2 \text{ for all } S \in \mathcal{O} , \tag{5.8}$$

$$y^*_v > 0 \text{ implies } x^*(\delta(v)) = 1 \text{ for all } v \in V . \tag{5.9}$$

We illustrate the use of these conditions by establishing the Edmonds–Gallai Theorem 4.9 from the output of the maximum cardinality matching algorithm (see section 3). Recall that when this algorithm terminates it has constructed a (maximum) matching with incidence vector x^* and also a number of Hungarian trees. Recall that D is the set of nodes of G which are even nodes, or contained in even pseudonodes, of the Hungarian trees; A is the set of odd nodes of the Hungarian trees and $C = V \backslash (A \cup D)$. We let $c_e = 1$ for all $e \in E$, and consider the following dual solution:

$$y_v^* = \begin{cases} 1 & \text{if } v \in A\,, \\ 0 & \text{if } v \in D\,, \\ \frac{1}{2} & \text{if } v \in C\,, \end{cases}$$

$$z_S^* = \begin{cases} 1 & \text{if } S \text{ is the nodeset of an even} \\ & \text{pseudonode of a Hungarian tree}\,, \\ 0 & \text{otherwise.} \end{cases}$$

It is easy to see that (5.1)–(5.9) hold, which establishes the optimality of y^*, z^* (and x^*). Moreover, since the incidence vector of every maximum matching must satisfy (5.7)–(5.9) with respect to this y^* and z^*, we deduce the following. First, by (5.9), only nodes of D can be left unsaturated by any maximum matching. It is easy to construct a maximum matching of G deficient at any desired node of D (using the Hungarian tree structure) so $D = D(G)$ (of the Gallai–Edmonds theorem). Since nodes of D are only adjacent to nodes of A, we have $A = A(G)$ and also $C = C(G)$. We saw in section 4 that a graph is critical if and only if it is shrinkable, implying (4.10). Since we have $y_u^* + y_v^* > 1 = c_{uv}$ for all $u \in A$ and $v \in A \cup C$, (5.7) implies (4.11a and c). Finally, (4.11b) follows from (5.9). Thus we have both proved Theorem 4.9 and shown how the maximum matching algorithm finds $D(G)$, $A(G)$ and $C(G)$.

Now we describe how to find a maximum weight perfect matching in a graph. From this, it is easy to find a maximum weight matching, which is described following the algorithm.

For the case of perfect matchings, we must make the following changes to our linear programming formulations. Condition (5.2) becomes the equation

$$x(\delta(v)) = 1 \text{ for all } v \in V\,. \tag{5.2'}$$

Consequently, dual feasibility condition (5.4) and complementary slackness condition (5.9) disappear.

At all times, the algorithm maintains a feasible dual solution y, z. It attempts to construct a perfect matching whose incidence vector $\hat{x}$ satisfies (5.7) and (5.8). If successful, it terminates. If unsuccessful, y and z are changed and the process repeats.

A family $\mathcal{F}$ of subsets of V is called *nested* if whenever S, $T \in \mathcal{F}$ satisfy $S \cap T \neq \emptyset$, either $S \subseteq T$ or $T \subseteq S$. Let y, z be a feasible dual solution. We let $\mathcal{O}' = \{S \in Q: z_S > 0\}$. At all times, the algorithm will ensure that $\mathcal{O}'$ is a nested family. Also, we let

$$E^{=} = \left\{uv \in E\colon y_u + y_v + \sum (z_S\colon u, v \in S \in \mathcal{O}') = c_{uv}\right\}. \tag{5.10}$$

For every $S \in \mathcal{O}'$, let G^S denote the graph obtained from $G[S]$ by shrinking all maximal members of $\mathcal{O}'$ properly contained in S. Then we will have

$$G^S \text{ is spanned by an odd cycle, all of whose edges are in } E^{=}\,. \tag{5.11}$$

Thus $G^S - (E \backslash E^=)$ is critical (cf. Theorem 4.2).

Weighted Perfect Matching Algorithm 5.12.

Input. A graph $G = (V, E)$, a vector $c = (c_e : e \in E)$ of edge weights.

Output. Either a set $X \subseteq V$ for which $|X| < \theta(G - X)$ (establishing the nonexistence of a perfect matching) or the incidence vector x^* of a maximum cost perfect matching and an optimum dual solution y^*, z^*.

Step 0. (*Construct feasible dual solution*). Let y satisfy $y_u + y_v \geqslant c_{uv}$ for all $uv \in E$, let $z_S = 0$ for all $S \in \mathcal{O}$ (i.e., $\mathcal{O}' = \emptyset$).

Step 1. (*Construct current equality subgraph*). Let $E^=$ be defined as in (5.10). Let $\bar{G} = (\bar{V}, \bar{E})$ be the graph obtained from $G^= = (V, E^=)$ by shrinking all maximal members of $\mathcal{O}' = \{S \in \mathcal{O}: z_S > 0\}$. (If $\mathcal{O}' = \emptyset$, then $\bar{G} = G^=$.)

Step 2. (*Solve maximum cardinality problem on* $\bar{G}$). Apply the maximum matching algorithm of section 3 to $\bar{G}$. If a perfect matching is found, go to Step 4. If not let D and A be the sets of all nodes of G which are equal to or contained in even and odd nodes respectively of the Hungarian trees in $\bar{G}$. Let C be the other nodes of G. Go to Step 3.

Step 3. (*Modify the dual solution*). Let

$\delta_1 = \min\{z_S\colon S \in \mathcal{O}'$ and S is the set of nodes belonging to an odd pseudonode of a Hungarian tree$\}$;

$\delta_2 = \min\{y_u + y_v - c_{uv}\colon uv \in E \backslash E^=, u \in D$ and $v \in C\}$.

$\delta_3 = \min\{y_u + y_v - c_{uv}\colon uv \in E \backslash E^=$ and u and v are nodes of D which do not belong to the same pseudonode of a Hungarian tree$\}$.

By convention, if any of these sets are empty, we let the corresponding $\delta_i = \infty$. Let $\delta = \min\{\delta_1/2, \delta_2, \delta_3/2\}$. If $\delta = \infty$ then G has no perfect matching as $\theta(G - A) > |A|$, and we terminate. Otherwise, let

$$y'_v = \begin{cases} y_v - \delta & \text{for } v \in D\,, \\ y_v + \delta & \text{for } v \in A\,, \\ y_v & \text{for } v \in C\,; \end{cases}$$

$$z'_S = \begin{cases} z_S + 2\delta & \text{if } S \text{ is the set of nodes contained in an even pseudonode}\,, \\ z_S - 2\delta & \text{if } S \in \mathcal{O}' \text{ is the set of nodes contained in an odd pseudonode}\,, \\ z_S & \text{otherwise}\,. \end{cases}$$

By our choice of δ, the dual solution y', z' is feasible. Moreover, any edge in the forest or in the current matching will still be in $E^=$, recomputed for y' and z'. Replace y, z with y', z' and go to Step 1.

Step 4. (*Extend matching in pseudonodes*). Using (5.11), it is easy to see that the incidence vector $\hat{x}$ of the perfect matching of $\bar{G}$ obtained can be extended to a perfect matching of G, whose incidence vector x^* satisfies (5.7) and

(5.8). Let $y^* = y$ and $z^* = z$. Then x^*, y^* and z^* are the desired optimal solutions. Terminate.

It is straightforward to verify that when the algorithm terminates, it has found either the required set $A \subseteq V$ such that $|A| < \theta(G - A)$ (Step 3) or has constructed optimal y^*, z^* and x^*. What remains to be shown is that the running time is finite, and in fact, polynomial in the size of the graph. To show this, we must establish an upper bound on the number of times Step 3 (dual change) can be performed.

Recall that $d(G)$ denotes the deficiency (minimum possible number of unsaturated nodes) of G. Let $\bar{G}$ and $\bar{G}'$ be the graphs obtained before and after a dual variable change. Let D and C be the sets of nodes of G belonging to $D(\bar{G})$ and $C(\bar{G})$ respectively. Let D' and C' be defined similarly for $\bar{G}'$. We show that

$$d(\bar{G}') \leqslant d(\bar{G})$$

and

$$\text{if } d(\bar{G}') = d(\bar{G}) \text{ then } |D'| \geqslant |D|$$

and

$$\text{if } |D'| = |D| \text{ then } |C'| > |C|\,.$$

From this it will follow that we can perform at most $|V|^3$ dual changes.

When Step 3 is performed, we have one or more of the following situations:

($\delta = \delta_1/2$) A pseudonode, belonging to A, is expanded.

($\delta = \delta_2$) An edge joining a node $u \in D$ and $v \in C$ enters $E^=$.

($\delta = \delta_3/2$) A node joining distinct even nodes u, v of the Hungarian trees enters $E^=$.

It is easy to see that none of these can increase $d(\bar{G})$, so suppose $d(\bar{G}') = d(\bar{G})$. Again, we can still find maximum matchings leaving any node of $D(\bar{G})$ unsaturated, so $|D'| \geqslant |D|$. We now try to construct a matching deficient at a node of $\bar{G}$ which does not belong to $D(\bar{G})$, which will show $|D'| > |D|$.

If $\delta = \delta_3/2$ we can start with a maximum matching of $\bar{G}$ deficient at u. If v is not saturated, then adding uv to the matching gives $d(\bar{G}') < d(G')$. If v is incident with an edge uw in the matching then add uv and remove vw. We now have a maximum matching deficient at $w \in A(\bar{G}')$ so $|D'| > |D|$.

If $\delta = \delta_2$ then start with a maximum matching of $\bar{G}$ which leaves u, or the pseudonode containing u, unsaturated. Some edge vw is in the matching, where v, $w \in C(\bar{G})$. Now remove vw and add uv. Then $w \in D'$, so $|D'| > |D|$.

If $\delta = \delta_1/2$ then either some node of the expanded pseudonode enters D' or else all go into C. Thus if $|D'| = |D|$ then $|C'| > |C|$, and we are done.

This shows that a maximum weight perfect matching can be found by solving $O(|V|^3)$ maximum matching problems. The algorithm as presented by Edmonds (1965b) was slightly different – when a Hungarian tree was found, a dual variable change would be made and the algorithm would continue with that tree, which would no longer be Hungarian. This resulted in an overall running time of

$O(|V|^4)$. Lawler (1976) showed that this could be reduced to $O(|V|^3)$ by the use of suitable data structures. This is still the best known bound for a weighted nonbipartite matching.

Each time we make a dual variable change, the value of the dual solution strictly decreases. All this requires is that we have identified a subset X of $V(\bar{G})$ such that $\theta(\bar{G}-X)>|X|$. However if we were to make such changes based on arbitrary X rather than $A(\bar{G})$, we could not even guarantee *finite* termination of the algorithm for real costs! This even applies in the bipartite case (Aráoz and Edmonds 1985).

5.2. Matching polyhedra, dual solutions and facets

The correctness of the preceding algorithm provides a proof of one of the fundamental theorems of polyhedral combinatorics. The *perfect matching polytope* $P(G)$ of a graph G is defined to be the convex hull of the incidence vectors of the perfect matchings of G.

Theorem 5.13 (Edmonds 1965b). *$P(G)$ is the set of all $x \in \mathbb{R}^E$ satisfying*

$$x_e \geq 0 \quad \text{for all } e \in E\,,$$
$$x(\delta(v)) = 1 \quad \text{for all } v \in V\,,$$
$$x(E(S)) \leq (|S|-1)/2 \quad \text{for all } S \in \mathcal{O}\,.$$

Proof. Let P denote the polyhedron defined by the above linear system. We have already noted that the incidence vector of any perfect matching satisfies these inequalities so $P(G) \subseteq P$. We complete the proof by showing that every vertex of P is in $P(G)$, i.e., is the incidence vector of a perfect matching. If $P(G) \neq \emptyset$ and if $\hat{x}$ is a vertex of P, then there is a vector c of edge weights such that $\hat{x}$ is the unique member of P maximizing $c\hat{x}$. Apply Algorithm 5.12 to G, with edge weights c. We have seen that it will terminate with the incidence vector x^* of a perfect matching, if one exists, plus an optimum dual solution y^*, z^*. Since they are feasible and satisfy complementary slackness, they are optimal solutions to the linear program maximize cx for $x \in P$. Therefore $x^* = \hat{x}$. If $P(G) = \emptyset$, then the algorithm terminates in Step 3, with $\delta = \infty$. This means that the objective function of the dual linear program can be made arbitrarily negative, so $P = \emptyset$ by the weak duality theorem of linear programming. □

See chapter 30 or Lovász and Plummer (1986) for alternate nonalgorithmic proofs.

It is interesting that Theorem 5.13 has provided the basis for a cutting plane approach to the maximum weight matching problem, which has proved to be successful on quite large problems (see Grötschel and Holland 1986). The idea is to begin with a small subset of the inequalities, in this case just the nonnegativity and degree constraints. First solve the relaxed problem, obtained by maximizing cx subject to these constraints. If the optimum solution is 0–1 valued, then we stop with the incidence vector of a maximum weight matching. If not, we find one or more violated constraints of the third type, add them to our constraint set so as to obtain a stronger relaxation, and resolve. This is then repeated until an optimum matching is obtained. (See chapter 30 for a thorough discussion of separation based algorithms.)

The crucial step in this process is to find a violated blossom inequality (5.3) if one exists. This problem was solved by Padberg and Rao (1982), who showed how the procedure of Gomory and Hu (1961) for constructing a flow equivalent tree (see chapter 2) could be adapted to find a minimum capacity odd cut.

The previous algorithm also makes it easy to deduce an important result concerning discreteness of optimal dual solutions.

Theorem 5.14 (Edmonds 1965a). *If G has a perfect matching and c is integral, then there exists an optimum dual solution y^*, z^* such that*

$$y_v^* \equiv 0 \ (\mathrm{mod}\ \tfrac{1}{2}) \quad \textit{for all } v \in V\,, \qquad (5.15)$$
$$z_S^* \equiv 0 \ (\mathrm{mod}\ 1) \quad \textit{for all } S \in \mathcal{O}\,.$$

Proof. Begin the algorithm with $z = 0$ and y integral. Then (5.15) holds, as well as the following:

$$y_u \equiv y_v \ (\mathrm{mod}\ 1) \quad \text{for all } u, v \in D(\bar{G})\,. \qquad (5.16)$$

(Since every $v \in A(\bar{G})$ is adjacent with some $u \in D(\bar{G})$, we have $y_u \equiv y_v$ (mod 1) for all u, $v \in D(\bar{G}) \cup A(\bar{G})$.) It is straightforward to see that these properties imply that δ_1 and δ_3 will be integer and δ_2 will be half-integer. Therefore δ will be half-integer, which ensures that (5.15) holds for the new solution. Finally, we can see that (5.16) continues to hold, for any new node of $D(\bar{G}')$ must be joined by a path of edges in $E^{=}$ to a node previously in $D(\bar{G})$, so since all costs are integer, this gives the result. □

This dual discreteness property has useful combinatorial consequences which we discuss later. First, however, we note that the problem of finding a maximum weight (not necessarily perfect) matching in G reduces easily to the perfect matching case. Let $G' = (V', E')$ be a second disjoint copy of $G = (V, E)$ and join each $v' \in V'$ to the corresponding $v \in V$. Let each new edge not in E have cost zero. Now a maximum weight perfect matching in the expanded graph will induce a maximum weight matching in G, since any matching of G can be extended to a perfect matching in the extended graph having the same weight. Moreover, suppose that y^*, z^* is an optimal dual solution. Define a new objective function $\bar{c}$ by interchanging the costs of the corresponding edges of E and E'. For any $S \subseteq V \cup V'$, we define $S' = \{v \in V\colon v' \in S\} \cup \{v' \in V'\colon v \in S\}$. Let y' and z' be the optimal dual solutions for $\bar{c}$ obtained by letting $y_i' = y_{i'}^*$ and $y_{i'}' = y_i^*$ for all $i \in V$, and $z_S' = z_{S'}^*$ for all $S \subseteq V \cup V'$. Then $\bar{y} = y^* + y'$, $\bar{z} = z^* + z'$ is an optimum dual solution to the perfect matching problem on the expanded graph, where

$c_{j'} = c_j$ for all $j \in E'$ and all new edges not in E' have weight zero. Moreover, $\bar{y}_v = \bar{y}_{v'}$ and $\bar{z}_S = \bar{z}_{S'}$ for all $v \in V$ and $S \subseteq V \cup V'$. Also, $\bar{y}$, $\bar{z}$ has the same discreteness properties as y^*, z^*.

Suppose there exists $S \subseteq V \cup V'$ such that $\bar{z}_S > 0$ and $V \cap S \neq \emptyset \neq V' \cap S$. Let $T = (V \cap S) \backslash (V' \cap S)'$ and let $\tilde{T} = (V' \cap S) \backslash (V \cap S)'$. One of T, $\tilde{T}$, say T, has odd cardinality. Let $W = (S \cap V) \cap (S \cap V')'$. Redefine $\bar{y}$ and $\bar{z}$ by letting $\bar{y}_v := \bar{y}_v + \bar{z}_S$ for all $v \in W \cup W'$, $\bar{y}_v := \bar{y}_v + \frac{1}{2}\bar{z}_S$ for all $v \in \tilde{T} \cup \tilde{T}'$, $\bar{z}_T := \bar{z}_{T'} := \bar{z}_S$ and $\bar{z}_S := \bar{z}_{S'} := 0$. Then the new $\bar{y}$, $\bar{z}$ is optimal and has all the discreteness properties of the old, but fewer components $\bar{z}_S > 0$ for which S intersects both V and V'. Repeating this we can remove all such components, at which point it is immediate that the restriction of $\bar{y}$, $\bar{z}$ to G is an optimal dual solution to the maximum weight matching problem, which satisfies (5.15). The *matching polytope* $M(G)$ is defined to be the convex hull of the incidence vectors of the matchings of G. We obtain the following generalization of Theorem 5.13.

Theorem 5.17 (Edmonds 1965a). *$M(G)$ is the set of all $x \in \mathbb{R}^E$ satisfying*

$$x_e \geq 0 \quad \text{for all } e \in E\,,$$

$$x(\delta(v)) \leq 1 \quad \text{for all } v \in V\,,$$

$$x(E(S)) \leq (|S| - 1)/2 \quad \text{for all } S \in \mathcal{O}\,.$$

This is a generalization since $P(G)$ is the face of $M(G)$ obtained by replacing the inequalities $x(\delta(v)) \leq 1$ for $v \in V$ by equations, and all faces of integral polyhedra have integral vertices.

We also obtain a dual discreteness result, analogous to Theorem 5.14, which can be strengthened as follows.

Theorem 5.18. *Let $G = (V, E)$ be a graph and let c be a vector of integral edge weights. Then there exists an integral optimal solution to the dual problem (5.4)–(5.6).*

Proof. Let y^*, z^* be an optimal dual solution which satisfies (5.15). Let $F = \{v \in V: y_v^* \equiv \frac{1}{2} \bmod 1\}$. Since the value of y^*, z^* is integer, $|F|$ is even. Choose any $\hat{v} \in F$ and let $\hat{F} = F \backslash \{\hat{v}\}$. Let $y_i^* := y_i^* - \frac{1}{2}$ for all $i \in \hat{F}$, $y_{\hat{v}}^* := y_{\hat{v}}^* + \frac{1}{2}$ and let $z_{\hat{F}}^* := z_{\hat{F}}^* + 1$. The new solution is the required integral optimal solution. □

An identical argument applies to $P(G)$. A linear system such as (5.1)–(5.3) which has the property that there exists an integral optimal dual solution, for every integral objective function c which has a maximum, is called *totally dual integral*. See chapter 30. Such theorems have useful combinatorial applications. For example, suppose we are only interested in finding a maximum cardinality matching. This is equivalent to maximizing cx for $x \in M(G)$, where $c_j = 1$ for all $j \in E$. Let y^*, z^* be an integral optimal dual solution. Then (see (5.6)) each component must have value 0 or 1. Let $\mathcal{K} = \{\{v\}: y_v^* = 1\} \cup \{S \subseteq V: z_S^* = 1\}$.

Then $\mathcal{H}$ is an odd set cover and $w(\mathcal{H}) = \max\{cx: x \in M(G)\} = \bar{\nu}(G)$. Thus we derive Theorem 1.5 as a special case.

In general, not all constraints (5.2) and (5.3) are required to define $M(G)$. For $S \in \mathcal{O}$, constraint (5.3) is essential if and only if $F_S = \{x \in M(G): x(E(S)) = (|S| - 1)/2\}$ induces a facet of $M(G)$, i.e., has dimension one less than that of $M(G)$. (See chapter 30 for definitions of these terms.) This is equivalent to F_S containing $|E|$ affinely independent points. The essential constraints (5.3) correspond precisely to those critical $S \subseteq V$ for which $G[S]$ contains no cutnode.

Theorem 5.19 (Pulleyblank and Edmonds 1974). *Let $S \in \mathcal{O}$ be such that the inequality $x(E(S)) \leq (|S| - 1)/2$ is not of the form* (5.2), *for some $v \in V$. Then this inequality is essential for $M(G)$, i.e., F_S is a facet of $M(G)$, if and only if $G[S]$ is critical and nonseparable.*

Proof. We describe how to construct $|E|$ affinely independent members of F_S. It will be sufficient to construct $|E(S)|$ independent members of F_S all of which are zero for all edges of $E \backslash E(S)$. For once this is done, we can consider each $e \in E \backslash E(S)$ in turn. Since at most one end of e is in S and $G[S]$ is critical, there exists a matching of G which contains e and some maximum matching of $G[S]$. The incidence vector of this matching belongs to F_S, and since it is the only vector x we construct for which $x_e \neq 0$, it is affinely independent of the others.

By Theorem 4.3, $G[S]$ has a critical ear decomposition $G_0, G_1, \ldots, G_p$ such that each G_i is 2-connected. We prove by induction that each G_i has $|E(G_i)|$ matchings, each of cardinality $(|V(G_i)| - 1)/2$, whose incidence vectors are affinely independent, which will complete the proof. Since G_0 is an odd cycle, the $|E(G_0)|$ maximum matchings of G_0 suffice for the case $i = 0$. Now suppose the result is true for $i < k$, and consider G_k. Let π be the (odd length) ear added to G_{k-1} to form G_k and let $u, v \in V(G_{k-1})$ be the ends of the ear.

First, by induction there are $|E(G_{k-1})|$ independent maximum matchings of G_{k-1}. Each can be extended to a maximum matching of G_k by adding the 2nd, 4th, ... edges of the ear. Then we can obtain $|E(\pi)| - 1$ additional matchings by considering each $w \in V(\pi) \backslash \{u, v\}$ and constructing a matching of the ear which leaves only w and one of u, v, say u, unsaturated. Since G_{k-1} is critical, we can extend it to a maximum matching of G_k by adding a perfect matching of $G_{k-1} - v$. This is the only matching constructed so far which does not saturate w, so it is independent of all others. Finally, we get our last matching by starting with a matching consisting of the 1st, 3rd, ... edges of π. We combine this with a perfect matching of $G - v$ from which the edge incident with u has been removed. This is the only matching constructed so far, which does not contain $(|V(G_{k-1})| - 1)/2$ edges of G_{k-1}, so it is independent of the others. We therefore have $|E(G_k)|$ maximum matchings of G_k whose incidence vectors are independent as required. The result follows by induction. □

For another proof, of a more polyhedral nature, see Lovász and Plummer (1986). Cunningham and Marsh (1978) showed that Theorem 5.18 remains true even if we restrict ourselves to a minimal defining system for $M(G)$. That is, such a system is totally dual integral.

The Birkhoff–Von Neumann Theorem 2.9 also follows directly from Theorem 5.19. As we remarked, no bipartite graphs are critical, so we have the following.

Corollary 5.20. *If G is bipartite, then $M(G)$ is given by* (5.1) *and* (5.2).

In general, not all degree constraints (5.2) are necessary either. Let $N(v)$ denote the set of all nodes adjacent to v (not including v). If $|N(v)| = 1$, then the inequality $x(\delta(v)) \leq 1$ is implied by the constraints $x \geq 0$ and $x(\delta(w)) \leq 1$, where $N(v) = \{w\}$. If $|N(v)| = 2$, then the inequality $x(\delta(v)) \leq 1$ is implied by the constraints $x \geq 0$ and $x(E(N(v) \cup \{v\})) \leq 1$, and this latter constraint is essential if the two nodes of $N(v)$ are adjacent. Pulleyblank and Edmonds (1974) show that in all other cases, the constraints (5.2) are essential.

5.3. Dimension and the matching lattice

The matching polytope $M(G)$ is of full dimension, since all unit vectors plus the zero vector belong to it. However, it is a nontrivial problem to compute the dimension of $P(G)$, the perfect matching polytope. For a bipartite graph $G = (V, E)$, an upper bound is $|E| - |V| - 1$ and, using the ear decomposition theorem 4.1 it is easy to show that this is the correct answer, if G is matching covered.

If G is nonbipartite, then let $\mathcal{O}' = \{S \in \mathcal{O}$: every perfect matching of G contains $(|S| - 1)/2$ edges of $G[S]\}$. We say that a nested subfamily $\mathcal{S}$ of $\mathcal{O}'$ has the *odd cycle property* if for each $S \in \mathcal{S} \cup \{V\}$, wen we shrink all maximal members of $\mathcal{S}$ contained in S, the resulting graph is nonbipartite. Naddef (1982) showed that all maximal such nested $\mathcal{S}$ with the odd cycle property have the same cardinality $r(G)$ and proved the following.

Theorem 5.21. *Let $G = (V, E)$ be a matching covered nonbipartite graph. The dimension of $P(G)$ is $|E| - |V| - r(G)$.*

Edmonds et al. (1982) showed that the dimension was determined by the number of bricks in the brick decomposition, described in section 4.

Theorem 5.22. *Let $G = (V, E)$ be a matching covered nonbipartite graph. The dimension of $P(G)$ is $|E| - |V| + 1 - k$, where k is the number of bricks in the brick decomposition of G.*

Another algebraic structure related to the perfect matching polyhedron is the *matching lattice* of a graph, introduced by Seymour (1979). It consists of all integer linear combinations of the incidence vectors of perfect matchings. A *tight cut* in a matching covered graph $G = (V, E)$ is a cut $\delta(S)$ for $\emptyset \neq S \subset V$, $|S|$ odd, such that every perfect matching contains exactly one edge of $\delta(S)$. For each $v \in V$, $\delta(v)$ is a trivial tight cut, but there are often others for which $3 \leq |S| \leq |V| -$

3. Clearly, for every vector x in the matching lattice, there exists an integer k such that $x(\delta(S)) = k$ for every tight cut. In the case of bipartite graphs this characterizes the matching lattice, even when we restrict ourselves to the trivial tight cuts. However, for nonbipartite graphs the situation is more difficult.

For example, suppose that G is the Petersen graph, and we define x_j to be 0 for the edges j belonging to two disjoint pentagons and 1 elsewhere. Then x satisfies the above property but is not in the matching lattice. Lovász showed that in a sense this was the only exception. (Recall that the brick decomposition procedure of section 4 constructs a nested family $\mathcal{S}$ of odd cardinality subsets of V such that $\delta(S)$ is a tight cut for every $S \in \mathcal{S}$.)

Theorem 5.23 (Lovász 1987). *Let G be a matching covered graph, w an integral vector defined on the edges of G and $\mathcal{S}$ the family constructed by the brick decomposition. Then w is in the matching lattice of G if and only if*

(a) *$w(C)$ is the same value for every $C = \delta(v)$ for $v \in V$ or $C = \delta(S)$ for $S \in \mathcal{S}$;*

(b) *for every brick which is a Petersen graph and for every pentagon in that brick, the sum of the entries corresponding to the edges of G mapped onto this pentagon is even.*

This also enabled Lovàsz to describe a polynomial algorithm for constructing a GF(2)-basis of the incidence vectors of the perfect matchings of G, and so determine the GF(2)-rank of this set.

6. Variations and extensions

A remarkable feature of matching is its "self-refining" property. There are many combinatorial problems which appear more general and yet can be reduced to a matching problem. At the same time, several innocent appearing extensions of matchings turn out to be NP-complete. We discuss a variety of such problems in this section.

6.1. f-factors

Let $f = (f_v : v \in V)$ be a vector of positive integers. An *f-factor* is a set F of edges of G such that $|\delta(v) \cap F| = f_v$ for fall $v \in V$. If $f_v = 1$ for all $v \in V$, then it is simply a perfect matching, sometimes called a *1-factor*. If $f_v = 2$ for all v, then F is the edge set of a family of node disjoint simple cycles on G which contain every node, commonly called a *2-factor*.

Tutte (1954) showed that by node splitting, an f-factor problem could be transformed into a perfect matching problem. Each node v of G is replaced by a set $B(v)$ of f_v new nodes, such that the sets $B(v)$ are pairwise disjoint. Each edge j is replaced by two new adjacent nodes u_j and v_j such that, if u, v are the nodes incident with j, then u_j is adjacent to all nodes of $B(u)$ and v_j is adjacent to all nodes of $B(v)$. See fig. 6.1. There is a natural correspondence between matchings in the new graph and f-factors in G. If we wish to solve a weighted f-factor

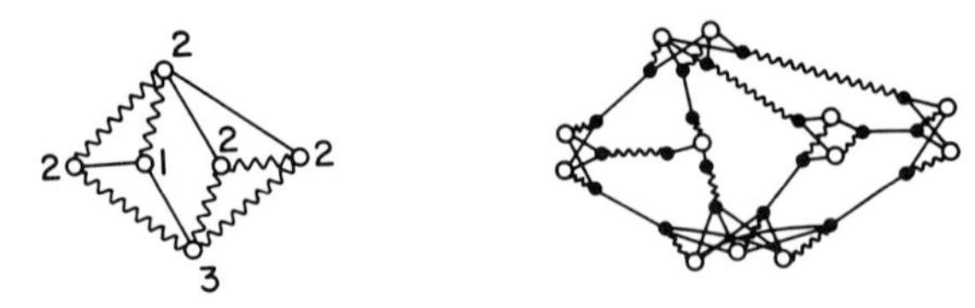

Figure 6.1. Transformation of f-factor problem to perfect matching problem.

problem, we define $c'_{uw} = c_{uw}$, for all $w \in B(u)$, for all $j \in E$, and $c_k = 0$ for all other new edges. (Here c is the original vector of edge costs.) The weights of the f-factors and the corresponding perfect matchings will agree.

Now we describe Tutte's f-factor theorem, which can be deduced, although not trivially, from the above construction. Let $G = (V, E)$ be a graph and let $f = (f_v: v \in V)$ be the vector of positive integral degree constraints. Let $S \subseteq V$ and let $T \subseteq V \backslash S$. For each $v \in T$, we let d_v^S denote the degree of v in $G - S$. We let $Q(S, T) = \sum (d_v^S - f_v: v \in T)$. Finally, we consider the graph $G - (S \cup T)$ and let $\mathscr{D}(S, T)$ denote the set of components of $G - (S \cup T)$ for which, if K is the nodeset, then $f(K) + |\delta(K) \cap \delta(T)|$ is odd. (Note that each such K, plus the edges joining K to T corresponds to an induced subgraph of G.)

Theorem 6.1 (Tutte 1952, see also Cook and Pulleyblank 1987 and Lovász and Plummer 1986). *The graph $G = (V, E)$ has an f-factor if and only if*

$$b(S) + Q(S, T) \geq |\mathscr{D}(S, T)| \quad \text{for all } S \subseteq V\,, \quad \text{for all } T \subseteq V \backslash S\,.$$

This result was further generalized by Lovász (1970). Instead of specifying a single "target" degree f_v for each node v, he permitted the specification of a set $F(v)$ of possible degrees. He defined a *gap* to be any maximal sequence of consecutive integers not in $F(v)$, but for which the lower and higher values are in $F(v)$. Then he generalized Theorem 6.1 to the case that no gap has length greater than one. This includes, for example, the f-factor problem and the case that a consecutive interval of degrees is permitted, for each $v \in V$. We will see later that if gaps of length two or more are permitted, then the problem of existence of an appropriate factor becomes NP-complete. Recently Cornuéjols (1988) developed a polynomially bounded extension of the blossom algorithm for this problem when no gap of size greater than one is permitted. However, the corresponding weighted problem has not yet been solved.

Edmonds and Johnson (1970) gave a linear system sufficient to define the convex hull of the incidence vectors of the f-factors. It is a special case of the more general Theorem 6.2 presented in the next section.

Suppose we have nonnegative integers $g_v \leq f_v$ defined for each node v. A *(g, f)-factor* is a set $F \subseteq E$ such that $g_v \leq |F \cap \delta(v)| \leq f_v$ for all $v \in V$. Hell and Kirkpatrick (1993) describe efficient methods for finding (g, f)-factors of minimum deficiency.

6.2. Capacitated and uncapacitated b-matchings

Another generalization of the f-factor problem is the capacitated perfect b-matching problem. In this case, each node v has a positive integral demand b_v and each edge j has capacity α_j. It is desired to assign an integer x_j satisfying $0 \leqslant x_j \leqslant \alpha_j$ to each edge j such that $x(\delta(v)) = b_v$ for all $v \in V$. (Alternatively, we may require $x(\delta(v)) \leqslant b_v$ for some nodes $v \in V$). The f-factor problem is just the special case that $\alpha_j = 1$ for all $j \in E$ and $f = b$. However, the b-matching problem can also be reduced to the f-factor problem by replacing each edge with α_j parallel edges and again letting $f = b$. Theorem 6.1 generalizes easily to this case; we define d_v^S to be the sum of the capacities of edges of $G - S$ incident with v and we let $\mathcal{D}(S, T)$ consists of those K for which $b(K) + \alpha(\delta(K) \cap \delta(T))$ is odd. Edmonds and Johnson described how to generalize the weighted matching algorithm to handle this problem and (see also Aráoz et al. 1983) showed that the matching polyhedron theorem can also be generalized to this case.

Theorem 6.2 (Edmonds and Johnson 1970). *The convex hull of the incidence vectors of the capacitated perfect b-matchings of G is the solution to the following system*:

$$0 \leqslant x_j \leqslant \alpha_j \quad \text{for all } j \in E \tag{6.3}$$

$$x(\delta(v)) = b_v \quad \text{for all } v \in V \tag{6.4}$$

$$x(E(S)) + x(J) \leqslant \frac{b(S) + \alpha(J) - 1}{2} \quad \text{for all } S \subseteq V \text{ and } J \subseteq \delta(S) \text{ such that } b(S) + \alpha(J) \text{ is odd}\,. \tag{6.5}$$

(It can also be shown that if we replace (6.4) with $x(\delta(v)) \leqslant b_v$, we get the convex hull of the "not necessarily perfect" capacitated b-matchings.)

If all capacities are infinite, we get a simpler generalization of the matching problem. We wish to assign a nonnegative integer x_j to each edge j of $G = (V, E)$ in such a way that $x(\delta(v)) = b_v$ for all $v \in V$, and cx is maximized, for a vector c of edge weights. This problem is called the maximum weight (uncapacitated) perfect b-matching problem. The characterization of when a perfect uncapacitated b-matching exists is much simpler than Theorem 6.1.

Theorem 6.6 (Tutte 1954). *A graph $G = (V, E)$ has a perfect uncapacitated b-matching if and only if*

$$b(S) \geqslant b(\theta_0(G - S)) + |\theta_1(G - S)| \quad \text{for all } S \subseteq V\,.$$

Here $\theta_0(G - S)$ is the set of nodes of G which are singleton components of $G - S$ and $\theta_1(G - S)$ is the set of nodesets K of components of $G - S$ which contain at least three nodes and for which $b(K)$ is odd.

Theorem 6.2 also has a simpler analogue in this case.

Theorem 6.7 (Edmonds and Johnson 1970, see also Pulleyblank 1973). *The convex hull of the uncapacitated perfect b-matchings of G is the solution to the following system*:

$$0 \leq x_j \quad \text{for all } j \in E ,$$

$$x(\delta(v)) = b_v \quad \text{for all } v \in V ,$$

$$x(E(S)) \leq \frac{b(S)-1}{2} \quad \text{for all } S \subseteq V \text{ such that } b(S) \text{ is odd.}$$

See Cook and Pulleyblank (1987) for a discussion of dual integrality results for b-matching polyhedra.

6.3. Bidirected matching problems

A still more general problem, which also reduces to this problem is the matching problem on a so-called bidirected graph. In this case, each edge has one or two "ends", each of which may be designated as a "head" or a "tail". Some edges, called *arcs*, have a head and a tail and others, called *links*, have two heads or two tails. A *slack* has one end, which may be either a head or a tail. For each node v, let $\delta^+(v)$ denote the set of edges having a tail incident with v and let $\delta^-(v)$ denote the set of edges having a head incident with v. A b-matching is a vector of nonnegative integers which satisfies

$$x(\delta^+(v)) - x(\delta^-(v)) = b_v \quad \text{for all } v \in V , \tag{6.8}$$

where b_v may be positive, negative or zero. A capacitated b-matching must also satisfy (6.3). Note that the presence of slacks which may have a head or a tail permits us to effectively replace the "=" in (6.8) with "$\leq$" or "$\geq$". Again we may ask whether a perfect (capacitated) b-matching exists, or ask for one which maximizes (or minimizes) $\sum_{j \in E} c_j x_j$, where c is a vector of edge weights.

A b-matching problem on a bidirected graph can be transformed to a b-matching problem on an undirected graph by node splitting. Basically, each node v is replaced with copies v^+ and v^-, and all edges having a head incident with v are made incident with v^-, and a tail with v^+. Then v^+ and v^- are joined with a new edge and we define $b_{v^-} = B_v$, $b_{v^+} = B_v + b_v$, where B_v is some value larger than the maximum possible values of $x(\delta^+(v))$, for all b-matchings. (If there are capacities, then B_v is trivially obtained. If not, it can be shown that setting $B_v = \sum (|b_i|: i \in V)$ for all $v \in V$ is sufficient.) Finally, a new node w^* is created, and all slacks are given a second end incident with w^*. We can define $b_{w^*} = \sum (|b_i|: i \in V)$, and if a perfect b-matching problem is desired, attach a loop to w^*. (See fig. 6.2.)

If capacities and/or weights are present they are transferred to the corresponding edges of the new graph, and all new edges are given zero cost and infinite capacity. (See Aráoz et al. 1983 or Lawler 1976 for more details.)

Theorem 6.9 (Edmonds and Johnson 1970, see Aráoz et al. 1983). *The convex*

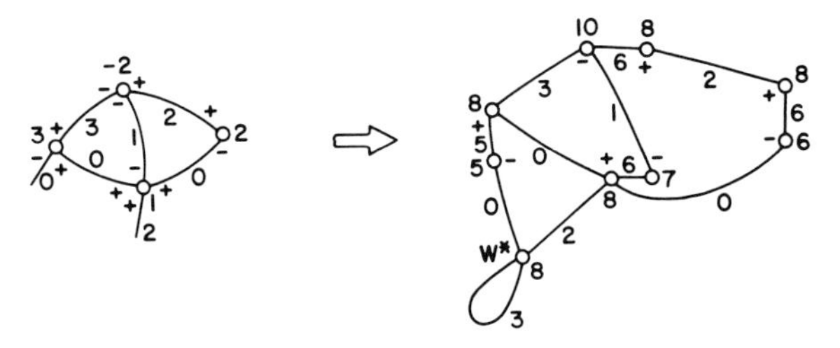

Figure 6.2. Transformation from bidirected to undirected.

hull of the b-matchings of a bidirected graph $G=(V,E)$ with integral capacities α, and integral degree constraints b is given by (6.3), (6.8) *and*

$$x(\delta(S)\backslash J) - x(J) \geqslant 1 - \alpha(J)\,, \quad \textit{for } S \subseteq V,\ J \subseteq \delta(S) \textit{ such that } b(S)+\alpha(J) \textit{ is odd}\,. \tag{6.10}$$

Note that the constraint (6.10) can be put in a form analogous to that of (6.5) by adding one half of the sum of the degree constraints (6.8). This theorem also permits edges to have infinite capacities, i.e., to be uncapacitated. In this case $\alpha_j = \infty$ and so such edges can never appear in sets J in the above inequalities.

6.4. *Edge covers*

A special case of the bidirected matching problem is the minimum edge cover problem. In this case we wish to select a minimum cardinality or weight subset of the edges of $G=(V,E)$ such that each node is incident with at least one. To reduce this to the bidirected b-matching problem we make each edge a link with two tails, and capacity 1, define $b_i = 1$ for all $i \in V$ and required "$\geqslant$" in (6.8).

Theorem 6.9 specializes to the following:

Theorem 6.11. *The convex hull of the incidence vectors of the edge covers of a graph $G=(V,E)$ is given by*

$$0 \leqslant x_j \leqslant 1 \quad \textit{for all } j \in E\,,$$

$$x(E(S)) + x(\delta(S)) \geqslant \frac{|S|+1}{2} \quad \textit{for all } S \subseteq V \textit{ such that } |S| \textit{ is odd}\,.$$

(It does require a simple argument to show that those constraints (6.10) with $J \neq \emptyset$ are redundant in this case.) See also Gamble and Pulleyblank (1989), Murty and Perin (1982) and White and Gillenson (1975).

6.5. *T-joins and the Chinese postman problem*

Let $T \subseteq V$ have even cardinality. A *T-join* is a set $F \subseteq E$ such that $|F \cap \delta(v)|$ is odd if $v \in T$ and even if $v \in V\backslash T$. If $|V|$ is even and $T = V$, then a minimum cardinality T-join is a perfect matching, if one exists. An application of T-joins is provided by the *Chinese postman problem*: Suppose we are given a graph $G=(V,E)$ and a vector c of edge weights. We wish to find a closed walk in G which passes through each edge at least once, and for which the sum of the edge costs is minimized. If G is Eulerian, then any Euler tour will suffice. Otherwise, we must find a minimum cost set of edges which when duplicated will yield an Eulerian graph. This is just a minimum cost T-join, in the case that we let T be the set of odd degree nodes of G.

Edmonds and Johnson (1973) proposed two different methods for solving the minimum cost T-join problem. In the first, which works when the sum of the costs is nonnegative for every cycle, a minimum cost path is computed between each pair (u,v) of nodes in T – let d_{uv} be the cost of this path. Then a minimum weight perfect matching is computed for the complete graph with nodeset T, where we let d_{uv} be the cost of the edge uv. The edge sets of the corresponding paths form a minimum weight T-join.

The second method reduces the problem to a single capacitated b-matching problem. For each $v\in V$, define b_v to be the largest odd integer less than or equal to $|\delta(v)|$ if $v\in T$, and the largest even integer less than or equal to $|\delta(v)|$ if $v\in V\setminus T$. Define the capacity α_j to be one for each edge j *of* E. Finally, add a loop l_v having both ends equal to v, for all $v\in V$. We let l_v have infinite capacity. (Purists may wish to subdivide each of these loops twice, introducing new nodes w where $b_w=\lfloor b_v/2\rfloor$, to remain simple.) There is a bijection between T-joins in G and perfect b-matchings in the expanded graph.

It is worth noting that by lowering b_v for some node v by some even amount, we effectively impose an upper bound on the degree of v in the T-join. By making the capacity of the loop l_v equal to k for some integer k, we are setting the minimum allowable degree of node v in the T-join to be b_v-2k. Note too that this reduction enables us to solve the minimum weight T-join problem for positive or negative edge weights. (This general case can also be reduced to the case of positive edge weights – see Lovász and Plummer 1986).

The shortest path problem in an undirected graph is a special case of the T-join problem. If we wish to find a shortest path joining nodes s, t, we define $T=\{s,t\}$ and construct a minimum weight T-join. Note that this will work even for negative weight edges. If there are no negative weight cycles, then the optimum solution will consist of a shortest s–t-path, plus, possibly, some zero weight cycles. If there are negative weight cycles, then at least one will show up in the minimum weight T-join. (Note that the simpler method of solving this problem – replace each edge with two parallel, oppositely directed arcs each having the same weight as the edge and use Dijkstra's directed graph algorithm – only works if all edge weights are nonnegative.)

A *T-cut* is a cut $\delta(S)$ in G, for $S\subseteq V$ such that $|S\cap T|$ is odd. Clearly $|F\cap\delta(S)|\geq 1$ for every T-join F, and T-cut $\delta(S)$ of G. A *double packing* of T-cuts is a family of T-cuts such that each edge is used at most twice.

Theorem 6.12 (Edmonds and Johnson 1973). *The minimum size of a T-join equals one half the maximum number of sets in a double packing of T-cuts.*

Suppose that each edge j of G has a nonnegative integral weight c_j. Replace j with a path of length c_j. (If $c_{uv} = 0$, then identify uv. The resulting node is in T if and only if both or neither of u, v were in T.) Then a minimum weight T-join in G corresponds to a minimum cardinality T-join in the new graph. Similarly, if we allow each edge of G to appear in at most $2c_j$ T-cuts, then a maximum such capacitated packing of T-cuts in G corresponds to a maximum double packing in the new graph. Thus we obtain the following.

Theorem 6.13 (Edmonds and Johnson 1973). *The convex hull of all vectors $x \in \mathbb{R}^E$ such that $x \geqslant u$, where u is the incidence vector of a T-join is given by the system*

$$x \geqslant 0\,, \tag{6.14}$$

$$x(J) \geqslant 1 \quad \textit{for every } T\text{-cut } J \textit{ of } G\,. \tag{6.15}$$

Moreover, if c is integral, then there exists an optimal solution to the dual linear program of minimizing cx subject to (6.14), (6.15) *for which each component is half-integer valued.*

Seymour (1981) showed that for bipartite graphs, a stronger result is true: The minimum size of a T-join is equal to the maximum number of T-cuts which can be packed. Moreover, in Theorem 6.13, we can find an integer optimal dual solution. Further, by subdividing edges, Seymour's extension can be seen to be valid whenever c is integral and has even sum around every cycle.

Seymour (1981) also showed that if we restrict the set of nodes which can comprise the set T, then we can obtain another integral result. We say that $G = (V, E)$ is *T-contractible* to K_4 if there exists a partition $V_1 \cup V_2 \cup V_3 \cup V_4$ of V such that, for each i, $G[V_i]$ is connected and $|T \cap V_i|$ is odd, and each pair of parts of the partition is joined by an edge.

Theorem 6.16. *If G is not T-contractible to K_4, then the minimum size of a T-join equals the maximum number of pairwise disjoint T-cuts.*

An easy corollary of this is that the integral result holds for series parallel graphs (graphs not contractible to K_4) for all choices of T. Gerards (1992) showed that the integral result also holds for all choices of T for a class of graphs which properly includes all bipartite graphs and all series-parallel graphs, namely those graphs which do not contain an odd subdivision of K_4. (An *odd subdivision* of a graph H is a subdivision G of H such that circuits in G coming from odd (even) circuits in H are odd (even) circuits in G.)

Sebő (1990) showed how the Gallai–Edmonds theorem can be extended to the case of T-joins. He also developed the connections between T-joins and so-called *conservative weightings* of graphs. A vector of edge weights is said to be conservative if the sum of the weights is nonnegative for the edgeset of every cycle. If we consider the special case that all weights are 1 or -1, then a weighting is conservative if and only if the edges assigned the value -1 form a minimum cardinality T-join for some set T.

Frank (1993) considered the problem of finding a conservative $(+1, -1)$-weighting for which the sum of the weights was minimized. This is equivalent to finding an even cardinality set T of nodes for which the size of a minimum cardinality T-join is maximized. He established the following surprising connection between this problem and the problem of finding an ear-decomposition of a 2-edge connected graph which now permits even length ears, but uses as few of them as possible. (It is well known that a graph has an ear-decomposition, using both even and odd length ears, if and only if it is 2-edge connected.)

Theorem 6.17 (Frank 1993). *The minimum of $w(E)$, taken over all conservative $(+1, -1)$ weightings of a graph $G = (V, E)$, equals the maximum number of odd length ears in an ear-decomposition of G.*

Let $G = (V, E)$ be a planar graph and let $G^D = (V^D, E^D)$ be the planar dual of G. Then $F \subseteq E$ is a cut in G (i.e., $F = \delta(S)$ for some $S \subseteq V$) if and only if the set F^D of corresponding edges of G^D decomposes into the union of circuits, i.e., has even degree at every node. The general problem of finding a minimum weight cut in a graph (when negative weights are permitted) is NP-hard, but this shows that for planar graphs it can be solved polynomially by solving a minimum weight T-join problem in the dual (Hadlock 1975). See Barahona (1990) for a comprehensive discussion of the connections between T-joins, the maximum cut problem and multicommodity flows.

6.6. Fractional matchings, 2-matchings and König–Egerváry graphs

One additional special case of b-matching is the uncapacitated 2-matching problem. In this case, each edge of G is to be assigned an integer $x_j \in \{0, 1, 2\}$ such that $x(\delta(v)) = 2$ (or $x(\delta(v)) \leq 2$) for all $v \in V$. In this case, it follows from Theorem 6.2 (take $\alpha_j = 2$ for all $j \in E$) that no constraints (6.5) exist, so any extreme solution $\hat{x}$ to $\{x \geq 0,\ x(\delta(v)) = 2$ for all $v \in V\}$ will be integer valued. (This is also easy to show directly.) Therefore the edges for which $\hat{x}_j = 1$ form pairwise disjoint circuits in G, and since $\hat{x}$ is extreme, each has an odd number of edges. Thus a perfect (uncapacitated) 2-matching in G corresponds precisely to a set of pairwise disjoint odd circuits and K_2's which contain every node. In this case Tutte's Theorem specializes to the following.

Theorem 6.18. *$G = (V, E)$ has a perfect 2-matching if and only if, for every $X \subseteq V$, $G - X$ has at most $|X|$ singleton components.*

We say that a graph $G = (V, E)$ is *2-bicritical* if $G - v$ has a perfect 2-matching for all $v \in V$. These graphs were introduced when studying a certain integer programming formulation of the maximum stable set problem, described in the next section. They can be characterized in terms of stable sets as follows.

Theorem 6.19 (Pulleyblank 1979). *For a graph* $G=(V,E)$ *the following are equivalent*:

(i) G *is* 2-*bicritical*;

(ii) $|I|<|N(I)|$ *for every stable set* I;

(iii) G *is nonbipartite, and for every* $j\in E$, *there is a perfect* 2-*matching* x *for which* $x_j>0$.

We let $N(I)$ denote the set of nodes not in I, but adjacent to one or more members of I.)

These graphs have ear decompositions similar to those described in section 4. See Bourjolly and Pulleyblank (1989).

Suppose we wish to study the stable sets of a graph from a polyhedral point of view, in a manner similar to the way we have considered matchings. For any graph $G=(V,E)$, we let $S(G)\subseteq\mathbb{R}^V$ denote the convex hull of the incidence vectors of the stable sets of G. We call $S(G)$ the *stable set polytope* of G. The incidence vector x of every stable set satisfies the following constraints:

$$x\geqslant 0\,, \tag{6.20}$$

$$x_u+x_v\leqslant 1 \quad \text{for every } uv\in E\,. \tag{6.21}$$

The following is easy to prove.

Proposition 6.22. $S(G)=\{x\in\mathbb{R}^V\colon x$ *satisfies* (6.20) *and* (6.21)$\}$ *if and only if* G *is bipartite*.

In general, a basic optimum solution $\hat{x}$ to the problem of maximizing cx, subject to (6.20) and (6.21) will have components receiving values 0, $\frac{1}{2}$ and 1. Nemhauser and Trotter (1974) showed that there always exists a stable set S in G, for which $c(S)$ is maximum and such that S contains all nodes v for which $\hat{x}_v=1$ and none for which $\hat{x}_v=0$. Pulleyblank (1979) showed that the 2-bicritical graphs are precisely those for which the unique optimum solution $\hat{x}$ to maximizing cx, subject to (6.20) and (6.21), has $\hat{x}_v=\frac{1}{2}$ for all $v\in V$. Further, almost all graphs (under the Erdős–Rényi model) are 2-bicritical. This was also shown to hold for a model of near regular random graphs by Grimmett and Pulleyblank (1985). See also Grimmett (1986).

Sometimes 2-matchings are studied as *fractional matchings* – vectors $x\in\mathbb{R}^E$ satisfying $x\geqslant 0$ and $x(\delta(v))=1$ for all $v\in V$. Since x is a fractional matching if and only if $2x$ is a 2-matching, these concepts are equivalent. See Balas (1981) and Uhry (1975) for the relationship between fractional matchings and matchings.

Problems involving 2-matchings are much easier than the corresponding problems for 1-matchings. In fact, by node-splitting, 2-matching problems can be transformed to 1-matching problems in bipartite graphs. It is also easy to modify the algorithms of section 5 to this case, see, e.g., Bourjolly and Pulleyblank (1989).

Recall that G was said to be a König–Egerváry graph if $\rho(G)=\nu(G)$. Such graphs were characterized by Deming (1979), Sterboul (1979) and Lovàsz (1983). See also Korach (1982). It is surprising though that they can be characterized in terms of 2-matchings. We say that a (not necessarily perfect) 2-matching x is maximum if $x(E)$ is maximum, over all 2-matchings. Let E^2 be the set of edges j of $G=(V, E)$ for which $x_j > 0$ in some maximum 2-matching.

Theorem 6.23 (Lovász and Plummer 1986). *G is a König–Egerváry graph if and only if* $G=(V, E^2)$ *is bipartite.*

For extensions of this, see Bourjolly and Pulleyblank (1989).

For further results concerning the structure of 2-matchings, see Mühlbacher et al. (1983) and Pulleyblank (1987).

Finally, note that the maximum 2-matching problem can be viewed as the problem of packing edges and odd circuits in a graph. Cornuéjols et al. (1982) and, independently, Hell and Kirkpatrick (1984) show that the problem of packing edges and any family of critical subgraphs can be solved polynomially. See also Cornuéjols and Pulleyblank (1983) and Cornuéjols and Hartvigsen (1986). Hell and Kirkpatrick (1983, 1984, 1986) show that in many other cases, the problem of determining whether G has a packing of edges plus subgraphs from a set $\mathscr{H}$ is NP-hard.

6.7. *Some* NP-*hard extensions*

We close this section by observing that many "natural" extensions of matching are NP-hard. One of the first problems to be shown to be NP-complete (Karp 1972) was the so-called 3-*dimensional matching problem*. Here we are given disjoint sets I, J, K such that $|I| = |J| = |K| = n$, and a set $\mathscr{J}$ of triples from $I \times J \times K$. We wish to determine whether there exists $\mathscr{J}' \subseteq \mathscr{J}$ such that each element of $I \cup J \cup K$ appears in exactly one triple.

This can be modelled as a matching problem, with one additional restriction. (See fig. 6.3.) Construct a node for each element of $I \cup J \cup K$; construct three nodes t_I, t_J, t_K for each $(t_I, t_J, t_K) \in \mathscr{J}$ and join each t_I to the node corresponding to the member of I, etc. Then there exists a 3-dimensional matching if and only if

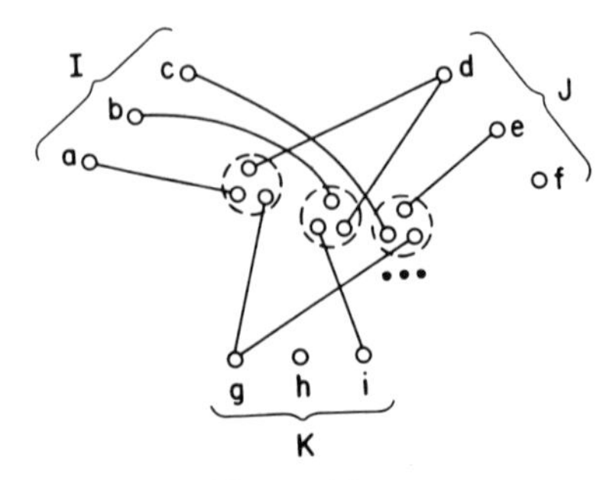

Figure 6.3.

the resulting graph has a matching which saturates all nodes corresponding to $I \cup J \cup K$, which also uses 0 or 3 edges of $\delta(\{t_I, t_J, t_K\})$ for each $(t_I, t_J, t_K) \in \mathcal{J}$. This construction shows that the following problems are NP-complete.

- Given $G = (V, E)$ and a set F_v of allowed degrees for each $v \in V$, does there exist $F \subseteq E$ such that $|F \cap \delta(v)| \in F_v$, for all $v \in V$? (Identify t_I, t_J, t_K to form a new node t and define $F_t = \{0, 3\}$, for all $(t_I, t_K, t_K \in \mathcal{J})$. (Recall that Cornuéjols 1988 showed that this could be solved polynomially if gaps of length two are not allowed.)
- Given (bipartite) $G = (V, E)$ and a partition of E, does G have a perfect matching which, for each set in the partition, uses all or no members?

Suppose we construct a bipartite graph with nodeset $I \cup J$ as follows. For each $(t_I, t_J, t_K) \in \mathcal{J}$, we join the corresponding nodes of I and J and "colour" the edge with colour t_K. Then a 3-dimensional matching corresponds to a perfect matching on this graph, which uses one edge of each colour. Thus the following is also NP-complete.

- Given (bipartite) $G = (V, E)$ and a colouring of E, does G contain a perfect matching with one edge of each colour?

7. Determinants, permanents and Pfaffians

Let $G = (V_1 \cup V_2, E)$ be a bipartite graph, with $|V_1| = |V_2|$. Construct the matrix $P(G) = (p_{uv}\colon u \in V_1, v \in V_2)$ as follows:

$$p_{uv} = \begin{cases} 1 & \text{if } uv \in E\,, \\ 0 & \text{if } uv \notin E\,. \end{cases}$$

The *permanent* of a matrix $A = (a_{ij}\colon 1 \leqslant i \leqslant n,\ 1 \leqslant j \leqslant n)$ is defined by

$$\text{per}(A) = \sum_{\pi \in \Pi_n} a_{1\pi(1)} a_{2\pi(2)} \cdots a_{n\pi(n)}\,,$$

where Π_n is the set of all permutations of $\{1, 2, \ldots, n\}$. Now consider per$(P(G))$. Each term in the summation will be 1 if all the factors p_{uv} in the term are 1, and zero otherwise. Therefore, the term is 1 if and only if the corresponding edges form a perfect matching in G. Hence per$(P(G)) \neq 0$ if and only if G has a perfect matching. Moreover, per$(P(G)) = \Phi(G)$, where $\Phi(G)$ denotes the number of perfect matchings of the graph G.

However, this fact does not help particularly in determining whether a graph has a perfect matching or in computing $\Phi(G)$, for computing per(A) is #P-complete (Valiant 1979), and, indeed, computing the number of perfect matchings of a bipartite graph is #P-complete. (This implies that the problem is at least as hard as any NP-complete problem. See chapter 29.) Nevertheless, upper and lower bounds obtained on the permanents of 0–1 matrices do yield important results bounding $\Phi(G)$ for the case of regular graphs. See Lovász and Plummer (1986, chapter 8) for details.

Recall that the determinant of matrix A is defined as

$$\det(A) = \sum_{\pi \in \Pi_n} \operatorname{sign}(\pi) a_{1\pi(1)} a_{2\pi(2)} \cdots a_{n\pi(n)} ,$$

where $\operatorname{sign}(\pi)$ is 1 if π is an even permutation and -1 if π is odd. We can compute $\det(A)$ efficiently and $|\det(A)| \leq |\operatorname{per}(A)|$. However, if cancellation occurs, it will be the case that $|\det(A)| < |\operatorname{per}(A)|$. If each nonzero entry in A is a distinct algebraic indeterminate, then no cancellation can occur, and $\det(A) \neq 0$ if and only if $\operatorname{per}(A) \neq 0$. But in this case, we do not know how to compute $\det(A)$ polynomially. Nevertheless, this observation provides the basis for so-called randomized algorithms, which we discuss later, in a more general setting.

Now we consider the case of general graphs. We require the following definition. Let $B = (b_{ij})$ be a skew symmetric matrix of dimension $p = 2n$. (That is, $b_{ij} = -b_{ji}$ for all i, j.) Let Λ be the family of all partitions $\{\{i_1, j_1\}, \{i_2, j_2\}, \ldots, \{i_n, j_n\}\}$ of the set $\{1, \ldots, 2n\}$ into pairs. For each member $\lambda = \{\{i_1, j_1\}, \{i_2, j_2\}, \ldots, \{i_n, j_n\}\}$ of Λ, define

$$b_\lambda = \operatorname{sign}((i_1, j_1, i_2, j_2, \ldots, i_n, j_n)) b_{i_1 j_1} b_{i_2 j_2} \cdots b_{i_n j_n} .$$

Note that since B is skew symmetric, the value of b_λ is independent of the order we write down the classes of the partition or the members of each class. The *Pfaffian* of B is defined by

$$\operatorname{pf}(B) = \sum_{\lambda \in \Lambda} b_\lambda .$$

We need the following identity from linear algebra.

Theorem 7.1. *If B is a skew symmetric matrix, then* $\det(B) = (\operatorname{pf}(B))^2$.

(If B is of odd order, then $\det(B) = 0$ and we take $\operatorname{pf}(B)$ to be 0.)

Let G be an arbitrary simple graph. Let $\vec{G}$ be any orientation of G, that is each edge uv of G is replaced with either arc (u, v) or (v, u). The *skew adjacency* matrix of $\vec{G}$ is defined as $A_S(\vec{G}) = (a_{uv}\colon u, v \in V)$ where

$$a_{uv} = \begin{cases} 1 & \text{if } (u, v) \in E(\vec{G}) , \\ -1 & \text{if } (v, u) \in E(\vec{G}) , \\ 0 & \text{otherwise} . \end{cases}$$

Now $\operatorname{pf}(A_S(\vec{G}))$ sums over each possible pairing of the nodes of G, and the corresponding term is nonzero if and only if the pairing corresponds to a perfect matching of G. Again, however, there is the possibility of cancellation, which can be handled in two ways. Let $A_S(\vec{G}, \boldsymbol{x})$ be obtained from $A_S(\vec{G})$ by replacing each 1 or -1 with a distinct real variable x or $-x$, respectively. Note that $\operatorname{pf}(A_S(\vec{G}, \boldsymbol{x}))$ is a polynomial of degree at most $|V|/2$, in at most $|V|^2$ variables. Therefore we obtain the following.

Theorem 7.2. *Let G be any graph and let $\vec{G}$ be any orientation of G. Then G has a perfect matching if and only if* $\mathrm{pf}(A_S(\vec{G}, \boldsymbol{x}))$ *is not identically zero.*

This criterion provided the basis for Tutte's original proof of Theorem 3.2.

Suppose that we could obtain and orientation $\vec{G}$ of G such that all terms in the expansion of $\mathrm{pf}(A_S(\vec{G}))$ had the same sign. Then no cancellation could occur, and $|\mathrm{pf}(A_S(\vec{G}))|$ would equal $\Phi(G)$. A graph is called *Pfaffian* if such an orientation, called a *Pfaffian orientation*, exists.

Kasteleyn's Theorem 7.3 (Kasteleyn 1963, 1967, see Lovász and Plummer 1986, chapter 8). *Every planar graph has a Pfaffian orientation. Such an orientation can be constructed in polynomial time.*

Proof (*Sketch*). Assume that G is matching covered. Let $\vec{G}$ be an orientation of G, let F be a perfect matching of G, let C be the edge set of an alternating circuit with respect to F and let $F' = F \triangle C$. Then the terms of $\mathrm{pf}(A_S(\vec{G}))$ corresponding to F and F' have the same sign if and only if C has an odd number of arcs oriented in each direction. So a sufficient condition of $\vec{G}$ to be a Pfaffian orientation is that every even circuit of $\vec{G}$ contain an odd number of arcs in each direction.

If G is planar, we can use an inductive construction to ensure that every face boundary has an odd number of arcs oriented in a clockwise direction. This implies that the orientation has the desired properties. □

Little (1974), see also Little (1973), showed that this could be extended to the case that G contains no subdivision of $K_{3,3}$. Barahona (1981) gave a method for computing $\Phi(G)$ for a toroidal graph, based on Theorem 7.3.

It is trivial to show that a given orientation is not Pfaffian – exhibit two perfect matchings whose symmetric difference has an even number of arcs in each direction. However, in spite of various characterizations known (see Lovász and Plummer 1986), none show that the problem is in NP, i.e., that there is a short way of verifying that a given orientation is Pfaffian. Lovász's procedure for constructing a GF(2)-basis of the perfect matchings does enable us to find an orientation $\vec{G}$ of G such that G is a Pfaffian graph if and only if $\vec{G}$ is a Pfaffian orientation. Recently Vazirani and Yannakakis (1989) showed that determining whether a given orientation is not Pfaffian is equivalent to determining whether a digraph has a (simple) dicircuit, containing an even number of arcs. This latter problem is a quite well-known problem whose complexity has not been determined. (See Klee et al. 1984 and Thomassen 1986.)

Theorem 7.3 also has interesting algorithmic consequences. First, if $\mathrm{pf}(A_S(\vec{G}, \boldsymbol{x}))$ is not identically zero, then there are real vectors $\boldsymbol{x}$ such that $\mathrm{pf}(A_S(G, \boldsymbol{x})) \neq 0$. Suppose we evaluated it, for a random value of $\boldsymbol{x}$. If the result was nonzero, we would know that G had a perfect matching. If the result was zero, then either G had no perfect matching, or we made an unlucky choice of $\boldsymbol{x}$. This is an example of a *randomized polynomial algorithm* (Lovász 1979) – an algorithm which runs in polynomial time, but may with "small" probability give a "NO" answer when the answer should be "YES". In this case, for example, if each component of $\boldsymbol{x}$ is chosen uniformly from $\{1, 2, \ldots, 20\}$, and $|E| = 100$, then the probability of accidental cancellation, i.e., an error, is less than 10^{-100}. (See Lovász and Plummer 1986.)

Moreover, this approach is well suited to parallel models of computation. (See chapter 29.) We can evaluate the determinants of $|V|/2 + 1$ matrices in parallel. (See Mulmuley et al. 1987 for a discussion of these issues.) This procedure only determines whether a perfect matching exists, it does not actually construct one. This latter problem could be solved by sequentially solving a perfect matching existence problem for each edge of G. If $G - e$ has a perfect matching delete e. If not, add e to the matching and delete all other edges incident with each end node. However, such a serial procedure is not acceptable when operating with a parallel model of computation. Mulmuley et al. (1987) show that this construction problem can be reduced to the problem of inverting a single matrix. They use a lemma which shows that if we independently assign to each edge of a graph a random weight between 1 and $2|E|$, then the probability is at least 0.5 that the minimum cost perfect matching is unique.

Suppose now we independently randomly orient the edges of a graph G so that each edge is equally likely to get either of the two possible orientations. Let $\vec{G}$ be the corresponding randomly oriented graph. If we were then to compute $|\mathrm{pf}(A_S(\vec{G}))|$, we would expect there to be large amounts of cancellation and so the value would be far from $\Phi(G)$. However, somewhat surprisingly, the expected value of the *determinant* of this matrix has magnitude exactly $\Phi(G)$.

Theorem 7.4 (Lovász and Plummer 1986). *Let $\vec{G}$ be a random orientation of a graph G, obtained by orienting each edge independently, with equal probability, in either direction. Then the expected value of $|\det(A_S(\vec{G}))|$ is $\Phi(G)$.*

It can also be shown that the variance of $|\det(A_S(\vec{G}))|$ is $\sum 2^{a(M_1, M_2)}$ where the sum is over all pairs M_1, M_2 of distinct perfect matchings of G and $a(M_1, M_2)$ is the number of (alternating) cycles contained in the symmetric difference of M_1 and M_2. If the variance divided by $\Phi^2(G)$ is a polynomial in the size of G, then we can efficiently estimate $\Phi(G)$ by repeatedly computing $|\det(A_S(\vec{G}))|$, for random orientations of G. This is true, for example, for complete graphs, but at present few more interesting classes are known.

The last topic of this section is the *exact matching problem*. We are given a graph $G = (V, E)$, an integer k and a set $R \subseteq E$ of "red" edges. We wish to determine whether G has a perfect matching with exactly k red edges. For general graphs, and even for bipartite graphs, the complexity of this problem is not settled. However Barahona and Pulleyblank (1987) show that it can be solved polynomially for Pfaffian graphs using the methods of this section. Replace the entries of the skew adjacency matrix $A_S(\vec{G})$ of a Pfaffian orientation which correspond to red edges with $\pm x$. Then the Pfaffian is a polynomial of degree at most $|V|/2$ in x, and the coefficient of x^k is just the number of perfect matchings with k red edges. This polynomial can be obtained by evaluating the Pfaffian for $|V|/2+1$ distinct values of x, and solving the resulting system of linear equations.

This relationship also enables the construction of a randomized algorithm for the exact matching problem. See Mulmuley et al. (1987).

8. Stable sets and claw free graphs

The problems of computing $\alpha(G)$, or of constructing a maximum cardinality stable set or a minimum cardinality node cover are NP-hard for general graphs (Karp 1972), however due to König's Theorem 2.1 and the Gallai identities (1.1) and (1.3), these can be solved efficiently for bipartite graphs.

Theorem 8.1. *Let $G=(V,E)$ be a bipartite graph with no isolated nodes. Then $\alpha(G)=\rho(G)$, i.e., the maximum cardinality of a stable set equals the minimum cardinality of an edge cover of the nodes.*

Moreover, in the bipartite case, the weighted stable set problem, wherein we have a vector $(c_v: v \in V)$ of node costs and require a stable set S for which $c(S)$ is maximized, is the dual of a minimum uncapacitated b-matching problem in a bipartite graph. Hence this can also be solved polynomially via matching theory.

For the rest of this section we focus on nonbipartite graphs. First, we note that an analogue of the augmenting path theorem 1.6 does hold for stable sets, but it is more complicated. For any stable set $S \subseteq V$, an *alternating sequence* relative to S is a sequence $\sigma=(t_1, s_1, t_2, s_2, \ldots)$ of distinct vertices alternately belong to $\bar{S}=V \backslash S$ and S such that

(i) $s_i \in S \backslash \{s_1, s_2, \ldots, s_{i-1}\}$ and $N(s_i) \cap \{t_1, \ldots, t_i\} \neq \emptyset$;
(ii) $t_{i+1} \in \bar{S} \backslash \{t_1, t_2, \ldots, t_i\}$ and $N(t_{i+1}) \cap \{s_1, \ldots, s_i\} \neq \emptyset$, $N(t_{i+1}) \cap \{t_1, \ldots, t_i\} = \emptyset$.

It is said to be *maximal*, if no vertices can be added without violating (i) or (ii). Notice that it has a tree like structure, rather than being a path. If we let $S(\sigma)$ and $\bar{S}(\sigma)$ denote the nodes of S and $\bar{S}$ respectively in σ, then $\bar{S}(\sigma)$ is a stable set, and if σ is maximal, then $S \cup \bar{S}(\sigma) \backslash S(\sigma)$ is also a stable set. If the length of σ is odd, then it will be larger than S.

Theorem 8.2 (Edmonds 1962, see also Berge 1985). *A stable set S is maximum if and only if there is no maximal alternating sequence having odd length.*

However, whereas alternating trees enable us to find augmenting paths in the case of matchings, no general method is known for finding these odd length maximal alternating sequences.

In one case, however, the situation simplifies. We say that G is *claw-free* if no induced subgraph is isomorphic to $K_{1,3}$. If σ is an odd maximal alternating

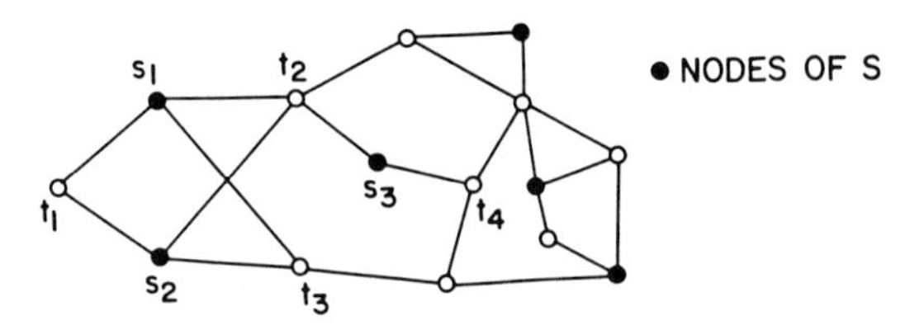

Figure 8.1. An odd, maximal, alternating sequence.

sequence in a claw-free graph G, then no node of $S(\sigma)$ can be adjacent to more than two nodes of $\bar{S}(\sigma)$, without creating a claw, and no node of $\bar{S}(\sigma)$ can be adjacent to more than two nodes of S. Therefore, every node of σ is adjacent to at most two other nodes of σ, and so σ is the node sequence of a path with an odd number of nodes or a circuit. But the circuit would have to have an even number of nodes, contradictory to the sequence being of odd length. Thus we obtain the following.

Theorem 8.3 (Minty 1980, Sbihi 1980). *Let S be a stable set in a claw-free graph $G = (V, E)$. Then S is maximum if and only if there is no induced simple path π with node sequence $t_1, s_1, t_2, s_2, \ldots, t_{n-1}, s_n, t_n$ such that each $s_i \in S$, each $t_i \in \bar{S}$, and no node t_i is adjacent to any node of S not in π.*

Thus the maximum stable set problem for a claw-free graph does appear much closer to matching than the general problem. As mentioned in section 1, it is a generalization of the maximum matching problem, since line graphs are a special class of claw-free graphs and the maximum matching problem in a graph is just the maximum stable set problem in its line graph.

Sbihi (1980) and Minty (1980) obtained quite different polynomially bounded algorithms to find a maximum stable set in a claw-free graph. The former directly generalized the blossom algorithm described in section 3. The latter reduced the problem of finding an alternating sequence to $O(|V|^2)$ problems of determining whether an augmenting path (with respect to a given matching) occurs in an auxiliary graph.

Recently, Lovász and Plummer (1986) showed that by means of a sequence of local replacements, a claw-free graph G could be transformed to a line graph G' in such a way that $\alpha(G)$ was reconstructible from $\alpha(G')$. Their method proceeds as follows. First, a node v of G is called *regular* if $N(v)$ can be partitioned into two complete subgraphs, and *irregular* otherwise. (All nodes in line graphs are regular.) We let $N_2(v)$ denote the set of all nodes of G at distance two from v, for all $v \in V$.

Lemma 8.4. *Let G be a claw-free graph and v an irregular node. Let Y be the set of nodes $y \in N_2(v)$ for which $N(v) \backslash N(y)$ induces a complete subgraph of G. Let G' be the graph obtained from G by deleting $\{v\} \cup N(v) \cup Y$ and then joining each* *nonadjacent pair of nodes of $N_2(v)\setminus Y$ with a new edge. Then G' is claw-free and $\alpha(G') = \alpha(G) - 2$.*

A clique Q of G is defined to be *reducible* if $\alpha(N(Q)) \leq 2$.

Lemma 8.5. *Let Q be any reducible clique in a claw-free graph G. Let G' be the graph obtained from G by deleting the nodes of Q and joining any nonadjacent nodes u and v of $N(Q)$ by an edge if and only if $Q \subseteq N(u) \cup N(v)$. Then G' is claw-free and $\alpha(G') = \alpha(G) - 1$.*

They next prove that if a graph contains neither irregular nodes nor reducible cliques, then it is a line-graph.

Theorem 8.6. *Let G be a graph such that every node v of G is contained in two irreducible cliques, which cover all members of $N(v)$. Then G is a line graph.*

They can now compute $\alpha(G)$ for a claw-free graph G as follows: Select any node v and see if it is regular. (This simply involves checking whether $N(v)$ induces a bipartite graph in the complement of G.) If v is irregular, we reduce G as described in Lemma 8.4. If v is regular, that is, $N(v)$ can be partitioned into complete subgraphs T_1 and T_2 of G, then we extend each T_i to a clique Q_i by adding node v. If either Q_1 or Q_2 is reducible, we apply Lemma 8.5. If not, we proceed to a new node v.

When this procedure terminates, by Theorem 8.6 we left with a graph G which is the line graph of a graph H. We use the maximum cardinality matching algorithm to compute $\nu(H) = \alpha(G)$.

Minty (1980) also solved the problem of finding a maximum weight stable set in a claw-free graph. However, his method did not lead to a polyhedral description of the convex hull of the incidence vectors of the stable sets. Indeed, the determination of such a system remains one of the more vexing open problems of polyhedral combinatorics. (To this date, no one has been successful in generalizing either the Sbihi or Lovász–Plummer method to the weighted case.)

Acknowledgement

I am grateful to Bruce Gamble, Bert Gerards, Lex Schrijver and the editors for comments and suggestions concerning a first draft of this chapter.

References

Anderson, I.
[1971] Perfect matchings of a graph, *J. Combin. Theory B* **10**, 183–186.
Aráoz, J., and J. Edmonds
[1985] A case of non-convergent pivoting in assignment problems, *Discrete Appl. Math.* **11**, 95–102.

Aráoz, J., W.H. Cunningham, J. Edmonds and J. Green-Krotki
[1983] Reductions to 1-matching polyhedra, *Networks* **13**, 455–473.
Balas, E.
[1981] Integer and fractional matchings, in: *Studies on Graphs and Discrete Programming*, ed. P. Hansen, *Ann. Discrete Math.* **11**, 1–13.
Barahona, F.
[1981] *Balancing signed toroidal graphs in polynomial time* (Depto. de Matemáticas, Universidad de Chile).
[1990] Planar multicommodity flows, max cut and the Chinese postman problem, in: *Polyhedral Combinatorics*, eds. W. Cook and P. Seymour, *DIMACS* **1**, 189–202.
Barahona, F., and W.R. Pulleyblank
[1987] Exact arborescences, matchings and cycles, *Discrete Appl. Math.* **16**, 91–99.
Berge, C.
[1957] Two theorems in graph theory, *Proc. Nat. Acad. Sci. U.S.A.* **43**, 842–844.
[1958] Sur le couplage maximum d'un graphe, *C.R. Acad. Sci. Paris Sér. I Math.* **247**, 258–259.
[1985] *Graphs*, 2nd revised Ed. (Elsevier, Amsterdam).
Biggs, N.L., E.K. Lloyd and R.J. Wilson
[1976] *Graph Theory 1736–1936* (Clarendon Press, Oxford).
Birkhoff, G.
[1946] Tres observaciones sobre el algebra lineal, *Rev. Fac. Ci. Exactas, Puras y Aplicadas Univ. Nac. Tucuman, Ser. A* **5**, 147–151.
Bourjolly, J.-M., and W.R. Pulleyblank
[1989] König–Egerváry graphs, 2-bicritical graphs and fractional matchings, *Discrete Appl. Math.* **24**, 63–82.
Burkard, R., and U. Derigs
[1980] *Assignment and Matching Problems: Solution Methods with FORTRAN Programs, Lecture Notes in Economics and Mathematical Systems*, Vol. 184 (Springer, Berlin).
Cook, W., and W.R. Pulleyblank
[1987] Linear systems for constrained matching problems, *Math. Oper. Res.* **12**, 97–120.
Cornuéjols, G.
[1988] General factors of graphs, *J. Combin. Theory B* **45**, 185–198.
Cornuéjols, G., and D. Hartvigsen
[1986] An extension of matching theory, *J. Combin. Theory B* **40**, 285–296.
Cornuéjols, G., and W.R. Pulleyblank
[1983] Critical graphs, matchings and tours or a hierarchy of relaxations for the travelling salesman problem, *Combinatorica* **3**, 35–52.
Cornuéjols, G., D. Hartvigsen and W.R. Pulleyblank
[1982] Packing subgraphs in a graph, *O.R. Lett.* **1**, 139–142.
Cunningham, W.H., and A.B. Marsh III
[1978] A primal algorithm for optimum matching, in: *Polyhedral Combinatorics (dedicated to the memory of D.R. Fulkerson)*, eds. M.L. Balinski and A.J. Hoffman, *Math. Programming Study* **8**.
Deming, R.W.
[1979] Independence numbers of graphs – an extension of the König–Egerváry theorem, *Discrete Math.* **27**, 23–33.
Derigs, U.
[1988] *Programming in Networks and Graphs, Lecture Notes in Economics and Mathematical Systems*, Vol. 300 (Springer, Berlin).
Edmonds, J.
[1962] Covers and packings in a family of sets, *Bull. Amer. Math. Soc.* **68**, 494–499.
[1965b] Maximum matching and a polyhedron with (0, 1) vertices, *J. Res. Nat. Bur. Standards Sect. B* **8**, 125–130.
Edmonds, J., and E.L. Johnson
[1970] Matching: a well-solved class of integer linear programs, in: *Proc. Calgary Int. Conf. on*

Combinatorial Structures and their Applications, eds. R.K. Guy et al. (Gordon and Breach, New York) pp. 89–92.
[1973] Matching, Euler tours and the Chinese postman, *Math. Programming* **5**, 88–124.

Edmonds, J., L. Lovász and W.R. Pulleyblank
[1982] Brick decompositions and the matching rank of graphs, *Combinatorica* **2**, 247–274.

Edmonds, J.R.
[1965a] Paths, trees, and flowers, *Canad. J. Math.* **17**, 449–467.
[1967] Systems of distinct representatives and linear algebra, *J. Res. Nat. Bur. Standards Sect. B* **71**, 241–245.

Edmonds, J.R., and D.R. Fulkerson
[1965] Transversals and matroid partitions, *J. Res. Nat. Bur. Standards B* **69**, 147–153.

Egerváry, E.
[1931] On combinatorial properties of matrices, *Mat. Lapok* **38**, 16–28.

Even, S., and O. Kariv
[1975] An $O(n^{5/2})$ algorithm for maximum matching in general graphs, in: *16th Annu. Symp. on Foundations of Computer Science (Berkeley)* (IEEE Computer Society Press, New York) pp. 100–112.

Frank, A.
[1993] Conservative weightings and ear-decompositions of graphs, *Combinatorica* **13**, 65–81.

Frobenius, G.
[1917] Über zerlegbare Determinanten, *Sitzungsber. König. Preuss. Akad. Wiss.* **XVIII**, 274–277.

Galil, Z.
[1986] Efficient algorithms for finding maximum matchings in graphs, *Comput. Surv.* **18**, 23–38.

Gallai, T.
[1959] Über extreme Punkt- und Kantenmengen, *Ann. Univ. Sci. Budapest Eötvös Sect. Math.* **2**, 133–138.
[1963] Kritische Graphen II, *Magyar. Tud. Akad. Mat. Kutató Int. Közl.* **8**, 373–395.
[1964] Maximale Systeme unabhängiger Kanten, *Magyar. Tud. Akad. Mat. Kutató Int. Közl.* **4**, 401–413.

Gamble, A.B., and W.R. Pulleyblank
[1989] Forest covers and a polyhedral intersection theorem, *Math. Programming* **45**, 49–58.

Gerards, A.M.H.
[1992] On shortest T-joins and packing T-cuts, *J. Combin. Theory B* **55**, 73–82.

Gomory, R.E., and T.C. Hu
[1961] Multi-terminal network flows, *SIAM J.* **9**, 551–570.

Grimmett, G.
[1986] An exact threshold theorem for random graphs and the nodepacking problem, *J. Combin. Theory B* **40**, 187–195.

Grimmett, G., and W.R. Pulleyblank
[1985] Random near-regular graphs and the node packing problem, *Oper. Res. Lett.* **4**, 169–174.

Grötschel, M., and O. Holland
[1986] Solving matching problems with linear programming, *Math. Programming* **33**, 243–259.

Hadlock, F.O.
[1975] Finding a maximum cut of a planar graph in polynomial time, *SIAM J. Comput.* **4**, 221–225.

Hall, P.
[1935] On representatives of subsets, *J. London Math. Soc.* **10**, 26–30.

Hell, P., and D. Kirkpatrick
[1983] On the complexity of general graph factor problems, *SIAM J. Comput.* **12**, 601–609.
[1984] Packing by cliques and by finite families of graphs, *Discrete Math.* **49**, 118–133.
[1986] Packing by complete bipartite graphs, *SIAM J. Alg. Discrete Methods* **7**, 199–209.
[1993] Algorithms for degree constrained graph factors of minimum deficiency, *J. Algorithms* **14**, 115–138.

Hetyei, G.
[1964] Rectangular configurations which can be covered by 2×1 rectangles, *Pécsi Tan. Föisk. Közl.* **8**, 351–367 (in Hungarian).

Hopcroft, J.E., and R.M. Karp
[1973] An $n^{5/2}$ algorithm for maximum matchings in bipartite graphs, *SIAM J. Comput.* **2**, 225–231.

Karp, R.M.
[1972] Reducibility among combinatorial problems, in: *Complexity of Computer Computations*, eds. R.E. Miller and J.W. Thatcher (Plenum Press, New York) pp. 85–102.

Kasteleyn, P.W.
[1963] Dimer statistics and phase transitions, *J. Math. Phys.* **4**, 287–293.
[1967] Graph theory and crystal physics, in: *Graph Theory and Theoretical Physics*, ed. F. Harary (Academic Press, New York) pp. 43–110.

Klee, V., R. Ladner and R. Manber
[1984] Sign solvability revisited, *Linear Algebra Appl.* **59**, 131–158.

König, D.
[1916a] Graphok és alkalmazśuk a determinánsok és halmazok elmlétŕe, *Math. Termész. Ert.* **34**, 104–119.
[1916b] Über Graphen und ihre Andwendung auf Determinantentheorie und Mengenlehre, *Math. Ann.* **77**, 453–465.

Korach, E.
[1982] On dual integrality, min–max equalities and algorithms in combinatorial programming, Ph.D. Thesis (University of Waterloo, Dept. of Combinatorics and Optimization).

Kotzig, A.
[1959a] On the theory of finite graphs with a linear factor II, *Mat.-Fyz. Casopis Slovensk. Akad. Vied* **9**, 136–159 (in Slovak).
[1959b] On the theory of finite graphs with a linear factor I, *Mat.-Fyz. Casopis Slovensk. Akad. Vied* **9**, 73–91 (in Slovak).
[1960] On the theory of finite graphs with a linear factor III, *Mat.-Fyz. Casopis Slovensk. Akad. Vied* **10**, 205–215 (in Slovak).

Lawler, E.L.
[1976] *Combinatorial Optimization: Networks and Matroids* (Holt, Rinehart & Winston, New York).

Little, C.H.C.
[1973] Kasteleyn's theorem and arbitrary graphs, *Canad. J. Math.* **25**, 758–764.
[1974] An extension of Kasteleyn's method of enumerating the 1-factors of planar graphs, in: *Combinatorial Mathematics, Proc. Second Australian Conference, Lecture Notes in Mathematics*, Vol. 403, ed. D. Holton (Springer, Berlin) pp. 63–72.

Little, C.H.C., and F. Rendl
[1989] An algorithm for the ear decomposition of a 1-factor covered graph, *J. Austral. Math. Soc. A* **46**, 296–301.

Lovász, L.
[1970] Generalized factors of graphs, in: *Combinatorial Theory and its Applications II*, eds. P. Erdős, A. Rényi and V.T. Sós, *Colloq. Math. Soc. János Bolyai* **4**, 773–781.
[1972a] On the structure of factorizable graphs, *Acta Math. Acad. Sci. Hung.* **23**, 179–195.
[1972b] A note on factor-critical graphs, *Studia Sci. Math. Hungar.* **7**, 279–280.
[1979] On determinants, matchings and random algorithms, in: *Fundamentals of Computation Theory, FCT '79, Proc. Conf. Algebraic, Arithmetic and Categorical Methods in Computation Theory*, Berlin/Wendisch-Reitz, 1979, ed. L. Budach, *Math. Res.* **2**, 565–574.
[1983] Ear-decompositions of matching-covered graphs, *Combinatorica* **2**, 395–407.
[1987] Matching structure and the matching lattice, *J. Combin. Theory B* **43**, 187–222.

Lovász, L., and M.D. Plummer
[1975] On bicritical graphs, in: *Infinite and Finite Sets (Colloq. Keszthely, Hungary, 1973)*, Vol. II, eds. A. Hajnal, R. Rado and V.T. Sós, *Colloq. Math. Soc. János Bolyai* **10**, 1051–1079.
[1986] *Matching Theory, Ann. Discrete Math.* **29**.

Menger, K.
[1927] Zur allgemeinen Kurventheorie, *Fund. Math.* **10**, 96–115.

Minty, G.J.
[1980] On maximal independent sets of vertices in claw-free graphs, *J. Combin. Theory B* **28**, 284–304.

Mühlbacher, J., F. Steinparz and G. Tinhofer
[1983] *On certain Classes of Fractional Matchings,* Preprint.
Mulmuley, K., U.V. Vazirani and V.V. Vazirani
[1987] Matching is as easy as matrix inversion, *Combinatorica* **7**, 105–113.
Murty, K.G., and C. Perin
[1982] A 1-matching blossom-type algorithm for edge covering problems, *Networks* **12**, 379–391.
Naddef, D.
[1982] Rank of maximum matchings in a graph, *Math. Programming* **22**, 52–70.
Nemhauser, G.L., and L.E. Trotter Jr
[1974] Properties of vertex packing and independence system polyhedra, *Math. Programming* **6**, 48–61.
Nemhauser, G.L., and L.A. Wolsey
[1988] *Integer and Combinatorial Optimization* (Wiley, New York).
Padberg, M.W., and M.R. Rao
[1982] Odd minimum cut-sets and b-matchings, *Math. Oper. Res.* **7**, 67–80.
Papadimitriou, C.H., and K. Steiglitz
[1982] *Combinatorial Optimization: Algorithms and Complexity* (Prentice-Hall, New York).
Petersen, J.
[1891] Die Theorie der regulären Graphen, *Acta Math.* **15**, 193–220.
Pulleyblank, W.R.
[1973] *Faces of matching polyhedra,* Ph.D. Thesis (University of Waterloo).
[1979] Minimum node covers and 2-bicritical graphs, *Math. Programming* **17**, 91–103.
[1987] Fractional matchings and the Edmonds–Gallai Theorem, *Discrete Appl. Math.* **16**, 51–58.
Pulleyblank, W.R., and J. Edmonds
[1974] Facets of 1-matching polyhedra, in: *Hypergraph Seminar, Lecture Notes in Mathematics,* Vol. 411, eds. C. Berge and D. Ray-Chaudhuri (Springer, Berlin) pp. 214–242.
Sbihi, N.
[1980] Algorithme de recherche d'un stable de cardinalité maximum dans un graphe sans étoile, *Discrete Math.* **29**, 53–76.
Sebő, A.
[1990] Undirected distances and the postman-structure of graphs, *J. Combin. Theory B* **49**, 10–39.
Seymour, P.D.
[1979] On multicolourings of cubic graphs, and conjectures of Fulkerson and Tutte, *Proc. London Math. Soc. Ser.* (3) **38**, 423–460.
[1981] On odd cuts and plane multicommodity flows, *Proc. London Math. Soc. Ser.* (3) **42**, 178–192.
Sterboul, F.
[1979] A characterization of the graphs in which the transversal number equals the matching number, *J. Combin. Theory B* **27**, 228–229.
Thomassen, C.
[1986] Sign-singular matrices and even cycles in directed graphs, *Linear Algebra Appl.* **75**, 27–41.
Tutte, W.T.
[1946] On Hamiltonian circuits, *J. London Math. Soc.* **21**, 98–101.
[1947] The factorisation of linear graphs, *J. London Math. Soc.* **22**, 107–111.
[1952] The factors of graphs, *Canad. J. Math.* **4**, 314–328.
[1954] A short proof of the factor theorem for finite graphs, *Canad. J. Math.* **6**, 347–352.
Uhry, J.-P.
[1975] Sur le problème du couplage maximal, *R.A.I.R.O. Rech. Opér.* **9**, 13–20.
Valiant, L.G.
[1979] The complexity of computing the permanent, *Theor. Comput. Sci.* **8**, 189–201.
Vazirani, V.
[1994] A theory of alternating paths and blossoms for proving correctness of the $O(\sqrt{v}E)$ general graph maximum matching algorithm, *Combinatorica* **14**, 71–109.

Vazirani, V., and M. Yannakakis
[1989] Pfaffian orientations, 0–1 permanents and even cycles in directed graphs, *Discrete Appl. Math.* **25**, 179–190.

Von Neumann, J.
[1953] A certain zero-sum two-person game equivalent to the optimal assignment problem, in: *Contributions to the Theory of Games,* Vol. II, ed. H.W. Kuhn, *Ann. of Math. Stud.* **28**, 5–12.

White, L.J., and M.L. Gillenson
[1975] An efficient algorithm for minimum k-covers in weighted graphs, *Math. Programming* **8**, 20–42.

CHAPTER 4

Colouring, Stable Sets and Perfect Graphs

Bjarne TOFT

Mathematics and Computer Science Department, Odense University, DK-5230 Odense, Denmark

Contents

HANDBOOK OF COMBINATORICS
Edited by R. Graham, M. Grötschel and L. Lovász

List of books and surveys

APPEL K. and W. HAKEN
[1989] *Every Planar Map is Four Colorable*, Contemporary Math. Series, Vol. 98 (American Mathematical Society, Providence, RI).

BERGE C. and V. CHVÁTAL (eds)
[1984] *Topics on Perfect Graphs*, *Annals of Discrete Math. 21* (North-Holland, Amsterdam).

BOLLOBÁS B.
[1978] *Extremal Graph Theory* (Academic Press, New York).

De WERRA D. and A. HERTZ (eds)
[1989] *Graph Colouring and Variations*, *Discrete Math. 74* and *Annals of Discrete Mathematics 39* (North-Holland, Amsterdam).

FIORINI S. and R.J. WILSON
[1977] *Edge-colourings of Graphs*, Research Notes in Mathematics 16 (Pitman, London).

GOLUMBIC M.C.
[1980] *Algorithmic Graph Theory and Perfect Graphs* (Academic Press, New York).

GRÖTSCHEL M., L. LOVÁSZ and A. SCHRIJVER
[1988] *Geometric Algorithms and Combinatorial Optimization* (Springer, Berlin).

GYÁRFÁS A.
[1987] Problems from the world surrounding perfect graphs, in: *Proceedings of the Int. Conf. on Combinatorial Analysis and its Applications*, Poland, Sept. 1985, ed. M.J. Syslo, *Zastosowania Matematyki (Applicationes Mathematicae) 19*, 413–441.

JENSEN T.R. and B. TOFT
[1995] *Graph Coloring Problems* (Wiley, New York).

KUBALE M.
[1991] Graph colouring, in: *Encyclopedia of Microcomputers*, Vol. 8, eds A. Kent and J.G. Williams (Marcel Dekker, New York), pp. 47–69.

LOVÁSZ L.
[1983] Perfect Graphs, in: *Selected Topics in Graph Theory*, Vol. 2, eds R.L. Wilson and L.W. Beineke (Academic Press, New York) pp. 55–87.

NELSON R. and R.J. WILSON (eds)
[1990] *Graph Colourings*, Pitman Research Notes in Mathematics Series, Vol. 218 (Longman, Harlow, UK).

READ R.C. and W.T. TUTTE
[1988] Chromatic polynomials, in: *Selected Topics in Graph Theory,* Vol. 3, eds R.L. Wilson and L.W. Beineke (Academic Press, New York) pp. 15–42.

SAATY T. and P. KAINEN
[1977] *The Four Color Problem. Assaults and Conquest* (McGraw-Hill, New York). Reprinted: Dover Publications, New York, 1986).

SACHS H. and M. STIEBITZ
[1989] On constructive methods in the theory of colour-critical graphs, *Discrete Math. 74*, 201–226.

STEINBERG R.
[1993] The state of the three color problem, in: *Quo Vadis, Graph Theory?*, eds J. Gimbel, J.W. Kennedy and L.V. Quintas, *Annals of Discrete Mathematics 55*, 211-248.

TUZA Z.
[1990] Problems and results on graph and hypergraph colourings, *Le Matematiche 45*, 219-238.

The book of Bollobás (1978) contains in chapter 5 an excellent survey of graph colouring theory, including critical graphs, graphs on surfaces, sparse graphs, perfect graphs and Ramsey theory.

Saaty and Kainen (1977) survey the four colour problem in its many variations and extensions, and Appel and Haken (1989) prove and discuss the four colour theorem.

Lovász (1983) and Grötschel et al. (1988) give excellent surveys of perfect graphs and stable sets in graphs, including combinatorial optimization aspects. Many types of perfect graphs and algorithmic aspects of them are treated by Golumbic (1980), and interesting problems by Gyárfás (1987).

Colouring by only three colours is surveyed by Steinberg (1993), algorithmic aspects in general by Kubale (1991), chromatic polynomials by Read and Tutte (1988) and edge-colourings up to 1977 by Fiorini and Wilson (1977). Sachs and Stiebitz (1989) include a comprehensive bibliography of colour-criticality. Tuza (1990) gives an interesting survey of unsolved problems. The monograph by Jensen and Toft (1995) contains a description of more than 200 unsolved graph colouring problems.

Finally, Berge and Chvátal (1984), De Werra and Hertz (1989), and Nelson and Wilson (1990) have edited interesting collections of research papers and surveys on perfect graphs and colouring theory in its many forms.

1. Basic definitions and motivation

Dividing a set of objects into classes according to certain rules is a fundamental process in mathematics. A simple set of rules determines for each pair of objects whether they are allowed in the same class or not. Graph colouring theory deals with exactly this situation. The objects form the vertex set $V(G)$ of a graph G, two vertices being joined by an edge in G whenever they are not allowed in the same class. To distinguish the classes we use a set of colours C, and the division into classes is given by a *colouring* $\varphi : V(G) \to C$, where $xy \in E(G) \Rightarrow \varphi(x) \neq \varphi(y)$. Thus vertices of the same colour form a stable set. The set C of colours can be any set; however, in this chapter we shall restrict C to be finite. Since the names and nature of the colours are irrelevant, most often we let C be equal to $\mathbb{N}_k = \{1, 2, \ldots, k\}$ for some $k \in \mathbb{N}$.

We shall assume that all graphs are simple, except if explicitly stated otherwise. If φ is a colouring of G with colours C and $|C| = k$, then φ is a *k-colouring* and G is *k-colourable*. We do not require all colours to be used, hence a k-colourable graph is also $(k+1)$-colourable. The smallest value of k for which G is k-colourable is the *chromatic number* $\chi(G)$ of G. If $\chi(G) = k$, but $\chi(G') < k$ for any proper subgraph of G, then G is *k-colour-critical*. If $\chi(G - t) < \chi(G)$ for $t \in V(G) \cup E(G)$, then t is a *critical element* of G. Hence G is critical iff all vertices and edges are critical. The colour-critical graphs were first defined, studied and used by Dirac (1952a,b,c, 1953). The only k-colour-critical graphs for $k = 1, 2$ are the complete graphs K_1 and K_2. The following result of König (1916, 1936, Satz X.12) is equivalent to the statement that the 3-colour-critical graphs are the odd circuits.

Theorem 1.1. *A graph G is 2-colourable iff G contains no odd circuit.*

Proof. If G contains an odd circuit C then $\chi(G) \geqslant \chi(C) = 3$. Conversely, if G contains no odd circuit, then the distance classes of G w.r.t. a vertex $x \in V(G)$ are all stable and hence can be coloured alternately by two colours. □

No reasonable characterization of the 4-colour-critical graphs, or equivalently of 3-colourability, seems possible, in fact the 4-colour-critical graphs form a very unruly class, as we shall see.

The following theorem by De Bruijn and Erdős (1951) shows that we may restrict our attention to finite graphs, in particular all colour-critical graphs are finite:

Theorem 1.2. *If all finite subgraphs of an infinite graph G are k-colourable then G is k-colourable.*

Proof. The theorem follows from compactness results in logic. A direct proof can be given as follows. Let G be a possible counterexample. By Zorn's lemma, G may be assumed to be maximal, i.e. the addition of any new edge e to G gives a finite subgraph G_e of $G + e$ which is not k-colourable. Suppose $xy \notin E(G)$, $yz \notin E(G)$, but $xz \in E(G)$. Then $(G_{xy} - xy) \cup (G_{yz} - yz) + xz$ would be a finite subgraph of G which is not k-colourable. Hence non-adjacency is an equivalence relation on $V(G)$. Since G is not k-colourable the number of equivalence classes is at least $k + 1$, but then $K_{k+1} \subseteq G$. Contradiction. □

Not only colouring theory, but much of graph theory, had its beginnings in work on the *four colour problem* of F. Guthrie, asking if any planar graph is 4-colourable. Well written accounts are contained in the monographs by Ringel (1959), Ore (1967), Saaty and Kainen (1977), Barnette (1983), and Aigner (1984). The four colour problem was mentioned in a letter from A. De Morgan to W.R. Hamilton in 1852, and seems first mentioned in print in 1860 by De Morgan. A proposed solution by Kempe (1879) stood for more than a decade until it was refuted by Heawood (1890) in his first paper. Heawood (1890) proved the five colour theorem, extended the problem to higher surfaces by giving a sufficient number of colours for each surface and showing the number 7 of colours to be necessary and sufficient for the torus. Moreover, Heawood considered maps with countries consisting of more than one connected part. Dirac (1963) gave a survey of Heawood's achievements.

After many attempts and partial results on the four colour problem throughout this century by many mathematicians, Appel and Haken in 1976 announced a complete proof, published in Appel and Haken (1977a) and Appel et al. (1977). The proof is based on the same basic idea as Kempe's proof, which is to find a set of unavoidable configurations, all being reducible in the sense that a 4-colouring of a planar graph containing one of the configurations can be obtained from a 4-colouring of a smaller reduced planar graph. But where Kempe's unavoidable configurations were nodes of degree 3, 4 or 5, Appel and Haken's initial set had 1936 configurations (in Appel and Haken 1977a it was announced that a proof with fewer configurations is possible). The detailed techniques of Appel and Haken are further developments of methods of Heesch (1969), who was the first to strongly emphasize a possible proof of the four colour theorem along these lines. Several surveys of the proof exist, for example Appel and Haken (1977b) and Woodall and Wilson (1978). Due to its length, its extensive use of computer verification, and its omission of details, the proof of Appel and Haken has been surrounded by some controversy. In any case it is safe to say that the last word on the four colour problem has not been said. Appel and Haken (1986; 1989) have answered the criticism raised against their proof and published an amended version of it.

Recently a new ingeniously improved proof of the four colour theorem has been presented by Robertson et al. (1994). The proof is similar to the proof of Appel and Haken, but it is simpler and more transparent in several ways. In particular it is based on a simpler procedure to obtain unavoidability (avoiding some of the more problematic aspects of Appel and Haken's proof with configurations that wrap around and meet themselves), and uses only 633 configurations. The reducibility proofs are however still based on computers.

But even if we accept that there is only little more to add about the four colour problem, there are still many attractive and easily formulated *unsolved problems* to consider, as exemplified in the book by Jensen and Toft (1995). Tutte (1978) says "The Four Colour Theorem is the tip of the iceberg, the thin end of the wedge and the first cuckoo of spring", and he describes several difficult conjectures generalizing the four colour theorem, among them the famous conjecture of Hadwiger (1943), which may be formulated as follows.

Conjecture 1.3. Let $\mathcal{G}$ be a class of graphs closed under deletions (of vertices and/or edges) and contractions of edges (removing possible loops and parallel edges that might arise). Then the maximum chromatic number χ of the graphs in $\mathcal{G}$ equals the size k of a largest complete graph in $\mathcal{G}$.

For $k = 3$ this was proved by Dirac (1952a), and the case $k = 4$ not only implies, but is equivalent to the four colour theorem, as proved by Wagner (1937). That Conjecture 1.3 is true when $\mathcal{G}$ is the class of all graphs embeddable on a surface S (different from the sphere), was proved by Ringel (1959).

As somewhat different examples of unsolved problems, here are two of the favourites of Erdős (1981).

Problem 1.4. Let G be the union of at most k complete k-graphs of which any two have at most one vertex in common. Is G k-colourable?

Problem 1.5. Let $V(G) = \mathbb{R}^2$, i.e. the vertices of G are all the points in the plane, where $xy \in E(G)$ iff x and y have distance 1. What is $\chi(G)$?

Problem 1.4. is due to Erdős, Faber and Lovász, and Erdős (1981) offers 500 dollars for a proof or a disproof. Problem 1.5. is due to Hadwiger and Nelson. By Theorem 1.2 it is sufficient to consider finite subgraphs of G. It is known that $4 \leqslant \chi(G) \leqslant 7$.

The chromatic number also plays a role outside pure colouring theory. A good illustration of this is the theorem of Erdős and Stone (1946) and Erdős and Simonovits (1966). Let G be a fixed graph with at least one edge. For $n \geqslant |V(G)|$ define $f(n,G)$ to be the maximum number of edges possible in a graph on n vertices not containing G as a subgraph. This extremal function, apparently with no connection to colouring, depends on $\chi(G)$.

Theorem 1.6. $f(n,G)/n^2 \to \frac{1}{2}(\chi(G)-2)/(\chi(G)-1)$ *for* $n \to \infty$.

Finally, although graph colouring theory has a recreational background, there are "real world" *applications*. For example, time tabling problems are basically graph colouring problems, where the activities to be scheduled are represented by vertices, two vertices being joined by an edge iff the corresponding activities are in conflict. Thus the times to be assigned to the activities correspond to colours. Radio frequency assignment is another example.

2. Constructions and examples

This section may be considered to be negative in nature. By exhibiting a variety of examples, the chromatic number of a graph G is seen not to have very strong implications for the structure and properties of G, even when G is colour-critical.

2.1. Sparse graphs

Obviously $\omega(G) \leqslant \chi(G)$, where $\omega(G)$ denotes the size of a largest complete subgraph of G. Moreover, $\omega(G)=1$ implies $\chi(G)=1$. Zykov (1949) showed by examples that apart from these there are no relations between ω and χ in general. In particular he proved the following.

Theorem 2.1. *For all k there exists k-chromatic triangle-free graphs G_k, i.e.* $\chi(G_k)=k$ *and* $G_k \not\supseteq K_3$.

Proof. By induction we assume that $G_1, G_2, \ldots, G_{k-1}$ have been obtained. Take disjoint copies of these. Let V be a set of $|V(G_1)| \cdot |V(G_2)| \cdots |V(G_{k-1})|$ new vertices corresponding to all selections of one vertex from each of $G_1, G_2, \ldots, G_{k-1}$. Then G_k is obtained from $G_1, G_2, \ldots, G_{k-1}$ and V by joining each vertex in V to its corresponding set of $k-1$ vertices, one vertex from each G_i.

G_k is k-colourable. On the other hand, if G_k had a $(k-1)$-colouring, then G_1 would have a vertex x_1, of some colour i_1, G_2 would have a vertex x_2 of a colour i_2 different from i_1, G_3 would have a vertex x_3 of a colour i_3 different from i_1 and i_2, etc. The vertex v of V joined to $x_1, x_2, \ldots, x_{k-1}$ would be joined to all colours of the $(k-1)$-colouring. Contradiction. Hence $\chi(G_k)=k$. □

Schäuble (1969) proved that the above construction, which is a refinement of Zykov's, produces a k-colour-critical G_k from critical $G_1, G_2, \ldots, G_{k-1}$. There are several other constructions of triangle-free k-chromatic graphs, the earliest ones by Descartes (1948a, 1954), Kelly and Kelly (1954) and Mycielski (1955). A survey was given by Sachs (1969).

Erdős (1957) proved by a geometric construction that the size of a smallest G_k is bounded by a polynomial in k (of degree 50). In fact, Erdős (1959, 1961) proved by probabilistic non-constructive arguments the extensions Theorems 2.2 and 2.3 of Theorem 2.1.

Theorem 2.2. *There is a constant c such that for all k there exists a k-chromatic triangle-free graph on* $\leqslant c \cdot k^2 \cdot (\log k)^2$ *vertices.*

Theorem 2.3. *For all k and* ℓ *there exist k-chromatic graphs without circuits of length* $\leqslant \ell$.

It is not known whether the upper bound in Theorem 2.2 is best possible, however Erdős and Hajnal (1985) obtained $c' \cdot k^2 \cdot \log k$ as a lower bound.

Extending the domain of operations to hypergraphs, Lovász (1968) gave a constructive proof of Theorem 2.3. This was later done more simply by Nešetřil and Rödl (1979).

There is a very simple way to obtain a k-chromatic graph in which all odd cycles are long. Let $X = \{1, 2, \ldots, 2n+k\}$ and define the *Kneser graph* $\mathrm{KG}_{n,k}$ by $V(\mathrm{KG}_{n,k}) = \{x \mid x \subseteq X \text{ and } x = n\}$ and $E(\mathrm{KG}_{n,k}) = \{xy \mid x \cap y = \emptyset\}$. The *Schrijver graph* $\mathrm{SG}_{n,k}$ is the subgraph of $\mathrm{KG}_{n,k}$ induced by $X' = \{x' \mid x'$ does not contain two consecutive elements in the cyclic order $(1, 2, \ldots, 2n+k, 1)\}$. If K_i denotes those vertices having least element i, then $K_1, K_2, \ldots, K_{k+1}$, $V(\mathrm{KG}_{n,k}) - K_1 - K_2 - \cdots - K_{k+1}$ defines a $(k+2)$-colouring of $\mathrm{KG}_{n,k}$. Kneser (1955) thus observed $\chi(\mathrm{KG}_{n,k}) \leqslant k+2$, and equality was proved by Lovász (1978a) using Borsuk's theorem, and subsequently more simply by Bárány (1978) using a further topological result of D. Gale. Schrijver (1978) proved that $\chi(\mathrm{SG}_{n,k}) = k+2$ and that all vertices of $\mathrm{SG}_{n,k}$ are critical.

Let C be an odd circuit of length ℓ in $\mathrm{KG}_{n,k}$. For each $xy \in E(C)$ consider the k-set $X - x - y$. Since C is odd these k-sets cover X, hence $\ell \geqslant (2n+k)/k$. By a well-known combinatorial result (Ryser 1963, Lemma 2.3), the graph $\mathrm{SG}_{n,k}$ has $((2n+k)/(n+k)) \cdot \binom{n+k}{k}$ nodes. For each $k \geqslant 2$ this implies the existence of infinitely many $(k+2)$-chromatic graphs G in which each odd circuit has length $\geqslant c_k \cdot |V(G)|^{1/k}$. For $k = 2$ the first such graphs were obtained by Gallai (1963a). Kierstead et al. (1984) proved the above order of magnitude to be best possible. For circuits in general the corresponding order of magnitude is only $c \cdot \log |V(G)| / \log k$, as proved by Erdős (1959, 1962).

Finally let us mention the following remarkable result due to Müller (1979). The proof of the second part uses Theorem 2.3. The case $r = 1$ of Theorem 2.4 implies the existence of uniquely k-colourable graphs without short circuits.

Theorem 2.4. *Let A be a set of vertices, and let* $P_1, P_2, \ldots, P_r$ *denote different partitions of A into at most k classes* $(k \geqslant 3)$. *Then there exists a k-chromatic graph G with* $A \subseteq V(G)$ *such that G has precisely r different k-colourings* $\varphi_1, \varphi_2, \ldots, \varphi_r$ *(where renaming or permuting the colours does not count as different colourings),* *and for all i the restriction of φ_i to A partitions A into colour classes like P_i. Moreover, for a given ℓ, there exists such a G in which all circuits have length $\geqslant \ell$ and any two vertices of A distance $\geqslant \ell$.*

2.2. Hypergraphs

In a hypergraph H the edges are subsets of the vertex set $V(H)$. The edges may be of arbitrary, not necessarily equal, size $\geqslant 2$. If all edges have the same size r the hypergraph is *r-uniform*. A *colouring* of H is a mapping $\varphi : V(H) \to C$, where $|\varphi(A)| \geqslant 2$ for all edges $A \in E(H)$. This definition is due to Erdős and Hajnal (1966). A 2-colour-critical hypergraph H has just one edge, i.e. $E(H) = \{V(H)\}$. No reasonable characterization of the 3-colour-critical hypergraphs, or equivalently the 2-colourable hypergraphs, seems possible. An example of a 3-uniform 3-colour-critical hypergraph is the finite projective plane of order 2, also called the Fano configuration, on seven vertices and seven edges. An example of an r-uniform k-colour-critical hypergraph is a set of $(r-1)(k-1)+1$ vertices together with all subsets of size r as edges. Other examples can be found, e.g. in Toft (1975).

Any k-chromatic or k-colour critical hypergraph with $k \geqslant 4$ can be transformed into a similar graph. The construction for Theorem 2.1 by Descartes (1948a, 1954) is an example of this transformation. From a graph G, a vertex x of G and a set X of vertices disjoint from G of size at most $\deg(x)$, a new graph G' can be obtained by *splitting* x into X, i.e. G' is obtained from X and $G - x$ by joining each vertex of X to one or more neighbours of x, such that each neighbour of x is joined to at least one vertex of X. If each neighbour of x is joined to exactly one vertex of X, then the splitting is *proper*. The following theorem is not deep, but it can be used to construct a variety of colour-critical graphs, because we can operate free of the constraint in graph theory that all edges have to be of size 2 (the above mentioned constructive proofs of Theorem 2.3 illustrates the same: even if we are only interested in graphs, the hypergraphs are indispensable in the constructions).

Theorem 2.5. *Let $k \geqslant 4$ and let H be a k-chromatic hypergraph (or graph) and A an edge of H. Let G be a k-chromatic graph disjoint from H with a node $x \in V(G)$ of degree $\geqslant |A|$. Obtain the hypergraph H' from H and from G by a proper splitting of x into A and the removal of A as an edge. Then:*

(a) $\chi(H') \geqslant k$.

(b) *If H and G are k-colour-critical, then H' is k-colour-critical in at least these cases:*

(i) $\deg(x) \leqslant 2k - 4$,

(ii) *$|A| = 2$ and the graph obtained from G by the proper splitting of x into A is $(k-1)$-colourable;*

(iii) *G consists of an odd circuit C completely joined to a K_{k-3} containing x.*

Proof. If H' had a $(k-1)$-colouring, then, since $\chi(H) = k$, the vertices of A would all have the same colour, and hence G would be $(k-1)$-colourable. Therefore $\chi(H') \geqslant k$. The proof of (b) is more tedious, but not difficult (see Toft 1974a). □

If H is k-critical with no other edge than possibly A having size $\geqslant 3$ and if G is k-critical with $\deg(x) = k-1 \geqslant |A|$, then H' is a k-colour-critical graph with a non-trivial cut (i.e. both sides of the cut have $\geqslant 2$ vertices) of $k-1$ edges. Any such H' can be obtained in this way, as proved by Gallai (unpublished) and Toft (1974a).

If H is a k-colour-critical graph and the condition in (b) (ii) is satisfied, then H' is a k-colour-critical graph containing a cutset of two vertices. Any such H' can be obtained in this way, as proved by Dirac (1953, 1964a) and Gallai (1963a).

2.3. Colour-critical graphs

Since the k-colour-critical graphs are minimal k-chromatic one might expect them to have relatively few edges. However, for $k \geqslant 4$ there exist positive constants a_k and infinitely many k-colour-critical graphs G with $\geqslant a_k|V(G)|^2$ edges. Dirac (1952a) observed the following.

Theorem 2.6. *Let G_1 and G_2 denote disjoint graphs, and let $G_1 * G_2$ denote the graph obtained from G_1 and G_2 by joining each vertex of G_1 to each vertex of G_2 by an edge. Then $\chi(G_1 * G_2) = \chi(G_1) + \chi(G_2)$. Moreover, $G_1 * G_2$ is colour-critical iff G_1 and G_2 are colour-critical.*

The proof of Theorem 2.6 is straightforward. Letting G_1 and G_2 be odd circuits of the same length n, Dirac (1952a) thus obtained 6-colour-critical graphs on $2n$ vertices and $n^2 + 2n$ edges, thereby getting $a_6 = \frac{1}{4}$ and answering a question of P. Erdős. For $k = 4$ one may similarly let G_1 and G_2 be 2-colour-critical hypergraphs of the same size n. Then $G_1 * G_2$ is a 4-colour-critical hypergraph with n^2 edges of size 2 and two edges of size n. Reducing $G_1 * G_2$ to a 4-colour-critical graph G by Theorem 2.5 (b) (iii), one may, for n odd, obtain a G on $4n$ vertices and $n^2 + 4n$ edges. These and other similar examples were obtained by Toft (1970). The best possible values of a_k are not known, and the above values $a_4 = \frac{1}{16}$ and $a_6 = \frac{1}{4}$ are the best known at present.

In the above examples most edges are concentrated in 2-colourable subgraphs, but V. Rödl, and Stiebitz (1987) proved that this need not be so.

Theorem 2.7. *For all $k \geqslant 4$ there exists a positive constant b_k and infinitely many k-colour-critical graphs G such that it requires the removal of $\geqslant b_k|V(G)|^2$ edges from G to reduce the chromatic number to $k-2$.*

Proof. We shall carry out the construction for $k = 4$ only. Let A_1, A_2, A_3, A_4, A_5 be disjoint sets of n vertices each, $n \geqslant 2$. Join A_i to A_{i+1}, $i = 1, 2, 3, 4$, and A_5 to A_1 by all possible edges. Add the five hyperedges $A_1 \cup A_3$, $A_2 \cup A_4$, $A_3 \cup A_5$, $A_4 \cup A_1$, $A_5 \cup A_2$ of size $2n$. The hypergraph H obtained is 4-colour-critical. Reducing H to a 4-colour-critical graph by Theorem 2.5 (b) (iii), one may obtain a G on $15n + 5$ vertices such that the removal of at least n^2 edges from G is required to reduce the chromatic number to 2. □

As a further surprise, the maximum number $\alpha(G)$ of vertices in a stable set in a k-colour-critical graph ($k \geqslant 4$), may exceed any fixed proportion of the total number of vertices, as proved by Brown and Moon (1969). Simonovits (1972) gave an alternative proof, based only on the existence of k-colour-critical graphs with many edges. He noticed that splitting a vertex of a k-colour-critical graph into new vertices corresponding to all subsets of size $k-1$ of the neighbours of x in G, and joining each new vertex to its corresponding subset, produces a new graph G' with $\chi(G') = k$. Moreover, a k-colour-critical subgraph G'' of G' is obtained by removing a subset of the new vertices. Since each neighbour y of x will still be joined to at least one new vertex (because $\chi(G - xy) = k - 1$), it follows that at least $\deg(x)/(k-1)$ new vertices remain in G''. If G has many edges, then splitting all vertices in a colour class incident with many edges produces a k-colour-critical graph with a large stable set consisting of new vertices. The order of magnitude is as in the construction of Brown and Moon. Lovász (1973a) proved this order of magnitude to be best possible for $k = 4$ and generalized it to all $k \geqslant 4$.

Theorem 2.8. *For $k \geqslant 4$, let $\alpha_k(n)$ denote the maximum number of vertices possible in a stable set of vertices in a k-colour-critical graph on n vertices. Then for infinitely many values of n*

(i) $\alpha_k(n) \geqslant n - k \cdot n^{1/(k-2)}$

and for all values of n

(ii) $\alpha_k(n) \leqslant n - \frac{1}{6}k \cdot n^{1/(k-2)}$.

Proof. Let G' be obtained from i-colour-critical graphs G_i, $i = 1, 2, \ldots, k-1$, as in the construction for Theorem 2.1 by Zykov and Schäuble, where each of $G_3, G_4, \ldots, G_{k-1}$ has N vertices, except that we shall let G_2 be a 2-colour-critical hypergraph, also on N vertices. Hence G' has precisely one hyperedge, and it may be reduced to a k-colour-critical graph G by Theorem 2.5 (b) (iii). This G has $n \geqslant \alpha(G) \geqslant N^{k-2}$, and hence we obtain (i):

$$n - a_k(n) \leqslant n - \alpha(G) \leqslant (k-1)N + 2 \leqslant k \cdot N \leqslant k \cdot n^{1/(k-2)}.$$

Lovász' proof of (ii) is an interesting application of a linear algebra technique. Let G be a k-colour-critical graph on n vertices with a stable set S of size $s = \alpha_k(n)$. By the splitting argument of Simonovits (1972) we may assume that all vertices of S have degree $k-1$ in G (otherwise split the vertices of S; the theorem for the new graph implies the theorem for G). Let T be the $t = n - \alpha_k(n)$ vertices of G outside S, and let A_i be the set of $k-1$ neighbours in T of vertex i from S for $i = 1, 2, \ldots, s$. Let $B_1, B_2, \ldots, B_q$ be all subsets of T of size $k-2$, i.e. $q = \binom{t}{k-2}$. Finally, let $M = \{m_{ij}\}$ be the $s \times q$-matrix with $m_{ij} = 1$ if $A_i \supseteq B_j$ and $m_{ij} = 0$ otherwise.

We shall prove below that the rank of M over GF[2] is s. Hence $s \leqslant q$, and therefore

$$n = s + t \leqslant \binom{t}{k-2} + t \leqslant \frac{2 \cdot t^{k-2}}{(k-2)!} \leqslant \left(\frac{\mathrm{e}t}{(k-2)}\right)^{k-2}$$

where we use that $i! \geqslant 2(i/\mathrm{e})^i$ by Stirling's formula. Then we obtain (ii):

$$n - \alpha_k(n) = t \geqslant \left(\frac{k-2}{3}\right) \cdot n^{1/(k-2)} \geqslant \frac{k}{6} n^{1/(k-2)}.$$

To see that the rank of M over GF[2] is s, colour the subgraph $G[T]$ of G induced by T with $k-1$ colours $1,2,\ldots,k-1$ in such a way that only A_1 of $A_1,A_2,\ldots,A_s$ get all $k-1$ colours. This is possible, since vertex 1 of S is critical in G. The total number of pairs (i,j), where B_j has colours $1,2,\ldots,k-2$, $A_i \supset B_j$, and $i \in \{1,2,\ldots,r\}$ for some r, $1 \leqslant r \leqslant s$, is an odd number, since there is exactly 1 pair for $i=1$, and 0 or 2 pairs for $i=2,\ldots,r$. Hence at least one B_j must be contained in an odd number of $A_1,A_2,\ldots,A_r$; but then the rows of M corresponding to $A_1,A_2,\ldots,A_r$ do not have sum 0. This argument shows in fact that no non-empty subset of the rows has sum 0, and hence the s rows are linearly independent over GF[2]. □

Using the splitting process of Simonovits, but splitting into vertices of large degree, Simonovits (1972) and Toft (1972a) independently proved the following.

Theorem 2.9. *There exist infinitely many 4-colour-critical graphs G for which the minimum degree $\delta(G)$ is at least $(\frac{1}{2}\cdot|V(G)|)^{1/3}$.*

Notice that Dirac's 6-colour-critical graphs G with many edges have $\delta(G) \geqslant c\cdot|V(G)|$. It is not known if a similar result holds for 4-colour-critical graphs.

Theorems 2.7, 2.8 and 2.9 are in sharp contrast to the situation for $k=3$, i.e. for the odd cycles. A further such result, based on the existence of 4-colour-critical graphs with many edges, was obtained by V. Rödl and published by Toft (1985).

Theorem 2.10. *There exists a constant $c>1$ such that the number of non-isomorphic 4-colour-critical graphs on n vertices is $> c^{(n^2)}$.*

3. Algorithmic aspects

Johnson (1978a) remarked that all the known graph colouring algorithms are horrible, and that all the unknown graph colouring algorithms are probably not much better. Much can be said to support these statements, as we shall see, the keywords being "worst case behaviour" and "NP-completeness". A *graph colouring algorithm A* is an algorithm that applied to any graph G produces a colouring of the vertices of G. The number of colours that A uses on G is denoted by $A(G)$. Ideally we would like A to be efficient and $A(G)$ to be equal to or close to $\chi(G)$. The *performance guarantee* $A(n)$ of A is the maximum of $A(G)/\chi(G)$ taken over all graphs G on n vertices.

3.1. Polynomial colouring algorithms

There are two basic types of polynomial graph colouring algorithms, *sequential colouring* and *maximal stable set colouring*. In a sequential colouring the vertices of the input graph G are ordered and then coloured in this order by colours $1,2,3,\ldots$ giving each vertex the least colour possible. In a maximal stable set colouring the vertices are also considered in some order, giving a vertex the colour 1 whenever it is not joined to a vertex already coloured 1. When no more vertices can be coloured 1, all the vertices of colour 1 are removed and the colouring continues likewise with colour 2 on the remaining graph, etc. The vertices of colour i form a maximal stable set in $G - V_1 - V_2 - \cdots - V_{i-1}$, where V_j are the vertices of colour j. The number of colours used will depend on the order in which the vertices are considered, in fact there will always be orderings giving a $\chi(G)$-colouring of G, but no polynomial algorithms to find such orderings are known. A possible polynomial ordering for sequential colouring is a *smallest last ordering* in which the vertex x_i in the order $x_1, x_2, \ldots, x_n$ is of smallest degree in $G - x_n - x_{n-1} - \cdots - x_{i+1}$, or a *largest first ordering* in which the vertices are ordered by non-increasing degrees. The philosophy behind these orderings is to have, by each colouring of a vertex, as few restrictions as possible. Polynomial graph colouring algorithms of the above types with various refinements, are described by Matula et al. (1972). More recent surveys were presented by Manvel (1985) and Kubale (1991).

For each of 13 different versions of such polynomial graph colouring algorithms, Johnson (1974) gave infinite classes of graphs, where the algorithms perform poorly. For example, let G be a bipartite graph with vertices $x_1, x_2, \ldots, x_n$ and $y_1, y_2, \ldots, y_n$ in the two sides and with all possible edges $x_i y_j$ with $i \neq j$. Then a sequential colouring with the ordering $x_n, y_n, x_{n-1}, \ldots, x_1, y_1$ uses n colours on the 2-colourable graph G. This is the largest first ordering, and a similar example for a smallest last ordering can be given. Such examples show the following.

Theorem 3.1. *For a largest first or smallest last sequential graph colouring algorithm A we have $A(G) \geqslant c_1 \cdot \chi(G) \cdot |V(G)|$ for a positive constant c_1 and infinitely many G. Moreover, $A(n) \geqslant c_2 \cdot n$ for a positive constant c_2.*

How bad the situation seems to be can best be explained by posing a question of Johnson (1978b).

Problem 3.2. Does there exist a polynomial graph colouring algorithm A for which $A(G) \leqslant c \cdot \chi(G) \cdot |V(G)|^{1-\varepsilon}$ for suitable positive constants c and ε?

The maximal stable set algorithms seem to behave slightly better than the sequential ones. As described by Johnson (1974), P. Erdős suggested the following.

Theorem 3.3. *There is a polynomial maximal stable set graph colouring algorithm A for which $A(G) \leqslant c \cdot (\log \chi(G)) \cdot |V(G)|/(\log |V(G)|)$ for all G with $\chi(G) \geqslant 2$ and some constant c.*

Proof. Let $|V(G)| = n$ and $\chi(G) = k \geqslant 2$. Then the size of a maximum stable set $\alpha(G)$ is $\geqslant n/k$, and therefore the minimum degree $\delta(G)$ is $\leqslant n - (n/k)$. We first colour a vertex x_1 of minimum degree by the colour 1. Then x_1 and all its neighbours are deleted. At least $n/k - 1$ vertices remain. Taking a vertex x_2 of minimum degree in the remaining graph, colouring x_2 by colour 1, and deleting x_2 and all its neighbours, at least $n/k^2 - 1/k - 1$ vertices remain. Continuing like this, after t vertices have been coloured 1, and as long as

$$\frac{n}{k^t} - \frac{1}{k^{t-1}} - \cdots - \frac{1}{k} - 1 > 0$$

at least one more vertex can be coloured 1. It follows that as long as $n \geqslant k^{t+1}$ at least one more vertex can be coloured 1. Therefore $|V_1| \geqslant \lfloor \log n / \log k \rfloor$, where V_1 are the vertices coloured 1. Continue in the same fashion with $G - V_1$ and the colour 2, etc. Let $G - V_1 - V_2 - \cdots - V_i$ have n_{i+1} vertices and let I be the first value of i for which $n_{i+1} < \sqrt{n}$. Then the number of vertices of colour i for $i = 1, 2, \ldots, I$ is $\geqslant \lfloor \log n_i / \log k \rfloor \geqslant \lfloor \frac{1}{2} \log n / \log k \rfloor$. Therefore $I \leqslant 2(\log k) \cdot n / (\log n - 2 \log k)$ and $A(G) \leqslant I + \sqrt{n}$. Hence $A(G) \leqslant c(\log k) \cdot n / \log n$. □

The performance guarantee of the above algorithm A is $\leqslant cn/\log n$. By refinements, Wigderson (1983) obtained a polynomial algorithm with performance guarantee $\leqslant cn(\log\log n)^2/(\log n)^2$ and in 1990 this was improved by M.M. Halldórsson with an extra factor $\log n$ in the denominator.

3.2. NP-completeness

As seen above, we are very far from a polynomial graph colouring algorithm A with $A(G)$ equal to $\chi(G)$. Probably no such algorithm exists. Karp (1972) showed the problem-type GRAPH k-COLOURABILITY, for a given graph G and number k to decide whether G is k-colourable, to be NP-complete. This means that a polynomial algorithm to solve it implies polynomial algorithms to solve any problem in the large class NP of problem-types, and hence NP would be equal to the class P of polynomially solvable problem-types. This is considered very unlikely. The monograph by Garey and Johnson (1979) is the authoritative text on NP-completeness.

Using the method in the proof of Theorem 1.1 it is easy to see that GRAPH 2-COLOURABILITY is in P. However, already the case $k = 3$ contains all the difficulties, even for rather special graphs.

Theorem 3.4. *The following problem-types are equivalent in the sense that if there is a polynomial algorithm to solve one of them, then there are polynomial algorithms for all of them.*

(a) HYPERGRAPH k-COLOURABILITY.
(b) GRAPH k-COLOURABILITY.
(c) GRAPH 3-COLOURABILITY.
(d) MAX DEGREE 4 PLANAR GRAPH 3-COLOURABILITY.
(e) HYPERGRAPH 2-COLOURABILITY.
(f) 3-UNIFORM HYPERGRAPH 2-COLOURABILITY.

Proofs of the various reductions establishing Theorem 3.4 can be found in Lovász (1973b), Garey et al. (1976) and Toft (1975).

Theorem 3.4 shows that we cannot expect ever to get a polynomial algorithm to $\chi(G)$-colour graphs G, and with the same degree of certainty one cannot expect even to get anywhere near to $\chi(G)$. Lund and Yannakakis (1993) proved the following.

Theorem 3.5. *For some constant $\varepsilon_0 > 0$, if there is a polynomial graph colouring algorithm A with $A(G) \leqslant \chi(G) \cdot |V(G)|^{\varepsilon_0}$, then there is a polynomial algorithm for GRAPH k-COLOURABILITY, and $\chi(G)$ can be found by a polynomial algorithm. In particular this is the case if there is a polynomial graph colouring algorithm A with $A(G) \leqslant r \cdot \chi(G) + d$ for some constants r and d.*

Proof in a very special case. We shall only prove the "in particular" case and only for $r < \frac{4}{3}$. Let G be any graph and let N be an integer such that $4N > r \cdot 3N + d$. Take N disjoint copies of G and join any two completely by all possible edges to obtain G^*. If G is 3-colourable, then $A(G^*) \leqslant r \cdot \chi(G^*) + d \leqslant r \cdot 3N + d < 4N$. If G is not 3-colourable, then $A(G^*) \geqslant \chi(G^*) \geqslant 4N$. Hence A applied to G^* solves (c) of Theorem 3.4. The result then follows from Theorem 3.4. □

The proof for $\frac{4}{3} \leqslant r < 2$ makes use of the Kneser graphs instead of the complete graphs K_N, but this is rather more complicated and was carried out by Garey and Johnson (1976). For $r \geqslant 2$ the result was obtained as a corollary of the first part of Theorem 3.5 by Lund and Yannakakis (1993).

3.3. Chromatic polynomials

For a graph G let $P(G,k)$ denote the number of k-colourings of G using colours $1, 2, \ldots, k$. Thus $P(K_n, k) = k \cdot (k-1) \cdot (k-2) \cdots (k-n+1)$ and $P(\bar{K}_n, k) = k^n$. For $e = xy \in E(G)$, let $G \cdot e$ denote the graph obtained from G by contracting e to a single vertex, leaving only one copy of each parallel edge that might arise. Since the k-colourings of $G - e$ are of two types, those with x and y coloured differently and those with x and y coloured alike, we have $P(G-e,k) = P(G,k) + P(G \cdot e, k)$. Both $G - e$ and $G \cdot e$ have fewer edges than G, hence applying the relation again to these graphs, and so on, $P(G,k)$ will be expressed as the sum of terms $P(G',k)$ for graphs G', without edges, the largest of which has $|V(G)|$ nodes. Therefore, the following holds.

Theorem 3.6. *$P(G,k)$ is polynomial in k with integral coefficients. If $|V(G)| = n$ then $P(G,k)$ is of degree n with leading term k^n. The coefficient of k^0 is 0.*

The above proof of Theorem 3.6 is constructive and can be used to compute $P(G,k)$, although this of course is not a polynomial algorithm. Theorem 3.6 can also be proved using the principle of inclusion and exclusion. One can easily prove that the coefficient of k^{n-1} is $-|E(G)|$, that the coefficients alternate in sign and that the smallest number r such that k^r has a nonzero coefficient is the number of connected components of G. Thus $P(G,k)$ contains several pieces of information about G in addition to the basic property that $\chi(G)$ is the minimal integer $k \geqslant 0$ for which $P(G,k) \neq 0$. However, chromatic polynomials are not well understood, for example it is not known what makes a polynomial chromatic.

The study of chromatic polynomials was initiated by Birkhoff (1912) and continued by Whitney (1932). An exposition of the early history of chromatic polynomials was given by Biggs et al. (1976). A good survey was presented by Read (1968), and more recently by Read and Tutte (1988). Many deep results have been obtained by W. T. Tutte in a sequence of papers. For a development of graph colouring theory with the chromatic polynomial as a basic concept, see Tutte (1984). One of Tutte's surprising and beautiful results is the golden identity.

Theorem 3.7. *Let M be a plane triangulation on n vertices. Then*

$$P(M,\tau+2)=(\tau+2)\cdot\tau^{3n-10}\cdot(P(M,\tau+1))^2,$$

where τ is the golden ratio $(1+\sqrt{5})/2$, i.e. $\tau+1=\tau^2$ and $\tau+2=\sqrt{5}\tau$.

Theorem 3.7 was obtained by Tutte (1970), who also noted that $P(M,\tau+1)\neq 0$. Hence we have the curious consequence that $P(M,3.618\ldots)>0$. The four colour theorem, of course, says $P(M,4)>0$.

4. Upper and lower bounds for the chromatic number

As explained in section 3 the chromatic number seems intractable, and it is therefore of importance to relate it to more tractable graph constants. We shall consider upper bounds in terms of the degrees in the graph. It seems more difficult to obtain reasonable lower bounds.

4.1. Brooks' theorem

A sequential colouring of the vertices of a graph G in any order, using for each vertex a least possible colour from $1,2,3\ldots$, produces a colouring with at most $\Delta(G)+1$ colours, i.e., $\chi(G)\leqslant\Delta(G)+1$. A deep theorem of Hajnal and Szemerédi (1970), see also Bollobás (1978), says that G can in fact be $(\Delta(G)+1)$-coloured with all colour-classes of equal or almost equal size. Another basic result in graph colouring theory, and the first non-trivial result relating χ to a tractable graph constant for graphs in general, is the theorem of Brooks (1941). It implies easily that the problem-type MAXIMUM DEGREE 3 GRAPH 3-COLOURABILITY can be solved by a polynomial algorithm (compare Theorem 3.4).

Theorem 4.1. *Let G be any graph. Then $\chi(G)\leqslant\Delta(G)+1$. If $\Delta(G)=2$ then equality holds iff G has an odd circuit as a connected component. If $\Delta(G)\neq 2$, then equality holds iff G has a complete graph on $\Delta(G)+1$ vertices as a connected component.*

Proof. For $\Delta(G)=1$ the theorem is obvious, and for $\Delta(G)=2$ it follows from Theorem 1.1. Hence we only need to prove that $\Delta(G)\leqslant s$ and $G\not\supseteq K_{s+1}$ for some $s\geqslant 3$, imply that G is s-colourable. The proof of this is by induction over $|V(G)|$. For $|V(G)|\leqslant 4$ the statement is obviously true.

If G is disconnected or has a cutvertex, then by induction each block of G is s-colourable, and hence G is. If G is 2-connected with a cutset of two vertices x and y, then $G=G_1\cup G_2$, where G_1 and G_2 each have fewer vertices than G and

$V(G_1 \cap G_2) = \{x, y\}$. By induction G_1 and G_2 are both s-colourable. If $xy \in E(G)$, then we may assume the two colours of x and y in the two s-colourings to be the same, hence the two s-colourings combine to an s-colouring of G. If $xy \notin E(G)$, then by induction $G_i' = G_i \cup xy$ is s-colourable or equal to a K_{s+1} for $i = 1, 2$. Then again G is s-colourable, except possibly when at least one of G_1' and G_2', say G_1', is a K_{s+1}. But in this case both x and y have degree 1 in G_2, since $\Delta(G) \leqslant s$. Hence G_2 has an s-colouring in which x and y have the same colour, since $s \geqslant 3$. Because $G_1 = K_{s+1} - xy$ this s-colouring of G_2 can be extended to an s-colouring of G. Thus we may assume G to be 3-connected.

Colouring the vertices of G in some order, using for each vertex a least possible colour from $1, 2, 3, \ldots$, produces an $(s+1)$-colouring of G. If the vertex x gets colour $s+1$, then $\deg(x) = s$ and the s neighbours of x have colours $1, 2, \ldots, s$. Moreover, in this situation we may recolour x by any colour i and recolour the unique neighbour of x of colour i by one of $1, 2, \ldots, s+1$. In this way we push the undesired colour $s+1$ from x to any neighbour of x or it disappears. Pushing along any path $x_1, x_2, x_3, \ldots, x_n$ starting in x_1, we obtain an $(s+1)$-colouring in which at most the vertex x_n of the path has colour $s+1$. Since there are paths from any vertex to x_n, we may push the colour $s+1$ so that it is present only at x_n. If $\deg(x_n) < s$ we may recolour x_n to obtain an s-colouring of G. Thus we may assume that all vertices of G have degree equal to s.

Let p and q be vertices of G not joined by an edge. Since there are paths from any vertex to q, we may push the colour $s+1$ so that it is present in G at most at q. In such a colouring the vertex p has two neighbours x and y of the same colour. Since G is 3-connected there is a path P from q to p avoiding x and y. Pushing the colour $s+1$ from q along P, either the colour $s+1$ disappears or it ends up at p. In the latter case, since x and y are coloured alike and $\deg(p) = s$, there is a colour i from $1, 2, \ldots, s$ that may be given to p, thus obtaining in any case an s-colouring of G. □

The above proof, based on sequential colouring and colour interchange, is the original argument of Brooks (1941). Melnikov and Vizing (1969) gave a simple proof based only on interchange, and Lovász (1975) gave one based on selecting the initial order of the vertices more carefully to obtain the s-colouring immediately.

For graphs G with $\Delta(G) \geqslant 3$ and not containing a $K_{\Delta(G)+1}$, Brooks' theorem improves the obvious bound $\chi(G) \leqslant \Delta(G) + 1$ by 1. Excluding the existence of smaller complete subgraphs, further improvements were obtained independently around the same time by Borodin and Kostochka (1977), Catlin (1978) and Lawrence (1978).

Theorem 4.2. *Let G be any graph. If $G \not\supseteq K_r$, where $4 \leqslant r \leqslant \Delta(G) + 1$, then $\chi(G) \leqslant \Delta(G) + 1 - \lfloor (Delta(G) + 1)/r \rfloor$, i.e. $\chi(G) \leqslant (r-1)/r \cdot (\Delta(G) + 2)$.*

Proof. The proof uses a result of Lovász (1966) that if $d_1 + d_2 + \ldots + d_q \geqslant \Delta(G) - q + 1$, then $V(G)$ can be decomposed into classes $V_1, V_2, \ldots, V_q$, such that the subgraph G_i induced by V_i has $\Delta(G_i) \leqslant d_i$. The general case follows immediately from the case $q = 2$. For $q = 2$ a partition for which $d_1 \cdot |E(G_2)| + d_2 \cdot |E(G_1)|$ is as small as possible has the desired property, as can easily be seen by moving a possible vertex of too high degree to the other side.

Now let $d_1 = d_2 = \cdots = d_{q-1} = r-1$ and $d_q \geqslant r-1$, so that $\Sigma d_i = \Delta(G) - q + 1$ and $q = \lfloor (\Delta(G)+1)/r \rfloor$. Then by Lovász' theorem there is a decomposition of G into $G_1, G_2, \ldots, G_q$ with $\Delta(G_i) \leqslant d_i$, and hence $\chi(G_i) \leqslant d_i$ by Brooks' theorem 4.1. Therefore $\chi(G) \leqslant \Sigma\chi(G_i) \leqslant \Sigma d_i = \Delta(G) + 1 - q$. □

For triangle-free graphs the theorem only gives the same as for K_4-free graphs; however, improvements in the triangle-free case have been obtained in 1982 by A. V. Kostochka: $\chi(G) \leqslant 2(\Delta(G)+3)/3$, and recently by J. H. Kim: $\chi(G) < c \cdot \Delta(G)/\log \Delta(G)$.

Brooks' theorem may be formulated and generalized in other directions, in particular in terms of colour-critical graphs. If x is a critical vertex of G, then $\deg(x) \geqslant \chi(G) - 1$. Using this and $\Sigma \deg(x) = 2 \cdot |E(G)|$ we may formulate Brooks' theorem as follows.

Theorem 4.3. *Let G be k-colour-critical. Then the minimum degree $\delta(G)$ is $\geqslant k-1$. In particular $2 \cdot |E(G)| \geqslant (k-1) \cdot |V(G)|$, and equality holds iff G is complete or an odd circuit.*

Extensions of this version of Brooks' theorem have been given by Dirac (1957a) and Gallai (1963a). An elegant proof of Dirac's theorem was given by Weinstein (1975).

Theorem 4.4. *Let G be k-colour-critical. If $k \geqslant 4$ and $G \neq K_k$, then $2 \cdot |E(G)| \geqslant (k-1) \cdot |V(G)| + (k-3)$.*

Theorem 4.5. *Let G be k-colour-critical. If $k \geqslant 4$ and $G \neq K_k$, then $2 \cdot |E(G)| > (k-1) \cdot |V(G)| + ((k-3)/(k^2-3)) \cdot |V(G)|$.*

4.2. The colouring number

Colouring the vertices of a graph G in the order $x_1, x_2, \ldots, x_n$, giving each vertex a smallest possible colour from $1, 2, \ldots$, produces a colouring using at most $\max_i \deg(x_i, G[x_1, x_2, \ldots, x_i]) + 1$ colours, since x_i is joined to only $\deg(x_i, G[x_1, x_2, \ldots, x_i])$ previously coloured vertices. $G[X]$ denotes the subgraph of G induced by X. Taking the minimum over all possible permutations P of the vertices, gives

$$\chi(G) \leqslant \min_P \max_i \deg(x_{P(i)}, G[x_{P(1)}, x_{P(2)}, \ldots, x_{P(i)}]) + 1.$$

The number on the right hand side is called the *colouring number* of G, denoted $\operatorname{col}(G)$. The name "colouring number" was first used by Erdős and Hajnal (1966). Obviously $\chi(G) \leqslant \operatorname{col}(G) \leqslant \Delta(G) + 1$, with $\operatorname{col}(G) = \Delta(G) + 1$ iff G has a $\Delta(G)$-regular connected component. The colouring number seems like another intractable graph constant as the minimum is taken over $n!$ permutations: however, it is not, as was first proved independently by Finck and Sachs (1969) and Matula

(1968). Before showing this, let us consider a very closely related bound, due to Szekeres and Wilf (1968).

Theorem 4.6. *Let f be a real valued function defined on all graphs, such that f is non-decreasing (i.e. $G_1 \subseteq G_2 \Rightarrow f(G_1) \leqslant f(G_2)$) and $f(G) \geqslant \delta(G)+1$ for all G. Then $\chi(G) \leqslant f(G)$. Moreover, the unique best such function f is*

$$SW(G) = \max \min_{x \in V'} \deg(x, G[V']) + 1 = \max \delta(G[V']) + 1,$$

where the maximum is taken over all subsets V' of $V(G)$.

Proof. Let G be k-chromatic and G' be a k-colour-critical subgraph of G with vertex set V'. By Theorem 4.3, $\delta(G') \geqslant k-1$, hence $k \leqslant \delta(G')+1 \leqslant \delta(G[V'])+1 \leqslant f(G[V']) \leqslant f(G)$.

For the second part, it is obvious that SW satisfies the two properties. And any G has a subset V' of $V(G)$ such that $\mathrm{SW}(G) = \delta(G[V']) + 1$. Moreover, $\delta(G[V'])+1 \leqslant f(G[V']) \leqslant f(G)$ for any f. Therefore $\mathrm{SW}(G) \leqslant f(G)$ for all G and all f satisfying the two properties. □

Again, since there are $2^{|V(G)|}$ possible sets V', the graph constant SW seems to be intractable, but it is not. The following surprising, but easy, theorem of Finck and Sachs (1969) shows that col(G) can be obtained from a smallest last ordering of the vertices of G.

Theorem 4.7. *For any graph G the numbers col(G) and SW(G) are equal. Moreover, if $x_1, x_2, \ldots, x_n$ is a smallest last ordering of the vertices of G, then*

$$\begin{aligned}\mathrm{col}(G) &= \max_i \deg(x_i, G[x_1, x_2, \ldots, x_i]) + 1 \\ &= \max_i \delta(G[x_1, x_2, \ldots, x_i]) + 1 = \mathrm{SW}(G).\end{aligned}$$

Proof. Let $x_1, x_2, \ldots, x_n$ be a smallest last ordering. It is obvious from the definitions of col and SW that

$$\begin{aligned}\mathrm{col}(G) &\leqslant \max_i \deg(x_i, G[x_1, x_2, \ldots, x_i]) + 1 \\ &= \max_i \delta(G[x_1, x_2, \ldots, x_i]) + 1 \leqslant \mathrm{SW}(G).\end{aligned}$$

On the other hand, it is a matter of routine to check that col has the two properties of Theorem 4.6 and therefore, by that theorem, $\mathrm{SW}(G) \leqslant \mathrm{col}(G)$. □

Graphs G for which $\mathrm{SW}(G) \leqslant k+1$ have also been called *k-degenerate* by Lick and White (1970). Thus the property to be of colouring number $k+1$ is often called to be of *degeneracy* k.

4.3. Lower bounds

The size $\omega(G)$ of a largest complete subgraph of G is an obvious lower bound for $\chi(G)$; however, Karp (1972) noted that ω is as intractable as χ. Moreover, ω is often not a very good lower bound, as indicated in Theorem 2.1. By taking into consideration a more general class of subgraphs than the complete ones, Hajós (1961) overcame this second deficiency of ω as a lower bound. A graph G is *k-constructible* if *either* $G = K_k$ *or* G is obtained from two disjoint k-constructible graphs G_1 and G_2 with edges $x_1y_1 \in E(G_1)$ and $x_2y_2 \in E(G_2)$ by removing x_1y_1 and x_2y_2, identifying x_1 and x_2 to one vertex x, and joining y_1 and y_2 by a new edge *or* G is obtained from a k-constructible graph G_1 by identifying two non-adjacent vertices x_1 and x_2, removing possible parallel edges that might arise. The two operations above we shall call *Hajós' construction* and *identification*, respectively.

It is easy to see that if G contains a k-constructible subgraph G' then $\chi(G) \geqslant \chi(G') \geqslant k$. Hajós' theorem states that this lower bound for $\chi(G)$ is best possible for any G, and thus the chromatic number is characterized in constructive terms.

Theorem 4.8. *Let G be any graph. Then $\chi(G) \geqslant k$ iff G has a k-constructible subgraph. In particular, if G is k-colour-critical then G is k-constructible.*

Proof. We shall assume that the theorem is false, i.e. there is a G with $\chi(G) \geqslant k$ and without a k-constructible subgraph. G may be assumed to be maximal in the sense that the addition of any new edge e to G gives rise to a k-constructible subgraph G_e of $G \cup e$. If non-adjacency is an equivalence relation on $V(G)$ then the number of equivalence classes is $\geqslant k$, since $\chi(G) \geqslant k$. But then G contains the k-constructible graph K_k. Therefore, there are three vertices x, y and z with $xy \notin E(G)$, $yz \notin E(G)$ and $xz \in E(G)$. Then disjoint copies of the k-constructible graphs G_{xy} and G_{yz} may be combined by Hajós' construction, removing the edges xy and yz, identifying the two copies of y and adding xz. For each vertex t both in G_{xy} and G_{yz} we then identify the two copies of t, thereby obtaining a k-constructible subgraph of G. Contradiction. □

Thus, to prove $\chi(G) \geqslant k$ we need only exhibit a k-constructible subgraph of G and by Theorem 4.8 there is at least one. The problem is that no reasonably sized upper bound on the length of the k-construction in terms of $|V(G)|$ is known, and probably no such polynomial bound exists. A detailed exposition of known results and unsolved problems related to Theorem 4.8 has been presented by Hanson et al. (1986).

A result reminiscent of Hajós' theorem 4.8 was obtained by D.J. Kleitman, and Lovász (1982, 1994) as a dual to an important corresponding result for the maximum size of a stable set by Li and Li (1981). For a graph G on n vertices $1,2,\ldots,n$ define the graph polynomial $f(G\colon x_1, x_2, \ldots, x_n) = \Pi(x_i - x_j)$, where the product is taken over all pairs (i,j) with $i < j$ and $ij \in E(G)$. The graph polynomial was first defined and used by J.J. Sylvester and J. Petersen in 1878 and 1891.

Theorem 4.9. *Let G be any graph. Then $\chi(G) \geqslant k$ iff there exist graphs $G_1, G_2, \ldots, G_m$ on $V(G)$, each containing a K_k as a subgraph, such that*

$$f(G\colon x_1, x_2, \ldots, x_n) = \sum_{i=1}^{m} \varepsilon_i f(G_i\colon x_1, x_2, \ldots, x_n),$$

where ε_i is a suitable sign (+1 or −1).

Thus, to prove $\chi(G) \geqslant k$ we need only exhibit the graphs G_i and the sum of their polynomials. A problem again is that no reasonably sized upper bound on m in terms of n is known; probably no such polynomial bound exists.

A different type of bound in terms of the eigenvalues λ of the adjacency matrix was obtained by Hoffman in 1970, see the survey by Hoffman (1975). The upper bound is a possible f in Theorem 4.6.

Theorem 4.10. $1 - (\lambda_{\max}/\lambda_{\min}) \leqslant \chi \leqslant 1 + \lambda_{\max}$.

Finally, let us mention a result of Lovász (1978a), used to prove $\chi(KG_{n,k}) \geqslant k + 2$ for the Kneser graphs, but whose graph theoretic meaning is not well understood. The *neighbourhood complex* of a graph G is a simplicial complex whose vertices are the vertices of G and whose simplices are those subsets of $V(G)$ which have a neighbour in common in G.

Theorem 4.11. *If the neighbourhood complex of G is a $(k-1)$-connected topological space, then $\chi(G) \geqslant k + 2$.*

5. Colour-critical graphs

The importance of the notion of a critical graph is that problems for k-chromatic graphs in general may often be reduced to k-colour-critical graphs, and these are less arbitrary. And even if the k-colour-critical graphs for $k \geqslant 4$ are not very restricted, as we saw in section 2, they do give rise to interesting results and problems.

5.1. Subgraphs of colour-critical graphs

The 4-colour-critical graphs with many edges described in section 2, and generalizations of them, show that any $(k-2)$-chromatic graph for $k \geqslant 4$ is a subgraph of some k-colour-critical graph. A $(k-1)$-chromatic graph G' with an edge xy, such that any $(k-1)$-colouring of $G' - xy$ give x and y different colours, is not a subgraph of any k-colour-critical graph G. The reason for this is that $G' - xy$ would be a subgraph of $G - xy$, and in a $(k-1)$-colouring of $G - xy$ the vertices x and y must be coloured the same. Greenwell and Lovász (1974) proved that this is exactly what can go wrong. A proof of the "if" part can be based on Theorem 2.4.

Theorem 5.1. *For $k \geqslant 4$, a graph G' is a proper subgraph of some k-colour-critical graph iff G' is $(k-1)$-colourable and for all edges xy of G' the graph $G'-xy$ has a $(k-1)$-colouring φ'_{xy} in which x and y are coloured the same.*

As far as more special subgraphs are concerned, there are of course many $(k-1)$-colour-critical subgraphs in a k-critical graph. Toft (1974b) proved the following.

Theorem 5.2. *Let G be k-colour-critical, $k \geqslant 3$, and let x_1y_1 and x_2y_2 be two different edges of G. Then there is a $(k-1)$-colour-critical subgraph G' of G containing x_1y_1, but not x_2y_2.*

Proof. Consider a $(k-1)$-colouring φ of $G-x_1y_1$ with $\varphi(x_1)=\varphi(y_1)=1$. One of x_2 and y_2 has a colour different from 1, say $\varphi(x_2)=k-1$. Let A be the set of all nodes of colour $k-1$. Then $G-x_1y_1-A$ is $(k-2)$-colourable, however, $G-A$ is $(k-1)$-chromatic, since A is stable. Any $(k-1)$-colour-critical subgraph of $G-A$ is a possible G'. □

Corollary 5.3. *Let G be k-colour-critical, $k \geqslant 3$. Then G is $(k-1)$-edge-connected.*

Proof. This is true for odd circuits, i.e. for $k=3$. By induction we may assume it to be true for $k-1$. Let F be a cut in G. Obviously $|F| \geqslant 2$. Let e_1 and e_2 be edges in F. Then by Theorem 5.2 there is a $(k-1)$-colour-critical subgraph G' containing e_1, but not e_2. Moreover, by induction $|F \cap E(G')| \geqslant k-2$. Then $|F| \geqslant k-1$ because of e_2. □

Corollary 5.4. *Let G be k-colour-critical, $k \geqslant 4$, with all $(k-1)$-colour-critical subgraphs isomorphic to K_{k-1}. Then $G=K_k$.*

Proof. Let x, y and z form a K_3 in G. By Theorem 5.2 and the assumption, there is a $G_1=K_{k-1}$ containing x and y, but not z, and a $G_2=K_{k-1}$ containing y and z, but not x. Then $(G_1-xy)\cup(G_2-yz)\cup xz$ is not $(k-2)$-colourable, hence contains a $(k-1)$-colour-critical subgraph G'. By assumption $G'=K_{k-1}$. This is only possible if $G_1\cup G_2\cup xz=K_k\subseteq G$, and hence $G=K_k$. □

Corollary 5.3 is a classical result of Dirac (1953) generalizing $\delta(G)\geqslant k-1$. Corollary 5.4 and its proof is due to Stiebitz (1987).

J. Nešetřil and V. Rödl, at the conference in Keszthely in Hungary in 1973, asked if a large k-colour-critical graph always contains a large $(k-1)$-colour-critical subgraph. For $k \geqslant 5$ this is still open.

Finally, let us mention a deep result of Gallai (1963b), proved ingeniously by factorization theory applied to the complement. By Theorem 2.6, it implies that if we want a catalogue of all k-colour-critical graphs on $\leqslant k+t$ vertices for a fixed t, then it is sufficient to consider the cases $k \leqslant t+1$. This has been done for $t \leqslant 4$. Let us in passing note the easy fact that no k-colour-critical graph can have exactly $k+1$ vertices, as first pointed out by Dirac (1952c).

Theorem 5.5. *If G is k-colour-critical and $|V(G)| \leqslant 2k-2$, then the complement of G is disconnected, i.e. G contains a complete bipartite subgraph H with $V(H)=V(G)$.*

5.2. Low and high vertices

As an important extension of Brooks' theorem 4.1, Gallai (1963a) characterized the subgraph $G[L]$ of a k-colour-critical graph G induced by the set L of *low vertices*, i.e. the vertices of degree $k-1$.

Theorem 5.6. *Let G be k-colour-critical, $k \geqslant 4$. Then the blocks of the subgraph $G[L]$ of G are odd circuits and/or complete graphs.*

Proof. The idea of the proof is similar to the proof of Brooks' theorem 4.1. Let C be an even circuit of $G[L]$ with vertices $x_1, x_2, \dots, x_{2n}$, and let us assume that x_1 is not incident with any chord of C. Let φ be a k-colouring of G with colours $1, 2, \dots, k$ such that only x_1 has the colour k, x_{2n} has colour 1, and x_2 has colour 2. Some x_j has a colour different from 1 and 2, say 3, since C is even. The $k-3$ neighbours of x outside C have colours $3, 4, \dots, k-1$.

Now, push the colour k around C, as the undesired colour $s+1$ was pushed in the proof of Theorem 4.1, until the colour k is back at x. This gives a new k-colouring of G with only x_1 of colour k, but with the other colours of C moved: x_i has the original colour of x_{i+1} for $2 \leqslant i \leqslant 2n-1$ and x_{2n} has the original colour of x_2. Pushing the colour k around in C a number of times, the vertex x_{2n} will eventually get the colour 3. But then x_1 is joined to two vertices of colour 3, and hence it may be recoloured by one of the colours $1, 2, \dots, k-1$ to which it is not joined. Thus we obtain a $(k-1)$-colouring of G, which is impossible.

The conclusion is that any vertex of any even circuit C of $G[L]$ must be incident with a chord of C. But then it is easy to prove by induction that an even circuit in $G[L]$ has all its chords. This implies the desired block structure by straightforward arguments. □

If G is k-colour-critical, $k \geqslant 4$ and $G \neq K_k$, then by Brooks' theorem 4.1 G has at least one *high vertex*, i.e. a vertex of degree $\geqslant k$. It is a curious fact, first noted by Borodin and Kostochka (1977), that no k-colour-critical graph with high vertices, all of degree equal to k, seems to be known for $k \geqslant 9$.

Gallai (1963a) proved constructively that Theorem 5.6 is best possible. G.A. Dirac and Sachs and Stiebitz (1983) proved constructively that any graph that is a proper subgraph of a k-colour-critical graph can be made the subgraph $G[H]$ induced by the high vertices H of some k-colour-critical G. Dirac (1957a) and Gallai (1963a) characterized the situation when $G[H] = K_1$. Further work in this direction by Sachs and Stiebitz (1989) suggests that as long as $\chi(G[H]) \leqslant k-2$ the structure of G is manageable. Finally, a proof of Theorem 4.5 can be based on Theorem 5.6.

5.3. Circuits, subdivisions and Kempe chains

The answer to the Keszthely problem of J. Nešetřil and V. Rödl mentioned above is affirmative for $k = 4$, since Kelly and Kelly (1954) proved the existence of a long circuit in a large k-colour-critical graph G, and therefore also a long odd circuit. An alternative argument has been presented by Voss (1977, 1991). The growth of the length of a longest circuit as a function of $|V(G)|$ may however be slow. Taking a connected graph in which each block is a K_{k-1} or a K_2, with all vertices of degree $k-2$ or $k-1$ and degree $k-2$ only in endblocks, together with an extra vertex x joined to all vertices of degree $k-2$, gives a k-colour-critical graph G in which a longest circuit has length less than $2(k-1)\cdot\log|V(G)|/\log(k-2)$, as proved by Gallai (1963a). It seems not to be known if this example is best possible. A result from the other side, due to Erdős and Hajnal (1966), says that a k-colour-critical G always has an odd circuit of length $\geqslant k-1$. A generalization of this result was proved in a very elegant way by Neumann-Lara (1982), using generalized Kempe chains.

Suppose φ is a k-colouring of a graph G with colours $1,2,\ldots,k$. A connected component in the subgraph induced by the vertices of two of the colours, say i and j, is called a *Kempe chain*. The importance of this concept lies in the fact that changing the two colours in a Kempe chain gives a new k-colouring of G. Such interchanges, originally used by Kempe (1879) in his attack on the four colour problem, is still the main tool for proving configurations in planar graphs to be reducible. A well written exposition on Kempe chains and their uses was given by Whitney and Tutte (1972).

More generally, let P be any permutation of the colours $1,2,\ldots,k$ of φ. Consider a vertex x, all neighbours N_1 of colour $P(\varphi(x))$ of x, all neighbours N_2 of colour $P(P(\varphi(x)))$ of vertices in N_1, all neighbours N_3 of colour $P^3(\varphi(x))$ of vertices in N_2, etc. Changing the colour $\varphi(y)$ of all vertices y in $N=x\cup N_1\cup N_2\cup\cdots$ from $\varphi(y)$ to $P(\varphi(y))$ gives a new k-colouring of G. We call N *a generalized Kempe chain from x w.r.t. P*. Letting P be cycle permutations of length j one can easily prove the following.

Theorem 5.7. *Let G be k-colour-critical. Then any edge xy of G is contained in at least as many circuits of length* 1 *modulo j for $j=2,3,\ldots,k-1$ as an edge in K_k. In particular, for k odd any edge is in at least* $(k-2)!$ *different odd circuits of length $\geqslant k$.*

A circuit may be thought of as a *subdivision* K_3S of K_3, i.e. it is obtained by inserting new vertices of degree 2 on the edges. G. Hajós suggested the following generalization of the four colour theorem.

Conjecture 5.8. Let G be 5-chromatic. Then G contains a subdivision K_5S of K_5.

Any 4-chromatic graph contains a K_4S as proved by Dirac (1952a), but this does not depend very strongly on the graph being 4-chromatic, since already $\delta(G)\geqslant 3$ implies the existence of a K_4S. For $k\geqslant 7$ the corresponding statement is not true, as proved by Catlin (1979), who exhibited the basic 8-colour-critical counterexamples of five complete 3-graphs joined completely to each other as a 5-cycle. Erdős and Fajtlowicz (1981) proved the following.

Theorem 5.9. *The statement "G contains a subdivision of a $K_{\chi(G)}$" is false for almost all graphs.*

Proof. Consider all graphs with vertex set $1, 2, \ldots, n$. If we consider each edge to appear with probability $\frac{1}{2}$, then each of the $2^{\binom{n}{2}}$ possible graphs G has probability $2^{-\binom{n}{2}}$. The average number of stable sets of size r is $\binom{n}{r} \cdot 2^{\binom{n}{2}-\binom{r}{2}}/2^{\binom{n}{2}}$. For $r = \lceil (2+\varepsilon)\log n/\log 2 \rceil$ this average number tends to 0 for $n \to \infty$. We express this by saying that almost all graphs have all stable sets of size at most $(2+\varepsilon)\log n/\log 2$ and hence chromatic number $\geqslant (\log 2/(2+\varepsilon)) \cdot (n/\log n)$.

Let $m \approx 3\sqrt{n}$. The average number of edges on any set of size m is $\frac{1}{2}\binom{m}{2}$, and it can be shown that almost all graphs have $< (\frac{1}{2}+\varepsilon) \cdot \binom{m}{2}$ edges on every set of m vertices. Therefore the existence of a $K_m S$ implies the existence of at least $(\frac{1}{2}-\varepsilon)\binom{m}{2}$ vertices to subdivide the edges of K_m. But $(\frac{1}{2}-\varepsilon)\binom{m}{2} > n$, so almost all graphs contain no $K_m S$. Since m is less than χ for almost all graphs the theorem follows. □

More precise estimates than those in the above proof were given by Bollobás and Catlin (1981).

5.4. Contraction-critical graphs

The existence of many different types of k-colour-critical graphs may tempt us to look for a less fine ordering than $\subseteq$ in the set of all graphs, to obtain more restrictive minimal graphs. Let $G_1 \geqslant G_2$ denote that G_2 can be obtained from G_1 by deletions of vertices and/or edges and contractions of edges, in any order. The terminology that G_2 is a *minor* of G_1 is often used. The k-chromatic graphs that are minimal w.r.t. $\geqslant$ are called *k-contraction-critical*. Hadwiger's conjecture 1.3 then becomes as follows.

Conjecture 5.10. $\chi(G) \geqslant k$ implies $G \geqslant K_k$, i.e. K_k is the only k-contraction-critical graph.

Dirac (1960, 1964a) obtained the basic properties of k-contraction-critical graphs.

Theorem 5.11. *Let G be k-contraction-critical $\neq K_k$. Then $\delta(G) \geqslant k \geqslant 5$. For each vertex x of G there are at most* $\deg(x) - k + 2$ *stable vertices among the neighbours of x. Moreover, G is 5-connected.*

K_6 is non-planar and $\delta(G) \geqslant 6$ implies non-planarity, hence the first statement of Theorem 5.11 for $k = 6$ implies the five colour theorem for planar graphs. A deeper generalization is the following result, due to Dirac (1964a).

Theorem 5.12. *If $\chi(G) \geqslant 6$ then $G \geqslant K_6^-$, where K_6^- denotes the complete 6-graph with one edge missing.*

Proof. $G \geqslant G^*$ where G^* is 6-contraction-critical. If $G^* = K_6$ we are done, hence assume $G^* \neq K_6$. A best possible extremal result of Dirac (1964b) proved by induction over $|V(H)|$ says that $2|E(H)| > 7|V(H)| - 15$ implies $H \geqslant K_6^-$. By this and Theorem 5.11, $\delta(G^*) = 6$. Let $\deg(x) = 6$ and let N be the six neighbours of

x in G^*. By Theorem 5.11 the graph $N^* = G^*[N]$ does not have a stable set of size 3, hence it must contain a triangle.

Suppose first that N^* contains two disjoint triangles on the vertices x_1, x_2, x_3 and x_4, x_5, x_6, respectively. Then by Theorem 5.11 $G - x$ contains three disjoint paths P_1, P_2, P_3, where P_i joins x_i and x_{i+3}. The removal of x, x_1, x_3 and x_5 leaves a connected graph by Theorem 5.11, hence it contains a path P between $P_2 - x_5$ and, say, $P_3 - x_3$. Contracting P into an edge and $P_1, P_2 - x_5$ and $P_3 - x_3$ into single vertices, we obtain a K_6^-.

If N^* does not have two disjoint triangles, then since there are no stable sets of three vertices, N^* contains a K_4^-. With x we get a K_5^- in G^*. Since G^* is 5-connected, any vertex y outside the K_5^- is joined to it by five paths, any pair of which have only y in common. But then again $G^* \geqslant K_6^-$. □

Further developments of the above results have been obtained by Mader (1968a,b), Jakobsen (1971) and Toft (1972b).

Finally, we turn our attention to Wagner's equivalence theorem. Wagner (1937) characterized those graphs not contractible to K_5. A simplified proof of Wagner's theorem was given by Halin (1964). If one is only interested in the equivalence theorem a shortcut is possible, since only the following corollary of Wagner's theorem is needed. This was pointed out by Young (1971), but the corollary was also explicitly mentioned by Ore (1967).

Theorem 5.13. *If G is 4-connected and contains a $K_{3,3}S$, then $G \geqslant K_5$.*

The proof is a rather simple case analysis. A slight extension of Theorem 5.13 is the basis of Halin's proof of the full Wagner theorem. The surprising consequence of Theorem 5.13 is the following.

Theorem 5.14. *If G is 5-contraction-critical $\neq K_5$, then G is planar. In particular Hadwiger's conjecture 5.10 for $k = 5$ is equivalent to the four colour theorem.*

Proof. A 5-contraction-critical G is 4-connected by Theorem 5.11. If G is non-planar then by Kuratowski's theorem G contains a K_5S or a $K_{3,3}S$. By Theorem 5.13 $G \geqslant K_5$, hence $G = K_5$.

Since $G \geqslant K_5$ implies G to be non-planar, Hadwiger's conjecture 5.10 for $k =$ 5 implies the four colour theorem. Conversely, a counterexample to Hadwiger's conjecture 5.10 for $k = 5$ implies a counterexample to the four colour theorem by the first part of Theorem (5.14). □

Recently Robertson et al. (1993) have obtained deep structural results on graphs not contractible to K_6. As a corollary they obtain that all such graphs are 5-colourable, assuming the four colour theorem, thus proving Hadwiger's conjecture 5.10 for $k = 6$.

6. Graphs on surfaces

Most of the early papers on colouring theory deal with maps M on surfaces and *face colourings* of M, i.e., colourings of the connected regions into which M divides the surface, such that faces with a common boundary line get different colours. By assigning a vertex to each face of M and joining any two of these by an edge whenever the corresponding two faces share a common boundary line, we obtain the *dual graph* G of M. The graph G can also be embedded on the surface and a colouring of the vertices of G corresponds to a face colouring of M. This translation of face colouring to vertex colouring was first pointed out by Kempe (1879). Kempe (1879) also mentioned the importance of colouring theory for abstract graphs, but until the papers by Whitney (1932) and Brooks (1941) the theory remained centered on maps and graphs on surfaces.

Colouring theory for graphs on surfaces is well developed. The monograph by Biggs et al. (1976) contains an account of the history until 1936.

6.1. Planar graphs

Instead of colouring faces or vertices, Tait (1878–80) considered colourings of the edges of a planar graph. A *plane triangulation* G is a graph embedded in the plane for which each face boundary is a K_3. Such a G is also called *maximal plane*; it has $3|V(G)|-6$ edges and is at least 3-connected for $|V(G)| \geqslant 4$. To prove the four colour theorem it is, of course, sufficient to prove that plane triangulations are 4-colourable. Tait (1878–80) stated the following theorem in a face-colouring formulation.

Theorem 6.1. *Let G be a plane triangulation. Then $\chi(G) \leqslant 4$ iff the edges of G can be labeled by three labels such that each face boundary has all three labels.*

Proof. Let φ be a 4-colouring of G with colours (0,0), (0,1), (1,0) and (1,1). These colours may be added coordinatewise modulo 2. For each edge xy define $\psi(xy) = \varphi(x) + \varphi(y)$. Then ψ is a desired labeling of the edges with labels (1,0), (0,1) and (1,1).

Conversely, let ψ be a labeling of the edges with labels (1,0), (0,1) and (1,1) such that each face has all three labels. Let C be any circuit of G. The interior of C is divided into triangular faces T_i. Hence, since we add modulo 2 and (1,0) + (0,1) + (1,1) = (0,0),

$$\sum_{e \in E(C)} \psi(e) = \sum_i \sum_{e \in T_i} \psi(e) = (0,0).$$

Colour an arbitrary vertex z of G by the colour $\varphi(z) = (0,0)$. For any vertex x let W be a walk from z to x. Define $\varphi(x) = \sum_{e \in E(W)} \psi(e)$. Then $\varphi(x)$ is well-defined, i.e. $\varphi(x)$ does not depend on the particular W chosen, since ψ sums to zero around any circuit. Moreover, φ is a 4-colouring with colours (0,0), (1,0), (0,1) and (1,1). □

The face-colouring version of Theorem 6.1 states that a plane 2-connected 3-regular graph is 4-face-colourable iff it is 3-edge-colourable, where any two adjacent edges in an *edge-colouring* have different colours. Tait (1878–80) suggested that any 2-connected 3-regular graph can be 3-edge-coloured, but the famous graph of Petersen (1898) disproves this.

The following theorem is due to Heawood (1898). It was also originally formulated for face colourings.

Theorem 6.2. *Let G be a plane triangulation. Then* $\chi(G) \leqslant 4$ *iff the faces can be labeled by two labels* $+1$ *and* -1, *such that the sum of the labels of the faces around any vertex is* 0 *modulo* 3.

There are, of course, other equivalents of the four colour theorem than those early ones implied by Theorems 6.1 and 6.2 above, for example two statements obtained from different characterizations of the chromatic number of an arbitrary graph in terms of orientations, one due to Minty (1962) and the other independently due to T. Gallai and B. Roy, see chapter 1.

As far as 2-colourings are concerned it follows from Theorem 1.1 that a 2-connected plane graph is 2-colourable iff all circuits are even iff all face boundaries are even. For 3-colourings of plane graphs we cannot expect any good characterization because of Theorem 3.4; however, for triangulations Heawood (1898) stated the following.

Theorem 6.3. *Let G be a plane triangulation. Then G is* 3*-colourable iff all vertices of G have even degree.*

A deep theorem obtained by Grötzsch (1959) says the following.

Theorem 6.4. *Let G be a plane graph without triangles. Then G is* 3*-colourable.*

Grünbaum (1963) suggested that Theorem 6.4 remains true, even if there are up to three triangles in G. This was proved by Aksionov (1974); see also Aksionov and Melnikov (1978) and Steinberg (1993), correcting a mistake discovered by T. Gallai in Grünbaum's original argument. A short proof of Grötzsch's theorem 6.4 has recently been given by Thomassen (1993b), who also proved that graphs embedded on the torus and having all circuits of length $\geqslant 5$ are 3-colourable.

6.2. Heawood's formula

Curiously, colouring theory for *higher surfaces*, i.e. connected compact 2-dimensional manifolds, is in general simpler than for the plane or sphere. We shall see why this is so. The topological prerequisite will be the generalized Euler formula

$$|F(G)| - |E(G)| + |V(G)| \geqslant 3 - h(S),$$

where G is a graph embedded on the surface S, $F(G)$ is the set of faces of G, and $h(S)$ is the *connectivity* of S. The number $3 - h(S)$ is also called the *Euler characteristic* of S. The possible S may be classified as follows: The *orientable surfaces* are the sphere S_0 with $h = 1$ and the sphere with g handles S_g with $h = 1 + 2g$. S_1 is the torus. The *non-orientable surfaces* are the projective plane P_0 with $h = 2$, the Klein bottle K_0 with $h = 3$, the projective plane with g handles P_g with $h = 2 + 2g$, and finally the Klein bottle with g handles K_g with $h = 3 + 2g$.

For a connected graph G with $|V(G)| \geqslant 3$ each face is bounded by at least 3 edges, hence $3|F(G)| \leqslant 2|E(G)|$. Then from the generalized Euler formula:

$$|E(G)| \leqslant 3|V(G)| - 9 + 3h. \tag{6.5}$$

The formula of Heawood (1890) is now easy to obtain.

Theorem 6.6. *Let G be embeddable on a surface S of connectivity $h \geqslant 2$. Then*

$$\chi(G) \leqslant \left\lfloor \frac{7 + \sqrt{24h - 23}}{2} \right\rfloor = H(h).$$

Proof. Suppose that G is k-chromatic with $k \geqslant 7$. Let G' be a k-colour-critical subgraph with n vertices and e edges. Since $\delta(G') \geqslant k - 1$ we have by (6.5)

$$(k - 1) \cdot n \leqslant 2e \leqslant 6n - 18 + 6h.$$

Since $n \geqslant k \geqslant 7$ this implies

$$(k - 7) \cdot k - 6(h - 3) \leqslant 0$$

or

$$k \leqslant \left\lfloor \frac{7 + \sqrt{24h - 23}}{2} \right\rfloor = H(h). \qquad \square$$

For the torus S_1 Heawood's formula gives $\chi(G) \leqslant 7$. Heawood (1890) noted that the 7-chromatic graph K_7 can be embedded on the torus, hence the colouring problem for the torus has the answer 7. For the Klein bottle the formula also gives $\chi(G) \leqslant 7$, but Franklin (1934) proved that the colouring problem for the Klein bottle has the answer 6. However, this is the only case for which the *Heawood number* $H(h)$ is not the correct answer to the colouring problem. This is the remarkable theorem of Ringel (1954) and Ringel and Youngs (1968). For orientable surfaces with $3 \leqslant h \leqslant 15$ this had been established already by Heffter (1891).

Theorem 6.7. *For a surface S of connectivity $h \geqslant 2$, where S is not the Klein bottle, the Heawood number $H(h)$ is the maximum chromatic number of graphs embeddable on S.*

The proof of this major result, completed in 1968, was obtained by ingenious embeddings of the complete $H(h)$ graphs on the surfaces of connectivity h as described also in the excellent monograph by Ringel (1974). This is, of course, sufficient for the proof of Theorem 6.7; however, Dirac (1952c) proved that it is also necessary, in fact in a very strong sense. The theorem of Dirac was perhaps the first breakthrough in colouring theory for higher surfaces after the paper of Heffter (1891). The idea of such a result and a proof in the case of the torus was first obtained by P. Ungar. We shall present the simple proof of Dirac (1957b).

Theorem 6.8. *For $h = 3$ and $h \geqslant 5$ any $H(h)$-chromatic graph G on a surface of connectivity h contains a $K_{H(h)}$ as a subgraph.*

Proof. Let $k = H(h)$ and G' be a k-colour-critical subgraph of G with n vertices and e edges. If $G' \neq K_k$ then by Theorems 4.4 and 6.5

$$(k-1) \cdot n + (k-3) \leqslant 2e \leqslant 6n + 6(h-3). \tag{6.9}$$

Since there are no k-colour-critical graph on $k+1$ vertices, as we remarked in section 5, we have $n \geqslant k+2$. Thus for $h = 3$ and $h \geqslant 5$ this gives a contradiction by an easy calculation. □

Dirac's Theorem 6.8 also holds in the missing cases $h = 2$ and 4 as proved by Albertson and Hutchinson (1979). The simple type of argument used above also shows the following.

Theorem 6.10. (a) *Let G be a k-colour-critical graph $\neq K_k$, $k \geqslant 8$, on a surface of connectivity $h \geqslant 5$. Then $|V(G)| \leqslant (6h - 15 - k)/(k-7)$.*

(b) *Let G be a k-contraction-critical graph $\neq K_k$, $k \geqslant 7$, on a surface of connectivity $h \geqslant 3$. Then $|V(G)| \leqslant (6h - 18)/(k-6)$.*

Dirac (1956, 1957c) obtained a number of results based on the above finiteness. For example Hadwiger's conjecture 5.10 is true for all $(H(h) - 1)$-chromatic graphs on a surface of connectivity $h \geqslant 5$. Albertson and Hutchinson (1980) proved the corresponding result for 6-chromatic graphs on the torus, using the four colour theorem, but a simpler argument has been obtained by Mayer (1989). Thomassen (1994a) proved that there are only four 6-critical graphs on the torus, and he observed that Gallai's theorem 5.6 implies that the number of 7-critical graphs on any fixed surface is finite. In fact, this follows more directly from Theorem 4.5.

It is tempting to suggest that any locally planar graph on any surface is 4-colourable (with some suitable definition of locally planar). However, Fisk (1978) constructed 5-chromatic graphs satisfying any reasonable local planarity definition on any surface except the sphere. Albertson and Stromquist (1982) proved that locally planar toroidal graphs are 5-colourable, a best possible result by Fisk's examples. Hutchinson (1984) defined "locally planar" to mean that all edges are "short" and proved that such locally planar graphs on any surface are 5-colourable. This was generalized by Thomassen (1993a), who proved that there exists a number $n(S)$ such that any graph embedded on S with any non-contractible circuit of length $> n(S)$ is 5-colourable (a circuit is called *contractible* if its deletion disconnects the surface in such a way that one of the resulting components is homeomorphic to a plane disc).

7. Stable sets

Colouring a graph G by k colours means covering the vertex set $V(G)$ by k stable sets. To ask for the existence in G of just one stable set of a given size seems simpler; however, the theory of stable sets in graphs is as complicated as colouring theory with many results of a similar nature. For example, the four colour theorem implies the existence of a stable set of size at least $|V(G)|/4$ in a planar graph, but no proof of this last result independent of the seemingly much stronger four colour theorem is known.

In this and the next section we look in particular at algorithmic aspects, critical graphs and stable set polytopes, although stable sets occur in many other contexts, for example in the theorems of Ramsey (1930), Turán (1941), Gallai and Milgram (1960) and Hajnal and Szemerédi (1970).

The monograph by Lovász and Plummer (1986) contains in chapter 12 an excellent survey of critical graphs and stable set polytopes.

7.1. *Algorithmic aspects*

To determine for a given graph G and a given integer k if G has a stable set of size k, is NP-complete as proved by Karp (1972). Thus $\alpha(G)$, the size of a maximum stable set in G, is as intractable as $\chi(G)$ is. For special types of graphs, however, $\alpha(G)$ is well-characterized and computable by polynomial algorithms. The following max–min theorem is due to König (1931) and Gallai (1958).

Theorem 7.1. *Let G be bipartite without isolated vertices. Then $\alpha(G)$ equals the size $\rho(G)$ of a minimum set of edges covering all vertices.*

Proof. $\alpha(G) \leqslant \rho(G)$ is obvious in any graph without isolated vertices. For the other direction, delete edges from G as long as α does not change. By Theorem 7.2 below, the obtained graph H consists of a set A of isolated vertices and a set B of disjoint edges. The edges of B together with one edge from each vertex of A cover all vertices, hence $\rho(G) \leqslant |A| + |B| = \alpha(H) = \alpha(G)$. □

The set B of edges in the proof of Theorem 7.1 is a maximum matching in G, since a larger matching M implies the contradiction $\alpha(G) \leqslant |V(G)| - |M| < |V(G)| - |B| = |A| + |B| = \alpha(G)$. Thus the well-known polynomial algorithms to determine a maximum matching in a bipartite graph G immediately also determine $\alpha(G)$, and in fact a stable set of size $\alpha(G)$. For line graphs $L(G)$ the polynomial matching algorithm of Edmonds (1965) applied to G determines $\alpha(L(G))$. More generally, a polynomial algorithm to determine $\alpha(G)$ for claw-free graphs G was obtained by Minty (1980) and Sbihi (1980).

Not being able to compute $\alpha(G)$ efficiently in general, one might look for upper and lower bounds as was done for $\chi(G)$ in section 4. Lovász (1982, 1994) gave excellent surveys of such bounds, including the theorem of Li and Li (1981) referred to in section 4.

7.2. *α-critical graphs*

In his elegant proof of Turán's theorem, Zykov (1949) introduced the notation of a *k-saturated* graph as a graph not containing a K_{k+1}, but containing a K_{k+1} whenever any new edge is added. If x is a vertex joined to all others in a graph G, then G

is k-saturated iff $G-x$ is $(k-1)$-saturated. Thus in a study of saturated graphs one might exclude such vertices x. Doing this and considering complements the class of *α-critical* graphs is obtained, i.e. a graph G is α-critical if it has no isolated vertices and $\alpha(G-xy) > \alpha(G)$ for all edges $xy \in E(G)$.

Because of the NP-completeness of $\alpha(G)$ we cannot hope that α-critical graphs have a simple structure; in fact, if there is a polynomial way to demonstrate that a graph is α-critical then NP would be equal to co-NP, which is considered unlikely. However, various structural properties and a deep classification result have been obtained, as we shall see. The following useful result is due to Berge (1972).

Theorem 7.2. *In an α-critical graph G any two adjacent edges uv and vw belong to an odd circuit without chords. In particular, a bipartite α-critical graph consists of a set of disjoint edges.*

Proof. Let T_u and T_w denote stable sets of size $\alpha(G)+1$ in $G-uv$ and $G-vw$, respectively. Let G' be the subgraph of G induced by the symmetric difference T of T_u and T_w, i.e. $T=(T_u \cup T_w)-(T_u \cap T_w)$. G' is bipartite with $T_u \backslash T_w$ and $T_w \backslash T_u$ forming the two sides. Of course, $u \in T_u \backslash T_w$ and $w \in T_w \backslash T_u$. The vertex v is joined by edges to only u and w in G', so if u and w are in the same connected component of G' then a shortest uw-path in G' and the edges uv and vw form the desired odd circuit. If u and w are in different connected components of G', then the graph induced by $T \cup v$ is bipartite. Thus $(T_u \cap T_w - v)$ together with at least half of the vertices of $T \cup v$ is a stable set in G. It has size $> \alpha(G)$. This contradiction proves the theorem. □

Obviously, a graph is α-critical iff all connected components are α-critical. A connected α-critical graph different from a single edge is 2-connected by Theorem 7.2. Moreover, the α-critical graphs that are 2-connected, but not 3-connected, can be characterized constructively. This was done independently by Gallai, Plummer and Wessel, see Wessel (1970). The construction corresponds to Theorem 2.5(b) (ii) for colour-critical graphs. A special case says that subdividing an edge of a 2-connected α-critical graph by an even number of new vertices preserves α-criticality. For later use let us note that this also preserves the value of $|V(G)|-2\alpha(G)$.

By induction over $|I|$ the following is easy to prove.

Theorem 7.3. *For any stable set I of an α-critical graph G the neighbours $N(I)$ of I, i.e. the vertices outside I joined to I by edges, satisfy $|N(I)| \geqslant |I|$. In particular, $|V(G)|-2\alpha(G) \geqslant 0$.*

The property of Theorem 7.3 is important in general, as Tutte (1953) proved it equivalent to having a *perfect 2-matching*, i.e. a system of disjoint circuits and edges covering all vertices.

The following three results were obtained by Hajnal (1965), Erdős et al. (1964) and Lovász (1977).

Theorem 7.4. *Let G be α-critical. Then*

(a) *All vertices have degree $\leqslant |V(G)|-2\alpha(G)+1$.*

(b) *G has at most* $\binom{|V(G)|-\alpha(G)+1}{2}$ *edges.*

(c) *For any subset* $T \subseteq V(G)$ *covering all edges of G the subgraph induced by T has at most* $\binom{|T|-\alpha(G)+1}{2}$ *edges.*

(b) follows from (a), whereas the proof of (c) is substantially more involved using multilinear algebra. (c) can be used to obtain a classification of all connected α-critical graphs G according to the value of $|V(G)| - 2\alpha(G)$. The classification was conjectured by Gallai and proved by Lovász (1978b).

Theorem 7.5. *For any fixed value* $v \geqslant 1$ *there exists a finite set of basis graphs such that all connected* α*-critical graphs G with* $|V(G)| - 2\alpha(G) = v$*, and only these, arise from the basis graphs by subdividing edges by an even number of vertices.*

In flavour this result is reminiscent of Gallai's theorem 5.5 for colour-critical graphs, implying for these graphs a finite basis for any fixed value t, where $t = |V(G)| - \chi(G)$.

8. Perfect graphs

The opposite extreme of the sparse graphs described in section 2 are the perfect graphs. A graph G is *perfect* if G and each of its induced subgraphs have the property that the chromatic number χ equals the size of a largest complete subgraph ω. The idea of perfect graphs, due to C. Berge, has been one of the most fruitful in graph theory, and it also relates to linear programming and to computational complexity. For perfect graphs both χ, ω and the size of a largest stable set α can be found by polynomial algorithms. This was proved by Grötschel et al. (1981) based on the ellipsoid method for linear programming. However, it is not known whether there is a polynomial algorithm to decide if a given graph is perfect or not.

8.1. Examples of perfect graphs

Berge (1961) considered several classes of perfect graphs, and he conjectured that complements of perfect graphs are perfect. For example, any bipartite graph is perfect, since $\chi = \omega = 2$ if there is at least one edge and $\chi = \omega = 1$ if there are none. Also the complements of bipartite graphs are perfect. This is equivalent to Theorem 7.1, since $\chi(\bar{G})$ is the minimum number of complete subgraphs of G covering all vertices of G, and thus equals $\rho(G)$ for G bipartite without isolated vertices. A reformulation of Theorem 7.1 is therefore as follows.

Theorem 8.1. *If G consists of two disjoint complete graphs and some edges between them, then* $\chi(G) = \omega(G)$.

Theorem 8.1 has been used as a lemma in graph colouring theory by Brown and Jung (1969), T. Gallai, and Toft (1972b). For example, the characterization of the

k-colour-critical graphs with a cut of $k-1$ edges, mentioned in section 2, follows from it.

A graph is called a *rigid-circuit graph* (or *triangulated* or *chordal*) if every circuit of length $\geqslant 4$ has a chord, i.e. an edge joining two non-adjacent vertices on the circuit.

Berge (1961) proved that rigid-circuit graphs are perfect, and Hajnal and Surányi (1958) and Dirac (1961) characterized them constructively and proved that their complements are perfect as well.

Theorem 8.2. *G is a rigid-circuit graph iff either G is complete or G can be obtained from two smaller disjoint rigid-circuit graphs G_1 and G_2 by identifying two complete subgraphs of the same size in G_1 and G_2. In particular a rigid-circuit graph is perfect.*

Proof. If G can be obtained as described then obviously G is a rigid-circuit graph. Conversely, let G be a non-complete rigid-circuit graph and let S be a minimal set of vertices in G such that $G-S$ is disconnected. Let $G=G_1\cup G_2$ with $G_1\cap G_2=G[S]$ and $G_1\neq G[S]\neq G_2$. If $G[S]$ contains two vertices x and y not joined by an edge, then let P_1 and P_2 be shortest paths from x to y in G_1 and G_2, respectively, with only the vertices x and y in S. Then $P_1\cup P_2$ is a circuit of length $\geqslant 4$ without a chord. This is a contradiction, and therefore $G_1\cap G_2$ is complete. Moreover, G_1 and G_2 are rigid-circuit graphs, since they are induced subgraphs of the rigid-circuit graph G. So the first part of the theorem follows. Since $\chi(G)=\max(\chi(G_1),\chi(G_2))$ the second part of the theorem follows by induction over $|V(G)|$. □

Theorem 8.3. *Let G be a rigid-circuit graph. Then for all proper complete subgraphs K of G there is a vertex x outside K for which the neighbours $N(x)$ of x in G induce a complete graph. In particular $\bar{G}$ is perfect.*

Proof. The existence of x is proved by induction over $|V(G)|$. If G is complete, the existence of x is obvious. Otherwise, by Theorem 8.2 $G=G_1\cup G_2$, where $G_1\cap G_2$ is complete, and G_1 and G_2 are smaller rigid-circuit graphs than G. In this case K is a subgraph of either G_1 or G_2, say G_1. Then by induction x can be chosen in $G_2-(G_1\cap G_2)$. This proves the existence of x.

That $\bar{G}$ is perfect is also obtained by induction over $|V(G)|$. From the first part we have

$$\omega(\bar{G})\geqslant\omega(\bar{G}-x-N(x))+1=\chi(\bar{G}-x-N(x))+1\geqslant\chi(\bar{G}),$$

where the middle equality holds by induction. Hence $\omega(\bar{G})=\chi(\bar{G})$ and $\bar{G}$ is perfect. □

There are several other classes of known perfect graphs, for example *comparability graphs*, where $V(G)$ is the set of elements of a partially ordered set and $xy\in E(G)$ iff x and y are comparable, *interval graphs*, where $V(G)$ is a set of intervals on a line, and $xy\in E(G)$ iff x and y overlap, and *split graphs*, where $V(G)$ can be partitioned into two classes, one inducing a complete graph and the other being stable. The interval graphs are those rigid-circuit graphs whose complements are comparability graphs, and the split graphs are those rigid-circuit graphs whose complements are also rigid-circuit graphs. These results are due to Gilmore and Hoffman (1964) and Foldes and Hammer (1977). Grötschel et al. (1988) have a more comprehensive list of known classes of perfect graphs.

8.2. The perfect graph theorem

The conjecture of Berge that the complement of a perfect graph is perfect was proved by Fulkerson (1971, 1972) for so-called pluperfect graphs, and by Lovász (1972a) in general. Fulkerson's proof that the complement of a pluperfect graph is pluperfect was based on the duality theory of linear programming. The following replacement theorem, proved by Lovász (1972a), implies that the classes of pluperfect and perfect graphs are identical.

Theorem 8.4. *Let G be perfect and let x be a vertex of G. Replace x by a complete graph K joined completely to all neighbours $N(x)$ of x in G. Then the obtained graph H is perfect.*

Proof. It is sufficient to prove the theorem for K being a complete 2-graph K_2. Moreover, it is sufficient to prove that $\chi(H) = \omega(H)$.

Obviously $\omega(G) \leqslant \omega(H) \leqslant \omega(G)+1$. If $\omega(H) = \omega(G)+1$, then $\chi(H) \leqslant \chi(G) + 1 = \omega(G)+1 = \omega(H)$, and hence $\chi(H) = \omega(H)$. Assume therefore that $\omega(H) = \omega(G)$. Then x is not contained in a complete $\omega(G)$-graph of G. Colour G by $\omega(G)$ colours so that x has colour 1. Let A denote the set of those vertices of colour 1 that are different from x. Then $\omega(G-A) < \omega(G)$, and hence a $(\omega(G)-1)$-colouring φ of $G-A$ is possible. Colouring the vertices of A and one of the vertices of the K_2 by a new colour, and the other vertex of the K_2 by $\varphi(x)$, we obtain from φ a $\omega(G)$-colouring of H, and $\omega(G) = \omega(H)$. Hence $\chi(H) = \omega(H)$. □

More generally, Lovász (1972a) proved that K in Theorem 8.4 can be any perfect graph. Lovász (1972a,b) also gave two further proofs of the perfect graph theorem, one based on hypergraphs and one characterizing the perfect graphs as follows.

Theorem 8.5. *A graph G is perfect iff every induced subgraph G' of G satisfies*

$$\alpha(G') \cdot \omega(G') \geqslant |V(G')|.$$

Lovász' ingenious hypergraph proof formulated for graphs gives a simple direct proof of the perfect graph theorem, as pointed out by Fulkerson (1973).

Theorem 8.6. *A graph G is perfect iff the complement $\bar{G}$ is perfect.*

Proof. Let G be perfect and suppose by induction that the theorem holds for all smaller graphs. Since the induced subgraphs of perfect graphs are perfect, it is sufficient to prove that $\chi(\bar{G}) = \omega(\bar{G})$. Let $K_1, K_2, \ldots, K_m$ be all the maximal complete graphs of G. If there is a K_j meeting all maximum stable sets of G, then $\omega(\bar{G} - K_j) < \omega(\bar{G})$, hence

$$\chi(\bar{G}) \leqslant \chi(\bar{G} - K_j) + 1 = \omega(\bar{G} - K_j) + 1 \leqslant \omega(\bar{G}),$$

and we are through.

Therefore, we may assume that $G - K_j$ contains a maximum stable set A_j of G for all j. For all vertices x_i in G, let ω_i denote the number of these sets A_j containing x_i. Let H be obtained from G by replacing each vertex x_i by a complete ω_i-graph. Then

$$\omega(H) = \max_j \sum_{x_i \in V(K_j)} \omega_i \leqslant m - 1,$$

since A_j does not count in the sum, and any other A_i counts at most once. Moreover,

$$\chi(H) \geqslant |V(H)|/\alpha(H) = (\sum \omega_i)/\alpha(G) = \alpha(G)m/\alpha(G) = m.$$

Hence $\omega(H) < \chi(H)$, but this contradicts H being perfect by Theorem 8.4. □

8.3. The strong perfect graph conjecture

Any graph containing an odd circuit of length $\geqslant 5$ or its complement as an induced subgraph is not perfect. Berge conjectured that the opposite is true as well. This implies Theorem 8.6 and is called *the strong perfect graph conjecture.* It may be formulated in terms of *critically imperfect graphs*, i.e. non-perfect graphs for which any proper induced subgraph is perfect.

Conjecture 8.7. The only critically imperfect graphs are the odd circuits of length $\geqslant 5$ and their complements.

The following is an equivalent formulation in terms of *k-vertex-critical graphs*, i.e. k-chromatic graphs for which $\chi(G - x) < k$ for all vertices x, due to Wessel (1977). Comparing with Corollary 5.4 indicates that it is the word "induced" that makes the conjecture difficult.

Conjecture 8.8. Let G be a k-vertex-critical graph, $k \geqslant 4$, for which any k'-vertex-critical induced subgraph is a complete k'-graph for all $k' < k$. Then G is a complete k-graph or the complement of a circuit of length $2k - 1$.

The strong perfect graph conjecture has been proved for special classes of graphs. For example, Tucker (1977) proved Conjecture 8.8 for $k = 4$, and Parthasarathy and Ravindra (1976) proved Conjecture 8.7 for graphs having no induced 3-star $K_{1,3}$. Thus the strong perfect graph conjecture is equivalent to the statement that no critically imperfect graph contains an induced 3-star.

By Theorem 8.5 a critically imperfect graph G has $\alpha(G) \cdot \omega(G) + 1$ vertices. Several other properties of critically imperfect graphs have been found by among others Padberg (1974), Bland et al. (1979) and Chvátal (1985).

Theorem 8.9. *A critically imperfect graph G on n vertices satisfies*

(a) $n = \alpha(G) \cdot \omega(G) + 1$.

(b) *G has exactly n complete* $\omega(G)$*-graphs.*

(c) *G has exactly n stable sets of size* $\alpha(G)$.

(d) *Every vertex of G belongs to exactly* $\omega(G)$ *complete* $\omega(G)$*-graphs and to exactly* $\alpha(G)$ *stable sets of size* $\alpha(G)$.

(e) *Every complete* $\omega(G)$*-graph is disjoint from exactly one stable set of size* $\alpha(G)$, *and vice versa.*

(f) *No set* $S \subseteq V(G)$ *with* $G - S$ *disconnected contains a vertex joined by edges to all other vertices of S, i.e. G does not have a star-cutset.*

8.4. Sufficient conditions for perfectness

Based on Theorem 8.9 (f), Hayward (1985) proved that weakly triangulated graphs, i.e. graphs for which neither it nor its complement contains a chordless circuit of length $\geqslant 5$, are perfect. There are other similar strengthenings of the condition in the strong perfect graph conjecture, sufficient to imply perfectness. Gallai (1962) proved that it is sufficient that each odd circuit of length $\geqslant 5$ has two non-crossing chords. Olaru (1969) and Sachs (1970) proved that it is sufficient that each odd circuit of length $\geqslant 5$ has two crossing chords. A common generalization was obtained by Meyniel (1976). The survey of Lovász (1983) contains an elegant proof.

Theorem 8.10. *Let G be a graph for which any odd circuit of length* $\geqslant 5$ *has at least two chords. Then G is perfect.*

From the definition of perfectness it immediately follows that in a perfect graph G any induced subgraph H has a stable set meeting all maximum complete subgraphs of H.

A graph G is called *very strong perfect* if any vertex of any induced subgraph H is contained in a stable set of H meeting all maximal (not just maximum) complete subgraphs of H. As proved by Hoàng (1987), these very strong perfect graphs are exactly the Meyniel-graphs of Theorem 8.10.

Finally we state the *semi-strong perfect graph theorem*, conjectured by Chvátal (1984) and proved by Reed (1987). Two graphs G and H on the same vertex set are called *P_4-isomorphic* if a set of four vertices induces a path in G iff it induces a path in H. Note that G and $\bar{G}$ are always P_4-isomorphic. Since, moreover, the only two graphs that are P_4-isomorphic with a chordless odd circuit of length $\geqslant 5$ are the circuit itself and its complement, Theorem 8.11 lies between the Perfect graph theorem 8.6 and the Strong perfect graph conjecture 8.7.

Theorem 8.11. *If G and H are P_4-isomorphic and G is perfect, then H is perfect.*

8.5. Stable set polytopes

For any graph G we let the *stable set polytope* $S(G)$ denote the convex hull in $\mathbb{R}^{|V(G)|}$ of the incidence vectors of the stable sets of G. A basic technique due to Grötschel et al. (1981, 1988) is to look for various classes of inequalities valid for $S(G)$ and then develop polynomial algorithms to maximize linear objective functions subject to these inequalities. Such algorithms provide upper bounds for the maximum weight of a stable set. If the inequalities in addition happen to describe the exact facets of $S(G)$ for certain types of graphs, then for such graphs $\alpha(G)$ can be determined in polynomial time. This turns out to be the case for interesting classes, most notably the perfect graphs.

Possible classes of inequalities satisfied by the incidence vectors of stable sets are

(a) $x \geqslant 0$,
(b) $x_i + x_j \leqslant 1$ for all edges ij,
(c) $x(K) \leqslant 1$ for all maximal complete subgraphs K,
(d) $x(C) \leqslant (|C| - 1)/2$ for all odd circuits C.

As an example of the basic technique described above, a weighted version of Theorem 7.1 implies that the inequalities (a) and (b) describe $S(G)$ iff G is bipartite without isolated vertices. Similarly, the following holds.

Theorem 8.12. *The inequalities* (a) *and* (c) *describe* $S(G)$ *iff* G *is perfect.*

Proof. Let $P(G)$ be the polytope described in (a) and (c). Obviously $S(G) \subseteq P(G)$. Assume first that G is perfect and let x be a rational vector in $P(G)$. Then for a common denominator q of the entries in x the vector qx is integer and it is $\geqslant 0$ by (a). For each vertex i of G replace i by a complete qx_i-graph. The obtained graph H is perfect by Theorem 8.4, hence $\chi(H) = \omega(H)$.

By (c) $\omega(H) \leqslant q$. Then H is q-colourable, implying the existence of stable sets $S_1, S_2, \ldots, S_q$ in G such that each vertex i of G is contained in exactly qx_i of them. The sum of the incidence vectors for $S_1, S_2, \ldots, S_q$ divided by q is thus x, and x is therefore a convex combination of vectors from G. Thus $x \in S(G)$, and $S(G) = P(G)$.

Assume conversely that $S(G) = P(G)$. By induction over $|V(G)|$ we shall first prove that $\chi(\bar{G}) = \omega(\bar{G})$. The face $\{x \in S(G) \mid \sum x_i = \alpha(G)\}$ of $S(G)$ is contained in a facet of $S(G)$, but since $S(G) = P(G)$ the facet must be of the form $\{x \in S(G) \mid x(K) = 1\}$ for some complete subgraph K. Thus K meets all stable sets of size $\alpha(G)$ in G. Then

$$\chi(\bar{G}) \leqslant \chi(\bar{G} - K) + 1 = \omega(\bar{G} - K) + 1 \leqslant \omega(\bar{G}),$$

where the middle equality is by induction, and we use $\omega(\bar{G} - K) = \alpha(G - K) < \alpha(G) = \omega(\bar{G})$.

$S(G) = P(G)$ implies easily that also for any induced subgraph H of G the polytope $S(H)$ is described by (a) and (c) Thus as above $\chi(\bar{H}) = \omega(\bar{H})$, and hence $\bar{G}$ is perfect.

We have so far in this proof obtained the implications (G perfect $\Rightarrow S(G) = P(G) \Rightarrow \bar{G}$ perfect). Applying this to $\bar{G}$ provides the opposite direction. □

Note that the above proof includes a proof of the perfect graph theorem based on some of the same ideas as the earlier proof. Contrary to what one might expect, optimization over the polytope $P(G)$ described by (a) and (c) is NP-hard, like over $S(G)$. This was proved by Grötschel et al. (1981). For the polytope described by (a), (b) and (d), however, optimization of linear objective functions can be carried out in polynomial time. The graphs G for which this polytope equals $S(G)$ are called *t-perfect*, studied first by Chvátal (1975).

Finally we mention briefly the subject of orthogonal representations due to Lovász (1979) and developed by Grötschel et al. (1988). For a graph G an *orthogonal representation* of G is a sequence $(u_i \mid i \in V(G))$ of vectors from $\mathbb{R}^n$ for some integer n, such that $\|u_i\| = 1$ for all i, and u_i and u_j are orthogonal for all pairs of non-adjacent vertices (for example, all the vectors could be mutually orthogonal). For any $c \in \mathbb{R}^n$ with $\|c\| = 1$ and any stable set S in G, the vectors u_i for $i \in S$ are mutually orthogonal, and so $\sum_{i \in S}(c \cdot u_i)^2 \leqslant 1$, or

(e) $\quad \displaystyle\sum_{i \in V}(c \cdot u_i)^2 \cdot x_i \leqslant 1$ for all incidence vectors x of stable sets.

TH(G) is now defined as $\{x \in \mathbb{R}^{|V(G)|} \mid x \geqslant 0$ and x satisfies (e) for all possible orthonormal representations of G and all possible unit vectors $c\}$. TH(G) is the intersection of infinitely many halfspaces, so it is convex, but it is not in general a polytope. However, $S(G) \subseteq \mathrm{TH}(G)$, and by taking $\{u_i \mid i \in V \backslash K\} \cup \{c\}$ to be mutually orthogonal unit vectors and $u_j = c$ for $j \in K$, the constraint (e) equals (c), and hence $\mathrm{TH}(G) \subseteq P(G)$, where $P(G)$ is described by (a) and (c). It follows by Theorem 8.12 that $S(G) = \mathrm{TH}(G)$ for perfect graphs G.

The most important property of TH(G) is that for arbitrary G, using the ellipsoid algorithm, one can optimize any linear objective function over TH(G) in polynomial time, i.e. $\theta(G, w) = \max\{wx \mid x \in \mathrm{TH}(G)\}$ can be found in polynomial time. In particular, $\alpha(G)$ can be polynomially computed for perfect graphs G, and hence also $\omega(G)$ and $\chi(G)$.

9. Edge-colourings

An *edge-colouring* of a simple graph or multigraph G is an assignment of colours to the edges of G such that no two adjacent edges are assigned the same colour. The *edge-chromatic* number $\gamma(G)$ is the minimum number of colours needed in an edge-colouring of G. The edge-chromatic number is called the *chromatic index* by some authors. If $L(G)$ denotes the line graph of G, then $\gamma(G) = \chi(L(G))$, and one may think of edge-colourings of graphs as vertex colourings of line graphs.

Compared with vertex- or face-colourings, edge-colourings have received less attention until relatively recently, although much studied concepts like matchings and latin squares have connections to edge-colourings. The first real breakthrough was the theorem of Vizing (1964, 1965a) that $\gamma(G)$ of a simple graph G is either $\Delta(G)$ or $\Delta(G) + 1$. Vizing considered multigraphs rather than just simple graphs, as also Shannon (1949) had done earlier. In much work on edge-colourings, however, authors restrict themselves to simple graphs, and multigraph edge-colouring is a topic where further work needs to be done.

Criticality plays an important role, as in the case of vertex-colouring. We shall call a simple graph or multigraph G *k-edge-critical* if $\gamma(G) = k$ and $\gamma(G') < k$ for all proper subgraphs G' of G. Edge-critical graphs were first considered by Vizing (1965b).

Fiorini and Wilson (1977, 1978) gave excellent surveys of edge-colourings and edge-critical graphs. More recently Hilton and Johnson (1987) described interesting conjectures and results on graphs whose vertices are critical with respect to the edge-chromatic number.

9.1. Bipartite graphs

After Tait's theorem 6.1 the next important observation on edge-colourings was the following theorem of König (1916). In a different formulation it says that the line graphs of bipartite graphs are perfect. The proof involves looking at a connected component in a subgraph consisting of the edges of two colours α and β, and changing the two colours there. Such a subgraph we shall call an (α, β)*-Kempe chain*. For edge-colourings a Kempe chain is either a path or a circuit.

Theorem 9.1. *If G is a bipartite simple graph or multigraph, then $\gamma(G) = \Delta(G)$.*

Proof. Obviously $\gamma(G) \geqslant \Delta(G)$. If $\gamma(G) > \Delta(G)$, then let G' be a $(\Delta(G)+1)$-edge-critical subgraph of G. Delete an edge xy from G' and $\Delta(G)$-edge-colour the graph $G' - xy$. Since x is incident with at most $\Delta(G) - 1$ coloured edges, there is at least one colour α of the $\Delta(G)$ colours missing at x, and likewise there is a colour β missing at y. If α and β can be chosen equal, then xy may be given that colour, and G' is $\Delta(G)$-edge-colourable. Since this is not the case, α and β are necessarily different. Therefore, the colour β is present at x and there is an (α, β)-Kempe-chain H in G' starting at x. Since G' is bipartite and β is missing at y, the vertex y is not in H. But then a $\Delta(G)$-edge colouring of G' can be obtained by changing the colours α and β on the edges of H and giving xy the colour β. This is a contradiction. □

An n-edge-colouring of the complete bipartite graph $K_{n,n}$ corresponds to a latin square. Answering a conjecture of T. Evans from 1960 on completing partial latin squares, Andersen and Hilton (1983) proved the following result, also characterizing the extremal cases.

Theorem 9.2. *A partial edge-colouring φ of at most n edges of $K_{n,n}$ can be extended to an n-edge-colouring of $K_{n,n}$, except when one of the following two cases occurs:*

(a) *For some uncoloured edge xy, there are n coloured edges of different colours, each incident with x or y.*

(b) *For some vertex x and some colour i the colour i is not present at x but is present at all vertices y for which xy is an uncoloured edge.*

9.2. The bounds of Shannon and Vizing

For a graph G obviously $\Delta(G) \leqslant \gamma(G)$. Vizing (1964, 1965a) proved that if $\mu(G)$ denotes the *maximum multiplicity* of G, i.e. the maximum number of edges joining the same pair of vertices, then $\gamma(G) \leqslant \Delta(G) + \mu(G)$. This same result was obtained independently by Gupta (1967). We shall consider the extension and elegant proof due to Kierstead (1984). An *s-triangle* is a graph with three vertices x, y and z and s xz-edges, $s-1$ xy-edges and one yz-edge.

Theorem 9.3. *Let G be a graph of multiplicity $\mu(G) \geqslant 1$. Then $\gamma(G) \leqslant \Delta(G) + \mu(G)$. Moreover, if $\mu(G) \geqslant 2$ and $\gamma(G) = \Delta(G) + \mu(G)$, then G contains a $\mu(G)$-triangle.*

Proof. Suppose that the theorem is false. Then there is an $s \geqslant 2$ such that $\mu(G) \leqslant s$, G does not contain an s-triangle and $\gamma(G) \geqslant \Delta(G) + s = \Delta + s$.

Then G contains a $(\Delta + s)$-edge-critical subgraph G'. Let $e_0 = x_0x_1$ be an edge of G' and let φ be a $(\Delta + s - 1)$-edge-colouring of $G' - e_0$. Let $\overline{\varphi}(x)$ denote the set of those of the $(\Delta + s - 1)$ colours not present at the vertex x. Then a path P in G with distinct vertices $x_0, x_1, x_2, \dots, x_n$ in this order and edges $e_0 = x_0x_1$, $e_1 = x_1x_2, \dots$, $e_{n-1} = x_{n-1}x_n$ is said to be *acceptable* if the first edge e_0 is the uncoloured edge of G', and $\varphi(e_i) \in \bigcup_{j<i} \overline{\varphi}(x_j)$ for all $i \geqslant 1$.

If P is acceptable in G' with respect to a $(\Delta + s - 1)$-edge-colouring φ of $G' - e_0$, and if there are i and j and a colour β such that $0 \leqslant i < j \leqslant n$ and $\beta \in \overline{\varphi}(x_i) \cap \overline{\varphi}(x_j)$, then G' can be $(\Delta + s - 1)$-edge-coloured. We shall prove this by double induction over j and over $j - i$. Since $\gamma(G') = \Delta + s$, the conclusion is that $\overline{\varphi}(x_i) \cap \overline{\varphi}(x_j)$ must be empty for any two vertices of an acceptable path.

If $j = 1$ then $i = 0$, and the colour β may be given to e_0, thus extending φ to a $(\Delta + s - 1)$-colouring of G'.

If $j \geqslant 2$ and $j - i = 1$, then β is missing at x_i and x_{i+1}. Let $\varphi(e_i) \in \overline{\varphi}(x_m)$, where $m < i$. We may recolour e_i by β and obtain the colouring φ'. The subpath P' of P from x_0 to x_i is an acceptable path w.r.t. φ'. Moreover, $\overline{\varphi}'(x_m) \cap \overline{\varphi}'(x_i)$ contains the colour $\varphi(e_i)$. Since $m < i < j$, the desired result follows by the induction over j.

So now consider the general case where $j - i \geqslant 2$. Let $\delta \in \overline{\varphi}(x_{i+1})$. If $\beta = \delta$, then j can be replaced by $i + 1$, and we are done by the induction over j. If $\beta \neq \delta$ necessarily, then the colour β is present at x_{i+1}, and there is a (β, δ)-Kempe-chain C starting at x_{i+1}. By the induction over j we may assume that the colour δ is present at all vertices $x_0, x_1, \dots, x_i$ and that β is present at all the vertices $x_0, x_1, \dots, x_{i-1}, x_{i+1}$. Thus none of the edges $e_1, e_2, \dots, e_i$ can have colours β or δ by the definition of an acceptable path. The chain C may or may not end at x_i. In any case change colours β and δ on C and obtain φ'.

If C does not end at x_i, then the subpath P' of P from x_0 to x_{i+1} is acceptable w.r.t. φ' and $\beta \in \overline{\varphi}'(x_i) \cap \overline{\varphi}'(x_{i+1})$. Hence we are done by the induction over j. If C ends at x_i, then P is acceptable w.r.t. φ' and $\beta \in \overline{\varphi}'(x_{i+1}) \cap \overline{\varphi}'(x_j)$. Hence we are done by the induction over $j - i$.

As mentioned, the conclusion is that $\overline{\varphi}(x_i) \cap \overline{\varphi}(x_j) = \emptyset$ for any two vertices on any acceptable path. But we shall prove that this is not so on a longest acceptable path P. This contradiction proves the theorem.

If $\overline{\varphi}(x_i) \cap \overline{\varphi}(x_j) = \emptyset$ for $i \neq j$, then $\sum_{0 \leqslant i < n} |\overline{\varphi}(x_i)| \geqslant (\sum(\Delta + s - 1 - \Delta)) + 2 \geqslant n(s-1) + 2$ different colours are missing at the vertices of the path $P - x_n$. (The additional 2 appears since x_0 and x_1 are each incident with at most $\Delta - 1$ coloured edges.) Moreover, all these $n(s-1)+2$ missing colours must be present at x_n, since otherwise $\overline{\varphi}(x_j) \cap \overline{\varphi}(x_n) \neq \emptyset$ for some j. Since P has maximal length, x_n is joined by at least $n(s-1)+2$ different edges to $P - x_n$. Since $\mu(G) \leqslant s$ and G does not contain an s-triangle, x_n is joined to x_0 and x_1 by $\leqslant 2s-2$ edges, to x_2 and x_3 by $\leqslant 2s-2$ edges, etc. This gives $\leqslant n(s-1)+1$ edges from x_n to $P - x_n$. Contradiction. □

For a simple graph G it was proved by Beineke (1968) that G is a line graph $L(H)$ of a simple graph H iff G does not contain any of nine specified graphs as an induced subgraph. Moreover, $\Delta(H) = \omega(G)$ where $\omega(G)$ is the size of a maximum complete subgraph of G, provided $\Delta(H) \geqslant 3$. Therefore, the case $\mu = 1$ of Theorem 9.3 may be formulated as the result that $\chi(G) \leqslant \omega(G)+1$ when G is a line graph of a simple graph. Based on Theorem 9.3, Kierstead and Schmerl (1983) were able (by a rather lengthy argument) to eliminate all but two of the forbidden nine subgraphs. This result is an important extension of Vizing's theorem for simple graphs.

Theorem 9.4. *Let G be a simple graph not containing K_5^- (the complete 5-graph with a missing edge) or $K_{1,3}$ as an induced subgraph. Then $\chi(G)$ is either $\omega(G)$ or $\omega(G)+1$.*

For graphs or multigraphs Shannon (1949) proved the following.

Theorem 9.5. $\gamma(G) \leqslant \frac{3}{2}\Delta(G)$.

Proof. Suppose $k = \gamma(G) \geqslant \frac{3}{2}\Delta(G)$ and let G' be a k-edge-critical subgraph of G. Since $k \leqslant \Delta(G) + \mu(G)$ by Theorem 9.3, there are two vertices x and y joined by $\geqslant \frac{1}{2}\Delta(G)$ edges. Delete an xy-edge e from G' and let φ be a $(k-1)$-edge-colouring of $G' - e$. In φ the number of the $k-1$ colours missing at x is $\geqslant (k-1) - (\Delta(G) - 1) = k - \Delta(G)$, and likewise at y. No colour is missing at both x and y, hence

$$2(k - \Delta(G)) + \frac{1}{2}\Delta(G) - 1 \leqslant k - 1$$

implying that $k \leqslant \frac{3}{2}\Delta(G)$. □

9.3. Various problems and results

Although Theorems 9.3 and 9.5 give tight bounds for γ, several interesting related problems can be posed. We shall briefly explain some of these.

Theorem 9.3 implies a way of classifying simple graphs into two classes. The graph G is said to be of *class 1* if $\gamma(G) = \Delta(G)$ and of *class 2* if $\gamma(G) = \Delta(G)+1$. It is an NP-complete problem for a given 3-regular simple graph G to decide if G is of class 1 or 2, as proved by Holyer (1981). A 3-regular 2-connected simple graph of class 2 has been conjectured by Tutte (1966, 1969) to contain a subdivision of the Petersen graph. Such graphs without any cutset of $\leqslant 3$ edges, except three edges at the same vertex, and without circuits of length $\leqslant 4$ have been called *snarks* by Gardner (1976). The Petersen graph is the smallest snark. Snark-hunting goes back to Descartes (1948b) and got a big push forward by Isaacs (1975). Very recently, snarks with arbitrarily long shortest circuits have been constructed by M. Kochol and M. Skoviera.

Shannon's theorem 9.5 is best possible as shown by a multigraph on three vertices, any two joined by the same number of edges. These graphs are examples of *ring-graphs*, i.e. odd circuits with parallel edges. For multigraphs, Jakobsen (1975) conjectured such ring-graphs to be extremal w.r.t. the number of vertices. More precisely, Jakobsen conjectured the following.

Conjecture 9.6. Let m be odd $\geqslant 3$ and let G be a k-edge-critical multigraph with

$$k > \frac{m}{m-1} \cdot \Delta(G) + \frac{m-3}{m-1}.$$

Then $|V(G)| \leqslant m-2$.

For $m = 3$ this is Shannon's theorem 9.5. For $m = 5, 7, 9$ and 11 the conjecture has been proved by Goldberg (1973), Andersen (1977), Goldberg (1984) and Nishizeki and Kashiwagi (1985).

Goldberg (1984) posed an interesting extension of Conjecture 9.6. Goldberg thought of this conjecture already around 1970, and equivalent conjectures have been posed by Andersen (1977) and Seymour (1979). Let

$$w(G) = \max_{H \subseteq G} \left\lceil \frac{|E(H)|}{\lfloor \frac{1}{2}|V(H)| \rfloor} \right\rceil.$$

It is easy to see that $w(G) \leqslant \gamma(G)$. Goldberg's conjecture states the following.

Conjecture 9.7. If $\gamma(G) \geqslant \Delta + 2$ then $\gamma(G) = w(G)$.

Conjecture 9.7 implies Conjecture 9.6, which in turn implies that there are only finitely many edge-critical multigraphs G with $\gamma(G) \geqslant \Delta(G) + 2$ for any fixed $\Delta(G)$. Whether this is so is not known. Conjecture 9.7 also implies all edge-critical multigraphs G with $\gamma(G) \geqslant \Delta(G) + 2$ to have an odd number of vertices. This *critical graph conjecture* was disproved by Goldberg (1979; 1981) for simple edge-critical graphs with $\gamma(G) = \Delta(G) + 1$.

10. Concluding remarks

This survey has demonstrated that graph colouring has come a long way since its beginning more than a century ago, but also that challenging unsolved problems remain.

We have concentrated on classical vertex- and edge-colouring. Many authors have considered extensions in various directions. We shall finish by briefly describing some of these.

(a) In a usual vertex-colouring the objective may be different from using as few colours as possible. For example, the maximum number of colours possible in a vertex-colouring, where any two different colours are present at vertices joined by an edge, is the *achromatic number*, introduced by Harary et al. (1967). The maximum number of colours from N for which any vertex of colour i is joined to vertices of all colours $1, 2, \ldots, i-1$ is the *Grundy number*, studied by Christen and Selkow (1979).

(b) A *graph homomorphism* of G into H is a mapping $\varphi : V(G) \to V(H)$ for which any edge xy of G is taken into an edge $\varphi(x)\varphi(y)$ of H. For $H = K_k$ a homomorphism is just a k-colouring and for general H it has been called an H-colouring. Albertson and Collins (1985) include a brief guide to graph homomorphism. Hell and Nešetřil (1990) proved that the existence problem for H-colouring is NP-complete iff H is non-bipartite. Also the directed case has been studied, for example Bang-Jensen et al. (1988) solved the complexity problem for tournaments.

(c) For any graph-property P an assignment of colours to the vertices (or edges) of G, such that each colour class induces a graph with property P, has been called a *conditional colouring* by Harary (1985). For example, P may be "k-degenerate", i.e. all subgraphs have vertices of degree $\leqslant k$, as studied by Lick and White (1970), or P may be "without paths of length k", as studied by Chartrand et al. (1968). The edge-case for the property "circuit-free" has a very satisfactory solution by W. T. Tutte and C. St. J. A. Nash–Williams. It is a special case of covering a matroid by a minimum number of independent sets, solved by J. Edmonds (for details, see chapter 8 of the monograph by Welsh 1976).

(d) In an *acyclic colouring* of a graph G any two colour classes together induce a subgraph of G without circuits. Borodin (1979) proved a conjecture of Grünbaum (1973) that any planar graph has an acyclic 5-colouring. Borodin's proof is reminiscent of the four colour theorem proof by K. Appel, W. Haken and J. Kock, involving some 450 reducible configurations (but no computers). No case of a surface S, except the sphere, where the maximum acyclic chromatic number for graphs on S is strictly greater than the maximum chromatic number, seems to be known. This was pointed out by M. O. Albertson and D. M. Berman, and independently by O. V. Borodin.

(e) Colourings may involve colouring vertices and/or edges and/or faces at the same time. For example, the *total chromatic number* is the smallest number of colours in a simultaneous colouring of vertices and edges, where any two incident or adjacent elements get different colours. For simple graphs, Behzad (1965) and independently Vizing (1968) conjectured that $\Delta(G) + 2$ colours always suffice in a total colouring. Another example is the best possible 6-colour-theorem of Borodin (1984, 1989), conjectured by G. Ringel: The vertices and faces of a planar graph may be 6-coloured with adjacent and incident elements getting different colours. More generally, Borodin (1984, 1989) proved: A 1-planar graph, i.e. a graph embedded in the plane such that each edge is crossed over by at most one other edge, is 6-colourable.

(f) Let for each vertex x [edge x] of a graph G a set list(x) of k colours be given. The *list chromatic number* $\chi_{\text{list}}(G)$ [*list edge chromatic number* $\gamma_{\text{list}}(G)$] is the minimum value of k for which a vertex [edge] colouring exists, where each x gets a colour from its list, no matter what the lists look like (but they must all be of size k).

This concept was first studied by Vizing (1976) and by Erdős et al. (1979) under the name *choosability*. Obviously the total chromatic number is at most $\gamma_{\text{list}}(G)+2$. Perhaps $\gamma_{\text{list}}(G)=\gamma(G)$ for all G; for bipartite G this question, known as Dinitz' problem was recently solved in the affirmative by F. Galvin with a surprisingly short proof. The obvious bound $\gamma_{\text{list}}(G)\leqslant 2\Delta-1$ was improved by Bollobás and Harris (1985), replacing the 2 by a smaller constant for large Δ.

Recently χ_{list} has received a considerable amount of attention following a new type of results by Alon and Tarsi (1992), proved by algebraic techniques applied to the graph polynomial, see also the recent survey by Alon (1993). A basic result of Alon and Tarsi is: If G is a directed graph with maximum outdegree d, and if the number of Eulerian (spanning) subgraphs of G (i.e. for all vertices $x\in V(G)$ the outdegree of x in the subgraph equals the indegree of x) with an even number of edges differs from the number of Eulerian (spanning) subgraphs of G with an odd number of edges, then $\chi_{\text{list}}(G)\leqslant d+1$.

In particular the above result of Alon and Tarsi (1992) was used by Fleischner and Stiebitz (1992) to prove that any 4-regular graph on $3n$ vertices having a decomposition into a Hamiltonian circuit and n pairwise edge-disjoint triangles, is 3-colourable. This was conjectured by P. Erdős and is equivalent to the statement that for any partition of the integers into triples there is another partition into three classes such that each class contains a member from each triple, but does not contain a consecutive pair of integers.

Recently, Thomassen (1994b) proved that the list chromatic number of a planar graph is $\leqslant 5$. In particular, Thomassen's proof provides a new proof of the century old five colour theorem, avoiding both the use of Euler's formula and the recolouring technique of Kempe, thus making it conceptually simpler than any previous proof.

(g) *Set-colourings* of graphs assign to each vertex a set of r colours, such that for any edge xy the vertices x and y are assigned disjoint sets. Under the name *set-chromatic number* the infimum over r of (the minimum number of colours divided by r) was studied by Bollobás and Thomason (1979).

(h) Finally, a *k-flow* is an assignment of a direction and an integer from the range $1,2,\ldots,k-1$ to each edge of a graph so that at each vertex the flow out equals the flow in. For k small, the solutions to k-colouring and k-flow problems are similar. A good survey was provided by Younger (1983), and Seymour (1981a) explored connections to colourings. The concept is due to W. T. Tutte. Tutte (1950, 1954, 1966) posed these attractive conjectures generalizing the five- and four-colour theorems. Seymour (1981b) proved the first conjecture with 5 replaced by 6. For further information, see the excellent appendix to this chapter by Seymour.

Conjecture 10.1. Let G be without a cut consisting of one edge. Then G has a 5-flow. If G contains no subdivision of the Petersen graph, then G has a 4-flow.

Acknowledgement

I wish to thank Tommy R. Jensen for much valuable help and advice, Ulla Pedersen for typing the manuscript nicely, and Margit Christiansen for effective library assistance.

References

Aigner, M.

[1984] *Graphentheorie – Eine Entwicklung aus dem 4-Farbenproblem* (B.G. Teubner, Stuttgart) [1987, English Translation (BCS Associates, Moscow, ID)].

Aksionov, V.A.

[1974] On continuation of 3-colourings of planar graphs (in Russian), *Diskret. Analiz.* **26**, 3–19.

Aksionov, V.A., and L.S. Melnikov

[1978] Essay on the theme: the three-color problem, in: *Combinatorics, Colloq. Math. Soc. János Bolyai* **18**, eds. A. Hajnal and V.T. Sós, pp. 23–34.

Albertson, M.O., and K.L. Collins

[1985] Homomorphisms of 3-chromatic graphs, *Discrete Math.* **54**, 127–132.

Albertson, M.O., and J.P. Hutchinson

[1979] The three excluded cases of Dirac's map-color theorem, *Ann. New York Acad. Sci.* **319**, 7–17.

[1980] On six-chromatic toroidal graphs, *Proc. London Math. Soc.* (3) **41**, 533–556.

Albertson, M.O., and W.R. Stromquist

[1982] Locally planar toroidal graphs are 5-colourable, *Proc. Amer. Math. Soc.* **84**, 449–456.

Alon, N.

[1993] Restricted colorings of graphs, in: *Surveys of Combinatorics, 1993, London Math. Soc. Lecture Note Series,* Vol. 187, ed. K. Walker (Cambridge University Press, Cambridge) pp. 1–33.

Alon, N., and M. Tarsi

[1992] Colorings and orientations of graphs, *Combinatorica* **12**, 125–134.

Andersen, L.D.

[1977] On edge-colourings of graphs, *Math. Scand.* **40**, 161–175.

Andersen, L.D., and A.J.W. Hilton

[1983] Thank Evans!, *Proc. London Math. Soc.* (3) **47**, 507–522.

Appel, K., and W. Haken

[1977a] Every planar map is four colorable: Part 1, Discharging, *Illinois J. Math.* **21**, 429–490.

[1977b] The solution of the four-color map problem, *Sci. Amer.* **237**, No. 4, Oct., 108–121.

[1986] The four colour proof suffices, *Math. Intelligencer* **8**, 10–20.

[1989] *Every Planar Map is Four Colorable* (AMS, Providence, RI).

Appel, K., W. Haken and J. Kock

[1977] Every planar map is four colourable: Part 2, Reducibility, *Illinois J. Math.* **21**, 491–567.

Bang-Jensen, J., P. Hell and G. MacGillivray

[1988] The complexity of colourings by semicomplete digraphs, *SIAM J. Discrete Math.* **1**, 281–298.

Bárány, I.

[1978] A short proof of Kneser's conjecture, *J. Combin. Theory A* **25**, 325–326.

Barnette, D.

[1983] *Map Coloring, Polyhedra, and the Four-Color Problem, Dolciani Mathematical Expositions,* No. 8 (Mathematical Association of America, Washington, DC).

Behzad, D.

[1965] *Graphs and their chromatic numbers,* Doctoral Thesis (Michigan University).

Beineke, L.W.
[1968] Derived graphs and digraphs, in: *Beiträge zur Graphentheorie*, eds. H. Sachs, H.-J. Voss and H. Walter (Teubner, Leipzig), pp. 17–33.

Berge, C.
[1961] Färbung von Graphen, deren sämtliche bzw. deren ungerade Kreise starr sind, *Wiss. Z. Martin-Luther-Univ. Halle-Wittenberg Math.-Nat. Reihe* **10**, 114–115.
[1972] Alternating chain methods: a survey, in: *Graph Theory and Computing*, ed. R.C. Read (Academic Press, New York), pp. 1–13.

Berge, C., and V. Chvátal
[1984] *Topics on Perfect Graphs, Ann. Discrete Math.* **21**.

Biggs, N., E.K. Lloyd and R.J. Wilson
[1976] *Graph Theory 1736–1936* (Clarendon Press, Oxford).

Birkhoff, G.D.
[1912] A determinant formula for the number of ways of colouring a map, *Ann. Math.* **14**, 42–46.

Bland, R.G., H.-C. Huang and L.E. Trotter Jr
[1979] Graphical properties related to minimal imperfection, *Discrete Math.* **27**, 11–22.

Bollobás, B.
[1978] *Extremal Graph Theory* (Academic Press, New York).

Bollobás, B., and P.A. Catlin
[1981] Topological cliques of random graphs, *J. Combin. Theory B* **30**, 224–227.

Bollobás, B., and A.J. Harris
[1985] List-colourings of graphs, *Graphs and Combinatorics* **1**, 115–127.

Bollobás, B., and A. Thomason
[1979] Set colourings of graphs, *Discrete Math.* **25**, 21–26.

Borodin, O.V.
[1979] On acyclic colorings of planar graphs, *Discrete Math.* **25**, 211–236.
[1984] Solution of Ringel's problem on vertex-face colouring of planar graphs and colouring 1-planar graphs (in Russian), *Met. Diskret. Anal.* **41**, 12–26.
[1989] A new proof of the 6 color theorem, *J. Graph Theory*, to appear.

Borodin, O.V., and A.V. Kostochka
[1977] On an upper bound of a graph's chromatic number, depending on the graph's degree and density, *J. Combin. Theory B* **23**, 247–250.

Brooks, R.L.
[1941] On colouring the nodes of a network, *Proc. Cambridge Phil. Soc.* **37**, 194–197.

Brown, W.G., and H.A. Jung
[1969] On odd circuits in chromatic graphs, *Acta Math. Acad. Sci Hungar.* **20**, 129–134.

Brown, W.G., and J.W. Moon
[1969] Sur les ensembles de sommets indépendants dans les graphes chromatiques minimaux, *Canad. J. Math.* **21**, 274–278.

Catlin, P.A.
[1978] A bound on the chromatic number of a graph, *Discrete Math.* **22**, 81–83.
[1979] Hajós' graph-coloring conjecture: variations and counterexamples, *J. Combin. Theory B* **26**, 268–274.

Chartrand, G., D.P. Geller and S. Hedetniemi
[1968] A generalization of the chromatic number, *Proc. Cambridge Philos. Soc.* **64**, 265–271.

Christen, C.A., and S.M. Selkow
[1979] Some perfect coloring properties of graphs, *J. Combin. Theory B* **27**, 49–59.

Chvátal, V.
[1975] On certain polytopes associated with graphs, *J. Combin. Theory B* **18**, 138–154.
[1984] A semi-strong perfect graph conjecture, in: *Topics on Perfect Graphs*, eds C. Berge and V. Chvátal, *Ann. Discrete Math.* **21**, 279–280.
[1985] Star-cutsets and perfect graphs, *J. Combin. Theory B* **39**, 189–199.

de Bruijn, N.G., and P. Erdős
[1951] A colour problem for infinite graphs and a problem in the theory of relations, *Nederl. Akad. Wetensch. Proc. Ser. A* **54** [= *Indag. Math.* **13**], 371–373.
de Werra, D., and A. Hertz
[1989] *Graph Colouring and Variations, Discrete Math.* **74**, *Ann. Discrete Math.* **39**.
Descartes, B.
[1948a] Solutions to problems in Eureka No. 9, *Eureka* **10**.
[1948b] Network-colourings, *Math. Gazette* **32**, 67–69.
[1954] Solution to advanced problem No. 4526, *Amer. Math. Monthly* **61**, 532.
Dirac, G.A.
[1952a] A property of 4-chromatic graphs and some remarks on critical graphs, *J. London Math. Soc.* **27**, 85–92.
[1952b] Some theorems on abstract graphs, *Proc. London Math. Soc. (3)* **2**, 69–81.
[1952c] Map colour theorems, *Canad. J. Math.* **4**, 480–490.
[1953] The structure of k-chromatic graphs, *Fund. Math.* **40**, 42–55.
[1956] Map colour theorems related to the Heawood colour formula, *J. London Math. Soc.* **31**, 460–471.
[1957a] A theorem of R.L. Brooks and a conjecture of H. Hadwiger, *Proc. London Math. Soc. (3)* **7**, 161–195.
[1957b] Short proof of a map-colour theorem, *Canad. J. Math.* **9**, 225–226.
[1957c] Map colour theorems related to the Heawood colour formula II, *J. London Math. Soc.* **32**, 436–455.
[1960] Trennende Knotenpunktmengen und Reduzibilität abstrakter Graphen mit Anwendung auf das Vierfarbenproblem, *J. Reine Angew. Math.* **204**, 116–131.
[1961] On rigid circuit graphs, *Abh. Math. Sem. Univ. Hamburg* **25**, 71–76.
[1963] Percy John Heawood, *J. London Math. Soc.* **38**, 263–277.
[1964a] On the structure of 5- and 6-chromatic abstract graphs, *J. Reine Angew. Math.* **214–215**, 43–52.
[1964b] Homomorphism theorems for graphs, *Math. Ann.* **153**, 69–80.
Edmonds, J.
[1965] Paths, trees and flowers, *Canad. J. Math.* **17**, 449–467.
Erdős, P.
[1957] Remarks on a theorem of Ramsey, *Bull. Res. Council Israel* **7F**, 21–24.
[1959] Graph theory and probability, *Canad. J. Math.* **11**, 34–38.
[1961] Graph theory and probability II, *Canad. J. Math.* **13**, 346–352.
[1962] On circuits and subgraphs of chromatic graphs, *Mathematika* **9**, 170–175.
[1981] On the combinatorial problems which I would most like to see solved, *Combinatorica* **1**, 25–42.
Erdős, P., and S. Fajtlowicz
[1981] On the conjecture of Hajós, *Combinatorica* **1**, 141–143.
Erdős, P., and A. Hajnal
[1966] On chromatic number of graphs and set-systems, *Acta Math. Acad. Sci. Hungar.* **17**, 61–99.
[1985] Chromatic number of finite and infinite graphs and hypergraphs, *Discrete Math.* **53**, 281–285.
Erdős, P., and M. Simonovits
[1966] A limit theorem in graph theory, *Studia Sci. Math. Hungar.* **1**, 51–57.
Erdős, P., and A.H. Stone
[1946] On the structure of linear graphs, *Bull. Amer. Math. Soc.* **52**, 1087–1091.
Erdős, P., A. Hajnal and J.W. Moon
[1964] A problem in graph theory, *Amer. Math. Monthly* **71**, 1107–1110.
Erdős, P., A.L. Rubin and H. Taylor
[1979] Choosability in graphs, in: *Proc. West Coast Conference in Combinatorics, Graph Theory and Computing, Congress. Numerantium* **26**, 125–157.
Finck, H.-J., and H. Sachs
[1969] Über eine von H.S. Wilf angegebene Schranke für die chromatische Zahl endlicher Graphen, *Math. Nachr.* **39**, 373–386.
Fiorini, S., and R.J. Wilson
[1977] *Edge-colourings of Graphs* (Pitman, London).

[1978] Edge-colourings of graphs, in: *Selected Topics in Graph Theory*, eds. L.W. Beineke and R.J. Wilson (Academic Press, New York) pp. 103–126.

Fisk, S.
[1978] The non-existence of colorings, *J. Combin. Theory B* **24**, 247–248.

Fleischner, H., and M. Stiebitz
[1992] A solution to a colouring problem by Erdős, *Discrete Math.* **101**, 39–48.

Foldes, S., and P.L. Hammer
[1977] Split graphs, in: *Proc. 8th S.E. Conf. on Combinatorics, Graph Theory and Computing, Congress. Numerantium* **19**, 311–315.

Franklin, P.
[1934] A six-color problem, *J. Math. Phys.* **13**, 363–369.

Fulkerson, D.R.
[1971] The perfect graph conjecture and pluperfect graph theorem, in: *Proc. 2nd Chapel Hill Conference on Combinatorial Mathematics and its Applications*, eds. R.C. Bose, I.M. Chakravarti, T.A. Dowling, D.G. Kelly and K.J.C. Smith (University of North Carolina, Chapel Hill, NC) pp. 171–175.
[1972] Anti-blocking polyhedra, *J. Combin. Theory B* **12**, 50–71.
[1973] On the perfect graph theorem, in: *Mathematical Programming*, eds. T.C. Hu and S. Robinson (Academic Press, New York) pp. 69–76.

Gallai, T.
[1958] Maximum–Minimum Sätze über Graphen, *Acta Math. Acad. Sci. Hungar.* **9**, 395–434.
[1962] Graphen mit triangulierbaren ungeraden Vielecken, *Publ. Math. Inst. Hung. Acad.* **7**, 3–37.
[1963a] Kritische Graphen I, *Publ. Math. Inst. Hung. Acad.* **8**, 165–192.
[1963b] Kritische Graphen II, *Publ. Math. Inst. Hung. Acad.* **8**, 373–395.
[1968] On directed paths and circuits, in: *Theory of Graphs*, eds. P. Erdős and G. Katona (Academic Press, New York) pp. 115–118.

Gallai, T., and A.N. Milgram
[1960] Verallgemeinerung eines graphentheoretischen Satzes von Rédei, *Acta Sci. Math. (Szeged)* **21**, 181–186.

Gardner, M.
[1976] Mathematical games, *Sci. Amer.* **234**(4), 126–130; **235**(3), 210–211.

Garey, M.R., and D.S. Johnson
[1976] The complexity of near-optimal graph colouring, *J. Assoc. Comput. Mach.* **23**, 43–49.
[1979] *Computers and Intractability. A Guide to the Theory of NP-Completeness* (W.H. Freeman, San Francisco, CA).

Garey, M.R., D.S. Johnson and L. Stockmeyer
[1976] Some simplified NP-complete graph problems, *Theor. Comput. Sci.* **1**, 237–267.

Gilmore, P.C., and A.J. Hoffman
[1964] A characterization of comparability graphs and interval graphs, *Canad. J. Math.* **16**, 539–548.

Goldberg, M.K.
[1973] On multigraphs with almost maximal chromatic class (in Russian) *Diskret. Analiz.* **23**, 3–7.
[1979] Critical graphs with even number of vertices (in Russian) *Bull. Acad. Sci. GSSR* **94**, 25–27.
[1981] Construction of class 2 graphs with maximum vertex degree 3, *J. Combin. Theory B* **31**, 282–291.
[1984] Edge-coloring of multigraphs: recoloring technique, *J. Graph Theory* **8**, 123–137.

Golumbic, M.C.
[1980] *Algorithmic Graph Theory and Perfect Graphs* (Academic Press, New York).

Greenwell, D., and L. Lovász
[1974] Applications of product colouring, *Acta Math. Acad. Sci. Hungar.* **25**, 335–340.

Grötschel, M., L. Lovász and A. Schrijver
[1981] The ellipsoid method and its consequences in combinatorial optimization, *Combinatorica* **1**, 169–197.
[1988] *Geometric Algorithms and Combinatorial Optimization* (Springer, Berlin).

Grötzsch, H.
[1959] Ein Dreifarbensatz für dreikreisfreie Netze auf der Kugel, *Wiss. Z. Martin-Luther-Univ. Halle-Wittenberg Math.-Nat. Reihe* **8**, 109–120.

Grünbaum, B.
[1963] Grötzsch's theorem on 3-colorings, *Michigan Math. J.* **10**, 303–310.
[1973] Acyclic colorings of planar graphs, *Israel J. Math.* **14**, 390–408.

Gupta, R.P.
[1967] *Studies in the theory of graphs,* Ph.D. Thesis (Tata Institute of Fundamental Research, Bombay).

Gyárfás, A.
[1987] Problems from the world surrounding perfect graphs, in: *Proc. Int. Conf. on Combinatorial Analysis and its Applications,* Poland, Sept. 1985, ed. M.J. Syslo, *Zastos. Mat. (Applicationes Mathematicae)* **19**, 413–441.

Hadwiger, H.
[1943] Über eine Klassifikation der Streckenkomplexe, *Vierteljschr. Naturforsch. Ges. Zurich* **88**, 133–142.

Hajnal, A.
[1965] A theorem on k-saturated graphs, *Canad. J. Math.* **17**, 720–724.

Hajnal, A., and J. Surányi
[1958] Über die Auflösung von Graphen in vollständige Teilgraphen, *Ann. Univ. Sci. Budapest Eötvös Sect. Math.* **1**, 113–121.

Hajnal, A., and E. Szemerédi
[1970] Proof of a conjecture of Erdős, in: *Combinatorial Theory and its Applications II,* eds. P. Erdős, A. Rényi and V.T. Sós, *Colloq. Math. Soc. János Bolyai* **4**, pp. 601–623.

Hajós, G.
[1961] Über eine Konstruktion nicht n-färbbarer Graphen, *Wiss. Z. Martin-Luther-Univ. Halle-Wittenberg Math.-Nat. Reihe* **10**, 116–117.

Halin, R.
[1964] Über einen Satz von K. Wagner zum Vierfarbenproblem, *Math. Ann.* **153**, 47–62.

Hanson, D., G.C. Robinson and B. Toft
[1986] Remarks on the graph colour theorem of Hajós, in: *Proc. 17th S.E. Conf. on Combinatorics, Graph Theory and Computing, Congress. Numerantium* **55**, 69–76.

Harary, F.
[1985] Conditional colorability in graphs, in: *Proc. 1st Colorado Symp. on Graph Theory,* eds. F. Harary and J.S. Maybee (Wiley, New York) pp. 127–136.

Harary, F., S. Hedetniemi and G. Prins
[1967] An interpolation theorem for graphical homomorphisms, *Portugal. Math.* **26**, 453–462.

Hayward, R.
[1985] Weakly triangulated graphs, *J. Combin. Theory B* **39**, 200–208.

Heawood, P.J.
[1890] Map colour theorem, *Quart. J. Pure Math.* **24**, 332–338.
[1898] On the four-colour map theorem, *Quart. J. Pure Math.* **29**, 270–285.

Heesch, H.
[1969] *Untersuchungen zum Vierfarbenproblem,* B.I. Hochschulscripten 810/810a/810b (Bibliographisches Institut Mannheim).

Heffter, L.
[1891] Über das Problem der Nachbargebiete, *Math. Ann.* **38**, 477–508.

Hell, P., and J. Nešetřil
[1990] On the complexity of H-coloring, *J. Combin. Theory B* **48**, 92–110.

Hilton, A.J.W., and P.D. Johnson
[1987] Graphs which are vertex-critical with respect to the edge-chromatic number, *Math. Proc. Cambridge Phil. Soc.* **102**, 211–221.

Hoàng, C.T.
[1987] On a conjecture of Meyniel, *J. Combin. Theory B* **42**, 302–312.

Hoffman, A.J.

[1975] Eigenvalues of graphs, in: *Studies in Graph Theory II,* ed. D.R. Fulkerson (Mathematical Association of America, Washington, DC) pp. 225–245.

Holyer, I.

[1981] The NP-completeness of edge-colouring, *SIAM J. Comput.* **10**, 718–720.

Hutchinson, J.P.

[1984] A five-color theorem for graphs on surfaces, *Proc. Amer. Math. Soc.* **90**, 497–504.

Isaacs, R.

[1975] Infinite families of non-trivial trivalent graphs which are not Tait colorable, *Amer. Math. Monthly* **82**, 221–239.

Jakobsen, I.T.

[1971] A homomorphism theorem with an application to the conjecture of Hadwiger, *Studia Sci. Math. Hungar.* **6**, 151–160.

[1975] On graphs critical with respect to edge-colourings, in: *Infinite and Finite Sets,* eds. A. Hajnal, R. Rado and V.T. Sós, *Colloq. Math. Soc. Janos Bolyai* **10**, 927–934.

Jensen, T.R., and B. Toft

[1995] *Graph Coloring Problems* (Wiley, New York).

Johnson, D.S.

[1974] Worst case behaviour of graph colouring algorithms, in: *Proc. 5th S.E. Conf. on Combinatorics, Graph Theory and Computing, Congress. Numerantium* **10**, 513–527.

[1978a] Graph colouring algorithm: between a rock and a hard place?, in: *Algorithmic Aspects of Combinatorics,* eds. B. Alspach, P. Hell and D.J. Miller, *Ann. Discrete Math.* **2**, 245.

[1978b] Research problem 13, in: *Algorithmic Aspects of Combinatorics,* eds. B. Alspach, P. Hell and D.J. Miller, *Ann. Discrete Math.* **2**, 243.

Karp, R.M.

[1972] Reducibility among combinatorial problems, in: *Complexity of Computer Computations,* eds. R.E. Miller and J.W. Thatcher (Plenum Press, New York) pp. 85–104.

Kelly, J.B., and L.M. Kelly

[1954] Paths and circuits in critical graphs, *Amer. J. Math.* **76**, 786–792.

Kempe, A.B.

[1879] On the geographical problem of four colours, *Amer. J. Math.* **2**, 193–200.

Kierstead, H.A.

[1984] On the chromatic index of multigraphs without large triangles, *J. Combin. Theory B* **36**, 156–160.

Kierstead, H.A., and J. Schmerl

[1983] Some applications of Vizing's theorem to vertex colourings of graphs, *Discrete Math.* **45**, 277–285.

Kierstead, H.A., E. Szemerédi and W.T. Trotter

[1984] On coloring graphs with locally small chromatic number, *Combinatorica* **4**, 183–185.

Kneser, M.

[1955] Aufgabe 360, *Jber. Deutsch. Math.- Verein.* **58**(2. Abt.), 27.

König, D.

[1916] Über Graphen und ihre Anwendung auf Determinantentheorie und Mengenlehre, *Math. Ann.* **77**, 453–465.

[1931] Graphen und Matrizen, *Mat. Fiz. Lapok* **38**, 116–119.

[1936] *Theorie der endlichen und unendlichen Graphen* (Leipzig). Reprinted: 1950 (Chelsea); 1986 (Teubner). English translation: 1989 (Birkhäuser, Basel).

Kubale, M.

[1991] Graph colouring, in: *Encyclopedia of Microcomputers,* Vol. 8, eds. A. Kent and J.G. Williams (Marcel Dekker, New York) pp. 47–69.

Lawrence, J.

[1978] Covering the vertex set of a graph with subgraphs of smaller degree, *Discrete Math.* **21**, 61–68.

Li, S.-Y.R., and W.-C.W. Li

[1981] Independence numbers of graphs and generators of ideals, *Combinatorica* **1**, 55–61.

Lick, D.R., and A.T. White
[1970] k-degenerate graphs, *Canad. J. Math.* **22**, 1082–1096.
Lovász, L.
[1966] On decomposition of graphs, *Studia Sci. Math. Hungar.* **1**, 237–238.
[1968] On chromatic number of finite set systems, *Acta Math. Sci. Hungar.* **19**, 59–67.
[1972a] Normal hypergraphs and the perfect graph conjecture, *Discrete Math.* **2**, 253–267.
[1972b] A characterization of perfect graphs, *J. Combin. Theory B* **13**, 95–98.
[1973a] Independent sets in critical chromatic graphs, *Studia Sci. Math. Hungar.* **8**, 165–168.
[1973b] Coverings and colourings of hypergraphs, in: *Proc. 4th S.E. Conf. on Combinatorics, Graph Theory and Computing, Congress. Numerantium* **8**, 3–12.
[1975] Three short proofs in graph theory, *J. Combin. Theory B* **19**, 269–271.
[1977] Flats in matroids and geometric graphs, in: *Combinatorial Surveys, Proc. Sixth British Combinatorics Conf.*, ed. P. Cameron (Academic Press, New York) pp. 45–86.
[1978a] Kneser's conjecture, chromatic number, and homotopy, *J. Combin. Theory A* **25**, 319–324.
[1978b] Some finite basis theorems in graph theory, in: *Combinatorics II*, eds A. Hajnal and V.T. Sós, *Colloq. Math. Soc. János Bolyai* **18**, 717–729.
[1979] On the Shannon capacity of a graph, *IEEE Trans. Inform. Theory* **IT-25**, 1–7.
[1982] Bounding the independence number of a graph, in: *Bonn Workshop on Combinatorial Optimization*, eds. A. Bachem, M. Grötschel and B. Korte, *Ann. Discrete Math.* **16**, 213–223.
[1983] Perfect graphs, in: *Selected Topics in Graph Theory 2*, eds. R.J. Wilson and L.W. Beineke (Academic Press, New York) pp. 55–87.
[1994] Stable sets and polynomials, *Discrete Math.* **124**, 137–153.
Lovász, L., and M.D. Plummer
[1986] *Matching Theory, Ann. Discrete Math.* **29**.
Lund, C., and M. Yannakakis
[1993] On the hardness of approximating minimization problems, in: *Proc. 25th ACM Symposium on Theory of Computing* (ACM, New York) pp. 286–293.
Mader, W.
[1968a] Über trennende Eckenmengen in homomorphiekritischen Graphen, *Math. Ann.* **175**, 243–252.
[1968b] Homomorphiesätze für Graphen, *Math. Ann.* **178**, 154–168.
Manvel, B.
[1985] Extremely greedy coloring algorithms, in: *Graphs and Applications, Proc. 1st Colorado Symp. on Graph Theory*, eds. F. Harary and J.S. Maybee (Wiley, New York) pp. 257–270.
Matula, D.W.
[1968] A max–min theorem for graphs with application to graph coloring, *SIAM Rev.* **10**, 481–482.
Matula, D.W., G. Marble and J.D. Isaacson
[1972] Graph coloring algorithms, in: *Graph Theory and Computing*, ed. R.C. Read (Academic Press, New York) pp. 109–122.
Mayer, J.
[1989] Conjecture de Hadwiger: un graphe k-chromatique contractioncritique n'est pas k-régulier, in: *Graph Theory in Memory of G.A. Dirac*, eds. L.D. Andersen, I.T. Jakobsen, C. Thomassen, B. Toft and P.D. Vestergaard, *Ann. Discrete Math.* **41**, 341–345.
Melnikov, L.S., and V.G. Vizing
[1969] New proof of Brooks' theorem, *J. Combin. Theory* **7**, 289–290.
Meyniel, H.
[1976] On the perfect graph conjecture, *Discrete Math.* **10**, 339–342.
Minty, G.J.
[1962] A theorem on n-coloring the points of a linear graph, *Amer. Math. Monthly* **69**, 623–624.
[1980] On maximal independent sets of vertices in claw-free graphs, *J. Combin. Theory B* **28**, 284–304.
Müller, V.
[1979] On colorings of graphs without short cycles, *Discrete Math.* **26**, 165–176.
Mycielski, J.
[1955] Sur le coloriage des graphes, *Colloq. Math.* **3**, 161–162.

Nelson, R., and R.J. Wilson

[1990] *Graph Colourings, Pitman Research Notes in Mathematics Series,* Vol. 218 (Longman, Harlow, UK).

Nešetřil, J., and V. Rödl

[1979] A short proof of the existence of highly chromatic hypergraphs without short cycles, *J. Combin. Theory B* **27**, 225–227.

Neumann-Lara, V.

[1982] The dichromatic number of a digraph, *J. Combin. Theory B* **33**, 265–270.

Nishizeki, T., and K. Kashiwagi

[1985] An upper bound on the chromatic index of multigraphs, in: *Graph Theory with Applications to Algorithms and Computer Science,* eds. Y. Alavi, G. Chartrand, L. Lesniak, D.R. Lick and C.E. Wall (Wiley, New York) pp. 595–604.

Olaru, E.

[1969] Über die Überdeckung von Graphen mit Cliquen, *Wiss. Z. TH Ilmenau* **15**, 115–121.

Ore, O.

[1967] *The Four-Color Problem* (Academic Press, New York).

Padberg, M.W.

[1974] Perfect zero–one matrices, *Math. Programming* **6**, 180–196.

Parthasarathy, K.R., and G. Ravindra

[1976] The strong perfect graph conjecture is true for $K_{1,3}$-free graphs, *J. Combin. Theory B* **21**, 212–223.

Petersen, J.

[1898] Sur le théorème de Tait, *Interméd. Math.* **5**, 225–227.

Ramsey, F.P.

[1930] On a problem of formal logic, *Proc. London Math. Soc. (2)* **30**, 264–286.

Read, R.C.

[1968] An introduction to chromatic polynomials, *J. Combin. Theory* **4**, 52–71.

Read, R.C., and W.T. Tutte

[1988] Chromatic polynomials, in: *Selected Topics in Graph Theory,* Vol. 3, eds. R.J. Wilson and L.W. Beineke (Academic Press, New York) pp. 15–42.

Reed, B.

[1987] A semi-strong perfect graph theorem, *J. Combin. Theory B* **43**, 223–240.

Ringel, G.

[1954] Bestimmung der Maximalzahl der Nachbargebiete von nichtorientierbaren Flächen, *Math. Ann.* **127**, 181–214.

[1959] *Färbungsprobleme auf Flächen und Graphen* (VEB Deutscher Verlag der Wissenschaften, Berlin).

[1974] *Map Color Theorem* (Springer, Berlin).

Ringel, G., and J.W.T. Youngs

[1968] Solution of the Heawood map-coloring problem, *Proc. Nat. Acad. Sci. U.S.A.* **60**, 438–445.

Robertson, N., P.D. Seymour and R. Thomas

[1993] Hadwiger's conjecture for K_6-free graphs, *Combinatorica* **13**, 279–361.

Robertson, N., D.P. Sanders, P. Seymour and R. Thomas

[1994] The four-colour theorem, Manuscript, February 1994; revised August 1994. 30pp + 17pp of figures.

Ryser, H.J.

[1963] *Combinatorial Mathematics* (Mathematical Association of America, Washington, DC).

Saaty, T.L., and P.G. Kainen

[1977] *The Four-Color Problem. Assaults and Conquest* (McGraw-Hill, New York). Reprinted: 1986 (Dover Publications, New York).

Sachs, H.

[1969] Finite graphs (Investigations and generalizations concerning the construction of finite graphs having given chromatic number and no triangles), in: *Recent Progress in Combinatorics,* ed. W.T. Tutte (Academic Press, New York) pp. 175–184.

[1970] On the Berge conjecture concerning perfect graphs, in: *Combinatorial Structures and their Applications,* eds. R. Guy, H. Hanani, N.W. Sauer and J. Schönheim (Gordon and Breach, New York) pp. 377–384.

Sachs, H., and M. Stiebitz

[1983] Construction of colour-critical graphs with given major-vertex subgraph, in: *Combinatorial Mathematics,* eds. C. Berge, D. Bresson, P. Camion, J.F. Maurras and F. Sterboul, *Ann. Discrete Math.* **17**, 581–598.

[1989] On constructive methods in the theory of colour-critical graphs, *Discrete Math.* **74**, 201–226.

Sbihi, N.

[1980] Algorithme de recherche d'un stable de cardinalité maximum dans un graphe sans étoile, *Discrete Math.* **29**, 53–76.

Schäuble, M.

[1969] Bemerkungen zur Konstruktion dreikreisfreier k-chromatischer Graphen, *Wiss. Z. TH Ilmenau* **15**, 59–63.

Schrijver, A.

[1978] Vertex-critical subgraphs of Kneser graphs, *Nieuw Archief voor Wiskunde* **26**, 454–461.

Seymour, P.D.

[1979] On multi-colourings of cubic graphs, and conjectures of Fulkerson and Tutte, *Proc. London Math. Soc. (3)* **38**, 423–460.

[1981a] On Tutte's extension of the four-colour problem, *J. Combin. Theory B* **31**, 82–94.

[1981b] Nowhere-zero 6-flows, *J. Combin. Theory B* **30**, 130–135.

Shannon, C.E.

[1949] A theorem on coloring the lines of a network, *J. Math. Phys.* **28**, 148–151.

Simonovits, M.

[1972] On colour-critical graphs, *Studia Sci. Math. Hungar.* **7**, 67–81.

Steinberg, R.

[1993] The state of the three color problem, in: *Quo Vadis, Graph Theory?* eds. J. Gimbel, J.W. Kennedy and L.V. Quintas, *Ann. Discrete Math.* **55**, 211–248.

Stiebitz, M.

[1987] Subgraphs of colour-critical graphs, *Combinatorica* **7**, 303–312.

Szekeres, G., and H.S. Wilf

[1968] An inequality for the chromatic number of a graph, *J. Combin. Theory* **4**, 1–3.

Tait, P.G.

[1878–80] On the colouring of maps, *Proc. Roy. Soc. Edinburgh Sect. A* **10**, 501–503.

Thomassen, C.

[1993a] Five-coloring maps on surfaces, *J. Combin. Theory B* **59**, 89–105.

[1993b] Grötzsch's 3-color theorem and its counterparts for the torus and the projective plane, Manuscript, The Technical University of Denmark, to appear in *J. Combin. Theory B.*

[1994a] Five-coloring graphs on the torus, *J. Combin. Theory B* **62**, 11–33.

[1994b] Every planar graph is 5-choosable, *J. Combin. Theory B* **62**, 180–181.

Toft, B.

[1970] On the maximal number of edges of critical k-chromatic graphs, *Studia Sci. Math. Hungar.* **5**, 461–470.

[1972a] Two theorems on critical 4-chromatic graphs, *Studia Sci. Math. Hungar.* **7**, 83–89.

[1972b] On separating sets of edges in contraction-critical graphs, *Math. Ann.* **196**, 129–147.

[1974a] Colour-critical graphs and hypergraphs, *J. Combin. Theory B* **16**, 145–161.

[1974b] On critical subgraphs of colour-critical graphs, *Discrete Math.* **7**, 377–392.

[1975] On colour-critical hypergraphs, in: *Infinite and Finite Sets,* eds. A. Hajnal, R. Rado and V.T. Sós, *Colloq. Math. Soc. János Bolyai* **10**, pp. 1445–1457.

[1985] Some problems and results related to subgraphs of colour-critical graphs, in: *Graphen in Forschung und Unterricht, Festschrift K. Wagner,* eds. R. Bodendiek, H. Schumacher and G. Walther (Verlag Barbara Franzbecker, Bad Salzdetfurth) pp. 178–186.

Tucker, A.C.

[1977] Critical perfect graphs and perfect 3-chromatic graphs, *J. Combin. Theory B* **23**, 143–149.

Turán, P.

[1941] An extremal problem in graph theory (in Hungarian), *Mat. Fiz. Lapok* **48**, 436–452 [English transl.: 1990, *Collected Papers by Paul Turán*, Vol. 1, ed. P. Erdős (Akadémiai Kiadó, Budapest) pp. 231–240].

Tutte, W.T.

[1950] On the imbedding of linear graphs in surfaces, *Proc. London Math. Soc. (2)* **51**, 474–483.

[1953] The 1-factors of oriented graphs, *Proc. Amer. Math. Soc.* **4**, 922–931.

[1954] A contribution to the theory of chromatic polynomials, *Canad. J. Math.* **6**, 80–91.

[1966] On the algebraic theory of graph colourings, *J. Combin. Theory* **1**, 15–50.

[1969] A geometrical version of the four color problem, in: *Combinatorial Mathematics and its Applications*, eds. R.C. Bose and T.A. Dowling (University of North Carolina Press, Chapel Hill, NC) pp. 553–560.

[1970] More about chromatic polynomials and the golden ratio, in: *Combinatorial Structures and their Applications*, eds. R. Guy, H. Hanani, N.W. Sauer and J. Schönheim (Gordon and Breach, New York) pp. 439–453.

[1978] Colouring problems, *Math. Intelligencer* **1**, 72–75.

[1984] *Graph Theory* (Addison-Wesley, Reading, MA).

Tuza, Z.

[1990] Problems and results on graph and hypergraph colourings, *Le Matematiche* **45**, 219–238.

Vizing, V.G.

[1964] On an estimate of the chromatic class of a p-graph (in Russian), *Diskret. Analiz.* **3**, 9–17.

[1965a] The chromatic class of a multigraph (in Russian), *Kibernetika* **3**, 29–39 [*Cybernetics* **1**, 32–41].

[1965b] Critical graphs with given chromatic class (in Russian), *Diskret. Analiz.* **5**, 9–17.

[1968] Some unsolved problems in graph theory (in Russian), *Uspekhi Math. Nauk.* **23**, 117–134 [*Russian Math. Surveys* **23**, 125–141].

[1976] Vertex colourings with given colours (in Russian), *Diskret. Analiz.* **29**, 3–10.

Voss, H.-J.

[1977] Graphs with prescribed maximal subgraphs and critical chromatic graphs, *Comment. Math. Univ. Carolin.* **18**, 129–142.

[1991] *Cycles and Bridges in Graphs* (Deutscher Verlag der Wissenschaften/Kluwer Academic Publishers, Berlin/Dordrecht).

Wagner, K.

[1937] Über eine Eigenschaft der ebenen Komplexe, *Math. Ann.* **114**, 570–590.

Weinstein, J.

[1975] Excess in critical graphs, *J. Combin. Theory B* **18**, 24–31.

Welsh, D.J.A.

[1976] *Matroid Theory* (Academic Press, New York).

Wessel, W.

[1970] Kanten-kritische Graphen mit der Zusammenhangszahl 2, *Manuscripta Math.* **2**, 309–334.

[1977] Some colour-critical equivalents of the strong perfect graph conjecture, in: *Beiträge zur Graphentheorie und deren Anwendungen vorgetragen auf dem internationalen Kolloquium in Oberhof, April 1977* (Mathematische Gesellschaft der DDR, Technische Hochschule Ilmenau) pp. 300–309.

Whitney, H.

[1932] A logical expansion in mathematics, *Bull. Amer. Math. Soc.* **38**, 572–579.

Whitney, H., and W.T. Tutte

[1972] Kempe chains and the four colour problem, *Utilitas Math.* **2**, 241–281. Reprinted in: *Studies in Graph Theory*, ed. D.R. Fulkerson (Mathematical Association of America, Washington, DC) pp. 378–413.

Wigderson, A.

[1983] Improving the performance guarantee for approximate graph coloring, *J. Assoc. Comp. Mach.* **30**, 729–735.

Woodall, D.R., and R.J. Wilson

[1978] The Appel–Haken proof of the four-color theorem, in: *Selected Topics in Graph Theory,* eds. L.W. Beineke and R.J. Wilson (Academic Press, New York) pp. 83–101.

Young, H.P.

[1971] A quick proof of Wagner's equivalence theorem, *J. London Math. Soc.* **3**, 661–664.

Younger, D.H.

[1983] Integer flows, *J. Graph Theory* **7**, 349–357.

Zykov, A.A.

[1949] On some properties of linear complexes (in Russian), *Mat. Sbornik N.S.* **24**, 163–188 [1952, *AMS Transl.* **79**]. Reprinted: 1962, *Translations Series 1,* Vol. 7, *Algebraic Topology* (American Mathematical Society, Providence, RI) pp. 418–449.

APPENDIX TO CHAPTER 4

Nowhere-Zero Flows

P.D. SEYMOUR

Bell Communications Research, 445 South Street, Morristown, NJ 07960-1910, USA

Contents

HANDBOOK OF COMBINATORICS
Edited by R. Graham, M. Grötschel and L. Lovász

1. Introduction

The theory of nowhere-zero flows was introduced by Tutte (1954), and provides an interesting way to generalize theorems about region-colouring planar graphs to general graphs. In this "mini-chapter" we shall survey the main ideas of the subject.

Let G be a digraph, and let Γ be an abelian group. A *Γ-circulation* in G is a function $\phi : E(G) \rightarrow \Gamma$ such that for every vertex v,

$$\sum (\phi(e)\colon e \in \delta^+(v)) = \sum (\phi(e)\colon e \in \delta^-(v)) ,$$

where $\delta^+(v)$, $\delta^-(v)$ are the sets of edges with tail v and head v respectively, and the summation is in Γ. An $\mathbb{R}$-circulation ($\mathbb{R}$ is the additive group of reals) is usually just called a *circulation*. If $k \geqslant 1$ is an integer, a *nowhere-zero k-flow* in G is a circulation ϕ such that for every edge e, $|\phi(e)|$ is one of $1, 2, \ldots, k-1$. We see that if an undirected graph G has a nowhere-zero k-flow for some assignment of directions to its edges, then it has one for every assignment (replace $\phi(e)$ by $-\phi(e)$ if the direction of e is changed). Thus, the problem raised by Tutte (1954) of determining for which values of k a graph has a nowhere-zero k-flow is a problem about graphs, not digraphs. Tutte's problem is motivated by the following observation.

Theorem 1.1. *Take a planar drawing of a planar graph G. The regions of this drawing may be coloured with k colours, such that for each edge the two regions bordering it have different colours, if and only if G has a nowhere-zero k-flow.*

Proof. Suppose for each region r, $\alpha(r) \in \{1, \ldots, k\}$ is a "colour", such that for each edge the regions bordering it have different colours. Assign a direction to each edge, and an orientation to the plane, and for each edge e let $\phi(e) = \alpha(r_1) - \alpha(r_2)$, where r_1, r_2 are the regions bordering e to its left and right respectively. Then ϕ is a nowhere-zero k-flow. For the converse, suppose that ϕ is a nowhere-zero k-flow. Since every integer-valued circulation in a planar graph is an integral combination of the "unit" circulations each of which runs around the perimeter of a region, it follows that for each region r there is an integer $\beta(r)$, such that for each edge e, $\phi(e) = \beta(r_1) - \beta(r_2)$, where again r_1, r_2 are the regions bordering e to its left and right respectively. Let $\alpha(r)$ be the residue of $\beta(r)$ modulo k; then α is a colouring of the regions. □

Theorem 1.1 allows us to state that a planar graph is k-region-colourable without mentioning regions, and hence in terms that make sense for general (non-planar) graphs. That suggests that we investigate how far the standard results about region-colouring planar graphs can be generalized to results about nowhere-zero flows in general graphs; and this is the main topic of the theory of nowhere-zero flows.

First, what about the four-colour theorem? By Theorem 1.1, it is equivalent to the assertion that every planar graph with no isthmus has a nowhere-zero 4-flow. (An *isthmus* is an edge not in any circuit. The isthmus condition is natural, for no graph with an isthmus has a nowhere-zero k-flow for any value of k.) Is this true for non-planar graphs as well? Unfortunately not, for the Petersen graph has no nowhere-zero 4-flow. Nevertheless, in contrast with vertex-colouring general graphs, there is a universal upper bound. In other words, the following holds.

Theorem 1.2. *There is a constant k such that every graph with no isthmus has a nowhere-zero k-flow.*

This was conjectured by Tutte (1954); indeed, Tutte conjectured that $k=5$ satisfied Theorem 1.2. This remains open, but Jaeger (1975) proved that $k=8$ satisfies Theorem 1.2, and later Seymour (1981a) showed it for $k=6$. We shall discuss these results later.

2. Group-valued flows

All the main results and problems about nowhere-zero flows concern nowhere-zero k-flows, that is, $\mathbb{R}$-circulations. But Tutte noticed that using Γ-circulations (where Γ is a finite abelian group) is often helpful in proving results about $\mathbb{R}$-circulations.

Thus, let Γ be a finite abelian group. A *nowhere-zero Γ-flow* in a digraph G is a Γ-circulation ϕ in G such that $\phi(e) \neq 0$ for all edges e. One can ask, given a digraph G and a finite abelian group Γ, does G have a nowhere-zero Γ-flow? This is relevant to the k-flow problem because Tutte noticed that G has a nowhere-zero Γ-flow if and only if G has a nowhere-zero $|\Gamma|$-flow. That is the main result of this section, and we shall prove this in two steps, as follows.

Theorem 2.1. *Let G be a digraph, and let Γ, Γ' be finite abelian groups with $|\Gamma| = |\Gamma'|$. Then the number of distinct nowhere-zero Γ-flows in G equals the number of distinct nowhere-zero Γ'-flows in G.*

Proof. Let $n(G)$ be the number of distinct nowhere-zero Γ-flows in G, and define $n'(G)$ similarly. We prove that $n(G) = n'(G)$ by induction on $|E(G)|$. If every edge of G is a loop, then

$$n(G) = (k-1)^{|E(G)|} = n'(G) ,$$

where $k = |\Gamma| = |\Gamma'|$, and the result holds. We assume then that some edge f is not a loop. Let A be the set of all Γ-circulations ϕ in G with $\phi(e) \neq 0$ ($e \in E(G)$, $e \neq f$) and $\phi(f) = 0$; and let B be the set of all nowhere-zero Γ-circulations in G. Let G_1 be obtained from G by contracting f. For every $\phi \in A \cup B$, the restriction of ϕ to $E(G_1)$ is a nowhere-zero Γ-flow in G_1, and conversely every nowhere-zero Γ-flow in G_1 is the restriction of a unique member of $A \cup B$ (since f is not a loop). Thus $n(G_1) = |A \cup B| = |A| + |B|$. But $n(G) = |B|$, and $n(G_2) = |A|$, where G_2 is obtained from G deleting f. Hence $n(G) = n(G_1) - n(G_2)$, and similarly $n'(G) = n'(G_1) - n'(G_2)$. But from the inductive hypothesis, $n(G_1) = n'(G_1)$ and $n(G_2) = n'(G_2)$; and so $n(G) = n'(G)$, as required. □

Theorem 2.2. *Let G be a digraph, and let $k \geq 1$ be an integer. Then G has a nowhere-zero $\mathbb{Z}_k$-flow if and only if G has a nowhere-zero k-flow.*

($\mathbb{Z}_k$ is the additive group of integers modulo k).

Proof. To prove "if", let ϕ be a nowhere-zero k-flow in G, and let $\phi'(e)$ be the residue of $\phi(e)$ modulo k. Since $\phi(e) \neq 0, \pm k$, it follows that $\phi'(e) \in \mathbb{Z}_k$ is non-zero, and hence ϕ' is a nowhere-zero $\mathbb{Z}_k$-flow.

The converse is less easy. Suppose that there is a nowhere-zero $\mathbb{Z}_k$-flow. Hence there is a function $\phi : E(G) \to R$ such that

(1) *for every edge e, $\phi(e)$ is one of* $\pm 1, \pm 2, \ldots, \pm(k-1)$;

and, denoting $\sum(\phi(e)\colon e \in \delta^+(v))$, $\sum(\phi(e)\colon e \in \delta^-(v))$ by $\phi(v^+)$, $\phi(v^-)$ respectively,

(2) *for every vertex v, $\phi(v^+) \equiv \phi(v^-)$ modulo k.*

For any such ϕ, let $D(\phi) = \sum_{v \in V(G)} |\phi(v^+) - \phi(v^-)|$, and choose ϕ satisfying (1) and (2) with $D(\phi)$ minimum. We shall show that ϕ is a nowhere-zero k-flow. If we reverse the direction of an edge e and replace $\phi(e)$ by $-\phi(e)$, then the new function ϕ still satisfies (1) and (2) in the new digraph, and $D(\phi)$ is unchanged, and so we may assume, to simplify notation, that

(3) *for every edge e, $\phi(e) \geq 0$, and so $\phi(e)$ is one of* $1, 2, \ldots, k-1$.

Let $S = \{v \in V(G)\colon \phi(v^+) > \phi(v^-)\}$, and $T = \{v \in V(G)\colon \phi(v^+) < \phi(v^-)\}$.

(4) *There is no directed path P from S to T.*

For if there is such a path, define $\phi'(e) = \phi(e) - k$ if $e \in E(P)$, and $\phi'(e) = \phi(e)$ otherwise; then ϕ' satisfies (1) and (2) (because of (3)) and $D(\phi') = D(\phi) - 2k$, contrary to our choice of ϕ.

For $X \subseteq V(G)$, we denote by $\delta^+(X)$ the set of edges with tail in X and head in $V(G) - X$, by $\delta^0(X)$ the set of edges with both ends in X, and by $\delta^-(X)$ the set of edges with head in X and tail in $V(G) - X$.

(5) $S = T = \emptyset$.

For by (4) there is a partition (A, B) of $V(G)$ with $S \subseteq A$, $T \subseteq B$ and $\delta^+(A) = \emptyset$. Then

$$\sum_{v \in A} \phi(v^+) - \phi(v^-) = \sum_{e \in \delta^0(A)} \phi(e) + \sum_{e \in \delta^+(A)} \phi(e) - \sum_{e \in \delta^0(A)} \phi(e) - \sum_{e \in \delta^-(A)} \phi(e) = - \sum_{e \in \delta^-(A)} \phi(e) \leq 0\,,$$

and since $\phi(v^+) - \phi(v^-) \geq 0$ for all $v \in A$ it follows that $S = \emptyset$. Similarly $T = \emptyset$.

From (5), ϕ is a nowhere-zero k-flow, as required. □

From Theorems 2.1 and 2.2 we deduce the following result of Tutte mentioned previously.

Theorem 2.3. *If Γ is a finite abelian group, then a digraph G has a nowhere-zero Γ-flow if and only if it has a nowhere-zero $|\Gamma|$-flow.*

Proof. Let $k = |\Gamma|$. By Theorem 2.1, G has a nowhere-zero Γ-flow if and only if it has a nowhere-zero $\mathbb{Z}_k$-flow. By Theorem 2.2, G has a nowhere-zero $\mathbb{Z}_k$-flow if and only if it has a nowhere-zero k-flow. □

3. Applications of Theorem 2.3

Theorem 2.3 is frequently useful, because Γ-flows are often easier to handle than k-flows. Let us see some of its applications.

There is a well-known result that any Eulerian planar triangulation is 3-vertex-colourable; let us cast this in terms of region-colourings and thence in terms of nowhere-zero flows, and thereby try to obtain an extension to non-planar graphs. In terms of region-colourings it asserts, via planar duality, that any bipartite cubic planar graph is 3-region-colourable, that is (by Theorem 1.1), has a nowhere-zero 3-flow. This can indeed be extended to general graphs, as follows.

Theorem 3.1. *Every bipartite cubic graph has a nowhere-zero 3-flow.*

Proof. Let (A, B) be a 2-colouring of a bipartite cubic graph G. Direct edge e from A to B and define $\phi(e) = 1$; then ϕ is a nowhere-zero $\mathbb{Z}_3$-flow. By Theorem 2.3, G has a nowhere-zero 3-flow, as required. □

Actually, a converse holds as well: a cubic graph has a nowhere-zero 3-flow if and only if it is bipartite. We omit the proof.

A second application is the following.

Theorem 3.2. *A graph G has a nowhere-zero 4-flow if and only if E(G) is the union of two cycles of G.*

(A *cycle* of G is a set of edges which may be partitioned into circuits.)

Proof. By Theorem 2.3, G has a nowhere-zero 4-flow if and only if it has a nowhere-zero $\mathbb{Z}_2 \times \mathbb{Z}_2$-flow. Let $C_1, C_2 \subseteq E(G)$, and for $e \in E(G)$ let $\phi(e) = (\phi_1(e), \phi_2(e))$, where

$$\phi_i(e) = \begin{cases} 1 & \text{if } e \in C_i\,, \\ 0 & \text{otherwise}\,, \end{cases}$$

for $i = 1, 2$. Then ϕ is a nowhere-zero $\mathbb{Z}_2 \times \mathbb{Z}_2$-flow if and only if C_1, C_2 are cycles with union $E(G)$. Since any function $\phi : E(G) \to \mathbb{Z}_2 \times \mathbb{Z}_2$ is describable this way, the result follows. □

Similarly, we have the following.

Theorem 3.3. *If G is cubic, then G has a nowhere-zero 4-flow if and only if G is 3-edge-colourable.*

Proof. By Theorem 2.3, G has a nowhere-zero 4-flow if and only if it has a nowhere-zero $\mathbb{Z}_2 \times \mathbb{Z}_2$-flow. Let the non-zero elements of $\mathbb{Z}_2 \times \mathbb{Z}_2$ be a, b, c. Then since $a+b+c=0$, it follows easily that a function $\phi : E(G) \to \{a, b, c\}$ is a nowhere-zero $\mathbb{Z}_2 \times \mathbb{Z}_2$-flow if and only if it is a 3-edge-colouring (with "colours" a, b, c). □

4. Nowhere-zero 4-flows

In view of Theorem 3.3, investigating which graphs have nowhere-zero 4-flows extends investigating which cubic graphs are 3-edge-colourable. Tutte (1966) proposed the following.

Conjecture 4.1. Every graph with no isthmus and no Petersen graph minor has a nowhere-zero 4-flow.

There has been a good deal of work done on Conjecture 4.1. For instance, in view of Theorem 3.2, Conjecture 4.1 asserts that if G has no isthmus and no Petersen graph minor then $E(G)$ is the union of two cycles. Seymour (1981b) showed that this conjecture is equivalent to a conjecture of Tutte (1966) about matroids, the "tangential 2-block" conjecture.

By Theorem 1.1, Conjecture 4.1 would extend the four-colour theorem, for planar graphs have no Petersen graph minor. A somewhat richer source of such graphs is those which can be obtained from a planar graph by adding one vertex, joined arbitrarily. Conjecture 4.1 implies that any such graph with no isthmus has a nowhere-zero 4-flow. This last conjecture is equivalent to the following conjecture of Grötzsch (unpublished).

Conjecture 4.2. Let G be a planar graph. Then G is 3-edge colourable if and only if

(i) every vertex of G has valency $\leqslant 3$, and
(ii) no subgraph of G has one vertex of valency 2 and all others of valency 3.

To see the equivalence, given G in Conjecture 4.2, add a new vertex adjacent to vertices of G such that every vertex of G has valency 3 in the enlarged graph. If the enlarged graph has a nowhere-zero 4-flow then (by the argument of Theorem 3.3) G is 3-edge colourable. The converse implication is a little more complicated, using "vertex-splitting", and we omit it.

Let us mention another extension of Conjecture 4.2 (Seymour 1979b).

Conjecture 4.3. Let G be a planar graph, and let $k \geqslant 0$ be an integer. Then G is k-edge-colourable if and only if

(i) every vertex of G has valency $\leqslant k$, and
(ii) for every $X \subseteq V(G)$ with $|X|$ odd, there are at most $\frac{1}{2}k(|X|-1)$ edges with both ends in X.

Conjecture 4.1 may be regarded as saying that in a sense, graphs with no Petersen graph minor are not much worse than planar graphs; and indeed there are now a couple of results of this kind. For instance, Alspach et al. (1994) have shown the following, generalizing a theorem of Seymour (1979a) for planar graphs.

Theorem 4.4. *Let G be a graph with no Petersen graph minor, and let $p: E(G) \to \mathbb{Z}^+$ be a function. Then there is a list of circuits of G such that every edge e occurs in precisely $p(e)$ of these circuits, if and only if for every $X \subseteq V(G)$*
(i) *$\sum (p(e)$: $e \in \delta(X))$ is even, and*
(ii) *$p(f) \leqslant \frac{1}{2} \sum (p(e)$: $e \in \delta(X)$ for every $f \in \delta(X)$.*
($\delta(X)$ is the set of edges with one end in X and the other in $V(G) - X$.)

Theorem 4.4 is related to the double cover conjecture (take $p = 2$), see chapter 5. It is also related to the "sums of circuits" property, discussed in chapter 10.
The Petersen graph is the only exception in an even stronger sense, in a theorem of Lovász (1987).

Theorem 4.5. *Let G be a 3-connected simple graph such that $G \backslash \{u, v\}$ has a perfect matching for every two vertices u, v. Let $w: E(G) \to \mathbb{Z}$ be such that $\sum (w(e)$: $e \in \delta(v))$ is the same for every vertex v. If G is not the Petersen graph, then w is expressible as an integral combination of the* (0, 1)*-characteristic functions of the perfect matchings of G.*

In the same paper Lovász obtained a characterization of the lattice of integral combinations of the 1-factors of a general graph G.

5. The 3-flow conjecture

The following famous theorem is due to Grötzsch (1958).

Theorem 5.1. *If G is planar and has no circuit of length $\leqslant 3$ then G is 3-vertex-colourable.*

Via planar duality and Theorem 1.1, this asserts that every 4-edge-connected planar graph has a nowhere-zero 3-flow. Tutte (1966) asked whether in this formulation planarity is needed, and conjectured not.

Conjecture 5.2. Every 4-edge-connected graph has a nowhere-zero 3-flow.

This remains open; indeed, it is not known if there is any k such that every k-edge-connected graph has a nowhere-zero 3-flow. Jaeger (1975) proved the following weakening of Conjecture 5.2.

Theorem 5.3. *Every 4-edge-connected graph has a nowhere-zero 4-flow.*

The proof of Theorem 5.3 uses the following lemma, a corollary of a theorem of Nash-Williams (1961).

Lemma 5.4. *For any $k \geq 1$, every 2k-edge-connected graph has k spanning trees, mutually edge-disjoint.*

Proof of Theorem 5.3. Let G be 4-edge-connected. By Lemma 5.4, G has two spanning trees T_1, T_2 with $E(T_1) \cap E(T_2) = \emptyset$. For each $e \in E(G) - E(T_1)$ let C_e be the unique set of edges in $E(T_1) \cup \{e\}$ forming a circuit. Let F be the set of all $f \in E(T_1)$ belonging to C_e for an odd number of edges $e \in E(G) - E(T_1)$, and let $C_1 = F \cup (E(G) - E(T_1))$. Then C_1 is a cycle with $E(G) - E(T_1) \subseteq C_1$. Similarly, there is a cycle C_2 with $E(G) - E(T_2) \subseteq C_2$. Then $C_1 \cup C_2 = E(G)$, and so G has a nowhere-zero 4-flow by Theorem 3.2. □

6. The 5-flow conjecture

As we stated earlier, Tutte (1954) proposed the following strengthening of Theorem 1.2, which would generalize the five-colour theorem for planar graphs.

Conjecture 6.1. Every graph with no isthmus has a nowhere-zero 5-flow.

Jaeger (1975) proved the following, which is now superceded but which had a very pretty proof that we would like to include here.

Theorem 6.2. *Every graph with no isthmus has a nowhere-zero 8-flow.*

Proof. It is easy to see that it suffices to prove that every 3-edge-connected graph has a nowhere-zero 8-flow. Thus, let G be 3-edge-connected. By applying Lemma 5.4 to the 6-edge-connected graph obtained from G by replacing each edge by two parallel edges, we deduce that there are three spanning trees T_1, T_2, T_3 of G with $E(T_1) \cap E(T_2) \cap E(T_3) = \emptyset$. As in Theorem 5.3, let $C_i \supseteq E(G) - E(T_i)$ be a cycle $(i = 1, 2, 3)$. Then $C_1 \cup C_2 \cup C_3 = E(G)$, and so (by an analogue of Theorem 3.2 for 8-flows) G has a nowhere-zero 8-flow. □

Theorem 6.2 was improved by the following result of Seymour (1981a).

Theorem 6.3. *Every graph with no isthmus has a nowhere-zero 6-flow.*

Proof. We proceed by induction on $|E(G)|$. We may assume that

(1) *G is simple, 3-connected, and cubic.*

For evidently, we may assume that G is connected and loopless, and hence 2-edge-connected since it has no isthmus. If $|\delta(X)| = 2$ for some $X \subseteq V(G)$, the result follows from the inductive hypothesis by contracting an edge in $\delta(X)$. Thus we may assume that G is 3-edge-connected. If some vertex v has valency $\geqslant 4$, then there are edges e_1, e_2 with ends u_1, v and u_2, v say, such that if we delete e_1, e_2 and add a new edge e_0 joining u_1, u_2, the resulting graph G' has no isthmus. By the inductive hypothesis G' has a nowhere-zero 6-flow ϕ', say; and hence so does G (define $|\phi(e_1)| = |\phi(e_2)| = |\phi(e)|$, and otherwise $\phi = \phi'$). Thus we may assume that there is no such v. Hence G is cubic, and so it is 3-connected.

For $X \subseteq E(G)$ we define X^2 to be the smallest set $Y \subseteq E(G)$ with $X \subseteq Y$ such that there is no circuit C with $0 < |E(C) - Y| \leqslant 2$. We say that X is *good* if there is a $\mathbb{Z}_3$-circulation ϕ with $\phi(e) \neq 0$ for all $e \in E(G) - X$.

(2) *If $X \subseteq E(G)$ and $X^2 = E(G)$ then X is good.*

For we proceed by induction on $|E(G) - X|$. We may assume that $X \neq E(G)$, and so there is a circuit C with $0 < |E(C) - X| \leqslant 2$. Let $X' = X \cup E(C)$. Then $X'^2 = E(G)$, and so from the inductive hypothesis X' is good; let ϕ' be the corresponding $\mathbb{Z}_3$-circulation. Let ϕ_0 be a $\mathbb{Z}_3$-circulation with $\phi_0(e) = \pm 1$ ($e \in E(C)$), and $\phi_0(e) = 0$ otherwise. Then one of $\phi', \phi' + \phi_0, \phi' - \phi_0$ is non-zero on all the (at most two) edges of $E(C) - X$, and hence is the required $\mathbb{Z}_3$-circulation.

(3) *There is a cycle C of G with $C^2 = E(G)$.*

For choose a cycle C with C^2 forming the edge-set of a connected subgraph H, and with C^2 maximal. Certainly H is induced, and no vertex of G not in $V(H)$ has $\geqslant 2$ neighbours in $V(H)$ (because H is connected). If $H \neq G$, there is a maximal 2-connected subgraph B of $G \backslash V(H)$ such that at most one vertex of B has a neighbour not in $V(B) \cup V(H)$. Now B must contain two vertices with neighbours in $V(H)$ since G is 3-connected, and hence there is a circuit D of B, two vertices of which have neighbours in $V(H)$. But $(C \cup D)^2$ is connected, contrary to the maximality of C. Thus $H = G$, as required.

From (2), there is a $\mathbb{Z}_3$-circulation ϕ_1 say such that $\phi_1(e) \neq 0$ for all $e \in E(G) - C$. Let ϕ_2 be a $\mathbb{Z}_2$-circulation such that $\phi_2(e) \neq 0$ for all $e \in C$. Let $\phi(e) = (\phi_1(e), \phi_2(e))$ for all e; then ϕ is a nowhere-zero $\mathbb{Z}_3 \times \mathbb{Z}_2$-flow in G. By Theorem 2.3, G has a nowhere-zero 6-flow, as required. □

References

Alspach, B., L.A. Goddyn and C.Q. Zhang
[1994] Graphs with the circuit cover property, *Trans. Amer. Math. Soc.* **344**, 131–154.

Grötzsch, H.
[1958] Ein Dreifarbensatz für Dreikreisfreie Netze auf der Kugel, *Wiss. Z. Martin-Luther Univ. Halle-Wittenberg, Math.-Naturwiss. Reihe* **8**, 109–120.

Jaeger, F.
[1975] On nowhere-zero flows in multigraphs, in: *Proc. 5th British Combinatorial Conference, Aberdeen, Congress. Numerantium* **XV**, 373–378.

Lovász, L.
[1987] Matching structure and the matching lattice, *J. Combin. Theory B* **43**, 187–222.
Nash-Williams, C.St.J.A.
[1961] Edge-disjoint spanning trees of finite graphs, *J. London Math. Soc.* **36**, 445–450.
Seymour, P.D.
[1979a] Sums of circuits, in: *Graph Theory and Related Topics,* eds. J.A. Bondy and U.S.R. Murty (Academic Press, New York) pp. 341–355.
[1979b] Unsolved problem, in: *Graph Theory and Related Topics,* eds. A. Bondy and U.S.R. Murty (Academic Press, New York) pp. 367–368.
[1981a] Nowhere-zero 6-flows, *J. Combin. Theory B* **30**, 130–135.
[1981b] On Tutte's extension of the four-colour problem, *J. Combin. Theory B* **31**, 82–94.
Tutte, W.T.
[1954] A contribution to the theory of chromatic polynomials, *Canad. J. Math.* **6**, 80–91.
[1966] On the algebraic theory of graph colorings, *J. Combin. Theory* **1**, 15–50.

CHAPTER 5

Embeddings and Minors

Carsten THOMASSEN

Mathematical Institute, The Technical University of Denmark, Building 303, DK-2800 Lyngby, Denmark

Contents

HANDBOOK OF COMBINATORICS
Edited by R. Graham, M. Grötschel and L. Lovász

1. Introduction

Various types of tilings of the Euclidean plane have been studied since ancient time. The problem of existence and classification of such tilings can be formulated as problems concerning infinite planar graphs with certain symmetry properties. In a series of recent papers Grünbaum and Shepard have investigated these problems extensively and written a beautiful book (Grünbaum and Shepard 1987) on the subject. The drawings in Grünbaum and Shepard (1987) show that these problems are not only interesting mathematically but also for aesthetic reasons. Another well-known classification problem of aesthetic nature involving planar graphs is that of classifying the Platonic solids, a problem which was solved by Euler's formula for planar graphs.

Planar graphs (or more generally graphs drawn in the plane with certain restrictions on how edges cross) are also of interest for practical reasons. The design of printed circuits has led to several problems on such graphs. Among such problems we may mention the problem of drawing a graph such that the total number of crossings is least possible and the problem of decomposing the graph into as few planar graphs as possible. The graph invariants thus arising are called the *crossing number* and *thickness*, respectively. These have been studied for their mathematical interest and we mention some of the results in this chapter.

Another practical reason for studying graph representation problems is that graph algorithms (when carried out in practice) may be very sensitive to the way in which the graphs are represented. One way is to represent the graphs as intersection graphs of set systems, where the sets may be some easily specified objects in a Euclidean space, for example d-intervals.

We say that a graph G can be *embedded* in a topological space X if the nodes of G can be represented by distinct elements in X and each edge of G can be represented by an arc in X, i.e., the image of a 1–1 continuous function from the unit interval $[0, 1]$ into X. Moreover, we require that two edges have at most one end in common. In this context many topological spaces X are uninteresting as the following observations show.

Proposition 1.1. *Every simple graph can be embedded into $\mathbb{R}^3$ such that all edges are straight line segments.*

Proof. For example, the nodes can all be represented on the curve consisting of all points (t, t^2, t^3) where $t \in \mathbb{R}$. □

We define a *k-book* as the topological space obtained from k disjoint unit squares by selecting one side of each and identifying all these sides into one line (called the *spine* of the k-book). Below is a short proof of a result of Atneosen (1968).

Proposition 1.2. *Every graph G can be embedded in a 3-book.*

Proof. Represent all nodes of G on page 1 of the 3-book and connect every vertex v by $\deg(v)$ straight line segments on page 1 to the spine. Now it is an easy exercise to extend the line segments to an embedding of G using only polygonal arcs on pages 2 and 3 (which may be thought of as a rectangle). □

Propositions 1.1 and 1.2 explain why almost all results on graph embeddings concern the plane $\mathbb{R}^2$ or compact 2-manifolds. The theory of knots in $\mathbb{R}^3$ is not considered as graph theory. One result by Conway and Gordon (1983) should, however, be mentioned in this context.

Theorem 1.3. *Let G be a graph embedded in the 3-dimensional space such that every edge is a polygonal arc. If G contains a K_6, then G contains a pair of disjoint cycles which are homologically linked. If G contains a K_7, then G contains a knotted cycle.*

Recently, Robertson et al. (1993b) have completely characterized the graphs which have linkless embeddings in the 3-dimensional space.

If we delete an edge e of a graph G embedded in X where X denotes the plane or a compact 2-manifold, then the resulting graph G-e is also embedded in X. Also, the contraction of the edge e results in a graph G/e embeddable in X, and hence the same holds for every *minor* of G, i.e., every graph obtained from G by successively deleting and/or contracting edges. The correlation between minors and embeddings is demonstrated by Kuratowski's (1930) classical planarity criterion that a graph is planar if and only if it contains no subdivision of any of the Kuratowski graphs K_5 and $K_{3,3}$ (see fig. 1.1). It is an easy exercise to see that a graph contains a subdivision of $K_{3,3}$ if and only if it has $K_{3,3}$ as a minor. Moreover, a graph G has K_5 as a minor, if and only if K either contains a subdivision of K_5 or has $K_{3,3}$ plus two independent edges as a minor. Hence Kuratowski's theorem is equivalent to the following result by Wagner (1937) (see also Harary and Tutte 1965).

Theorem 1.4. *A graph is planar if and only if it has none of K_5 or $K_{3,3}$ as a minor.*

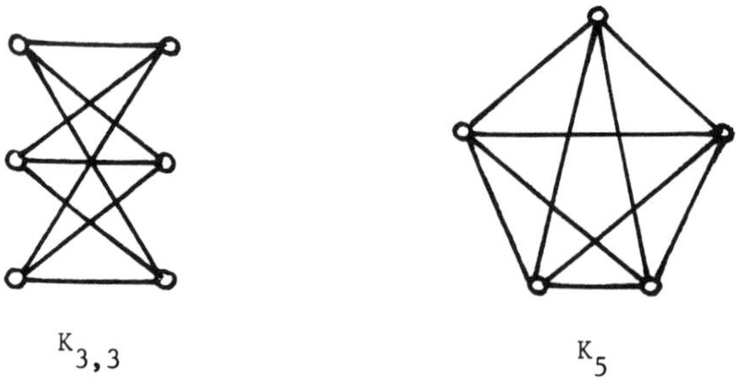

Figure 1.1. The Kuratowski graphs.

Kuratowski's theorem is not only the most fundemental planarity criterion as we point out in section 2.1. It also is a prototype of characterization in terms of forbidden minors, a type of characterization which is not only important in connection with graph embeddings, but also in other parts of graph theory and matroid theory. The study of minors has led to some of the deepest results in graph theory, some of which will be mentioned in this chapter.

In this chapter we first treat planarity and related invariants such as thickness and crossing numbers. Also, we consider more special representations of planar graphs and graphs in the plane. Then we proceed to graphs on higher surfaces. The investigations in this area were for several years centered around the Heawood conjecture on the minimum genus of a 2-manifold into which the complete graph K_n can be embedded. Ringel solved the problem in special cases and the solution was completed by Ringel and Youngs (1968), (see the survey of White 1978b). In section 3.3 we reproduce a complete solution when $n \equiv 7 \pmod{12}$. The method is based on the *rotation embedding scheme*, a simple idea which turns many embedding problems into combinatorial problems.

One of the most important unsolved problems in graph theory is *Hadwiger's conjecture* (see also the chapter by Toft) that every graph of chromatic number at least k has K_k as a minor. In section 4 we present the classical result of Wagner (1937) which gives a complete description of the graphs having no K_5 as a minor. As a corollary (called Wagner's equivalence theorem), Hadwiger's conjecture for $k = 5$ is equivalent to the 4-colour-theorem. We also mention methods and results related to the theorem of Wagner.

A well-known theorem of Kruskal (1960) asserts that, given an infinite sequence of trees there is always one that contains a subdivision of another. This does not hold for graphs in general, but if "subdivision" is replaced by "minor" then the statement above becomes true with "trees" replaced by "graphs". This has been proved by Robertson and Seymour in a series of "graph minor" papers beginning with Robertson and Seymour (1983). This remarkable results has far reaching consequences. Consider graph property p such that any minor of a graph with property p also has property p. For example, p might be the property of being embeddable in a given surface. Then the Robertson–Seymour theorem implies the existence of a finite set of graphs $G_1, \ldots, G_m$ such that a graph G has property p iff it has no minor isomorphic to any of $G_1, \ldots, G_m$. Combined with the solution of the k-path problem (also due to Robertson and Seymour) it gives a polynomial time algorithm for testing whether a graph has property p. Section 4 is devoted to a discussion of the Robertson–Seymour theory.

The literature on graph embeddings is immense and therefore many important results and methods have not been included in this chapter. Also, there are already excellent books or survey articles on the subject. White's (1973) book covers embeddings on higher surfaces and emphasizes also the algebraic aspects. White and Beineke (1978) is a more recent survey and the most recent book in the same spirit is the one by Gross and Tucker (1987). The Heawood conjecture is surveyed in White (1978b) and is the subject of Ringel's (1974) book. Crossing numbers are treated in surveys of Guy, for example Guy (1971). Furthermore, many text books on graph theory contain information on these subjects. However, we hope that the present chapter may be of use to those who wish to become acquainted with some of the important results and methods in a relatively fast way. Thus we demonstrate the power of the rotation scheme by including a short proof of one of the cases in the Heawood conjecture.

We treat various embeddings in the plane in some detail and describe the basic methods in connection with minors exemplified by a short proof of Wagner's characterization of the graphs that do not have K_5 as a minor. Finally, we emphasize the interaction between minors and embeddings as best demonstrated by the Robertson–Seymour theory.

2. Graphs in the plane

2.1. Planarity criteria

We distinguish between a *plane* graph (a graph embedded in the plane) and a *planar graph* (i.e., an abstract graph isomorphic to a plane graph). It is easy to see that a plane graph can be redrawn such that all edges are polygonal arcs and from now on we assume that plane graphs are embedded in this way. A *face* of a plane graph G is a (polygonal) arcwise connected point set of $\mathbb{R}^2 \backslash G$ and the *boundary* of a face is defined in the obvious way. The topological prerequisites needed for handling plane graphs are summarized in the next proposition.

Proposition 2.1. (a) *A cycle embedded in the plane has precisely two faces (called the interior and exterior of the cycle, respectively) each of which has the cycle as boundary.*

(b) *A plane isomorphic to $K_{2,3}$ has precisely three faces. The boundaries of these faces are the three cycles of $K_{2,3}$.*

(c) *Every face of a 2-connected plane graph G has a cycle of G as boundary.*

Proof. (a) is a variant of the *Jordan curve theorem*; a short proof of this together with (b) is given in Thomassen (1981). (c) follows easily from (b) with the aid of the following observation of H. Whitney: Every 2-connected graph can be obtained from a cycle by successively selecting two distinct nodes and adding a path between them. □

We also point out that a simple graph operation reduces many problems on graphs (in particular planar graphs) to the 3-connected case. We consider a 2-connected graph G which has a separating set $\{x, y\}$ of two nodes and form the two graphs G_1 and G_2 indicated in fig. 2.1. In these graphs x and y are adjacent while x and y may or may not be adjacent in G. It is an easy exercise to show that G is planar if and only if both G_1 and G_2 are planar. Also, it is easy to see that a graph is planar if and only if each block is planar and from these observations we may conclude that Kuratowski's theorem (and several other results on planar graphs) is only of interest for 3-connected graphs. If one (or both) of G_1 and G_2 in

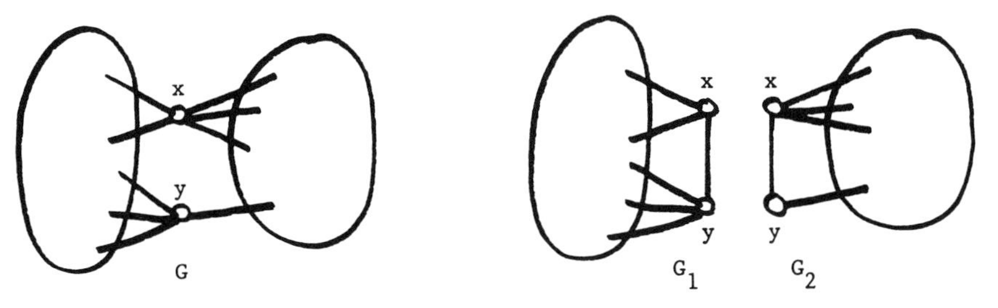

Figure 2.1. Splitting a graph of connectivity 2.

fig. 2.1 has a separating set of two nodes we may repeat the operation of fig. 2.1 until we obtain a collection of 3-connected graphs. These graphs turn out to be independent of the order in which we perform the operation of fig. 2.1. Hopcroft and Tarjan (1974) call them the *3-connected components* of the graph. They showed that the 3-connected components can be found in linear time and used that to describe a linear planarity test algorithm (see section 3.5).

Here we describe another planarity test algorithm which also proves Kuratowski's theorem. It is based on the following contraction lemma which is a 3-connected counterpart to the observation of Whitney mentioned in the proof of Proposition 2.1.

Lemma 2.2. *Every simple 3-connected graph other than K_4 has an edge whose contraction results in a 3-connected graph.*

Lemma 2.2 can be obtained from Tutte's characterization of the 3-connected graphs, but Lemma 2.2 also has a short independent proof (see Thomassen 1980). Based on Lemma 2.2 our planarity algorithm is as follows: Let G be any simple 3-connected graph. By Lemma 2.2, we form a sequence of graphs H_1, $H_2, \ldots, H_m$ where $H_1 = G$, $H_m = K_4$ and H_{i+1} is a 3-connected graph obtained from H_i by contracting an edge (and replacing any multiple edge that might occur by a single edge) for $i = 1, 2, \ldots, m-1$. Now draw H_m in the plane and, if H_{i+1} is plane, then we modify it, if possible, to a plane drawing of H_i by splitting a vertex z up into two adjacent vertices x and y keeping all edges not incident with z unchanged. In particular, the modification of H_{i+1} only affects the edges inside (or outside) the cycle C of $H_{i+1} - z$ which is the boundary of the face of $H_{i+1} - z$ containing z. So, if H_i cannot be drawn as a plane graph, then the reason is that C together with x, y and the edges incident with x and y is nonplanar. Then it is easy to find a subdivision of K_5 or $K_{3,3}$ in this subgraph of H_i (for details, see Thomassen 1981). In that case H_i (and hence also G) contains K_5 or $K_{3,3}$ as a minor.

An edge xy satisfies the conclusion of Lemma 2.2 iff the graph obtained by deleting x and y is 2-connected. This can be tested in $O(q)$ time where q is the number of edges of the graph. Thus an edge satisfying the conclusion of Lemma 2.2 can be found in $O(q^2)$ time. Hence the above sequence $H_1, H_2, \ldots, H_m$ can be found in $O(q^3)$ time. The reverse operating (going from H_m to H_1) can be done in fewer steps. Thus the algorithm is an $O(q^3)$ time algorithm.

Other short proofs and applications of Kuratowski's theorem are discussed in Thomassen (1981), and recently Tverberg (1987) obtained yet another short proof. The history of the theorem is the subject of Kennedy et al. (1985), see also the chapter by Biggs et al. in this volume. Here we just mention a few implications.

Tutte (1963) (see also Thomassen 1980) pointed out that Kuratowski's theorem implies MacLane's (1937) planarity criterion Theorem 2.3 below. For the definition of the *cycle space* the reader is referred to chapter 1 by Bondy.

Theorem 2.3. *A graph is planar if and only if its cycle space has a basis of cycles such that every edge is in at most two of these cycles.*

We also mention the characterization due to Chartrand and Harary (1967) of the *outplanar graphs*, i.e., the graphs that can be embedded in the plane such that all nodes are on the outer face boundary. We express the result in terms of minors (see also the chapter by Seymour).

Theorem 2.4. *A graph G is outerplanar if and only if it contains no minor isomorphic to K_4 or $K_{2,3}$.*

Proof. G is outerplanar if and only if the graph G' obtained from G by adding a new node joined to all other nodes is planar. Hence Theorem 2.4 follows from Theorem 1.4. □

2.2. The dual graph

If G and H are connected graphs we say that H is a *dual graph* of G, and write $H = G^*$, if there exists a function $\varphi: E(G) \to E(H)$ which is 1–1 and onto such that a subset $E \subseteq E(G)$ forms a cycle in G iff $\varphi(E)$ is a minimal edge cut in H. One can prove a number of fundamental and interesting facts concerning (abstract) dual graphs (for example that G is a dual of G^*). Here we mention only one such example which emphasizes the "duality" of the two operations used in the definition of minors.

Proposition 2.5. *If G^* is a dual of G, and e is an edge of G, and e^* is the corresponding edge in G^*, then $G^* - e^*$ is a dual of G/e and G^*/e^* is a dual of $G - e$.*

In particular, if G has a dual graph, then every minor of G has a dual graph. One can show that none of K_5 and $K_{3,3}$ has a dual graph. (This is a finite problem.) Hence Kuratowski's theorem implies the nontrivial part of Whitney's (1932) planarity criterion.

Theorem 2.6. *A graph G has a dual graph if and only if G is planar.*

The easy part of Theorem 2.6 is established by the *geometric dual graph* of a plane graph G. In every face of G we insert a node and for every edge e of G, we draw an edge e^* such that e^* crosses e (and no other edge of G or the geometric dual graph) and joins the two nodes in the faces adjacent to e. If e is an isthmus of G, then only one face is adjacent to e and in that case e^* becomes a loop.

The importance of duality becomes perhaps more clear when going from graphs to matroids (see the chapters by Recski and Welsh, respectively). For planar graphs the geometric dual graph is often a useful tool (primarily because of Proposition 2.7 below) and some results which hold for planar graphs and not for graphs in general can sometimes by "dualized" such that they give rise to theorems or problems for general graphs. The dual graph (or matroid) is of interest in electrical networks as indicated by the following simple example: Let N be an electrical network whose underlying graph is plane and whose edges are resistances, voltage generators and current generators. Let N^* be the dual network, i.e., the underlying graph of N^* is the geometric dual of that of N, and the resistances of N^* are the reciprocals of those of N, and voltage (resp. current) generators in N^* correspond to current (resp. voltage) generators of N. Then the solution of N (i.e., the current and voltage vectors) becomes the solution of N^* simply by interchanging between the current and voltage in every edge.

Theorem 2.6 can be extended as follows: If G, H is a pair of (abstract) dual graphs then they have a plane embedding as geometric dual graphs. Suppose now that G is 2-connected. (It is then easy to see that every dual graph of G is also 2-connected.) If G_1 and G_2 are plane graphs isomorphic to G, and the facial cycles (i.e., the face boundaries) of G_1 and G_2 are the same cycles in G and, furthermore, the outer cycle in G_1 corresponds to that in G_2, then it is not difficult to prove that the two plane representations G_1 and G_2 are *equivalent* in the sense that there exists a homeomorphism of the plane taking G_1 onto G_2. If we regard G_1 and G_2 as embeddings of G on the sphere, then the same conclusion holds without the assumption on the outer cycles. Hence we get the following.

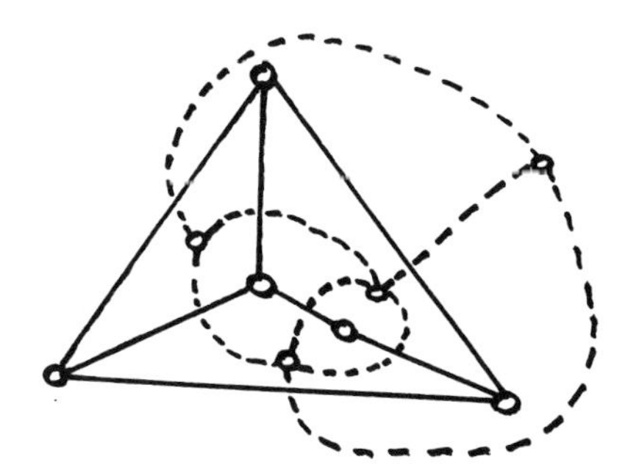

Figure 2.2. A plane graph and its dual graph.

Proposition 2.7. *Let G be a 2-connected plane graph. There is a 1–1 correspondence between the nonequivalent embeddings of G into the sphere and the nonisomorphic dual graphs of G.*

Let H be a 2-connected graph and let G be a dual graph as the leftmost graph in fig. 2.1. A *2-switching* of G is obtained by deleting every edge zx (where z is in G_1) and replacing it by zy. The resulting graph G' clearly has the same cycles as G (in the sense that an edge set forms a cycle in G iff the corresponding edge set in G' also forms a cycle). Hence H is a dual of G' and so G' is a dual of H. Whitney (1933) proved that every dual graph of H can be obtained from G by a sequence of 2-switchings. Combined with Proposition 2.7 this gives a combinatorial description of the nonequivalent embeddings of H into the sphere. In particular, if H is 3-connected, then H has only one such embedding (up to equivalence). Another combinatorial explanation of the last fact was also given by Tutte (1963).

Proposition 2.8. *If H is a simple 3-connected planar graph, then the facial cycles in any plane embedding of H are those cycles in H which are chordless and nonseparating* (*in the sense that the deletion of the nodes of the cycle leaves a connected graph*).

The nonequivalent embeddings into the plane (or sphere) can also be described algebraically based on extensions of Theorem 2.3.

2.3. Convex representations

The graph $G(P)$ of d-polytope P (i.e., the convex hull of a finite point set in $\mathbb{R}^d$) is the graph whose nodes are the vertices of P and whose edges are the edges of P. It can be shown (see Grünbaum 1967) that $G(P)$ is d-connected. Moreover, if $d = 3$, then an appropriate projection of P onto $\mathbb{R}^2$ shows that $G(P)$ has a *convex representation*, i.e., a plane representation such that all facial cycles are convex polygons. In particular, $G(P)$ is planar and 3-connected. Steinitz' theorem (see Grünbaum 1967) below shows that the converse holds.

Theorem 2.9. *A graph G is the graph of a 3-polytope if and only if G is simple planar and 3-connected.*

In particular, every simple 3-connected planar graph has a convex representation. Tutte (1963) considered planar straight line representations such that every node not on the outer cycle is the mass center of its neighbours. This is a special type of convex representations and finding such a representation amounts to solving certain system of linear equations. Tutte proved that this can always be solved for 3-connected planar graphs.

Tutte (1960) also proved that any convex polygon can play the role of the outer cycle (provided, of course, that it has the right number of corners). Tutte's result was in fact more general and we state here the generalization of Thomassen (1980) of Tutte's result.

Theorem 2.10. *Let G be a 2-connected plane graph with outer cycle S. Redraw S as a convex polygon Σ.* Then Σ can be extended to a convex representation of G (inside Σ) if and only if G satisfies the following conditions:

(a) *If H is a connected component of $G - V(S)$, then the nodes of S joined to H are not all on the same straight line segment of Σ.*

(b) *If e is an edge of $G - E(S)$, then the ends of e are not on the same straight line segment of S.*

(c) *If G has a separating node set $\{x, y\}$ as in fig.* 2.1, *and S is in G_1, then G_2 is just a path from x to y.*

The afore-mentioned result of Tutte (1960) is the special case of Theorem 2.10 when Σ is strictly convex. In this case (b) can be omitted, and (a) reduces to the statement that no two consecutive nodes on S form a separating set. The extension Theorem 2.10 of Tutte's result is convenient when going from finite to infinite graphs as pointed out in Thomassen (1984). For example, every infinite, locally finite 3-connected simple planar graph with no node accumulation points has a convex representation, even such that every bounded face is a convex polygon with at most 8 corners. (For finite graphs we can replace 8 by 6 which is best possible.) Moreover, if the graph is doubly-periodic (i.e., its automorphism group includes two translations in different directions), as is the case for many of the tilings mentioned in the introduction, then the graph has a doubly-periodic convex representation. This last result was proved first by complicated methods by Mani-Levitska et al. (1979) and a short proof based on Theorem 2.10 appears in Grünbaum and Shepard (1981) and Thomassen (1984).

The following application of Theorem 2.10 appears in Thomassen (1989b).

Theorem 2.11. *Let G, S and Σ be as in Theorem* 2.10 *and suppose* (a), (b), (c) *are satisfied. Suppose further that Σ has no horizontal edge. Now orient each edge of G*

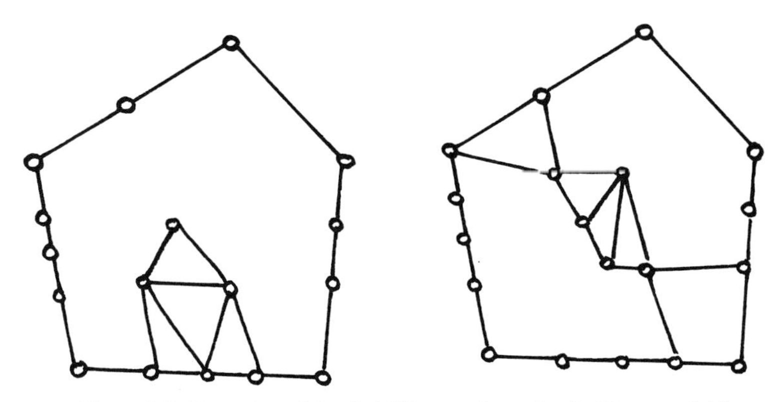

Figure 2.3. Examples of the forbidden configuration in Theorem 2.10.

such that the top (respectively bottom) vertex of Σ is the only sink (respectively source). Then Σ can be extended to a convex representation of G such that all edges are directed upwards.

It is easy to see that a maximal plane graph is 3-connected and has therefore by Theorem 2.10 a convex representation. In particular, every planar graph has a straight line representation. This is usually called Fary's theorem although it was first discovered by Wagner (1936). Similarly, Theorem 2.11 implies the following result which extends a result of Platt (1976). (Here, an *oriented graph* is obtained from a graph by assigning an orientation to every edge.)

Corollary 2.12. *Let G be an oriented simple graph with only one source x and one sink y and assume further that* $G \cup \{xy\}$ *is planar. Then G has a plane straight line representation such that all edges are directed upwards.*

It is an interesting unsolved problem in connection with Hasse diagrams to characterize those oriented graphs which have plane straight line representations with all edges directed upwards. For oriented graphs with a node dominating all other nodes (a directed analogue of outerplanar graphs) there is a Kuratowski type theorem by Thomassen (1989b) with more than 20 types of forbidden subgraphs, so the general problem is probably difficult. N. Alon and L. Lovász (and possibly others) have generalized the above-mentioned result of Wagner and Fary to general surfaces: If a graph embeds in the surface it embeds with geodesics.

A plane graph is *rectangular* if all faces are bounded by rectangles whose sides are either vertical or horizontal straight line segments. Ungar (1953) proved the following result on rectangular representations.

Theorem 2.13. *Every planar, cubic, 3-connected, cyclically 4-edge-connected graph has a rectangular representation.*

2.4. Other special representations of planar graphs

A *d-interval graph G* is the intersection graph of a collection of d-intervals (boxes) in $\mathbb{R}^d$. If the intervals can be chosen such that no two have an interior point in common and any two intersecting intervals have a $(d-1)$-interval in common, then we say that G is a *strict d-interval graph*. The d-interval graphs have been characterized for $d = 1$ only. Thomassen (1986) extended Theorem 2.13 to give a necessary and sufficient condition for rectangular representations in terms of forbidden configurations in the same spirit as Theorem 2.10 and that was used to prove the following.

Theorem 2.14. *A graph G has a strict 2-interval representation if and only if G is a proper subgraph of a 4-connected planar graph.*

Moreover, Thomassen (1986) obtained a characterization of the graphs in Theorem 2.14 in terms of forbidden subgraphs and proved also the following.

Theorem 2.15. *Every planar graph has a strict 3-interval representation.*

Theorem 2.15 was conjectured by Scheinerman and West inspired by their counterpart of Theorem 2.15 proved in (Scheinerman and West 1983).

Theorem 2.16. *Every planar graph is the intersection graph of a set system where each set is the union of three real intervals.*

Melnikov (1981) introduced an analogous type of representation (which we here call a *Melnikov representation*: The nodes of the graph are disjoint horizontal straight line segments in $\mathbb{R}^2$ and the edges are all possible vertical straight line segments which connect two nodes and intersect no other node or edge. Duchet et al. (1983) proved that every maximal planar graph has a Melnikov representations. Thomassen (1984) extended this to all 3-connected graphs. Tamassia and Tollis (1986) proved the following.

Theorem 2.17. *A graph G has a Melnikov representation if and only if it has a plane representation such that all cutnodes are on the boundary of the outer face.*

We can test, in linear time, if a graph G satisfies the conclusion of Theorem 2.17: Just add a new node v and join v to all cutnodes of G. Then apply a planarity test algorithm to the resulting graph G'. Applying Kuratowski's theorem to G' yields the following.

Corollary 2.18. *A planar graph G has no Melnikov representation iff either*

(i) *G contains a subdivision of K_4 such that two "opposite" paths contain cutnodes of G, or*

(ii) *G contains a subdivision of $K_{2,3}$ such that each of the three paths between the nodes of degree 3 contains a cutnode of G, or*

(iii) *G contains a subdivision of K_5 minus an edge such that the two nodes of degree 3 are cutnodes in G, or*

(iv) *G contains a subdivision of K_4 such that the nodes of degree 3 are cutnodes in G.*

The following result is due to Koebe (1936) and Andreev (1970).

Theorem 2.19. *Every planar graph can be represented as the intersection graph of closed discs in the plane such that no two discs have an interior point in common.*

Sinden raised the problem of characterizing those graphs which are intersection graphs of open curves in the plane. Such graphs are investigated carefully by Kratochvíl et al. (1986). Bouchet (1987) obtained an excluded minor characterization of those string graphs where the curves are all chords of a fixed circle. Kratochvíl (private communication) has proved that recognizing string graphs in general is NP-hard.

2.5. *Plane drawings with crossings*

Every graph can be drawn in the plane such that any two edges have at most one point in common (and such that any such point is an end or a crossing point) and, furthermore, such that no three edges have a crossing point in common. Such a drawing will be called *normal*.

The *crossing number* $\mathrm{cr}(G)$ and the *rectilinear crossing number* $\overline{\mathrm{cr}}(G)$ is the smallest number of crossings in a normal drawing (respectively normal drawing with straight edges) of G. $\mathrm{cr}(G)$ has been investigated for several classes of graphs (see, e.g., Guy 1971), but only few exact results are known. Here we consider complete graphs and complete bipartite graphs only.

The crossing number $\mathrm{cr}(K_n)$ was considered first by P. Erdős in 1940, as mentioned by Guy (1969b) who was the first to obtain the following upper bound.

Theorem 2.20.

$$\nu(K_n) \leqslant \tfrac{1}{4}\lfloor n/2\rfloor\, \lfloor (n-1)/2\rfloor\, \lfloor (n-2)/2\rfloor\, \lfloor (n-3)/2\rfloor\,.$$

Proof. Guy (1969b) gave the following elegant proof: Let the nodes of K_n be the vertices of a regular n-gon. Joint two nodes by straight line segments if these are in directions within a given quadrand. Otherwise join them outside. □

Another simple proof is described by White and Beineke (1978). The big unsolved problem on crossing numbers is whether the inequality of Theorem 2.20 is in fact an equality. This has been verified for $n \leqslant 10$ (see Erdős and Guy 1973).

Guy (private communication) has shown that $\overline{\mathrm{cr}}(K_9) = \mathrm{cr}(K_9) = 36$ but $\overline{\mathrm{cr}}(K_8) > 18 = \mathrm{cr}(K_8)$ and $\overline{\mathrm{cr}}(K_{10}) > \mathrm{cr}(K_{10})$. It is believed that

$$\overline{\mathrm{cr}}(K_n) > \mathrm{cr}(K_n) \quad \text{for all } n \geqslant 10\,.$$

The best known upper bound for $\overline{\mathrm{cr}}(n)$ has been obtained by Jensen (1971).

Theorem 2.21.

$$\overline{\mathrm{cr}}(K_n) \leqslant \lfloor (7n^4 - 56n^3 + 128n^2 + 48n\lfloor (n-7)/3\rfloor + 108)/432\rfloor\,.$$

It was conjectured by Erdős and Guy (1973) that Theorem 2.21 is an equality. However, Jensen (private communication) has found several better drawings. Jensen can prove that $\overline{\mathrm{cr}}(K_n) < \frac{1}{63}n^2$ for n large (which is better than Theorem 2.21) and believes that $\overline{\mathrm{cr}}(K_n)/\mathrm{cr}(K_n) \to 1$ as $n \to \infty$.

If K_{n+1} is drawn with $\mathrm{cr}(K_{n+1})$ crossings, then the average number of crossings in the $n+1$ K_n's in this K_{n+1} is $(n-3)\,\mathrm{cr}(K_{n+1})/(n+1)$ and hence this is an upper bound for $\mathrm{cr}(K_n)$. In other words, $\mathrm{cr}(K_n)/\binom{n}{4} \leqslant \mathrm{cr}(K_{n+1})/\binom{n+1}{4}$. Hence

$\lim_{n\to\infty} \mathrm{cr}(K_n)/n^4$ exists. Similarly, $\lim_{n\to\infty} \overline{\mathrm{cr}}(K_n)/n^4$ exists. Since $\overline{\mathrm{cr}}(K_5) = \mathrm{cr}(K_5) = 1$ we conclude that

$$\overline{\mathrm{cr}}(K_n) \geq \mathrm{cr}(K_n) \geq \tfrac{1}{5}\binom{n}{4}.$$

Kleitman (1970) obtained the better inequality

$$\overline{\mathrm{cr}}(K_n) \geq \mathrm{cr}(K_n) \geq \tfrac{3}{10}\binom{n}{4}$$

for n sufficiently large.

The crossing number problem for complete bipartite graphs is also known as Turán's brick-factory problem and it was believed to be solved by Zarankiewicz (1954). This is discussed by Guy (1969a) who notes that Zarankiewicz proved only the upper bound.

Theorem 2.22.

$$\mathrm{cr}(K_{m,n}) \leq \lfloor m/2 \rfloor \lfloor (m-1)/2 \rfloor \lfloor n/2 \rfloor \lfloor (n-1)/2 \rfloor.$$

It is believed that Theorem 2.22 is in fact an equality. (This has been verified by Kleitman (1970) for $\min(m, n) \leq 6$). If true, then $\overline{\mathrm{cr}}(K_{m,n}) = \mathrm{cr}(K_{m,n})$. It was conjectured by Eggleton (1986) and proved by Thomassen (1988a), that if G is a normal drawing such that every edge crosses at most one other edge, then there exists a homeomorphism of $\mathbb{R}^2$ onto $\mathbb{R}^2$ taking G into a graph where every edge is a straight line segment unless G contains a so-called B- or W-configuration. (Such a configuration has at most 4 edges and is therefore easy to detect.) As a consequence, every 2-connected 3-edge-connected planar graph G and its dual can be drawn as geometric dual graphs such that all edges (except one) are straight line segments. As another easy consequence, $\mathrm{cr}(G) \leq 1$ implies $\overline{\mathrm{cr}}(G) \leq 1$. Bienstock and Dean (1993) proved that $\mathrm{cr}(G) \leq 3$ implies $\overline{\mathrm{cr}}(G) = \mathrm{cr}(G)$ and that there are infinitely many graphs H such that

$$\overline{\mathrm{cr}}(G) > 4 = \mathrm{cr}(G).$$

Garey and Johnson (1982) proved that it is NP-complete to settle the following question: Given a graph G and a natural number k, is $\mathrm{cr}(G) \leq k$?

There are other problems on plane representations of graphs with certain crossings allowed. For example, Woodall (1971) gave a complete characterization of the graphs which have a normal drawing in the plane such that every two nonadjacent edges cross under the assumption that no such graph can have more edges than nodes. This assumption, which is the so-called thrackle conjecture by Conway, has not been verified, though.

2.6. Graph thickness

The *thickness* $t(G)$ of a graph G is the smallest number k such that G can be decomposed into k planar graphs. A lower bound for $t(G)$ is easily obtained from

Euler's formula

$$n - e + f = 2 , \tag{2.23}$$

where n, e, f are the number of nodes, edges, and faces (respectively) of a connected plane graph G. Euler's formula follows easily, by induction on e, from Proposition 2.1 (a) and (b). If G if simple, then every face has at least three edges on its boundary, and every edge is on the boundary on at most two faces. Hence, for G simple,

$$e \leqslant 3n - 6 . \tag{2.24}$$

Now, (2.24) implies

$$t(G) \geqslant e/(3n - 6) \tag{2.25}$$

and hence

$$\begin{aligned} t(K_n) &\geqslant \lceil (1/2n(n-1))/(3n-6) \rceil \\ &= \lfloor (1/2n(n-1) + 3n - 7)/3n - 6 \rfloor = \lfloor (n+7)/6 \rfloor . \end{aligned} \tag{2.26}$$

Beineke and Harary (1969) showed that the inequality of (2.26) is an equality for almost all n. They first showed that the inequality of (2.25) is an equality for the graph obtained from K_{6k} by deleting a perfect matching. So, this graph has thickness k. Then they showed that the inequality for $t(K_n)$ is an equality for all n, $n \neq 4 \pmod 6$. Others modified the construction of Beineke and Harary (1969) (see White and Beineke 1978) to dispose of the cases $n \equiv 4 \pmod 6$ except the case $n = 16$ which was finally settled by Mayer (1972) as shown in fig. 2.4.

Battle et al. (1962b) showed that $t(K_9) = 3$. Combining the above results we get.

Theorem 2.27. $t(K_9) = t(K_{10}) = 3$, *and*

$$t(K_n) = \lfloor (n+7)/6 \rfloor$$

for all $n \neq 9$, 10.

The *book-thickness* of a graph G is the smallest number k such that G can be embedded into the k-book (defined in the introduction) such that all nodes are on the spine and no edge intersects more than one page (except at the spine). It is easy to see that a graph G has book-thickness 1 iff it is outerplanar. If G has book-thickness $\leqslant 2$, then clearly G is planar. Furthermore, we can add edges to G (in the 2-book) so that we obtain a Hamiltonian cycle of the resulting graph G' such that all edges of $E(G') \backslash E(G)$ except one are on the spine. Conversely, a Hamiltonian planar graph clearly has book-thickness $\leqslant 2$. So, a graph G has book-thickness $\leqslant 2$ iff G is a subgraph of a Hamiltonian planar graph. Chvátal (private communication) has shown that it is NP-complete to decide if a maximal planar graph is Hamiltonian. Hence it is NP-complete to decide if a graph has

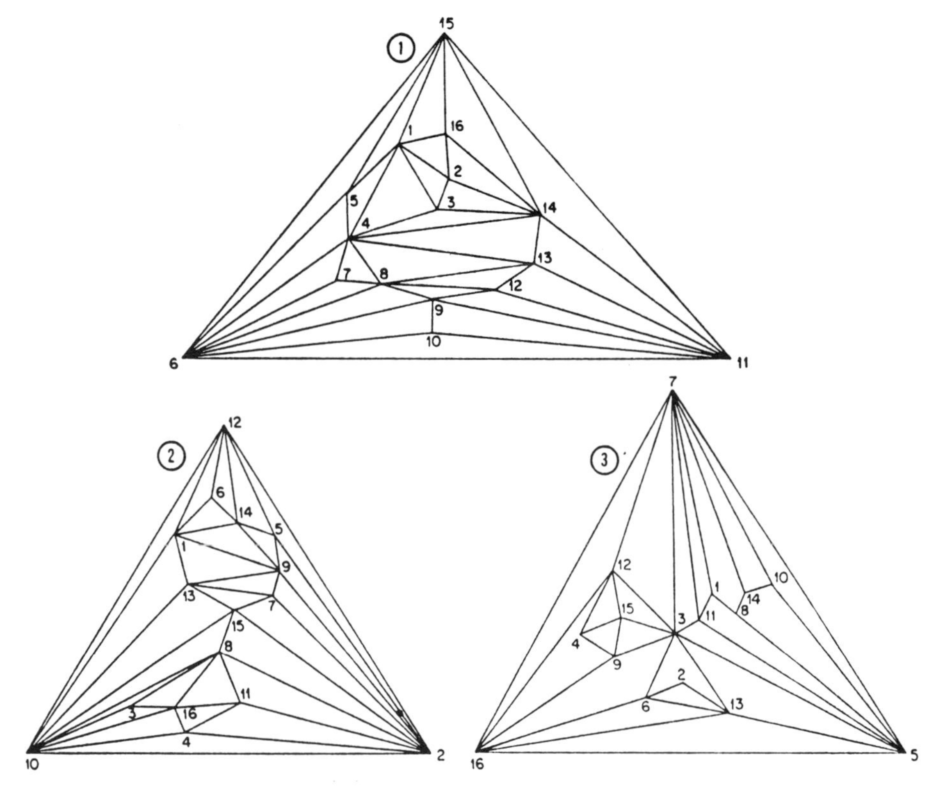

Figure 2.4. $t(K_{16}) = 3$.

book-thickness ≤2. Mansfield (1983) proved that it is NP-complete to decide if a graph has thickness ≤2.

Yannakakis (1989) proved that every planar graph has book-thickness ≤4. Every triangle free planar graph is a subgraph of a 4-connected (and hence Hamiltonian) planar graph (see, e.g., Thomassen 1986). So, every triangle free planar graph has book-thickness 1 if it is outerplanar and book-thickness 2 otherwise. Malitz (1988) proved that graphs of genus g (defined in section 3.2 below) have book-thickness at most $O(\sqrt{g})$ which is essentially best possible.

3. Graphs on higher surfaces

3.1. *The classification of surfaces*

As noted earlier, a graph embedded in the plane may also be regarded as a graph on the sphere. In this section we consider graphs on other surfaces. The topological prerequisites are the basic definitions such as compact and arcwise connected Hausdorff topological space, homeomorphism, open and closed sets and continuous deformation of a curve (which will not be repeated here).

In what follows a *surface* S is a compact arcwise connected Hausdorff topological space which is locally homeomorphic to a disc (which means that, for every point p on S, there is an open set in S containing p which is homeomorphic to an open disc in the plane). A surface can be obtained in the following way: Take a collection of pairwise disjoint convex polygons (and their interior) of side length 1 in the plane. Identify each side with precisely one other side (possibly in the same polygon). This results in a topological space S and a simple graph G (whose nodes are the corners of the polygons and whose edges are the sides). If G is connected, then also S is arcwise connected. If, in addition, S is locally homeomorphic to a disc at every node of G, then S is a surface. We say that G is a *2-cell embedding* of S. If all the polygons are triangles (and no side of a triangle is identified with a side of the same triangle) then G is a *triangulation* of S and S is a *triangulated surface*. The simplest triangulated surface is the (boundary of the) tetrahedron. A *face* of a surface with a 2-cell embedding is the interior of one of the polygons.

We shall now define the (triangulated) surfaces S_g and N_h.

S_0 is the sphere. If we cut out two disjoint open discs in S_0 and identify their boundaries such that the clockwise orientations of these boundaries disagree, then we have added a *handle* to S_0. If we add g handles we obtain S_g. If instead we cut out an open disc in S_0 and identify any two opposite points on the boundary, then we have added a *crosscap* to S_0. If we add h crosscaps ($h \geq 1$), then we obtain N_h. S_1, N_1, N_2 are the *torus*, the *projective plane* and the *Klein bottle*, respectively. It is easy to see that all surfaces S_g and N_h are triangulated. (They can be obtained from the boundary of the tetrahedron by cutting out triangles instead of discs). We shall use the following version of the classification theorem.

Theorem 3.1. *Every triangulated surface is homeomorphic to* S_g ($g \geq 0$) *or* N_h ($h \geq 1$).

A short graph theoretic proof of Theorem 3.1 is given by Thomassen (1992). Theorem 3.1 also holds when "triangulated" is omitted, since one can prove that every surface can be triangulated (see, e.g., Kerékjártó 1923). Thomassen (1992) gave a simple graph theoretic proof of the Jordan–Schönfliess theorem and used that to give a very short proof of the fact that every surface can be triangulated.

The proof of Theorem 3.1 in Thomassen (1992) also includes Euler's formula.

Theorem 3.2. *Let G be a 2-cell embedding of a surface S with n nodes and e edges and f faces. If S is homeomorphic to S_g, then*

$$n - e + f = 2 - 2g .$$

If S is homeomorphic to N_h, then

$$n - e + f = 2 - h .$$

So, for fixed S, the number $n - e + f$ does not depend on G. It is called the *Euler characteristic of* S. Euler's formula implies the following.

Theorem 3.3. *If G is a simple graph with n nodes and e edges embedded in S_g, respectively N_h, then*

$$e \leqslant 3n - 6 + 6g ,$$

respectively

$$e \leqslant 3n - 6 + 3h .$$

Proof. We consider only S_g (the proof for N_h is similar). Let H be a triangulation of S_g. Now we can redraw G, if necessary, such that all edges are polygonal arcs (when we think of S_g as a union of triangles in the plane). We may think of $G \cup H$ as a graph M embedded in S_g with say m nodes and q edges. We can assume that G is connected since otherwise we can add edges to G and still have an embedding in S_g. So, M is a 2-cell embedding of S_g and by Theorem 3.2 M has $2 - 2g - m + q$ faces. We now delete successively from M edges and nodes of degree 1 such that, at each stage, the current graph is connected and such that we end up with G. Whenever we delete an edge the number of faces is either unchanged or decreased by 1. Thus G has at most $2 - 2g - n + e$ faces. Since every edge is on the boundary of at most two faces and every face has at least three faces on its boundary, an easy count shows that

$$e \leqslant 3n - 6 + 6g . \quad \square$$

An easy count also shows that Theorem 3.3 is an equality for each triangulation of S_g or N_h. Thus a triangulation of S_g has too many edges to be embedded in $S_{g'}$ for $g' < g$. Hence all of $S_0, S_1, \ldots$ are pairwise nonhomeomorphic. Also $N_1, N_2, \ldots$ are pairwise nonhomeomorphic. Clearly, N_1 (and hence each $N_h, h \geqslant 1$) contains a Möbius strip. S_g does not, see, e.g., Thomassen (1992). Hence we get the following.

Theorem 3.4. *All the surfaces $S_0, S_1, \ldots$ and $N_1, N_2, \ldots$ are pairwise nonhomeomorphic.*

The surfaces S_g are called *orientable* while the surfaces N_h are called *nonorientable*.

Rather than starting with a surface S_g or N_h and drawing a graph G on it, we are going to start with G and then "extend" G to a surface S in which G is a 2-cell embedding and in which the number f of faces is easy to determine. We then decide which surface we get by computing the Euler characteristic and then decide whether or not the surface contains a Möbius strip.

3.2. The rotation scheme

The basic combinatorial tool for graph embeddings in orientable surfaces is a simple but powerful observation attributed to Heffter (1891) and Edmonds (1960). Consider a graph G in S_g and let p be a node. The embedding defines a cyclic permutation π_p of the edges incident with p where $\pi_p(a)$ is the successor of a in the clockwise orientation around p. Let us now consider an edge $a = pq$ and the following sequence: $paqbuc \cdots$ where $b = \pi_q(a)$, u is the end of b distinct from q, $c = \pi_u(b)$ etc. Since G is finite we will traverse some edge twice in the same direction. This will happen first for the edge a. The set of edges (considered now as *ordered* pairs of vertices) in this sequence is called the *orbit* containing a (and $b, c, \ldots$). Note that every edge $a = pq$ belongs to two orbits (which may be identical), one for each direction of a. The orbits depend only on the permutations π_p. Returning to the embedding it is clear that

$$f \leq r , \tag{3.5}$$

where f is the number of faces and r the number of orbits. So, by Euler's formula

$$n - e + r \geq 2 - 2g , \tag{3.6}$$

where n and e are the number of nodes and edges, respectively, and if equality holds, then the embedding is a 2-cell embedding. We now formulate the rotation scheme.

Theorem 3.7. *Let G be a connected graph such that, for each node p, π_p is a cyclic permutation of the edges incident with G. Then there exists a 2-cell embedding of G in some S_g, $g \geq 0$, such that π_p is the clockwise ordering of the edges incident with p. For any such embedding, $n - e + r = n - e + f = 2 - 2g$, where r is the number of orbits and f is the number of faces. In particular, g is unique.*

Proof. Consider any orbit $paqbuc \cdots$ of G. We draw a convex polygon in the plane such that the sides are pq, $qu, \ldots$ and such that the walk $paqbuc \cdots$ is anticlockwise around the polygon. We do this for each orbit such that the r polygons (and their interior) are disjoint. We now form a topological space as follows. The edge a occurs in two polygons (which may be identical) namely as pq and qp. We identify these sides. Doing this for each edge results clearly in a surface S. It only remains to be proved that S contains no Möbius strip. We shall only indicate the proof. First one can see that, if a surface with a 2-cell embedding contains a Möbius strip, then it contains a closed polygonal arc, such that left and right interchange as we walk along it (see Thomassen 1992). Then we can show that S has no such closed polygonal arc. The last part of Theorem 3.7 follows from Theorem 3.2. □

We now define the *genus* $\gamma(G)$ and the maximum genus $\gamma_M(G)$ of a graph G as the minimum (respectively maximum) g such that G has an embedding (respectively 2-cell embedding) in S_g. By Theorem 3.7, it is a finite problem to determine

$\gamma(G)$ and $\gamma_M(G)$ when G is fixed. We might also consider the minimal genus for 2-cell embeddings but this leads to $\gamma(G)$ by the following observation of Youngs (1963).

Theorem 3.8. *Every embedding of a connected graph G into $S_{\gamma(G)}$ is a 2-cell embedding.*

Proof. If Theorem 3.8 were false, then by Theorem 3.2 there is an embedding of G into $S_{\gamma(G)}$ such that

$$n - e + f > 2 - 2\gamma(G) .$$

This embedding defines a permutation π_p of the edges incident with p, for every node p. If r is the number of orbits, then $r \geqslant f$, i.e.,

$$n - e + r > 2 - 2\gamma(G) .$$

By Theorem 3.7, there exists an embedding of G into $S_{g'}$ where

$$n - e + r = 2 - 2g' .$$

This implies $g' < \gamma(G)$, a contradiction. □

We also mention Duke's (1966) interpolation theorem.

Theorem 3.9. *If $\gamma(G) \leqslant g \leqslant \gamma_M(G)$, then there exists a* 2-cell embedding of G into S_g.

Proof. Every cyclic permutation π'_p can be obtained from every other cyclic permutation π_p by successively shifting two consecutive edges. Such a shift affects at most three orbits (in the sense that three orbits may be transformed into one orbit and vice versa). Since the genus and the number of orbits are related as described in Theorem 3.7, the genus is changed by at most one by a shift of two consecutive edges. hence Theorem 3.9 is a consequence of Theorem 3.7. □

Theorem 3.7 can also be used to give a purely combinatorial proof of the following result of Battle et al. (1962a) (see Gross and Tucker 1987).

Theorem 3.10. *The genus of a graph is the sum of the genera of its blocks.*

An analogue of Theorem 3.10 for nonorientable surfaces was obtained by Stahl and Beineke [1977]. Their formula is a slight modification of Theorem 3.10. The different behaviour of orientable and nonorientable surfaces in connection with genus additivity becomes more striking when we consider more complicated graph amalgamations as demonstrated by Archdeacon (1986).

We conclude this section with an application of the rotation scheme to the maximum genus. By Theorem 3.7,

$$\gamma_M(G) \leq \lfloor (e-n+1)/2 \rfloor$$

and equality holds if and only if there exist cyclic permutations π_p such that G has at most two orbits. The graphs with this property were described elegantly by Xuong (1979) and, independently, by Jungerman (see Behzad et al. 1979).

Theorem 3.11. *If G is a connected graph with e edges and n nodes, then*

$$\gamma_M(G) \leq \lfloor (e-n+1)/2 \rfloor$$

with equality holding if and only if G has a spanning tree T such that at most one component of $G - E(T)$ has an odd number of edges.

Sketch of proof. Assume first that G has a spanning tree T as described in Theorem 3.11. We prove, by induction on e, that G has a cyclic permutation π_p (for each node p) such that G has at most two orbits (and such that at least one of these meets all nodes). This is obvious if $e = n - 1$ or n (because in that case G can be drawn in the plane with at most two faces). So assume that $e \geq n + 1$. Let H be a connected component of $G - E(T)$ with at least two edges. Since H has a node v such that $H - v$ is connected, it is easy to find two edges a_1, a_2 of H incident with v (if v has degree ≥ 2) or the neighbour of v (if v has degree 1) such that at most one component of $H - \{a_1, a_2\}$ has an odd number of edges. Hence at most one component of $G - \{a_1, a_2\} - E(T)$ has an odd number of edges. By induction, $G - \{a_1, a_2\}$ has cyclic permutations π_p such that $G - \{a_1, a_2\}$ has at most two orbits. Let $a_i = qv_i$ for $i = 1, 2$. We now modify $\pi_q, \pi_{v_1}, \pi_{v_2}$ into cyclic permutations in G as indicated in fig. 3.1 where we have also indicated (in dotted lines) the orbit of $G - \{a_1, a_2\}$ that meets all nodes.

Suppose conversely that G has cyclic permutations π_p such that G has at most two orbits. We shall find a spanning tree as described in Theorem 3.11. If $e \leq n$, there is nothing to prove, so assume that $e \geq n + 1$. If G has two orbits, then some edge $a = uv$ is contained in both orbits. (Since G is connected, some vertex v

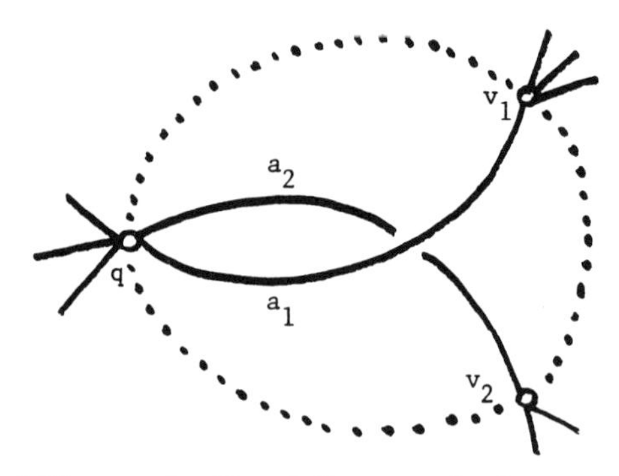

Figure 3.1. Adding a path of length 2 to a graph with at most two orbits.

must be in distinct orbits, and then some edge incident with v must be in distinct orbits.) If we delete edge a and modify π_u and π_v just by ignoring a (i.e., the successor of a around u in G is the successor of the predecessor of a in the new cyclic permutation), then $G - a$ has only one orbit and the proof is completed by induction. On the other hand, if G has only one orbit, then we select an edge a such that the walk from a to a in the orbit is smallest possible. If a is incident with a vertex u of degree 1, then we complete the proof by applying the induction hypothesis to $G - u$. If a is succeeded by b in the above walk, where $b \neq a$, then the minimality property of a implies that $G - \{a, b\}$ is connected and it is easy to see that it has only one orbit (when we ignore a and b as above). The proof is now completed by induction. □

Xuong (1979) extended the ideas of Theorem 3.11 to show that

$$\gamma_{\mathrm{M}}(G) = \max\lfloor(e - n + 1 - c_0(G - T))/2\rfloor, \tag{3.12}$$

where $c_0(H)$ is the number of components of H with an odd number of edges and the maximum is taken over all spanning trees T of G.

By a result of Tutte (1961), every $2k$-edge-connected graph has k edge-disjoint spanning trees. Combining this with Theorem 3.11 we obtain the following.

Corollary 3.13. *If G is 4-edge-connected, and has n nodes and e edges, then*

$$\gamma_{\mathrm{M}}(G) = \lfloor(e - n + 1)/2\rfloor .$$

Furst et al. (1988) showed that the problem of determining the maximum in (3.12) can be reduced (in polynomial time) to the following problem which was solved by Giles (1982): Given a graph whose edge set is partitioned into pairs, find a largest forest whose edge set is a union of some of the above pairs.

3.3. The Heawood conjecture

Theorem 3.14. *For each $n \geq 13$, $\gamma(K_n) = \lceil(n-3)(n-4)/12\rceil$.*

Theorem 3.14 was announced in 1890 by Heawood (see White 1978b). The lower bound $\gamma(K_n) \geq (n-3)(n-4)/12$ follows immediately from Theorem 3.3. But the upper bound is far from trivial.

In a series of papers, particularly by G. Ringel, special cases of Theorem 3.14, which became known as the *Heawood conjecture*, were settled, and in 1968 Ringel and Youngs (1968) announced the final solution.

In this section we describe a proof of Theorem 3.14 in the case $n \equiv 7 \pmod{12}$. In other words, K_{12k+7} can be embedded into S_{12k^2+7k+1}. Since Theorem 3.3 is an equality for such an embedding every face is bounded by a K_3. In other words, S_{12k^2+7k+1} is triangulated by the embedding of K_{12k+7}. The number of faces is $48k^2 + 52k + 14$ by Theorem 3.2. So, by Theorem 3.7, this problem is equivalent with that of describing a cyclic permutation π_p for each node of K_{12k+7} such that the number of orbits is $48k^2 + 52k + 14$, or equivalently, each orbit has three

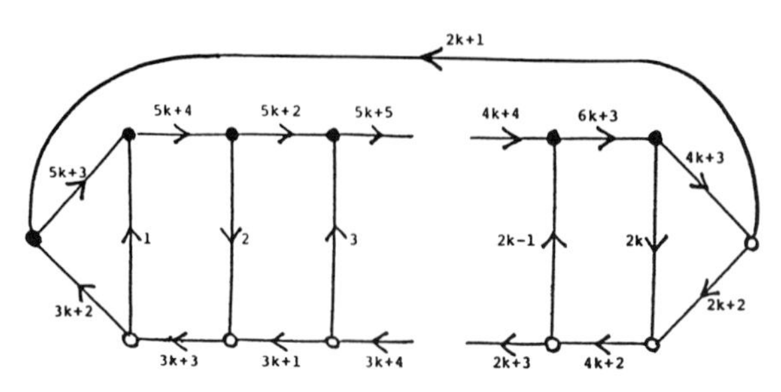

Figure 3.2. Currents from $\mathbb{Z}_{12k+7}$.

edges. Before proving this, we stress the nontrivial character of the problem by pointing out that in general, it is not known when a graph G with n nodes and e edges can be embedded into the surface S_g such that every face is bounded by a triangle. A necessary condition is that Theorem 3.3 is an equality which implies that g is uniquely determined by G. Another necessary condition is that G is locally a wheel in the sense that the subgraph of G induced by the neighbours of each node has a Hamiltonian cycle. These two conditions are sufficient when $g = 0$ as shown by Skupien (1966). But, they are not sufficient in general as demonstrated by Ringel (1978).

Consider now the cubic graph G_k of fig. 3.2. G_k has $4k + 2$ nodes which are partitioned into black and white vertices. Now we define a cyclic permutation of the edges incident with a fixed node by the clockwise (or anti-clockwise) ordering of its incident edges when the vertex is black (respectively white). It is easy to see that the graph gets exactly one orbit in this way. (In particular, this determines a 2-cell embedding of G_k into an orientable surface of maximum possible genus, by Theorem 3.7. But this will not be used.) This orbit is a cyclic sequence of edges such that every edge appears twice. Replacing each edge by the element of $\mathbb{Z}_{12k+7}$ as indicated in fig. 3.2 gives a cyclic permutation of $\mathbb{Z}_{12k+7}\setminus\{0\}$. Here, $\mathbb{Z}_{12k+7}$ are the integers reduced modulo $12k + 7$ and, when an edge labelled j is traversed in the negative direction, the corresponding element in the sequence is $-j$ (modulo $12k + 7$). If we start with the edge labelled $3k + 4$ we get the cyclic permutation of $\mathbb{Z}_{12k+7}\setminus\{0\}$:

$$\pi_0\colon 3k+4, 3k+1, 3k+3, 3k+2, 5k+3, 12k+6, 9k+4, 12k+5, \\ 5k+2, \ldots .$$

Now we regard $\mathbb{Z}_{12k+7}$ as the node set of K_{12k+7}. For each node we consider the cyclic orientation π_j of the edges incident with j (or, equivalently, of the neighbouring nodes) which is obtained from π_0 by adding j to each term. That is,

$$\pi_j\colon 3k+4+j, 3k+1+j, \ldots .$$

The proof is now completed when we show that each orbit in K_{12k+7} consists of three edges. But this is straightforward to verify using the following property of G_k: For each node v the sum of labels on arcs entering v equals the sum of labels on arcs leaving v.

The graph G_k of fig. 3.2 is called a *current graph*. Instead of using $\mathbb{Z}_{12k+7}$ as labels one may use any abelian group. Current graphs, which were introduced by Gustin (1963), played an important role for the solution of the Heawood conjecture. For more information on current graphs and the dual concept of *voltage graphs*, and their use in embedding theory (in particular the Heawood conjecture), the reader is referred to White (1973, 1978b). Voltage graphs were introduced by Gross and are discussed in detail by Gross and Tucker (1987).

Ringel (1965a) also used the rotation technique to establish the following.

Theorem 3.15. *For all natural numbers* $k, m \geqslant 2$,

$$\gamma(K_{k,m}) = \lceil (k-2)(m-2)/4 \rceil .$$

The genus has been determined or estimated for various other classes of graphs. For example, White (1970) and Pisanski (1980) have calculated the genus of several cartesian products of graphs. For details we refer again to White (1973, 1978b) and Gross and Tucker (1987).

3.4. Graphs on nonorientable surfaces

We define the *nonorientable genus* $\tilde{\gamma}(G)$ of G as the smallest h such that G can be embedded into N_h. Clearly,

$$\tilde{\gamma}(G) \leqslant 2\gamma(G) + 1$$

since N_{2h+1} can be obtained from S_h by adding a crosscap. As observed by Auslander et al. (1963) there is no upper bound for $\gamma(G)$ in terms of $\tilde{\gamma}(G)$. This is illustrated in fig. 3.3.

The graph H_m in fig. 3.3 has $n = m^2$ vertices $e = 2m^2 - 2$ edges and $f = m^2 - 1$

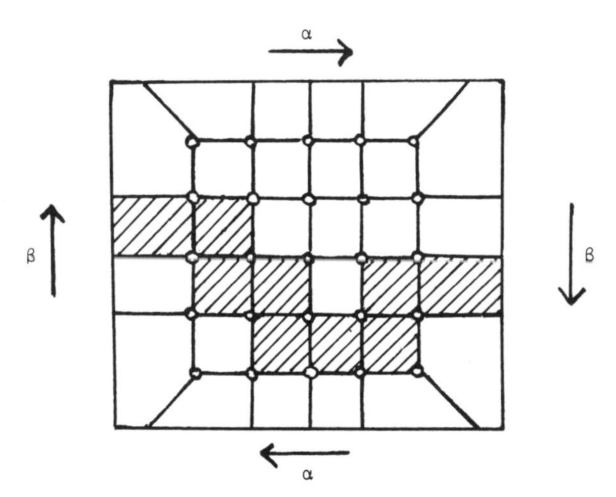

Figure 3.3. Graphs of large genus in the projective plane.

faces each of which is bounded by a 4-cycle. Suppose an embedding of H_m into $S_{\gamma(H_m)}$ has f' faces. The shaded regions in fig. 3.3 form a Möbius strip which is not contained in $S_{\gamma(H_m)}$. Hence one of the 4-cycles in the Möbius strip is not a facial cycle in the embedding of H_m in $S_{\gamma(H_m)}$. So, there are many edges in the $S_{\gamma(H_m)}$-embedding which are on faces whose boundaries have more than four edges. This implies that $f-f'$ is large (if m is large). Since

$$2\gamma(H_m) = 2 - n + e - f' = 1 + f - f' ,$$

it follows that $\gamma(H_m)$ is large when m is large. Hence, it seems, in some sense, easier to embed graphs in a surface N_h than in S_g.

Ringel (1954, 1965b) obtained the following results.

Theorem 3.16. *For all natural numbers n, s, t where $n>3$, $s \geqslant t \geqslant 2$,*

$$\bar{\gamma}(K_n) = \lceil (n-3)(n-4)/6 \rceil \quad \textit{and} \quad \bar{\gamma}(K_{s,t}) = \lceil (s-2)(t-2)/2 \rceil .$$

Bouchet (1978) has obtained general results on regular complete k-partite graphs. Ringel (1977) and Stahl (1978) showed that the maximum h for which a connected graph with n nodes and e edges has a 2-cell embedding into N_h is simply $e-n+1$ which is simpler than (3.12). Stahl (1975) also established the nonorientable version of Duke's interpolation theorem 3.9. As pointed out by White and Beineke (1978) the analogue of Theorem 3.8 does not hold for nonorientable surfaces. Also, it is not known if every 2-connected graph G can be embedded into S_g or N_h in such a way that every face is bounded by a cycle of G. This would imply the *double cover conjecture* formulated independently by W.T. Tutte and P.D. Seymour.

Conjecture 3.17. Every 2-connected graph has a collection of cycles such that every edge is in precisely two of these cycles.

One can develop a rotation scheme for nonorientable surfaces. As we are not going to use it here we refer the reader to Ringel (1974).

3.5. Algorithms for embeddings in orientable surfaces

We have already mentioned that many embedding problems are difficult: It is NP-hard to find the thickness, the book-thickness, and the crossing number of a graph, respectively, and it is NP-hard to recognize string graphs. The following unsolved problem in Garey and Johnson (1979) was solved by Thomassen (1989a).

Theorem 3.18. *The following problem is* NP*-complete: Given a graph G and a natural number k, is $\gamma(G) \leqslant k$?*

The proof of Theorem 3.18 can be extended to the nonorientable genus as well.

It also shows that it is NP-complete to decide if a given embedding is of minimum genus. For fixed genus g, however, the situation is different. The fastest algorithm for $g = 0$ is due to Hopcroft and Tarjan (1974).

Theorem 3.19. *There exists a linear time algorithm for deciding if a graph is planar.*

The first polynomially bounded algorithm for fixed genus was found by Filotti et al. (1979).

Theorem 3.20. *The genus of a graph with n nodes can be determined in* $O(n^{O(g)})$ *steps.*

The genus has been determined for some graphs of a very special type including the complete graphs, as we have seen. Also, Theorem 3.20 shows that the genus can be found for graphs of bounded genus. We shall describe a method for finding the genus for a large general class of graph with unbounded genus. Consider an embedding (which we think of as a rotation scheme) of a connected graph G. Let C be a cycle in G. If we choose a positive orientation of G we can speak of edges incident with C which go to the left (respectively right) side of C. We define the right side R (respectively left side L) of C as C together with all paths that start with an edge on the right (respectively left) side and which has no intermediate vertex in common with C. Clearly R and L can be found in polynomial time. Now C is *contractible* if $R \cap L = C$ and one of R or L is embedded in the sphere (by the induced embedding). To verify this is just a matter of using Euler's formula. If we think of G as drawn on a surface, then C is contractible iff C can be continuously deformed into a point. We say that an embedding is an *LEW-embedding* (large-edge-width-embedding) if all noncontractible cycles have length greater than all orbits (that is, face boundaries). Thomassen (1990) proved the following.

Theorem 3.21. *There exists a polynomially bounded algorithm for finding a shortest noncontractible cycle in an embedded graph. In particular, there exists a polynomially bounded algorithm for deciding if an embedding is an LEW-embedding. Moreover, there exists a polynomially bounded algorithm that finds an LEW-embedding for any 2-connected graph G that has such an embedding. If G is 3-connected and has an LEW-embedding, then that is the unique minimum genus embedding of G.*

The genus of another large class of graphs can be computed fast by the following result of Fiedler et al. (1994).

Theorem 3.22. *There exists a polynomially bounded algorithm that computes the genus* $\gamma(G)$ *of every graph G satisfying* $\bar{\gamma}(G) \leqslant 1$.

Once a graph has been embedded in S_g, where g is fixed, many problems concerning G can be solved fast (or nearly solved) simply by using the fact that G is either small or has a vertex of small degree. We shall illustrate this by considering the chromatic number $\chi(G)$. Although it is difficult (in fact NP-complete) to find $\chi(G)$ even if $g = 0$ (see Toft's chapter) we have the following: For fixed g, there exists a polynomially bounded algorithm for colouring the nodes of a graph G of genus $\leqslant g$ in $\max\{7, \chi(G)\}$ colours such that no two adjacent nodes receive the same colour. The algorithm is easy: If G has a node v_1 of degree $\leqslant 6$, delete it. If $G - v_1$ has a node of degree $\leqslant 6$, delete it. This results in a sequence $v_1, v_2, \ldots, v_m$ of nodes such that $H = G - \{v_1, v_2, \ldots, v_m\}$ is either empty or has minimum degree $\geqslant 7$. By Theorem 3.3, H has at most $12(g-1)$ nodes. In constant time we can colour the nodes of H with colours $1, 2, \ldots, \chi(H)$ such that no two neighbours receive the same colour. Now we put back $v_m, v_{m-1}, \ldots, v_1$ giving each of them one of the colours $1, 2, \ldots, 7$.

Other embedding algorithms following from the Robertson–Seymour theory are discussed in sections 5.3 and 5.4.

4. Graph minors

In the previous sections we concentrated on the rotation scheme as a tool for embedding a given graph into a given orientable surface. It is equally natural to seek an explanation of why a given graph cannot be embedded into a given surface. Euler's formula 3.2 can sometimes be used for that. But, as the graphs H_m of fig. 3.3 show, it is possible that a graph satisfies Euler's formula for S_g without being embeddable into S_g.

Kuratowski's classical Theorem 1.4 shows that nonplanarity can be explained in terms of excluded minors and, the work of Robertson and Seymour (which we review in section 5) shows that the same holds for embedding (and other) problems in general. In this section we describe some basic results on graph minors.

4.1. Simplicial decompositions and tree width

Suppose G is a connected graph and A a minimal separating node set of G. Then we can write $G = G_1' \cup G_2'$ where G_1' and G_2' are connected and $G_1' \cap G_2' = G(A)$. Now suppose further, that $G(A)$ is a complete graph. If G_1' or G_2' (say G_1') has a separating node set which induces a complete graph, then we can write $G_1' = G_3' \cup G_4'$ such that G_3' and G_4' are connected and $G_3' \cap G_4'$ is a complete subgraph of G. We proceed like this until none of the resulting subgraphs $G_1, G_2, \ldots, G_k$ has a complete separating subgraph. With this notation we have the following.

Proposition 4.1. *The subgraphs $G_1, G_2, \ldots, G_k$ are independent of the order in which the decomposition is carried out.*

Proof (*by induction on the order of* G). If the above graphs G_1', G_2' are chosen such that G_2' has a few vertices as possible, then any separating complete subgraph of G is contained in G_1'. Hence G_2' is in the sequence $G_1, G_2, \ldots, G_k$ regardless of the order in which the decomposition is performed. The proof is now completed by applying the induction hypothesis to G_1'. □

The graphs $G_1, G_2, \ldots, G_k$ are called the *simplicial summands* of G. Tarjan (1985) has described a polynomially bounded algorithm for finding a separating complete graph, and hence the simplicial summands can be found in polynomial time. If all the separating node sets in the above decomposition have only one vertex, then the simplicial summands are simply the blocks of the graph. The blocks of a graph form a tree-like structure. More generally, any graph can be built up from its simplicial summands in a tree-like fashion by successively identifying complete subgraphs.

One of the first results on simplicial decompositions is the following of Dirac (1961).

Theorem 4.2. *A graph G has the property that every cycle of length at least 4 has a chord if and only if every simplicial summand of G is a complete graph.*

The graphs in Theorem 4.2 are called *chordal* or *rigid circuit* graphs. A subclass of these are the K_w-cockade, i.e., the connected graphs in which every simplicial summand is a complete graph K_w. If every minimal separating complete graph in a K_w-cockade has m nodes, then we say that the K_w-cockade has *strength m*.

We can also now introduce the tree width of a graph. We say that G has *tree width* $\leqslant w$, and write $\mathrm{tw}(G) \leqslant w$, if G is a subgraph of a K_{w+1}-cockade. With this notation we have the following.

Proposition 4.3. $\mathrm{tw}(G) \leqslant 1$ *iff G does not contain K_3 as a minor (i.e., G is a forest).* $\mathrm{tw}(G) \leqslant 2$ *iff G does not contain K_4 as a minor.*

The first of these statements is trivial. The second is equivalent to Dirac's (1960) characterization of the graphs which contain no subdivision of K_4 combined with the following (which is an easy exercise).

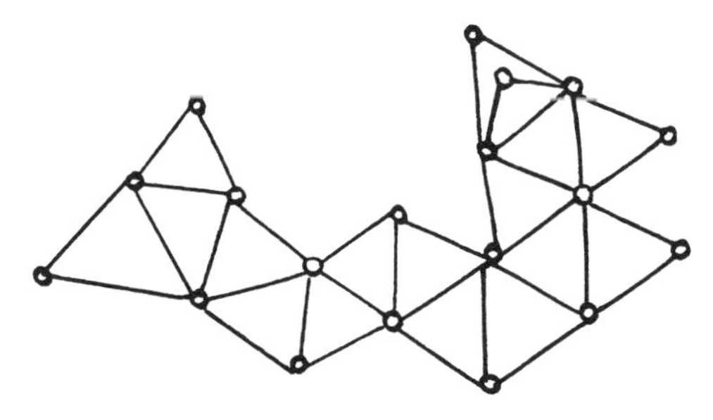

Figure 4.1. A K_3-cockade of strength 2.

Proposition 4.4. *Let G_0 be a graph of maximum degree at most 3. Then G contains a subdivision of G_0 iff G has G_0 as a minor.*

The graphs of tree width $\leqslant 2$ are known as *series–parallel graphs*. Proposition 4.3 leads to the problems of characterizing the graphs of tree width $\leqslant k$ and the graphs containing no K_k (or, more generally, an arbitrary graph G_0) as a minor. For k fixed, polynomial algorithms for these problems follow from the Robertson-Seymour theory as we shall see. But, there are still important unsolved problems in this area, the most important being Hadwiger's conjecture (see section 4.3).

4.2. Wagner's characterization of graphs with no K_5-minor

The following observation shows how simplicial decompositions are useful for characterizing the graphs which contain no (fixed) graph G_0 as a minor.

Lemma 4.5. *Let G_0 be a graph which is either $(k+1)$-connected or a K_{k+1}. Let G be a graph such that $G = G_1' \cup G_2'$ where G_1' and G_2' are connected, $G_1' \cap G_2' = K_k$, and $V(G_1') \cap V(G_2')$ is a minimal separating set of nodes in G. Then G has a minor isomorphic to G_0 if and only if one of G_1', G_2' has a minor isomorphic to G_0. Suppose, furthermore, that $G_1' \cap G_2'$ is not contained in a K_{k+1} in G. Then G is edge-maximal with respect to not having G_0 as a minor if and only if both of G_1', G_2' are edge-maximal with respect to not having G_0 as a minor.*

Proof. Suppose G has G_0 as a minor. That is, it is possible to contract edges in G such that we get a graph containing G_0 as a subgraph. The connectivity condition on G_0 ensures that this subgraph cannot have vertices separated by $G_1' \cap G_2'$. Hence this subgraph is in (a minor of) G_1' or G_2'. The same argument shows that, if G is edge-maximal with respect to not having G_0 as a minor then so are G_1' and G_2'. Finally, if both G_1' and G_2' are edge-maximal with respect to not having G_0 as a minor, then we show that $G \cup \{uv\}$ has G_0 as a minor for any two nonadjacent nodes u and v. If u, v are in the same G_i', this is obvious. If $u \in V(G_1') \backslash V(G_2')$ and $v \in V(G_2') \backslash V(G_1')$, then there is a v' in $V(G_1') \cap V(G_2')$ which is not adjacent to u. Now $G \cup \{uv\}$ has $G_1' \cup \{uv'\}$ and hence G_0 as a minor. □

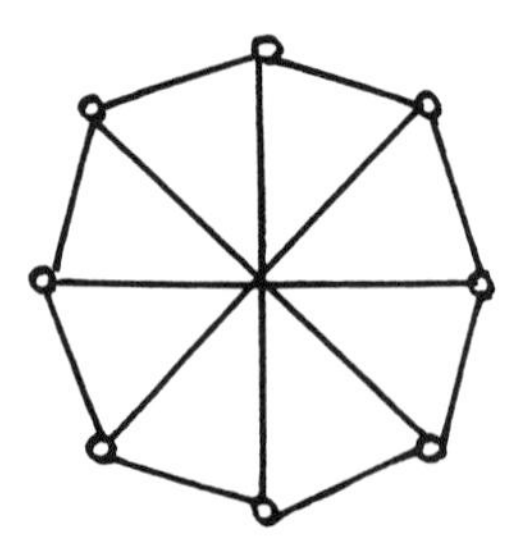

Figure 4.2. The graph L.

Lemma 4.5 shows that in order to characterize the edge-maximal graphs not having G_0 as a minor it is sufficient to characterize their simplicial summands (perhaps with some additional information of how they are "pasted together" along complete subgraphs). Thus Wagner's (1937) fundamental characterization of the graphs having no K_5 as a minor can be formulated as follows.

Theorem 4.6. *Let G be an edge-maximal graph having no K_5 as a minor and having only one simplicial summand (i.e., G has no separating complete graph). Then G is either a maximal planar graph or isomorphic to the graph L of fig.* 4.2.

Proof. Since G has no separating complete graph, G is 2-connected. We claim that G is 3-connected. For suppose that $G - \{u, v\}$ is disconnected, i.e., $G = G_1' \cup G_2'$ where G_1', G_2' are connected and $G_1' \cap G_2' = \{u, v\}$. Since $G \cup \{uv\}$ has K_5 as a minor, Lemma 4.5 implies that $G_1' \cup \{uv\}$, say, has K_5 as a minor. But G has $G_1' \cup \{uv\}$ as a minor. This contradiction shows that G is 3-connected.

G need not be 4-connected (because L is not 4-connected) but G is "almost 4-connected" in the following sense: If $G - \{x, y, z\}$ is disconnected, then it has precisely two components one of which has only one node. To prove this we assume that $G = G_1 \cup G_2$ where $V(G_1 \cap V(G_2) = \{x, y, z\}$ and $|V(G_i)| \geqslant 5$ for $i = 1, 2$. For any node $u \in V(G_1) \backslash \{x, y, z\}$, G has three paths P_1, P_2, P_3 from u to $\{x, y, z\}$ such that $V(P_i) \cap V(P_j) = \{u\}$ whenever $i \neq j$. Since $|V(G_1)| \geqslant 5$ and G is 3-connected, it is easy to see that $G_1 - u$ has a path P_4 connecting two of $P_1 - u$, $P_2 - u$, $P_3 - u$. Using P_1, P_2, P_3, P_4 we see that G has a minor G_2' which is obtained from G_2 by adding all edges between two of $\{x, y, z\}$. Since G has no K_5-minor, G_2' has no K_5-minor. Similarly, $G_1' = G_1 \cup \{xy, xz, yz\}$ has no K_5-minor. By Lemma 4.5, $G \cup \{xy, xz, yz\}$ has no K_4-minor. By the maximality of G, $G = G \cup \{xy, xz, yz\}$. But, this contradicts the assumption that G has no separating complete graph. Hence G is "almost 4-connected".

If G is planar, then the maximality of G implies that G is a maximal planar graph. So assume that G is nonplanar. By Kuratowski's theorem 1.4, G contains a subgraph H which is a subdivision of $K_{3,3}$. Let A, B be the two node sets of H corresponding to the partite sets of $K_{3,3}$. Since G is almost 4-connected, the three components of $H - A$ are not in distinct components of G. Hence G has a path P_1 connecting two of the distinct components of $H - A$. If the nodes of A belong to three distinct components of $(H \cup P_1) - B$, then G has a path P_2 connecting two of these components.

Now it is easy to see that $H \cup (P_1 \cup P_2)$ contains either a path P_3 connecting two disjoint $A - B$ paths of H or else $H \cup (P_1 \cup P_2)$ has K_5 as a minor. Since G has no K_5 as a minor we can assume the former. Then $H' = H \cup P_3$ is a subdivision of L.

We claim that $G = H' = L$. Note that L has no K_3 and is maximal with respect to not having K_5 as a minor. Now if $G \neq H' = L$, then we pick a vertex $u \in V(G) \backslash V(H')$ and three paths from u to H' which are disjoint (except at u). This gives a subgraph having K_5 as a minor (because L is K_3-free). On the other hand, if $H' \neq L$, then H' has a node of v of degree 3 such that the three nodes v_1,

v_2, v_3 which correspond to the neighbours of v in L are not all neighbours of v in H'. Hence G has a path connecting the distinct components of $H' - \{v_1, v_2, v_3\}$ and this gives a subgraph of G having K_5 as a minor. This contradiction proves that $G = L$. □

Wagner's theorem 4.6 combined with Proposition 4.1 and the discussion preceding Proposition 4.1 shows that the edge-maximal graphs having no K_5-minor are obtained by successively pasting maximal planar graphs and copies of L together along complete subgraphs. Conversely, if we form a graph of this type by always choosing the complete subgraphs such that they are maximal in one of the graphs we paste together, then the resulting graph can easily be shown to be edge-maximal with respect to not having K_5 as a minor.

Theorem 4.6 is sometimes called *Wagner's equivalence theorem* because it shows that the 4-colour-theorem is equivalent to Hadwiger's conjectures for $k = 5$. This follows because the chromatic number of a graph is the maximum chromatic number of its simplicial summands.

It is easy to see that if a graph has K_5 as a minor then either it contains a K_5-subdivision or else it has $K_{3,3}$ as a minor. Moreover, every 3-connected graph with at least six nodes which contains a K_5-subdivision has $K_{3,3}$ as a minor. Hence Theorem 4.6 together with its method of proof implies the following.

Corollary 4.7. *If G is an edge-maximal graph having no $K_{3,3}$ as a minor, then the simplicial summands of G are either maximal planar graphs or copies of K_5.*

Note that the proof method of Theorem 4.6 also quickly gives the part of Proposition 4.3 concerning K_4.

Theorem 4.6 and Corollary 4.7 describe the "building stones" for the graphs having no minor isomorphic to K_5, respectively $K_{3,3}$. These building stones have also been described for other small graphs (see Bollobás 1978 and Wagner 1970). Recently, Robertson (private communication) has solved the problem for the graph L of fig. 4.2.

4.3. Hadwiger's conjecture

Hadwiger's conjecture (which is also discussed in Toft's chapter) states the following.

Conjecture 4.8. If the graph G has chromatic number at least k, then it has K_k as a minor.

Wagner (1964) proved the following weakening of Hadwiger's conjecture.

Theorem 4.9. *There exists a function $f(k)$ such that every $f(k)$-chromatic graph has K_k as a minor.*

Since every k-chromatic graph contains a subgraph of minimum degree at lest $k-1$ (see Toft's chapter), Theorem 4.9 also follows from the following result of Mader (1967).

Theorem 4.10. *For each k there exists a smallest natural number $g(k)$ such that every graph of minimum degree at least $g(k)$ contains K_k as a minor.*

Proof. We show by induction on k, that $g(k) \leqslant 2^k$. For $k = 1$ this is trivial. So assume that G is a graph of minimum degree $\geqslant 2^k$. Then G has at least $2^{k-1}|V(G)|$ edges. Let H be a maximal connected subgraph of G such that contracting H to a single vertex results in a graph H' with at least $2^{k-1}|V(H')|$ edges. Let H'' denote the subgraph of $G - V(H)$ induced by the nodes joined to H. Let u be any vertex of H'' and let m be the degree of u in H''. Let H'' be the graph obtained from G by contracting $V(H) \cup \{u\}$ into a single vertex. The maximality of H implies that

$$|E(H''')| < 2^{k-1}|V(H''')| = 2^{k-1}(|V(H')| - 1)\,.$$

Clearly,

$$|E(H''')| = |E(H')| - (m+1) \geqslant 2^{k-1}|V(H')| - (m+1)\,.$$

Hence $m \geqslant 2^{k-1}$. By the induction hypothesis, H'' has K_{k-1} as a minor and, since each node of H'' is joined to H which is connected, G has K_k as a minor. □

The function $g(k)$ was determined for small complete (or nearly complete) graphs up to eight vertices by Dirac (1964) and Jakobsen (1972). These investigations indicated that $g(k)$ is linear (see Wagner 1970, p. 156 and Bollobás 1978, p. 378). But the following holds.

Theorem 4.11. *With the terminology of Theorem* 4.10, *there exist positive constants c_1, c_2 such that*

$$c_1 k\sqrt{\log k} < g(k) < c_2 k\sqrt{\log k}\,.$$

The upper bound was found by Mader (1967) and the lower bound was derived from random graphs by Kostochka (1984) and De la Vega (1983).

One of Dirac's (1964) results states the following.

Theorem 4.12. *If G is a graph with n nodes and at least $(7n-15)/2$ edges, then G has K_6 (less an edge) as a minor unless G is a K_5-cockade of strength* 3.

Using Theorem 4.12 Dirac then proved the following generalization of the 5-colour-theorem.

Theorem 4.13. *If G has chromatic number at least* 6, *then G has K_6 (less an edge) as a minor.*

Recently, Hadwiger's conjecture for $k=6$ was proved by Robertson et al. (1993a).

5. Embeddings and well-quasi-orderings of graphs

König (1936) asked if there exists a Kuratowski type characterization of the graphs not embeddable in a given surface S. Glover et al. (1979) and Archdeacon (1980) obtained such a characterization for the projective plane. For no surface other than the sphere and the projective plane a complete characterization is known. A penetrating result of Archdeacon and Huneke (1989) asserts that, for any nonorientable surface N_h, there exists a finite family of graphs $\mathscr{F}_h$ such that a graph G is embeddable in N_h if and only if G has no minor (isomorphic to a graph) in $\mathscr{F}_h$. This general result is a special case of the following powerful result of N. Robertson and P. Seymour which is central in the theory of minors and embeddings. It provides in particular an affirmative answer to König's question.

Theorem 5.1. *Let p be any graph property such that, for any graph G with property p, every minor of G has property p. Then there exists a finite family* $\mathscr{F}_p$ *of graphs such that a graph G has property p if and only if G has no minor in* $\mathscr{F}_p$.

This theorem is the culmination of the series of graph minor papers of Robertson and Seymour. The rest of this chapter is devoted to a discussion of graph structure results and algorithms related to Theorem 5.1.

5.1. *Well-quasi-orderings*

A partially ordered set $(M, \leqslant)$ is *well-quasi-ordered* (abbreviated wqo) if for each infinite sequence $a_0, a_1, a_2, \ldots$ of elements from M, there are natural numbers i, j such that $i<j$ and $a_i \leqslant a_j$. Clearly, M has no infinite strictly decreasing sequence. Ramsey's theorem for infinite graphs can be formulated as follows: Every infinite oriented graph contains an infinite set of nodes which induce a subgraph with no edges or a subgraph isomorphic to the positive (respectively negative) numbers where there is a directed edge from m to p when $m<p$. So, if $(M, \leqslant)$ is wqo, and $a_1, a_2, \ldots$ is any infinite sequence, then there exists an infinite subsequence $a_{i_1}, a_{i_2}, \ldots$ such that $a_{i_1} \leqslant a_{i_2} \leqslant \cdots$.

If $\boldsymbol{a} = a_1, a_2, \ldots, a_n$ and $\boldsymbol{b} = b_1, b_2, \ldots, b_m$ are finite sequences from M we write $\boldsymbol{a} \leqslant \boldsymbol{b}$ if there exists a subsequence $b_{i_1}, b_{i_2}, \ldots, b_{i_n}$ $(1 \leqslant i_1 < i_2 < \cdots < i_n \leqslant m)$ such that $a_j \leqslant b_{i_j}$ for $j = 1, 2, \ldots, n$. With this notation Higman's (1952) theorem states the following.

Theorem 5.2. *If M is wqo, then the set of finite sequences of M is wqo.*

Proof (*by contradiction*). Assume $\boldsymbol{a}_1, \boldsymbol{a}_2, \ldots$ is an infinite sequence of finite sequences such that $\boldsymbol{a}_i \not\leqslant \boldsymbol{a}_j$ whenever $i<j$. Assume that the sequence has been

chosen such that $\boldsymbol{a}_1$ has minimum length; subject to this condition $\boldsymbol{a}_2$ has minimum length; subject to these conditions $\boldsymbol{a}_3$ has minimum length, etc. Let a_i be the first element of $\boldsymbol{a}_i$ and let $\boldsymbol{a}_i'$ be obtained from $\boldsymbol{a}_i$ by deleting a_i. Now, there is an infinite sequence $a_{i_1}, a_{i_2}, \ldots$ such that $i_1 < i_2 < \cdots$ and $a_{i_p} \leq a_{i_{p+1}}$ for $p = 1, 2, \ldots$. The minimality of $\boldsymbol{a}_1, \boldsymbol{a}_2, \ldots$ implies that the sequence $\boldsymbol{a}_1, \boldsymbol{a}_2, \ldots, \boldsymbol{a}_{i_1-1}, \boldsymbol{a}_{i_1}', \boldsymbol{a}_{i_2}', \ldots$ is not a counterexample to Theorem 5.2. So either $\boldsymbol{a}_i \leq \boldsymbol{a}_j$ for some i, j where $1 \leq i < j \leq i_1 - 1$, or else $\boldsymbol{a}_i \leq \boldsymbol{a}_{i_j}'$ for some i, j where $1 \leq i \leq i_1 - 1$ and $j \geq 1$, or else $\boldsymbol{a}_{i_p}' \leq \boldsymbol{a}_{i_q}'$ for some p, q where $1 \leq p < q$. In each case we obtain a contradiction to the assumption that $\boldsymbol{a}_1, \boldsymbol{a}_2, \ldots$ is a counterexample to Theorem 5.2. □

Now consider two finite trees T_1, T_2 such that each node in T_1 and T_2 is labelled by an element of M and such that a node v_i in T_i is distinguished as a root. We write $T_1 \leq T_2$ if T_2 contains a tree T_1' such that T_1' is a subdivision of T_1, the vertex in T_1' corresponding to v_1 is v_2, and if a is the label of a vertex v in T_1 and b is the label of the corresponding vertex in T_2, then $a \leq b$. With this terminology we have the following general version of Kruskal's (1960) theorem.

Theorem 5.3. *The collection of finite rooted trees labelled by elements of a wqo set M is wqo.*

If T is a tree with root v and $T_1, T_2, \ldots, T_p$ are the trees of $T - v$ each rooted at the node adjacent to v, then we may think of T as the sequence T_1, $T_2, \ldots, T_p$. This observation enables us to prove Theorem 5.3 in almost the same way as Theorem 5.2.

Kruskal's theorem (without reference to M) is sometimes formulated as follows: The collection of finite trees is wqo *under topological containment*. Wagner (1970) observed that this does not hold for graphs in general: just consider the sequence of double cycles, i.e., the graphs obtained from a cycle of length n by replacing each edge by a double edge. Also, the collection of trees is not wqo under inclusion. However, Theorems 5.2 and 5.3 imply the following results observed by Robertson and Seymour, and Mader (1972), respectively.

Theorem 5.4. *Let k be a natural number. The collection of graphs with no path of length k is wqo under inclusion.*

Proof. We prove by induction on k the following stronger statement: If $H_1, H_2, \ldots$ is an infinite sequence of graphs with no path of length k and each node of each H_i is labelled by an element of a prescribed finite set M, then there exist i, j $(i < j)$ such that H_j contains an induced subgraph which is isomorphic (considered as a labelled graph) to H_i. For $k = 1$ the statement is an easy exercise, so assume that $k \geq 2$. By Theorem 5.2 it is sufficient to prove the statement for connected graphs. We can assume that each H_i has a path $x_1^i x_2^i \cdots x_k^i$ of length $k - 1$ since otherwise we can consider an appropriate subsequence and apply the induction hypothesis. Without loss of generality we can assume that the subgraphs

$H_i(\{x_1^i, x_2^i, \ldots, x_k^i\})$ are label-isomorphic for $i = 1, 2, \ldots$. Now $H_i' = H_i - \{x_1^i, x_2^i, \ldots, x_k^i\}$ has no path of length $k - 1$ (because no connected graph has two disjoint longest paths). Every node v of H_i' has a label from M. We now give it a label from the cartesian product of M and the set of subsets of $\{1, 2, \ldots, k\}$; the second coordinate is the set of those j such that v is adjacent to x_j^i. By the induction hypothesis, some H_j' has an induced subgraph which is label-isomorphic to some H_i' such that $i < j$. Then H_j contains an induced subgraph which is label-isomorphic to H_i. □

For the next result we need the following result of Erdős and Pósa (1965).

Theorem 5.5. *There exists a function $f(k)$ such that every graph G with no k disjoint cycles contains a set S of at most $f(k)$ nodes such that $G - S$ is a forest.*

Theorem 5.6. *Let k be a natural number. The collection of graphs having no k disjoint cycles are wqo under topological containment.*

Proof. Let $H_1, H_2, \ldots$ be an infinite sequence of graph with no k disjoint cycles. By Theorem 5.5, let $S_i \subseteq V(H_i)$ such that $|S_i| \leqslant f(k)$ and $H_i - S_i$ is a forest for $i = 1, 2, \ldots$. Without loss of generality we can assume that all the graphs $H_i(S_i)$ are isomorphic. Let us assume that $S_1 = S_2 = \cdots$ and let us label each node of $H_i - S_i$ by the subset of $S_i = S_1$ to which it is joined. By Theorem 5.3, there exist i, j such that $i < j$ and $H_j - S_j$ contains a forest which is a subdivision of $H_i - S_i$ with "greater labels". This completes the proof. □

5.2. Tree width and minors

Recall that tree width is defined in section 4.1. The graphs of tree width $\leqslant w$, where w is a fixed natural number, behave in many respects like trees although they are not wqo under topological containment as shown by (subdivisions of) the double cycles. By Theorem 5.1 they are wqo by minors as shown first by Robertson and Seymour (1990a). This is not surprising in view of Theorem 5.3, but the proof involves many technical details.

A further partial result towards Theorem 5.1 is that, if an infinite sequence of graphs contains just one planar graph, then one of the graphs in the sequence has another as a minor. This follows from the above-mentioned result of wqo of graphs of bounded tree width combined with the fundamental *tree width theorem* of Robertson and Seymour which we shall now describe.

Consider the grid G_k consisting of k disjoint paths $P_i\colon x_{1,i}x_{2,i} \cdots x_{k,i}$ and the path $P_1'\colon x_{1,1}x_{1,2} \cdots x_{1,k}$ and the path $P_2'\colon x_{k,1}x_{k,2} \cdots x_{k,k}$ and all edges $x_{i,j}x_{i+1,j}$ where $i + j$ is odd. (See fig. 5.1.)

If a graph G contains a minor of G_k (and hence a subdivision of G_k by Proposition 4.4) for k large, then G has large tree width. This follows because any subgraph of a graph of tree width $\leqslant w$ has tree width $\leqslant w$. And, G_k does not have small tree width when k is large. The easiest way of seeing this is probably to

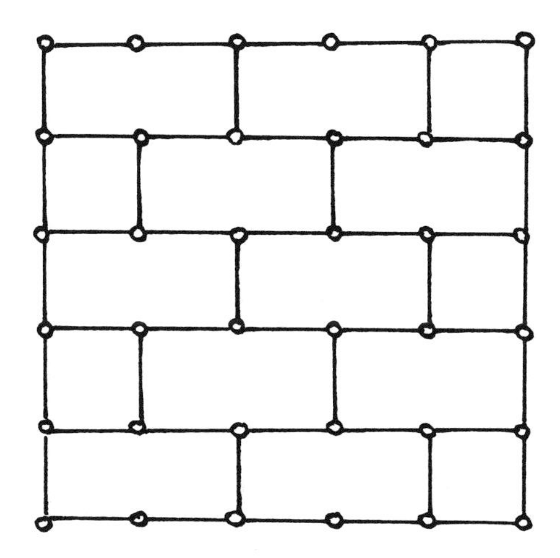

Figure 5.1. The grid G_6.

verify that G_k does not have the separation property of Proposition 5.11 when k is large and w is small. The tree width theorem 5.7 shows therefore that a necessary and sufficient condition for a graph to have large tree width is that it contains a G_k-subdivision for some large k.

Theorem 5.7. *There exists a function $f(k)$ such that every graph of tree width $>f(k)$ contains a subdivision of G_k.*

For any fixed planar graph H_0 there exists a natural number k such that G_k has H_0 as a minor. To see this we first observe that there exists a cubic planar graph H_1 having H_0 as a minor. Now H_1 can be drawn in the plane such that all edges are straight line segments and every such edge can be modified such that it becomes a polygonal arc whose segments are all vertical or horizontal. Then it is easy to see that, for some large k, G_k contains a subdivision of H_1. So, Theorem 5.7 implies that every graph with sufficiently large tree width has H_0 as a minor. This statement is best possible in the sense that it becomes false if H_0 is nonplanar. This follows from the fact that the grids G_k are planar and of large tree width when k is large.

Besides its afore-mentioned applications to wqo by minors, Theorem 5.7 has other applications. Let us define a *K_w-path-cockade* as a graph which is the union of graphs $H_1, H_2, \dots, H_q$ each of which is a K_w such that, for each $i = 2, 3, \dots, q$, $(H_1 \cup H_2 \cup \cdots \cup H_{i-1}) \cap H_i$ is a (complete) subgraph of H_{i-1}. We say that a graph G has *path width* $\leq w$ if G is contained in a K_{w+1}-path-cockade. Also this concept was introduced by Robertson and Seymour (1983) who proved, roughly speaking, that a graph has large path width if and only if it has any large tree as a minor. More precisely, let T_k be the tree containing a vertex x_0 such that x_0 has degree 2, all other vertices have degree 1 or 3, and all maximal paths starting at x_0 have length k. It is not difficult to show that graphs of bounded path width cannot contain subdivisions of T_k for arbitrarily large k. This shows that, in a sense, the following analogue of Theorem 5.7 is best possible.

Theorem 5.8. *There exists a function $f'(k)$ such that every graph of path width $>f'(k)$ contains a subdivision of T_k.*

Theorem 5.8 was proved before Theorem 5.7. But Theorem 5.8 is not difficult to derive from Theorem 5.7 because a G_k with k large contains a subdivision of $T_{k'}$ for k' large. Hence it is sufficient to prove Theorem 5.8 for graphs of bounded treewidth which is not difficult (although some technical considerations are needed).

Theorem 5.7 can also be applied to prove a considerable extension of Theorem 5.5 of Erdős and Pósa. Motivated by Theorem 5.5 let us say that a family $\mathscr{F}$ of graphs has the *E–P-property* if there exists, for every natural number k, a natural number $f(k, \mathscr{F})$ such that the following holds: If G is any graph, then either G contains k disjoint subgraphs each isomorphic to a member of $\mathscr{F}$ or else G contains a set S of at most $f(k, \mathscr{F})$ nodes such that $G - S$ has no subgraph isomorphic to a member of $\mathscr{F}$. Theorem 5.5 says that the collection of cycles has the E–P-property. Graphs in the projective plane (of the type indicated in fig. 3.3) show that the collection of odd cycles does not have the E–P-property. Thomassen (1988b) showed that every family $\mathscr{F}$ of connected graphs has the E–P-property if we only consider graphs of bounded tree width. It was also shown that, for every natural number d, k, there exists a natural number $h(d, k)$ such that every subdivision of the grid $G_{h(d, k)}$ contains a subdivision of G_k such that every "edge" of G_k is a path of length divisible by d. Then these results combined with Theorem 5.7 give the following general result.

Theorem 5.9. *Let G_0 be a planar graph of maximum degree at most 3 and let d be any fixed natural number. If $\mathscr{F}$ is a family of connected graphs containing all those subdivisions of G_0 in which every "edge" is a path of length divisible by d, then $\mathscr{F}$ has the E–P-property.*

In particular, if $\mathscr{F}$ contains all cycles of length divisible by a fixed number d, then $\mathscr{F}$ has the E–P-property by Theorem 5.9.

We say that $\mathscr{F}$ is closed *under subdivision* if every subdivision of a graph in $\mathscr{F}$ is also in $\mathscr{F}$.

Theorem 5.10. *Let $\mathscr{F}$ be a family of connected graphs closed under subdivision. A sufficient condition for $\mathscr{F}$ to have the E–P-property is that $\mathscr{F}$ contains a planar graph of maximum degree 3. A necessary condition is that $\mathscr{F}$ contains a planar graph which can be drawn in the plane (without crossings of edges) such that all nodes of degree ≥ 4 are on the boundary of the outer face.*

The first part of Theorem 5.10 follows from Theorem 5.9. The necessity of Theorem 5.10 was proved by Thomassen (1988b). Robertson and Seymour (1990a) proved previously that $\mathscr{F}$ must contain a planar graph. For, if all graphs

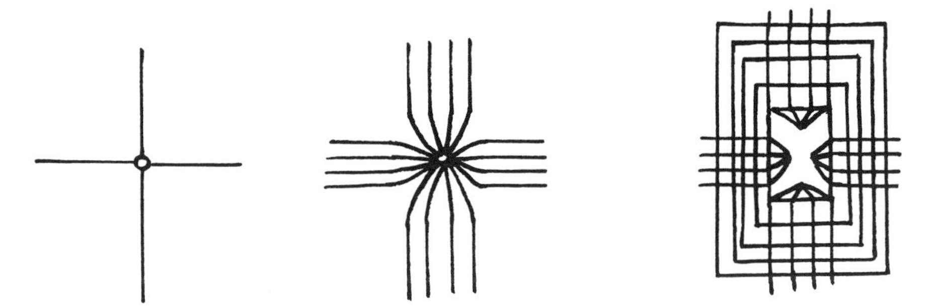

Figure 5.2. Modifying a graph to a "dense" graph.

in $\mathscr{F}$ are nonplanar, then we choose a graph H_0 in $\mathscr{F}$ of minimum genus g. Now let H_0 be embedded in S_g. We replace every edge of H_0 by a large multiple edge and modify the embedding locally as indicated in fig. 5.2.

The resulting graph will contain a subdivision of H_0 even after the removal of a large node set. On the other hand, it contains no two disjoint graphs of $\mathscr{F}$ since the disjoint union of such graphs has genus $\geqslant 2g > g$ by the genus additivity theorem 3.10 and the definition of g.

5.3. Tree width and algorithms

From an algorithmic point of view the graphs whose tree width are bounded above by a fixed number are convenient since they have a tree-like structure which we formalize below. Many problems which are intractible for graphs in general are solvable in polynomial time for graphs of bounded tree width, a fact which has been discovered independently in applied mathematics, for example in so-called expert systems and other areas in statistics (S. Lauritzen, private communication).

With any graph G of tree width $\leqslant w$ we can associate a tree as follows: Assume G' is a connected K_{w+1}-cockade containing G as a spanning subgraph. Then G' can be described as a union of its simplicial summands $H_1, H_2, \ldots, H_q$ such that, for each i, $H_1 \cup H_2 \cup \cdots \cup H_i$ is connected and $(H_1 \cup H_2 \cup \cdots \cup H_i) \cap H_{i+1}$ is a complete subgraph contained in some H_{j_i}, $1 \leqslant j_i \leqslant i$. Let T be the tree with node set $x_1, x_2, \ldots, x_q$ such that, with the above notation, x_{i+1} is joined to x_{j_i} and no other node in $\{x_1, x_2, \ldots, x_i\}$. We shall call T a *w-tree associated with* G. Note that T is not unique. If $G = K_{1,4}$, then there are two 1-trees associated with G, namely $K_{1,3}$ and the path of length 3. If T is a tree with q nodes, then it is easy to find a node v such that no component of $T - v$ has more than $q/2$ nodes. Hence we can partition the components of $T - v$ into two parts none of which has more than $2(q-1)/3$ nodes. This leads to the following partition property of graphs of tree width $\leqslant w$ as noted by Robertson and Seymour (1986a).

Proposition 5.11. *Every graph G of tree width $\leqslant w$ has a node set S and two subgraphs G_1, G_2 such that $G = G_1 \cup G_2$, $V(G_1) \cap V(G_2) = S$, $|S| \leqslant w + 1$, and $|V(G_i)| \leqslant w + 1 + 2(|V(G)| - w - 1)/3$ for $i = 1, 2$.*

Proposition 5.11 implies the existence of polynomially bounded algorithms for several NP-complete graph problems when these are restricted to graphs of bounded tree width. To illustrate this, we describe polynomially bounded algorithms for k-colouring a graph G or proving that it has tree width $>w$ (where k and w are fixed natural numbers). We first consider all $(w+1)$-element subsets S of $V(G)$ in order to find the decomposition in Proposition 5.11. This can be done in $O(n^{w+2})$ steps. If such a decomposition does not exist, then G has tree width $>w$. If it does exist we consider all possible k-colourings of $G(S)$. (There are at most k^{w+1} such colourings). For each colouring of $G(S)$ and for each of G_1, G_2 we contract each colour class into a single vertex (and joint any two of these vertices by an edge), and repeat the algorithm for the resulting graphs. An easy count shows that this algorithm is polynomially bounded. Note that this algorithm does not involve a polynomially bounded algorithm for deciding if G has tree width $\leqslant w$. Such an algorithm follows from the solution of the k-path problem which we discuss in the next section. The k-path problem for the graphs of bounded tree width was solved in Robertson and Seymour (1986a) and that method has interesting consequences, too. Thomassen (1988b) pointed out that it yields the following.

Proposition 5.12. *For fixed natural numbers* $w, k, d, d_1, d_2, \ldots, d_k$ *there exists a polynomially bounded algorithm for deciding whether a graph* G *has tree width* $>w$ *or contains, for any prescribed nodes* $x_1, y_1, x_2, y_2, \ldots, x_k, y_k$, k *disjoint paths* $P_1, P_2, \ldots, P_k$ *such that* P_i *connects* x_i *and* y_i *and has length* $d_i \pmod d$ *for* $i=1,2,\ldots,k$.

The algorithm of Proposition 5.12 is similar to the afore-mentioned colouring algorithm. We split the graph into G_1 and G_2 as in Proposition 5.11 and then we translate the k-path problem in G into a (fixed) number of k'-path problems in G_1 and G_2. We can assume that $k \geqslant w$ and then $k' \leqslant 2k$. If necessary, we use (a modification of) Proposition 5.11 to split G_1 and G_2 such that we get $k' \leqslant k$ for each of the graphs into which G has been split. Then we repeat the algorithm for each of these graphs.

Corollary 5.13. *For each fixed natural number* d *there exists a polynomially bounded algorithm for deciding whether a graph* G *has a cycle of length divisible by* d.

Proof. If G has sufficiently large tree width then it has such a cycle by Theorem 5.7 and the remark preceding Theorem 5.9. If G has bounded tree width, then we apply Proposition 5.12 to each pair of adjacent nodes x_1, y_1 letting $k=1$ and $d_1 = d-1$. □

Arnborg and Proskurowski (1989) have shown that many difficult graph problems (for example the k-colouring problem and that of finding a Hamiltonian cycle) can be solved even in linear time for graphs of bounded tree width.

5.4. The k-path problem with applications to embeddings: Excluded minor theorems and algorithms

Robertson and Seymour (1994b) proved the following.

Theorem 5.14. *For each fixed natural number k, there exists a polynomially bounded algorithm for deciding whether a graph G with prescribed nodes x_1, y_1, x_2, y_2, ..., x_k, y_k contains k disjoint paths P_1, P_2, ..., P_k such that P_i joins x_i and y_i for $i = 1, 2, \ldots, k$.*

Corollary 5.15. *For each fixed graph G_0, there exists a polynomially bounded algorithm for deciding whether a graph G contains a subdivision of G_0.*

Proof. Assume that G_0 has p nodes and k edges. Now consider a graph G with n nodes. If G' is a subdivision of G_0 in G, then we define the *skeleton* of G' as the subgraph of G' obtained by deleting all those nodes of G' which have distance (in G') at least 2 to those nodes of G' that correspond to $V(G_0)$. The skeleton of G' has at most $p + 2\binom{p}{2}$ nodes. Thus the number of subgraphs of G which can play the role of a G_0-skeleton is polynomially bounded, and, for each such subgraph, the problem of deciding whether it really is a skeleton of a G_0-subdivision is the k-path problem. □

Proposition 5.16. *For each graph G_0, there exists a finite collection G_1, G_2, ..., G_p of graphs such that an arbitrary graph G has G_0 as a minor if and only if G contains a subdivision of one of G_1, G_2, ..., G_p.*

Proof. Let G_0' be the graph obtained from G_0 by subdividing each edge of G_0 twice. Consider any node v in G_0' corresponding to a node in G_0 and let $N(v)$ denote the neighbours of v in G_0'. Now delete v, partition $N(v)$ into classes and identify each class into a single vertex. Then add new nodes and edges such that these together with (the classes of) $N(v)$ form a tree where none of the new nodes have degree 1. Thus the number of new vertices we add is less than $|N(v)| < |V(G_0)|$. We perform this construction for every node v of G_0 and let G_1, G_2, ..., G_p denote all the graphs which can arise in this way. Clearly each G_i has G_0 as a minor. Conversely, if G has G_0 as a minor, then, for every vertex v of G_0, the vertex set in G which has been contracted into v induces a connected subgraph $G(v)$ of G. We consider all the connected subgraphs $G(v)$ and for any pair $G(v)$, $G(u)$ we select an edge $e(u, v)$ in G between them provided v and u are adjacent in G_0. Then we select in each $G(v)$ a minimal connected subgraph $G'(v)$ containing all ends of the edges $e(u, v)$, $u \in V(G_0)$. The union of the edges $e(u, v)$ and the trees $G'(v)$ form a subdivision of one of G_1, G_2, ..., G_p. □

Theorem 5.1, Corollary 5.15 and Proposition 5.16 imply the following which is probably the most general result on graph algorithms.

Theorem 5.17. *Let p be any graph property such that, for every graph G with property p, every minor of G has property p. Then there exists a polynomially bounded algorithm for deciding whether an arbitrary graph has property p.*

Robertson and Seymour proved that the algorithm in Theorem 5.17 can be chosen such that it is bounded by $C_p n^3$ where n is the number of nodes of G and C_p is a constant depending on p.

An important consequence of Theorems 5.1 and 5.17 is the following.

Theorem 5.18. *Let S be any 2-dimensional surface. There exists a finite collection* $G_1, G_2, \ldots, G_q$ *of graphs such that an arbitrary graph G can be embedded into S if and only if G has none of* $G_1, G_2, \ldots, G_q$ *as a minor. Furthermore, there exists a polynomially bounded algorithm for deciding this.*

In particular, there exists a polynomially bounded algorithm for deciding if G can be embedded into S_g or N_h (when g and h are fixed).

If a graph G has tree width $\leq w$, then every minor of G has tree width $\leq w$. Thus Theorem 5.17 implies the existence of a polynomially bounded algorithm for deciding if a graph has tree width $\leq w$. A corresponding excluded minor theorem follows from Theorem 5.1.

The excluded minor characterizations that follow from Theorem 5.1 and hence also the algorithms of Theorem 5.17 are implicit in the sense that, in general, we do not know the family $\mathcal{F}_p$ in Theorem 5.1. As mentioned earlier, the sphere and the projective plane are the only compact 2-dimensional surfaces for which we know a quantitative version of Theorem 5.18. The afore-mentioned result of Archdeacon (1980) and Glover et al. (1979) states the following.

Theorem 5.19. *There exists a collection of* 35 *graphs* $G_1, G_2, \ldots, G_{35}$ *such that an arbitrary graph G can be embedded in the projective plane if and only if it has none of the graphs* $G_1, G_2, \ldots, G_{35}$ *as a minor.*

5.5. Concluding remarks

As we have seen, the well-quasi-ordering result Theorem 5.1 and the connectivity result Theorem 5.14 are of central importance to minors and embeddings. Conversely, their proofs are based on investigations of embeddings. Previously, embeddings on higher surfaces were studied for their own sake. (White's application 1978a of embeddings to design theory is an exception.) The Robertson–Seymour theory shows that graph embeddings have a natural place in general graph theory. Also, the proofs of Theorem 5.1 and Theorem 5.14 give much additional insight in embeddings, minors and connectivity which is worth mentioning. We have indicated how the solution of the k-path problem for graphs of bounded tree width depends on nice separation properties of such graphs. Similar (but more complicated) separation properties are used for the k-path problem for graphs of bounded genus in Robertson and Seymour (1988). Suppose

G is a graph on a surface S and suppose $s_1, s_2, \ldots, s_k, t_1, \ldots, t_k$ are nodes of G and we want to find disjoint paths $P_1, P_2, \ldots, P_k$ such that P_i connects s_i and t_i for $i = 1, 2, \ldots, k$. Robertson and Seymour showed that these paths exist if G has the following properties (which we describe informally).

(i) the vertices $s_1, \ldots, t_k$ are "far apart" in the sense that every closed curve on S which separates $\{s_1, \ldots, t_k\}$ in the topological and sense and every open curve which connects two of $\{s_1, \ldots, t_k\}$ must have many points in common with G and

(ii) G is "dense" on S in the sense that every non-null-homotopic curve on S has many points in common with G.

This result was also applied to the following result on minors (again described informally).

If H is a fixed graph on S and G is any graph on S which is dense on S (in the sense of (ii) above), then G has H as a minor.

This result was then used to prove the restriction of Theorem 5.1 to graphs on a fixed surface and thereby obtain the excluded minor theorem (which is a special case of Theorem 5.1) for graphs on a fixed surface.

In the general k-path problem also the tree width theorem plays an important role since the presence of a large grid can be useful for connecting prescribed nodes with disjoint paths.

Although Theorem 5.17 gives a so-called good characterization for describing the graphs G that have no fixed graph H as a minor, it would still be desirable to have a more precise structural characterization of these graphs of the same type as Wagner's characterization Corollary 4.7 of the graphs having no K_5 as a minor. Robertson and Seymour (1984a) suggested that, for any fixed graph H, there might exist natural numbers k and g (depending on H) such that any graph G with no minor isomorphic to H contains a set S of at most k nodes such that $G - S$ is contained in a graph G' with the property that each of the simplicial summands of G' can be embedded into S_g. However, Seese and Wessel (1985) answered this in the negative by the following interesting grid like graphs: Consider the graph whose nodes are all pairs (i, j), where $1 \leqslant i \leqslant n$, $1 \leqslant j \leqslant n$, and whose edges are all $((i, j), (i', j'))$ such that $|i - i'| + |j - j'| = 1$. Then add a path $x_1 x_2 \cdots x_n$ and join x_i to $(2, i)$. The resulting graphs have no K_8 as a minor and yet they do not have the above embedding property. Robertson and Seymour (1994a) got around this problem by a structural characterization (more complicated than suggested above) of the graphs which have no (fixed) graph H as a minor. Roughly speaking, those graphs are obtained by pasting graphs of genus $<\gamma(H)$ together in a tree-like fashion, except for a few nodes and a few "bad" areas.

The restriction of Theorem 5.1 to graphs of maximum degree $\leqslant 3$ is known as Vazsonyi's conjecture. Nash-Williams (1965) suggested the following extensions of Vazsonyi's conjecture, known as the *immersion conjecture*.

Theorem 5.20. *Suppose $G_1, G_2, \ldots$ is an infinite sequence of graphs. Then there are integers i, j $(i < j)$ and an injective map $f: V(G_i) \rightarrow V(G_j)$ and a map g from*

$E(G_i)$ into a set of pairwise edge-disjoint paths of G_j such that, for any edge $e = xy$ of G_i, $g(e)$ connects $f(x)$ and $f(y)$.

Robertson and Seymour (private communication) have extended Theorem 5.1 to hypergraphs in such a way that the immersion conjecture follows as a "dual version" of Theorem 5.1.

Another problem related to Theorem 5.1 and motivated by König's problem mentioned at the beginning of this section is the *intertwining conjecture* of Lovász and Milgram (see Ungar 1979).

Theorem 5.21. *For any two graphs G_1, G_2 of minimum degree at least 3, there are only finitely many graphs G of minimum degree at least 3 which are edge-minimal with respect to the property of containing subdivisions of both G_1 and G_2.*

Robertson and Seymour (private communication) have recently obtained a labelled version of Theorem 5.1 from which the intertwining conjecture follows.

Finally we mention that Theorem 5.1 is not true for infinite graphs, as proved by Thomas (1988). It is unsolved whether or not Theorem 5.1 holds for countable graphs.

References

Andreev, E.M.

[1970] On convex polyhedra in Lobačevskiĭ spaces, *Mat. Sb. (N.S.)* **81**, 445–478 [*Math. USSR Sb.* **10**, 413–440].

Archdeacon, D.

[1980] *A Kuratowski theorem for the projective plane,* Thesis (Ohio State University).

[1986] The orientable genus is nonadditive, *J. Graph Theory* **10**, 385–401.

Archdeacon, D., and P. Huneke

[1989] A Kuratowski theorem for nonorientable surfaces, *J. Combin. Theory B* **46**, 173–231.

Arnborg, S., and A. Proskurowski

[1989] Linear time algorithms for NP-hard problems on graphs embedded in k-trees, *Discrete Appl. Math.* **23**, 11–24.

Atneosen, G.

[1968] *On the embeddability of compacta in n-books: intrinsic and extrinsic properties,* Ph.D. Thesis (Michigan State University).

Auslander, L., I.A. Brown and J.W.T. Youngs

[1963] The imbedding of graphs in manifolds, *J. Math. Mech.* **12**, 629–634.

Battle, J., F. Harary, Y. Kodama and J.W.T. Youngs

[1962a] Additivity of the genus of a graph, *Bull. Amer. Math. Soc.* **68**, 565–568.

Battle, J., F. Harary and Y. Kodama

[1962b] Every planar graph with nine points has a nonplanar complement, *Bull. Amer. Math. Soc.* **68**, 569–571.

Behzad, M., G. Chartrand and L. Lesniak-Foster

[1979] *Graphs and Digraphs* (Prindle, Weber and Schmidt, Boston).

Beineke, L.W., and F. Harary

[1969] The thickness of the complete graph, *Canad. J. Math.* **21**, 850–859.

Bienstock, D., and N. Dean
[1993] Bounds for rectilinear crossing numbers, *J. Graph Theory* **17**, 333–348.
Bollobás, B.
[1978] *Extremal Graph Theory* (Academic Press, New York).
Bouchet, A.
[1978] Triangular imbeddings into surfaces of a join of equicardinal independent sets following an Eulerian graph, in: *Theory and Applications of Graphs,* eds. Y. Alavi and D.R. Lick (Springer, Berlin), 86–115.
[1987] Reducing prime graphs and recognizing circle graphs, *Combinatorica* **7**, 243–254.
Chartrand, G., and F. Harary
[1967] Planar permutation graphs, *Ann. Inst. H. Poincaré Sect. B* **3**, 433–438.
Conway, J.H., and C.McA. Gordon
[1983] Knots and links in spatial graphs, *J. Graph Theory* **7**, 445–453.
de la Vega, W.F.
[1983] On the maximum density of graphs which have no subcontraction to K^s, *Discrete Math.* **46**, 109–110.
Dirac, G.A.
[1960] In abstrakten Graphen vorhandene vollständige 4-Graphen und ihre Unterteilungen, *Math. Nachr.* **22**, 61–85.
[1961] On rigid circuit graphs, *Abh. Math. Sem. Hamburg* **25**, 71–76.
[1964] Homorphism theorems for graphs, *Math. Ann.* **153**, 69–80.
Duchet, P., Y. Hamidoune, M. Las Vergnas and H. Meyniel
[1983] Representing a planar graph by vertical lines joining different levels, *Discrete Math.* **46**, 221–332.
Duke, R.A.
[1966] The genus, regional number, and Betti number of a graph, *Canad. J. Math.* **18**, 817–822.
Edmonds, J.R.
[1960] A combinatorial representation for polyhedral surfaces, *Notices Amer. Math. Soc.* **7**, 646.
Eggleton, R.B.
[1986] Rectilinear drawings of graphs, *Utilitas Math.* **29**, 149–172.
Erdős, P., and R.K. Guy
[1973] Crossing number problems, *Amer. Math. Monthly* **80**, 52–58.
Erdős, P., and L. Pósa
[1965] On the independent circuits contained in a graph, *Canad. J. Math.* **17**, 347–352.
Fiedler, J.R., J.P. Huneke, R.B. Richter and N. Robertson
[1994] Computing the orientable genus of projective graphs, *J. Graph Theory,* to appear.
Filotti, L.S., G.L. Miller and J. Reif
[1979] On determining the genus of a graph in $O(v^{O(g)})$ steps, in: *Proc. 11th Annu. ACM Symp. on Theory of Computing,* pp. 27–37.
Furst, M.L., J.L. Gross and L.A. McGeoch
[1988] Finding a maximum-genus graph imbedding, *J. Assoc. Comput. Mach.* **35**, 523–534.
Garey, M.R., and D.S. Johnson
[1979] *Computers and Intractability, A Guide to the Theory of NP-Completeness* (W.H. Freeman, San Francisco, CA).
[1982] The NP-completeness column: An ongoing guide, *J. Algorithms* **3**, 89–99.
Giles, R.
[1982] Optimum matching forests I: special weights, *Math. Programming* **22**, 1–11.
Glover, H.H., J.P. Huneke and C.S. Wang
[1979] 103 graphs that are irreducible for the projective plane, *J. Combin. Theory B* **27**, 332–370.
Gross, J.L., and T.W. Tucker
[1987] *Topological Graph Theory* (Wiley, New York).
Grünbaum, B.
[1967] *Convex Polytopes* (Wiley, London).

Grünbaum, B., and G.C. Shephard
[1981] The geometry of planar graphs, in: *Proc. 8th British Combinatorial Conf.*, ed. H.N.V. Temperley (Cambridge University Press, Cambridge) pp. 124–150.
[1987] *Tilings and Patterns* (W.H. Freeman, New York).
Gustin, W.
[1963] Orientable embedding of Cayley graphs, *Bull. Amer. Math. Soc.* **69**, 272–275.
Guy, R.K.
[1969a] The decline and fall of Zarankiewicz's theorem, in: *Proof Techniques in Graph Theory*, ed. F. Harary (Academic Press, New York) pp. 63–69.
[1969b] A combinatorial problem, *Nabla, Bull. Malayan Math. Soc.* **7**, 68–72.
[1971] Latest results on crossing numbers, in: *Lecture Notes in Mathematics*, Vol. 186, eds. M. Capabianco, J.M. Frechen and M. Krolik (Springer, Berlin) pp. 143–156.
Harary, F., and W.T. Tutte
[1965] A dual form of Kuratowski's Theorem, *Canad. Math. Bull.* **8**, 17–20.
Heffter, L.
[1891] Über das Problem der Nachbargebiete, *Math. Ann.* **38**, 477–508.
Higman, G.
[1952] Ordering by divisibility in abstract algebras, *Proc. London Math. Soc.* **2**, 326–336.
Hopcroft, J.E., and R.E. Tarjan
[1974] Efficient planarity testing, *J. Assoc. Comput. Math.* **21**, 549–568.
Jakobsen, I.T.
[1972] On certain homomorphism properties, I and II, *Math. Scand.* **31**, 379–404; **52**, 229–261.
Jensen, H.F.
[1971] An upper bound for the rectilinear crossing number of the complete graph, *J. Combin. Theory B* **10**, 212–216.
Kennedy, J.W., L.V. Quintas and M.M. Syslo
[1985] The theorem on planar graphs, *Historia Math.* **12**, 356–368.
Kerékjártó, B.
[1923] *Vorlesungen über Topologie* (Springer, Berlin).
Kleitman, D.J.
[1970] The crossing number of $K_{5,n}$, *J. Combin. Theory* **9**, 315–323.
Koebe, P.
[1936] Kontaktprobleme der konformen Abbildung, *Ber. Verh. Sächs. Akad. Wiss. Leipzig, Math.-Phys. Kl.* **88**, 141–164.
König, D.
[1936] *Theorie der endlichen und unendlichen Graphen* (Akademische Verlagsgesellschaft, Leipzig).
Kostochka, A.V.
[1984] Lower bound on the Hadwiger number of graphs by their average degree, *Combinatorica* **4**, 307–316.
Kratochvíl, J., M. Goljan and P. Kučera
[1986] *String Graphs* (Rozpravy Československé Akademie Věd).
Kruskal, J.B.
[1960] Well-quasi-ordering, the tree theorem, and Vázsonyi's conjecture, *Trans. Amer. Math. Soc.* **95**, 210–225.
Kuratowski, K.
[1930] Sur le problème des courbes gauches en topologie, *Fund. Math.* **15**, 271–283.
MacLane, S.
[1937] A combinatorial condition for planar graphs, *Fund. Math.* **28**, 22–32.
Mader, W.
[1967] Homomorphieeigenschaften und mittlere Kantendichte von Graphen, *Math. Ann.* **174**, 265–268.
[1972] Wohlquasigeordnete Klassen endlicher Graphen, *J. Combin. Theory B* **12**, 105–122.
Malitz, S.M.
[1988] Genus g graphs have page number $O(\sqrt{g})$, in: *Proc. 29th Annu. Symp. on Foundations of Computer Science* (IEEE Computer Soc. Press, New York) pp. 458–468.

Mani-Levitska, P., B. Guigas and V. Klee
[1979] Rectifiable n-periodic maps, *Geom. Dedicata* **8**, 127–137.

Mansfield, A.
[1983] Determining the thickness of graphs is NP-hard, *Math. Proc. Cambridge Philos. Soc.* **93**, 9–23.

Mayer, J.
[1972] Décomposition de K_{16} en trois graphes planaires, *J. Combin. Theory B* **13**, 71.

Melnikov, L.A.
[1981] Problem, *Sixth Hung. Coll. on Combinatorics, Eger.*

Nash-Williams, C.St.J.A.
[1965] On well-quasi-ordering infinite trees, *Proc. Cambridge Philos. Soc.* **61**, 697–720.

Pisanski, T.
[1980] Genus of cartesian products of regular bipartite graphs, *J. Graph Theory* **4**, 31–42.

Platt, C.R.
[1976] Planar lattices and planar graphs, *J. Combin. Theory B* **21**, 30–39.

Ringel, G.
[1954] Bestimmung der Maximalzahl der Nachbargebiete auf nichtorientierbaren Flächen, *Math. Ann.* **127**, 181–214.
[1965a] Das Geschlecht des vollständigen paaren Graphen, *Abh. Math. Sem. Univ. Hamburg* **38**, 139–150.
[1965b] Der vollständige paare Graph auf nichtorientierbaren Flächen, *J. Reine Angew. Math.* **220**, 89–93.
[1974] *Map Color Theorem* (Springer, Berlin).
[1977] The combinatorial map color theorem, *J. Graph Theory* **1**, 141–155.
[1978] Non-existence of graph embeddings, in: *Theory and Applications of Graphs,* eds. Y. Alavi and D.R. Lick (Springer, Berlin) pp. 455–475.
[1985] 250 Jahre Graphentheorie, in: *Graphen in Forschung und Unterricht, Festschrift K. Wagner,* eds. R. Bodendiek, H. Schumacher and G. Walter (Franzbecker, Bad Salzdetfurth) pp. 136–152.

Ringel, G., and J.W.T. Youngs
[1968] Solution of the Heawood map-coloring problem, *Proc. Nat. Acad. Sci. U.S.A.* **60**, 438–445.

Robertson, N., and P.D. Seymour
[1983] Graph minors. I. Excluding a forest, *J. Combin. Theory B* **35**, 39–61.
[1984a] Graph width and well-quasi-ordering: a Survey, in: *Progress in Graph Theory,* eds. J.A. Bondy and U.S.R. Murty (Academic Press, Toronto) pp. 399–406.
[1984b] Graph minors. III. Planar tree-width, *J. Combin. Theory B* **36**, 49–64.
[1986a] Graph minors. II. Algorithmic aspect of tree-width, *J. Algorithms* **7**, 309–322.
[1986b] Graph minors. V. Excluding a planar graph, *J. Combin. Theory B* **41**, 92–114.
[1986c] Graph minors. VI. Disjoint paths across a disc, *J. Combin. Theory B* **41**, 115–138.
[1988] Graph minors. VII. Disjoint paths on a surface, *J. Combin. Theory B* **45**, 212–254.
[1990a] Graph minors. IV. Tree-width and well-quasi-ordering, *J. Combin. Theory B* **48**, 227–254.
[1990b] Graph minors. VIII. A Kuratowski theorem for general surfaces, *J. Combin. Theory B* **48**, 255–288.
[1990c] Graph minors. IX. Disjoint crossing paths, *J. Combin. Theory B* **49**, 40–77.
[1991] Graph minors. X. Obstructions to tree-decomposition, *J. Combin. Theory B* **52**, 153–190.
[1994a] Graph minors. XII. Excluding a non-planar graph, submitted.
[1994b] Graph minors. XIII. The disjoint paths problem, submitted.
[1994c] Graph minors. XIV. Embedding structures on a surface, in preparation.

Robertson, N., P.D. Seymour and R. Thomas
[1993a] Hadwiger's conjecture for K_6-free graphs, *Combinatorica* **13**, 279–361.
[1993b] Linkless embeddings of graphs in 3-space, *Bull. Amer. Math. Soc.* **28**, 84–89.

Scheinermann, E.R., and D.B. West
[1983] The interval number of a planar graph: Three intervals suffice, *J. Combin. Theory B* **35**, 224–239.

Seese, D.G., and W. Wessel
[1989] Grids and their minors, *J. Combin. Theory B* **47**, 349–360.

Skupien, Z.
[1966] Locally Hamiltonian and planar graphs, *Fund. Math.* **58**, 193–200.

Stahl, S.
[1975] *Self-dual embedding of graphs,* Ph.D. Thesis (Western Michigan University).

[1978] Generalized embedding schemes, *J. Graph Theory* **2**, 41–52.

Stahl, S., and L. Beineke

[1977] Blocks and the nonorientable genus of a graph, *J. Graph Theory* **1**, 75–78.

Tamassia, R., and I.G. Tollis

[1986] A unified approach to visibility representations of planar graphs, *Discrete Comput. Geom.* **1**, 321–341.

Tarjan, R.E.

[1985] Decomposition by clique separators, *Discrete Math.* **55**, 221–232.

Thomas, R.

[1988] A counterexample to Wagner's conjecture for infinite graphs, *Math. Proc. Cambridge Philos. Soc.* **103**, 55–57.

Thomassen, C.

[1980] Planarity and duality of finite and infinite graphs, *J. Combin. Theory B* **29**, 244–271.

[1981] Kuratowski's theorem, *J. Graph Theory* **5**, 225–241.

[1984] Plane representations of graphs, in: *Progress in Graph Theory,* eds. J.A. Bondy and U.S.R. Murty (Academic Press, New York) pp. 43–69.

[1986] Interval representations of planar graphs, *J. Combin. Theory B* **40**, 9–20.

[1988a] Rectilinear drawings of graphs, *J. Graph Theory* **12**, 335–342.

[1988b] On the presence of disjoint subgraphs of a specified type, *J. Graph Theory* **12**, 101–112.

[1989a] The graph genus problem is NP-complete, *J. Algorithms* **10**, 568–576.

[1989b] Planar acyclic oriented graphs, *Order* **5**, 349–361.

[1990] Embeddings of graphs with no short noncontractible cycles, *J. Combin. Theory B* **48**, 155–177.

[1992] The Jordan–Schönfliess theorem and the classification of surfaces, *Amer. Math. Monthly* **99**, 116–130.

Tutte, W.T.

[1960] Convex representations of graphs, *Proc. London Math. Soc.* **10**, 304–320.

[1961] On the problem of decomposing a graph into connected factors, *J. London Math. Soc.* **36**, 221–230.

[1963] How to draw a graph, *Proc. London Math. Soc.* **13**, 743–768.

Tverberg, H.

[1987] A proof of Kuratowski's theorem, in: *Graph Theory in Memory of G.A. Dirac,* eds. L.D. Andersen et al. (North-Holland, Amsterdam).

Ungar, P.

[1953] On diagrams representing maps, *J. London Math. Soc.* **28**, 336–342.

[1979] A dissection problem and an intertwining problem, in: *Graph Theory and Related Topics,* eds. J.A. Bondy and U.S.R. Murty (Academic Press, New York) pp. 370–371.

Wagner, K.

[1936] Bemerkungen zum Vierfarbenproblem, *Jber. Deutsch. Math.-Verein.* **46**, 26–32.

[1937] Über eine Eigenschaft der Ebenen Komplexe, *Math. Ann.* **114**, 570–590.

[1964] Beweis einer Abschwächung der Hadwiger-Vermutung, *Math. Ann.* **153**, 139–141.

[1970] *Graphentheorie* (BI-Hochschultaschenbücher-Verlag, Mannheim).

White, A.T.

[1970] The genus of repeated cartesian products of bipartite graphs, *Trans. Amer. Math. Soc.* **151**, 393–404.

[1973] *Graphs, Groups and Surfaces* (North-Holland, Amsterdam).

[1978a] Block designs and graph imbeddings, *J. Combin. Theory B* **25**, 166–183.

[1978b] The proof of the Heawood Conjecture, in: *Selected Topics in Graph Theory,* eds. L.W. Beineke and R.J. Wilson (Academic Press, New York) pp. 51–81.

White, A.T., and L.W. Beineke

[1978] Topological graph theory, in: *Selected Topics in Graph Theory,* eds. L.W. Beineke and R.J. Wilson (Academic Press), 15–50.

Whitney, H.

[1932] Non-separable and planar graphs, *Trans. Amer. Math. Soc.* **34**, 339–362.

[1933] 2-isomorphic graphs, *Amer. J. Math.* **55**, 245–254.

Woodall, D.R.
[1971] Thrackles and deadlock, in: *Combinatorial Mathematics and its Applications,* ed. D.J.A. Welsh (Academic Press, New York) pp. 335–347.

Xuong, N.H.
[1979] How to determine the maximum genus of a graph, *J. Combin. Theory B* **26**, 217–225.

Yannakakis, M.
[1989] Embedding planar graphs in four pages, in: *Proc. 18th Annu. ACM Symp. on Theory of Computing, J. Comput. System Sci.* **38**, 36–67.

Youngs, J.W.T.
[1963] Minimal imbeddings and the genus of a graph, *J. Math. Mech.* **12**, 303–315.

Zarankiewicz, K.
[1954] On a problem of P. Turán concerning graphs, *Fund. Math.* **41**, 137–145.

CHAPTER 6

Random Graphs

Michał KAROŃSKI

Faculty of Mathematics and Computer Science, Adam Mickiewicz University, Matejki 48/49 60-769 Poznań, Poland

and

Department of Mathematics and Computer Science, Emory University, Atlanta, GA 30322, USA

Contents

HANDBOOK OF COMBINATORICS
Edited by R. Graham, M. Grötschel and L. Lovász

1. Introduction

In the most general terms, a *random graph* is a pair $(\mathcal{G}, P)$ where $\mathcal{G}$ is a family of graphs and P is a probability distribution over $\mathcal{G}$.

More specifically take $\mathcal{G}$ to be the family of all graphs on n labeled vertices and assign to each $G \in \mathcal{G}$ a probability measure

$$\mathrm{P}(G) = p^{e(G)}(1-p)^{\binom{n}{2}-e(G)} ,$$

where $0 \leqslant p \leqslant 1$ and $e(G)$ denotes the number of edges of G.

We can describe this model in terms of independent Bernoulli experiments performed on the edges of a complete graph K_n with n vertices. Namely, delete each of the $\binom{n}{2}$ edges with probability $1-p$, independently of all other edges. Equivalently, one may start with an empty graph on the vertex set and insert edges independently, with the same probability p. In the sequel such a random graph is denoted $G(n, p)$ and is often called a *binomial random graph*. Note, that in $G(n, p)$ the number of edges is a binomially distributed random variable with expectation $\binom{n}{2}p$. From now on we often assume that the edge probability p is a function of the number of vertices, i.e., $p = p(n)$.

Another fundamental model, having a "counting" flavor, is obtained by considering $\mathcal{G}$ to be the family of all graphs with n labeled vertices and with exactly N edges, $0 \leqslant N \leqslant \binom{n}{2}$. To every $G \in \mathcal{G}$, assign a probability measure

$$\mathrm{P}(G) = |\mathcal{G}|^{-1} = \binom{\binom{n}{2}}{N}^{-1} .$$

Intuitively, we sample at random a graph from the family $\mathcal{G}$. Equivalently, one can get such a random graph by picking at random a given number N of edges, in such a way that all possible $\binom{\binom{n}{2}}{N}$ choices are equally likely, and removing the remaining ones from K_n, or by simply inserting N edges at random into an empty graph on n vertices. We denote such a random graph by $G(n, N)$ and call it a *uniform random graph*.

Many natural questions of deterministic graph theory can now be translated into a probabilistic language. In random graph theory we are mainly concerned with establishing the probability that a random graph possesses a given property, i.e., the probability that the random graph belongs to the class of graphs possessing it. Interesting properties include connectedness, hamiltonicity or containment of a perfect matching, to name only a few. Most often we are interested in when the probability of such an event tends to one as the number of vertices of a random graph goes to infinity. In this case we say that the random graph has this property almost surely (a.s., for short). Moreover, various numerical invariants, for instance vertex degree, vertex or edge connectivity, chromatic number and girth, can be considered random variables when the graph in question is random. We then ask for the probability distribution of such random variables. Usually we are interested in asymptotic distributions as $n \to \infty$.

In this chapter we are concerned with problems of the above type and we restrict our attention to binomial and uniform random graphs. We aim to give the reader an idea of the nature of typical problems in the field, basic methods used in proofs, as well as to lead the reader from the origins to most recent results.

It is widely accepted that the birth of the theory of random graphs is the publication of the seminal paper on the evolution of random graphs by Erdős and Rényi (1960), followed by a sequence of their other fundamental contributions (for the entire collection see Erdős 1973 or Rényi 1976). Since then several hundred papers have been published in various areas of random graph theory. The growth of the field and its impact on other branches of combinatorics, is extensively presented in the excellent monograph of Bollobás (1985). This book is highly recommended to all who want to undertake a serious study of the field. One may, however, begin with an introductory book by Palmer (1985), as well as to consult review papers by Grimmett (1983), Karoński (1982) or Stepanov (1973). (For recent developments in research on random graphs we refer the reader to *Random Structures & Algorithms*, the journal solely devoted to that topic.)

2. Evolving graphs

Let us adopt a dynamic view in which a graph, acquiring edges, grows from empty to complete. More precisely, start with n labelled vertices $\{1, 2, \ldots, n\}$ and add edges at random, one at a time, until the resulting graph is complete. This simple model of a graph evolving in a random manner was first investigated in detail by Erdős and Rényi (1960). The key question they posed may be stated in general terms as follows: what is a typical structure of a random graph during its evolution? We say that a structure (or a graph property) is *typical* if a random graph possesses it almost surely.

Denote by $\mathcal{G}$ the set of all labelled graphs on the vertex set $\{1, 2, \ldots, n\}$. Define a *graph process* $(G_t)_0^{\binom{n}{2}}$ as a sequence of graphs from $\mathcal{G}$, such that $e(G_t) = t$ and $G_0 \subset G_1 \subset \cdots \subset G_{\binom{n}{2}}$. A *graph process* is *random* if G_{t+1} arises from G_t by adding to it a single edge chosen at random from the $\binom{n}{2} - t$ remaining edges, $0 \leqslant t \leqslant \binom{n}{2} - 1$. Equivalently, we may choose at random one particular process $(G_t)_0^{\binom{n}{2}}$ from all $\binom{n}{2}!$ graph processes. Clearly, since G_{t+1} depends only on G_t, this random graph process has the Markov property.

By a *graph property* we mean a subset of the set $\mathcal{G}$. We say that a graph has a property $\mathcal{A}$ if it is an element of $\mathcal{A}$. For example, $\mathcal{A}$ may be the set of all connected graphs, Hamiltonian graphs, graphs with isolated vertices, graphs with a given chromatic number, graphs without a forbidden subgraph, etc. A graph property $\mathcal{A}$ is *increasing* (*decreasing*) if the fact that $G \in \mathcal{A}$ and G is a spanning subgraph (supergraph) of the graph H implies that $H \in \mathcal{A}$. The property $\mathcal{A} = \{G \in \mathcal{G}\colon G \text{ is connected}\}$ for example, is increasing, $\mathcal{A} = \{G \in \mathcal{G}\colon G \text{ has isolated vertices}\}$ is decreasing while $\mathcal{A} = \{G \in \mathcal{G}\colon \chi(G) = k, k \in \mathbb{N}\}$ is neither increasing or decreasing, i.e., is not *monotone*. Notice also that the complement of an increasing property $\mathcal{A}$ ("not $\mathcal{A}$") is decreasing. if $\mathcal{A}$ is an increasing property and

$(G_t)_0^{\binom{n}{2}}$ is a graph process, then a natural question arises: when is the first time G_t acquires the property $\mathcal{A}$, i.e., what is the *hitting time* $\tau = \min\{t: G_t \in \mathcal{A}\}$ of $\mathcal{A}$? Obviously, when the process is random then the problem is to find the limit of $\Pr(\tau = N)$ for $N = 0, 1, \ldots, \binom{n}{2}$. Although such a "dynamic" discrete time approach to the study of the evolution leads to very precise results, we shall take an easier "static" approach, looking at the state of the process at a given time $t = N$. Notice that the set of all possible $\binom{\binom{n}{2}}{N}$ states (graphs) of the process $(G_t)_0^{\binom{n}{2}}$ at $t = N$ consists of all labelled graphs on n vertices with N edges, and that each particular state is equally likely. Hence, G_N can be identified with a random graph $G(n, N)$, and instead of finding the hitting time of $\mathcal{A}$ in $(G_t)_0^{\binom{n}{2}}$ we try to specify the smallest value of N for which $\mathcal{A}$ is likely to occur in $G(n, N)$.

One can also view a random graph process as a continuous time process. Let $\{T_e\}$, $e \in \{\{i, j\}: i, j = 1, 2, \ldots, n, i < j\}$, be a sequence of independent random variables with a common continuous distribution on $[0, \infty]$. Each T_e can be viewed as the time it takes to "grow" edge e. Let $G_n(t)$ denote the graph with vertex set $\{1, 2, \ldots, n\}$ and any edge e for which $T_e \leq t$, $t \in [0, \infty)$. Hence, we treat T_e as the time of appearance of the edge e in a graph $G_n(t)$. Furthermore, notice that $G_n(t)$ can be identified with a random graph $G(n, p)$ when $p = \Pr(T_e \leq t)$, i.e., $G(n, p)$ can be treated as a state of the process $G_n(t)$ at the time t. (In fact $G(n, N)$ can be recovered from this process as well by taking $G_n(T_{[N]})$, where $T_{[N]}$ denotes the Nth smallest order statistic of $\{T_e\}$.) Hence, we see that random graphs $G(n, N)$ and $G(n, p)$ may provide information about the asymptotic properties of random graph processes, both with discrete and continuous time at fixed moments of the evolution. It turns out however, that in many cases $G(n, N)$ and $G(n, p)$ are asymptotically equivalent. Intuitively, one may expect that such equivalence should take place when the number of edges of $G(n, N)$ is equal to the expected number of edges of $G(n, p)$, i.e., when $N = \lfloor \binom{n}{2} p \rfloor \sim n^2 p/2$.

This natural relationship was described by Erdős and Rényi (1960) but a formal proof was given by Bollobás (see Bollobás 1985, Theorem 2, p. 34). Here, we present a slight generalization of Bollobás' equivalence theorem due to Łuczak (1990a).

Theorem 2.1. *Let* $0 \leq p_0 \leq 1$, $s(n) = (n^2 p(1-p))^{1/2} \to \infty$ *and* $\omega(n) \to \infty$ *as* $n \to \infty$.

(1) *If* $\mathcal{A}$ *is a graph property and for all* $N \in \mathbb{N}$ *such that*

$$\binom{n}{2} p - \omega(n) s(n) < N < \binom{n}{2} p + \omega(n) s(n) ,$$

$$\Pr(G(n, N) \in \mathcal{A}) \to p_0 , \quad \textit{then} \quad \Pr(G(n, p) \in \mathcal{A}) \to p_0 .$$

(2) *If* $\mathcal{A}$ *is a monotone property and* $p^+ = p + \omega(n)s(n)/n^2$, $p^- = p - \omega(n)s(n)/n^2$, *then the facts that*

$$\Pr(G(n, p^-) \in \mathcal{A}) \to p_0 \quad \textit{and} \quad \Pr(G(n, p^+) \in \mathcal{A}) \to p_0$$

imply that $\Pr(G(n, N) \in \mathcal{A}) \to p_0$ *as* $n \to \infty$, *where* $N = \lfloor \binom{n}{2} p \rfloor$.

Notice that the results for $G(n, N)$ are sharper than results about $G(n, p)$, since for any graph property $\mathscr{A}$,

$$\Pr(G(n, p) \in \mathscr{A}) = \sum_{N=0}^{\binom{n}{2}} \Pr(G(n, N) \in \mathscr{A}) \Pr(e(G(n, p)) = N) .$$

However, $G(n, N)$, unlike $G(n, p)$, suffers from the fact that the edges are not independent of each other. Since the calculations are often easier in $G(n, p)$, this model is more popular though it is less intuitive than $G(n, N)$. From now on we shall study the evolution of $G(n, p)$ as the edge occurrence probability grows from zero to one. The reader more used to the graph-theoretic than the probabilistic approach is encouraged to "translate" the results by substituting $2N/n^2$ for p, whenever Theorem 2.1 allows.

3. Evolution – Main epochs

The process of the evolution of a random graph (or rg for short) consists of clearly distinguishable epochs, decisive events and turning points. This subsection is a "historical sketch", which will later serve as a guide, leading us in our "travel in time" through particular characteristic periods.

At the beginning ($p \equiv 0$) our universe is empty, i.e., it consists of isolated vertices only. Next, for some time, we stay in a forest, where trees grow larger and larger. The following theorem shows how long this "forest epoch" lasts.

Theorem 3.1. *If $np \to 0$ as $n \to \infty$, then $G(n, p)$ is a.s. a forest.*

Proof. Denote by $\mathscr{A}$ the set of all graphs on n vertices, with at least one cycle. Then

$$\Pr(G(n, p) \in \mathscr{A}) \leq \sum_{k=3}^{n} \Pr(X_{n,k} > 0) \leq \sum_{k=3}^{n} \mathrm{E}X_{n,k} ,$$

where the random variable $X_{n,k}$ counts cycles of length k in $G(n, p)$. But,

$$\mathrm{E}X_{n,k} = \binom{n}{k} \tfrac{1}{2}(k-1)!p^k < (np)^k$$

and the theorem follows. □

This simple proof illustrates the usefulness of the so-called *first moment method*. This name is attributed to the immediate consequence of Markov's inequality, that for a non-negative valued random variable X, $\Pr(X > 0) \leq \mathrm{E}X$. It should be stressed that the knowledge of the asymptotic behavior of the expectation of a given numerical characteristic of a random graph is usually of a critical importance. A careful analysis of when the expectation tends to zero, to a constant, or to infinity as $n \to \infty$, often explains the typical structure of a random graph.

Let us now take a closer look at that period of the evolution of $G(n, p)$ when

$np \to 0$. Following the observation above, consider the expectation of the number $Y_{n,k}$ of trees on k vertices, contained in $G(n, p)$ as subgraphs. Obviously, $\mathrm{E}Y_{n,k} = \binom{n}{k} k^{k-2} p^{k-1}$. Hence, by the first moment method, it follows that, for example, when $pn^2 \to 0$, $G(n, p)$ consists a.s. of isolated vertices only, if $pn^{3/2} \to 0$ there are a.s. no trees on three vertices, and in general, $G(n, p)$ a.s. does not contain trees on k vertices as subgraphs when $pn^{k/(k-1)} \to 0$. Furthermore, as noted by Erdős and Rényi (1960), when $pn^{k/(k-1)} \to \infty$ then $G(n, p)$ a.s. contains a tree on k vertices. However, this cannot be deduced from the behavior of the expectation itself and one has to employ a more sophisticated approach known as the second moment method (as it makes use of the second moment of a random variable by way of Chebyshev's inequality).

The "forest epoch" ends when $np \to c$, $c > 0$, however as long as $c < 1$, as the next result shows, the structure of a random graph does not change too much, compared with the period when $np \to 0$.

Theorem 3.2. *If $np \to c$, as $n \to \infty$, and $0 < c < 1$, then $G(n, p)$ a.s. consists of components with at most one cycle.*

Not surprisingly, in that period of the evolution, trees dominate the structure of a rg because the probability that $G(n, p)$ is a forest tends to $(1 - c)^{1/2} \exp(c/2 + c^2/4)$. The expected number of vertices which belong to components with one cycle tends to a constant. Thus a.s. all but a negligible fraction of vertices are contained in isolated trees. Moreover, in that brief epoch, all components are rather small since they are $\mathrm{O}(\log n)$ in size.

The structure of a rg changes dramatically when the constant c exceeds 1.

Theorem 3.3. *If $np \to c$, $c > 1$, then $G(n, p)$ a.s. consists of a unique giant component and small components with at most one cycle.*

The unique component is "giant" indeed, since it contains approximately $(1 - x(c)/c + \mathrm{o}(1))n$ vertices of $G(n, p)$, where $x(c)$ is the only root of the equation $x\,\mathrm{e}^{-x} = c\,\mathrm{e}^{-c}$, satisfying $0 < x(c) < 1$, while the second largest component is of size $\mathrm{O}(\log n)$.

Theorems 3.2 and 3.3 indicate that $c = 1$ is a critical point in the sense that, when $np \to 1$, $G(n, p)$ very quickly changes from a scattered collection of small trees and unicyclic components to a coagulated lump of components that dominates the graph in size. This short period when the giant component emerges is called the phase transition because of its resemblance to the physical phenomenon of that name. In a subsequent section, we shall provide more details about this turning point in the history of the evolution of $G(n, p)$.

After the phase transition, the giant component begins to consume the small components not yet attached to it, and finally $G(n, p)$ achieves the structure described below.

Theorem 3.4. *If* $np \to \infty$, *but* $np - \log n \to -\infty$ *as* $n \to \infty$, *then* $G(n, p)$ *is a.s. disconnected and all components, with the exception of the giant one, are trees.*

When we look closer into the period of evolution characterized by Theorem 3.4, we see (a simple first moment argument) that the giant component gradually "swallows" the remaining tree-components in the reverse order (with respect to their size) in which they have been emerging in $G(n, p)$ (when $np \to 0$). Indeed, if $Z_{n,k}$, $k \in \mathbb{N}$, denotes the number of tree-components on k vertices, then

$$\mathrm{E}Z_{n,k} = \binom{n}{k} k^{k-2} p^{k-1} (1-p)^{\binom{k}{2}-k+1+k(n-k)}$$

and one may easily check that $\mathrm{E}Z_{n,k} \to 0$ as $n \to \infty$ as soon as $np - (\log n - (k-1)\log\log n)/k \to \infty$. So, the asymptotic behavior of $\mathrm{E}Z_{n,k}$ suggests that, when $np - \log n \to c$, $-\infty < c < \infty$, then $G(n, p)$ a.s. consists of the giant component and a number of isolated vertices only. This is in fact true, which further implies that a rg becomes connected as soon as the last vertex ceases to be isolated.

Theorem 3.5. *If* $np - \log n \to \infty$, *as* $n \to \infty$ *then* $G(n, p)$ *is a.s. connected.*

Notice that the expected vertex degree in $G(n, p)$ is equal to np. This means that the before connectedness (b.c.) era ends and the after connectedness (a.c.) era begins when each vertex has on the average, $\log n$ neighbors. Also, as soon as $np \to \infty$, $G(n, p)$ a.s. contains a triangle K_3, but one must wait until the expected vertex degree is of the order $n^{1/3}$ to be almost sure that it contains K_4 as a subgraph. In general, complete graphs appear in $G(n, p)$ gradually in a manner similar to trees. Namely, when $pn^{2/(k-1)} \to 0$ then $G(n, p)$ a.s. does not contain K_k, while for $pn^{2/(k-1)} \to \infty$, $G(n, p)$ has a.s. at least one subgraph isomorphic to K_k.

Increasing the edge probability p further, we make $G(n, p)$ more and more dense to such an extent that when p is a constant (i.e., each vertex is, roughly speaking, joined to a positive fraction of the remaining vertices) then, for example, its diameter is a.s. equal to two. The process of the evolution of $G(n, p)$ comes to its end when for $p \equiv 1$, it becomes a "dull" complete graph K_n.

4. Phase transition

We now take a closer look at the phase transition in a random graph, the most fascinating, but still somewhat mysterious, period of its evolution. A distant view, offered by Theorems 3.2 and 3.3, is quite clear: when $p = c/n$, $0 < c < 1$, then $G(n, p)$ is composed of many small components, mainly trees, while for $c > 1$, it has one large component which dominates the whole graph. Early attempts to better understand the nature of the phase transition, concentrated on the analysis of the typical structure of $G(n, p)$ at the critical point $c = 1$, i.e., when $p = 1/n$. For example, Erdős and Rényi (1960) proved that for such p, the largest component a.s. has size $O(n^{2/3})$. Recall that its size is $O(\log n)$ when $c<1$ and is comparable with n for $c>1$. At first glance, this suggests that there is a "double jump" in the order of the largest component as c passes 1. But it is only a matter of the lens through which we observe $G(n, p)$ at the critical point of its evolution. In order to get an undistorted picture of the phase transition, we have to significantly magnify the neighborhood of the point $c=1$, assuming that $c=1\pm\varepsilon$, where $\varepsilon=\varepsilon(n)\to 0$ as $n\to\infty$. Then, as we shall see from the next theorem, the "double jump" somehow vanishes. Also, while a distant view suggests the existence of the critical point, $c=1$, the closer look offered by this theorem shows that we should speak about a critical "interval" $c=1\pm\varepsilon$, where $\varepsilon^3 n=O(1)$, instead.

Theorem 4.1. *Let $\varepsilon\to 0$ but $\varepsilon^3 n\to\infty$ as $n\to\infty$. Then*

(i) *$G(n,(1-\varepsilon)/n)$ a.s. consists of components with at most one cycle. The largest component is a.s. a tree on $(2+o(1))\varepsilon^{-2}\log(\varepsilon^3 n)$ vertices;*

(ii) *$G(n,(1+\varepsilon)/n)$ a.s. has the unique large component and a number of small components with at most one cycle. The giant one has $(2+o(1))\varepsilon n$ vertices and $(\frac{2}{3}+o(1))\varepsilon^3 n$ edges more than vertices, while small components have less than $n^{2/3}$ vertices each.*

To get an idea of how the giant component emerges in $G(n, p)$, it is convenient to return to the original "dynamic" interpretation of the evolutionary process. Therefore, let $p_2>p_1$ and assume that $G(n, p_2)$ arises from $G(n, p_1)$ by adding new edges to the latter graph with the probability $\Delta p=(p_2-p_1)/(1-p_1)$ (formally, $G(n, p_2)$ can be interpreted as the sum of $G(n, p_1)$ and $G(n, \Delta p)$).

Following Theorem 4.1, we start to observe the process just before the critical interval begins, by looking at two particular instances, when $p_1=(1-2\varepsilon)/n$ and $p_2=(1-\varepsilon)/n$, where $\varepsilon^3 n\to\infty$ as $n\to\infty$. Then, one can prove that in $G(n, p_1)$ there are many components of the same size as the largest one, which, recall, consists roughly of $\varepsilon^{-2}\log\varepsilon^3 n$ vertices. Let us take two such components and insert additional edges with probability $\Delta p\sim\varepsilon/n$ in order to obtain $G(n, p_2)$ and ask about the chance that those components will merge together in $G(n, p_2)$. By the first moment, the probability of such an event is at most $(\varepsilon^{-2}\log\varepsilon^3 n)^2(\varepsilon/n)=(\log\varepsilon^3 n)/(\varepsilon^3 n)\to 0$ as $n\to\infty$. Thus, when $p=(1-\varepsilon)/n$, $\varepsilon^3 n\to\infty$, the large components are unable to "swallow" each other and therefore are forced to hunt for smaller quarry. Hence large components grow absorbing only small ones and no clear favorite to win the race to become the giant emerges. As the edge probability p increases, the number of contenders for the role of giant decreases. When $p=(1-\varepsilon)/n$, $\varepsilon^3 n=O(1)$, the probability that two specified large components will form a new component is bounded away from zero, but still is too small to ensure the creation of the unique giant component. At the same time, a big gap between the orders of large and small components arises which prevents a new large component being formed from the small ones. Next, as soon as $p=(1+\varepsilon)/n$, $\varepsilon\to 0$ but $\varepsilon^3 n\to\infty$, all large components a.s. merge together and the unique giant component emerges from "chaos" as the unquestioned victor.

The critical interval $c = 1 \pm \varepsilon$, $\varepsilon^3 n = O(1)$, which sharply differs from other epochs due to its many peculiarities, is the most specific and hard-to-study period of the evolution. For example, this is the only time when $G(n, p)$ may have more than one component which is neither a tree nor a unicyclic graph. This is also the only moment when the size of the largest component is "fuzzy", (i.e., cannot be a.s. determined up to a factor of one, etc.). The critical interval divides the whole b.c. (before connectedness) epoch into two phases: *subcritical* – when components compete to be the largest; and *supercritical* in which the largest component rules. In addition to this fundamental difference, there is a striking similarity between the typical structure of $G(n, p)$ in its subcritical phase and the structure of the part of $G(n, p)$, in its supercritical phase, which is induced by vertices not belonging to the giant component. Indeed, Theorems 3.1–3.4 and 4.1 show that in both cases all components are rather small and each contain at most one cycle. In fact, ties connecting those two phases are much stronger since the evolution of the part of $G(n, p)$ which remains outside the giant component during the supercritical phase, looks like a backward replay of the evolution of the whole of $G(n, p)$ during the subcritical phase. The supercritical phase, restricted to that part of a rg which is not absorbed by the giant component, is a symmetric reflection of the subcritical phase only when $p = (1 \pm \varepsilon)/n$ and $\varepsilon \to 0$. In general, roughly speaking, all structural events which take place when $np \to c$, $c > 1$ have counterparts located in a relatively short period when $np \to c'$, $0 < c' < 1$ (in the sense, that a given c corresponds to c' defined by $c' \mathrm{e}^{-c'} = c\,\mathrm{e}^{-c}$. The same "relationship" connects two other epochs: The one when $np \to \infty$ (but $np - \log n \to -\infty$) with that when $np \to 0$ (compare, for example, how tree-components of a given size emerge and how they disappear). One can say that outside the giant component, in the supercritical phase, "time" flows more slowly than in the whole rg during the subcritical phase.

Let us present a result due to Łuczak et al. (1994), which adds substantial support to the claim that the phase transition leads to a significant qualitative change in the structure of a random graph.

Theorem 4.2. *Let $\varepsilon = \varepsilon(n) \to 0$ as $n \to \infty$. Then $G(n, p)$ is*

(i) *a.s. planar, when $p = (1 - \varepsilon)/n$, $\varepsilon^3 n \to \infty$;*

(ii) *planar with probability tending to $a(\lambda)$, $0 < a(\lambda) < 1$, as $n \to \infty$, when $p = (1 + \varepsilon)/n$, where $\varepsilon^3 n \to \lambda$ and $-\infty < \lambda < \infty$ is a constant;*

(iii) *a.s. non-planar, when $p = (1 + \varepsilon)/n$, $\varepsilon^3 n \to \infty$.*

Notice that planarity is a decreasing property of graphs which, by the above theorem, implies that $G(n, p)$ is a.s. planar during the subcritical phase while after the phase transition it is a.s. non-planar.

A systematic study of the phase transition was originated by Bollobás who revealed the mechanism of the formation of the giant component. For an extensive account of those results the reader is referred to chapter 6 of Bollobás' (1985) monograph. The reader interested in recent developments in this direction is referred to papers by Kolchin (1986), Stepanov (1987), Łuczak (1990b) and, first of all, to a seminal paper by Janson et al. (1993).

Obviously, investigations of the critical "interval", when the phase transition takes place, requires very sophisticated and delicate tools. Janson et al. (1993) applied machinery of generating functions with a great precision. They were able to study the structure of evolving graphs (and multigraphs) when edges are added one at a time and at random, mainly looking at so-called excess and deficiency of a graph. To give the reader a taste of their results let us quote the following theorem.

Theorem 4.3. *The probability that a random graph of multigraph with n vertices and $\frac{1}{2}n + O(n^{1/3})$ edges has exactly r bicyclic components (i.e., components with exactly two cycles), and no components of higher cyclic order, is*

$$\left(\frac{5}{18}\right)^r \sqrt{\frac{2}{3}} \frac{1}{(2r)!} + O(n^{-1/3}) .$$

They also consider the following fascinating problem: What is the probability that the component which during the evolution becomes the first "complex" component (i.e., the first component with more than one cycle) is the only complex component which emerges during the whole process. So they ask what is the probability that the first bicyclic component is the "seed" for the giant one. They prove that it happens quite often indeed.

Theorem 4.4. *The probability that an evolving graph or multigraph on n vertices never has more than one complex component throughout its evolution approaches* $5\pi/18 \approx 0.8727$ *as* $n \to \infty$.

5. Thresholds, threshold spectra and 0–1 laws

Erdős and Rényi (1960) were the first to discover that when a graph evolves acquiring edges in a random manner, an abrupt change in its typical structure occurs as the order of magnitude of the number of edges allocated hits some critical values. In the specific case of $G(n, p)$, this means that many properties are associated with critical edge probabilities, called thresholds, around which the probability that $G(n, p)$ possess a given property jumps from 0 to 1 in the limit.

Let $\mathscr{A} \subset \mathscr{G}$ be an increasing property of graphs on n labeled vertices. We say that $p_{\mathscr{A}} = p_{\mathscr{A}}(n)$ is the *threshold* for $\mathscr{A}$ if $\Pr(G(n, p) \in \mathscr{A}) \to 0$ when $p \ll p_{\mathscr{A}}$ ($p = o(p_{\mathscr{A}})$), while $\Pr(G(n, p) \in \mathscr{A}) \to 1$ if $p \gg p_{\mathscr{A}}$ ($p_{\mathscr{A}} = o(p)$). Bollobás and Thomason (1986) proved, in a general setting, the following existential statement.

Theorem 5.1. *Every non-trivial increasing property* $\mathscr{A}$ *has a threshold in* $G(n, p)$.

First, consider the following basic increasing property: $G(n, p)$ has a subgraph isomorphic to a given graph G. Erdős and Rényi (1960) established the threshold for this property under the condition that G must be balanced. This simply means that the density of G: $d(G) = e(G)/|G|$, where $|G|$ denotes the number of vertices of G, is equal to its maximum subgraph density $m(G) = \max\{d(H)\colon H \subseteq G\}$ (i.e., $d(H) \leqslant d(G)$ for every subgraph H of G). It is not hard to check that the class of balanced graphs includes trees, cycles as well as unicyclic and complete graphs.

The threshold for balanced graphs played a crucial role in the explanation of the nature of the evolution first given by Erdős and Rényi. Twenty years later, Bollobás (1981a) extended this fundamental result by, removing the condition that G be balanced using a sophisticated method of grading. Here we shall present an elementary proof of this result due to Ruciński and Vince (1986). To the surprise of many, their argument is similar to the one which Erdős and Rényi used to solve the case of balanced graphs.

Theorem 5.2. *The threshold for the property that a rg $G(n, p)$ contains a subgraph isomorphic to an arbitrary graph G is $n^{-1/m(G)}$.*

Proof. For the sake of simplicity, denote "$G(n, p) \supset G$" the property that $G(n, p)$ contains a subgraph isomorphic to G. Moreover, let $X_n(G)$ be the number of copies of G in $G(n, p)$. To prove that $\Pr(G(n, p) \supset G) \to 0$, when $p \ll n^{-1/m(G)}$ choose a subgraph H of G such that $d(H) = m(G)$. Then

$$\begin{aligned}\Pr(G(n, p) \supset G) &\leqslant \Pr(G(n, p) \supset H) \leqslant \Pr(X_n(H) > 0) \\ &\leqslant \mathrm{E}X_n(H) = \binom{n}{|H|} \frac{|H|!}{\mathrm{aut}(H)}\, p^{e(H)} \\ &\asymp n^{|H|} p^{e(H)} = (pn^{1/m(G)})^{e(H)} ,\end{aligned}$$

where $\mathrm{aut}(H)$ is the number of automorphisms of H. Hence $n^{-1/m(G)}$ satisfies the first condition to be the threshold. To show that $\Pr(G(n, p) \supset G) \to 1$ when $p \gg n^{-1/m(G)}$ the first moment argument does not suffice. Now we have to apply the following bound:

$$\Pr(X_n(G) = 0) \leqslant \frac{\mathrm{Var}\, X_n(G)}{\{\mathrm{E}X_n(G)\}^2} .$$

This bound is a simple consequence of the Chebyshev inequality and is widely known in the literature as the *second moment method* (for other applications, see chapter 33).

Denote by $G_1, G_2, \ldots, G_t$, $t = \binom{n}{|G|} |G|!/\mathrm{aut}(G)$, all copies of the graph G on $\{1, 2, \ldots, n\}$. Let $I_n^{(i)}$, $i = 1, 2, \ldots, t$, be the indicator of the event "$G(n, p) \supset G_i$" i.e.,

$$I_n^{(i)} = \begin{cases} 1 & \text{if } G(n, p) \supset G_i , \\ 0 & \text{otherwise} . \end{cases}$$

Then

$$\operatorname{Var} X_n(G) = \sum_{1\le i,\, j\le t}\sum \operatorname{Cov}(I_n^{(k)}, I_n^{(j)}) = \sum\sum \{\mathrm{E}(I_n^{(i)}I_n^{(j)}) - (\mathrm{E}I_n^{(i)})(\mathrm{E}I_n^{(j)})\}$$
$$= \sum\sum \{\Pr(I_n^{(i)} = 1, I_n^{(j)} = 1) - p^{2e(G)}\}\ .$$

Observe that the random variables $I_n^{(i)}$ and $I_n^{(j)}$ are independent if and only if G_i and G_j are edge-disjoint. Consequently, in those cases the covariance vanishes and therefore

$$\operatorname{Var} X_n(G) \asymp \sum_{H\subset G, e(H)>0} n^{2|G|-|H|} p^{2e(G)-e(H)}\ .$$

On the other hand $\mathrm{E}X_n(G) \asymp n^{|G|}p^{e(G)}$, thus

$$\Pr(G(n, p) \not\supset G) = \Pr(X_n(G) = 0) = \mathrm{O}\Big(\sum_{H\subseteq G, e(H)>0} n^{-|H|}p^{-e(H)}\Big)\ .$$

Hence, $\Pr(G(n, p) \not\supset G) \to 0$ as $pn^{1/m(G)} \to \infty$ since $d(H) \le m(G)$, so $n^{-1/m(G)}$ is the threshold. □

The above theorem shows why trees and complete graphs appear gradually in an evolving graph. Remember that trees and complete graphs are balanced, hence the threshold for a k vertex tree is $n^{-k/(k-1)}$ while for a complete graph of order k is $n^{-2/(k-1)}$. The behavior of trees and complete graphs contrasts with that of unicyclic graphs whose thresholds, regardless of their sizes, are always equal to n^{-1}.

Theorem 5.2 discusses only the asymptotic behavior of $\Pr(G(n, p) \supset G)$. However, in many random graph problems it is important to have a good estimate for this probability. Janson et al. (1990) were able to show that for every graph G with at least one edge there are constants $c_1, c_2 > 0$ (depending on G only), such that for all p and n

$$\exp(-c_1 M) \le \Pr(G(n, p) \supset G) \le \exp(-c_2 M)\ ,$$

where $M = M(n, p, G) = \min\{n^{|H|}p^{e(H)}\colon H \subseteq G,\ e(H) > 0\}$.

From the nature of the thresholds, it follows that if $\mathscr{A}$ is increasing then, in the limit, $\Pr(G(n, p) \in \mathscr{A})$ jumps from 0 to 1 and stays equal to 1 up to the end of the evolutionary process. Obviously, when a property is not monotone, its behavior in $G(n, p)$ can be much more complicated. For example, let $\mathscr{A}$ be the property that the largest component has exactly k vertices. Our current knowledge of rg evolution suggests that we should look then at a very early period of the evolution, namely at its "forest epoch." Indeed, $\Pr(G(n, p) \in \mathscr{A})$ remains equal to 0 until we reach the threshold $n^{-k/(k-1)}$ for a k-vertex tree. Next it jumps to 1 and remains unchanged as long as larger trees do not appear. This a.s. happens when edge probability $p \gg n^{-(k+1)/k}$ and therefore, in the limit, $\Pr(G(n, p) \in \mathscr{A})$ jumps again, now from 1 to 0. In general, one can easily construct properties with finitely many such jumps but, in order to study when they happen, it is necessary to extend the notion of the threshold.

Call $a > 0$ a point of continuity if there is an $\varepsilon > 0$ and $\delta \in \{0, 1\}$ so that if $n^{-a-\varepsilon} < p < n^{-a+\varepsilon}$ then $\Pr(G(n, p) \in \mathcal{A}) \to \delta$ as $n \to \infty$. The *threshold spectrum* $\mathrm{Spec}(\mathcal{A})$ is the complement (in $[0, \infty)$) of all points of continuity. The elements of $\mathrm{Spec}(\mathcal{A})$ are called points of evolutionary discontinuity. So for the property $\mathcal{A}$ that $G(n, p)$ contains a graph G as a subgraph we have $\mathrm{Spec}(\mathcal{A}) = \{1/m(G)\}$, while $\mathrm{Spec}(\mathcal{A}) = \{k/(k-1), (k+1)/k\}$ when $\mathcal{A}$ states that the largest component has exactly k vertices.

Theorem 5.1 says that any increasing graph property has a threshold spectrum consisting of a single point. Shelah and Spencer (1988) gave a characterization of $\mathrm{Spec}(\mathcal{A})$ for properties $\mathcal{A}$ expressible in the first order theory of graphs (FO) which means that the fact "$G \in \mathcal{A}$" can be stated in the language consisting of Boolean connectivities, two binary predicates: equality and adjacency, universal quantifiers and variables identified with vertices of G (for example, the property that G has no isolated vertices is expressible in FO since it can be written as $\forall_v \exists_u (v \leftrightarrow u)$, where $\leftrightarrow$ denotes adjacency).

Theorem 5.3. *If $\mathcal{A}$ is expressible in FO then* $\mathrm{Spec}(\mathcal{A})$ *consists only of rational numbers and is a well-ordered set under $>$ and has order type less than ω^{ω}.*

The first statement of Theorem 5.3 follows from another result of Shelah and Spencer, which is itself concerned with 0–1 laws, another "hotspot" in current research in random graphs (as well as in logic and combinatorics; see Compton 1989). We say that edge probability $p = p(n)$ satisfies the 0–1 *law* if for all graph properties $\mathcal{A}$ expressible in FO, $\Pr(G(n, p) \in \mathcal{A})$ tends to 0 or 1 as $n \to \infty$.

Theorem 5.4. *If a is an irrational number between* 0 *and* 1 *then $p = n^{-a}$ satisfies the* 0–1 *law.*

The reader may expect that, in the limit, $\Pr(G((n, p) \in \mathcal{A})$ is bounded away from 0 and 1 if the edge occurrence probability $p \sim cp_{\mathcal{A}}$, $-\infty < c < \infty$, where $p_{\mathcal{A}}$ is the threshold for $\mathcal{A}$. Indeed, as we shall show in subsequent sections, this observation does hold and therefore the 0–1 law can be satisfied as long as p falls "between the cracks" of the spectrum of threshold functions. Shelah and Spencer (1988) showed the following.

Theorem 5.5 *If any of the following conditions holds then p satisfies the* 0–1 *law.*

(i) $p \ll n^{-2}$,
(ii) $n^{-k/(k-1)} \ll p \ll n^{-(k+1)/k}$ *for some positive integer k,*
(iii) $n^{-1-\varepsilon} \ll p \ll n^{-1}$ *for all $\varepsilon > 0$,*
(iv) $n^{-1} \ll p \ll n^{-1} \log n$,
(v) $n^{-1} \log n \ll p \ll n^{-1+\varepsilon}$ *for all $\varepsilon > 0$,*
(vi) $n^{-\varepsilon} \ll p$ *for all $\varepsilon > 0$.*

The statement (vi) is due to Glebskii et al. and Fagin and its elegant proof, combining graphs, logic and probability, is presented in Spencer's lectures on the probabilistic method (Spencer 1987, p. 23).

The results established for threshold spectra and 0–1 laws are restricted to properties which are expressible in the first order language. The class is rather narrow and does not contain such important properties as, for example, connectedness, hamiltonicity and planarity. In view of Theorem 4.2, the planarity of $G(n, p)$ enjoys a much sharper threshold than n^{-1}, which is the threshold satisfying the definition from the beginning of this section.

The same is true in the case of two other properties mentioned above. Indeed, Erdős and Rényi (1959) proved the following.

Theorem 5.6. *If* $np - \log n \to c$, $-\infty < c < \infty$, *then*

$$\Pr(G(n, p) \text{ is connected}) \to \exp(-e^{-c}) \quad \text{as } n \to \infty .$$

The threshold for a Hamiltonian cycle, a long standing open problem of Erdős and Rényi, was found by Bollobás (1983) and Komlós and Szemerédi (1983).

Theorem 5.7. *If* $np - \log n - \log\log n \to c$, $-\infty < c < \infty$, *then*

$$\Pr(G(n, p) \text{ is Hamiltonian}) \to \exp(-e^{-c}) .$$

Hence, in both instances, a critical range in which the probability jumps from 0 to 1 is very narrow and reduces to the case when $c = c(n)$ and $c(n)$ tends to $-\infty$ or $+\infty$, respectively. Obviously, the usual thresholds for those properties are identical and equal $n^{-1}\log n$. However, notice that the threshold spectrum is even less informative since it extracts from the threshold its main factor (n^{-a}) and ignores the secondary ones. Therefore, connectedness and Hamiltonicity (but also planarity) all have the spectrum $\{1\}$.

6. Distributions

Another important branch of the theory of random graphs with a pronounced probabilistic flavor is concerned with limit distributions of numerical characteristics associated with graphs. In previous sections we reduced the analysis of the evolution to the problem of whether a given property is typical or not. However, in many cases we are able to give a more detailed picture. For example, suppose we know that $G(n, p)$ contains a copy of a graph G with a probability bounded away from 0 as $n \to \infty$. Then, one is led naturally to the following question: how many copies of G can be found in large $G(n, p)$? Clearly, this number is a non-negative integer valued random variable (rv), therefore the answer involves its asymptotic probability distribution. Most often, we encounter two types of asymptotic behavior of random variables (rvs) associated with random graphs: Poisson and normal, especially when those rvs are themselves sums of indicators (Bernoulli rvs). The problem is that, as a rule, those indicators are not independent. Hence, generally speaking, one is forced then to prove results which have the form of the Poisson and central limit theorems for sums of random variables with a given, quite often weak, dependence structure. In random graphs, we deal with various modes of convergence of a sequence of rvs $\{X_n\}$ to a rv X. Usually, convergence in distribution, denoted as $X_n \xrightarrow{\mathcal{D}} X$, is studied and if $X_n \xrightarrow{\mathcal{D}} X$ we often say instead that X_n has asymptotically the same distribution as X. For convenience, $X_n \xrightarrow{\mathcal{D}} 0$ means here that $\{X_n\}$ converges to a rv degenerate at 0. Moreover, $\mathrm{Po}(\lambda)$ and $\mathrm{N}(0, 1)$ represent random variables having the Poisson distribution with the expectation λ and the standard normal distribution, respectively. Finally, if X is a rv, then $\tilde{X} = (X - \mathrm{E}X)/(\mathrm{Var}\, X)^{1/2}$.

We begin our discussion with a fairly complete answer to the following question: What is the asymptotic distribution of the number $X_n(G)$ of subgraphs of a rg $G(n, p)$ isomorphic to a given graph G (the number of copies of G).

Theorem 6.1. *Let G be a graph with the density $d(G)$ and the maximum subgraph density $m(G)$. Then*

(i) $X_n(G) \xrightarrow{\mathcal{D}} 0$ *if* $p \ll n^{-1/m(G)}$

(ii) $X_n(G) \xrightarrow{\mathcal{D}} \mathrm{Po}(\lambda)$, $\lambda = c^{e(G)}/\mathrm{aut}(G)$ *if* $p \sim cn^{-1/d(G)}$, $c > 0$ *and G is strictly balanced, i.e., such that $d(H) < d(G)$ for every proper subgraph H of G.*

(iii) $\tilde{X}_n(G) \xrightarrow{\mathcal{D}} \mathrm{N}(0, 1)$ *if* $p \gg n^{-1/m(G)}$ *and* $1 - p \gg n^{-2}$.

Sketch of proof of (ii). The assertion (ii) was proved independently by Bollobás (1981a) and Karoński and Ruciński (1983). The most useful tool in proving such statements is the following fact, called the *method of moments*, which says that if the probability distribution of a rv X is uniquely determined by its moments and for every $r = 1, 2, \ldots, \mathrm{E}X_n^r \to \mathrm{E}X^r$ then $X_n \xrightarrow{\mathcal{D}} X$. To prove (ii) it is more convenient to consider a variant of the method of moments and to investigate convergence of the factorial moments of $X_n(G)$ (see Bollobás 1985, p. 23). It is not hard to notice that the rth factorial moment of $X_n(G)$, $\mathrm{E}_r = \mathrm{E}_r X_n(G) = \mathrm{E}[X_n(G)(X_n(G) - 1) \cdots (X_n(G) - r + 1)]$, $r = 1, 2, \ldots$ is equal to the expected number of ordered r-tuples $(G_1, G_2, \ldots, G_r)$ of graphs isomorphic to G and contained in $G(n, p)$ as subgraphs. Having in mind the underlying dependence structure (see the proof of Theorem 5.2), the usual trick is to split E_r into two separate parts E_r' and E_r'', where E_r' is defined over those r-tuples which consist of vertex-disjoint copies of G while E_r'' takes care of the other cases. Next, one shows that, under the assumptions of (ii), E_r'' is negligible. Indeed

$$\mathrm{E}_r'' \leqslant \sum_{|G| \leqslant k \leqslant r|G|-1} \binom{n}{k} \left\{ \binom{k}{|G|} \frac{|G|!}{\mathrm{aut}(G)} \right\}^r p^{e_r},$$

where e_r is a lower bound for the number of edges in $\bigcup_{i=1}^r G_i$ while $k = |\bigcup_{i=1}^r G_i|$. By a purely combinatorial argument one can show that when G is strictly balanced then $e_r \geqslant kd(G) + 1/|G|$. Hence

$$\mathrm{E}_r'' = \mathrm{O}((np^{d(G)})^k p^{1/|G|}) = \mathrm{O}(p^{1/|G|}) = \mathrm{o}(1).$$

To complete the proof, note that

$$E_r' \asymp \{n^{|G|} p^{e(G)}/\mathrm{aut}(G)\}^r ,$$

which implies that $E_r X_n(G) = (1 + o(1))\lambda^r$, for $p \sim cn^{-1/d(G)}$ and $r = 1, 2, \ldots$. It means that the factorial moments of $X_n(G)$ converge to the factorial moments of the Poisson rv with the expectation λ, which completes the proof of (ii). □

The statement (i) of Theorem 6.1 follows trivially from Theorem 5.2 while the central limit theorem given in (iii) was proved by Ruciński (1988) also by the method of moments (though his proof involves more subtle combinatorial arguments). It is not hard to verify that (ii) holds if and only if when $p \sim cn^{-1/d(G)}$ and G is strictly balanced (examples of strictly balanced graphs are trees, cycles and complete graphs). If we assume only that G is balanced, but not strictly balanced (for example, G is connected unicyclic graph different than a cycle), then it is quite hard to establish the asymptotic distribution of $X_n(G)$ and no "closed form" solution can be expected. Indeed even in the simplest case when G is balanced and has exactly one subgraph $H \neq G$ such that $d(H) = m(G)$ (for instance G is a cycle with a pendant vertex attached to it), the limit distribution of $X_n(G)$ when $p \sim cn^{-1/d(G)}$, is the same as the distribution of a rv $\Sigma_{i=1}^{Z} Y_i$, where $Z, Y_1, Y_2, \ldots,$ are independent Poisson rvs with $EZ = c^{|H|}/\mathrm{aut}(H)$ and $EY_i = c^{|G|-|H|}\mathrm{aut}(H)/\mathrm{aut}(G)$ for $i = 1, 2, \ldots$ (see Janson 1987).

In 1970, Stein introduced a powerful new technique for obtaining estimates of the rate of convergence to the standard normal distribution. His approach was subsequently extended to cover convergence to the Poisson distribution by Chen in 1975. Finally, Barbour (1982) ingeniously adapted both methods to random graphs. The Stein–Chen approach has several advantages over the more popular method of moments. The principal advantage is that a rate of convergence is automatically obtained. Also the computations are often easier and fewer moment assumptions are required. Moreover, it frequently leads to conditions for convergence weaker than those obtainable by the method of moments.

Let us consider the following mode of convergence of a sequence of non-negative integer-valued rvs $\{X_n\}$ to the Poisson distribution. We say that $\{X_n\}$ is *Poisson convergent* if the total variation distance between the distribution $\mathcal{P}(X_n)$ of X_n and the Poisson distribution with the expectation $\lambda_n = EX_n$, tends to 0 as $n \to \infty$. So, we ask when

$$d(\mathcal{P}(X_n), \mathrm{Po}(\lambda_n)) = \sup_{A \subset \mathbb{Z}^+} |\Pr(X_n \in A) - \sum_{i \in A} \frac{\lambda_n^i}{i!} e^{-\lambda_n}| \to 0$$

as $n \to \infty$, $\mathbb{Z}^+ = \{0, 1, \ldots\}$.

It is important to notice that if $\{X_n\}$ is Poisson convergent and $\lambda_n \to 0$, $\lambda_n \to \lambda$ or $\lambda_n \to \infty$, then $X_n \xrightarrow{\mathcal{D}} 0$, $X_n \xrightarrow{\mathcal{D}} \mathrm{Po}(\lambda)$ and $\tilde{X}_n \xrightarrow{\mathcal{D}} N(0, 1)$, respectively.

The original Stein method investigates the convergence to the standard normal distribution in the following metric:

$$d_1(\mathscr{P}(\tilde{X}_n), \mathrm{N}(0,1)) = \sup_h \|h\|^{-1} \left| \int h(x)\, \mathrm{d}F_n(x) - \int h(x)\, \mathrm{d}\Phi(x) \right| ,$$

the supremum is taken over all bounded test functions h with bounded derivative, $\|h\| = \sup|h(x)| + \sup|h'(x)|$, F_n is the distribution function of X_n, while Φ denotes the distribution function of $\mathrm{N}(0,1)$. One can also show that if $d_1(\mathscr{P}(X_n), \mathrm{N}(0,1)) \to 0$ as $n \to \infty$ then $\tilde{X}_n \xrightarrow{\mathscr{D}} \mathrm{N}(0,1)$.

The basic feature and advantage of Stein–Chen approach is that it gives computationally tractable upper bounds for distances d and d_1 even when the random variables in question are sums of indicators with a fairly general dependence structure. For example, if $Z_{n,k} = Z_n(T_k)$ is the number of k-vertex tree components then $Z_n(T_k) = \Sigma_{\alpha \in [n]^k} I_\alpha$, where $I_\alpha = 1$ if the k-vertex subset α of the vertex set of $G(n,p)$ spans a component being a tree and $I_\alpha = 0$ otherwise. Note that the dependence of those indicators is no longer "local" as in the case of $X_n(G)$ (see Theorems 5.2 and 6.1) since now *all* I_α are pairwise dependent. The next theorem combines results of Barbour (1982) and Barbour et al. (1989) and gives a better insight into the character of the approximation obtained via the Stein–Chen approach.

Theorem 6.2. *If* $k \in \mathbb{N}$, $k \geqslant 2$, *then*
 (i) $d(\mathscr{P}(Z_{n,k}), \mathrm{Po}(\mathrm{E}Z_{n,k})) = \mathrm{O}(n^{-1}\mathrm{E}Z_{n,k})$
 (ii) $d_1(\mathscr{P}(\tilde{Z}_{n,k}), \mathrm{N}(0,1)) = \mathrm{O}((\mathrm{E}Z_{n,k})^{-1/2})$.

Notice that for a fixed k, $\mathrm{E}Z_{n,k} \sim n^k p^{k-1} \mathrm{e}^{-knp}$, the theorem above implies that $Z_{n,k} \xrightarrow{\mathscr{D}} 0$ when $p \ll n^{-k/(k-1)}$ or $knp - \log n - (k-1)\log\log n \to \infty$. $Z_{n,k}$ enjoys the Poisson distribution at both "ends" of the "lifetime interval" and $Z_{n,k}$ has the standard normal distribution "in between". Knowing the fate of tree-components during the evolution from previous sections, one might expect such asymptotic behavior of $Z_{n,k}$.

Another example of an important numerical characteristic of $G(n,p)$ which acts in a similar way as $Z_{n,k}$ is the number $Y_{n,d}$ of vertices of degree d, $d \in \mathbb{N}$. Note that neither the property that a graph has a tree component nor the property that it contains a vertex of fixed degree is monotone. Although a stronger result, analogous to Theorem 6.2, can be proved for $Y_{n,d}$ also (see Barbour et al. 1989), we shall describe its limit distribution in a weaker but more transparent form.

Theorem 6.3. *If* $d \in \mathbb{N}$, *and* $n \to \infty$, *then*
 (i) $Y_{n,d} \xrightarrow{\mathscr{D}} 0$ *if* $p \ll n^{-(d+1)/d}$,
 (ii) $Y_{n,d} \xrightarrow{\mathscr{D}} \mathrm{Po}(d^d/d!)$ *if* $p \sim cn^{-(d+1)/d}$, $c > 0$,
 (iii) $\tilde{Y}_{n,d} \xrightarrow{\mathscr{D}} \mathrm{N}(0,1)$ *if* $p \gg n^{-(d+1)/d}$, *and* $np - \log n - d\log\log n \to -\infty$,
 (iv) $Y_{n,d} \xrightarrow{\mathscr{D}} \mathrm{Po}(e^{-c}/d!)$ *if* $np - \log n - d\log\log n \to c$, $-\infty < c < \infty$,
 (v) $Y_{n,d} \xrightarrow{\mathscr{D}} 0$ *if* $np - \log n - d\log\log n \to \infty$.

The reader may notice that we excluded the case $k = 1$ in Theorem 6.2 and the case $d = 0$ in Theorem 6.3. Obviously, $Z_{n,1} = Y_{n,0}$ and both rvs count simply the number of isolated vertices $G(n, p)$. Denote this number by S_n and observe that the reason why we have to treat this case separately is that the property that a graph has an isolated vertex is decreasing. Therefore, as we shall see, S_n behaves in an opposite manner to $X_n(G)$, the rv related to an increasing property (see Theorem 6.1).

Theorem 6.4. *If* $n \to \infty$, *then*
(i) $\tilde{S}_n \xrightarrow{\mathcal{D}} \mathrm{N}(0, 1)$ *if* $p \gg n^{-2}$ *and* $np - \log n \to -\infty$.
(ii) $S_n \xrightarrow{\mathcal{D}} \mathrm{Po}(\mathrm{e}^{-c})$ *if* $np - \log n \to c, c > 0$.
(iii) $S_n \xrightarrow{\mathcal{D}} 0$ *if* $np - \log n \to \infty$.

The Poisson and Normal are the most common limit distributions in random graphs since the majority of numerical graph characteristics can be represented as sums of weakly dependent indicators, each with a small probability of being non-zero. However, especially when $G(n, p)$ is in the phase transition period, other types of limit distributions occur. For example, in the case when $\varepsilon^3 n \to \infty$, Kolchin (1986) showed that the number U_n of unicyclic components and the number W_n of vertices in such components in the subcritical phase have the following limit distributions.

Theorem 6.5. *If* $p = (1 - \varepsilon)/n$, $\varepsilon = \varepsilon(n)$ *and* $\varepsilon^2 n \to \infty$ *as* $n \to \infty$ *then,*
(i) $(U_n + \frac{1}{2}\log \varepsilon)/(-\frac{1}{2}\log \varepsilon)^{1/2} \xrightarrow{\mathcal{D}} \mathrm{N}(0, 1)$,
(ii) $\varepsilon^2 W_n/2 \xrightarrow{\mathcal{D}} \Gamma(\frac{1}{4})$, *where* $\Gamma(\beta)$ *denotes the standard gamma distribution with the expectation* β.

We should mention, that inside the phase transition interval, i.e., when $p = (1 \pm \varepsilon)/n$, $\varepsilon = \varepsilon(n)$ and $\varepsilon^3 n = \mathrm{O}(1)$, the limiting results are extremely hard to study (see Stepanov 1987).

Let us conclude this section with another example of the limit theorem, which provides the answer to the following question of Paul Erdős: "What is the length L_n of the first cycle in an evolving random graph?" The distribution of L_n was found by Janson (1987) while the expectation of L_n was determined by Flajolet et al. (1989).

Theorem 6.6. (i) $L_n \xrightarrow{\mathcal{D}} L$ *as* $n \to \infty$, *where* L *is a random variable with the distribution*

$$\Pr(L = l) = \tfrac{1}{2} \int_0^1 t^{l-1}(1 - t)^{1/2}\, \mathrm{e}^{t/2 + t^2/4}\, \mathrm{d}t\,, \quad \textit{for } l = 3, 4, \ldots,$$

(ii) $\mathrm{E}L_n = Cn^{1/6} + \mathrm{O}(n^{1/8})$ *where* $C \approx 2.033$.

7. Extreme characteristics

Another interesting class of rvs defined on a rg $G(n, p)$ consists of extreme graph characteristics such as the minimum and maximum vertex degree, the length of the shortest or longest path between two specified vertices, the diameter and girth, the length of the longest cycle, the size of the smallest or largest subset of the vertex set having a given property, etc. From the point of view of probability theory, all the random variables listed above are examples of extreme order statistics.

Let us consider the minimum and maximum vertex degree of $G(n, p)$ denoted by δ and Δ, respectively, and let X_i, $i = 1, 2, \ldots, n$, be the degree of the ith vertex of $G(n, p)$. Clearly, the X_i' are identically distributed (but not independent) rvs each having the binomial distribution with parameters $n-1$ and p. Moreover, if $X_{[1]}, X_{[2]}, \ldots, X_{[n]}$ stand for the degrees arranged in the ascending order, then $\delta = X_{[1]}$ and $\Delta = X_{[n]}$ are the two extreme order statistics of the degree sequence. For a detailed study of the asymptotic behavior of those rvs, the reader is referred to chapter 3 of the book by Bollobás (1985).

Let us concentrate on the limit distribution of the rv δ. Theorem 6.4(i) implies that if the edge probability p is such that $pn - \log n \to -\infty$ then the minimum vertex degree of $G(n, p)$ is a.s. equal to 0. If we raise p slightly, then the following result of Erdős and Rényi (1961) holds.

Theorem 7.1. *Let* $d \geqslant 0$, $c_d = c_d(n) = np - \log n - d \log\log n$ *and* $n \to \infty$.

(i) $\Pr(\delta = d) \to 1 - \exp(-e^{-c}/d!)$ *and* $\Pr(\delta = d+1) \to \exp(-e^{-c}/d!)$ *if* $c_d \to c$, *where* $-\infty < c < \infty$.

(ii) $\Pr(\delta = d+1) \to 1$ *if* $c_d \to \infty$ *but* $c_{d+1} \to -\infty$.

Ivchenko (1973) proved that the above theorem is true in a more general setting, namely δ can be replaced by $X_{[i]}$, $i \geqslant 1$, and $\exp(-e^{-c}/d!) = \lambda$ by $\sum_{k=0}^{i-1} (\lambda^i/i!)\, e^{-\lambda}$. When $p \ll \log n/n$ then $G(n, p)$ has many vertices of minimum and maximum degree. However, Bollobás (1982) showed that if $p \gg \log n/n$ (but $1-p \ll \log n/n$) then it a.s. has a unique vertex of minimum degree and a unique vertex of maximum degree. In fact even more is true. One can show (see Palka 1987) that for a fixed i both sequences $X_{[1]}, \ldots, X_{[i]}$ and $X_{[n-i]}, \ldots, X_{[n]}$ are a.s. strictly increasing in $G(n, p)$.

Recall now the well-known theorem of Whitney which states that for any graph G, $\kappa(G) \leqslant \lambda(G) \leqslant \delta(G)$, where $\kappa(G)$ and $\lambda(G)$ denote the vertex and edge connectivity of G, respectively. Erdős and Rényi (1961) asked about the value of $\kappa(G)$ and $\lambda(G)$ when G is random. They showed that, at least for the edge probability p specified in the assumptions of Theorem 7.1, a minimal vertex cut in $G(n, p)$ is a.s. composed of the neighbors of any vertex of minimum degree. Bollobás and Thomason (1985) were able to prove that this holds during the whole evolution. Hence we have the following random graph version of Whitney's theorem.

Theorem 7.2. *Let $p = p(n)$, $0 < p < 1$. Then in $G(n, p)$*

$$\Pr(\kappa = \lambda = \delta) \to 1 \quad \text{as } n \to \infty .$$

The above result indicates a remarkable fact that in a rg trivial necessary conditions expressible in terms of the minimum vertex degree δ are often a.s. sufficient. Indeed, the evolving random graph becomes a.s. k-connected as soon as the last vertex of degree $k-1$ vanishes. Moreover, one can show that "$\delta \geq 1$" and "$\delta \geq 2$" are a.s. sufficient conditions for a perfect matching and Hamiltonian cycle, respectively (compare Theorems 5.6 and 5.7 with Theorems 6.3 and 6.4). In fact, the following very precise and general result, due to Bollobás and Thomason (1985) and Bollobás and Frieze (1985) hold.

Theorem 7.3. *Let $k \in \mathbb{N}$ and $\tau(\mathcal{A})$ denote the hitting time for an increasing property $\mathcal{A}$. Then almost every random graph process $(G_t)_0^{\binom{n}{2}}$ is such that $\tau((\delta \geq k) = \tau(\kappa \geq k) = \tau(M_k)$, where M_k is the property that a graph contains $\lfloor k/2 \rfloor$ disjoint Hamiltonian cycles plus a disjoint perfect matching when k is odd.*

The proof of the second equality in Theorem 7.3 is based on a sophisticated "coloring method " of Fenner and Frieze (1983). The introduction of this tool to random graphs was a turning point in the study of matchings and Hamiltonian cycles. For a closer look at how the method works, the reader is encouraged to consult the paper by Frieze (1989).

The diameter of a rg is another example of an extreme order statistic since, $\operatorname{diam}(G(n, p)) = Y_{[\binom{n}{2}]}$ where Y_i, $i = 1, 2, \ldots, \binom{n}{2}$ denotes the length of a shortest path between the ith pair of vertices in $G(n, p)$. Notice that the problem of the limit distribution of the diameter is more complicated as the Y_i themselves are extreme order statistics. Nevertheless, Burtin (1975) was able to establish the asymptotic behavior of this characteristic, but only for relatively large edge probabilities p.

Theorem 7.4. *Let $d \in N$, $c_d = p^d n^{d-1} - 2 \log n$ and $n \to \infty$. Then*

(i) $\Pr(\operatorname{diam}(G(n, p)) = d) \to \exp(-e^{-c}/2)$, $\Pr(\operatorname{diam}(G(n, p) = d + 1) \to 1 - \exp(-e^{-c}/2)$ *if $d \geq 2$ and $c_d \to c$, $-\infty < c < \infty$,*

(ii) $\Pr(\operatorname{diam} G((n, p) = d) \to 1$ *if $d \geq 3$ and $c_d \to \infty$ but $c_{d-1} \to -\infty$, while* $\Pr(\operatorname{diam}(G(n, p) = 2) \to 1$ *if $c_2 \to \infty$ but $n^2(1-p) \to \infty$.*

The results of Bollobás (1981b) indicate that the diameter of $G(n, p)$ is one of at most two values when p is just about large enough to guarantee that $G(n, p)$ is a.s. connected. Furthermore, he shows that when $np = c$, and c is sufficiently large, then the diameter of the giant component is $o(\log n)$.

A related problem of the length $L = L(G(n, p))$ of the longest path in $G(n, p)$, for a similar range of p, was considered by Ajtái et al. (1981).

Theorem 7.5. *If $np = c$, then there is a function $f: (1, \infty) \to (0, 1)$, $f(c) \to 1$ as $c \to \infty$, such that* $\Pr(n^{-1}L \geq f(c)) \to 1$ *as $n \to \infty$.*

Frieze (1986), solving the more general problem of the length of the longest cycle in $G(n, c/n)$, gave essentially the best possible upper bound on $f(c)$ for large c. A natural generalization of these questions is: what is the size of the largest induced subgraph of a given type (path, tree, cycle, etc.)?

We conclude this section with two important results concerning the size $\alpha = \alpha(G(n, p))$ of the largest independent set in $G(n, p)$. Obviously, α is yet another example of the extreme order statistics since $\alpha = U_{[2^n]}$, where for $i = 1, 2, \ldots, 2^n$, $U_i = |V_i|$ if V_i is an independent set in $G(n, p)$ and $U_i = 0$ otherwise. The reader may expect that, as in all previous cases, the limit distribution of α should be concentrated in a small range of possible values. Indeed, Frieze (1990) gave an astonishingly precise estimate for the independence number α, when the edge probability $p = p(n) \to 0$ as $n \to \infty$.

Theorem 7.6. *Let $np = c$ and $\varepsilon > 0$ be fixed. Suppose $c_\varepsilon \leq c = o(n)$ for sufficiently large fixed constant c_ε. Then*

$$\Pr\left(\left|\alpha(G(n, p)) - \frac{2n}{c}(\log c - \log\log c - \log 2 + 1)\right| \leq \frac{\varepsilon n}{c}\right) \to 1$$

as $n \to \infty$.

Systematic studies of the independence number of a rg were originated by Matula (1972) who proved that when p is fixed, the limit distribution of α is concentrated on at most two possible values.

Theorem 7.7. *For any constant p, $0 < p < 1$ and $\varepsilon > 0$,*

$$\Pr(\lfloor z((n, p) - \varepsilon \rfloor \leq \alpha(G(n, p)) \leq \lfloor z(n, p) + \varepsilon \rfloor) \to 1 \quad \text{as } n \to \infty,$$

where $z(n, p) = 2\log_b n - 2\log_b \log_b n + 2\log_b(\mathrm{e}/2) + 1$ and $b = 1/(1-p)$.

In the next section we shall find both above results very useful in determining the chromatic number of a rg.

8. Coloring

Over the last decade the chromatic number $\chi(G(n, p))$ of a rg $G(n, p)$ has attracted the interest of many people (cf. chapter 4). We shall present, briefly, the milestones in the history of the struggle to determine $\chi(G(n, p))$, showing how those efforts have contributed to the most important methodological breakthrough in random graphs in recent years. Observe first that the picture before the phase transition is quite clear.

Theorem 8.1. *Let* $n \to \infty$ *and* $\chi = \chi(G(n, p))$. *Then*
(i) $\Pr(\chi = 1) \to 1 - e^{-c/2}$ *and* $\Pr(\chi = 2) \to e^{-c/2}$ *if* $p \sim cn^{-2}$,
(ii) $\Pr(\chi = 2) \to 1$ *if* $p \gg n^{-2}$ *but* $p \ll n^{-1}$
(iii) $\Pr((\chi = 2) \to a = e^{c/2}((1-c)/(1+c))^{1/4}$ *and* $\Pr(\chi = 3) \to 1 - a$ *if* $p \sim cn^{-1}$, $0 < c < 1$.

Indeed, statements (i) and (ii) trivially follow from Theorems 3.1 and 6.1, while the proof of (iii) reduces to the determination of the probability that $G(n, p)$ contains an odd cycle. The question, what is the chromatic number of $G(n, p)$ when $p = cn$, $c \geqslant 1$, was raised by Erdős and Rényi (1960). They noticed that for such edge probabilities $\chi \geqslant 3$, since $G(n, p)$ contains an odd cycle a.s. (now we know that $\chi = 3$ a.s. if $p = 1/n$). For large c an answer is given by the following result of Łuczak (1990c).

Theorem 8.2. *Let* $np = c$ *and* $\varepsilon > 0$ *be fixed. Suppose* $c_\varepsilon \leqslant c = o(n)$ *for sufficiently large constant* c_ε. *Then*

$$\Pr\left(\frac{c}{2\log c} < \chi(G(n, p)) < (1+\varepsilon)\frac{c}{2\log c}\right) \to 1 \quad \text{as } n \to \infty .$$

Notice that a trivial lower bound for the chromatic number of $G(n, p)$ *is* n/α, where $\alpha = \alpha(G(n, p))$ denotes, as before, the size of the largest independent set. Hence, the results of Frieze and Matula concerning α (see Theorems 7.6 and 7.7) yield the respective lower bounds on $\chi(G(n, p))$. The hard part is to establish the upper bound. For a constant p, Grimmett and McDiarmid (1975) were able to prove that for any $\varepsilon > 0$, $\chi(G(n, p)) < (1+\varepsilon)n/\log_b n$ a.s. where $b = 1/(1-p)$. They showed that a greedy algorithm, which assigns colors to vertices of a rg sequentially, in such a way that a vertex gets the earliest available color, a.s. needs, approximately, $n/\log_b n$ colors to produce a proper coloring of $G(n, p)$. Grimmett and McDiarmid conjectured that this algorithm uses twice as many colors as necessary. An important step towards a resolution was made by Matula (1987). He constructed a refined version of the greedy algorithm, called the "expose and merge" algorithm, which uses roughly $(2n/3\log_b n)$ colors. Next Shamir and Spencer (1987) proved that χ is sharply concentrated in an interval of length of order $n^{1/2}$, and in their paper, for the first time, a new powerful technique based on concentration measure for martingales was applied to random graphs. (Let us restrict our attention to discrete rvs and recall that a sequence $\{X_i: i \geqslant 0\}$ is a *martingale* with respect to the sequence $\{Y_i \geqslant 0\}$ if for all $i \geqslant 1$, $E|X_i| < \infty$ and, with probability one, $E(X_i \mid Y_1, \ldots, Y_{i-1}) = X_{i-1}$, where $E(\cdot \mid \cdot)$ denotes the conditional expectation.) In particular, Shamir and Spencer (1987) used the Azuma inequality, which states that if $X_0, X_1, \ldots, X_n$ is a martingale and $|X_i - X_{i-1}| \leqslant 1$ for all i, then for any positive λ, $\Pr(X_0 - X_n \geqslant \lambda) < e^{-\lambda^2/2n}$.

The chromatic number problem, for dense graphs, was solved by Bollobás (1988) who showed how the potential of this large deviation theorem could be realized. We shall present the main ideas of his ingenious and elegant proof.

Theorem 8.3. *Let $0<p<1$ be fixed and $b=1/(1-p)$. Then for every $\varepsilon>0$*

$$\Pr\left(\frac{n}{2\log_b n}<\chi(G(n,p))<(1+\varepsilon)\frac{n}{2\log_b n}\right)\to 1 \quad \textit{as } n\to\infty.$$

Sketch of proof. It seems to be clear that to get the upper bound on χ when p is fixed, one has to employ primarily independent sets of essentially optimal order $(2+o(1))\log_b n$.

Indeed, Bollobás uses independent sets of the cardinality $s_0=\lfloor 2\log_b n - 8\log_b\log_b n+O(1)\rfloor$, which is only a little bit smaller than the maximal order of the largest independent set in $G(n,p)$ (see Theorem 7.7). Next, he sets $n_0=\lfloor n/(\log n)^2\rfloor$ and shows that *every* set of n_0 vertices of $G(n,p)$ contains an independent set of order s_0, with probability tending to one as n tends to infinity. To prove this last statement, one has to show that

$$\binom{n}{n_0}\Pr(X=0)<2^n\Pr(X=0)\to 0 \quad \text{as } n\to\infty,$$

where X is the number of independent sets of order s_0 in $G(n_0,p)$. To meet this requirement, $\Pr(X=0)$ has to be exponentially small, which means that the expectation $\mathrm{E}X$ should be large and the distribution of X sharply concentrated around $\mathrm{E}X$. This is, in fact, the key moment of the proof and we shall need martingales to proceed further. Notice that a random graph $G(n,p)$ can be viewed as a vector of indicators rvs $(I_1,I_2,\ldots,I_{\binom{n}{2}})$, where $\Pr(I_i=1)=p$ and $\Pr(I_i=0)=1-p$, independently of other I_i. Moreover, if an rv $Z=Z(G(n,p))=Z(x_1,\ldots,x_{\binom{n}{2}})$, $x_i\in\{0,1\}$, satisfies

$$|Z(x_1,\ldots,x_{\binom{n}{2}})-Z(x_1',\ldots,x_{\binom{n}{2}}')|\leq 1$$

whenever $(x_1,\ldots,x_{\binom{n}{2}})$ and $(x_1',\ldots,x_{\binom{n}{2}}')$ differ in exactly one coordinate, then the same is true for the martingale

$$X_i=\mathrm{E}(Z\mid Y_i), \qquad Y_i=(I_1,\ldots,I_i), \qquad X_0\equiv \mathrm{E}Z, \qquad X_{\binom{n}{2}}=Z,$$
$$\text{i.e.,} \quad |X_i-X_{i-1}|\leq 1.$$

Unfortunately, X does not satisfy the above condition because when two sequences (graphs) differ on exactly one place (edge) the number of independent sets can differ substantially (in our case by as much as $\binom{n_0-2}{s_0-2}$). Therefore, Bollobás switches to the random variable X^*, the cardinality of a maximal collection of independent sets of order s_0, no two of them sharing more than one vertex in $G(n_0,p)$. Then $\mathrm{E}X^*\geq O(n_0^2/(\log n_0)^4)=n_0^{2-o(1)}$. Furthermore, notice that a single edge can alter the size of such a family by at most one, so Azuma's inequality gives

$$\begin{aligned}\Pr(X=0)&\leq\Pr(X^*=0)\leq\Pr(\mathrm{E}X^*-X^*\geq\mathrm{E}X^*)\\&<\exp(-(\mathrm{E}X^*)^2/(n_0(n_0-1)))\leq\exp(-n_0^{2-o(1)})=o(2^{-n}).\end{aligned}$$

Hence, in every subset of n_0 vertices of $G(n,p)$ there is an independent set of order s_0 a.s. Now, assign the first color to such a set and peel it off from $G(n, p)$. Next choose another independent set of order s_0 in the remainder and again peel it off. Repeat the procedure until at most n_0 vertices remain. Obviously, $\chi(G(n, p)) \leq n/s_0 + n_0$, which gives the required upper bound. $\square$

Let us conclude our discussion of the chromatic number of a random graph with a few comments. Matula and Kučera (1990) gave an alternative proof of Theorem 8.3, using the second moment method and "expose-and-merge" approach. We should remark that Łuczak's proof of Theorem 8.2 is an ingenious blend of the martingale and "expose-and-merge" techniques. The reader may also notice that the chromatic number is yet another extreme order statistic. So, we should be able to show that, at least for some range of p, the limit distribution of χ is concentrated on at most two points. Theorem 8.1 shows that it is true when $p \leq 1/n$. Shamir and Spencer (1987) proved that if $p \leq n^{-((5/6)+\varepsilon)}$, when $\varepsilon > 0$, then the chromatic number takes on a.s. at most five different values. Łuczak (1991) was able to show that for such sparse random graphs, the chromatic number is, as expected, a.s. two-point concentrated.

Now assume that instead of vertices of $G(n, p)$ we color edges. Recall that Vizing's theorem says that the chromatic index $\gamma \in \{\Delta, \Delta + 1\}$. Furthermore, from the proof of this theorem, it follows that if $\gamma = \Delta + 1$ then a graph has two adjacent vertices of maximum degree Δ. Since in a rg $G(n, p)$ such an event is unlikely (remember that for large p a rg $G(n, p)$ has only a unique vertex of degree Δ), therefore we have the following result.

Theorem 8.4. *Let* $p = p(n)$, $0 < p < 1$. *Then*

$$\Pr(\gamma = \Delta) \to 1 \quad \text{as } n \to \infty .$$

Another interesting class of coloring problems is concerned with Ramsey properties of random graphs. There are only some preliminary results in this area, although random graphs are a quite common tool used in proofs of Ramsey theorems for deterministic graphs (see chapter 33). Here we ask about thresholds for the vertex and edge Ramsey properties: $G(n, p) \to (G)_r^v$ and $G(n, p) \to (G)_r^e$, which formally denotes the fact that for every r-coloring of vertices (edges) an rg $G(n, p)$ contains a monochromatic copy of a graph G. The vertex-coloring case is much easier than the edge-coloring, in contrast to the previously considered problem of the chromatic number and the chromatic index of $G(n, p)$. Łuczak et al. (1992) were able to establish a sharp threshold for the vertex Ramsey property for any r and G.

Theorem 8.5. *For every integer* $r \geq 2$ *and every graph* G *with at least one edge* (G *non-trivial when* $r = 2$, *i.e., containing a path on three vertices*), *there exist constants* $c_2 \geq c_1 > 0$ *such that, as* $n \to \infty$,

$$\Pr(G(n, p) \to (G)_r^v) \to \begin{cases} 0 & \text{if } p \leq c_1 n^{-1/m^*(G)}, \\ 1 & \text{if } p \geq c_2 n^{-1/m^*(G)}, \end{cases}$$

where $m^*(G) = \max_{H \subseteq G, |H| \geq 3} e(H)/(|H| - 1)$.

Edge-coloring Ramsey properties of random graphs appeared to be definitely more challenging. In the papers by Frankl and Rödl (1986) and Łuczak et al. (1992) the threshold was established in the restricted case when $G = K_3$ and $r = 2$, i.e., when G is a triangle and one colors the edges of a rg by two colors. Only very recently, using sophisticated probabilistic and graph theoretic tools such as Janson's inequality and Szemerédi's regularity lemma, Rödl and Ruciński (1993, 1995) found the threshold for any graph G.

Theorem 8.6. *For every integer* $r \geq 2$ *and every graph* G *which is not a star forest, there exist constants* $c_2 \geq c_1 > 0$ *such that, as* $n \to \infty$,

$$\Pr(G(n, p) \to (G)_r^e) \to \begin{cases} 0 & \text{if } p \leq c_1 n^{-1/m^{**}(G)}, \\ 1 & \text{if } p \geq c_2 n^{-1/m^{**}(G)}, \end{cases}$$

where $m^{**}(G) = \max_{H \subseteq G, |H| \geq 3} (e(H) - 1)/(|H| - 2)$.

The results discussed above paved the way for further research on similar properties of other random discrete structures, such as random hypergraphs, random subsets of the set of integers, etc.

Appendix A. Quasi-random graphs

As has been noted earlier, there is a large collection of properties which the random graph $G = G(n, 1/2)$ a.s. has. Some of these are as follows.

$P_1(t)$: For a fixed $t \geq 4$, each t-vertex graph $H(t)$ occurs $(1 + o(1))n^t/2^{\binom{t}{2}}$ times as an (ordered) induced subgraph of G.

P_2: G has at least $(1 + o(1))n^2/4$ edges and at most $(1 + o(1))n^4/16$ 4-cycles.

P_3: $\sum_{u,v \in V(G)} \big| |nd(u) \cap nd(v)| - n/4 \big| = o(n^3)$ where $nd(x)$ denotes the neighborhood of vertex x in G, i.e., the set of all y adjacent to x.

P_4: For all $S \subset V(G)$ with $|S| = \lfloor n/2 \rfloor$, S spans $(1 + o(1))n^2/16$ edges in G.

P_5: G has $(1 + o(1))n^2/4$ edges; $\lambda_1(G) = (1 + o(1))n/2$ and $\lambda_2(G) = o(n)$, where $\lambda_1(G), \lambda_2(G), \ldots$ denote the eigenvalues of the adjacency matrix of G with $|\lambda_1(G)| \geq |\lambda_2(G)| \geq \cdots$.

It turns out that these properties are part of a rather large equivalence class of properties, termed *quasi-random* properties in Chung et al. (1989) (see also Thomason 1987a). Any graphs $G = G(n)$ satisfying *any one* of the quasi-random properties must in fact satisfy *all* of the quasi-random properties. This is perhaps somewhat surprising since these properties appear to have rather different strengths, e.g., P_2 looks a lot weaker than $P_1(1000)$. Moreover, since it is often easy to prove that a particular class of graphs satisfies a certain quasi-random property then the graphs in this class must possess all the quasi-random properties and so, behave like random graphs in many ways. For example, since it is trivial to show that the Paley graph $P(p)$ with vertex set $\mathbb{Z}/p$ (where p is a prime $\equiv 1 \pmod 4$) and edges $\{i, j\}$ for $i-j$ a quadratic residue modulo p, satisfies P_3, then in fact it must contain $(1+o(1))p^{10}/2^{45}$ induced copies of the Petersen graph, etc. Similar remarks apply to the *even intersection* graph $I(n)$ with all $X \subset \{1, 2, \ldots, n\}$ as vertices, and all $\{X, Y\}$ with $|X \cap Y| \equiv 0 \pmod 2$ as edges.

In fact, a more quantitative form of the theory is available which is based on the concept of the *deviation* of a graph $G = G(n)$, denoted by $\operatorname{dev} G$. This is defined to be

$$\operatorname{dev} G = \frac{1}{n^4} \sum_{v_1, v_2, v_3, v_4} \mu(v_1, v_2)\mu(v_2, v_3)\mu(v_3, v_4)\mu(v_4, v_1),$$

where for vertices x, y of G,

$$\mu(x, y) = \begin{cases} -1 & \text{if } \{x, y\} \text{ is an edge of } G, \\ 1 & \text{otherwise}. \end{cases}$$

Quasi-randomness is equivalent to $\operatorname{dev} G$ going to 0 (as $n \to \infty$), and so, all quasi-random properties can be expressed in terms of $\operatorname{dev} G$ (see Chung and Graham 1991). For example it can be shown in the case of property P_2 that for any graph G with n vertices and $e(G)$ edges:

(i) $e(G) \geq (n^2/4)(1 - (\operatorname{dev} G)^{1/4})$,

(ii) G has at most $(n^4/16)(1 + (\operatorname{dev} G)^{1/4})^4$ 4-cycles.

The theory of quasi-randomness can be carried over to hypergraphs as well (see Chung and Graham 1990, 1991 for details). It turns out that the following "threshold" result holds for k-uniform hypergraphs (= k-graph). If $G^{(k)}(n)$ is a k-graph which contains all $2k$-vertex k-graphs as (ordered) induced sub-k-graphs asymptotically the same number of times (i.e., $(1+o(1))n^{2k}/2^{\binom{2k}{2}}$ times) then for any fixed t, $G^{(k)}(n)$ will contain any t-vertex k-graph $(1+o(1))n^t/2^{\binom{t}{k}}$ times as an induced sub-k-graph. The value $2k$ is sharp here; the corresponding statement is not true if $2k$ is replaced by $2k-1$.

In fact, this philosophy can be applied to a large class of structures (e.g., integer sequences, (0, 1)-matrices, tournaments, directed graphs, etc.) although many of the details still remain to be worked out. We remark that the threads of quasi-randomness can be traced back to early work of Wilson (1972, 1974) on block designs, Erdős and Sós (1982) in Ramsey–Turán problems for graphs and hypergraphs, Rödl (1986) and Graham and Spencer (1971) (both on certain universality properties of graphs) and more recently Thomason (1987a,b, 1989), Haviland (1989) and Haviland and Thomason (1989), who deal with a related concept called (p, α)-jumbledness.

References

Ajtái, M., J. Komlós and E. Szemerédi

[1981] The longest path in a random graph, *Combinatorica* **1**, 1–12.

Barbour, A.D.

[1982] Poisson convergence and random graphs, *Math. Proc. Cambridge Philos. Soc.* **92**, 349, 359.

Barbour, A.D., M. Karoński and A. Ruciński
[1989] A central limit theorem for decomposable random variables with applications to random graphs, *J. Combin. Theory B* **47**, 125–145.
Bollobás, B.
[1981a] Threshold functions for small subgraphs, *Math. Proc. Cambridge Philos. Soc.* **90**, 197–206.
[1981b] The diameter of random graphs, *Trans. Amer. Math. Soc.* **267**, 41–52.
[1982] Distinguishing vertices of random graphs, *Ann. Discrete Math.* **13**, 33–50.
[1983] Almost all regular graphs are Hamiltonian, *European J. Combin.* **4**, 97–106.
[1985] *Random Graphs* (Academic Press, London).
[1988] The chromatic number of random graphs, *Combinatorica* **8**, 49–55.
Bollobás, B., and A.M. Frieze
[1985] On matchings and Hamiltonian cycles in random graphs, *Ann. Discrete Math.* **28**, 23–46.
Bollobás, B., and A. Thomason
[1985] Random graphs of small order, *Ann. Discrete Math.* **28**, 47–97.
[1986] Threshold functions, *Combinatorica* **7**, 35–38.
Burtin, Yu.D.
[1975] On the extreme metric characteristics of a random graph II. Limit distributions, *Theory Probab. Appl.* **20**, 83–101.
Chung, F.R.K., and R.L. Graham
[1990] Quasi-random hypergraphs, *Random Structures and Algorithms* **1**, 105–124.
[1991] Quasi-random set systems, *J. Amer. Math. Soc.* **4**, 151–196.
Chung, F.R.K., R.L. Graham and R.M. Wilson
[1989] Quasi-random graphs, *Combinatorica* **9**, 345–362.
Compton, K.J.
[1989] 0–1 laws in logic and combinatorics, in: *Algorithms and Order (Ottawa, 1987)* (Kluwer Academic Publishers, Dordrecht) pp. 353–383.
Erdős, P.
[1973] The art of counting, in: *Selected Writings,* ed. J.E. Spencer (MIT Press, Cambridge, MA).
Erdős, P., and A. Rényi
[1959] On random graphs I, *Publ. Math. Debrecen* **6**, 290–297.
[1960] On the evolution of random graphs, *Publ. Math. Inst. Hung. Acad. Sci.* **5**, 17–61.
[1961] On the strength of connectedness of a random graph, *Acta Math. Sci. Hung.* **12**, 261–267.
Erdős, P., and V.T. Sós
[1982] On Ramsey–Turán type theorems for hypergraphs, *Combinatorica* **2**, 289–295.
Fenner, T.I., and A. Frieze
[1983] On the existence of Hamilton cycles in a class of random graphs, *Discrete Math.* **45**, 301–305.
Flajolet, Ph., D.E. Knuth and B. Pittel
[1989] The first cycle in an evolving graph, *Discrete Math.* **75**, 167–215.
Frankl, P., and V. Rödl
[1986] Large triangle-free subgraphs in graphs without K_4, *Graphs and Combinatorics* **2**, 135–144.
Frieze, A.
[1986] On large matchings and cycles in sparse random graphs, *Discrete Math.* **59**, 243–256.
[1989] On matchings and Hamilton cycles in random graphs, in: *Surveys in Combinatorics, Invited Papers at the 12th British Combinatorial Conf.,* ed. J. Siemond (Norwich) pp. 84–114.
[1990] On the independence number of random graphs, *Discrete Math.* **81**, 171–176.
Graham, R.L., and J.E. Spencer
[1971] A constructive solution to a tournament problem, *Canad. Math. Bull.* **14**, 45–58.
Grimmett, G.R.
[1983] Random graphs, in: *Selected Topics in Graph Theory,* Vol. 2, eds. L. Beineke and R. Wilson (Academic Press, London) pp. 201–235.
Grimmett, G.R., and C.J.H. McDiarmid
[1975] On colouring random graphs, *Math. Proc. Cambridge Philos. Soc.* **77**, 313–324.

Haviland, J.
[1989] *Cliques and independent sets,* Ph.D. Thesis (University of Cambridge).

Haviland, J., and A. Thomason
[1989] Pseudo-random hypergraphs, *Discrete Math.* **75**, 255–278.

Ivchenko, G.I.
[1973] On the asymptotic behaviour of the degrees of vertices in a random graph, *Theory Probability Appl.* **18**, 188–195.

Janson, S.
[1987] Poisson convergens and Poisson processes with applications to random graphs, *Stochastic Process. Appl.* **26**, 1–30.

Janson, S., T. Łuczak and A. Ruciński
[1990] An exponential bound for the probability of nonexistence of a specified subgraph in a random graph, in: *Random Graphs, 1987,* eds J. Jaworski, M. Karoński and A. Ruciński (Wiley, Chichester).

Janson, S., D.E. Knuth, T. Łuczak and B. Pittel
[1993] The birth of the giant component, *Random Structures and Algorithms* **4**, 233–358.

Karoński, M.
[1982] A review of random graphs, *J. Graph Theory* **6**, 349–389.

Karoński, M., and A. Ruciński
[1983] On the number of strictly balanced subgraphs of a random graph, in: *Graph Theory, Lagów 1981, Lecture Notes in Mathematics,* Vol. 1018 (Springer, Berlin) pp. 799–830.

Kolchin, V.F.
[1986] On the limit behaviour of a random graph near the critical point, *Theory Probability Appl.* **31**, 439–451.

Komlós, J., and A. Szemerédi
[1983] Limit distribution for existence of Hamilton cycles in a random graph, *Discrete Math.* **43**, 55–63.

Łuczak, T.
[1990a] On the equivalence of two basic models of random graphs, in: *Random Graphs, 1987,* eds. J. Jaworski, M. Karoński and A. Ruciński (Wiley, Chichester) pp. 151–158.
[1990b] Component behavior near the critical point of the random graph process, *Random Structures and Algorithms* **1**, 287–310.
[1990c] On the chromatic number of spare random graphs, *Combinatorica* **10**, 377–385.
[1991] A note on the sharp concentration of the chromatic number of random graph, *Combinatorica* **11**, 295–297.

Łuczak, T., A. Ruciński and B. Voigt
[1992] Ramsey properties of random graphs, *J. Combin. Theory B* **56**, 55–68.

Łuczak, T., B. Pittel and J.C. Wierman
[1994] The structure of a random graph at the point of the phase transition, *Trans. Amer. Math. Soc.* **341**, 721–748.

Matula, D.
[1972] The employee party problem, *Notices Amer. Math. Soc.* **19**, A-382.
[1987] Expose-and-merge exploration and the chromatic number of a random graph, *Combinatorica* **7**, 275–284.

Matula, D., and L. Kučera
[1990] An expose-and-merge algorithm and the chromatic number of random graph, in: *Random Graphs, 1987,* eds. J. Jaworski, M. Karoński and A. Ruciński (Wiley, Chichester) pp. 175–188.

Palka, Z.
[1987] On the degree sequence in a random graph, *J. Graph Theory* **11**, 121–134.

Palmer, E.M.
[1985] *Graphical Evolution: An Introduction to the Theory of Random Graphs* (Wiley, New York).

Rényi, A.
[1976] *Selected Papers,* ed. P. Turán (Akadémiai Kiadó, Budapest).

Rödl, V.
[1986] On universality of graphs with uniformly distributed edges, *Discrete Math.* **59**, 125–134.

Rödl, V., and A. Ruciński

[1993] Lower bounds on probability thresholds for Ramsey properties, in: *Combinatorics, Paul Erdős is Eighty, 1993, Bolyai Math. Study* (Akadémia Kiadó, Budapest) pp. 317–346.

[1995] Threshold functions for Ramsey properties, *J. Amer. Math. Soc.*, to appear.

Ruciński, A.

[1988] When are small subgraphs of a random graph normally distributed?, *Probab. Theory Related Fields* **78**, 1–10.

Ruciński, A., and A. Vince

[1986] Strongly balanced graphs and random graphs, *J. Graph Theory* **10**, 251–264.

Shamir, E., and J. Spencer

[1987] Sharp concentration of the chromatic number of random graphs $G_{n,0}$, *Combinatorica* **7**, 121–129.

Shelah, S., and J. Spencer

[1988] Zero–one laws for sparse random graphs, *J. Amer. Math. Soc.* **1**, 97–115.

Spencer, J.

[1987] *Ten Lectures on the Probabilistic Method* (SIAM, Philadelphia).

Stepanov, V.E.

[1973] *Random graphs, Voprosy Kibernetiki* (in Russian) pp. 164–185.

[1987] Some features of the structure of random graphs in the neighborhood of the critical point, *Theory Probab. Appl.* **32**, 573–594.

Thomason, A.

[1987a] Random graphs, strongly regular graphs and pseudo-random graphs, in: *Surveys of Combinatorics 1987, London Mathematical Society Lecture Note Series,* Vol. 123, ed. C. Whitehead (Cambridge University Press, Cambridge) pp. 173–196.

[1987b] Pseudo-random graphs, in: *Proc. Random Graphs, Poznán, 1985,* ed. M. Karoński, *Ann. Discrete Math.* **33**, 307–331.

[1989] Dense expanders and pseudo-random bipartite graphs, *Discrete Math.* **75**, 381–386.

Wilson, R.M.

[1972] Cyclotomy and difference families on abelian groups, *J. Number Theory* **4**, 17–47.

[1974] Constructions and uses of pairwise balanced designs, in: *Combinatorics, Mathematical Centre Tracts,* Vol. 55, eds. M. Hall Jr and J.H. van Lint (Mathematisch Centrum, Amsterdam) pp. 18–41.

CHAPTER 7

Hypergraphs

Pierre DUCHET

Laboratoire de Structures Discrètes et de Didactique, UA CNRS 393, BP 53X, F-38041 Grenoble Cedex, France

Contents

HANDBOOK OF COMBINATORICS
Edited by R. Graham, M. Grötschel and L. Lovász

1. Hypergraphs and set systems

Hypergraph theory is a part of the general study of combinatorial properties of (finite) families of (finite) sets. Hypergraphs are systems of sets which are conceived as natural extensions of graphs: elements correspond to nodes, sets correspond to edges which are allowed to connect more than two nodes. Usually, no distinction is made between isomorphic set systems; for instance, the *complete r-uniform hypergraph* on n nodes, denoted by K_n^r, is the pair consisting of a ground n-set and the collection of all its r-subsets.

1.1. Historical background

Influenced by the axiomatic tendencies of the beginning of the 20th century, and faced with the fast-growing economical interest of discrete mathematics, combinatorialists of the 1960s were urged to develop a systematic combinatorial approach of finite sets.

Earlier fundamental theorems, already known in the 1930s, like the Sperner theorem (in 1928), the Ramsey theorem (in 1930) and the Marriage theorem of König [from Frobenius in 1912. . . to Hall in 1935: see chapter 3 and Lovász and Plummer (1986)], took progressively their place, with the efforts of the Hungarian school, into a larger field of investigations focusing on "*set systems*"; chapters 24, 25 and 33 are mostly illustrative of this approach.

During the same period, set systems also appeared to the French school as efficient generalizations of graphs. A theory of *hyper*graphs would extend, simplify, and unify the results of the theory of graphs; in return, the theory of graphs would supply a convenient language for a theory of set systems. With the term "*hypergraph*", suggested by Berge at the Tihany Colloquium (in 1966), and with the forerunning paper of Erdős and Hajnal "On the chromatic number of graphs and set systems" (1966), the premises of a hypergraph theory were settled.

The reader interested in the first steps of the theory is referred to Berge (1970, 1973, 1976), Berge and Ray-Chaudhuri (1974), and Zykov (1974). Recent general references are Lovász (1979), and Berge's textbook (1989). Packing and covering problems for hypergraphs are extensively treated in Schrijver (1979a,b), and in a survey by Füredi (1988).

1.2. Hypergraphs nowadays

Inheriting the usual terminology, and hence the usual tools, of graph theory, hypergraph theory gains two major advantages: (1) hypergraph structure is preserved under the basic set-theoretic operations, and (2) in a set system the elements (= the *nodes* or the *vertices*) and the sets (the *edges*) play dual roles. The generalization to set systems of graph parameters (such as $\Delta, \omega, \alpha, \tau, \rho, \nu, \chi, \gamma$), constitute, at the present time, the distinctive feature of hypergraph theory. Actually, they have led to new fields of investigation, which, more powerful and more unified than the original ones, are of independent interest.

Some outstanding graph-theoretic results have been obtained via hypergraphs. A celebrated example is Lovász's use of "normal hypergraphs" to solve Berge's "weak perfect graph conjecture": the complement of a perfect graph is perfect (see Theorem 3.19 and chapter 4). Hypergraphs also appear explicitly in constructions of Ramsey graphs and graphs with large chromatic number and girth (see section 5). More recently, Robertson and Seymour proved a conjecture by Nash-Williams on "graph immersions" by the means of an extension to hypergraphs of their "Graph minor theorem" (see Theorem 2.6 and chapter 5).

Hypergraphs also turn out to be remarkably efficient in discrete optimization problems: offering a convenient setting to numerous minimax-type theorems (see section 3), they lead to a pertinent formulation of general inequalities on packing and covering parameters (section 4).

1.3. Structural-type properties

Section 2 introduces the reader to difficulties and advantages in generalizing graph problems to hypergraphs. There, some hypergraph analogues of important graph-theoretic results involving basic notions are mentioned: degrees, cycles, edge contractions, cliques, etc. An undeniable success is the above mentioned result on hypergraph embeddings. A typical difficulty arises with edge intersection problems: a (connected, simple) graph on more than three nodes is fully determined by edge-to-edge incidences but a hypergraph is not. The point is that families of pairwise intersecting edges in a hypergraph can have a very intricate structure. Even for simplicial complexes (hypergraphs in which every nonempty subset of an edge is an edge), the maximum size of an intersecting family of edges is unknown (see Chvátal's conjecture 2.18 and chapter 24). Among possible simplifying hypotheses, the Helly property (i.e., "pairwise intersecting edges share a common node") provides an efficient way of investigating graphs via set systems.

1.4. Hypergraphs and min–max relations

A variety of hypergraph classes is presented in section 3. Those set systems generalize either trees or bipartite graphs; they frequently arise in the polyhedral approach of Combinatorial Optimization problems (cf. chapters 3 and 30): certain hypotheses – embeddability in a highly structured set, coloring properties, forbidden configurations – imply general min–max type relations, namely when assumptions warrant the integrability of optima in some related linear programs. The interest in "tree-like" hypergraphs (e.g., *arboreal hypergraphs*, *totally balanced hypergraphs*, . . .) has been largely motivated by greedy heuristics: the idea is to handle a structure which is preserved under some *elimination scheme* (like successive removals of leaves of a tree).

Balanced hypergraphs and further generalizations of bipartite graphs (such as *normal hypergraphs*, *Mengerian hypergraphs*) share the *König property*: the minimum number of nodes needed to cover all the edges (i.e., the *blocking number*) equals the maximum number of pairwise disjoint edges (i.e., the *matching number*). Actually, numerous min–max type properties amount to say that a certain hypergraph (such as paths or cliques of a graph, pieces of a pattern, intervals, chains of a partially ordered set, . . .) possesses the König property.

1.5. Packing and covering

The world of matching and blocking numbers (denoted by ν and τ) forms the cornerstone of discrete packing and covering problems, discussed in section 4. This field is the most active part of the theory of hypergraphs: almost all combinatorial optimization problems can indeed be reformulated as the determination of τ or ν in a certain hypergraph.

Deep results in this area, inequalities or estimates, have been obtained by a powerful combination of classical "hypergraph-style" reasonings (extensions of König's theorem, averaging methods, set theoretic operations and incidence properties) with linear algebra, probabilistic methods and polyhedral combinatorics. For instance, with the linear programming approach of optimization problems, parameters τ and ν are viewed as integral optima of *dual linear programs* whose rational optima coincide. Extremal situations frequently involve block designs.

1.6. Colorings

Hypergraph coloring problems constitute, because of their intrinsic difficulty, a very attractive subject, with natural connections with other fields (e.g., Ramsey theory): less unity, more invention. The central question, treated in section 5, is to find the *chromatic number* of a set system: how many colors do we need in order to color the nodes so that at least two different colors appear in every edge of cardinality 2 or more.

As a clear demonstration of "French style" incidence methods, we give in full the proof of a bicolorability theorem, 5.8, showing that a nonbicolorable hypergraph must contain a special type of odd cycles.

For uniform hypergraphs, the best upper bounds to date on the chromatic number (in term, for instance, of their size) were obtained by "Hungarian style" probabilistic methods (e.g., see Theorem 5.17).

A famous coloring problem, originating with Kneser's conjecture in 1955, was to determine the minimum number of classes in which the h-subsets of a given n-set can be partitioned such that no class contains t pairwise disjoint members. A general solution has recently been obtained with the help of topological methods, inspired by Lovász's solution of Kneser's conjecture. See Theorem 5.24.

Some partitioning problems, occurring in design theory and in discrepancy theory (cf. chapters 14 and 26), as the "factorization problem" for complete uniform hypergraphs, have been successfully treated by hypergraph methods on *edge colorings*: see Theorem 5.29 and Corollaries 5.30–5.32.

1.7. General terminology

(1.1) Let $V = \{v_1, \ldots, v_n\}$ be a finite set. A *hypergraph on* V is a pair $\mathcal{H} = (V, \mathcal{E})$, where $\mathcal{E}$ is a family $(E_i)_{i \in I}$ of subsets of V. Elements of V (respectively members of $\mathscr{E}$) are called the *nodes* (or *vertices*) of $\mathscr{H}$ (respectively the *edges* of $\mathscr{H}$). The numbers n and $|I|$ are frequently referred to as the *order* and the *size* of $\mathscr{H}$, respectively. Notice that repeated edges, empty edges and isolated nodes (= in no edge) are permitted. One-element edges are called *loops*.

In fact, *when* $\mathscr{H}$ *is defined as a family of sets*, $\mathscr{H}=(E_i)_{i\in I}$, *or by the list of its edges*, $\mathscr{H}=(E_1, E_2, \ldots, E_m)$, *we implicitly assume* (in accordance with Berge 1989) *that the vertex set* $V(\mathscr{H})$ *is the union of the edges*.

(1.2) We may view a general hypergraph on V as a multiset of V-subsets: every V-subset appears in $\mathscr{H}$ with a certain *multiplicity*, a *nonedge* corresponding to multiplicity 0.

Hypergraphs correspond as well to 0–1 matrices. With hypergraph $\mathscr{H}$ is associated in a canonical way its *edge–vertex incidence matrix* $M(\mathscr{H})$: rows correspond to edges, columns correspond to vertices; entry $m_{i,j}$ equals 1 or 0, depending on whether E_i contains v_j or not. The pairs $[v, i]$ such that $v\in E_i$, form the edges of the *bipartite incidence graph* of $\mathscr{H}$.

(1.3) The *dual* of $\mathscr{H}$, denoted by $\mathscr{H}^{\top}$, corresponds to the transposed matrix. We have $\mathscr{H}^{\top}=(I, (\mathscr{H}(v))_{v\in V})$, where $\mathscr{H}(v)$, the *star of* v, denotes the set of all $i\in I$ such that $v\in E_i$.

(1.4) The letter $\mathscr{H}$ refers, if no other specification is made, to the *generic hypergraph* $\mathscr{H}=(V,\mathscr{E})$, with $\mathscr{E}=(E_i)_{i\in I}$; n and m denote its order and size, respectively. Note that, in general, n and $|\bigcup_i E_i|$ may be different.

(1.5) The *image* $\phi(\mathscr{H})$ of $\mathscr{H}$ by a mapping $\phi: V\rightarrow W$ is the hypergraph $(\phi(V), (\phi(E_i))_{i\in I})$. If ϕ is one-to-one, hypergraphs $\mathscr{H}$ and $\phi(\mathscr{H})$ are *isomorphic* (we write $\mathscr{H}\simeq\phi(\mathscr{H})$). Throughout the chapter, we do not distinguish between isomorphic hypergraphs, unless otherwise specified.

(1.6) The *product* (*direct product in* Lovász (1979)) of hypergraphs $\mathscr{H}=(V,(E_i)_{i\in I})$ and $\mathscr{K}=(W,(F_j)_{j\in J})$ is the hypergraph $\mathscr{H}\otimes\mathscr{K}:=(V\times W, (E_i\times F_j)_{(i,j)\in I\times J})$.

(1.7) Subhypergraphs correspond to submatrices: a hypergraph $\mathscr{H}'=(V',\mathscr{E}')$ is a *subhypergraph* of $\mathscr{H}=(V,\mathscr{E})$ if $V'\subseteq V$ and if every $\mathscr{H}'$-edge is the intersection with V' of some $\mathscr{H}$-edge. A subhypergraph may have empty edges or isolated nodes. Let $A\subseteq V$ and $J\subseteq I$. The subhypergraph of $\mathscr{H}$ *induced by* A [or the *restriction* of $\mathscr{H}$ to A, Lovász (1979)] is the hypergraph $\mathscr{H}[A]=(A,(E_i\cap A: E_i\cap A\neq\emptyset)_{i\in I})$ [an alternative notation is $\mathscr{H}_A$, Berge (1973, 1989)]. Notice that an induced hypergraph has no empty edge. The notation $\mathscr{H}[V\backslash x]$ is abridged into $\mathscr{H}\backslash x$. A *partial* (*sub*)*hypergraph of* $\mathscr{H}$ is a hypergraph that corresponds to a subfamily of edges. A partial hypergraph has no isolated vertex. The family $(E_j)_{j\in J}$ constitutes the partial hypergraph *determined by* J (or *generated by* J, Berge 1989). The partial hypergraph determined by $I\backslash\{i\}$ is denoted by $\mathscr{H}\backslash E_i$.

(1.8) The partial hypergraph formed with all edges of $\mathscr{H}$ that are contained in A is called the *trace of* $\mathscr{H}$ *by* A, denoted by $\mathscr{H}[\![A]\!]$ (the term *section* was used in Berge 1989, with notation $\mathscr{H}\times A$).

(1.9) The *k-section* of $\mathscr{H}$ (or the kth *shadow*, cf. chapter 24), denoted by $\mathscr{H}_{[k]}$, is the hypergraph on $V(\mathscr{H})$ formed with all k-sets F such that $F\subseteq E$ for some edge E of $\mathscr{H}$.

(1.10) The *line-graph* $L(\mathscr{H})$ of $\mathscr{H}$ [or *representative graph* (Berge 1989), *intersection graph*] of $\mathscr{H}$ is the 2-section of $\mathscr{H}^{\top}$: nodes of $L(\mathscr{H})$ are $\mathscr{H}$-edges, edges of $L(\mathscr{H})$ are pairs of intersecting $\mathscr{H}$-edges.
(1.11) A sequence $\sigma = v_1 E_1 v_2 E_2 \cdots v_p E_p v_{p+1}$ (where the E_i are distinct $\mathscr{H}$-edges, the v_i are distinct $\mathscr{H}$-vertices for $1 \leqslant i \leqslant p$ such that $v_i, v_{i+1} \in E_i$ for $1 \leqslant i \leqslant p$) is called a *path of length p* (joining v_1 to v_{p+1}) *in* $\mathscr{H}$ if v_{p+1} is different from the other v_i's. The hypergraph $\mathscr{H}$ is *connected* if any two vertices are joined by a path. If $v_{p+1} = v_1$, the sequence σ is a *cycle of length p in* $\mathscr{H}$. It is sometimes useful to consider a more restrictive notion: if, in addition, the sets $E_{i+1} \backslash E_i$ (with $p+1 \equiv 1$) form a partition of $E_1 \cup E_2 \cup \cdots \cup E_p$, the cycle will be called a *hypercycle*. A graph-theoretic cycle will be referred to as a *polygon*.
(1.12) $\mathscr{H}$ is *simple* if there are no repeated edges. If, moreover $E_i \not\subseteq E_j$ for $i \neq j$, the hypergraph $\mathscr{H}$ is called a *clutter* or an *antichain* or a *Sperner system*. $\mathscr{H}$ is *separating* if $\mathscr{H}^{\top}$ is a clutter or, equivalently, if, for every node v, the intersection of all $\mathscr{H}$-edges containing v is $\{v\}$.
(1.13) The *rank* (respectively *corank*) of $\mathscr{H}$ is the maximum (minimum) cardinality of an edge. $\mathscr{H}$ is *r-uniform* if $|E_i| = r$ for all i. An *r-graph* is a *simple* r-uniform hypergraph.
(1.14) $\mathscr{H}$ is *linear* if $|E_i \cap E_j| \leqslant 1$ for $i \neq j$.
(1.15) $\mathscr{H}$ is *intersecting* if its edges pairwise meet. For positive integers k, t, with $k \geqslant 2$, $\mathscr{H}$ is *k-wise t-intersecting* whenever any k distinct edges have at least t common nodes. If $t = 1$, then t is omitted. Intersecting partial hypergraphs of $\mathscr{H}$ (or *intersecting families*, Berge 1989) are referred to as "cliques" of $\mathscr{H}$ by some authors (compare with (1.27)).
(1.16) $\mathscr{H}$ is a *star* if all its edges have a common node. The *star centered at a node* v, denoted by $\mathscr{H}(v)$, is the family of all $\mathscr{H}$-edges that contain v.
(1.17) $\mathscr{H}$ has the *Helly property* if every intersecting family of edges forms a star. If $\mathscr{H}^{\top}$ has the Helly property, $\mathscr{H}$ is said to be *conformal*.
(1.18) The *degree* in $\mathscr{H}$ of a node v, denoted by $\deg_{\mathscr{H}}(v)$, is the size of the star $H(v)$. The size of the largest *linear* star containing v is the *linear-degree* of v. The maximum degree of $\mathscr{H}$-nodes (respectively the maximum linear-degree) is denoted by $\Delta(\mathscr{H})$ ($\Delta_l(\mathscr{H})$).
(1.19) A set $S \subseteq V(\mathscr{H})$ is *stable* (respectively *strongly stable*) if no edge is contained in S (respectively if $|S \cap E| \leqslant 1$ for every edge E). A hypergraph is *r-partite* if its vertex set can be partitioned into r strongly stable sets. The *stability number* $\alpha(\mathscr{H})$ is the maximum cardinality of a stable set. The maximum cardinality of a strongly stable set is denoted by $\alpha_{[2]}(\mathscr{H})$. We have $\alpha_{[2]}(\mathscr{H}) = \alpha(\mathscr{H}_{[2]})$.
(1.20) A set $B \subseteq V(\mathscr{H})$ is a *blocking set of* $\mathscr{H}$ [or a *node-cover*, Lovász (1979), Füredi (1988), or a *transversal*, Berge (1973, 1989)] if it meets every edge. Inclusion-minimal blocking sets are the edges of the *blocker* of $\mathscr{H}$, denoted by $\mathrm{Bl}(\mathscr{H})$. The minimum cardinality of a blocking set is referred to as the *blocking number* and is denoted by $\tau(\mathscr{H})$. Alternative terms and notation are *transversal hypergraph* $\mathrm{Tr}(\mathscr{H})$ and *transversal number*, Berge (1989), or *node-covering number*, Lovász (1979), Füredi (1988).

(1.21) An *edge-cover of* $\mathscr{H}$ is a node-cover of $\mathscr{H}^\top$. The *edge-covering number* $\rho(\mathscr{H})$ is the minimum size of an edge-cover.
(1.22) A *matching of* $\mathscr{H}$ is a set of pairwise disjoint $\mathscr{H}$-edges. The *matching number* $v(\mathscr{H})$ is the maximum number of edges in a matching.
(1.23) A *fractional blocking set* of $\mathscr{H}$ (or a *fractional* (*node-*)*cover*, a *fractional transversal*) is a weight function $p: V \to \mathbb{R}_+$ such that $\sum_{v \in E} p(v) \geq 1$ for every edge E. Dually a *fractional matching* is a function $q: I \to \mathbb{R}_+$ such that $\sum_{v \in E_i} q(i) \leq 1$ for every vertex v. The numbers $\sum_{v \in V} p(v)$ and $\sum_{i \in I} q(i)$ are the respective *weights* of p and q. The *fractional blocking number* $\tau^*(\mathscr{H})$ is the maximum weight of a fractional transversal. It turns out (see section 3) that $\tau^*(\mathscr{H})$ coincides with the *fractional matching number* $\nu^*(\mathscr{H})$, i.e., the minimum weight of a fractional matching.
(1.24) A (*proper*) *k-coloring* of $\mathscr{H}$ (or *k-coloration*, Lovász 1979) is an assignment of "colors" to nodes so that (i) each node receives exactly one color, (ii) at most k colors are used, and (iii) no nontrivial edge is monochrome (thus loops and empty edges do not play any role). The *chromatic number* $\chi(\mathscr{H})$ is the smallest integer k such that $\mathscr{H}$ admits a k-coloring. If $\chi(\mathscr{H}) = k$ (respectively $\chi(\mathscr{H}) \leq k$, $\chi(\mathscr{H}) \leq 2$), then $\mathscr{H}$ is said to be *k-chromatic* (*k-colorable*, *bicolorable*).
(1.25) A (*proper*) *edge-coloring* of $\mathscr{H}$ is an assignment of colors to $\mathscr{H}$-edges so that distinct intersecting edges receive different colors. The *chromatic index* $\gamma(\mathscr{H})$ (or *edge-coloring* number) is the smallest number of colors needed. Note that $\gamma(\mathscr{H}) = \chi(L(\mathscr{H}))$.
(1.26) A *hereditary* hypergraph is a *simplicial complex*, i.e., a simple hypergraph in which every nonempty subset of an edge is an edge. The *hereditary closure* of $\mathscr{H}$, denoted by $\hat{\mathscr{H}}$, is the smallest hereditary hypergraph containing $\mathscr{H}$.
(1.27) If a hypergraph $\mathscr{H}$ is endowed with a partition $V_i, V_2, \ldots, V_p$ of its node set, we say $\mathscr{H}$ is *partitioned* (or *p-partitioned* to be more precise). Given a mapping $M: \mathbb{N}^p \to \mathbb{N}$ and integers $n_i = |V_i|$, the *complete* p-partitioned hypergraph $K^M_{n_1,n_2,\ldots,n_p}$ is the multiset where every subset E of V occurs with multiplicity $M(|E \cap V_1|, |E \cap V_2|, \ldots, |E \cap V_p|)$. If M takes value 1 for a single p-tuple $(r_1, r_2, \ldots, r_p)$ and is 0 otherwise, the standard notation is $K^{r_1,r_2,\ldots,r_p}_{n_1,n_2,\ldots,n_p}$, abridged to $K^p_{n_1,n_2,\ldots,n_p}$ if $r_1 = r_2 = \cdots = r_p = 1$ (*complete p-partite p-graph*) and to K^r_n if $p = 1$ (*complete r-graph* of order n). A *clique* of an r-graph G is any subset A of nodes such that $G[A] \simeq K^r_{|A|}$.

1.8. Notation

$E \in \mathscr{H}$	E is an edge of $\mathscr{H}$.
$\mathscr{K} \subseteq \mathscr{H}$	$\mathscr{K}$ is a *partial* subhypergraph of $\mathscr{H}$.
$V(\mathscr{H})$	vertex set.
$M(\mathscr{H})$	$m \times n$ edge-vertex incidence matrix.
$\mathscr{H}^\top$	dual of $\mathscr{H}$.
$\phi(\mathscr{H})$	image hypergraph of $\mathscr{H}$ under ϕ.
$\mathscr{H} \simeq \mathscr{K}$	isomorphic hypergraphs.
$\mathscr{H} \otimes \mathscr{K}$	product hypergraph of $\mathscr{H}$ and $\mathscr{K}$.

$\deg_{\mathscr{H}}(x)$	degree of node x in $\mathscr{H}$.
$\Delta(\mathscr{H})$	maximum degree.
$\Delta_l(\mathscr{H})$	maximum linear-degree.
$\mathscr{H}[A]$	subhypergraph of $\mathscr{H}$ induced by A ($\mathscr{H}_A$ in Berge 1989, Lovász 1979).
$\mathscr{H}[\![A]\!]$	trace of $\mathscr{H}$ by A.
$\mathscr{H}(x)$	star centered at node x.
$\mathscr{H}\backslash E$	partial subhypergraph obtained after deletion of the edge E.
$\mathscr{H}\backslash x$	induced subhypergraph obtained by deletion of x: $\mathscr{H}\backslash x = \mathscr{H}[V\backslash x]$.
$\mathscr{H}_{[k]}$	k-section of $\mathscr{H}$.
$L(\mathscr{H})$	line-graph of $\mathscr{H}$.
$\alpha(\mathscr{H})$	stability number.
$\alpha_{[2]}(\mathscr{H})$	maximum cardinality of a strongly-stable set.
$\tau(\mathscr{H})$	blocking number (or transversal number, node-covering number).
$\nu(\mathscr{H})$	matching number.
$\rho(\mathscr{H})$	edge-covering number.
$\chi(\mathscr{H})$	chromatic number.
$\gamma(\mathscr{H})$	chromatic index (or edge-coloring number).
K_n^r	complete r-graph of order n.

2. Hypergraphs versus graphs

This section focuses on results which are directly inspired by graph theory and deals essentially with edge–node incidences. The generalizations concerning cycles (cyclomatic number, acyclicity) or hypergraph minors will sound familiar to graph theorists. They will be probably more puzzled by innocent looking questions involving degree sequences, line-graphs or intersecting families. A way of modelizing graphs in hypergraph theory is shown in section 2.5, where the role of the Helly property is emphasized.

2.1. *Trees, cycles and their relatives*

Hypergraphs with no cycles constitute an obvious generalization of the concept of a tree. As easily seen, a necessary and sufficient condition for a hypergraph $\mathscr{H} = (E_j)_{i\in I}$ to be cycle-free is that, for every nonempty subset $J \subset I$, the following inequality holds:

$$|J| > \sum_{i\in J} |E_i| - \left|\bigcup_{i\in J} E_i\right| . \tag{2.1}$$

More generally, Acharya and Las Vergnas (1982) have extended to hypergraphs the notion of a cyclomatic number by taking into account the cardinality of the intersections $E_i \cap E_j$. Let us assign to every edge $\{E_i, E_j\}$ of the line-graph $L(\mathscr{H})$ the weight $|E_i \cap E_j|$ and denote by $w(\mathscr{H})$ *the maximum weight of a forest* in $L(\mathscr{H})$.

Then the parameter

$$\mu(\mathcal{H}) = \sum_{i \in I} |E_i| - \left|\bigcup_{i \in I} E_i\right| - w(\mathcal{H}) \tag{2.2}$$

is nonnegative. It is named the *cyclomatic number* of $\mathcal{H}$. Hypergraphs with cyclomatic number 0 are called *acyclic hypergraphs*. An inequality, proved by Lovász in 1968, shows that hypergraphs with no cycle of length ≥ 3 are acyclic. More generally, we have the following characterization.

Theorem 2.3 (Acharya and Las Vergnas 1982). *A hypergraph $\mathcal{H}$ is acyclic if and only if $\mathcal{H}$ is conformal and for every cycle of length ≥ 3 in $\mathcal{H}$, some edge of $\mathcal{H}$ contains 3 vertices of that cycle.*

Equivalently, the edges of an acyclic clutter represent the maximal cliques of some chordal graph (i.e., a graph in which every cycle of length ≥ 4 has a chord, see chapter 4). The edges of an acyclic hypergraph can be represented by the nodes of some tree T such that the star of every node induces a subtree of T (see Theorem 3.8).

(2.4) Let $a \geq 1, b$ be fixed integers. Given an arbitrary hypergraph $\mathcal{H} = (E_i)_{i \in I}$, the set function $f: 2^I \to \mathbb{Z}$ defined by $f(X) = a|\bigcup_{i \in X} E_i| + b$ for $X \subset I$ is submodular. Such functions induce matroidal structures (see chapters 9 and 11 for details) that generalizes graphic matroids (the case $a = 1, b = -1$, $|E_i| = 2$). Then standard methods of matroid theory lead to the following generalization of (2.1).

Theorem 2.5 (Las Vergnas 1970, see Berge 1973). *Let $r \geq 2$ be an integer and $(E_i)_{i \in I}$ be a hypergraph. There exists a cycle-free r-graph $\mathcal{H}' = (E'_i)_{i \in I}$ such that $E'_i \subset E_i$ for every $i \in I$ if and only if one has*

$$\left|\bigcup_{j \in J} E_j\right| > (r-1)|J| \quad \textit{for every } J \subset I\,.$$

This yields a sufficient condition for a hypergraph to admit a partition of its vertex-set into r *blocking sets* [a cycle-free hypergraph is *balanced*, hence Theorem 3.5 (vi) applies; for $r = 2$ cf. Theorem 5.12].

2.2. Hypergraph minors

A graph G' is a minor of a graph G if (up to isomorphism) G' is obtained from G by successive node-deletions, edge-deletions and edge-contractions (in any order). In a series of outstanding papers, Robertson and Seymour proved the following conjecture of Wagner: In *every infinite sequence of (finite) graphs there is a graph which is a minor of another* (see chapter 5). Robertson and Seymour recently obtained a generalization of their result to hypergraphs.

Without entering into details, let us say that a hypergraph is a *minor* of another hypergraph $\mathcal{H}$ if it arises from $\mathcal{H}$ as the result of successive elementary operations, performed in any order. Elementary operations include the deletion of a node, of an edge, the identification of two nodes of an edge (elementary "*collapsing*"), and the replacement of an edge by any subset of this edge ("*shrinking*").

Theorem 2.6 (Robertson and Seymour 1987). *Let $\mathcal{F}$ be a family of hypergraphs such that no hypergraph in $\mathcal{F}$ is isomorphic to a minor of another hypergraph in $\mathcal{F}$. Then $\mathcal{F}$ is finite.*

Such a generalization was highly motivated by its application to the following conjecture of Nash-Williams on graph embeddings (see chapter 5 for details and references): *Any infinite list of graphs contains a graph that can be immersed in another graph of the list.* Here, an *immersion* of a graph G *into* a graph H consists of a pair (α, β) of mappings with the following properties:

(a) $\alpha : V(G) \rightarrow V(H)$ is one-to-one.

(b) To every edge $e = \{x, y\}$ of G corresponds a path $\beta(e)$ of H with endpoints $\alpha(x)$ and $\alpha(y)$.

(c) Paths of the form $\beta(e)$ are pairwise edge-disjoint.

A proof of the Nash-Williams conjecture is promptly derived from Theorem 2.6, by considering hypergraph *duals* of the graphs of the list.

2.3. Degree sequences and edge cardinalities

(2.7) Degree-sequences of graphs are easily characterized by a system of inequalities (see chapter 1). From the fundamental theorem on the existence of a compatible integral flow in a network (see chapter 2) follows a generalization to hypergraphs.

Theorem 2.8 (Gale 1957, Ryser 1957; see Berge 1989). *Given $n + m$ nonnegative integers $r_1 \geqslant r_2 \geqslant \cdots \geqslant r_m$, $d_1 \geqslant d_2 \geqslant \cdots \geqslant d_n$, there exists a hypergraph $(E_1, E_2, \ldots, E_m)$ on a set $\{x_1, x_2, \ldots, x_n\}$ such that $|E_j| = r_j$ for $1 \leqslant j \leqslant m$ and $\deg_{\mathcal{H}}(x_i) = d_i$ for $1 \leqslant i \leqslant n$ if and only if*

$$\sum_{i=1}^{k} d_i \leqslant \sum_{i=1}^{k} r_i^* \quad \textit{for } k < n , \tag{1}$$

where r_i^* denotes the number of r_j's such that $r_j \geqslant i$, and

$$\sum_{i=1}^{n} d_i = \sum_{i=1}^{n} r_1^* = \sum_{i=1}^{m} r_j . \tag{2}$$

The Kruskal–Katona theorem (see chapter 24) yields a characterization of sequences $(|E_i|)_{i=1,\ldots,m}$ for simplicial complexes and, with little more effort, for clutters (Clements, Daykin, Godfrey, Hilton) and intersecting clutters of rank at most $\frac{1}{2}n$ (Greene, Katona, Kleitman): see chapter 24 for details and references. Characterizations of degree sequences have been obtained for certain finite set systems arising in combinatorial convex geometry: face complexes of simple polytopes (Stanley, Billera and Lee); finite families of convex sets in $\mathbb{R}^d$ (Kalai): see chapters 34 and 18.

Open problem 2.9 (Berge 1989). Determine, for $r \geq 3$, the possible degree-sequences of r-graphs (= simple r-uniform hypergraphs).

2.4. Line-graphs and edge intersections

To what extent is the structure of a hypergraph reflected by its line-graph or more generally by a quantitative informative information on edge intersections?
(2.10) Every graph G is the line-graph of some hypergraph (see (2.17) for a canonical construction). Let $\Omega(G)$ denote the smallest order of a hypergraph with line-graph G. If G has n vertices, the edges of G can be covered by at most $[\frac{1}{4}n^2]$ cliques (Erdős, Goodman, Pósa, see chapter 1); hence $\Omega(G) \leq [\frac{1}{4}n^2]$. The exact value of $\Omega(G)$ can be interpreted as the chromatic number of some associated graph, Berge (1989).
(2.11) While the line-graphs of graphs and multigraphs can be characterized by a finite number of forbidden induced subgraphs (see chapter 1), no such characterization is possible for the line-graphs of r-uniform hypergraphs for any $r \geq 3$. Constructions of infinitely many forbidden subgraphs are known in the case $r = 3$ (Nickel 1973 (unpublished), Gardner 1977 (unpublished), Bermond et al. 1977, Berge 1989.) Nevertheless, Naik et al. (1982) obtained the following result.

Theorem 2.12. *For any integer $r \geq 3$, there exists a smallest integer $f(r)$ such that, among the graphs of minimum degree $\geq f(r)$, the line-graphs of linear r-graph can be characterized by a finite list of forbidden induced subgraphs. Moreover, $f(r) = O(r^3)$ and $f(3) \leq 69$.*

(2.13) In 1932, Whitney proved that the "claw" $K_{1,3}$ and the triangle K_3 are the only nonisomorphic connected simple graphs with isomorphic line-graphs. Moreover, if G and G' are connected simple graphs of order greater than 4, any isomorphism $L(G) \rightarrow L(G')$ comes from an underlying isomorphism $G \rightarrow G'$. First generalizations of Whitney's theorem were obtained by Berge and Rado who constructed, for every integer $p \geq 2$, a pair of "nearly isomorphic" hypergraphs $\mathcal{B}_p$ and $\mathcal{D}_p$ as follows. Let $P = 1, \ldots, p$. Denote by B (respectively D) the set of P-subsets with even (odd) cardinality. Define $\mathcal{B}_p$ and $\mathcal{D}_p$ as the dual hypergraphs of (P, B) and (P, D), respectively. The next theorem is a further generalization of Berge and Rado's results.

Theorem 2.14 (Fournier 1980, Gardner 1984). *Let $\mathcal{H} = (V, E_1, \ldots, E_m)$ and $\mathcal{H}' = (V, E_1', \ldots, E_m')$ be two hypergraphs with the same vertex set and the same size. Suppose p is an integer such that both $\mathcal{H}$ and $\mathcal{H}'$ have rank less than 2^{p-1}. Then there exists an isomorphism $\varphi: v \mapsto v'$ mapping every E_i to E_i' if and only if the following two conditions are fulfilled:*

(1) *For all* $J \subset \{1, \ldots, m\}$ *with* $|J| < p$, *we have* $|\bigcap_{i \in J} E_i| = |\bigcap_{i \in J} E'_i|$.
(2) $\mathscr{H}$ *and* $\mathscr{H}'$ *do not contain subhypergraphs* $\mathscr{H}_J[X]$ *and* $\mathscr{H}_J[X']$ (*with* $|J| = p$ *and* $|X| = |X'| = 2^{p-1}$), *respectively isomorphic to* $\mathscr{B}_p$ *and* $\mathscr{D}_p$, *or vice-versa.*

Theorem 2.14 implies that certain hypergraphs $(E_i)_{i \in I}$ are determined (up to isomorphism) by their "intersection matrix" (i.e., with entries $|E_i \cap E_j|$). Interval hypergraphs and, more generally, totally balanced hypergraphs (cf. section 3) do have that property.
(2.15) The famous Ulam–Kelly *graph reconstruction conjecture* (see chapter 27) can be extended to hypergraphs in an obvious way. Within the context of hypergraphs, edge-reconstruction and vertex reconstruction become dual problems. The example of pairs $\mathscr{B}_p$ and $\mathscr{D}_p$, as defined above, shows that additional hypotheses are necessary to imply the reconstructibility of hypergraphs. More sophisticated nonreconstructible hypergraphs are known.

2.5. Helly property

A family of sets has the *k-Helly property* (we omit k for $k = 2$) if every k-wise intersecting finite subfamily is a star. Helly's celebrated theorem asserts that convex sets in $\mathbb{R}^{k-1}$ have the k-Helly property. Actually, the Helly property occurs in numerous mathematical fields. In arithmetics, the "Chinese remainder theorem" amounts to say that arithmetical progressions have the Helly property. Other classical examples of Helly families are families of intervals in a lattice (sets of the form $\{x : a \leq x \leq b\}$) and families of connected subsets of a (topological) tree (cf. arboreal hypergraphs in section 3). Notice that the k-Helly property is preserved under the product of hypergraphs. For instance, Euclidean boxes (i.e., parallelepipeds with faces parallel to the coordinate axes) possess the Helly property (see Theorem 4.40 for more).

Theorem 2.16 (Berge and Duchet 1975). *Let* $\mathscr{H}$ *be a hypergraph. Then*
(1) $\mathscr{H}$ *has the k-Helly property if and only if, for every set* A *of* $k + 1$ *nodes, all the edges* E *such that* $|E \cap A| \geq k$ *share a common node.*
(2) $\mathscr{H}^{\top}$, *the dual of* $\mathscr{H}$, *has the k-Helly property if and only if the* $\subset$*-maximal edges of* $\mathscr{H}$ *are precisely the* $\subset$*-maximal cliques of some k-graph.*

(2.17) In particular, with any graph G can be associated a separating Helly hypergraph $\mathscr{H}^{\top}(G)$ such that $L(\mathscr{H}^{\top}(G)) \simeq G$: hypergraph $\mathscr{H}^{\top}(G)$ is the dual of the clutter formed with the maximal cliques of G. That one-to-one correspondence constitutes a bridge between graphs and hypergraphs and appears to be very useful. Properties of a given class of graphs, say $\mathfrak{G}$, may be derived, with the tools of hypergraph theory, from more general properties of *Helly classes of hypergraphs*. By a Helly class we mean (Duchet 1978) a class of Helly hypergraphs whose line-graph lies in $\mathfrak{G}$ and which is closed with respect to certain operations. Numerous examples of Helly classes are provided in section 3.
Various problems on set systems are drastically simplified when restricted to

Helly hypergraphs. Theorem 4.40 gives an example of a covering problem; let us conclude this section with a celebrated intersection problem.

Conjecture 2.18 (Chvátal 1972). The hereditary closure $\hat{\mathscr{H}}$ of any hypergraph $\mathscr{H}$ contains at most $\Delta(\hat{\mathscr{H}})$ pairwise intersecting edges.

From a result of Schönheim, it follows that the conjecture holds true when $\mathscr{H}$ is a Helly hypergraph. For further information on Chvátal's conjecture, we refer to chapter 24 (see also the discussion on the "edge-coloring property" in section 5).

3. Remarkable hypergraphs and min–max properties

In many cases, a satisfactory solution of a combinatorial optimization problem, such as a min–max relation, amounts to saying that certain 0–1 matrices (= hypergraphs) have particular structural properties. Such matrices may, for instance, express constraints in a linear program, and some special properties imply the integrality of rational optima. It is often useful to consider those 0–1 matrices as hypergraphs, although, in a large part of the extensive literature on the subject, the hypergraph structure is not explicitly mentioned.

The same classes of hypergraphs, and related min–max type properties, may be seen in a different way. Either they generalize remarkable classes of graphs (bipartite graphs, trees), or they correspond to interesting patterns of some combinatorial structure (paths of a graph, branchings, . . .).

3.1. A hierarchy of hypergraph properties

The synoptic diagram in fig. 3.1 visualizes remarkable properties generalizing either those of trees or those of bipartite graphs. In the diagram, all possible unrefinable implications are marked by arrows.

Balanced hypergraphs (section 3.3) form a very natural generalization of bipartite graphs. They are defined as hypergraphs for which no subhypergraph is an odd polygon. Balanced hypergraphs are bicolorable (i.e., they admit a proper 2-coloring (1.24)) and have the Helly property (i.e., pairwise intersecting edges must share a common point (2.16)–(2.17)). Moreover, they possess the main min–max type properties of bipartite graphs (viz. chapters 3 and 4):
– The *König property*: the blocking number τ coincides with the matching number ν (see definitions (1.20) and (1.22);
– The *dual* König property: the edge-covering number ρ coincides with the maximum cardinality of a strongly stable set $\alpha_{[2]}$ (see (1.21) and (1.19));
– The *edge-coloring property*: the chromatic index γ coincides with the maximum degree Δ (see definitions (1.25) and (1.18));
– The *Gupta property*: the edges can be partitioned into δ edge-covers, where δ denotes the minimum degree.

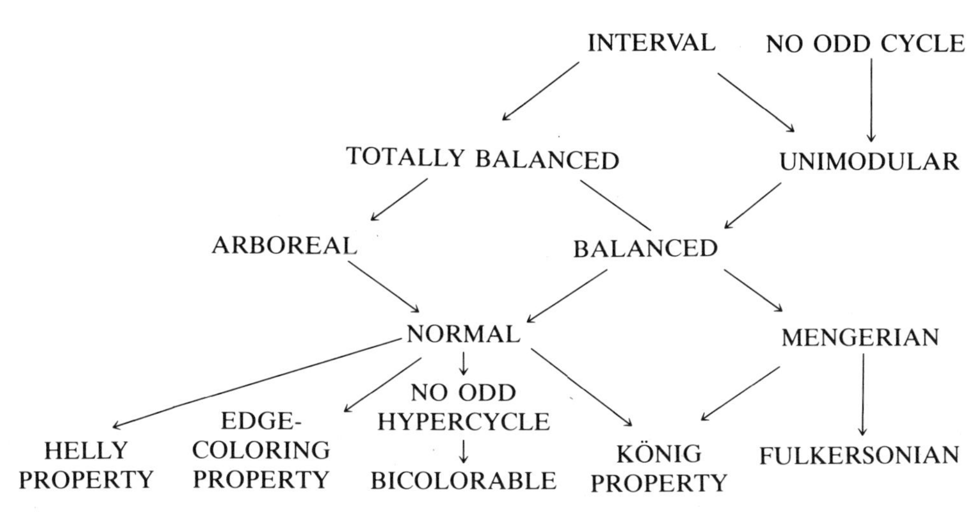

Figure 3.1.

Investigation on tree-like structures was motivated by practical applications. Families of intervals of $\mathbb{Z}$ (*Interval hypergraphs*, see section 3.4) and systems of paths in a tree occur in numerous seriation problems (e.g., location of genes in a DNA molecule, phylogenetic trees, . . .). *Arboreal hypergraphs* (section 3.4), i.e., families of connected subsets of a tree, generalize the preceding classes. They are used to organize information retrieval systems in relational data bases. They also offer a natural setting to some "greedy-type" algorithms, based on the existence of "removable nodes". *Totally-balanced hypergraphs* (section 3.4), defined by the absence of *any* polygon as a subhypergraph, were encountered in the polyhedral approach of some location problems. They get together the advantages of arboreal structures and of balanced matrices.

Normal hypergraphs (section 3.5) were introduced as Helly hypergraphs corresponding (viz. section 2.5) to perfect graphs (see chapter 4). A hypergraph $\mathscr{H}$ is normal iff every partial hypergraph $\mathscr{H}' \subseteq \mathscr{H}$ has the edge-coloring property. The "perfect graph theorem" amounts to saying that normal hypergraphs have the König property.

Mengerian hypergraphs and *Fulkersonian hypergraphs* (section 3.6) provide a general setting for important minimax theorems encountered in the polyhedral approach of Network Theory (Menger theorem, two-commodity flows, . . .). They are defined by integrality conditions on *fractional blocking* and *matching numbers*. It is more convenient here to extend definition (1.23) using the polyhedral approach.

3.2. A glimpse of the polyhedral approach

Given a hypergraph $\mathscr{H} = (V, \mathscr{E})$, with incidence matrix H, and vectors $\boldsymbol{a} \in \mathbb{Q}_+^{\mathscr{E}}$, $\boldsymbol{b} \in \mathbb{Q}_+^V$, the *fractional blocking-set polyhedron* $\mathbf{FB}(\boldsymbol{a})$ and the *fractional matching polyhedron* $\mathbf{FM}(\boldsymbol{b})$ associated with $\mathscr{H}$ are respectively defined by:

$$\mathbf{FB}(\boldsymbol{a}) = \{\boldsymbol{x} \in \mathbb{Q}_+^V \mid H\boldsymbol{x} \geq \boldsymbol{a}\} , \qquad \mathbf{FM}(\boldsymbol{b}) = \{\boldsymbol{y} \in \mathbb{Q}_+^{\mathscr{E}} \mid \boldsymbol{y}H \leq \boldsymbol{b}\} . \tag{3.1}$$

Vectors in $\mathbf{FB}(\boldsymbol{a})$, $\mathbf{FM}(\boldsymbol{b})$ are called *fractional* $\boldsymbol{a}$*-blocking sets* and *fractional* $\boldsymbol{b}$*-matchings*, respectively. The adjective "fractional" is omitted if we restrict ourselves to integral vectors. We omit $\boldsymbol{a}, \boldsymbol{b}$ in the cases $\boldsymbol{a} = \mathbf{1}, \boldsymbol{b} = \mathbf{1}$. Integral $\mathbf{1}$-blocking sets and $\mathbf{1}$-matchings are respective incidence vectors of usual matchings and blocking sets. We recall that the clutter consisting of $\subset$-minimal blocking sets is the *blocker* $\mathrm{Bl}(\mathscr{H})$ of $\mathscr{H}$.

By linear programming duality, we have

$$\min\{\boldsymbol{b} \cdot \boldsymbol{x} \mid \boldsymbol{x} \in \mathbf{FB}(\boldsymbol{a})\} = \max\{\boldsymbol{y} \cdot \boldsymbol{a} \mid \boldsymbol{y} \in \mathbf{FM}(\boldsymbol{b})\} . \tag{3.2}$$

Hence, the *fractional blocking number* τ^* and the *fractional matching number* ν^* coincide:

$$\tau^* = \min\{\mathbf{1} \cdot \boldsymbol{x} \mid \boldsymbol{x} \in \mathrm{FB}\} = \max\{\boldsymbol{y} \cdot \mathbf{1} \mid \boldsymbol{y} \in \mathrm{FM}\} = \nu^* .$$

The blocking number τ (respectively the matching number ν) corresponds to the integral optimum of the "min" ("max") in (3.2), with $\boldsymbol{a} = \mathbf{1}$ and $\boldsymbol{b} = \mathbf{1}$. More generally, the integral optima

$$\tau_k(\mathscr{H}) = \min\{\mathbf{1} \cdot \boldsymbol{x} \mid \boldsymbol{x} \in \mathbb{Z}_+^V \cap k(\mathbf{FB})\}$$

and

$$\nu_k(\mathscr{H}) = \max\{\boldsymbol{y} \cdot \mathbf{1} \mid \boldsymbol{y} \in \mathbb{Z}_+^{\mathscr{E}} \cap k(\mathbf{FM})\}$$

are the *k-blocking number* and the *k-matching number* of $\mathscr{H}$ respectively. A k-matching corresponds to a set of edges such that each node belongs to at most k of them. A k-blocking set is a set of nodes such that each edge contains at least k of them. Remark that repetition of nodes and edges is permitted. We have fundamental inequalities (obtained in 1968 by Berge, Simonovits and Lovász, see Berge 1989, Lovász 1979 or Schrijver 1979a):

$$\begin{aligned}\nu(\mathscr{H}) = \nu_1(\mathscr{H}) = \min_{k \in \mathbb{N}} \frac{\nu_k(\mathscr{H})}{k} \leq \sup_{k \in \mathbb{N}} \frac{\nu_k(\mathscr{H})}{k} = \max_{k \in \mathbb{N}} \frac{\nu_k(\mathscr{H})}{k} = \nu^*(\mathscr{H}) \\ = \tau^*(\mathscr{H}) = \min_{k \in \mathbb{N}} \frac{\tau_k(\mathscr{H})}{k} = \inf_{k \in \mathbb{N}} \frac{\tau_k(\mathscr{H})}{k} \leq \max_{k \in \mathbb{N}} \frac{\tau_k(\mathscr{H})}{k} = \tau_1(\mathscr{H}) = \tau(\mathscr{H}) .\end{aligned} \tag{3.3}$$

The aim of the hypergraph "fractional theory" is to find the integers k that force min–max relations of the type $\nu^* = \nu_k/k$ or $\tau^* = \tau_k/k$. (More on the subject in section 4; viz. chapters 28 and 30.)

3.3. Balanced hypergraphs

Since balanced graphs are defined by the absence of odd polygons, they form a self-dual class. Berge introduced this important class of hypergraphs in 1969 and proved bicolorability under the stronger following form (existence of a "good k-coloring").

Lemma 3.4. *For every integer $k \geq 2$, a balanced hypergraph admits a k-coloring of its nodes such that the number of colors appearing in any edge E is exactly* $\min(k, |E|)$.

The next theorem summarizes the minimax types properties of balanced hypergraphs (see Berge 1989, Lovász and Plummer 1986 or Schrijver 1979a for proofs).

Theorem 3.5 (Berge 1970, Berge and Las Vergnas 1970). *For a hypergraph $\mathcal{H}$, the following properties are equivalent:*

(i) *$\mathcal{H}$ is balanced;*
(ii) *$\mathcal{H}^{\top}$, the dual of $\mathcal{H}$, is balanced;*
(iii) *Every subhypergraph of $\mathcal{H}$ is bicolorable;*
(iv) *Every subhypergraph of $\mathcal{H}$ has the König property;*
(v) *Every subhypergraph of $\mathcal{H}$ has the edge-coloring property;*
(vi) *The blocker of any subhypergraph of $\mathcal{H}$ has the König property.*

Applying, for instance, (vi) to duals of bipartite graphs, we obtain Gupta's theorem. Theorem 3.5, together with a result of Fulkerson et al. (1974), see Berge (1989) and Schrijver (1979a), implies that either in case $\boldsymbol{a} \in \{1, +\infty\}^{\mathscr{E}}$, $\boldsymbol{b} \in \mathbb{Z}_{+}^{V}$ or in case $\boldsymbol{a} \in \mathbb{Q}_{+}^{\mathscr{E}}$, $\boldsymbol{b} \in \{1, +\infty)^{V}$, both dual programs in (3.2) have integral optimum solutions (hence, balanced hypergraphs are Mengerian: see (3.21)). These results can be formulated equivalently in terms of total dual integrality, blockers, antiblockers, etc. (see chapters 30 and 28).

No polynomial recognition algorithm for balanced hypergraphs was known until Conforti et al. (1990) succeeded, by the use of a decomposition theorem relative to bipartite incidence graphs (as defined in (1.2)).

In a bipartite graph B, a *2-complete cut* is a cut (= an edge set whose removal disconnects B) which consists of the union of the edge sets of two node-disjoint complete bipartite subgraphs. A *2-star cutset* is a pair of adjacent nodes whose removal disconnects B.

Theorem 3.6 (Conforti et al. 1990, cf. Conforti et al. 1994). *The incidence bipartite graph of a balanced hypergraph admits a 2-complete cut or a 2-star cutset.*

(3.7) *Unimodular hypergraphs* correspond to totally-unimodular 0–1 matrices, treated in depth in chapter 30. Unimodular hypergraphs are balanced, moreover both polyhedra $\mathbf{FB}(\boldsymbol{a})$ and $\mathbf{FM}(\boldsymbol{b})$ have integral vertices for every integral vectors

$\boldsymbol{a}, \boldsymbol{b}$. Ghouila-Houri's characterization theorem which asserts the existence of an "equitable 2-coloring", can be extended as follows (De Werra 1971, see Berge 1989): *For any integer k, the nodes of an unimodular hypergraph can be partitioned into k stable sets S_i such that, for each edge E, the cardinalities $|E \cap S_i|$ differ by at most* 1.

Hypergraphs with no odd cycles are unimodular (Berge 1989): examples are bipartite multigraphs, and subhypergraphs of any "grid-hypergraph" consisting of rows and columns of a two-dimensional array.

Interval hypergraphs (section 3.4) and more generally systems of oriented paths in an oriented tree (3.18) are unimodular hypergraphs. See chapter 30 for further examples. A polynomial-time algorithm to test whether a matrix is totally unimodular results from the work of Seymour on regular matroids (see chapter 10).

3.4. Arboreal hypergraphs

A hypergraph $\mathcal{H}$ is said to be *arboreal* when there exists a tree T on node set $V(\mathcal{H})$ such that every edge of $\mathcal{H}$ induces a T-subtree. Such trees are called *representative trees* for $\mathcal{H}$. As shown by the structural characterization which follows, a hypergraph is arboreal iff its dual hypergraph is acyclic (viz. Theorem 2.3).

Theorem 3.8 (Duchet 1978, Flament 1978, Slater 1978). *A hypergraph $\mathcal{H}$ is an arboreal hypergraph if and only if $\mathcal{H}$ has the Helly property and every cycle of length $\geqslant 3$ possesses three intersecting edges.*

Arboreal hypergraphs form a Helly class (2.17) corresponding to chordal graphs. Since chordal graphs are perfect (see chapter 4), arboreal hypergraphs share the min–max type properties of normal hypergraphs (Theorem 3.19, see also Theorem 4.40 for related results). Algorithms concerning arboreal structures frequently use an "elimination scheme": the nodes of an arboreal hypergraph can be ordered as $v_1, \ldots, v_n$ such that every subhypergraph $\mathcal{H}[v_i, v_{i+1}, \ldots, v_n]$ is arboreal. To handle representative trees, Duchet (1985) introduced an efficient decomposition method; see Theorem 3.16 and (3.18) for some applications.

(3.9) *Totally balanced hypergraphs*. A hypergraph $\mathcal{H}$ (equivalently, its incidence matrix H) is said to be *totally balanced* if no subhypergraph is a polygon. Observe that, by Theorem 3.8, totally balanced hypergraphs are arboreal [see Lehel (1985) for a construction of the representative trees] and form a self-dual subclass of balanced hypergraphs. If H is an $m \times n$ totally balanced matrix, the general *weighted $\boldsymbol{b}$-matching* problem

$$\max\{\boldsymbol{y} \cdot \boldsymbol{a} \mid \boldsymbol{y} \in \mathbb{Q}_+^m, \boldsymbol{y} \leqslant \boldsymbol{y}_0 \text{ and } H\boldsymbol{y} \leqslant \boldsymbol{b}\} \tag{3.10}$$

has an integral optimal solution for every $\boldsymbol{a}, \boldsymbol{y}_0, \boldsymbol{b} \geqslant 0$ (such problems arise as duals of some location problems, for instance, if, on a tree, distributing centers are to be chosen in order to serve clients at a minimum cost: see (3.14)).

Moreover, the solutions of (3.10) can be found by a "greedy" algorithm: coordinates are optimized one by one according to a certain reordering of rows and columns of $\boldsymbol{M}$, based on the following theorem.

Theorem 3.11 (Lubiw 1985, Hoffman et al. 1985). *A* 0–1 *matrix is totally balanced if and only if rows and columns can be reordered such that the resulting matrix has no submatrix of the form*

$$\Gamma = \begin{pmatrix} 1 & 1 \\ 1 & 0 \end{pmatrix}.$$

An elegant way of proving Theorem 3.11 is to consider the colexicographic order on vectors: $(a_1, a_2, \ldots) <_{\text{colex}} (b_1, b_2, \ldots)$ if $a_k < b_k$ for the *largest* index k such that $a_k \neq b_k$. By suitable permutations of rows and columns any matrix can be arranged in "*colexical form*" (i.e., both rows and columns of the transformed matrix are in colexicographic order). For a totally balanced matrix, every colexical form is Γ-free! Hence the method also provides a fast algorithm to test whether a given matrix is totally balanced. A straightforward consequence of Theorem 3.11 is the following property, due to Brouwer and Kolen: any totally balanced hypergraph contains a "nest point", i.e., a node v such that the edges containing v are totally ordered under inclusion.

To prove that a hypergraph is totally balanced, the following criterion, due to Lubiw, is useful.

Proposition 3.12. *The* 0–1 *intersection matrix of the edges of a totally balanced hypergraph is totally balanced.*

Line-graphs of totally balanced hypergraphs ("*strongly chordal graphs*") have been extensively studied. The above results were obtained independently by Farber. For instance, Farber showed that a graph is totally balanced if and only if its adjacency matrix is totally balanced (compare with Proposition 3.12) and he characterized those graphs by a simple list of excluded induced subgraphs.

(3.13) A *chordal bipartite graph* is a bipartite graph for which every cycle of length >4 has a chord. Chordal bipartite graphs were discussed by Golumbic (1980) in relation with perfect Gaussian elimination for systems of linear equations. Actually, chordal bipartite graphs coincide with the bipartite incidence graphs of totally balanced hypergraphs. Perfect elimination schemes correspond to Γ-free orderings; to the first non-zero entry of the first row of a Γ-free form of the matrix corresponds a *bisimplicial edge* of the graph, i.e., an edge $\{i, j\}$ such that all vertices adjacent to i or j induce a complete bipartite graph.

(3.14) Consider a tree T and let $S \subseteq V(T)$. For every node x let $0 \leq d_0^x \leq d_1^x \leq d_2^x \leq \cdots$ denote the sequence of distances from x to elements of S and let T_k denote the minimal subtree containing x and the elements $s \in S$ with $d(x, s) \leq d_k^x$. For $p \in \mathbb{N}$ denote by $S(x, p)$ the set of all nodes y of T_k such that $d(x, y) \leq p$ where k is chosen so that $d_{k-1}^x < p \leq d_k^x$. Tamir (1985) showed that the hypergraph $(E_i)_{i \in I} := \{S(x, p) \mid x \in V(T), p \in \mathbb{N}\}$ is totally balanced. By Proposition

3.12 the hypergraph $(\bar{E}_i)_{i\in I}$ on I, where $\bar{E}_i = \{j \in I \mid E_i \cap E_j \neq \emptyset\}$, is totally balanced too. For $S = V$, we obtain a "neighborhood system" (Giles). For $|S| = 1$, hypergraph (E_i) consists of paths of an arborescence while the sets $\bar{E}_i$ form a "bipath-system" (Frank).

For more information on totally balanced structures, see Anstee and Farber (1984), Lubiw (1985), Hoffman et al. (1985) and their references.

Definition 3.15. $\mathcal{H}$ is an *interval hypergraph* if, for some total ordering of its nodes, every edge forms an interval. Interval hypergraphs constitute a Helly class (2.17) which is closed under passing to subhypergraphs and corresponds to *interval graphs* (see chapter 4). More precisely, a hypergraph $\mathcal{H}$ is an interval hypergraph if and only if it has the Helly property and the line-graph of the augmented hypergraph $\mathcal{H}^+$, resulting from $\mathcal{H}$ by adding all singletons as new edges, is an interval graph.

A characterization of interval hypergraphs by forbidden configuration (Tucker 1972) can be promptly derived from the following "betweenness theorem" (in hypergraph $\mathcal{H}$, a node x lies *between* two other nodes y, z if every y–z path has an edge containing x).

Theorem 3.16 (Duchet 1978). *$\mathcal{H}$ is an interval hypergraph iff every set of* 3 *nodes contains a node which lies between the two others.*

Investigating the König property for polyominoes (see (3.20)), Györi (1984) found a nice min–max property of interval hypergraphs.

Theorem 3.17. *Let $\mathcal{H} = (E_i)_{i\in I}$ be a family of intervals of a totally ordered set V. Let $\alpha_\cup(\mathcal{H})$ denote the maximum length of a sequence $i_1, \ldots, i_p$ such that $E_{i_{k+1}} \not\subseteq E_{i_1} \cup E_{i_2} \cup \cdots \cup E_{i_k}$ for $1 \leqslant k < p$. Let $\theta_\cup(\mathcal{H})$ denote the minimal number of V-intervals we can choose so that every edge of $\mathcal{H}$ is expressible as a union of chosen intervals. Then*

$$\alpha_\cup(\mathcal{H}) = \theta_\cup(\mathcal{H}) .$$

(3.18) *Paths in trees.* Families of node-sets of paths in a tree (*TP-hypergraphs*), families of oriented paths in an oriented tree (*OTP-hypergraphs*) or in a rooted tree (*RTP-hypergraphs*) generalize interval hypergraphs. The problem of characterizing such structures was raised by Renz in 1970. By (3.7), OTP-hypergraphs are unimodular. RTP-hypergraphs are preserved under taking subhypergraphs [hence they have the additional property of being totally balanced, compare with (3.14)], but this is not true for OTP-hypergraphs and TP-hypergraphs. Nevertheless, in each case, a characterization by a (huge) list of excluded configurations can be given: a hypergraph is in the class iff it is arboreal and has no subhypergraph in the list (Arami and Duchet 1988). It also appears that a TP-hypergraph is an OTP-hypergraph if and only if it is balanced. Polynomial recognition algorithms are easily derived. Similar results on the corresponding line-graphs follow from a suitable "Helly class representation method" (see 2.17).

3.5. Normal hypergraphs

Following Lovász (1972), let us call a hypergraph $\mathcal{H}$ *normal* if (i) it has the Helly property, and (ii) its line-graph $L(\mathcal{H})$ is perfect (i.e., $\chi(G') = \omega(G')$ for every induced subgraph G'; see chapter 4)). Arborcal hypergraphs are normal, since, by a theorem of Gallaï, chordal graphs are perfect.

For $\mathcal{H}' \subseteq \mathcal{H}$, $L(\mathcal{H}')$ is an induced subgraph of $L(\mathcal{H})$. Since $\chi(L(\mathcal{H}')) = \gamma(\mathcal{H}')$ and $\omega(L(\mathcal{H}')) = \Delta(\mathcal{H}')$, we see that $\mathcal{H}$ is normal iff every partial hypergraph $\mathcal{H}' \subseteq \mathcal{H}$ has the edge-coloring property: $\gamma(\mathcal{H}') = \Delta(\mathcal{H}')$. Hence, balanced hypergraphs are normal.

By (2.17), to every perfect graph G corresponds (canonically) a normal hypergraph $\mathcal{H} = \mathcal{K}^{\mathrm{T}}(\mathrm{G})$; clutters of the form $\mathcal{K}(G)$ are called *conormal* clutters. Matchings of $\mathcal{H}$ correspond to stable sets of $L(\mathcal{H})$; blocking sets correspond (because of the Helly property) to clique-covers of G. Equivalently, $\nu(\mathcal{H}) = \alpha(G) = \omega(\bar{G})$ and $\tau(\mathcal{H}) = \chi(\bar{G})$. Thus, the implication "G perfect $\Rightarrow$ $\bar{G}$ perfect" (Lovász's perfect graph theorem) is a part of the following general theorem.

Theorem 3.19 (Lovász, see Lovász 1979, Berge 1989, Schrijver 1979a). *For any hypergraph $\mathcal{H}$, the following conditions are equivalent*:

(i) *$\mathcal{H}$ is normal*;
(ii) *Every partial hypergraph $\mathcal{H}' \subseteq \mathcal{H}$ has the König property*: $\tau(\mathcal{H}') = \nu(\mathcal{H}')$;
(iii) $\tau^*(\mathcal{H}') = \tau(\mathcal{H}')$ *for every* $\mathcal{H}' \subseteq \mathcal{H}$;
(iv) $\tau_2(\mathcal{H}') = 2\tau(\mathcal{H}')$ *for every* $\mathcal{H}' \subseteq \mathcal{H}$;
(v) $\nu(\mathcal{H}')\Delta(\mathcal{H}') \leqslant |\mathcal{H}'|$ *for every every* $\mathcal{H}' \subseteq \mathcal{H}$.

See chapters 3 and 30 for polyhedral proofs of this theorem. An easy (but crucial!) point is that multiplication of edges in a normal hypergraph preserves the edge coloring property. It follows, that $\mathcal{H}$ is normal iff the maximum in (3.2) has an integral optimal solution for $\boldsymbol{b} = \mathbf{1}$ and every $\mathbf{a} \in \mathbb{Z}_+^{\mathscr{E}}$ (equivalently the fractional matching polyhedron is integral). Actually, the system defining FM is *totally dual integral* (Lovász 1972, Chvátal 1975, Fulkerson 1972). Inequality 3.19 (v) may be viewed as the "width–length inequality" of *antiblocking pairs* of polyhedra (the duals of normal clutters corresponding to complementary graphs form an antiblocking pair: viz. chapter 30). See chapter 4 for more information on perfect graphs.

(3.20) *Polyominoes*. Let S denote the set of unit squares (= *cells*) with integer vertices in $\mathbb{R}^2$. A *polyomino* is a finite subset of S (here, a polyomino need not be connected). A *box* is a rectangular polyomino. A hypergraph $\mathcal{H}(P)$ is associated with any polyomino P as follows: the nodes are the cells of P, and the edges are the $\subset$-maximal boxes contained in P. Polyomino-hypergraphs have the Helly property and are conform (Theorem 2.16). If P is simply connected, then $\mathcal{H}(P)$ is normal (Shearer 1982, Berge 1989). That is the case when P is horizontally convex, i.e., when cuts of P by horizontal lines are intervals (Berge et al. 1982). Theorem 3.5 has the following consequence: hypergraphs corresponding to horizontally convex polyominoes have the dual König property: $\rho(\mathcal{H}(P)) = \alpha_{[2]}(\mathcal{H}(P))$. For simply connected polyominoes, Györi proved: $\rho(\mathcal{H}(P)) \leq 2\alpha_{[2]}(\mathcal{H}(P)) - 1$.

3.6. *Mengerian and Fulkersonian hypergraphs*

Let $\mathcal{H} = (V, \mathscr{E})$ be a hypergraph. *Expanding* a node $v \in V$ by an integer k means replacing v by k new vertices $v_1, \ldots, v_k$ and each edge E containing v by k new edges $E - v \cup v_1, \ldots, E - v \cup v_k$. For instance, if v is the set of arcs of a directed graph D with two given nodes s, t, and $\mathcal{H}$ is the collection of s–t paths, then expanding v by k corresponds to replacing, in D, arc v by k parallel arcs. Expanding a vertex by 0 is equivalent to considering the partial hypergraph $\mathcal{H} - v$.

More generally, for $\boldsymbol{w} \in \mathbb{Z}_+^V$, the *expanded* hypergraph $\mathcal{H}^{\boldsymbol{w}}$ arises from $\mathcal{H}$ by expanding, successively, every node by the corresponding component of $\boldsymbol{w}$. Let us consider the following properties:

$$\begin{aligned} &\text{(a)} \quad \nu(\mathcal{H}^{\boldsymbol{w}}) = \nu^*(\mathcal{H}^{\boldsymbol{w}}) \quad \text{for every } \boldsymbol{w} \in \mathbb{Z}_+^V, \\ &\text{(b)} \quad \tau(\mathcal{H}^{\boldsymbol{w}}) = \tau^*(\mathcal{H}^{\boldsymbol{w}}) \quad \text{for every } \boldsymbol{w} \in \mathbb{Z}_+^V. \end{aligned} \tag{3.21}$$

As easily seen, (a) (respectively (b)) amounts to saying that the maximum (minimum) in (3.2) has integral optimal solution for $\boldsymbol{a} = \mathbf{1}$ and every $\boldsymbol{b} = \boldsymbol{w}$. Thus (see chapter 30), condition (a) expresses that the system defining the fractional blocking set polyhedron **FB** is totally dual integral, while condition (b) means that **FB** is integral. Hence ν and τ do not play a symmetric role: (a) is stronger than (b).

Hypergraphs which satisfy (3.21) (a) are said to be *Mengerian* (or hypergraphs with the $\mathbb{Z}_+$*-max-flow min-cut property*, Seymour (1977); the term "Mengerian" refers to the flow version of Menger's theorem: see Example 3.24(1)). Mengerian hypergraphs are characterized as follows.

Theorem 3.22 (Lovász 1976, see Schrijver 1979a, Berge 1989). *The following conditions are equivalent*:

(i) $\mathcal{H}$ *is Mengerian*;
(ii) *Every expanded hypergraph* $\mathcal{H}^{\boldsymbol{w}}$ *has the König property*;
(iii) $\nu_2(\mathcal{H}^{\boldsymbol{w}}) = 2\nu(\mathcal{H}^{\boldsymbol{w}})$ *for every* $\boldsymbol{w} \in \mathbb{Z}_+^V$;
(iv) *For some integer* $k \geq 2, \nu_k(\mathcal{H}^{\boldsymbol{w}}) = k\nu(\mathcal{H}^{\boldsymbol{w}})$ *holds for every* $\boldsymbol{w} \in \mathbb{Z}_+^V$.

Hypergraphs which satisfy (3.21)(b) are said to be *Fulkersonian* [= hypergraphs with the $\mathbb{Q}_+$*-max-flow min-cut property* in Seymour (1977), *paranormal hypergraphs in Berge* (1989)]. We underlined above the importance of anti-blocking polyhedra for normal hypergraphs; for Fulkersonian hypergraphs, the theory of blocking-polyhedra, developed by Fulkerson (1971) and Lehman (1979) (see Schrijver 1979b, 1983, Berge 1989, and chapter 30), plays a similar role.

Theorem 3.23 (Fulkerson 1971, Lehman 1979). *Let $\mathcal{H}$ be a hypergraph and let $\mathcal{K} = \mathrm{Bl}(\mathcal{H})$ denote its blocker. The following conditions are equivalent:*

(i) *$\mathcal{H}$ is Fulkersonian;*
(ii) *$\tau^*(\mathcal{H}^w)$ is an integer for every $w \in \mathbb{Z}_+^V$;*
(iii) *$\tau^*(\mathcal{H}^w) = \tau(\mathcal{H}^w)$ for every $w \in \mathbb{Z}_+^V$;*
(iv) *$\tau(\mathcal{H}^w)\tau(\mathcal{K}^l) \leq w \cdot l$ for every $w, l \in \mathbb{Z}_+^V$;*
(i)′–(iii)′ *obtained from* (i)–(iii) *with $\mathcal{K}$ instead of $\mathcal{H}$.*

Observe that condition (iv) of this theorem (the "*width–length inequality*") is preserved under taking blockers, since $\mathrm{Bl}(\mathrm{Bl}(\mathcal{H}))$ consists of the $\subset$-minimal edges of $\mathcal{H}$ (see (4.4)).

Seymour showed that if a clutter $\mathcal{C}$ is Mengerian or Fulkersonian, then any *contraction–deletion minor* of $\mathcal{C}$ has the same property. Here, a contraction–deletion minor $\mathcal{C}'$ arises from $\mathcal{C}$ by taking disjoint subsets $V_1, V_2 \subseteq V(\mathcal{C})$ and defining $\mathcal{C}'$ to be the collection of minimal sets in

$$\{E \setminus V_1 \mid E \in \mathcal{C}, E \cap V_2 = \emptyset\} .$$

The clutter $Q_6 = K_4^{\top}$ is an example of a minimal non-Mengerian clutter. Its Blocker $\mathrm{Bl}(Q_6)$ is Mengerian. It was conjectured that if $\mathcal{C}$ has no Q_6 minor and if $\mathrm{Bl}(\mathcal{C})$ is Mengerian, then $\mathcal{C}$ is Mengerian. This is, however, contradicted by an example due to Schrijver (1980, 1983). To characterize Mengerian or Fulkersonian hypergraphs by excluded minors seems to be a difficult problem; see Theorem 3.25 for a partial answer.

Examples 3.24. By Theorem 3.5, balanced hypergraphs and their blockers are Mengerian. Several min–max theorems in Graph Theory provide interesting blocking pairs of Fulkersonian clutters $\mathcal{C}$, $\mathrm{Bl}(\mathcal{C})$ [see chapter 30, and Schrijver (1979a, 1983) for details and proofs].

(1) Given a directed graph $D = (V, A)$ and $s, t \in V$, let $V(\mathcal{C}) = A$; edges of $\mathcal{C}$ are minimal s–t directed cuts, edges of $\mathrm{Bl}(\mathcal{C})$ are s–t minimal directed paths. Clutters $\mathcal{C}$ and $\mathrm{Bl}(\mathcal{C})$ are Mengerian (max-potential min-work and max-flow min-cut theorems).

(2) Given a digraph $D = (V, A)$ with root r, let $V(\mathcal{C}) = A$; edges of $\mathcal{C}$ are minimal r-cuts, edges of $\mathrm{Bl}(\mathcal{C})$ are r-arborescences. Again, $\mathcal{C}$ and $\mathrm{Bl}(\mathcal{C})$ are Mengerian (Fulkerson's optimum branching theorem and Edmonds's disjoint branching theorem).

(3) Given a graph $G = (V, E)$ and $s_1, t_1, s_2, t_2 \in V$, let $V(\mathcal{C}) = E$; edges of $\mathcal{C}$ are minimal "two-commodity cuts" which separate each of the pairs $\{s_1, t_1\}$, $\{s_2, t_2\}$, edges of $\mathrm{Bl}(\mathcal{C})$ are minimal s_1–t_1 paths and minimal s_2–t_2 paths. Seymour and Hu proved that $\mathcal{C}$ and $\mathrm{Bl}(\mathcal{C})$ are $\frac{1}{2}$-Mengerian, i.e., in both cases, the maximum in (3.2) has half-integral optimal solution for $\boldsymbol{a} = \mathbf{1}, \boldsymbol{b} \in \mathbb{Z}_+^E$.

(4) Given $G = (V, E)$ and $S \subseteq V$, let $V(\mathcal{C}) = E$; edges of $\mathcal{C}$ are minimal S-cuts, edges of $\mathrm{Bl}(\mathcal{C})$ are minimal S-joins. The clutter $\mathcal{C}$ is $\frac{1}{2}$-Mengerian (Edmonds and Johnson 1973).

(5) Given $D=(V,A)$, let $V(\mathscr{C})=A$; edges of $\mathscr{C}$ are minimal directed cuts (= sets of arcs entering a nonempty proper subset S such that no arc of D leaves S), edges of $\mathrm{Bl}(\mathscr{C})$ are minimal coverings by directed cuts. The Lucchesi–Younger theorem asserts that $\mathscr{C}$ is Mengerian; generally $\mathrm{Bl}(\mathscr{C})$ is not Mengerian (Schrijver 1980, 1983, Berge 1989).

Seymour was able to characterize Mengerian hypergraphs for an interesting class of clutters. A clutter $\mathscr{C}$ is called *binary* if for all $E_1, E_2, \ldots, E_k$ in $\mathscr{C}$ with k odd, the set $E_1 \Delta E_2 \Delta \cdots \Delta E_k$ includes some edge of $\mathscr{C}$. The blocker of a binary clutter is binary again. Examples 3.24 (1), (3) and (4) are examples of binary clutters.

Theorem 3.25 (Seymour 1977). *A binary clutter is Mengerian if and only if it has no Q_6 minor.*

No characterization of Fulkersonian binary clutters is known. The binary clutter $\mathscr{O}(G)$ formed with the edge-sets of odd circuits of a graph G is Fulkersonian if G is planar, but $\mathscr{O}(K_5)$ is not Fulkersonian.

3.7. Further refinements

The following refinements of Theorems 3.19, 3.22 and 3.23 provide general conditions under which the k-matching number ν_k coincides with the k-blocking number.

Theorem 3.26 (Lovász 1975a, 1977, see Schrijver 1979a, Füredi 1988). *Let $k \in \{1,2,3\}$ be given. If $k\tau^*(\mathscr{H}')$ is an integer for each partial hypergraph $\mathscr{H}' \subseteq \mathscr{H}$, then $\tau_k(\mathscr{H}) = \nu_k(\mathscr{H})$.*

Theorem 3.27 (Schrijver and Seymour 1979, Schrijver 1979a, Füredi 1988). *Let k be a positive integer. Each of the following conditions implies $\tau_k(\mathscr{H}) = \nu_k(\mathscr{H})$:*
(i) *$k\tau^*(\mathscr{H}^w)$ is an integer for every $w \in \mathbb{Z}_+^{V(\mathscr{H})}$;*
(ii) *$\nu_{2k}(\mathscr{H}^w) = 2\nu_k(\mathscr{H}^w)$ for every $w \in \mathbb{Z}_+^{V(\mathscr{H})}$.*

Theorem 3.26 does not hold for every k. Schrijver and Seymour gave an example for which $20\tau^*(\mathscr{H}') \in \mathbb{N}$ for every $\mathscr{H}' \subseteq \mathscr{H}$ but $\tau_{20}(\mathscr{H}) \neq 20\tau^*(\mathscr{H})$ and $\nu_{60}(\mathscr{H}) \neq 60\nu^*(\mathscr{H})$.

4. Stability, transversals and matchings

For a general hypergraph, the problem of determining its packing and covering parameters such as α, $\alpha_{[2]}$, ν, ρ, and τ (viz. definitions in section 1.7) is NP-hard (practical algorithms may be found in Nemhauser and Wolsey 1988); hence estimates are useful. Basic inequalities are

$$
\begin{array}{ll}
\text{(i)} & \nu \leqslant \tau \text{ and, for } r\text{-graphs, } \tau \leqslant r\nu\,, \\
\text{(ii)} & m/\Delta \leqslant \tau\,, \\
\text{(iii)} & \nu \leqslant \rho \text{ for uniform hypergraphs}\,, \\
\text{(iv)} & \nu + (\rho - \nu)/r \leqslant n/r \leqslant \rho - (\rho - \nu)/r \text{ for } r\text{-graphs}\,, \\
\text{(v)} & \tau \leqslant r\tau^* - r + 1 \text{ for } r\text{-graphs}\,.
\end{array}
\tag{4.1}
$$

Equality is possible in each case.

Gallai observed that the identity $\alpha + \tau = n$ holds for every graph. That trivially extends to hypergraphs, since stable sets correspond to blocking sets under taking complements. Considering dual hypergraphs, we have the additional identities:

$$\rho(\mathcal{H}) = \tau(\mathcal{H}^\top)\,, \qquad \alpha_{[2]}(\mathcal{H}) = \nu(\mathcal{H}^\top) = \alpha(\mathcal{H}_{[2]})\,. \tag{4.2}$$

Consequently, we may focus without loss of generality on *blocking number* (node-covering problems) and *matching numbers* (edge-packing problems), except for few cases in which other parameters appear to be more natural. The König–Hall theorem and its relatives (systems of distinct representatives, "transversal theory", cf. chapters 3 and 9) has widely inspired the earliest results on blocking and matching numbers. Although this point of view is still active (e.g., Theorem 4.15), important inequalities were obtained by linear algebra, averaging methods and probabilistic arguments. Polyhedral combinatorics [see (3.1)–(3.3) and chapter 30] are of a particular efficiency since the *fractional packing problem* (finding a maximum matching with weighted edges) and the *fractional covering problem* (finding a minimum blocking set with weighted nodes) appear as dual linear programs. Recalling (3.3), we have

$$
\begin{aligned}
\nu(\mathcal{H}) = \nu_1(\mathcal{H}) &= \min_{k\in\mathbb{N}} \frac{\nu_k(\mathcal{H})}{k} \leqslant \max_{k\in N} \frac{\nu_k(\mathcal{H})}{k} = \nu^*(\mathcal{H}) \\
&= \tau^*(\mathcal{H}) = \min_{k\in\mathbb{N}} \frac{\tau_k(\mathcal{H})}{k} \leqslant \max_{k\in\mathbb{N}} \frac{\tau_k(\mathcal{H})}{k} = \tau_1(\mathcal{H}) = \tau(\mathcal{H})
\end{aligned}
\tag{4.3}
$$

Section 2 was mainly devoted to the equality cases in (4.3). Here we wish to estimate how good the inequalities are, so the central question becomes: how far is τ from ν?

4.1. Examples of packing and covering problems

(4.4) *Blocking clutters*. Let $\mathcal{H}$ and $\mathcal{K}$ be two clutters on the same node-set V. Then $\mathcal{K} = \mathrm{Bl}(\mathcal{H})$ if and only if, for every bicoloring of the nodes, say in pink and green, *either* $\mathcal{H}$ has a pink edge *or* (exclusive) $\mathcal{H}'$ has a green edge. This property, which implies $\mathrm{Bl}(\mathrm{Bl}(\mathcal{H})) = \mathcal{H}$, is the discrete version of the "width–length inequality" which characterizes blocking pairs of polyhedra (see Theorem 3.23 and chapter 30). In a matroid, for instance, bases and cocircuits form a pair of blocking clutters and the above property is usually referred to as Minty's Lemma (see chapter 9). Various other examples were given in section 3.6.

Let $\mathcal{H}, \mathcal{K} = \mathrm{Bl}(\mathcal{H})$ be a pair of blocking clutters and f be any real-valued function on V. The following duality formula ("bottleneck extrema") was obtained by Edmonds and Fulkerson (1970) in the context of facility location problems (see chapter 35):

$$\max_{E \in \mathcal{H}}(\min_{v \in E} f(v)) = \min_{F \in \mathcal{K}}(\max_{v \in F} f(v)) . \tag{4.5}$$

(4.6) *Self-blocking clutters* (i.e., $\mathcal{H} = \mathrm{Bl}(\mathcal{H})$ holds) coincide with intersecting 3-chromatic clutters (cf. section 5). Obvious examples are K^r_{2r-1} (for $\mathrm{Bl}(K^r_n) \simeq K^{n-r+1}_n$), the Fano plane) and the *fan*, on ground-set $1, \ldots, n$ with edges $2, \ldots, n$ and $1, i$ for $i = 2, \ldots, n$. The following less trivial example is due to Erdős and Lovász: as node-set, let us take the union of r pairwise disjoint sets V_i with $|V_i| = i$ for $1 \leq i \leq r$; the edges have the form $V_i \cup T$, where $1 \leq i \leq r$, and T is any set such that $|T| = r - i$ and $|T \cap V_j| = 1$ for $i < j \leq r$. See (5.27).

(4.7) Let $\mathcal{H}$ be the hypergraph whose edges are the hyperlanes of the k-dimensional affine space over $\mathrm{GF}(q)$. The trivial lower bound on the blocking number is in fact the right value: $\tau(\mathcal{H}) = k(q-1) + 1$ (Jamison 1977, see chapter 32 for a proof). If lines are considered instead of hyperplanes, the blocking number is not known, even for $q = 3$.

(4.8) The Ramsey number $R(p, q)$ (see chapter 25) may be viewed as the optimum of a matching problem (Erdős et al. 1971, see Berge 1989):

$$R(p, q) = 1 + \max(\nu(\mathcal{H} \otimes \mathcal{H}') \mid \nu(\mathcal{H}) < p, \nu(\mathcal{H}') < q) .$$

Turán's problem 4.9. For $1 < r \leq n$, let $T(n, p, r)$ denote the minimum number of edges of r-graph $\mathcal{H}$ of order n with $\alpha(\mathcal{H}) < p$. $T(n, p, r)$ is the blocking number of the hypergraph $K^r_p \mid K^r_n$ whose edges are the copies of K^r_p in K^r_n. Inequality (4.1) (ii) gives $\binom{n}{r}\binom{p}{r}^{-1} \leq T(n, p, r)$.

In fact, the ratio $T(n, p, r)\binom{n}{r}^{-1}$ has a limit $t(p, r)$ as n goes to infinity (Katona et al. 1964). The determination of $t(p, r)$ is one of the most challenging problems in extremal set theory (see chapter 24). Let $\alpha(n, r, p) = \binom{n}{r} - T(n, p, r)$ denote the stability number of $K^r_p \mid K^r_n$. Erdős and Sauer (see Erdős 1981) made the following conjecture.

Conjecture 4.10. The edges of any r-uniform hypergraph can be partitioned into at most $\alpha(n, p, r)$ edges and copies of K^r_p.

For partial results see Theorem 4.15 and its consequences.

Zarankiewicz's Problem 4.11. The following problem, raised in 1951 is not yet completely solved: what is the smallest integer $z(p, q)$ such that every boolean $p \times q$-matrix with $z(p, q)$ entries equal to 1 necessarily contains an $r \times s$-submatrix of which all entries equal 1? Again, this is equivalent to determining a blocking number: $z(p, q) = \alpha(K_p^r \otimes K_q^s) + 1 = pq + 1 - \tau(K_p^r \otimes K_q^s)$. See chapter 24 for further information.

4.2. Stability number

The stability number appears in situations where a given pattern is to be avoided in a large part of some structure. The lower bound on the Turán number given above, easily implies the following inequality:

$$\alpha \geqslant nm^{-1/r} \quad \text{for any } r\text{-uniform hypergraph}. \tag{4.12}$$

For loop-free hypergraphs, inequality $\alpha \geqslant n/(\Delta + 1)$ is obvious. If $\mathcal{H}$ is loop-free, the minimum number of stable sets needed to cover $V(\mathcal{H})$ is precisely the chromatic number $\chi(\mathcal{H})$. Hence, from inequalities (5.14) and (5.15), we have the following strengthening:

$$\alpha \geqslant \frac{n}{\Delta_l + 1} \quad \text{for any loopless hypergraph}. \tag{4.13}$$

Here Δ_l stands for the maximum *linear degree*, i.e., the maximum size of a linear star. Various similar bounds can be obtained by simple counting arguments, Berge (1989). For instance, if α' denotes the minimum cardinality of an inclusion-maximal stable set, Meyer (1975) and Lorea (1972) (see Berge 1989) obtained the following inequality, where h denotes the minimum cardinality of an edge:

$$\alpha' \geqslant \frac{n(h-1)}{\Delta + h - 1}. \tag{4.14}$$

Parameter α' is intimately related to the Helly property: let $\mathcal{H} = (E_1, \ldots, E_m)$ be a hypergraph on a set V of $p + q$ elements. Put $\mathcal{H}^c = (V - E, \ldots, V - E_m)$. By Theorem 2.16, we have $\alpha'(\mathcal{H}) \geqslant p$ if and only if $\mathcal{H}^c$ has the q-Helly property.

The following interesting extension of the König–Egervary theorem for bipartite graphs was proved by Lehel (1982).

Theorem 4.15. *Suppose $\mathcal{H}$ has no isolated node and every $A \subseteq V(\mathcal{H})$ contains a stable set of $\mathcal{H}$ of cardinality at least $\frac{1}{2}|A|$. Then $\rho(\mathcal{H}) \leqslant \alpha(\mathcal{H})$.*

Applying this theorem to suitable hypergraphs, we obtain:

Corollary 4.16. *For $1 < r < p$, any r-graph G can be edge-covered by $\alpha(G, p, r)$ edges and copies of K_p^r, where $\alpha(G, p, r)$ is the maximum number of edges of a K_p^r-free* partial hypergraph of G.

Corollary 4.17 (Lehel and Tuza 1982). *Let F be a nonbipartite graph and G an arbitrary graph. Then the edge-set of G can be covered by $\alpha_F(G)$ edges and copies of F, where $\alpha_F(G)$ is the maximum number of edges of an F-free subgraph of G.*

Corollary 4.17 answered a conjecture of Bollobás: with the notation of Turán's Problem and Conjecture 4.10, the edges of any r-uniform hypergraph can be covered by $\alpha(n, p, r)$ edges or copies of K_p^r. Lehel conjectures that in fact we can pack: the copies of K_p^r could be chosen to be pairwise edge-disjoint. This would imply the Erdős–Sauer Conjecture 4.10.

More precise lower bounds on α are known under additional assumptions.

Theorem 4.18 (Ajtai et al. 1982). *Suppose an r-graph $\mathcal{H}$ of order n has no cycle of length less than 5 and has average degree t with $r \ll t \ll n$.* Then, *for some constant c_r depending only on r, we have*

$$\alpha(\mathcal{H}) \geqslant c_r n t^{-1} (\log t)^{1/r} .$$

4.3. Fractional optima

The series of inequalities (4.3), obtained by the linear programming approach of packing and covering problems, yields basic bounds on the common fractional optimum $\tau^* = \nu^*$ of the blocking number and the matching number. Let X be a blocking-set of hypergraph $\mathcal{H}$ and $E \in \mathcal{H}$. Then, for any partial hypergraph $\mathcal{H}' \subset \mathcal{H}$, we have

$$\frac{m(H')}{\Delta(\mathcal{H}')} \leqslant \nu^* = \tau^* \leqslant \frac{|X|}{|E \cap X|} . \tag{4.19}$$

As Lovász observed, equality occurs on the left-hand side (respectively the right-hand side) if the automorphism group of $\mathcal{H}$ is transitive over the edges (nodes). Moreover, for regular r-uniform hypergraphs, we have

$$\frac{m}{\Delta} = \nu^* = \tau^* = \frac{n}{r} . \tag{4.20}$$

An r-uniform hypergraph $\mathcal{H}$ is said to be *quasi-regularizable* (respectively *regularizable*) if, for some natural integers $k(E)$ with positive sum (respectively if, for some positive integers $k(E)$), a regular r-uniform hypergraph $\mathcal{R}'$ can be obtained by multiplying each edge E of $\mathcal{H}$ by $k(E)$. Quasi-regularizable r-uniform hypergraphs are precisely those for which τ^* equals n/r (see Berge 1989). The structure of regularizable hypergraphs is not known.

Berge and Simonovits (Berge 1989, Lovász 1979, Füredi 1988) gave further corollaries of (4.3), relative to hypergraph products (as defined in (1.6)).

Corollary 4.21. *For any pair $\mathcal{H}$, $\mathcal{G}$ of hypergraphs, we have*

$$\begin{aligned}\nu(\mathcal{H})\nu(\mathcal{G}) &\leqslant \nu(\mathcal{H} \otimes \mathcal{G}) \leqslant \nu(\mathcal{H})\nu^*(\mathcal{G}) \leqslant \nu^*(\mathcal{H})\nu^*(\mathcal{G}) = \nu^*(\mathcal{H} \otimes \mathcal{G}) \\ &= \tau^*(\mathcal{H} \otimes \mathcal{G}) = \tau^*(\mathcal{H})\tau^*(\mathcal{G}) \leqslant \tau(\mathcal{H})\tau^*(\mathcal{G}) \leqslant \tau(\mathcal{H} \otimes \mathcal{G}) \\ &\leqslant \tau(\mathcal{H})\tau(\mathcal{G}) .\end{aligned}$$

Corollary 4.22. *For every hypergraph $\mathcal{H}$, we have*

$$\tau^*(\mathcal{H}) = \min_{\mathcal{G}} \frac{\tau(\mathcal{H} \otimes \mathcal{G})}{\tau(\mathcal{G})} = \max_{\mathcal{G}} \frac{\nu(\mathcal{H} \otimes \mathcal{G})}{\nu(\mathcal{G})} = \nu^* ,$$

where $\mathcal{G}$ runs over all hypergraphs.

From the left-hand equality in this corollary it follows that $\tau(\mathcal{H}) = \tau^*(\mathcal{H})$ holds if and only if we have $\tau(\mathcal{H} \otimes \mathcal{G}) = \tau(\mathcal{H})\tau(\mathcal{G})$ for every hypergraph $\mathcal{G}$. The right-hand equality was first obtained in implicit form by Rosenfeld (Füredi 1988).

What are the possible values of τ^*? Lovász (1975b) proved that any rational number ($\geqslant 1$) is the fractional blocking number of some hypergraph. More precisely, given positive integers a, b, c and d such that $a \leqslant b/c \leqslant d$, $b/c > 1$, there exist a hypergraph for which $\nu = a$, $\tau^* = b/c$ and $\tau = d$.

Chung et al. (1988) brought a new insight. Given any rational $t, 0 \leqslant t < 1$, a rank 3 hypergraph exists with $\tau^* - [\tau^*] = t$. Nevertheless, if p/q (with $\gcd(p, q) = 1$) is the fractional blocking number of a hypergraph of rank at most r, then

$$\frac{p}{q} \geqslant \frac{2 \log q}{r \log r} . \tag{4.23}$$

Hence the set N_r of possible values of τ^* for hypergraphs of rank at most r forms a discrete sequence. For instance,

$$N_3 = \{1, \tfrac{4}{3}, \tfrac{3}{2}, \tfrac{5}{3}, \tfrac{7}{4}, \tfrac{9}{5}, 2, \ldots\} .$$

We shall see below that further information on $\tau^* = \nu^*$ can be obtained by standard tools of linear programming. Let us mention here an interesting consequence of Farkas's Lemma of linear programming.

Lemma 4.24 (Füredi 1981). *Any hypergraph $\mathcal{H}$ of rank at least 2 contains a partial subhypergraph $\mathcal{H}' = (V', \mathcal{E}')$ such that $\tau^*(\mathcal{H}') = \tau^*(\mathcal{H})$ and $|\mathcal{E}'| < |V'|$.*

4.4. Comparison between fractional and integral optima

The first general result that bounds the ratio τ/τ^* in terms of maximum degree was obtained by Lovász (1975a), see Berge (1989), Lovász (1979) and Füredi (1988), and in a slightly different form, by Stein (1974), see Füredi (1988) [we refer to Lovász (1975b) and Füredi (1988) for other basic bounds on τ that involve τ^* and the ratios $\tau_k/k\tau^*$]:

$$\tau \leqslant \left(1 + \tfrac{1}{2} + \tfrac{1}{3} + \cdots + \frac{1}{\Delta}\right)\tau^* \leqslant (1 + \log \Delta)\tau^* . \tag{4.25}$$

Inequality is sharp, up to a constant factor.

Proof. In eq. (4.25), τ can be in fact replaced by $|T|$, where T is any minimal blocking set obtained by a greedy algorithm. Let $T = \{x_1, \ldots, x_t\}$ be a blocking set defined as follows: node x_1 has maximum degree Δ_1 in hypergraph $\mathcal{H}_1 = \mathcal{H}$; node x_{i+1} has maximum degree Δ_{i+1} in the partial hypergraph $\mathcal{H}_{i+1} = \mathcal{H}_i - x_i$. Observe that $\Delta_i = m_i - m_{i+1}$, where m_i is the size of $\mathcal{H}_i$ (with $m_{t+1} = 0$). Writing

$$t = \sum_{i=1}^{t} \frac{\Delta_i}{\Delta_i} = \sum_{i=1}^{t} m_i \alpha_i \quad \text{with } \alpha_1 = \frac{1}{\Delta_1} \quad \alpha_i = \frac{1}{\Delta_i} - \frac{1}{\Delta_{i-1}}, \ i = 2, \ldots, t,$$

we have $\alpha_i \geqslant 0$. Applying (4.19), we obtain $m_i \leqslant \tau^*(\mathcal{H}_i)\Delta_i \leqslant \tau^*(\mathcal{H})\Delta$; hence

$$t \leqslant \tau^* \sum_{i=1}^{t} \Delta_i \alpha_i ,$$

which easily implies (4.25). □

Using (4.25) and Corollary 4.21, we easily obtain the following.

Corollary 4.26 (McEliece and Posner 1971, see Berge 1989, Lovász 1979). *Setting $\mathcal{H}^{\otimes p} = \mathcal{H} \otimes \cdots \otimes \mathcal{H}$ (p factors), we have*

$$\tau^*(\mathcal{H}) = \lim_{p \to \infty} \sqrt[p]{\tau_p(\mathcal{H}^{\otimes p})} .$$

The corresponding statement about ν is not true: numbers $\sqrt[p]{\nu_p(\mathcal{H}^{\otimes p})}$ go to a limit $S(\mathcal{H})$, called the *Shannon capacity* of $\mathcal{H}$, but, in general, $S(\mathcal{H}) \neq \nu^*(\mathcal{H})$ (see chapter 1).

For Δ-regular r-graphs, inequality (4.25) can be improved under rather weak "sparseness" assumptions. It is more convenient from now on to deal with the dual problem about the edge-covering number ρ and the corresponding fractional optimum ρ^* which, by (4.19), is $\rho^* = n/r = m/\Delta = \nu^*$. Let an arbitrary s-graph $\mathcal{P}$ be given and denote by $\mathcal{P} \mid K_N^s$ the "factor hypergraph" whose edges correspond to copies of $\mathcal{P}$ in K_N^s (hence, the nodes are the s-subsets of a N-set). A celebrated conjecture formulated by Erdős and Hanani in 1963 (cf. chapter 14) amounts to saying that the ratio $\nu(K_p^s \mid K_N^s)/\rho(K_p^s \mid K_N^s)$ tends to 1 when $N \to \infty$ (p, s fixed). Roughly speaking, near perfect matchings and coverings exist (i.e., $\nu \cong \nu^*$ and $\rho \cong \rho^*$). Rödl found in 1985 an ingenious probabilistic method which succeeded in proving the Erdős–Hanani conjecture. Frankl and Rödl (1985) generalized Rödl's method and result in a hypergraph setting. They proved that near perfect edge-coverings exist in nearly regular r-graphs if the number of edges containing an arbitrary pair of nodes is not too large; the existence of near perfect matching follows since, in view of (4.1)(iv), any inequality of the form $\rho \leqslant (1+\varepsilon)^{n/r}$ implies $\nu \geqslant (1 - \varepsilon(r-1))^{n/r}$. A similar result was previously obtained for random r-graphs (De La Vega 1982). Recently, Pippenger and Spencer obtained a simpler and more powerful version.

Theorem 4.27 (Pippenger et al. 1989). *For all integers $r \geqslant 2$, reals $\lambda \geqslant 1$, and $\varepsilon > 0$*

there exist $\xi > 0$ *so that: if an r-graph of order n has, for some d, the following properties*:

(i) $\Delta(\mathcal{H}) \leq \lambda d$,

(ii) $(1-\xi)d < \deg_{\mathcal{H}}(v) < (1+\xi)d$ *holds for all but at most* ξn *nodes*,

(iii) *any pair of different nodes lies in less than* ξd *edges*,

then $\rho(\mathcal{H}) < (1+\varepsilon)n/r$.

Corollary 4.28 (Frankl and Rödl 1985). *Suppose* $\mathcal{P}$ *is a fixed r-graph of size M and N tends to infinity. Then*

$$\nu(\mathcal{P} \mid K_N^s) = (1 - o(1))\binom{N}{s}/M .$$

For further information and applications, see chapter 33 and Füredi (1988). Perfect decompositions of K_N^s are treated in chapter 14, see also Theorems 5.28 and 5.29 and Corollaries 5.30–5.32.

Returning to general r-graphs, we now compare τ^* with ν. The trivial inequality $\tau^* \leq r\nu$ is not sharp: as shown by the hypergraph formed with disjoint copies of a projective plane of order $r-1$ (lines have r points), the ratio τ^*/ν can reach $r-1+1/r$ (if such a plane exists). Actually, that is precisely the extremal case.

Theorem 4.29 (Füredi 1981). *Let* $\mathcal{H}$ *have rank* $r \geq 2$. *Then*

$$\text{(i)} \quad \tau^*(\mathcal{H}) = \nu^*(\mathcal{H}) \leq \frac{r^2 - r + 1}{r} \nu(\mathcal{H})$$

holds with equality if and only if $\mathcal{H}$ *consists of* $\nu(\mathcal{H})$ *disjoint copies of a projective plane* $\mathcal{P}_r$ *of order* $r-1$. *More precisely, if* $r \geq 3$ *and if* $\mathcal{H}$ *does not contain* $p+1$ *copies of a* $\mathcal{P}_r$, *then we have*

$$\text{(ii)} \quad \tau^*(\mathcal{H}) = \nu^*(\mathcal{H}) \leq (r-1)\nu(\mathcal{H}) + \frac{p}{r} .$$

Theorem 4.29 answered questions which had been considered at their time difficult; the case of intersecting regular r-graphs had been solved by Lovász (1975b) and the case of regular graphs by Bóllobas and Eldridge (chapter 23). Nevertheless, Füredi's proof (mainly by induction on ν) is surprisingly simple. We show below the (crucial) case $\nu = 1$.

Proof (*for* $\nu = 1$). Let $\mathcal{H}$ be an intersecting r-graph with n nodes and m vertices. We prove $\tau^* \leq r-1$. By Lemma 4.24 we may suppose $m \leq n$.

Case 1. There exists an $x_0 \in V(\mathcal{H})$ with degree $k < r$. Set $\mathcal{H}(x_0) = (E_1, \ldots, E_k)$ and

$$t(x) = \begin{cases} 0 & \text{if } x = x_0 , \\ \frac{1}{k} d_0(x) & \text{if } x \neq x_0 , \end{cases}$$

where $d_0(x)$ stands for the degree of x in $\mathcal{H}(x_0)$. As easily seen, t is a fractional blocking. Hence

$$\tau^*(\mathcal{H}) \leqslant \frac{1}{k} \sum_{x \neq x_0} d_0(x) = \frac{1}{k}(r-1)k = r-1 .$$

Case 2. $\mathcal{H}$ has minimum degree $\delta \geqslant r$. We have

$$\delta n \leqslant \sum_{x \in V} d(x) = rm \leqslant rn ,$$

hence $\mathcal{H}$ is regular and m equals n. By (4.19), $\tau^* = m/r$. We prove $m \leqslant r^2 - r$. Let $E_1 \in \mathcal{H}$. Since $\mathcal{H}$ is r-regular, we obtain

$$m(\mathcal{H}) \leqslant 1 + r(r-1) ,$$

with equality only if $|E \cap E_1| = 1$ for any other edge E. Hence $m \leqslant r^2 - r$ holds except when $\mathcal{H}$ is a linear intersecting r-graph with $m = n = r^2 - r + 1$; these requirements force $\mathcal{H}$ to be a projective plane. □

Füredi's theorem was generalized in two directions.

Theorem 4.30 (Frankl and Füredi 1986). *Let $\mathcal{H}$ be a (2-wise) s-intersecting hypergraph of rank $r \geqslant 2$. Then, either*

(i) $\tau^* \leqslant \dfrac{r-1}{s} + \dfrac{1}{r} - \dfrac{r-s}{r(r-1)s}$, or

(ii) *$\mathcal{H}$ is a symmetric (r, s)-design and* $\tau^* = \dfrac{r-1}{s} + \dfrac{1}{r}$.

It is conjectured that Theorem 4.30 still holds if (i) is changed to $\tau^* \leqslant (r-1)/s$. The next theorem concerns the fractional matching polyhedron **FM** associated with a hypergraph, as defined by (3.1), and the (integral) matching polyhedron **IM** which is the convex hull of the set of integral vectors in **FM**.

Theorem 4.31 (Füredi et al. 1993, Füredi 1988). (a) *Any hypergraph $\mathcal{H}$ admits a matching $\mathcal{M} \subset \mathcal{H}$ such that*

$$\tau^*(\mathcal{H}) \leqslant \sum_{E \in \mathcal{M}} \left(|E| - 1 + \frac{1}{|E|}\right).$$

(b) *Suppose $\mathcal{H}$ is r-uniform; if I and F are the matching polytope and the fractional matching polytope, respectively, then*

$$\mathbf{FM} \subseteq \left(r - 1 + \frac{1}{r}\right)\mathbf{IM} .$$

Using a different approach, Aharoni et al. (1985) obtained the following powerful inequality involving ν and ν^*.

Theorem 4.32. *Suppose hypergraph $\mathscr{H}$ has corank h. Then we have:*

$$\nu \geqslant f(\nu^*, m, n) = \frac{(\nu^*)^2}{n - (h-1)(\nu^*)^2/m} \geqslant \frac{(\nu^*)^2}{n} .$$

There are infinitely many cases where equality holds.

This theorem is effective in the sense that we can find a matching of size $f(\nu^*, m, n)$ in time polynomial in m, n.

Proof (*Sketch*). One starts with the following straightforward inequality:

$$\boldsymbol{p}^\top M^\top M\boldsymbol{p} \leqslant n , \tag{4.32a}$$

where $\boldsymbol{p}$ is an optimal fractional matching. The matrix $M^\top M$ is estimated as follows:

$$M^\top M \geqslant \operatorname{diag}(|E_i| - 1) + J - B , \tag{4.32b}$$

where J is the $m \times m$ matrix with all entries equal 1 and B is the adjacency matrix of the graph $G = L(H)$. Consider separately the three terms in (4.32b). Then

$$\boldsymbol{p}^\top J\boldsymbol{p} = (\nu^*)^2 . \tag{4.32c}$$

The Cauchy–Schwartz inequality, joined to the fact that $|E_i| \leqslant h$, yields

$$\boldsymbol{p}^\top \operatorname{diag}(|E_i| - 1)\boldsymbol{p} \geqslant (h-1)(\nu^*)^2 . \tag{4.32d}$$

To estimate the term $\boldsymbol{p}^\top B\boldsymbol{p}$, remark that cliques of G correspond to matchings of $\mathscr{H}$ and apply to this graph an idea developed by Motzkin and Strauss in their proof of Turán's Theorem (see chapter 23): The *maximum of* $\boldsymbol{x}^\top A(G)\boldsymbol{x}$, *subject to conditions* $\boldsymbol{x} \geqslant 0$ *and* $\sum x_i = 1$, *is achieved when* $\boldsymbol{x}$ *is the incidence vector of a largest clique of* G. One obtains

$$\boldsymbol{p}^\top B\boldsymbol{p} \leqslant \left(1 - \frac{1}{\nu}\right)(\upsilon^*)^2 . \tag{4.32e}$$

The inequality of Theorem 4.32 follows. □

4.5. Blocking number versus matching number

Of course, we may compare τ with ν by combining inequalities that involves the fractional optima. But direct comparisons are better in most cases. A possible method consists in keeping one of the parameters τ and ν fixed, then estimating separately the other one. For instance, a greedy heuristic to find a k-matching yields:

Theorem 4.33 (Lovász 1977, 1979). *For every hypergraph of rank r we have*

$$r\nu_k \geqslant k\tau + (k-1)(r-1)\,.$$

It follows that, if $\mathcal{H}$ is assumed to be t-wise intersecting, we have

$$\tau(\mathcal{H}) \leqslant \frac{r-1}{t-1} + 1\,. \tag{4.34}$$

See chapter 24 for further results of that kind.

Following the lines of Theorem 4.32, Aharoni et al. (1985) made a direct comparison between τ and ν.

Theorem 4.35. *Set* $\xi = \mathrm{e}\sqrt{n\nu}$ *(with* $\mathrm{e} = 2.71828\ldots$*). If* $\mathcal{H}$ *has at least* $\mathrm{e}\xi$ *edges, then*

$$\tau \leqslant \min(n, 3\xi\sqrt{\log n/\xi})\,,$$

moreover, this bound is, up to a constant factor, best possible whenever $m > en$. *If* $\mathcal{H}$ *has less than* $\mathrm{e}\xi$ *edges, then* τ *can be greater than* $m/5$.

Proof (*Sketch*). The theorem is effective in that a blocking of the required cardinality is exhibited in polynomial time. Three arguments are involved.

(1) If "small" edges exist, one considers a maximum matching among them; the set of nodes in the union meets all small edges.

(2) The generic step of the construction is a greedy one (as in (4.25)): one picks a node of maximum degree, remove it and the edges containing it.

(3) Towards the end of the process, one estimates the number of steps by saying that it does not exceed the number of remaining edges. □

The comparison of blocking numbers with matching numbers in *multipartite hypergraphs* has received the most attention, since extensions of the König theorem for bipartite graphs are eagerly expected. The following striking conjecture of Ryser (1970, unpublished) would be a remarkable generalization of the König–Egervary theorem on bipartite graphs.

Conjecture 4.36. Every r-partite r-graph satisfies $\tau \leqslant (r-1)\nu$.

Let $\rho_0(\mathcal{H})$ denote the minimum number of pairwise disjoint edges and 1-element sets whose union covers $V(\mathcal{H})$. Lehel (1982) showed that Conjecture 4.36 may be restated as follows.

Conjecture 4.37. Every r-partite r-graph satisfies $\rho_0 \leqslant \alpha$.

A weaker form of Conjecture 4.36 was independently stated stated by Lovász (1975b). A stronger form is due to Meyniel (1984, unpublished):

Conjecture 4.38. If $\mathscr{H}$, and every subhypergraph of $\mathscr{H}$, is k-colorable, then $\tau(\mathscr{H}) \leqslant (k-1)\nu(\mathscr{H})$.

Very little is known on Ryser's conjecture. Szemerédi and Tuza settled some particular cases: $r=3$, $\nu \leqslant 4$; $r=4$, $\nu \leqslant 2$; $r=5$, $\nu=1$. For $r=3$, the current best bound is $\tau \leqslant \frac{8}{3}\nu$. See Tuza (1987).

Gyárfás proved the "fractional version": $\tau^* \leqslant (r-1)\nu$. In fact, this is a corollary of Theorem 4.29. Ryser's conjecture would imply $\tau \leqslant (r-1)\tau^*$. Actually, a much stronger statement holds.

Theorem 4.39 (Lovász 1975b). *For any r-partite r-graph,* $\tau \leqslant \frac{1}{2}r\tau^*$.

The Helly property (2.16)–(2.17) suggests another natural direction towards interesting relationships between τ and ν, for a hypergraph $\mathscr{H}$ has the Helly property if and only if implication $\nu(\mathscr{H}') = 1 \Rightarrow \tau(\mathscr{H}') = 1$ holds true for every $\mathscr{H}' \subseteq \mathscr{H}$. A class of hypergraphs is said to be *τ-bounded* if there exists a "binding function" $f(x)$ such that $\tau(\mathscr{H}) \leqslant f(\nu(\mathscr{H}))$ for every $\mathscr{H}$ in the family. Two interesting classes of hypergraphs, denoted here by $\mathfrak{Q}_k$ and $\mathfrak{F}_k$, were considered by Gyárfás and Lehel (1983).

(a) $\mathscr{H} \in \mathfrak{Q}_k$ iff $\mathscr{H}$ has the Helly property and its line-graph has no induced subgraph isomorphic to the k-dimensional *octahedron graph* Q_k (obtained from the complete graph K_{2k} by deleting k independent edges).

(b) $\mathscr{H} \in \mathfrak{F}_k$ iff there exists a tree T on $V(\mathscr{H})$ such that every edge of $\mathscr{H}$ induces a subforest of T with at most k components.

Theorem 4.40. *Classes* $\mathfrak{Q}_k$ *and* $\mathfrak{F}_k$ *are τ-bounded.*

Surprisingly, the presence of octahedron graphs in statements of that kind is, in some sense, unavoidable. If $\mathfrak{B}$ denotes the Helly class associated to "Berge graphs" (in a Berge graph, neither odd polygons nor their complements arise as induced subgraphs), the Strong Perfect Graph Conjecture (see chapter 4) would imply that members of $\mathfrak{B}$ are normal hypergraphs (hence $\tau = \nu$, by Theorem 3.19). But it is not even known whether $\mathfrak{B}$ is τ-bounded.

4.6. Other estimates of blocking and matching numbers

Most upperbounds for blocking numbers and lower bounds for matching numbers are derived from the preceding results. For instance, the following inequalities on the matching number are trivial consequences of Füredi's Theorem 4.29:

$$\text{(i)}\quad m(\mathscr{H}) \leqslant \frac{r^2-r+1}{r}\Delta(\mathscr{H})\nu(\mathscr{H}) \quad \text{if } \mathscr{H} \text{ has rank } r\,,$$

$$\text{(ii)}\quad n(\mathscr{H}) \leqslant (r^2-r+1)\nu(\mathscr{H}) \quad \text{if } \mathscr{H} \text{ is regular } r\text{-uniform}\,. \qquad (4.41)$$

Equalities are possible only for disjoint copies of projective planes.

The matching problem appears to be more easily handled than the covering problem. For graphs, a min–max formula and a fast algorithm for the matching number are known (chapter 3). For uniform hypergraphs, some interesting lower bounds are obtained by counting arguments. For r-graphs of order n, the inequality

$$m \leqslant \binom{n-1}{r-1} + \cdots + \binom{n-\nu}{r-1} \tag{4.42}$$

holds for sufficiently large n (Hajnal and Rotschild 1973, Lovász 1979). The case $\nu = 1$ of (4.42) is the celebrated Erdős–Chao–Ko–Rado theorem (chapter 24). For *linear* hypergraphs more can be said:

Theorem 4.43 (Seymour 1982). *Suppose $\mathcal{H}$ is a linear hypergraph without multiple loops; then*

$$\nu \geqslant \frac{m}{n}.$$

The case $\nu = 1$ is the famous De Bruijn–Erdős–Ryser inequality for intersecting linear set-systems (chapter 14). Compare with Theorem 5.12. It had been conjectured that the inequality $m < n$ holds for any intersecting r-graph which is maximal in the following sense: the addition of any new edge (with r elements) causes the matching number to increase. This is however contradicted by a construction due to Blokhuis, see Füredi (1988). The construction was improved by Boros, Füredi and Kahn who exhibited a maximal intersecting $(q+1)$-graph of size $\frac{1}{2}q^2 + O(q)$ for any prime power q, $q \equiv 5 \pmod 6$, $q > 20$.

To obtain bounds in the other direction (*upper* estimates of ν, *lower* estimates of τ) we may look for a small part of the hypergraph which forces the matching number to be small or which shows that the blocking number is big. That is the aim of the notions of ν-critical or τ-critical hypergraphs.

A hypergraph $\mathcal{H}$ is *ν-critical* when any substitution of an edge by a smaller new one causes the matching number to increase. A hypergraph $\mathcal{H}$ is *τ-critical* whenever $\tau(\mathcal{H} \setminus E) < \tau(\mathcal{H})$ for every edge E. Main problems about such structures are of extremal nature and we refer to chapter 24 and Füredi (1988) for a detailed account on the subject. We just mention here a few basic methods and results.

The investigation of τ-critical graphs was initiated by Erdős and Gallai in 1961. Lovász (Berge and Ray-Chaudhuri 1974) observed that in a τ-critical hypergraph we have strict inequality $\tau^* < \tau$.

Given an integer r, there are only a finite number of τ-critical r-graphs; An upper bound on the size of a τ-critical hypergraph is given by the following result.

Theorem 4.44 (Bollobás 1965, Jaeger and Payan 1971, see Berge 1989, Lovász 1979). *A τ-critical r-graph has at most $\binom{\tau(\mathcal{H})+r-1}{\tau(\mathcal{H})-1}$ edges.*

For a proof and extensions, see chapter 24.

Rather good estimates of the maximum number of nodes of a τ-critical hypergraph with given blocking number are derived from the next lemma originated in Gallai's work.

Lemma 4.45 (Gyárfás et al. 1982). *If S is a strongly stable set in a τ-critical hypergraph $\mathscr{H}$, then $|\Gamma(S)| \geq |S| + d_{\mathscr{H}}(x) - 1$, where $\Gamma(S)$ is the set of all subsets $E - s$ for $E \in \mathscr{H}, s \in S$.*

In the case of graphs this was proved by Lovász and by Surányi. Recently, Tuza determined the maximum order of a τ-critical r-graph, apart from a constant factor. Notice that determining the minimum size of a τ-critical r-graph on n nodes is precisely Turán's Problem.

It is also true that for a given r, there are only a finite number of ν-critical r-graphs. The earliest result is due to Calczinska–Karlowicz who proved in 1964 the existence of some function $f(r)$ such that, for every intersecting hypergraph of rank r, there is a set S with no more than $f(r)$ elements which induces still an intersecting subhypergraph. With other terms, the function $f(r)$ is the maximum order of a ν-critical intersecting r-graph. The current best bounds for arbitrary ν are due to Tuza who found astonishing connections with bounds on intersecting set–pair systems.

5. Coloring problems

We already encountered in section 3 some important properties relative to hypergraph node-colorings (bicolorability) or to edge-colorings (the "edge-coloring property"). We now turn to a more comprehensive approach of those hypergraph coloring problems. This field has various applications (timetabling and scheduling problems, planning of experiments, multi-user source coding, . . .) and offers rich connections with other combinatorial areas: probabilistic methods, Extremal Set Theory, Ramsey Theory, Discrepancy Theory, etc.

When speaking about hypergraph node-coloring problems, loops play no role (hence they are tacitly deleted). The chromatic number $\chi(\mathscr{H})$ of a hypergraph $\mathscr{H}$ is the smallest number of colors needed to color the nodes so that no monochrome edge occurs. $\mathscr{H}$ is *k-colorable* (respectively *k-chromatic*) if $\chi(\mathscr{H}) \leq k$ ($\chi(\mathscr{H}) = k$); if $\mathscr{H}$ is k-chromatic but any edge-removal causes the chromatic number to drop, $\mathscr{H}$ is said to be *k-chromatic critical* (or *k-critical* for short). These concepts are close analogues of graph theoretical concepts (chapter 4) but they differ in spirit; frequently, they have been at the origin of new ideas or methods.

Any assignment of "colors" to the edges of a hypergraph $\mathscr{H}$ is referred to as an *edge-coloring*. An edge-coloring is *proper* if any two intersecting edges receive different colors. The *chromatic index* of $\mathscr{H}$ [i.e., the least number of colors used in a proper edge-coloring, denoted by $\gamma(\mathscr{H})$] coincides with the usual chromatic number of the line-graph $L(\mathscr{H})$. Parameter $\gamma(\mathscr{H})$ might be thought more relevant to Graph Theory. In fact, important results and conjectures, related to the "edge-coloring property" (i.e., $\gamma(\mathscr{H}) = \Delta(\mathscr{H})$) refer fundamentally to the hypergraph structure; we close the section with them.

5.1. *Some instances of hypergraph coloring problems*

Some examples enlighten connections and contrasts with Graph Theory; others show how hypergraph coloring problems arise in various domains.

(5.1) *The* 4-*color problem and hypergraph bicoloring*. Associate with any graph G a hypergraph $\mathscr{H}$ as follows: nodes of $\mathscr{H}$ correspond to edges of G; edges of $\mathscr{H}$ correspond to odd cycles of G. Then, G is 4-colorable if and only if $\mathscr{H}$ is 2-colorable (Stein, and Woodall 1972).
(5.2) *The bicolorability problem for hypergraphs is* NP-complete (cf. chapter 29). Let F be a boolean formula in conjuctive normal form with V_F as set of variables and $\mathscr{C}_F$ as set of clauses. Define a hypergraph $\mathscr{H} = (V, \mathscr{E})$ as follows:

$$V = \{x \mid x \in V_F\} \cup \{\bar{x} \mid x \in V_F\} \cup \{\#\},$$

where # is a new symbol. For each clause C in $\mathscr{C}_F$ let $e(C)$ be the set containing # and every variable appearing in C, taking with its sign (thus, for example, if $C = x_1 \vee \bar{x}_2$ then $e(C) = \{\#, x_1, \bar{x}_2\}$). Set

$$\mathscr{E} = \{e(C) \mid C \in \mathscr{C}_F\} \cup \{\{x, \bar{x}\} \mid x \in V_F\}.$$

Then F is satisfiable if and only if $\mathscr{H}$ is 2-colorable (Linial and Tarsi 1985).
(5.3) *Hypergraphs do not jump*. A real λ, $0 \leqslant \lambda \leqslant 1$, is a *jump* for an integer $r \geqslant 2$ if, for every positive ε and every integer $p \geqslant r$, any r-graph with order $n > n_0(\varepsilon, p)$ and with size not less than $(\lambda + \varepsilon)\binom{n}{r}$ contains necessarily a partial r-graph of order p with at least $(\lambda + c)\binom{p}{r}$ edges, where c is a positive constant depending only on λ. Graphs "jump", i.e., for $r = 2$ every λ is a jump; an explanation for this phenomenon is that, for any 2-graph G, the ratio $\alpha(G \mid K_n^2)/n^2$ (where $\alpha(G \mid K_n^2)$ denotes the maximum number of edge-disjoint copies of G in K_n^2) asymptotically tends to

$$\frac{1}{2}\left(\frac{\chi(G) - 2}{\chi(G) - 1}\right).$$

Frankl and Rödl (1984) proved that hypergraphs do not jump: for $r \geqslant 3$, $k \geqslant 2r$, the number $(k^{r-1} - 1)/k^{r-1}$ is not a jump. For details and connections with the Erdős–Stone density theorem and Turán's Problem, the reader is referred to chapters 23 and 24.
(5.4) *Ramsey graphs*. Roughly speaking, a Ramsey graph is a graph with neither a large clique nor a large independent set. Let $R_k^r(p)$ denote the smallest integer n such that, for any partition of the edges of K_n^r into k parts (the "colors"), some part contains a (monochrome) copy of K_p^r. To estimate $R_k^r(p)$ amounts to estimating the chromatic number of the *factor hypergraph* $K_p^r \mid K_n^r$ (where nodes are the r-subsets of an n-set and edges correspond to copies of K_p^r): we have

$R_k^r(p) > n$ iff $\chi(K_p^r \mid K_n^r) \leq k$. Lower bounds for R_k^r thus follow from upper bounds on the chromatic numbers of hypergraphs; for $k = 2$, the existence of large Ramsey graphs follows from the bicolorability of r-graphs with few edges. Actually, numerous results in Ramsey Theory use hypergraphs. Since the pioneering work of Erdős on Ramsey graphs, existence theorems and estimates are frequently obtained by probabilistic methods: see eq. (5.16), Theorem 5.17, Corollary 5.18 and eq. (5.19) and chapter 25.

(5.5) Consider a $p \times q$-matrix M. Erdős and Rado asked in 1956 for the largest integer k such that, for every k-coloring of the entries of M, there exists an $r \times s$ submatrix all of whose entries have the same color (see chapter 25). Again, this is a Ramsey-type problem, related to Zarankiewicz's Problem (see chapter 23). The required number is $\chi(K_p^r \otimes K_q^r) - 1$ whose precise value is unknown. Berge and Simonovits (1974), see Berge (1989), proved that $K_p^r \otimes K_q^r$ has maximum chromatic number among products of a p-colorable r-graph by a q-colorable r-graph.

(5.6) In a *positional game* (such as HEX) two players, say Red and Blue, mark alternatively the vertices of a hypergraph $\mathscr{H}$. The first who obtains a monochrome edge of his own color is the winner, Berge (1989). The possibility of a draw depends upon the existence of a *uniform* bicoloring (the numbers of red and blue nodes differ by at most 1). Erdős and Selfridge (1973) showed that the second player has a strategy warranting a draw if $\mathscr{H}$ satisfies the inequality $m(\mathscr{H}) + \Delta(\mathscr{H}) < 2^h$, where h is the minimum cardinality of an edge; compare with (5.16).

(5.7) Let $\mathrm{st}(\mathscr{H})$ denote the family of stable sets of a hypergraph $\mathscr{H}$ (with no loops). Then $\chi(\mathscr{H}) = \rho(\mathrm{st}(\mathscr{H}))$. For instance, if $\mathscr{C}_M$ and $\mathscr{I}_M$ denote the family of circuits and of independent sets of a matroid $M(E)$, respectively, we have, by the Edmonds–Nash-Williams theorem on unions of matroids (see chapter 9),

$$\chi(\mathscr{C}_M) = \rho(\mathscr{I}_M) = \max(\lceil |A| / \mathrm{rank}_M(A) \rceil ; \emptyset \neq A \subset E) .$$

5.2. Bicolorable hypergraphs

What properties of bipartite graphs do bicolorable hypergraphs share? We already noticed that excluding odd polygons as induced subhypergraphs insures bicolorability (Theorem 3.5) on balanced hypergraphs). The problem, raised by Lovász, of deciding whether normal hypergraphs (see section 3.5) are bicolorable or not was answered into the affirmative by the following deep result, due to Fournier and Las Vergnas (1972, 1974), which gives some precise information on the obstructions to bicolorability.

Theorem 5.8. *Suppose a hypergraph $\mathscr{H}$ does not contain any odd cycle $E_1, \ldots, E_{2q+1}, E_1$ with the following properties*:

(a) $|E_i \cap E_{i+1}| = 1$ *for* $i = 1, \ldots, 2q$,

(b) $|E_{2q+1} \cap E_1| \geq 1$,

(c) $E_i \cap E_j \cap E_k = \emptyset$ *for every distinct indices* i, j, k.

Then H is bicolorable.

The hypotheses of this theorem are satisfied by normal hypergraphs and more generally by hypergraphs having *no odd hypercycle* (also named "*pseudobalanced hypergraphs*", Berge 1989).

Proof. The idea is to perform color interchanges on disjoint subsets until either a bicoloring or an odd cycle of the forbidden type is obtained. We suppose $\mathscr{H}$ is a 3-critical hypergraph. We choose an edge E_0 and a vertex $b_0 \in E_0$. By the hypotheses, $\mathscr{H} \backslash E_0$ admits a proper 2-coloring (X_0, Y_0) and we may suppose $E_0 \subset X_0$. Setting $B_0 = \{b_0\}$, we form a sequence of triples (X_i, Y_i, B_i) for $i = 0, 1, \ldots, p$, *as long as possible*, with the following properties (let us recall that $\mathscr{H}[\![A]\!]$ denotes the *trace* of $\mathscr{H}$ on a subset $A \subseteq V(\mathscr{H})$, formed with $\mathscr{H}$-edges included in A):

(i) (X_i, Y_i) is a partition of $V(\mathscr{H})$;
(ii) $X_i = Y_{i-1} \cup B_{i-1}$ and $Y_i = X_{i-1} - B_{i-1}$ (for $i \geqslant 1$);
(iii) Y_i is a stable set of $\mathscr{H}$;
(iv) B_i is a minimal blocking set of $\mathscr{H}[\![X_i]\!]$;
(v) $B_i \subseteq X_0$ for i even, and $B_i \subseteq Y_0$ for i odd;
(vi) $B_0, B_1, \ldots, B_i$ are pairwise disjoint.

All conditions are satisfied by (X_0, Y_0, B_0). No X_i is stable, otherwise (X_i, Y_i) would be a 2-coloring. Hence $B_i \neq \emptyset$ and condition (vi) insures the existence of a largest p. Conditions (5.9) imply:

(vii) $X_i = X_0 - B_0 \cup B_1 \cdots - B_{i-2} \cup B_{i-1}$ for i even;
(viii) $X_i = Y_0 \cup B_0 - B_1 \cdots - B_{i-2} \cup B_{i-1}$ for i odd;
(ix) $X_i \cap X_{i+1} = B_i$.

By the definition of a minimal blocking set, for any $b_i \in B_i$ some edge E_i of $\mathscr{H}[\![X_i]\!]$ is such that $E_i \cap B_i = \{b_i\}$. Since $E_i \subseteq X_i = Y_{i-1} \cup B_{i-1}$ there is a vertex b_{i-1} in $E_i \cap B_{i-1}$. Again, some edge E_{i-1} of $\mathscr{H}[\![X_{i-1}]\!]$ is such that $E_{i-1} \cap B_{i-1} = \{b_{i-1}\}$. Hence by (ix) we have $E_i \cap E_{i-1} = \{b_{i-1}\}$. Iterating the argument, we obtain a sequence

$$b_i E_i b_{i-1} E_{i-1} b_{i-2} \cdots b_1 E_1 b_0 E_0$$

such that $|E_j \cap E_{j+1}| = 1$ for $i \geqslant j \geqslant 1$. Such a sequence is named a *cascade*. The following point is crucial:

Lemma 5.10. *Edges of any cascade are distinct. Moreover, two edges of a cascade are disjoint when their indices have the same parity.*

It is not difficult to see that if some cascade contradicts this lemma, then it contains the edges of an odd cycle with properties (a)–(c). Hence, to finish the proof of Theorem (5.8), it suffices to obtain a contradiction when assuming

Lemma (5.10). The contradiction follows from the construction of a new triple $(X_{p+1}, Y_{p+1}, B_{p+1})$ that satisfies (5.9).

We put $X_{p+1} = Y_p \cup B_p$ and $Y_{p+1} = X_p - B_p$. By (iii), the set Y_{p+1} is stable, hence the trace $\mathcal{H}[\![X_{p+1}]\!]$ is not empty.

– If p is odd, every edge of $\mathcal{H}[\![X_{p+1}]\!]$ meets X_0 (since Y_0 is stable), hence X_0 contains a minimal blocking set B_{p+1} of $\mathcal{H}[\![X_{p+1}]\!]$. By (v), B_{p+1} is disjoint from $B_1, B_3, \ldots, B_p$. Since $B_{p+1} \subseteq X_{p+1} = X_{p-1} - B_p$, the set B_{p+1} is also disjoint from $B_2, B_4, \ldots, B_{p-1}$. Then, conditions (5.9) are fulfilled.

– If p is even, we have $E_0 \cap B_p = \emptyset$ (otherwise, any vertex $b_p \in E_0 \cap B_p$ would be the beginning of a cascade with $E_p \cap E_0 \neq 0$, a contradiction with Lemma 5.10). Hence $E_0 \not\subseteq X_{p+1}$ (since $X_{p+1} = Y_p \cup B_p$, and Y_p is stable). Since E_0 is the only edge of $\mathcal{H}$ to be contained in X_0, every edge of $\mathcal{H}[\![X_{p+1}]\!]$ meets Y_0, hence Y_0 contains a minimal blocking set B_{p+1} of $\mathcal{H}[\![X_{p+1}]\!]$. As for p odd, we check that the sets $X_{p+1}, Y_{p+1}, B_{p+1}$ satisfy (5.9). □

An interesting generalization has been conjectured by Sterboul in 1972 (unpublished).

Conjecture 5.11. A nonbicolorable hypergraph contains an odd hypercycle $E_1, \ldots, E_{2k+1}$ such that $|E_i \cap E_{i+1}| = 1$ for $i = 1, \ldots, 2k$.

As implied by Theorem 5.8 the conjecture holds for 3-graphs.

We shall see later on some sufficient conditions for bicolorability as specializations of general upper bounds on the chromatic number. Let us mention here some other results on bicolorable hypergraphs which mainly rely on the existence of a matching in some associated bipartite graph.

Lovász (1979) observed in 1967 that a hypergraph whose every partial subhypergraph has fewer edges than nodes is bicolorable. This proposition, which is easily derived from Theorem 2.5, led Khatchatryan (1982) to give a general upper bound on $\chi(\mathcal{H})$ in terms of the parameter $\max_{\mathcal{H}' \subset \mathcal{H}}(m(\mathcal{H}') - kn(\mathcal{H}'))$, where k is any fixed integer. Woodall (Berge 1989) exhibited a 3-chromatic hypergraph $\mathcal{H}$ such that $m(\mathcal{H}) = n(\mathcal{H})$ but $m(\mathcal{H}') < n(\mathcal{H}')$ holds for every $\mathcal{H}' \subsetneq \mathcal{H}$. Seymour (1974) gave a characterization of 3-chromatic critical hypergraphs such that $m(\mathcal{H}) = n(\mathcal{H})$ and strengthened Lovász's result as follows.

Theorem 5.12. *Every 3-critical hypergraph satisfies* $m(\mathcal{H}) \geq n(\mathcal{H})$. *Moreover, the bipartite incidence graph admits a node-to-edge matching.*

Seymour's proof is by linear algebra (see chapter 33 and compare with Theorem 5.26). Aharoni and Linial (1986) extended Theorem 5.12 to the infinite case and gave a purely combinatorial proof, using matchings and transversals. Theorem 4.15 of Lehel (1982) is obtained with similar tools and has the following corollary (which extends the König–Egervary theorem on bipartite graphs to bicolorable hypergraphs):

Theorem 5.13. *Every bicolorable hypergraph $\mathcal{H}$ satisfies $\rho(\mathcal{H}) \leq \alpha(\mathcal{H})$.*

5.3. General upper bounds

A simple examination of the greedy coloring algorithms leads to the following inequality, as observed in 1968 by Tomescu and Lovász, see Berge (1973, 1989):

$$\chi(\mathcal{H}) \leq \Delta_l(\mathcal{H}) + 1 , \tag{5.14}$$

where Δ_l denotes the maximum linear degree, i.e., the maximum size of a linear star. Inequality (5.14) has various applications (see Berge 1989); for instance, it leads to estimates of the stability number (such as (4.13)), in conjunction with the following trivial inequalities:

$$\chi(\mathcal{H})\alpha(\mathcal{H}) \geq n , \qquad \chi(\mathcal{H}) + \alpha(\mathcal{H}) \leq n + 1 . \tag{5.15}$$

Brooks's graph-coloring theorem (see chapter 4) can be extended to *linear* hypergraphs: a connected linear hypergraph $\mathcal{H}$ forces equality in (5.14) if and only if $\mathcal{H}$ is an odd polygon or a complete graph (Lepp and Gardner 1973). No exact analogue is known for general hypergraphs: equality in (5.14) is obtained with complete r-graphs but also with nontrivial structures (see, e.g., Lovász 1968, Berge 1973, p. 145).

For uniform hypergraphs, substantially stronger results have been obtained by a probabilistic approach of the following fundamental question: estimate $m_k(r)$, the minimum size of a non k-colorable r-graph.

The basic result amounts to Erdős (1964), see Erdős and Spencer (1974) and chapter 33. Let us color randomly the nodes of an r-graph $\mathcal{H}$, using k colors (each color has probability $1/k$ to be assigned to any given node). If $\mathcal{H}$ has no more than k^{r-1} edges, then, with *positive* probability, a proper k-coloring is obtained. Hence a proper k-coloring does exist! With other words, we have

$$m_k(r) \geq k^{r-1} . \tag{5.16}$$

With each edge E may be associated the event "E is monochrome". To disjoint edges correspond independent events, hence the line-graph $L(\mathcal{H})$ is the dependency graph for those events. Applying Lovász's "local lemma" (see chapter 33), we obtain (with $\mathrm{e} = 2.71828\ldots$):

Theorem 5.17 (Erdős and Lovász 1975). *A non k-colorable r-graph $\mathcal{H}$ must contain an edge which meets at least k^{r-1}/e other edges. Moreover, if $\mathcal{H}$ is linear, it contains at least $k^{r-2}/\mathrm{e}(r-1)$ nodes with degree at least $k^{r-2}/\mathrm{e}(r-1)$ and contains at least $k^{r-2}/\mathrm{e}r(r-1)$ pairwise disjoint edges.*

Corollary 5.18. *For any r-graph, $\chi(\mathcal{H}) \leq (\mathrm{e}r\Delta)^{1/(r-1)}$.*

The lower bound on $m_k(r)$ was improved by Beck (1978) [for a simple proof, see Spencer (1981) and chapter 33]:

$$m_k(r) \geqslant ck^{r-1}r^{1/3} \text{ for some constant } c\,. \tag{5.19}$$

An upper bound for $m_k(r)$ can be obtained as follows. For every k-coloring Γ of a given set S, let us denote by $\mathscr{E}_\Gamma$ the collection of all monochrome r-subsets of S. Then an r-graph $(S, \mathscr{E})$ has no proper k-coloring if and only if $\mathscr{E}$ meets every $\mathscr{E}_\Gamma$. Applying inequality (4.25) on blocking numbers, we obtain (Erdős and Schmidt 1964):

$$m_k(r) \leqslant cr^2(\log k)k^2\,. \tag{5.20}$$

Other interesting estimates (weaker, in most cases, than those above) have been given.

Theorem 5.21.

$$\frac{rk^r}{r + k(k-1)} \leqslant m_k(r) \leqslant \binom{kr - k + 1}{r}\,.$$

The upper bound is due to Herzog and Schönheim (1972). The lower bound (Johnson 1976) can be deduced from a more involved formula due to Hansen and Lorea (1978), see Berge (1989):

Theorem 5.22. *The following inequality implies the k-colorability of $\mathscr{H}$:*

$$\sum_{E \in \mathscr{H}} k^{-|E|}\left(1 + \frac{k^2 - k}{|E|}\right) \cdot \sum_{E \in \mathscr{H}} k^{-|E|} < 1\,.$$

5.4. Lower bounds via topology

To obtain lower estimates of chromatic numbers a powerful topological method was initiated in 1978 by Lovász. To any r-graph $\mathscr{H}$, associate a simplicial complex $C(\mathscr{H})$ as follows: the vertices of $C(\mathscr{H})$ are all the $n!m(\mathscr{H})$ ordered r-tuples $(v_1, v_2, \ldots, v_r)$ of vertices of $\mathscr{H}$, where $\{v_1, \ldots, v_r\} \in \mathscr{H}$. A set of vertices $(v_1^i, \ldots, v_r^i)_{i \in I}$ of $C(\mathscr{H})$ forms a face if there is a complete r-partite subhypergraph of $\mathscr{H}$ on the (pairwise disjoint) sets of vertices $V_1, \ldots, V_r$ such that $v_j^i \in V_j$ for all $i \in I$ and $1 \leqslant j \leqslant r$.

Recall that for $s \geqslant 0$ a topological space T is *s-connected* if for all $0 \leqslant p \leqslant s$ every continuous mapping from the p-dimensional sphere S^p into T can be extended to a continuous mapping from the $(p+1)$-dimensional ball B^{p+1}, with boundary S^p, into T.

Theorem 5.23 (Alon et al. 1986). *For any r-graph $\mathscr{H}$, where r is a prime, we have $\chi(\mathscr{H}) > k$ if $C(\mathscr{H})$ is $(k-1)(r-1)-1$-connected.*

This theorem extends a previous result of Lovász (case $r = 2$) and is conjectured to hold for every positive integer r. A (far nontrivial) consequence of Theorem

5.23 is the solution of a conjecture of Erdős that generalized Kneser's famous conjecture (see below and chapter 4). As a matter of fact, a self contained 3-page proof of a stronger result was recently obtained by a different use of simplicial complexes:

Theorem 5.24 (Sarkaria 1990). *Let n, h, j, k, r be positive integers. If $n(j-1) \geqslant (k-1)(r-1) + rh$, then for every k-coloring of the h-subsets of an n-set there is at least one r-tuple of h-sets having the same color such that any j of them have empty intersection.*

Erdős's conjecture was the case $j = 2$, while Kneser's was the subcase $j = 2$, $r = 2$. The r-graphs whose nodes are the h-subsets of a given set and whose edges correspond to r-tuples of pairwise disjoint h-subsets are often referred to as *Kneser hypergraphs*. The key in the proof of Theorem 5.24 – as in every proof of the special cases mentioned above – is a variant of the Borsuk–Ulam theorem (on the existence of pairs of antipodal points in some member of an arbitrary covering of S^p by $p+1$ open sets, see chapters 4 and 34). Indeed, following Sarkaria, Theorem 5.24 can be seen itself as a "Borsuk–Ulam result".

5.5. Highly chromatic hypergraphs

As we have seen above, to prove that a chromatic number is large is by no means easy. The question is widely related to Ramsey theory. As indicated by (5.4), any Ramsey type result implies that the chromatic number of a certain hypergraph is large. For instance the r-dimensional subspaces of the affine space $\mathrm{GF}(q)^n$ form the edges of a hypergraph with large chromatic number if n is sufficiently large. Most of the constructive or existential proofs of Ramsey Theory rely on highly chromatic graphs or hypergraphs. Of course, for estimates on Ramsey numbers, it is important to "minimize" the construction, i.e., to obtain k-chromatic critical hypergraphs with as few edges as possible (see section 5.6).

Lovász (1968), and then Nešetril and Rödl (1979), gave a constructive proof of the following important result of Erdős and Hajnal (1966), see Berge (1989).

Theorem 5.25. *For all integers r, k, s with $r \geqslant 2$, there exists an r-graph $\mathscr{H}$ with $\chi(\mathscr{H}) \geqslant k$ in which no cycle is shorter than p (i.e., the girth is at least p).*

Interestingly, the earliest constructive proof of this theorem uses a double induction on p, k: the construction of a graph of girth p with chromatic number $\geqslant k$ depends on a previous construction of r-graphs (with r large) of girth $p-1$.

From Theorem 5.25 can be derived for instance the following result: given a graph G of order n and a positive integer k, there exists a graph $R(G)$ with the *same clique number* as G such that for every k-coloring of the nodes of $R(G)$, some monochrome *induced* copy of G occurs (see chapter 25).

Highly chromatic Steiner systems are known (see chapter 14). For highly chromatic infinite hypergraphs, see chapter 42.

5.6. Chromatic critical hypergraphs

Let us recall that a hypergraph $\mathscr{H}$ is *k-chromatic critical*, or *k-critical* for short, when $\chi(\mathscr{H}) = k$ and removing any edge causes the chromatic number to drop. As mentioned above with regard to Theorem 5.23, constructions of χ-critical r-graphs often use r'-graphs with $r' > r$: a consequence of various constructions of this type (cf. chapter 4) is the existence of χ-critical r-graphs with arbitrarily large minimum linear degree, for $\chi \geqslant 4$ and $r \geqslant 2$ (Toft, see chapter 4). Moreover, in contrast with the case of graphs, Müller, Rödl and Turzik exhibited, for any integers $r \geqslant 3$ and h, a 3-critical r-graph with all linear-degrees at least h and prove the existence of a *linear* hypergraph with the same properties. Notice that a k-critical r-graph on n vertices exists for $n \geqslant (k-1)(r-1)+1$ (Abbott and Hanson for $r=3$, Toft (op. cit.) for $r \geqslant 4$).

By use of linear algebra, Lovász found an upper bound for the size of 3-critical r-graphs:

Theorem 5.26. *For any 3-critical r-graph $\mathscr{H}$ on n vertices, we have*

$$m(\mathscr{H}) \leqslant \binom{n}{r-1} .$$

This bound has the correct order of magnitude, as justified by Toft's constructions.

There are only a finite number of 3-critical intersecting r-graphs. Bounds on their maximum size $N(r)$ were given by Erdős and Lovász (1975):

$$\sum_{s=1}^{r} \frac{r!}{s!} = \lfloor (e-1)r! \rfloor \leqslant N(r) \leqslant r^r . \tag{5.27}$$

The lower bound is conjectured to be exact.

5.7. Edge-colorings

The chromatic index $\gamma(\mathscr{H})$ of a hypergraph $\mathscr{H}$ cannot be less than the largest size of an intersecting subhypergraph and a fortiori cannot be less than the maximum degree $\Delta(\mathscr{H})$. The "edge-coloring property", i.e., $\gamma(\mathscr{H}) = \Delta(\mathscr{H})$, holds for bipartite graphs (König's theorem) and for some much more general classes of hypergraphs described in section 3.

For regular hypergraphs, to find a proper edge-coloring in $\Delta(\mathscr{H})$ colors amounts to finding a decomposition of the edge set into perfect matchings, i.e., a "1-*factorization*". Factorization problems appear in the theory of block designs (a 1-factorization of K_{15}^3 is a solution of the famous "Fifteen Schoolgirls Problem" of Kirkman: see chapter 14) and in Discrepancy Theory (see chapter 26). As shown by Pippenger and Spencer (1989), near 1-factorizations exist for near regular hypergraphs (compare with Theorem 4.27):

Theorem 5.28. *Let $\mathscr{H}$ be an r-graph of order n such that any node belongs to*

$d(1+o(1))$ edges and any two different nodes are in $o(d)$ edges (r is fixed, $d=d(n)\to\infty$). Then
(i) *$\gamma(\mathcal{H})$–$d(1+o(1))$, and*
(ii) *$\mathcal{H}$ edges can be partitioned into $d(1+o(1))$ edge covers.*

A major contribution *to factorization problems* was given by Baranyi who showed how network flows and "integer making lemmas" can be used to find the *chromatic index* of complete hypergraphs K_n^r and of their hereditary closures [see chapter 26 and Berge (1989) for more information on Baranyai's method]. Häggkvist and Hellgren recently developed a purely combinatorial approach unifying Baranyai's results with natural hypergraph extensions of König's edge-coloring theorem. They obtained a far-reaching generalization to hypergraphs of a theorem of Ryser on when a proper n-edge-coloring with n colors of the complete bipartite graph $K_{r,s}$ can be extended to a proper n-edge-coloring of $K_{n,n}$.

Let us be given a *complete p-partitioned* hypergraph $\mathcal{H}$ with partition $V_1, V_2, \ldots, V_p$, as defined in (1.27). The *tincture* of a subset $E \subset V(\mathcal{H})$ is the p-tuple with components $|E \cap V_k|$ for $1 \leqslant k \leqslant p$. Notice that, since $\mathcal{H}$ is complete, the multiplicity of an edge only depends on its tincture. To any (possibly improper) edge-coloring of $\mathcal{H}$ corresponds a *color-chart* Λ which matches to every color c the multiset $\Lambda(c)$ formed with the tinctures of the edges colored c; for $1 \leqslant k \leqslant p$, the sum of the kth components of $\Lambda(c)$ is denoted by $\Lambda_k(c)$. The following theorem means that a proper edge-coloring of a subhypergraph $\mathcal{H}'$ can be *extended* – a term to be taken in a self-explanatory way – to a proper edge-coloring of $\mathcal{H}$ with given specifications, provided that each color can be extended separately. (Remark the role of the empty edges: in a proper coloring, any empty edge can share the color of an other edge.)

Theorem 5.29 (Häggkvist and Hellgren 1993). *Let $\mathcal{H}=(V,(E_i)_{i\in I})$ be a complete p-partitioned hypergraph with partition $V=V_1\cup V_2\cup\cdots\cup V_p$. Let $V'\subset V$ and $V_k'=V'\cap V_k$. Let λ' denote a proper edge-coloring, with color chart Λ', of the subhypergraph $\mathcal{H}'=(V',(E_i\cap V')_{i\in I})$ and assume that λ' extends to some (possibly improper) edge-coloring of $\mathcal{H}$, with color chart Λ. Then, a necessary and sufficient condition for λ' to be extendable to a* proper *edge-coloring of $\mathcal{H}$ is that, for every color c, we have*

$$\Lambda_k(c)-\Lambda_k'(c)\leqslant |V_k|-|V_k'| \quad \text{for } 1\leqslant k\leqslant p\,.$$

This theorem has many interesting consequences, such as completion theorems for incomplete Latin squares (Cruse, Hoffman, Andersen, etc. See chapter 14). Below are some applications in hypergraph theory.

Corollary 5.30 (Baranyai 1975). *The complete hypergraph K_n^r has the edge-coloring property if and only if $r\,|\,n$.*

Proof. Choose $p=1$ (each edge has tincture r), $V'=\emptyset$ and color chart Λ with $\Delta=\binom{n-1}{r-1}$ colors, each color to be used n/r times. □

Corollary 5.31 (Baranyai 1979). *Let $\mathcal{H}$ be a complete hypergraph of order n in which every edge of cardinality i has multiplicity m(i). Then $\mathcal{H}$ has the edge coloring property if and only if there exist positive integers t_{ij} ($i = 1, 2, \ldots, n$, $j = 1, 2, \ldots$) such that*

$$\sum_j t_{ij} = m(i)\binom{n}{i} \quad \text{and} \quad \sum_i it_{ij} = n\,.$$

Proof. Apply Theorem 5.29 with $p = 1$ (so tincture i occurs exactly $m(i)\binom{n}{i}$ times), $V' = \emptyset$ and color chart Λ such that exactly t_{ij} i-edges receive color j. □

Corollary 5.32 (Generalization of Ryser's theorem). *Let $\mathcal{H} = K^p_{q,q,\ldots,q}$ be the complete p-partite p-graph on pq vertices. Let $A \subset V(\mathcal{H})$ and $\mathcal{H}_A = (E \cap A)_{A \in \mathcal{H}}$. Then every edge coloring of $\mathcal{H}_A$ where each color occurs q times can be extended to all of $\mathcal{H}$.*

It follows from Theorem 5.29 that complete p-partite p-graphs and also their hereditary closure have the edge-coloring property (Berge, and Berge and Johnson, see Berge 1989). The equality between $\gamma(\hat{\mathcal{H}})$ and $\Delta(\hat{\mathcal{H}})$, relative to the hereditary closure of a hypergraph $\mathcal{H}$, is of special interest: it implies that Chvátal's conjecture 2.18 is valid, in a strong sense, for $\hat{\mathcal{H}}$. Equality holds true when $\mathcal{H}$ is a star (Berge 1989) and, as Berge observed, is equivalent to the famous Vizing theorem (see chapter 4) when $\mathcal{H}$ is a graph.

Attempts to extend Vizing's theorem have inspired interesting open problems.

Conjecture 5.33 (Berge 1989). If $\mathcal{H}$ is a linear hypergraph, then

$$\gamma(\hat{\mathcal{H}}) = \Delta(\hat{\mathcal{H}})\,.$$

Conjecture 5.34 (Berge 1985, Füredi 1986, Meyniel, unpublished). If $\mathcal{H}$ is a simple linear hypergraph, then

$$\gamma(\mathcal{H}) \leq \Delta(\mathcal{H}_{[2]}) + 1\,.$$

Conjecture 5.34 would imply the following much publicized problem.

Conjecture 5.35 (Erdős, Faber and Lovász, *see* Erdős 1976). Let $\mathcal{H}$ be a linear hypergraph $\mathcal{H}$ of order n vertices. Then $\gamma(\mathcal{H}) \leq n$.

The original formulation of this conjecture asserts that if the edge set of a graph is the edge-disjoint of n complete graphs on n vertices, then its chromatic number equals n [the equivalence was pointed out by Hindman (1981) and Seymour (1982)]. Conjecture 5.35 was proved up to $n = 10$ (Hindman 1981), also in case $\mathcal{H}$ is a cyclic Steiner system (Colbourn and Colbourn 1982) and in case $\mathcal{H}$ is intersecting (Füredi 1986). Berge and Hilton (1990) proved Conjecture 5.34 if the edges of $\mathcal{H}$ with more than 2 elements are assumed pairwise disjoint; if, in addition, no edge has more than 3 elements, Conjecture 5.32 holds. Seymour's inequality (Theorem 4.43) is a step towards Conjecture 5.33 and suggests the following related problems.

Conjecture 5.36 [(i) Seymour 1982, (ii) Füredi 1986]. Let $\omega: \mathcal{H} \to \mathbb{R}^+$ be a nonnegative function on the edges of a linear hypergraph $\mathcal{H}$, then there exists a matching $\mathcal{M} \subseteq \mathcal{H}$ such that

(i) $\sum_{\mathcal{M}} \omega(E) \geqslant \sum_{\mathcal{H}} \omega(E)/n$,

(ii) $\sum_{\mathcal{M}} \omega(E) \geqslant \sum_{\mathcal{H}} \omega(E)/(\Delta(\mathcal{H}_{[2]}) + 1)$.

Conjecture 5.37 (*Origin unknown*). Suppose any partial hypergraph $\mathcal{H}' \subseteq \mathcal{H}$ satisfies $\nu(\mathcal{H}') \geqslant m(\mathcal{H}')/[\Delta(\mathcal{H}') + 1]$. Then $\gamma(\mathcal{H}) \leqslant \Delta(\mathcal{H}) + 1$.

References

Acharya, B.D., and M. Las Vergnas
[1992] Hypergraphs with cyclomatic numher zero, triangulated graphs and an inequality, *J. Combin. Theory B* **33**, 52–56.

Aharoni, R., and N. Linial
[1986] Minimal non two-colorable hypergraphs and minimal unsatisfiable formulas, *J. Combin. Theory A* **43**, 196–204.

Aharoni, R., P. Erdős and N. Linial
[1985] Dual Integer Linear Programs and the relationship between their optima, in: *Proc. 17th Annu. ACM-SIGACT Symp. on Theory of Computing,* pp. 476–483.

Ajtai, M., J. Komlós, J. Pintz, J.H. Spencer and E. Szemerédi
[1982] Extremal uncrowded hypergraphs, *J. Combin. Theory A* **32**, 321–325.

Alon, N., P. Frankl and L. Lovász
[1986] The chromatic number of Kneser hypergraphs, *Trans. Amer. Math. Soc.* **298**, 359–370.

Anstee, R.P., and M. Farber
[1984] Characterizations of totally balanced matrices, *J. Algorithms* **5**, 215–230.

Arami, Z., and P. Duchet
[1988] Arboreal structures, II: paths in trees, in preparation.

Baranyai, Zs.
[1975] On the factorization of the complete uniform hypergraph, in: *Infinite and finite sets, Colloq. Math. Soc. János Bolyai, Vol. 10, Keszthely, 1973* (North-Holland, Amsterdam) pp. 91–107.
[1979] The edge coloring of complete hypergraphs I, *J. Combin. Theory B* **26**, 276–294.

Beck, J.
[1978] On 3-chromatic hypergraphs, *Discrete Math.* **24**, 127–137.

Berge, C.
[1970] Sur certains hypergraphes généralisant les graphes bipartis, in: *Combinatorial Theory and its Applications,* eds. P. Erdős, A. Rényi and V.T. Sós (North-Holland, Amsterdam) pp. 119–133.
[1973] *Graphs and Hypergraphs,* 1st Ed. (North-Holland, Amsterdam) [translation of *Graphes et Hypergraphes* (Dunod, Paris, 1970)].
[1976] *Graphs and Hypergraphs,* 2nd rev. Ed. (North-Holland, Amsterdam).
[1985] On the chromatic index of a linear hypergraph and the Chvátal conjecture, *Ann. New York Acad. Sci.,* Proc. of 1985 meeting, to appear.
[1989] *Hypergraphs, Combinatorics of Finite Sets* (North-Holland, Amsterdam) [translation of *Hypergraphes, Combinatoire des Ensembles Finis* (Gauthier-Villars, Paris, 1987)]. 240pp.

Berge, C., and P. Duchet
[1975] A generalisation of Gilmore's theorem, in: *Recent Advances in Graph Theory,* ed. M. Fiedler (Acad. Praha, Prague) pp. 49–55.

Berge, C., and A.J.W. Hilton
[1990] On two conjectures about edge-coloring of hypergraphs, in preparation.

Berge, C., and M. Las Vergnas
[1970] Sur un théoreme du type König pour hypergraphes, *Ann. New York Acad. Sci.* **175**, 0.

Berge, C., and D. Ray-Chaudhuri
[1974] eds., *Hypergraph Seminar, Lecture Notes in Mathematics,* Vol. 411 (Springer, Berlin).

Berge, C., and M. Simonovits
[1974] The coloring numbers of the direct product of two hypergraphs, in: *Hypergraph Seminar, Springer Lecture Notes,* Vol. 411, eds. C. Berge and D. Ray-Chaudhury (Springer, Berlin) pp. 21–33.

Berge, C., C.C. Chen, V. Chvátal and C.S. Seow
[1982] Combinatorial properties of polyominoes, *Combinatorica* **2**, 217–224.

Bermond, J.-C., M.-C. Heydemann and D. Sotteau
[1977] Line graphs of hypergraphs, I, *Discrete Math.* **18**, 235–241.

Bollobás, B.
[1965] On generalized graphs, *Acta Math. Acad. Sci. Hungar.* **16**, 447–452.

Chung, F.R.K., Z. Füredi, M.R. Carey and R.L. Graham
[1988] On the fractional covering number of hypergraphs, *SIAM J. Discrete Math.* **1**, to appear.

Chvátal, V.
[1972] Problem 7, in: *Hypergraph Seminar, Lecture Notes in Mathematics,* Vol. 411, eds. C. Berge and D. Ray-Chaudhuri (Springer, Berlin) p. 280.
[1975] On certain polytopes associated with graphs, *J. Combin. Theory B* **18**, 138–154.

Colbourn, J., and M. Colbourn
[1982] The chromatic number of cyclic Steiner 2-designs, *Int. J. Math. Math. Sci.* **5**, 823–825.

Conforti, M., G. Cornuejols and M.R. Rao
[1990] A decomposition theorem for balanced matrices, *J. Combin. Theory,* submitted.

Conforti, M., G. Cornuéjols, A. Kapoor and K. Vuskovic
[1994] Recognizing balanced matrices, in: *Proc. 5th Annu. ACM–SIAM Symp. on Discrete Algorithms, Arlington, VA, 1994* (ACM, New York) pp. 103–111.

de la Vega, W.F.
[1982] Sur la cardinalité maximum des couplages d'hypergraphes aléatoires uniformes, *Discrete Math.* **40**, 315–318.

de Werra, D.
[1971] Equitable coloration of graphs, *Rev. Fr. Informatique et R.O.* **3**, 3–8.

Duchet, P.
[1978] Propriété de Helly et problèmes de représentations, in: *Problèmes Combinatoires et Théorie des Graphes, Coll. Orsay 1976* (Edition du CNRS, Paris) pp. 117–118.
[1985] *Tree-hypergraphs and their Representative Trees,* Mimeograph (University of Paris 6). 35pp.

Edmonds, J., and D.R. Fulkerson
[1970] Bottleneck extrema, *J. Combin. Theory* **8**, 299–306.

Edmonds, J., and E.L. Johnson
[1973] Matching, Euler tours and the Chinese postman, *Math. Programming* **5**, 88–124.

Erdős, P.
[1964] On a combinatorial problem II, *Acta Math. Acad. Sci. Hungar.* **15**, 445–447.
[1975] Problems and results in graph theory and combinatorial analysis, in: *Proc. 5th British Combinatorial Conference, Aberdeen, Congress. Numerantium* **XV**, 169–192.
[1981] On combinatorial problems which I would like to see solved, *Combinatorica* **1**, 25–42.

Erdős, P., and A. Hajnal
[1966] On chromatic number of graphs and set-systems, *Acta Math. Acad. Sci. Hungar.* **17**, 61–99.

Erdős, P., and L. Lovász
[1975] Problems and results on 3-chromatic hypergraphs and some related questions, in: *Infinite and finite sets, Colloq. Math. Soc. János Bolyai,* Vol. 10, Keszthely, 1973 (North-Holland, Amsterdam) pp. 609–627.
Erdős, P., and J.L. Selfridge
[1973] On a combinatorial game, *J. Combin. Theory B* **14**, 298–301.
Erdős, P., and J. Spencer
[1974] *Probabilistic Methods in Combinatorics* (Akadémiai Kiadó, Budapest).
Erdős, P., R.J. McEliece and H. Taylor
[1971] About Ramsey numbers, *Pacific J. Math.* **37**, 45–46.
Flament, C.
[1978] Hypergraphes arborés, *Discrete Math.* **21**, 223–226.
Fournier, J.-C.
[1980] Isomorphismes d'hypergraphes par intersections équicardinales d'arêtes et configurations exclues, *J. Combin. Theory B* **29**, 321–3.
Fournier, J.-C., and M. Las Vergnas
[1972] Une classe d'hypergraphes bichromatiques, *Discrete Math.* **2**, 407–410.
[1974] Une classe d'hypergraphes bichromatiques II, *Discrete Math.* **7**, 99–106.
Frankl, P., and Z. Füredi
[1986] Finite projective spaces and intersecting hypergraphs, *Combinatorica* **6**, 335–354.
Frankl, P., and V. Rödl
[1984] Hypergraphs do not jump, *Combinatorica* **4**, 149–159.
[1985] Near perfect coverings in graphs and hypergraphs, *European J. Combin.* **6**, 317–326. MR #88a:05116.
Fulkerson, D.R.
[1971] Blocking and anti-blocking pairs of polyhedra, *Math. Programming* **1**, 168–194.
[1972] Antiblocking polyhedra, *J. Combin. Theory B* **12**, 50–71.
Fulkerson, D.R., A.J. Hoffman and R. Oppenheim
[1974] On balanced matrices, *Math. Programming Stud.* **1**, 120–132.
Füredi, Z.
[1981] Maximum degrees and fractional matchings in uniform hypergraphs, *Combinatorica* **1** 154–162.
[1986] The chromatic index of simple hypergraphs, *Graphs and Combinatorics* **2**, 89–92.
[1988] Matchings and covers in hypergraphs, *Graphs and Combinatorics* **4**, 115–206.
Füredi, Z., J. Kahn and P.D. Seymour
[1993] On the fractional matching polytope of a hypergraph, *Combinatorica* **13**, 167–180.
Gale, D.
[1957] A theorem on flows in networks, *Pacific J. Math.* **7**, 1073–1082.
Gardner, M.L.
[1984] Hypergraphs and Whitney's theorem on edge-isomorphism of graphs, *Discrete Math.* **51**, 1–9.
Golumbic, M.
[1980] *Algorithmic Graph Theory and Perfect Graphs* (Academic Press, New York).
Gyárfás, A., and J. Lehel
[1983] Hypergraph families with bounded edge cover or transversal number, *Combinatorica* **3**, 351–358.
Gyárfás, A., J. Lehel and Z. Tuza
[1982] Upper bound on the order of τ-critical hypergraphs, *J. Combin. Theory B* **33**, 161–165.
Györi, E.
[1984] A minimax theorem for intervals, *J. Combin. Theory* B **37**, 1–9.
Häggkvist, R., and T.B. Hellgren
[1993] Extensions of edge-colorings in hypergraphs I, in: *Combinatorics, Paul Erdős is Eighty,* Vol. 1 (János Bolyai Math. Soc., Budapest) pp. 215–238.
Hajnal, A., and B. Rothschild
[1973] A generalisation of the Erdős–Ko–Rado theorem on finite set systems, *J. Combin. Theory A* **15**, 359–362.

Hansen, P., and M. Lorea
[1978] Deux conditions de colorabilité des hypergraphes, *Cahiers du C.E.R.O., Bruxelles* **20**, 3–4; 405–410.
Herzog, M., and J. Schönheim
[1972] The B_r property and chromatic numbers of generalized graphs, *J. Combin. Theory B* **12**, 41–49.
Hindman, N.
[1981] On a conjecture of Erdős, Faber, and Lovász about n-colorings, *Canad. J. Math.* **33**, 563–570.
Hoffman, A.J., M. Sakarovitch and A.W.J. Kolen
[1985] *Totally balanced and greedy matrices,* Report 8293/0 (Erasmus University, Rotterdam).
Jaeger, F., and C. Payan
[1971] Détermination du nombre maximum d'arêtes d'un hypergraphe τ-critique, *C.R. Acad. Sci. Paris* **271**, 221–223.
Jamison, R.
[1977] Covering finite fields with cosets of subspaces, *J. Combin. Theory B* **22**, 253–266.
Johnson, D.S.
[1976] On property B_r, *J. Combin. Theory B* **20**, 64–66.
Katona, G.O.H., T. Nemetz and M. Simonovits
[1964] On a problem of Turán in the theory of graphs (in Hungarian, English summary), *Mathematikai Lapok* **15**, 228–238.
Khatchartryan, M.A.
[1982] Generalisation of a deficit and certain applications (in Russian), *Operation Research and Programming* (Chtiintsa, Kishiniev), pp. 162–168. MR #85i:05177. (See also MR #83g:05057, 83g:05058).
Las Vergnas, M.
[1970] Sur un théorème de Rado, *C.R. Acad. Sci. Paris A–B* **270**, A733–A735.
Lehel, J.
[1982] Covers in hypergraphs, *Combinatorica* **2**, 301–309.
[1985] A characterization of totally-balanced hypergraphs, *Discrete Math.* **57**, 59–65.
Lehel, J., and Z. Tuza
[1982] Triangle-free partial graphs and edge covering theorems, *Discrete Math.* **39**, 59–65.
Lehman, A.
[1979] On the width–length inequality, *Math. Programming* **16**, 245–259.
Linial, N., and M. Tarsi
[1985] Deciding hypergraph 2-colorability by H-resolution, *Theor. Comput. Sci.* **38**, 265–268.
Lorea, M.
[1972] Ensembles stables dans les hypergraphes, *C.R. Acad. Sci. Paris A–B* **275**, A163–A165.
Lovász, L.
[1968] On chromatic number of finite set-systems, *Acta Math. Hung.* **19**, 59–67.
[1972] Normal hypergraphs and the perfect graph conjecture, *Discrete Math.* **2**, 253–267.
[1975a] On the ratio of optimal integral and fractional covers, *Discrete Math.* **13**, 383–390.
[1975b] On minimax theorems of combinatorics (Doctoral Thesis) (in Hungarian), *Mathematikai Lapok* **26**, 209–264.
[1976] On two minimax theorems in graph theory, *J. Combin. Theory B* **21**, 96–103.
[1977] Certain duality principles in integer programming, *Ann. Discrete Math.* **1**, 363–374.
[1978] Kneser's conjecture, chromatic number and homotype, *J. Combin. Theory A* **25**, 319–324.
[1979] *Combinatorial Problems and Exercises* (Akadémiai Kiadó/North-Holland, Budapest/Amsterdam).
Lovász, L., and M.D. Plummer
[1986] *Matching Theory, Ann. Discrete Math.* **29**.
Lubiw, A.
[1985] Doubly lexical orderings of matrices, in: *Proc. 17th ACM Symp. on Theory of Computing* (ACM, New York) pp. 396–403.
McEliece, R.J., and E.C. Posner
[1971] Hide and seek, data storage and entropy, *Amer. Statist.* **42**, 1706–1716.

Meyer, J.-C.
[1972] Ensembles stables maximaux dans les hypergraphes, *C.R. Acad. Sci. Paris A–B* **274**, 144–147.
Naik, R.N., S.B. Rao, S.S. Shrikhande and N.W. Singhi
[1982] Intersection graphs of k-uniform linear hypergraphs, *European J. Combin.* **3**, 159–172.
Nemhauser, G.L., and L.A. Wosley
[1988] *Integer and Combinatorial Optimization* (Wiley, New York).
Nešetřil, J., and V. Rödl
[1979] A short proof of the existence of highly chromatic hypergraphs without short cycles, *J. Combin. Theory B* **27**, 225–227.
Pippenger, N., and J.H. Spencer
[1989] Asymptotic behavior of the chromatic index for hypergraphs, *J. Combin. Theory A* **51**, 24–42.
Robertson, N., and P. Seymour
[1987] Hypergraph minors, lecture at the British Combinatorial Conference, Manuscript in preparation.
Ryser, H.J.
[1957] Combinatorial properties of matrices of zeros and ones, *Canad. J. Math.* **9**, 371–377.
Sarkaria, K.S.
[1990] A generalized Kneser conjecture, *J. Combin. Theory B* **49**, 239–240.
Schrijver, A.
[1979a] ed., *Packing and Covering in Combinatorics, Mathematical Centre Tracts,* Vol. 106 (Mathematisch Centrum, Amsterdam). 313pp.
[1979b] Fractional packing and covering, in: *Packing and Covering in Combinatorics, Mathematical Centre Tracts,* Vol. 106 (Mathematisch Centrum, Amsterdam) 201–274.
[1980] A counterexample to a conjecture by Edmonds and Giles, *Discrete Math.* **32**, 213–214.
[1983] Min–max results in combinatorial optimization, in: *Mathematical Programming – The State of the Art,* eds. A. Bachem et al. (Springer, Berlin) 439–500.
Schrijver, A., and P.D. Seymour
[1979] Solution of two fractional packing problems of Lovász, *Discrete Math.* **26**, 177–184.
Seymour, P.D.
[1972] On the two-colouring of hypergraphs, *Quart J. Oxford (3)* **25**, 303–312.
[1974] On the two-colouring of hypergraphs, *Quart. J. Math. Oxford* **25**, 303–312.
[1977] The matroids with the max-flow min-cut property, *J. Combin. Theory B* **23**, 189–222.
[1982] Packing nearly-disjoint sets, *Combinatorica* 91–97.
Shearer, J.B.
[1982] A class of perfect graphs, *SIAM J. Algebraic Discrete Methods* **3**, 281–284.
Slater, P.J.
[1978] A characterization of soft hypergraphs, *Canad. Math. Bull.* **21**, 335–337.
Spencer, J.
[1981] Coloring n-sets red and blue, *J. Combin. Theory A* **30**, 112–113.
Stein, S.K.
[1974] Two combinatorial covering theorems, *J. Combin. Theory A* **16**, 391–397.
Tamir, A.
[1985] *Totally Balanced and Totally Unimodular Matrices Defined by Center Location Problems,* Mimeograph (New York University).
Tucker, A.
[1972] A structure theorem for the consecutive 1's property, *J. Combin. Theory* **12**, 361–381.
Tuza, Z.
[1987] On the order of vertex sets meeting all edges of a 3-partite hypergraph, *Ars Combin.* **24**, 59–63.
Woodall, D.R.
[1972] Property B and the four color problem, in: *Combinatorics, Proc. Oxford Conf., 1972* (Inst. Math. Appl., Southend-on Sea) pp. 323–340.
Zykov, A.A.
[1974] Hypergraphs (in Russian), *Uspekhi Mat. Nauk* **29**, 89–154.

CHAPTER 8

Partially Ordered Sets

William T. TROTTER

Department of Mathematics, Arizona State University, Tempe AZ 85287, USA

Contents

HANDBOOK OF COMBINATORICS
Edited by R. Graham, M. Grötschel and L. Lovász

Introduction

Interest in finite partially ordered sets has been heightened in recent years by a steady stream of theorems combining clever ad hoc arguments with powerful techniques from other areas of mathematics. In this chapter, we present a sampling of results exhibiting these characteristics. In those instances where we do not present a complete proof, we outline enough of the general contours of the argument to allow the reader to supply the missing details with little difficulty. We also outline anticipated research directions in the combinatorics of partially ordered sets, and we discuss briefly some of the most interesting open problems in this field.

Since this Handbook contains chapters on Extremal Set Theory and Enumeration, we have limited our discussion to results on general partially ordered sets. Still some difficult choices had to be made concerning results to be included – especially in view of our emphasis on proof techniques. West's survey articles (West 1982, 1985) offer more of a catalogue of theorems in the area and have extensive bibliographies. Also, we recommend the recent books by Anderson (1987), Fishburn (1986), Stanley (1986), and Trotter (1992) as well as the conference volumes (Rival 1982, 1985) for additional material on partially ordered sets and related topics.

1. Notation and terminology

Formally, a *partially ordered set* is a pair (X, P) where X is a set, and P is a reflexive, antisymmetric, and transitive binary relation on X. The set X is called the *ground set* and P is called a *partial order*. Throughout this chapter, we use the short form *poset* for a partially ordered set. Many researchers choose to drop the adjective "partially" and use *ordered set* to mean a poset. A poset (X, P) is *finite* if the ground set X is finite. In this chapter, we will be concerned primarily with finite posets.

In some settings, we find it convenient to use a single symbol such as $\mathbf{P}$ to denote a poset (X, P). This notation is particularly handy when both the ground set X and the partial order P remain fixed. In other settings, especially when we have several partial orders on the same ground set, we will use the ordered pair notation for posets.

The notations $(x, y) \in P$, xPy, $x \leqslant y$ in P, and $y \geqslant x$ in P are used interchangeably. The notation $x < y$ in P means $x \leqslant y$ in P and $x \neq y$. Distinct points x, y are *comparable* when either $x < y$ or $y < x$ in P. Otherwise, we say x and y are *incomparable* and write $x \| y$ in P. When using a single symbol like $\mathbf{P}$ for a poset, we will write $x < y$ in $\mathbf{P}$, $x \| y$ in $\mathbf{P}$, etc.

A poset $\mathbf{P} = (X, P)$ is a *chain* (also a *totally ordered set* or a *linearly ordered set*) if each pair of distinct points is comparable. We will use the symbols $\mathbb{R}$, $\mathbb{Q}$, $\mathbb{Z}$ and $\mathbb{N}$ to denote the *reals*, *rationals*, *integers* and *positive integers*, respectively. Each of these posets is a chain.

Dually, $\mathbf{P} = (X, P)$ is an *antichain* if each pair of distinct points is incomparable.

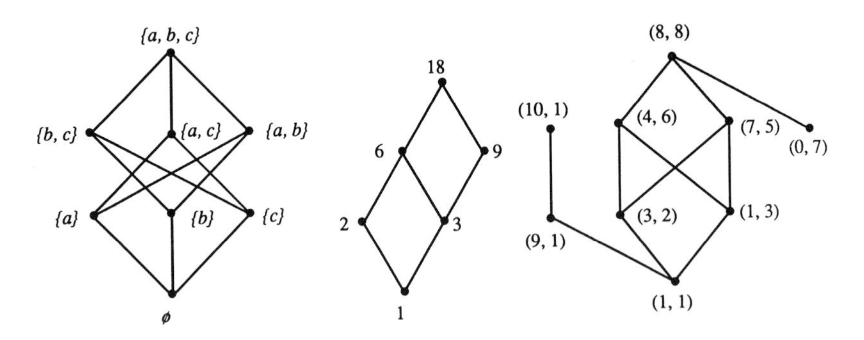

Figure 1.1.

If $Y \subset X$ and Q is the restriction of P to Y, then the poset $\mathbf{Q} = (Y, Q)$ is called a *subposet* of (X, P). A subset $Y \subset X$ is also called a *chain* (*antichain*) if the subposet (Y, Q) is a chain (antichain). The *height* of a poset is the maximum cardinality of a chain, and the *width* is the maximum cardinality of an antichain.

When $\mathbf{P} = (X, P)$ and $\mathbf{Q} = (Y, Q)$ are posets, a map $f : X \to Y$ is called an *embedding* (of $\mathbf{P}$ into $\mathbf{Q}$) if $x_1 \leqslant x_2$ in $P \iff f(x_1) \leqslant f(x_2)$ in Q. An embedding $f : X \to Y$ is an *isomorphism* when $f(X) = Y$. In this chapter, we prefer not to distinguish between isomorphic posets and to write $\mathbf{P} = \mathbf{Q}$ to indicate that the two posets are isomorphic. Similarly we say that $\mathbf{P}$ is *contained* in $\mathbf{Q}$ (also $\mathbf{P}$ is a *subposet* of $\mathbf{Q}$) when there exists an embedding of $\mathbf{P}$ in $\mathbf{Q}$.

We say y *covers* x in P and write $x <: y$ in P when there is no z for which both $x < z$ and $z < y$ in P. The *cover graph* associated with the poset $\mathbf{P} = (X, P)$ is the graph $\mathbf{G} = (X, E)$ whose edge set E consists of the pairs xy for which $x <: y$ in P. A drawing of the cover graph $\mathbf{G} = (X, E)$ in the Euclidean plane is called a *Hasse diagram* (or *order diagram*) of the poset $\mathbf{P} = (X, P)$ if x is lower in the plane than y whenever $x <: y$ in P.

Here are some frequently encountered examples of posets. Any family of sets is partially ordered by set inclusion; a set of positive integers is partially ordered by division; and a subset of $\mathbb{R}^n$ is partially ordered by $(a_1, a_2, \ldots, a_n) \leqslant (b_1, b_2, \ldots, b_n) \iff a_i \leqslant b_i$ in $\mathbb{R}$ for $i = 1, 2, \ldots, n$. In fig. 1.1, we show particular instances of these examples. Each has height 4, and their respective widths are 3, 2, and 4.

If P and Q are partial orders on the same ground set X, Q is called an *extension* of P when $P \subseteq Q$. The partial order Q is called a *linear extension* of P if Q is an extension of P and (X, Q) is a chain.

When $\mathbf{P} = (X, P)$ is a poset, an element $x \in X$ is called a *maximal* (*minimal*) element if there is no $y \in X$ for which $x < y$ in P ($y < x$ in P). The set of maximal (minimal) elements is denoted $\mathrm{MAX}(X, P)$ ($\mathrm{MIN}(X, P)$). The subsets $\mathrm{MAX}(X, P)$

and $\text{MIN}(X,P)$ always determine antichains, although neither may be as large as the width of (X,P).

When $Y \subset X$, the set $\{z \in X: y \leqslant z \text{ in } P \text{ for every } y \in Y\}$ is called the set of *upper bounds* for Y. Note that this set may be empty. When the set of upper bounds of Y is nonempty and has a least element, this unique point is called the *least upper bound* of Y and is denoted l.u.b.(Y). Dually, the *greatest lower bound* (if it exists) of Y is denoted g.l.b.(Y).

A poset $\mathbf{P} = (X,P)$ is called a *lattice* when each nonempty subset $Y \subset X$ has both a least upper bound and a greatest lower bound. When $\mathbf{P} = (X,P)$ is a lattice and $x, y \in X$, we write $x \vee y$ for l.u.b.$\{x,y\}$ and $x \wedge y$ for g.l.b.$\{x,y\}$. The binary operations $\vee$ (join) and $\wedge$ (meet) are commutative and associative. The lattice is *distributive* if $x \wedge (y \vee z) = (x \wedge y) \vee (x \wedge z)$ for all $x, y, z \in X$.

When $\mathbf{P}$ and $\mathbf{Q}$ are posets, the *disjoint sum* of $\mathbf{P}$ and $\mathbf{Q}$, denoted $\mathbf{P}+\mathbf{Q}$, is obtained by taking the union of disjoint copies of the two posets with no comparabilities between the points in one and points in the other. A poset is *disconnected* if it is the disjoint sum of two proper subposets; otherwise it is *connected*. The maximal connected subposets of a disconnected poset are *components*.

The *cartesian product* of $\mathbf{P} = (X,P)$ and $\mathbf{Q} = (Y,Q)$, denoted $\mathbf{P} \times \mathbf{Q}$, consists of the ordered pairs (x,y) where $x \in X$ and $y \in Y$ with partial ordering $(x_1,y_1) \leqslant (x_2,y_2) \Longleftrightarrow x_1 \leqslant x_2$ in $\mathbf{P}$ and $y_1 \leqslant y_2$ in $\mathbf{Q}$. The cartesian product of n copies of $\mathbf{P}$ is denoted $\mathbf{P}^n$.

Given posets $\mathbf{P} = (X,P)$ and $\mathbf{Q} = (Y,Q)$, a function $f: X \to Y$ is an *order preserving* (or *monotone*) map from $\mathbf{P}$ to $\mathbf{Q}$ if $x_1 \leqslant x_2$ in $P \Rightarrow f(x_1) \leqslant f(x_2)$ in Q. The set of all order preserving maps from $\mathbf{P}$ to $\mathbf{Q}$ is partially ordered by $f_1 \leqslant f_2 \Longleftrightarrow f_1(x) \leqslant f_2(x)$ in Q for every $x \in X$. This poset is denoted $\mathbf{Q}^{\mathbf{P}}$.

Throughout the chapter, we use $\mathbf{k}$ to denote a k-element chain $0 < 1 < 2 < \cdots < k-1$. The poset $\mathbf{2}^n$ is isomorphic to the set of subsets of an n-element set partially ordered by inclusion. A poset $\mathbf{P}$ is a distributive lattice if and only if there is a poset $\mathbf{Q}$ so that $\mathbf{P}$ is isomorphic to $\mathbf{2}^{\mathbf{Q}}$ [see chapter 3 in Birkhoff (1973)].

When $\mathbf{P} = (X,P)$ is a poset and $\mathcal{F} = \{\mathbf{P}_x = (Y_x, Q_x): x \in X\}$ is a family of posets indexed by the ground set of $\mathbf{P}$, the *lexicographic sum* of $\mathcal{F}$ over $\mathbf{P}$ is the poset whose ground set is $\{(x,y): x \in X, y \in Y_x\}$. The partial ordering is defined by $(x_1,y_1) \leqslant (x_2,y_2) \Longleftrightarrow (x_1 < x_2 \text{ in } P)$ or $(x_1 = x_2$ and $y_1 \leqslant y_2$ in $Q_{x_1})$. A lexicographic sum is nontrivial if $|X| \geqslant 2$ and if at least one Y_x satisfies $|Y_x| \geqslant 2$. A poset $\mathbf{P}$ is *decomposable* if it is isomorphic to a nontrivial lexicographic sum; otherwise $\mathbf{P}$ is *indecomposable.* Note that the disjoint sum of two posets is a lexicographic sum over a 2-element antichain.

2. Dilworth's theorem and the Greene–Kleitman theorem

Dilworth's decomposition theorem (Dilworth 1950) has played an important role in motivating research in posets, as evidenced by results discussed in this section as well as in sections 3, 6, 7 and 8. Also, Dilworth's theorem surfaces in a variety of extremal problems (see, for example, Duffus et al. 1991). There are several elementary proofs; the one we present is patterned after Perles (1963).

Theorem 2.1. *If $\mathbf{P} = (X, P)$ is a poset of width n, then there exists a partition $X = C_1 \cup C_2 \cup \cdots \cup C_n$ where each C_i is a chain.*

Proof. We proceed by induction on $|X|$ and note that the result is trivial when $|X| = 1$. Assume validity when $|X| < k$ and consider a poset $\mathbf{P}$ with $|X| = k$. We may assume that the width n of $\mathbf{P}$ is larger than 1.

Choose $x \in \mathrm{MAX}(\mathbf{P})$ and $y \in \mathrm{MIN}(\mathbf{P})$ with $y \leqslant x$. Let $\mathbf{Q}$ be the poset obtained by removing x and y from $\mathbf{P}$. If the width of $\mathbf{Q}$ is less than n, then we can partition $\mathbf{Q}$ into fewer than n chains which together with the chain $\{x, y\}$ form a partition of X into (at most) n chains. So we may assume that $\mathbf{Q}$ has width n. Thus $y < x$ in $\mathbf{P}$. Choose an n-element antichain $A = \{a_1, a_2, \ldots, a_n\}$ in $\mathbf{Q}$.

Then let $U = \{u \in X\colon u \geqslant a_i$ for some $a_i \in A\}$ and $D = \{d \in X\colon d \leqslant a_j$ for some $a_j \in A\}$. Evidently $x \in U - D$ and $y \in D - U$. Thus there are chain partitions $U = C'_1 \cup C'_2 \cup \cdots \cup C'_n$ and $D = C''_1 \cup C''_2 \cup \cdots \cup C''_n$. We may label these chains so that $a_i \in C'_i \cap C''_i$ for $i = 1, 2, \ldots, n$. Then $C_i = C'_i \cup C''_i$ is a chain for each i and the desired partition is $X = C_1 \cup C_2 \cup \cdots \cup C_n$. □

In introductory combinatorics texts, Dilworth's theorem is grouped with other max–min theorems having a common theme: P. Hall's marriage theorem, the König–Egervary theorem, Menger's theorem, and the max flow–min cut theorem for network flows. This last result most clearly captures the linear programming core common to all. (See chapters 2 and 3 by Frank and Pulleyblank for additional material.)

Dilworth's theorem has a trivial dual version for antichains.

Theorem 2.2. *If $\mathbf{P} = (X, P)$ is a poset of height n, then there exists a partition $X = A_1 \cup A_2 \cup \cdots \cup A_n$ where each A_i is an antichain.*

Proof. Set $A_1 = \mathrm{MAX}(\mathbf{P})$. Thereafter set $A_{i+1} = \mathrm{MAX}(\mathbf{P}_i)$ where $\mathbf{P}_i$ is the subposet obtained by removing the antichains $A_1, A_2, \ldots, A_i$ from $\mathbf{P}$. □

The first major result in this chapter is an important generalization of Dilworth's chain partitioning theorem due to Greene and Kleitman (1976). The proof we give here is patterned after algorithmic proofs given by Saks (1979) and Perfect (1984). An alternative proof using network flows is given in this volume in chapter 2.

We need some preliminary notation and terminology. Let $\mathbf{P} = (X, P)$ be a poset and k a positive integer. A subset $S \subset X$ is called a *Sperner k-family* if S does not contain a chain of $(k+1)$-elements. The maximum cardinality of a Sperner k-family is denoted $d_k(\mathbf{P})$. When $\mathscr{C} = \{C_1, C_2, \ldots, C_t\}$ is a family of chains forming a partition of X, we define $e_k(\mathscr{C}) = \sum_{i=1}^{t} \min\{k, |C_i|\}$. If S is any Sperner k-family and $\mathscr{C} = \{C_1, C_2, \ldots, C_t\}$ is any chain partition of X, we note that $|S \cap C_i| \leqslant \min\{k, |C_i|\}$. Thus $|S| \leqslant e_k(\mathscr{C})$, so that $d_k(\mathbf{P}) \leqslant e_k(\mathscr{C})$. The chain partition $\mathscr{C}$ is said to be *k-saturated* if $d_k(\mathbf{P}) = e_k(\mathscr{C})$.

We also need a preliminary lemma whose elementary proof is omitted. Let $\mathcal{M}(\mathbf{P})$ denote the set of all maximum antichains of $\mathbf{P}$. Define a partial order on $\mathcal{M}(\mathbf{P})$ by $A \leqslant B \iff$ for every $a \in A$, there exists $b \in B$ with $a \leqslant b$.

Lemma 2.3. *The set $\mathcal{M}(\mathbf{P})$ of maximum antichains of a poset $\mathbf{P} = (X, P)$ has a unique greatest element.*

With this background, here is the Greene–Kleitman theorem.

Theorem 2.4. *Let $\mathbf{P}$ be a poset and k a positive integer. Then there exists a chain partition $\mathscr{C}$ of $\mathbf{P}$ which is simultaneously k-saturated and $(k+1)$-saturated, i.e., $d_k(\mathbf{P}) = e_k(\mathscr{C})$ and $d_{k+1}(\mathbf{P}) = e_{k+1}(\mathscr{C})$.*

Proof. We first show that $d_1(\mathbf{P} \times \mathbf{k}) = d_k(\mathbf{P})$ for every $k \geqslant 1$. Let A be a maximum antichain in $\mathbf{P} \times \mathbf{k}$, and let $A_i = \{x \in X\colon (x, i) \in A\}$. Then each A_i is an antichain in $\mathbf{P}$, so the set $S = A_1 \cup A_2 \cup \cdots \cup A_k$ is a Sperner k-family. Furthermore, $|S| = |A|$ since $A_i \cap A_j = \emptyset$ when $i \neq j$. Thus $d_1(\mathbf{P} \times \mathbf{k}) \leqslant d_k(\mathbf{P})$.

Conversely, let S be a maximum Sperner k-family in $\mathbf{P}$. Partition S into k antichains by setting $A_1 = \mathrm{MAX}(S)$ and $A_{i+1} - \mathrm{MAX}(S - (A_1 \cup A_2 \cup \cdots \cup A_i))$. Then $A = \{(a, i)\colon a \in A_i\}$ is an antichain in $\mathbf{P} \times \mathbf{k}$ with $|A| = |S|$. This shows $d_1(\mathbf{P} \times \mathbf{k}) \geqslant d_k(\mathbf{P})$. Thus $d_1(\mathbf{P} \times \mathbf{k}) = d_k(\mathbf{P})$.

For the remainder of the proof, we fix a positive integer k. Then we make several definitions concerning chain partitions of $\mathbf{P} \times (\mathbf{k+1})$. When $\mathscr{C} = \{C_1, C_2, \ldots, C_t\}$ is a chain partition of $\mathbf{P} \times (\mathbf{k+1})$, we let $M(\mathscr{C}) = \{\mathrm{MAX}(C_i)\colon 1 \leqslant i \leqslant t\}$. We say *$y$ covers x* in $\mathscr{C}$ if there is some $C_i \in \mathscr{C}$ so that y covers x in the chain C_i. When $S \subset \mathbf{P} \times (\mathbf{k+1})$, the set $\{x \in X\colon (x, i) \in S$ for some $i\}$ is called the *projection* of S on $\mathbf{P}$. For each i, the subset $S \cap (\mathbf{P} \times \{i\})$ is called *level i* of S. The projection on $\mathbf{P}$ of level i of $M(\mathscr{C})$ is denoted by $M_i(\mathscr{C})$.

A chain partition $\mathscr{C}$ of $\mathbf{P} \times (\mathbf{k+1})$ is *special* if the following two conditions hold:

(i) $M_0(\mathscr{C}) \supset M_1(\mathscr{C}) \supset M_2(\mathscr{C}) \supset \cdots \supset M_{k-1}(\mathscr{C})$;

(ii) If $x \in M_k(\mathscr{C}) - M_{k-1}(\mathscr{C})$, then (x, k) covers $(x, k-1)$ in $\mathscr{C}$.

A special chain partition of $\mathbf{P} \times \mathbf{k+1}$ is *very special* if it also satisfies the following two conditions:

(iii) Exactly $d_{k+1}(\mathbf{P}) - d_k(\mathbf{P})$ of the chains in $\mathscr{C}$ are subsets of level 0; and

(iv) $|\mathscr{C}| = d_1(\mathbf{P} \times (\mathbf{k+1}))$.

When $\mathscr{C}$ is special, it follows from the second condition in this definition that $M_k(\mathscr{C}) = N_1(\mathscr{C}) \cup N_2(\mathscr{C})$ where $N_1(\mathscr{C}) = M_k(\mathscr{C}) \cap M_{k-1}(\mathscr{C})$. If $x \in N_2(\mathscr{C})$, then (x, k) covers $(x, k-1)$ in $\mathscr{C}$.

We now show that the theorem follows whenever $\mathbf{P} \times (\mathbf{k+1})$ has a very special chain partition. To see this, let $\mathscr{C}_{k+1} = \{C_1, C_2, \ldots, C_t\}$ be a very special chain partition of $\mathbf{P} \times (\mathbf{k+1})$ where $t = d_1(\mathbf{P} \times (\mathbf{k+1}))$. Set $s = d_{k+1}(\mathbf{P}) - d_k(\mathbf{P})$. We assume that $C_1, C_2, \ldots, C_s$ are subsets of level 0. For each $j = 1, 2, \ldots, t$, let $D_j = \{(x, i)\colon (x, i+1) \in C_j\}$. Of course, $D_1, D_2, \ldots, D_s$ are all empty. Let $\mathscr{C}_k$ be the collection of all nonempty D_j's. Then $\mathscr{C}_k$ is a chain partition of $\mathbf{P} \times \mathbf{k}$ and $|\mathscr{C}_k| \leqslant t - s = d_k(\mathbf{P}) = d_1(\mathbf{P} \times \mathbf{k})$. Thus $|\mathscr{C}_k| = d_k(\mathbf{P})$. Furthermore, it is easy to see that $\mathscr{C}_k$ is a special chain partition of $\mathbf{P} \times \mathbf{k}$.

Now level k in $\mathbf{P} \times (\mathbf{k}+\mathbf{1})$ forms a copy of $\mathbf{P}$ and the $|M_k(\mathscr{C}_{k+1})|$ chains in $\mathscr{C}_{k+1}$ which intersect level k determine a chain partition of $\mathbf{P}$ which we denote by $\mathscr{C}$. We show that $\mathscr{C}$ is both k-saturated and $(k+1)$-saturated.

Now let $j \in \{k, k+1\}$. Then

$$\begin{aligned}
d_j(\mathbf{P}) = e_1(\mathscr{C}_j) = |M(\mathscr{C}_j)| &= \sum_{i=0}^{j-1} |M_i(\mathscr{C}_j)| \\
&\geqslant (j-1)|M_{j-2}(\mathscr{C}_j)| + |M_{j-1}(\mathscr{C}_j)| \\
&= (j-1)|M_{j-2}(\mathscr{C}_j)| + |M_{j-1}(\mathscr{C}_j) \cap M_{j-2}(\mathscr{C}_j)| + |N_2(\mathscr{C}_j)| \\
&\geqslant j|M_{j-1}(\mathscr{C}_j) \cap M_{j-2}(\mathscr{C}_j)| + |N_2(\mathscr{C}_j)| \\
&= j|N_1(\mathscr{C}_j)| + |N_2(\mathscr{C}_j)| \\
&\geqslant \sum_{E \in \mathscr{E}} \min\{j, |E|\} \\
&= e_j(\mathscr{E}) \geqslant d_j(\mathbf{P}).
\end{aligned}$$

Thus $\mathscr{E}$ is both k-saturated and $(k+1)$-saturated as claimed. To complete the proof, we need only show the existence of a very special chain partition of $\mathbf{P} \times (\mathbf{k}+\mathbf{1})$. Set $t = d_{k+1}(\mathbf{P}) = d_1(\mathbf{P} \times (\mathbf{k}+\mathbf{1}))$. Of all partitions of $\mathbf{P}(k+1)$ into t chains, choose one having as many chains as possible as subsets of level 0. Call this partition $\mathscr{C} = \{C_1, C_2, \ldots, C_t\}$ and label the chains in $\mathscr{C}$ so that $C_1, C_2, \ldots, C_s$ are subsets of level 0 but $C_{s+1}, C_{s+2}, \ldots, C_t$ are not. Since the last $t-s$ chains in $\mathscr{C}$ cover a copy of $\mathbf{P} \times \mathbf{k}$, we know $t - s \geqslant d_1(\mathbf{P} \times \mathbf{k}) = d_k(\mathbf{P})$, so $s \leqslant d_{k+1}(\mathbf{P}) - d_k(\mathbf{P})$. We show that $s = d_{k+1}(\mathbf{P}) - d_k(\mathbf{P})$. Suppose to the contrary that $s < d_{k+1}(\mathbf{P}) - d_k(\mathbf{P})$.

Let $\mathbf{Q} = \mathbf{P} \times (\mathbf{k}+\mathbf{1}) - (C_1 \cup C_2 \cup \cdots \cup C_s)$. Clearly, the width of $\mathbf{Q}$ is $t-s$. Let A be the unique greatest element in the poset $\mathscr{M}(\mathbf{Q})$ of maximum antichains in $\mathbf{Q}$. A contains at least $d_{k+1}(\mathbf{P}) - d_k(\mathbf{P}) - s$ elements from level 0 since the width of the top k levels is only $d_k(\mathbf{P})$. Choose an element $a_0 \in A$ which comes from level 0. Without loss of generality $a_0 \in C_{s+1}$. Let $\mathbf{Q}' = \mathbf{Q} - \{c \in C_{s+1} \colon c \leqslant a_0\}$.

We claim that the width of $\mathbf{Q}'$ is less than $t-s$, for if $\mathbf{Q}'$ contains a $(t-s)$-element antichain B, then B contains an element b with $a_0 < b$ in $\mathbf{Q} \times (\mathbf{k}+\mathbf{1})$. This contradicts our choice of A. It follows that we can partition Y' into $t-s-1$ chains which together with $C_1, C_2, \ldots, C_s$ and $\{c \in C_{s+1} \colon c \leqslant a_0\}$ form a partition of $\mathbf{P} \times (\mathbf{k}+\mathbf{1})$ into t chains. In this partition, there are $s+1$ chains which are subsets of level 0. The contradiction shows $s = d_{k+1}(\mathbf{P}) - d_k(\mathbf{P})$.

We now proceed to transform $\mathscr{C}$ into a very special partition by a series of operations called *insertions* and *switches*. At this moment $\mathscr{C}$ satisfies properties (iii) and (iv), and both operations preserve these properties.

We first perform a series of insertions. Choose points (x, i), (y, j) so that (x, i) covers (y, j) in $\mathscr{C}$ and $i > j$. If $i \neq j+1$ or $x \neq y$, remove $(y, j+1)$ from the chain to which it currently belongs and insert it in the chain containing (x, i) and (y, j). Repeat until no further insertions are possible.

Next, we perform a series of switches. For an integer $j \geqslant 1$, locate a point $(x, j) \in M(\mathscr{C})$ so that either: (1) $j < k$ and $(x, j-1) \notin M(\mathscr{C})$; or (2) $j = k$ and (x, j) does not cover $(x, j-1)$ in $\mathscr{C}$, and $(x, j-1) \notin M(\mathscr{C})$.

Let (y, i) be the point covering $(x, j-1)$ and let C be the chain containing $(y, i+1)$. Let C'' consist of those points in $\mathscr{C}$ which are less than $(y, i+1)$ and let $C'' = C - C'$. Then let D be the chain containing (x, j) and set $D' = D \cup C''$. Replace C and D in $\mathscr{C}$ by C' and D'. Repeat until no further switches are possible.

It is obvious that the series of insertions must stop, but it takes a moment's reflection to see that this is also true for the series of switches. For $j = 1, 2, \ldots, k$, let v_j count the number of points $x \in X$ for which (x, j) covers (y, i) in $\mathscr{C}$ and $(x, j-1)$ covers $(y, i-1)$ in $\mathscr{C}$. Each time we perform a switch, the vector $(v_1, v_2, \ldots, v_k)$ increases lexicographically. Since $v_j \leqslant |X|$ for each j, the procedure stops. □

This theorem has many significant applications. The following corollary follows easily, but we know of no simple proof avoiding the use of Theorem 2.4.

Corollary 2.5. *Let* $\mathbf{P}$ *be a finite poset. Then for each* $k \geqslant 1$, $d_k(\mathbf{P}) - d_{k-1}(\mathbf{P}) \geqslant d_{k+1}(\mathbf{P}) - d_k(\mathbf{P})$.

3. Kierstead's chain partitioning theorem

In this section we outline the proof of a theorem of Kierstead (1981) which asserts that for each $n \geqslant 1$, there is a $t = t(n)$ for which there exists an on-line algorithm which will partition any poset $\mathbf{P}$ of width at most n into t chains. By an on-line partition, we mean that the poset and the partition are constructed one point at a time. An adversary (infinitely clever) constructs the poset and we must devise the partition. At each round, the adversary presents the new point and describes its comparabilities and incomparabilities to all preceding points. We must then add the new point to one of the sets making up the partition. Both players' moves are permanent.

As a warm-up, we first present the on-line version of the dual to Dilworth's theorem. The result is an unpublished theorem of Schmerl, although a short proof is given in Kierstead (1986).

Theorem 3.1. *For each* $n \geqslant 1$, *there exists an algorithm which will construct an on-line partition of a poset of height at most* n *into* $n(n+1)/2$ *antichains.*

Proof. When the new point x is added to the poset, let $r = r(x)$ be the maximum number of points in a chain having x as least element, and let $s = s(x)$ be the maximum number of points in a chain having x as greatest element. Assign x to the set $A(r, s)$. Clearly, each $A(r, s)$ is an antichain. Since $r + s - 1 \leqslant n$, there are $n(n+1)/2$ such sets. □

Szemerédi produced a simple argument to show that Theorem 3.1 is best possible, and we invite the reader to reconstruct his proof. Full details are given in

Kierstead (1986). As a first step, show that there exists a strategy for constructing a poset **P** of height at most n which will force any opponent producing an on-line partition into antichains to use at least $n(n+1)/2$ antichains in covering **P** and at least n antichains in covering MAX(**P**).

Here is Kierstead's on-line chain partitioning theorem (Kierstead 1981).

Theorem 3.2. *For each $n \geqslant 1$, there exists an algorithm which will construct an on-line partition of a poset of width at most n into $(5^n - 1)/4$ chains.*

Proof. The argument proceeds by induction on n with the case $n = 1$ being trivial. The heart of the argument is the case $n = 2$ where we have to partition a width-2 poset into 6 chains.

We first construct a greedy chain C_1. As a new point enters the poset, we insert it in C_1 whenever it is comparable to all other points previously placed in C_1. Thus for every $x \in X - C_1$, there is a nonempty set $I(x)$ of points from C_1 which are incomparable to x. Although $I(x)$ may grow with time, it is always a set of consecutive points from C_1. When $x, y \in X - C_1$, we write $I(x) < I(y)$ when $u < v$ for every $u \in I(x)$ and every $v \in I(y)$. Note that if x and y are incomparable points in $X - C_1$, then the following condition holds:

(K). When the latter of x and y enters, either

$$I(x) < I(y) \text{ or } I(y) < I(x).$$

In fact, when $n = 2$, the qualifying phrase "when the latter of x and y enters" can be dropped since $I(x) \cap I(y) = \emptyset$ whenever $x \parallel y$. Regardless, we choose the weaker statement since it is crucial to the inductive step.

We define a partial order, called the $*$-order, on $X - C_1$ as follows. When the new point x enters, we set $x * y$ if

(1) $x < y$ in $\mathbf{P} - C_1$, or

(2) $x \parallel y$ and $I(x) < I(y)$.

Similarly, we set $y * x$ if

(3) $y < x$ in $\mathbf{P} - C_1$, or

(4) $x \parallel y$ and $I(y) < I(x)$.

With this definition, it is straightforward to verify that $(X - C_1, *)$ is a chain, i.e., $*$ is a linear extension of the original partial order on $X - C_1$. Next, we define an equivalence relation on $\mathbf{P} - C_1$. Just as is the case with the $*$-order, the definition of this equivalence relation is on-line. The relation will satisfy:

(a) each equivalence class is a set of consecutive elements of $X - C_1$ in the $*$-order, and

(b) if x and y are consecutive elements belonging to the same equivalence class, then $I(x) \cap I(y) \neq \emptyset$.

When a new point x enters $X - C_1$, we put x in the same equivalence class as y if $x <: y$ in $*$ and $I(x) \cap I(y) \neq \emptyset$. If no such y exists, we put x in the same class as z if $z <: x$ in $*$ and $I(x) \cap I(z) \neq \emptyset$. If neither of these results in the assignment of x to an existing class, start a new equivalence class whose only element (at this moment) is x.

Note that if x enters between two consecutive points y and z belonging to the same class, then $I(x) \cap I(y) \neq \emptyset \neq I(x) \cap I(z)$. This insures that property (b) will be preserved when x is added to this class.

To complete the proof of the width-2 case, we need to verify the following claim whose proof we leave as an exercise.

Claim. *If S_1 and S_2 are equivalence classes of $X - C_1$, and there are at least two other equivalence classes between them in the $*$-order, then $S_1 \cup S_2$ is a chain in* $\mathbf{P}$.

Once the chain is verified, we may use it to devise a simple strategy for constructing an on-line partition $X - C_1 = C_2 \cup C_3 \cup \cdots \cup C_6$. Each of these five chains is the union of equivalence classes, and any two classes which are subsets of the same chain have at least two other classes between them in the $*$-order. When we start a new class for a point x, we assign it to a chain which does not contain any of the four classes – two above x and two below x in the $*$-order.

To obtain the general result when the width of $\mathbf{P}$ is $n \geqslant 3$, start by constructing a greedy chain C_1 just as before. Then define a partial order $*$ on $X - C_1$ so that $*$ is an extension of the original order on $X - C_1$ and the width of $(X - C_1, *)$ is $n - 1$. When the new point x enters $X - C_1$, we set $x * y$ if either (1) or (2) holds. However, we also set $x * y$ if:

(1′) there exists $u \in X - C_1$ so that $x < u$ in $\mathbf{P}$ and $u * y$, or

(2′) there exists $v \in X - C_1$ so that $x * v$ and $v < y$ in $\mathbf{P}$.

It is easy to see that this more general definition of $*$ is necessary in order to insure that $*$ is transitive. The definition of when $y * x$ must be expanded analogously. With these observations, it is clear that the width of $(X - C_1, *)$ is at most $n - 1$, for if $A = \{a_1, a_2, \ldots, a_m\}$ is an antichain in $(X - C_1, *)$, then $\bigcap_{i=1}^{m} I(a_i) \neq \emptyset$. If $x \in C_1$ and $x \parallel a_i$ for $i = 1, 2, \ldots, m$, then $\{x\} \cup A$ is an antichain in $\mathbf{P}$. It follows that there is a strategy for partitioning $X - C_1$ into $(5^{n-1} - 1)/4$ subsets each of which is a chain in the $*$-order. Observe that a $*$-chain satisfies property (K), and the algorithm described for the width-2 case will then partition a $*$-chain into five chains in $\mathbf{P}$. The theorem follows since $(5^n - 1)/4 = 1 + 5(5^{n-1} - 1)/4$. □

It is apparently a very difficult problem to determine just how good the upper bound in Theorem 3.2 actually is. For $n = 2$, Theorem 3.2 gives an upper bound of 6, but Felsner (1995) has just shown that the correct answer is 5. Saks observed that the techniques used to show that Theorem 3.1 is best possible can be dualized to produce a lower bound of the form $n(n+1)/2$ for Theorem 3.2. This bound is probably a very weak result, and it would be interesting to determine whether there is an algorithm which will partition on-line a poset of width at most n into n^c chains for some absolute constant c.

Recently, Kierstead et al. (1994) have shown that for every radius-two tree T, there exists a function $f_T : \mathbb{N} \to \mathbb{N}$, so that if G is any graph which does not contain T as an induced subgraph and does not contain a complete subgraph on k vertices, then G can be colored on-line with $f_T(k)$ colors. The complement of a comparability graph does not contain the subdivision of $K_{1,3}$, so it follows that there exists a function $g : \mathbb{N} \to \mathbb{N}$ so that a comparability graph with independence number n can be partitioned on-line into $g(n)$ complete subgraphs. This result does not follow from a straightforward extension of the ideas presented in this section. The difficulty is that the argument presented here makes specific use of the order relation between points – not just the information as to which pairs of points are comparable.

4. Sperner's lemma and the cross cut conjecture

A poset $\mathbf{P}$ is said to be *ranked* if every maximal chain in $\mathbf{P}$ has the same number of points. When $\mathbf{P}$ is ranked and $x \in X$, we let $r(x)$ be the largest i so that there exists a chain of i points having x as its least element. The value $r(x)$ is called the *rank* of x, and the antichains $A_i = \{x\colon r(x) = i\}$ are called *ranks*. The poset $\mathbf{P}$ is said to be a *Sperner* poset if the width of $\mathbf{P}$ equals the maximum cardinality of its ranks. The following now classic result is due to Sperner (1928).

Theorem 4.1. *For each $n \geqslant 1$, $\mathbf{2}^n$ is a Sperner poset. In particular, the width of $\mathbf{2}^n$ is $\binom{n}{\lfloor n/2 \rfloor}$ (cf. chapter 24).*

Proof. We consider $\mathbf{2}^n$ as the set of all subsets of $\{1, 2, \ldots, n\}$ ordered by inclusion. It is easy to see that the maximum cardinality of a rank of $\mathbf{2}^n$ is the binomial coefficient $\binom{n}{\lfloor n/2 \rfloor}$. Also there are $n!$ maximal chains in $\mathbf{2}^n$. Now suppose the width of $\mathbf{2}^n$ is t, and let $\mathscr{A} = \{A_1, A_2, \ldots, A_t\}$ be a maximum antichain in $\mathbf{2}^n$. If $A \in \mathscr{A}$ and $|A| = k$, then there are $k!(n-k)!$ maximal chains in $\mathbf{2}^n$ which contain A. It follows that $\sum_{i=1}^{t} k_i!(n-k_i)! \leqslant n!$ where $k_i = |A_i|$. Thus $t/\binom{n}{\lfloor n/2 \rfloor} \leqslant \sum_{i=1}^{t} k_i!(n-k_i)!/n! \leqslant 1$, so that $t \leqslant \binom{n}{\lfloor n/2 \rfloor}$ as claimed. □

An enormous amount of research has been done on generalizations of this elementary result, and we encourage the reader to consult chapter 24 or the book Anderson (1987) which concentrates on this subject. Also, chapter 32 contains an important result from Sperner theory. In view of our space limitations, we include here only an outline of the theorem of Canfield (1978) which asserts that sufficiently large partition lattices are not Sperner posets. The argument we give is patterned after the argument given subsequently by Shearer (1979).

The *partition lattice* $\mathbf{\Pi}_n$ is the poset whose elements are the partitions (into equivalence classes) of the set $\{1, 2, \ldots, n\}$. In $\mathbf{\Pi}_n$, we set $\pi_1 \leqslant \pi_2 \iff$ each class in π_1 is a subset of a class in π_2. Partition lattices are natural combinatorial objects and have been studied extensively both in combinatorial mathematics and in related areas. For example, an important theorem in lattice theory due to Pudlak and Tuma (1980) asserts that every lattice is a sublattice of a partition lattice. The problem of investigating the Sperner property for partition lattices was popularized by Rota.

The rank sizes of the partition lattice $\mathbf{\Pi}_n$ are the Stirling numbers of the second kind $S(n, k)$ for $k = 0, 1, 2, \ldots, n$. These numbers form a unimodal sequence achieving maximum value when $k = k_n \sim n/\log n$. If $\pi \in \mathbf{\Pi}_n$ and π has k_i classes of size i for each i, then π is called a partition of type $1^{k_1}2^{k_2}3^{k_3}\cdots n^{k_n}$.

The following lemma is an easy exercise.

Lemma 4.2. *The number of partitions in Π_n of type $1^{k_1}2^{k_2}\cdots n^{k_n}$ is*

$$n!/(1!)^{k_1}(2!)^{k_2}\cdots(n!)^{k_n}k_1!k_2!\cdots k_n!.$$

Here is Canfield's (1978) solution to Rota's conjecture.

Theorem 4.3. *If n is sufficiently large, the partition lattice Π_n is not a Sperner poset.*

Proof. We actually prove a slightly weaker result. We show that for certain large values of n, the partition lattice Π_n is not a Sperner poset. It is a relatively straightforward extension to obtain the general result. Let $\mathcal{P}$ consist of all partitions of type $m^{h_1}(2m)^{h_2}(3m)^{h_3}$ where $h_1+h_2+h_3=k+1$, and $k=k_n$ is chosen to maximize the Stirling number $S(n,k)$. Note that $mh_1+2mh_2+3mh_3=n$. Furthermore, h_1, h_2, and h_3 satisfy:

$$h_1=[5(k+1)-2n/m-\theta]/3,$$
$$h_2=[n/m-(k+1)+2\theta]/3,$$
$$h_3=[n/m-(k+1)-\theta]/3.$$

The value θ is taken from $\{-1,0,+1\}$ so that each h_i is an integer. However, for the time being m is unspecified, although we assume that m divides n.

Any partition into k classes which is comparable to a partition in $\mathcal{P}$ must belong to one of the following six types:

Type 1: $m^{h_1-2}(2m)^{h_2+1}(3m)^{h_3}$

Type 2: $m^{h_1-1}(2m)^{h_2-1}(3m)^{h_3+1}$

Type 3: $m^{h_1-1}(2m)^{h_2}(3m)^{h_3-1}(4m)^1$

Type 4: $m^{h_1}(2m)^{h_2-2}(3m)^{h_3}(4m)^1$

Type 5: $m^{h_1}(2m)^{h_2-1}(3m)^{h_3-1}(5m)^1$

Type 6: $m^{h_1}(2m)^{h_2}(3m)^{h_3-2}(6m)^1$

We will show that for sufficiently large n, there are fewer than $|\mathcal{P}|$ partitions of these six types. The remaining partitions into k classes together with the partitions in $\mathcal{P}$ will then form an antichain of more than $S(n,k)$ elements which shows that Π_n is not a Sperner poset.

By Lemma 4.2, the ratio of the number of partitions of Type 1 divided by the number of partitions in $\mathcal{P}$ is $h_1(h_1-1)/(h_2+1)\binom{2m}{m}$. Then it is an easy (although tedious) calculation to show that if we set $m=\lfloor n/1.06k\rfloor$, then this ratio goes to 0 as n tends to infinity. An analogous computation shows that for this value of m, the ratio of the number of partitions of Type i divided by the number of partitions in $\mathcal{P}$ goes to 0 for each of the other five types. It may be shown that when $n>4\times10^9$, all six ratios are sufficiently small that their sum is less than 1 and the theorem follows. $\square$

Some progress has been made in reducing the value of n for which it can be shown that $\mathbf{\Pi}_n$ is not a Sperner poset. Jichang and Kleitman (1984) have lowered the estimate to 3.4×10^6. However, the enormity of these estimates and the width of the corresponding partition lattices are striking testimony to the adage well known to researchers in combinatorical mathematics: Woe be to those who make conclusions based on detailed examinations of small examples. Sometimes we are startled to learn just how large small can be.

5. Linear extensions and correlation

Let $\mathbf{P}$ be a poset, and let $\mathscr{E}$ denote the set of all linear extensions of $\mathbf{P}$. It is natural to consider the elements of $\mathscr{E}$ as equally likely outcomes in a finite probability space. When x and y are distinct points in $\mathbf{P}$, we let $\text{Prob}[x < y]$ denote the probability of the event consisting of all linear extensions in which $x < y$. Observe that $\text{Prob}[x < y]$ is the ratio of the number of linear extensions with $x < y$ divided by $|\mathscr{E}|$. Note that $\text{Prob}[x < y] = 1 \iff x < y$ in $\mathbf{P}$, and $\text{Prob}[x < y] = 0 \iff y < x$ in $\mathbf{P}$.

Similarly, if x, y, z are three distinct points in $\mathbf{P}$, we let $\text{Prob}[x < y | x < z]$ denote the conditional probability that in a random selection of a linear extension, the relation $x < y$ holds given that $x < z$ holds. In 1980, Rival and Sands made the following conjecture, which quickly became known as the xyz-conjecture.

Conjecture 5.1. If x, y, z are distinct points in a poset $\mathbf{P}$, then

$$\text{Prob}[x < y] \leqslant \text{Prob}[x < y | x < z].$$

It is easy to see that the xyz-conjecture holds except possibly when $\{x, y, z\}$ is a three-element antichain. In this case, Rival and Sands conjectured that the inequality in Conjecture 5.1 was strict, and this stronger version became known as the strict xyz-conjecture. The original conjecture was settled in the affirmative by Shepp (1980, 1982) using the FKG-inequality from statistical mechanics. The strict xyz-conjecture was proved by Fishburn (1984) using an important generalization of the FKG-inequality proved by Ahlswede and Daykin (1978), which we call the AD-inequality.

Although we do not include its proof here, the AD-inequality is a marvelous device with a growing list of applications in combinatorics. We encourage the reader to study its elementary proof carefully.

Let $\mathbf{P}$ be a distributive lattice with meet and join denoted $\wedge$ and $\vee$ respectively. When A and B are subsets of X, we let $A \wedge B = \{a \wedge b\colon a \in A \text{ and } b \in B\}$ and $A \vee B = \{a \vee b\colon a \in A \text{ and } b \in B\}$. When $f : X \to \mathbb{R}$ and $A \subset X$, let $f(A) = \sum_{x \in A} f(x)$. Let $\mathbb{R}_0$ denote the nonnegative real numbers.

Here is the AD-inequality.

Theorem 5.2. *Let* $\mathbf{P}$ *be a distributive lattice and let* α, β, γ, *and* δ *be functions from* X *to* $\mathbb{R}_0$ *satisfying:*

(i) $\alpha(x)\beta(y) \leqslant \gamma(x \wedge y)\delta(x \vee y)$ *for all* $x, y \in X$.

Then the following inequality holds for every $A, B \subset X$:

(ii) $\alpha(A)\beta(B) \leqslant \gamma(A \wedge B)\delta(A \vee B)$.

Corollary 5.3 (Fortuin et al. 1971). *Let* $\mathbf{P}$ *be a distributive lattice and let* $\mu : X \to [0, 1]$ *satisfy:*

(i) $\mu(x)\mu(y) \leqslant \mu(x \wedge y)\mu(x \vee y)$, *for all* $x, y \in X$.

If f *and* g *are monotonic functions from* X *to* $\mathbb{R}$, *then the following inequality holds:*

(ii) $[(f\mu)(X)][(g\mu)(X)] \leqslant [\mu(X)][(fg\mu)(X)]$.

Proof. Assume first that f and g map X to $\mathbb{R}_0$. For each $x \in X$, define $\alpha(x) = f(x)\mu(x)$, $\beta(x) = g(x)\mu(x)$, $\gamma(x) = \mu(x)$, and $\delta(x) = f(x)g(x)\mu(x)$. Then for an arbitrary $x, y, \in X$ we have:

$$\begin{aligned}
\alpha(x)\beta(y) &= f(x)\mu(x)g(y)\mu(y) \\
&= f(x)g(y)\mu(x)\mu(y) \\
&\leqslant f(x)g(y)\mu(x \wedge y)\mu(x \vee y) \\
&\leqslant \mu(x \wedge y)f(x \vee y)g(x \vee y)\mu(x \vee y) \\
&= \gamma(x \wedge y)\delta(x \vee y).
\end{aligned}$$

Therefore $\alpha(X)\beta(X) \leqslant \gamma(X)\delta(X)$ as claimed. When the range of f or g includes negative reals, we increase both functions by some suitably large constant. □

A subset S of a poset $\mathbf{P}$ is called a *down set* (*up set*) if $x \in S$ and $y \leqslant x$ in $\mathbf{P}$ ($x \leqslant y$ in $\mathbf{P}$) always implies $y \in S$.

Corollary 5.4. *Let* U_1 *and* U_2 *be up sets in a distributive lattice* $\mathbf{P}$. *Then* $|U_1||U_2| \leqslant |U_1 \cap U_2||X|$.

Proof. Set $\alpha(x) = \beta(x) = \gamma(x) = \delta(x) = 1$ for every $x \in X$. Observe that $U_1 \vee U_2 = U_1 \cap U_2$ and $U_1 \wedge U_2 \subset X$. □

The special case of Corollary 5.4 when $\mathbf{P} = \mathbf{2}^n$ was first proved by Kleitman (1966) and in dual form by Seymour (1973).

We close this section by outlining Shepp's (1982) proof of the xyz-conjecture.

Theorem 5.5. *Let* $\mathbf{P}$ *be a poset and let* x, y, *and* z *be three distinct points in* X. *Then*

$$\text{Prob}[x < y] \leqslant \text{Prob}[x < y | x < z].$$

Proof. We assume $\{x, y, z\}$ is a three-element antichain in $\mathbf{P}$. Let k be a positive integer, and let Y_k denote the set of all order preserving functions from $\mathbf{P}$ to $\mathbf{k}$. Define a partial order P_k on Y_k by setting $f \leqslant g$ in $P_k \iff f(x) \geqslant g(x)$ and

$f(u) - f(x) \leqslant g(u) - g(x)$ for every $u \in X$. It is straightforward to verify that the poset (Y_k, P_k) is in fact a distributive lattice.

Now let $U_1(k) = \{f \in Y_k\colon f(x) < f(y)\}$ and $U_2(k) = \{f \in Y_k\colon f(x) < f(z)\}$. Then $U_1(k)$ and $U_2(k)$ are up sets in the distributive lattice (Y_k, P_k). Therefore

$$|U_1(k)||U_2(k)| \leqslant |U_1(k) \cap U_2(k)||Y_k|,$$

so that:

$$\frac{|U_1(k)|}{|Y_k|} \leqslant \frac{|U_1(k) \cap U_2(k)|/|Y_k|}{|U_2(k)|/|Y_k|}.$$

However, it is easy to see that as k tends to ∞, the left-hand side of this inequality approaches $\mathrm{Prob}[x < y]$ while the right-hand side approaches $\mathrm{Prob}[x < y|x < z]$. □

The reader should note that the truly clever part of this proof is the nonstandard definition of the partial order P_k so that (Y_k, P_k) is a distributive lattice having $U_1(k)$ and $U_2(k)$ as up sets. Fishburn's (1984) proof of the strict xyz-inequality requires two applications of the AD-inequality. Winkler's (1986) survey article is a good starting point for an overview of work on correlation.

6. Balancing pairs and the $\frac{1}{3}$-$\frac{2}{3}$ conjecture

The following conjecture is due to Kislitysn (1968), although it was also made independently by Fredman (1979) and Linial (1984):

Conjecture 6.1 (*The* $\frac{1}{3}$–$\frac{2}{3}$ *conjecture*). If $\mathbf{P}$ is not a chain, then $\mathbf{P}$ contains distinct points x and y for which $\frac{1}{3} \leqslant \mathrm{Prob}[x > y] \leqslant \frac{2}{3}$.

If true, this conjecture is best possible as is evidenced by the three-point poset $\mathbf{2}+\mathbf{1}$. Given a poset $\mathbf{P} = (X, P)$ which is not a chain, let $\delta(\mathbf{P})$ denote the largest positive number for which there exists a pair $x, y \in X$ with $\delta(\mathbf{P}) \leqslant \mathrm{Prob}[x > y] \leqslant 1 - \delta(\mathbf{P})$. Using this terminology, we can restate the $\frac{1}{3}$–$\frac{2}{3}$ conjecture as follows.

Conjecture 6.2. If $\mathbf{P}$ is not a chain, then $\delta(\mathbf{P}) \geqslant \frac{1}{3}$.

The original motivation for studying balancing pairs in posets was the connection with sorting. The problem was to answer whether it is always possible to determine an unknown linear extension of a poset $\mathbf{P}$ with $O(\log t)$ rounds (questions) where t is the number of linear extensions of P. The answer would be "yes" if one could prove that there exists an absolute constant δ_0 so that $\delta(\mathbf{P}) \geqslant \delta_0$ for any $\mathbf{P}$ which is not a chain.

Linial (1984) has shown that the $\frac{1}{3}$–$\frac{2}{3}$ conjecture holds for posets of width two. Fishburn et al. (1992) show that it holds for posets of height at most two. Although the conjecture remains open for general posets, we present a partial result, due to Kahn and Saks (1984), which is particularly appealing in view of its nontrivial application of the Alexandrov–Fenchel inequalities for mixed volumes.

Theorem 6.3. *If* **P** *is not a chain, then X contains a distinct pair* x, y *so that*

$$\tfrac{3}{11} \leqslant \text{Prob}[x > y] \leqslant \tfrac{8}{11},$$

and thus $\delta(\mathbf{P}) \geqslant \frac{3}{11}$.

Proof. Clearly, we may assume that **P** does not have a least element. Let $\mathscr{E}$ denote the set of all linear extensions of **P** and let $n = |X|$. For each $L \in \mathscr{E}$, let $h_L : X \to \mathbf{n}$ be the order preserving injection determined by L. Let $|\mathscr{E}| = t$ and then define $h : X \to \mathbb{R}_0$ by $h(x) = (\sum_{L \in \mathscr{E}} h_L(x))/t$. The value $h(x)$ is the average height of x among the linear extensions in $\mathscr{E}$. Since no element satisfies $h(x) = 0$, it follows that there exist distinct elements $x, y \in X$ with $0 \leqslant h(y) - h(x) < 1$. Note that such a pair must be incomparable. We will show that $\frac{3}{11} \leqslant \text{Prob}[x > y] \leqslant \frac{8}{11}$. The argument depends on a series of lemmas which hold for an arbitrary incomparable pair in **P**.

Fix an arbitrary incomparable pair $x, y \in X$. For a positive (negative) integer i, let e_i be the number of linear extensions of **P** in which x is below (above) y by exactly i positions.

Lemma 1. $e_1 = e_{-1}$.

Proof. Since x and y occur consecutively, they may be interchanged. □

Lemma 2. $e_2 + e_{-2} \leqslant e_1 + e_{-1}$.

Proof. Suppose $L \in \mathscr{E}$ and $|h_L(y) - h_L(x)| = 2$. Let u be the unique point between x and y. If $u \parallel x$, exchange u and x. If u is comparable with x, then $u \parallel y$. In this case, the cyclic permutation (uyx) converts L into an extension in which x and y are consecutive. The mapping is easily seen to be an injection. □

Lemma 3. *If* $|i| \geqslant 2$, *then* $e_i \leqslant e_{i-1} + e_{i+1}$.

Proof. Without loss of generality $i \geqslant 2$. Suppose the lemma is false and of all counterexamples, choose one for which $|X| = n$ is minimum. For this value of n, we choose **P** so that the number of comparable pairs is maximum. We first show that for every incomparable pair $u, v \in X$, one of u and v is greater than or equal to x and the other is less than or equal to y.

Suppose to the contrary that $u \parallel v$ and that the above statement does not hold. Let $\mathbf{P}'$ be the poset obtained by adding the relation $u < v$ to the partial order on X and taking the transitive closure. Also let $\mathbf{P}''$ be the poset obtained by adding $v < u$. In $\mathbf{P}'$, we let e'_j denote the number of linear extensions with x below y by exactly j positions. Also, let e''_j denote the corresponding number for $\mathbf{P}''$. Then $e_i = e'_j + e''_j$ for each $j \geqslant 1$. Since $\mathbf{P}'$ and $\mathbf{P}''$ have more comparable pairs than **P**, we know that $e'_i \leqslant e'_{i-1} + e'_{i+1}$ and $e''_i \leqslant e''_{i-1} + e''_{i+1}$, so that $e_i \leqslant e_{i-1} + e_{i+1}$ as claimed.

We next claim that $x \in \text{MIN}(\mathbf{P})$ and $y \in \text{MAX}(\mathbf{P})$. For suppose $u \in \text{MIN}(\mathbf{P})$ and that $u < x$. If $u < y$ in **P**, then n must be comparable with every point in X, i.e.,

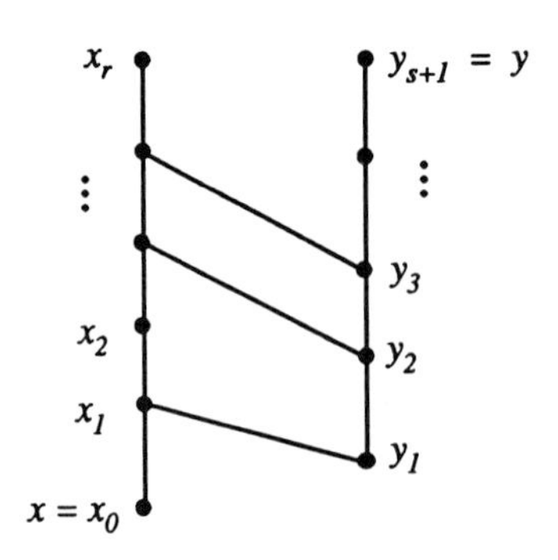

Figure 6.1.

u is the least element of $\mathbf{P}$. In this case, we consider the poset $\mathbf{P}' = \mathbf{P} - \{u\}$ and the numbers e'_j defined in the natural way for $\mathbf{P}'$. Since $e_j = e'_j$, we would conclude $e_i \leqslant e_{i-1} + e_{i+1}$. The contradiction shows $x \in \mathrm{MIN}(\mathbf{P})$. Dually, $y \in \mathrm{MAX}(\mathbf{P})$.

These remarks show that X is the union of two chains $x = x_0 < x_1 < x_2 < \cdots < x_r$ and $y_1 < y_2 < \cdots < y_s < y_{s+1} = y$. Furthermore, no y_i is larger than any x_j in $\mathbf{P}$. (See fig. 6.1.)

We now distinguish two cases.

Case 1. $i < n - 1$. Choose $L \in \mathscr{E}$ with $h_L(y) - h_L(x) = k$. If x is not the least element in L, exchange x with the element immediately under it. If x is the least element in L, then y is not the greatest. Let z be the element immediately over x. If $x \parallel z$, exchange them; otherwise exchange y with the element immediately over it. This procedure is an injection which transforms L into a linear extension in which x is below y by either $i - 1$ or $i + 1$ positions.

Case 2. $i = n - 1$. For each subposet $\mathbf{Q} \subset \mathbf{P}$, let $e(\mathbf{Q})$ denote the number of linear extensions of $\mathbf{Q}$. Let $\mathbf{Z} = \mathbf{P} - \{x, y\}$. Then $e_{i-1} = e(\mathbf{Z} - \{x_r\}) + e(\mathbf{Z} - \{y_1\})$, $e_i = e(\mathbf{Z})$ and $e_{i+1} = 0$. Now $e(\mathbf{Z}) = e(\mathbf{Z} - \{x_1\}) + e(\mathbf{Z} - \{y_1\})$ so, to complete the argument, we need only show that $e(\mathbf{Z} - \{x_r\}) \geqslant e(\mathbf{Z} - \{x_1\})$. However, this inequality follows immediately since the mapping $x_j \to x_{j-1}$ transforms a linear extension of $\mathbf{Z} - \{x_1\}$ into a linear extension of $\mathbf{Z} - \{x_r\}$. This completes the proof of Lemma 3. □

The next lemma is a special case of the Alexandrov–Fenchel inequalities for mixed volumes. This method was pioneered by Stanley (1981).

Lemma 4. *Let K_0 and K_1 be convex subsets of $\mathbb{R}^n$. For each λ with $0 < \lambda < 1$, let $K_\lambda = \{(1 - \lambda)v_0 + \lambda v_1\colon\ v_0 \in K_0,\ v_1 \in K_1\}$. Let d be the dimension of the affine hull* *of K_λ for $\lambda \in (0,1)$. Then there exist unique numbers $a_0, a_1, \ldots, a_d$ so that for all λ, the d-dimensional volume of K_λ satisfies:*

$$\mathrm{Vol}(K_\lambda) = \sum_{k=0}^{d} \binom{d}{k} a_k (1-\lambda)^{d-k} \lambda^k.$$

Furthermore, the sequence $a_0, a_1, a_2, \ldots, a_d$ is logarithmically concave, i.e.,

$$a_k^2 \geqslant a_{k-1}a_{k+1}, \quad \textit{for } k = 1, 2, \ldots, d-1.$$

Lemma 5. *The sequences $e_1, e_2, e_3, \ldots, e_n$ and $e_{-n}, e_{-n+1}, \ldots, e_{-1}$ are logarithmically concave.*

Proof. Let $\mathbb{R}^{\mathbf{P}}$ denote the vector space of all order preserving functions from $\mathbf{P}$ to $\mathbb{R}$. Also let $C(\mathbf{P}) = \{f \in \mathbb{R}^{\mathbf{P}}\colon 0 \leqslant f(x) \leqslant 1 \text{ for all } x \in \mathbf{P}\}$. For λ with $0 \leqslant \lambda \leqslant 1$, let $K_\lambda = \{f \in C(\mathbf{P})\colon f(y) - f(x) = \lambda\}$. Note that $K_\lambda = (1-\lambda)K_0 + \lambda K_1$. Also note that when $0 < \lambda < 1$, K_λ has dimension $n-1$.

For each $L \in \mathscr{E}$ with $x < y$ in L, let $\Delta_\lambda(L) = \{f \in K_\lambda\colon v < w \text{ in } L \Rightarrow f(v) \leqslant f(w)\}$. Then K_λ is the union of the $\Delta_\lambda(L)$'s taken over all L with $x < y$ in L. Since the $\Delta_\lambda(L)$'s have disjoint interiors, we see that

$$\mathrm{Vol}(K_\lambda) = \sum_{\substack{L \in \mathscr{E} \\ x<y \text{ in } L}} \mathrm{Vol}(\Delta_\lambda(L)).$$

Now consider some particular $L \in \mathscr{E}$ with $x < y$ in L. Suppose L orders X as $v_0 < v_1 < \cdots < v_{n-1}$ with $x = v_i$ and $y = v_j$. Then $\Delta_\lambda(L)$ consists of those $f \in C(\mathbf{P})$ for which $0 \leqslant f(v_0) \leqslant f(v_1) \leqslant \cdots \leqslant f(v_{n-1}) \leqslant 1$ and $f(v_j) = f(v_i) + \lambda$. Now consider the volume preserving mapping defined by:

$$\begin{aligned} &f(v_k) \to f(v_k) - \lambda && \text{when } k > j; \\ &f(v_k) \to f(v_k) - f(v_i) && \text{when } j \geqslant k > i; \quad \text{and} \\ &f(v_k) \to f(v_k) && \text{when } i \geqslant k. \end{aligned}$$

Then the image of $\Delta_\lambda(L)$ under this mapping is the set of $f \in C(\mathbf{P})$ for which

$$0 \leqslant f(v_0) \leqslant f(v_1) \leqslant \cdots \leqslant f(v_i) \leqslant f(v_{j+1}) \leqslant \cdots \leqslant f(v_{n-1}) \leqslant 1 - \lambda$$

and $0 \leqslant f(v_i) \leqslant f(v_{i+1}) \leqslant \cdots \leqslant f(v_j) = \lambda$. However, this set is the product of two simplices and its volume is therefore

$$\frac{(1-\lambda)^{n-j+1}}{(n-j+i)!} \; \frac{\lambda^{j-i-1}}{(j-i-1)!}.$$

It follows that

$$\begin{aligned} \mathrm{Vol}(K_\lambda) &= \sum_{0 \leqslant i < j \leqslant n-1} \; \sum_{\substack{L \in \mathscr{E} \\ h_L(x)=i \\ h_L(y)=j}} \frac{(1-\lambda)^{n-j+i}}{(n-j+i)!} \; \frac{\lambda^{j-i-1}}{(j-i-1)!} \\ &= \sum_{i \geqslant 1} e_i \frac{(1-\lambda)^{n-i}}{(n-i)!} \; \frac{\lambda^{i-1}}{(i-1)!}. \end{aligned}$$

From Lemma 4, we conclude that $a_i = e_{i-1}/(n-1)!$, and thus the sequence $e_1, e_2, e_3, \ldots$ is logarithmically concave. The argument for the other sequence is dual. □

For the remainder of the proof, we assume that x and y are an incomparable pair with $0 \leqslant h(y) - h(x) \leqslant 1$. For each $i \geqslant 1$, we let $b_i = e_i/t$ and $a_i = e_{-i}/t$. We then know that the following conditions hold:

(1) $a_1, a_2, a_3, \ldots$ and $b_1, b_2, b_3, \ldots$ are sequences of nonnegative real numbers so that $\sum_{i\geqslant 1} a_i + \sum_{i\geqslant 1} b_i = 1$.

(2) $a_1 = b_1$.

(3) $a_2 + b_2 \leqslant a_1 + b_1$.

(4) $a_i \leqslant a_{i-1} + a_{i+1}$ and $b_i \leqslant b_{i-1} + b_{i+1}$ for all $i \geqslant 2$.

(5) $a_i^2 \geqslant a_{i-1}a_{i+1}$ and $b_i^2 \geqslant b_{i-1}b_{i+1}$ for all $i \geqslant 2$.

(6) If $a_i = 0$, then $a_{i+1} = 0$, for all $i \geqslant 1$.

(7) If $b_i = 0$, then $b_{i+1} = 0$, for all $i \geqslant 1$.

It remains only to show that whenever these seven properties are satisfied, we always have $\sum_{i\geqslant 1} ia_i - \sum_{i\geqslant 1} ib_i \geqslant 1$ whenever $\sum_{i\geqslant 1} b_i \leqslant \frac{3}{11}$. To accomplish this, we fix a value $b = b_1 = a_1$, and the value $B = \sum_{i\geqslant 1} b_i$. For the pair (b, B), we determine the unique sequences $\{a_i\colon i \geqslant 1\}$, $\{b_i\colon i \geqslant 1\}$ satisfying these seven conditions which minimize $\sum ia_i - \sum ib_i$.

The infinite geometric sequence obtained by setting $\varepsilon = b/B$ and $b_i = b(1-\varepsilon)^{i-1}$ is easily seen to maximize $\sum ib_i$ among all sequences $\{b_i\colon i \geqslant 1\}$ satisfying $b_1 = b$, $\sum b_i = B$, and $b_i^2 \geqslant b_{i-1}b_{i+1}$ for all $i \geqslant 2$. The argument to minimize $\sum ia_i$ is somewhat more complicated. We know $a_2 + b_2 \leqslant a_1 + b_1 = 2b$ so $a_2 \leqslant 2b - b_2 = b(1+\varepsilon)$. Therefore $a_{i+1} \leqslant (1+\varepsilon)a_i$ for every $i \geqslant 1$. On the other hand, set $A = 1 - B$ and for each $j \geqslant 1$, let $s_j = A - \sum_{i=1}^{j} a_i$. Then $a_{j+1} \leqslant s_j$. Also $a_{j+1} \leqslant a_j + a_{j+2} \leqslant a_j + s_{j+1} = a_j + s_j - a_{j+1}$, so that $a_{j+1} \leqslant (a_j + s_j)/2$. It is then easy to verify that $\sum ia_i$ is minimized when there is some $k \geqslant 1$ so that $a_i = b(1+\varepsilon)^{i-1}$ for all $i = 1, 2, \ldots, k$ and either:

Type 1: $a_{k+1} = s_k$, $a_i = 0$ for all $i \geqslant k+2$; or

Type 2: $a_{k+1} = (s_k + a_k)/2$ where $s_k > a_k$, $a_{k+2} = s_{k+1} = (s_k - a_k)/2$, and $a_i = 0$ for all $i \geqslant k+3$.

Set $\alpha = a_{k+1}/a_k$ and $\beta = a_{k+2}/a_k$. We verify that $\sum ia_i - \sum ib_i \geqslant 1$ for a Type 2 sequence $\{a_i\colon i \geqslant 1\}$. The reader may enjoy the challenge of handling the Type 1 case – it is somewhat easier. Now for a Type 2 sequence, we know:

$$1 + \varepsilon \geqslant \alpha \geqslant 1, \quad \beta = \alpha - 1, \text{ and} \tag{1}$$

$$\alpha + \beta = \frac{(1+\varepsilon)^{1-k}}{\varepsilon B} - 1 - 1/\varepsilon. \tag{2}$$

Using the definitions of $\{a_i\colon i \geqslant 1\}$ and $\{b_i\colon i \geqslant 1\}$, we find that

$$\sum ia_i - \sum ib_i = \frac{[k\varepsilon - \varepsilon - 1 + \varepsilon^2 k(\alpha + \beta + 1) + \varepsilon^2(\alpha + 2\beta)]}{\varepsilon + \varepsilon^2(\alpha + \beta + 1)}.$$

The inequality $\sum ia_i - \sum ib_i \geqslant 1$ is then equivalent (using (1)) to

$$\frac{(1+\varepsilon)^{k+1}}{k\varepsilon} \leqslant \frac{1}{B} + \frac{\varepsilon\beta}{k}(1+\varepsilon)^{k-1}. \tag{3}$$

This in turn is equivalent to

$$\frac{(4\varepsilon^2+5\varepsilon+2)(1+\varepsilon)^{k-1}}{(2k+1)\varepsilon} \leqslant \frac{1}{\beta}. \tag{4}$$

Now $\beta = \frac{1}{2B\varepsilon(1+\varepsilon)^{k-1}} - \frac{1}{2\varepsilon} - 1$ so the inequality $0 \leqslant \beta \leqslant \varepsilon$ converts to

$$(2\varepsilon+1)(1+\varepsilon)^{k-1} \leqslant \frac{1}{B} \tag{5}$$

and

$$(2\varepsilon^2+2\varepsilon+1)(1+\varepsilon)^{k-1} \geqslant \frac{1}{B}. \tag{6}$$

We may assume $\varepsilon < 2/(2k-1)$; otherwise $k \geqslant 1/\varepsilon + \frac{1}{2}$ and

$$\begin{aligned}\frac{(4\varepsilon^2+5\varepsilon+2)(1+\varepsilon)^{k-1}}{(2k+1)\varepsilon} &\leqslant \frac{(4\varepsilon^2+5\varepsilon+2)(1+\varepsilon)^{k-1}}{2+2\varepsilon}\\ &\leqslant (2\varepsilon+1)(1+\varepsilon)^{k-1}\\ &\leqslant 1/B.\end{aligned}$$

We may also assume $\varepsilon > 3/(3k+1)$, for if $\varepsilon \leqslant 3/(3k+1)$, inequality (5) implies

$$\left(1+\frac{3}{3k+1}\right)^{k-1}\left[\frac{18}{(3k+1)^2}+\frac{6}{3k+1}+1\right] \geqslant \frac{11}{3}. \tag{7}$$

However, inequality (6) is false for all $k \geqslant 1$.

To complete the proof, we observe that the left-hand side of inequality (3) is an increasing function of ε when $\varepsilon > 3/(3k+1)$. It suffices to test the validity of (3) at an upper bound for ε. When $k = 1$, the trivial bound $\varepsilon \leqslant 1$ works. For $k \geqslant 2$, use the bound $\varepsilon \leqslant 2/(2k-1)$. This completes the proof. □

In a certain sense, the approach taken by Kahn and Saks cannot be improved. In fig. 6.2, we show a poset **P** containing two points x, y satisfying $h(y) - h(x) = 1$ and $\text{Prob}[x > y] = \frac{3}{11}$.

Other proofs bounding $\delta(\mathbf{P})$ away from zero have been given. Khachiyan (1989) uses geometric techniques to show $\delta(\mathbf{P}) \geqslant 1/e^2$. Kahn and Linial (1991) provide a short and elegant argument using the Brunn–Minkowski theorem to show that $\delta(\mathbf{P}) \geqslant 1/2e$. Friedman (1993) also applies geometric techniques to obtain even better constants when the poset satisfies certain additional properties. Kahn and Saks (1984) conjectured that $\delta(\mathbf{P})$ approaches $\frac{1}{2}$ as the width of **P** tends to infinity. Komlós (1990) provides support for this conjecture by showing that for every

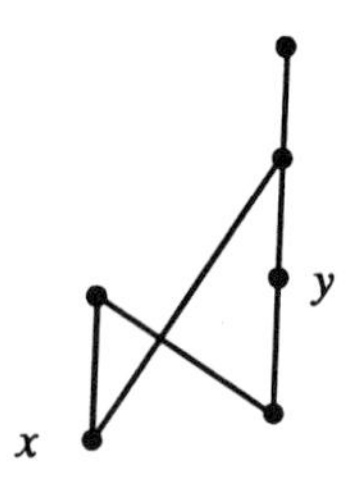

Figure 6.2.

$\varepsilon > 0$, there exists a function $f_\varepsilon(n) = o(n)$ so that if $\mathbf{P} = (X, P)$ is a poset with $|X| = n$ and at least $f_\varepsilon(n)$ minimal points, then $\delta(\mathbf{P}) > \frac{1}{2} - \varepsilon$.

As Kahn and Saks (1984) point out, the value of the constant in Theorem 6.3 could be improved if we could show that there exists a positive absolute constant γ so that if $\mathbf{P}$ is not a chain, then it is always possible to find an ordered pair (x, y) with $0 \leqslant h(y) - h(x) \leqslant 1 - \gamma$. However, nobody has yet been able to settle whether such a γ exists. If it does, then as shown by Saks (1985), it must satisfy $\gamma \leqslant 0.133$. Even this value would not be enough to prove $\delta(\mathbf{P}) \geqslant \frac{1}{3}$. However, it is of interest to determine the maximum value of $|h(y) - h(x)|$ which allows one to conclude that $\frac{1}{3} \leqslant \text{Prob}[x > y] \leqslant \frac{2}{3}$. Felsner and Trotter (1993) discuss how to modify the Kahn–Saks proof technique to obtain the next result, which is clearly best possible.

Theorem 6.4. *Let (x, y) be distinct points in a poset $\mathbf{P}$, and suppose that $0 \leqslant h(y) - h(x) \leqslant \frac{2}{3}$. Then $\frac{1}{3} \leqslant \text{Prob}[x > y] \leqslant \frac{2}{3}$.*

Felsner and Trotter (1993) obtain a slight improvement in the Kahn–Saks bound by considering subposets in which the points are relatively close in average height.

Theorem 6.5. *There exists an absolute constant $\varepsilon > 0$ so that if $\mathbf{P}$ is a poset which is not a chain, then $\delta(\mathbf{P}) > \varepsilon + \frac{3}{11}$.*

In developing this theorem, Felsner and Trotter made a correlation conjecture which is of independent interest. Let x, y and z be distinct points in a poset $\mathbf{P} = (X, P)$. For each $i, j \in \mathbb{Z}$, let $p(i, j)$ denote the probabilty that $h_L(y) - h_L(x) = i$ and $h_L(z) - h_L(y) = j$ in a random linear extension L of P. Felsner and Trotter then made the following *cross product* conjecture.

Conjecture 6.6. Let $x <: y <: z$ in a poset $\mathbf{P} = (X, P)$. Then for all $i, j \in \mathbb{N}$,

$$p(i, j)p(i + 1, j + 1) \leqslant p(i, j + 1)p(i + 1, j).$$

Brightwell, Felsner and Trotter (1995) prove the cross product conjecture in the special case $i = j = 1$. They then prove the following lower bound.

Theorem 6.7. *If* **P** *is a finite poset which is not a chain then*

$$\delta(\mathbf{P}) > \frac{5-\sqrt{5}}{10}.$$

The results of this section have emphasized existence questions – disregarding the issue as to how one actually goes about finding an incomparable pair of points which are balanced. Brightwell and Winkler (1991) showed that the problem of computing the number of linear extensions of a poset is #P-complete. However, if one is willing to use randomized algorithms, then a good approximation to the volume of a polytope can be efficiently computed, so that the theorems presented here can form the basis for a sorting algorithm.

On the other hand, if we limit our attention to deterministic algorithms, then an alternative approach is necessary. Kahn and Kim (1994) use a concept of entropy for posets and show the existence of a polynomial-time deterministic algorithm for sorting in $O(\log t)$ rounds. Their algorithm shows how to efficiently locate pairs to use in queries so that, regardless of the responses, the determination of the unknown linear extension is made in $O(\log t)$ rounds. However, at individual rounds, the pairs need not be balanced in the sense that for a given pair (x, y) used in the algorithm, $\text{Prob}[x > y]$ may be arbitrarily close to zero.

7. Dimension and posets of bounded degree

The *dimension* of a poset $\mathbf{P} = (X, P)$ is the least t for which there exists a family $R = \{L_1, L_2, \ldots, L_t\}$ of linear extensions of P so that $P = L_1 \cap L_2 \cap \cdots \cap L_t$. In fact, the dimension of $\mathbf{P}$ is the least t for which there exists a family $R = \{L_1, L_2, \ldots, L_t\}$ of linear orders (not necessarily linear extensions of P) on X so that whenever $x \not\leqslant y$ in P, there exists at least one i with $y < \{z \colon x \leqslant z \text{ in } P\}$ in L_i. The concept of dimension was introduced by Dushnik and Miller (1941).

In section 2, we presented Dilworth's decomposition theorem, but we did not give much of an explanation for the role this theorem plays in research on posets – other than to motivate the Greene–Kleitman theorem. However, Dilworth's theorem plays a major role in dimension theory. For example, Hiraguchi (1951) used it to prove that the dimension of a poset never exceeds its width. Hiraguchi (1951) also proved that the dimension of a poset $\mathbf{P} = (X, P)$ is at most $|X|/2$, provided $|X| \geqslant 4$. On the other hand, it is still unknown whether every poset $\mathbf{P}$ with three or more points always contains a pair whose removal decreases the dimension at most one (see Kelly 1984, Reuter 1989, Kierstead and Trotter 1991).

Dilworth's theorem is critical to dimension-theoretic ineqalities appearing in Bogart and Trotter (1973), Trotter (1974b, 1975a,c, 1976a). It also plays a major role in the variants of dimension investigated in Bogart and Trotter (1976a,b), Kierstead et al. (1987), Kierstead and Trotter (1985, 1989). For additional background material on dimension, the reader is encouraged to consult the monograph Trotter (1992) and the survey articles Kelly and Trotter (1982) and Trotter (1982).

Also, discussions of open problems in dimension theory are given in Trotter (1989, 1992, 1994).

In the remainder of this section, we discuss one important problem for which there is an interesting partial solution – utilizing the probabilistic method on posets. Brightwell's survey article (Brightwell 1993) highlights the recent work in this rapidly growing area of research.

For integers $n \geqslant 3$, $k \geqslant 0$, define the *crown* $\mathbf{S}_n^k$ as the poset of height 2 having $n+k$ minimal elements $a_1, a_2, \ldots, a_{n+k}$ and $n+k$ maximal elements $b_1, b_2, \ldots, b_{n+k}$ with $a_i < b_{i-1}, b_{i-2}, b_{i-3}, \ldots, b_{i-n+1}$ for each i (cyclically). Trotter (1974a) showed that $\dim(\mathbf{S}_n^k) = \lceil 2(n+k)/(k+2) \rceil$. As noted previously, Hiraguchi (1951) proved that the dimension of a poset on m points does not exceed $m/2$, when $m \geqslant 4$, so the crowns $\{\mathbf{S}_n^0 \colon n \geqslant 3\}$ show that this inequality is best possible. In fact, the crown $\mathbf{S}_n^0$ is called the *standard* example of an n-dimensional poset. However, it is of interest to investigate conditions which force the dimension of a poset to be small in comparison to the number of points.

We now proceed to study one such condition which surfaced in the investigation of crowns. For a point x in a poset $\mathbf{P}$, let $\deg(x)$ count the number of points comparable (but not equal) to x in $\mathbf{P}$. Then let $\Delta(\mathbf{P})$ denote the maximum value of $\deg(x)$ for $x \in \mathbf{P}$. Rödl and Trotter proved that if $\Delta(\mathbf{P}) \leqslant k$, then $\dim(\mathbf{P}) \leqslant 2k^2 + 2$. For each $x \in \mathbf{P}$, let $U(x) = \{y \colon x < y \text{ in } \mathbf{P}\}$ and let $u = \max\{|U(x)| \colon x \in \mathbf{P}\}$. Then $u \leqslant \Delta(\mathbf{P})$.

We now present a strengthening of this result due to Füredi and Kahn (1986). Lemmas 7.1, 7.2, 7.5, Corollary 7.4 and Theorem 7.6 all come from that paper.

Lemma 7.1. $\dim(\mathbf{P}) < 2(u+1)(\log|X|) + 1$.

Proof. Let $|X| = n$ and set $t = \lceil 2(u+1)\log n \rceil$. Let $L_1, L_2, \ldots, L_t$ be linear orders of X chosen at random. Then for each $i = 1, 2, \ldots, t$ and for each $x \in \mathbf{P}$, the probability that $x < y$ in L_i for all $y \in U(x)$ is at least $1/(u+1)$. Hence the probability that no L_i satisfies $x < y$ in L_i for all $y \in U(x)$ is at most $[1 - 1/(u+1)]^t < 1/n^2$. Thus the probability that there exists a pair $x, y \in X$ with $y \not\leqslant x$ in X but there is no L_i with $x < y$ in L_i for all $y \in U(x)$ is less than 1. Thus $\dim(\mathbf{P}) \leqslant t$ as claimed. □

For integers n, k with $1 < k \leqslant n$, let $\dim(1, k; n)$ denote the dimension of the poset formed by the 1-element and k-element subsets of $\{1, 2, \ldots, n\}$ ordered by inclusion. Then $\dim(1, k; n)$ is the least t for which there exist t linear orders $L_1, L_2, \ldots, L_t$ on $\{1, 2, \ldots, n\}$ so that for each $(k+1)$-element subset $S \subset \{1, 2, \ldots, n\}$ and for each $x \in S$, there is at least one L_i in which x is the least element of S. The following lemma follows along the same lines as Lemma 7.1.

Lemma 7.2. *For all integers n, k with* $1 < k \leqslant n$, $\dim(1, k; n) \leqslant k^2(1 + \log(n/k))$.

The Füredi–Kahn argument also depends on the following result, which is known as the Lovász local lemma (Erdős and Lovász 1973). Other applications of this lemma are given in chapter 33 by Spencer.

Lemma 7.3. *Let G be a graph on $\{1,2,\ldots,m\}$ and let $k = \Delta(G)$ denote the maximum degree in G. Suppose $A_1, A_2, \ldots, A_m$ are events in a probability space so that for each $i = 1,2,\ldots,m$:*
(1) $\mathrm{Prob}[A_i] \leqslant 1/4k$, *and*
(2) A_i *is jointly independent of the events* $\{A_j\colon ij$ *is not an edge in* $G\}$.
Then $\mathrm{Prob}[\bar{A}_i \bar{A}_2 \cdots \bar{A}_m] > 0$.

Corollary 7.4. *Let $b \geqslant 500$, $s = \lceil b/\log b \rceil$, and $v = \lceil 4.7 \log b \rceil$. Let $\mathcal{H}$ be a hypergraph whose edges are subsets of sizes at most b from a set X. Suppose further that no point of X belongs to more than b edges in $\mathcal{H}$. Then there is a partition $X = X_1 \cup X_2 \cup \cdots \cup X_s$ so that $|H \cap X_i| \leqslant v$ for every edge $H \in \mathcal{H}$ and every $i = 1,2,\ldots,s$.*

Proof. Let $X = X_1 \cup X_2 \cup \cdots \cup X_s$ be a random partition of X. We denote by $A(H,i)$ the event $|H \cap X_i| > v$. Then let G be the graph whose vertex set consists of the pairs (H,i) where H is an edge in $\mathcal{H}$ and $1 \leqslant i \leqslant s$. The edges of G are the pairs $(H,i)(H',i')$ for which $H \cap H' \neq \emptyset$. Therefore $\Delta(G) \leqslant (1 + b(b-1))s \leqslant b^3$.

However, since $|H| \leqslant b$,

$$\begin{aligned}
\mathrm{Prob}[A(H,i)] &= \mathrm{Prob}[|H \cap X_i| > v] \\
&\leqslant \sum_{t>v} \binom{b}{t} \left(\frac{1}{s}\right)^t \left(1 - \frac{1}{s}\right)^{b-t} \\
&< \frac{1}{3} \binom{b}{v} \left(\frac{1}{s}\right)^v \left(1 - \frac{1}{s}\right)^{b-v} \\
&< \frac{1}{3} \left(\frac{be}{vs}\right)^v \frac{1}{\sqrt{2\pi v}} \left(1 - \frac{1}{s}\right)^{b-v} \\
&< \frac{1}{4b^3}.
\end{aligned}$$

The conclusion then follows from the Lovász local lemma. □

We also need the following lemma whose elementary proof we leave as an exercise.

Lemma 7.5. *Let a and b be positive integers. Let $\mathcal{H}$ be a hypergraph on the vertex set Y so that each edge in $\mathcal{H}$ has at most a elements and no vertex of Y belongs to more than b edges. Then there exists a partition $Y = Y_1 \cup Y_2 \cup \cdots \cup Y_r$ with $r = (a-1)b + 1$ so that $|H \cap Y_i| \leqslant 1$ for all edges $H \in \mathcal{H}$ and all i with $1 \leqslant i \leqslant r$.*

We are now ready to present the upper bound on dimension established by Füredi and Kahn (1986).

Theorem 7.6. *Let $\mathbf{P}$ be any poset with $\Delta(\mathbf{P}) \leqslant k$. Then*

$$\dim(\mathbf{P}) \leqslant 50k(\log k)^2.$$

Proof. Let $n = |X|$. When $k < 500$, the result follows from the Rödl–Trotter inequality $\dim(\mathbf{P}) \leqslant 2k^2 + 2$, so we assume $k \geqslant 500$. Let $\mathcal{H}$ be the hypergraph whose vertex set is X and whose edges are the up sets $U(x) = \{y \in X: x < y \text{ in } \mathbf{P}\}$. By Corollary 7.4, we obtain a partition $X = X_1 \cup X_2 \cup \cdots \cup X_s$ where $s = \lceil (k+1)/\log(k+1) \rceil$ so that $|U(x) \cap X_i| \leqslant v = \lceil 4.7 \log(k+1) \rceil$ for every $x \in X$ and every $i = 1, 2, \ldots, s$.

Then let $\mathcal{H}_i$ be the hypergraph obtained by restricting $\mathcal{H}$ to X_i. Each edge in $\mathcal{H}_i$ has size at most v and no point of X_i belongs to more than $k+1$ edges. By Lemma 7.5, we obtain a partition $X_i = X_{i1} \cup X_{i2} \cup \cdots \cup X_{ir}$ where $r = (v-1)(k+1)+1$ so that $|U(x) \cap X_{ij}| \leqslant 1$ for all x, i, j. Now we know by Lemma 7.2 that

$$d = \dim(1, v+1; r) \leqslant (v+1)^2 \left(1 + \log \frac{(v-1)(k+1)+1}{v+1}\right).$$

Let $M_1, M_2, \ldots, M_d$ be a family of linear orders on $\{1, 2, \ldots, r\}$ so that for each $(v+1)$-element subset $S \subset \{1, 2, \ldots, r\}$ and each $y \in S$, there is some i with y the least element of S in M_i. For each ij, let N_{ij} be an arbitrary linear order on X_{ij} and let $\hat{N}_{ij}$ denote the dual of N_{ij}, i.e., $u < v$ in $\hat{N}_{ij} \iff v < u$ in N_{ij}. Finally, for each $i = 1, 2, \ldots, s$ and $j = 1, 2, \ldots, d$, we define two linear orders L_{ij} and L'_{ij} on X by

$$L_{ij} = N_{iM_j(1)} < N_{iM_j(2)} < \cdots < N_{iM_j(r)} < X - X_i, \text{ and}$$

$$L'_{ij} = \hat{N}_{iM_j(1)} < \hat{N}_{iM_j(2)} < \cdots < \hat{N}_{iM_j(r)} < X - X_i.$$

In both L_{ij} and L'_{ij}, the ordering on $X - X_i$ is arbitrary. Furthermore, the subscripts are interpreted so that M_j orders $\{1, 2, \ldots, r\}$ by $M_j(1) < M_j(2) < \cdots < M_j(r)$.

Now suppose $x, y \in \mathbf{P}$ with $y \nleq x$ in $\mathbf{P}$. Choose α so that $x \in X_{i\alpha}$. Set $T = \{j: U(y) \cap X_{ij} \neq \emptyset\}$. Since $|T| \leqslant v$, there is some β for which α is the least element of $T \cup \{\alpha\}$ in M_β. If $y \notin X_{i\alpha}$, then $x < y$ in both $L_{i\beta}$ and $L'_{i\beta}$. If $y \in X_{i\alpha}$, then $x < y$ in exactly one of $L_{i\beta}$ and $L'_{i\beta}$.

This shows that $\dim(\mathbf{P}) \leqslant 2sd < 50k(\log k)^2$ as claimed. □

Until recently it was not known whether there exists an absolute constant c so that $\dim(\mathbf{P}) < ck$ whenever $\Delta(\mathbf{P}) \leqslant k$. Note that for each $k \geqslant 2$, the crown $\mathbf{S}^0_{k+1}$ satisfies $\Delta(\mathbf{S}^0_{k+1}) = k$ and $\dim(\mathbf{S}^0_{k+1}) = k+1$. However, Erdős et al. (1991) have substantially improved this lower bound by investigating the dimension of a height-two random poset. They define the sample space $\Omega(n, p)$ as consisting of all height-two posets containing n minimal points $a_1, a_2, \ldots, a_n$ and n maximal points $b_1, b_2, \ldots, b_n$. For a poset $\mathbf{P} \in \Omega(n, p)$, the probability that $a_i < b_j$ in $\mathbf{P}$ is equal to p (which in general is a function of n). Events corresponding to distinct pairs of points in $\mathbf{P}$ are independent.

Erdős et al. (1991) develop upper and lower bounds on the expected value of the dimension of a random poset for values of p in the range

$$\frac{(\log n)^{1+\varepsilon}}{n} < p < 1 - n^{-1+\varepsilon}.$$

However, taking the particular value $p = 1/\log n$, their results imply that there exist absolute positive constants δ_1 and δ_2 so that the following inequalities almost surely hold for the random poset $\mathbf{P}$:

$$\Delta(\mathbf{P}) < \delta_1 n/\log n \quad \text{and} \quad \dim(\mathbf{P}) > \delta_2 n.$$

This shows that if we define the function $f : \mathbb{N} \to \mathbb{N}$ by "$f(k)$ is the largest positive integer for which there exists a poset $\mathbf{P} = (X, P)$ with $\Delta(\mathbf{P}) \leqslant k$ and $\dim(\mathbf{P}) \leqslant f(k)$", then there exist absolute positive constants c_1 and c_2 so that

$$c_1 k(\log k) \leqslant f(k) \leqslant c_2 k(\log k)^2.$$

It is probably a very difficult problem to determine the correct exponent on the logarithmic factor in the preceding inequality. Perhaps the answer will come with improved bounds on the expected value of the dimension of a random poset in the sparse case, i.e., for values of p satisfying $p \leqslant (\log n)^{1+\varepsilon}/n$. Other dimension-related questions for random posets are posed in Erdős et al. (1991); further applications of dimension-theoretic concepts for random posets are given in Brightwell and Trotter (1994b).

Many other interesting dimension-theoretic questions center around families of subsets ordered by inclusion. For example, the proof of Theorem 7.6 requires an estimate on the dimension of the poset formed by the 1-element and $(\log n)$-element subsets of an n-element set. Lemma 7.2 gives an upper bound of the form $c(\log n)^3$, and Kierstead (1995) has just shown a lower bound of the form $c \log^3 n/\log\log n$. So some genuinely new idea is needed to determine more accurate estimates for $f(k)$.

Dushnik (1950) developed upper and lower bounds for $\dim(1, k; n)$ which result in an exact formula when $k > 2\sqrt{n}$, and Spencer (1972) gave asymptotic results when k is relatively small in comparison to n. Füredi et al. (1991) give the following asymptotic formula:

$$\dim(1, 2 : n) = \lg\lg n + \left(\frac{1}{2} + o(1)\right) \lg\lg\lg n.$$

Hurlbert et al. (1994) show that $\dim(2, n-2; n) = n - 1$, when $n \geqslant 5$; and Brightwell et al. (1994) show that $\dim(s, s+k; n) = O(k^2 \log n)$.

8. Interval orders and semiorders

Let $\mathcal{I}$ be a collection of closed intervals of $\mathbb{R}$. Define a partial order P on $\mathcal{I}$ by $I_1 < I_2$ in $P \iff x < y$ in $\mathbb{R}$ for every $x \in I_1$, and $y \in I_2$. Posets obtained from this construction are called *interval orders*. The following theorem of Fishburn (1970) provides a forbidden subposet characterization of interval orders. We leave the proof as an exercise.

Theorem 8.1. *A poset* $\mathbf{P}$ *is an interval order* $\iff$ $\mathbf{P}$ *does not contain* $\mathbf{2} + \mathbf{2}$.

A poset $\mathbf{P}$ is called a *semiorder* if there exists a function $f: X \to \mathbb{R}$ so that $x < y$ in $\mathbf{P} \iff f(y) > f(x) + 1$. Evidently, a semiorder is an interval order having a representation in which all intervals have length 1. The following theorem due to Scott and Suppes (1958) provides a forbidden subposet characterization of semiorders. Again the proof is omitted.

Theorem 8.2. *A poset* $\mathbf{P}$ *is a semiorder* $\iff$ $\mathbf{P}$ *does not contain either* $\mathbf{2}+\mathbf{2}$ *or* $\mathbf{3}+\mathbf{1}$.

Rather than present the proofs of these two important, but by now well-known, theorems, we choose instead to discuss some recent work in which interval orders and semiorders surface in a surprising manner.

Let $\mathcal{P}$ be a class of posets. We say that the on-line dimension of $\mathcal{P}$ is at most t if there exists a strategy for constructing a realizer $L_1, L_2, \ldots, L_t$ for any poset from $\mathcal{P}$ constructed in an on-line fashion. As was discussed in section 4, the poset and the realizer are to be constructed one point at a time. At each step, a new point is added to the poset. This point is then inserted into each of the existing linear extensions in a manner such that they remain a realizer.

The reader may enjoy showing that if $\mathcal{P}$ is the set of posets of dimension at most 2, then the on-line dimension of $\mathcal{P}$ is infinite, i.e., for each t, there is no algorithm which will construct on-line realizers of size t for posets from $\mathcal{P}$.

The members of family $\{\mathbf{S}_3^k\colon k \geqslant 0\}$ of 3-dimensional crowns are called 3-*crowns*. The following result is due to Kierstead et al. (1984).

Theorem 8.3. *Let* $\mathcal{P}_n$ *denote the class of posets of width at most n which do not contain any* 3-*crowns. Then the on-line dimension of* $\mathcal{P}_n$ *is at most* $((5^n - 1)/4)!$.

Proof. Set $t = (5^n - 1)/4$. Use Kierstead's theorem (4.2) to construct an on-line partition of a poset $\mathbf{P}$ in $\mathcal{P}$ into t chains $C_1, C_2, \ldots, C_t$. For each $x \in X$, let $C(x)$ denote the unique α for which $x \in C_\alpha$. Let M be any of the $t!$ linear orders on $\{1, 2, \ldots, t\}$. We construct on-line a linear extension L_M of a poset from $\mathcal{P}_n$. When the new point x enters $\mathbf{P}$, let $S_1 = \{y \in \mathbf{P}\colon x < y \text{ in } \mathbf{P}\}$ and $S_2 = \{y \in \mathbf{P}\colon x \parallel y \text{ in } \mathbf{P} \text{ and } C(x) < C(y) \text{ in } M\}$. If $S_1 \cup S_2 = \emptyset$, insert x at the top of L_M. If $S_1 \cup S_2 \neq \emptyset$, and y is the lowest element of $S_1 \cup S_2$ in L_M, insert x immediately under y.

We now show that the set of all $t!$ linear extensions of $\mathbf{P}$ determined by this procedure form a realizer of $\mathbf{P}$. To accomplish this, choose an arbitrary incomparable pair $x \parallel y$ from $\mathbf{P}$. We show that there is at least one M for which $x > y$ in L_M. In fact, we will show that $x > y$ in some L_M so that $C(y)$ is the least element in M and $C(x)$ is the greatest element in M.

A *fence* F starting up at x and ending at y is a sequence $x = x_0, x_1, x_2, \ldots, x_i = y$ for which the only comparabilities between these points are $x_0 < x_1 > x_2 < x_3 > \cdots$. Note that such a fence starts up from $x_0 = x$, but can end either up or down at $x_i = y$.

Let $Y = \{1, 2, \ldots, t\} - \{C(x), C(y)\}$. Define a binary relation Q on Y by "$\alpha Q \beta \iff$ there exist points u, v and a fence F starting up at x and ending at v so that $\alpha = C(u), \beta = C(v), u < y$ and $\{u, y\} \parallel F$ in $\mathbf{P}$".

Claim. *If $\alpha Q\beta$ and $\gamma Q\delta$, then either $\alpha Q\delta$ or $\gamma Q\beta$.*

Proof. Choose points u, v and a fence $F = \{x = x_0, x_1, x_2, \ldots, x_i = v\}$ which witness $\alpha Q\beta$. Similarly, choose points w, z and a fence $G = \{x = u_0, u_1, \ldots, u_j = z\}$ which witness $\gamma Q\delta$. Suppose that neither $\alpha Q\delta$ nor $\gamma Q\beta$. Then u is less than one or more points in G. Choose the least m so that $u < u_m$. Then m is odd (and positive). Similarly, let p be the least integer so that $w < x_p$. The p is odd and positive. Then the set $H = \{x, x_1, x_2, \ldots, x_p, u_1, u_2, \ldots, u_m\}$ is connected and has x_p and u_m as distinct maximal elements. Let $\mathbf{K}$ be a minimum-size connected subposet of $\mathbf{H}$ containing x_p and u_m. Then $\mathbf{K} \cup \{y, u, w\}$ is a 3-crown. The contradiction completes the proof of the claim. □

From its definition, it is clear that Q is irreflexive, i.e., we never have $\alpha Q\alpha$ for any $\alpha \in Y$. By taking $\beta = \gamma$ in the claim, we conclude that either $\alpha Q\gamma$ or $\beta Q\beta$. Hence $\alpha Q\delta$. This shows Q is transitive, and is therefore a strict partial order on Y. Furthermore, in view of Theorem 8.1, (Y, Q) is an interval order!

Choose an interval representation of (Y, Q) in which all end points are distinct. Then let M' be the linear extension of Q determined by the left end points in this representation. Form M from M' by adding $C(y)$ as least element and $C(x)$ as greatest element. We claim that $x > y$ in L_M.

Suppose to the contrary that $x < y$ in L_M. When the latter of these two points enters the poset, let $x = u_0 < u_1 < u_2 < \cdots < u_s < u_{s+1} = y$ be the sequence of points between x and y in L_σ. Note that for all $i = 0, 1, 2, \ldots, s$, either ($u_i < u_{i+1}$ in $\mathbf{P}$) or ($u_i \parallel u_{i+1}$ and $C(u_i) < C(u_{i+1})$ in M). We call any sequence $x = v_0, v_1, v_2, \ldots, v_m, v_{m+1} = y$ a *blocking chain* if

(1) $v_0 < v_1 < \cdots < v_m < v_{m+1}$ in L_M;

(2) for $i = 0, 1, \ldots, m$, either ($v_i < v_{i+1}$ in $\mathbf{P}$) or ($v_i \parallel v_{i+1}$ and $C(v_i) < C(v_{i+1})$ in M).

Of all blocking chains, we choose one, say $\{v_0, v_1, \ldots, v_{m+1}\}$, for which m is as small as possible. For each $i = 1, 2, \ldots, m$, let $\alpha_i = C(v_i)$. The minimality of m implies that $v_i \parallel v_j$ and $\alpha_i > a_j$ in M whenever $0 \leqslant i, j \leqslant m+1$ and $|j - i| \geqslant 2$. It follows that m is even and that for even i, $vi < v_{i+1}$ in X. For odd i, $v_i \parallel v_{i+1}$ and $\alpha_i < \alpha_{i+1}$ in M. Thus $C(y) < \alpha_{m-1} < \alpha_m < \alpha_{m-3} < \alpha_{m-2} < \cdots < \alpha_5 < \alpha_6 < \alpha_3 < \alpha_4 < \alpha_1 < \alpha_2 < C(x)$ in M.

Next we observe that the set $F = \{v_0, v_1\}$ is a fence starting up at x and ending at v_1. Also $\{y, v_m\} \parallel F$. It follows that $\alpha_m < \alpha_1$ in Q. Choose the largest integer i so that $\alpha_m < \alpha_i$ in Q. Suppose first that i is odd. Then $\alpha_i < \alpha_{i+1}$ in M, which implies that the left end point of the interval corresponding to α_i is less than the left end of the interval corresponding to α_{i+1}. However, the inequality $\alpha_m < \alpha_i$ in Q implies that the interval for α_m lies entirely to the left of the interval for α_i. This in turn implies $\alpha_m < \alpha_{i+1}$ in Q, contradicting our choice of i.

Now suppose that i is even. Choose points $u \in C_{\alpha_m}$, $z \in C_{\alpha_i}$ and a fence $F = \{x = z_0, z_1, z_2, \ldots, z_r = z\}$ starting up at x and ending at z so that $u < y$ and $\{u, y\} \parallel F$. Let $u' = \mathrm{MAX}\{u, v_m\}$ and let G be a minimum-size connected subposet of $F \cup \{v_i, v_{i+1}\}$ so that G contains both x and v_{i+1}. Then G is a fence starting up at x and ending at v_{i+1}. Furthermore, $\{y, u'\} \parallel G$ which implies $\alpha_m < \alpha_{i+1}$ in Q. This is also a contradiction. □

It is mildly irritating that we do not know whether it is necessary to exclude all 3-crowns from the posets in order to have finite on-line dimension. It is certainly necessary to exclude $\mathbf{S}_3^0$, but perhaps this is enough.

We next discuss an extremal problem for interval orders with a surprising connection to hamiltonian circuit problems. It is well known that there exist posets whose cover graphs have large chromatic number [see Kříž and Nešetřil (1991), for example]. It is easy to see that such graphs exist as cover graphs of interval orders. In connection with this topic, Felsner and Trotter (1995) made the following conjecture.

Conjecture 8.4. Let $n \geqslant 1$, and let $t = 2^n$. Then there exists a permutation $A_1, \ldots, A_t$ of the subsets of $\{1, \ldots, n\}$ so that:

(1) $A_1 = \emptyset$, and

(2) For each $i = 1, 2, \ldots, t-1$, either $A_i \subset A_{i+1}$ or $A_{i+1} \subset A_i$. Furthermore, $|A_i \Delta A_{i+1}| = 1$.

(3) For each $i, j = 1, 2, \ldots, t$, if $A_i \subset A_j$ and $i > j$, then $i = j + 1$.

Here is a re-formulation of the preceding conjecture as an extremal problem.

Conjecture 8.5. For each $n \geqslant 1$, let $f(n) = s$ be the largest integer for which there exists a sequence $B_1, B_2, \ldots, B_s$ of distinct subsets of $\{1, 2, \ldots, n\}$ so that:

(1) $B_1 = \emptyset$, and

(2) for each $i = 1, 2, \ldots, s-1$, $B_{i+1} \not\subset B_i$, and

(3) for each $i = 1, 2, \ldots, s-2$, $B_{i+2} \not\subseteq B_i \cup B_{i+1}$.

Then $f(n) = 2^{n-1} + \lfloor (n+1)/2 \rfloor$.

Trotter and Felsner show that $f(n) \leqslant 2^{n-1} + \lfloor (n+1)/2 \rfloor$ and that equality holds if and only if Conjecture 8.4 is valid. These conjectures are related to the following (surprisingly difficult) problem.

Conjecture 8.6. Let $\mathbf{G}$ denote the comparabilty graph of the poset formed by the k-element and $(k+1)$-element subsets of a $(2k+1)$-element set partially ordered by inclusion. Then $\mathbf{G}$ has a hamiltonian cycle.

Although Conjecture 8.6 is known to be true for small values of k, most of the results thus far are negative, i.e., a hamiltonian cycle *cannot* be formed by combining certain types of matchings [see Duffus et al. (1988) and Kierstead and Trotter (1988), for example]. On the other hand, it is shown in Felsner and Trotter (1995) that there exists a cycle whose size is at least one-fourth of the total number of vertices. This fraction has subsequently been raised by C. Savage and P. Winkler (pers. comm).

We next discuss an elementary extremal problem for posets. For integers n, k with $0 \leqslant k \leqslant \binom{n}{2}$, let $Q(n,k)$ denote the class of all posets having n points and

k comparable pairs. For each poset $\mathbf{P} \in Q(n,k)$, let $e(\mathbf{P})$ count the number of linear extensions of $\mathbf{P}$. Then set $e(n,k) = \max\{e(\mathbf{P})\colon \mathbf{P} \in Q(n,k)\}$. Fishburn and Trotter (1992) then show that the extremal posets are semiorders.

Theorem 8.7. *Every poset $\mathbf{P} \in Q(n,k)$ with $e(\mathbf{P}) = e(n,k)$ is a semiorder.*

Proof. Let $\mathbf{P}$ be such a poset. We first show that $\mathbf{P}$ does not contain $\mathbf{2}+\mathbf{2}$. Suppose to the contrary that the chains $u < x$ and $v < y$ are incomparable. Of all such pairs of chains, we choose two for which $|U(x)| + |U(y)|$ is minimum. We may therefore assume that $U(x) \subset U(y)$. Let $\mathbf{P}'$ be the poset obtained by replacing the relations $z < y$ by $z < x$ for all $z \in D(y) - D(x)$. Then $\mathbf{P}' \in Q(n,k)$. We now show that $e(\mathbf{P}') > e(\mathbf{P})$.

Exchanging x and y maps $\mathscr{E}(\mathbf{P}) - \mathscr{E}(\mathbf{P}')$ to $\mathscr{E}(\mathbf{P}') - \mathscr{E}(\mathbf{P})$. Although the map is 1–1, it is not onto since any linear extension in which $y < u < v < x$ is not in the image of the map. The contradiction shows $\mathbf{P}$ is an interval order.

If $\mathbf{P}$ contains a 3-element chain $x < y < z$ with all three points incomparable to w, form $\mathbf{P}''$ from $\mathbf{P}$ by replacing the relations $y < z$ by $w < z$ for all $z \in U(y) - U(w)$. As before, $e(\mathbf{P}'') > e(\mathbf{P})$. The contradiction shows $\mathbf{P}$ is a semiorder. □

At first glance, Theorem 8.7 seems to be very helpful in determining $e(n,k)$, and some progress is made in Fishburn and Trotter (1992). However, the general problem remains open.

Recently, P. Winkler (pers. comm.) has proposed another extremal problem involving semiorders. For a poset $\mathbf{P} = (X, P)$ and a point $x \in X$, let $\deg(x) = |\{y \in X\colon y < x \text{ in } P, \text{ or } y > x \text{ in } P\}|$. Then define the *flexibility* of $\mathbf{P}$, denoted flex($\mathbf{P}$), by

$$\operatorname{flex}(\mathbf{P}) = \sum_{x \in X} (\deg(x))^2.$$

Now let n and k be fixed integers with $0 \leqslant k \leqslant \binom{n}{2}$. It is an easy exercise to show that among all posets containing n points and k comparable pairs, any poset with maximum flexibility is a semiorder.

Problem 8.8. For fixed n and k, find all semiorders with n points and k comparable pairs for which the flexibility is maximum.

Problem 8.9. For a poset $\mathbf{P} = (X, P)$ with n points and k comparable pairs, let i be an integer with $0 \leqslant i \leqslant k$ and let a_i denote the number of permutations (linear orders, not necessarily linear extensions) of the ground set X so that exactly i of the k comparable pairs are in the same order as in P. Then $\sum_{i=0}^{k} a_i = n!$, and $a_i = a_{k-i}$, for each $i = 0, 1, \ldots, k$. Is the sequence $a_0, a_1, \ldots, a_k$ unimodal?

Winkler noted that the sequence need not be log-concave, so the mixed volumes approach used by Stanley (see the discussion in section 6) will not apply.

The next result is a recent theorem of Brightwell (1989) establishing the $\frac{1}{3}$–$\frac{2}{3}$ conjecture for semiorders.

Theorem 8.10. *Let $\mathbf{P} = (X, P)$ be a semiorder which is not a chain. Then X contains a pair x, y of incomparable points with $\frac{1}{3} \leqslant \text{Prob}[x < y] \leqslant \frac{2}{3}$.*

Proof. Suppose the result is false and choose a counterexample with $|X|$ minimum. Define a linear extension L by $x < y \iff \text{Prob}[x < y] > \frac{2}{3}$. Let $|X| = n$ and label the points in X so that $x_1 < x_2 < \cdots < x_n$ in L.

Since $\mathbf{P}$ is a semiorder, there exists a function $f : X \to \mathbb{R}$ so that $x < y$ in $\mathbf{P} \iff f(y) > f(x) + 1$. We now show that $f(x_1) < f(x_2) < \cdots < f(x_n)$. Suppose to the contrary that $1 \leqslant i < j \leqslant n$, but that $f(x_j) < f(x_i)$. Then $x_i \not< x_j$ in $\mathbf{P}$. However, $x_i < x_j$ in L, so $x_j \not< x_i$ in $\mathbf{P}$. Thus $x_i \parallel x_j$ in $\mathbf{P}$. However, the inequality $f(x_j) < f(x_i)$ implies that $u > x_j$ in $\mathbf{P}$ whenever $u > x_i$ in $\mathbf{P}$. Dually $v < x_i$ in $\mathbf{P}$ whenever $v < x_j$ in $\mathbf{P}$.

It follows that every $L \in \mathscr{E}(\mathbf{P})$ with $x_i < x_j$ in L can be transformed into a linear extension L' with $x_j < x_i$ in L' just by interchanging these two points. The mapping is 1–1 which shows $\text{Prob}[x_j < x_i] \geqslant \frac{1}{2}$. Thus $\text{Prob}[x_j < x_i] > \frac{2}{3}$ and $x_j < x_i$ in L. The contradiction shows $f(x_1) < f(x_2) < \cdots < f(x_n)$ as claimed.

We now show that $x_i \parallel x_{i+1}$ for all $i = 1, 2, \ldots, n-1$. Suppose to the contrary that $x_i < x_{i+1}$ in $\mathbf{P}$. Then every point of $X' = \{x_1, x_2, \ldots, x_i\}$ is less than every point of $X'' = \{x_{i+1}, x_{i+1}, \ldots, x_n\}$. At least one of $\mathbf{P}'$ and $\mathbf{P}''$ is not a chain, and we can restrict our attention to that subposet to locate x and y. The contradiction shows $x_i \parallel x_{i+1}$ for $i = 1, 2, \ldots, n-1$ as claimed.

We say x_j *separates* x_i *and* x_{i+1} *from above* if $x_j :> x_i$ and $x_j \parallel x_{i+1}$. Dually we say x_j *separates* x_i *and* x_{i+1} *from below* if $x_j <: x_{i+1}$ and $x_j \parallel x_i$. We say x_j *separates* x_i *and* x_{i+1} if it either separates them from above or it separates them from below.

If x_j separates x_i and x_{i+1} from above, then $x_k < x_j$ in $\mathbf{P}$ for $k = 1, 2, \ldots, i$. This implies that x_j does not separate x_k and x_{k+1} when $1 \leqslant k < i$. Dually, if x_j separates x_i and x_{i+1} from below, then x_j does not separate x_k and x_{k+1} when $i < k < n$. So each x_j separates at most one pair from below and at most one pair from above.

However, x_1 and x_2 cannot separate pairs from above and x_{n-1} and x_n cannot separate any pair from below. If follows that there are at most $2(n-4) + 4 = 2n - 4$ ordered pairs (i, j) so that x_j separates x_i and x_{i+1}. Hence there is at least one (in fact at least two) values of i with $1 \leqslant i < n$ for which there is at most one j so that x_j separates x_i and x_{i+1}.

For such a value of i, partition $\mathscr{E}(\mathbf{P})$ into three sets by letting $\mathscr{E}_1 = \{L \in \mathscr{E}(\mathbf{P})$: $x_i < x_{i+1}$ in L, and no element separating x_i and x_{i+1} is between them in $L\}$; $\mathscr{E}_2 = \{L \in \mathscr{E}(\mathbf{P})$: $x_i < x_{i+1}$ and $L \notin \mathscr{E}_1\}$, and $\mathscr{E}_3 = \{L \in \mathscr{E}(\mathbf{P})$: $x_{i+1} < x_i$ in $L\}$. Now let $t = |\mathscr{E}(\mathbf{P})|$. Then $|\mathscr{E}_1| + |\mathscr{E}_2| > 2t/3$.

If $|\mathscr{E}_2| \geqslant t/3$, let j be the unique integer so that x_j separates x_i and x_{i+1}. Then x_j is between x_i and x_{i+1} in every $L \in \mathscr{E}_3$. If $j > i + 1$, this implies $\text{Prob}[x_j < x_{i+1}] \geqslant \frac{1}{3}$, and if $j < i$, it implies $\text{Prob}[x_i < x_j] \geqslant \frac{1}{3}$. Both of these implications are false, so we know $|\mathscr{E}_2| < t/3$. Thus $|\mathscr{E}_1| > t/3$.

Now let $L \in \mathscr{E}_1$. Form L' from L by interchanging x_i and x_{i+1}. This interchange is possible since any point between x_i and x_{i+1} in L is incomparable with both. This procedure determines a 1–1 map from $\mathscr{E}_1$ to $\mathscr{E}_3$. However, no such map exists because $|\mathscr{E}_3| < t/3$. The contradiction completes the proof. $\square$

The connection between semiorders and the $\frac{1}{3}$–$\frac{2}{3}$ conjecture is even more complex than suggested by the preceding result. Consider the infinite poset $\mathbf{P} = (X, P)$, where $X = \{x_i\colon i \in \mathbb{Z}\}$. Furthermore, we define $x_i < x_j$ in P if and only if $i < j - 1$ in $\mathbb{Z}$. Then $\mathbf{P}$ is a width-two semiorder. Also observe that any incomparable pair in $\mathbf{P}$ is of the form (x_i, x_{i+1}), for some $i \in \mathbb{Z}$. For a positive integer n, let $\mathbf{P}_n$ denote the subposet of $\mathbf{P}$ determined by the points whose subscripts in absolute value are at most n. Then it is an easy exercise to show that

$$\lim_{n\to\infty} \mathrm{Prob}[x_0 > x_1] = (5 - \sqrt{5})/10.$$

Note that $(5 - \sqrt{5})/10 \approx 0.2764 < \frac{1}{3}$. So the $\frac{1}{3}$–$\frac{2}{3}$ conjecture is *false* for infinite width-two semiorders, even though it is true for any finite poset which is either a width-two poset or is a semiorder! Also note that this example shows that the inequality in Theorem 6.7 is best possible when considering infinite posets of bounded width. Further results on infinite posets and balanced pairs are given in the recent papers Brightwell (1988, 1993).

Rabinovitch (1978) showed that the dimension of a semiorder is at most 3, but that an interval order may have arbitrarily large dimension. Füredi et al. (1991) showed that if $\mathbf{P} = (X, P)$ is an interval order of height n, then

$$\dim(\mathbf{P}) \leqslant \lg\lg n + \left(\tfrac{1}{2} + \mathrm{o}(1)\right) \lg\lg\lg n.$$

This inequality is best possible. For an integer $n \geqslant 2$, let $\mathbf{I}_n$ denote the *canonical interval order* consisting of all intervals with integer end points from $\{1, 2, \ldots, n\}$. Then Füredi et al. (1991) showed that

$$\dim(\mathbf{I}_n) = \lg\lg n + \left(\tfrac{1}{2} + \mathrm{o}(1)\right) \lg\lg\lg n.$$

This formula is closely related to the asymptotic formula given in section 7 for the dimension of the poset formed by the 1-element and 2-element subsets of $\{1, 2, \ldots, n\}$.

9. Degrees of freedom

Given a family $\mathscr{F}$ of sets, a poset $\mathbf{P}$ is called an *$\mathscr{F}$–inclusion order* if there exists a mapping which assigns to each $x \in X$ a set $S(x) \in \mathscr{F}$ so that $x \leqslant y$ in $\mathbf{P} \iff S(x) \subset S(y)$. As an example, if $\mathscr{F}$ is the collection of all closed intervals of $\mathbb{R}$, then the $\mathscr{F}$-inclusion orders are exactly the posets with dimension at most 2.

Fishburn and Trotter (1990) studied the class of *angle orders*. These are the posets which arise when $\mathscr{F}$ is the set of angular regions in the Euclidean plane, i.e., convex regions bounded by two rays emanating from a common point. They proved that every interval order is an angle order, as is every poset of dimension at most four. Both results admit elementary proofs, but while the first result may be mildly surprising, the second is certainly not. In a certain sense, to specify an angle requires four coordinates – two to locate the corner point and one for each ray to specify the angle from $[0, 2\pi)$ at which it leaves the corner point.

Fishburn and Trotter conjectured that not all 5-dimensional posets are angle orders, but were only able to prove the existence of a 7-dimensional poset which is not an angle order. R. Jamison (pers. comm.) settled this conjecture in the affirmative with an intricate ad hoc argument. However, Alon and Scheinerman (1988) have produced a much more general result using a powerful theorem of Warren (1968). We now outline their approach.

For $x \in \mathbb{R}$, let $\text{sgn}(x) = +$ if $x \geqslant 0$ and $\text{sgn}(x) = -$ if $x < 0$. For a vector $\mathbf{x} = (x_1, x_2, \ldots, x_t)$, $\text{sgn}(\mathbf{x})$ denotes the vector $(\text{sgn}(x_1), \text{sgn}(x_2), \ldots, \text{sgn}(x_t))$. The vector $\text{sgn}(\mathbf{x})$ is called the *sign pattern* of $\mathbf{x}$ and t is the *length* of the pattern. We say $\mathcal{F}$ has *at most k degrees of freedom*, and write $\deg(\mathcal{F}) \leqslant k$, if the following conditions are satisfied:

(1) there exists a mapping F which assigns to each $S \in \mathcal{F}$ a k-tuple $F(S) = (S(1), S(2), \ldots, S(k))$ from $\mathbb{R}^k$;

(2) there exists a finite set $P_1, P_2, \ldots, P_t$ of polynomials in $2k$ variables $x_1, x_2, \ldots, x_{2k}$; and

(3) there exists a set J of sign patterns of length t so that for every pair S, T of sets from $\mathcal{F}$, $S \subset T \iff \text{sgn}(\mathbf{y}(S,T)) \in J$ where $\mathbf{y}(S,T)$ is the vector of length t whose jth coordinate is given by $P_j(S(1), S(2), \ldots, S(k), T(1), T(2), \ldots, T(k))$.

To illustrate this definition, let $\mathcal{F}$ denote the set of closed disks in $\mathbb{R}^2$. With each set (disk) S, we take $(S(1), S(2))$ as the coordinates of the center of S and $S(3)$ as the radius. Take $P_1 = x_6 - x_3$ and $P_2 = (x_6 - x_3)^2 - (x_5 - x_2)^2 - (x_4 - x_1)^2$. Then take $J = \{(+,+)\}$. Let S and T be disks. Then $S \subset T \iff$ the distance from the center of S to the center of T plus the radius of S is less than or equal to $T \iff \text{sgn}(\mathbf{y}(S,T)) \in J$. This shows $\deg(\mathcal{F}) \leqslant 3$, i.e., $\mathcal{F}$ has at most three degrees of freedom. As a second example, the set $\mathcal{A}$ of angular regions in $\mathbb{R}^2$ has at most four degrees of freedom. Here is Warren's theorem (Warren 1968).

Theorem 9.1. *Let $P_1, P_2, \ldots, P_t$ be polynomials in m variables and let d denote the maximum degree among these polynomials. Then there are at most $(4edt/m)^m$ sign patterns of the form $\text{sgn}(P_1(\mathbf{x}), P_2(\mathbf{x}), \ldots, P_t(\mathbf{x}))$ where $\mathbf{x} = (x_1, x_2, \ldots, x_m)$ ranges over $\mathbb{R}^m$.*

Let $P(n,k)$ denote the number of labelled posets on n points having dimension at most k. Clearly $P(n,k) \leqslant (n!)^k \leqslant n^{kn}$. Subsequent arguments will require the following lower bound on $P(n,k)$ due to Alon and Scheinerman (1988). We leave the proof as an exercise.

Theorem 9.2. *The number $P(n,k)$ of labelled posets on n points having dimension at most k satisfies:*

$$P(n,k) \geqslant (n/\log n)^{nk - 2k^2 n/\log n}.$$

The preceding two results combine easily to prove the following striking result of Alon and Scheinerman (1988).

Theorem 9.3. *Let $\mathcal{F}$ be any family of sets having at most k degrees of freedom. Then there exists a poset* $\mathbf{P}$ *with* $\dim(\mathbf{P}) \leqslant k+1$ *which is not an* $\mathcal{F}$*-order.*

Proof. Suppose $\deg(\mathcal{F}) \leqslant k$ is witnessed by the set $P_1, P_2, \ldots, P_t$ of polynomials and the set J of test patterns. Let d denote the maximum degree of these polynomials. We show that when n is sufficiently large, there is a labelled poset on n points having dimension at most $k+1$ which is not an $\mathcal{F}$-order.

When $\mathbf{P}$ is a labelled poset on the ground set $X = \{1, 2, \ldots, n\}$ and $\mathbf{P}$ is an $\mathcal{F}$-order, we designate the sets in $\mathcal{F}$ corresponding to the points in X by $(S_1, S_2, \ldots, S_n)$. Each S_i is associated with a vector $\mathbf{x}_i = (x_{i1}, x_{i2}, \ldots, x_{ik})$. For each ordered pair (i, j) with $1 \leqslant i,\ j \leqslant n$ and $i \neq j$, we know $i < j$ in $\mathbf{P} \Longleftrightarrow S_i \subset S_j \Longleftrightarrow \operatorname{sgn}(\mathbf{y}(S_i, S_j)) \in J$. Recall that the αth coordinate of the vector $\mathbf{y}(S_i, S_j)$ is $P_\alpha(x_{i1}, x_{i2}, \ldots, x_{ik}, x_{j1}, x_{j2}, \ldots, x_{jk})$.

Concatenate in lexicographic order the $n(n-1)$ vectors $\mathbf{y}(S_i, S_j)$ into a single vector $\mathbf{y}$ of length $n(n-1)t$. Then $\operatorname{sgn}(\mathbf{y})$ is the sign pattern of a vector whose entries are determined by a family of $n(n-1)t$ polynomials in nk variables $x_{i\beta}$ where $1 \leqslant i \leqslant n$ and $1 \leqslant \beta \leqslant k$.

If $(T_1, T_2, \ldots, T_n)$ is another n-tuple of sets from $\mathcal{F}$ which yields the same sign pattern $\operatorname{sgn}(\mathbf{y})$ as $(S_1, S_2, \ldots, S_n)$, then these two n-tuples correspond to the same labelled poset $\mathbf{P}$. By Warren's theorem, we conclude that there are at most $(4edn(n-1)t/nk)^{nk}$ possible sign patterns. However, this number is clearly less than $P(n, k+1)$ when n is sufficiently large. $\square$

There are a number of perplexing open problems involving the representation of posets as a family of sets ordered by inclusion. Here are two of the most appealing. Given a point x in d-dimensional Euclidean space and a positive number r, let $B_d(\mathbf{x}, r)$ denote the ball of radius r centered at $\mathbf{x}$, i.e., the set of points at distance at most r from $\mathbf{x}$. It is customary to call $B_d(\mathbf{x}, r)$ a *d-dimensional sphere.* A poset $\mathbf{P} = (X, P)$ is called a *d-dimensional sphere order* if for each $x \in X$, there exists a d-dimensional sphere B_x so that $x \leqslant y$ in P if and only if $B_x \subseteq B_y$, for all $x, y \in X$.

Problem 9.4. If $\mathbf{P}$ is a finite poset, does there always exist a positive integer d so that $\mathbf{P}$ is a d-dimensional sphere order?

For historical reasons, a 2-dimensional sphere order is called a *circle order*, although it might be better to call it a *disk order*. Fishburn (1988) proved that every interval order is a circle order. Also, it is easy to see that every 2-dimensional poset is a circle order; in fact, we may require that the centers of the circles used in the representation be collinear. On the other hand, by Theorem 9.3, there exists a 4-dimensional poset which is not a circle order.

Problem 9.5. If $\mathbf{P}$ is a finite poset and $\dim(\mathbf{P}) \leqslant 3$, is $\mathbf{P}$ a circle order?

Problem 9.5 is intriguing because Scheinerman and Weirman (1989) showed that the countably infinite 3-dimensional poset $\mathbb{Z}^3$ is not a circle order. On the other hand, it is a relatively easy exercise to show that if $\mathcal{F}_n$ is the family of all regular n-gons in the plane (with bottom side horizontal) and $\mathbf{P}$ is any finite poset with $\dim(\mathbf{P}) \leqslant 3$, then $\mathbf{P}$ is a $\mathcal{F}_n$-inclusion order, for all $n \geqslant 3$.

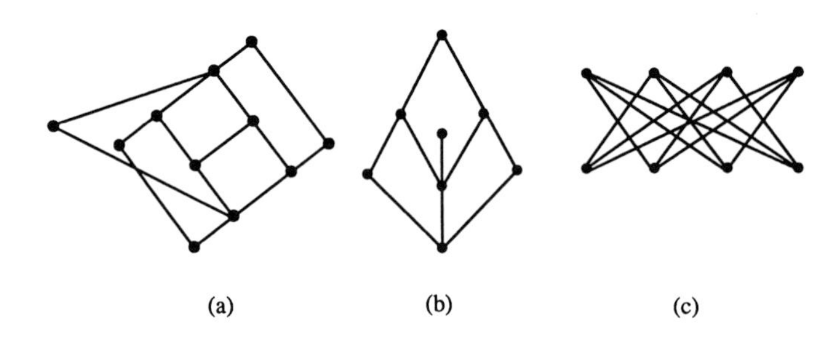

Figure 10.1.

10. Dimension and planarity

A poset **P** is said to be *planar* if it has a planar Hasse diagram. The poset shown in fig. 10.1a is nonplanar, but the posets in figs. 10.1b and 10.1c are planar. Note that the diagram for the last example can be redrawn without edge crossings.

As is well known, a planar poset **P** having a greatest and least element has dimension at most 2. Kelly and Rival (1975) provide a forbidden subposet characterization of nonplanar lattices by providing a minimum list $\mathscr{L}$ of lattices so that a lattice **P** is planar if and only if **P** contains a lattice from $\mathscr{L}$ as a subposet. One lattice from $\mathscr{L}$ is shown in fig. 10.1a. The lengthy argument for this theorem must be cleverly organized just so it can be written down on a finite number of pages. There are several other theorems in dimension theory for posets which exhibit these same characteristics: Kelly's (1977) determination of all 3-irreducible posets, Trotter's (1981) determination of all 3-interval irreducible posets of height 2, and Kimble's (1973) proof that if $n \geqslant 4$ and $|X| \leqslant 2n+1$, then $\dim(\mathbf{P}) < n$ unless **P** contains the standard n-dimensional poset $\mathbf{S}_n^0$. In fact, Gallai's (1967) forbidden subgraph characterization of comparability graphs belongs in this same grouping – especially in view of its value in obtaining a list of all 3-irreducible posets (see Trotter 1992, Trotter and Moore 1976).

Planar posets can have dimension exceeding 2: the planar posets in figs. 10.1b and 10.1c have dimensions 3 and 4 respectively. Trotter and Moore (1977) proved that a planar poset having either a greatest or least element has dimension at most 3. Kelly (1981) then constructed planar posets of arbitrary dimension by the device of embedding $\mathbf{S}_n^0$ in a planar poset. Kelly's construction is illustrated in fig. 10.2.

Three interesting problems remain. Do there exist irreducible planar posets of arbitrarily large dimension? Provide a characterization of planar posets in terms of forbidden subdiagrams. Develop a fast algorithm which will produce a planar drawing of the Hasse diagram of a poset if such a drawing exists.

Recently, Schnyder (1989) produced a striking theorem relating dimension and planarity in a different manner. Let $\mathbf{G} = (V, E)$ be an ordinary undirected

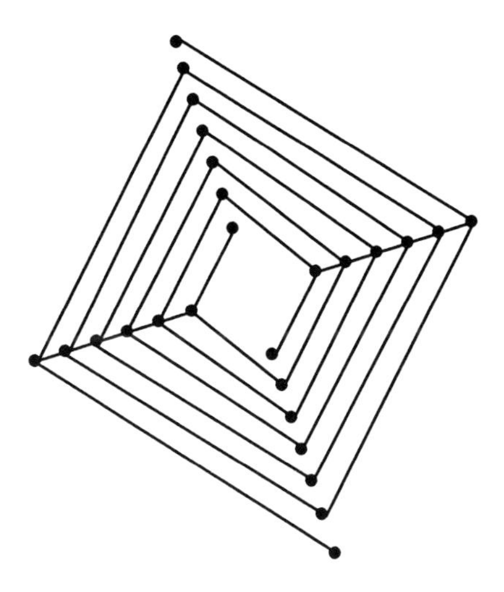

Figure 10.2.

graph. We associate with **G** a poset $\mathbf{P} = \mathbf{P}(\mathbf{G})$ of height 2. In **P**, $\text{MIN}(\mathbf{P}) = V$ and $\text{MAX}(\mathbf{P}) = E$. Also vertex x is less than edge e in $\mathbf{P} \Longleftrightarrow x$ is an end point of e. We call **P** the *incidence poset* of **G**. Here is Schnyder's theorem.

Theorem 10.1. *Let* **G** *be a graph. Then G is planar $\Longleftrightarrow$ the dimension of its incidence poset is at most* 3.

Proof. Let $\mathbf{P} = \mathbf{P}(\mathbf{G})$ be the incidence poset of **G**. Suppose first that $\dim(\mathbf{P}) \leqslant 3$. We show that **G** is planar. Suppose to the contrary that **G** is nonplanar. [The argument we give for this part is patterned after a proof of Babai and Duffus (1981).]

Choose an embedding of **P** in $\mathbb{R}^3$ which associates with each $y \in V \cup E$ a vector $\mathbf{y} = (y_1, y_2, y_3) \in \mathbb{R}^3$ so that $u \leqslant v$ in $\mathbf{P} \Longleftrightarrow u_i \leqslant v_i$ in $\mathbb{R}$ for $i = 1, 2, 3$. For each $y \in X \cup E$, let $\pi(y)$ be the orthogonal projection of $\mathbf{y}$ on the plane $x_1 + x_2 + x_3 = 0$ in $\mathbb{R}^3$. Without loss of generality, all points in $X \cup E$ project to distinct points on the plane $x_1 + x_2 + x_3 = 0$, and these points are in general position.

For each $u \in X$ and each $e \in E$ containing u as an end point, join $\pi(u)$ and $\pi(e)$ with a straight line segment. Since **G** is nonplanar, there exist distinct vertices $u, v \in V$ and distinct edges $e, f \in E$ so that u is an end point of e but not of f, v is an end point of f but not of e, and the line segment $\pi(u)\pi(e)$ crosses the line segment $\pi(v)\pi(f)$ at a point $\mathbf{p}$ interior to both. Let $\mathbf{z}$ be the point on the line segment $\mathbf{u}e$ in $\mathbb{R}^3$ so that $\pi(\mathbf{z}) = \mathbf{p}$. Also let $\mathbf{w}$ be the point on the line segment $\mathbf{v}f$ in $\mathbb{R}^3$ so that $\pi(\mathbf{w}) = \mathbf{p}$. Then either $\mathbf{z} \leqslant \mathbf{w}$ in $\mathbb{R}^3$ or $\mathbf{w} \leqslant \mathbf{z}$ in $\mathbb{R}^3$. However, $\mathbf{z} \leqslant \mathbf{w}$ implies $\mathbf{u} \leqslant \mathbf{z} \leqslant \mathbf{w} \leqslant \mathbf{f}$, which is false since u is not an end point of f. Similarly $\mathbf{w} \leqslant \mathbf{z}$ implies $\mathbf{v} \leqslant \mathbf{e}$ which is also false. The contradiction shows that **G** is planar.

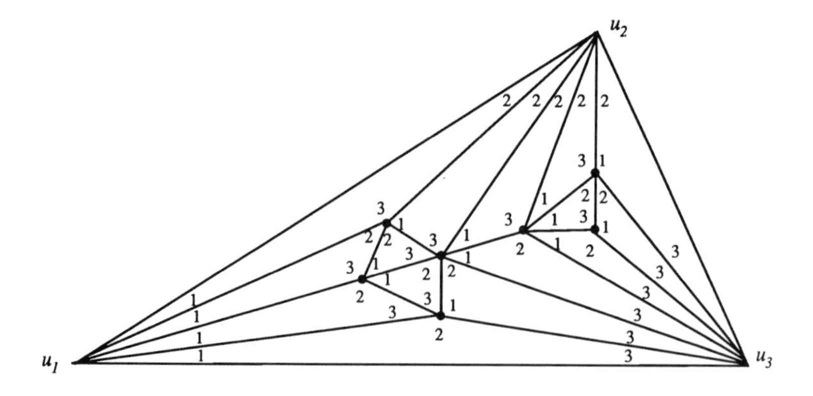

Figure 10.3.

Now suppose that **G** is planar. We show that **P** has dimension at most 3. Without loss of generality, we assume that **G** is maximal planar. Choose a planar diagram of **G** using straight line segments for the edges. This diagram is a triangulation T of the plane. Each interior region is a triangle, and T has three exterior vertices which we label in clockwise order u_1, u_2, and u_3.

Now consider a function f which assigns to each angle of each interior triangle of T a color selected from $\{1, 2, 3\}$. The function f is called a *normal* coloring of $T \iff$

(1) all angles incident with exterior vertex u_i are mapped by f to color i for $i = 1, 2, 3$;

(2) at each interior vertex u of T, there is an angle mapped by f to color i for $i = 1, 2, 3$;

(3) at each interior vertex u of T, all angles mapped by f to color i are consecutive for $i = 1, 2, 3$;

(4) at each interior vertex u of T the block of angles mapped by f to color 2 appears immediately after the block of angles mapped by f to color 1; and

(5) for each elementary triangle of T, f assigns the three angles to colors 1, 2, and 3 in clockwise order.

We illustrate this definition in fig. 10.3 with a normal coloring of a triangulation.

The following claim yields to a straightforward inductive argument and its proof is left as an exercise.

Claim 1. *Every planar triangulation has a normal coloring.*

Let C be a cycle in a planar triangulation T which has been colored normally. A vertex x belonging to C is called a *Type* i vertex on C if all angles incident with x and interior to C are colored i. When C is exterior triangle, u_i is a Type i vertex on C.

Claim 2. *If C is a cycle in T, then C contains a Type i vertex, for each $i = 1, 2, 3$.*

Proof. Suppose the claim is false. Choose a counterexample C containing the minimum number of elementary triangles. Clearly C is not the boundary of an elementary triangle. Now suppose C does not have a Type 1 vertex.

Suppose that C has two nonconsecutive vertices x and y which are adjacent via an edge $e = xy$ interior to C. Then the region bounded by C can be partitioned into regions bounded by cycles C' and C'' having e as a common edge. Now C' and C'' both have a Type 1 vertex. If x is a Type 1 vertex for C' and for C'', then x is a Type 1 vertex for C. An analogous statement holds for y. We conclude that one of x and y is a Type 1 vertex for C' and the other is a Type 1 vertex for C''. Consideration of the two elementary triangles sharing the edge shows this is impossible.

Now let $C = \{x_1, x_2, \ldots, x_s\}$ and let x_i and x_{i+1} be any two consecutive vertices of C and let z_i be the vertex so that $x_i x_{i+1} z_i$ is an elementary triangle interior to C. Let C_i be the cycle obtained by deleting the edge $x_i x_{i+1}$ and adding the edges $x_i z_i$ and $z_i x_{i+1}$. Then C_i has a Type 1 vertex because it contains fewer elementary triangles than C. Clearly z_i cannot be a Type 1 vertex on C_i because z_i is an interior vertex of T.

It follows that one of x_i and x_{i+1} is a Type 1 vertex on C_i. If x_i is Type 1 on C_i, then the angle of triangle $x_i x_{i+1} z_i$ incident with x_i must be colored 3; else x_i is Type 1 on C. Thus the angle of $x_i x_{i+1} z_i$ incident with x_{i+1} is colored 1. This implies that x_{i+1} is not Type 1 for C_i. Dually, if x_{i+1} is Type 1 for C_i, then the angle of $x_i x_{i+1} z_i$ incident with x_{i+1} is colored 2, the angle of $x_i x_{i+1} z_i$ incident with x_i is colored 1, and x_i is not Type 1 for C_i.

If some vertex x_{i+1} is Type 1 for both C_i and C_{i+1}, then x_i is Type 1 for C. So either x_i is Type 1 for C_i for $i = 1, 2, \ldots, s$, or x_{i+1} is Type 1 for C_i for $i = 1, 2, \ldots, s$. In the first case, there is no Type 2 vertex on C_1; in the second, there is no Type 3 vertex on C_2. The contradiction completes the proof. □

Claim 3. *Let P_i be the binary relation on the set V of vertices of* **G** *defined by $xP_iy \iff$ there exists an elementary triangle T having x and y as vertices in which the angle incident at y is colored i. Then the transitive closure $Q_i = \bar{P}_i$ is a partial order on X.*

Proof. It suffices to show Q_i has no directed cycles. This follows from Claim 2 since a directed cycle in Q_i could not have either a Type $i+1$ or a Type $i+2$ vertex. □

For each $i = 1, 2, 3$, let M_i be a linear extension of Q_i. Then let L_i be any linear extension of **P** so that:

(1) The restriction of L_i to V is M_i.

(2) For each $e \in E$, the M_i-largest element of V which is less than e in M_i is less than e in **P**.

Alternatively, L_i is obtained from M_i by inserting the elements of E as low as possible. To complete the proof, it suffices to show that $P = L_1 \cap L_2 \cap L_3$. To

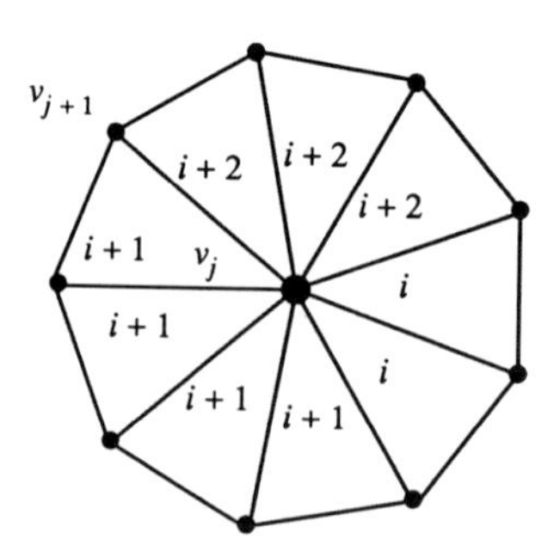

Figure 10.4.

accomplish this, it is enough to show that for each edge $e = xy$ and each vertex z not an end point of e, there exists some i so that $z > e$ in L_i. This means that we must find some M_i in which z is above both x and y in M_i. In fact we show that there is some i for which $z > x$ and $z > y$ in Q_i.

If z is an exterior vertex, say $z = u_i$, then z is the largest element in Q_i. Now suppose z is an interior vertex. Then for each $i = 1, 2, 3$, there is a path $S_i(z)$ from z to the ith exterior vertex u_i. The starting point of $S_i(z)$ is $v_0 = z$. If v_j has been determined, and v_j is an interior vertex, then v_{j+1} is the unique vertex so that the angles at v_j on either side of the edge $v_j v_{j+1}$ are colored $i+1$ and $i+2$.

The paths $S_1(z), S_2(z)$, and $S_3(z)$ are pairwise disjoint and partition T into three regions R_1, R_2, and R_3 as shown in fig. 10.5.

If the edge $e = xy$ joins two vertices in the region R_i, then z is greater than both x and y in Q_i. This completes the proof. □

It is well known that the problem of deciding whether a poset **P** satisfies $\dim(\mathbf{P}) \leqslant 2$ belongs to the class P of problems admitting a polynomial-time solution. For fixed $t \geqslant 3$, Yannakakis (1982) proved that the problem of deciding whether a poset **P** satisfies $\dim(\mathbf{P}) \leqslant t$ is NP-complete. For these reasons, Schnyder's theorem (10.3) is all the more striking since it equates a well-known polynomial-time problem, planarity testing, with an apparently NP-complete problem, deciding whether a particular poset has dimension at most 3. However, the poset being tested has a special form. The maximal elements all have degree two in the comparability graph. Also, it is not known whether it is NP-complete to answer whether the dimension of a height-two poset is at most 3. The answer is "yes" for dimension 4 or more.

Schnyder's theorem has been applied to find efficient algorithms for laying out a planar graph on a grid (see Kant 1992, Schnyder 1990, Schnyder and Trotter 1995). Recently, Brightwell and Trotter (1994a) have extended Schnyder's theorem to arbitrary planar maps.

Theorem 10.2. *Let* **G** *be a planar multigraph and let* D *be a drawing of* **G** *in*

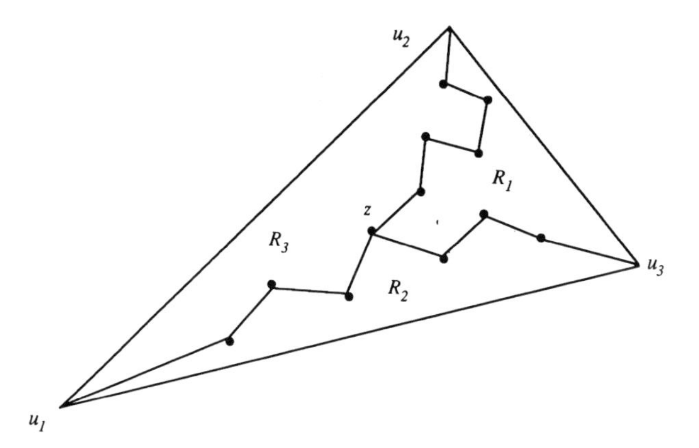

Figure 10.5.

the plane so that no edges cross. Then let $\mathbf{P}(D)$ *be the poset consisting of the vertices, edges and faces of the drawing D partially ordered by inclusion. Then* $\dim(\mathbf{P}(D)) \leqslant 4$.

The proof of Theorem 10.4 depends on the development of special graph-theoretic techniques applied to ordinary planar graphs satisfying a property somewhat weaker than 3-connectedness. The argument is inductive and required Brightwell and Trotter (1993) to first establish the following theorem.

Theorem 10.3. *Let* $\mathbf{M}$ *be a convex polytope in* $\mathbb{R}^3$, *and let* $\mathbf{P}_M$ *denote the poset consisting of the vertices, edges and faces of* $\mathbf{M}$ *partially ordered by inclusion. Then* $\dim(\mathbf{P}_M) = 4$.

In fact, the proof of Theorem 10.3 yields the even stronger conclusion that the subposet of $\mathbf{P}_M$ determined by the vertices and the faces of $\mathbf{M}$ is 4-irreducible. Theorem 10.3 cannot be extended to yield a bound of the dimension of the face lattice of a convex polytope in $\mathbb{R}^n$ for $n \geqslant 4$. This is due to the existence of cyclic polytopes [see the discussion in Brightwell and Trotter (1993)].

However, Theorem 10.3 can be extended to surfaces of higher genus since it is easy to prove by induction on n the existence of a function $f(n)$ so that the dimension of the poset of vertices, edges and faces of a multigraph drawn without edge crossings on a surface of genus n has order dimension at most $f(n)$. The only difficulty encountered in establishing the existence of $f(n)$ is the case $n = 0$. Here we know of no elementary proof of any finite bound, although of course Theorem 10.3 yields an upper bound of 4 in this case.

11. Regressions and monotone chains

If $\mathbf{P} = (X, P)$ is a poset, we call a map $f: X \to X$ a *regression* if $f(x) \leqslant x$ for every $x \in X$. When $C = \{x_1 < x_2 < \cdots < x_k\}$ is a k-element chain in $\mathbf{P}$, we say that a regression f is *monotone* on C if $f(x_1) \leqslant f(x_2) \leqslant \cdots \leqslant f(x_k)$. By convention, a regression is monotonic on any 1-element chain.

For $k \geqslant 1$, there are several interesting conditions on a poset which guarantee that every regression is monotonic on some k-element chain. Here is an important example due to Rado (1971).

Theorem 11.1. *For every $k \geqslant 1$, there exists an integer $n_0 = n_0(k)$ so that if $n \geqslant n_0$ and f is a regression on the subset lattice $\mathbf{2}^n$, then f is monotonic on some k-element chain.*

An alternative proof of Theorem 11.1 has been provided by Harzheim (1982) and this argument extends to a wider class of posets. However, neither argument gives much information about how large n_0 must be in terms of k. This is not surprising in view of the arguments' dependence on Ramsey-theoretic tools emphasizing existence.

By way of contrast, we present in this section a sharp result for posets of bounded width. The result is due to Peck et al. (1984).

Theorem 11.2. *Let w and k be positive integers and let $\mathbf{P} = (X, P)$ be a poset of width at most w. If $|X| \geqslant (w+1)^{k-1}$ then every regression is monotonic on some k-element chain.*

Proof. We proceed by induction on k, noting that the case $k = 1$ is trivial. Now assume $k \geqslant 2$ and that the theorem holds for smaller values of k. Let $\mathbf{P} = (X, P)$ be any poset of width at most w and let f be any regression on $\mathbf{P}$. We show f is monotonic on some k-element chain.

For each $x \in X$, let $H(x)$ be the largest t for which there is a t-element chain $x_1 < x_2 < \cdots < x_t = x$ on which f is monotonic. Without loss of generality $H(x) \leqslant k - 1$ for all $x \in X$.

Then let $Y = \{x \in X\colon H(x) < k-1\}$, $F = \{x \in X\colon H(x) = k-1, f(x) = x\}$ and $M = \{x \in X\colon H(x) = k-1, f(x) \neq x\}$. Evidently, $X = Y \cup F \cup M$ is a partition.

Now suppose that F is not an antichain. Choose $x, x' \in F$ with $x < x'$ in P. Then choose a $(k-1)$-element chain $x_1 < x_2 < \cdots < x_{k-1} = x$ on which f is monotonic. Then adding x' to this chain yields a k-element chain on which f is monotonic since $f(x_1) \leqslant f(x_2) \leqslant \cdots \leqslant f(x_{k-1}) = f(x) = x < x' = f(x')$. The contradiction shows that F is an antichain and thus $|F| \leqslant w$.

Next suppose that $x \in M$ and that $H(f(x)) = k-1$. Then we may choose a $(k-1)$-element chain $x_1 < x_2 < \cdots < x_{k-1} = f(x)$ on which f is monotonic. Since $f(x) < x$, we may add x to this chain to obtain a k-element chain on which f is monotonic. The contradiction shows $H(f(x)) \leqslant k-2$ for every $x \in M$, i.e., $k \geqslant 3$ and $f(M) \subset Y$.

Now let $y \in Y$. It is easy to see that $f(y)$ also belongs to y. Thus the restriction of f to Y is a regression. Since this restriction is not monotonic on any chain of $k-1$ points, it follows from the inductive hypothesis that $|Y| < (w+1)^{k-2}$.

Since $|X| \geqslant (w+1)^{k-1}$, $|Y| < (w+1)^{k-2}$ and $|F| \leqslant w$, we conclude that

$$\begin{aligned}|M| &= |X| - |Y| - |F| \\ &> (w+1)^{k-1} - (w+1)^{k-2} - w \\ &= w[(w+1)^{k-2} - 1] \\ &\geqslant w|Y|\end{aligned}$$

Since the width of **P** is at most w, it follows that there is some $y_0 \in Y$ for which the inverse image $f^{-1}(y_0)$ is not an antichain. We may then choose distinct points $x, x' \in M$ for which $f(x) = f(x') = y_0$ and $x < x'$ in P. As before, we choose a $(k-1)$-element chain $x_1 < x_2 < \cdots < x_{k-1} = x$ on which f is monotonic and add x' to form the desired k-element chain. □

The reader may enjoy the challenge of showing that thc inequality $|X| \geqslant (w+1)^{k-1}$ in Theorem 11.2 is best possible. The basic idea is to fix w and then construct a poset $\mathbf{P}_k = (X_k, P_k)$ and a regression f on $\mathbf{P}_k$ by induction on k. The poset $\mathbf{P}_1$ is a w-element antichain, and $\mathbf{P}_k$ is constructed by placing w disjoint chains, each containing $1 + (w+1)^{k-2}$ points, on top of $\mathbf{P}_{k-1}$. We refer the reader to Peck et al. (1984) for further details.

There appears to be some intrinsic connection between regressions and arithmetic progressions. Following Trotter and Winkler (1987), we define an *arithmetic progression* in a poset $\mathbf{P} = (X, P)$ as a chain $x_1 < x_2 < \cdots < x_t$ for which there is a constant d so that there are exactly d points in each of the intervals $\{y \in X\colon x_i \leqslant y < x_{i+1}\}$ for $i = 1, 2, \ldots, t-1$. The following result is due to Trotter and Winkler (1987).

Theorem 11.3. *Let k and w be positive integers and let $\varepsilon > 0$. Then there exists a number $n_0 = n_0(k, w, \varepsilon)$ so that if $\mathbf{P} = (X, P)$ is a poset of width at most w and $|X| \geqslant n_0$, then for every subset $S \subset X$ with $|S| > \varepsilon|X|$, there is a k-element chain $x_1 < x_2 < \cdots < x_k$ contained in S which is also a k-term arithmetic progression in* **P**.

The proof of Theorem 11.3 proceeds by induction on w with the case $w = 1$ being a restatement of Szemerédi's (1975) celebrated theorem on arithmetic progressions in subsets of $\mathbb{N}$ having positive upper density. It is reasonable to conjecture that for each $k \geqslant 1$, there is some $n_0 = n_0(k)$ so that if $\mathbf{L} = (X, P)$ is any distributive lattice with $|X| \geqslant n_0$, then every regression on **L** is monotonic on some k-element chain. This is supported by Theorems 11.1 and 11.2.

Also, we believe that for every $k \geqslant 1$ and every $\varepsilon > 0$, there is some $n_0 = n_0(k, \varepsilon)$ so that if $\mathbf{L} = (X, P)$ is a distributive lattice with $|X| \geqslant n_0$ and S is any subset of X with $|S| > \varepsilon|X|$, then S contains a k-term arithmetic progression. It is an easy exercise to show that this conjecture holds in the case where $\mathbf{L}$ is a subset lattice of the form $\mathbf{2}^n$.

Some modest progress has been made on these conjectures. Alon et al. (1987) study regressions on up sets in $\mathbf{n}^2$, while Kahn and Saks (1988) show that for each $\varepsilon > 0$, there exists an integer n_0 so that if $\mathbf{L} = (X, P)$ is a distributive lattice and $|X| \geqslant n_0$, then any antichain in $\mathbf{L}$ has less than $\varepsilon|X|$ points.

References

Ahlswede, R., and D. Daykin
[1978] An inequality for the weights of two families of sets, their unions and intersections, *Z. Wahrsch. V. Geb.* **43**, 183–185.
Alon, N., and E.R. Scheinerman
[1988] Degrees of freedom versus dimension for containment orders, *Order* **5**, 11–16.
Alon, N., W.T. Trotter and D. West
[1987] Regressions and monotone chains II: The poset of integer intervals, *Order* **4**, 155–164.
Anderson, I.
[1987] *Combinatorics of Finite Sets* (Clarendon Press, Oxford).
Babai, L., and D. Duffus
[1981] Dimension and automorphism groups of lattices, *Algebra Universalis* **12**, 279–289.
Birkhoff, G.
[1973] *Lattice Theory, Amer. Math. Soc. Colloq. Publ.* **25**.
Bogart, K.P., and W.T. Trotter
[1973] Maximal dimensional partially ordered sets II, *Discrete Math.* **5**, 33–44.
[1976a] On the complexity of posets, *Discrete Math.* **16**, 71–82.
[1976b] Maximal dimensional partially ordered sets III: A characterization of Hiraguchi's inequality for interval dimension, *Discrete Math.* **15**, 389–400.
Brightwell, G.
[1988] Linear extensions of infinite posets, *Discrete Math.* **70**, 113–136.
[1989] Semiorders and the 1/3–2/3 conjecture, *Order* **5**, 369–380.
[1993] Models of random partially ordered sets, in: *Surveys in Combinatorics 1993, London Math. Soc. Lecture Notes,* Vol. 187, ed. K. Walker (Cambridge University Press, Cambridge) pp. 53–83.
Brightwell, G., and E. Scheinerman
[1992] Fractional dimension of partial orders, *Order* **9**, 139–152.
Brightwell, G., and W.T. Trotter
[1993] The order dimension of convex polytopes, *SIAM J. Discrete Math.* **6**, 230–245.
[1994a] The order dimension of planar maps, *SIAM J. Discrete Math.,* to appear.
[1994b] Incidence posets of trees in posets of large dimension, *Order* **11**, 159–168.
Brightwell, G., and P.W. Winkler
[1991] Counting linear extensions, *Order* **8**, 225–242.
Brightwell, G., and C.D. Wright
[1992] The 1/3–2/3 conjecture for 5-thin posets, *SIAM J. Discrete Math.* **5**, 467–474.
Brightwell, G., H. Kierstead, A. Kostochka and W.T. Trotter
[1994] The dimension of suborders of the Boolean lattice, *Order* **11**, 127–134.
Canfield, E.R.
[1978] On a problem of Rota, *Adv. in Math.* **29**, 1–10.
Dilworth, R.P.
[1950] A decomposition theorem for partially ordered sets, *Ann. Math.* **51**, 161–165.

Duffus, D., B. Sands and R. Woodrow
[1988] Lexicographical matchings cannot form hamiltonian cycles, *Order* **5**, 149–161.
Duffus, D., H. Kierstead and W.T. Trotter
[1991] Fibres and ordered set coloring, *J. Combin. Theory A* **58**, 158–164.
Dushnik, B.
[1950] Concerning a certain set of arrangements, *Proc. Amer. Math. Soc.* **1**, 788–796.
Dushnik, B., and E.W. Miller
[1941] Partially ordered sets, *Amer. J. Math.* **63**, 600–610.
Erdős, P., and L. Lovász
[1973] Problems and results on 3-chromatic hypergraphs and some related questions, *Colloq. Math. Soc. János Bolyai* **10**, 609–627.
Erdős, P., H. Kierstead and W.T. Trotter
[1991] The dimension of random ordered sets, *Random Structures and Algorithms* **2**, 253–275.
Felsner, S.
[1995] *On-line chain partitions of orders,* Preliminary Manuscript.
Felsner, S., and W.T. Trotter
[1993] Balancing pairs in partially ordered sets, in: *Combinatorics, Paul Erdős is Eighty, Bolyai Soc. Math. Study* (Akadémia Kiadó, Budapest) pp. 145–157.
[1995] Colorings of diagrams of interval orders and α-sequences of sets, *Discrete Math.,* to appear.
Fishburn, P.C.
[1970] Intransitive indifference with unequal indifference intervals, *J. Math. Psych.* **7**, 144–149.
[1984] A correlational inequality for linear extensions of posets, *Order* **1**, 127–137.
[1986] *Interval Orders and Interval Graphs* (Wiley, New York).
[1988] Circle orders and interval orders, *Order* **5**, 225–234.
Fishburn, P.C., and W.T. Trotter
[1990] Angle orders, *Order* **1**, 333–343.
[1992] Linear extensions of semiorders: A maximization problem, *Discrete Math.* **103**, 25–40.
[1993] Posets with large dimension and relatively few critical pairs, *Order* **10**, 317–328.
Fishburn, P.C., W.G. Gehrlein and W.T. Trotter
[1992] Balance theorems for height-2 posets, *Order* **9**, 43–53.
Fortuin, C.M., P.W. Kasteleyn and J. Ginibre
[1994] Correlation inequalities on some partially ordered sets, *Comm. Math. Phys.* **22**, 89–103.
Fredman, M.
[1979] How good is the information theory bound in sorting?, *Theor. Comput. Sci.* **1**, 355–361.
Friedman, J.
[1993] A note on poset geometries, *SIAM J. Comput.* **22**, 72–78.
Füredi, Z., and J. Kahn
[1986] On the dimension of ordered sets of bounded degree, *Order* **3**, 15–20.
Füredi, Z., P. Hajnal, V. Rödl and W.T. Trotter
[1991] Interval orders and shift graphs, in: *Sets, Graphs and Numbers,* eds. A. Hajnal and V.T. Sós, *Colloq. Math. Soc. János Bolyai* **60**, 297–313.
Gallai, T.
[1967] Transitiv orientierbare Graphen, *Acta Math. Acad. Sci. Hungar.* **18**, 25–66.
Greene, C., and D.J. Kleitman
[1976] The structure of Sperner k-families, *J. Combin. Theory A* **20**, 41–68.
Harzheim, E.
[1982] Combinatorial theorems on contractive mappings in power sets, *Discrete Math.* **40**, 193–201.
Hiraguchi, T.
[1951] On the dimension of partially ordered sets, *Sci. Rep. Kanazawa Univ.* **1**, 77–94.
Hurlbert, G., A.V. Kostochka and L.A. Talysheva
[1994] On dimension of $P(2,k;n)$ for large k, *Order*, to appear.

Jamison, R.
[personal communication]
Jichang, S., and D.J. Kleitman
[1984] Superantichains in the lattice of partitions of a set, *Studia Appl. Math.* **71**, 207–241.
Kahn, J., and J. Kim
[1994] Entropy and sorting, *JACM*, to appear.
Kahn, J., and N. Linial
[1991] Balancing extensions via Brunn–Minkowski, *Combinatorica* **11**, 363–368.
Kahn, J., and M. Saks
[1984] Balancing poset extensions, *Order* **1**, 113–126.
[1988] On the widths of finite distributive lattices, *Order* **5**.
Kant, G.
[1992] *Drawing planar graphs using the lmc-ordering*, Technical Report RUU-CS-92–93 (Utrecht University).
Kelly, D.
[1977] The 3-irreducible partially ordered sets, *Canad. J. Math.* **29**, 367–383.
[1981] On the dimension of partially ordered sets, *Discrete Math.* **35**, 135–156.
[1984] Removable pairs in dimension theory, *Order* **1**, 217–218.
Kelly, D., and I. Rival
[1975] Planar lattices, *Canad. J. Math.* **27**, 636–665.
Kelly, D., and W.T. Trotter
[1982] Dimension theory for ordered sets, in: *Ordered Sets*, ed. I. Rival (Reidel, Dordrecht) pp. 172–211.
Khachiyan, L.
[1989] Optimal algorithms in convex programming, decomposition and sorting, in: *Computers and Decision Problems*, ed. J. Jaravlev (Nauka, Moscow) pp. 161–205 (in Russian).
Kierstead, H.
[1981] An effective version of Dilworth's theorem, *Trans. Amer. Soc.* **268**, 63–77.
[1986] Recursive ordered sets, in: *Contemp. Math.* **57**, 75–102.
[1995] *The dimension of 1 versus k element subsets of [n]*, Preliminary Manuscript.
Kierstead, H., and W.T. Trotter
[1985] Inequalities for the greedy dimension of ordered sets, *Order* **2**, 145–164.
[1988] Explicit matchings in the middle two levels of a boolean algebra, *Order* **5**, 163–171.
[1989] Super-greedy linear extensions of ordered sets, in: *Combinatorial Mathematics*, eds. G. Bloom et al., *Ann. New York Acad. Sci.* **555**, 262–271.
[1991] A note on removable pairs, in: *Graph Theory, Combinatorics and Applications*, Vol. 2, eds. Y. Alavi, G. Chartrand, O.R. Dellermann and A.J. Schwenk (Wiley, New York) pp. 739–742.
Kierstead, H., G. McNulty and W.T. Trotter
[1984] Recursive dimension for partially ordered sets, *Order* **1**, 67–82.
Kierstead, H., Z. Bing and W.T. Trotter
[1987] Representing an ordered set as the intersection of supergreedy linear extensions, *Order* **4**, 293–311.
Kierstead, H., S. Penrice and W.T. Trotter
[1994] On-line coloring and recursive graph theory, *SIAM J. Discrete Math.* **7**, 72–89.
Kimble, R.
[1973] *Extremal problems in dimension theory for partially ordered sets*, Ph.D. Thesis (MIT, Cambridge, MA).
Kislitsyn, S.S.
[1968] Finite partially ordered sets and their associated sets of permutations, *Mat. Zametki* **4**, 511–518.
Kleitman, D.J.
[1966] Families of non-disjoint sets, *J. Combin. Theory* **1**, 153–155.
Komlós, J.
[1990] A strange pigeon-hole principle, *Order* **7**, 107–113.

Kříž, I., and J. Nešetřil
[1991] Chromatic number of Hasse diagrams, eyebrows and dimension, *Order* **8**, 41–48.
Linial, N.
[1984] The information theoretic bound is good for merging, *SIAM J. Comput.* **13**, 795–801.
Peck, G.W., P. Schor, W.T. Trotter and D. West
[1984] Regressions and monotone chains, *Combinatorica* **4**, 117–119.
Perfect, H.
[1984] Addendum to a paper of M. Saks, *Glasgow Math. J.* **25**, 31–33.
Perles, M.A.
[1963] A proof of Dilworth's decomposition theorem for partially ordered sets, *Israel J. Math.* **1**, 105–107.
Pudlak, P., and J. Tuma
[1980] Every finite lattice can be embedded in a finite partition lattice, *Algebra Universalis* **10**, 74–95.
Rabinovitch, I.
[1978] The dimension of semiorders, *J. Combin. Theory A* **25**, 50–61.
Rado, R.
[1971] A theorem on chains of finite sets II, *Acta Arithmetica* **43**, 257–261.
Reuter, K.
[1989] Removing critical pairs, *Order* **6**, 107–118.
Rival, I.
[1982] ed., *Ordered Sets* (Reidel, Dordrecht).
[1985] *Graphs and Order, NATO ASI Series* (Reidel, Dordrecht).
Saks, M.
[1979] A short proof of the existence of k-saturated partitions of partially ordered sets, *Adv. in Math.* **33**, 207–211.
[1985] Balancing linear extensions of ordered sets, *Order* **2**, 323–330.
Savage, C., and P. Winkler
[personal communication]
Scheinerman, E., and J. Weirman
[1989] On circle containment orders, *Order* **4**, 315–318.
Schnyder, W.
[1989] Planar graphs and poset dimension, *Order* **9**, 45–473.
[1990] Embedding planar graphs on the grid, in: *Proc. ACM-SIAM Symposium on Discrete Algorithms, San Francisco,* pp. 138–147.
Schnyder, W., and W.T. Trotter
[1995] *Convex embeddings of 3-connected planar graphs,* Preliminary Manuscript.
Scott, D., and P. Suppes
[1958] Foundational aspects of theories of measurement, *J. Symbol. Logic* **23,** 113–128.
Seymour, P.
[1973] On incomparable collections of sets, *Mathematika* **20**, 208–209.
Shearer, J.B.
[1979] A simple counterexample to a conjecture of Rota, *Discrete Math.* **28**, 327–330.
Shepp, L.
[1980] The FKG inequality and some monotonicity properties of partial orders, *SIAM J. Algebraic Discrete Methods* **1**, 295–299.
[1982] The XYZ conjecture and the FKG inequality, *Ann. Probab.* **10**, 824–827.
Spencer, J.
[1972] Minimal scrambling sets of simple orders, *Acta Math. Hungar.* **22**, 349–352.
Sperner, E.
[1928] Ein Satz über Untermengen einer endlichen Menge, *Math. Z.* **27**, 544–548.

Stanley, R.P.

[1981] Two combinatorial applications of the Aleksandrov–Fenchel inequalities, *J. Combin. Theory A* **31**, 56–65.

[1986] *Enumerative Combinatorics*, Vol. 1 (Wadsworth and Brooks/Cole, Monterey, CA).

Szemerédi, E.

[1975] On sets of integers containing no k elements in arithmetic progression, *Acta Arithmetica* **27**, 199–245.

Trotter, W.T.

[1974a] The dimension of the crown S_n^k, *Discrete Math.* **8**, 85–103.

[1974b] Irreducible posets with arbitrarily large height exist, *J. Combin. Theory A* **17**, 337–344.

[1975a] Inequalities in dimension theory for posets, *Proc. Amer. Math. Soc.* **47**, 311–316.

[1975b] Embedding finite posets in cubes, *Discrete Math.* **12**, 165–172.

[1975c] A note on Dilworth's embedding theorem, *Proc. Amer. Math. Soc.* **52**, 33–39.

[1976a] A forbidden subposet characterization of an order dimension inequality, *Math. Systems Theory* **10**, 91–96.

[1976b] A generalization of Hiraguchi's inequality for posets, *J. Combin. Theory A* **20**, 114–123.

[1981] Stacks and splits of partially ordered sets, *Discrete Math.* **35**, 229–256.

[1982] Graphs and partially ordered sets, in: *More Selected Topics in Graph Theory*, eds. R. Wilson and L. Beineke (Academic Press, New York).

[1989] Problems and conjectures in the combinatorial theory of ordered sets, *Ann. Discrete Math.* **41**, 401–416.

[1992] *Combinatorics and Partially Ordered Sets: Dimension Theory* (Johns Hopkins University Press, Baltimore, MD).

[1994] Progress and new directions in dimension theory for finite partially ordered sets, in: *Proc. Visegrad Conf. on Extremal Set Theory*, eds. G. Katona et al., to appear.

Trotter, W.T., and J.I. Moore

[1976] Characterization problems for graphs, partially ordered sets, lattices and families of sets, *Discrete Math.* **16**, 361–381.

[1977] The dimension of planar posets, *J. Combin. Theory B* **22**, 54–67.

Trotter, W.T., and P. Winkler

[1987] Arithmetic progressions in partially ordered sets, *Order* **4**, 37–42.

Warren, H.E.

[1968] Lower bounds for approximation by nonlinear manifolds, *Trans. Amer. Math. Soc.* **133**, 167–178.

West, D.B.

[1982] Extremal problems in partially ordered sets, in: *Ordered Sets*, ed. I. Rival (Reidel, Dordrecht).

[1985] Parameters of partial orders and graphs: packing, covering, and representation, in: *Graphs and Orders*, ed. I. Rival (Reidel, Dordrecht).

Winkler, P.

[personal communication]

[1986] Correlation and order, in: *Combinatorics and Ordered Sets*, ed. I. Rival, *Contemp. Math.* **57**, 151–174.

Yannakakis, M.

[1982] On the complexity of the partial order dimension problem, *SIAM J. Algebraic Discrete Methods* **3**, 351–358.

CHAPTER 9

Matroids: Fundamental Concepts

D.J.A. WELSH

Merton College and Mathematical Institute, University of Oxford, Oxford, UK

Contents

HANDBOOK OF COMBINATORICS
Edited by R. Graham, M. Grötschel and L. Lovász

1. Introduction/History

Matroid theory dates from the 1930s when Whitney (1935) produced his seminal paper in which the term *matroid* first appeared. As the word suggests, Whitney conceived a matroid as an abstract generalisation of a matrix, and much of the language of the theory is based on that of linear algebra. However, Whitney's approach was also motivated by his work in graph theory and as a result some of the matroid terminology has a distinct graphical flavour. Some time later Van der Waerden also used the concept of abstract dependence in his "Modern Algebra".

Apart from isolated papers by Birkhoff (1935), Maclane (1936) and Dilworth (1941–1944), on the lattice theoretic and geometric aspects of matroid theory, and two important papers by Rado (1942, 1949) on the combinatorial applications of matroids and infinite matroids, respectively, the subject lay virtually dormant until the late fifties when Tutte (1958–1959), published his fundamental papers on matroids and graphs and Rado (1957) studied the representability problem for matroids. Since then interest in matroids and their applications in combinatorial theory has accelerated rapidly. Indeed it was realised that matroids have important applications in the field of combinatorial optimisation and also that they unify and simplify apparently diverse areas of pure combinatorics.

2. Axiom systems

A matroid has many different but equivalent definitions, several of which were described in Whitney's original paper. Deciding which set of axioms to use is difficult. To some extent it depends on the background of the reader or user. However, if we take the concept of a vector space as one of the most basic in mathematics we can regard matroid theory as having exactly the same relationship to linear algebra as does point set topology to the theory of real variables. That is, it *postulates* certain sets to be "independent" (linearly independent) and develops a fruitful theory from certain axioms which it demands hold for this collection of independent sets.

Accordingly we define a *matroid* M to be a finite set S and a collection $\mathcal{I}$ of subsets of S (called *independent* sets) such that (I1)–(I3) are satisfied.

(I1) $\phi \in \mathcal{I}$.

(I2) If $X \in \mathcal{I}$ and $Y \subseteq X$ then $Y \in \mathcal{I}$.

(I3) If U, V are members of $\mathcal{I}$ with $|U| = |V| + 1$ there exists $x \in U \backslash V$ such that $V \cup x \in \mathcal{I}$.

A subset of S not belonging to $\mathcal{I}$ is called *dependent*.

Example. Let S be a finite subset of a vector space V and let $\mathcal{I}$ be the collection of linearly independent subsets of vectors of S. Then $(S, \mathcal{I})$ is a matroid.

Following the analogy with vector spaces we make the following definitions. A *base* of M is a maximal independent subset of S, the collection of bases is denoted by $\mathscr{B}$ or $\mathscr{B}(M)$.

The *rank function* of a matroid is a function $r: 2^s \to \mathbb{Z}$ defined by

$$r(A) = \max(|X|: X \subseteq A,\ X \in \mathscr{I}) \quad (A \subseteq S).$$

The *rank of the matroid* M, sometimes denoted by $r(M)$, is the rank of the set S. A subset $A \subseteq S$ is *closed* or a *flat* or a *subspace* of the matroid M if for all $x \in S \backslash A$

$$r(A \cup x) = r(A) + 1.$$

In other words A is closed if no element can be added to it without increasing its rank. If for $x \in S$, $A \subseteq S$, $r(A \cup x) = r(A)$ we say that x *depends* on A. We define the *closure operator* of the matroid to be a function $\sigma: 2^s \to 2_s$ such that $\sigma(A)$ is the set of elements which depend on A. It is easy to prove from the axioms that $\sigma(A)$ is the smallest closed set containing A. A subset X is *spanning* in M if and only if it contains a base.

All the above concepts are very familiar from vector space theory.

To illustrate what can be done starting just with the basic axioms (I1)–(I3) we now obtain some elementary properties of independent sets and bases of a matroid M on S.

It is clear that if A is independent there exists a base B such that $A \subseteq B$.

The following stronger result is used extensively.

Proposition 2.1 (The augmentation theorem). *Suppose that X, Y are independent in M and that $|X| < |Y|$. Then there exists $Z \subseteq Y \backslash X$ such that $|X \cup Z| = |Y|$ and $X \cup Z$ is independent in M.*

Proof. Let Z_0 be a set such that over all $Z \subseteq Y \backslash X$ such that $X \cup Z$ is independent, $|X \cup Z_0|$ is a maximum. If $|X \cup Z_0| < |Y|$ then $Y_0 \subseteq Y$, $|Y_0| = |X \cup Z_0| + 1$, and since Y_0 is independent by (I3), $\exists y \in Y_0 \backslash (X \cup Z_0)$ such that $X \cup Z_0 \cup y$ is independent. The set $Z_0 \cup y$ contradicts the choice of Z_0. $\square$

An immediate consequence of this is the following result, which extends the well-known property of bases of a vector space.

Corollary 2.2. *All bases of a matroid on S have the same cardinality.*

It is fairly obvious that knowledge of the bases, or rank function, or closure operator is sufficient to uniquely determine the matroid. Hence it is not surprising that there exist axiom systems for a matroid in terms of each of these concepts. We list some of these axioms in the following theorems.

Proposition 2.3 (Base axioms). *A non-empty collection $\mathscr{B}$ of subsets of S is the set of bases of a matroid on S if and only if it satisfies the following condition:*

(B1) *If B_1, $B_2 \in \mathscr{B}$ and $x \in B_1 \backslash B_2$, $\exists y \in B_2 \backslash B_1$ such that $(B_1 \cup y) \backslash x \in \mathscr{B}$.*

Proposition 2.4 (Rank axioms). *A function* $r: 2^s \to \mathbb{Z}$ *is the rank function of a matroid on S if and only if for* $X \subseteq S$, $y, z \in S$;
(R1) $r(\emptyset) = 0$;
(R2) $r(X) \leq r(X \cup y) \leq r(X) + 1$;
(R3) *if* $r(X \cup y) = r(X \cup z) = r(X)$ *then* $r(X \cup y \cup z) = r(X)$.

Proposition 2.5 (Alternative rank axioms). *A function* $r: 2^s \to \mathbb{Z}$ *is the rank function of a matroid on S if and only if for any subsets X, Y of S*
(R1′) $0 \leq r(X) \leq |X|$;
(R2′) $X \subseteq Y \Rightarrow r(X) \leq r(Y)$;
(R3′) $r(X \cup Y) + r(X \cap Y) \leq r(X) + r(Y)$.

The condition (R3′) is a statement of the important property that the rank function of a matroid is a submodular function. This submodularity is at the heart of this area of combinatorics.
We can also easily obtain another axiomatisation of a matroid by its independent sets.

Proposition 2.6. *A collection* $\mathcal{I}$ *of subsets of S is the collection of independent sets of a matroid on S if and only if* $\mathcal{I}$ *satisfies conditions* (I1), (I2) *and the following statement*:
(I3′) *If A is any subset of S all maximal subsets Y of A with* $Y \in \mathcal{I}$ *have the same cardinality.*

An alternative and attractive approach is to axiomatise a matroid via its closure operator. It needs little more than definition chasing to prove that such a characterisation is as follows.

Proposition 2.7. *A function* σ *on the subsets of S is the closure operator of a matroid on S if it satisfies the following conditions for all* $A, B \subseteq S$, *and elements x, y of S*:
(S1) $A \subseteq \sigma(A)$;
(S2) $A \subseteq B \Rightarrow \sigma(A) \subseteq \sigma(B)$;
(S3) $\sigma(\sigma(A)) = \sigma(A)$;
(S4) *if* $y \notin \sigma(A)$ *but* $y \in \sigma(A \cup \{x\})$ *then* $x \in \sigma(A \cup \{y\})$.

The conditions (S1)–(S3) are the usual conditions satisfied by any set function with pretensions to being a closure function. But condition (S4) is special to matroids. It is sometimes referred to as the *Steinitz–Maclane exchange axiom*.
The proofs of all the above assertions are fairly routine (see Welsh 1976).

We shall call a set S with any of the above systems ($\mathcal{I}, r, \sigma, \mathcal{B}$) a matroid and will usually just denote it by M. Unless otherwise specified, S will always be the set supporting M and we sometimes signify this by writing S(M). Throughout r and σ will denote a rank function and closure operator of a general matroid.

Two matroids M_1 and M_2 on S_1 and S_2 respectively are *isomorphic* if there is a bijection $\phi: S_1 \to S_2$ which preserves independence. It is clear that equivalently ϕ is an isomorphism if and only if it preserves the rank function, bases, and so on. We write $M_1 \simeq M_2$ if M_1 and M_2 are isomorphic.

3. Some examples

The uniform matroid U_n^k. Let S be a set of cardinality n and let $\mathscr{I}$ be all subsets of S of cardinality $\leqslant k$. This is a matroid on S, called the uniform matroid of rank k and denoted by U_n^k.

Its rank and closure functions are defined by

$$r(A) = \begin{cases} |A| & |A| \leqslant k, \\ k & |A| \geqslant k, \end{cases} \quad (A \subseteq S),$$

$$\sigma(A) = \begin{cases} A & |A| < k, \\ S & |A| \leqslant k, \end{cases} \quad (A \subseteq S).$$

The special matroid in which every set is independent is the *free matroid*.

Vectorial matroids. Let S be any finite subset of a vector space V. Let a set $X = \{\underline{x}_1, \ldots, \underline{x}_k\} \in \mathscr{I}$ if and only if the vectors $\underline{x}_1, \underline{x}_2, \ldots, \underline{x}_k$ are linearly independent in V. Then $\mathscr{I}$ is the collection of independent sets of a matroid M. The rank function is just the rank (or dimension) function of V restricted to the set S. Any matroid obtained this way is called *vectorial*, and a matroid is *representable* if it is isomorphic to some vectorial matroid.

Cycle matroids of graphs. Let G be a graph, let S be its set of edges $E(G)$ and let $X \in \mathscr{I}$ if and only if X does not contain a circuit if G. Then $\mathscr{I}$ is the collection of independent sets of a matroid on S, called the *cycle matroid* or *polygon matroid* of the graph G and denoted by $M(G)$.

Algebraic matroids. Let F be a field and let K be an extension of F.

Proposition 3.1. *Let S be a finite subset of K and let $X \in \mathscr{I}$ if and only if $X \subseteq S$ and the elements of X are algebraically independent over F. Then $\mathscr{I}$ is the collection of independent sets of a matroid on S.*

The proof of this is a straightforward piece of field extension theory. Any matroid obtained this way is called *algebraic*.

Affine dependence. Let S be a finite set of points in d-dimensional Euclidean space $\mathbb{R}^d$. An element $\underset{\sim}{x} \in \mathbb{R}^d$ is *affinely dependent* on the set $\{\underset{\sim}{x}_1, \ldots, \underset{\sim}{x}_r\}$ of $\mathbb{R}^d$ if there exist real numbers λ_i $(1 \leqslant i \leqslant r)$ such that

$$\underset{\sim}{x} = \sum_{i=1}^{r} \lambda_i \underset{\sim}{x}_i,$$

and

$$\sum_{i=1}^{r} \lambda_i = 1 .$$

A subset $X \subseteq \mathbb{R}^d$ is *affinely dependent* if no element $x \in X$ is affinely dependent on $X \backslash x$. It is easy to prove that if $\mathcal{I}$ is the collection of subsets of S which are affinely independent then $\mathcal{I}$ is a matroid on S.

A Euclidean representation of matroids of low rank. Matroids of rank $\leqslant 3$ have the following very useful geometric description in the plane.

Let M be a matroid of rank 3 on the set $S = \{x_1, \ldots, x_n\}$. Place the n points $x_1, \ldots, x_n$ on the plane and draw a line through each closed set A such that $|A| \geqslant 3$, $r(A) = 2$. Then the bases of M are all subsets of S of cardinal 3 which are not collinear in this diagram.

Example. Let M be the matroid on $\{x_1, x_2, \ldots, x_6\}$ whose bases are all 3-sets except

$$\{x_1, x_3, x_6\}, \quad \{x_1, x_4, x_5\}, \quad \{x_2, x_5, x_6\}, \quad \{x_2, x_3, x_4\} .$$

Then a Euclidean representation of M is shown in fig. 3.1.

Conversely it is easy to check that any diagram of points and lines in the plane in which a pair of lines meet in at most one point represents a matroid whose bases are those 3-sets of points which are not collinear in the diagram.

Example. The diagram D of fig. 3.2 is a Euclidean representation of the matroid on $\{x_1, \ldots, x_7\}$ with bases all 3-sets except $\{x_1, x_2, x_6\}$, $\{x_1, x_4, x_7\}$, $\{x_1, x_3, x_5\}$, $\{x_2, x_3, x_4\}$, $\{x_2, x_5, x_7\}$, $\{x_3, x_6, x_7\}$, $\{x_4, x_5, x_6\}$.

The reader familiar with projective geometry will recognise this matroid as the well-known Fano plane which is vectorial over the field GF(2) under the map

$$x_1 \to (1,0,0), \quad x_2 \to (0,1,0), \quad x_3 \to (0,0,1),$$
$$x_4 \to (0,1,1), \quad x_5 \to (1,0,1), \quad x_6 \to (1,1,0),$$
$$x_7 \to (1,1,1).$$

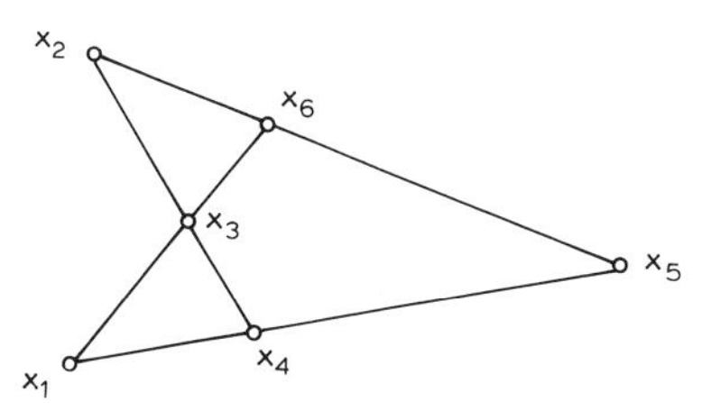

Figure 3.1.

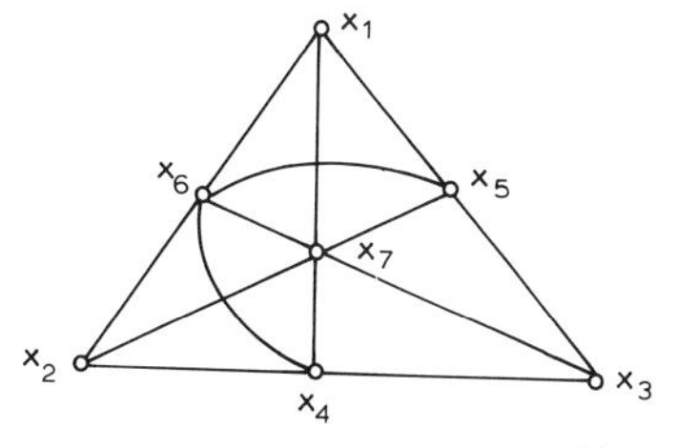

Figure 3.2. The Fano matroid.

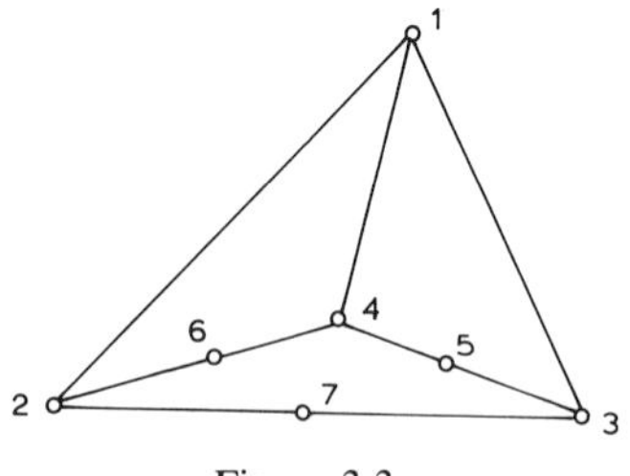

Figure 3.3.

This matroid, called the *Fano matroid*, and denoted by F_7 is more well known as the projective geometry PG(2, 2) (see chapter 13).

In both the above examples there are no two points x, y which form a dependent set. When such points are present they can be signified by "double points".

Example. The following diagram represents a matroid of rank 2 on $\{1, \ldots, 6\}$ in which $r(\{2, 3\}) = r(\{4, 5, 6\}) = 1$.

$\{1\}$ $\{2, 3\}$ $\{4, 5, 6\}$

This idea can be extended to rank 4 matroids by using 3-dimensional configurations as shown in the following example.

Example. Consider the tetrahedron in fig. 3.3. This is a matroid on $\{1, \ldots, 7\}$ in which the bases are the union of 1 with any 3 non-collinear points on the base.

4. The polygon/cycle matroid of a graph, circuits, connectedness

As mentioned earlier, Whitney introduced matroids to deal with graph properties in an abstract setting. The abstraction rest on the following easy observation.

Proposition 4.1. *The edge sets of the forests of a graph G are the independent sets of a matroid on $E(G)$.*

This matroid is called the *cycle* or *polygon matroid* of G and is denoted by $M(G)$.

Example. Let G be the graph of fig. 4.1. The polygon matroid of G has as its independent sets all subsets of $\{a, b, c, d, e\}$ which do not contain one of the sets $\{a, b, c\}$, $\{c, d, e\}$, $\{a, b, d, e\}$.

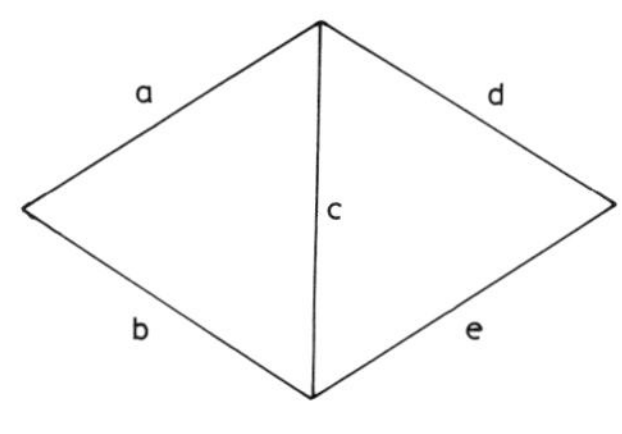

Figure 4.1.

From very elementary graph theory we can list the following basic properties of $M(G)$.

4.2. *If G is a connected graph the bases of $M(G)$ are the spanning trees of G.*

4.3. *If G is a disconnected graph the bases of $M(G)$ are the maximum forests of G.*

Another elementary observation is that the polygon matroid $M(G)$ can just as well be defined by listing the circuits of G. This clearly carries over to general matroids and continuing this analogy Whitney defined a *circuit* of an arbitrary matroid M to be a minimal dependent subset of $S(M)$.

Example. In the uniform matroid U_n^k on S in which all subsets of S of k or fewer elements are independent, the circuits are all $(k+1)$-subsets of S.

Example. If G is a graph, the circuits of $M(G)$ are just the edge sets which comprise the circuits of G.

Clearly, knowledge of the set of circuits of a matroid completely specifies the matroid. It is not surprising therefore that a matroid can be axiomatised in terms of its circuits, and for those readers motivated by graph theory, this may well be the most useful approach.

Obviously every dependent set contains a circuit and it is easy to see the following.

4.7. *If C is a circuit then $r(C) = |C| - 1$, $|C| \leq r(S) + 1$, and every proper subset of C is independent.*

Proposition 4.8. *A collection $\mathscr{C} \neq \{\emptyset\}$ of sets is the set of circuits of a matroid if conditions* (C1) *and* (C2) *hold.*

(C1) *If C_1, C_2 are distinct circuits then C_1 is not contained in C_2.*

(C2) *If C_1, C_2 are distinct circuits and $z \in C_1 \cap C_2$ there exists a circuit C_3 such that $C_3 \subseteq (C_1 \cup C_2)\backslash\{z\}$.*

In fact Whitney used a stronger axiomatisation replacing (C2) by what is now called the *strong* circuit axiom (C3) given by:

(C3) *If C_1, C_2 are distinct members of $\mathscr{C}$ and $y \in C_1 \backslash C_2$ then for each $x \in C_1 \cap C_2$ there exists $C_3 \in \mathscr{C}$ such that*

$$y \in C_3 \subseteq (C_1 \cup C_2) \backslash \{x\} .$$

Proving the equivalence of (C1)–(C2) with (C1) and (C3) is not trivial.

Continuing the analogy with graphs we note that in a graph G an edge e is a loop iff $\{e\}$ is a circuit of $M(G)$, while distinct edges e, f, are parallel iff $\{e, f\}$ is a circuit of $M(G)$. Accordingly we make the following definitions for general matroids.

An element $x \in S$ is a *loop* of M on S if and only if $\{x\}$ is a circuit of M. A pair of distinct elements x, y of S are *parallel* in M iff $\{x, y\}$ is a circuit of M. Trivially, we have

$$x \text{ is a loop} \Leftrightarrow r(\{x\}) = 0 , \qquad (4.9)$$

$$x \text{ and } y \text{ are parallel} \Leftrightarrow r(\{x\}) = r(\{y\}) = r(\{x, y\}) = 1 . \qquad (4.10)$$

For many purposes one can "forget about" loops and parallel elements in matroid theory in the same way as many authors in graph theory only consider graphs with no loops or parallel edges. However, as we will see, if we adopt this course the intrinsic simplicity and symmetry of duality theory (to be studied shortly) vanishes.

In the same way as the adjective "simple" is used to describe a graph which has no loops or parallel edges we define a *simple matroid* to be a matroid with no loops or parallel elements.

When G is a connected graph any spanning tree has $|V(G)| - 1$ edges. Thus counting up the connected components we get the following.

4.11. *The rank of the matroid $M(G)$ is $|V(G)| - k(G)$ where $k(G)$ is the number of connected components of G.*

More generally if A is any subset of edges of G let $k(A)$ denote the number of components in the subgraph of G with the edge set A and vertex set those vertices of G incident with a member of A.

4.12. *For any subset $A \subseteq E(G)$ the rank of A in $M(G)$ is $r_G(A) = |V(A)| - k(A)$.*

It is clear from this how to characterise the closure operator and hence implicitly the flats of $M(G)$.

4.13. *Let A be a set of edges of G, let $e \in E(G) \backslash A$. Then e belongs to $\sigma(A)$, the closure of A in $M(G)$, if and only if there is a circuit C of G with $e \in C \subseteq A \cup \{e\}$.*

However, despite the power of matroid theory as a tool in the clarification of certain graphical ideas we warn that many problems of graph theory cannot even be posed in matroid language. Basically this is because there is nothing in a matroid which corresponds exactly to a vertex in a graph.

For example since any tree T on n edges has a cycle matroid which is isomorphic to that of *any* forest on the same number of edges, it is clear that there can be no concept in matroid theory corresponding to ordinary graphical connectedness. It is easy however to introduce a notion of separability which corresponds exactly to 2-connection in graph theory.

A matroid M is *connected* or *non-separable* if for any pair of distinct elements e, f of $S(M)$, there exists a circuit C containing e and f. The following observations are easy to check.

4.14. *A matroid M on S is connected if and only if for any proper subset A of S*

$$r(A) + r(S \setminus A) > r(S) .$$

4.15. *Write $e \sim f$ if there is a circuit C containing both e and f. Then $\sim$ is an equivalence relation on the groundset S and the equivalence classes are called the connected components of M.*

In the same way as M can be naturally broken up into its connected components, the reverse operation of taking the "direct sum" of the components resurrects the original matroid.

The *direct sum* of matroids M_1 on S_1 and M_2 on S_2, with $S_1 \cap S_2 = \emptyset$ is the matroid $M_1 \oplus M_2$ on $S_1 \cup S_2$ which has as its independent sets all sets $\{I_1 \cup I_2$: $I_1 \in \mathcal{I}(M_1),\ I_2 \in \mathcal{I}(M_2)\}$. Equivalently the set of circuits of $M_1 \oplus M_2$ is just the union of the sets of circuits of M_1 and M_2.

Clearly

$$r(M_1 \oplus M_2) = r(M_1) + r(M_2)$$

and a disconnected matroid is just the direct sum of its connected components.

Finally we address the problem of characterising those matroids isomorphic to the cycle matroid of some graph. Call such matroids *graphic*. It is easy to see that most matroids are not graphic.

Example. The uniform matroid U_4^2 is not graphic. (Try to draw a graph with 4 edges such as its circuits are all edge sets of size 3.)

A complete characterisation of graphic matroids was given by Tutte (1959) in one of the really deep theorems on the subject; for further details of this and more on the subject of higher connectivity and Seymour's theory of splitters see Welsh (1982), Oxley (1992) and chapters 10 and 11.

5. Duality

The concept of matroid duality (which we emphasise has nothing to do with vector space duality) is of fundamental importance in the applications of matroids to combinatorial theory.

The basic result due to Whitney is easily proved.

Theorem 5.1. *If* $\{B_i: i \in I\}$ *is the set of bases of a matroid* M *on* S *then* $\{S \setminus B_i: i \in I\}$ *is the set of bases of a matroid on* S, *called the dual matroid of* M.

We denote the *dual matroid* of M by M^*. Obviously M and M^* are related by

$$(M^*)^* = M . \tag{5.2}$$

The following properties of M^* are straightforward deductions from its definition.

5.3. *A subset* $X \subseteq S$ *is independent in* M^* *if and only if* $S \setminus X$ *is spanning in* M.

5.4. *An element* $x \in S$ *is a loop of* M *if and only if* x *belongs to every base of* M^*.

5.5. *The rank of* M^* *is* $|S| - r(M)$.

This last result is a special case of the following more general fact.

5.6. *The rank functions* r, r^* *of* M, M^* *respectively are related by*

$$r^*(A) = |A| - r(S) + r(S \setminus A) \quad (A \subseteq S) .$$

The function $r^*: 2^S \to \mathbb{Z}$ we call the *corank* function of M. A *cobase* of M is a base of M^*: a *cocircuit* of M is a circuit of M^*, if x is a loop of M^* it is called a *coloop* of M and so on.

Since M^* determines M uniquely we have the obvious result that a matroid is uniquely determined by its cobases, cocircuits or corank function. Moreover, each of the earlier axiomatisations can be dualised to give an axiomatisation of a matroid in terms of these dual concepts.

For example if we define a *hyperplane* of M to be a maximal proper flat it is routine to check the following.

5.7. H *is a hyperplane of* M *on* S *iff* $S \setminus H$ *is a circuit of* M^*.

As a consequence, since the circuits of a matroid determine it uniquely, we have the following.

5.8. *A matroid is uniquely determined by its collection of hyperplanes.*

In general there is a simple but useful *duality principle* – that to every statement about a matroid there is a dual statement. For example a dualisation of the statement

(α) An element x is a loop of M if and only if x does not belong to any base of M.

is the statement

$(\alpha)^*$ An element x is a coloop of M if and only if x does not belong to any cobase of M.

As another example, because of 5.7 and 5.8 the reader can obtain axioms for a matroid in terms of its hyperplanes merely by complementing the circuit axioms (C1)–(C2) of section 4.

More interesting is the light duality sheds on the concept of planarity in graph theory. To understand this first consider the following class of matroids.

A *cocircuit* (minimum cut) of a graph G is a minimal set of edges whose deletion from G increases the number of connected components. A fundamental observation is the following.

Proposition 5.9. *The collection of cocircuits of a graph G is the set of circuits of a matroid $M^*(G)$ on the edge set E of G.*

This matroid is called the *cocycle* or *bond* matroid of G and as the notation suggests we have the following.

Proposition 5.10. *For any graph G, $M^*(G)$ and $M(G)$ are dual matroids, that is*

$$M^*(G) = (M(G))^* .$$

The reader familiar with graph theory will not find it difficult to verify 5.9 by showing that the cocircuits of a graph satisfy the circuit axioms (C1)–(C2) of a matroid.

A consequence of 5.10 is that any statement about circuits in a graph must have an exact dual counterpart in terms of cocircuits. For example try and verify directly that the cocircuits of a graph satisfy the strong circuit axiom (C3).

Another useful property of duality is the following.

Proposition 5.11. *M is a connected matroid if and only if M^* is.*

Proof. Use criterion 4.14 in terms of the rank function. □

As an application, we note that a graph is 2-connected iff any pair of edges belongs to a cocircuit.

Finally we relate duality to planarity. A matroid M is *cographic* if it is isomorphic to the cocycle matroid of some graph G. Again it is easy to verify the following.

5.12. *U_4^2 is the smallest matroid which is not cographic.*

The concept of abstract duality allowed Whitney to characterise planar graphs in the following way.

Theorem 5.13. *A graph is planar if and only if its cycle matroid is cographic.*

Equivalently, by dualising the above statement, we have the following.

Corollary 5.14. *A graph is planar if and only if its cocycle matroid is graphic.*

As in the case of graphic matroids Tutte's theorem completely characterises cographic matroids.

An alternative reinterpretation of Whitney's planarity condition is as follows.

Proposition 5.15. *A matroid is both graphic and cographic if and only if it is the cycle matroid of a planar graph (equivalently the cocycle matroid of a planar graph).*

Thus the statement that M_1 and M_2 are dual matroids can be regarded as the natural generalisation of the statement that G and H are dual planar graphs.

6. Submatroids and minors

We now look at the different ways in which a matroid M on S induces "smaller" matroids which in some sense preserve some of the structure of M. The simplest operation is defined by the following.

6.1. *Let M be a matroid on S and Let $0 \leq k \leq r(M)$. Then if $\mathcal{I}_k = \{X\colon X \subseteq \mathcal{I}(M);$ $|X| \leq k\}$, $\mathcal{I}_k$ is the collection of independent sets of a matroid M_k on S.*

We call M_k the *k-truncation* of M. It is trivial to verify 6.1 and to note that M_k has the rank function r_k given by $r_k(A) = \min(k, r(A))$ $(A \subseteq S)$ and the bases of M_k are the independent sets of M which have cardinality k.

Example. The uniform matroid U_n^k is the k-truncation of the free matroid on n elements.

We next show how a matroid M on S induces two matroids on a subset T of S which correspond in the natural way to subgraphs of a graph obtained by the operations of deleting and contracting of edges.

If $\mathcal{I}(M)$ is the set of independent sets of M on S and $T \subseteq S$, let

$$\mathcal{I}(M \mid T) = \{X\colon X \subseteq T,\ X \in \mathcal{I}(M)\}\ .$$

6.2. *$\mathcal{I}(M \mid T)$ is the set of independent sets of a matroid on T.*

We denote this matroid by $M \mid T$ and call it the *restriction* of M to T.

The proof of 6.2 is again trivial. When M is the cycle matroid $M(G)$ of a graph G and $T \subseteq E(G)$, recall that $G \mid T$ denotes the subgraph with edge set consisting of those edges of G not belonging to $E(G) \setminus T$. Then it is easy to verify

$$M(G \mid T) = M(G) \mid T . \tag{6.3}$$

Because of this analogy between $M \mid T$ and the effect of removing the elements of $S \setminus T$ the matroid $M \mid T$ is sometimes called the matroid obtained from M by *deleting* $S \setminus T$.

An alternative and commonly used notation is to write $M \setminus X$ to denote the matroid obtained from M by deleting the subset X of its groundset $S(M)$. Thus $M \setminus X = M \mid (S \setminus X)$.

Obviously the rank function of $M \mid T$ is just the restriction of r to T. The following basic properties of $M \mid T$ are easily checked.

6.4. *If X is dependent in M and $X \subseteq T$ then X is dependent in $M \mid T$.*

6.5. *$M \mid T$ has as its circuits all circuits of M which are contained in T.*

Having carried out a restriction of M to the set T it is natural to examine the dual operation. This turns out to be the exact counterpart of the concept of contraction in graphs. There are several equivalent ways of defining it as can be seen from the following statements.

Proposition 6.6. *Let M be a matroid on S and let $T \subseteq S$. Define $\mathcal{I}(M \cdot T)$ to be the matroid on T with rank function λ given in terms of the rank function r of M by*

$$\lambda(A) = r(A \cup (S \setminus T)) - r(S \setminus T) \quad (A \subseteq T) .$$

Then

$$(M \cdot T)^* = M^* | T , \tag{6.7}$$

$$(M \mid T)^* = M^* \cdot T . \tag{6.8}$$

The circuits of $M \cdot T$ are the minimal non-empty sets of the form $C \cap T$, C a circuit of M. (6.9)

X is independent in $M \cdot T$ if there exists a maximal independent set Y of $M \mid (S \setminus T)$ such that $X \cup Y$ is independent in M. (6.10)

We call $M \cdot T$ the *contraction* of M to T because when M is the cycle matroid of a graph G and $T \subseteq E(G)$, $M \cdot T$ is the cycle matroid of the graph $G \cdot T$ obtained from T by contracting out of G the edges in $E(G) \setminus T$.

For $X \subseteq S(M)$, we often us M/X to denote the matroid obtained from M by

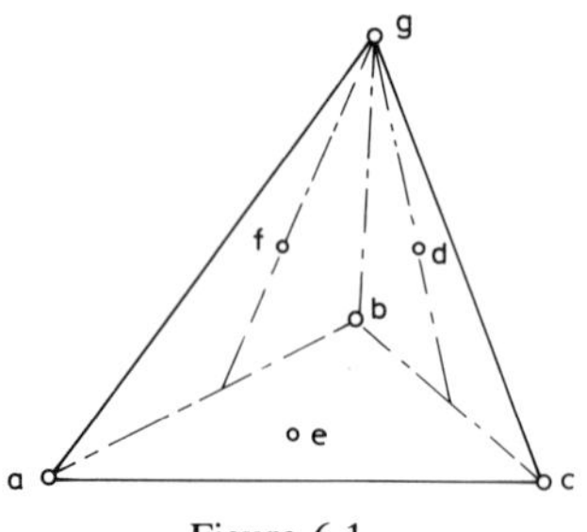

Figure 6.1.

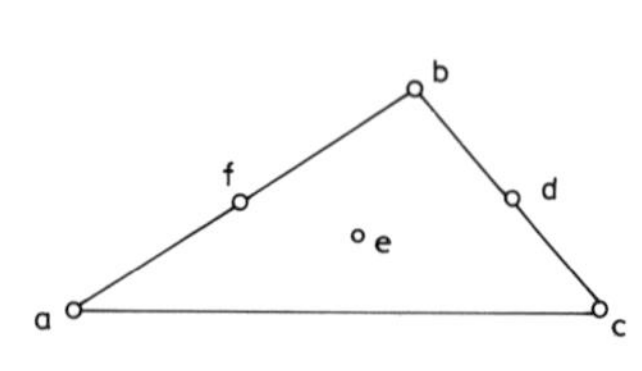

Figure 6.2.

contracting out X. Thus

$$M/X = M \cdot (S/X) .$$

Note by (6.7), (6.8) that deletion and contraction are dual concepts in matroid theory.

Contraction also has an attractive geometric interpretation. Consider for example M on $S = \{a, b, c, d, e, f, g\}$ where a, b, c, g are the vertices of a tetrahedron in $\mathbb{R}^3$ and d, e, f are points in general position in the faces $\{g, b, c\}$, $\{a, b, c\}$, $\{a, b, g\}$ as shown in fig. 6.1. Then the contraction of g out of M gives the matroid $M \cdot T$ of rank 3 obtained by just *projecting* down from g and represented geometrically by fig. 6.2.

Any matroid N which can be obtained from a matroid M by some sequence of contractions and deletions is called a *minor* of M. Since the operations of contraction and deletion commute the following is equivalent.

6.11. *N is a minor of M if and only if M can be obtained from M by a restriction (contraction) followed by a contraction (restriction).*

Moreover, since the operations are dual the following holds.

6.12. *N is a minor of M if and only if N^* is a minor of M^*.*

The crucial importance of minors in matroid theory is because many of the interesting properties seem to be preserved under the taking of minors and a natural way to obtain a good characterisation of such a property π is to find some finite class of matroids $\{N_1, \ldots, N_t\}$ such that having property π is equivalent to not having one of the N_i as a minor. In recent years, due mainly to the research efforts of Tutte and Seymour, many deep characterisations of this type have been obtained – we refer to chapter 10 for details.

7. Geometric lattices

A completely different approach to matroid theory was initiated by Birkhoff (1935).

Recall that a *flat* or *closed* set of a matroid M is just any subset A of $E(M)$ such that $r(A \cup \{e\}) = r(A) + 1$ for any $e \not\in A$. It corresponds exactly to a subspace in a vector space. Clearly the flats of M, ordered by inclusion form a partially ordered set (see chapter 8). However, more can be said. It is easy to prove that if A, B are any two flats then $A \cap B$ is also a flat. This leads to the following easy observation.

7.1. *For any matroid M, the set of flats of M ordered by inclusion forms a lattice $\mathscr{L}(M)$.*

In particular the lattice $\mathscr{L}(M)$ has the following special properties.

7.2. *$\mathscr{L}(M)$ has a minimal element* 0 *and a maximal element I together with a height function h such that for any pair of elements x, y*

$$h(x) + h(y) \geqslant h(x \wedge y) + h(x \vee y) .$$

The proofs of 7.1 and 7.2 are easy. An element x of $\mathscr{L}(M)$ corresponds to a flat X of M. The zero element 0 is just the closure of the empty set, namely the set of loops of M. The height $h(x)$ of x is then just the rank $r(X)$ of the corresponding flat X and 7.2 follows from the submodularity of the rank function.

From 7.2 it follows that $\mathscr{L}(M)$ is a semimodular lattice which has the additional property that any element of $\mathscr{L}(M)$ can be written as the join of atoms. But this is exactly the definition of what is known as a *geometric lattice*.

It is natural therefore to consider the reverse proposition that every finite geometric lattice $\mathscr{L}$ can be obtained by the above construction from some matroid M. This is the case. To prove it just construct a matroid M on the set A of atoms (elements of height 1) of $\mathscr{L}$ in which a subset $X = \{x_1, \ldots, x_t\}$ is called independent if and only if

$$h\{x_1 \vee \cdots \vee x_t\} = t .$$

Verifying that this construction gives a matroid M with the property that $\mathscr{L}(M) \simeq \mathscr{L}$ is just a routine check.

Example 1. If M is the free matroid in which every subset is independent, $\mathscr{L}(M)$ is the distributive lattice of all subsets of $S(M)$.

Example 2. If M is the uniform matroid of rank 3 on $S = \{a, b, c, d\}$, $\mathscr{L}(M)$ is the lattice of fig. 7.1.

Example 3. Semimodular lattices which are not the lattices of flats of a matroid are shown in fig. 7.2. In both cases the element g is not the join of atoms.

We can therefore sum up the previous discussion by stating the following.

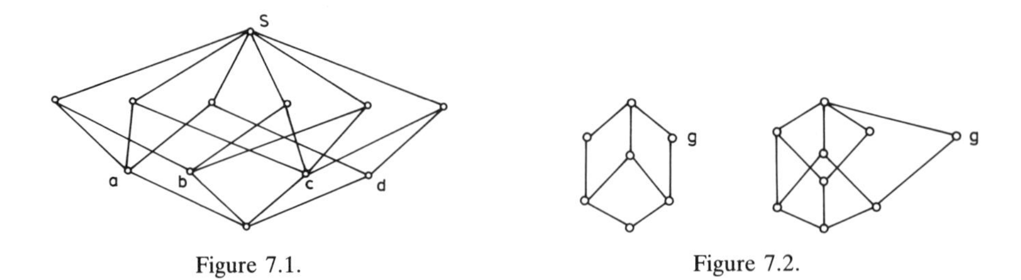

Figure 7.1.

Figure 7.2.

Theorem 7.3. *A finite lattice is isomorphic to the lattice of flats of a matroid if and only if it is a geometric lattice.*

Thus in a sense matroid theory can be regarded as just a small subset of lattice theory. However, the flaw in this approach is that it does not seem to give any insight into duality and it is from duality that much of the richness of the theory comes. As a test of this observation consider the following problem. Given a geometric lattice $\mathscr{L}$ construct the geometric lattice of the matroid dual to that of $\mathscr{L}$. There seems to be no natural lattice theoretic answer to this.

Another point in connection with Theorem 7.3 is that the correspondence between matroids and geometric lattices is only a bijection between the class of *simple* matroids and geometric lattices. This is because if M_1 can be got from M by deleting loops or elements in parallel then the geometric lattices $\mathscr{L}(M)$ and $\mathscr{L}(M_1)$ are indistinguishable.

The relation between matroids and geometric lattices parallels the formation of projective spaces from vector spaces, 1-dimensional subspaces in the vector space become points of the projective space.

However, provided we are concerned only with simple matroids, much can be gained by a geometric lattice approach and it is this approach which is taken in the influential work of Crapo and Rota (1970). (They use the term *combinatorial geometry* for what we call simple matroid and *pregeometry* for matroid.)

Finally we remark that minors of M are clearly visible in the geometric lattice of M. That is to say, if we consider M on S with geometric lattice $\mathscr{L}(M)$ as shown, then if T is a flat of M, the minors $M|T$ and $M \cdot (S \backslash T)$ are the sublattices indicated in fig. 7.3. Similarly if $A \subseteq B$ are any two flats the interval sublattice $[A, B]$ is the geometric lattice of the minor M obtained by contracting $S \backslash B$ out and then restricting to $S \backslash A$.

8. Pavings, transversals and linkages

We observed in 5.8 that a matroid is uniquely determined by its hyperplanes. By dualising the circuit axioms and recalling that H is a hyperplane of M only if its complement is a circuit we have the following.

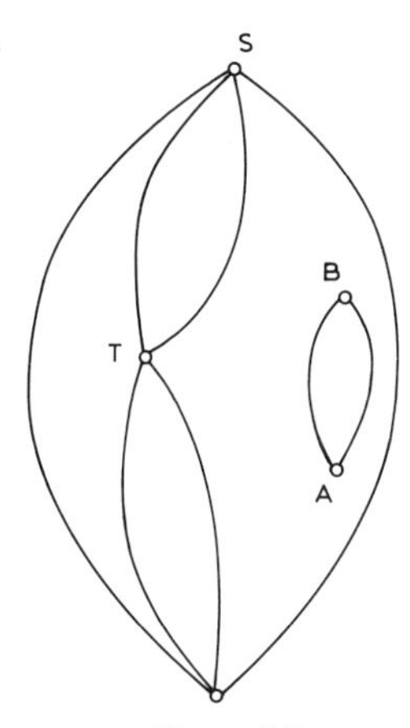

Figure 7.3.

Proposition 8.1. *A collection* $\mathscr{H}$ *of subsets of* S *is the set of hyperplanes of a matroid on* S *if and only if*

(H1) *No member of* $\mathscr{H}$ *properly contains another.*

(H2) *If* H_1, H_2 *are distinct members of* $\mathscr{H}$ *and* $x \notin H_1 \cup H_2$ *there exists* $H_3 \in \mathscr{H}$ *such that* $(H_1 \cap H_2) \cup x \subseteq H_3$.

From this basic result it is easy to recognise the following large class of matroids.

A *d-paving* of S is a collection $\mathscr{P}$ of subsets of S each of cardinality $\geq d$, no one of which is contained in another and with the property that each d-subset of S is contained in a unique member of $\mathscr{P}$. A trivial consequence of this definition and 8.1 is the following.

Proposition 8.2. *If* $\mathscr{P}$ *is a d-paving of* S *then it is the set of hyperplanes of a matroid on* S *of rank* $(d+1)$.

Any matroid isomorphic to one obtained in this way we call a *paving matroid*. Although they have a very simple structure (every set of cardinality $<d$ is independent and every circuit has cardinality $d+1$ or $d+2$), they often occur as counterexamples, possibly because they are far away from the linear case.

A particularly regular class of paving matroid is when all the hyperplanes have the same cardinality. We then have the very special structure, well known in the theory of block designs as *Steiner systems* see chapter 14. Explicitly we have the following.

Corollary 8.3. *The blocks of a Steiner system are the hyperplanes of a matroid.*

Despite their rather limited structure there are surprisingly many paving matroids. For example if $f(n)$, $p(n)$ denote respectively the number of matroids and paving matroids with groundset of cardinality n then we have

$$2^{2^{n-(3/2)\log(n)+O(\log\log n)}} \leq p(n) \leq f(n) \leq 2^{2^{n-\log n+O(\log\log n)}} . \tag{8.4}$$

Apart from the ways in which we have already met matroids in connections with graphs and Steiner systems they seem to have a natural place in the theory of linkages.

The simplest way to see this is to consider the following class of matroids occurring in transversal theory. Let $\mathcal{A} = (A_i: i \in I)$ be a finite family of subsets of S. A *partial transversal* of $\mathcal{A}$ is a set $X = \{x_1, x_2, \ldots, x_t\}$ such that there is an injection $\psi\{1, \ldots, t\} \to I$ with the property that $x_i \in A_{\psi(i)}$ $(1 \leq i \leq t)$.

The following observation, first made by Edmonds and Fulkerson (1965) is fundamental.

Proposition 8.5. *The collection $P(\mathcal{A})$ of partial transversals of a finite family $\mathcal{A}$ of subsets of S is the set of independent sets of a matroid on S.*

Any matroid which is isomorphic to one which is obtainable in this way is called a *transversal matroid*.

In a sense this is just a special case of the following more general construction. Take an arbitrary graph G and a distinguished subset A of vertices of G. We say that a set $\{x_1, \ldots, x_t\}$ of vertices of G can be *linked* into A if there is a sequence $P_1, \ldots, P_t$ of paths of G which have the following properties:

(a) P_i, P_j are vertex disjoint if $i \neq j$,

(b) the endpoints of P_i are x_i and a_i where $a_i \in A$.

An obvious extension of 8.5 is the following.

Proposition 8.6. *For any graph G and any distinguished subset A of vertices of G the collection $\mathcal{P}(A)$ of sets of vertices which can be linked into A form the collection of independent sets of a matroid on A.*

Any matroid which can be obtained in this way is called a *gammoid*.

Clearly every transversal matroid is a gammoid obtained by taking G to be a bipartite graph.

A closed related class of matroids occurs by taking a set X of vertices of a graph G to be independent if there exists some *matching* of G which contains X among its endpoints. This defines the class of *matching matroids*.

Induced matroids

A natural extension of these matroids obtained from linkages is the idea of inducing matroids.

For any graph G let M be a matroid with groundset S a subset of the vertex set $V(G)$. If $T \subseteq V$, G *induces* a matroid $G(M)$ on T by defining $X \subseteq T$ to be independent in $G(M)$ if and only if there are vertex disjoint paths in G linking X to a subset Y of $V(G)$ which is independent in M.

A particular case of this occurs when f is a function on $S \to T$. By taking G to be the obvious associated bipartite graph with vertex set $S \cup T$ and edges $(s, f(s))$ we may define $f(M)$ to be the induced matroid $G(M)$. This construction was first introduced by Nash-Williams (1966), the extensions to arbitrary linkages was carried out some years later, for details see Mirsky (1971), Welsh (1976) or Oxley (1992).

9. Submodular set functions

A function $\mu : S \to \mathbb{R}$ is *submodular* if

$$\mu(X) + \mu(Y) \geqslant \mu(X \cup Y) + \mu(X \cap Y) \quad (X, Y \subseteq S) .$$

There is an intrinsic relationship between matroids and submodular set functions; not only because of the submodularity of the rank function, but because of the following result.

Theorem 9.1. *Let μ be an integer valued, submodular, non-negative set function on S with $\mu(\emptyset) = 0$. Then μ induces on S the matroid $M(\mu)$ which has as its independent sets the collection of sets X, such that for all $A \subseteq S$*

$$\mu(A) \geqslant |X \cap A| .$$

The rank function r of $M(\mu)$ is given by

$$r(X) = \inf\{\mu(A) + |X \backslash A|): A \subseteq S\} .$$

Obviously any matroid can be obtained by choosing μ to be the "right" function. What is interesting is to use the above theorem in a most natural way so that the class of matroid drops out.

Example. As a very simple example, let $(A_i : i \in I)$ be a family of subsets of S and let $\mu : I \to \mathbb{Z}^+$ be defined by

$$\mu(J) = |A(J)| \quad J \subseteq I$$

and where $A(J) = \bigcup \{A_i : i = J\}$.

It is easily checked that μ is submodular and the resulting matroid has as its independent sets those J for which the family $(A_i : i \in J)$ has a partial transversal.

Interchanging the roles of points and sets in the above (sometimes rather grandly called *point–set duality*) gives the class of transversal matroids.

More interesting is the following construction which has an easy interpretation in terms of submodular functions. The sum of two rank functions on the same set is a submodular function and leads to the following construction.

Given two matroids M_1 on S_1 and M_2 on S_2. Define $M_1 \vee M_2$, the *union* of M_1 and M_2 to be the matroid on $S_1 \cup S_2$ in which a set X is independent if it can be expressed in the form $X_1 \cup X_2$ where X_1 is independent in M_1 and X_2 is independent in M_2.

Clearly when $S_1 \cap S_2 = \emptyset$ the union of M_1 and M_2 is just the direct sum $M_1 \oplus M_2$. In general, however, it is a difficult operation to unravel. For example the following question first posed in Welsh (1971) is still unresolved.

Problem. Call a matroid M *irreducible* or *prime* if there do not exist non-trivial M_1, M_2 such that $M = M_1 \vee M_2$. Classify, or find necessary and sufficient conditions for a matroid to be irreducible.

An almost immediate consequence of its definition is the following.

Proposition 9.2. *M is transversal if and only if M is the union of matroids of rank one.*

Much less obvious is the following fundamental calculation often called the matroid partition theorem (see also chapter 11).

Theorem 9.3. *If $M = M_1 \vee \cdots \vee M_k$ with M_i having rank function r_i then M has rank r given by*

$$r(S) = \min_{A \subseteq S} (r_1(A) + \cdots + r_k(A) + |S \backslash A|) \quad (A \subseteq S) .$$

This follows immediately from Theorem 9.1 and the observation that since each r_i is submodular the sum $r_1 + \cdots + r_k$ is a submodular set function.

From 9.3 we have what are often described as the two covering/packing theorems about matroids.

Corollary 9.4. *A matroid M has k pairwise disjoint bases iff for all $A \subseteq S$*

$$kr(A) + |S \backslash A| \geq kr(S) .$$

Corollary 9.5. *The ground set S of a matroid M can be covered by k or fewer independent sets if and only if*

$$kr(A) \geq |A| \quad (A \subseteq S) .$$

In both cases 9.4 and 9.5 the conditions are easily shown to be necessary. They have many applications in different areas of combinatorics and versions of these results restricted to graphs or vector spaces or transversals were originally proved by quite complicated and ad hoc arguments which were relevant to the particular specialisations see for example chapter 2. The realisation that matroids gave a unified and simple treatment of this general area was a main impetus to interest in the subject. For more on this we refer to chapter 11.

By taking the union of M_1 with the dual of M_2 we get an easy condition for M_1 and M_2 to have an independent set of prescribed size. More precisely, we get the following result called the matroid intersection theorem due to Edmonds (1965) (for a proof and algorithms see chapter 11).

Theorem 9.6. *If M_1 and M_2 are matroids on S and $k \in \mathbb{Z}^+$ then M_1 and M_2 have a common independent set of size k if and only if for any subset A of S*

$$r_1(A) + r_2(S \setminus A) \geq k .$$

A curiosity is that the problem of deciding whether three or more matroids have a common independent set of size k is NP-*hard* (see chapter 29) and appears to be hopelessly difficult.

Both 9.3 and 9.6 can be easily deduced from, or lead to a very nice result of Rado (1942) which generalises in a very natural way the classic Hall theorem (see chapter 3). This can be stated as follows.

Rado–Hall theorem 9.7. *If M is a matroid on S and $\mathbf{A} = (A_i\colon i \in I)$ is a family of subsets of S then $\mathbf{A}$ has a transversal which is independent in M if and only if for all $J \subseteq I$*

$$r(A(J)) \geq |J| .$$

For a proof of this, an account of its many applications in transversal theory and its relation with Theorems 9.3–9.7 we refer to the text by Mirsky (1971).

10. Linear representability

One of the oldest problems in matroid theory is the question of finding necessary and sufficient criteria for a given matroid to be embedded in a vector space over a given field.

There certainly exist matroids which are not representable over any field.

A very simple example of such a matroid is the following.

Example. Consider the rank 3 matroid M on the 14 elements $\{a, b, c, d, e, f, g, a', b', c', d', e', f', g'\}$ which has a Euclidean representation of the form of fig. 10.1. M is just $F_7 \oplus F_7^0$ where F_7^0 is the Fano matroid "less a line". F_7 is only representable over fields of characteristic 2, F_7^0 is only representable over fields

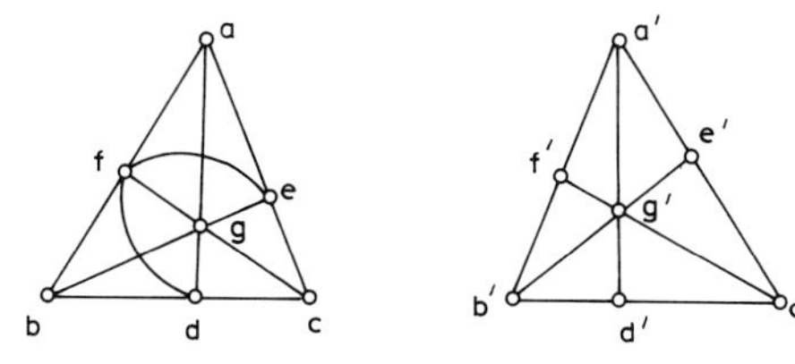

Figure 10.1.

not of characteristic 2. As a result this direct sum is not representable over any field.

Similar examples of any rank can be constructed by taking the direct sum of projective spaces,

$$M = \mathrm{PG}(n, q) \oplus \mathrm{PG}(m, q'),$$

where q and q' are powers of different primes.

However, in one sense these are unconvincing examples since the lack of connectivity is crucial for the non-representability. For example suppose we say that a flat Z is *modular* in M if

$$r(Z) + r(Y) = r(Z \cup Y) + r(Z \cap Y)$$

for all flats Y of M.

Call a matroid *modular* if every flat is modular.

Examples of modular matroids are the free matroids and the projective geometries, and any matroid obtained from these by taking direct sums.

It turns out (and this is really a theorem about lattices, see Birkhoff (1967, pp. 90–93) that this is in fact a complete characterisation: more precisely we have the following.

Theorem 10.1. *A connected simple matroid is modular if and only if it is either a free matroid or a projective space.*

Corollary 10.2. *A connected modular matroid is representable.*

For more on these sort of questions we refer to chapter 13.

The first "interesting" non trivial example of a non-representable matroid was found by Maclane (1936) who noticed that the nine element example whose Euclidean representation is given by fig. 10.2 is only representable over a field if the points a, b, c are collinear. That is $\{a, b, c\}$ has rank 2 rather than rank 3 as in fig. 10.2. This is because the classic theorem of Pappus demands that in such a geometric configuration of points coordinatisable over a field these points of intersection a, b and c are collinear.

For some time the above "*non-Pappus*" *matroid* was believed to be the

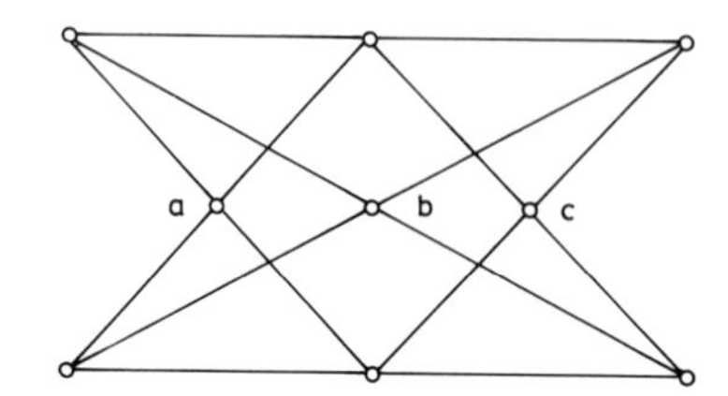

Figure 10.2. The non-Pappus configuration.

smallest non-representable matroid. This turns out to be false as is shown by the following example of Vamos (1971).

Example 10.3. V_8 is a self-dual matroid on 8 elements which is not representable over any field, see fig. 10.3.

We now list the basic properties of representability, for proofs see Welsh (1976) or Oxley (1992).

10.4. *If M is representable over F then so is any minor of M.*

10.5. *If M is representable over F so is its dual M^*, moreover, if the representation of M is by the columns of the r by n matrix* ($r = r(M)$, $n = S(M)$)

$$M \simeq [I_r, A]$$

then M^* is representable by the columns of the matrix

$$M^* \simeq [-A^{\mathrm{T}}, I_{n-r}] .$$

10.6. *If M is representable over F then any matroid induced from M is representable over some extension of F.*

10.7. *If M_1 and M_2 are representable over F then $M_1 \vee M_2$ is representable over any sufficiently large extension of F. (In other words it is the size of the extension not its algebraic structure which matters.)*

Turning now to the representability of classes of matroids, we first have the following.

Proposition 10.8. *Graphic matroids are representable over every field and hence by duality and* (10.5) *so are cographic matroids.*

Matroids which are representable over *every* field form an important subclass of matroids. They are known as *regular* matroids and a complete characterisation (or more precisely) synthesis of these matroids is contained in a remarkable theorem of Seymour (1980). This shows that any regular matroid is effectively obtained by composing graphic matroids, cographic matroids and one special 10-element

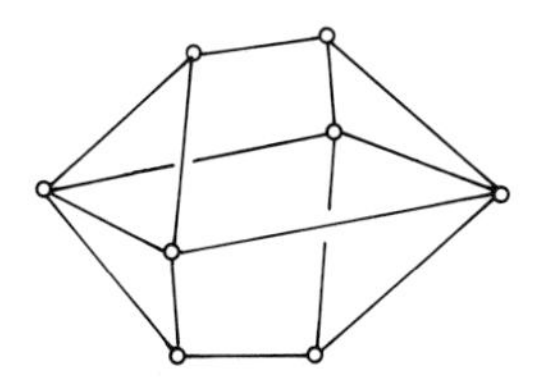

Figure 10.3.

matroid. For more on this and a characterisation of regular matroids in terms of their forbidden minors we refer to chapter 10.

10.9. *Transversal matroids, being the union of matroids of rank one, are representable over all sufficiently large fields.*

As far as representability over specific fields goes only a few cases have been settled.

A matroid is *binary* if it can be represented over the field GF(2). The smallest non-binary matroid is the uniform matroid U_4^2. There are many good characterisations of this property which we illustrate in the following statement.

Theorem 10.10. *The following statements about a matroid are equivalent.*

(a) *M is binary.*

(b) *M has no minor isomorphic to U_4^2.*

(c) *The symmetric difference of any two circuits is the union of disjoint circuits.*

The only other field for which a complete characterisation of representability is known is the ternary case. This is a deep result (see chapter 10).

Finally we briefly consider the *characteristic set* problem. Let P be the set of all primes together with 0. Given a matroid M let $C(M)$ be the set of integers $p \in P$ such that M is representable over a field of characteristic p. $C(M)$ is the *characteristic set* of M and K is a *characteristic set* if $K = C(M)$ for some matroid M. The problem is to characterise the subsets K of P which are characteristic sets of a matroid.

Clearly from our earlier results we know the following.

10.11. *$\emptyset$ and P are characteristic sets, as is $\{p\}$ for any prime $p \in P$.*

An early result of Rado (1957) subsequently extended by Vamos (1971) is the following.

10.12. *If $0 \in C(M)$ then all sufficiently large primes belong to P and conversely if $0 \not\in C(M)$ then $C(M)$ is a finite set.*

In other words, the following holds.

10.13. *If K is a characteristic set then either*

(i) $0 \in K$, $|P \backslash K| < \infty$, *or*

(ii) $0 \not\in K$, $|K| < \infty$.

Reid (1970) showed that all sets satisfying (i) are characteristic sets. He also produced some exotic finite sets of primes which are characteristic sets, for example $\{1103, 2089\}$, for details see Brylawski and Kelly (1980). A definitive solution to the characteristic set problem was completed when Kahn (1982) proved the following.

10.14. *All finite subsets of* $P \backslash \{0\}$ *are characteristic sets.*

Combining all these results we obtain the complete answer to the characteristic set problem.

Theorem 10.15. *K is a characteristic set if and only if it satisfies* (i) *or* (ii) *of* (10.13).

11. Algebraic matroids

The idea of treating algebraic dependence over a field axiomatically is implicit in the treatment by Van der Waerden (1937) in his classic text "Moderne Algebra".

First recall the definitions. Let K be an extension field of the field F. Elements $e_1, \ldots, e_n$ of K are *algebraically dependent* over F when there is a non-zero polynomial $p(x_1, \ldots, x_n)$ with coefficients from F such that $p(e_1, \ldots, e_n) = 0$. The collection of algebraically independent subsets of a subset S of K are the independent sets of a matroid on S and any matroid isomorphic to one which is obtained in this way we say is *algebraic over F*.

A matroid is *algebraic* if it is algebraic over some field, and the sort of questions which are interesting are similar to those asked about linear representability, namely: find conditions for M to be algebraic over a specific field or over some field. What is the "algebraic characteristic set" of a given matroid and the like.

Some progress was made on these in the early seventies by Piff (1972), Ingleton, and Main (1975) and more recently by Lindström (1985–1988) and Dress and Lovász (1987).

First we clarify the relation with linear representability by stating some of the basic results.

11.1. *If M is linear over F then M is algebraic over F.*

The converse of this is not true; the non-Fano matroid is algebraic over GF(2) but is not linearly representable over a field of characteristic 2.

11.2. *If M is algebraic over F then any minor of M and any truncation of M is algebraic over some extension of F.*

11.3. *If M on S is algebraic over F and $f: S \to T$ then $f(M)$ is algebraic over F.*

This leads to easy proofs of the following results

11.4. *If M_i: $1 \leq i \leq n$, are algebraic over F then so is $M_1 \vee \cdots \vee M_n$.*

11.5. *Transversal matroids are algebraic over every field.*

A recent result of Lindström (1985b) is the following.

11.6. *If M is linearly representable over the rationals then M is algebraic over every field.*

From this the following can be shown.

11.7. *For any field F there is no finite set of excluded minors for algebraic representability over F.*

Another interesting contrast with the situation vis a vis linear representability as given is that algebraic characteristic sets not containing 0 are not forced to be finite.

Example (Lindström 1986). The non-Pappus matroid is algebraic over GF(p^2) for every prime p but it is not algebraic over any field of characteristic zero.

A recent review paper by Lindström (1988) gives a very good account of the many new results obtained about possible algebraic characteristic sets and algebraic matroids. Whether or not any fixed finite set of primes can be algebraic characteristic set of a matroid seems to be a difficult problem.

We close this brief account by restating a problem first posed in Welsh (1971) and on which there seems to have been only limited progress.

Problem. If M is algebraic is its dual M^* also algebraic?

A more specific form of this to which it may be easier to find a counterexample is as follows.

Problem. If M is algebraic over F is M^* algebraic over F?

One of the difficulties in this sort of problem is that there are not "large stocks" of non-algebraic matroids. Ingleton and Main (1975) were the first to exhibit a non-algebraic matroid by showing that the rank 4 matroid V_8 of section 10 was not algebraic and Lindström (1985a) shows that the non-Desargues matroid consisting of the standard Desargues configuration with one line missing is not algebraic. He has also constructed an infinite family of such non-algebraic matroids. For more on this see Oxley (1992).

12. Structural properties

There are a rich variety of constructions which give a deeper understanding of the underlying structure of a matroid. At this level of exposition it is impossible to do more than uncover the tip of an iceberg by mentioning a few of the more important constructions and problems.

Strong maps

The concept of a linear mapping is so important in linear algebra that it is natural to try and develop an analogue in a matroid setting. This is achieved by defining a *strong map* between two geometric lattices $\mathscr{L}_1$ and $\mathscr{L}_2$ to be a function $f: \mathscr{L}_1 \to \mathscr{L}_2$ satisfying

(a) $a, b \in \mathscr{L}_1 \Rightarrow f(a \vee b) = f(a) \vee f(b)$,
(b) if x is an atom of $\mathscr{L}_1$, $f(x)$ is either an atom or the zero element of $\mathscr{L}_2$,
(c) $f(0) = 0$.

Although this definition can be formulated in terms of matroids it is less complicated to describe it lattice-theoretically as above. Geometric lattices and strong maps form a category and it seems clear that this appears to be the most natural category of maps to use.

Some elegant factorisation theorems for strong maps exist and a fairly rich theory has been developed, see for example Kung (1986). However, it should be emphasised that if $\mathscr{L}(M_2)$ is the image of $\mathscr{L}(M_1)$ under a strong map very few properties of M_1 are shared by M_2. This is because any simple matroid M on a set S is the image of the free matroid under the canonical map ϕ defined by

$$\phi(x) = x \quad x \in S ,$$

$$\phi(A) = \sigma\Big(\bigcup x : x \in A\Big) \quad A \subseteq S ,$$

where σ is the closure function of M.

Extensions and lifts

A matroid N is an *extension* of M on S if N restricted to S is M. It is a *single element extension* if N has ground set $S \cup e$, $e \notin S$.

Two very trivial ways of extending M are by adding either a loop or an element in parallel with an existing element of M. Almost, but not quite, as trivial is to add a coloop, that is an element so that $r(N) = r(M) + 1$.

We call an extension non-trivial if it is not of any of these three types, and henceforth only consider such extensions.

A *lift* of M is a matroid N such that N^* is an extension of M^*.

Geometrically we think of M as being obtained from a single element lift N by projection.

Quotients

If M and Q are matroids which share a common groundset, Q is a *quotient* of M if every flat of Q is also a flat of M. It is clear that the identity map on the groundset induces a strong map between M and Q. Conversely any strong map between lattices whose atoms are in one to one correspondence induces a quotient between the corresponding matroids. When $r(Q) = r(M) - 1$, Q is an *elementary quotient* of M.

As might be expected, elementary quotients are intimately related to single element extensions.

12.1. *If M, Q have groundset S, Q is an elementary quotient of M iff there exists a single element extension N of M on say $T = S \cup e$ such that $N \cdot S = Q$. In other words we have*

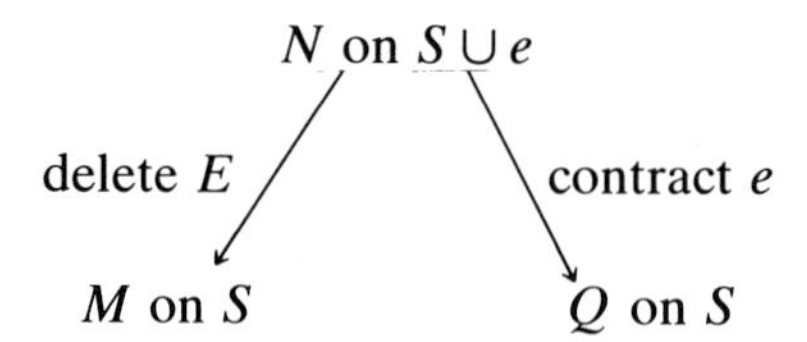

More generally, we have the following.

12.2. *If $r(M) = r(Q) + k$ and E is disjoint from S with $|E| = k$ then Q is a quotient of M if and only if there exists an extension N of M by E such that E is independent in N and*

$$N | S = M , \qquad N \cdot S = Q .$$

The Dilworth truncation of a matroid

A more interesting construction is what is known as the *Dilworth truncation* $D_k(M)$ of M. This is the matroid on the set $\mathscr{F}^k$ of k-flats of M and which has independent sets defined to be

$$\mathscr{I} = \left\{ I: r\left(\bigcup_{i \in J} F_i \right) \geq |J| + k - 1 \ \forall J \subseteq I \right\} .$$

In other words, $D_k(M)$ is the matroid induced on the set $\mathscr{F}^k$ by the submodular function $\mu : \mathscr{F}^k \to \mathbb{Z}^+$ defined by

$$\mu(J) = r\left(\bigcup F_i \colon i \in J \right) - k + 1 .$$

Somewhat surprisingly the Dilworth truncation of a matroid inherits many of the properties of its underlying matroid. For example Mason (1977) shows the following.

12.3. *M is representable over a field of characteristic p if and only if $D_k(M)$ is representable over the same characteristic.*

12.4. *If M is transversal then so is $D_k(M)$.*

For a more detailed and comprehensive account of these and other constructions we refer to the article of Kung (1986). We close with a brief discussion of one of the most intriguing structural problems in matroid theory.

The *Whitney number* W_k of a matroid M is defined to be the number of k-flats of M. Possibly inspired by the intuition that geometric lattices are "eggshaped" when regarded as posets, a long standing conjecture about matroids is that for any matroid the sequence $(W_k\colon 1 \leq k \leq r(M))$ is unimodal. This is certainly true for the cases where the numbers are known explicitly, projective and affine geometries, boolean algebras and the like. Stronger forms of the unimodal conjecture are that the sequence W_k is log concave, that is

$$W_k^2 \geq W_{k-1}W_{k+1}\,, \quad 2 \leq k \leq r-1$$

or even that the ratio $W_k^2/W_{k-1}W_{k+1}$ is a minimum when the underlying matroid is free. For more on this for cases where the W_k are known explicitly see chapter 31.

Not much progress has been made on any of these problems over the last 15 years apart from a very beautiful paper by Dowling and Wilson (1975) who gave a very elegant proof of the result that for any simple M

$$W_1 + W_2 + \cdots + W_k \leq W_{r-k} + \cdots + W_{r-2} + W_{r-1}\,, \tag{12.5}$$

where r is the rank of M and $1 \leq k \leq r-1$.

13. Colourings, flows and the critical problem

The chromatic polynomial of a graph is a very familiar concept in graph theory. Indeed, it has become so familiar that one tends to overlook the significance of the fact that, if we calculate the chromatic polynomial of a given graph by the well-known technique of successively contracting and deleting edges, then *the resulting polynomial is independent of the order in which we consider the edges.*

This basic observation is the key to the Tutte–Grothendieck theory developed by Brylawski (1972) and having applications in fields as diverse as percolation theory and codes.

Brylawski's idea was to define a "Tutte–Grothendieck" invariant for matroids as follows. A *matroid invariant* is a function f from the set of matroids to a commutative ring, such that, if M is isomorphic to N, then $f(M) = f(N)$. It is a *Tutte–Grothendieck invariant* if, in addition, it satisfies the following two conditions: for any e which is not a loop or coloop, then

(i) $f(M) = f(M \backslash e) + f(m/e)$

(ii) $f(M_1 \oplus M_2) = f(M_1)f(M_2)$.

It is easy to check that examples of Tutte–Grothendieck invariants are the numbers of bases of M, the number of independent sets of M, and the number of spanning sets of M.

Brylawski's main result is the following.

Theorem 13.1. *Any Tutte–Grothendieck invariant is uniquely determined by its values on the two single-element matroids consisting of a loop or coloop. More precisely, there exists a polynomial $T(M; x, y)$ in two variables x, y such that, if f is a Tutte–Grothendieck invariant, and if $f(M_0) = x$ and $f(M_0^*) = y$ where M_0 is the single-element matroid of rank* 1, *then $f(M) = T(M; x, y)$.*

The polynomial $T(M; x, y)$ is known as the *Tutte polynomial* of the matroid M. It is most easily defined by its relation to the (Whitney) *rank generating function* $R(M; x, y)$ which is given by

$$T(M; x, y) = R(M; x-1, y-1) , \qquad (13.2)$$

where R is defined by

$$R(M; x, y) = \sum_{A \subseteq S} x^{r(S)-r(A)} y^{|A|-r(A)} . \qquad (13.3)$$

Example. The reader is invited to verify that if M is the cycle matroid of the graph G of fig. 13.1 then by successively deleting and contracting the elements, the Tutte polynomial of M is given by

$$T(M; x, y) = x^3 + 2x^2 + 2xy + y^2 + x + y .$$

The remarkable thing is that, no matter in what order we carry out our contractions and deletions, we always get the same polynomial. Furthermore, any function which can be calculated by this contraction–deletion method must be an evaluation of this polynomial for suitable x and y. For example, $b(M)$, the number of bases of M, is given by $b(M) = T(M; 1, 1)$, since the value of b on a coloop and a loop is in both cases 1. Similarly, $i(M)$, the number of independent sets of M, is given by $i(M) = T(M; 2, 1)$, since a coloop has two independent sets whereas a loop has only one.

The prototype Tutte–Grothendieck invariant is the chromatic polynomial of a

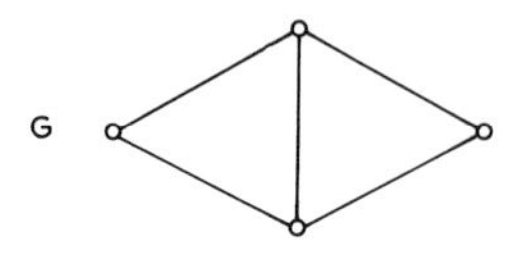

Figure 13.1.

graph, see chapter 1. If we define the *chromatic* or *characteristic polynomial* of a matroid M by

$$P(M; \lambda) = (-1)^{r(M)} T(M; 1-\lambda, 0) , \qquad (13.4)$$

where M is the cycle matroid of a graph G, the chromatic polynomial $P(G; \lambda)$ of the graph G is given by the equation

$$\lambda^{k(G)} P(M(G); \lambda) = P(G; \lambda) , \qquad (13.5)$$

where $k(G)$ is the number of connected components of G. Incidentally, this shows that two graphs with isomorphic cycle matroids must have the same chromatic number.

Let G be a connected graph having no isthmuses and suppose that the edges of G are oriented, giving a directed graph $\vec{G}$. For any finite abelian group H an *H-flow* on $\vec{G}$ is a map $\phi : E(G) \to H$ such that Kirchhoff's conservation laws are obeyed at each vertex of G. Clearly whether or not ϕ is an H-flow depends on the orientation given to G. A *nowhere-zero* H-flow is one which assigns a non-zero element of H to each edge of G.

A crucial observation is that for any abelian group H, if $F(G; H)$ denotes the number of nowhere-zero H-flows on G then

$$F(G; H) = F(G/e; H) - F(G\backslash e; H)$$

for any $e \in G$ which is not a loop or coloop. Thus F is essentially a Tutte–Grothendieck invariant of $M(G)$ and we have the somewhat surprising consequence.

Proposition 13.6. *The number of nowhere-zero H-flows on G is given by $T(M(G); 0, 1-\lambda)$ where λ is the order of the abelian group H.*

Thus we may sensibly speak of G possessing a *nowhere-zero n-flow* to mean that G has a nowhere-zero H-flow for some (and hence every) abelian group H of order n.

Two longstanding conjectures of Tutte (1966) are the following.

Conjecture 13.7 (*Tutte's 5-flow conjecture*)**.** Every bridgeless graph has a 5-flow.

Conjecture 13.8 (*Tutte's 4-flow conjecture*)**.** A cubic bridgeless graph has a 4-flow if it has no subgraph contractible to the Petersen graph P_{10}.

Both of these conjectures have attracted a lot of attention over the years. The 4-flow conjecture is intimately related to edge colourings, while Seymour (1981a) has improved an earlier result of Jaeger (1976) by showing that every bridgeless graph possesses a 6-flow. For more on these problems we refer to chapters 1 and 4.

For a general matroid M which is representable over a finite field of order q the chromatic polynomial has an intriguing interpretation in terms of a fundamental blocking problem of projective geometry. This goes back at least as far as Veblen (1912).

Consider the projective space $PG(r-1, q)$. For any positive integer t, a *t-block* in $PG(r-1, q)$ is a set X of points such that $X \cap F \neq \emptyset$ for each flat F of rank $r-t$. In particular, the 1-blocks are the sets which have non-empty intersection with every hyperplane of $PG(r-1, q)$. We call X a *minimal t-block* if X is a t-block but $X \backslash p$ is not a t-block for each $p \in X$.

Standard vector space arguments can be used to show that the projective space $PG(t, q)$ as a subspace of $PG(r-1, q)$ is a minimal t-block for each integer $t \geqslant 2$ and each prime power q. The following remarkable theorem gives the relationship between blocking sets and the chromatic polynomial of the underlying matroid.

Theorem 13.9. *Let M be a matroid of rank r on S which is embedded in* $PG(r-1, q)$. *Then S is a t-block if and only if $P(M; q^t) = 0$. More generally, $P(M; q^t)$ enumerates the collection of hyperplanes whose intersection is an $(r-t)$-flat avoiding S.*

For a class of matroids which are representable over $GF(q)$ Crapo and Rota (1970) define $c(M; q)$ to be the minimum t such that $P(M; q^t) > 0$. Determining $c(M; q)$ is called the *critical problem* of combinatorial geometry.

Note also that Theorem 13.9 is a justification for the slight abuse of language in the statement "M is a t-block". What we mean is that *any* of the various vector representations of M in $PG(r-1, q)$ is a t-block; in other words, it is not their coordinatisation which is important, but their geometrical structure.

As an example, consider the Fano matroid F_7. Since

$$P(F_7; \lambda) = (\lambda - 1)(\lambda - 2)(\lambda - 4),$$

F_7 is a 2-block over $GF(2)$ and a 1-block over $GF(4)$. More generally we have the following result.

Theorem 13.10. *For any prime p, a t-block over* $GF(p)$ *is a* 1*-block over* $GF(p^t)$ *for each positive integer t.*

The converse of Theorem 13.10 is not always true since there exist 1-blocks over $GF(p^t)$ which, are not representable over $GF(p)$. However, if a 1-block over $GF(p^t)$ is representable over $GF(p)$, then it is a t-block over $GF(p)$. Similarly, since the uniform matroid U_n^2 has chromatic polynomial $\lambda^2 - n\lambda + n - 1$, we know that the $(q+1)$-point line is a 1-block over the field $GF(q)$ for each prime power q.

For complete graphs K_n, we have

$$P(M(K_n); \lambda) = (\lambda - 1)(\lambda - 2) \cdots (\lambda - n + 1).$$

It follows that for any prime power $q = p^t$, the cycle matroid of K_{q+1} is a

1-block over GF(q), and is thus also a t-block over GF(p). Now the cycle matroid $M(G)$ of any graph G is representable over every field, and so $M(G)$ is a t-block over GF(q) if and only if G has chromatic number $\chi(G) > q^t$. Thus the graphic 1-blocks over GF(2) are just the cycle matroids of non-bipartite graphs.

When we come to 2-blocks over GF(2), however, the situation is much more complicated. Since K_5 is a minimal graph which is not 4-colourable we deduce that $M(K_5)$ is a minimal 2-block. However, there are many other minimal 2-blocks, since if G is any edge-critical graph for which $\chi(G) = 5$, but $\chi(G \backslash e) = 4$ for each edge e, then $M(G)$ is a minimal 2-block.

If we let P_{10} denote the Petersen graph then Tutte (1966) noticed that each of the matroids $M(K_5)$, F_7 and $M^*(P_{10})$ has the additional property that no loop-free minor is also a 2-block. A 2-block with this property is called a *tangential 2-block*. In 1966, Tutte proved that these three matroids are the only tangential 2-blocks with rank at most 6, and Datta (1976) used a complicated geometrical argument to show that there is no tangential 2-block of rank 7.

Tutte's tangential 2-block conjecture, originally made in 1966 and still unsettled, can be stated in the following form.

Conjecture 13.11 (*Tutte's tangential block conjecture*). The only tangential 2-blocks are $M(K_5)$, F_7 and $M^*(P_{10})$.

Seymour's theory of splitters discussed in chapter 10 is a major step towards proving this conjecture. It is related to Hadwiger's conjecture which, in its full form, reads as follows.

Conjecture 13.12 (*Hadwiger*). If a loopless graph G is not n-colourable it contains a subgraph contractible to K_{n+1}.

Dirac (1957) showed that Hadwiger's conjecture is true for $n = 3$, and Wagner (1964) showed that for $n = 4$ it was equivalent to the four-colour conjecture, and thus it holds for $n = 4$. Thus we know that there can be no new tangential 2-block which is the cycle matroid of a graph. Seymour (1980) used his characterisation of regular matroids to prove the following striking result.

Theorem 13.13. *Any new tangential 2-block must be the cocycle matroid of a graph.*

Seymour's result shows that Tutte's tangential 2-block conjecture is exactly equivalent to the 4-flow conjecture 13.8. Thus he has reduced this seemingly intractable geometrical problem to the conceptually much simpler problem of characterising those graphs which have no 4-flow. More precisely, there exists a tangential 2-block other than $M(K_5)$, F_7 and $M^*(P_{10})$ if and only if there is a bridgeless graph G which does not contain a subgraph contractible to the Petersen graph and which has no 4-flow.

There are many other applications of these polynomials to diverse fields including coding theory (Greene 1976), arrangements of hyperplanes in space (Zaslavsky 1975), percolation and reliability (Oxley and Welsh 1979) and more recently to knot polynomials (Jaeger et al. 1990). For a much fuller discussion we refer to the review article of Brylawski and Oxley (1992).

14. Varieties and universal models

Apart from Seymour's theory of splitters discussed in chapter 10 one of the two major developments in the subject over the last decade stems from the paper of Kahn and Kung (1982) and can be summarised as follows.

A *hereditary class* of geometries (simple matroids) is a collection of geometries which is closed under taking minors and direct sums. Thus they are the analogues of varieties in universal algebra. A *sequence of universal* models for a hereditary class $\mathscr{H}$ is a sequence (T_n) of geometries in $\mathscr{H}$ with $r(T_n) = n$ and satisfying the following property.

> If G is a geometry in $\mathscr{H}$ of rank n then G is a subgeometry (restriction minor) of T_n. (14.1)

A *variety* of geometries is a hereditary class with a sequence of universal models. The achievement of Kahn and Kung was to give a complete characterisation of the possible varieties.

First some examples.

Example 14.2. *The collection of geometries coordinatisable over a fixed finite field is a variety. The universal models are the projective geometries* PG(n, q).

Example 14.3. *The collection of graphic matroids is a variety, the universal models are the cycle matroids of the complete graphs.*

Example 14.4. *The trivial variety consisting of the free geometries, has universal models the Boolean algebras.*

Two other simple varieties are the following:

(i) Let M_1 be the rank one geometry and M_2 be the line with $q+1$ points. Let $M_{2n}(q) = M_2 \oplus \cdots \oplus M_2$ and $M_{2n+1}(q) = M_2 \oplus \cdots \oplus M_2 \oplus M_1$. The subgeometries of these geometries form a variety called the *variety of matchstick geometries of order* q.

(ii) If B_n is the Boolean algebra on $\{1, \ldots, n\}$ and on each of the lines $\overline{12}$, $\overline{23}, \ldots, \overline{(n-1)n}$ we add $q-1$ points in general position let $o_n(q)$ denote the resulting geometry. Then subgeometries of these geometries form a variety called the variety of *origami geometries of order* q.

(iii) *A partial partition* of $N = \{1, \ldots, n\}$ is a collection of non-empty, disjoint subsets of N. Ordered by refinement the partial partitions of N form a geometric lattice which is isomorphic to the lattice of flats of the cycle matroid $M(K_n)$ of the complete graph. Now take finite group A and define a *partial A-partition* of N to be a partial partition π of N together with a labelling of the elements of each block of π with elements from A. The partial partitions can be equated in a "projective manner" and if the resulting equivalence classes are ordered by refinement we again get a geometric lattice of rank n and which is denoted by $Q_n(A)$. These lattices were discovered by Dowling (1973), they give a large stock of non-representable matroids since he showed that $Q_n(A)$ is representable if and only if A is a subgroup of the group of units of a finite field. Moreover, $Q_n(A) \simeq Q_n(A')$ iff A and A' are isomorphic. These lattices known as *Dowling lattices* were demystified to some extent in a series of papers by T. Zaslavsky on voltage-graphic matroids (see for example Whittle 1989). What is surprising about Dowling geometries is their prominence in Kahn and Kung's major classification theorem.

Theorem 14.5. *Let $\mathscr{H}$ be a variety of geometries. Then $\mathscr{H}$ is one of the following collections*:

(1) *the variety of free geometries*;
(2) *the variety of matchstick geometries of order q*;
(3) *the variety of origami geometries of order q*;
(4) *the variety of geometries coordinatisable over* GF(q);
(5) *the variety of subgeometries of the Dowling geometry $Q_n(A)$ for some fixed finite group A.*

One consequence of Theorem 14.5 is the following statement which seems hard to prove directly.

Corollary 14.6. *The only hereditary classes which are also closed under the taking of duals are the varieties of geometries coordinatisable over a fixed finite field.*

Theorem 14.5 has led to renewed interest in the voltage graphic geometries of Zaslavsky and their universal model, the Dowling geometries. As an example of another extremely beautiful and surprising theory prompted by the paper of Kahn and Kung, Whittle (1989) has completely extended the theory of the critical problem for projective spaces to the Dowling geometries $Q_n(A)$ for any finite group A.

15. Oriented matroids

A feature of graphs which is difficult to capture in a matroid is the notion of direction. Minty (1966) seems to have been the first to attempt to extend the notion of orientability from graphs to matroids but it turned out that his notion of orientability was exactly the concept of regularity discussed in chapter 10. A more general and interesting concept of orientability which extended that of Minty was proposed by Rockafellar (1969). Rockafellar's ideas were motivated more by considerations from convexity and linear programming and have certainly had a strong influence on the theory of oriented matroids as it exists today.

In what follows I will attempt to give some idea of the different approaches which led to the independent discovery of oriented matroids in the middle seventies by Las Vergnas (1975), Bland (1977), Folkman and Lawrence (1978).

We start first with the graphically motivated approach proposed in Bland and Las Vergnas (1978).

Given a matroid M on S an *orientation* of M is the partitioning of each circuit C into a *positive* part C^+ and a negative part C^-, and we say x is *positive* (*negative*) in C if $x \in C^+$ (respectively C^-). The *sign* function is defined for any circuit C and $x \in C$ by

$$\operatorname{sign}_c(x) = \begin{cases} + & x \in C^+ , \\ - & x \in C^- . \end{cases}$$

Given three circuits C_1, C_2, C_3 with $C_3 \subseteq C_1 \cup C_2$ we say C_3 satisfies the *rule of signs* with respect to C_1 and C_2 if the following conditions hold:

$$x \in C_3 \cap (C_1 \backslash C_2) \Rightarrow \operatorname{sign}_{c_3}(x) = \operatorname{sign}_{c_1}(X) , \tag{a}$$

$$x \in C_3 \cup (C_2 \backslash C_1) \Rightarrow \operatorname{sign}_{c_3}(x) = \operatorname{sign}_{c_2}(x) , \tag{b}$$

$$\begin{aligned} &x \in C_3 \cap C_1 \cap C_2 \quad \text{and} \quad \operatorname{sign}_{c_1}(x) = \operatorname{sign}_{c_2}(x) \\ &\quad \Rightarrow \operatorname{sign}_{c_3}(x) = \operatorname{sign}_{c_1}(x) = \operatorname{sign}_{c_2}(x) . \end{aligned} \tag{c}$$

Now it is easy to see that given any graphic matroid all the rule of signs means is that when the graph's edges are directed the resulting orientation satisfies the rule of signs.

However, more than this is needed. An orientation of a matroid M is *compatible* with its structure if for any pair of circuits C, D, and $x \in C \cap D$ such that

$$\operatorname{sign}_C(x) = -\operatorname{sign}_D(x)$$

and $y \in D \backslash C$ then there exists a circuit E such that E obeys the rule of signs with respect to C and D and $y \in E \subseteq (C \cup D) \backslash x$.

Finally we say M is *orientable* if there exists some orientation which is compatible with its structure.

As should be expected from the origin of the definition we have the following basic result.

Proposition 15.1. *Every graphic matroid is orientable.*

Proof. Use any orientation of the associated graph. □

Less obvious is the following.

Proposition 15.2. *A binary matroid is orientable if and only if it is regular.*

Proposition 15.3. *If M is orientable so is its dual M^* and so is any minor of M.*

These results 15.2 and 15.3 become more apparent if one takes the alternative approach to orientability proposed by Folkman and Lawrence (1978). This is motivated more by ideas from convexity. Their definition of an oriented matroid (which turns out to be cryptomorphic with that of Bland and Las Vergnas treated above) is as follows.

An *involution* on a set E is a function $*: E \to E$ such that $(x^*)^* = x$ for each element $x \in E$; it is *fixed point free* if $x^* \neq x$ for each $x \in E$.

An *oriented matroid* is a triple $(E, \mathscr{C}, *)$ where E is a finite set, $\mathscr{C}$ is a collection of non-empty subsets of E and $*$ is a fixed point free involution on E such that

(a) $\mathscr{C}$ is a *clutter* (that is no member properly contains another);
(b) If $A \in \mathscr{C}$ then $A^* \in \mathscr{C}$ and $A \cap A^* = \emptyset$;
(c) If $A, B \in \mathscr{C}$ and $x \in A \cap B^*$ and $A \neq B^*$ then there exists $C \in \mathscr{C}$ with

$$C \subseteq (A \cup B) \backslash \{x, x^*\} .$$

The members of $\mathscr{C}$ are called the *circuits* of the oriented matroid.

Example. Suppose that E is a finite set of non-zero vectors in a vector space over an ordered field and $E = -E$.

For $x \in E$ let $x^* = -x$.

Let $\mathscr{C}$ be the minimal subsets C of E such that

(i) $C \cap C^* = \emptyset$,
(ii) there exist *positive* real numbers α_v $(v \in C)$ such that

$$\sum_{v \in C} \alpha_v v = 0 .$$

Then $(E, \mathscr{C}, *)$ is an oriented matroid. In other words $C \in \mathscr{C}$ if and only if it is the vertex set of a simplex containing the origin in its interior and is not of the form $\{u, -u\}$.

The connection between the two definitions of oriented matroid which we have given is seen from the following.

Consider an oriented matroid defined by a triple $(E, \mathscr{C}, *)$.

For each element $x \in E$ and subset $A \subseteq E$ define

$$\bar{x} = \{x, x^*\} , \qquad \bar{A} = \{\bar{x}: x \in A\} .$$

Let $\bar{\mathscr{C}} = \{\bar{C}: C \in \mathscr{C}\}$. Then we have the following.

Proposition 15.4. *If $(E, \mathscr{C}, *)$ is an oriented matroid, then the collection $\bar{\mathscr{C}}$ of subsets of $\bar{E}$ is the set of circuits of a matroid on $\bar{E}$.*

Conversely given a matroid M on S which is oriented it is an easy exercise to reverse the above process and construct a triple $(E, \mathscr{C}, *)$ satisfying the axioms (a)–(c).

Let $E = S \cup S'$ where S' is a disjoint copy of S. If $C = C^+ \cup C^-$ is an oriented circuit of M the corresponding circuit of $\mathscr{C}$ is in the obvious notation $C^+ \cup (C^-)'$.

A theory of duality and minors, closely analogous to the unoriented case exists for oriented matroids. However, as an indication of the diverse nature of orientability consider the following examples.

Proposition 15.5. *Any matroid which is linearly representable over an ordered field is orientable.*

Example 15.6. *The matroid V_8 described in section* 10 *is a matroid which is not coordinatisable over any field but it is orientable.*

Proposition 15.7. *There exists an infinite collection of non-orientable matroids, none of which is a minor of the other. Hence there can be no finite excluded minor criterion for orientability.*

Because the structure of oriented matroids retains many properties of vector spaces over ordered fields they are of considerable interest in the study of abstract ideas of linear programming and convex polyhedra, for further details we refer to the seminal papers of Bland and Las Vergnas (1978), Folkman and Lawrence (1978), Las Vergnas (1980), and especially to the recent monograph of Björner et al. (1993).

16. Extensions of matroids

Over the last twenty years the ideas and techniques of matroid theory have been generalised in several directions. In this final section we can do no more than give very brief pointers to these extensions.

Historically one of the earliest and certainly one of the most important is the theory of polymatroids introduced by Edmonds (1970). Polymatroids are convex polyhedra in $\mathbb{R}^n$ associated in a very natural way with submodular set functions. They are treated in greater detail in chapter 11.

An alternative generalisation of matroids introduced by Dunstan et al. (1972) are supermatroids which are defined as follows.

Let P be a finite partial order with a zero and let $\mathscr{I}$ be a collection of elements of P satisfying

(S1) $0 \in \mathscr{I}$,

(S2) if $x \in \mathscr{I}$ and $y \leqslant x$ in P then $y \in \mathscr{I}$,

(S3) for any element $a \in P$ if $P(a) = \{x: x \leqslant a\}$ then all maximal elements of $P(a) \cap \mathscr{I}$ have the same height in P.

The condition (S3) means that rank is well defined and when P is a distributive lattice it turns out to be a submodular function on the lattice. This leads to a generalisation of many combinatorial aspects of matroids, for example duality, a Rado–Hall theorem and the like.

Supermatroids are closely connected with F-geometries introduced by Faigle (1980) and to the notion of a greedoid introduced by Korte and Lovász (1981, 1983) and discussed below. For a discussion of their interrelationship and details of the first matroid intersection theorem for these classes we refer to Tardos (1990).

A *greedoid* $(S, \mathcal{G})$ is a set system such that

(G1) $\emptyset \in \mathcal{G}$,

(G2) $A \in \mathcal{G}$, $A \neq \emptyset \Rightarrow \exists x \in A$ such that $A \backslash \{x\} \in \mathcal{G}$,

(G3) $X, Y \in \mathcal{G}$, $|X| > |Y| \Rightarrow \exists x \in X \backslash Y$ such that $Y \cup \{x\} \in \mathcal{G}$.

This concept was introduced by Korte and Lovász (1981) as a characterisation of structures for which a greedy type algorithm "works" for a wide class of objective functions. As with matroids there are many different axiomatisations, the above is the most natural in the present setting but unlike matroids, greedoids have alternative formulations in terms of strings and sequences. As a class they strictly include matroids and can be characterised (in the obvious way) as set-systems having a rank function r satisfying

$$r(\emptyset) = 0\,, \qquad r(A) \leq |A|\,, \qquad X \subseteq Y \Rightarrow r(X) \leq r(Y)\,,$$

together with the following substantially weakened form of submodularity

$$r(X) = r(X \cup \{a\}) = r(X \cup \{b\}) \Rightarrow r(X \cup \{a\} \cup \{b\}) = r(X)\,. \tag{16.1}$$

A particularly important class of greedoid is that of *interval greedoids* which have the additional property

If A, B, $C \in \mathcal{G}$ with $A \subseteq B \subseteq C$ and $x \in S$ is such that $A \cup \{x\}$ and $C \cup \{x\} \in \mathcal{G}$ then $B \cup \{x\} \in \mathcal{G}$. (IG)

As an example of an interval greedoid which is not a matroid and which appears not to be in any of the earlier classes consider the following.

Example (*Directed branching greedoid*). Let G be a directed graph with $r \in V(G)$ a distinguished vertex to be called the *root*. If $\mathcal{G}$ is the set of arborescences in G rooted at r then $(E(G), \mathcal{G})$ is a greedoid.

This example highlights the difference between greedoids and matroids and at the same time gives the reader an understanding of the way greedoids might be defined in terms of ordered strings. For an account of the now quite well-developed theory of greedoids we refer to the monograph of Korte et al. (1991).

Greedoids include matroids and what are known as *antimatroids*. These are interval greedoids $(S, \mathcal{G})$ which also satisfy the condition that for all $A, B \in \mathcal{G}$

with $B \subseteq A$ and all $x \in S \backslash A$,

$$B \cup \{x\} \in \mathcal{G} \Rightarrow A \cup \{x\} \in \mathcal{G} .$$

Antimatroids were introduced by Edelman and Jamison (1985) as abstract models of convexity. The classic example of an antimatroid is the following.

Example. Let S be a finite subset of $\mathbb{R}^n$ and let $X = \{x_1, \ldots, x_k\} \in \mathcal{G}$ if x_i does not belong to the convex hull of $S \backslash \{x_1, \ldots, x_i\}$ for $1 \leq i \leq k$. Then $(S, \mathcal{G})$ is an antimatroid.

The above example motivates us to call the complements of members of $\mathcal{G}$ a *convex set* in the antimatroid $(S, \mathcal{G})$. Various combinatorial properties of convex sets can be studied in this more abstract framework. For example, define the *Helly number* of an antimatroid as the smallest integer h with the property (H):

> If $\mathcal{H}$ is a collection of convex sets such that every $\mathcal{H}' \subseteq \mathcal{H}$ with $|\mathcal{H}'| \leq h$ has a non-empty intersection then the members of $\mathcal{H}$ have a non-empty intersection. (H)

The following theorem characterises the Helly-number of an antimatroid.

Theorem 16.2. *The Helly number of an antimatroid* $(S, \mathcal{G})$ *is the maximum size of a set* $X \subseteq S$ *such that every subset of* X *is convex.*

For more details on this and related topics we refer to Edelman and Jamison (1985) or Korte et al. (1991).

Lastly we remark that throughout this article we have assumed the underlying groundset of M is finite. There are many different plausible extensions of the finite theory to the infinite case, we refer the reader to Oxley (1978) for an indication of possible approaches.

17. Conclusion

Over the last twenty years or so there has been a substantial increase in the applications and theory of matroids. In this brief account it has been impossible to do more than cover basic concepts and highlight some topics. Others will be treated in chapters 10 and 11, but a much fuller picture can be obtained from a wide range of texts devoted primarily to the subject, namely Crapo and Rota (1970), Tutte (1971), Welsh (1976), Bryant and Perfect (1980), Recski (1989),

Truemper (1992), the volumes edited by White (1986, 1987 and 1992) and especially the monograph of Oxley (1992) where a comprehensive treatment of the fundamentals of the subject can be found.

References

Birkhoff, G.
[1935] Abstract linear dependence in lattices, *Amer. J. Math.* **57**, 800–801.
[1967] *Lattice Theory,* 3rd Ed. (American Mathematical Society, Providence, RI).

Björner, A., M. Las Vergnas, B. Sturmfels, N. White and G. Ziegler
[1993] *Oriented Matroids* (Cambridge University Press, Cambridge).

Bland, R., and M. Las Vergnas
[1978] Orientability of matroids, *J. Combin. Theory B* **24**, 94–123.

Bland, R.G.
[1977] A combinatorial abstraction of linear programming, *J. Combin. Theory B* **23**, 33–57.

Bryant, V., and H. Perfect
[1980] *Independence Theory in Combinatorics* (Chapman and Hall, London).

Brylawski, T.H.
[1972] A decomposition for combinatorial geometries, *Trans. Amer. Math. Soc.* **171**, 235–282.

Brylawski, T.H., and D.G. Kelly
[1980] *Matroids and Combinatorial Geometries* (University of North Carolina Press, Chapel Hill, NC).

Brylawski, T.H., and J.G. Oxley
[1992] The Tutte polynomial and its applications, in: *Matroid Applications II,* ed. N. White (Cambridge University Press, Cambridge) pp. 123–225.

Crapo, H.H., and G.-C. Rota
[1970] *On The Foundations of Combinatorial Theory: Combinatorial Geometries* (MIT Press, Cambridge, MA).

Datta, B.T.
[1976] Nonexistence of six-dimensional tangential 2-blocks, *J. Combin. Theory B* **21**, 171–193.

Dilworth, R.P.
[1941a] Ideals in Birkhoff lattices, *Trans. Amer. Math. Soc.* **49**, 325–353.
[1941b] Arithmetic theory of Birkhoff lattices, *Duke Math. J.* **8**, 286–299.
[1944] Dependence relations in a semimodular lattice, *Duke Math. J.* **11**, 575–587.

Dirac, G.A.
[1957] A theorem of R.L. Brooks and a conjecture of H. Hadwiger, *Proc. London Math. Soc.* **7**, 161–195.

Dowling, T.A.
[1973] A class of geometric lattices based on finite groups, *J. Combin. Theory B* **14**, 61–86. Erratum: **15**, 211.

Dowling, T.A., and R.M. Wilson
[1975] Whitney number inequalities for geometric lattices, *Proc. Amer. Math. Soc.* **47**, 504–512.

Dress, A., and L. Lovász
[1987] On some combinatorial properties of algebraic matroids, *Combinatorica* **7**, 39–48.

Dunstan, F.D.J., A.W. Ingleton and D.J.A. Welsh
[1972] Supermatroids, in: *Combinatorics,* eds. D.J.A. Welsh and D.R. Woodall (Institute of Mathematics and its Applications, London) pp. 72–122.

Edelman, P.H., and R. Jamison
[1985] The theory of convex geometries, *Geom. Dedicata* **19**, 247–271.

Edmonds, J.
[1965] Minimum partition of a matroid into independent subsets, *J. Res. Nat. Bur. Standards B* **69**, 67–72.
[1970] Submodular functions, matroids, and certain polyhedra. In: *Combinatorial Structures and their Applications,* eds. R. Guv, H. Hanani, N. Sauer and J. Schonheim (Gordon and Breach, New York) pp. 69–87.

Edmonds, J.R., and D.R. Fulkerson
[1965] Transversals and matroid partition, *J. Res. Nat. Bur. Standards B* **69**, 147–153.
Faigle, U.
[1980] Geometries on partially ordered sets, *J. Combin. Theory B* **28**, 26–51.
Folkman, J., and J. Lawrence
[1978] Oriented matroids, *J. Combin. Theory B* **25**, 199–236.
Greene, C.
[1976] Weight enumeration and the geometry of linear codes, *Studia Appl. Math.* **55**, 119–128.
Ingleton, A.W., and R.A. Main
[1975] Non-algebraic matroids exist, *Bull. London Math. Soc.* **7**, 144–146.
Jaeger, F.
[1976] On nowhere-zero flows in multigraphs, in: *Proc. 5th British Combinatorial Conf.*, eds. C.St.J.A. Nash-Williams and J. Sheehan, *Congress. Numerantium* **XV**, 373–378.
Jaeger, F., D.L. Vertigan and D.J.A. Welsh
[1990] On the computational complexity of the Jones and Tutte polynomials, *Math. Proc. Cambridge Philos. Soc.* **108**, 35–53.
Kahn, J.
[1982] Characteristic sets of matroids, *J. London Math. Soc. (2)* **26**, 207–217.
Kahn, J., and J.P.S. Kung
[1982] Varieties of combinatorial geometries, *Trans. Amer. Math. Soc.* **271**, 485–499.
Korte, B., and L. Lovász
[1981] Mathematical properties underlying the greedy algorithm, in: *Fundamentals of Computational Theory, Lecture Notes in Computer Sciences*, Vol. 117, ed. F. Grecseg (Springer, New York) pp. 205–209.
[1983] Structural properties of greedoids, *Combinatorica* **3–4**, 359–374.
Korte, B., L. Lovász and R. Schrader
[1991] *Greedoids, Algorithms and Combinatorics* (Springer, Berlin).
Kung, J.P.S.
[1986] Strong maps, in: *Theory of Matroids*, Vol. 1, ed. N.L. White (Cambridge University Press, Cambridge) pp. 254–271.
Las Vergnas, M.
[1975] Matroides orientables, *C.R. Acad. Sci. Paris Ser. A* **280**, 61–64.
[1980] Convexity in oriented matroids, *J. Combin. Theory B* **29**, 231–241.
Lindström, B.
[1985a] A Desarguesian theorem for algebraic combinatorial geometries, *Combinatorica* **5**, 237–239.
[1985b] A class of algebraic matroids with simple characteristic set, *Proc. Amer. Math. Soc. (1)* **95**, 147–151.
[1986] The non-Pappus matroid is algebraic over any finite field, *Utilitas Math.* **30**, 53–55.
[1988] Matroids, algebraic and non-algebraic, in: *Algebraic, Extremal and Metric Combinatorics, London Mathematical Society Lecture Note Series*, Vol. 131, eds. M.-M. Deza, P. Frankl and I.G. Rosenberg (Cambridge University Press, Cambridge) pp. 166–174.
Maclane, S.
[1936] Some interpretations of abstract linear dependence in terms of projective geometry, *Amer. J. Math.* **58**, 236–240.
Mason, J.
[1977] Matroids as the study of geometrical configurations, in: *Higher Combinatorics*, ed. M. Aigner (Reidel, Dordrecht) pp. 133–176.
Minty, G.J.
[1966] On the axiomatic foundations of the theories of directed linear graphs, electrical networks and network programming, *J. Math. Mech.* **15**, 485–520.
Mirsky, L.
[1971] *Transversal Theory* (Academic Press, London).

Nash-Williams, C.St.J.A.
[1967] An application of matroids to graph theory, in: *Theory of Graphs, Proc. Int. Symp., Rome, 1966* ed. P. Rosenstiehl (Gordon and Breach, New York), pp. 263–265.
Oxley, J.G.
[1978] Infinite matroids, *Proc. London Math. Soc.* **83**, 123–130.
[1992] *Matroid Theory* (Oxford University Press, Oxford).
Oxley, J.G., and D.J.A. Welsh
[1979] The Tutte polynomial and percolation, in: *Graph Theory and Related Topics,* eds. J.A. Bondy and U.S.R. Murty (Academic Press, New York) pp. 329–339.
Piff, M.J.
[1972] *Some problems in combinatorial theory,* Ph.D. Thesis (Oxford).
Rado, R.
[1942] A theorem on independence relations, *Quart. J. Math. (Oxford) (2)* **13**, 83–89.
[1949] Axiomatic treatment of rank in infinite sets, *Canad. J. Math.* **1**, 337–343.
[1957] Note on independence functions, *Proc. London. Math. Soc. (3)* **7**, 300–320.
Recski, A.
[1989] *Matroid Theory and Applications in Electric Network Theory and Statics* (Springer, Berlin).
Reid, R.
[1970] *Obstructions to representations of combinatorial geometries,* Unpublished Manuscript.
Rockafellar, R.T.
[1969] The elementary vectors of a subspace of $\mathbb{R}^N$, in: *Combinatorial Mathematics and its Applications,* eds. R.C. Bose and T.A. Dowling (Chapel Hill, NC) pp. 104–127.
Seymour, P.D.
[1980] Decomposition of regular matroids, *J. Combin. Theory B* **28**, 305–359.
[1981a] Nowhere-zero 6-flows, *J. Combin. Theory B* **30**, 130–135.
[1981b] On Tutte's extension of the four-colour problem, *J. Combin. Theory B* **31**, 82–94.
Tardos, E.
[1990] An intersection theorem for supermatroids, *J. Combin. Theory B* **50**, 150–159.
Truemper, K.
[1992] *Matroid Decomposition* (Academic Press, Boston).
Tutte, W.T.
[1958] A homotopy theory for matroids I, II, *Trans. Amer. Math. Soc.* **88**, 144–174.
[1959] Matroids and graphs, *Trans. Amer. Math. Soc.* **90**, 527–552.
[1966] On the algebraic theory of graph coloring, *J. Combin. Theory* **1**, 15–50.
[1971] *Introduction to the Theory of Matroids* (American Elsevier, New York).
Vamos, P.
[1971] A necessary and sufficient condition for a matroid to be linear, in: *Conference on Möbius Algebras,* eds. H. Crapo and G. Roulet (University of Waterloo) pp. 162–169.
van der Waerden, B.L.
[1937] *Moderne Algebra,* 2nd Ed. (Springer, Berlin).
Veblen, O.
[1912] An application of modular equations in analysis situs, *Ann. of Math.* **14**, 86–94.
Wagner, K.
[1961] Beweis einer Abschwächung der Hadwiger-Vermutung, *Math. Ann.* **153**, 139–141.
Welsh, D.J.A.
[1971] Combinatorial problems in matroid theory, in: *Combinatorial Mathematics and its Applications* (Academic Press, New York) pp. 291–307.
[1976] *Matroid Theory, London Mathematical Society Monographs,* Vol. 8 (Academic Press, London).
[1982] Matroids and combinatorial optimisation, in: *Matroid Theory and its Applications,* ed. A. Barlotti, *Centr. Int. Mat. Estivo* **3**, 323–417.
White, N.
[1992] ed., *Matroid Applications* (Cambridge University Press, Cambridge).

White, N.L.

[1986] *Theory of Matroids* (Encycl. of Math. and Its Applications) (Cambridge University Press, Cambridge).

[1987] *Combinatorial Geometries* (Encycl. of Math. and Its Applications) (Cambridge University Press, Cambridge).

Whitney, H.

[1935] On the abstract properties of linear dependence, *Amer. J. Math.* **57**, 509–533.

Whittle, G.P.

[1989] Dowling group geometries and the critical problem, *J. Combin. Theory B* **47**, 80–92.

Zaslavsky, T.

[1975] Facing up to arrangements: face count formulas for partitions of space by hyperplanes, *Mem. Amer. Math. Soc.* **154**.

CHAPTER 10

Matroid Minors

P.D. SEYMOUR

Bell Communications Research, 445 South Street, Morristown, NJ 07960-1910, USA

Contents

HANDBOOK OF COMBINATORICS
Edited by R. Graham, M. Grötschel and L. Lovász

1. Introduction

A flourishing branch of matroid theory concerns results of the type "for any matroid M, either M contains an object of type T, or M has the structure Σ, and not both". In this chapter, we survey some such results.

Although what we mean here by "structure Σ" can vary widely, the "object of type T" is generally a minor. We begin then with a discussion of minors. Let M be a matroid on a set S. For $X \subseteq S$, $M \backslash X$ denotes the matroid N on $S - X$, in which $Y \subseteq S - X$ is independent if and only if it is independent in M. We define M/X to be $(M^* \backslash X)^*$ where M^* denotes the dual of M. If $e \in S$, we abbreviate $M \backslash \{e\}$ to $M \backslash e$, etc. Any matroid expressible in form $M \backslash X / Y$ for some X, $Y \subseteq S$ with $X \cap Y = \emptyset$ is a *minor* of M. Since $(M \backslash X)/Y = (M/Y) \backslash X$, the following holds.

Theorem 1.1. (i) *Any minor of a minor of M is a minor of M.*
(ii) *If N is a minor of M then N^* is a minor of M^*.*

The prime example of the kind of theorem we are concerned with is the following result of Tutte (1958).

Theorem 1.2. *A matroid is binary if and only if it has no U_4^2 minor.*

(M is *representable* over a field F if M is isomorphic to the linear independence matroid of the columns of some matrix with entries in F, and M is *binary* if it is representable over GF(2). The matroid U_n^k has n elements, any k of which form a base. In general, an *N minor of M* means a minor of M isomorphic to N.)

We shall sketch a proof of Theorem 1.2 due essentially to Truemper. Let M be a matroid on S, and for each base B let $A_M(B)$ be the $(0,1)$-matrix $(a_{ef}$: $e \in B, f \in S - B)$ with rows and columns indexed by B and $S - B$ respectively, where for $e \in B$ and $f \in S - B$, $a_{ef} = 1$ if and only if e belongs to the unique circuit included in $B \cup \{f\}$. Now a knowledge of $A_M(B)$ for one base B does not determine M uniquely; but if M_1, M_2 are both matroids on S, and every base B of M_1 is also a base of M_2 and $A_{M_1}(B) = A_{M_2}(B)$, then $M_1 = M_2$. (For, if possible, choose bases B, B' of M_2 such that B is a base of M_1, and B' is not, with $|B' - B| = 1$. Let $A_{M_1}(B) = (a_{ef}$: $e \in B, f \in S - B)$, and let $B' - B = \{f\}$, $B - B' = \{e\}$. If $a_{ef} = 1$ then B' is a base of M_1, a contradiction; while if $a_{ef} = 0$ then B' is not a base of M_2, since $A_{M_1}(B) = A_{M_2}(B)$, again a contradiction.) Consequently we have the following lemma.

Lemma 1.3. *Let M_1, M_2 be distinct matroids on S, and let B be a base of both matroids with $A_{M_1}(B) = A_{M_2}(B)$. Then there are bases B', B'' of both matroids such that $|B'' - B'| = 1$, $A_{M_1}(B') = A_{M_2}(B')$, and $A_{M_1}(B'') \neq A_{M_2}(B'')$; and M_1, M_2 are not both binary.*

Proof. Choose a base B' of both matroids with $A_{M_1}(B') = A_{M_2}(B')$, such that for some base B'' of M_1 with $|B''-B'|=1$, either B'' is not a base of M_2 or $A_{M_1}(B'') \neq A_{M_2}(B'')$. The first is impossible since $A_{M_1}(B')=A_{M_2}(B')$, and the second is what we want. M_1 and M_2 are not both binary because a binary matroid M is uniquely determined by a knowledge of $A_M(B)$ for one base B (for a representation of M may be obtained by adding an identity matrix to $A_M(B)$). □

Proof of Theorem 1.2. Since U_4^2 is not binary and minors of binary matroids are binary, the "only if" part is clear. To prove the "if" part, let M_1 be a non-binary matroid on S. For each base B of M_1, let $A^+_{M_1}(B)$ be the matrix with rows indexed by B and columns indexed by B and columns indexed by S, consisting of $A_{M_1}(B)$ augmented by a $B \times B$ identity matrix. Now choose some base B, and let M_2 be the binary matroid represented by $A^+_{M_1}(B)$ over GF(2). Since M_1 is not binary it follows that $M_1 \neq M_2$, and $A_{M_1}(B)=A_{M_2}(B)$. Choose B', B'' as in Lemma 1.3. Since M_2 is binary it is represented by both $A^+_{M_2}(B')$ and $A^+_{M_2}(B'')$. Let $B'-B''=\{e_1\}$, $B''-B'=\{f_1\}$, and let $A_{M_1}(B'')$, $A_{M_2}(B'')$ differ in the (e_2, f_2)-entry, where $e_2 \in B''$, $f_2 \in S-B''$. Let $Z=\{e_1, f_1, e_2, f_2\}$, let $X=S-(B' \cup B'' \bigcup Z)$ and let $Y=(B' \cup B'')-Z$. Now $B' \cap Z$, $B'' \cap Z$ are both bases of $M_1 \backslash X/Y$ and $M_2 \backslash X/Y$, and they are related as in Lemma 1.3, that is $|(B'' \cup Z)-(B' \cup Z)|=1$, $A_{M_1\backslash X/Y}(B' \cup Z)=A_{M_2\backslash X/Y}(B' \cup Z)$, $A_{M_1\backslash X/Y}(B'' \cup Z) \neq A_{M_2\backslash X/Y}(B'' \cup Z)$, (for these matrices are submatrices of $A_{M_1}(B')$, $A_{M_2}(B')$, $A_{M_1}(B'')$, $A_{M_2}(B'')$), and in particular $M_1\backslash X/Y \neq M_2 \backslash X/Y$; and so one of $M_1\backslash X/Y$, $M_2\backslash X/Y$ is not binary. Since M_2 is binary, it must be the former. We have shown than that M_1 has a non-binary minor with at most four elements, and now the result follows easily. □

A characterization of the form of Theorem 1.2 can exist for binary matroids since any minor of a binary matroid is also binary. On the other hand, it is surprising that the characterization is so simple. For one might expect there to be many, even infinitely many, minor-minimal non-binary matroids, yet Theorem 1.2 asserts that there is, up to isomorphism, only one. In general, it happens quite frequently that if we calculate the minor-minimal matroids not possessing a certain natural structure, we obtain a more elegant answer than we have any right to expect, a small, interesting set of matroids. This is the main motivation behind the theory of matroid minors; that the exclusion of simple sets of matroids often corresponds quite closely with the imposition of natural structures. Of course, it does not always work; for instance, there are infinitely many minor-minimal matroids not representable over the reals – see Ingleton (1971).

As for the "structure Σ", this can vary widely, but generally what we have in mind is a way to describe the matroid with a description of length bounded by a polynomial in the number of elements. Most matroids have no such description, because there are too many matroids on a given set for them all to have "short" descriptions, and so finding such a description is usually very helpful. Our structures are all constructions of one kind or another – for instance, we could take as the structure Σ the properties of being graphic, being representable over a given field, being a gammoid, being uniform, or being made by piecing together matroids with a structure Σ' in a prescribed way.

Results of this sort can be divided naturally into two classes, because their motivation derives either from the object excluded or from the structure imposed. In other words, we might want to obtain a short description of the matroids without certain minors, or we might be interested in a certain structure and wish to find a characterization of the matroids with this structure by excluded minors. We begin with a survey of "structure-driven" results like Theorem 1.2. The following was first proved by R. Reid (unpublished), and then independently by Bixby (1979) and Seymour (1979a). For a short proof see Kahn and Seymour (1988).

Theorem 1.4. *A matroid is representable over* GF(3) *if and only if it has no* U_5^2, U_5^3, F_7 *or* F_7^* *minor.*

(F_7 is the matroid represented over GF(2) by the seven non-zero 3-tuples.)

The excluded minors for representability over any other fields are still unknown, although results of Kahn (1988) give some hope for GF(4). The problem for larger fields is probably too difficult, however. Even G.C. Rota's popular conjecture (Rota 1970) that there are only finitely many minor-minimal matroids not representable over each finite field is still open. The general problem "which matroids are representable over which fields?" is very complicated; see chapter 9 for further details.

Next we look at matroids representable over *all* fields. A matrix of real numbers is *totally unimodular* if every square submatrix has determinant ± 1 or 0. (Such matrices, investigated by Hoffman and Kruskal (1956), are important in linear programming – see chapter 11.) A matroid is *regular* if it can be represented over the reals by the columns of a totally unimodular matrix. As a corollary of Theorems 1.2 and 1.4 we have the following result of Tutte (1958), which was provided with an elegant proof by Gerards (1989).

Theorem 1.5. *For a matroid* M, *the following are equivalent*:
(i) M *is regular*,
(ii) M *can be represented over every field*,
(iii) M *can be represented over* GF(2) *and* GF(3),
(iv) M *has no* U_4^2, F_7 *or* F_7^* *minor.*

Proof. The equivalence of (iii) and (iv) follows from Theorems 1.2 and 1.4. But (i)$\Rightarrow$(ii) (the same totally unimodular matrix represents M over any field) and (ii)$\Rightarrow$(iii) is trivial. Let us show (iii)$\Rightarrow$(i). Choose a representation $I\,|\,A$ of M over GF(3), and view A as a real matrix of 0s and ± 1s. By $I\,|\,A$ we mean a matrix whose columns are partitioned into two sets, the first of which forms an identity submatrix. Any representation of a matroid can be converted to one in this so-called *standard* form by row operations.) Since M is binary it follows that $I\,|\,A$

also represents M over GF(2) (setting $-1=+1$); and since there is a bijection between bases of M and non-singular square submatrices of A, it follows that for every square submatrix B of A,

$$\det(B)\equiv 0 \pmod 2 \Leftrightarrow \det(B)\equiv 0 \pmod 3 .$$

Since it is easy to show (by induction on the size – pivot on some non-zero entry) that a minimal matrix of 0, ± 1s which is not totally unimodular has determinant ± 2, it follows that A is totally unimodular, as required. □

Being regular is not exactly a structural condition, since it is not apparent how to construct regular matroids; we shall return to this later.

There is a third major result in a similar vein, the following, due to Tutte (1959).

Theorem 1.6. *A matroid is graphic if and only if it has no* U_4^2, F_7, F_7^*, $M^*(K_5)$ *or* $M^*(K_{3,3})$ *minor.*

(For a graph G, $M(G)$ denotes its *polygon matroid*, on the set $E(G)$ and with circuits the edge-sets of circuits of G. The dual matroid is denoted by $M^*(G)$. If $M=M(G)$ for some graph G, M is said to be *graphic*. K_5 and $K_{3,3}$ are the usual "Kuratowski" graphs.)

Theorem 1.6 is connected with the following well-known theorem of Kuratowski (1930) and Wagner (1937).

Theorem 1.7. *A graph is planar if and only if no subgraph is contractable to* K_5 *or to* $K_{3,3}$.

Proof. Theorem 1.6 implies that for a graph G, $M^*(G)$ is graphic if and only if $M(G)$ has no $M(K_5)$ or $M(K_{3,3})$ minor, that is, if and only if no subgraph of G is contractible to K_5 or to $K_{3,3}$. But G is planar if and only if $M^*(G)$ is graphic; for if $M^*(G)=M(H)$, the "vertex stars" of H give us the face boundaries of an embedding of G in the plane. The result follows. □

This completes the list of the main "structure-driven" excluded minor theorems in matroid theory.

2. Connectivity

For most of the remainder of this chapter we study the oppositely motivated theorem. Characteristically, such results involved connectivity, and we begin by discussing that.

A *separation* of a matroid M on a set S is a partition (A, B) of S. The *order* of

(A, B) is

$$r(A) + r(B) - r(S) + 1 ,$$

where $r(X)$ denotes the rank in M of $X \subseteq S$. Thus every separation has order $\geqslant 1$. If $k \geqslant 1$ is an integer, a *k-separation* of M is a separation (A, B) of order $\leqslant k$ with $|A|, |B| \geqslant k$. We say M is *k-connected* if it has no k'-separation for $1 \leqslant k' < k$. (We usually abbreviate "2-connected" to "connected".)

For example, we have the following.

Theorem 2.1. *For a connected graph G,*

(i) *if $|V(G)| \geqslant 2$, $M(G)$ is connected $\Leftrightarrow G$ is 2-connected and loopless,*

(ii) if $|V(G)| \geqslant 3$, *$M(G)$ is 3-connected $\Leftrightarrow G$ is 3-connected and simple,*

(iii) *for every separation (A, B) of $M(G)$ of order $<k$ one of A, B spans $M \Leftrightarrow G$ is k-connected.*

Connectivity may enter into our theorems in at least three ways, and this gives a convenient classification of the results. The simplest way is when the structure predicted is just a low-order separation. For example, from graph theory we have the following form of a theorem of Menger (1927) (see chapters 1, 2).

Theorem 2.2. *Let G be a graph, let $X, Y \subseteq V(G)$, with $|X| = |Y| = k$. Then exactly one of the following holds:*

(i) *there are k paths of G from X to Y, mutually vertex-disjoint,*

(ii) *there is a separation (A, B) of G with $X \subseteq V(A)$, $Y \subseteq V(B)$ and $|V(A) \cap V(B)| < k$.*

(A separation of a graph G is a pair (A, B) of subgraphs with $V(A) \cup V(B) = V(G)$, $E(A) \cup E(B) = E(G)$, and $E(A) \cap E(B) = \emptyset$.)

The following elementary result generalizes the $k = 2$ case of Theorem 2.2 to matroids.

Theorem 2.3. *Let x, y be distinct elements of a matroid M. Then either there is a circuit of M containing both x and y, or there is a 1-separation (A, B) with $x \in A$ and $y \in B$, and not both.*

For given $X, Y \subseteq V(G)$ with $|X| = |Y| = 2$, we may add new edges x, y joining the elements of X and of Y respectively, and apply Theorem 2.3 to deduce Theorem 2.2 with $k = 2$. A similar device due to Tutte (1965) allows us to reformulate Theorem 2.2 in matroid terms for general k, and thereby generalize Theorem 2.2 to matroids. Given X, Y as in Theorem 2.2, we add two trees to G, both with $k - 1$ edges, connecting together the vertices in X and in Y respectively. The paths of Theorem 2.2 (i) exist if and only if we may contract a subgraph of G such that the vertex set of one tree is contracted onto the vertex set of the other, the vertices of each tree remaining distinct from one another. Thus the following (Tutte 1965) generalizes Theorem (2.2).

Theorem 2.4. *Let M be a matroid on a set S, and let X,* $Y \subseteq S$ *with* $|X| = |Y| = k - 1$ *and* $X \cap Y = \emptyset$. *Then exactly one of the following holds*:

(i) *M has a minor N on* $X \cup Y$ *such that X and Y are both bases of N,*

(ii) *there is a separation* (A, B) *of M of order* $<k$ *with* $X \subseteq A$ *and* $Y \subseteq B$.

In spite of the fact that edges of a graph are "matroidal objects" while vertices are not, it is more difficult to find a matroid generalization of the version of Menger's theorem for edge-disjoint paths. One partially successful way to do so is as follows. If M is a matroid on a set S and $e \in S$, let us say e is a *free-flowing element* of M if for all integers $k \geq 0$ and for all non-negative integer functions c on $S - \{e\}$, exactly one of the following holds:

(i) there are k circuits of M, each containing e and at most $c(f)$ containing any other element f

(ii) there is a cocircuit D of M with $e \in D$ and $\sum_{f \in D - \{e\}} c(f) < k$.

Then the edge-disjoint form of Menger's theorem may be expressed in matroid language, as follows.

Theorem 2.5. *Every element of a graphic matroid is free-flowing.*

This was extended by Minty (1966) to the following.

Theorem 2.6. *Every element of a regular matroid is free-flowing.*

Unfortunately for our attempted generalization, not all elements of all elements of all matroids are free-flowing. The following was proved (with a long and cumbersome proof) by Seymour (1977b).

Theorem 2.7. *An element e of a matroid M is free-flowing if and only if there is no* U_4^2 *or* F_7^* *minor of M using e.*

Another method of proving Theorem 2.7 was given by Truemper (1987), using a result of Tseng and Truemper (1986); the latter explicitly find all matroids not possessing these two minors. In the next section we shall sketch their proof, because it illustrates the use of a number of results that we wish to survey.

3. Connectivity to ensure the spread of information

Suppose that we wish to understand the structure of matroids which, for some special element e, do not have a certain minor containing e. If the minor we are excluding is connected, the absence of this minor imposes no restriction whatsoever on components which do not contain e, and so we would only hope to get a structural description of the component containing e. (A *component* of a matroid is a connected minor with maximal element set.) The same kind of thing happens with higher connectivity, as we shall see in this section.

How then to prove Theorem 2.7? One half is easy; if e is free-flowing in M, then it is free-flowing in every minor N of M which contains e, and yet no element of U_4^2 or F_7^* is free-flowing. This proves the "only if" direction of Theorem 2.7.

For the "if" direction, we wish to analyze the matroids which have an element not in any U_4^2 or F_7^* minor. The first step is the following seminal result of Bixby (1974).

Theorem 3.1. *Let e be an element of a matroid M. There is a U_4^2 minor of M using e if and only if the component of M containing e is non-binary.*

In other words, we have the following.

Theorem 3.2. *Let e be an element of a connected matroid M. If M has a U_4^2 minor then it has such a minor using e.*

To prove Theorem 3.2, we first verify it for matroids with 5 elements. Then we let N be a U_4^2 minor of M (which we may assume does not use e) and the result follows from the following easy lemma of Seymour (1977a).

Lemma 3.3. *Let N be a connected minor of a connected matroid M, and let e be an element of M which is not an element of N. There is a connected minor N' of M using e such that $N = N'\backslash e$ or N'/e.*

Oxley (1984) showed that the curious property of U_4^2 in Theorem 3.2 is not possessed by any other matroid except tiny ones; nevertheless, Theorem 3.2 does have certain generalizations. For instance, one can again use Lemma 3.3 to show the following

Theorem 3.4. *Let e be an element of a connected matroid M. If M has a U_2^2, U_5^3, F_7 or F_7^* minor, then it has such a minor (i.e., one of the four) using e.*

Another relative of Theorem 3.2, due to Seymour (1985), is the following.

Theorem 3.5. *Let e, f be distinct elements of a 3-connected matroid M. If M has a U_4^2 minor then it has such a minor using both e and f.*

There is a parallel relative of Lemma 3.3 which can be used to prove Theorem 3.5, due to Bixby and Coullard (1987).

Theorem 3.6. *Let N be a 3-connected minor of a 3-connected matroid M, and let e be an element of M which is not an element of N. There is a 3-connected minor N'*

of M which uses e, such that N is a minor of N′ and N′ has at most four more elements than N.

A similar result of Truemper (1984) is the following.

Theorem 3.7. *Let N be a 3-connected minor of a 3-connected matroid $M \neq N$. There is a 3-connected minor N' of M such that $N' \neq N$, N is a minor of N' and N' has at most three more elements than N.*

Let us return to our proof of Theorem 2.7. We are given an element e of a matroid M, not in any U_4^2 or F_7^* minor, and we wish to prove that e is free-flowing. By induction we may assume that M is connected, and then Theorem 3.2 implies that M is binary. Again by induction (although the details are a little awkward – see Truemper 1987) we may assume that M is 3-connected and there is no 2-separation (A, B) of M/e with $|A|, |B| \geq 3$. But then we may apply the following (difficult) theorem of Tseng and Truemper (1986) (see also Bixby and Rajan 1989).

Theorem 3.8. *Let e be an element of a 3-connected binary matroid M, such that there is no 2-separation (A, B) of M/e with $|A|, |B| \geq 3$. Then one of the following holds:*

(i) *there is a F_7^* minor of M using e,*
(ii) *M is regular,*
(iii) $M \simeq F_7$.

By means of Theorem 3.8 we may assume that our matroid is regular; but by Theorem 2.6, every element of a regular matroid is free-flowing. This completes the proof of Theorem 2.7.

Theorems 2.4 and 2.7 arose then as matroid generalizations of versions of Menger's theorem (or the max-flow min-cut theorem). There is another theorem about disjoint paths in graphs which is capable of matroid generalization (Robertson and Seymour 1990, Seymour 1980b, Shiloach 1980, Thomassen 1980, and see also Jung 1970).

Theorem 3.9. *Let s_1, s_2, t_1, t_2 be distinct vertices of a graph G such that there is no separation (A, B) of G with $s_1, s_2, t_1, t_2 \in V(A)$, $|E(B)| \geq 4$ and $|V(A) \cap V(B)| \leq 3$. Then exactly one of the following holds:*

(i) *there are paths P_1, P_2 of G such that P_i has ends s_i, t_i $(i = 1, 2)$ and $V(P_1) \cap V(P_2) = \emptyset$,*

(ii) *G is planar, and can be drawn in the plane with s_1, s_2, t_1, t_2 incident with the infinite region, occurring on its boundary in that order.*

The condition about there being no separation (A, B) in Theorem 3.9 is important, for if there were such a separation (A, B), we could obtain no information about what is inside B from assuming the falsity of (i). For even if B

is highly non-planar, the paths in (i) cannot utilize this non-planarity because they cannot both get into B and out again. This is another instance of high connectivity being needed to "ensure the spread of information".

To obtain a formulation of Theorem 3.9 in matroid language, we use an idea of Chakravarti and Robertson (unpublished), as follows. Given G as in Theorem 3.9, we form a new graph G^+ by adding three edges e, f, g to G, joining s_1s_2, s_2t_1 and t_1t_2 respectively. Then paths as in Theorem 3.9 (i) exist if and only if there is a circuit of $M^*(G^+)$ containing e, f and g. We would therefore have a matroid generalization of Theorem 3.9 if we could answer in general when three given elements of a matroid are in a circuit. This is still open, but for binary matroids is solved by the following, due to Seymour (1986), for it is easy to reduce the general binary problem to the case when M is 3-connected and internally 4-connected. (A matroid M is *internally 4-connected* if there is no 3-separation (A, B) with $|A|, |B| \geq 4$.)

Theorem 3.10. *Let e, f, g be distinct elements of a 3-connected, internally 4-connected binary matroid M. Then one of the following holds:*

(i) *there is a circuit D of M with $e, f, g \in D$,*

(ii) *$M = M(G)$ for some graph G, and e, f, g are incident in G with a common vertex,*

(iii) *$\{e, f, g\}$ is a cocircuit of M.*

4. Connectivity to eliminate hybrid structure

In the previous section, we studied the second way in which connectivity may enter into our results; there we were looking for minors with some elements prescribed, and as we might expect, the absence of such minors gives us information about the structure of the whole matroid provided that no part of the matroid is "insulated" from our prescribed elements by a low-order separation. In this section, we study the third way that connectivity may enter, which is rather more unexpected. In some situations when we exclude certain minors, there is more than one structure possible; a sufficiently connected matroid without our minors will have either this structure or that, while a matroid of low connectivity may have some hybrid structure. We begin with an easy example from graph theory.

Let us say a graph H is a *minor* of a graph G if H can be obtained from a subgraph of G by contracting edges. (If H is a minor of G then $M(H)$ is a minor of $M(G)$, but the converse need not hold. In our applications, however, the converse will always hold, since H and G will be sufficiently connected, and so we hope no confusion will arise.) A graph is *outerplanar* if it can be drawn in the plane with all its vertices incident with the infinite region. Chartrand and Harary (1967) showed the following.

Theorem 4.1. *G has no K_4 or $K_{2,3}$ minor if and only if G is outerplanar.*

(This may be derived from Theorem 1.7 by adding a vertex adjacent to all others.) This result, and Theorems 4.2–4.6, are discussed more fully in chapter 5.

If we just exclude K_4 we get quite a different structure. A graph is *series–parallel* if it can be constructed by starting with a forest, adding loops, and repeatedly replacing edges by pairs of edges, either "in series" or "in parallel". Dirac (1952) and Duffin (1965) showed the following.

Theorem 4.2. *G has no K_4 minor if and only if G is series–parallel.*

On the other hand, if we just exclude $K_{2,3}$ we get a structure not much different from outerplanarity.

Theorem 4.3. *G has no $K_{2,3}$ minor if and only if every block of G is either outerplanar or has ≤ 4 vertices.*

(A *block* of G is a maximal 2-connected subgraph.) In other words, we have the following.

Theorem 4.4. *If G is 2-connected, then G has no $K_{2,3}$ minor if and only if either G is outerplanar or $|V(G)| \leq 4$.*

Thus, if we exclude $K_{2,3}$, we get two possible structures for 2-connected graphs, one non-trivial and one trivial. The same phenomenon happens quite frequently, although in most interesting cases we need to assume 3-connectedness instead of 2-connectedness. For instance, Hall (1943) showed the following.

Theorem 4.5. *If G is 3-connected, then G has no $K_{3,3}$ minor if and only if either G is planar or $|V(G)| \leq 5$.*

Similarly, Wagner (1937) showed the following.

Theorem 4.6. *If G is a simple 3-connected graph with no K_5 minor, then either G has no V_8 minor or $G \simeq V_8$.*

(V_8 is constructed from a circuit of length 8 by adding 4 edges joining the opposite pairs of vertices, and $\simeq$ denotes isomorphism.)

Theorem 4.7. *If M is a 3-connected binary matroid, then M has no F_7^* minor if and only if either M is regular or $M \simeq F_7$.*

Theorem 4.8. *If M is a 3-connected regular matroid, then either M has no R_{10} minor or $M \simeq R_{10}$.*

(R_{10} is the matroid represented over GF(2) by the ten 5-tuples with three 1s and two 0s.)

These four results are all corollaries of the following, due to Seymour (1980a).

Theorem 4.9. *Let $\mathscr{F}$ be a class of matroids closed under isomorphism and under taking minors. Let $N \in \mathscr{F}$ be 3-connected with at least four elements, and such that there is no 3-connected $M \in \mathscr{F}$ with one more element than N and with N as a minor. Suppose also that if N is a wheel, $\mathscr{F}$ contains no larger wheel, and if N is a whirl, $\mathscr{F}$ contains no larger whirl. Then for every 3-connected $M \in \mathscr{F}$, either M has no N minor or $M \simeq N$.*

(A *wheel* is a polygon matroid of the graph obtained from a circuit by adding one vertex joined (by one edge) to all others. A *whirl* is obtained from a wheel by declaring the "rim" circuit independent, while leaving the dependence of all other sets unchanged.)

For instance, to derive Theorem 4.7 we let $\mathscr{F}$ be the class of binary matroids with no F_7^* minor, and we let $N = F_7$. We verify that no 3-connected 8-element member of $\mathscr{F}$ has N as a minor, and then Theorem 4.7 follows from Theorem 4.9. The other three results follow in the same way.

There is an attractive formulation of Theorem 4.9 which was independently discovered (for graphs) by Negami (1982), as follows.

Theorem 4.10. *Let N be a 3-connected minor of a 3-connected matroid M with at least four elements. Suppose that if N is a wheel, no larger wheel is a minor of M, and similarly for whirls. Then either $M = N$ or there is a 3-connected minor N' of M with an element e such that one of $N' \backslash e$, N'/e is isomorphic to N.*

(It is important that we only ask that our matroid be isomorphic to N; if we ask that it *is* N, the best we do is Theorem 3.7.)

To see the equivalence of Theorems 4.9 and 4.10, we proceed as follows. It is clear that Theorem 4.10 implies Theorem 4.9. For the converse, given M and N as in Theorem 4.10, let $\mathscr{F}$ be the class of matroids isomorphic to minors of M. Then Theorem 4.10 follows by applying Theorem 4.9.

Another corollary of Theorem 4.10 is the following "wheels and whirls" theorem of Tutte (1966a).

Corollary 4.11. *Let M be a 3-connected matroid with at least four elements. Then either M is a wheel or whirl, or for some element e one of $M \backslash e$, M/e is 3-connected.*

Proof. Certainly M has a 3-connected wheel or whirl minor with $\geqslant 4$ elements (for example, from Theorem 1.6 and Theorem 4.2, because U_4^2 is a whirl and $M(K_4)$ is a wheel). Thus we may choose a 3-connected minor N of M such that $N \neq M$ and N has at least four elements, since we may assume that M is not a wheel or whirl. Choose N with as many elements as possible. By Theorem 4.10 there is a 3-connected minor N' of M such that for some element e of N', one of $N' \backslash e$, N'/e

is isomorphic to N. From the maximality of N, $N' = M$, and the result follows. □

Let us illustrate the use of Theorem 4.10 further by a more complicated application, the following theorem of Wagner (1960).

Theorem 4.12. *Let G be a 3-connected simple graph with $|V(G)| \geq 4$, with no $K_5 - e$ minor. Then either $M(G)$ is a wheel or $G \simeq K_{3,3}$ or $G \simeq \bar{C}_6$.*

($K_5 - e$ is the graph obtained from K_5 by deleting one edge. $\bar{C}_6$ is the complement of a circuit of length 6.)

Proof. We assume G is not isomorphic to $K_{3,3}$ or $\bar{C}_6$. For any simple 3-connected graph H, if one of $M(H)\backslash e$, $M(H)/e$ is isomorphic to one of $M(K_{3,3})$, $M(\bar{C}_6)$ for some edge e, then H has a $K_5 - e$ minor (as we see by checking cases). By Theorem 4.10 it follows that G has no $K_{3,3}$ or $\bar{C}_6$ minor. But it has a K_4 minor by Theorem 4.2, and so we may choose a minor H of G such that $M(H)$ is a wheel, as large as possible. For any simple 3-connected graph J, if $M(J)\backslash e$ is a wheel for some $e \in E(J)$ then J has a $K_5 - e$ minor; while if $M(J)/e$ is a wheel then J has a $K_{3,3}$ or $\bar{C}_6$ minor. In either case no such J is a minor of G. By Theorem 4.10 we deduce that $G = H$, as required. □

Two more difficult theorems of the same type were proved by Oxley (1987a,b).

Theorem 4.13. *Let M be a 3-connected binary matroid with at least four elements. Then M has no W_4 minor if and only if M is a minor of A_n for some $n \geq 3$.*

(W_4 is the "four-spoke wheel". A_n is the matroid represented over GF(2) by the n n-tuples with $n-1$ 0s and one 1, together with the n n-tuples with one 0 and $n-1$ 1s.)

Theorem 4.14. *Let M be a 3-connected matroid representable over* GF(3), *with no $M(K_4)$ minor. Then either M is a whirl or M has at most twelve elements.*

(Indeed, there are only two maximal (non-whirl) such matroids, one with 8 elements and the other with 12, related to the (5, 6, 12) Steiner system.)

So far in this section the results have only used 3-connectedness. There are two results of the same kind which use varieties of 4-connectedness. The first is due to Wagner (1937).

Theorem 4.15. *Let G be a simple 3-connected graph with no K_5 minor. Then either G is planar, or $G \simeq V_8$, or there are three vertices u, v, w of G such that $G\backslash\{u, v, w\}$ has at least three components.*

In particular, every 4-connected graph with no K_5 minor is planar. Wagner used this to prove his "equivalence theorem", the equivalence of the four-colour conjecture with the four-colouring case of Hadwiger's conjecture.

The second is the following, due to Seymour (1980a).

Theorem 4.16. *Let M be a 3-connected regular matroid. Then either M is graphic, or M is cographic, or $M \simeq R_{10}$, or there is a 3-separation (A, B) of M with $|A|$, $|B| \geqslant 6$.*

(M is *cographic* if M^* is graphic.)

In particular every 4-connected regular matroid except R_{10} is graphic or cographic. We shall give some applications of Theorem 4.16 in section 6. The following lemma is used to prove Theorem 4.16, and is sometimes of use in its own right.

Lemma 4.17. *Let M be a 3-connected regular matroid. Then M is either graphic or cographic if and only if M has no R_{10} or R_{12} minor.*

(R_{12} is the 3-sum of $M^*(K_{3,3})$ and $M(K_5 - e)$, using the 3-element circuit of $K_5 - e$ through the three vertices of valency 4. "3-sum" is defined in the next section.)

5. Decomposition

While Theorem 4.4 applies only to 2-connected graphs, it is equivalent to Theorem 4.3 which applies to all graphs. In this section we describe a decomposition method for matroids which allows us similarly to extend our results from highly-connected matroids to all matroids.

If M_1 and M_2 are matroids on S_1, S_2 respectively, where $S_1 \cap S_2 = \emptyset$, we define M to be the matroid on $S_1 \cup S_2$ in which a set is a circuit if and only if it is a circuit of one of M_1, M_2. If S_1 and S_2 are not empty, we call M the 1-*sum* (or *direct sum* or *disjoint union*) of M_1 and M_2. The following is clear.

Lemma 5.1. *If M is the 1-sum of M_1 and M_2 then (S_1, S_2) is a 1-separation of M. Conversely, if (A, B) is a 1-separation of M, then M is the 1-sum of $M \backslash B$ and $M \backslash A$.*

Now let M_1, M_2be matroids on S_1, S_2 with $S_1 \cap S_2 = \{z\}$, where z is not a loop or coloop of either M_1 or M_2, and with $|S_1|$, $|S_2| \geqslant 3$. Let M be the matroid on $S = S_1 \cup S_2 - \{z\}$, in which $X \subseteq S$ is a circuit if and only if either

(i) X is a circuit of one of M_1, M_2, or

(ii) $(X \cap S_i) \cup \{z\}$ is a circuit of M_i $(i = 1, 2)$.

M is indeed a matroid, called the 2-*sum* of M_1 and M_2. The next two lemmas are due to Seymour (1980a).

Lemma 5.2. *If M is the 2-sum of M_1 and M_2 then $(S_1 - S_2, S_2 - S_1)$ is a 2-separation of M. Conversely, if (A, B) is a 2-separation of M of order 2, and z is a new element, then there are matroids M_1, M_2 on $A \cup \{z\}$, $B \cup \{z\}$ respectively such that M is the 2-sum of M_1 and M_2.*

Lemma 5.3. *If M is the 1- or 2-sum of M_1 and M_2 then M_1, M_2 are both isomorphic to proper minors of M.*

(A minor of M is *proper* if it is not M itself.)

Armed with these three lemmas, let us return to the results of the last section. From Theorem 4.7, for instance, we deduce the following.

Theorem 5.4. *Every binary matroid with no F_7^* minor may be constructed by 1- and 2-sums, starting from regular matroids and copies of F_7.*

Proof. Let M be binary, with no F_7^* minor. We proceed by induction on the number of elements of M. If M is 3-connected then it satisfies the theorem by Theorem 4.7. If not, then it is expressible as a 1- or 2-sum of two smaller matroids by Lemmas 5.1 and 5.2. By Lemma 5.3 these smaller matroids are isomorphic to minors of M and hence themselves have no F_7^* minor. From our inductive hypothesis, these smaller matroids can be constructed in the required way, and hence so can M, as required. □

We remark that Theorem 5.4 has an easy converse, that *only* binary matroids with no F_7^* minor may be constructed as in Theorem 5.4.

Similarly, Theorem 4.13 yields a construction for all binary matroids with no W_4 minor, and so on. To go further, we need some composition corresponding to 3-separations. This is difficult in general (but see Truemper 1985), and we shall only attempt it for binary matroids. If M is a binary matroid on S, we say $X \subseteq S$ is a *cycle* of M if X may be partitioned into circuits. Let M_1, M_2 be binary matroids on S_1, S_2, with $S_1 \cap S_2 = Z$, where $|Z| = 3$ and Z is a circuit of M_1 and of M_2, and Z includes no cocircuit of either matroid, and $|S_1|$, $|S_2| \geq 7$. Let M be the binary matroid on $S = S_1 \triangle S_2$ ($X \triangle Y$ means $(X - Y) \cap (Y - X)$), in which $X \subseteq S$ is a cycle if and only if there are cycles X_1 of M_1 and X_2 of M_2 with $X = X_1 \triangle X_2$. Then M is said to be the 3-*sum* of M_1 and M_2. Seymour (1980a) showed the following.

Theorem 5.5. *If M is the 3-sum of binary matroids M_1 and M_2 then $(S_1 - S_2, S_2 - S_1)$ is a 3-separation of M, and $|S_1 - S_2|$, $|S_2 - S_1| \geq 4$. Conversely, if (A, B) is a 3-separation of a binary matroid M with $|A|$, $|B| \geq 4$, then there are binary matroids M_1, M_2 on sets S_1, S_2 with $S_1 - S_2 = A$, $S_2 - S_1 = B$ such that M is the 3-sum of M_1 and M_2.*

Theorem 5.6. *If M is 3-connected and is the 3-sum of M_1 and M_2 then M_1, M_2 are both isomorphic to proper minors of M.*

Using Theorems 5.5 and 5.6 we deduce the following from Theorem 4.16.

Theorem 5.7. *Every regular matroid may be constructed by* 1-, 2- *and* 3-*sums, starting from graphic and cographic matroids and copies of* R_{10}.

Brylawski (1975) proved the converse.

Theorem 5.8. *If* M *is the* 1-, 2- *or* 3-*sum of two regular matroids then* M *is regular.*

A warning; our 3-sum decomposition is not invariant under duality (that is, if M is the 3-sum of M_1 and M_2 then M^* need not be the 3-sum of M_1^* and M_2^* – indeed, it never is). Since by Theorem 5.5 we can express M as a 3-sum as soon as we find a 3-separation (A, B) with $|A|$, $|B| \geqslant 4$, and (A, B) is also a 3-separation of M^*, it follows that we could equally well express M^* as a 3-sum, and what we do is our choice. This choice makes a significant difference, as illustrated by the following, which is an extension of Theorem 4.15 to regular matroids, due to Seymour (1981a).

Theorem 5.9. *For a 3-connected regular matroid* M, *the following are equivalent:*

(i) M *has no* R_{10}, $M(V_8)$ *or* $M(K_5)$ *minor*,

(ii) M *is not isomorphic to* R_{10} *or* $M(V_8)$, *and* M *has no* $M(K_5)$ *minor*,

(iii) *either* $M \simeq M(K_{3,3})$, *or* M *is constructible by* 3-*sums starting from cographic matroids*,

(iv) M^* *is constructible by* 3-*sums starting from graphic matroids and copies of* $M^*(K_{3,3})$.

Proof. (ii)$\Rightarrow$(i) follows from Theorem 4.9. That (iii), (iv)$\Rightarrow$(ii) we omit. Let us see that (i)$\Rightarrow$(iii), (iv). To prove (i)$\Rightarrow$(iii) we proceed by induction on the number of elements of M. We may assume that M is not a 3-sum of two matroids M_1, M_2; for if so then M_1, M_2 are not isomorphic to $M(K_{3,3})$ since the latter has no 3-element circuits, and so by induction M_1, M_2 and hence M are 3-sums of cographic matroids. By Theorem 5.7, M is graphic or cographic, and we assume that it is graphic, and $M = M(G)$ say, where G is simple and 3-connected. By Theorem 4.15, either G is planar or $G \simeq V_8$ or $K_{3,3}$, or there is a 3-separation (A, B) of G with $|E(A)|$, $|E(B)| \geqslant 4$. The last cannot occur because M is not expressible as a 3-sum; and certainly G is not isomorphic to V_8; and so either M is cographic, or $M \simeq M(K_{3,3})$, as required. To see (i)$\Rightarrow$(iv), again we proceed by induction, and we may assume that M^* is not a 3-sum. By Theorem 5.7, M^* is graphic or cographic, and we may assume the latter, and $M^* = M^*(G)$ say, where G is simple and 3-connected. As before, either G is planar or $G \simeq K_{3,3}$, and so either M^* is graphic or $M^*(G) \simeq M^*(K_{3,3})$, as required. □

Although (iii) and (iv) are equivalent structures by Theorem 5.9, they do not seem to be obviously equivalent. In particular, they are not just reformulations of one another under duality. For instance, the 3-sum of two graphic matroids is graphic, while the 3-sum of two cographic matroids need not be cographic. Also, $K_{3,3}$ occurs in (iii) and (iv) in completely different ways. In (iii) it is just an exceptional case, while in (iv) it is crucial; without it all the matroids constructed would be graphic. The difference between (iii) and (iv) is the more remarkable because their derivations from (i) are so similar.

6. Applications of the regular matroid decomposition

Next, we sketch several applications of Theorems 4.16, 5.7 and 5.9. The first is to totally unimodular matrices; Theorem 5.7 yields a construction for them. To obtain this, we have to express the 1-, 2- and 3-sum constructions in matrix terms. They become basically "diagonal" constructions, as follows. (Our pictures in the following are not intended to imply any particular ordering of rows and columns. u, v, etc. will be column vectors, and vertical and horizontal lines represent partitions of the rows and columns.)

Construction 6.1. *If A and B are totally unimodular matrices, then so is*

$$\begin{array}{c|c} A & 0 \\ \hline 0 & B \end{array}.$$

Construction 6.2. *If $A\,|\,u$ and $B\,|\,v$ are totally unimodular, then so is*

$$\begin{array}{c|c} A & u \cdot v^{\mathrm{T}} \\ \hline 0 & B^{\mathrm{T}} \end{array},$$

where ^{T} *denotes transpose.*

Construction 6.3. *If $A\,|\,u_1\,|\,u_2\,|\,u_3$ and $B\,|\,v_1\,|\,v_2\,|\,v_3$ are totally unimodular, and $u_1 + u_2 + u_3 = 0$ and $v_1 + v_2 + v_3 = 0$, and every column of C is one of $\pm u_1$, $\pm u_2$, $\pm u_3$, 0, and every row of C is one of $\pm v_{12}^{\mathrm{T}}$, $\pm v_2^{\mathrm{T}}$, $\pm v_3^{\mathrm{T}}$, 0, then*

$$\begin{array}{c|c} A & C \\ \hline 0 & B^{\mathrm{T}} \end{array}$$

is totally unimodular.

Construction 6.4. *If $x = \pm 1$, and*

$$\begin{array}{c|c|c} A & u_1 & u_1 \\ \hline u_2^{\mathrm{T}} & 0 & x \end{array} \qquad \begin{array}{c|c|c} B & v_1 & v_1 \\ \hline v_2^{\mathrm{T}} & 0 & x \end{array}$$

are both totally unimodular, then so is

$$\begin{array}{c|c} A & u_1 \cdot v_2^{\mathrm{T}} \\ \hline v_1 \cdot u_2^{\mathrm{T}} & B \end{array}.$$

In each case we combine two totally unimodular matrices to produce a third. Let us define the *size* of an $m \times n$ matrix to be $m + n$. If a matrix is produced from two other matrices of strictly smaller size by one of these four constructions, we say that the large matrix is the *composition* of the two smaller ones.

These then will be our constructions; now we need the building blocks. Let T be a tree with its edges directed. A path P of T provides a *path vector from* T, a vector of 0, ± 1s indexed by $E(T)$, where the coordinate corresponding to $e \in E(T)$ is $+1$, -1 or 0 depending on whether P uses e forwards, backwards, or not at all. A matrix is a *tree matrix* if there is some directed tree such that every column of the matrix is a path vector from T. Tree matrices are totally unimodular, and so are their transposes.

Let us say a matrix is *sporadic* if it is 5×5 and can be obtained from one of the matrices of fig. 1 by permuting rows and columns and scaling some rows and columns by -1.

$$\begin{matrix} -1 & 1 & 0 & 0 & 1 \\ 1 & -1 & 1 & 0 & 0 \\ 0 & 1 & -1 & 1 & 0 \\ 0 & 0 & 1 & -1 & 1 \\ 1 & 0 & 0 & 1 & -1 \end{matrix} \qquad \begin{matrix} 1 & 1 & 0 & 0 & 1 \\ 0 & 1 & 1 & 0 & 1 \\ 0 & 0 & 1 & 1 & 1 \\ 1 & 0 & 0 & 1 & 1 \\ 1 & 1 & 1 & 1 & 1 \end{matrix}$$

Figure 1. Sporadic matrices.

Sporadic matrices are also totally unimodular. Our structure theorem for totally unimodular matrices is the following.

Theorem 6.5. *Every totally unimodular matrix can be constructed by repeated composition, starting from tree matrices, transposes of tree matrices, and sporadic matrices.*

Proof. Let A be a totally unimodular matrix. We prove, by induction on the size of A, that it can be constructed as claimed in Theorem 6.5. Prefix A with an identity matrix, to obtain $I \mid A$, and let M be the matroid represented by the columns of $I \mid A$. Then M is regular and so Theorem 4.16 applies. But it can be shown that

$M \simeq R_{10} \Leftrightarrow A$ is sporadic ,

M is graphic $\Leftrightarrow A$ is a tree matrix ,

M is cographic $\Leftrightarrow A^{\mathrm{T}}$ is a trcc matrix ,

M has a 1-separation $\Leftrightarrow A$ is a composition as in Construction 6.1 ,

M has a 2-separation $\Leftrightarrow A$ is a composition as in Construction 6.2 .

Moreover, if none of these apply, then

M has a 3-separation (X, Y) with $|X|, |Y| \geqslant 4$

$\Leftrightarrow A$ is a composition as in Constructions 6.3 or 6.4.

The result follows. $\square$

The argument of Theorem 6.5 may be adapted to yield a polynomial algorithm to test if a matrix is totally unimodular, using an algorithm of Cunningham and Edmonds (1980) to test if M has a 1-, 2- or 3-separation as above.

Our second application is due to Gerards et al. (1990). It concerns the structure of graphs not containing an "odd-K_4". An odd-K_4 is a graph obtained from K_4 by replacing its edges by internally disjoint paths, in such a way that the parity of each circuit is unchanged. Some examples of graphs with no odd-K_4 subgraph are

(i) graphs G such that $G \backslash v$ is bipartite for some vertex v ($G \backslash v$ is obtained from G by deleting v),

(ii) planar graphs G which can be drawn so that at most two regions have an odd number of sides.

Gerards et al. showed that every graph with no odd-K_4 can be constructed from these two special kinds by piecing them together at small cutsets. The composition rules are quite complicated, however, and we restrict ourselves here to the adequately connected case.

Theorem 6.6. *Let G be a simple, 3-connected graph with no 3-separation (A, B) with $|E(A)|, |E(B)| \geq 5$. Then G has no odd-K_4 if and only if G is of type* (i) *or* (ii) *above.*

Proof. Take a new element f, and let M be the binary matroid on $E(G) \cup \{f\}$ such that $M/f = M(G)$ and every circuit of M has even cardinality. Now it is easily shown that M has an F_7^* minor using f if and only if G has an odd-K_4. Thus we may apply Theorems 3.8 and 4.16. The result follows since

$$M \text{ is graphic} \Leftrightarrow G \text{ is of type (i)},$$

$$M \text{ is cographic} \Leftrightarrow G \text{ is of type (ii)}.$$

(Actually, the paper gives a proof which bypasses the use of Theorem 3.8.) □

Our third application is similar, and is due to Lovász – see Gerards et al. (1990). It concerns the structure of graphs such that any two odd circuits have a common vertex. Some examples of such graphs are

(i) graphs G such that $G \backslash v$ is bipartite for some vertex v,

(ii) graphs G with three edges e, f, g forming a circuit, such that the graph obtained by deleting e, f, g is bipartite,

(iii) simple graphs with ≤ 5 vertices,

(iv) graphs which can be drawn in the projective plane such that every region has an even number of sides.

Lovász showed the following.

Theorem 6.7. *Let G be a simple, 3-connected, internally 4-connected graph. Then every two odd circuits of G have a common vertex if and only if G is of one of the four types* (i), . . . , (iv) *above.*

Proof. Let M be the binary matroid on $E(G)$, with cycles the cycles of $M(G)$ of even cardinality. If M has an F_7 or F_7^* minor one can show that two odd circuits of G are disjoint. Thus by Theorem 1.5 we may assume that M is regular. Except in easy cases M is 3-connected and internally 4-connected. If $M \simeq R_{10}$ then $G \simeq K_5$, and if M is graphic some detailed analysis shows that G is of types (i) or (ii). Suppose then that M is cographic, and $M = M^*(H)$ for some graph H with $E(H) = E(G)$. There are $|V(H)|$ circuits of M corresponding to the vertex stars of H, and every edge is in exactly two of them. Now we know the rank of M (assuming that G is not bipartite), and hence $|V(H)| = |E(G)| - |V(G)| + 1$. Every circuit of M is a disjoint union of circuits of $M(G)$, and so we obtain a "2-covering" of the edge-set of G by circuits of G, at least $|E(G)| - |V(G)| + 1$ of them. But any 2-covering may be viewed as the face-boundaries of an embedding of G in some pseudo-surface, and so Euler's formula implies that the number of circuits in a 2-covering is at most $|E(G)| - |V(G)| + 2$, with equality only if the pseudo-surface is a sphere, and missing equality by 1 only for the projective plane and for the sphere with two points identified. The result follows from this and the observation that the stars of H have even cardinality (since every circuit of M is even). □

Our text application is to the "sums of circuits" property. Let M be a matroid on a set S. With each circuit C of M let us associate a (0, 1)-function $\chi^C : S \to \mathbb{R}$, where $\chi^C(e) = 1$ if and only if $e \in C$ ($\mathbb{R}$ is the set of real numbers). The *cone of circuits* of M is the set of all non-negative linear combinations of these functions χ^C. We say that M has the *sums of circuits property* if for every $p : S \to \mathbb{R}$, p belongs to the cone of circuits if and only if

(i) $p(e) \geq 0$ for all $e \in S$,

(ii) for every cocircuit D and $f \in D$, $p(f) \leq \sum(p(e) : e \in D - \{f\})$.

("Only if" is always true.) The following was proved by Seymour (1979b).

Theorem 6.8. *Graphic matroids have the sums of circuits property.*

Which other matroids have this property? The answer was given by Seymour (1981a), as follows.

Theorem 6.9. *For a matroid M, the following are equivalent*:

(i) *M has the sums of circuits property,*

(ii) *M has no U_4^2, F_7^*, R_{10} or $M^*(K_5)$ minor,*

(iii) *M^* is constructible by 1-, 2- and 3-sums, starting from cographic matroids and copies of F_7^*, $M(K_{3,3})$ and $M(V_8)$.*

Proof. If M is the 1-, 2- or 3-sum of M_1 and M_2, and M_1^*, M_2^* have the sums of circuits property, one can show that so does M^*. Thus (iii)$\Rightarrow$(i), for graphic matroids have the property by Theorem 6.8 and one verifies directly that F_7, $M^*(K_{3,3})$, $M^*(V_8)$ have the property. (Incidentally, it is not true that if M is the 3-sum of M_1 and M_2, and M_1, M_2 both have the property then so does M; this explains the dual formulation of (iii).) Now (i)$\Rightarrow$(ii) is easy, and it remains to show that (ii)$\Rightarrow$(iii). Let M satisfy (ii). Then M is binary, by Theorem 1.2, and we may assume by induction that M^* is not expressible as a 1-, 2- or 3-sum. By Theorem 4.7, either $M \simeq F_7$ or M is regular; and if the latter, then by Theorem 5.9 applied to M^*, either $M^* \simeq M(V_8)$ or $M(K_{3,3})$ or M^* is cographic. The result follows. □

There are many similar results in the same paper of Seymour (1981a), formulating a graph optimization theorem (particularly multicommodity flow theorems) in terms of matroids, and finding an excluded minor characterization of the matroids with the property by applying a structure theorem. We omit further details.

Our final application is to a problem of Tutte. A *cocycle* in a binary matroid is a set which can be partitioned into cocircuits. Let us say that a binary matroid M is a *tangential* 2-*block* if M is minor-minimal with the property that it has no loops and no two cocycles have as union the element set of M. This is connected with the four-colour problem (see chapters 4 and 5). Tutte (1966b) proposed the following.

Conjecture 6.10. The only tangential 2-blocks are F_7, $M(K_5)$, and $M^*(P)$, where P is the Petersen graph.

Using the structure theorems, Seymour (1981b) showed the following.

Theorem 6.11. *Every tangential* 2-*block is cographic, except for* F_7 *and* $M(K_5)$.

Proof. Let M be a tangential 2-block, not isomorphic to F_7 or $M(K_5)$. Then it has no F_7 or $M(K_5)$ minor. But if M^* is the 1-, 2- or 3-sum of M_1^* and M_2^*, it is easy to see that since M_1, M_2 both are covered by two cocycles, so is M, a contradiction. Thus M^* is not a 1-, 2- or 3-sum, and so, by Theorem 5.4, either $M \simeq F_7^*$, or M is regular; and in the latter case, by Theorem 5.9, either $M \simeq R_{10}$, or $M \simeq M(V_8)$, or $M \simeq M(K_{3,3})$, or M is cographic. Now none of F_7^*, R_{10}, $M(V_8)$ and $M(K_{3,3})$ are tangential 2-blocks, and so M is cographic as required. □

In view of Theorem 6.11, Conjecture 6.10 is equivalent to the conjecture that $M^*(P)$ is the only cographic tangential 2-block; or equivalently,

Conjecture 6.12. If G is a graph with no isthmus and with no Petersen graph minor, then $E(G)$ may be expressed as the union of two cycles of G.

(A *cycle* of G is a subset of $E(G)$ which may be partitioned into circuits.)

It is shown in the Appendix to chapter 4 that Conjecture 6.12 is equivalent to Tutte's well-known "nowhere-zero 4-flow" conjecture.

References

Bixby, R.E.
[1974] l-matrices and a characterization of binary matroids, *Discrete Math.* **8**, 139–145.
[1979] On Reid's characterization of the ternary matroids, *J. Combin. Theory B* **26**, 174–204.

Bixby, R.E., and C.R. Coullard
[1987] Finding a small 3-connected minor maintaining a fixed minor and a fixed element, *Combinatorica* **7**, 231–242.

Bixby, R.E., and A. Rajan
[1989] A short proof of the Truemper–Tseng theorem on max-flow min-cut matroids, *Linear Algebra Appl.* **114**, 277–292.

Brylawski, T.H.
[1975] Modular constructions for combinatorial geometries, *Trans. Amer. Math. Soc.* **203**, 1–44.

Chartrand, G., and F. Harary
[1967] Planar permutation graphs, *Ann. Inst. Henri Poincaré B* **3**, 433–438.

Cunningham, W.H., and J. Edmonds
[1980] A combinatorial decomposition theory, *Canad. J. Math.* **32**, 734–765.

Dirac, G.A.
[1952] A property of 4-chromatic graphs and remarks on critical graphs, *J. London Math. Soc.* **27**, 85–92.

Duffin, R.J.
[1965] Topology of series-parallel networks, *J. Math. Anal. Appl.* **10**, 303–318.

Gerards, B.
[1989] A short proof of Tutte's characterization of totally unimodular matrices, *Linear Algebra Appl.* **114**, 207–212.

Gerards, B., L. Lovász, A. Schrijver, P.D. Seymour and K. Truemper
[1990] in preparation.

Hall, D.W.
[1943] A note on primitive skew curves, *Bull. Amer. Math. Soc.* **49**, 935–937.

Hoffman, A.J., and J.B. Kruskal
[1956] Integral boundary points of convex polyhedra, in: *Linear Inequalities and Related Systems,* eds. H.W. Kuhn and A.W. Tucker, *Ann. of Math. Stud.* **38**, 223–246.

Ingleton, A.W.
[1971] Representation of matroids, in: *Combinatorial Mathematics and Its Applications,* ed. D.J.A. Welsh (Academic Press, New York) pp. 149–167.

Jung, H.A.
[1970] Eine Verallgemeinerung des n-fachen Zusammenhangs für Graphen, *Math. Ann.* **187**, 95–103.

Kahn, J.
[1988] On the uniqueness of matroid representations over GF(4), *Bull. London Math. Soc.* **20**, 5–10.

Kahn, J., and P.D. Seymour
[1988] On forbidden minors for GF(3), *Proc. Amer. Math. Soc.* **102**, 437–440.

Kuratowski, K.
[1930] Sur le problème des courbes gauches en topologie, *Fund. Math.* **156**, 271–283.

Menger, K.
[1927] Zur allgemeinen Kurventheorie, *Fund. Math.* **10**, 96–115.

Minty, G.J.
[1966] On the axiomatic foundations of the theories of directed linear graphs, electrical networks and network programming, *J. Math. Mech.* **15**, 485–520.

Negami, S.
[1982] A characterization of 3-connected graphs containing a given graph, *J. Combin. Theory B* **32**, 69–74.

Oxley, J.G.
[1984] On singleton 1-rounded sets of matroids, *J. Combin. Theory B* **37**, 189–197.
[1987a] A characterization of the ternary matroids with no $M(K_4)$-minor, *J. Combin. Theory B* **42**, 212–249.
[1987b] The binary matroids with no 4-wheel minor, *Trans. Amer. Math. Soc.* **301**, 63–75.
Robertson, N., and P.D. Seymour
[1990] Graph minors. IX. Disjoint crossed paths, *J. Combin. Theory B* **49**, 40–77.
Rota, G.-C.
[1970] Combinatorial theory, old and new, in: *Actes du Congrès International des Mathématiciens (Proc. Int. Congress of Mathematicians), Nice, 1970* (Gauthier-Villars, Paris) pp. 229–233.
Seymour, P.D.
[1977a] A note on the production of matroid minors, *J. Combin. Theory* **22**, 289–295.
[1977b] The matroids with the max-flow min-cut property, *J. Combin. Theory B* **23**, 189–222.
[1979a] Matroid representation over GF(3), *J. Combin. Theory B* **26**, 159–173.
[1979b] Sums of circuits, in: *Graph Theory and Related Topics,* eds. J.A. Bondy and U.S.R. Murty (Academic Press, New York).
[1980a] Decomposition of regular matroids, *J. Combin. Theory B* **28**, 305–359.
[1980b] Disjoint paths in graphs, *Discrete Math.* **29**, 293–309.
[1981a] Matroids and multicommodity flows, *European J. Combin.* **2**, 257–290.
[1981b] On Tutte's extension of the four-colour problem, *J. Combin. Theory B* **31**, 82–94.
[1985] On minors of 3-connected matroids, *European J. Combin.* **6**, 375–382.
[1986] Triples in matroid circuits, *European J. Combin.* **7**, 177–185.
Shiloach, Y.
[1980] A polynomial solution to the undirected two paths problem, *J. Assoc. Comput. Mach.* **27**, 445–456.
Thomassen, C.
[1980] 2-linked graphs, *European J. Combin.* **1**, 371–378.
Truemper, K.
[1984] Partial matroid representations, *European J. Combin.* **5**, 377–394.
[1985] A decomposition theory for matroids. I. General results, *J. Combin. Theory B* **39**, 43–76.
[1987] Max-flow min-cut matroids: polynomial testing and polynomial algorithms for maximum flow and shortest routes, *Math. Oper. Res.* **12**, 72–96.
Tseng, F.T., and K. Truemper
[1986] A decomposition of the matroids with the max-flow min-cut property, *Discrete Appl. Math.* **15**, 329–364.
Tutte, W.T.
[1958] A homotopy theorem for matroids, I, II, *Trans. Amer. Math. Soc.* **88**, 144–174.
[1959] Matroids and graphs, *Trans. Amer. Math. Soc.* **90**, 527–552.
[1965] Menger's theorem for matroids, *J. Res. Nat. Bur. Standards B* **69**, 49–53.
[1966a] Connectivity in matroids, *Canad. J. Math.* **18**, 1301–1324.
[1966b] On the algebraic theory of graph colorings, *J. Combin. Theory* **1**, 15–50.
Wagner, K.
[1937] Über eine Eigenschaft der ebenen Komplexe, *Math. Ann.* **114**, 570–590.
[1960] Bemerkungen zu Hadwigers Vermutung, *Math. Ann.* **141**, 433–451.

CHAPTER 11

Matroid Optimization and Algorithms

Robert E. BIXBY

Rice University, Department of Computational and Applied Mathematics, Brown School of Engineering, Houston, TX 77251, USA

William H. CUNNINGHAM

Faculty of Mathematics, Department of Combinatorics and Optimization, University of Waterloo, Waterloo, Ont. N2L 3G1, Canada

Contents

HANDBOOK OF COMBINATORICS
Edited by R. Graham, M. Grötschel and L. Lovász

1. Introduction

This chapter considers matroid theory from a constructive and algorithmic viewpoint. A substantial part of the developments in this direction have been motivated by optimization. Matroid theory has led to a unification of fundamental ideas of combinatorial optimization as well as to the solution of significant open problems in the subject. In addition to its influence on this larger subject, matroid optimization is itself a beautiful part of matroid theory.

The most basic optimizational property of matroids is that for any subset every maximal independent set contained in it is maximum. Alternatively, a trivial algorithm maximizes any $\{0,1\}$-valued weight function over the independent sets. Most of matroid optimization consists of attempts to solve successive generalizations of this problem. In one direction it is generalized to the problem of finding a largest common independent set of two matroids: the matroid intersection problem. This problem includes the matching problem for bipartite graphs, and several other combinatorial problems. In Edmonds' solution of it and the equivalent matroid partition problem, he introduced the notions of good characterization (intimately related to the NP class of problems) and matroid (oracle) algorithm.

A second direction of generalization is to maximizing any weight function over the independent sets. Here a greedy algorithm also works: Consider the elements in order of non-increasing weight, at each step accepting an element if its weight is positive and it is independent of the set of previously-accepted elements. A most significant step was Edmonds' recognition of a polyhedral interpretation of this fact. He used the fact that the greedy algorithm optimizes any linear function over the convex hull of characteristic vectors of independent sets to establish a linear-inequality description of that polyhedron. He then showed that a greedy algorithm also works for the much larger class of polymatroids: polyhedra that are defined from functions that, like matroid rank functions, are submodular. This polyhedral approach led to the solution of the weighted matroid intersection problem and extensive further generalizations, culminating in the optimal submodular flow problem. Other important related problems are that of minimizing an arbitrary submodular function and a polymatroid generalization of nonbipartite matching. Complete solutions of these problems remain to be discovered, but substantial progress has been made.

There is a second aspect of matroid theory, other than the optimizational one, for which a constructive approach is desirable. This might be called the structural aspect. We seek constructive answers to such fundamental questions as the connectivity, graphicness, or linear representability of a given matroid. It is pleasant to realize how many of the classical structural results of Tutte and Seymour are essentially constructive in nature. In fact, the most important structural result to date, Seymour's characterization of regular matroids, yields an algorithm to recognize regularity by reducing that problem to recognizing graphicness and connectivity, two problems to which Tutte contributed substantially. On the other hand many structural questions can be proved to be unsolvable by matroid algorithms. Indeed, recognizing regularity turns out to be essentially the only question on linear representability that is solvable.

We use the matroid terminology introduced in chapter 9 by Welsh. The following additional notation is used. For a subset A of the set S, we use $\bar{A}$ to denote $S \backslash A$. Where J is an independent set of matroid $M = (S, \mathscr{I})$, and $e \in \bar{J}$ with $J \cup \{e\} \notin \mathscr{I}$, we use $C(J, e)$ to denote the unique circuit (*fundamental circuit*) contained in $J \cup \{e\}$. Use of this notation will imply that $J \cup \{e\} \notin \mathscr{I}$.

2. Matroid optimization

Two of the most natural optimization problems for a matroid $M = (S, \mathscr{I})$ with weight vector $c \in \mathbb{R}^S$ are to find an independent set of maximum weight, and to find a circuit of minimum weight. These generalize standard optimization problems on graphs. We show in this section that the classical greedy algorithm solves the independent set problem. While this problem is easy, it leads to very useful polyhedral methods. Considering the greedy algorithm requires discussion of the efficiency of matroid algorithms. It turns out that, with respect to the resulting notion of algorithmic solvability, the circuit problem above is intractable.

2.1. The optimal independent set problem

If we are asked to find a maximum-*cardinality* independent set, we know from the independent-set axioms (chapter 9 by Welsh) that any maximal independent set is a solution. Hence the following trivial algorithm works: Where $S = \{e_1, e_2, \ldots, e_n\}$, start with $J = \emptyset$ and treat the e_i sequentially, adding e_i to J if and only if $J \cup \{e_i\}$ is independent. The *maximum-weight independent set problem*

$$\max(c(J) \colon J \in \mathscr{I}) \tag{2.1}$$

includes the above problem as a special case (take each $c_j = 1$). Moreover, (2.1) is solved by making two simple modifications to the above algorithm. First, we treat the elements of S in order of non-increasing weight, and second we do not add to J any negative-weight elements. The resulting method is the *greedy algorithm* (GA).

Greedy algorithm for the maximum-weight independent set problem.

Order $S = \{e_1, e_2, \ldots, e_n\}$ so that $c_{e_1} \geqslant c_{e_2} \geqslant \cdots \geqslant c_{e_m} \geqslant 0 \geqslant c_{e_{m+1}} \geqslant \cdots \geqslant c_{e_n}$
$J := \emptyset$
For $i = 1$ to m do
  If $J \cup \{e_i\} \in \mathscr{I}$ then $J := J \cup \{e_i\}$

Theorem 2.2. *For any matroid $M = (S, \mathscr{I})$ and any $c \in \mathbb{R}^S$, GA solves* (2.1).

Proof. Suppose that $J = \{j_1, \ldots, j_k\}$ is found by GA, but $Q = \{q_1, \ldots, q_\ell\}$ has larger weight. Assume that the j_i are in the order in which GA added them, and that $c_{q_1} \geqslant c_{q_2} \geqslant \cdots \geqslant c_{q_\ell}$. There is a least index i such that $c_{q_i} > c_{j_i}$ or $c_{q_i} > 0$ and $i > k$. Then $\{j_1, \ldots, j_{i-1}\}$ is a basis of $A = \{j_1, \ldots, j_{i-1}, q_1, \ldots, q_i\}$, for otherwise GA would choose one of $q_1, \ldots, q_i$ as j_i. But $\{q_1, \ldots, q_i\}$ is a larger independent subset of A, a contradiction. □

A pair $(S, \mathcal{I})$ satisfying only that $\mathcal{I} \neq \emptyset$ and $A \subseteq B \in \mathcal{I}$ implies $A \in \mathcal{I}$, is sometimes called an *independence system*. Both the problem (2.1) and the greedy algorithm can be stated for any independence system, and one wonders whether other independence systems are similarly nice. However, the matroid axioms say that $(S, \mathcal{I})$ is a matroid if and only if GA solves (2.1) for every $c \in \{0,1\}^S$. In view of this observation, Theorem 2.2 can be restated as follows.

Theorem 2.3. *An independence system* $(S, \mathcal{I})$ *is a matroid if and only if GA solves* (2.1) *for every* $c \in \mathbb{R}^S$.

A second proof of Theorem 2.2, almost as short and much more useful, leads us to the polyhedral method.

Proof of Theorem 2.2. Let $\bar{x}$ be the characteristic vector of the set J produced by GA, and let x be the characteristic vector of any independent set J'. Then $c(J') = \sum c_{e_i} x_{e_i} = c \cdot x$. Let $T_i = \{e_1, \ldots, e_i\}$ for $0 \leqslant i \leqslant n$. Notice that $\bar{x}(T_i) \geqslant x(T_i)$ for $1 \leqslant i \leqslant m$, because $J \cap T_i$ is a maximal independent subset of T_i. Then

$$\begin{aligned}
c \cdot x &= \sum_{i=1}^{m} c_{e_i} x_{e_i} + \sum_{i=m+1}^{n} c_{e_i} x_{e_i} \\
&= \sum_{i=1}^{m} c_{e_i} (x(T_i) - x(T_{i-1})) + \sum_{i=m+1}^{n} c_{e_i} x_{e_i} \\
&= \sum_{i=1}^{m-1} (c_{e_i} - c_{e_{i+1}}) x(T_i) + c_m x(T_m) + \sum_{i=m+1}^{n} c_{e_i} x_{e_i} \\
&\leqslant \sum_{i=1}^{m-1} (c_{e_i} - c_{e_{i+1}}) \bar{x}(T_i) + c_m \bar{x}(T_m) + \sum_{i=m+1}^{n} c_{e_i} \bar{x}_{e_i}.
\end{aligned}$$

But the last line is $c \cdot \bar{x}$, since the inequality holds with equality for $x = \bar{x}$. □

Notice that the only properties of $x, \bar{x}$ used in the second proof of Theorem 2.2 were that $x \geqslant 0$ and $x(T_i) \leqslant \bar{x}(T_i)(= r(T_i))$, $1 \leqslant i \leqslant m$. So GA actually solves the following linear programming problem, since $\bar{x}(T_i) = r(T_i)$ implies $x(T_i) \leqslant \bar{x}(T_i)$:

$$\begin{aligned}
&\text{maximize } c \cdot x \\
&\text{subject to } \quad x(A) \leqslant r(A), \quad A \subseteq S; \\
&\qquad\qquad\quad\;\; x \geqslant 0.
\end{aligned} \tag{2.4}$$

This observation implies the following *Matroid Polytope Theorem* of Edmonds (1970).

Theorem 2.5. *For any matroid $M = (S, \mathscr{I})$, the extreme points of the polytope $P(M) = \{x \in \mathbb{R}_+^S\colon x(A) \leqslant r(A)$ for all $A \subseteq S\}$ are precisely the characteristic vectors of independent sets of M.*

Proof. It is easy to see that the characteristic vector of any independent set is an extreme point of $P(M)$. Now let x' be an extreme point of $P(M)$. Then there is $c \in \mathbb{R}^S$ such that x' is the unique optimal solution of $\max(c \cdot x\colon x \in P(M))$. Applying GA to M and this c, we obtain $\bar{x}$, the characteristic vector of an independent set, and $\bar{x}$ solves the same linear programming problem. Hence $x' = \bar{x}$, as required. □

The dual linear program to (2.4) is:

$$\begin{aligned} &\text{minimize } \textstyle\sum(r(A)y_A : A \subseteq S) \\ &\text{subject to } \textstyle\sum(y_A : j \in A) \geqslant c_j, \quad \text{for } j \in S; \\ &\qquad\qquad y_A \geqslant 0, \quad \text{for } A \subseteq S. \end{aligned} \tag{2.6}$$

Analysis of the second proof of Theorem 2.2 shows that the following formula, sometimes called the *dual greedy algorithm*, gives an optimal solution y' to (2.6):

$$\begin{aligned} &y'_{T_i} = c_{e_i} - c_{e_{i+1}}, \quad 1 \leqslant i < m; \\ &y'_{T_m} = c_{e_m}; \\ &y'_A = 0 \quad \text{for all other } A \subseteq S. \end{aligned}$$

It follows that (2.6) has an optimal solution that is integer-valued whenever c is integer-valued, that is, that the constraint system of (2.4) is *totally dual integral.* (This important notion is developed in chapter 30 by Schrijver. One fundamental fact that we do mention, is that if $Ax \leqslant b$ is totally dual integral, b is integer-valued, and $\max(cx : Ax \leqslant b)$ has an optimal solution, then it has one that is integer-valued.) One can also use y' and the linear-programming optimality conditions to prove the following converse to Theorem 2.2. (The point here is that there can be many choices of the ordering of S, because of equal weights or zero weights.)

Theorem 2.7. *$J \in \mathscr{I}$ has maximum c-weight if and only if it can be found by* GA.

2.2. *Efficient matroid algorithms*

How efficient is GA? Because sorting can be done in polynomial time, GA runs in polynomial time if and only if there is a polynomial-time algorithm to answer the question: "Is $J \in \mathscr{I}$?" (In fact, we need to answer at most n such questions.) It is usual for matroids to be represented, for purposes of algorithms, by such an (independence-testing) oracle. An abstract matroid algorithm is *good* if both the number of calls to the oracle and the amount of additional computation is bounded by a polynomial in n and the size of any additional input (such as the weights). Hence GA is a good matroid algorithm. For concrete classes, such as matroids represented by matrices, for which there exists a subroutine for independence-testing that is polynomial-time in the usual sense (that is, relative to input size), a good matroid algorithm does yield a polynomial-time algorithm.

Having defined what is meant by "good matroid algorithm" and having described an important example, we digress briefly on some related issues. First, let us explain why the oracle representation is necessary. A natural alternative would be to have a general encoding for matroids, as we do, say, for graphs, and to measure algorithm efficiency relative to the encoding-size of the input matroid. The difficulty is that the number of matroids on an n-element set (see chapter 9 by Welsh) is so large that any general encoding scheme would require space exponential in n, which would affect the meaning of polynomial time.

Since we are not using the usual model of computation, the theory of NP-completeness does not play the same role, and it is natural to ask whether there is an analogous theory of difficult matroid problems. We call a matroid problem *intractable* if there is no good matroid algorithm solving the problem. In contrast to the situation in ordinary complexity theory, we can *prove* that certain natural problems are intractable. To illustrate a typical proof method, we consider the problem of finding the girth (minimum size of a circuit) of a matroid. Let $|S| = n = 2m$, let M be the uniform matroid of rank m on S, and, for each $A \subseteq S$ with $|A| = m$, let M_A denote the matroid obtained from M by making A a circuit. If an algorithm concludes that the girth of M is $m+1$ after making fewer than $\binom{2m}{m}$ oracle calls, then there is some A such that the same (incorrect) conclusion would have been reached if the input had been M_A. (Notice that the argument uses the fact that all calls other than A receive the same answer for both M and M_A.) Hence a correct algorithm requires at least $\binom{2m}{m}$ calls, which is not bounded by a polynomial in n, and so computing the girth is intractable. (We mention that it is an open problem whether there is a polynomial-time algorithm to compute the girth of the matroid of a given binary matrix; on the other hand, finding a smallest circuit containing a fixed element in such a matroid is known to be NP-hard.) Other examples of intractable problems are described in later sections. Proofs can be more intricate than this example, but use similar "adversary" arguments.

Finally, we discuss why independence is an appropriate choice of oracle on which to base a theory of matroid algorithms. Consider the following alternatives: (i) "What is $r(B)$?"; (ii) "Is B a circuit?"; (iii) "What is the minimum size of a circuit contained in B?" It is easy to see that (i) is equivalent to the independence-testing oracle, since there is a good matroid algorithm for each with respect to the other. On the other hand (ii) is weaker than independence-testing: $|B|+1$ calls to the independence-testing oracle will be enough to determine whether B is a circuit, but there can be no polynomial-time simulation of independence-testing by an oracle for (ii). (To see this, use an adversary approach, considering matroids on S having at most one circuit.) So (ii) would lead to a theory of matroid algorithms with respect to which independence-testing is intractable–in contrast to what the standard classes of matroids would lead us to expect. On the other hand, we have seen that (iii) would lead to a theory in which the existence of a good matroid algorithm would not imply the existence of a polynomial-time algorithm for the corresponding matrix problem.

3. Matroid intersection

We are given matroids M_1, M_2 on the same set S. We want to find a maximum weight (or maximum cardinality) common independent set. Obviously this problem generalizes the optimal independent set problem (2.1), taking $M_1 = M_2$. Some examples of applications are the following:

Problem 3.1. finding a maximum-weight matching in a bipartite graph;

Problem 3.2. finding a maximum-weight branching (forest in which each node has indegree at most 1) in a digraph;

Problem 3.3. finding a maximum-weight subgraph that is the union of k forests in a graph;

Problem 3.4. given bases B_1, B_2 of a matroid M and $X_1 \subseteq B_1$, finding $X_2 \subseteq B_2$ such that $(B_1 \backslash X_1) \cup X_2$, $(B_2 \backslash X_2) \cup X_1$ are both bases.

Problem (3.1) is a direct special case of weighted matroid intersection. Where $\{V_1, V_2\}$ is a bipartition of $G = (V, E)$, take $S = E$ and $\mathcal{I}_i = \{J \subseteq S$: each $v \in V_i$ is incident with at most one element of $J\}$. Problem (3.2) is also a special case; (3.3) and (3.4) will be treated later. We begin the discussion with the maximum-cardinality case.

3.1. Matroid Intersection Theorem

It is obvious, for any $J \in \mathcal{I}_1 \cap \mathcal{I}_2$ and any $A \subseteq S$, that

$$|J| = |J \cap A| + |J \cap \overline{A}| \leqslant r_1(A) + r_2(\overline{A}).$$

So if we can find J and A satisfying this relationship with equality, then we know that this J is maximum. In fact, such a pair (J, A) always exists; this is the content of the Matroid Intersection Theorem of Edmonds (1970). We remark that applying this theorem to maximum-cardinality bipartite matching as above immediately yields König's Theorem (chapter 3 by Pulleyblank). While the theorem follows from the algorithm described later, we first give a short non-constructive proof.

Theorem 3.5. *For matroids M_1, M_2 on S,*

$$\max(|J|: J \in \mathcal{I}_1 \cap \mathcal{I}_2) = \min(r_1(A) + r_2(\overline{A}): A \subseteq S).$$

Proof. The proof is by induction on $|S|$. Let k be the minimum of $r_1(A) + r_2(\overline{A})$, and choose $e \in S$ with $\{e\} \in \mathcal{I}_1 \cap \mathcal{I}_2$. (If none exists, $k = 0$, and we are finished.) If the minimum of $r_1(A) + r_2(S' \backslash A)$ over subsets A of $S' = S \backslash \{e\}$ is k, then we are finished, by induction. If M_i' denotes $M_i / \{e\}$ and the minimum of $r_1'(B) + r_2'(S' \backslash B)$ over subsets B of S' is at least $k - 1$, then induction gives a common independent set of M_1', M_2' of size $k - 1$; adding e to it gives the desired J. We conclude that, if there is no common independent set of size k, then there exist subsets A, B of S' such that

$$r_1(A) + r_2(S' \backslash A) \leqslant k - 1$$

and

$$r_1(B \cup \{e\}) - 1 + r_2\left((S' \backslash B) \cup \{e\}\right) - 1 \leqslant k - 2.$$

Adding and applying submodularity, we have

$$r_1(A \cup B \cup \{e\}) + r_1(A \cap B) + r_2\left(S \backslash (A \cap B)\right) + r_2\left(S \backslash (A \cup B \cup \{e\})\right) \leqslant 2k - 1;$$

it follows that the sum of the middle two terms, or the sum of the other two terms, is at most $k - 1$, a contradiction. □

(Notice that this proof gives no hint of how to find the desired subsets J and A efficiently.)

3.2. The Matroid Intersection Algorithm

The Matroid Intersection Algorithm (MIA) uses an augmenting path approach that generalizes a common method for bipartite matching. It maintains $J \in \mathcal{I}_1 \cap \mathcal{I}_2$, at each step either finding a larger such J, or finding A giving equality with J in the min–max formula, proving that J is maximum. Given $J \in \mathcal{I}_1 \cap \mathcal{I}_2$, an auxiliary digraph $G = G(M_1, M_2, J)$ having node-set $S \cup \{s, t\}$ is constructed. It contains:

an arc (e, t) for every $e \in S \backslash J$ such that $J \cup \{e\} \in \mathcal{I}_1$;

an arc (s, e) for every $e \in S \backslash J$ such that $J \cup \{e\} \in \mathcal{I}_2$;

an arc (e, f) for every $e \in S \backslash J, f \in J$ such that $(J \cup \{e\}) \backslash \{f\} \in \mathcal{I}_1$;

an arc (f, e) for every $e \in S \backslash J, f \in J$ such that $(J \cup \{e\}) \backslash \{f\} \in \mathcal{I}_2$.

The following "augmenting path" theorem is the basis for the algorithm.

Theorem 3.6. (a) *If there exists an (s, t)-dipath in G, then J is not maximum; in fact, if $s, e_1, f_1, \ldots, e_k, f_k, e_{k+1}, t$ is a chordless (s, t)-dipath, then* $J' = (J \cup \{e_1, \ldots, e_{k+1}\}) \backslash \{f_1, \ldots, f_k\} \in \mathcal{I}_1 \cap \mathcal{I}_2$. (b) *If there exists no (s, t)-dipath in G, then J is maximum; in fact, if $A \subseteq S$ and $\delta^+(A \cup \{s\}) = \emptyset$, then $|J| = r_1(A) + r_2(\overline{A})$.*

We remark that, while the restriction to chordless dipaths is not necessary for the special case of bipartite matching, it is needed in general. The proof of (b) is easy. Consider $e \in A \backslash J$. Since (e, t) is not an arc, $J \cup \{e\}$ contains an M_1-circuit C. Since there is no arc (e, f) with $f \in S \backslash A$, we have $C \subseteq (A \cap J) \cup \{e\}$. Therefore, $J \cap A$ M_1-spans A. Similarly, $J \cap \overline{A}$ M_2-spans $\overline{A}$. Hence

$$|J| = |J \cap A| + |J \cap \overline{A}| = r_1(A) + r_2(\overline{A}).$$

We can deduce (a) from the following result, whose proof is a straightforward induction. (Recall that $\sigma(A)$ denotes the closure of the subset A.)

Lemma 3.7. *Let $M = (S, \mathcal{I})$ be a matroid, let $J \in \mathcal{I}$ and let $x_1, y_1, \ldots, x_k, y_k$ be a sequence of distinct elements of S such that*
(a) $x_i \notin J$, $y_i \in J$ *for* $1 \leqslant i \leqslant k$;
(b) $y_i \in C(J, x_i)$ *for* $1 \leqslant i \leqslant k$;
(c) $y_i \notin C(J, x_j)$ *for* $1 \leqslant i < j \leqslant k$.
Then where $J' = (J \cup \{x_1, \ldots, x_k\}) \backslash \{y_1, \ldots, y_k\}$, $J' \in \mathcal{I}$ *and* $\sigma(J') = \sigma(J)$.

3.3. Efficiency of the Matroid Intersection Algorithm

Since MIA will terminate after at most $n = |S|$ augmentations, since the dipath or set A as in Theorem 3.6 can be found with standard methods, and since each auxiliary digraph can be constructed with $O(n^2)$ independence tests, it follows that MIA is a good matroid algorithm. Here we mention the complexity for some refinements and special cases. Just as MIA generalizes a basic bipartite matching algorithm, more efficient versions of MIA generalize some of the ideas used to speed up matching and network flow algorithms. (See chapter 2 by Frank.) The most important refinement is a natural one: At each step augment J using a shortest (s, t)-dipath of the auxiliary digraph. Of course, any such dipath will automatically be chordless. The next result implies that a large proportion of the resulting augmentations will be on very short augmenting paths. It is from Cunningham (1986), and part (a) is also due to Gabow and Stallmann (1985).

Theorem 3.8. *If MIA using shortest augmenting paths is applied to matroids M_1, M_2 on S, $|S| = n$, then:*
(a) *The length of a shortest augmenting path never decreases, and the number of different lengths is* $O(\sqrt{n})$;
(b) *The sum of lengths of augmenting paths used is* $O(n \log n)$.

It follows from Theorem 3.6(a) that the work of the algorithm can be divided into $O(\sqrt{n})$ stages; during each stage all augmenting paths have the same length. It is possible to find and perform all of the augmentations of a stage more efficiently than if the auxiliary digraph were reconstructed after each stage. This leads to a version of MIA (Cunningham 1986) that requires $O(n^{2.5})$ independence tests rather than the $O(n^3)$ of the basic algorithm. How close is this bound to being best possible? One of the few results on this theme is that the greedy algorithm for finding a basis of a single matroid is optimal; it requires exactly n independence tests, and an easy argument shows that no algorithm uses fewer. For matroid intersection nothing more is known, that is, it is an open problem to find a nonlinear lower bound on the number of independence tests required.

For concrete classes of matroids, one can usually obtain better bounds than arise from simply multiplying the number of oracle calls by the oracle complexity.

For example, for matroids arising from two matrices each having at most n rows, Theorem 3.8(b) can be used to show that the complexity of MIA is $O(n^3 \log n)$, assuming that arithmetic operations are counted as single steps. As another example, Gabow and Stallmann (1985) have given an $O(p^{2.5})$ time bound for MIA on the cycle matroids of two graphs, each having at most p nodes.

3.4. Matroid partitioning

Many of the applications of matroid intersection are most easily derived through the theory of matroid partitioning. In fact this theory is equivalent to that for matroid intersection and actually was discovered earlier by Edmonds.

The matroid partitioning problem is, given matroids $M_i = (S, \mathcal{I}_i)$, $1 \leqslant i \leqslant k$, to find a maximum cardinality subset $J \subseteq S$ that is *partitionable*, that is, $J = \bigcup (J_i : 1 \leqslant i \leqslant k)$ where $J_i \in \mathcal{I}_i$, $1 \leqslant i \leqslant k$. Obviously, we may assume that the J_i are disjoint. Moreover, the assumption that all M_i have underlying set S is made only for convenience. The main result of the theory is the following Matroid Partition Theorem. It is implicit in Edmonds and Fulkerson (1965). The special case in which the M_i are all equal is the subject of Edmonds (1965a). We mention that a theorem of Rado (1942) (see chapter 9 by Welsh) can be shown to be equivalent.

Theorem 3.9. *Let J be a maximum-cardinality partitionable subset with respect to $M_i = (S, \mathcal{I}_i)$, $1 \leqslant i \leqslant k$. Then $|J| = \min_{A \subseteq S} \left(\sum_{i=1}^{k} r_i(A) + |\overline{A}| \right)$.*

As usual we can observe that for any such J and A, we have

$$\begin{aligned} |J| &= |J \backslash A| + |J \cap A| \\ &\leqslant |S \backslash A| + \Sigma\, |J_i \cap A| \\ &\leqslant |\overline{A}| + \Sigma r_i(A). \end{aligned}$$

The matroid partitioning problem is reduced to a matroid intersection problem as follows. [This construction, and the reverse one described later, are due to Edmonds (1970).] Make k disjoint copies $S_1, S_2, \ldots, S_k$ of S, and imagine M_i as being defined on S_i rather than S. Let N_a be the direct sum of the M_i and let N_b be the matroid on $S' = \cup S_i$ in which a set is independent if and only if it contains at most one copy of e for each $e \in S$. It is easy to see that there is a correspondence between partitionable sets with respect to $M_1, \ldots, M_k$ and common independent sets of N_a, N_b. It is also easy to see that a set $B \subseteq S'$ that minimizes $r_a(B) + r_b(S' \backslash B)$ can be chosen to consist of all the copies of elements of A, for some $A \subseteq S$. It follows from the Matroid Intersection Theorem that the maximum size of a partitionable set is $\min(|\overline{A}| + \Sigma r_i(A)\colon A \subseteq S)$, proving Theorem 3.9.

There is an important strengthening of Theorem 3.9, namely that, in its statement, "maximum-cardinality" can be replaced by "maximal". It follows from the observation that whenever a copy of e is deleted from the common independent set of N_a, N_b by the intersection algorithm, it is replaced by another copy of e, so that no element is ever deleted from the partitionable set. Since the same argument could be applied to maximal partitionable subsets of an arbitrary subset B, we conclude that every maximal partitionable subset of B has the same cardinality. A consequence is the following result, which appeared explicitly for the first time in Nash-Williams (1967).

Theorem 3.10. *The subsets of S partitionable with respect to the M_i, form the independent sets of a matroid. Its rank function r is given by $r(B) = \min(|B\backslash A| + \Sigma r_i(A)\colon A \subseteq B)$.*

There is a neater description of the partitioning algorithm, obtained by identifying all copies of each element e of S in the auxiliary digraph for the intersection algorithm. The resulting digraph has node-set $S \cup \{s,t\}$ and has

an arc (s,e) for each $e \in S\backslash J$;

an arc (e,t) for each $e \in S$ such that $J_i \cup \{e\} \in \mathcal{I}_i$ for some i;

an arc (e,f) for each $e,f \in S$ such that $f \in C_i(J_i,\ e)$ for some i.

At termination of the algorithm, any set $A \subseteq S$ such that $\delta^+(A \cup \{s\}) = \emptyset$ has the property that $J_i \cap A$ M_i-spans A, and so $|J| = |\overline{A}| + \Sigma r_i(A)$. It is worthwhile also to observe that *every* set A that minimizes $|\overline{A}| + \Sigma r_i(A)$ must have this property and so must satisfy $\delta^+(A \cup \{s\}) = \emptyset$.

Now recall Problem 3.3 at the beginning of the section. By Theorem 3.10 the feasible solutions form the independent sets of a matroid. Hence Problem 3.3 can be solved by the greedy algorithm, with independence tests requiring applications of the matroid partition algorithm to k copies of the cycle matroid of the graph. The Matroid Partition Theorem applied to this example yields standard graph results of Nash-Williams and Tutte on the existence of disjoint spanning trees and the covering of edges by forests. (See chapter 2 by Frank.) There are also beautiful combinatorial applications in transversal theory; see chapter 9 (by Welsh) and Mirsky (1971).

Finally, let us describe Edmonds' reduction of intersection to partitioning. The proof is easy.

Theorem 3.11. *Let B be a maximal partitionable subset with respect to M_1 and M_2^*, and let J_1, J_2 be an associated partitioning. Extend J_2 to a basis B_2 of M_2^*. Then $B\backslash B_2$ is a maximum-cardinality common independent set of M_1 and M_2.*

3.5. Basis exchange

Recall Problem 3.4. If X_2 has the required properties, then $B_2\backslash X_2$, X_2 provides a partitioning of B_2 with respect to the matroids $M_1 = M/X_1$ and $M_2 = M/(B_1\backslash X_1)$. Thus X_2 can be found with the Matroid Partition Algorithm. Moreover, applying the Matroid Partition Theorem one gets, after a short calculation, the following result of Brylawski and Greene.

Theorem 3.12. *Let B_1, B_2 be bases of a matroid M and let $X_1 \subseteq B_1$. Then there exists $X_2 \subseteq B_2$ such that $(B_1 \backslash X_1) \cup X_2$ and $(B_2 \backslash X_2) \cup X_1$ are also bases.*

3.6. Solution of the Shannon game

The Shannon game, proposed by Shannon and generalized to matroids and solved by Lehman (1964), is a game played on a matroid with a single distinguished element e. (See also chapter 43 by Guy.) The two players, Short and Cut, alternately choose elements of $S \backslash \{e\}$, with elements chosen by Cut deleted from M and elements chosen by Short contracted. Short's (Cut's) objective is to reach a minor in which e is a loop (coloop).

A game (M, e) is called *short* (*cut*) if there is a winning strategy for Short (Cut) playing second (and hence also playing first). The game is *neutral* if it is neither cut nor short, that is, if the first player, whether Cut or Short, has a winning strategy. It is easy to see that (M, e) is a cut game if and only if (M^*, e) is short, and that (M, e) is neutral if and only if (M, e) is not short and (M', e) *is* short, where M' is obtained from M by adding an element parallel to e. Hence it is enough to characterize short games.

Theorem 3.13. *(M, e) is short if and only if there exist disjoint independent sets I_1, I_2 of M such that $e \in \sigma(I_1) = \sigma(I_2)$ and $e \notin I_1 \cup I_2$.*

The "if" part of the theorem can be proved by checking that the following strategy, applied iteratively, works. If Cut plays $f \in I_1$ then Short plays $e \in I_2$, where (with respect to the current minor just before f is deleted) $(I_1 \cup \{e\}) \backslash \{f\}$ is independent, and similarly for $f \in I_2$. (If Cut does not play an element of $I_1 \cup I_2$, then this only makes life easier for Short.) The "only if" part can be proved (this is harder) by showing that I_1, I_2 exist for one of $(M, e), (M^*, e), (M', e)$.

The condition of Theorem 3.13 is easily recognized by the partitioning algorithm applied with $M_1 = M_2 = M$. (In fact, Lehman's work was one of the motivations for Edmonds' development of matroid partitioning.) We know that the minimizers of g, where $g(A) = |S \backslash A| + 2r(A)$, are precisely the sets $A \subseteq S$ such that, in the auxiliary digraph at termination of the algorithm, $\delta^+(A \cup \{s\}) = \emptyset$. There is a unique smallest such A (easily found); call it A'. If $e \in A'$, then the minimum of g remains the same when e is deleted, so there is a maximum partitionable set $J = J_1 \cup J_2$ with $e \notin J$. Then $e \in A' = \sigma(J_1 \cap A') = \sigma(J_2 \cap A')$, so $J_1 \cap A', J_2 \cap A'$ are the required sets. On the other hand, we claim that if such sets I_1, I_2 exist, then necessarily $e \in A'$.

For suppose we start the partition algorithm with $J_1 = I_1$, $J_2 = I_2$. Then in the auxiliary digraph, we have $\delta^+(\sigma(I_1)) = \emptyset$, so no augmenting path can use any element of $\sigma(I_1)$ and so $I_1 \subseteq J_1$, $I_2 \subseteq J_2$, $e \notin J_1 \cup J_2$ will be maintained throughout execution of the algorithm. At termination, $\delta^+(A' \cup \{s\}) = \emptyset$, and so $e \in A'$, as required. Finally, we point out that the sets I_1, I_2, can be found by applying the algorithm to $M \backslash e, M \backslash e$ after first checking that $e \in A'$.

Further analysis of the game and interesting extensions can be found in Edmonds (1965b), Bruno and Weinberg (1971), and Hamidoune and Las Vergnas (1986).

3.7. Weighted matroid intersection

Just as $r_1(A) + r_2(\overline{A})$ provides an upper bound for the size of a common independent set J, where $c \in \mathbb{R}^S$ we can define a simple upper bound for $c(J)$. Namely let (c^1, c^2) be a "weight-splitting", that is, $c^1 + c^2 = c$. Then

$$c(J) = c^1(J) + c^2(J) \leqslant \max_{J_1 \in \mathcal{I}_1} c^1(J_1) + \max_{J_2 \in \mathcal{I}_2} c^2(J_2). \tag{3.14}$$

Hence, if we find $J \in \mathcal{I}_1 \cap \mathcal{I}_2$ and a weight-splitting (c^1, c^2) such that equality holds in (3.14), we know that J has maximum weight. In fact, this is always possible.

Theorem 3.15. *For matroids M_1, M_2 on S and $c \in \mathbb{R}^S$, there exists a weight-splitting (c^1, c^2) and a set J such that J has maximum c^i-weight among independent sets of M_i, for $i = 1$ and 2.*

We shall also see that, if c is integer-valued, then there is an integral weight-splitting in Theorem 3.15. Actually, Theorem 3.15 is equivalent to Theorem 3.16, the *Matroid Intersection Polytope Theorem* of Edmonds (1970). [Notice that the converse of (3.16) is trivially true.]

Theorem 3.16. *For matroids M_1, M_2 on S, every extreme point of $P(M_1) \cap P(M_2)$ is the characteristic vector of a common independent set.*

Proof of equivalence of Theorem 3.15 and 3.16. Suppose that Theorem 3.15 holds and let $\bar{x}$ be an extreme point of $P(M_1) \cap P(M_2)$. Choose $c \in \mathbb{R}^S$ such that $\bar{x}$ is the unique optimal solution of $\max(c \cdot x\colon x \in P(M_1) \cap P(M_2))$. Then by Theorem 3.15 we have $J \in \mathcal{I}_1 \cap \mathcal{I}_2$ and (c^1, c^2) such that $c(J) = c^1(J) + c^2(J) \geqslant c^1 \cdot \bar{x} + c^2 \cdot \bar{x} = c \cdot \bar{x}$. It follows that $\bar{x}$ is the incidence vector of J.

Now suppose that every extreme point of $P(M_1) \cap P(M_2) = \{x \in \mathbb{R}^S_+ : x(A) \leqslant r_1(A),\ x(A) \leqslant r_2(A), \text{ for } A \subseteq S\}$ is the characteristic vector of a common independent set. Given $c \in \mathbb{R}^S$, let J be a maximum-weight common independent set, and let $\bar{x}$ be its characteristic vector. Let (y^1, y^2) be an optimal solution of the linear program dual to $\max(c \cdot x\colon x \in P(M_1) \cap P(M_2))$. Then the optimality conditions imply that $|J \cap A| = r_i(A)$ whenever $y^i(A) > 0$. For $i = 1$ and 2 and $j \in S$, let $c^i_j = \sum(y^i(A)\colon j \in A)$; then $c \leqslant c^1 + c^2$. Thus $\bar{x}, y^i$ satisfy the optimality conditions for $\max(c^i \cdot x\colon x \in P(M_i))$ and its dual, so J is c^i-optimal in M_i, for $i = 1$ and 2. Moreover, $c(J) = c^1(J) + c^2(J)$, so $c^1_j + c^2_j > c_j$ implies $j \notin J$, so c^2_j can be lowered to $c_j - c^1_j$ without affecting the c^2-optimality of J. □

Although Edmonds (1970) used the idea of weight-splitting in a non-constructive proof of Theorem 3.16, it was Frank (1981) who showed how to use the optimality conditions based on Theorem 3.15 to simplify the weighted Matroid Intersection Algorithms of Edmonds (1979) and Lawler (1975). We describe here an algorithm based essentially on Frank's, but with an additional simplification.

The basic idea is to generalize the unweighted MIA by using a weight-splitting to assign costs to the arcs of the auxiliary digraph. Where c_0^i denotes $\max(c_e^i\colon e \notin J, J \cup \{e\} \in \mathcal{I}_i)$, the arc costs w_{uv} are defined by:

$$w_{et} = c_0^1 - c_e^1;$$

$$w_{se} = c_0^2 - c_e^2;$$

$$w_{ef} = -c_e^1 + c_f^1;$$

$$w_{fe} = -c_e^2 + c_f^2.$$

We shall require that $(c^1, c^2), J$ satisfy the properties $c_0^2 = 0$, and $w_{uv} \geqslant 0$ for each arc uv. If in addition, we have $c_0^1 \leqslant 0$, then the conditions of Theorem 3.15 are satisfied (by Theorem 2.7) and we are finished. We can begin with $c^1 = c$, $c^2 = 0$, $J = \emptyset$. (Notice that, if we augment on an $(s,\ t)$ dipath P to obtain J' from J, then the cost of P is $c_0^1 + c_0^2 + c(J) - c(J')$. This motivates choosing P to have least cost. In fact, solving a least-cost dipath problem gives a way to update the weight-splitting too. This observation makes possible the following simpler presentation of Frank's algorithm. We remark that Lawler also used a shortest-path calculation, but without weight-splitting.)

Iteration of weighted MIA.
If $c_0^1 \leqslant 0$, stop;
Form $G = G(M_1, M_2, J, c^1, c^2)$;
Compute a dipath from s to v of least-cost d_v for each $v \in S \cup \{s,t\}$;
For each $v \in S$, let $\alpha_v = \min(d_v, d_t, c_0^1)$, and replace c_v^1 by $c_v^1 - \alpha_v$, c_v^2 by $c_v^2 + \alpha_v$;
If $c_0^1 \leqslant 0$, stop;
Augment J on a zero-cost (s,t) dipath having as few arcs as possible.

Notice that the resulting algorithm has essentially the same complexity as its unweighted version, since the non-negative-cost shortest-path calculation can be done in time $O(n^2)$ (see chapter 2 by Frank). We outline a proof of validity of the algorithm. There are three things to check: (i) that the change in (c^1, c^2) preserves the properties required of it; (ii) that J remains common independent after an augmentation; (iii) that an augmentation does not violate the properties required of (c^1, c^2). It is straightforward to check (i), using the fact that the d_v satisfy $d_u + w_{uv} \geqslant d_v$. Notice that (ii) is not obvious, since the dipath may have chords (but not zero-cost ones). One actually shows that the subgraph induced by the zero-cost arcs is $G(M_1',\ M_2', J)$ for new matroids M_1', M_2' for which J is common independent, and every common independent set is also independent in both M_1 and M_2. Then the result follows from the validity of the unweighted MIA. To define M_1', let $p_1 > p_2 > \cdots > p_k$ be the distinct values of c_e^1 that occur in J. Let $T_0 = \emptyset$, let $T_i = \{e \in S\colon c_e^1 \geqslant p_i\}$ for $1 \leqslant i \leqslant k-1$, and let $T_k = \sigma(J)$. Where $N_i = (M/T_i) \backslash \bar{T}_{i+1}, 0 \leqslant i \leqslant k$, put $M_1' = N_0 \oplus \cdots \oplus N_k$. For (iii), it is easy to show that the change in J preserves the property that $c_0^2 = 0$; to show that it also preserves $w \geqslant 0$, we use the fact that the latter condition is equivalent to J being c^i-optimal of its cardinality in $\mathcal{I}_i$, $i = 1$

and 2. Consider $i = 1$, and let e_{m+1} be the second-from-last node of the augmenting path P. Then $c^1_{m+1} = c^1_0$, so $J \cup \{e_{m+1}\}$ is c^1-optimal of cardinality $|J| + 1$ in $\mathcal{I}_1$. But $c^1(J') = c^1(J \cup \{e_{m+1}\})$, since each element of J on P has the same c^1-weight as the preceding node on P, and we are done.

A good deal of work has been done on faster implementations of an MIA for special classes of matroids. See Gabow and Tarjan (1984), Brezovec et al. (1988), and Gabow and Xu (1989).

4. Submodular functions and polymatroids

Let f be a function defined on subsets of S, with values in $\mathbb{R}$; f is *submodular* if

$$f(A) + f(B) \geqslant f(A \cup B) + f(A \cap B) \quad \text{for all } A, B \subseteq S.$$

Some examples:

Example 4.1. *Let M be a matroid on S, and let f be the rank function of M.*

Example 4.2. *Let $G = (V, E)$ be a digraph, let $s \in V$, let $S = V \backslash \{s\}$, and let $f(A) = |\delta^-(A)|$. (Specifying s is not necessary here, but will be useful later.)*

Given a set function f on S we let $P(f)$ denote the polyhedron $\{x \in \mathbb{R}^S_+ : x(A) \leqslant f(A)$ for all $A \subseteq S\}$. A *polymatroid* is a polyhedron of the form $P(f)$ where f is submodular and non-negative. It is said to be *integral* if f is integral. It follows from Theorem 2.5 that the polymatroid defined by f of Example 4.1 is the convex hull of characteristic vectors of independent sets of M. In Example 4.2, if there exists in G a family of arc-disjoint directed paths, each beginning at s and ending in S, and we let x_v denote the number of dipaths ending at v, then it is easy to see that $x \in P(f)$. It is a consequence of a form of Menger's Theorem that every integer-valued element of $P(f)$ arises in this way.

The first result (Theorem 4.3) shows that our polymatroids do satisfy the central condition of the original geometric definition of Edmonds (1970): For any $u \in \mathbb{R}^S_+$, all maximal (with respect to component-wise order) vectors $x \in P$ with $x \leqslant u$, have the same component-sum. Another consequence of this result is a construction for matroids: The $\{0,1\}$-valued vectors in an integral polymatroid are the characteristic vectors of the independent sets of a matroid, and Theorem 4.3 gives a formula for its rank function. In particular, the fact that r of Theorem 3.10 is a matroid rank function follows from applying this to $f = \Sigma r_i$.

Theorem 4.3. *Let f be submodular on S, let $u \in \mathbb{R}^S_+$, and let x be any maximal vector satisfying $x \leqslant u, x \in P(f)$. Then $x(S) = \min_{A \subseteq S}(f(A) + u(\overline{A}))$. Moreover, if f and u are integer-valued and x is required to be integer-valued, the conclusion is still satisfied.*

Proof. First we observe that, for any $A \subseteq S$ and any x (maximal or not), we have $x(S) = x(A) + x(\overline{A}) \leqslant f(A) + u(\overline{A})$. Therefore, it will be enough to prove that there is some A for which equality holds. Obviously, for each $j \in S$, if $x_j \neq u_j$, then by the maximality of x, there is a set $A_j \subseteq S$ with $j \in A_j$ such that $x(A_j) = f(A_j)$. Call such a set x-*tight*, or just *tight*. A fundamental fact is:

Claim 4.4. *The intersection and union of tight sets are also tight.*

Proof of Claim 4.4. If A, B are tight, we have

$$\begin{aligned} x(A \cup B) + x(A \cap B) &\leqslant f(A \cup B) + f(A \cap B) \\ &\leqslant f(A) + f(B) \\ &= x(A) + x(B) \\ &= x(A \cup B) + x(A \cap B), \end{aligned}$$

so $x(A \cup B) = f(A \cup B)$ and $x(A \cap B) = f(A \cap B)$. □

Now if we choose A to be the union of the A_j, then A is tight and $x(\overline{A}) = u(\overline{A})$, so we are finished. The same proof applies to the integer-restricted version. □

It is a consequence of Theorem 4.3 that a greedy algorithm maximizes $x(S)$ over $P(f)$ (or more generally over $\{x: x \in P(f),\ x \leqslant u\}$). The algorithm begins with $x = 0$, and for each $j \in S$ increases x_j as much as possible subject to the restriction that $x \in P(f)$. Just as in the special case of matroid polytopes, we generalize to arbitrary weight-vectors by treating the elements in order of non-increasing weight.

Polymatroid Greedy Algorithm (PGA).
Order $S = \{e_1, \ldots, e_n\}$ so that $c_{e_1} \geqslant \cdots \geqslant c_{e_m} \geqslant 0 \geqslant c_{e_{m+1}} \geqslant \cdots \geqslant c_{e_n}$
$x := 0$
For $i = 1$ to m do
  Choose x_{e_i} as large as possible so that $x \in P(f)$.

Theorem 4.5. *For any non-negative submodular function f on S and any $c \in \mathbb{R}^S$, PGA optimizes $c \cdot x$ over $P(f)$. Moreover, if f is integer-valued, the output of PGA is integer-valued.*

Where $T_i = \{e_1, \ldots, e_i\}$ and $\bar{x}$ is the output of PGA, one can apply Theorem 4.3 to deduce that $\bar{x}(T_i) = \max(x(T_i): x \in P(f))$. With this observation, Theorem 4.5 can be proved in the same way as Theorem 2.2. Other results for matroid polytopes immediately generalize. These include the fact that the extreme points of $P(f)$ are precisely the vectors that can be the output of PGA, the dual greedy algorithm, and the total dual integrality of the linear system for $P(f)$. One also can prove a converse of Theorem 4.5 similar to Theorem 2.3, characterizing polymatroids as the compact subsets of $\mathbb{R}^S_+$ that are closed below and for which a greedy algorithm always works.

Call a function f a *polymatroid function* if f is submodular, normalized ($f(\emptyset) = 0$), and monotone ($f(A) \geqslant f(B)$ if $A \supseteq B$). For an arbitrary submodular function f, the function f' defined by $f'(A) = \min(f(B): B \supseteq A)$, is obviously monotone, and is easily seen to be submodular. Notice that, for the f of Example 4.2, $f(A)$ is the maximum number of edge-disjoint directed paths beginning at s and ending in A.

More generally, applying Theorem 4.3 with $u_j = 0, j \notin A$ and u_j large, $j \in A$, yields

$$f'(A) = \max(x(A): x \in P(f)), \text{ for } A \neq \emptyset. \tag{4.6}$$

It follows that $P(f) \subseteq P(f')$. But $f' \leqslant f$ by definition, so $P(f) = P(f')$. Notice also that if f is a polymatroid function, then $f = f'$ and PGA reduces to a formula (which works in the slightly more general case when f is non-negative and monotone); namely, the output $\bar{x}$ of PGA satisfies

$$\bar{x}_{e_i} = \begin{cases} f(T_i), & i = 1; \\ f(T_i) - f(T_{i-1}), & 2 \leqslant i \leqslant m; \\ 0, & i > m. \end{cases} \tag{4.7}$$

It is easy to derive from these observations the following result of Edmonds (1970).

Theorem 4.8. *Every polymatroid is determined by a (unique) polymatroid function.*

4.1. Submodular function minimization

We have seen from (4.7) that when f is non-negative and monotone, the greedy algorithm is especially simple. Its efficiency depends on the ease with which we can obtain function values, and the size (number of digits) of the values. There is no difficulty with the first of these, since we assume that the function is given via an evaluation oracle. On the other hand, the maximum size of function values must, like $n = |S|$, be treated as a measure of input size. (We must do the same for the element weights c_j.) With these ground rules, the greedy algorithm can be regarded as a polynomial-time oracle algorithm, when f is monotone.

On the other hand, if no additional assumption on f is made, computing component e of $\bar{x}$ in the greedy algorithm requires minimizing $f(A) - \bar{x}(A)$ over subsets A containing e, a problem easily seen to be equivalent to that of finding the minimum of an arbitrary submodular function. This latter problem is fundamental; it includes as special cases both the minimum cut problem and, by Theorem 3.9, the problem of finding the maximum size of a partitionable set. Using some of the above results, we shall show that the minimum of a submodular function can be well-characterized. (The obverse problem of maximizing a submodular function, on the other hand, includes NP-hard special cases, and can be easily proved intractable in the oracle context.)

It is useful to reduce the general problem of submodular function minimization to that of minimizing a function of the form $f(A) - u(A)$, where f is a polymatroid function and $u \in \mathbb{R}^S_+$. Let g be a submodular function on S and let $u_j = g(S \backslash \{j\}) - g(S)$, $j \in S$. If $u_j < 0$, it is easy to see that no minimizer of g will include j, so the problem could be restricted to subsets of $S \backslash \{j\}$. Hence we may assume that $u \geqslant 0$. The function f defined by $f(A) = g(A) + u(A) - g(\emptyset)$ is easily shown to be a polymatroid function. Hence minimizing g is equivalent to minimizing $f(A) + u(\overline{A})$, since this differs from g by the constant $u(S) - g(\emptyset)$. Then Theorem 4.3 characterizes the minimum. That this is a useful characterization is not completely obvious, since the maximizing x must be certifiably in $P(f)$. But x can be expressed as a convex combination of at most $n+1$ extreme points of $P(f)$, by a standard result in polyhedral theory, and these can be generated by PGA, since f is a polymatroid function.

In fact, a minimization algorithm can be based on the above ideas: Maintain $x \in P(f)$ with $x \leqslant u$ explicitly as a convex combination of extreme points of $P(f)$, and at each step either find A giving equality in Theorem 4.3, or find a new x with $x(S)$ larger. This combinatorial approach was first developed (Cunningham 1984) for the special case in which f is a matroid rank function. The resulting algorithm, which can be viewed as a generalization of the matroid partition algorithm, runs in polynomial time. For general f, a finite algorithm occurs in Bixby et al. (1985), and it was modified to run in "pseudo-polynomial" time (Cunningham 1985).

Grötschel et al. (1981) *did* find a polynomial-time algorithm for submodular function minimization. It is based on the equivalence, via the ellipsoid method, of the optimization and separation problems for polyhedra. (See chapter 30 by Schrijver.) Since we wish to minimize $f(A) - u(A)$, it is enough to be able to determine, given $K \in \mathbb{R}$, either a set $A \subseteq S$ for which $f(A) - u(A) < K$, or the information that no such A exists. (For then one could search over K for a sufficiently small K for which A does exist.) The function f_K defined by $f_K(B) = f(B) - K$ is submodular and monotone, and such A exists if and only if $u \notin P(f_K)$. But this is the separation problem for $P(f_K)$, and since f_K is monotone, the optimization problem for $P(f_K)$ is solvable, and we are done. The resulting algorithm, while theoretically acceptable, is not computationally useful. An important open question is the existence of a polynomial-time *combinatorial* minimization algorithm.

A set function is symmetric if $f(A) = f(\overline{A})$ for all $A \subseteq S$. It is easy to see that a symmetric submodular function is minimized by $\emptyset$ and by S. However, it is an interesting problem to minimize such a function over $\{A\colon \emptyset \subset S \subset V\}$. This problem includes the "global minimum cut" problem for undirected graphs. It is solved by a combinatorial algorithm requiring $O(|S|^3)$ function evaluations in Queyranne (1994).

4.2. Polymatroid intersection

The Matroid Intersection Theorem (3.5) is a special case of a result of Edmonds (1970) on polymatroids.

Theorem 4.9. *Let f_1, f_2 be polymatroid functions on S. Then*

$$\max\,(x(S) : x \in P(f_1) \cap P(f_2)) = \min_{A \subseteq S}\,(f_1(A) + f_2(S \backslash A))\,.$$

Moreover, if f_1, f_2 are integer-valued, then the maximizing x can be chosen to be integer-valued.

We remark that, if f_1, f_2 are not required to be monotone, then by monotonization arguments the same result holds, except that the right-hand side becomes $\min(f_1(A) + f_2(B) : A \cup B = S)$. The inductive proof of Theorem 3.5 outlined in section 2 generalizes to a proof of the integral version of Theorem 4.9. (The induction is now on $\Sigma(\min(f_1(\{j\}), f_2(\{j\}))\colon j \in S)$, and the appropriate analogue of contracting $e \in S$ is to form the function f_i^e by $f_i^e(A) = \min(f_i(A), f_i(A \cup \{e\}) - 1)$. The non-integral version can be deduced from the integral version in a straightforward way. Later we shall see other proofs and algorithmic aspects of (generalizations of) Theorem 4.9.

The following sandwich theorem of Frank (1982) is a useful and attractive restatement of Theorem 4.9. Its resemblance to classical results on separation of convex and concave functions is evident; we shall see that the relationship is more than an analogy. A set function f is *supermodular* if $-f$ is submodular, and is *modular* if it is both sub- and supermodular. It is easy to see that a function m is modular on subsets of S if and only if $m(A) = x(A) + k$ for some $x \in \mathbb{R}^S, k \in \mathbb{R}$.

Theorem 4.10. *Let g, h be defined on subsets of S such that g is supermodular, h is submodular, and $g \leqslant h$. Then there exists a modular function m satisfying $g \leqslant m \leqslant h$. Moreover, if f and g are integer-valued, then m may be chosen integer-valued.*

To derive Theorem 4.10 from Theorem 4.9 one proceeds as follows. Add a constant to g and h to make $g(\emptyset) = 0$. Lower $h(\emptyset)$ to 0. Raise $g(S)$ to $h(S)$. Add a function p of the form $p(A) = M\,|A|$ to make f and g monotone. Now take $f_1 = h$, define f_2 by $f_2(A) = g(S) - g(S \backslash A)$, apply Theorem 4.9 to find $x \in P(f_1) \cap P(f_2)$ with $x(S) = g(S)$, and define m by $m(A) = x(A)$. A similar construction allows the derivation of Theorem 4.9 from Theorem 4.10.

4.3. Optimization over the intersection of polymatroids

The problem of optimizing a linear function over the intersection of two polymatroids may be stated as a linear program:

$$\begin{aligned} &\text{maximize } c \cdot x \\ &\text{subject to } \quad x(A) \leqslant f_1(A), \quad A \subseteq S; \\ &\qquad\qquad\quad\; x(A) \leqslant f_2(A), \quad A \subseteq S; \\ &\qquad\qquad\quad\; x_j \geqslant 0, \quad j \in S. \end{aligned} \tag{4.11}$$

The dual linear program is

$$\begin{aligned} &\text{minimize } \sum (f_1(A) y_A^1 + f_2(A) y_A^2 : A \subseteq S) \\ &\text{subject to } \quad \sum (y_A^1 + y_A^2 : j \in A \subseteq S) \geqslant c_j, \quad j \in S; \\ &\qquad\qquad\quad\; y_A^1, y_A^2 \geqslant 0, \quad A \subseteq S. \end{aligned} \tag{4.12}$$

The main result (Edmonds 1970) on this topic may be stated as follows.

Theorem 4.13. *If f_1, f_2 are integer-valued, then* (4.11) *has an optimal solution that is integer-valued. If c is integer-valued, then* (4.12) *has an optimal solution that is integer-valued.*

A number of important results are consequences of Theorem 4.13. For example, taking f_1, f_2 to be matroid rank functions, we can conclude that the intersection of two matroid polyhedra is a polyhedron with $\{0,1\}$-valued extreme points, and thus derive the Matroid Intersection Polyhedron Theorem. A second consequence is the Polymatroid Intersection Theorem (4.9), obtained by taking each $c_j = 1$ and observing that y^1, y^2 can be required to take a very special form.

A proof of Theorem 4.13 based on the theory of total dual integrality and total unimodularity can be found in chapter 30 by Schrijver. In the next section we treat a generalization of Theorem 4.13.

It is worthwhile to identify the results that extend to the intersection of three (or more) polymatroids. It is possible to optimize any linear function over the intersection of three polymatroids, using the ellipsoid method, since the separation problem can be solved efficiently. However, the Integrality Theorem (4.13) does not generalize, and optimizing over the integral vectors in three polymatroids, even over common independent sets of three matroids, is difficult. Although we are not aware of a proof that this problem is unsolvable in the oracle context, it is well known to contain NP-hard problems.

4.4. Some extensions

It is frequently useful in applications to relax some of the assumptions in the definition of polymatroid. A first such variant is to drop non-negativity, considering $Q(f) = \{x \in \mathbb{R}^S\colon x(A) \leqslant f(A) \text{ for all } A \subseteq S\}$. For such *submodular polyhedra* Theorem 4.3 still holds with the same proof. In addition the greedy algorithm works (for any $c \in \mathbb{R}^S_+$) with the same proof. Moreover, we do not need monotonicity for the formula (4.7) to be correct.

This greedy algorithm for $Q(f)$ motivates the definition (Lovász 1983) of an extension of a set function. For $c \in \mathbb{R}^S_+$ let $\hat{f}(c)$ denote $\max(c \cdot x\colon x \in Q(f))$. It is easy to see (essentially from the proof of the greedy algorithm) that if f is submodular, $\hat{f}(c)$ can also be calculated as follows: c can be expressed (uniquely) as $\sum_{i=1}^{k} \lambda_i \chi^{T_i}$ for $\lambda_i > 0$ and $T_i \subseteq S$ with $T_1 \supset T_2 \supset \cdots \supset T_k$. Then $\hat{f}(c) = \sum_{i=1}^{k} \lambda_i f(T_i)$. This can be taken as the definition of the extension of any set function, submodular or not, to a function on $\mathbb{R}^S_+$. Now we can make the connection between submodularity and convexity more explicit.

Theorem 4.14. *f is submodular on S if and only if $\hat{f}$ is convex on $\mathbb{R}^S_+$.*

This result, from Lovász (1983), has a straightforward proof. When combined with a standard result on separation of convex and concave functions, it implies the first part of Frank's Theorem (4.10).

Another useful extension is to allow a function to take value ∞ on some of the subsets of S. With this extension essentially all of the previous results still obtain, with obvious exceptions caused by unboundedness of $P(f)$. (It *is* true that if $P(f)$ is bounded and non-empty, then it is a polymatroid.) This idea is often combined with another important extension, namely, requiring the submodular inequality to hold only for certain pairs of sets. We say that f is *intersecting* (*crossing*) *submodular* if $f(A)+f(B) \geqslant f(A\cup B)+f(A\cap B)$ whenever $A\cap B\neq\emptyset$ ($A\cap B\neq\emptyset$ and $A\cup B\neq S$). (When considering such weaker notions, we sometimes refer to ordinary submodular functions as *fully* submodular.) Edmonds (1970) considered intersecting submodular functions and proved extensions of most of our earlier results. For example, the next result generalizes Theorem 4.3; the same idea underlies the proof.

Theorem 4.15. *Theorem 4.3 is true if f is intersecting submodular, with* $\min_{A\subseteq S}(f(A) + u(\overline{A}))$ *replaced by* $\min(\Sigma f(A_i)+u(\overline{\cup A_i})\colon \emptyset\neq A_i\subset S,\ A_i$ *pairwise disjoint).*

As for Theorem 4.3, we can conclude from Theorem 4.15 that if f is integer-valued and intersecting submodular, then the $\{0,1\}$-valued vectors in $P(f)$ correspond to the independent sets of a matroid. A classical example of this construction results in the forest matroid of a graph $G(V,E)$. Here we take $S=E$ and $f(A)=|V(A)|-1$ for $A\neq\emptyset$, with $f(\emptyset)=0$.

The following consequence of Theorem 4.15 is also useful.

Theorem 4.16. *If f is intersecting submodular on S and f' is defined by* $f'(A) = \min(\Sigma f(A_i)\colon A=\bigcup A_i,\ \emptyset\neq A_i\subseteq S,\ A_i$ *pairwise disjoint), then f' is submodular on S, and $Q(f')=Q(f)$.*

There is also a construction that produces an intersecting submodular function beginning with a crossing submodular function; it is from Frank (1982) and also implicitly Fujishige (1984).

Theorem 4.17. *If f is crossing submodular on S with $f(S)$ finite, and f' is defined by* $f'(A)=\min(\Sigma f(A_i)\colon \overline{A}=\bigcup\overline{A}_i,\ A_i\subset S,\ \overline{A}_i$ *pairwise disjoint), then f' is intersecting submodular; moreover,* $Q(f')\cap\{x\in\mathbb{R}^S\colon x(S)=f(S)\}=Q(f)\cap\{x\in\mathbb{R}^S\colon x(S)=f(S)\}$.

Notice the essential difference between Theorems 4.16 and 4.17. If f is crossing submodular, it need not be true that $Q(f)=Q(f')$; in fact, $Q(f)$ need not be a submodular polyhedron. It is the *base polyhedron* $B(f)=Q(f)\cap\{x\in\mathbb{R}^S\colon x(S)=f(S)\}$ that is preserved. However, we can still construct a matroid from a crossing submodular function (Frank and Tardos 1984).

Theorem 4.18. *Let f be integer-valued and crossing submodular on S and let $k\in\mathbb{Z}_+$. Then $\{B\subseteq S\colon |B|=k,\ \chi^B\in Q(f)\}$, if non-empty, is the basis family of a matroid.*

5. Submodular flows and other general models

In this section we describe several more general models. We give considerable attention to submodular flows, a generalization of polymatroid intersection. In particular, we describe the basic ideas behind solution algorithms. (This has not been done for polymatroid intersection.) We also report on the polymatroid matching problem, a submodular model that includes graph matching as a special case. Finally, we describe two of the many additional generalizations of matroids, delta-matroids and greedoids.

5.1. Submodular flows: Models and applications

The optimal submodular flow problem is a generalization of the problem of optimizing over the intersection of two polymatroids, that keeps the important integrality and algorithmic properties of the latter problem. In addition, it contains several other fundamental problems. There are a number of closely related models, introduced under other names, and many of the important contributions have been made in these differing contexts. [Schrijver (1984) explains the connections between these models.] However, we define just two of the models, and state the main results in the language of one of them.

First we describe the *polymatroidal network flow* model of Lawler and Martel (1982); it was introduced independently by Hassin (1982). Let $G=(V,E)$ be a digraph, let s,t be distinct elements of V, and for each $v \in V$ let f_v, g_v be polymatroid functions on $\delta^-(v), \delta^+(v)$. A *feasible flow* is a vector $x=(x_j\colon j \in E)$ satisfying

$$\begin{aligned}
&x(\delta^-(v)) - x(\delta^+(v)) = 0, \quad \text{for } v \in V\backslash\{s,t\};\\
&x(A) \leqslant f_v(A), \quad \text{for all } v \in V, \text{ all } A \subseteq \delta^-(v);\\
&x(A) \leqslant g_v(A), \quad \text{for all } v \in V, \text{ all } A \subseteq \delta^+(v);\\
&x_j \geqslant 0, \quad \text{for all } j \in E.
\end{aligned}$$

If $u \in \mathbb{R}^E_+$ and we take $f_v(A)=u(A)$, $g_v(A)=u(A)$, then the feasible flows are the feasible flows of an ordinary (single-source, single-sink) flow network. If we take $V=\{s,t\}$ and allow g_s, f_t to be arbitrary polymatroid functions, the feasible flows are the elements of $P(g_s) \cap P(f_t)$.

The model for which we shall present the main results is one introduced by Edmonds and Giles (1977). Let $G=(V,E)$ be a directed graph, let b be a crossing submodular function on V, and let $\ell, u, c \in \mathbb{R}^E$. We allow values of u and b to be ∞, and we allow values of ℓ to be $-\infty$. The *optimal submodular flow problem* is

$$\begin{aligned}
&\text{maximize } \sum(c_j x_j\colon j \in E)\\
&\text{subject to } \; x(\delta^-(A)) - x(\delta^+(A)) \leqslant b(A), \quad A \subseteq V; \qquad (5.1)\\
&\qquad\qquad\quad \ell_j \leqslant x_j \leqslant u_j, \quad j \in E.
\end{aligned}$$

We shall use the term *feasible flow* to refer to a vector x satisfying the constraints of (5.1). The special case where b is identically 0, is the well-known optimal circulation problem of network flow theory. Let us also show that the polymatroid intersection problem (4.11) can be cast in this form. For each $j \in S$, let j_1, j_2 be copies of j, let A_i denote $\{j_i\colon j \in A\}$ for any $A \subseteq S$, and let $G=(V,E)$ be defined by $V = S_1 \cup S_2$ and $E = S$ with $j = (j_1, j_2)$ for each $j \in S$. Define $\ell_j = 0,\ u_j = \infty$ for each j. Define $b(A_2) = f_1(A)$ and $b((S_1 \backslash A_1) \cup S_2) = f_2(A)$ for all $A \subseteq S$, $b(A) = \infty$ for all other $A \subseteq V$. The resulting (5.1) is exactly (4.11). We encourage the reader to check that b is crossing submodular.

The main integrality result for (5.1) is due to Edmonds and Giles (1977).

Theorem 5.2. *If ℓ, u, b are integer-valued and* (5.1) *has an optimal solution, then it has one that is integer-valued. If c is integer-valued and the dual of* (5.1) *has an optimal solution, then it has one that is integer-valued.*

The proof of Theorem 5.2 uses an idea similar to that of (4.13) in chapter 30 by Schrijver: the dual of (5.1) has an optimal solution whose non-zero components form an optimal solution to a linear program having a totally unimodular constraint matrix. Such an optimal solution is obtained from any optimal solution y by successive "uncrossings", that is, given sets A, B with $A \cap B$, $A \backslash B$, $B \backslash A$, $V \backslash (A \cup B)$ all non-empty and $y_A, y_B > 0$, decreasing y_A, y_B by $\varepsilon = \min(y_A, y_B)$ and increasing $y_{A \cup B}, y_{A \cap B}$ by ε.

As yet another illustration of the power of the submodular flow model, we show how the Lucchesi–Younger dicut-covering result (chapter 2 by Frank) can be derived from Theorem 5.2. Given $G = (V, E)$, we put $\ell_j = 0,\ c_j = -1$, and $u_j = \infty$ for each $j \in E$. For each $A \subseteq V$ such that $\delta^-(A) = \emptyset$ and $\emptyset \neq A \neq V$ we put $b(A) = -1$, and $b(A) = \infty$ for all other A. Then it is easy to check that an optimal integer-valued feasible flow is the characteristic vector of a minimum-cardinality cover of directed cuts, and an optimal integer-valued solution of the dual problem picks out a collection of arc-disjoint directed cuts. Hence the Lucchesi–Younger result follows from Theorem 5.2 and the linear programming duality theorem.

5.2. *Submodular flow algorithms*

Suppose we are given $x = (x_j \colon j \in E)$ and want to determine whether x is a feasible flow. Then it will be enough to be able to minimize $g(A) = b(A) - x(\delta^-(A)) + x(\delta^+(A))$ over $A \subseteq V$. Since g is a (crossing) submodular function, there exists a polynomial-time (ellipsoid) algorithm to minimize it. Hence by the equivalence of separation and optimization, there exists a polynomial-time algorithm for (5.1). The resulting algorithm uses the ellipsoid method on two different levels. The search for better algorithms has succeeded in decreasing this reliance on the ellipsoid method. We shall outline an efficient combinatorial algorithm for (5.1), assuming the availability of a subroutine for minimizing a submodular function. It follows that, if an efficient combinatorial algorithm for the latter problem is discovered, then (5.1) is also solved in a satisfactory way. In addition, there exist instances of (5.1) for which the submodular functions arising can be minimized by efficient combinatorial algorithms. An example is the problem of re-orienting the arcs of a digraph at minimum cost so as to make it k-arc-diconnected (Frank 1982).

Notice that x is a feasible flow if and only if $\ell \leqslant x \leqslant u$ and $Bx \in Q_0(b) = Q(b) \cap \{z \in \mathbb{R}^V \colon z(V) = 0\}$, where B is an appropriately defined matrix. (Recall that $Q(b) = \{x \in \mathbb{R}^S \colon x(A) \leqslant b(A),\ A \subseteq S\}$.) The results (4.16) and (4.17) tell us that $Q_0(b) = Q_0(b')$ for some fully submodular function b', since there is no harm in assuming that $b(V) = 0$. Hence the feasible flows remain the same when b is replaced by b'; this is a result of Fujishige (1984). In addition the submodular oracle that is used has the same output for b and for b'. Therefore, we may pretend that b is fully submodular even if it is not. (There are two exceptions to this statement; for purposes of this exposition we ignore them.)

5.3. Submodular flow algorithms: Maximum flows and consistent BFS

The *maximum (submodular) flow problem* is to find a feasible flow that maximizes x_f for a fixed arc $f \in E$. (Notice that it is an equivalent problem to minimize x_f, since f's direction could be reversed.) This is a special case of (5.1), and includes as special cases the ordinary network maximum-flow problem (chapter 2 by Frank), and the problem of finding a maximum component-sum vector in the intersection of two polymatroids. (The first is easy to see; the second requires modifying the previously-described submodular flow representation of polymatroid intersection.) The shortest augmenting path technique of maximum flow theory (see chapter 2 by Frank) generalizes to this context, but we also need an important further refinement of breadth-first search.

The algorithm for the maximum flow problem generalizes the cardinality Matroid Intersection Algorithm, as well as the usual network maximum-flow algorithm. To motivate the augmentation used, we first describe two special cases. Suppose that we find a circuit in G including f and such that $x_e < u_e$ for all arcs having the same orientation as f and $x_e > \ell_e$ for all arcs having opposite orientation to f. Then x_f could be increased by sending flow around the circuit, increasing x_e by ε for arcs of the first kind and decreasing x_e by ε for arcs of the second kind; ε must not exceed $u_e - x_e$ for any arc of the first kind, or $x_e - \ell_e$ for any arc of the second kind. Next, suppose that we have a path with the same properties, say from q to p, and we attempt to send flow along the path. Now there is an additional limitation on ε; for the new flow to satisfy the constraints of (5.1), ε cannot exceed $\varepsilon_x(p,q)$, defined to be $\min(b(A) - x(\delta^-(A)) + x(\delta^+(A))$: $p \in A, q \notin A)$. This limitation could be represented as an upper bound for the flow on a fictitious arc (p, q). An actual augmentation in the algorithm will consist of a sequence of augmentations on paths linked into a circuit by the addition of fictitious arcs, and is best described via an auxiliary digraph.

Given a feasible flow x, let $G' = G'(G, b, \ell, u, x)$ have node-set V and:

for each $e = (p,q) \in E$ with $x_e < u_e$, an arc (p,q) with capacity $u_e - x_e$;

for each $e = (p,q) \in E$ with $x_e > \ell_e$, an arc (q,p) with capacity $x_e - \ell_e$;

for each $p,q \in V$ with $\varepsilon_x(p,q) > 0$, an arc (p,q) with capacity $\varepsilon_x(p,q)$.

We call the arcs *forward, backward*, and *jumping*, respectively. Suppose that $f = (t,s)$. A dipath P in G' from s to t together with (t,s) yields a directed circuit C in G'. Let ε be the minimum capacity of its arcs. The augmentation corresponding to P increases x_e by ε if a forward arc of C arises from e, and decreases x_e by ε

if a backward arc of C arises from e. The next lemma occurs in Frank (1984a); similar results for other models can be found in Fujishige (1978), Hassin (1982), Lawler and Martel (1982), and Schönsleben (1980).

Lemma 5.3. *If P is a chordless (s,t)-dipath in G', then the augmentation corresponding to P results in a feasible flow.*

On the other hand, if there is no (s,t)-dipath in G', then x_f is maximum, and the algorithm may terminate. To see this we observe that in this case there is a set $A \subseteq V$ with $s \in A$, $t \notin A$ such that $x_e = u_e$ for all $e \in \delta^+(A)$, $x_e = \ell_e$ for all $e \in \delta^-(A)\backslash\{f\}$, and for every $p \in A$, $q \notin A$ there is a tight set containing p and not q. [A set is *tight* if its inequality in (5.1) holds with equality.] Since we may assume that b is fully submodular, it is easy to prove that the intersection and union of tight sets is tight. It follows that A is tight. Therefore,

$$x_f = b(A) - \ell(\delta^-(A)\backslash\{f\}) + u(\delta^+(A)).$$

But obviously no feasible flow can have x_f exceeding this, so x_f is maximum. The resulting min–max theorem is the following.

Theorem 5.4. *If there is a maximum flow and b is fully submodular, then*

$$\begin{aligned}&\max(x_f\colon x \text{ a feasible flow})\\ &\quad = \min(u_f,\ \min(b(A) - \ell(\delta^-(A)\backslash\{f\}) + u(\delta^+(A))\colon s \in A \subseteq V\backslash\{t\})).\end{aligned}$$

There is a more general version of Theorem 5.4 for crossing submodular functions. It can be derived from Theorem 5.4 by applying Theorems 4.16 and 4.17. In addition, feasibility can be tested by applying the maximum submodular flow algorithm to a certain auxiliary problem and so the following feasibility characterization is also a consequence. Again a more general version (Frank (1984a)) is available.

Theorem 5.5. *If b is fully submodular then there exists a feasible flow if and only if, for all $A \subseteq V$, $\ell(\delta^+(A)) - u(\delta^-(A)) \leqslant b(A)$.*

Now we discuss the efficiency of the algorithm. First, we observe that if b, u, ℓ and the initial x are integer-valued, then x remains integer-valued, and the algorithm terminates (assuming there exists a maximum flow). However, we would like to have a bound on the number of augmentations that is not so dependent on the size and form of the input numbers. A similar difficulty arises in ordinary network flows, where the classical solution is to find shortest augmenting paths, found by "breadth-first search": scanning nodes in the order in which they are labelled. The analysis of that method is based on the following facts. For a flow x, let $k(x)$ denote the length of a shortest augmenting path with respect to x, and let $E(x)$ denote the set of arcs contained in some shortest augmenting path with respect to x. Suppose that an augmentation replaces flow x by flow x'. Then:

(a) $k(x') \geqslant k(x)$;

(b) if $k(x') = k(x)$, then $E(x') \subset E(x)$.

It follows from (a) that the computation is divided into at most $n = |V|$ stages and it follows from (b) that each stage takes at most n^2 augmentations. In the more general situation of submodular flows, (a) still holds but (b) fails. However, careful examination of *how* it can fail, leads to a further refinement. In addition to scanning nodes in the order labelled, we label nodes (from a node being scanned) in an order consistent with a fixed ordering of V. The resulting path has a node-sequence that is lexicographically least among node-sequences of shortest augmenting paths. This important technique was introduced independently by Schönsleben (1980) and Lawler and Martel (1982), who used it in contexts similar to the present one. Cunningham (1984) used it in his algorithm for testing membership in matroid polyhedra. He also labelled the technique "consistent breadth-first search", and described its essential properties in a context-free way. For the submodular flow model we are using, the following result is due to Frank (1984a).

Theorem 5.6. *If consistent breadth-first search is used, the maximum flow algorithm terminates after* $O(n^3)$ *augmentations.*

5.4. Submodular flow algorithms: Optimization

Recall that the weighted MIA used the following three ideas: (i) simpler optimality conditions (using weight-splitting) than those coming from a straightforward use of linear programming duality; (ii) use of the same auxiliary digraph as the unweighted algorithm, except that arcs were assigned costs; (iii) use of a least-cost dipath computation to update the dual solution, followed by use of the unweighted algorithm on arcs of cost zero. These three ideas will be used in extending the work of the last section to an algorithm for the optimal submodular flow problem.

The first important idea, suggested by Frank (1982), is that an optimal dual solution for (5.1) can be represented by a vector of *potentials* π_v, $v \in V$. Given such a vector π and an arc $e = (p,q) \in E$, $\bar{c}_e$ denotes $c_e + \pi_p - \pi_q$.

Theorem 5.7. *Suppose that* x *is a feasible flow and* π, x *satisfy*

(a) *if* $\bar{c}_e > 0$ *then* $x_e = u_e$, *for* $e \in E$;

(b) *if* $\bar{c}_e < 0$ *then* $x_e = \ell_e$, *for* $e \in E$;

(c) *if* $\pi_p > \pi_q$, *then* $\varepsilon_x(p,q) = 0$, *for* $p, q \in V$.

Then x *is optimal.*

This result is easily proved, by showing that linear-programming optimality conditions are satisfied by x and the dual solution (y, f, g) constructed from π as follows. Let $\pi_0 < \pi_1 < \cdots < \pi_k$ be the distinct values of π, and let A_i denote $\{v \in V\colon \pi_v \geqslant \pi_i\}$, $1 \leqslant i \leqslant k$. Define y_A to be $\pi_i - \pi_{i-1}$ if $A = A_i$ and to be 0 otherwise. Define f_e to be max $(\bar{c}_e, 0)$ and g_e to be max $(-\bar{c}_e, 0), e \in E$. (The dual variable f_e corresponds to the constraint $x_e \leqslant u_e$; g_e corresponds to $-x_e \leqslant -\ell_e$.) Notice that (y, f, g) is integer-valued if π is.

The algorithm maintains a feasible flow x and a potential π satisfying Theorem 5.7(c). Let $f = (t,s)$ be an arc violating condition (a); the other case is similar.

It is convenient to assume that f is the only arc violating (a) or (b). This can be accomplished by temporarily changing the appropriate bound of any other offending arc e to x_e. Now we form the auxiliary digraph G' of the last section and assign arc-costs as follows: A forward arc (p, q) has cost $w_{pq} = -\bar{c}_e$; a backward arc (q, p) has cost $w_{qp} = \bar{c}_e$; a jumping arc (p, q) has cost $\pi_q - \pi_p$. Now each arc of G' (except the forward arc (t, s), which we delete) has non-negative cost, and indeed, this is equivalent to the condition that (a)–(c) are violated only by f.

Next we use an $O(n^2)$ shortest-path algorithm to find a least-cost dipath in G' from s to v for each $v \in V$; let d_v be its cost. For $v \in V$ we replace π_v by $\pi_v - \min(d_v, \bar{c}_f)$. It is quite easy to check that the arc-costs remain non-negative. Now either $\bar{c}_f = 0$, in which case we have ended f's violation of the optimality conditions, or $\bar{c}_f > 0$, in which case there exists in G' an (s, t)-dipath consisting of arcs having weight 0. In this case we use consistent breadth-first search to find such a dipath and perform an augmentation. [It is possible that this dipath yields a directed circuit of G' having no finite capacity; in this case (5.1) is unbounded, and the algorithm terminates.] That such an augmentation preserves (a)–(c) is obvious. That it delivers a feasible flow can be proved as for the weighted MIA, by showing that it is an augmentation of the maximum submodular flow algorithm applied to a more restricted submodular flow problem; see Cunningham and Frank (1985). From this observation it follows also that there will be at most $O(n^3)$ augmentations before π changes again. The number of potential changes can be shown to be finite, and better bounds hold when c is nice. Refinements based on scaling c lead to a polynomial bound (Cunningham and Frank 1985) and to a bound polynomial in n alone (Fujishige et al. 1989) for the number of augmentations.

5.5. *Polymatroid matching*

An important common generalization of graph matching and matroid intersection may be formulated as a problem on *2-polymatroids*: polymatroids whose polymatroid function f is integer-valued and satisfies $f(\{j\}) \leqslant 2$ for each $j \in S$. A *matching* is a set $J \subseteq S$ such that $f(J) = 2|J|$. Ordinary matching on a graph $G = (V, E)$ arises when we take $S = E$ and $f(A) = |V(A)|$ for $A \subseteq S$. Matroid intersection for matroids M_1, M_2 on S arises when we take $f = r_1 + r_2$. Historically the first common generalizations of these two important problems were equivalent models called matroid parity (Lawler), matchoids (Edmonds), and matroid matching. The latter is the case of polymatroid matching in which we are given a graph $G = (V, E)$ and a matroid M on V and take $S = E$ with $f(A) = r(V(A))$. However, the maximum matching problem, even in this special case, is intractable from the oracle viewpoint (Lovász (1980b), Korte and Jensen 1982), and also contains NP-hard problems.

Nevertheless, a great deal of progress has been made on the 2-polymatroid matching problem, mainly due to the work of Lovász. He has described (Lovász 1980b) general reduction techniques for computing a maximum matching. For several important special classes these lead to efficient algorithms and min-max results. Among the applications are finding a maximum-cardinality forest in a 3-

uniform hypergraph, and finding a maximum family of openly disjoint A-paths in a graph. (See chapter 2 by Frank for more on the latter problem.)

The most important special case is that of *linear* 2-polymatroids. Here S is a set of lines (equivalently, pairs of points) in a vector space, and $f(A)$ is $\mathrm{ar}(\bigcup\{e\colon e \in A\})$, where ar denotes affine rank. Graph matching and matroid intersection for two matroids linearly represented over the same field are both special cases, and their min-max formulas are generalized by the next result from Lovász (1980a).

Theorem 5.8. *The maximum size of a matching in a linear 2-polymatroid is*

$$\min \mathrm{ar}(A_0) + \sum_{i=1}^{k} \lfloor \frac{\mathrm{ar}(A_i)}{2} \rfloor,$$

where the minimum is over sets $A_0, A_1, \ldots, A_k$ of points of the space such that, for every $e \in S$, either $\mathrm{ar}(A_0 \cup e) \leqslant \mathrm{ar}(A_0) + 1$ or, for some i, $\mathrm{ar}(A_i \cup e) = \mathrm{ar}(A_i)$.

Lovász has also given a (complex) polynomial-time algorithm for linear 2-polymatroid matching. Gabow and Stallmann (1986) contains a different, much more efficient, algorithm. Its running time is $O(n^4)$, where $n = |S|$, surprisingly close to the best bound known for the special case of linear matroid intersection. An outstanding question is the solvability of the corresponding weighted problem, about which little is known.

5.6. Delta-matroids and bisubmodular polyhedra

Many of the optimizational properties of matroids are preserved in an interesting generalization introduced under various names by Dress and Havel (1986), Bouchet (1987), and Chandrasekaran and Kabadi (1988). A *delta-matroid* is a pair $(S, \mathscr{F})$ where S is a finite set and $\mathscr{F}$ is a family of subsets of S, called the *feasible* sets, satisfying the following *symmetric exchange axiom*:

$$\text{If } F_1, F_2 \in \mathscr{F} \text{ and } a \in F_1 \triangle F_2, \text{ there exists } b \in F_1 \triangle F_2 \text{ such that } F_1 \triangle \{a, b\} \in \mathscr{F}. \tag{5.9}$$

(Here $\triangle$ denotes symmetric difference.) Matroids, defined by their basis families, are precisely the delta-matroids in which the feasible sets all have the same cardinality. They form the fundamental examples, although the independent sets of a matroid also form the feasible sets of a delta-matroid. Other interesting examples are *matching* delta-matroids (S is the node-set of a graph and $F \subseteq S$ is feasible if and only if it is the set of end-nodes of some matching), and *linear* delta-matroids (where $A_{S,S}$ is a skew-symmetric (or symmetric) matrix over a field, $F \subseteq S$ is feasible if and only if $A_{F,F}$ is non-singular).

The maximum-weight feasible set problem can be solved by a type of greedy algorithm. This algorithm needs an oracle that, for disjoint subsets A, B of S, determines whether there is a feasible set F satisfying $A \subseteq F \subseteq S \backslash B$.

Symmetric greedy algorithm for the maximum-weight feasible set problem.
Order $S = \{e_1, e_2, \ldots, e_n\}$ so that $|c_{e_1}| \geqslant |c_{e_2}| \geqslant \cdots \geqslant |c_{e_n}|$
For each i, let R_i denote $\{e_i, \ldots, e_n\}$
$J := \emptyset$
For $i = 1$ to n do
  If $c_{e_i} \geqslant 0$ and there exists $F \in \mathcal{F}$ with $J \cup \{e_i\} \subseteq F \subseteq J \cup R_i$ then $J := J \cup \{e_i\}$
  If $c_{e_i} < 0$ and there exists no $F \in \mathcal{F}$ with $J \subseteq F \subseteq J \cup R_{i+1}$ then $J := J \cup \{e_i\}$

Theorem 5.10. *$(S, \mathcal{F})$ is a delta-matroid if and only if for any $c \in \mathbb{R}^S$, the symmetric greedy algorithm finds a maximum-weight feasible set.*

Again the "if" part is an easy consequence of the definition. The other part can be proved by methods similar to the ones used for matroids. However, it is also possible to derive many of the results for delta-matroids from corresponding matroid facts. One useful technique is the following. (The reader should consider the effect of choosing N to be the set of negative-weight elements of S.)

Proposition 5.11. *$(S, \mathcal{F})$ is a delta-matroid if and only if for every $N \subseteq S$, the maximal members of $\{F \triangle N \colon F \in \mathcal{F}\}$ are the bases of a matroid.*

We define a rank function f for a delta-matroid by $f(A, B) = \max(|F \cap A| - |F \cap B| \colon F \in \mathcal{F})$ for subsets A, B of S. Clearly the characteristic vector of any feasible set, and therefore any convex combination of such vectors, satisfies the inequalities

$$x(A) - x(B) \leqslant f(A, B), \qquad A, B \subseteq S,\ A \cap B = \emptyset. \tag{5.12}$$

Theorem 5.13. *The convex hull of characteristic vectors of feasible sets of a delta-matroid is precisely the set of solutions of* (5.12).

As for matroids, there is one property of the rank function that is the essential one for proving such results. Let f be a function defined on ordered pairs of disjoint subsets of S. We say that f is *bisubmodular* if it satisfies

$$f(A, B) + f(A', B') \geqslant f(A \cap A', B \cap B') + f((A \cup A') \backslash (B \cup B'), (B \cup B') \backslash (A \cup A')). \tag{5.14}$$

Kabadi and Chandrasekaran (1990), Nakamura (1988), and Qi (1988) have shown that for any bisubmodular f, the system (5.12) is totally dual integral. In particular, this implies that if f is also integer-valued, then every extreme point of the corresponding *bisubmodular polytope*, that is, the solution set $P(f)$ of (5.12), is also integer-valued. This result easily implies Theorem 5.13. The total dual integrality of (5.12) can again be proved by a greedy algorithm for maximizing $c \cdot x$ over the bisubmodular polyhedron, a generalization of the one for delta-matroids. In fact, this algorithm already appears in Dunstan and Welsh (1973). For $P \subseteq \mathbb{R}^S$, $A \subseteq S$, and $x \in \mathbb{R}^A$, we say that $x \hat{\in} P$ if there exists $y \in P$ such that $y_j = x_j$ for all $j \in A$.

Symmetric greedy algorithm for bisubmodular polyhedra.
Order $S = \{e_1, e_2, \ldots, e_n\}$ so that $|c_{e_1}| \geqslant |c_{e_2}| \geqslant \cdots \geqslant |c_{e_n}|$
For each i, let R_i denote $\{e_i, \ldots, e_n\}$
$J := \emptyset$
For $i = 1$ to n do
  If $c_{e_i} \geqslant 0$ choose x_{e_i} as large as possible so that $(x_{e_1}, \ldots, x_{e_i}) \hat{\in} P$
  If $c_{e_i} < 0$ choose x_{e_i} as small as possible so that $(x_{e_1}, \ldots, x_{e_i}) \hat{\in} P$

Dunstan and Welsh call a compact set $P \subseteq \mathbb{R}^S$ *greedy* if the above version of the greedy algorithm maximizes $c \cdot x$ over P for every $c \in \mathbb{R}^S$. To describe their characterization of greedy sets, we need to define some terms. For a point $x \in \mathbb{R}^S$ and a subset $A \subseteq S$, we denote by $x \triangle A$ the point obtained by replacing x_j by its negative for each $j \in A$. For a subset P of $\mathbb{R}^S$, we write $P \triangle A$ to denote $\{x \triangle A\colon x \in P\}$. Finally, the hereditary closure $\{y \in \mathbb{R}^S\colon y \leqslant x$ for some $x \in P\}$ of P is denoted by $\mathrm{dn}(P)$. The equivalence of (a) and (c) below is due to Dunstan and Welsh (1973); that of (a) and (b) is due to Kabadi and Chandrasekaran (1990) and Nakamura (1988). It is an interesting puzzle to understand why the recent interesting work on delta-matroids and bisubmodular polyhedra occurred so long after the initial work of Dunstan and Welsh.

Theorem 5.15. *For a polytope $P \subseteq \mathbb{R}^S$, the following statements are equivalent:*
(a) *P is greedy;*
(b) *P is a bisubmodular polyhedron;*
(c) *for every $A \subseteq S$, $\mathrm{dn}(P \triangle A)$ is a submodular polyhedron.*

The bad news about delta-matroids and bisubmodular polyhedra is the absence of an intersection theorem. In fact, the matroid matching problem is easily seen to be the intersection of a matroid with a matching delta-matroid, so the delta-matroid intersection problem is intractable. Frank (1984b) introduced a special class of bisubmodular polyhedra that is better behaved in this respect, but still includes polymatroids and base polyhedra. A *generalized polymatroid* is a bisubmodular polyhedron determined by a bisubmodular function f satisfying $f(A, B) = g(A) - h(B)$, where g is submodular, h is supermodular, and they satisfy

$$g(A) - h(B) \leqslant g(A \backslash B) - h(B \backslash A), \qquad A, B \subseteq S. \tag{5.16}$$

Actually, the definition of Frank is slightly more general. The paper of Frank and Tardos (1988) is an excellent reference, giving a wealth of results on generalized polymatroids and related topics. One result that helps to explain their good behaviour is that every such polyhedron can be obtained by projection from a base polyhedron. The main integrality result, generalizing the Polymatroid Intersection Theorem 4.9, may be stated as follows.

Theorem 5.17. *The union of the defining systems of two generalized polymatroids in $\mathbb{R}^S$ is totally dual integral.*

5.7. Greedoids and independence systems

A pair $(S, \mathcal{F})$, where $\mathcal{F}$ is a family of subsets of S containing $\emptyset$, is a matroid if it satisfies:

$$\text{if } A \subseteq B \in \mathcal{F}, \text{then } A \in \mathcal{F}; \tag{5.18}$$

$$\text{if } A,\ B \in \mathcal{F},\ |A| > |B|, \text{ then there is } a \in A \backslash B \text{ with } B \cup \{a\} \in \mathcal{F}. \tag{5.19}$$

Requiring only one of (5.18), (5.19) yields two different generalizations. The first, of course, is independence systems. These objects seem to be too general to have an interesting theory. We do mention one result on the effectiveness of the greedy algorithm. For any $A \subseteq S$, let $r^{+}(A)$ denote $\max(|F|\colon A \supseteq F \in \mathcal{F})$, and let $r^{-}(A)$ denote $\min(|F|\colon A \supseteq F \in \mathcal{F},\ F \text{ maximal})$. Of course, an independence system is a matroid precisely when $r^{+}(A) = r^{-}(A)$ for all $A \subseteq S$. That is, the quantity $\min(r^{-}(A)/r^{+}(A)\colon A \subseteq S,\ r^{+}(A) > 0) = p(S,\ \mathcal{F})$ is 1 for matroids, and $p(S,\ \mathcal{F})$ is a measure of how close $(S,\ \mathcal{F})$ is to being a matroid. We know from Theorem 2.2 that the maximum-weight independent set problem is solved optimally by the greedy algorithm when $p(S,\ \mathcal{F}) = 1$. Jenkyns (1976), and independently Korte and Hausmann (1978), have proved more generally that the greedy algorithm works well if $p(S,\ \mathcal{F})$ is not too small.

Theorem 5.20. *Let $(S,\ \mathcal{F})$ be an independence system. For any $c \in \mathbb{R}^S$ the greedy algorithm delivers a feasible set of weight at least $p(S,\ \mathcal{F})\max(c(F)\colon F \in \mathcal{F})$.*

Now let us consider the objects obtained when we drop (5.18) and keep (5.19). The resulting structures are called *greedoids*. They were introduced by Korte and Lovász and, while very general, have a surprising amount of structure. (In fact, enough to justify a book; see Korte et al. (1991).) Here we mention a few of the connections with optimization.

For a greedoid $(S,\ \mathcal{F})$ a *basis* of $(S,\ \mathcal{F})$ is a maximal feasible set. Given a weight function c such that $c(F) \in \mathbb{R}$ for each $F \in \mathcal{F}$, we consider the problem of finding a basis F maximizing $c(F)$. We do *not* assume that c is linear, that is, determined by $c(F) = \Sigma(c_j\colon j \in F)$ for an element-weighting $c \in \mathbb{R}^S$. We state a greedy algorithm for the maximum-weight basis problem.

Greedy algorithm for a greedoid.

$J := \emptyset$
While there exists $e \in S \backslash J$ with $J \cup \{e\} \in \mathcal{F}$ do
  Choose such e with $c(J \cup \{e\})$ maximum
  $J := J \cup \{e\}$

Notice that this algorithm, for the case when c is linear and $(S,\ \mathcal{F})$ is a matroid, is equivalent to the greedy algorithm for finding a matroid basis of maximum weight. In that case the algorithm finds an optimal solution. The situation for greedoids is more complicated. We illustrate two of its aspects by considering a class of greedoids arising from a digraph $G = (V, E)$ with a fixed node $s \in V$. We take $S = E$ and take $\mathscr{F}$ to be the arc-sets of arborescences rooted at s. Notice that (5.19) is preserved under the truncation operation $\mathscr{F}^k = \{F \in \mathscr{F}: |F| \leqslant k\}$. Let $T \subseteq V \backslash \{s\}$ be a set of target nodes and assign (linear) weights of 1 to arcs having head in T and 0 to the others. By finding a maximum-weight basis of $(S, \mathscr{F}^k)$, we can decide whether there is an s-rooted arborescence having at most k edges and including all the target nodes. This latter problem is a version of the "Steiner tree problem" and is NP-hard. So optimizing linear functions over greedoid bases is, in general, difficult.

On the other hand there are interesting *nonlinear* functions that the greedy algorithm optimizes. For an example, consider again $(S, \mathscr{F})$ from above. Given $d \in \mathbb{R}^S_+$ and $F \in \mathscr{F}$, we let $c(F)$ denote $\sum_{v \in V(F)} (\sum_{j \in P_v} -d_j)$ where P_v is the arc-set of the unique dipath in F from s to v. Then a basis F of $(S, \mathscr{F})$ maximizing $c(F)$ provides least-cost dipaths (with respect to d) from s to v for all $v \in V$, and the greedy algorithm will find such a basis. (These facts are consequences of standard results on shortest path problems; see chapter 2 by Frank.) More generally, there is a large class of functions that the greedy algorithm optimizes over the bases of any greedoid. There is also a characterization of greedoids in terms of the functions that the greedy algorithm optimizes, and a characterization of the greedoids for which it optimizes every linear function. For these results and others, we refer the reader to Korte et al. (1991).

Finally, we describe a generalization of the matroid intersection theorem to a larger class of greedoids. A *distributive supermatroid* (we do not attempt to explain the name) is a greedoid $(S, \mathscr{F})$ together with a partial order on S satisfying

$$\text{if } A \subseteq B \in \mathscr{F} \text{ and } A \text{ is an ideal, then } A \in \mathscr{F}. \tag{5.21}$$

Notice that every matroid satisfies (5.21), by taking the partial order to be trivial. It is a result of Tardos (1990) that, for two distributive supermatroids $(S, \mathscr{F}_1)$, $(S, \mathscr{F}_2)$ on the same set *with the same partial order*, the quantity $\max(|F|: F \in \mathscr{F}_1 \cap \mathscr{F}_2)$ is well-characterized. If the partial orders are not assumed to agree, then the intersection problem can be shown to be intractable in general. We state a simple version of Tardos's min–max theorem. This is a bit misleading since it involves quantities that are difficult to compute; her paper provides a more complicated formula that does not have this drawback. For $i = 1$ and 2 and $A \subseteq S$, let $\beta_i(A)$ denote $\max(|F \cap A|: F \in \mathscr{F}_i)$. (Of course, if $(S, \mathscr{F}_i)$ is a matroid, then β_i is just the rank function.)

Theorem 5.22. *Let* $(S, \mathscr{F}_1)$, $(S, \mathscr{F}_2)$ *be distributive supermatroids with the same partial order. Then* $\max(|F|: F \in \mathscr{F}_1 \cap \mathscr{F}_2) = \min(\beta_1(A) + \beta_2(\overline{A}): A \subseteq S)$.

6. Matroid connectivity algorithms

Connectivity is a fundamental structural property of matroids. Algorithmically this notion took on a central role with Seymour's work on regular matroids. The connectivity-based decompositions applied there have since been further applied and significantly extended by Seymour (1981a), Truemper (1986), and Truemper (1992).

Let M be a matroid on S with rank function r. A partition $\{S_1, S_2\}$ of S is an *m-separation* of M for $m \geqslant 1$ if

$$|S_1| \geqslant m \leqslant |S_2|, \tag{6.1}$$

and

$$r(S_1) + r(S_2) - r(S) \leqslant m - 1. \tag{6.2}$$

Define M to be *k-connected* for $k \geqslant 2$ if M has no m-separation for $m < k$ (Tutte 1966); in this case we say that M has *connectivity* k. 2-connected matroids are called *connected* or *nonseparable*. It is easy to see that $\{S_1, S_2\}$ is a 1-separation of M if and only if every circuit of M is either a subset of S_1 or a subset of S_2.

The above definition is motivated by graph connectivity. Tutte proved that the polygon-matroid $M(G)$ of a graph G is k-connected if and only if G is k-connected in the usual sense (that is, connected and remains so upon the deletion of any $k-1$ or fewer nodes), and either $|S| \leqslant 2k-1$ or G has no cycle of size less than k. To obtain an exact analogue of node k-connectivity in graphs we can replace the cardinality condition (6.1) by $r(S_1) \geqslant m \leqslant r(S_2)$. This definition, however, is not invariant under duality, whereas Tutte's definition is: it is easy to show that $r(S_1) + r(S_2) - r(S) = r^*(S_1) + r^*(S_2) - r^*(S)$ for any pair $\{S_1, S_2\}$.

The following table summarizes the best algorithms available for testing k-connectivity for various values of k. Note that for general k, the complexity does grow exponentially with $|S|$, as it inevitably must since computing the connectivity of a matroid specified by an oracle is easily shown to be intractable (see the argument used for girth in section 2).

Table 1
Algorithms for testing k-connectivity

k	Complexity (oracle calls)	Method
2	$O(\|S\|^2)$	Shifting
3	$O(\|S\|^3)$	Bridges or shifting
4	$O(\|S\|^{4.5}\sqrt{\log \|S\|})$	Shifting
$\geqslant 5$	$O(\|S\|^{k+1})$	Matroid intersection

6.1. *Partial representations*

Partial representations (Truemper 1984) will be used in several places in this section. Let X be a basis of a matroid M. The *partial representation*, abbreviated PR, of M with respect to X is the $\{0,1\}$-matrix R with rows indexed on elements $x \in X$ and columns indexed on elements $y \in Y = S \backslash X$ such that the (x, y) entry is 1 if and only if $x \in C(X,y)$. If M is binary, (I,R) is an actual representing matrix for M over GF(2), where I is an identity matrix of appropriate dimension.

For $X' \subseteq X$ and $Y' \subseteq Y$, let $R(X',Y')$ denote the submatrix of R with row index set X' and column index set Y'. $R(X',Y')$ is then a PR of $M\backslash(Y\backslash Y')/(X\backslash X')$ with respect to $X\backslash X'$. Define the *R-rank* of $R(X',Y')$ by

$$\mathrm{rk}(R(X',Y')) = r(Y' \cup (X\backslash X')) - |X\backslash X'|,$$

where r is the usual matroid rank function for M. If $\mathrm{rk}(R(X',Y')) = |X'| = |Y'|$, then $R(X',Y')$ is called *nonsingular*.

6.2. 2-Connectivity

By definition, a matroid is disconnected (separable) if there is a partition $\{S_1,S_2\}$ of S with $S_1 \neq \emptyset \neq S_2$ and $r(S_1)+r(S_2) = r(S)$ ($r(S_1)+r(S_2) \geqslant r(S)$ holds by submodularity). There is an easy algorithm for testing this condition using PRs.

Let X be any basis of M and let R be a corresponding PR. R has a naturally associated bipartite graph $G(R)$ in which there is one node for each row and column of R and one edge for each nonzero entry.

Theorem 6.3. *Let M be matroid, X a basis of M, R the* PR *corresponding to X and $G(R)$ the associated bipartite graph. Then M is connected if and only if $G(R)$ is connected. (Thus, M is connected if and only if R has no nontrivial block decomposition.)*

Proof. We prove one half of the theorem. Suppose that M is not connected, let $\{S_1,S_2\}$ be a separator, and let X be a basis of M. Define $X_i = S_i \cap X$ $(i=1,2)$. Now $X_i \subseteq S_i$ is independent and so

$$|X| = r(S) = r(S_1)+r(S_2) \geqslant |X_1|+|X_2| = |X|.$$

It follows that X_i is a basis of S_i $(i=1,2)$. Let $Y_i = S_i\backslash X_i$. We conclude that every fundamental circuit $C(X,e)$ for $e \in Y_i$ is contained in S_i. That is, the submatrices $R(X_i,Y_j)$, $i \neq j$, are identically zero. It follows that $G(R)$ is not connected. $\square$

Theorem 6.3 yields an algorithm for testing the connectivity of M, as follows. First construct a basis and the corresponding PR R. This step requires at most $|S| + r(S)(|S|-r(S)) \leqslant |S|^2$ calls to an independence oracle. Now test the connectivity of the graph $G(R)$. The time for this latter computation is dominated by that for the first, and so the total is $O(|S|^2)$ oracle calls.

6.3. Computing minors containing a fixed element

The relationship introduced in the last subsection between 2-connectivity and bipartite graph connectivity is the basis for an algorithmic proof of the following result. This approach is typical of several proof techniques in the subject. It plays a fundamental role in Truemper (1992).

Theorem 6.4 (Seymour 1977a). *If N and M are connected matroids, N is a minor of M, and $e \in S(M)\backslash S(N)$, then there exists a connected minor N' of M such that $N = N'\backslash e$ or $N = N'/e$ (these are actual equalities – no isomorphism is involved).*

Proof. Let M and N be as in the statement of the theorem. Let X' be independent and Y' coindependent in M such that $N = M\backslash Y'/X'$. Take a PR R corresponding to a basis X such that $X' \subseteq X \subseteq S\backslash Y'$. Let $e \in S(M)\backslash S(N) = X' \cup Y'$. By duality assume $e \in Y'$. Now if column e of R has a nonzero in the rows of N, we are done: $N' = M\backslash(Y'\backslash\{e\})/X'$ satisfies the requirements of the theorem (by Theorem 6.3). Otherwise, we find a chordless path in the bipartite graph $G(R)$ from e to any element in N. Deleting the column elements not in this path, and contracting the row elements not in the path yields the following picture (or a slight variant), where all entries not in N and not on the path are zero:

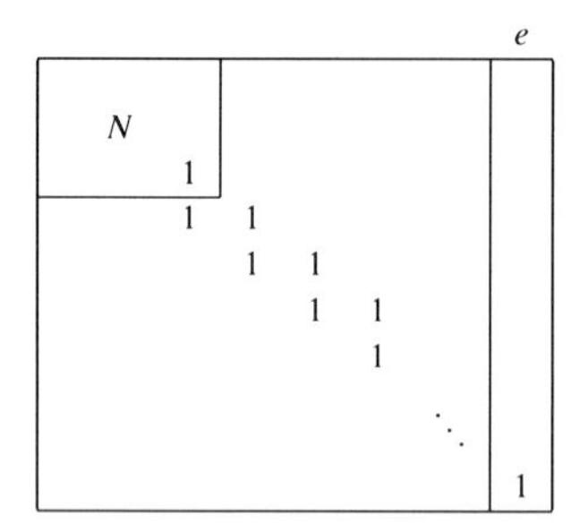

Denote the set of elements corresponding to non-N columns in this picture by Y'' and the set corresponding to non-N rows by X''. One can verify that the minor obtained by deleting X'' and contracting $Y''\backslash\{e\}$ has the property asserted in the theorem. □

Theorem 6.4 may be applied in the following way. Suppose that we know an excluded-minor characterization for some class of matroids and ask the question, whether one of the excluded minors, when it occurs, can be isolated to a part of the ground set of a given matroid, or must be present throughout that matroid. This question leads to the notion of 1-roundedness: We say that a subfamily $\mathcal{P}'$ of a family $\mathcal{P}$ of connected matroids is 1-*rounded* if whenever $M \in \mathcal{P}$ has a minor N isomorphic to a member of $\mathcal{P}'$, then for every element $e \in S(M)$, there is a minor N' of M such that N' is isomorphic to a member of $\mathcal{P}'$ and $e \in S(N')$. An example is provided by the binary matroids. In that case we take $\mathcal{P}$ to be the class of connected non-binary matroids, and $\mathcal{P}' = \{U_{2,4}\}$, where $U_{2,4}$ denotes a matroid on four elements every two of which form a basis. By Theorem 6.4, to prove that $\{U_{2,4}\}$ is 1-rounded, we need only prove that for every connected 1-element extension of $U_{2,4}$, the new element in the extension is an element of a minor isomorphic to $U_{2,4}$. This result together with Tutte's theorem (Tutte (1965a)) that every non-binary matroid has a minor isomorphic to $U_{2,4}$ implies a result of Bixby (1974): If M is a connected non-binary matroid, then for every element $e \in M$, there is a minor of M isomorphic to $U_{2,4}$ and containing e. Note that the proof of Theorem 6.4 gives an algorithm to find such a minor, given any single $U_{2,4}$ minor of M. This idea is used later in computing the special 3-separations needed for testing the regularity of general matroids not known to be binary.

To use the algorithm suggested in the proof of Theorem 6.4 we need to know N in the form $N = M\backslash Y''/X''$, for disjoint sets X'' and Y''. In $O(|S|)$ oracle calls we can find a maximal independent subset of X'', extend it to a maximal independent set in $X'' \cup S(N)$ and then further extend to a basis X of M. Let $X' = X \cap (X'' \cup Y'')$ and $Y' = (X'' \cup Y'')\backslash X'$. Then X' is independent, Y' is coindependent and $N = M\backslash Y'/X'$. Now constructing the partial representation R corresponding to X and finding the chordless path described in the above proof takes time $O(|S|^2)$. Finally, computing the necessary contractions and deletions is no more than the work for computing a PR for N. Thus, the complete algorithm runs in $O(|S|^2)$ oracle calls.

6.4. 3-Connectivity and bridges

Connectivity (i.e. 2-connectivity) is relatively easy to analyze. 3-connectivity is thus the first difficult instance of k-connectivity. It is particularly important in the theory of regular matroids and in decomposition theory in general. In this section we present a special-purpose recursive algorithm for testing 3-connectivity in connected matroids. Some of the ideas in this algorithm will be used later in a graph-realization algorithm.

Define the *elementary separators* of a matroid to be the minimal nonempty sets S' such that $\{S', S\backslash S'\}$ is a 1-separation. Let M be a matroid, and let D be a cocircuit of M. The elementary separators of $M\backslash D$ are called the *bridges* of D. D is called a *separating cocircuit* if it has more than one bridge. The corresponding *D-components* are the matroids $M/(S\backslash(D \cup A))$ where A is a bridge. The *A-segments* of D, for a bridge A, are the parallel classes of $M/(S\backslash(D \cup A))$ in D.

The above quantities are easily computed using PRs. Take a PR of M such that D is a fundamental cocircuit. Then D minus some single element is a row of this matrix. Deleting this row and the incident columns gives a PR for $M\backslash D$. The bridges of D can then be computed using the connectivity algorithm given by Theorem 6.3. The corresponding D-component for a particular bridge A is obtained by deleting the columns and rows not in $D \cup A$. The computation of the A-segments is then straightforward.

One more definition is required. Say that two D-bridges A_1 and A_2 *avoid* if there are A_i-segments Y_i $(i = 1, 2)$ such that $Y_1 \cup Y_2 = D$. Define the *bridge graph* of D to have nodes corresponding to the bridges of D and an edge joining two nodes if and only if they *do not* avoid.

Theorem 6.5 (Bixby and Cunningham 1979). *Let M be a connected matroid that is both simple and cosimple, and let D be a cocircuit of M. Then*

(a) *M has a 2-separation $\{S_1, S_2\}$ such that $D \subseteq S_2$ if and only if the matroid obtained from some D-component by identifying parallel elements has a 2-separation;*

(b) *M has a 2-separation $\{S_1, S_2\}$ such that D meets both S_1 and S_2 if and only if the bridge graph of D is not connected.*

The above theorem suggests a recursive procedure for testing 3-connectivity, given a method for finding separating cocircuits. But this latter task is easy, because of the following lemma.

Lemma 6.6. *Let M be a simple, connected matroid, and let X be a basis of M. If every fundamental cocircuit with respect to X is nonseparating, then M is 3-connected.*

The algorithm based on the above results proceeds as follows. First construct a PR. Check its rows to see whether any is separating. If none is, the matroid is 3-connected by Lemma 6.6; otherwise, take a separating cocircuit D and compute its D-components. Compute the corresponding bridge graph. By Theorem 6.5 we can determine whether M is 3-connected by examining the connectivity of the bridge graph and determining whether the D-components are 3-connected. Since the latter matroids are smaller than the original matroid, the suggested procedure can be applied recursively. An appropriate implementation gives a bound of $|S|^3$ oracle calls.

6.5. *k*-Connectivity

We describe two algorithms for testing k-connectivity for general k. The first is based on an elementary shifting idea in partial representations. For 2-connectivity, this algorithm reduces to the one given by Theorem 6.3. The second algorithm uses the cardinality Matroid Intersection Algorithm.

6.5.1. *A shifting algorithm*

This algorithm was suggested by work of Cunningham (1982) and Truemper (1985). To describe it we need an interpretation of m-separation in PRs. Let M be a matroid, and let R be a PR of M determined by a basis X. Let $\{S_1, S_2\}$ be a partition of M and define $X_i = S_i \cap X$ and $Y_i = S_i \cap Y$ $(i = 1, 2)$, where $Y = S \backslash X$. Define $R_{ij} = R(X_i, Y_j)$. This situation is depicted below:

$$\begin{array}{c|cc|} & Y_1 & Y_2 \\ \hline X_1 & R_{11} & R_{12} \\ X_2 & R_{21} & R_{22} \\ \hline \end{array} \qquad (6.7)$$

PR for M

Using the above notation, an equivalent definition of m-separation in terms of PRs is obtained from the computation:

$$\begin{aligned} \operatorname{rk}(R_{21}) + \operatorname{rk}(R_{12}) &= r((X \backslash X_2) \cup Y_1) - |X \backslash X_2| \\ &\quad + r((X \backslash X_1) \cup Y_2) - |X \backslash X_1| \\ &= r(S_1) + r(S_2) - r(S). \end{aligned} \qquad (6.8)$$

Thus, a partition of the elements S of a matroid into two sets is an m-separation if and only if its blocks are sufficiently large and, in any partial representation R, the corresponding "off-diagonal" submatrices determined by this partition have total R-rank at most $m-1$.

We now present the shifting algorithm. It is convenient to assume that M is known to be m-connected (we could first apply the algorithm for smaller values of m). Consider a basis X and the corresponding PR R, and suppose R'_{12} and R'_{21} are submatrices of R having disjoint row and column sets and total R-rank $m-1$ ($m \geqslant 1$). These matrices induce a partitioning of R, as illustrated below:

	Y'_1	Y'_2
X'_1	R'_{11}	R'_{12}
X'_2	R'_{21}	R'_{22}

PR for M

We also assume that $|X'_1 \cup Y'_1| \geqslant m$. The basic routine of the shifting algorithm determines whether R'_{12}, R'_{21} can be extended to a pair R_{12}, R_{21} determining an m-separation as in (6.7). (We call R_{12}, R_{21} a *legal* extension of R'_{12}, R'_{21}.) It consists of the following two operations.

Row shifting: Given $x \in X$ such that $\text{rk}(R(X'_2 \cup \{x\}, Y'_1)) > \text{rk}(R'_{21})$, set $X'_1 = X'_1 \cup \{x\}$.

Column shifting: Given $y \in Y$ such that $\text{rk}(R(X'_1, Y'_2 \cup \{y\})) > \text{rk}(R'_{12})$, set $Y'_1 = Y'_1 \cup \{y\}$.

It is easy to see that these operations are valid, in the sense that any legal extension R_{12}, R_{21} of R'_{12}, R'_{21} must also be an extension of the new R'_{12}, R'_{21}. Now suppose that we repeatedly apply these operations. If $\text{rk}(R'_{12}) + \text{rk}(R'_{21})$ ever increases, then we can stop; no legal extension exists. On the other hand, if $\text{rk}(R'_{12}) + \text{rk}(R'_{21}) = m-1$, but no shifting operation is possible, then $X_1 = X'_1$, $X_2 = X'_2 \cup (X \backslash X'_1)$, $Y_1 = Y'_1$, $Y_2 = Y'_2 \cup (Y \backslash Y'_1)$ defines a legal extension, unless $|X'_2 \cup Y'_2| < m$, in which case no legal extension exists.

Next we explain how to use the basic routine to test M for the existence of an m-separation. For matrices R_{12}, R_{21} as in (6.7) and determining an m-separation of M, there exist square non-singular submatrices P of one and Q of the other whose total R-rank is $m-1$. Suppose that we are given P and Q, and we want to find R_{12}, R_{21}. Let $z \in Y$ such that z does not index a column of P or of Q. We run the basic routine twice. First we initialize R'_{12} to be P and R'_{21} to be Q with column z appended, and second we initialize R'_{12} to be Q and R'_{21} to be P with column z appended. (We are taking advantage of the fact that we may assume z indexes a column of R'_{21}.) The shifting algorithm applies this procedure for all choices of P, Q.

6.5.2. Complexity of shifting

For $m = 1$, testing 2-connectivity, the shifting algorithm reduces to the algorithm given by Theorem 6.3. First, since $m - 1 = 0$, the matrices P, Q must be 0×0 matrices, implying there is only one pair to consider. There is also only one case to consider for the special element z – being appended to P is now equivalent to being appended to Q. Finally, the shifting procedure is equivalent to using a standard graph algorithm to compute the component containing z in $G(R)$.

In general, the shifting algorithm can be shown to have a complexity of $\mathrm{O}(|S|^{2m})$ oracle calls for computing m-separations. This involves an $\mathrm{O}(|S|^2)$ implementation of the basic routine for a given pair P, Q and the observation that there are at most $\mathrm{O}(|S|^{2m-2})$ pairs of nonsingular matrices with total R-rank $m - 1$. A defect in this approach is that, for general m, we do not know any device to decrease the estimate of the number of pairs considered. Such a device is known for the matroid intersection approach given in the next subsection.

In a quite different setting, Cunningham (1982) suggested the following idea, which can be applied to the special case $m = 2$. Assume M is connected and select a spanning tree of $G(R)$. Since $m - 1 = 1$, it follows that one of the matrices R_{12}, R_{21} in (6.7) must be a zero matrix, so that the other must then contain one of the elements from the spanning tree. Using this observation, we see that only $|S| - 1$ pairs P, Q must be considered, yielding an overall bound of $\mathrm{O}(|S|^3)$ oracle calls, the same as that derived from Theorem 6.5. For the binary case, this bound can be further improved to $\mathrm{O}(|S|^2)$ running time using a graph decomposition algorithm of Spinrad (1989).

Finally, an improved bound is also known for testing 4-connectivity. The details are too involved to present here, but using a graph-theory lemma of Szegedy, Rajan (1986) has shown how to reduce the bound to $\mathrm{O}(|S|^{4.5}\sqrt{\log|S|})$ oracle calls.

6.5.3. An algorithm using matroid intersection

This algorithm is due to Cunningham and Edmonds; see Cunningham (1973) for an early version. Let M be a matroid on S and suppose that we wish to test whether M has an m-separation for some $m \geqslant 1$. To do this it suffices to test for each pair of disjoint sets U_1, U_2, both of cardinality m, whether there exists an m-separation $\{V_1, V_2\}$ such that $U_i \subseteq V_i$. For a particular choice of U_1 and U_2, define matroids $M_1 = M/U_1 \backslash U_2$ and $M_2 = M/U_2 \backslash U_1$. These are matroids on $S' = S \backslash (U_1 \cup U_2)$, and for any partition $\{V_1', V_2'\}$ of S' we have

$$r_1(V_1') + r_2(V_2') = r(V_1' \cup U_1) - r(U_1) + r(V_2' \cup U_2) - r(U_2).$$

Since $r(U_1) + r(U_2)$ is a constant, minimizing this quantity over partitions $\{V_1', V_2'\}$ of S' is equivalent to minimizing $r(V_1) + r(V_2)$ over partitions $\{V_1, V_2\}$ of S such that $U_i \subseteq V_i$ $(i = 1, 2)$. Thus, using Theorem 3.5 and the Matroid Intersection Algorithm we can determine whether a given pair $\{U_1, U_2\}$ induces an m-separation. It follows that with $\mathrm{O}(|S|^{2m})$ applications of the matroid intersection algorithm we can test $(m+1)$-connectivity.

This bound can be significantly improved using the following observation. Fix some set $Q \subseteq S$ with $|Q| = m$. Fix a partition $\{Q_1, Q_2\}$ of Q. There are $\mathrm{O}(|S|^m)$

ways to complete this partition to a pair of disjoint sets $\{U_1, U_2\}$ such that $|U_1| = |U_2| = m$. Since there are 2^m partitions of Q into two sets, and this number is a constant relative to $|S|$, we see that $O(|S|^m)$ applications of matroid intersection will do. Since matroid intersection takes times $O(|S|^{2.5})$ we obtain an overall bound of $O(|S|^{m+2.5})$ oracle calls. By taking into account the similarity of the matroid intersection instances being solved, the bound can be reduced to $O(|S|^{m+2})$ oracle calls, and in the linear case, to $O(|S|^{m+2})$ total work.

6.5.4. Menger's Theorem for matroids

Tutte has generalized Menger's Theorem to matroids (Tutte 1965b). Let $M = M(G)$ for a graph G, and let S be the edge-set of G. Pick two nodes u, v of G, let P and Q be the stars of these nodes and assume that $P \cap Q = \emptyset$. Menger's Theorem for graphs (see chapter 2 by Frank) asserts that the minimum number of nodes, distinct from u and v, the deletion of which separates u and v equals the maximum number of internally node-disjoint paths joining them. If G is 2-connected, this minimum can be expressed in matroid terms as

$$\min_{P \subseteq A \subseteq S \setminus Q} r(A) + r(S \setminus A) - r(S) + 1.$$

To express the maximum imagine that we have found a family of m node-disjoint paths joining u and v. Deleting all edges not on these paths, and contracting all remaining edges other than those in P and Q, yields a graph in which u and v are still joined by m paths. This minor of G corresponds to a minor M' of M, and for this minor we have $m = r'(P) + r'(Q) - r'(P \cup Q) + 1$, where r' is the rank function of M'. Now it is easy to prove that for any such minor, $r'(P) + r'(Q) - r'(P \cup Q) + 1$ is no bigger than the minimum above. Tutte proved that equality can always be achieved.

Theorem 6.9 (Tutte). *Let M be a matroid on S and let P and Q be disjoint subsets of S. Let $\mathcal{M}'$ be the family of minors of M on the element set $P \cup Q$. Then*

$$\max_{M' \in \mathcal{M}'} r'(P) + r'(Q) - r'(P \cup Q) = \min_{P \subseteq A \subseteq S \setminus Q} r(A) + r(S \setminus A) - r(S).$$

Note that the quantity on the right can be computed using the matroid intersection algorithm. This fact was the basis for the connectivity algorithm in the subsection 6.5.3. It works by defining $M_1 = M \setminus P / Q$, $M_2 = M / P \setminus Q$ and finding a maximum-cardinality set J jointly independent in M_1 and M_2. It can then be proved that for this J, the matroid $M' = M / J \setminus (S \setminus (P \cup Q \cup J))$ achieves the maximum in Theorem 6.9. This idea, due to Edmonds, yields a proof of Theorem 6.9.

7. Recognition of representability

7.1. Graph realization for general matroids

A matroid M is *graphic* if there is a graph G such that $S = E(G)$ and the circuits of M are exactly the edge-sets of simple cycles in G. In this case we write $M = M(G)$.

Graph realization (GR) is the problem, given a matroid M, of determining that M is not graphic or finding a graph G such that $M = M(G)$. Seymour (1981b) first solved GR. We use here a slight variation of Seymour's result due to Truemper (1984).

Theorem 7.1. *Let M and M' be matroids on S, and let G be a graph with edge-set S. Suppose that*

(1) *M and M' are connected,*

(2) *M and M' have a common basis X such that the corresponding partial representations are identical,*

(3) *$M' = M(G)$, and*

(4) *for every node v of G, the star of v contains a cocircuit of M.*

Then $M = M'$, which implies that M is graphic.

We remark that (1) is necessary. Let $S = \{a,b,c,d,e\}$ and define two matroids M and M' on S as follows: M is the direct sum of U_4^2 and U_1^1 where $S(U_1^1) = \{e\}$, and $M' = M(G)$ where G has four nodes the stars of which are $\{a,c,d\}$, $\{b,c,d\}$, $\{a,b,e\}$ and $\{e\}$. Then the pair M,M' satisfies (2)–(4) with $X = \{a,b,e\}$, but $M \neq M'$.

Lemma 7.2. *Assume that M is a connected matroid, X is a basis, C is a circuit, and $S\backslash(X \cup C) \neq \emptyset$. Then there is an element $e \in S\backslash(X \cup C)$ such that either $M\backslash e$ is connected or e is in series with some element of X.*

Lemma 7.3. *Assume that (M, M') satisfy* (2)–(4). *Let $e \in S$, and assume e is a coloop of neither M nor M'.*

(a) *For $e \in S\backslash X$, $(M\backslash e, M'\backslash e)$ satisfy* (2)–(4) *with G replaced by $G\backslash e$.*

(b) *For $e \in X$, if e is parallel to no edge in G, then $(M/e, M'/e)$ satisfy* (2)–(4) *with G replaced by G/e and X replaced by $X\backslash\{e\}$.*

Lemma 7.4. *Assume that (M, M') satisfy* (1)–(4).

(a) *If $e \in S\backslash X$ is parallel to some element of X (in either M or M'), then* (1)–(4) *hold for $(M\backslash e, M'\backslash e)$ with G replaced by $G\backslash e$.*

(b) *If $e \in X$ is in series with some element of $S\backslash X$ (in either M or M'), then* (1)–(4) *hold for $(M/e, M'/e)$ with G replaced by G/e and X replaced by $X\backslash\{e\}$.*

Proof of Theorem 7.1. Suppose that $M \neq M'$, and that S has been chosen minimal subject to this condition. Let C be a circuit of one of M, M', independent in the other. Then by Lemmas 7.2, 7.3(a), and 7.4(b), $C \supseteq S\backslash X$, and so $r(S) \geqslant r^*(S)$. Let D be a cocircuit of one of M, M', coindependent in the other. Then, by the dual of Lemmas 7.2, and by 7.3(b) and 7.4(a), $D \supseteq X$, and so $r^*(S) \geqslant r(S)$. It follows that $r(S) = r^*(S)$, $C = S\backslash X$ and $D = X$.

Now suppose that C is a circuit of M and that D is a cocircuit of M. Then for $e \in X$, adding e to C creates no circuit in M, because of D. Hence, by the minimality of S, $C \cup \{e\}$ is a circuit in M'. But if $|X| \geqslant 2$, this implies X contains a circuit of M', since M' is binary. Moreover, $|X| \neq 1$ (otherwise $M = M'$), and so we conclude that C is a circuit of M' and D is a cocircuit of M'. But then deleting D from M', that is, G, leaves M' connected, because of C, which implies that D is the star of a node in G. On the other hand, C is a basis of M, and so D contains no cocircuit of M, contrary to (4). This proves the theorem. □

In the next subsection we will describe two algorithms for recognizing when a binary matroid is graphic; given such an algorithm, Theorem 7.1 yields an algorithm for GR, as follows. Let M be a matroid and assume that M is connected (if not, apply the algorithm to its connected components). Construct a partial representation R for M and determine whether the associated binary matroid $M((I,R))$ is graphic. If not, stop – M is not graphic; otherwise, taking G to be a graph with representation (I,R), use Theorem 7.1 to test whether $M = M(G)$.

The computational complexity of this algorithm may be derived as follows. The construction of a partial representation R requires $O(r(M)|S|)$ calls to an independence oracle. To determine whether (I,R) is the binary representation of a graphic matroid, we may use the algorithm of Bixby and Wagner described in the next subsection. The work for this step is bounded by $O(z\alpha(z,r(M)))$, where $\alpha(\cdot,\cdot)$ is an inverse of the Ackermann function (see Theorem 7.8) and z is the number of nonzero entries in R. Finally, to apply Theorem 7.1 one must check the star of each node in G to see whether it contains a cocircuit. This checking requires a total of $O(r(M)|S|)$ calls to an independence oracle. Thus, the entire algorithm uses $O(r(M)|S|)$ oracle calls, plus $O(z\alpha(z,r(M)))$ other work.

In closing we mention another result that can be used to solve GR. This result generalizes a result of Tutte for binary matroids. It uses concepts defined in the discussion preceding Theorem 6.5.

Theorem 7.5 (Bixby). *Let M be a matroid and let D be a cocircuit of M. Then M is graphic if and only if*

(a) *the D-components of M are graphic, and*
(b) *the bridge graph of D is bipartite.*

This theorem suggests a recursive algorithm for GR, very similar to the first algorithm discussed in the next subsection. The bound for this algorithm is $O(r(M)z)$, where z is the number of nonzero elements in some partial representation of M.

7.2. Graph realization for binary matroids

Numerous algorithms have been proposed for GR on binary matroids. We describe two. The first is based on Theorem 7.5 for the case of binary matroids. Let M be a binary matroid and let R be a partial representation of M. The steps in the algorithm, due to Tutte (1960), are the following:

Step 1. If (I,R) has at most 2 ones in each column, M is graphic; otherwise, choose a column having a one in each of three rows, corresponding to cocircuits D_1,D_2 and D_3.

Step 2. If each of D_1,D_2 and D_3 has just one bridge, then M is not graphic (because edges of graphs can be in at most two nonseparating cocircuits, corresponding to their end vertices); otherwise, let D be a separating cocircuit. (For definitions of terms used here, see the discussion preceding Theorem 6.5.)

Step 3. Compute the bridge graph of D. If it is not bipartite, M is not graphic; otherwise, apply the above steps recursively to the D-components of M. M is graphic if and only if each of these is graphic.

In Bixby and Cunningham (1980) it is shown that this procedure can be implemented to run in time $O(r(M)z)$ where z is the number of nonzero elements in the representation (I,R).

As our second algorithm, we describe one (Bixby and Wagner 1988) based on an idea of Löfgren. The algorithm has the same computational complexity as an algorithm outlined by Fujishige, which is also based on Löfgren's procedure. This complexity is the best for any known GR algorithm on binary matroids. It is conjectured in Bixby and Wagner (1988) to be best possible.

Let G be a 2-connected graph with edge-set E. For $E' \subseteq E$, let $V(E')$ denote the set of nodes in G incident to some edge in E'. Let $\{E_1, E_2\}$ be a partition of the edge-set of G such that $V(E_1) \cap V(E_2) = \{u,v\}$. Let G_1 be the subgraph of G induced by E_1. Define G' to be the graph obtained by interchanging in G_1 the incidences of the nodes u and v. Then G' is said to be obtainable from G by *reversing* G_1. In general, G'' is *2-isomorphic* to G if G'' is obtainable from G by a sequence of subgraph reversals.

Theorem 7.6 (Whitney 1933). *Let G and G' be 2-connected graphs on the same edge-set. Then G and G' are 2-isomorphic if and only if they have the same matroid.*

Let R be a partial representation of a binary matroid M. Define R to be *graphic* if M is graphic. Let R_k denote the matrix made up of the first k columns of R, where rows consisting entirely of zeros have been deleted. R is called *totally nonseparable* if R_k is nonseparable for $1 \leqslant k \leqslant |S| - r(M)$. For any matrix A such that $G(A)$ is connected, it is easy to compute a permutation of the columns such that the resulting matrix is totally nonseparable.

Where C_k is the fundamental circuit defined by column k of R, define $P_k = C_k \cap \{\bigcup_{j<k} C_j\}$. A set of edges P of a graph G is a *hypopath* of G if P is a path in some graph 2-isomorphic to G. The following statement, provable directly from Theorem 7.6, is the "subrearrangement theorem" of Löfgren (1959).

Theorem 7.7. *Let R be a totally nonseparable $\{0,1\}$-matrix with c columns. Assume that for some $1 \leqslant k \leqslant c$, R_k is graphic with realization G_k. Then R_{k+1} is graphic if and only if P_{k+1} is a hypopath of G_k.*

Löfgren suggested the following procedure for testing whether R is graphic. Assume R is totally nonseparable. Clearly R_1 is graphic. Suppose there exists a graph G_k that realizes R_k. Further, suppose P_{k+1} is a hypopath of G_k. Then there exists a graph G'_k 2-isomorphic to G_k such that P_{k+1} is a path in G'_k. Add the edges of $C_{k+1} \backslash P_{k+1}$ to G'_k so that they form a path between the ends of P_{k+1} but are not incident to any other vertices of G'_k. It is straightforward to verify that the resulting graph G_{k+1} is a realization of R_{k+1}. If the above procedure breaks down at any point, it follows that R is not graphic.

To implement the above idea requires a polynomial-time method for constructing G'_k from G_k. A natural approach is to invoke some representation of G_k that "displays" all graphs 2-isomorphic to G_k. For this representation we use a graph-decomposition theory developed by Tutte in which a 2-connected graph is uniquely decomposed into polygons, bonds and 3-connected graphs. Using this decomposition, we can determine in polynomial time whether a given subset of edges is a hypopath.

The complexity of the algorithm is stated in the following theorem. The function $\alpha(\cdot,\cdot)$ is an inverse of the Ackermann function, and is very slow growing, being for all practical purposes never bigger than 4.

Theorem 7.8 (Bixby and Wagner). *Given an $r \times c$ $\{0,1\}$-matrix (I,R) with z nonzero entries, there is an algorithm that runs in time $O(z\alpha(z,r))$ and uses space $O(z)$ to determine whether the binary matroid $M((I,R))$ is graphic.*

7.3. An application to linear programming

A *(linear) network (flow) problem* is a linear programming problem $\min\{c^{\mathrm{T}}x\colon Nx = b, x \geqslant 0\}$, where is N is a $\{0,\pm 1\}$-matrix with no column having two equal nonzero entries. We call a matrix with this property a *network matrix*. These are exactly the matrices that occur as submatrices of node–arc incidence matrices of digraphs.

Two matrices A and R are *projectively equivalent* if one can be obtained from the other by elementary row operations and nonzero column scaling. A linear program $\min\{c^{\mathrm{T}}x\colon Ax = b, x \geqslant 0\}$ is called a *hidden network* if its constraint matrix A is projectively equivalent to a network matrix N. In this case, given explicit knowledge of N, one can easily produce an equivalent network problem with constraint matrix N. The motivation for doing so is primarily computational: It is well established that network linear programs can be solved much more efficiently than general linear programs.

The relationship between GR and hidden networks is summarized in the following theorem. In particular, it follows from this result that any polynomial-time algorithm for GR on binary matroids implies a polynomial-time algorithm for testing whether a given linear program is a hidden network.

Theorem 7.9 (Iri 1968). *Let $A = (I,R)$ be a real-valued matrix, where I is an identity matrix, and let $A' = (I,R')$ where R' is obtained from R by replacing nonzero entries with 1's. Then A is projectively equivalent to a network matrix if and only if the following two conditions hold:*

(a) *R' is the partial representation of a graphic matroid M; and*

(b) *where G is any graph whose matroid is M, D is any orientation of G, and N is its corresponding network matrix, A is projectively equivalent to N.*

7.4. Recognizing total unimodularity

A $\{0,1\}$-matrix is *totally unimodular* if every square submatrix has determinant ± 1 or 0. The significance of these matrices in optimization was pointed out by A. J. Hoffman and J. B. Kruskal (see chapter 30 by Schrijver for an extensive discussion of total unimodularity), who observed that linear programming problems $\min\{c^{\mathrm{T}}x\colon Ax = b, x \geqslant 0\}$ with integral b and totally unimodular constraint matrix A have integral basic feasible solutions (a simple consequence of Cramer's Rule). It is well known that network matrices are totally unimodular, and Hoffman and Kruskal asked whether other interesting classes could be found. Seymour gave an answer to this question:

Theorem 7.10 (Seymour 1980). *Let $M = M(A)$ where A is totally unimodular, and assume that M is 3-connected and has no 3-separation $\{S_1, S_2\}$ with $|S_1| \geqslant 4 \leqslant |S_2|$. Then M is either graphic, cographic (the dual of a graphic matroid), or isomorphic to R_{10}. (See chapter* 10 *by Seymour for a definition of R_{10}.)*

The above theorem is perhaps best viewed as a "decomposition" result. In chapter 10 by Seymour, the connectivity-based notions of 1-, 2- and 3-sum are defined. In terms of these sums, Theorem 7.10 can be stated as follows: Every matroid arising from a totally unimodular matrix can be constructed using 1-, 2- and 3-sums starting with only graphic and cographic matroids and copies of R_{10}. Thus, apart from the matrices representing R_{10}, all "indecomposable" totally unimodular matrices arise from graphs, or duals of graphs. This view suggests an algorithm for testing whether a matrix is totally unimodular. More generally, it can be used to test whether a given matroid M is regular (i.e. $M = M(A)$ where A is totally unimodular).

To see that a solution to the second problem above yields a solution to the first, suppose that $M = M((I, A))$ has been shown to be regular where A is a given real matrix with $\{0, \pm 1\}$ entries, and I is an identity matrix of appropriate dimension. Let A_k denote the submatrix of A made up of the first k columns. A_1 is clearly totally unimodular. In general, assume that A_k ($k \geqslant 1$) is known to be totally unimodular. Find the components of the bipartite graph $G(A_k)$, and let i' and i'' be two row indices in the same component of $G(A_k)$ and such that $a_{i',k+1} \neq 0 \neq a_{i'',k+1}$. Find a shortest path from i' to i'' in $G(A_k)$. The row and column indices of this path together with the column index $k+1$ determine a square submatrix of A_{k+1}, since the path was chosen to be shortest. If this submatrix has determinant other than ± 1 or 0, then A_{k+1} is not totally unimodular, and so neither is A; otherwise, after checking all such pairs (i', i''), it follows that A_{k+1} is totally unimodular. The validity of this last assertion can be deduced by showing that if (I, A') is a totally unimodular matrix with the same zero, nonzero pattern as (I, A) ((I, A') exists because M is regular), then the success of the checking procedure for A implies that A can be scaled to A' by multiplying some rows and columns of A by -1.

We now sketch an algorithm for determining whether a general matroid is regular, using Seymour's theorem above. The description proceeds inductively on $|S(M)|$. Thus, we assume that a matroid M is given, specified by an independence oracle, and that we can test for regularity any matroid that has a smaller number of elements.

Obviously we can test whether M is isomorphic to R_{10}, and using the results on graph realization given earlier in this section, we can test whether M is graphic or cographic. If any of these tests is positive, then M is regular and we are done; otherwise, using results of section 6, we test whether M has a 1-separation $\{S_1, S_2\}$. If so, let $M_i = M \backslash S_i$ $(i = 1, 2)$; it is easy to see that M is regular if and only if M_1 and M_2 are regular. Hence, we may assume that M is connected. Now, test for a 2-separation $\{S_1, S_2\}$. If one exists, find a circuit C of M such that $C_i = C \cap S_i \neq \emptyset$ $(i = 1, 2)$. Select $e_i \in C_i$ $(i = 1, 2)$ and let $M_i = M \backslash (S_i \backslash C_i)/(C_i \backslash \{e_i\})$ $(i = 1, 2)$. Then both M_1 and M_2 are smaller than M, and it can be shown that M is regular if and only if M_1 and M_2 are regular.

Finally, it remains to consider the case when M is 3-connected, neither graphic nor cographic, and not R_{10}. This case is difficult because we must, in effect, consider representing M as a 3-sum, 3-sums are defined only for binary matroids, and M is not known to be binary (moreover, if M is not regular we have no way to test whether it is binary, short of determining that it is regular, since testing whether a general matroid is binary is intractable). The method given below for dealing with this difficulty is due to Truemper (1982).

Using the connectivity algorithm based on matroid intersection (see section 6), we can easily determine whether M has a 3-separation $\{S_1, S_2\}$ such that $|S_1| \geqslant 4 \leqslant |S_2|$. Suppose this is the case. Let X_2 be a basis of S_2, and extend X_2 to a basis $X = X_1 \cup X_2$ of M, where $X_1 \cap X_2 = \emptyset$. Let $Y_i = S_i \backslash X_i$ $(i = 1, 2)$. Let R be the PR of M determined by X:

$$R = \boxed{\begin{matrix} A_1 & 0 \\ D & A_2 \end{matrix}}$$

where $A_i = R(X_i, Y_i)$ $(i = 1, 2)$ and $\text{rk}(D) = 2$. Suppose that D has the structure

$$D - \boxed{\begin{matrix} J_1 & 0 \\ 0 & J_2 \end{matrix}}$$

where J_1 and J_2 are matrices of all 1's. If D does not have this structure, we take $R' = R$, $X_2' = X_2$ and $Y_2' = Y_2$; otherwise, since M is 3-connected, it follows that there is a shortest path in A_2 joining some row of J_1 and some row of J_2. A *pivot* on a nonzero element r_{xy} of R is a sequence of elementary row operations (over GF(2)) that remove all nonzeros in column y, except r_{xy}, followed by a resetting of all entries in this column to their original values. Performing GF(2) pivots on appropriate 1's along this path in A_2 we obtain a matrix

$$R' = \boxed{\begin{matrix} A_1 & 0 \\ D' & A_2' \end{matrix}}$$

where $D' = R'(X_2', Y_1)$ and $A_2' = R'(X_2', Y_2')$, and where D' has some row of all 1's and $X_2' \cup Y_2' = S_2$. Note that if R' is not a PR of M, then M is not binary, and hence not regular.

Now define $M_1 = M/X_1$ and $M_2 = M_2\backslash Y_2'$. Let N_1 be the binary matroid with PR (D', A_2), and let N_2 be the binary matroid with PR A_1. Remove loops, coloops, series and parallel elements from N_1 and N_2. If these elements are not loops, coloops, series and parallel in M_1 and M_2, respectively, then M is not regular; otherwise, let M_1' and M_2' be the corresponding reduced versions of M_1 and M_2, respectively. Now M_1 and M_2 are smaller than M, and Truemper has proved that they are both regular if and only if M is regular.

The above procedure clearly runs in polynomial time. We will not attempt to estimate its complexity. Truemper (1990) has given a more complicated, direct algorithm with complexity $O(|S|^3)$.

7.5. *Intractable problems*

In section 2 it was shown that computing the girth of a matroid is intractable (i.e. that there is no oracle polynomial-time algorithm for this problem), and, as noted at the beginning of section 6, the same argument applies to show that computing the connectivity of a matroid is intractable.

In this section we have given polynomial-time algorithms for a small set of "representability" questions: Testing whether a matroid is graphic, testing whether a binary matroid is graphic, and testing whether a matroid is regular. The surprising fact is that these are essentially the *only* "interesting" representability questions for which polynomial-time algorithms are *possible.* Thus, at one extreme, testing whether there exists a field over which a given matroid is representable is intractable, as is the other extreme case of testing whether a given matroid is representable over a given field.

The most basic result along these lines was proved by Seymour (1981b), who showed that testing whether a matroid is binary is intractable. His proof runs as follows. Let $S = \{x_1, \ldots, x_k, y_1, \ldots, y_k\}$ be a 2k-element set, define $Y = \{y_1, \ldots, y_k\}$ and define two families of subsets $\mathcal{A}$ and $\mathcal{B}$ of S as follows:

$$\mathcal{A} = \{\{x_i, y_i, x_j, y_j\} \subseteq S\colon i < j\}$$

$$\mathcal{B} = \{Z = \{z_1, \ldots, z_k\} \subseteq S\colon z_i = x_i \text{ or } y_i, \text{ and } |Z \cap Y| \text{ is even}\}$$

Then $\mathcal{C} = \mathcal{A} \cup \mathcal{B}$ is the family of circuits of a binary matroid M on S. Now, for each $Z \in \mathcal{B}$, let M_Z denote the nonbinary matroid that is identical to M except that Z is independent (M_Z is a matroid since Z is a circuit and hyperplane of M). The existence of the matroids M_Z implies that any algorithm proving that M is binary must have made at least $|\mathcal{B}| = 2^{k-1}$ calls to the independence oracle.

8. Matroid flows and linear programming

8.1. Maximum flows

Let M be a matroid on S. Fix $l \in S$, and let $\mathscr{C}_l$ be the family of circuits of M containing l. Let A be the $\{0,1\}$-matrix with columns indexed on elements $e \in S\backslash\{l\}$ and rows indexed on circuits $C \in \mathscr{C}_l$, such that $a_{Ce} = 1$ if and only if $e \in C$. Let $\mathscr{D}_l$ be the family of cocircuits of M containing l. We say that M is an l-MFMC matroid, that is, has the (*integral*) *max-flow min-cut property* with respect to l, if for every choice of nonnegative integral vector w defined on $S\backslash\{l\}$,

$$\min_{D\in\mathscr{D}_l} w(D\backslash\{l\}) = \max(\mathbf{1}^{\mathrm{T}}y\colon y^{\mathrm{T}}A \leqslant w^{\mathrm{T}}, y \geqslant 0 \text{ and integral}),$$

where $\mathbf{1}$ is a vector of 1's.

A nonnegative vector y satisfying $y^{\mathrm{T}}A \leqslant w^{\mathrm{T}}$ is called a *flow*, or *l-flow*, and $\mathbf{1}^{\mathrm{T}}y$ is its *value*. A flow of maximum value is a *maximum flow*. The special element l is called the *demand* element. Define M to be a MFMC matroid if it is an l-MFMC matroid with respect to every choice of demand element l. (In chapter 10 by Seymour MFMC matroids are called *free flowing*.)

Theorem 8.1 (Seymour 1977b). *A matroid M is an MFMC matroid if and only if it is binary and contains no F_7^* minor.*

This theorem can be proved using "splitter theory" (Seymour 1980), from which it follows that a 3-connected matroid M is binary and has no F_7^* minor if and only if it is either regular or isomorphic to F_7. This structural result gives an algorithm for computing maximum flows in MFMC matroids and for testing whether a matroid is an MFMC matroid. We describe this algorithm for computing flows. Truemper (1992) contains an extensive treatment of this topic.

Suppose M is not 2-connected. Then it is easy to see that M is an MFMC matroid if and only if each of its 2-connected components is. In particular, for any choice of demand element, the computation of a maximum flow for this demand element can be restricted to the component containing it.

Assume M is 2-connected but not 3-connected. Let $\{S_1, S_2\}$ be a 2-separation of M and let C be a circuit of M such that $C_i = C \cap S_i \neq \emptyset$ $(i = 1,2)$. Select $e_i \in C_i$ $(i = 1,2)$ and let $M_i = M\backslash(S_i\backslash C_i)/(C_i\backslash\{e_i\})$ $(i = 1,2)$. Then M is an MFMC matroid if and only if M_1 and M_2 are. Assuming that M_1 and M_2 satisfy this property, we may compute maximum flows for M as follows. Select an element $l \in S$ and assume $l \in S(M_1)$. Let w be an integral vector defined on S. Restricting w to S_2, compute a maximum e_1-flow in M_2. Let r be the value of this flow. Now define w' on S_1 by $w'(S_1\backslash\{e_2\}) = w$ and $w'(e_2) = r$. Now compute the value of a maximum l-flow on M_1 with respect to w'. It is easy to prove that this is the maximum flow value for M; moreover, it is not difficult to construct the actual flow for M from the flows for M_1 and M_2.

Finally, suppose that M is 3-connected. Then M is either isomorphic to F_7 or regular. Both of these properties can be checked, the latter using the algorithm given in section 7. To complete the description of the algorithm for finding maximum flows in MFMC matroids, we describe a method for computing maximum flows in regular matroids. This can be done in a variety of ways. The simplest is via linear programming. Let M be regular and assume K is a totally unimodular matrix such that $M = M(K)$. Then it can be proved that the optimal value of the following linear program is the value of a maximum l-flow in M:

$$\max(x_l\colon Kx = 0, 0 \leqslant x_e \leqslant w_e \ (e \neq l))$$

Since K is totally unimodular, this LP will have an integral optimal solution x. By decomposing this x into multiples of rows of A, we obtain the desired flow vector y.

The characterization of the l-MFMC matroids, for a fixed l, is very similar to that for the MFMC matroids, but the proof is much more difficult.

Theorem 8.2 (Seymour 1977b). *A matroid M is an l-MFMC matroid with respect to some fixed demand element l if and only if it is binary and contains no F_7^* minor containing l.*

Let us denote the class of binary matroids that are l-MFMC matroids with respect to some fixed l by $\mathcal{M}_l$. Truemper (1987) has given a polynomial-time algorithm for testing membership in $\mathcal{M}_l$ and has shown that the maximum flow problem on this class always has an integral solution by giving an algorithm for computing such a solution. We describe Truemper's algorithm for computing maximum flows on $\mathcal{M}_l$. To simplify the presentation, we do not concern ourselves with finding integral flows. We also do not treat membership, although the development given below can easily be modified to do so.

The problem we will actually solve is the problem of finding shortest l-paths. An *l-path* is a set P of the form $C\backslash\{l\}$ where $C \in \mathscr{C}_l$. Given a real-valued "length" function d defined on S, the *length* of P is $d(P)$. P is *shortest* if its length is minimum. Now, to see that we can solve maximum flow problems by computing shortest paths, note first that to solve a given maximum flow problem, it is sufficient to solve the corresponding dual:

$$\min(w^{\mathrm{T}}x\colon Ax \geqslant \mathbf{1},\ x \geqslant 0). \tag{8.3}$$

Of course, (8.3) generally has an enormous number of constraints, and it is no more obvious how to solve it than to solve the primal. Note, however, that for a given $x \geqslant 0$, checking $Ax \geqslant \mathbf{1}$ is nothing but the problem of checking whether

$$\min_{C \in \mathscr{C}_l} x(C\backslash\{l\}) \geqslant 1.$$

Thus, if we can compute shortest l-paths in polynomial time, then we can check for violated inequalities in (8.3) in polynomial time, and so we can solve (8.3) in polynomial time using the ellipsoid method.

Let $M \in \mathcal{M}_l$ and consider the problem of computing a shortest l-path with respect to some nonnegative weight function w defined on S. We begin by discussing several special cases.

If M is not 3-connected, this shortest-path problem on M can be solved, or, more precisely, reduced to smaller shortest-path problems, using exactly the same methods we used to compute maximum flows in the non-3-connected MFMC matroids.

Suppose M is either regular or isomorphic to F_7. The F_7 case obviously presents no difficulty. For the case when M is regular, we can use the following computation, where $M^* = M(K^*)$ and K^* is totally unimodular:

$$\max(x_l\colon K^*x = 0,\ 0 \leqslant x_e \leqslant w_e\ (e \neq l)). \tag{8.4}$$

Since regular matroids are MFMC matroids, the optimal value in (8.4) equals the minimum weight of a circuit containing l.

Suppose that there is a triad $\{x, y, z\}$ of M, not containing l. (A *triad* is a cocircuit of cardinality 3.) In this case we make use of the following result:

Lemma 8.5. *Suppose that $M \in \mathcal{M}_l$, $\{x, y, z\}$ is a triad of M not containing l, and $\{e, f, g\}$ is disjoint from S. Then $M' \in \mathcal{M}_l$, where M' is obtained from M by creating circuits $\{e, x, y\}$, $\{f, y, z\}$ and $\{g, z, x\}$.*

This lemma justifies the following construction. Assign weights $w_e = w_x + w_y$, $w_f = w_y + w_z$ and $w_g = w_z + w_x$ to the elements e, f and g, respectively. Let $M'' = M' \backslash \{x, y, z\}$. Now, using any shortest l-path in M'', we easily construct a shortest l-path in M. This procedure reduces by one the number of triads not containing l.

We are now reduced to considering the case when M is 3-connected, not isomorphic to F_7, not regular, and contains no triad missing l. In this final case, M can be appropriately decomposed, and the shortest-path problem solved on it by solving appropriate problems on the smaller matroids that result from the decomposition.

Assume that $\{S_1, S_2\}$ is a 3-separation of M with the properties that $|S_1| \geqslant 4 \leqslant |S_2|$, $l \in S_2$ and $S_2 \backslash \{l\}$ spans l. Let X_2 be a basis of $S_2 \backslash \{l\}$ and extend X_2 to a basis $X = X_1 \cup X_2$ of M, where $X_1 \cap X_2 = \emptyset$. Let $Y_i = S_i \backslash X_i$ $(i = 1, 2)$. Let R be the PR determined by X (in fact, R is a *representation* of M since M is binary). Let $X_2' \subseteq X_2$ be such that $X_2' \cup \{l\}$ is the fundamental X-circuit determined by l, and let $X_2'' = X_2 \backslash X_2'$. Then we may display R as

	Y_1	l	Y_2 (rest)
X_1	A_1	$\mathbf{0}$	
X_2'	D'	$\mathbf{1}$	A_2'
X_2''	D''	$\mathbf{0}$	A_2''

where **1** and **0** are matrices of 1's and 0's, respectively, of appropriate dimensions. Now if D' is identically 0, then there must be an element in Y_2 such that the corresponding column of R has a 1 in both A_2' and A_2''; otherwise, M is not connected. By pivoting on a 1 in the intersection of such a column with A_2'', D' becomes nonzero. We would also like to arrange that $\mathrm{rk}(D') = 2$. Suppose this is not the case. Let $\bar{X}_2'$ be the subset of elements in X_2' determined by the nonzero rows in D', and let $\bar{X}_2''$ be the subset of elements in X_2'' determined by the nonzero rows in D'', different from those in $R(\bar{X}_2', Y_1)$. Then there is a path in $G = G(R(X_2, Y_2))$ between some element of $\bar{X}_2'$ and some element of $\bar{X}_2''$, for in the alternative case M is not 3-connected (where T is the vertex-set of the component of G containing $\bar{X}_2'$, $\{T, S\backslash T\}$ is either a 1- or 2-separation of M if $S\backslash T \neq \emptyset$). Taking a shortest such path, and pivoting on appropriate 1's in the path, results in $\mathrm{rk}(D') \geqslant 2$. Now, applying essentially the same argument to A_1, letting any nonzero row in A_1 play the role of l, it follows that we may assume R has the following form, for appropriate elements e, f, x, y and z:

$$
\begin{array}{c|c|cc|c|c|}
 & & y & z & l & \\
\hline
 & A_1 & & & & \mathbf{0} \\
x & & 1 & 1 & & \\
\hline
e & D_1 & 1 & 0 & 1 & \\
f & & 0 & 1 & 1 & \\
\hline
 & & D_2 & & & A_2 \\
\hline
\end{array}
$$

From the above representation for M, we construct two matroids M_1 and M_2 with PRs given by R_1 and R_2:

$$
R_1 = \begin{array}{c|c|cc|c|}
 & & y & z & l \\
\hline
 & A_1 & & & \mathbf{0} \\
x & & 1 & 1 & \\
\hline
e & D_1 & 1 & 0 & 1 \\
f & & 0 & 1 & 1 \\
\hline
\end{array}
\quad \text{and} \quad
R_2 = \begin{array}{c|cc|c|c|}
 & y & z & l & \\
\hline
x & 1 & 1 & & \mathbf{0} \\
\hline
e & 1 & 0 & 1 & \\
f & 0 & 1 & 1 & \\
\hline
 & D_2 & & & A_2 \\
\hline
\end{array}
$$

We can now state

Theorem 8.6. *Let $M \in \mathcal{M}_l$ be 3-connected. Assume that M is not regular, not isomorphic to F_7 and has no triad missing l. Then there is a 3-separation $\{S_1, S_2\}$ of M such that $|S_1| \geqslant 4 \leqslant |S_2|$, $l \in S_2$, $S_2\backslash\{l\}$ spans l and M_1 is either isomorphic to F_7 or is regular.*

There are now two steps left. We must show how to compute a 3-separation of the form guaranteed by Theorem 8.6, and we must show how the resulting decomposition into M_1 and M_2 can be used to compute shortest l-paths. For the computation of the 3-separation we use the following expedient algorithm. (A much more efficient, direct algorithm is given in Tseng and Truemper 1986). For each disjoint pair $\{U_1, U_2\}$, where $l \in U_2$ and $|U_1| = 4 = |U_2|$, we apply the Matroid Intersection Algorithm to the pair of matroids $M' = M/U_1\backslash U_2$ and $M'' = M/U_2\backslash U_1$ to determine the unique minimal set $A \subseteq S' = S\backslash(U_1 \cup U_2)$ that minimizes $r_{M'}(A) + r_{M''}(S'\backslash A)$.[1] Then $\{U_1 \cup A, S\backslash(U_1 \cup A)\}$ is a 3-separation if and only if there exists

[1] To see that we can compute A with this property, note that A here is the same as the A in Theorem 3.6(b), where M' plays the role of M_1 and M'' the role of M_2. The minimal A is just the set of vertices $v \neq s$ of G such that there exists an (s, v)-dipath.

some 3-separation $\{V_1, V_2\}$ such that $U_i \subseteq V_i$ $(i = 1, 2)$. Now, by the minimality of the A's, for any 3-separation $\{S_1, S_2\}$ that satisfies the conditions of Theorem 8.6, we will find some A determining a 3-separation and such that $U_1 \cup A \subseteq S_1$. It follows that this 3-separation also satisfies the conclusion of the theorem.

We are now in a position to complete the description of a method for finding a shortest l-path in M. By the computation of the previous paragraph, there is a 3-separation giving rise to matroids M_1 and M_2 satisfying the conditions of Theorem 8.6. The method for finding shortest paths is then straightforward. Set $w_e = w_f = w_l = 0$ in M_1. Find the shortest e-, f- and l-paths in $M_1\backslash\{f, l\}$, $M_1\backslash\{e, l\}$ and $M_1\backslash\{e, f\}$, respectively, and let their corresponding lengths be d_e, d_f and d_l. (These computations are possible since M_1 is either regular or isomorphic to F_7.) Now add new elements e', f' and g' to M_2 so that $\{e', x, y\}$, $\{f', y, z\}$ and $\{g', z, x\}$ are circuits; then delete x, y and z, denoting the resulting matroid by M_2'. By Lemma 8.5, $M_2' \in \mathcal{M}_l$; moreover, M_2' has fewer elements than M, and so we may assume (by induction on $|S|$) that an algorithm is available to compute shortest l-paths in M_2'. Set $w_i = d_i$ for $i = e', f', g'$. It is straightforward to see that a shortest l-path in M_2', together with the three shortest paths constructed in M_1 can be used to build a shortest l-path in M.

8.2. Oriented matroids and linear programming

Matroid theory arises as a combinatorial abstraction of properties of linear dependence. From this viewpoint, oriented matroid theory arises when attention is restricted to ordered fields. Hence, it allows interpretation and generalization of ideas of real linear algebra. We indicate here how this can be done with linear programming. It is a very attractive theory, which has also produced new insights and methods in linear programming. Good references for this subject are Bachem and Kern (1992) and Björner et al. (1993).

Oriented matroids are treated in chapter 9 by Welsh. We give a definition that is useful for our purposes, and explain the relationship with the earlier definition. A *sign vector* on S is an element of $\{0, 1, -1\}^S$. To each vector in $\mathbb{R}^S$ we associate a sign vector in the obvious way. For any vector $x \in \mathbb{R}^S$, the *support* of x is $\{j \in S\colon x_j \neq 0\}$. For a subspace L of $\mathbb{R}^S$, a vector $x \in L$ is *elementary* if $x \neq 0$ and the support of x is minimal. The supports of elementary vectors of L are the circuits of a matroid M, and those of $L^\perp$, the orthogonal complement of L, are the circuits of M^*. In particular, where A is a matrix with columns indexed by S, and $L = \{x\colon Ax = 0\}$, M is the matroid whose independent sets correspond to linearly independent sets of columns of A. One can easily check that the sets $\mathcal{F}$, $\mathcal{F}^*$ of sign vectors corresponding to elementary vectors of L, $L^\perp$ satisfy:

the supports of members of $\mathcal{F}$ are the circuits of a matroid M, and those of $\mathcal{F}^*$ are the circuits of M^*; (8.7)

for every circuit C of $M(M^*)$, there are exactly two elements x, y of $\mathcal{F}(\mathcal{F}^*)$ such that C is the support of both, and $x = -y$; (8.8)

$$\text{if } x \in \mathscr{F}, y \in \mathscr{F}^*, \text{ and } x_i y_i \neq 0 \text{ for some } i, \text{ then there exists } j, k \in S \text{ with } x_j y_j = 1 \text{ and } x_k y_k = -1. \tag{8.9}$$

For a triple $(S, \mathscr{F}, \mathscr{F}^*)$ such that $\mathscr{F}$, $\mathscr{F}^*$ are sets of signed vectors on S, we take (8.7)–(8.9) as the definition of a *dual pair of oriented matroids.* It is not difficult to see that $(S, \mathscr{F})$ determines $\mathscr{F}^*$, so oriented matroids do not need to be defined in dual pairs. (See chapter 9 by Welsh.) It is also unnecessary (and a little inconvenient) to deal with *elementary* sign vectors. We have done so because the corresponding matroidal objects (circuits) are more familiar. The price we pay is that, in applications, we often have "elementary" as a qualifier when we do not really want it. In many cases, the following result can be used to get rid of it.

Lemma 8.10. *Let L be a subspace of $\mathbb{R}^S$ and let $x \in L$, $x \neq 0$. Then there exist elementary vectors $x^1, x^2, \ldots, x^k$ of L such that $\sum_{i=1}^k x^i = x$ and $x^i_j \neq 0$ implies $x^i_j x_j > 0$.*

As an example, suppose that we are interested in the existence of a non-negative solution of $A'x' = b$. By taking $A = (A', -b)$, this is equivalent to the existence of a non-negative solution of $Ax = 0$ with a particular component x_e positive. By Lemma 8.10 the latter is equivalent to the existence of such a solution that is also elementary. So the following oriented matroid theorem (due independently to Bland and Las Vergnas, see Bland 1977) gives a characterization. (Notice that both situations in Theorem 8.11 cannot occur, because of (8.9).)

Theorem 8.11. *Let $(S, \mathscr{F}, \mathscr{F}^*)$ be a dual pair of oriented matroids, and let $e \in S$. Then either there exists $x \in \mathscr{F}$ with $x \geqslant 0$ and $x_e = 1$, or there exists $y \in \mathscr{F}^*$ with $y \geqslant 0$ and $y_e = 1$.*

Applying Lemma 8.10 and Theorem 8.11, we have that $A'x' = b$ has a non-negative solution if and only if there does not exist a vector $z \geqslant 0$ in the row space of $(A', -b)$ that is positive in the last component, that is, if and only if there does not exist y such that $y^{\mathrm{T}}A' \geqslant 0$, $y^{\mathrm{T}}b < 0$. This is the classical Farkas Lemma; see chapter 30 by Schrijver.

Somewhat similar techniques allow a generalization of the duality theorem of linear programming. One form of that theorem says that if the problems $\max(c^{\mathrm{T}}x' \colon A'x' = b, x' \geqslant 0)$ and $\min(y^{\mathrm{T}}b \colon y^{\mathrm{T}}A' \geqslant c^{\mathrm{T}})$ have feasible solutions, then they have ones for which $c^{\mathrm{T}}x' = y^{\mathrm{T}}b$. An equivalent statement of the latter condition is that $(y^{\mathrm{T}}A' - c^{\mathrm{T}})x' = 0$. We handle the objective function $c^{\mathrm{T}}x'$ by adding a new variable x_f and a new equation $x_f - c^{\mathrm{T}}x' = 0$ as the last equation in the system. Thus, where the components of x' and the columns of A' are indexed by elements of $S \backslash \{e, f\}$, we get a new problem with unknown vector $x \in \mathbb{R}^S$. So the existence of x' such that $A'x' = b$ and $x' \geqslant 0$ is equivalent to the existence of x such that $Ax = 0$, $x_e = 1$ and $x_j \geqslant 0$ for $j \in S \backslash \{f\}$. The existence of a vector y such that $y^{\mathrm{T}}A' \geqslant c^{\mathrm{T}}$ is equivalent to the existence of a vector $z = (y^{\mathrm{T}}, 1)A$ in the row space of A, such that $z_j \geqslant 0$ for $j \in S \backslash \{e\}$ and $z_f = 1$. If x, y and z have the above properties, then the condition $(y^{\mathrm{T}}A' - c^{\mathrm{T}})x' = 0$ is equivalent to $x_j z_j = 0$ for $j \in S \backslash \{e, f\}$. Hence, the following result of Lawrence generalizes the duality theorem.

Theorem 8.12. *Let $(S, \mathcal{F}, \mathcal{F}^*)$ be a dual pair of oriented matroids and let $e, f \in S$ with $e \neq f$. If there exists $x \in \mathcal{F}$ with $x_e = 1$ and $x_j \geqslant 0$ for $j \in S \backslash \{f\}$, and there exists $z \in \mathcal{F}^*$ with $z_f = 1$ and $z_j \geqslant 0$ for $j \in S \backslash \{e\}$, then there exist such x, z with $x_j z_j = 0$, $j \in S \backslash \{e, f\}$.*

Considerable effort has gone into finding constructive proofs of results such as Theorems 8.11 and 8.12. One would like to have, for example, an oriented matroid algorithm that constructs the x and z of Theorem 8.12. Bland (1977) found an extension to oriented matroids of the simplex method of linear programming, although the extension is not as complete as one might hope. [More recently other algorithms for "optimizing" in oriented matroids have been introduced. See Bachem and Kern (1992) for references.]

We briefly indicate how the simplex method can be imitated. Consider a basis B of M, such that $e \notin B$ and $f \in B$. We define a tableau $(a_{ij}: i \in B, j \in S)$ determined by B as follows. For each $i \in B$, $(a_{ij}: j \in S)$ is the vector $z \in \mathcal{F}^*$ whose support is the fundamental circuit $C(S \backslash B, i)$ of M^*, and that also satisfies $z_i = 1$. It is easy to show that for each $j \in B$, the vector x defined by $x_j = -1$, $x_k = a_{kj}$ for $k \in B$, and $x_k = 0$ for $k \in (S \backslash B) \backslash \{j\}$, is in $\mathcal{F}$, and its support is the fundamental circuit $C(B, j)$ of M. In particular, the $z \in \mathcal{F}^*$ determined by choosing $i = f$ above, and the $x \in \mathcal{F}$ determined by choosing $j = e$, satisfy the condition $x_j z_j = 0$ for all $j \in S \backslash \{e, f\}$ of Theorem 8.12. If this x satisfies $x_j \geqslant 0$, $j \in S \backslash \{f\}$, then the tableau is *primal feasible*; if this z satisfies $z_j \geqslant 0$, $j \in S \backslash \{e\}$, it is *dual feasible.*

A strengthening of Theorem 8.12 is that (under the same hypotheses) there exists a tableau that is both primal and dual feasible. A primal simplex method is one that begins from a primal feasible tableau and performs pivot operations, replacing B by a basis $B' = (B \cup \{j\}) \backslash \{k\}$ where $j \in S \backslash (B \cup \{e\})$ violates dual feasibility. There is a simple rule in ordinary linear programming for choosing k, but it is based on numerical comparisons, and so is unavailable in the oriented matroid setting. However, it can be proved that there does exist a choice for k, if the hypotheses of Theorem 8.12 are satisfied. A more serious difficulty is that cycling (repeating a basis) can occur in a different way from that in ordinary linear programming, making rules for finite termination considerably more complicated.

Acknowledgement

We are grateful to Kazuo Murota and András Sebő for their comments on earlier versions of this paper.

References

Bachem, A., and W. Kern
[1992] *Linear Programming Duality: An Introduction to Oriented Matroids* (Springer, Berlin).

Bixby, R.E.
[1974] l-matrices and a characterization of binary matroids, *Discrete Math.* **8**, 139–145.

Bixby, R.E., and W.H. Cunningham

[1979] Matroids, graphs and 3-connectivity, in: *Graph Theory and Related Topics,* eds. J.A. Bondy and U.S.R. Murty (Academic Press, New York) pp. 91–103.

[1980] Converting linear programs to network problems, *Math. Oper. Res.* **5**, 321–357.

Bixby, R.E., and D.K. Wagner

[1988] An almost linear-time algorithm for graph realization, *Math. Oper. Res.* **13**, 99–123.

Bixby, R.E., W.H. Cunningham and D.M. Topkis

[1985] The poset of a polymatroid extreme point, *Math. Oper. Res.* **10**, 367–378.

Björner, A., M. Las Vergnas, B. Sturmfels, N. White and G.M. Ziegler

[1993] *Oriented Matroids* (Cambridge University Press, Cambridge).

Bland, R.G.

[1977] A combinatorial abstraction of linear programming, *J. Combin. Theory B* **23**, 33–57.

Bouchet, A.

[1987] Greedy algorithm and symmetric matroids, *Math. Programming* **38**, 147–159.

Brezovec, C., G. Cornuéjols and F. Glover

[1988] A matroid algorithm and its application to the solution of two optimization problems on graphs, *Math. Programming* **42**, 471–488.

Bruno, J., and L. Weinberg

[1971] The principal partition of a matroid, *Linear Algebra Appl.* **4**, 17–54.

Chandrasekaran, R., and S.N. Kabadi

[1988] Pseudomatroids, *Discrete Math.* **71**, 205–217.

Cunningham, W.H.

[1973] *A combinatorial decomposition theory,* Thesis (University of Waterloo).

[1982] Decomposition of directed graphs, *SIAM J. Algebra Discrete Methods* **3**, 214–228.

[1984] Testing membership in matroid polyhedra, *J. Combin. Theory B* **36**, 161–188.

[1985] On submodular function minimization, *Combinatorica* **5**, 185–192.

[1986] Improved bounds for matroid partition and intersection algorithms, *SIAM J. Comput.* **15**, 948–957.

Cunningham, W.H., and A. Frank

[1985] A primal dual algorithm for submodular flows, *Math. Oper. Res.* **10**, 251–262.

Dress, A., and T. Havel

[1986] Some combinatorial properties of discriminants in metric vector spaces, *Adv. in Math.* **62**, 285–312.

Dunstan, F.D.J., and D.J.A. Welsh

[1973] A greedy algorithm for solving a certain class of linear programmes, *Math. Programming* **5**, 338–353.

Edmonds, J.

[1965a] Minimum partition of a matroid into independent subsets, *J. Res. Nat. Bur. Standards B* **69**, 67–72.

[1965b] Lehman's switching game and a theorem of Tutte and Nash-Williams, *J. Res. Nat. Bur. Standards B* **69**, 73–77.

[1970] Submodular functions, matroids, and certain polyhedra, in: *Combinatorial Structures and their Applications,* eds. R.K. Guy, H. Hanani, N. Sauer and J. Schönheim (Gordon and Breach, New York) pp. 69–87.

[1979] Matroid intersection, *Ann. Discrete Math.* **4**, 39–49.

Edmonds, J., and R. Giles

[1977] A min–max relation for submodular functions on directed graphs, *Ann. Discrete Math.* **1**, 185–204.

Edmonds, J.R., and D.R. Fulkerson

[1965] Transversals and matroid partition, *J. Res. Nat. Bur. Standards B* **69**, 147–153.

Frank, A.

[1981] A weighted matroid intersection algorithm, *J. Algorithms* **2**, 328–336.

[1982] An algorithm for submodular functions on graphs, *Ann. Discrete Math.* **16**, 97–120.

[1984a] Finding feasible vectors of Edmonds–Giles polyhedra, *J. Combin. Theory B* **36**, 221–239.

[1984b] Generalized polymatroids, in: *Finite and Infinite Sets, Eger 1981,* ed. A. Hajnal (North-Holland, Amsterdam) pp. 295–294.

Frank, A., and E. Tardos
[1984] Matroids from crossing families, in: *Finite and Infinite Sets, Eger 1981,* ed. A. Hajnal (North-Holland, Amsterdam), pp. 295–304.
[1988] Generalized polymatroids and submodular flows, *Math. Programming* **42**, 489–563.

Fujishige, S.
[1978] Algorithms for the independent flow problems, *J. Oper. Res. Soc. Japan* **21**, 189–203.
[1980] An efficient PQ-graph algorithm for solving the graph realization problem, *J. Comput. System Sci.* **21**, 63–86.
[1984] Structures of polytopes determined by submodular functions on crossing families, *Math. Programming* **29**, 125–141.

Fujishige, S., H. Röck and U. Zimmermann
[1989] A strongly polynomial algorithm for minimum cost submodular flow problems, *Math. Oper. Res.* **14**, 60–69.

Gabow, H., and M. Stallmann
[1985] Efficient algorithms for graphic matroid intersection and parity, in: *Automata, Languages and Programming: 12th Colloquium, Lecture Notes in Computer Sciences,* Vol. 194, ed. W. Brauer (Springer, Berlin) pp. 210–220.
[1986] An augmenting path algorithm for the linear matroid parity problem, *Combinatorica* **6**, 123–150.

Gabow, H., and R.E. Tarjan
[1984] Efficient algorithms for a family of matroid intersection problems, *J. Algorithms* **5**, 80–131.

Gabow, H., and Y. Xu
[1989] *Efficient Theoretic and Practical Algorithms for Linear Matroid Intersection Problems,* Technical Report CU-CS-424–89 (University of Colorado).

Grötschel, M., L. Lovász and A. Schrijver
[1981] The ellipsoid method and its consequences in combinatorial optimization, *Combinatorica* **1**, 169–197.

Hamidoune, Y.O., and M. Las Vergnas
[1986] Directed switching games on graphs and matroids, *J. Combin. Theory B* **40**, 237–269.

Hassin, R.
[1982] Minimum cost flow with set constraints, *Networks* **12**, 1–21.

Hoffman, A.J., and J.B. Kruskal
[1956] Integral boundary points of convex polyhedra, in: *Linear Inequalities and Related Systems,* eds. H.W. Kuhn and A.W. Tucker (Princeton University Press, Princeton, NJ) pp. 223–246.

Iri, M.
[1968] On the synthesis of loop and cutset matrices and the related problems, *RAAG Memoirs 4* **A-XIII**, 4–38.

Jenkyns, T.
[1976] The efficacy of the "greedy" algorithm, in: *Proc. 7th Southeastern Conf. on Combinatorics, Graph Theory and Computing,* eds. F. Hoffman, L. Lesniak, R. Mullin, K.B. Reid and R. Stanton (Utilitas Math., Winnipeg) pp. 341–350.

Kabadi, S.N., and R. Chandrasekaran
[1990] On totally dual integral systems, *Discrete Appl. Math.* **26**, 87–104.

Korte, B., and D. Hausmann
[1978] An analysis of the greedy algorithm for independence systems, *Ann. Discrete Math.* **2**, 65–74.

Korte, B., and P. Jensen
[1982] Complexity of matroid property algorithms, *SIAM J. Comput.* **11**, 184–190.

Korte, B., L. Lovász and R. Schrader
[1991] *Greedoid Theory* (Springer, Berlin).

Lawler, E.L.
[1975] Matroid intersection algorithms, *Math. Programming* **9**, 31–56.

Lawler, E.L., and C. Martel
[1982] Finding maximal polymatroidal network flows, *Math. Oper. Res.* **7**, 334–347.

Lehman, A.
[1964] A solution of the Shannon switching game, *SIAM J. Appl. Math.* **12**, 687–725.
Löfgren, L.
[1959] Irredundant and redundant Boolean branch-networks, *IRE Trans. Circuit Theory* **CT-6**(Spec. Suppl.), 158–175.
Lovász, L.
[1980a] Selecting independent lines from a family of lines in a space, *Acta Sci. Univ. Szeged* **42**, 121–131.
[1980b] Matroid matching and some applications, *J. Combin. Theory B* **28**, 208–236.
[1983] Submodular functions and convexity, in: *Mathematical Programming: The State of the Art*, eds. A. Bachem, M. Grötschel and B. Korte (Springer, Berlin) pp. 235–257.
Mirsky, L.
[1971] *Transversal Theory* (Academic Press, New York).
Nakamura, M.
[1988] A characterization of those polytopes in which the greedy algorithm works (abstract), *13th Int. Symp. on Mathematical Programming, Tokyo.*
Nash-Williams, C.St.J.A.
[1967] An application of matroids to graph theory, in: *Theory of Graphs, Proc. Int. Symp., Rome, 1966* ed. P. Rosenstiehl (Gordon and Breach, New York), pp. 263–265.
Qi, L.
[1988] Directed submodularity, ditroids, and directed submodular flows, *Math. Programming* **42**, 579–599.
Queyranne, M.
[1994] *A combinatorial algorithm for minimizing symmetric submodular functions*, Report 393/1994, Mathematics (Technical University of Berlin).
Rado, R.
[1942] A theorem on independence relations, *Quart. J. Math. Oxford* **13**, 83–89.
Rajan, A.
[1986] *Algorithmic applications of connectivity and related topics in matroid theory*, Ph.D. Dissertation (Northwestern University, Evanston, IL).
Schönsleben, P.
[1980] *Ganzzahlige Polymatroid-Intersektions-Algorithmen*, Thesis (ETH Zürich).
Schrijver, A.
[1984] Total dual integrality from directed graphs, crossing families and sub- and supermodular functions, in: *Progress in Combinatorial Optimization*, ed. W.R. Pulleyblank (Academic Press, New York) pp. 315–362.
Seymour, P.D.
[1977a] A note on the production of matroid minors, *J. Combin. Theory B* **22**, 289–295.
[1977b] The matroids with the max-flow min-cut property, *J. Combin. Theory B* **23**, 189–222.
[1980] Decomposition of regular matroids, *J. Combin. Theory B* **28**, 305–359.
[1981a] Matroids and multicommodity flows, *European J. Combin.* **2**, 257–290.
[1981b] Recognizing graphic matroids, *Combinatorica* **1**, 75–78.
Sprinrad, J.
[1989] Prime testing for the split decomposition of a graph, *SIAM J. Discrete Math.* **2**, 590–599.
Tardos, E.
[1990] An intersection theorem for supermatroids, *J. Combin. Theory B* **50**, 150–159.
Truemper, K.
[1982] On the efficiency of representability tests for matroids, *European J. Combin.* **3**, 275–291.
[1984] Partial matroid representations, *European J. Combin.* **5**, 377–394.
[1985] A decomposition theory for matroids, I: General results, *J. Combin. Theory B* **39**, 43–76.
[1986] A decomposition theory for matroids, III: Decomposition conditions, *J. Combin. Theory B* **41**, 275–305.
[1987] Max-flow min-cut matroids: Polynomial testing and polynomial algorithms for maximum flow and shortest routes, *Math. Oper. Res.* **12**, 72–96.

[1990] A decomposition theory for matroids: V. Testing matrix total unimodularity, *J. Combin. Theory B* **49**, 241–281.

[1992] *Matroid Decomposition* (Academic Press, Boston).

Tseng, F.T., and K. Truemper

[1986] A decomposition of matroids with the max-flow min-cut property, *Discrete Appl. Math.* **15**, 329–364.

Tutte, W.T.

[1960] An algorithm for determining whether a given binary matroid is graphic, *Proc. Amer. Math. Soc.* **11**, 905–917.

[1965a] Lectures on matroids, *J. Res. Nat. Bur. Standards B* **69**, 1–47.

[1965b] Menger's theorem for matroids, *J. Res. Nat. Bur. Standards B* **69**, 49–53.

[1966] Connectivity in matroids, *Canad. J. Math.* **18**, 1301–1324.

Whitney, H.

[1933] 2-isomorphic graphs, *Amer. J. Math.* **55**, 245–254.

CHAPTER 12

Permutation Groups

Peter J. CAMERON

School of Mathematical Sciences, Queen Mary and Westfield College, Mile End Road, London E1 4NS, UK

Contents

HANDBOOK OF COMBINATORICS
Edited by R. Graham, M. Grötschel and L. Lovász

1. Introduction

The theory of permutation groups can be traced to the work of Galois on the solubility of equations, if not earlier. It became an independent discipline with the work of the Paris school, notably Jordan and Mathieu, after the mid-19th century. Frobenius, Burnside and Schur linked it to representation theory, while Zassenhaus and Witt obtained the first general classification theorems. In the flourishing of finite simple group theory in the period from 1955 to 1980, permutation group theory both used and contributed to the new techniques; some of the contributions were Fischer's work on 3-transposition groups and the construction of several sporadic groups. Throughout this period, the subject had clear goals, though these appeared to be out of reach. Perhaps the most important problem was: are there any 6-transitive groups apart from the symmetric and alternating groups?

Also, throughout this period, permutation groups went hand-in-hand with parts of finite geometry and graph theory. This was also a two-way process: Dembowski's book on finite geometries shows clearly the impact of the new group-theoretic techniques, while the sporadic groups appeared as automorphism groups of combinatorial objects.

Then, in 1980, the classification of the finite simple groups was completed. Overnight, the balance changed; many of the long-term problems about permutation groups were settled, using this new and powerful machine. Thus, group theory became an even more powerful tool for finite geometers, but the key rôle of combinatorial argument in permutation group theory gave way to increasing use of representation theory to understand better the finite simple groups. A typical theorem about permutation groups in the 1980s asserts that a group with some specified properties is one of a long list (perhaps taking a page or two to state) of known groups.

In the first part of this chapter, I have taken a post-1980 view of the subject. After an introductory section giving the basic theory, I turn to the theorem of O'Nan and Scott, which reduces many problems to specific questions about simple groups. Then there is a brief discussion of the finite simple groups. (Fortunately, it is not necessary to understand how the classification is obtained in order to apply it – the proof is estimated to be 10000 pages long.) Then I turn to some results on primitive groups: classification theorems (including the list of multiply transitive groups), results about degrees and orders, the Sims conjecture and its application to distance-transitive graphs, and so on.

As mentioned earlier, representation theory (and, from a slightly different point of view, character theory) has been associated with permutation groups for a long time. In combinatorial terms, we are led to study "coherent configurations", which generalise association schemes and Schur rings. There are applications to enumeration theory, statistics, and other fields. A specific topic having very strong links with combinatorics is the character theory of the symmetric group.

One part of the theory which has become more important recently is the study of algorithms for permutation groups. At the most basic level, if we are given a set of permutations, we want to obtain information about the group they generate, such as its order, or a membership test for arbitrary permutations. I describe these techniques in detail, and give a summary of results about more recondite properties.

The final section concerns infinite permutation groups. This has seen very rapid developments during the last decade, due in large part to its very close links with model theory.

2. Transitivity and primitivity

This section contains the basic theory of permutation groups: transitivity and primitivity, coset spaces, the orbit-counting lemma, direct and wreath products, and so on. Much of this material can be found in Wielandt (1964).

A permutation group is simply a subgroup of the symmetric group, the group of all permutations of a set. It is convenient, however, to have a more flexible concept available. An *action* of a group G on a set Ω is a homomorphism from G into the symmetric group $\mathrm{Sym}(\Omega)$ on Ω. This allows us to consider actions of the same group on different sets (such as the points and lines of some geometry). Two group elements may induce the same permutation in a given action. Nevertheless, it is convenient to suppress the homomorphism in the notation for an action. Thus, writing permutations on the right, we let αg denote the image of the point $\alpha \in \Omega$ under the permutation (corresponding to) $g \in G$.

Thus, if G is a permutation group, then the identity map on G is an action. (This will be the usual situation.) Also, if G acts on Ω, then the image of the action homomorphism is a permutation group, which is called the group *induced* on Ω by G (and denoted by G^{Ω}).

If G acts on Ω, we often refer to Ω as a *G-space* or *G-set*.

Actions of G on sets Ω_1 and Ω_2 are *isomorphic* if there is a bijection $\theta : \Omega_1 \to \Omega_2$ such that

$$(\alpha g)\theta = (\alpha\theta)g$$

for all $g \in G$, $\alpha \in \Omega_1$.

Let G act on Ω. Define a relation $\sim$ on Ω by the rule that $\alpha \sim \beta$ if and only if $\alpha g = \beta$ for some $g \in G$. Then $\sim$ is an equivalence relation on Ω. (The reflexive, symmetric and transitive laws for $\sim$ follow directly from the group axioms asserting the existence of an identity, inverses, and closure respectively.) The equivalence classes are the *orbits* of G on Ω. We say that G is *transitive* on Ω, or that Ω is a *transitive* G-space, if there is just one orbit. In general, a *transitive constituent* of G is the permutation group induced by G on one of its orbits. Thus, we have the following

Theorem 2.1. *Any G-space can be expressed in a unique way as a disjoint union of transitive G-spaces.*

Things are not quite so simple, however, if we consider permutation groups rather than G-spaces. Any permutation group is a subgroup of the cartesian product of its transitive constituents. More specifically, it is a "subcartesian product", which is to say that it projects onto each factor. (The cartesian product is the set of all those permutations of Ω whose action on each orbit is the same as that of some element of G.) We could say that the problem of describing all permutation groups has been reduced to that of describing transitive groups modulo the problem of constructing subcartesian products.

In the case of finite permutation groups (or, more generally, groups with finitely many orbits), the cartesian product coincides with the direct product, and the latter term is more commonly used.

A special class of transitive actions of a group is given by the *coset spaces*. Let H be a subgroup of G, and $H\backslash G$ the set of right cosets Hx for $x \in G$. Then G acts on $H\backslash G$ by right multiplication: the permutation of $H\backslash G$ induced by the group element $g \in G$ is

$$Hx \mapsto Hxg.$$

This action is clearly transitive. And it is not so special after all.

Theorem 2.2.

(i) *Any transitive G-space Ω is isomorphic to the coset space $H\backslash G$, where H is the* stabiliser *of a point $\alpha \in \Omega$ (the subgroup*

$$\{g \in G : \alpha g = \alpha\}).$$

(ii) *The actions of G on the coset spaces $H\backslash G$ and $K\backslash G$ are isomorphic if and only if H and K are conjugate subgroups of G.*

The stabiliser of α is denoted by G_α or $\mathrm{Stab}_G(\alpha)$. This notation can be extended to stabilisers of sequences of points, or of subgroups of Ω. (The notation is not quite standard; but, if Δ is a subset of Ω, it is fairly common to use G_Δ or $G_{\{\Delta\}}$ to denote the setwise stabiliser of Δ, and $G_{(\Delta)}$ for the pointwise stabiliser.)

Corollary 2.3 (Lagrange's theorem)**.** *If G acts transitively on Ω, then*

$$|\Omega| \cdot |G_\alpha| = |G|.$$

Corollary 2.4 (Cayley's theorem)**.** *Every group is isomorphic to a transitive permutation group.*

To see this, we take the action of G on itself by right multiplication (that is, on the coset space $1\backslash G$), which is obviously faithful. An action (or a permutation group) is *regular* if it is transitive and the stabiliser of a point is the identity. By Theorem 2.2, any regular action is isomorphic to the action described in Cayley's theorem.

For finite permutation groups, the *orbit-counting lemma* (often called Burnside's lemma: it is given without attribution in Burnside's 1911 book but has been traced back to Cauchy by Neumann 1979) gives a formula for the number of orbits.

Theorem 2.5 (Orbit-counting lemma). *Let the finite group G act on the finite set Ω. For $g \in G$, let* $\mathrm{fix}(g)$ *be the number of points $\alpha \in \Omega$ for which $\alpha g = \alpha$. Then the number of orbits of G in Ω is the average, over G, of* $\mathrm{fix}(g)$.

Proof. Suppose first that G is transitive. Count in two ways the number of pairs (α, g) with $\alpha \in \Omega$, $g \in G$, and $\alpha g = \alpha$, to obtain

$$\sum_{g \in G} \mathrm{fix}(g) = |\Omega| \cdot |G_\alpha|.$$

By Corollary 2.3, we obtain

$$\frac{1}{|G|} \sum_{g \in G} \mathrm{fix}(g) = 1,$$

as required.

In general, let G have t orbits $\Omega_1, \ldots, \Omega_t$, and let $\mathrm{fix}_i(g)$ be the number of fixed points of g in Ω_i for $1 \leqslant i \leqslant t$. Then

$$\mathrm{fix}(g) = \sum_{i=1}^{t} \mathrm{fix}_i(g);$$

summing over i and using the result for transitive groups we obtain

$$\frac{1}{|G|} \sum_{g \in G} \mathrm{fix}(g) = t$$

as required. □

This result stands at the root of two major areas, the Redfield–Pólya theory of enumeration under group action, and the character theory of permutation groups. The second of these is discussed in section 5; for the first, see chapter 21.

Let G act transitively on Ω. A *congruence* on Ω is an equivalence relation which is invariant under G. Its equivalence classes are called *blocks of imprimitivity*, and the set of blocks is a *system of imprimitivity*. A non-empty subset Δ of Ω is a block of imprimitivity if and only if

$$\Delta \cap \Delta g = \emptyset \text{ or } \Delta$$

for all $g \in G$. We say that G is *primitive* if the only congruences are equality and the universal relation which holds between all pairs of points.

Proposition 2.6. *The transitive group G is primitive if and only if the stabiliser of a point is a maximal subgroup of G.*

Proof. If Δ is a non-trivial block of imprimitivity (that is, neither a singleton nor all of Ω), then $G_\alpha < G_{\{\Delta\}} < G$. Conversely, if $G_\alpha < H < G$, then the H-orbit containing α is a block. □

Just as an intransitive action can be "reduced" to transitive ones, an imprimitive action can also be reduced to smaller actions, in a manner which I now describe.

Let H and K be permutation groups on sets Γ and Δ respectively. Let $\Omega = \Gamma \times \Delta$, regarded as a covering of Δ with fibres $B_\delta = \{(\gamma, \delta) : \gamma \in \Gamma\}$ bijective with Γ. Now let N be the group of permutations of Ω which fix each fibre set-wise and act on it as an element of H acting on Γ; and let K° be the group of permutations of Ω induced by K acting on the index set of the set of fibres. Then the group generated by N and K° (which is their semi-direct product) is called the *wreath product* of the permutation groups H and K (written $H \operatorname{Wr} K$). If neither Γ nor Δ is a singleton, then the wreath product is imprimitive; the fibres are blocks of imprimitivity.

Theorem 2.7. *Let G be a transitive permutation group on Ω with system of imprimitivity Δ containing a block Γ. Let H be the permutation group induced on Γ by its set-wise stabiliser, and K the group induced on Δ by G. Then G is embeddable in $H \operatorname{Wr} K$ in such a way that its actions on Ω and on $\Gamma \times \Delta$ are isomorphic.*

For finite permutation groups, this reduction reaches primitive groups after a finite number of steps. Thus, information about primitive groups tells us a great deal about transitive groups, though there is still a non-trivial problem involved in re-assembling the imprimitive group from its components. For infinite groups, this is no longer true, but there is a more general concept of wreath product of an arbitrary partially ordered set of groups which does the same job, defined by Hall (1962) and others.

3. The O'Nan–Scott theorem

While every abstract group is isomorphic to a transitive permutation group (Cayley's theorem), the structure of primitive groups is more restricted. The normal subgroup structure was considered by O'Nan and by Scott (1980), who proposed this as a means for using the classification of finite simple groups (which was imminent when they wrote) to study maximal subgroups of symmetric groups. We will see that it does this and much more. In fact, their result is only a little stronger than was obtained over a century earlier by Jordan (1861). I will give the result in a fairly combinatorial form. To state it, a little terminology is required.

The *socle* of a finite group is the product of its minimal normal subgroups. (Each minimal normal subgroup is itself the direct product of isomorphic simple groups.)

A *power structure* on a set Ω (also called a *hypercube* or *Hamming scheme*) is a bijection between Ω and the set Γ^Δ of all functions from Δ to Γ, for some sets Γ and Δ, neither of which is a singleton. If H and K are permutation groups on Γ and Δ respectively, then the wreath product $H \operatorname{Wr} K$ defined previously has a natural action on Γ^Δ. (If Γ and Δ are finite, say $|\Gamma| = m$, $|\Delta| = k$, then Γ^Δ can be identified with the set of all k-tuples of symbols from the "alphabet" Γ. Now the factors of N act on the symbols in each coordinate, while K° permutes the coordinates. This is called the *product action* of the wreath product.) It can be shown that a group G preserving a power structure can be embedded in $H \operatorname{Wr} K$ so that its actions on Ω and on Γ^Δ are isomorphic, where H and K are obtained from the action of G. Furthermore, $H \operatorname{Wr} K$ acts primitively on Γ^Δ if and only if H is primitive but not regular on Γ and K is transitive on Δ.

We call a primitive group *basic* if there is no power structure on Ω preserved by G. So we have a "reduction" for non-basic groups similar to that for imprimitive groups.

Next, we define some special classes of permutation groups.

Let V be a finite vector space, H a group of linear transformations of V. Let G be the group

$$\{v \mapsto vA + c \colon A \in H, c \in V\}$$

of permutations of V. It is a *semi-direct product* of the translation group of V by H. Then G is transitive on V; it is primitive if and only if H is an irreducible linear group; and it is basic if and only if H is a primitive linear group (that is, preserves no non-trivial direct sum decomposition of V). We call G a group of *affine type*. If H is the general linear group $\mathrm{GL}(n,q)$ (all non-singular linear transformations), then G is the affine general linear group $\mathrm{AGL}(n,q)$. (Here V is n-dimensional over $\mathrm{GF}(q)$.)

Let T be a non-abelian finite simple group, G the direct product of k copies of T, and H the diagonal of G (the subgroup $\{(t,t,\dots,t) \colon t \in T\}$). Let G act on the coset space $H\backslash G$. The normaliser of G in the symmetric group is generated by G together with

(i) the automorphism group $\mathrm{Aut}(T)$ of T, acting in the same way on each factor of G (note that inner automorphisms are induced by elements of H); and

(ii) the group of permutations of the factors of G.

Any group lying between H and its normaliser is said to be of *diagonal type*.

A group G is called *almost simple* if it has a minimal normal subgroup T which is a non-abelian simple group and has trivial centralizer in G (equivalently, if

$$T \leqslant G \leqslant \mathrm{Aut}(T)$$

for some non-abelian simple group T).

The O'Nan–Scott theorem can be stated in two parts as follows.

Theorem 3.1. *A basic primitive permutation group is either of affine or diagonal type, or almost simple.*

Theorem 3.2. *Let G be a primitive group preserving a power structure Γ^{Δ}, and embedded in $H \operatorname{Wr} K$ in the natural way. Let S be the socle of G. Then $S \cong N^k$, where $k = |\Delta|$ and N is either the socle or a regular minimal normal subgroup of H.*

Several proofs have appeared, though some of the early versions have a small error – the second case in Theorem 3.2, where S is a non-abelian regular normal subgroup of G, is omitted. See Scott (1980), Cameron (1981), Aschbacher and Scott (1985), Kovács (1986), Liebeck et al. (1988).

A corollary of the theorem, which is easily proved directly (and was known much earlier), is the following.

Corollary 3.3. *A primitive permutation group has at most two minimal normal subgroups; if there are two, then they are regular and non-abelian.*

4. Finite simple groups

In 1980, the classification of the finite simple groups was completed (apart from a couple of details about some of the groups which were resolved subsequently). This result has probably the longest proof of any single theorem, and hundreds of mathematicians contributed to its achievement. The techniques needed to apply the theorem, however, have little to do with those in its proof, and it is the former that concern us here.

The result is as follows.

Theorem 4.1 *A finite simple group is one of the following:*

(i) *a cyclic group of prime order;*
(ii) *an alternating group A_n, $n \geqslant 5$;*
(iii) *a group of Lie type over a finite field;*
(iv) *one of* 26 *"sporadic" groups.*

For an overview of the proof, see Gorenstein (1982).

I will now attempt to give a brief description of the groups in the conclusion of the theorem. The *cyclic groups of prime order* require no comment.

The *alternating group A_n* consists of all even permutations of $\{1,\ldots,n\}$. Its simplicity for all $n \geqslant 5$ was known to Galois, as an ingredient in his proof of the unsolvability by radicals of the general polynomial equation.

Groups of Lie type form the most important class, and are also the most difficult to define. They can be approached in different ways. A large subclass consists of the *Chevalley groups*, defined in a uniform way by Chevalley (1955), roughly as follows. For each of the simple Lie algebras over $\mathbb{C}$ (of types A_n, $n \geqslant 1$; B_n, $n \geqslant 2$; C_n, $n \geqslant 3$; D_n, $n \geqslant 4$; E_6, E_7, E_8, F_4 and G_2), it is possible to choose a *Chevalley basis*, relative to which the multiplication constants are integers. Then the matrices of the adjoint mappings corresponding to these basis vectors have entries which are polynomials with integer coefficients. These matrices can then be used to define a group over any field K, the Chevalley group over K. *Twisted groups*, modifications of the Chevalley groups, were defined by Steinberg (1959) and Ree (1961). They exist whenever the Dynkin diagram of the Lie algebra has an automorphism, and the field has an automorphism of the same order; for diagrams with multiple bonds, the characteristic of the field is restricted as well. The types are ${}^2A_n, {}^2D_n, {}^3D_4, {}^2E_6$ (Steinberg), ${}^2F_4, {}^2G_2$ (Ree), ${}^2B_2$ (Suzuki); the superscript denotes the order of the graph automorphism. For further discussion, a good general reference is Carter (1972). See also chapter 13 on buildings.

However, for everyday purposes, a different approach is often more useful, involving a different subdivision. The first class consists of the *classical groups*. These are the Chevalley and most twisted groups associated with the Lie algebras A_n, B_n, C_n and D_n. They have been identified with the projective special linear, symplectic, orthogonal and unitary groups. (Take a finite-dimensional vector space V, possibly carrying a non-degenerate sesquilinear or quadratic form f; let G be the group of all invertible linear transformations of V, or all those which are isometries of f; in all but a few small cases, G has a unique non-abelian composition factor.) See Dickson (1900), Dieudonné (1963) or Artin (1957) for these groups, and Carter (1972) for their identification with groups of Lie type.

One particularly important class is that consisting of the *projective special linear groups* $\mathrm{PSL}(d,q) = \mathrm{SL}(d,q)/Z$, where $\mathrm{SL}(d,q)$ is the group of $d \times d$ matrices of determinant 1 over $\mathrm{GF}(q)$, and Z the subgroup of $\mathrm{SL}(d,q)$ consisting of scalar matrices. In Lie notation, $\mathrm{PSL}(n+1,q) = A_n(q)$. It is known that $\mathrm{PSL}(d,q)$ is simple except in the cases $(d,q) = (2,2)$ or $(2,3)$. (In fact some of these groups occur in other guises: $\mathrm{PSL}(2,2) \cong S_3$, $\mathrm{PSL}(2,3) \cong A_4$, $\mathrm{PSL}(2,4) \cong \mathrm{PSL}(2,5) \cong A_5$, $\mathrm{PSL}(3,2) \cong \mathrm{PSL}(2,7)$, and $\mathrm{PSL}(4,2) \cong A_8$.) The group $\mathrm{PSL}(d,q)$ is a permutation group on the set Ω of 1-dimensional subspaces of the d-dimensional vector space over $\mathrm{GF}(q)$ (that is, the points of the projective space – see chapter 13), since Z is the kernel of the action of $\mathrm{SL}(d,q)$ on this set. We have

$$|\mathrm{PSL}(d,q)| = \frac{(q^d-1)(q^d-q)\cdots(q^d-q^{d-1})}{(q-1)(q-1,d)},$$

which is roughly a polynomial in q of degree d^2-1.

(As a matter of notation, GL refers to the *general linear* group of all invertible matrices; replacing G by S denotes determinant 1; replacing G or S by Γ or Σ denotes adjoining field automorphisms; prefacing by P denotes taking the quotient by the group of scalar matrices; and prefacing with A denotes taking the semi-direct product with the translation group of the vector space.)

classical groups are $C_n(q)$ (the symplectic group $\mathrm{PSp}(2n,q)$), ${}^2A_n(q^2)$ (the unitary group $\mathrm{PSU}(n+1,q)$), and $D_n(q), B_n(q)$ and ${}^2D_n(q^2)$ (the orthogonal groups $\mathrm{P\Omega}^+(2n,q), \mathrm{P\Omega}(2n+1,q)$ and $\mathrm{P\Omega}^-(2n,q)$).

The remaining groups of Lie type are referred to as the *exceptional groups*. The most significant fact about them is that an exceptional group over a field K has a matrix representation over K of bounded degree; and so, if $K = \mathrm{GF}(q)$, its order is bounded by a fixed polynomial in q.

Table 4.1.
The finite groups of Lie type

Group	Condition	N	d
Chevalley groups:			
$A_n(q)$	$n \geqslant 1$	$q^{n(n+1)/2}\prod_{i=1}^{n}(q^{i+1}-1)$	$(n+1, q-1)$
$B_n(q)$	$n \geqslant 2$	$q^{n^2}\prod_{i=1}^{n}(q^{2i}-1)$	$(2, q-1)$
$C_n(q)$	$n \geqslant 3$ q odd	$q^{n^2}\prod_{i=1}^{n}(q^{2i}-1)$	$(2, q-1)$
$D_n(q)$	$n \geqslant 4$	$q^{n(n-1)}(q^n-1)\prod_{i=1}^{n-1}(q^{2i}-1)$	$(4, q^n-1)$
$E_6(q)$		$q^{36}(q^{12}-1)(q^9-1)(q^8-1)(q^6-1)(q^5-1)(q^2-1)$	$(3, q-1)$
$E_7(q)$		$q^{63}(q^{18}-1)(q^{14}-1)(q^{12}-1)(q^{10}-1)(q^8-1)(q^6-1)(q^2-1)$	$(2, q-1)$
$E_8(q)$		$q^{120}(q^{30}-1)(q^{24}-1)(q^{20}-1)(q^{18}-1)(q^{14}-1)$	1
		$\times(q^{12}-1)(q^8-1)(q^2-1)$	1
$F_4(q)$		$q^{24}(q^{12}-1)(q^8-1)(q^6-1)(q^2-1)$	1
$G_2(q)$		$q^6(q^6-1)(q^2-1)$	1
Steinberg groups:			
${}^2A_n(q^2)$	$n \geqslant 2$	$q^{n(n+1)/2}\prod_{i=1}^{n}(q^{i+1}-(-1)^{i+1})$	$(n+1, q+1)$
${}^2D_n(q^2)$	$n \geqslant 4$	$q^{n(n-1)}(q^n+1)\prod_{i=1}^{n-1}(q^{2i}-1)$	$(4, q^n+1)$
${}^2E_6(q^2)$		$q^{36}(q^{12}-1)(q^9+1)(q^8-1)(q^6-1)(q^5+1)(q^2-1)$	$(3, q+1)$
${}^3D_4(q^3)$		$q^{12}(q^8+q^4+1)(q^6-1)(q^2-1)$	1
Suzuki and Ree groups:			
${}^2B_2(q)$	$q=2^{2m+1}$	$q^2(q^2+1)(q-1)$	1
${}^2G_2(q)$	$q=3^{2m+1}$	$q^3(q^3+1)(q-1)$	1
${}^2F_4(q)$	$q=2^{2m+1}$	$q^{12}(q^6+1)(q^4-1)(q^3+1)(q-1)$	1

Here n is a positive integer, m a non-negative integer, q a prime power. The group occurs as a quotient of a group of order N by a central subgroup of order d (and so has order N/d). All groups in the table are simple except for $A_1(2) \cong S_3$, $A_1(3) \cong A_4$, $B_2(2) \cong S_6$, $G_2(2)$ (which has a subgroup of index 2 isomorphic to ${}^2A_2(3^2)$), ${}^2A_2(2^2)$ (a Frobenius group of order 72), ${}^2B_2(2)$ (a Frobenius group of order 20), ${}^2G_2(3)$ (which has a subgroup of index 3 isomorphic to $A_1(2^3)$), and ${}^2F_4(2)$ (which has a simple subgroup of index 2, the *Tits group*, not occurring elsewhere in the table). Other coincidences between groups of Lie type and alternating groups are $A_1(2^2) \cong A_1(5) \cong A_5$, $A_1(7) \cong A_2(2)$, $A_1(3^2) \cong A_6$, $A_3(2) \cong A_8$, $B_2(3) \cong {}^2A_3(2^2)$.

Note that $C_n(q)$ is defined for $n=2$ and for q even but is isomorphic to $B_n(q)$ in these cases; similarly $D_3(q)$ and ${}^2D_3(q^2)$ are isomorphic to $A_3(q)$ and ${}^2A_3(q^2)$ respectively.

Some of these groups are isomorphic to classical groups, as follows: $A_n(q) \cong \mathrm{PSL}(n+1, q)$, $B_n(q) \cong \mathrm{P}\Omega(2n+1, q)$, $C_n(q) \cong \mathrm{PSp}(2n, q)$, $D_n(q) \cong \mathrm{P}\Omega^+(2n, q)$, ${}^2A_n(q^2) \cong \mathrm{PSU}(n+1, q)$, ${}^2D_n(q^2) \cong \mathrm{P}\Omega^-(2n, q)$. Other common names are $Sz(q)$ for the Suzuki group ${}^2B_2(q)$, and $R_1(q)$ and $R_2(q)$ for the Ree groups ${}^2G_2(q)$ and ${}^2F_4(q)$ respectively.

The groups of Lie type are listed in table 4.1.

No uniform description of the sporadic groups is known; it is necessary to give 26 separate constructions. Each sporadic group is the unique non-abelian composition factor in the automorphism group of some rather special combinatorial or algebraic object (design, graph, code, lattice or non-associative algebra). We refer to the "ATLAS of Finite Groups" (Conway et al. 1985) for more details and references.

The sporadic groups are listed in table 4.2.

Table 4.2.
The sporadic simple groups

Group	Name	Order
M_{11}	Mathieu	$2^4 \cdot 3^2 \cdot 5 \cdot 11$
M_{12}	Mathieu	$2^6 \cdot 3^3 \cdot 5 \cdot 11$
M_{22}	Mathieu	$2^7 \cdot 3^2 \cdot 5 \cdot 7 \cdot 11$
M_{23}	Mathieu	$2^7 \cdot 3^2 \cdot 5 \cdot 7 \cdot 11 \cdot 23$
M_{24}	Mathieu	$2^{10} \cdot 3^3 \cdot 5 \cdot 7 \cdot 11 \cdot 23$
J_1	Janko	$2^3 \cdot 3 \cdot 5 \cdot 7 \cdot 11 \cdot 19$
J_2	Hall–Janko	$2^7 \cdot 3^3 \cdot 5^2 \cdot 7$
J_3	Janko	$2^7 \cdot 3^5 \cdot 5 \cdot 17 \cdot 19$
J_4	Janko	$2^{21} \cdot 3^3 \cdot 5 \cdot 7 \cdot 11^3 \cdot 23 \cdot 29 \cdot 31 \cdot 37 \cdot 43$
Co_1	Conway	$2^{21} \cdot 3^9 \cdot 5^4 \cdot 7^2 \cdot 11 \cdot 13 \cdot 23$
Co_2	Conway	$2^{18} \cdot 3^6 \cdot 5^3 \cdot 7 \cdot 11 \cdot 23$
Co_3	Conway	$2^{10} \cdot 3^7 \cdot 5^3 \cdot 7 \cdot 11 \cdot 23$
Fi_{22}	Fischer	$2^{17} \cdot 3^9 \cdot 5^2 \cdot 7 \cdot 11 \cdot 13$
Fi_{23}	Fischer	$2^{18} \cdot 3^{13} \cdot 5^2 \cdot 7 \cdot 11 \cdot 13 \cdot 17 \cdot 23$
Fi_{24}'	Fischer	$2^{21} \cdot 3^{16} \cdot 5^2 \cdot 7^3 \cdot 11 \cdot 13 \cdot 17 \cdot 23 \cdot 29$
F_1	Fischer–Griess	$2^{46} \cdot 3^{20} \cdot 5^9 \cdot 7^6 \cdot 11^2 \cdot 13^3 \cdot 17 \cdot 19 \cdot 23 \cdot 29 \cdot 31 \cdot 41 \cdot 47 \cdot 59 \cdot 71$
F_2	Fischer	$2^{41} \cdot 3^{13} \cdot 5^6 \cdot 7^2 \cdot 11 \cdot 13 \cdot 17 \cdot 19 \cdot 23 \cdot 31 \cdot 47$
F_3	Thompson	$2^{15} \cdot 3^{10} \cdot 5^3 \cdot 7^2 \cdot 13 \cdot 19 \cdot 31$
F_5	Harada–Norton	$2^{14} \cdot 3^6 \cdot 5^6 \cdot 7 \cdot 11 \cdot 19$
HS	Higman–Sims	$2^9 \cdot 3^2 \cdot 5^3 \cdot 7 \cdot 11$
Mc	McLaughlin	$2^7 \cdot 3^6 \cdot 5^3 \cdot 7 \cdot 11$
Suz	Suzuki	$2^{13} \cdot 3^7 \cdot 5^2 \cdot 7 \cdot 11 \cdot 13$
Ly	Lyons	$2^8 \cdot 3^7 \cdot 5^6 \cdot 7 \cdot 11 \cdot 31 \cdot 37 \cdot 67$
He	Held	$2^{10} \cdot 3^3 \cdot 5^2 \cdot 7^3 \cdot 17$
Ru	Rudvalis	$2^{14} \cdot 3^3 \cdot 5^3 \cdot 7 \cdot 13 \cdot 29$
O'N	O'Nan	$2^9 \cdot 3^4 \cdot 5 \cdot 7^3 \cdot 11 \cdot 19 \cdot 31$

Some of these groups have alternative names: for example, $\mathrm{Co}_n = .n$ $(n = 1,2,3)$; $\mathrm{Fi}_n = M(n)$ $(n = 22,23,24)$; $F_1 = \mathrm{FG} = M$ (the *friendly giant* or *monster*); $F_2 = \mathrm{BM} = B$ (the *baby monster*); $F_3 = \mathrm{Th}$; $F_5 = \mathrm{HN}$; $\mathrm{Mc} = \mathrm{McL}$; $J_2 = \mathrm{HJ}$; etc.

The main information we need about these groups is their maximal subgroups; or, more precisely (in view of the O'Nan–Scott theorem), maximal subgroups of groups G with $T \leqslant G \leqslant \mathrm{Aut}(T)$ with T simple. The O'Nan–Scott theorem was first used for this very purpose in connection with the symmetric and alternating groups (an application of the so-called "bootstrap principle"); see also Aschbacher (1984), Kleidman and Liebeck (1988), and the ATLAS.

5. Applications of the classification

5.1. Multiple transitivity and small rank

Let t be a positive integer not exceeding $|\Omega|$. We say that G acts *t-transitively* on Ω if, given any two t-tuples $(\alpha_1, \ldots, \alpha_t)$ and $(\beta_1, \ldots, \beta_t)$ of distinct elements of Ω, there is an element $g \in G$ carrying α_i to β_i for $i = 1, \ldots, t$. Note that t-transitivity implies $(t-1)$-transitivity; the condition gets stronger as t increases. The symmetric and alternating groups of degree n are n- and $(n-2)$-transitive, respectively.

Theorem 5.1. *All finite 2-transitive groups are known. In particular, the only 6-transitive groups are the symmetric and alternating groups.*

Table 5.1 gives the list of 2-transitive groups.

Table 5.1(a)
Finite 2-transitive groups with abelian socle

Degree	$H = G_0$	Condition	No. of actions
q^d	$\mathrm{SL}(d,q) \leqslant H \leqslant \Gamma\mathrm{L}(d,q)$		up to q if q even, $d=2$ 2 if $(d,q)=(3,2)$
q^{2d}	$\mathrm{Sp}(d,q) \trianglelefteq H$	$d \geqslant 2$	up to q if q even
q^6	$G_2(q) \trianglelefteq H$	q even	up to q
q	$2^{1+2}.3 = \mathrm{SL}(2,3) \trianglelefteq H$	$q = 5^2, 7^2, 11^2, 23^2$	1
q	$2^{1+4} \trianglelefteq H$	$q = 3^4$	1
q	$\mathrm{SL}(2,5) \trianglelefteq H$	$q = 11^2, 19^2, 29^2, 59^2$	1
2^4	A_6		2
2^4	A_7		1
2^6	$\mathrm{PSU}(3,3)$		2
3^6	$\mathrm{SL}(2,13)$		1

The degree n is a prime power p^m, and the socle is elementary abelian of order n and regular. The stabiliser $G_0 = H$ of the origin is a subgroup of $\mathrm{GL}(m,p)$. 2^{1+2k} denotes an extraspecial group of this order.

The maximum degree of transitivity is 2 in all cases except the first with $q=2$ (4-transitive if $d=2$, 3-transitive if $d>2$) or $(d,q)=(1,3)$ or $(1,4)$ (3- or 4-transitive respectively), and the case $n=2^4$, $H=A_7$ (3-transitive). In the first, fourth, fifth and sixth rows of the table, the conditions are not sufficient for 2-transitivity of G. For example, in the case $n=3^4$, $P=2^{1+4} \trianglelefteq H$, we have $H/P \leqslant S_5$, and G is 2-transitive if and only if H/P contains an element of order 5. For precise conditions in all cases, see Huppert (1957), Hering (1985).

The number of actions is the order of the cohomology group $H^1(G_0, N)$.

In outline, the proof runs as follows. It was shown by Burnside (1911, p. 202), that a 2-transitive group G has a unique minimal normal subgroup N, which is either elementary abelian and regular, or primitive and simple. (This can be seen using the O'Nan–Scott theorem as follows. First, G is certainly primitive, since it preserves no non-trivial binary relations. Similarly, it preserves no power structure.

Table 5.1(b)
Finite 2-transitive groups with non-abelian socle

n	Condition	N	$\max\|G/N\|$	$\min(t)$	$\max(t)$	No. of actions
n	$n \geqslant 5$	A_n	2	$n-2$	n	2 if $n=6$ 1 otherwise
$\frac{q^d-1}{q-1}$	$d \geqslant 2$ $(d,q) \neq (2,2)$, $(2,3)$	PSL(d,q)	$(d,q-1)e$	3 if $d=2, q$ even 2 otherwise	3 if $d=2$ 2 otherwise	2 if $d>2$ 1 otherwise
$2^{2d-1}+2^{d-1}$	$d \geqslant 3$	Sp$(2d,2)$	1	2	2	1
$2^{2d-1}-2^{d-1}$	$d \geqslant 3$	Sp$(2d,2)$	1	2	2	1
q^3+1	$q \geqslant 3$	PSU$(3,q)$	$(3,q+1)e$	2	2	1
q^2+1	$q=2^{2d+1}>2$	Sz(q)	$2d+1$	2	2	1
q^3+1	$q=3^{2d+1}>3$	$R_1(q)$	$2d+1$	2	2	1
11		PSL(2, 11)	1	2	2	2
11		M_{11}	1	4	4	1
12		M_{11}	1	3	3	1
12		M_{12}	1	5	5	2
15		A_7	1	2	2	2
22		M_{22}	2	3	3	1
23		M_{23}	1	4	4	1
24		M_{24}	1	5	5	1
28		PSL(2, 8)	3	1	2	1
176		HS	1	2	2	2
276		Co_3	1	2	2	1

Here q is a prime power $q=p^e$; n and d are positive integers, n being the degree. N is the socle of G. The next column gives the index of N in the largest possible G, viz., its normaliser in S_n. Groups with the same socle may have different degrees of transitivity; the least and greatest such degrees are given. Note that the socle is 2-transitive in all cases except one. The final column gives the number of non-isomorphic actions. (For example, for PSL(d,q), $d>2$, these actions are on points and hyperplanes of the projective space.)

It is straightforward to check that no diagonal group can be 2-transitive. So G is either of affine type or almost simple. In the latter case, if the simple socle N of G is imprimitive, then a geometric argument shows that G acts on a linear space (whose lines are the G-translates of an N-block) admitting a parallelism, and N fixes all the parallel classes, and hence that the stabiliser of two points in N is trivial. A theorem of Frobenius now shows that N has a regular normal subgroup, which is impossible.)

Suppose that G is of affine type. Then the linear group $H=G_0$ acts transitively on the non-zero vectors of the vector space V. Huppert (1957) determined all solvable linear groups with this property. Non-solvable groups were considered by Hering (1985), who showed that there can be at most one non-abelian composition factor, and then considered the possible simple groups to establish which could occur in this situation.

Suppose that G is almost simple, that is, $T \leqslant G \leqslant \mathrm{Aut}(T)$ for some simple group

T. The proof is completed by considering the possibilities for T. As an illustration, I will discuss the case where T is an alternating group A_m, $m \geqslant 5$. Since $\mathrm{Aut}(A_m) = S_m$ for $m \neq 6$, we may (by treating one case separately) assume that $G = S_m$ or $G = A_m$. Suppose that $G = S_m$. Then G contains a conjugacy class $\mathscr{C}$ of $1/2m(m-1)$ "transpositions" (in its natural action of degree m; we do not know how they act in the unknown action!). A point α is exchanged with each further point β by equally many members of $\mathscr{C}$; so $n - 1 \leqslant 1/2m(m-1)$. On the other hand, G_α is a subgroup of index n in G. There are classical lower bounds (in terms of m) for the index of a subgroup of S_m. Comparing these bounds, we find that either the unknown permutation representation coincides with the known one, or m is bounded. Then only a finite amount of work remains. (The lower bounds referred to have a long history. It is straightforward to write down the index of a maximal intransitive or imprimitive subgroup of S_m. For primitive subgroups, a result more than good enough for this application has been available since the turn of the century; but it has been dramatically improved by use of the classification, as we will see later. This is another instance of the "bootstrap principle".) If $G = A_m$, we use the class of 3-cycles instead. This argument is due to Maillet (1895).

Suppose that G is a group of Lie type. The pattern of argument is similar; there are a distinguished small conjugacy class (consisting of transvections in the case of most classical groups), and a lower bound for the index of a subgroup. However, the argument can be shortened by using some knowledge of the structure and character theory of the groups of Lie type. This was done by Howlett (1974) and Curtis et al. (1976).

Finally, for the sporadic groups, the problem is a finite one.

For $t > 2$, the t-transitive groups form a subclass of the 2-transitive groups, which is easily determined by inspection.

Theorem 5.2. *Let G be t-transitive of degree n, but not S_n or A_n.*

(i) *If $t = 5$ then G is a Mathieu group M_{12} or M_{24} ($n = 12, 24$).*

(ii) *If $t = 4$, then either G is one of the above, or G is M_{11} or M_{23} ($n = 11, 23$).*

(iii) *If $t = 3$, then either G is one of the above, or G is $\mathrm{AGL}(d,2)$ ($n = 2^d$), or G is $V_{16}.A_7$ ($n = 16$), or G is M_{11} ($n = 12$), or G is M_{22} or $\mathrm{Aut}(M_{22})$ ($n = 22$), or*

$$\mathrm{PSL}(2,q) \leqslant G \leqslant \mathrm{P\Gamma L}(2,q)$$

($n = q + 1$, q a prime power).

The *Mathieu groups* M_n, $n = 11, 12, 22, 23, 24$ were the first sporadic simple groups to be discovered. They are closely connected to many important combinatorial configurations such as designs, geometries, codes and Hadamard matrices (see chapter 14).

The result has obvious applications in the classification of designs, codes, etc. with 2-transitive automorphism groups. Sometimes, the 2-transitivity arises naturally and does not have to be assumed. To take an example: A *Jordan group* is a permutation group containing a subgroup fixing at least one point and acting transitively on the points it does not fix; it is *proper* if it is not $(t+1)$-transitive, where t is the number of points fixed by the subgroup. Jordan (1871) showed that a primitive Jordan group is 2-transitive. The argument has a geometric interpretation. A Jordan group is an automorphism group of a matroid, whose flats are the fixed point sets of subgroups with the property of the definition. Now if the group is primitive, then the matroid is geometric (flats of rank 1 have cardinality 1), whence the 2-transitivity is clear. See chapter 9 for more about matroids. From Theorem 5.1, the complete classification of proper Jordan groups follows.

Theorem 5.3. *A proper primitive Jordan group G satisfies one of the following:*
(i) $\mathrm{PSL}(d,q) \leqslant G \leqslant \mathrm{P\Gamma L}(d,q), d \geqslant 3$;
(ii) $\mathrm{ASL}(d,q) \leqslant G \leqslant \mathrm{A\Gamma L}(d,q), d \geqslant 2$;
(iii) $G = M_{22}, \mathrm{Aut}(M_{22}), M_{23}$, *or* M_{24};
(iv) $G = A_7$ *(degree 15) or* $V_{16}.A_7$ *(degree 16).*

In particular, Jordan groups of sufficiently large rank are necessarily projective or affine groups. The latter assertion has important applications in model theory, in the work of Cherlin et al. (1985) on $\aleph_0$-categorical, ω-stable theories. It has also been given a purely combinatorial proof (not using the classification) by Evans (1986) and Zil'ber (1988). (These "elementary" proofs were found after the classification, despite much earlier effort in this direction.)

Another application is Kantor's (1985c) determination of the 2-transitive linear spaces and basis-transitive matroids.

Next we weaken the assumption of 2-transitivity. The *rank* of a transitive permutation group G on Ω is the number of orbits of G on Ω^2. The diagonal is always an orbit; so the rank of a group of degree greater than 1 is at least 2, with equality if and only if it is 2-transitive.

Theorem 5.4.
(i) *All primitive permutation groups of rank* 3 *are known.*
(ii) *For any given r, all but finitely many primitive almost simple permutation groups of rank r are known.*

Part (i) was completed by Liebeck (1987), building on work of Bannai (1972), Kantor and Liebler (1982), and Liebeck and Saxl (1986). The strategy is similar to that for 2-transitive groups. Part (ii) is due to Kantor (1979) using a result of Seitz (1974).

Theorem 5.5. *All almost simple primitive permutation groups of odd degree are known.*

This is due to Kantor (1987) and Liebeck and Saxl (1985). Note that if G is primitive of odd degree but not basic, then $G \leqslant H \,\mathrm{Wr}\, K$, where H is basic of odd degree; and no group of diagonal type has odd degree (since, by the Feit–Thompson 1963 theorem, non-abelian simple groups have even order). The unknown factor here is thus the list of groups of affine type; determining this is equivalent to finding all irreducible linear groups in odd characteristic (and so somewhat out of reach yet.)

A result similar in spirit to Theorem 5.4(ii) was proved by Kantor et al. (1989), also in the context of model theory.

Theorem 5.6. *There is a function f on the natural numbers, tending to infinity with n, with the property that any primitive group of degree n having fewer than $f(n)$ orbits on 5-tuples of points is explicitly known.*

5.2. Degree and order

It was known to Mathieu (1861, see the footnote on p. 274) that 2-transitive groups of degree n (other than S_n and A_n) exist for $5 \leqslant n \leqslant 33$. However, this pattern is not typical.

Theorem 5.7. *For almost all n, the only primitive groups of degree n are S_n and A_n. More precisely, if*

$$\mathscr{E} = \{n\colon \text{ there exists a primitive group of degree } n, \text{ not } S_n \text{ or } A_n\},$$

and $e(x) = |\mathscr{E} \cap [1, x]|$, then

$$e(x) = 2\pi(x) + (1 + \sqrt{2})x^{1/2} + \mathrm{O}(x^{1/2}/\log x),$$

where $\pi(x)$ is the number of primes not exceeding x.

This is due to Cameron et al. (1982). In outline: degrees of non-basic groups are proper powers, of which squares contribute $x^{1/2}$ and the others $\mathrm{O}(x^{1/3+\varepsilon})$. Degrees of groups of affine type are prime powers, so either prime (contributing $\pi(x)$) or proper powers. Degrees of diagonal groups are either orders of simple groups or proper powers. For the density of the former we turn to the classification. There are $\mathrm{o}(\log x)$ alternating groups, $\mathrm{o}(x^{1/3} \log x)$ groups of Lie type (since the orders are roughly polynomials of degree at least 3 in a prime power argument), and $\mathrm{O}(1)$ sporadic groups, of order less than x. The most difficult estimation is that of almost simple groups. The crucial results are the lower bounds for indices of subgroups; these show that not too many simple groups can contribute to $e(x)$. The number of values contributed by T is at most the number of divisors of $|T|$, by Lagrange's theorem. It turns out that the dominant contributions are $\pi(x-1)$ (from $\mathrm{PSL}(2, p)$, in its representation of degree $p+1$, p prime) and $(2x)^{1/2}$ (from the triangular numbers $\binom{n}{2}$, the degree of S_n acting on pairs). The asymptotic expansion could be continued to higher precision.

I have already mentioned classical results asserting that primitive groups of degree n cannot be too large. The best bound obtained without using the classification is $n^{4\sqrt{n}\log n}$ for primitive, not 2-transitive groups, by Babai (1981). (The 2-transitive groups are now known to be much smaller, at most $n^{c \log n}$. The original "elementary" result of Babai (1982) was much weaker; but Pyber (1993a) has proved a bound $n^{c\log^2 n}$ by elementary means.) However, using the classification, one can show the following.

Theorem 5.8. *With an infinite family and finitely many more exceptions, a primitive group of degree n has order at most $n^{c\log n}$, where c is a constant.*

The infinitely many exceptions satisfy

$$(A_m)^l \leqslant G \leqslant S_m \operatorname{Wr} S_l$$

with the product action, where S_m acts on k-sets, k and l are bounded and m arbitrary. By allowing further exceptions (among basic groups, just symmetric groups acting on partitions of fixed shape, classical groups acting on an orbit of subspaces in their natural modules, and certain groups of affine type), the bound can be improved to $n^{c\log\log n}$. For almost simple groups, the order is at most n^c with known exceptions. For applications of these methods to graph isomorphism problem, see chapter 27.

An up-to-date survey of "asymptotic permutation group theory" is given by Pyber (1993b).

5.3. Fixed-point-free elements

It is elementary that a transitive finite group with degree greater than 1 contains an element with no fixed points. (This can be seen from the Orbit-counting lemma 2.3: the average number of fixed points is 1, and the identity fixes more than one point, so some element fixes less than one.) The next result, then, is not unexpected.

Proposition 5.9. *A transitive group of degree greater than* 1 *has a fixed-point-free element of prime power order.*

There are three surprising things about this result. First, that it is so difficult. The proof (Fein et al. 1981) involves first reducing the problem to consideration of primitive simple groups, and going through the list of simple groups, using quite detailed information about each class. Second is the applicability of the result (to a question about relative Brauer groups of finite extensions of global fields). Third is the fact that the following plausible extension is still open.

Conjecture. For any prime p, there is a function f_p such that a transitive group of degree n, where $n = p^a \cdot b$ and $a \geqslant f_p(b)$, contains a fixed-point-free element of p-power order.

As well as several applications in number theory, this problem has arisen in game theory (Isbell 1960) and extremal set theory (Cameron et al. 1989a).

5.4. Distance-transitive graphs

Cameron et al. (1983) confirmed a conjecture of Sims by showing the following.

Theorem 5.10. *There is a function f such that, if G is primitive and G_α has an orbit (other than $\{\alpha\}$) of length d, then $|G_\alpha| \leqslant f(d)$.*

(Of course, the group induced by G_α on the orbit has order at most $d!$; the point is that the kernel of this action has bounded order.)

Using this, they showed the following. A connected graph is *distance-transitive* if its automorphism group acts transitively on ordered pairs of vertices at any given distance.

Theorem 5.11. *There are only finitely many finite distance-transitive graphs of given valency $d > 2$.*

Proof (*sketch*)**.** Using techniques of Smith (1971), one reduces to the case where the automorphism group acts primitively. By the Sims conjecture, the number of vertices at given distance from a fixed vertex is bounded (all these sets are orbits of the vertex stabiliser). By the Compactness theorem of first-order logic, an infinite distance-transitive graph with the same properties exists. But this contradicts Macpherson's (1982) classification of these graphs. □

Further work on distance-transitive graphs has concentrated on applying methods of group theory to the full automorphism group (rather than to the point stabiliser as above). Smith's result cited above asserts that any (finite) imprimitive distance-transitive graph other than a cycle is either a "bipartite double" or an "antipodal cover" of a smaller distance-transitive graph; in at most two such reductions, we reach a primitive graph. So, naturally, most attention has been given to this case. Application of the O'Nan–Scott theorem (with some further work) reduces to the affine and almost simple cases. (The Hamming graphs, whose automorphism groups preserve an obvious power structure, are characterised along the way. See Praeger et al. 1987.) Then these cases must be dealt with – this is the hard part! At the time of writing, it appears that the classification of primitive distance-transitive graphs is almost complete. Brouwer et al. (1989) is the standard reference on distance-transitive graphs, which are also discussed in chapter 15.

Closely related is the concept of arc-transitivity. A graph is *s-arc transitive* if its automorphism group acts transitively on the set of arcs (walks with no two consecutive edges equal) of length s. Weiss (1981) showed the following.

Theorem 5.12. *A finite graph (other than a cycle) cannot be more than 7-arc transitive.*

By contrast, a countable regular tree is s-arc transitive for all s. There are close connections with the flourishing theory of group amalgams. It is also known that finite s-arc-transitive digraphs of valency d exist for every s and d (Praeger 1989).

6. Characters and configurations

The conclusion of the orbit-counting lemma can be phrased in terms of permutation characters. Further information can be obtained by considering other actions, especially on Ω^2. A combinatorial version of these matters is Higman's theory of "coherent configurations". General references on character theory are Serre (1977) and Curtis and Reiner (1962); for a survey of Higman's ideas, see Higman (1975).

A *matrix representation* M of a group G is a homomorphism from G to the group of invertible $n \times n$ matrices over a field. I will always take this field to be $\mathbb{C}$, but important results are obtained by considering subfields of $\mathbb{C}$ (such as $\mathbb{R}$ or $\mathbb{Q}$), p-adic fields, or fields of non-zero characteristic.

The *character* of G associated with a representation is the function

$$\chi(g) = \operatorname{trace}(M(g)).$$

It can be shown that two representations are similar if and only if their characters are equal.

A representation M is *irreducible* if the space $V(n, \mathbb{C})$ has no proper subspace invariant under $\{M(g): g \in G\}$; it is *indecomposable* if it is not the direct sum of two non-trivial invariant subspaces. *Maschke's theorem* asserts that any invariant subspace has an invariant complement, so these concepts coincide. (This is true generally over fields of characteristic zero, or whose characteristic does not divide the group order.) We call a character *irreducible* if the corresponding representation is.

Any character of G is constant on the conjugacy classes of G. The vector space of *class functions* (functions from G to $\mathbb{C}$ which are constant on the conjugacy classes) admits an inner product given by

$$(f_1, f_2) = \frac{1}{|G|} \sum_{g \in G} f_1(g) \overline{f_2(g)}.$$

Theorem 6.1.

(i) *The irreducible characters form an orthonormal basis for the space of complex-valued class functions (with respect to the above inner product).*

(ii) *The number of irreducible characters of* G *is equal to the number of conjugacy classes.*

(iii) *The sum of squares of the degrees of the irreducible characters is equal to* $|G|$.

Note that arbitrary characters are linear combinations of irreducible characters with non-negative integer coefficients.

The *character table* of a group G is the matrix whose (i, j) entry is equal to the value of the ith character on elements of the jth conjugacy class. It is a square matrix which is orthogonal (with respect to suitable weight functions on the classes and characters).

Any group has a *principal character*, the identically-1 function, which is clearly irreducible. Also, if G acts on Ω, then G has a matrix representation by permutation matrices. Now the trace of a permutation matrix is the number of 1s on the diagonal, that is, the number of fixed points. Combining all these remarks, we obtain the following translation of the Orbit-counting lemma 1.3.

Theorem 6.2. *Let the finite group G act on the finite set Ω. Then the number of orbits of G is equal to the inner product of the permutation character with the principal character, that is, to the multiplicity of the principal character in the permutation character.*

This observation allows techniques of character theory to be used. It also has some combinatorial consequences. One consequence is Block's lemma (see chapter 31): If G is a group of automorphisms of an incidence structure whose incidence matrix has rank equal to the number of rows, then it can be shown that the permutation character of G on points is contained in the permutation character on blocks; taking inner product with the principal character, we see that G has at least as many block-orbits as point-orbits.

For another example, I give an argument due to O'Nan (1985). A set S of permutations of Ω is called *uniformly transitive* if, for any $\alpha, \beta \in \Omega$, the number of elements $g \in S$ satisfying $\alpha g = \beta$ is a constant, say λ. If so, then we have $\lambda \cdot |\Omega| = |S|$. We call S *sharply transitive* if it is uniformly transitive with $\lambda = 1$.

Lemma 6.3. *Let G act transitively on Ω_1 and Ω_2. Suppose that every irreducible constituent of the permutation character on Ω_2 also occurs in the permutation character on Ω_1. If a subset S of G is uniformly transitive on Ω_1, then S is uniformly transitive on Ω_2. In particular, if S is sharply transitive on Ω_1, then $|\Omega_2|$ divides $|\Omega_1|$.*

Corollary 6.4. *For $d > 2$, the group $\mathrm{P\Gamma L}(d,q)$ contains no sharply 2-transitive subset.*

Proof. It can be shown that the character condition applies, with Ω_1 the set of ordered pairs of distinct points and Ω_2 the set of antiflags (non-incident point-hyperplane pairs); but $|\Omega_2|$ does not divide $|\Omega_1|$. □

For an extension, see Cameron et al. (1987).

Further results can be obtained by considering different actions of G. The basic observation is the following.

Lemma 6.5. *Let the group G act transitively on sets Γ and Δ. Then the numbers of G-orbits on $\Gamma \times \Delta$ and of G_γ-orbits on Δ ($\gamma \in \Gamma$) are equal, and each is the inner product of the permutation characters on Γ and Δ.*

Proof. A bijection between the specified orbits is given by

$$O \mapsto O(\gamma) = \{\delta : (\gamma, \delta) \in O\},$$

for a G-orbit O on $\Gamma \times \Delta$. If π_Γ and π_Δ are the permutation characters on Γ and Δ, then $\pi_\Gamma \pi_\Delta$ is the permutation character on $\Gamma \times \Delta$; and we have

$$(\pi_\Gamma \pi_\Delta, 1) = (\pi_\Gamma, \pi_\Delta)$$

where 1 is the principal character. □

The following illustration of this result is due to Livingstone and Wagner (1965).

Corollary 6.6. *If $k \leqslant l \leqslant n/2$, then a permutation group of degree n has at least as many orbits on l-sets as on k-sets.*

For, under these hypotheses, if Ω_k and Ω_l are the sets of k-sets and l-sets respectively, then the symmetric group S_n has $k+1$ orbits on $\Omega_k \times \Omega_l$; then an easy induction shows that the permutation character of S_n on Ω_k is the sum of $k+1$ irreducible characters, all of which appear in the character on Ω_l. So the permutation character on k-sets is contained in that on l-sets. Now the proof proceeds as in Block's lemma.

Corollary 6.7. *If G is transitive on Ω, then the rank of G is the inner product of the permutation character with itself.*

This is immediate from the definition of rank as the number of orbits of the stabiliser of a point. In particular, G is 2-transitive if and only if the permutation character is the sum of the principal character and one further irreducible.

It is true in general that (π_Ω, π_Ω) is the number of G-orbits on Ω^2 (though, if G is intransitive, this number cannot be equated with the number of G_α-orbits). D. G. Higman, in a number of papers dating from the mid-1960s, has developed a combinatorial setting for this observation. Regard each orbit O_i of G on Ω^2 as a binary relation R_i on Ω. Now:

(i) the given relations partition Ω^2;

(ii) a subset of the given relations partitions the diagonal of Ω^2;

(iii) the set of relations is closed under taking converses;

(iv) for relations R_i, R_j, R_k and a pair $(\alpha, \beta) \in R_k$, the number of points $\gamma \in \Omega$ such that $(\alpha, \gamma) \in R_i$ and $(\gamma, \beta) \in R_j$ depends only on i, j and k, not on α and β.

A family of binary relations satisfying these conditions is called a *coherent configuration*. These structures, which include strongly regular and distance-regular graphs, association schemes, symmetric and quasi-symmetric designs, and regular two-graphs, can be studied by representation-theoretic methods like those used for groups. These matters are considered further in chapter 15; see also chapter 41, section 4.

We will also need a generalisation of the concept of permutation characters, that of *induced characters*. Let G be a group, H a subgroup, χ a character of H, afforded by a matrix representation M. Now let $|G:H| = n$, and let $x_1, \ldots, x_n$ be right coset representatives for H in G. For $g \in G$, let $\overline{g}$ denote the coset representative of g. Now let $\Omega = H\backslash G$, and $\alpha = H$, so that $\Omega = \{\alpha x_1, \ldots, \alpha x_n\}$.

As before, we consider the representation P of G by permutation matrices corresponding to the action on Ω. Now define a new representation of G of degree nm as follows: g is represented by block matrices with $(\alpha x_i, \alpha x_j)$ block equal to O if the $(\alpha x_i, \alpha x_j)$ entry of $P(g)$ is zero, and to $M(x_i g x_j^{-1})$ otherwise. (Note that, if

$\alpha x_i g = \alpha x_j$, then $\overline{x_i g} = x_j$, and so the argument of M really is an element of H.) It is easily checked that we really have a representation of G; it is called the *induced representation*, and its character is the *induced character* χ^G.

If χ is the principal character of H, then χ^G is simply the permutation character of G on the space $H\backslash G$ of cosets of H.

7. The characters of the symmetric group

There is one family of groups whose character theory is particularly important to combinatorics, namely the symmetric groups. This theory has connections with partitions, symmetric functions, Schur functions, etc.; these matters are discussed further in chapter 21. General references here are Robinson (1961), James and Kerber (1981), and the first chapter of Macdonald (1979).

The initial observation is the following.

Proposition 7.1. *There is a natural bijection between the characters of the symmetric group S_n and the partitions of n.*

Proof. Given any permutation g on Ω, there is a unique decomposition of Ω into cycles of g (orbits of the group generated by g):

$$g = (\alpha_1\ \alpha_2 \cdots)(\beta_1\ \beta_2 \cdots)\cdots.$$

Since

$$h^{-1}gh = (\alpha_1 h\ \alpha_2 h \cdots)(\beta_1 h\ \beta_2 h \cdots)\cdots,$$

conjugate permutations have cycles of the same lengths, and conversely; so the partition of n determined by the cycles of a permutation (its *cycle structure*) characterises it up to conjugacy. □

Hence the number of conjugacy classes is equal to the number $p(n)$ of partitions of n. This is a well-studied function; see chapter 21.

In view of the preceding section, we know that the irreducible characters of S_n are also bijective with the partitions of n. I will give a brief description of this bijection, and mention a few consequences.

A partition of n is represented by a *Ferrers diagram*: if the partition λ has c parts with lengths $l_1, \ldots, l_c$, with $l_1 \geqslant \cdots \geqslant l_c$, the diagram for λ has c rows with l_i cells in the ith row, aligned on the left.

Corresponding to a partition as above, there is a well-defined conjugacy class of subgroups of S_n, one for each partition of Ω with parts of the appropriate sizes, consisting of all permutations fixing every part of the partition. These are called *Young subgroups*. We use S^λ to denote a typical Young subgroup corresponding to the partition λ. We will be interested in the action of S_n on all partitions of n of the appropriate shape (with labelled parts); the corresponding Young subgroups are just the stabilisers in this action.

Now we define characters π^λ and τ^λ of S_n as follows:

- π^λ is the permutation character of S_n acting on partitions of shape λ (that is, the character induced from the trivial character of S^λ);
- τ^λ is the character of S_n induced from the sign character of S^λ (the character of degree 1 whose value is +1 on even permutations and −1 on odd permutations).

A partition λ has a *dual* λ', whose ith part is equal to the number of parts of λ which exceed i. Its Ferrers diagram is the transpose of that of λ. Hence $\lambda'' = \lambda$ for any partition λ.

Finally, there is a total order $\prec$ on partitions, as follows: if λ and μ have parts $l_1, \ldots, l_c$ and $m_1, \ldots, m_d$, then we write $\lambda \prec \mu$ if, for some i, we have $l_j = m_j$ for $j < i$ while $l_i > m_i$ (where undefined parts are taken to be zero).

Now the main result is as follows.

Theorem 7.2. *For any partition λ, there is a unique irreducible χ^λ which is a constituent of π^λ but not of π^μ for any $\mu \prec \lambda$. Furthermore, χ^λ is the unique common constituent (with multiplicity* 1*) of π^λ and $\tau^{\lambda'}$.*

For example, the first few characters π^λ, with their decompositions into irreducibles, are as follows:

$$\begin{aligned}
\pi^{(n)} &= \chi^{(n)}, \quad \text{the principal character;} \\
\pi^{(n-1,1)} &= \chi^{(n)} + \chi^{(n-1,1)}; \\
\pi^{(n-2,2)} &= \chi^{(n)} + \chi^{(n-1,1)} + \chi^{(n-2,2)}; \\
\pi^{(n-2,1,1)} &= \chi^{(n)} + 2\chi^{(n-1,1)} + \chi^{(n-2,2)} + \chi^{(n-2,1,1)}.
\end{aligned}$$

In particular, for $n > 2k$, the permutation character $\pi^{(n-k,k)}$ on k-sets is the sum of the irreducibles $\chi^{(n-i,i)}$ for $i \leqslant k$, in accordance with our observation in the proof of Corollary 6.6.

Theorem 7.3.

(i) *The degree of π^λ is equal to the number of ways of writing the numbers $\{1, \ldots, n\}$ into the cells of λ so that the rows are strictly increasing.*

(ii) *The degree of χ^λ is equal to the number of ways of writing $\{1, \ldots, n\}$ into the cells of λ so that the rows and columns are strictly increasing.*

The first statement holds because, if $\Omega = \{1, \ldots, n\}$, then any partition of shape λ gives rise to a tableau with increasing rows (writing the elements of the ith part into the ith row in increasing order. For the second, see any of the books referred to earlier.

Note that, just as χ^λ is the "intersection" of π^λ and $\tau^{\lambda'}$, so the arrangements counted to obtain its degree are those satisfying both the constraint for π^λ and (in transposed form) that for $\tau^{\lambda'}$. The objects counted in Theorem 7.3(ii) are called *standard tableaux of shape* λ, and their number (the degree of χ^λ) is denoted by f_λ.

For example, if $n = 2m$ and the partition λ has two parts of size m, then $\chi^\lambda(1)$ is equal to the number of ways that m votes can be cast for each of two candidates in an election so that the first candidate is never behind in the count. This is a familiar interpretation of the mth *Catalan number* C_m (see chapter 21).

From Theorem 6.1(iii) we see that $\sum(f_\lambda)^2 = n!$ where the summation is over all partitions of n. In other words, the number of ordered pairs of standard tableaux of the same shape is equal to the number of permutations of $\{1,\ldots,n\}$. A bijection between these two sets is given by the *Schensted correspondence.* In this correspondence, the two tableaux are equal if and only if the permutation satisfies $g^2 = 1$; so

$$\sum f_\lambda = s(n),$$

where $s(n)$ is the number of involutions on $\{1,\ldots,n\}$. (This is also a special case of a general result about characters.)

In principle, the entries of the character table can now be calculated: the values of $\pi^\lambda(g)$ are easily established, and the matrix of the system of equations for the $\chi^\lambda(g)$ in terms of the $\pi^\lambda(g)$ is lower triangular.

The *dimension* of a character χ^λ is defined to be $n - d$, where d is the largest part of λ. Now consider partitions of n, for $n > 2d$ with fixed d, having dimension d and such that the parts other than the largest constitute a fixed partition of d. It can be shown that the degrees of such characters are given by a polynomial in n of degree d, with leading term $n^d/d!$. Moreover, Schur proved the following.

Proposition 7.4. *For $n > 2d$, the irreducible constituents of the permutation character of S_n on d-tuples of distinct elements are the characters of dimension d or less. Moreover, a permutation group G is d-transitive if and only if the restrictions to G of all these characters (except the principal character) are orthogonal to the principal character of G.*

An application of these ideas is given by Cameron et al. (1987).

Theorem 7.5. *Let S be a set of permutations of an n-set such that the Hamming distances*

$$d(g,h) = \mathrm{fix}(gh^{-1})$$

for $g,h \in S, g \neq h$, take just d distinct values. Then S is bounded above by the sum of squares of the characters of S_n of dimension d or smaller.

(This bound is a polynomial in n of degree $2d$, with leading term $n^{2d} \cdot p(n)/(d!)^2$, where $p(d)$ is the number of partitions of d.) The bound is not too far from best possible: examples of size roughly $(n/2d)^{2d}$ are known.

8. Computing in permutation groups

A good example to keep in mind for this section is Rubik's cube. This celebrated puzzle can be thought of as a cube divided into $3 \times 3 \times 3$ smaller cubes. In its initial configuration, the nine squares on each face have the same colour. Each face (more precisely, the set of nine small cubes with faces lying in a single face of the large cube) can be rotated through any multiple of $\pi/2$. A few rotations will destroy the original arrangement and leave the cube in an apparently random configuration. The problem is: given the cube in a disordered state, decide whether or not the initial state can be restored by a sequence of moves (rotations of faces) and, if so, find a sequence which does this. We can regard the six squares at the centres of faces as fixed, so that the remaining 48 squares are permuted by any move. The six face rotations through $\pi/2$ in the positive sense generate a group of permutations. The problem can now be formulated and generalized as follows.

`Input`: permutations $g_1, \ldots, g_s$ of a finite set Ω.

`Output`: information about the group G generated by $g_1, \ldots, g_s$. (In the example, the required information is a membership test for arbitrary permutations; in other situations, we may want the order of G, whether G is simple, even a "name" for G, etc.)

I will give an outline of some of the basic algorithms, up to the point where we have the membership test and the group order; and then summarise more quickly some further techniques.

The first step involves finding the orbit containing a point α. This is the equivalence class containing α of the smallest equivalence relation $\sim$ in which $\beta \sim \gamma$ if $\beta g_i = \gamma$ for some i. It can be found by essentially the same algorithm that computes connected components in a digraph. For later use, this is refined so that with each point β of the orbit comes a "witness", a word w in the generators $g_1, \ldots, g_s$ such that $\alpha w = \beta$. Our discussion of coset spaces in section 2 shows that these witnesses form a complete set of right coset representatives for the stabiliser G_α. Also, the algorithm has the property that any initial segment of a witness is also a witness (for a point of the orbit found earlier); so instead of the actual witness, we only need to remember which generator was used last. This involves a considerable saving in space. The appropriate data structure is called a *Schreier vector*.

Next, we need a set of generators for G_α. (In Rubik's cube, once we have moved one square, or a set of squares, into their correct positions, we want to do all subsequent manipulations without disturbing these squares.) The following lemma is due to Schreier.

Lemma 8.1. *Let $g_1, \ldots, g_s$ be generators of a group G, and $x_1, \ldots, x_t$ right coset representatives for a subgroup H of index t. If $\overline{g}$ denotes the coset representative of an element $g \in G$, then H is generated by the ns elements*

$$\{x_i g_j (\overline{x_i g_j})^{-1} : 1 \leqslant i \leqslant t, 1 \leqslant j \leqslant s\}.$$

This principle (in words, "move α to β, using the witness for β; apply a generator to β, moving it (say) to γ; then return γ to α with the inverse of the witness for γ"), is surely familiar to every solver of Rubik's cube.

The disadvantage here is the number of generators. The index of G_α in G (the size of the orbit) may be as large as the degree n of G; so we may be left with ns generators for G_α. As we stabilise further points, the numbers of generators will grow explosively, and we will quickly run out of space, even though the groups being considered are getting smaller.

This is dealt with by a "filter" to reduce the number of generators. The best algorithm is due to Jerrum (1982). Its input is a sequence of permutations of Ω, given one at a time "on-line". At any stage, the output is a list of permutations of Ω such that

(i) the input and output permutations generate the same group;
(ii) the output list contains at most $n-1$ entries, where $n = |\Omega|$.

This gives an upper bound for the number of generators we have to store; the output from Schreier's algorithm can be fed straight into Jerrum's.

Jerrum's filter has a theoretical consequence.

Theorem 8.2. *Any subgroup of S_n can be generated by at most $n-1$ elements.*

McIver and Neumann (1987) (see also Cameron et al. 1989b) have announced without proof that this bound can be improved to $\max(2, \lfloor n/2 \rfloor)$; this is best possible. However, the proof does not provide an algorithm to find these generators.

Now we proceed to the important step, the Schreier–Sims algorithm for a base and strong generating set. A *base* for a permutation group G consists of a sequence $(\alpha_1, \ldots, \alpha_d)$ of elements of Ω such that only the identity fixes $\alpha_1, \ldots, \alpha_d$. Thus every permutation in G is uniquely determined by its effect on a base. (This means that, once we have a base, in subsequent computations only d values of a permutation need be stored, rather than n; and d is usually much smaller than n, as we shall see.) A base can be found by successively fixing points until the stabiliser consists of the identity only. Clearly we may assume that no base point is fixed by the stabiliser of the preceding points. (A base with this property is called *irredundant.*) A sequence $\boldsymbol{g}$ of elements of G is called a *strong generating set* (relative to the given base) if, for $0 \leqslant i \leqslant d$, the stabiliser of $(\alpha_1, \ldots, \alpha_i)$ is generated by some terminal subsequence of $\boldsymbol{g}$.

We can now find a base and strong generating set as follows:

Start with the empty sequence of elements of Ω, and the empty sequence of elements of G. While the generating set for G is non-empty, do the following. Choose a point α not fixed by all generators of G; adjoin it to the first list. Now compute the orbit of α, with witnesses; add the witnesses to the second list. Then compute generators for G_α, and replace G by G_α.

On termination, the first list is a base, and the second is a strong generating set of a special kind: it has the form $\boldsymbol{g} = (\boldsymbol{g}_0, \ldots, \boldsymbol{g}_{d-1})$, where $\boldsymbol{g}_i$ is a list of coset representatives for G_i in G_{i-1}, where G_j is the stabiliser of $(\alpha_1, \ldots, \alpha_j)$. Thus

$$|\boldsymbol{g}_i| = |G_{i-1} : G_i|.$$

Now the membership test works as follows. Given an arbitrary permutation g, check first whether $\alpha_1 g$ is in the G-orbit of α_1 (using the witnesses in $\mathbf{g}_0$). If not, then $g \notin G$; otherwise, $\alpha_1 g = \alpha_1 g_{0i_0}$ for some i_0, so that $gg_{0i_0}^{-1} \in G_1$. Proceed inductively. At the conclusion, either we have shown that $g \notin G$, or we have

$$gg_{0i_0}^{-1} \cdots g_{d-1i_{d-1}}^{-1} = 1,$$

so that

$$g = g_{d-1i_{d-1}} \cdots g_{0i_0} \in G.$$

As a consequence, we see that each element $g \in G$ has a *unique* expression of the above form, so that

$$|G| = |\mathbf{g}_0| \cdots |\mathbf{g}_{d-1}|.$$

Many further algorithms for studying a permutation group are known. It is possible to find centralisers and normalisers of subsets, Sylow subgroups, normal subgroups, composition factors, cores; test simplicity; and so on. See Butler and Cannon (1982), Kantor (1985b), Neumann (1987). These algorithms usually start with a base and strong generating set.

Some theoretical issues are raised by this point of view. Clearly it is important to find as small a base as possible, and to do so quickly. Blaha (1992) has shown the following.

Theorem 8.3. *If the smallest base for G has size d, then any irredundant base has size at most $d \log n$, and any base found by the greedy algorithm has size at most $2d \log \log n$; these bounds are best possible, up to constant factors.*

(The *greedy algorithm* chooses each base point in a largest orbit of the stabiliser of the preceding base points.)

The examples Blaha gives to show that his bounds are best possible are intransitive. It seems likely that much stronger results hold for primitive groups. It is *conjectured* that

(i) there is an absolute constant c such that, if the primitive group G has a base of size d, the greedy algorithm finds one of size at most cd;

(ii) there is an absolute constant c such that, if G is primitive and almost simple, then with known exceptions, G has a base of size at most c.

(A group with a base of size c has order at most n^c; the exceptions in (ii) should be the primitive almost simple groups whose order is not polynomially bounded, see section 5.)

Maund (1989) has proved some results in the direction of (ii). She has also found all permutation groups which permute their irredundant bases transitively. (Such a group G is a Jordan group, and its bases are the bases of a matroid; this is far from true in general.)

Another issue concerns the extent to which computations in a permutation group can be done in polynomial time. The Schreier–Sims algorithm with a filter like Jerrum's is polynomial, so we can always assume that we are given a base and strong generating set for the group in question. Many structural properties are in P (Kantor 1985a). A striking example is finding a Sylow subgroup, see Kantor (1985b). (The classical proofs of this important theorem are constructive, but yield exponential algorithms; the polynomial algorithm is much more complicated.)

In practice, randomised algorithms are often used. For example, often we need to find an element of prime order p. This can be done quickly by choosing random elements of G until we find one with order divisible by p, and raising it to an appropriate power.

Many of these algorithms are implemented in group-theoretic programming environments, of which the most comprehensive are MAGMA, a development of CAYLEY (Cannon 1984) with new features for doing combinatorics (among other things), and GAP, an open system developed by Neubüser.

For further detail on computational group theory, see chapter 27.

9. Infinite permutation groups

Until the middle 1970s, the subject "infinite permutation groups" scarcely existed. (In the Mathematical Reviews, permutation groups were explicitly finite.) Now, this is a rapidly expanding area, with close links to model theory, Ramsey theory, enumeration and topology, among others. In this section I will sketch briefly some of these developments.

9.1. Groups of finitary permutations

A finitary permutation is one which moves only finitely many points. The set of such permutations on an infinite set forms the *finitary symmetric group*; the even permutations in it form the *alternating group*. Wielandt (1959), extending a result of Jordan about finite permutation groups, showed the following.

Theorem 9.1. *A primitive permutation group of infinite degree containing a finitary permutation contains the alternating group.*

The general theory of groups of finitary permutations was developed by Neumann (1976). The first observation is that all non-trivial blocks of imprimitivity are finite. Such a group of uncountable degree has a unique system of maximal blocks, and is an extension of a subdirect product of finite groups by the finitary symmetric or alternating group. For countable degree, there are additional possibilities, and examples including transitive p-groups.

9.2. Groups of cofinitary permutations

The dual situation is where all non-identity permutations in a group have only finitely many fixed points. Barlotti and Strambach (1983) showed the following.

Theorem 9.2. *For any finite k, there is a k-transitive group in which any non-identity permutation fixes at most k points.*

The group can be chosen free of countable rank. A generalisation of this result has applications in the theory of measurement in mathematical psychology (Cameron 1989). A survey of this topic is in preparation.

9.3. Normal subgroups

No simple analogue of the O'Nan–Scott theorem can hold for infinite groups. There are two reasons for this. The more serious is the absence of minimal normal subgroups in general (e.g., in the free group of countable rank in the last subsection). The other is illustrated by the following fact.

Proposition 9.3. *There is a 2-transitive group which is the direct product of two simple regular subgroups.*

This follows from the existence of a (simple) group S in which all non-identity elements are conjugate (Higman et al. 1949). Now $G = S \times S$, acting on $\Omega = S$ by the rule

$$(g, h) : s \mapsto g^{-1}sh,$$

is 2-transitive. (This group is of diagonal type.)

9.4. Jordan groups

Apart from some very recent examples due to Hrushovski (1993), the known infinite Jordan groups fall into three types:

(i) Automorphism groups of matroids (projective or affine spaces or the geometry of algebraically closed subfields of a field).

(ii) Automorphism groups of "order-like" relations. The prototype is the group of order-preserving permutations of $\mathbb{Q}$.

(iii) Homeomorphism groups of topological spaces (e.g., manifolds).

Some characterisations of the first two classes exist. For example, a primitive Jordan group with a cofinite Jordan set preserves a projective or affine space over a finite field; this fact has very important implications in model theory (see Cherlin et al. 1985). Adeleke and Macpherson (1994) have characterised all primitive, not highly transitive Jordan groups as automorphism groups of certain relations; see Macpherson's article in Kaye and Macpherson (1994) for discussion. Also, Wagner's article in that volume discusses Hrushovski's constructions.

9.5. Oligomorphic permutation groups

The permutation group G on Ω is *oligomorphic* (="few shapes") if it has only finitely many orbits on Ω^n for all n. The importance of this class of groups derives from the following theorem of Engeler (1959), Ryll–Nardzewski (1959) and Svenonius (1959).

Theorem 9.4. *A countable first-order structure is characterised, up to isomorphism, by first-order axioms (and the countability condition) if and only if its automorphism group is oligomorphic.*

Oligomorphic permutation groups are also closely connected with enumeration theory: counting orbits on n-sets or n-tuples is equivalent to counting unlabelled or labelled finite structures in certain classes. A version of the cycle index can be defined for any oligomorphic group.

The earliest theorem on these groups was due to Cameron (1976).

Theorem 9.5. *If the infinite permutation group G on Ω is transitive on n-subsets of Ω for all n, but not n-transitive for some n, then there is a linear or circular order on Ω preserved or reversed by G.*

A more detailed account of oligomorpic permutation groups is given by Cameron (1990). See also several articles in Kaye and Macpherson (1994).

References

Adeleke, S.A., and H.D. Macpherson
[1994] A classification of infinite Jordan groups, to appear.

Artin, E.
[1957] *Geometric Algebra* (Interscience, New York).

Aschbacher, M.
[1984] On the maximal subgroups of the finite classical groups, *Invent. Math.* **76**, 469–514.

Aschbacher, M., and L.L. Scott
[1985] Maximal subgroups of finite groups, *J. Algebra* **92**, 44–80.

Babai, L.
[1981] On the orders of uniprimitive permutation groups, *Ann. of Math. (2)* **113**, 553–568.
[1982] On the orders of doubly transitive permutation groups, *Invent. Math.* **65**, 473–484.

Babai, L., P.J. Cameron and P.P. Palfy
[1982] On the orders of primitive groups with restricted non-abelian composition factors, *J. Algebra* **79**, 161–168.

Bannai, E.E.
[1972] Maximal subgroups of low rank of finite symmetric and alternating groups, *J. Fac. Sci. Univ. Tokyo* **18**, 475–486.

Barlotti, A., and K. Strambach
[1983] The geometry of binary systems, *Adv. in Math.* **49**, 1–105.

Blaha, K.
[1992] Minimum bases for permutation groups: the greedy approximation, *J. Algorithms* **13**, 297–306.

Brouwer, A.E., A.M. Cohen and A. Neumaier
[1989] *Distance-Regular Graphs* (Springer, Berlin).

Burnside, W.
[1911] *Theory of Groups of Finite Order* (Cambridge University Press, Cambridge). Reprinted: 1955 (Dover, New York).

Butler, G.A., and J.J. Cannon
[1982] Computing in permutation and matrix groups, I, II, *Math. Comput.* **39**, 663–670, 671–680.

Cameron, P.J.
[1976] Transitivity of permutation groups on unordered sets, *Math. Z.* **148**, 127–139.
[1981] Finite permutation groups and finite simple groups, *Bull. London Math. Soc.* **13**, 1–22.
[1989] Groups of order-automorphisms of the rationals with prescribed scale type, *J. Math. Psychol.* **33**, 163–171.
[1990] *Oligomorphic Permutation Groups, London Mathematical Society Lecture Note Series,* Vol. 152 (Cambridge University Press, Cambridge).

Cameron, P.J., P.M. Neumann and D.N. Teague
[1982] On the degrees of primitive permutation groups, *Math. Z.* **180**, 141–149.

Cameron, P.J., C.E. Praeger, J. Saxl and G.M. Seitz
[1983] On the Sims conjecture and distance-transitive graphs, *Bull. London Math. Soc.* **15**, 499–506.

Cameron, P.J., M. Deza and P. Frankl
[1987] Sharp sets of permutations, *J. Algebra* **111**, 220–247.

Cameron, P.J., P. Frankl and W.M. Kantor
[1989a] Intersecting families of finite sets and fixed-point-free 2-elements, *European J. Combin.* **10**, 149–160.

Cameron, P.J., R. Solomon and A. Turull
[1989b] Chains of subgroups in symmetric groups, *J. Algebra* **127**, 340–352.

Cannon, J.J.
[1984] An introduction to the group theory language CAYLEY, in: *Computational Group Theory,* ed. M.D. Atkinson (Academic Press, London) pp. 145–183.

Carter, R.W.
[1972] *Simple Groups of Lie Type* (Wiley, London).

Cherlin, G.L., L. Harrington and A.H. Lachlan
[1985] $\aleph_0$-categorical, $\aleph_0$-stable structures, *Ann. Pure Appl. Logic* **28**, 103–135.

Chevalley, C.
[1955] Sur certains groupes simples, *Tôhoku J. Math* **7**, 14–66.

Conway, J.H., R.T. Curtis, S.P. Norton, R.A. Parker and R.A. Wilson
[1985] ATLAS *of Finite Groups* (Oxford University Press, Oxford).

Curtis, C.W., and I. Reiner
[1962] *Representation Theory of Finite Groups and Associative Algebras* (Interscience, New York).

Curtis, C.W., W.M. Kantor and G.M. Seitz
[1976] The 2-transitive representations of the finite Chevalley groups, *Trans. Amer. Math. Soc.* **218**, 1–57.

Dickson, L.E.
[1900] *Linear Groups with an Exposition of the Galois Field Theory* (University of Chicago Press, Chicago). Reprinted: 1958 (Dover, New York).

Dieudonné, J.
[1963] *La Géometrie des Groupes Classiques* (Springer, Berlin).

Engeler, E.
[1959] Äquivalenzklassen von *n*-Tupeln, *Z. Math. Logik Grundl. Math.* **5**, 340–345.

Evans, D.M.
[1986] Homogeneous geometries, *Proc. London Math. Soc. (3)* **52**, 305–327.

Fein, B., W.M. Kantor and M. Schacher
[1981] Relative Brauer groups II, *J. Reine Angew. Math.* **328**, 39–57.

Feit, W., and J.G. Thompson
[1963] Solvability of groups of odd order, *Pacific J. Math.* **13**, 775–1029.

Gorenstein, D.
[1982] *Finite Simple Groups: An Introduction to their Classification* (Plenum Press, New York).

Hall, P.
[1962] Wreath powers and characteristically simple groups, *Proc. Cambridge Philos. Soc.* **58**, 170–184.

Hering, C.
[1985] Transitive linear groups and linear groups which contain irreducible subgroups of prime order, II, *J. Algebra* **93**, 191–198.

Higman, D.G.
[1975] Invariant relations, coherent configurations and generalized polygons, in: *Combinatorics,* eds. M. Hall Jr and J.H. van Lint (Reidel, Dordrecht) pp. 347–363.

Higman, G., B.H. Neumann and H. Neumann
[1949] Embedding theorems for groups, *J. London Math. Soc.* **24**, 247–254.

Howlett, R.B.
[1974] On the degrees of Steinberg characters of Chevalley groups, *Math. Z.* **135**, 125–135.

Hrushovski, E.
[1993] A new strongly minimal set, *Ann. of Pure & Appl. Logic* **62**, 147–166.

Huppert, B.
[1957] Zweifach transitive, auflösbare Permutationsgruppen, *Math. Z.* **68**, 126–150.

Isbell, J.
[1960] Homogeneous games II, *Proc. Amer. Math. Soc.* **11**, 159–161.

James, G.D., and A. Kerber
[1981] *The Representation Theory of the Symmetric Group* (Addison-Wesley, Reading, MA).

Jerrum, M.R.
[1982] A compact representation for permutation groups, in: *23rd Annu. Symp. on Foundations of Computer Science, Chicago, 1982* (IEEE, New York) pp. 126–133.

Jordan, C.
[1861] Mémoire sur le nombre des valeurs des fonctions, *J. École Poly.* **22**, 113–194.
[1871] Théorèmes sur les groupes primitifs, *J. Math. Pures Appl. (Liouville) (2)* **16**, 383–408.

Kantor, W.M.
[1979] Permutation representations of the finite classical groups of small degree or rank, *J. Algebra* **60**, 158–168.
[1985a] Sylow's theorem in polynomial time, *J. Comput. System Sci.* **30**, 359–394.
[1985b] Notes on polynomial-time group theory, in: *Proc. Rutgers Group Theory Year 1983–1984,* eds. M. Aschbacher et al. (Cambridge University Press, Cambridge) pp. 19–21.
[1985c] Homogeneous designs and geometric lattices, *J. Combin. Theory A* 38, 66–74.
[1987] Primitive permutation groups of odd degree, and an application to finite projective planes, *J. Algebra* **106**, 15–45.

Kantor, W.M., and R.A. Liebler
[1982] The rank 3 permutation representations of the finite classical groups, *Trans. Amer. Math. Soc.* **271**, 1–71.

Kantor, W.M., M.W. Liebeck and H.D. Macpherson
[1989] $\aleph_0$-categorical structures smoothly approximated by finite substructures, *Proc. London Math. Soc. (3)* **59**, 439–463.

Kaye, R., and H.D. Macpherson
[1994] eds., *Automorphisms of First-Order Structures* (Oxford University Press, Oxford).

Kleidman, P.B., and M.W. Liebeck
[1988] A survey of the maximal subgroups of the finite simple groups, *Geom. Dedicata* **25**, 375–389.

Kóvacs, L.G.
[1986] Maximal subgroups in composite finite groups, *J. Algebra* **119**, 114–131.

Liebeck, M.W.
[1987] The affine permutation groups of rank 3, *Proc. London Math. Soc. (3)* **54**, 477–516.

Liebeck, M.W., and J. Saxl
[1985] The primitive permutation groups of odd degree, *J. London Math. Soc. (2)* **31**, 250–264.
[1986] The finite primitive permutation groups of rank three, *Bull. London Math. Soc.* **18**, 165–172.

Liebeck, M.W., C.E. Praeger and J. Saxl
[1988] On the 2-closures of finite permutation groups, *J. London Math. Soc. (2)* **37**, 241–252.

Livingstone, D., and A. Wagner
[1965] Transitivity of finite permutation groups on unordered sets, *Math. Z.* **90**, 393–403.

Luks, E.M.
[1980] Isomorphism of graphs of bounded valence can be tested in polynomial time, in: *Proc. 21st IEEE Symp. on Foundations of Computer Science, Syracuse, 1980* (IEEE, New York) pp. 42–49.

Macdonald, I.G.
[1979] *Symmetric Functions and Hall Polynomials* (Oxford University Press, Oxford).

Macpherson, H.D.
[1982] Infinite distance-transitive graphs of finite valency, *Combinatorica* **2**, 63–69.

Maillet, E.
[1895] Sur les isomorphes holoédriques transitifs des groupes symétriques ou alternés, *J. Math. Pures Appl.* **1**, 5–34.

Mathieu, É.
[1861] Mémoire sur l'étude des fonctions des plusiers quantités, sur le maniére de les former et sur les substitutions qui les laissent invariables, *J. Math. Pure Appl. (Liouville) (2)* **6**, 241–323.

Maund, T.
[1989] Thesis (Oxford University).

McIver, A., and P.M. Neumann
[1987] Enumerating finite groups, *Quart. J. Math Oxford (2)* **38**, 473–488.

Neumann, P.M.
[1976] The structure of finitary permutation groups, *Arch. Math. (Basel)* **27**, 3–17.
[1979] A lemma that is not Burnside's, *Math. Sci.* **4**, 133–141.
[1987] Some algorithms for computing with finite permutation groups, in: *Groups – St. Andrews, 1985, London Mathematical Society Lecture Note Series*, Vol. 121, eds. C.M. Campbell and E.F. Robertson (Cambridge University Press, Cambridge) pp. 59–92.

O'Nan, M.E.
[1985] Sharply 2-transitive sets of permutations, in: *Proc. Rutgers Group Theory Year 1983–1984*, eds. M. Aschbacher et al. (Cambridge University Press, Cambridge).

Praeger, C.E., J. Saxl and K. Yokoyama
[1987] Distance-transitive graphs and finite simple groups, *Proc. London Math. Soc. (3)* **55**, 1–21.

Pyber, L.
[1993a] On the orders of doubly transitive groups: elementary estimates, *J. Combin. Theory A* **62**, 361–366.
[1993b] Asymptotic results for permutation groups, in: *Proc. DIMACS Conf. on Computational Group Theory*, eds. L. Finkelstein and W.M. Kantor (American Mathematical Society, Providence, RI) pp 197–219.

Ree, R.
[1961] A family of simple groups associated with the simple Lie algebra of type (F4), *Amer. J. Math.* **83**, 401–420.

Robinson, G. de B.
[1961] *Representation Theory of the Symmetric Group* (Edinburgh University Press, Edinburgh).

Ryll-Nardzewski, C.
[1959] On category in power $\leqslant \aleph_0$, *Bull. Acad. Polon. Sci. Sér. Math. Astron. Phys.* **7**, 545–548.

Scott, L.L.
[1980] Representations in characteristic *p*, in: *The Santa Cruz Conference on Finite Groups*, eds. B. Cooperstein and G. Mason, *Proc. Symp. Pure Math.* **37**, 319–331.

Seitz, G.M.
[1974] Small rank permutation representations of finite Chevalley groups, *J. Algebra* **28**, 508–517.

Serre, J.-P.
[1977] *Linear Representations of Finite Groups* (Springer, New York).

Smith, D.H.
[1971] Primitive and imprimitive graphs, *Quart. J. Math. Oxford (2)* **22**, 551–557.

Steinberg, R.
[1959] Variations on a theme of Chevalley, *Pacific J. Math.* **9**, 875–891.

Svenonius, L.
[1959] $\aleph_0$-categoricity in first-order predicate calculus, *Theoria (Lund)* **25**, 82–94.
Weiss, R.M.
[1981] The non-existence of 8-transitive graphs, *Combinatorica* **1**, 309–311.
Wielandt, H.
[1959] *Unendliche Permutationsyruppen,* (Universität Tübingen).
[1964] *Finite Permutation Groups* (Academic Press, New York).
Zil'ber, B.I.
[1988] Finite homogeneous geometries, in: *Proc. 6th Easter Conf. on Model Theory,* eds. B. Dahn and H. Wolter (Humboldt Universität, Berlin) pp. 186–208.

CHAPTER 13

Finite Geometries

Peter J. CAMERON

School of Mathematical Sciences, Queen Mary and Westfield College, Mile End Road, London E1 4NS, UK

Contents

HANDBOOK OF COMBINATORICS
Edited by R. Graham, M. Grötschel and L. Lovász

1. Introduction

The chief difficulty in surveying finite geometries is the huge number of kinds of geometries which have been studied. A survey must steer between the extremes of omitting a large part of the subject, or becoming little more than a list of lists of axioms. I have tried to do this by a combination of selectivity and vagueness.

The central actors in this story are undoubtedly the projective and affine spaces. Section 2 is devoted to a survey of these, giving descriptions, axiomatisations, and some of their properties. The next three sections give an account of their collineation groups, a study of configurations in projective spaces, and a brief pointer to a connection with coding theory. Projective and affine planes are important in view of their status as low-dimensional exceptions; some features of their theory appear in section 6.

In the next few sections, the theme is generalization. Buildings form an important and natural extension of projective spaces. Most finite examples, apart from projective spaces themselves, are the polar spaces formed by the isotropic varieties with respect to polarities, or the singular varieties with respect to quadratic forms, in projective spaces. Section 8 treats buildings, following an account of rank 2 buildings (generalized polygons) in section 7. The next section treats a wider generalization to the class of Buekenhout geometries. Finally, I discuss partial geometries.

2. Projective and affine spaces

Projective spaces are defined, in general, over possibly non-commutative fields (division rings or skew fields). However, *Wedderburn's theorem* asserts that any finite field is commutative; and *Galois' theorem* asserts that there is a unique field $\mathrm{GF}(q)$ of any given prime power order q, and no field of non-prime-power order.

Accordingly, let V be a vector space of dimension $n+1$ over $\mathrm{GF}(q)$. The *projective geometry* $\mathrm{PG}(n,q)$ is the geometry whose points, lines, planes, ... are the 1-, 2-, 3-, ... dimensional subspaces of V. (We use the general term *variety* for a subspace of arbitrary dimension. A $k+1$-dimensional subspace will be called a *k-flat*, so that a point is a 0-flat, etc.) We say that two varieties are *incident* if one contains the other. We use familiar geometric language; a point lies on a line if the point and line are incident; two lines are concurrent if some point is incident with both; and so on.

Now some familiar geometric properties hold: two points lie on a unique line; a non-incident point and line lie on a unique plane; and so on. (These properties can be summarized by saying that the projective geometry is a *matroid*. More precisely, if we identify an arbitrary variety with the set of points incident with it, then the varieties are the flats of a matroid. See chapter 9.) In addition various non-matroid properties hold: two coplanar lines are concurrent, and so on.

Proposition 2.1. *The number of k-dimensional subspaces of an n-dimensional vec-*

tor space over GF(q) *is the* Gaussian coefficient

$$\begin{bmatrix} n \\ k \end{bmatrix}_q = \prod_{i=0}^{k-1} \frac{q^n - q^i}{q^k - q^i}.$$

In particular, the number of points of $\mathrm{PG}(n,q)$ is $(q^{n+1}-1)/(q-1)$; each line contains $q+1$ points, each plane contains q^2+q+1, etc.

The Gaussian coefficient can be defined by the formula of Proposition 2.1 for any positive real value of q except 1; its limit, as $q \to 1$, is the binomial coefficient $\binom{n}{k}$. It has a combinatorial interpretation, and is of great importance; see chapter 21.

Proposition 2.2. *For* $0 < d < n$, *the points and k-flats in* $\mathrm{PG}(n,q)$ *are the points and blocks of a* 2-$\left(\begin{bmatrix} n+1 \\ 1 \end{bmatrix}_q, \begin{bmatrix} k+1 \\ 1 \end{bmatrix}_q, \begin{bmatrix} n-1 \\ k-1 \end{bmatrix}_q\right)$ *design.*

See chapter 14 for the concepts of design theory. Here, as earlier, a variety is identified with the set of points incident with it. Note that the design is square (i.e. has equally many points and blocks) if and only if the blocks are *hyperplanes* (subspaces of codimension 1). Dembowski (1968), in his influential book, used the term "projective design" for what is here called a "square design", but the term did not catch on; these objects are almost universally referred to as "symmetric designs".

How do you recognise a projective geometry? The result here is stated in the finite case only, but does not really require finiteness. It is due to Veblen and Young (1916), building on the work of Hilbert (1899) on the foundations of geometry.

A *projective plane* of order n is a 2-$(n^2+n+1, n+1, 1)$ design (whose blocks are called "lines"); it has the property that any two lines are concurrent. (See chapter 14.) $\mathrm{PG}(2,q)$ is a projective plane of order q; but there are others, as we shall see. It can alternatively be defined as a square design with $\lambda = 1$, or as a connected matroid of rank 3 in which any two lines meet.

Theorem 2.3. *Let* $\mathcal{P}$ *be a finite set of points, and* $\mathcal{L}$ *a collection of subsets of* $\mathcal{P}$ *called* lines. *Suppose that the following conditions hold:*

(a) *any line contains at least three points;*

(b) *any two points lie on a unique line;*

(c) *if a line meets two sides of a triangle, not at their intersection, then it meets the third side also.*

Then exactly one of the following conclusions holds:

(i) $\mathcal{P} = \mathcal{L} = \emptyset$;

(ii) $|\mathcal{P}| = 1$, $\mathcal{L} = \emptyset$;

(iii) $|\mathcal{P}| \geqslant 3$, $\mathcal{L} = \{\mathcal{P}\}$;

(iv) $(\mathcal{P}, \mathcal{L})$ *is a projective plane;*

(v) $\mathcal{P}$ *and* $\mathcal{L}$ *can be identified with the point and line sets of* $\mathrm{PG}(n,q)$ *for some* $n \geqslant 3$ *and some prime power* q.

Axiom (c) above is sometimes called the *Veblen–Young axiom* or *Veblen's axiom* or (with less justification) *Pasch's axiom*. An equivalent formulation is that any three non-collinear points are contained in a subspace which is a projective plane.

Note that the whole geometry can be recovered from the points and lines; for a set of points is a variety if and only if it contains the line through any two of its points.

Condition (a) of Theorem 2.3 can be weakened. Given a collection of *linear spaces* (point-line geometries satisfying condition (b)), we define their direct sum as follows. The point set is the disjoint union of the point sets of its summands; every line of every summand, and every set consisting of two points from different summands, is a line of the sum. Then the sum is a linear space. Its varieties are all the sets of points which contain the line through any two of their points. (This is just the matroid direct sum defined in chapter 9.) It is now trivial to show the following.

Proposition 2.4. *A point-line geometry satisfying* (a)–(c) *of Theorem* 2.3, *with "three" replaced by "two" in* (a), *is a direct sum of a collection of geometries each satisfying the conclusion of that theorem.*

For example, given a family of (at least two) subsets (called lines) of a finite set, each line containing at least two points, such that any two points lie on a unique line and any two lines meet in a unique point, then the family is a projective plane, or the direct sum of a point and a line, or a triangle (the direct sum of three points). This is the *De Bruijn–Erdős* (1948) theorem (see chapter 14).

The geometries described by Proposition 2.4 are sometimes called "generalized projective geometries". Another terminology is to call a geometry satisfying Theorem 2.3(a) *thick*, and one satisfying the weakened form in Proposition 2.4 *firm*. This terminology will be extended later.

Another point of view is to regard the projective geometry as a partially ordered set: the ordering is just inclusion. This poset is a lattice.

Birkhoff (1935) characterized the generalized projective geometries in lattice-theoretic terms:

Theorem 2.5. *A finite lattice is the lattice of varieties of a generalized projective geometry if and only if it is atomic and modular.*

(An *atom* in a lattice is an element covering 0; the lattice is *atomic* if every element is a join of finitely many atoms. A lattice is *modular* if $a \vee (b \wedge c) = (a \vee b) \wedge c$ holds whenever $a \leqslant c$.) An equivalent formulation is as follows.

Theorem 2.6. *A geometric (i.e. simple) matroid is a generalized projective geometry if and only if its rank function ρ is* modular, *i.e.*

$$\rho(a \wedge b) + \rho(a \vee b) = \rho(a) + \rho(b)$$

for any flats a, b.

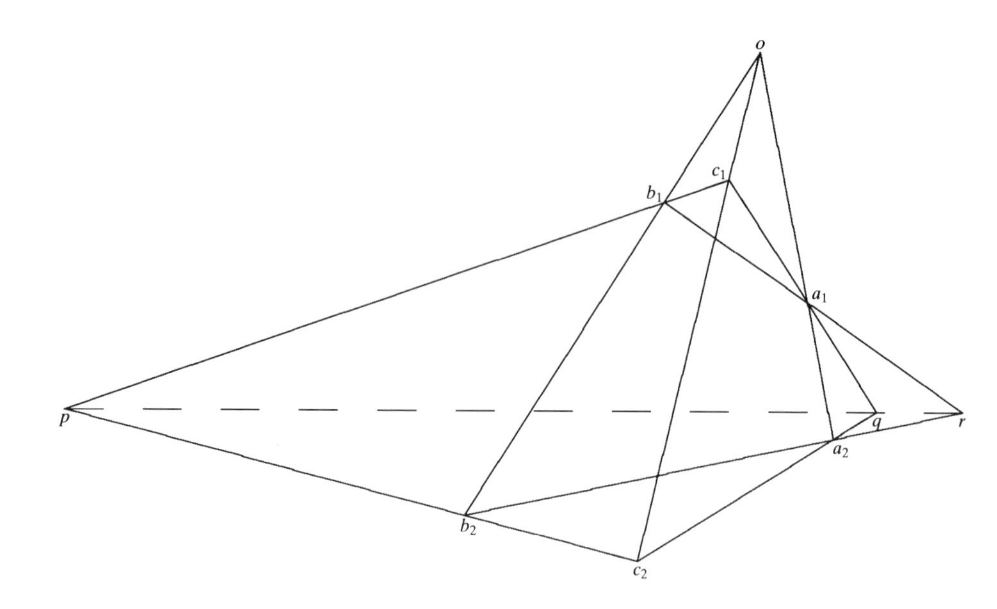

Figure 2.1. The Desargues' configuration.

The next result characterizes $\mathrm{PG}(2,q)$ among projective planes.

Theorem 2.7. *For a projective plane Π, the following are equivalent:*
(i) *Π is isomorphic to* $\mathrm{PG}(2,q)$ *for some prime power q;*
(ii) *Π satisfies Desargues' theorem (fig.* 2.1*);*
(iii) *Π satisfies Pappus' theorem (fig.* 2.2*).*

Here and subsequently, a configuration theorem described by a geometric figure is the assertion that, if all the points and all but one of the lines with the required incidences are given, then the last line exists also. The figure may degenerate, i.e. there may be additional incidences which are not among those specified.

The status of Theorem 2.7 for possibly infinite projective planes is interesting. A geometric argument shows that Pappus' theorem implies Desargues'. (See Pickert 1955, for example.) Furthermore, Desargues' theorem is equivalent to the plane's being coordinatized by a division ring, and Pappus' theorem to the commutativity of this division ring. (So, for example, the projective plane over the real quaternions satisfies Desargues' theorem but not Pappus'.) Thus, the fact that Desargues' theorem implies Pappus' theorem for finite projective planes is equivalent to Wedderburn's theorem. No purely geometric proof of this implication is known.

This result is used in the proof of Theorem 2.3. A geometric argument shows that any projective space of dimension at least three satisfies Desargues' theorem. (If we view fig. 2.1 as the projection of a 3-dimensional configuration, we see that

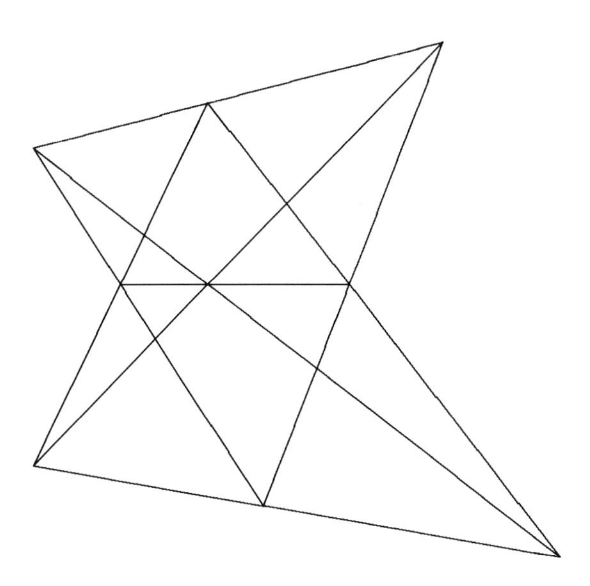

Figure 2.2. Pappus' configuration.

the theorem must hold in the "generic" case in which the configuration does not lie in a plane.) Then Theorem 2.7 shows that the planes can be coordinatized, and these coordinatizations can be glued together to prove the theorem.

A projective space is determined by the design of proper subspaces of any fixed positive dimension. For the line through two points is the intersection of all the subspaces containing them; and then the other varieties are characterized as before.

More generally, in any 2-design, the *line* through any two points is defined to be the intersection of all blocks containing those points. The points and lines then form a linear space (that is to say, two points lie on a unique line); and any block is a linear subspace. The lines do not, in general, all have the same cardinality.

Dembowski and Wagner (1960) characterized the point-hyperplane designs of projective spaces in terms of these concepts. A *triangle* will mean a set of three non-collinear points.

Theorem 2.8. *The following conditions are equivalent for a square design* $\mathcal{D}$ *with* $\lambda > 1$.

(i) $\mathcal{D}$ *is the point-hyperplane design of* $\mathrm{PG}(n, q)$ *for some* $n \geqslant 2$ *and some prime power* q;

(ii) *every line meets every block*;

(iii) *the number of blocks containing a triangle is constant*;

(iv) *the automorphism group of* $\mathcal{D}$ *acts transitively on (ordered) triangles.*

Condition (ii) of the Dembowski–Wagner theorem is often expressed in dual form: if E is the intersection of two blocks, and p a point not in E, then there is a (unique) block containing E and p. This condition implies that the points and blocks are the points and hyperplanes of a simple matroid; since there are equally many points and blocks, the matroid is modular, and hence a projective space (since it is easily seen to be connected). (cf. chapter 9). Notice that the full force of the fact that $\mathcal{D}$ is a design is not needed.

It is also possible to characterize projective spaces under a weakening of the group-theoretic hypothesis (iv). For projective planes, Ostrom and Wagner (1959) showed that 2-transitivity on points characterizes $\mathrm{PG}(2,q)$; and since the classification of finite simple groups was completed, all 2-transitive square designs have been known (cf. chapter 12).

The *affine geometry* $\mathrm{AG}(n,q)$ can be defined in either of two ways:

(1) Remove a hyperplane from $\mathrm{PG}(n,q)$ (that is, remove its points from all varieties, and remove those varieties which are contained in the hyperplane).

(2) Take an n-dimensional vector space V over $\mathrm{GF}(q)$; the points of the affine space are the vectors, and the k-flats are all the cosets of k-dimensional vector subspaces.

To match up the descriptions, let $x_0,\ldots,x_n$ be coordinates for the $(n+1)$-dimensional vector space from which $\mathrm{PG}(n,q)$ is built. If the hyperplane $x_0 = 0$ is removed, every point (1-dimensional subspace) remaining contains a unique vector with $x_0 = 1$, and the bijection

$$\langle(1,x_1,\ldots,x_n)\rangle \mapsto (x_1,\ldots,x_n)$$

between the point sets is an isomorphism between the two descriptions.

All the results described for projective spaces have analogues for affine spaces. For example, we have the following.

Proposition 2.9. *The number of k-flats in* $\mathrm{AG}(n,q)$ *is* $q^{n-k}\begin{bmatrix} n \\ k \end{bmatrix}_q$.

Proposition 2.10. *The points and k-flats of* $\mathrm{AG}(n,q)$ *form a* 2-$(q^n, q^k, \begin{bmatrix} n-1 \\ k-1 \end{bmatrix}_q)$ *design. If* $q = 2$, *it is even a* 3-*design.*

An *affine plane* is a 2-$(n^2,n,1)$ design, for some $n > 1$; in other words, an affine design with $\lambda = 1$. (See chapter 14.)

Theorem 2.11 (Buekenhout 1969). *A non-trivial linear space with more than three points on some line is either an affine plane or the point-line design of* $\mathrm{AG}(n,q)$ *for some n and q if and only if every triangle is contained in an affine plane.*

The conclusion is trivially false if lines have two points (then the structure is trivial — every point pair is a line). Hall (1967) first gave a counterexample with three points on every line. Such objects are Steiner triple systems in which every triangle lies in a 9-point subsystem. They are closely connected with commutative Moufang loops of exponent 3 (see Bruck 1958), and have been studied by Young (1973), Bénéteau (1986), Fischer (1964), Hall (1974), and others.

Theorem 2.12 (Dembowski 1964b). *For an affine design $\mathscr{D}$ with $\lambda > 1$, the following are equivalent:*

(i) *$\mathscr{D}$ is either the design of points and hyperplanes of* $\mathrm{AG}(n,q)$ *for some $n \geqslant 3$ and prime power q, or a Hadamard* 3*-design*;

(ii) *every line meets every block*;

(iii) *the number of blocks containing a triangle is constant.*

Kimberley (1971) has studied Hadamard 3-designs with a view to quantifying the extent to which they resemble affine spaces over GF(2).

For $q = 2$, every pair of points is a line, and the geometry cannot be recovered from its lines alone. However, lines and *parallelism* suffice for any q, where parallelism is an equivalence relation on lines satisfying *Euclid's parallel postulate*: each equivalence class contains a unique line through every point. See Lenz (1954) for axiomatisations, and Cameron (1976) for more on the case $q = 2$. Note that any affine plane admits a parallelism. Hence, under the hypotheses of Buekenhout's Theorem 2.11, two lines can be defined to be parallel if they are contained in an affine plane (and parallel there); the difficulty is to show that this relation is transitive.

3. Collineation groups

The so-called *Fundamental Theorem of Projective Geometry* is a description of the collineations of $\mathrm{PG}(n,q)$. A *collineation* is a permutation of the set of varieties which preserves the dimension and the relation of incidence. Any non-singular linear transformation of V acts in a natural way on the set of subspaces of V, and clearly induces a collineation. The set of such collineations is a group, the *general linear group* $\mathrm{GL}(n,q)$. If we coordinatize V as $\mathrm{GF}(q)^n$, then the group of automorphisms of $\mathrm{GF}(q)$ also acts coordinate-wise as a collineation group. (This group is cyclic of order e, where $q = p^e$, p prime.) The product of these two groups (which is their semi-direct product) is denoted by $\Gamma\mathrm{L}(n,q)$; its elements are called *semilinear transformations*. The kernel of the action of this group on the projective space is easily seen to be the group Z of non-zero scalar transformations. We let $\mathrm{P\Gamma L}(n,q)$ denote the quotient, which is isomorphic to the group of collineations induced by $\Gamma\mathrm{L}(n,q)$.

Theorem 3.1 (Fundamental Theorem of Projective Geometry). *For $n \geqslant 2$, any collineation of* $\mathrm{PG}(n,q)$ *is induced by a semilinear transformation of the underlying vector space; in other words, the full collineation group is* $\mathrm{P\Gamma L}(n,q)$.

Theorem 3.2. *For $n \geqslant 2$, the collineation group of* $\mathrm{AG}(n,q)$ *is the semidirect product of the translation group of V by the group* $\Gamma\mathrm{L}(n,q)$ *of semilinear transformations.*

(This group is denoted by $\mathrm{A\Gamma L}(n,q)$.)

A projective line, as we have defined it, has no structure; its collineation group is the symmetric group. It is possible to give structure to a projective line in various ways, either by using cross-ratio or as an algebraic curve; but I will not discuss this.

Theorem 3.3 (Singer's theorem). *There is a collineation of* $\mathrm{PG}(n,q)$ *which permutes the points in a single cycle.*

This result is trivial to prove — if we regard the underlying vector space as having the structure of $\mathrm{GF}(q^{n+1})$, the required collineation is induced by multiplication by a primitive element of the field — but is the origin of the theory of difference sets; see chapter 14.

There is a large body of work on collineation groups of projective spaces. The assumptions made on the groups fall into three classes: generation, transitivity, and low dimension.

The first type of problem concerns collineation groups generated by particular kinds of elements, such as transvections (whose fixed points form a hyperplane, corresponding to linear transformations of the form $I+E$, where E has rank 1 and $E^2=0$), or reflections (induced by diagonalizable transformations with all but one eigenvalue equal to 1). Some of this material dates from the beginning of this century. A sample result is McLaughlin's (1967) determination of irreducible groups generated by full transvection subgroups. See Kantor (1982) for a survey.

Typical of the second type of problem is a result of Cameron and Kantor (1979) and Orchel (1978).

Theorem 3.4. *Let G be a group of collineations of* $\mathrm{PG}(n,q)$ *($n \geqslant 2$) which acts 2-transitively on points. Then either G contains* $\mathrm{PSL}(n+1,q)$, *or $q=2$, $n=3$, and G is isomorphic to the alternating group A_7.*

Of course, this result is now superseded by the classification of all finite 2-transitive permutation groups; nevertheless, its proof is elementary and involves some nice geometry. In fact, the classification of the 2-transitive groups with elementary abelian regular normal subgroups by Huppert (1957) and Hering (1974, 1985) is equivalent to a determination of all collineation groups which are transitive on points. (These are the groups $H=G_0$ listed in table 5.1(a) of chapter 12.) Many similar problems are resolved by the classification of finite simple groups.

Block's lemma (see chapters 14, 31) implies that a collineation group of $\mathrm{PG}(n,q)$ ($n \geqslant 2$) has at least as many orbits on lines as on points. When does equality hold? This problem was studied by Cameron and Liebler (1982) and Penttila (1984), but is not yet settled.

The classification of subgroups of low-dimensional projective groups is almost as old as group theory, having been considered by Galois in his letter to Chevalier in 1832. Dickson (1901) gave a list of all the subgroups of $\mathrm{PSL}(2,q)$. Throughout this century, work continued on finding all (or at least all maximal) subgroups of $\mathrm{PSL}(n,q)$ for small n. This work is part of the general attempt to find all maximal subgroups of "almost simple" groups. The relevance of this can be seen in part from the O'Nan–Scott theorem (chapter 12), which gives a good description of those primitive permutation groups which are not almost simple.

An important theorem of Aschbacher (1984) is the analogue of the O'Nan–Scott theorem (in its original form concerning maximal subgroups of symmetric groups) for collineation groups. Aschbacher defines certain classes of large subgroups: reducible (fixing a subspace); imprimitive (fixing a direct sum decomposition); the stabilizer of a tensor product decomposition; the projective group over a subfield of $\mathrm{GF}(q)$; a group of smaller dimension over an extension field of $\mathrm{GF}(q)$; a classical group (the stabiliser of a non-degenerate sesquilinear or quadratic form); or the normalizer of an irreducible extraspecial group. His theorem asserts that a maximal subgroup of $\mathrm{PSL}(n,q)$ which is not in any of these classes is almost simple and absolutely irreducible. There is an analogous result for the other classical groups. In addition, we expect that these "exceptional" subgroups will be small. Liebeck (1985) gives an upper bound of q^{3n} for the order of such a subgroup, and improved bounds with "known" exceptions have been obtained subsequently. A typical example is the occurrence of A_5 as a maximal subgroup of $\mathrm{PSL}(2,p)$ whenever p is a prime congruent to $\pm 1 \pmod 5$. These matters, and much more detail about subgroups of classical (and other) simple groups, are discussed by Kleidman and Liebeck (1988, 1990).

4. Galois geometry

To Euclid, the plane and space in which he worked were familiar objects. His axioms were not intended to characterize these structures, but as self-evident truths on which to base propositions about configurations in the plane or space. Of course, since Descartes, algebraic or analytic proofs have become more common than Euclid's geometric ones. In a similar way, the classical results of section 2 (due to Hilbert, Veblen and Young, Birkhoff, etc.) have made $\mathrm{PG}(n,q)$ a familiar object. The study of configurations in $\mathrm{PG}(n,q)$ is largely the creation of B. Segre (1961). Part of his legacy is the pre-eminence of Italian mathematicians in this area.

The study of $\mathrm{PG}(n,q)$ has two aspects, not always sharply differentiated: the study of configurations within the space, for example by their intersections with subspaces; and the study of intrinsic properties of subgeometries (including, for example, conditions for representability of matroids).

Before beginning, I draw the reader's attention to Hirschfeld (1979, 1985), Hirschfeld and Thas (1991) a very detailed account of material which can at most be sketched here.

A *cap* is a set of points in $\mathrm{PG}(n,q)$ with no three collinear. (For $n=2$, a cap is called an *arc*.) The general problem here, which is yet unsolved, is to determine the size of the largest cap in $\mathrm{PG}(n,q)$. Consider first the case $n=2$. The lines joining one point on an arc to the others are all distinct; since a point lies on $q+1$ lines, there are at most $q+2$ points in an arc. If this bound is attained, then any line which meets the arc does so in two points; then the arc is partitioned into sets of size 2 by the lines through any outside point, and so q must be even. Moreover, if q is even, then a $(q+1)$-arc has a *knot* or *nucleus*, each line through which meets the arc once; adjoining this point gives a $(q+2)$-arc. (This paragraph applies to any projective plane of order q.)

A *conic* in $\mathrm{PG}(2,q)$ is a set of points equivalent under $\mathrm{PGL}(3,q)$ to

$$\{\langle(x,y,z)\rangle : xz = y^2\} = \{\langle(1,t,t^2)\rangle : t \in \mathrm{GF}(q)\} \cup \langle(0,0,1)\rangle\}.$$

It is easily seen to be a $(q+1)$-arc. If q is even, its nucleus is $\langle(0,1,0)\rangle$. Hence, the following holds.

Proposition 4.1. *The maximum size of an arc in* $\mathrm{PG}(2,q)$ *is*

$$\begin{cases} q+1, & \text{if } q \text{ is odd;} \\ q+2, & \text{if } q \text{ is even.} \end{cases}$$

An arc containing this bound is called an *oval.* (Sometimes the term “oval” is used for a $q+1$-arc, a $q+2$-arc being called a *hyperoval.*)

A celebrated theorem of Segre (1955) describes all ovals in the case where q is odd. It was proved as a result of a speculation in cosmology (Järnefelt 1951).

Theorem 4.2 (Segre’s theorem)**.** *Any oval in* $\mathrm{PG}(2,q)$, *for* q *odd, is a conic.*

No such result is known for even q. For $q \leqslant 8$, any oval consists of a conic and its nucleus; but this is false for $q = 16$ (Lunelli and Sce 1958), and no classification is known without strong additional hypotheses. An example of a “non-classical” oval is the set

$$\{\langle(1,t^m,t^{m+1})\rangle\} \cup \{\langle(0,0,1)\rangle\}$$

in $\mathrm{PG}(2,2^e)$, where $m = 2^d$, with $1 < d < e$. (Some of these arcs are equivalent to conic plus nucleus; but for $e = 5$ or $e \geqslant 7$, there are some which are not.)

This leaves the problem of arcs maximal under inclusion but not of maximum cardinality. A complete classification is too much to hope for; so define functions f, g such that

(i) an arc of cardinality less than $f(q)$ is not maximal (and $f(q)$ is the smallest number with this property);

(ii) an arc of cardinality greater than $g(q)$ is contained in an oval (and $g(q)$ is the greatest number with this property).

These functions have been much studied. The next theorem surveys some of the results.

Theorem 4.3. (i) $cq^{1/2} \leqslant f(q) \leqslant cq^{3/4}$.

(ii)

$$g(q) \leqslant \begin{cases} q - q^{1/2} + 1, & \text{if } q \text{ is even;} \\ q - \frac{1}{4}q^{1/2}, & \text{if } q \text{ is even;} \\ \frac{44}{45}q + 1, & \text{if } q \text{ is prime.} \end{cases}$$

(iii)

$$g(q) \geqslant \begin{cases} \frac{1}{2}q + q^{1/2}, & \textit{in general;} \\ q - q^{1/2} + 1, & \textit{if } q \textit{ is a square.} \end{cases}$$

The lower bound in (i) is trivial: an arc with fewer than $(2q)^{1/2}$ points has at most q secants, which cannot cover all the points of the plane. The rest is a compilation of results by Boros, Fisher, Hirschfeld, Segre, Szőnyi, Thas, Voloch and Zirilli. Note that (iii) depends in part on a result of Schoof and Waterhouse on the tightness of Weil's theorem (see chapter 32). See also Thas (1983) for another proof of Segre's theorem.

More generally, given a set S of $q+1$ points in $\mathrm{PG}(2,q)$, a point p is called a *nucleus* of S if every line through p meets S. (Note that if S is a line, then every point off S is a nucleus.) Some recent results characterize various sets by the number or structure of their set of nuclei. For example, Blokhuis and Wilbrink (1987) showed the following.

Proposition 4.4. *If S is not a line, then it has at most q 1 nuclei.*

The largest cap in $\mathrm{PG}(3,q)$ has size q^2+1 (Barlotti 1955). A cap of this size is called an *ovoid*. Ovoids have such a rich structure theory that I will digress to describe them further.

Let O be an ovoid. Then, for any point $p \in O$, the lines tangent to O at p (i.e. meeting O only in p) form a plane, the *tangent plane* at p. Any other plane intersects O in a set of $q+1$ points forming an arc in that plane. These arcs are called *circles*. Thus, any three points lie in a unique circle, and the points and circles form a 3-$(q^2+1, q+1, 1)$ design.

Any 3-$(n^2+1, n+1, 1)$ design is called an *inversive plane* or *Möbius plane*. An inversive plane which arises from an ovoid is called *egglike*.

The classical example of an ovoid is the *elliptic quadric*, the set

$$\{\langle(x_0, x_1, x_2, x_3)\rangle : x_0x_1 + f(x_2, x_3) = 0\},$$

where f is an irreducible quadratic form in two variables over $\mathrm{GF}(q)$.

Theorem 4.5. (i) *An inversive plane is egglike if and only if it satisfies the Bundle theorem (fig.* 4.1*) (Kahn).*

(ii) *An inversive plane arises from an elliptic quadric if and only if it satisfies Miquel's theorem (fig.* 4.2*) (Van der Waerden–Smid).*

(iii) *Every inversive plane of even order is egglike (Dembowski).*

See Dembowski (1964a, 1968) and Kahn (1980). Note that the Bundle theorem is the assertion that the rank 4 matroid associated with the design (whose hyperplanes are the circles) does not contain the Vamos matroid as a minor: see chapter 9.

Apart from the elliptic quadrics, the only known ovoids are the *Suzuki–Tits*

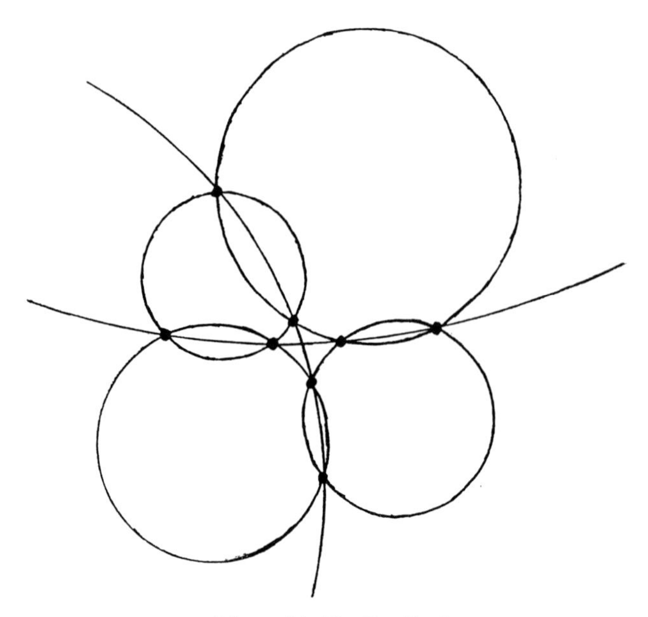

Figure 4.1. The Bundle theorem.

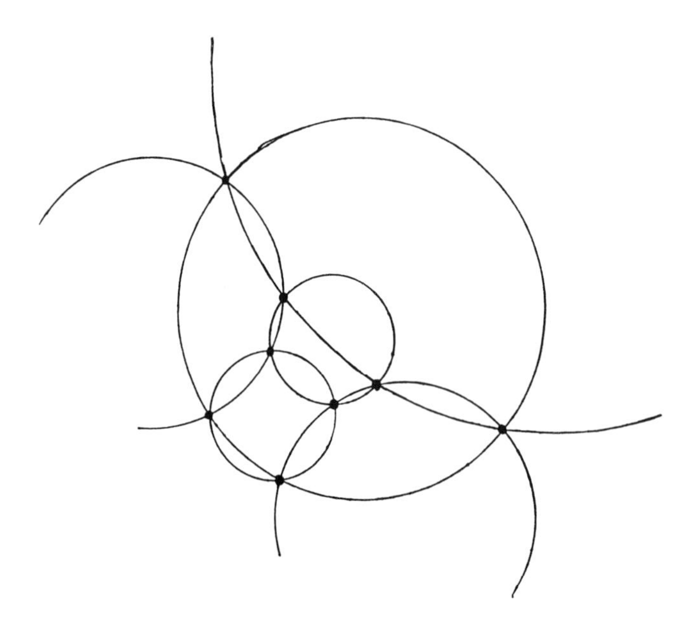

Figure 4.2. Miquel's theorem.

ovoids, which live in PG$(3,q)$ ($q=2^{2d+1}$, $d\geqslant 1$) and admit the Suzuki groups as automorphism groups (see chapter 12). Moreover, no non-egglike inversive plane is known. There are results asserting that an inversive plane admitting a sufficiently large collineation group must be known; see Lüneburg (1965), for example.

In the same way that Möbius planes model conics on an elliptic quadric, the other quadrics in 3-space (the hyperbolic or ruled quadric, and the cone) are modelled by *Minkowski* and *Laguerre planes*. These have also been intensively studied, and there is a unified treatment of matters related to Theorem 4.5 (Kahn 1980). More specifically, the classification of *flocks* (sets of pairwise disjoint circles in Möbius, Minkowski, and Laguerre planes, or of plane sections of quadrics in 3-space) has now been essentially completed (see Bader and Lunardon 1989, Thas 1992, for example). This work has many applications to inversive planes, translation planes, "maximum exterior sets" of quadrics, etc.

The largest cap in PG$(n,2)$ is easily seen to be the complement of a hyperplane. Apart from this, only isolated results are known. For example, the largest caps in PG$(4,3)$ and PG$(5,3)$ have size 20 and 56 respectively (Pellegrino 1971, Hill 1978). The unique cap in the latter case contains half of the points of a quadric and admits the simple group PSL$(3,4)$.

A related problem (identical in the plane) is to determine the largest set of points in PG(n,q) with the property that no $n+1$ are contained in a hyperplane. Such a set is called an *arc*. Two examples of arcs are:

(a) the *simplex*, the points spanned by the basis vectors and the all-1 vector, with cardinality $n+2$;

(b) the *normal rational curve* consisting of the points spanned by the vectors $(1,t,t^2,\ldots,t^n)$ for $t\in$ GF(q) and $(0,0,\ldots,1)$, with cardinality $q+1$.

These examples show that the answer will depend on the relative magnitudes of n and q. The existence of an arc with k points is equivalent to the representability of the uniform matroid U^k_{n+1} over GF(q). For $n\geqslant q-1$ every maximal arc is a simplex; this follows from the observation that if an arc with $n+3$ points exists, then the matroid U^{n+3}_{n+2} and so its dual U^{n+3}_{2} are representable over GF(q), and hence $n+3\leqslant q+1$.

Arcs are closely connected with maximum distance separable codes. (This is an instance of a general connection between projective geometry and coding theory which is discussed in section 5 and in chapter 16.) See Bruen et al. (1988).

Another variant asks for sets meeting any line in at most k points, for some k. This question is less well developed, except in the plane. The analogue of a $q+2$-arc is a set which meets every line in 0 or k points. Such a set has cardinality $qk-q+k$. A trivial necessary condition is that k divides q (proved as for the evenness of q when $q+2$-arcs exist. These sets have been given the misnomer "maximal arcs"; possibly this is because a set meeting every line in at most k points has cardinality at most $qk-q+k$, with equality if and only if it is a "maximal arc".

Proposition 4.6. (i) *A "maximal arc" exists whenever k and q are powers of* 2 *and* $q\geqslant k$.

(ii) *There is no "maximal arc" with* $k=3$, $q>3$ (*or, dually, with* $k=q/3>1$).

These results are due to Denniston (1969a) and Thas (1975) respectively.

For the maximum of generality, let h and k be integers and L a set of integers. Which sets S of h-flats of $\mathrm{PG}(n,q)$ have the property that, for any k-flat E, the number of h-flats contained in E which are members of S belongs to L? This is so general that it is impossible to survey the results; but see Hirschfeld (1985).

Specialising slightly, a set of points is called an *h-blocking set* if it meets every h-flat but contains none. The term (its original form was "blocking coalition") comes from game theory, in which the original work was done (see Di Paola 1969). (Roughly: if the h-flats are the minimal winning coalitions in a game, then a blocking coalition is one which can prevent any other group from winning, but cannot itself win.) The subject has been revitalized by connections with coding theory, to be discussed in section 5. For its connections to matroid theory, see the treatment of t-blocks in chapter 9. A 1-blocking (i.e. line-blocking) set in $\mathrm{PG}(2,q)$ is simply called a *blocking set*.

Proposition 4.7. (i) *A blocking set in* $\mathrm{PG}(2,q)$ *has cardinality at least* $q+q^{1/2}+1$, *with equality if and only if it is a* Baer subplane, *the set of points with coordinates in* $\mathrm{GF}(q^{1/2})$.

(ii) *A blocking set in* $\mathrm{PG}(2,q)$ *which is minimal (under inclusion) has cardinality at most* $q^{3/2}+1$.

These results are due to Bruen (1971a), Bruen and Thas (1977).

The lower bound is met in any Desarguesian plane of square order; but it has been improved to $q+(2q)^{1/2}+1$ for sufficiently large odd non-square q by Blokhuis and Brouwer (1986) (for a proof, see chapter 32). An example meeting the upper bound is given by a *hermitian unital*, the set of zeros of the form

$$x_0\bar{x}_2+x_1\bar{x}_1+x_2\bar{x}_0,$$

where $\bar{x}=x^{q^{1/2}}$, the image of x under the field automorphism of order 2. Note that the two extreme cases in the theorem are characterized by the fact that any line meets the blocking set in 1 or k points for some k (and in fact $k=q^{1/2}+1$).

For affine planes, the value is much larger (Jamison 1977), see also Brouwer and Schrijver 1977)

Proposition 4.8. *A blocking set in* $\mathrm{AG}(2,q)$ *has cardinality at least* $2q-1$.

For a proof, see chapter 32.

There are analogues for larger values of h, and for other situations. For an up-to-date account of blocking sets, see Jungnickel (1989).

A version of Ramsey's theorem was proved for projective spaces by Graham et al. (1972).

Theorem 4.9. *Let integers k, l, r and a prime power q be given. Then there is an integer n such that, if the k-flats of* $\mathrm{PG}(n,q)$ *are partitioned into r classes, then there is an l-flat, all of whose k-flats are contained in the same class.*

For further discussion of this important theorem and its context, see chapter 25. Here it has the following corollary.

Corollary 4.10. *For fixed l and n, there exists q_0 such that* PG(n,q) *possesses l-blocking sets only if $q \geqslant q_0$.*

For, if a blocking set exists, partition the points into the blocking set and its complement; there is no monochromatic l-flat. Lower and upper bounds for q_0 are known (Mazzocca and Tallini 1985, Cameron and Mazzocca 1986), but there is a wide gap between them!

An important theorem from algebraic geometry, which concerns finite geometry but has not been exploited to the full by finite geometers, is the Hasse–Weil theorem, see chapter 32.

Turning now to intrinsic characterisations, the most important class of configurations consists of polar spaces. A *duality* or *correlation* of PG(n,q) is a lattice isomorphism from PG(n,q) to its dual; it is a *polarity* if it has period 2. The Fundamental Theorem of Projective Geometry implies that any duality arises from a *sesquilinear form* on V, a function of two variables which is linear in the first variable and semilinear in the second. Further analysis shows that polarities arise from forms f which are either *σ-hermitian*, for some field automorphism σ whose square is the identity (this means that $f(v,w) = f(w,v)^{\sigma}$), or *alternating* (this means that $f(v,v) = 0$ for all v). Note that σ may be the identity, in which case a σ-hermitian form is symmetric. A variety is *totally isotropic* for a polarity if it is contained in its image; equivalently, if the form f vanishes identically on it. The *polar space* associated with a polarity is the geometry consisting of the totally isotropic varieties.

In odd characteristic, if f is a symmetric bilinear form, then $Q(v) = f(v,v)$ defines a *quadratic form*, and the totally isotropic varieties are just those which are *totally singular* for Q, i.e. on which Q vanishes. Moreover, f can be recovered from Q by *polarization*, i.e.

$$f(v,w) = \tfrac{1}{2}(Q(v+w) - Q(v) - Q(w)).$$

In even characteristic, this is no longer true: a quadratic form polarises to an alternating, not a symmetric, form. Since the geometry of totally singular varieties of a non-degenerate quadratic form possesses properties similar to those of polar spaces of polarities, it has been found best to extend the definition to include them, and to dispense with the symmetric bilinear forms altogether.

Thus there are three types of polar spaces, which are commonly named after the associated classical groups (see chapter 12):

- *unitary*, from σ-hermitian forms with $\sigma \neq 1$;
- *symplectic*, from alternating bilinear forms;
- *orthogonal*, from quadratic forms.

The *rank* of a polar space is defined to be one more than the projective dimension of the largest variety it contains. (In the orthogonal case, this is the Witt index of the quadratic form.) The major result here is due to Tits (1974), extending earlier work by Veldkamp (1959):

Theorem 4.11. *For $r \geqslant 3$, finite polar spaces of rank r are characterized by the following axioms:*

(a) *Any variety, together with the varieties it contains, is a projective space of dimension at most $r-1$.*

(b) *The intersection of any set of varieties is a variety.*

(c) *If E is an $(r-1)$-flat and p a point not in E, then the set of points of E joined to p by lines of the polar space is a hyperplane of E, and the union of these lines is an $(r-1)$-flat.*

(d) *There exist two disjoint $(r-1)$-flats.*

Tits' theorem is really more general; the infinite geometries satisfying the axioms are classified as well. There are analogues of the finite ones, as well as some new types. See Tits (1974) for details.

Tits' axioms in Theorem 4.11 were considerably simplified by Buekenhout and Shult (1974) as follows. In a point-line geometry, we call two points *adjacent* if they are incident with a common line. A *subspace* is a set of points containing every line through any two of its points; it is *singular* if any two of its points are adjacent.

Theorem 4.12 (Buekenhout–Shult theorem). *Suppose that a finite point-line geometry satisfies the following conditions:*

(a) *If a point p is not on a line L, then p is adjacent to either one or all points of L.*

(b) *Every line is incident with at least three points, and no two lines are incident with the same set of points.*

(c) *No point is adjacent to all other points.*

Then the points and singular subspaces satisfy Tits' axioms in Theorem 4.11 *for a polar space.*

Notes: It follows from the axioms of Buekenhout and Shult (though it is by no means obvious) that two points lie on at most one line. Hence every line is a singular subspace! Again, finiteness of the geometry is not really needed; it suffices that chains of singular subspaces should be finite.

Many numerical properties of the polar spaces can be simply expressed. For example, the following holds.

Proposition 4.13. *The number of points of a polar space of rank r over* GF(q) *is*

$$(q^r - 1)(q^{r+\varepsilon} + 1)/(q-1),$$

and the number of $r-1$-flats is

$$\prod_{k=1}^{r}(q^{k+\varepsilon} + 1),$$

where ε is given in table 4.1.

Table 4.1
Polar spaces

Type	dim(V)	ε
Symplectic	$2r$	0
Orthogonal	$2r$	-1
Orthogonal	$2r+1$	0
Orthogonal	$2r+2$	1
Unitary	$2r$	$-\frac{1}{2}$
Unitary	$2r+1$	$\frac{1}{2}$

The quadrics corresponding to polar spaces with $\varepsilon = 1, -1$ are sometimes called *elliptic* and *hyperbolic* respectively. The formulae are also valid for $r = 1, 2$; we have seen examples with $r = 1$ (conics, elliptic quadrics in projective 3-space, and hermitian unitals in the plane). Note that every point of the projective space is a point of the symplectic polar space. Note also that, in even characteristic, the symplectic and orthogonal polar spaces with $\varepsilon = 0$ are isomorphic.

There are also characterisations of polar spaces in terms of their embeddings. For example, Tallini–Scafati (1967) showed the following.

Theorem 4.14. *Let q be an odd square, and $n > 2$. Then a set of points in* PG(n,q) *which meets every line in* 1, k *or* $q+1$ *points is the point set of a unitary polar space, with* $k = q^{1/2} + 1$.

A characterisation of sets satisfying the same condition with q even and $q > 4$ is given by Hirschfeld and Thas (1980a,b). As well as hermitian varieties, certain projections of quadrics can arise.

The result of (4.11) is similar to the axiomatisation of projective spaces in Theorem 2.3, in that low-dimensional exceptions occur. In the case $r = 1$, a polar space consists of points only. For $r = 2$, however, it is more interesting. The hypotheses of Theorem 4.11 assert that we have a geometry of points and lines, with two points on at most one line; if a point p is not on a line L, then p is collinear with exactly one point of L; and no point is collinear with all others. A geometry satisfying these conditions is called a *generalized quadrangle* or *GQ*. There will be more about these geometries in general later. The important result in the present context is the theorem of Buekenhout and Lefèvre (1974).

Theorem 4.15. *If a generalized quadrangle $\mathcal{Q}$ is embedded as a spanning subset of a projective space $\mathcal{P}$, so that every point of $\mathcal{P}$ on a line of $\mathcal{Q}$ is a point of $\mathcal{Q}$, then $\mathcal{Q}$ is a classical polar space in $\mathcal{P}$.*

Note that both the isomorphism type of $\mathcal{Q}$ and its embedding in $\mathcal{P}$ are determined.

The configuration of h-flats in PG(n,q) has intrinsic characterisations. We can structure it in two possible ways. One is to take it as the point set of an incidence structure whose blocks are the $(h+1)$-flats. The other is to define lines to be sets of the form

$$\{X : U \subset X \subset W\},$$

where U is an $(h-1)$-flat and W an $(h+1)$-flat with $U \subset W$; in this case, a line has $q+1$ points, and two points lie on at most one line. By taking the $(h+1)$st exterior power of the underlying vector space, the set of h-flats of $\mathrm{PG}(n,q)$ is embedded in the point set of $\mathrm{PG}(\binom{n+1}{h+1}-1,q)$; the lines contained within this set are precisely the lines described above. See Sprague (1985), Tallini (1981).

Another class of embedding theorems is due to Kantor (1974, 1975) and Percsy (1981). Kantor's theorems assert that, if a matroid has the property that all the "top intervals" of dimension at most 3 are embeddable as sufficiently large parts of projective spaces, then the whole matroid is so embeddable. The canonical application is the following.

Proposition 4.16. *A* 4-$(q^2+2, q+2, 1)$ *design exists only for* $q=2$, 3 *or possibly* 13.

For standard divisibility arguments restrict q to finitely many values of which all except 3 and 13 are even. If q is even, then the derived design (an inversive plane) is egglike, by Dembowski's theorem 4.5(iii); by Kantor's theorem, the whole design is embeddable in $\mathrm{PG}(4,q)$, and then an easy counting argument shows that $q=2$. (The divisibility conditions alone do not preclude the existence of a 12-(74, 18, 1) design!)

The rank bound in Kantor's theorem is necessary. For example, the 5-(12, 6, 1) design, a 2-point extension of the egglike inversive plane of order 3, is embeddable in $\mathrm{PG}(5,3)$ (Coxeter 1958); but the 5-(24,8,1) Witt-design, a 3-point extension of $\mathrm{PG}(2,4)$, is not embeddable in $\mathrm{PG}(5,4)$. (For the constructions of Witt-designs, see chapter 14.)

Percsy (1981) proved an analogous result for embeddings in polar spaces, but there is some doubt about the correctness of his results.

5. The coding connection

There are many connections between finite geometry and coding theory. This section is a brief account of one of these; for the 2-dimensional case, see also chapter 16.

Let A be an $n \times N$ matrix over $\mathrm{GF}(q)$ having the following properties:

(a) the rows of A are linearly independent;

(b) any two columns of A are linearly independent.

From A, we can construct two objects:

(1) A set of N points in $\mathrm{PG}(n-1,q)$, those spanned by the N columns of A. By (b), the points are all distinct; by (a), they span the space. This set is unaltered by multiplying A on the right by a monomial matrix: permutations of the columns just

permute the points in the set, while multiplying a column by a non-zero scalar does not change the 1-dimensional subspace it spans. On the other hand, multiplying A on the left by a non-singular matrix has the effect of applying a collineation (an element of $\mathrm{PGL}(n,q)$) to the projective space, and so replaces A by an equivalent set of points.

(2) A code of length N and dimension n, the subspace of $\mathrm{GF}(q)^N$ spanned by the rows of A. (The dimension is guaranteed by (a).) Condition (b) asserts that the dual code has minimum weight at least 3, that is, it is 1-error-correcting. (See chapter 16 for concepts of coding theory.) Multiplying A on the left by a non-singular matrix does not change the code, merely changes the basis. On the other hand, multiplication on the right by a monomial transformation replaces the code by an equivalent code (under coordinate permutations and scalar multiplications), not changing coding-theoretic parameters of interest.

Under these circumstances, we would expect that information about the code and the configuration would transfer back and forth. For example, words of weight 3 in the dual code correspond to collinear triples of points; complements of codewords in the original code are hyperplane sections of the set; etc.

More strikingly, Greene (1976) showed that the weight enumerator of the code is a specialisation of the Tutte polynomial of the matroid represented by the point set. In this way, he was able to relate coding and matroid duality.

For a simple example, consider the set of all points of $\mathrm{PG}(n-1,q)$. The dual of the corresponding code has length $(q^n-1)/(q-1)$ and dimension $(q^n-1)/(q-1)-n$, and is 1-error-correcting, and hence perfect; it is a Hamming code.

Again, consider a set of N points with the property that no n of them are contained in a hyperplane, as discussed in section 4. The corresponding code has length N, dimension n, and minimum weight at least $N-n+1$ (by the remark on hyperplane sections); so it is maximum distance separable (i.e. it attains the Singleton bound). Conversely, a linear MDS code gives a set of points with this property. See Bruen et al. (1988). For a similar discussion of caps, see Hill (1978).

A far-reaching generalization of this construction produces the so-called *algebraic geometry codes*, see Goppa (1984), Drinfel'd and Vladut (1983), Manin and Vladut (1985), Van Lint and Van der Geer (1987). These codes are constructed from divisors on an algebraic curve; the Riemann–Roch theorem gives lower bounds for the dimension and minimum distance in terms of the genus of the curve.

6. Projective and affine planes

The importance of projective and affine planes is underlined by their appearance in Theorem 2.3. In Dembowski's influential book "Finite Geometries", published in 1968, these planes account for over half the total content (though the proportion may be less today, because of an upsurge of interest in other topics).

We obtain an affine plane by removing a line and all of its points from a projective plane. Conversely, given an affine plane, we reconstruct the projective plane by adding a "point at infinity" to all the lines in each parallel class, and adding one line incident with all the new points. The canonical nature of the construction shows that the collineation group of the affine plane is the subgroup of the collineation group of the projective plane fixing the chosen line; and that two affine planes are isomorphic if and only if the projective planes are isomorphic and the lines lie in corresponding orbits of the collineation group.

Wilker (1977), following Mendelsohn (1972), showed the next result.

Proposition 6.1. *Given a group G, there is a projective plane Π with the properties*
(i) $\mathrm{Aut}(\Pi) = G$;
(ii) *for any subgroup H of G, there is a line L of Π such that H is the stabiliser of L (and hence the collineation group of the corresponding affine plane).*

It is not known whether, if the group is finite, the plane can also be taken to be finite. It is not even known whether every finite group is a subgroup of the collineation group of some finite projective plane.

(A related observation concerns substructures. Any set of points and lines form a *partial linear space*, that is, any two points lie on at most one line and dually. A well-known free construction embeds any partial linear space in a projective plane. Is every finite partial linear space embeddable in a finite projective plane? The question may be specialized to *linear spaces* or simple matroids of rank 3 (two points on exactly one line).)

It is possible to coordinatize any projective plane by an algebraic structure known as a *planar ternary ring*. (In the case of $\mathrm{PG}(2,q)$, this structure is $\mathrm{GF}(q)$ equipped with the ternary operation $(x,m,c) \mapsto x \cdot m + c$.) There is a considerable amount of theory about the interplay between algebraic properties of the PTR, configuration theorems in the plane, and the existence of certain collineations.

A collineation θ of a projective plane Π is called *central* if there is a point p (its *centre*) such that θ fixes every line through p. Dually, θ is *axial* if it has an *axis*, a line L such that θ fixes every point on L. A non-identity collineation has at most one centre and at most one axis, and is central if and only if it is axial. A central collineation is called an *elation* if its centre is incident with its axis, a *homology* otherwise.

Desargues' theorem is equivalent to the statement that Π has "all possible" central collineations; that is, for every point p and line L, the group of central collineations with centre p and axis L acts transitively on the points of M other than p and $L \cap M$, where M is any line on p different from L. We abbreviate this condition by saying that Π is *(L,p)-transitive*. If Π is (L,p)-transitive for all L and p, then the group of elations (resp. homologies) with given centre and axis can be identified with the additive (resp. multiplicative) group of the coordinatising field; hence the PTR satisfies the axioms for a field. This is the prototype of the connections alluded to above.

The most ambitious classification of finite projective planes is the *Lenz–Barlotti classification*, according to the configuration formed by the pairs (L,p) for which the plane is (L,p)-transitive. (Lenz considered elations, i.e. incident pairs, and distinguished seven types; by including homologies, Barlotti refined this to 53 types. The Desarguesian planes stand at one extremity; at the other are the planes which are not (L,p)-transitive for any (L,p). See Dembowski (1968), Hughes and Piper (1973) for an account of this.

One particularly important case is that where there is a line L such that Π is (L,p)-transitive for all p incident with L. Equivalently, in the affine plane obtained by removing L, the group of *translations* (collineations fixing every parallel class of lines but no points) is transitive on points. Such an affine plane is called a *translation plane*. The translation group must be elementary abelian, and hence a vector space V over GF(p) for some prime p; its dimension must be even, say $2m$. The points of the plane can be identified with the non-zero vectors. Then the lines containing 0 are m-dimensional vector subspaces, and the remaining lines are cosets of these subspaces.

The *kernel* of a translation plane is the set of all those endomorphisms of the vector space which map the lines through 0 to themselves. It is a field k containing GF(p), and V and the lines through 0 are k-vector spaces. So we can replace GF(p) in the above by $k = \mathrm{GF}(q)$, say, and redefine m so that $\dim_k V = 2m$.

The lines through the origin form a set of $q^m + 1$ subspaces of dimension $m-1$ in PG($2m-1, q$) which partition the point set. Such a collection of subspaces is called a *spread*. Many examples of translation planes have been constructed by taking a linear group G over GF(q) and looking for a G-invariant spread; see Ostrom (1970), for example.

Translation planes of order q^2 with kernel containing GF(q) have received most attention, and their theory is best seen in relation to the Klein quadric.

The set W of skew-symmetric 4×4 matrices over a field k is a 6-dimensional vector space. The determinant of a skew-symmetric matrix is a square, $\det(A) = \mathrm{Pf}(A)^2$, where Pf($A$), the *Pfaffian* of A, is (in this case) a quadratic form on W. Thus the singular matrices are precisely those on which the Pfaffian vanishes. Moreover, the rank of a skew-symmetric matrix is necessarily even, so (in this case) 0, 2 or 4.

Any skew symmetric matrix of rank 2 can be written as $v^{\mathrm{T}}w - w^{\mathrm{T}}v$, where v and w are row vectors. The vectors themselves are not uniquely determined, but their span is; so we have a bijection from the set of lines of PG($3,k$) to the points of PG($5,k$) which are singular for the quadratic form given by the Pfaffian. This set $\mathcal{Q}$ of points is called the *Klein quadric*.

Now it is easy to see that two lines L, L' of PG($3,k$) intersect if and only if the corresponding points of the Klein quadric are joined by a line lying in the quadric. (The points on this line translate back into the lines of the plane pencil in PG($3,k$) determined by L and L'.) So a spread of lines in PG($3,k$) (and hence a translation plane of the type we are considering) corresponds to an *ovoid* on the quadric, a maximal set of pairwise non-collinear points of $\mathcal{Q}$. (If $k = \mathrm{GF}(q)$, the number of points on an ovoid is q^2+1.)

The Pappian plane (the plane AG($2, q^2$) in the finite case) is obtained when the ovoid is an elliptic quadric (a 3-dimensional section of $\mathcal{Q}$). The corresponding spread is called *regular*.

As we saw in section 4, this quadric has the structure of an inversive plane. Its circles are conics, i.e. plane sections of $\mathcal{Q}$. Translating back, the set of lines corresponding to a conic is a *regulus*, one ruling of lines on a hyperbolic quadric.

Any regulus $\mathcal{R}$ has an *opposite regulus* $\mathcal{R}^{\mathrm{op}}$, the set of lines which are transversals to every line through $\mathcal{R}$. (On the Klein quadric $\mathcal{Q}$, $\mathcal{R}^{\mathrm{op}}$ is the section by the plane perpendicular to the plane defining $\mathcal{R}$.)

For an intrinsic definition, recall that three pairwise skew lines in 3-space have a unique common transversal through any point of one of them; a regulus is the set of all common transversals to the three lines. Any three lines of a regulus have the same set of transversals; this set is the opposite regulus.

Each point on a line of $\mathcal{R}$ lies on a unique line of $\mathcal{R}^{\mathrm{op}}$. This means that we can replace the lines of $\mathcal{R}$ in a regular spread by those of $\mathcal{R}^{\mathrm{op}}$, obtaining another spread and hence another translation plane. This process is known as *derivation*.

More generally, if any spread contains a regulus, then it can be derived; in particular, the regular spread can be derived with respect to any set of pairwise disjoint reguli (circles in the Miquelian inversive plane).

This can be still further generalized. More general sets of points of the Klein quadric can be used. For example, suppose that we have a set of circles of the inversive plane such that every point lies in 0 or k of them. The corresponding reguli of $\mathrm{PG}(3,k)$ contain each line of the regular spread 0 or k times; if they are all replaced by their opposites, another spread results. See Bruen (1971b).

More generally, the connection with the Klein quadric has led to a body of work relating translation planes, configurations on the quadric, and coding theory (Kerdock codes), and has led to the construction of large numbers of planes. See Kantor (1983).

Another idea, due to Ostrom (1966), does not even require a translation plane; some lines (forming a so-called *replaceable net*) of an arbitrary plane of square order m^2 are replaced by a suitable collection of *Baer subplanes* (of order m).

A different classification of projective planes by collineation groups was proposed by Hering (1979). I will take the easiest of Hering's sixteen cases. We say that G acts *irreducibly* on Π if G fixes no point, line, triangle, or proper subplane setwise. Suppose that G is irreducible, and that G contains a central collineation. Then Hering shows that a minimal normal subgroup of G is either elementary abelian of order 9 (and equal to its centralizer), or non-abelian simple (and having trivial centralizer). Thus G is contained in either $\mathrm{AGL}(2,3)$ or $\mathrm{Aut}(S)$, where S is non-abelian simple. (This result is in the same spirit as the O'Nan–Scott theorem; see chapter 12.) The combined efforts of Hering, Walker, Reifart and Stroth have determined which possible simple groups can occur, up to one unknown, the sporadic group J_2. (It is known that J_2 can act irreducibly on an infinite projective plane.)

There have been many other characterizations of projective planes by symmetry properties. Perhaps the most famous is the *Ostrom–Wagner theorem*, mentioned earlier: a finite projective plane admitting a 2-transitive collineation group is Desarguesian. This result can be strengthened now, using the classification of finite simple groups. Kantor (1987) showed the following.

Theorem 6.2. *Let G be a collineation group of the finite projective plane Π of order n, which is either transitive on flags, or primitive on points. Then either*
(i) *Π is Desarguesian; or*
(ii) *n^2+n+1 is prime, and G is solvable.*

Kantor obtained this as a corollary of his classification of almost simple primitive groups of odd degree (see chapter 12), since the number of points in a projective plane is odd. The implication from flag-transitivity to point-primitivity was proved much earlier, by Higman and McLaughlin (1961). In case (ii), the plane is given by a difference set in the cyclic group of order n^2+n+1. This occurs in the Desarguesian planes (but only for $n=2$ and $n=8$ can G be flag-transitive); no non-Desarguesian examples are known.

The next two results summarize some general combinatorics of projective and affine planes.

Proposition 6.3. (i) *If a projective plane of order n has a proper subplane of order m, then either $n=m^2$ or $n \geqslant m^2+m$; the first alternative occurs if and only if every line of the plane is incident with a point of the subplane (and dually).*

(ii) *An involution of a projective plane of order n has $n+1$, $n+2$ or $n+n^{1/2}+1$ fixed points, forming a line, a line and a point off it, or a subplane respectively.*

(iii) *A polarity of a projective plane of order n has at least $n+1$ absolute points; if there are $n+1$, they form a line if n is even, an oval if n is odd.*

Notes: (i) A subplane of order $n^{1/2}$ of a plane of order n is called a *Baer subplane*. No example with $n=m^2+m$ is known. The theorem is false for affine planes; AG$(2,3)$ is isomorphic to the configuration of inflection points of a cubic curve, and is embedded in AG$(2,q)$ for all $q \equiv 1 \pmod 3$, $q>4$ (and, indeed, in PG$(2,4)$, as the hermitian unital). However, if a projective plane of order n has an affine subplane of order $m<n$, then either $(n,m)=(4,3)$, or $n \geqslant m^2-\frac{1}{2}m-\frac{1}{2}$; see Cameron (1980).

(ii) An *involution* is a collineation of order 2. It is an elation, a homology, or a *Baer involution* respectively in the three cases. The result is proved by observing that the fixed points and lines form a (possibly degenerate) subplane, and that any non-fixed line is incident with a fixed point and dually.

(iii) A *polarity* is an incidence-preserving map of order 2 interchanging points and lines; a point or line is *absolute* if it is incident with its image. In PG$(2,q)$, any polarity has either $q+1$ or $q^{3/2}+1$ absolute points, as we saw in section 4.

Gleason (1956) characterized the Desarguesian planes of even order by the "Fano configuration" (fig. 6.1), so-called because it was excluded by the Axiom of Fano (1892) (!)

Theorem 6.4. *A projective plane is isomorphic to* PG$(2,2^d)$ *for some d if and only if the diagonal points of any complete quadrangle are collinear.*

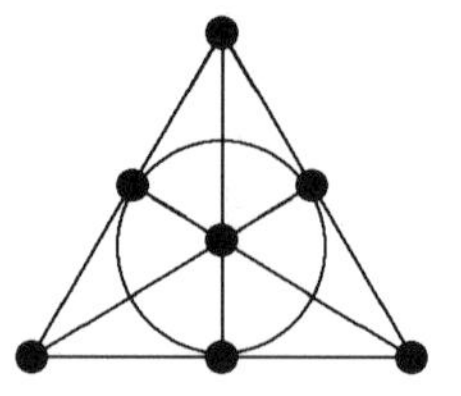

Figure 6.1. The Fano configuration.

Table 6.1
Small projective planes

Order	2	3	4	5	6	7	8	9	10
Number	1	1	1	1	0	1	1	4	0

Proposition 6.5. *The following objects are equivalent:*
(i) *an affine plane of order n with two distinguished parallel classes;*
(ii) *a set of* $n-1$ *mutually orthogonal Latin squares of order n;*
(iii) *a sharply 2-transitive set of permutations of an n-set.*

For let $\{H_1,\dots,H_n\}$ and $\{V_1,\dots,V_n\}$ be the two given parallel classes of lines, and set $\{p_{ij}\} = H_i \cap V_j$. For each of the $n-1$ further parallel classes $\{L_1,\dots,L_n\}$, take the Latin square with (i,j) entry k if and only if $p_{ij} \in L_k$. The "equation" of each line L_k is a permutation: $i \mapsto j$ if and only if $p_{ij} \in L_k$. The constructions reverse.

Thus the maximum number of mutually orthogonal Latin squares of order n is a measure of "how close a plane of order n is to existing". This function has received much attention since the question was first posed by Euler in 1782.

If we take fewer than $n-1$ mutually orthogonal Latin squares, the construction yields a *net*. Nets are discussed further in section 10. (See also chapter 14, section 7.)

All known projective planes have prime power order. In the other direction, the *Bruck–Ryser* (1949) *theorem* was, until very recently, the strongest necessary condition for the existence of a projective plane.

Theorem 6.6. *If a projective plane of order n exists, where n is congruent to* 1 *or* 2 (mod 4), *then n is the sum of two squares.*

(Cf. chapter 14, section 3.) For small orders, the number of non-isomorphic planes is given in table 6.1.

The non-existence of planes of order 6 follows from the Bruck–Ryser theorem, though this special case had been settled much earlier (by Tarry 1901). For order 10, however, this is the end-product of a massive computation by Lam et al. (1989). I would like to sketch briefly the history of this problem.

Over the course of time, attempts to construct a plane of order 10 gave way to attempts to prove non-existence. The first phase involved looking at collineations. If we are trying to construct an object, the larger collineation group we assume, the easier our job will be (if the object exists!). Hughes, Hall, Whitesides, Janko and many others analysed the possible collineations of prime order. Gradually, the possible prime divisors of the order of the collineation group were whittled away. Some general results were found by Hughes, using Hasse–Minkowski theory and other techniques. For example, we have the following (Hughes 1957).

Proposition 6.7. *A projective plane whose order is congruent to* 2 (mod 4) *but greater than* 2 *cannot have a collineation of order* 2.

(Such a collineation would necessarily be an elation, by Proposition 6.3(ii).)

The other line of attack was using coding theory. This provides a technique for showing that the plane must contain one of a fairly small number of configurations. It is then necessary to try with bare hands to extend each configuration to a plane.

Let Π be a projective plane of order n. Set $N = n^2 + n + 1$. In the vector space $\mathrm{GF}(2)^N$, let C be the row space of the incidence matrix of Π. (We take the convention that points index columns of the matrix.) Now C is a code; the *support* of any codeword (the set of coordinates where it has a non-zero entry) is a set of points in the plane which can be expected to have interesting properties. It is elementary that, if n is even, then the minimum weight of the code is $n+1$; the support of any word of weight $n+1$ (resp. $n+2$) is a line (resp. oval). (Cf. chapter 16 (9.2).)

If $n \equiv 2 \pmod 4$, then $C^{\perp}$ is a subcode of C with codimension 1. This means that a set S of points supports a codeword if and only if

$$|S \cap L| \equiv |S| \pmod 2$$

for every line L. It is also true that all weights in C are congruent to 0 or 3 (mod 4). Now we specialise to the case $n = 10$. Then, by the MacWilliams identities, the weight enumerator of C is completely determined by the numbers of words of weight 12, 15 and 16. In particular, C must contain a word of weight 12, 15, 16 or 19. Now, as described, all configurations of these sizes can be determined, and each tested for embeddability in a plane of order 10.

A word of weight 15 gives a 15-set meeting every line in an odd number of points, at most 5 (its symmetric difference with a line has cardinality at least 16, since it clearly cannot be an oval). This case was excluded by MacWilliams et al. (1973) in their pioneering application of the technique. (In fact, an easy counting argument shows that there are exactly six 5-secants, and that these form a complete arc in the dual plane; this possibility had been excluded by an earlier computation by Denniston 1969b.) As remarked earlier, words of weight 12 are ovals; this case was settled by Lam et al. (1983). Two subsequent computations by Lam et al. (1986, 1989) excluded the other two possible weights.

These techniques can in principle be applied to planes of other orders; but the scale of the problem, even for the next admissible order 12, puts any further applications out of reach for some time.

The classification of planes of order 9 is due to Lam et al. (1991).

The code of a projective plane can be computed over GF(p) for any prime p. It is non-trivial if and only if p divides the order n, and contains its dual with codimension 1 if and only if p divides n to the first power only. For example, the code of the Desarguesian plane of order p^m has dimension $\binom{p+1}{2}^m + 1$. It has been conjectured by Hamada (1973) that the code of any non-Desarguesian plane of the same order has strictly larger dimension. For example, the code (mod 3) of the Desarguesian plane of order 9 has dimension 37; for each of the three non-Desarguesian planes, the dimension is 41.

Lander (1981) associated a chain of r codes with a plane whose order is exactly divisible by p^r. If r is odd, the middle code in the chain contains its dual with codimension 1, so techniques like those discussed earlier can be applied.

Despite our lack of knowledge of planes of non-prime-power order, there are enormous numbers of planes of prime power order p^r (for fixed p and large r), including many non-isomorphic translation planes. In particular, non-Desarguesian planes of all prime power orders except 4, 8 and primes are known. The existence of non-Desarguesian planes of prime order is one of the most important open problems at present.

7. Generalized polygons

Generalized polygons form an important class of geometries including both projective planes and GQs; they also serve as ingredients in the construction of buildings, to be defined in the next section.

A geometry has varieties of d different types, with an incidence relation between them. It can be represented by a d-partite *incidence graph*, whose vertices are the varieties, two vertices adjacent if they are incident. Now let us specialise to the case $d = 2$. A connected bipartite graph of diameter m which is not a tree has girth at most $2m$. Generalized polygons are extremal graphs for this bound: a *generalized m-gon* is a geometry with two types of variety, whose (bipartite) incidence graph has diameter m and girth $2m$, with the property that, for any variety, there is a variety at distance m from it.

Proposition 7.1. (i) *A generalized 2-gon is a geometry (with at least two points and at least two lines) in which any point is incident with any line.*

(ii) *A generalized 3-gon is a generalized projective plane, i.e. a projective plane, the direct sum of a point and a line, or a triangle.*

(iii) *A generalized 4-gon is a GQ.*

A generalized polygon is called *thick* if every vertex in the incidence graph has valency at least 3.

Proposition 7.2. (i) *In a thick generalized m-gon, vertices of the same type have the same valency; if m is odd, all vertices have the same valency.*

(ii) *A non-thick generalized m-gon with m even is the flag geometry of a generalized m/2-gon.*

(The *flag geometry* of a geometry $\mathcal{G}$ has as points the varieties of $\mathcal{G}$ and as lines the incident pairs, or dually.)

In a thick generalized m-gon, we denote by $s+1$ the number of points on a line, and by $t+1$ the number of lines on a point. Then $s=t$ if m is odd. The most important result is the *Feit–Higman* (1964) *theorem.*

Theorem 7.3. *A finite thick generalized m-gon has* $m=2$, 3, 4, 6 *or* 8. *Moreover, if* $m=6$ *then st is a square; if* $m=8$ *then* $2st$ *is a square.*

Further necessary conditions are known, derived from the representation theory of coherent configurations (cf. chapters 12, 15). These include divisibility conditions (expressing the integrality of the multiplicity of eigenvalues, see Higman 1975), and (from the Krein bounds) some inequalities due to Higman (1975) and Haemers and Roos (1981).

Proposition 7.4. (i) *A thick generalized* 4-*gon or* 8-*gon satisfies* $t \leqslant s^2$ *and* $s \leqslant t^2$.

(ii) *A thick generalized* 6-*gon satisfies* $t \leqslant s^3$ *and* $s \leqslant t^3$.

A curious question related to Proposition 7.4 asks whether there exists a thick generalized polygon with s finite and t infinite. All that is known is that there is no GQ with $s=2$ or 3 and t infinite.

Quite a large number of GQs are known (though they are not as prolific as projective planes). By Theorem 7.3(ii), the non-thick ones are complete bipartite graphs or rectangular grids. The parameters of the known thick GQs, up to duality, are (q,q), (q,q^2), (q^2,q^3), or $(q-1,q+1)$, where q is any prime power. The first three include the polar spaces of rank 2, except for the hyperbolic (ruled) quadric in $\mathrm{PG}(3,q)$, which is a square grid. (In the notation of table 4.1, we have $s=q$, $t=q^{1+\varepsilon}$.) A simple construction for the first non-trivial example, the symplectic polar space in $\mathrm{PG}(3,2)$, is as follows: the points and lines are the edges and 1-factors of the complete graph K_6; incidence is membership.

A construction due to Tits (1962) ties in several concepts we have met already. Let $\mathcal{O}$ be an ovoid in $\mathrm{PG}(3,q)$, C the code constructed from $\mathcal{O}$ as in section 5 (so that C has length q^2+1, dimension 4, and weights q^2 and q^2-q). The points are of three types: a special point ∞; $\mathrm{GF}(q)\times\mathcal{O}$; and C. Form a graph by joining ∞ to all points of the second type; joining $(\alpha,p)\in\mathrm{GF}(q)\times\mathcal{O}$ to $v\in C$ if and only if $v_p=\alpha$; and joining $v,v'\in C$ if and only if their distance is q^2. This graph is strongly regular; its $(q+1)$-cliques are the lines of a GQ with $s=q$, $t=q^2$, which is classical if and only if $\mathcal{O}$ is an elliptic quadric.

The simplest class of examples of the fourth type, due to Ahrens and Szekeres (1969), is constructed from an oval $\mathcal{O}$ in $\mathrm{PG}(2,q)$, q even. Points are the vectors in the underlying 3-dimensional vector space; lines are cosets of the 1-dimensional subspaces in $\mathcal{O}$.

There are many results on GQs; the theory parallels that of projective planes. See Payne and Thas (1984) for a survey. Among many results, I choose just one to mention: Walker's (1977) analogue of Hering's result on irreducible collineation groups.

The only known finite thick generalized m-gons with $m > 4$ are constructed from certain exceptional or twisted groups of Lie type. Up to duality, $G_2(q)$ gives a 6-gon with $s = t = q$; ${}^3D_4(q^3)$ gives a 6-gon with $s = q$, $t = q^3$; and ${}^2F_4(q)$ gives an 8-gon with $s = q$, $t = q^2$, $q = 2^{2d+1}$. (See chapter 12 for these groups.)

The most important general result about generalized m-gons is the classification of thick Moufang polygons: all are classical, i.e. (in the finite case), $\mathrm{PG}(2,q)$, a rank 2 polar space, or one of the 6-gons or 8-gons just described. (The Moufang condition extends, in a natural way, the condition of (L,p)-transitivity for all incident pairs p, L (see section 6).) This result is due to Fong and Seitz (1973, 1974) in the finite case, and to Faulkner, Tits and Weiss in general (see Tits 1976, Weiss 1979).

8. Buildings

The definition of a building exists in several versions, none of which is simple to explain. I will attempt to provide a definition here. Keep in mind that buildings form a class of geometries including projective and polar spaces, and standing in the same relation to generalized polygons as projective spaces do to projective planes. Two recent books (Brown 1989 and Ronan 1989) give good accounts of the theory.

We need yet another reformulation of the notion of geometry. A *chamber system* of rank d is a set $\mathscr{C}$ carrying d equivalence relations $\sim_0, \ldots, \sim_{d-1}$. It is usual to assume that these relations intersect pairwise in the relation of equality.

From a geometry with d types of varieties, we construct a chamber system as follows. The chambers are the complete flags (i.e. sets of d pairwise incident varieties, one of each type.) We set $F \sim_i F'$ if the varieties in F and F' of all types *except* the ith are the same. (Thus, in a point-line geometry, two flags are "point-related" if they involve the same line but different points, and dually.)

Not all interesting chamber systems come from geometries. For example, a Latin square can be conveniently regarded as a chamber system, where the cells of the square are the chambers, and there are three equivalence relations: "same row", "same column", and "same entry". This is the correct picture of a Latin square in experimental design, where the cells might correspond to experimental plots and the equivalence relations to significant inter-plot effects.

A path in a chamber system is a sequence of chambers in which each consecutive pair of chambers is distinct and satisfies one of the relations $\sim_i$. Since these relations are equivalences, we may (by omitting vertices if necessary) assume that, if $C_m \sim_i C_{m+1}$ and $C_{m+1} \sim_j C_{m+2}$, then $i \neq j$. A path satisfying this condition will be called *reduced*. A shortest path is necessarily reduced; and in a *thin* chamber complex (in which every equivalence class has size 2), a path (in which successive vertices are distinct) is necessarily reduced.

In the chamber system of a generalized m-gon, an *apartment* is a subsystem isomorphic to the chamber system of an ordinary m-gon. Then:

(a) any two chambers are contained in an apartment; and

(b) the shortest path between two chambers is unique, unless they are opposite in every apartment containing them.

Moreover, these properties characterize generalized m-gons among chamber systems of rank 2.

The definition of a building is of the same form; only the apartments need to be changed, and (b) reformulated in such a way that it can be generalized.

A *Coxeter group* is a group defined by a presentation with generators $x_0, \ldots, x_{d-1}$ and relations asserting that $x_i^2 = 1$ for all i and that $(x_i x_j)^{m_{ij}} = 1$ for suitable integers m_{ij}. (It is sometimes convenient to allow some relations of the second type to be absent.) A Coxeter group is described by a *Coxeter diagram* in which the nodes correspond to the generators, and x_i and x_j are joined by $m_{ij} - 2$ edges (or, if m_{ij} is large, by a single edge with label (m_{ij})). Note that non-edges ($m_{ij} = 2$) correspond to commuting generators; so the Coxeter group of a disconnected diagram is the direct product of the Coxeter groups of the components. The dihedral group of order $2m$ has a presentation

$$\langle x_0, x_1 : x_0^2 = x_1^2 = (x_0 x_1)^m = 1 \rangle,$$

and so has the diagram $I_2^{(m)}$ below. *Coxeter's theorem* (see Coxeter and Moser 1957) asserts the following.

Theorem 8.1. *The finite Coxeter groups are those given by disjoint unions of the diagrams in table* 8.1. *Each is a group of euclidean transformations generated by reflections fixing the origin.*

Now a *Coxeter complex* is defined to be a chamber system whose chambers are the elements of a Coxeter group, with $g \sim_i h$ if and only if either $g = h$ or $gx_i = h$. (This is the Cayley diagram of the group with respect to the distinguished generators, with edges labelled by generators in the usual way.) It is a thin chamber complex, in the sense described earlier.

The Coxeter group of type A_n is the symmetric group on the set $\{0, 1, \ldots, n\}$, the distinguished generators being the "adjacent transpositions" $x_i = (i \quad i+1)$ for $i = 0, \ldots, n-1$. It is straightforward to check that the Coxeter complex of this group is isomorphic to the flag complex of the free matroid on $\{0, 1, \ldots, n\}$ (whose flats are all the subsets).

In any chamber complex whose relations are parametrized by the generators of a Coxeter group, we can "translate" any simple path into a group element, by multiplying together the generators corresponding to the equivalence relations used on the path. Condition (b) in the definition of a generalized m-gon can now be stated as follows:

(bb) two shortest paths between the same pair of chambers translate into the same element of the Coxeter group.

Table 8.1
The finite Coxeter groups

Type	Diagram	Order
A_n	 (n nodes)	$(n+1)!$
C_n	 (n nodes)	$2^n n!$
D_n	 (n nodes)	$2^{n-1} n!$
E_6	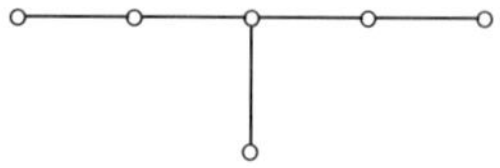	51840
E_7	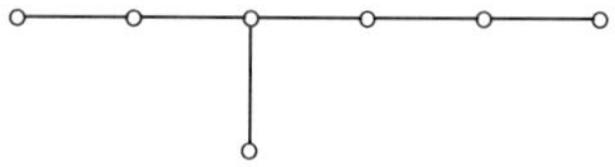	2903040
E_8	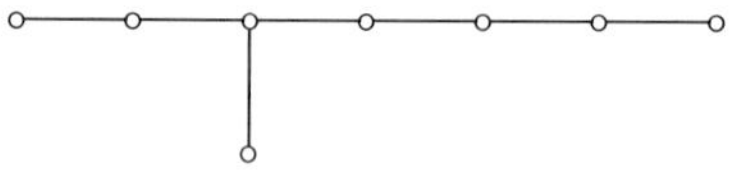	696729600
F_4		1152
H_3	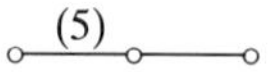 (5)	120
H_4	 (5)	14400
$I_2^{(m)}$	(m)	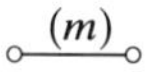 $2m$

(In the case of a generalized m-gon, if two chambers are "opposite", then the elements of the Coxeter group corresponding to shortest paths between them are either $x_0 x_1 \cdots$ (m terms) or $x_1 x_0 \cdots$ (m terms). But these are equal, as a consequence of the relations

$$x_0^2 = x_1^2 = (x_0 x_1)^m = 1$$

which hold in the dihedral group of order $2m$.)

Now we define a *building* to be a chamber complex which satisfies (a) and (bb), where the apartments are Coxeter complexes for an arbitrary (but fixed) Coxeter group. The *type* of the building is that of the Coxeter group. The building is *thick* if each equivalence class of each $\sim_i$ has size at least 3.

It is an instructive exercise to verify that buildings of type A_n are precisely the flag complexes of generalized projective spaces; their apartments correspond to bases. Thick buildings of this type are just the projective spaces.

Thick buildings of types C_n and D_n are polar spaces, and conversely. (The Coxeter complex is the flag complex of the geometry with point set $\{u_i, v_i : 1 \leqslant i \leqslant n\}$, the flats being the sets which contain at most one of u_i and v_i for each i.

This is realized as an apartment in a polar space by a set of vectors satisfying $f(u_i, v_j) = \delta_{ij}, f(u_i, u_j) = f(v_i, v_j) = 0$, where f is the sesquilinear form.)

A building is called *spherical* (or *of spherical type*) if its apartments are finite. Clearly any finite building is spherical! The types of spherical buildings are just those given in table 8.1.

Let J be a subset of the index set of the equivalence relations in a chamber system $\mathscr{C}$. Then a connected component for $\{\sim_j : j \in J\}$ is itself a chamber system, and is a building (resp. thick building) if $\mathscr{C}$ is. This, together with the Feit–Higman theorem, shows that there is no finite thick building of type H_3 or H_4.

Let G be a group of Lie type (see chapter 12), defined over a field of characteristic p. A *Borel subgroup* B of G is the normalizer of a Sylow p-subgroup. The *minimal parabolics* $P_0, \ldots, P_{d-1}$ of G (relative to B) are the atoms of the lattice of subgroups between B and G; the *rank* d of G is their number. Now define a chamber system $\mathscr{C}(G)$, as follows: The chambers are the right cosets of B in G; and $Bx \sim_i By$ if and only if $P_i x = P_i y$.

Now, at last, we can state the finite version of the main theorem of Tits (1974).

Theorem 8.2. *The finite thick buildings with rank at least* 3 *are precisely the chamber systems* $\mathscr{C}(G)$ *arising from finite groups* G *of Lie type and rank at least* 3.

It should be stressed that this major theorem includes the axiomatisations of both projective spaces of Theorem 2.3 and polar spaces of Theorem 4.12, though the proof of the latter result takes up the bulk of Tits (1974). As in both of those results, finiteness of the geometry can be weakened, to the assumption that the building is of spherical type; some more examples have to be added to the list.

Tits (1986) achieved the next stage by classifying the so-called "affine buildings" of rank at least 4. (These have Coxeter groups which are generated by euclidean reflections not all having a common fixed point. Their diagrams are suitable one-point extensions of those of the spherical buildings.) For buildings with more general diagrams, there are too many for such a classification to be possible: see Ronan and Tits (1987).

Table 8.2 lists the groups of Lie type and the types of the corresponding buildings. See table 4.1 of chapter 12 for more information about these groups.

9. Buekenhout geometries

In the last section, we had to resort to chamber systems in order to obtain the correct definition of a building. There is no doubt that this adds a level of conceptual difficulty. It turns out that there is a very natural way to attempt to define buildings as geometries; it does not quite yield the right definition because too many geometries are obtained. Buekenhout made a virtue of this by showing that, if we widen the definition just a little more, we get a class of geometries including operands for many of the sporadic simple groups as well as those of Lie type.

Recall that a geometry has a set X of varieties of d different types — this can be described by a *type map* $\mathrm{tp}: X \to \Delta$, where Δ is a d-set of types — with an

Table 8.2
Buildings of Lie type

Group	Type of building
$A_n(q)$	A_n
$B_n(q)$	C_n
$C_n(q)$	C_n
$D_n(q)$	D_n
$E_n(q)$ $(n = 6, 7, 8)$	E_n
$F_4(q)$	F_4
$G_2(q)$	$I_2^{(6)}$
${}^2A_n(q^2)$	$C_{\lfloor n/2 \rfloor}$
${}^2D_n(q^2)$	C_{n-1}
${}^3D_4(q^3)$	$I_2^{(6)}$
${}^2E_6(q^2)$	F_4
${}^2B_2(q)$, $q = 2^{2d+1}$	A_1
${}^2G_2(q)$, $q = 3^{2d+1}$	A_1
${}^2F_4(q)$, $q = 2^{2d+1}$	$I_2^{(8)}$

incidence relation holding between some pairs of varieties. For convenience we assume that any variety is incident with itself and with no other variety of the same type.

A *flag* is a set of mutually incident varieties. Thus a flag has cardinality at most d, with equality if and only if it is *complete* (contains one variety of each type). A geometry is called *firm* (resp. *thick*) if any incomplete flag is contained in at least two (resp. three) complete flags. A geometry is *connected* if its incidence graph is connected. The *rank* of a geometry is the number of types.

Let F be a flag of *cotype* J (that is, $\Delta \setminus \mathrm{tp}(F) = J$). The *residue* of F is the set of varieties not in F which are incident with every variety in F. It is a geometry with type set J, and is firm (resp. thick) if the original geometry is. A geometry is *residually connected* if every residue of rank at least 2 is connected.

Buekenhout's idea is that interesting classes of geometries can be axiomatized by simple global assumptions (e.g. residual connectedness) together with detailed hypotheses about the rank 2 residues. These hypotheses can be expressed by a diagram. We identify various interesting classes of rank two geometries by "edge labels". A *diagram* is a set Δ with a class $\mathcal{K}_{ij}$ of connected point-line geometries for each distinct $i, j \in \Delta$ (with the convention that the members of $\mathcal{K}_{ji}$ are the duals of those of $\mathcal{K}_{ij}$). Now a *Buekenhout geometry* with diagram Δ is a residually connected firm geometry with type set Δ, in which the residue of any flag of cotype $\{i, j\}$ is a member of $\mathcal{K}_{ij}$ (the points and lines having type i and j respectively).

We adapt the convention of the last section: the class of generalized m-gons is denoted by $m - 2$ parallel edges (or by an edge with label (m), if m is large).

Proposition 9.1. *A building with Coxeter diagram Δ is a Buekenhout geometry with type Δ.*

To illustrate, consider a projective space $PG(n,q)$: the varieties of type i are the i-flats, for $0 \leqslant i \leqslant n-1$. Consider a residue of rank 2 and type $\{i,j\}$, with $i<j$.

(a) Suppose that $j=i+1$. Let the flag be

$$U_0 \subset \cdots \subset U_{i-1} \subset U_{i+2} \subset \cdots \subset U_{n-1}.$$

Its residue consists of all i-flats and $(i+1)$-flats X satisfying $U_{i-1} \subset X \subset U_{i+2}$, and is a projective plane $PG(2,q)$ (a generalized 3-gon).

(b) Suppose that $j>i+1$. Let the flag be

$$U_0 \subset \cdots \subset U_{i-1} \subset U_{i+1} \subset \cdots \subset U_{j-1} \subset U_{j+1} \subset \cdots \subset U_{n-1}.$$

The residue consists of the i-flats X with $U_{i-1} \subset X \subset U_{i+1}$, and the j-flats Y with $U_{j-1} \subset Y \subset U_{j+1}$. Clearly any such X and Y are incident. So the residue is a generalized 2-gon.

Stay with geometries over diagrams involving generalized polygons for a moment. These are called *Tits geometries*, or *geometries that are almost buildings* (GABs). Much of the language of buildings can be re-used here; for example, a GAB is *spherical* if its type is one of those listed in table 8.1. There is no finite thick GAB of type H_3 or H_4: the same argument that excludes buildings of these types applies.

There are two ways in which a GAB can fail to be a building:

(i) It may not be "simply connected". A *covering map* between geometries with the same diagram is a surjection preserving type and incidence and inducing an isomorphism on any rank 2 residue. (Strictly, this is what is known as a 2-covering, to distinguish it from a topological covering of the simplicial complex whose simplexes are the flags.) A geometry $\mathcal{G}$ has a *universal cover*, which covers every cover of $\mathcal{G}$. Now buildings are simply connected, i.e. equal to their universal covers. The most we can hope for in a general characterization is to conclude that our geometries are covered by buildings. But in the finite spherical case, things are nicer, because of the theorem of Brouwer and Cohen (1986).

Theorem 9.2. *A finite thick spherical GAB which is covered by a building is a building.*

(ii) It may have nothing to do with a building. A celebrated example is Neumaier's *A_7-geometry* (Neumaier 1984), defined as follows.

The point set is $X=\{1,2,3,4,5,6,7\}$.

The lines are the 3-subsets of X.

There are 30 ways of giving X the structure of $PG(2,2)$, falling into two orbits of 15 under A_7. The planes are the objects in one of these orbits.

Every point is incident with every plane; incidence between the other pairs of types is membership.

This geometry has diagram $C_3 =$ o——o══o. (The 2-gon and 3-gon are clear. Consider the residue of the point 7. This consists of 15 lines, identified with the

2-subsets of $\{1,\ldots,6\}$; and 15 planes, identified with the partitions of $\{1,\ldots,6\}$ into three 2-sets. (Given lines 712, 734, 756, for example, there are just two ways to complete to a PG(2,2), and these lie in different A_7-orbits.) This is a GQ we met earlier.)

This geometry is simply connected, and so is not covered by a building. But Tits (1981) showed that things like this are the only obstructions.

Theorem 9.3. *Let $\mathcal{G}$ be a GAB in which every rank 3 residue of type C_3 or H_3 is covered by a building. Then $\mathcal{G}$ is covered by a building.*

We observed that there are no finite thick GABs of type H_3. Combining this with Theorems 9.2 and 9.3, we see that a finite GAB with no C_3 residues (or, indeed, no "bad" C_3 residues) must be a building.

Once we leave spherical GABs, the infinite cannot be avoided: many finite GABs have covers which are infinite buildings (see Kantor 1981). Much recent work has concentrated on GABs whose universal cover is an affine building. A theorem of Kantor et al. (1987) describes all finite flag-transitive GABs whose universal cover is an algebraic affine building defined by a simple algebraic group of relative rank at least 2 over a locally compact local field. There are only finitely many universal covers, but passage modulo primes leads to an infinite number of finite examples.

The Buekenhout–Shult theorem 4.13 gives a characterization of polar spaces in terms of points and lines only. This has been extended to many other buildings and GABs; after work by Cohen and Cooperstein (1983) and many others, the best result to date is that of Hanssens (1988).

Buekenhout exploited the freedom given by his definition in another way, by introducing some new classes of rank 2 geometries. The most commonly used, in increasing generality, are:

$\circ\overset{c}{\text{———}}\circ$, circle or complete graph: lines are all pairs of points;

$\circ\overset{L}{\text{———}}\circ$, linear space: two points lie on a unique line;

$\circ\overset{\Pi}{\text{———}}\circ$, partial linear space: two points lie on at most one line.

If a class of geometries has symbol $\circ\overset{S}{\text{———}}\circ$, the class of duals is denoted by $\circ\overset{S^*}{\text{———}}\circ$.

A simple result illustrates these.

Proposition 9.4. (i) *A geometry with diagram*

$$\circ\overset{L}{\text{———}}\circ\overset{L}{\text{———}}\circ \cdots \circ\overset{L}{\text{———}}\circ$$

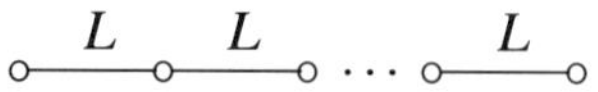

is the same as a geometric (simple) matroid.

(ii) *A geometry with diagram*

$$\circ\overset{c}{\text{———}}\circ\text{———}\circ$$

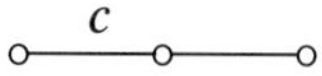

is an extension of a projective plane of order n. Such a geometry exists only for $n = 1$, 2 or 4; there is a unique example in each case.

The last part of (ii) is a classical result of Hughes (1965) on extensions of projective planes, afforced by the computer proof by Lam et al. (1983) of the nonexistence of a plane of order 10 containing an oval.

Many specific diagrams have been explored. Buekenhout collected large numbers of examples, including (as noted) geometries connected with many sporadic simple groups; see his paper (Buekenhout 1979) for some of these. A striking theorem is due to Sprague (1985).

Theorem 9.5. *A finite geometry with diagram*

$$\circ\overset{L^*}{\text{———}}\circ\overset{L}{\text{———}}\circ$$

consists of the $(i-1)$-, i- and $(i+1)$-flats of a generalized projective geometry.

Geometries with diagram

$$\circ\overset{c}{\text{———}}\circ\overset{c^*}{\text{———}}\circ$$

are semibiplanes; see Wild (1980). Some theory has been developed for

$$\circ\overset{c}{\text{———}}\circ{=\!=\!=}\circ$$

(extended generalized quadrangles: Cameron et al. 1990). More generally, Buekenhout and Hubaut (1977) considered *locally polar spaces*

$$\circ\overset{c}{\text{———}}\circ\text{———}\circ\text{———}\circ\ \cdots\ \circ\text{———}\circ{=\!=\!=}\circ,$$

and Del Fra et al. (1991) found the flag-transitive geometries in this class (under some extra hypotheses). These include geometries for several of the sporadic simple groups (Fischer, McLaughlin and Higman–Sims).

An important recent development concerns the work of Ivanov and his collaborators on the so-called *P-geometries* and *tilde-geometries*. For these, the diagram is linear, and all strokes are projective planes (of order 2) except for the stroke at one end, which is either $\circ\overset{P}{\text{———}}\circ$ (for the points and lines of the Petersen graph) or $\circ\overset{\sim}{=\!=\!=}\circ$ (for a particular triple cover of the generalized quadrangle of order 2 having 45 points). Many of the sporadic simple groups, including the Monster, the Baby Monster, and Janko's fourth group, act flag-transitively on such geometries. The classification of flag-transitive geometries in these classes has recently been completed (see Ivanov 1993 for a summary), and has a number of applications in other areas.

10. Partial geometries

There are many classes of finite geometries which have not been covered in this survey. Some, such as designs and association schemes, find their place elsewhere in this volume. But one which deserves a mention is the class of partial geometries. This class includes both projective planes and GQs, but generalizes them in quite a different direction from generalized polygons.

A *partial geometry* with parameters (s,t,α), or $\mathrm{pg}(s,t,\alpha)$, is a point-line geometry having the properties:

(a) any line is incident with $s+1$ points, and any two lines with at most one point;

(b) any point is incident with $t+1$ lines, and any two points with at most one line;

(c) if the point p and line L are not incident, then exactly α points of L are collinear with p (and so also exactly α lines on p are concurrent with L).

It is clear that the dual of a $\mathrm{pg}(s,t,\alpha)$ is a $\mathrm{pg}(t,s,\alpha)$. A projective plane of order n is a $\mathrm{pg}(n,n,n+1)$, while a GQ is a $\mathrm{pg}(s,t,1)$. Bose (1963), who introduced partial geometries, used the notation (r,k,t). Partial geometries are conveniently divided into four classes:

(1) 2-designs with $\lambda = 1$ (the case $\alpha = s+1$) and their duals (the case $\alpha = t+1$);

(2) transversal 2-designs with $\lambda = 1$ (the case $\alpha = s$) and their duals, viz. nets (the case $\alpha = t$);

(3) GQs (the case $\alpha = 1$);

(4) sporadic partial geometries (the remaining cases).

A net, or $\mathrm{pg}(s,t,t)$, is equivalent to a set of $t-1$ mutually orthogonal Latin squares of order $s+1$. (The points are the cells, and the lines correspond to rows, columns, and entries in each square, as in Theorem 6.4.)

Bose's purpose was to prove characterization theorems for some strongly regular graphs. The *point graph* of a partial geometry (whose vertices are the points, adjacent if they are collinear), is easily seen to be strongly regular; the intersection numbers are rational functions of s,t and α. A strongly regular graph is said to be *pseudo-geometric* if its intersection numbers are of this form (for some integral s,t,α; it is *geometric* if it is the point graph of a partial geometry. Building on previous ideas of Bruck for nets, Bose showed the following.

Theorem 10.1. *A pseudo-geometric strongly regular graph whose parameters satisfy*

$$s > \tfrac{1}{2}(t+2)(t-1+\alpha(t^2+1))$$

is geometric.

This important result has several corollaries.

Corollary 10.2. *The triangular graph $T(n)$ and the square lattice graph $L_2(n)$ are characterized by their intersection numbers for $n > 8$, $n > 4$ respectively.*

Here $T(n)$ and $L_2(n)$ are the point graphs of the (trivially unique) partial geometries $\mathrm{pg}(n-2,1,2)$ and $\mathrm{pg}(n-1,1,1)$ respectively. The characterizations are due to Hoffman (1960) and Shrikhande (1959) respectively. The inequalities are best possible.

Corollary 10.3. *A net of order n and deficiency d can be completed to an affine plane if*

$$n > \tfrac{1}{2}(d^4 - 2d^3 + 2d^2 + d - 2).$$

As we saw earlier, a net is a $\mathrm{pg}(n-1,t,t)$, and corresponds to $t-1$ MOLS; its *deficiency* is $d = n - t$ (the number of extra squares required for an affine plane). The complement of its point graph is a pseudo-geometric $(n-1, d-1, d-1)$ graph. If the inequality holds, it is geometric, and hence the required squares exist. This result is due to Bruck (1963).

Bose's result was refined by Neumaier (1979), who derived the same conclusion from an inequality on the intersection numbers. This gives a non-existence criterion for strongly regular graphs: no such graph whose intersection numbers satisfy Neumaier's inequality but fail to be pseudo-geometric can exist.

Several results about projective planes or GQs have been extended to partial geometries, including bounds on subgeometries. Thas and De Clerck (1978) determined all partial geometries embeddable in projective spaces, extending the Buekenhout–Lefèvre theorem 4.15.

Finally, partial geometries provide us with a new "stroke" which can be used to enlarge the class of Buekenhout geometries. Examples of its use include the 3-nets of Laskar (1974), Dunbar and Laskar (1978), and the EPGs of Hobart and Hughes (1990), with diagrams o—pg—o——o and o—c—o—pg—o respectively.

A number of generalizations have been considered (semi-partial geometries, partial geometric designs, $1\frac{1}{2}$-designs, partially balanced designs, and so on). One further type is the class of *near polygons*, partial linear spaces in which every line contains a unique nearest neighbour to every point. This class includes the generalized polygons; near polygons with point diameter 2 are the same thing as GQs. Shult and Yanushka (1980) showed that a near polygon with thick lines containing a quadrangle has diagram o===o—L—o. Among such geometries are the "dual polar spaces", characterized by Cameron (1982).

References

Ahrens, R.W., and G. Szekeres
[1969] On a combinatorial generalization of 27 lines associated with a cubic surface, *J. Austral. Math. Soc.* 485–492.

Aschbacher, M.
[1984] On the maximal subgroups of the finite classical groups, *Invent. Math.* **76**, 469–514.

Bader, L., and G. Lunardon
[1989] On the flocks of $Q^{+}(3,q)$, *Geom. Dedicata* **29**, 177–183.
Barlotti, A.
[1955] Un'estensione del teorema di Segre–Kustaanheimo, *Boll. Un. Mat. Ital.* **10**, 498–506.
Bénéteau, L.
[1986] Symplectic geometry, quasigroups and Steiner systems, in: *Combinatorics '84, Ann. Discrete Math.* **30**, 9–13.
Birkhoff, G.
[1935] Combinatorial relations in projective geometries, *Ann. of Math.* **37**, 823–843.
Blokhuis, A., and A.E. Brouwer
[1986] Blocking sets in Desarguesian projective planes, *Bull. London Math. Soc.* **18**, 132–134.
Blokhuis, A., and H. Wilbrink
[1987] A characterization of exterior lines of certain sets of points in PG$(2,q)$, *Geom. Dedicata* **23**, 253–254.
Bose, R.C.
[1963] Strongly regular graphs, partial geometries and partially balanced designs, *Pacific J. Math.* **13**, 389–419.
Brouwer, A.E., and A.M. Cohen
[1986] Local recognition of Tits geometries of classical type, *Geom. Dedicata* **20**, 181–199.
Brouwer, A.E., and A. Schrijver
[1977] The blocking number of an affine space, *J. Combin. Theory A* **24**, 251–253.
Brown, K.
[1989] *Buildings* (Springer, New York).
Bruck, R.H.
[1958] *A Survey of Binary Systems* (Springer, Berlin).
[1963] Finite nets II: Uniqueness and embedding, *Pacific J. Math.* **13**, 421–457.
Bruck, R.H., and H.J. Ryser
[1949] The nonexistence of certain finite projective planes, *Canad. J. Math.* **1**, 88–93.
Bruen, A.A.
[1971a] Blocking sets in finite projective planes, *SIAM J. Appl. Math.* **21**, 380–392.
[1971b] Partial spreads and replaceable nets, *Canad. J. Math.* **23**, 389–391.
Bruen, A.A., and J.A. Thas
[1977] Blocking sets, *Geom. Dedicata* **6**, 193–203.
Bruen, A.A., J.A. Thas and A. Blokhuis
[1988] M.D.S. codes, arcs in PG(n,q) with q even, and a solution of three fundamental problems of B. Segre, *Invent. Math.* **92**, 441–459.
Buekenhout, F.
[1969] Une caractérisation des espaces affines basée sur la notion de droite, *Math. Z.* **111**, 367–371.
[1979] Diagrams for geometries and groups, *J. Combin. Theory A* **27**, 121–151.
Buekenhout, F., and X. Hubaut
[1977] Locally polar spaces and related rank three groups, *J. Algebra* **45**, 391–434.
Buekenhout, F., and C. Lefèvre
[1974] Generalized quadrangles in projective spaces, *Arch. Math. (Basel)* **25**, 540–552.
Buekenhout, F., and E.E. Shult
[1974] On the foundations of polar geometry, *Geom. Dedicata* **3**, 155–170.
Cameron, P.J.
[1976] *Parallelisms of Complete Designs, London Mathematical Society Lecture Note Series,* Vol. 23 (Cambridge University Press, Cambridge).
[1980] Extremal results and configuration theorems for Steiner systems, in: *Topics in Steiner Systems,* eds. C.C. Lindner and A. Rosa, *Ann. Discrete Math.* **7**, 43–63.
[1982] Dual polar spaces, *Geom. Dedicata* **12**, 75–85.
Cameron, P.J., and W.M. Kantor
[1979] 2-transitive and antiflag transitive collineation groups of finite projective spaces, *J. Algebra* **60**, 384–422.

Cameron, P.J., and R.A. Liebler
[1982] Tactical decompositions and orbits of projective groups, *Linear Algebra Appl.* **46**, 91–102.
Cameron, P.J., and F. Mazzocca
[1986] Bijections which preserve blocking sets, *Geom. Dedicata* **21**, 219–229.
Cameron, P.J., D.R. Hughes and A. Pasini
[1990] Extended generalized quadrangles, *Geom. Dedicata* **35**, 193–228.
Cohen, A.M., and B.N. Cooperstein
[1983] A characterization of some geometries of exceptional Lie type, *Geom. Dedicata* **15**, 73–105.
Coxeter, H.S.M.
[1958] Twelve points in PG(5, 3) with 95 040 self-transformations, *Proc. Roy. Soc. (A)* **247**, 279–293.
Coxeter, H.S.M., and W.O.J. Moser
[1957] *Generators and Relations for Discrete Groups* (Springer, Berlin).
de Bruijn, N.G., and P. Erdős
[1948] On a combinatorial problem, *Indag. Math.* **10**, 421–423.
Del Fra, A., D. Ghinelli, T. Meixner and A. Pasini
[1991] Flag-transitive extensions of C_n geometries, *Geom. Dedicata* **37**, 253–273.
Dembowski, P.
[1964a] Möbiusebenen gerader Ordnung, *Math. Ann.* **157**, 179–205.
[1964b] Eine Kennzeichnung der endlichen affinen Raüme, *Arch. Math.* **15**, 146–154.
[1968] *Finite Geometries* (Springer, Berlin).
Dembowski, P., and A. Wagner
[1960] Some characterizations of finite projective spaces, *Arch. Math.* **11**, 465–469.
Denniston, R.H.F.
[1969a] Some maximal arcs in finite projective planes, *J. Combin. Theory* **6**, 317–319.
[1969b] Non-existence of a certain projective plane, *J. Austral. Math. Soc.* **10**, 214–218.
di Paola, J.
[1969] On minimum blocking coalitions in small projective plane games, *SIAM J. Appl. Math.* **17**, 378–392.
Dickson, L.E.
[1901] *Linear Groups, with an Exposition of the Galois Field Theory* (Teubner, Leipzig). Reprinted: 1958 (Dover, New York.).
Drinfel'd, V.G., and S.G. Vladut
[1983] Number of points of an algebraic curve, *Funct. Anal. Appl.* **17**, 53–54.
Dunbar, J., and R. Laskar
[1978] Partial geometry of dimension 3, *J. Combin. Theory A* **24**, 187–201.
Fano, G.
[1892] Sui postulati fondamentali della geometria proiettiva, *Giorn. Mat.* **30**, 106–132.
Feit, W., and G. Higman
[1964] The nonexistence of certain generalized polygons, *J. Algebra* **1**, 114–131.
Fischer, B.
[1964] Distributive Quasigruppen endlicher Ordnung, *Math. Z.* **83**, 267–303.
Fong, P., and G.M. Seitz
[1973] Groups with a BN-pair of rank 2, I, *Invent. Math.* **21**, 1–57.
[1974] Groups with a BN-pair of rank 2, II, *Invent. Math.* **24**, 237–292.
Gleason, A.M.
[1956] Finite Fano planes, *Amer. J. Math* **78**, 797–807.
Goppa, V.D.
[1984] Codes and information, *Russian Math. Surveys* **39**, 87–141.
Graham, R.L., K. Leeb and B.L. Rothschild
[1972] Ramsey's theorem for a class of categories, *Adv. in Math.* **8**, 417–433.
Greene, C.
[1976] Weight enumerators and the geometry of linear codes, *Studia Appl. Math.* **55**, 119–128.

Haemers, W.H., and C. Roos
[1981] An inequality for generalized hexagons, *Geom. Dedicata* **10**, 219–222.
Hall, J.I.
[1974] Steiner triple systems with geometric minimally generated subsystems, *Quart. J. Math. Oxford (2)* **25**, 41–50.
Hall Jr, M.
[1967] Group theory and block designs, in: *Proc. Int. Conf. on Theory of Groups* (Gordon and Breach, New York) pp. 460–472.
Hamada, N.
[1973] On the p-rank of the incidence matrix of a balanced or partially balanced incomplete block design and its application to error-correcting codes, *Hiroshima Math. J.* **3**, 154–226.
Hanssens, G.
[1988] A characterization of point-line geometries for finite buildings, *Geom. Dedicata* **25**, 297–315.
Hering, C.
[1974] Transitive linear groups and linear groups which contain irreducible subgroups of prime order, I, *Geom. Dedicata* **2**, 425–460.
[1979] On the structure of collineation groups of finite projective planes, *Abh. Math. Sem. Univ. Hamburg* **49**, 155–182.
[1985] Transitive linear groups and linear groups which contain irreducible subgroups of prime order, II, *J. Algebra* **93**, 151–164.
Higman, D.G.
[1975] Invariant relations, coherent configurations and generalized polygons, in: *Combinatorics,* eds. M. Hall Jr and J.H. van Lint (Reidel, Dordrecht) pp. 347–363.
Higman, D.G., and J.E. McLaughlin
[1961] Geometric ABA-groups, *Illinois J. Math.* **5**, 382–397.
Hilbert, D.
[1899] *Grundlagen der Geometrie* (Teubner, Leipzig).
Hill, R.
[1978] Caps and codes, *Discrete Math.* **22**, 111–137.
Hirschfeld, J.W.P.
[1979] *Projective Geometries over Finite Fields* (Oxford University Press, Oxford).
[1985] *Finite Projective Spaces of Three Dimensions* (Oxford University Press, Oxford).
Hirschfeld, J.W.P., and J.A. Thas
[1980a] The characterization of projections of quadrics over finite fields of even order, *J. London Math. Soc. (2)* **22**, 226–238.
[1980b] Sets of type $\{1, n, q+1\}$ in PG(d, q), *Proc. London Math. Soc. (3)* **41**, 254–278.
[1991] *General Galois Geometries* (Oxford University Press, Oxford).
Hobart, S.A., and D.R. Hughes
[1990] Extended partial geometries, *European J. Combin.* **11**, 357–372.
Hoffman, A.J.
[1960] On the uniqueness of the triangular association scheme, *Ann. Math. Statist.* **31**, 492–497.
Hughes, D.R.
[1957] Generalized incidence matrices over group algebras, *Illinois J. Math.* **1**, 545–551.
[1965] On t-designs and groups, *Amer. J. Math.* **87**, 761–778.
Hughes, D.R., and F.C. Piper
[1973] *Projective Planes* (Springer, New York).
Huppert, B.
[1957] Zweifach transitive, auflösbare Permutationsgruppen, *Math. Z.* **68**, 126–150.
Ivanov, A.A.
[1993] Graphs with projective subconstituents which contain short cycles, in: *Surveys in Combinatorics, London Math. Soc. Lecture Notes,* Vol. 187, ed. K. Walker (Cambridge University Press, Cambridge) pp. 173–190.

Jamison, R.E.
[1977] Covering finite fields with cosets of subspaces, *J. Combin. Theory A* **22**, 253–266.

Järnefelt, G.
[1951] Reflections on a finite approximation to Euclidean geometry, Physical and astronomical prospects, *Ann. Acad. Sci. Fenn.* **96**.

Jungnickel, D.
[1989] *Blocking Sets: Proc. Gießen Conf., 1989* (Universität Giessen, Giessen).

Kahn, J.
[1980] Locally projective-planar lattices which satisfy the Bundle Theorem, *Math. Z.* **175**, 219–247.

Kantor, W.M.
[1974] Dimension and embedding theorems for geometric lattices, *J. Combin. Theory A* **17**, 173–195.
[1975] Envelopes for geometric lattices, *J. Combin. Theory A* **18**, 12–26.
[1981] Some geometries that are almost buildings, *European. J. Combin.* **2**, 239–247.
[1982] Generation of linear groups, in: *The Geometric Vein: The Coxeter Festschrift,* eds. C. Davis, B. Grünbaum and F.A. Sherk (Springer, Berlin) pp. 479–509.
[1983] Spreads, translation planes and Kerdock sets, I, II, *SIAM J. Algebraic and Discrete Methods* **3**, 151–165, 308–318.
[1987] Primitive permutation groups of odd degree, with an application to finite projective planes, *J. Algebra* **106**, 15–45.
[1988] Reflections on concrete buildings, *Geom. Dedicata* **25**, 121–145.

Kantor, W.M., R.A. Liebler and J. Tits
[1987] On discrete chamber-transitive automorphism groups of affine buildings, *Bull. Amer. Math. Soc.* **16**, 129–133.

Kimberley, M.E.
[1971] On the construction of certain Hadamard designs, *Math. Z.* **119**, 41–59.

Kleidman, P.B., and M.W. Liebeck
[1988] A survey of the maximal subgroups of the finite simple groups, *Geom. Dedicata* **25**, 375–389.
[1990] *The Subgroup Structure of the Finite Classical Groups, London Mathematical Society Lecture Note Series,* Vol. 129 (Cambridge University Press, Cambridge).

Lam, C.W., J. McKay, S. Swiercz and L. Thiel
[1983] The nonexistence of ovals in a projective plane of order 10, *Discrete Math.* **45**, 319–321.

Lam, C.W., S. Swiercz and L. Thiel
[1986] The nonexistence of codewords of weight 16 in a projective plane of order 10, *J. Combin. Theory A* **42**, 207–214.
[1989] The nonexistence of finite projective planes of order 10, *Canad. J. Math.* **41**, 1117–1123.

Lam, C.W., G. Kolesova and L. Thiel
[1991] A computer search for finite projective planes of order 9, *Discrete Math.* **92**, 187–195.

Lander, E.S.
[1981] Symmetric designs and self-dual codes, *J. London Math. Soc. (2)* **24**, 193–204.

Laskar, R.
[1974] Finite nets of dimension three, *J. Algebra* **32**, 8–25.

Lenz, H.
[1954] Zur Begründung der analytischen Geometrie, *Sitz.-Ber. Bayer. Akad. Wiss.* 17–72.

Liebeck, M.W.
[1985] On the orders of maximal subgroups of the finite classical groups, *Proc. London Math. Soc. (3)* **50**, 426–446.

Lüneburg, H.
[1965] *Die Suzukigruppen und ihre Geometrien, Lecture Notes in Mathematics,* Vol. 10 (Springer, Berlin).
[1967] Kreishomogene endliche Möbius-Ebenen, *Math. Z.* **101**, 68–70.

Lunelli, L., and M. Sce
[1958] *k-archi completi nei piani proiettivi desarguesiani di rango 8 e 16* (Centro Calcol. Numerici, Politecnico Milano).

MacWilliams, F.J., N.J.A. Sloane and J.G. Thompson

[1973] On the existence of a projective plane of order 10, *J. Combin. Theory A* **14**, 66–78.

Manin, Y.A., and S.G. Vladut

[1985] Linear codes and modular curves, *J. Soviet Math.* **30**, 2611–2643.

Mazzocca, F., and G. Tallini

[1985] On the non-existence of blocking sets in PG(n, q) and AG(n, q) for all large enough n, *Simon Stevin* **59**, 43–50.

McLaughlin, J.E.

[1967] Some groups generated by transvections, *Arch. Math.* **18**, 364–368.

Mendelsohn, E.

[1972] Every group is the collineation group of a projective plane, *J. Geometry* **2**, 97–106.

Neumaier, A.

[1979] Strongly regular graphs with smallest eigenvalue $-m$, *Arch. Math.* **33**, 392–400.

[1984] Some sporadic geometries related to PG(3, 2), *Arch. Math.* **42**, 89–96.

Orchel, A.W.

[1978] Thesis (University of London).

Ostrom, T.G.

[1966] Replaceable nets, net collineations and net extensions, *Canad. J. Math.* **18**, 666–672.

[1970] *Finite Translation Planes, Lecture Notes in Mathematics,* Vol. 158 (Springer, Berlin).

Ostrom, T.G., and A. Wagner

[1959] On projective and affine planes with transitive collineation groups, *Math. Z.* **71**, 186–199.

Payne, S.E., and J.A. Thas

[1984] *Finite Generalized Quadrangles, Research Notes in Mathematics* (Pitman, Boston).

Pellegrino, G.

[1971] Sul massimo ordine delle calotte in $S_{4,3}$, *Matematiche* **25**, 149–157.

Penttila, T.

[1984] Thesis (Oxford University).

Percsy, N.

[1981] Locally embeddable geometries, *Arch. Math. (Basel)* **37**, 184–192.

Pickert, G.

[1955] *Projektive Ebenen* (Springer, Berlin).

Ronan, M.A.

[1989] *Lectures on Buildings* (Academic Press, Boston).

Ronan, M.A., and S.D. Smith

[1985] Sheaves on buildings and modular representations of Chevalley groups, *J. Algebra* **96**, 319–346.

Ronan, M.A., and J. Tits

[1987] Building buildings, *Math. Ann.* **278**, 291–306.

Segre, B.

[1955] Sulle ovali nei piani lineari finiti, *Atti Accad. Naz. Lincei Rendic.* **17**, 141–142.

[1961] *Lectures on Modern Geometry* (Cremonese, Roma).

Shafarevich, I.R.

[1974] *Basic Algebraic Geometry* (Springer, Berlin).

Shrikhande, S.S.

[1959] The uniqueness of the L_2 association scheme, *Ann. Math. Statist.* **30**, 781–798.

Shult, E.E., and A. Yanushka

[1980] Near n-gons and line systems, *Geom. Dedicata* **9**, 1–72.

Sprague, A.

[1985] Rank 3 incidence structures admitting dual-linear, linear diagrams, *J. Combin. Theory A* **38**, 254–259.

Tallini, G.

[1981] On a characterization of the Grassman manifold representing the lines in a projective space, in: *Finite Geometries and Designs, London Mathematical Society Lecture Note Series,* Vol. 49, eds. P.J. Cameron, J.W.P. Hirschfeld and D.R. Hughes (Cambridge University Press, Cambridge) pp. 354–358.

Tallini-Scafati, M.
[1967] Caraterizzazione grafica delle forme hermitiane di un $S_{r,q}$, *Rend. Mat. e Appl.* **26**, 273–303.

Tarry, G.
[1901] Le problème des 36 officiers, *C.R. Acad. Franc. Avanc. Sci. Nat.* **2**, 170–203.

Thas, J.A.
[1975] Some results concerning $\{(q+1)(n-1), n\}$-arcs and $\{(q+1)(n-1)+1, n\}$-arcs in finite projective planes of order q, *J. Combin. Theory A* **19**, 228–232.
[1983] Elementary proofs of two fundamental theorems of B. Segre without using the Hasse–Weil theorem, *J. Combin. Theory A* **35**, 58–66.
[1992] Recent results on flocks, maximal exterior sets, and inversive planes, in: *Combinatorics '88*, eds. A. Barlotti et al. (Mediterranean Press, Roma) pp. 95–108.

Thas, J.A., and F. de Clerck
[1978] Partial geometries in finite projective spaces, *Arch. Math. (Basel)* **30**, 537–540.

Tits, J.
[1962] Géométries polyédriques et groupes simples, in: *2e Reunion Math. d'expression latine, Firenze/Bologna, 1961* (Cremonese, Roma) pp. 66–68.
[1974] *Buildings of Spherical Type and Finite BN-pairs, Lecture Notes in Mathematics*, Vol. 386 (Springer, Berlin).
[1976] Non-existence de certains polygones généralisés, *Invent. Math.* **36**, 275–284.
[1981] A local approach to buildings, in: *The Geometric Vein: The Coxeter Festschrift*, eds. C.C. Davis, B. Grünbaum and F.A. Sherk (Springer, Berlin) pp. 519–547.
[1986] Immeubles de type affine, in: *Buildings and the Geometry of Diagrams, Lecture Notes in Mathematics*, Vol. 1181 (Springer, Berlin) pp. 157–191.

van Lint, J.H., and G. van der Geer
[1987] *Introduction to Coding Theory and Algebraic Geometry, DMV Sem.*, Vol. 12 (Birkhäuser, Basel).

Veblen, O., and J.W. Young
[1916] *Projective Geometry*, 2 volumes (Ginn & Co., Boston).

Veldkamp, F.D.
[1959] Polar geometry I–V, *Proc. Kon. Nederl. Akad. Wetensch. A* **62**, 512–551; *A* **63**, 207–212.

Walker, M.
[1977] On the structure of collineation groups containing symmetries of generalized quadrangles, *Invent. Math.* **40**, 245–265.

Weiss, R.
[1979] The nonexistence of certain Moufang polygons, *Invent. Math.* **51**, 261–266.

Wild, P.R.
[1980] Thesis (University of London).

Wilker, J.B.
[1977] On Mendelsohn's pathology for projective and affine planes, *Bull. London Math. Soc.* **9**, 163–164.

Young, H.P.
[1973] Affine triple systems and matroid designs, *Math. Z.* **132**, 343–359.

CHAPTER 14

Block Designs

Andries E. BROUWER

Department of Mathematics, Eindhoven University of Technology, 5600 MB Eindhoven, The Netherlands

Contents

HANDBOOK OF COMBINATORICS
Edited by R. Graham, M. Grötschel and L. Lovász

Introduction

The theory of finite incidence structures can be divided into three parts: finite geometry, theory of block designs and hypergraph theory. It is difficult to tell these apart, but roughly speaking finite geometry concerns itself with incidence structures that satisfy geometrical, that is, structural requirements, the theory of block designs concerns itself with incidence structures that satisfy numerical requirements, and hypergraph theory concerns itself with completely arbitrary systems of sets. Typically, for designs with a small number of points, or designs that satisfy certain inequalities (almost) with equality, it is possible to extract structural information from the numerical data; on the other hand, in the general case pure chaos reigns, and not much more can be said about designs than about arbitrary hypergraphs. A relatively high structured part of design theory is the theory of association schemes. These are discussed in chapter 15. In this chapter we very briefly discuss the main facts concerning t-designs and a few related subjects. Only in the section on Witt designs and Golay codes, the longest of this chapter, do we attempt a somewhat fuller treatment.

1. Generalities

A *(block) design* is a pair $(X, \mathscr{B})$ where X is a set (the *point set*) and $\mathscr{B}$ (the family of *blocks*) is a family of subsets of X (not necessarily pairwise distinct) such that some regularity condition is satisfied. Thus, there are many types of (block) design; often however, "block design" is taken to be an abbreviation for "balanced incomplete block design" (BIBD) (see below).

If two members of $\mathscr{B}$ are incident with the same points then we say that the design has *repeated blocks*. A design without repeated blocks is called *simple*.

The (point–block) *incidence matrix* of a design is the 0–1 matrix with rows indexed by the points and columns indexed by the blocks of the design, where the (p, B) entry is 1 if $p \in B$ and 0 otherwise. We shall use I for the identity matrix and J for the all-1 matrix of any suitable size.

The *dual* of a design $(X, \mathscr{B})$ is the design $(\mathscr{B}, X)$; the incidence matrix of the dual is the transpose of that of the original design. (Note that we here identify the point x with the set of blocks containing x; in general we shall not worry when the incidence relation between points and blocks is different from the ordinary membership relation.)

A design is called *square* whenever its incidence matrix is square. A design is said to have a *polarity* whenever its incidence matrix can be written as a symmetric matrix. (Note that what are called "symmetric designs" in the literature, are square BIBDs, and do not in general have a polarity.)

When the design is finite, some numbers have conventional denotations. The number of points is called v, the number of blocks b, and the size of the blocks (when constant) k. The *replication number* $r(x)$ or r_x of a point x is the number of blocks on x. We write just r if it does not depend on x.

Let us give names to some important types of design.

A t-(v, k, λ) *design* is a block design on v points with blocks of size k such that any set of t points is covered by (i.e., is a subset of) precisely λ blocks. We shall always assume that $k \geqslant t$; now a t-(v, k, λ) design is also an i-(v, k, λ_i) design for $0 \leqslant i \leqslant t$, where $\lambda_t = \lambda$ and $\lambda_i = (v-i)\lambda_{i+1}/(k-i)$ for $i < t$. Note that $\lambda_0 = b$ and $\lambda_1 = r$.

A *Steiner system* $S(t, k, v)$ is a t-$(v, k, 1)$ design. One also uses $S_\lambda(t, k, v)$ as synonym for t-(v, k, λ).

A *balanced incomplete block design* (*BIBD*) is a 2-(v, k, λ) design (where the "incomplete" originally required that $k < v$, but that requirement is now often forgotten). This concept came from statistics, and some statistical terminology and notation is still current; the points are sometimes called "varieties" (that is why there are v of them) and the blocks "treatments".

A *pairwise balanced design* (*with index* λ) is a design such that any pair of points is in precisely λ blocks. Clearly, a BIBD is a pairwise balanced design (PBD) with the additional requirement that all blocks be of the same size k. (We shall not need t-wise balanced designs for $t \neq 2$. For some results on these, cf. Kramer 1983.)

An (r, λ)-*design* is a pairwise balanced design with index λ in which each point lies in r blocks.

A *partial linear space* (sometimes called "semilinear space") is a design $(X, \mathscr{L})$ with point set X and set of *lines* $\mathscr{L}$ such that no two lines have two points in common. The dual of a partial linear space is again a partial linear space.

A *linear space* is a pairwise balanced design of index unity, i.e., a partial linear space such that any two of its points are joined by a line. (Do not confuse these linear spaces with those from functional analysis.)

A *projective plane* $\mathrm{PG}(2, n)$ is a Steiner system $S(2, n+1, n^2+n+1)$. An *affine plane* $\mathrm{AG}(2, n)$ is a Steiner system $S(2, n, n^2)$. More generally, $\mathrm{PG}(d, q)$ and $\mathrm{AG}(d, q)$ denote the projective and affine space of order q and dimension d. (For $d \geqslant 3$, these are necessarily coordinatized by a field (everything is finite in this chapter), and q is a prime power.)

A subset of the point set of a partial linear space is called a *subspace* if any line that meets it in at least two points, is entirely contained in it.

2. De Bruijn–Erdős and Fisher inequalities and variations

Often it is possible to show that a design has at least as many blocks as it has points. One standard argument goes as follows: let the design have incidence matrix A, and suppose that AA^{T} is nonsingular. Then A does not have more rows than columns. For instance, when the design is a BIBD then $AA^{\mathrm{T}} = (r-\lambda)I + \lambda J$ with determinant $kr(r-\lambda)^{v-1}$ so that if $v > k$ (and $b > 0$) then $b \geqslant v$. This is called Fisher's inequality. Many generalizations exist; e.g., for $2s$-(v, k, λ) designs one has the Petrenjuk–Ray-Chaudhuri–Wilson inequality (Petrenjuk 1968, Ray-Chaudhuri and Wilson 1975) which states that if $v \geqslant k + s$, then $b \geqslant \binom{v}{s}$.

A beautiful argument due to Conway for the case of linear spaces is the following. Suppose $b \leqslant v$. Observe that if the point x is not on the line L then $r_x \geqslant k_L$ where k_L denotes the size of the line L. Now we have

$$b = \sum_L 1 = \sum_L \sum_{x \notin L} \frac{1}{v - k_L} \leqslant \frac{b}{v} \sum_x \sum_{L \not\ni x} \frac{1}{b - r_x} = \frac{b}{v} \sum_x 1 = b$$

and hence all inequalities are equalities so that $b = v$ and $r_x = k_L$ whenever $x \notin L$. This shows that in a linear space $b \geqslant v$ with equality iff the linear space is a (possibly degenerate) projective plane, a result first published by De Bruijn and Erdős (1948) (Hanani 1951 says he found this in 1938; see also Motzkin 1951.)

Ryser (1968) proved $b \geqslant v$ for pairwise balanced designs (that are nondegenerate: $\lambda > 0$, $v > 1$ and $r_x > \lambda$ for each x), thus generalizing both Fisher and De Bruijn–Erdős. (Proof: Again AA^{T} is nonsingular.)

In case a group action is given we have analogous results: Let $\bar{v}$ and $\bar{b}$ be the number of point and block orbits of some design under a group G. If the design is square, with nonsingular incidence matrix then $\bar{v} = \bar{b}$ (Brauer 1941, cf. Parker 1957, Hughes 1957, Dembowski 1958). (Proof: In case $G = \langle g \rangle$, let the permutation matrices P, Q represent the action of g on points and blocks, so that $PAQ = A$. Now $\operatorname{tr} Q = \operatorname{tr} A^{-1}P^{-1}A = \operatorname{tr} P^{-1} = \operatorname{tr} P$, so that g fixes as many points as blocks. In the general case apply Burnside's lemma.) In the general case, where the incidence matrix need not be square and has rank ρ we have (Block 1967) $\bar{v} \leqslant \bar{b} + v - \rho$ and $\bar{b} \leqslant \bar{v} + b - \rho$. (This is a purely combinatorial result: it holds when $\bar{v}$ and $\bar{b}$ are the number of point and block classes in a tactical decomposition.) When A has rank v (as is the case for 2-designs) we find $\bar{v} \leqslant \bar{b}$. Similarly, generalizing the Wilson–Petrenjuk inequality, Kreher (1986) showed for a $2s$-(v, k, λ) design with $v \geqslant k + s$ that the number of block orbits is at least the number of orbits of G on s-sets.

A *planar space* is a linear space in which certain subspaces are called *planes*, and one requires that any three noncollinear points determine a unique plane. Hanani (1954/5) shows that $p \geqslant v$ if p is the number of planes, and that equality holds iff the space is a projective 3-space. More generally, in any matroid with v points and h hyperplanes (cf. chapter 9) we have $h \geqslant v$ (cf. Motzkin 1951); Conway's proof generalizes to this situation since $r_x \geqslant k_H$ holds by induction. Equality holds if and only if we have a (possibly degenerate) projective space.

3. Square 2-designs

Cases of equality in one of the above inequalities are especially interesting.

Ito calls a $2s$-design *tight* when equality holds in the Wilson–Petrenjuk inequality, and Ito (1975, 1978), Enomoto et al. (1970) and Bremner (1979) determine all tight 4-designs. (The only example up to complementation is the unique Steiner system $S(4, 7, 23)$ with $b = 253$.) Peterson (1977) shows that no tight 6-designs exist. It should be feasible to show that no tight $2s$-designs exist with $2s > 6$. (Cf. Deza 1975, Bannai 1977.)

Ryser (1968) shows that a nondegenerate pairwise balanced design that is square (i.e., has $b = v$) either has constant k and r (i.e., is a square BIBD) or has precisely two distinct block sizes k_1 and k_2, and $k_1 + k_2 = v + 1$. The dual of a design of the latter kind is called a *λ-design*. (But note that specifying λ will cause confusion with the concept of t-design.) Examples may be obtained by starting with a square 2-$(v, k, k - \lambda)$ design and replacing all blocks B containing a fixed point x_0 by $\{x_0\} \cup X \backslash B$. The λ-design conjecture says that all examples are obtained in this way. See, e.g., Bridges (1983), Woodall (1970), Shrikhande and Singhi (1976) and Seress (1987).

There exists a large body of literature on square BIBDs; they are usually called "symmetric designs", Dembowski uses "projective designs", here we shall call them SBIBDs. The most popular examples are projective planes PG$(2, q)$ and, more generally, the designs of points and hyperplanes in a projective space PG(d, q). For a list of all known SBIBDs, see the appendix. Let us list some important properties here. By definition $b = v$, and it follows that $r = k$ and $k(k-1) = (v-1)\lambda$. The dual of an SBIBD is again one (since $AJ = JA = kJ$ and hence $AA^{\mathrm{T}} = (r - \lambda)I + \lambda J = A^{\mathrm{T}}A$), and it follows that any two blocks have λ points in common. For an SBIBD the following parameter restriction is known.

Theorem (Bruck–Ryser–Chowla). *If a SBIBD has parameters* 2-(v, k, λ) *then the following holds.*

(i) (Bruck and Ryser 1949, Schutzenberger 1949, Shrikhande 1950). *If v is even then $k - \lambda$ is a square.*

(ii) (Chowla and Ryser 1950). *If v is odd then the equation $z^2 = (k - \lambda)x^2 + (-1)^{(v-1)/2}\lambda y^2$ has a nontrivial integer solution.*

Proof. (i) follows from $(\det A)^2 = k^2(r - \lambda)^{v-1}$; (ii) expresses the fact that the quadratic forms $\sum_{j=1}^{v} (\sum_{i=1}^{v} a_{ij}x_i)^2$ and $(k - \lambda) \sum_{i=1}^{v} x_i^2 + \lambda(\sum_{i=1}^{v} x_i)^2$ are equivalent. In fact (ii) is necessary and sufficient for the existence of a rational matrix A satisfying the abovementioned equations. □

Condition (i) e.g., rules out 2-(22, 7, 2); (ii), e.g., rules out PG(2, 6), i.e., 2-(43, 7, 1). No other restrictions are known, i.e., no case is known of a parameter set for a SBIBD passing the Bruck–Ryser–Chowla criterion (and Fisher's inequality), while it is known that no corresponding design exists. (However, just recently Lam has announced that exhaustive computer search shows that no projective plane of order 10 exists.)

There is an analogue of this in case a group action is given (Hughes 1957): *Let α be an automorphism of prime order p of a SBIBD, with f fixed points. Then α has $m = f + (v - f)/p$ point and block orbits, and if we put $n = k - \lambda$ then the equation $z^2 = nx^2 + (-1)^{(m-1)/2}p^{f+1}\lambda y^2$ has a nontrivial integer solution.*

Lander (1983, Theorem 3.20) gives the following rather strong condition on p and f: *Suppose that p is odd and that we have $q^j \equiv -1 \pmod p$ for some j and for some prime q dividing the square free part of n. Then f is odd.*

A bound on f is given by Wilbrink (cf. Lander 1983, (3.7)) (using Haemers' interlacing result, see Haemers 1980, (3.1.1)): *Let α be a nonidentity automorphism of a SBIBD with f fixed points. Then $f \leq v - 2n$ and $f \leq k + \sqrt{n}$, and if equality holds in either inequality then α must be an involution and every nonfixed block contains precisely λ fixed points.*

Corollary (Feit 1970). *$f \leq v/2$ and if equality holds then $v = 4n$ and α is an involution.*

More precisely, if k_0 is the average number of fixed points on a nonfixed block, then $\lambda f \leq k_0(k + \sqrt{n})$. Trivially $k_0 \leq \lambda$, but sometimes sharper bounds follow from the observation that each orbit of a nonfixed block yields an equidistant constant weight code with word length $v - f$, weight $k - k_0$ and Hamming distance $2(k - \lambda)$. Another trivial bound is: if $f > 0$ then $p \leq k$.

A *difference set* D in a group G (written additively) is a set such that $(G, \{D + g \mid g \in G\})$ is a SBIBD; equivalently, $|G| = v$, $|D| = k$, and in the list of differences $d_1 - d_2$ $(d_1, d_2 \in D)$ each nonzero group element occurs precisely λ times. (More generally, one considers *difference families* and designs where the block set is the union of several G-orbits. This is called Bose's (1939) method of symmetrically repeated differences. One calls the given representatives for the orbits *base blocks* or *initial blocks* and constructing the G-orbits *developing* (mod G).)

An important example was given by Singer (1938) who showed that the design of points and hyperplanes in the projective space PG(d, q) has a cyclic difference set (i.e., a difference set in $G = \mathbb{Z}_v$).

The most important result on difference sets is the multiplier theorem.

Theorem (Hall and Ryser 1951). *Let D be a difference set in $G = \mathbb{Z}_v$ and suppose that p is a prime with $p \mid k - \lambda$, $(p, v) = 1$, $p > \lambda$. Then p is a multiplier of D, i.e., $pD = D + e$ for some $e \in G$.*

Many generalizations exist. For a discussion of multiplier theorems see Ryser (1963, chapter 9), Hall (1967, chapter 11), Dembowski (1968, pp. 87–90), Lander (1983), Beth et al. (1985, chapter 6). One may always choose D so as to be fixed by a given multiplier p (Mann 1965) (for: p induces an automorphism of the design and fixed the point 0 hence must also fix some block D), and, in case G is abelian, one may even find a D which is fixed by all (numerical) multipliers of D (McFarland and Rice 1978).

A lot has been done in the special case where $\lambda = 1$ ("planar difference sets"), see, for example, Jungnickel and Vedder (1984) and Wilbrink (1985); for a survey, see the books mentioned.

For a table of parameters of the known symmetric designs, see the appendix. Dembowski (1968, pp. 105–108), Baumert (1971), Hall (1974), Kibler (1978), Lander (1983), and Beth et al. (1985, chapter 6) survey the known difference sets. A good discussion of difference sets, and an update to Lander's tables can be found in Jungnickel (1992).

4. Inequalities in 2-designs

Consider the incidence matrix A of a BIBD, so that $AA^{\mathrm{T}} = (r - \lambda)I + \lambda J$. Since AA^{T} has spectrum $(kr)^1(r - \lambda)^{v-1}$ (with multiplicities written as exponents), it follows that $A^{\mathrm{T}}A$ has spectrum $(kr)^1(r - \lambda)^{v-1}0^{b-v}$, and hence the matrix $Q = (r - \lambda)I + (\lambda k/r)J - A^{\mathrm{T}}A$ is positive semidefinite (it has spectrum $0^v(r - \lambda)^{b-v}$). It follows that if Q_m is a principal submatrix of Q of order m, then $\det Q_m \geqslant 0$ (and $\det Q_m = 0$ for $m > b - v$). This result is known as *Connor's inequalities*, cf. Connor (1953). The case $m = 1$ of these inequalities gives us Fisher's inequality again. The case $m = 2$ of these inequalities says that for any two blocks B, C we have

$$k + \lambda - r \leqslant |B \cap C| \leqslant r - k - \lambda + \frac{2\lambda k}{r}.$$

These inequalities are sometimes called *Majumdar's inequalities*, although Connor already states them explicitly. Majumdar (1953) observed that we have equality in the left-hand inequality if and only if $|B \cap D| = |C \cap D|$ for all blocks $D \neq B$, C (provided B and C are not incident with the same set of points), and that equality in the right-hand inequality holds if and only if $|B \cap D| + |C \cap D| = 2\lambda k/r$ for all blocks $D \neq B$, C. (Indeed, let e_B be the Bth unit vector. Then, since Q is positive semidefinite, we have for blocks B, C:

$$0 \leqslant (e_B - e_C)^{\mathrm{T}} Q (e_B - e_C) = 2(|B \cap C| - k - \lambda + r)$$

with equality if and only if $Qe_B = Qe_C$. Thus, meeting in $k + \lambda - r$ points is an equivalence relation, and equivalent blocks meet other blocks in the same number of points. Conversely, if B and C meet all other blocks in the same number of points, then $Q(e_B - e_C) = (|B \cap C| - k - \lambda + r)(e_B - e_C)$ so that $|B \cap C| - k - \lambda + r$ is an eigenvalue of Q, i.e., 0 or $r - \lambda$, and hence $|B \cap C|$ is either k or $k + \lambda - r$. Similarly, we have equality in the right-hand inequality above if and only if $Qe_B = -Qe_C$.) See also Bekers and Haemers (1980). It follows that if a BIBD contains two disjoint blocks, then $r \geqslant k + \lambda$, i.e., $b \geqslant v + r - 1$, a strengthening of Fisher's inequality. A bound on the maximum number of blocks disjoint from a given block was given by Majindar (1962).

Another direct consequence of Connor's inequalities is *Mann's inequality* (Mann 1969): If in a BIBD with $v > k$ a block is repeated m times (i.e., if m blocks are incident with the same set of points), then $b \geqslant mv$. (Indeed, for the corresponding principal submatrix we find $Q_m = (r - \lambda)I - (k(r - \lambda)/r)J$ with eigenvalue $(r - \lambda)(1 - km/r) \geqslant 0$.)

A very good discussion of these results is contained in Wilson (1984); generalizations of these results to t-designs are also given there.

5. Derived and residual designs, extensions

Given a design $\mathscr{D} = (X, \mathscr{B})$ and a point $x \in X$, the *derived design* (at x) is defined as $\mathscr{D}_x = (X \backslash \{x\}, \{B \backslash \{x\} \mid x \in B \in \mathscr{B}\})$ and the *residual design* (at x) as $\mathscr{D}^x =$

$(X\backslash\{x\}, \{B \mid x \notin B \in \mathcal{B}\})$. Conversely, $\mathcal{D}$ is called an *extension* of $\mathcal{D}_x$, and $\mathcal{D}_x$ is called *extendable*. In case $\mathcal{D}$ is a t-(v, k, λ) design we find that $\mathcal{D}_x$ is a $(t-1)$-$(v-1, k-1, \lambda)$ design, while $\mathcal{D}^x$ is a $(t-1)$-$(v-1, k, \lambda_{t-1} - \lambda)$ design (where $\lambda_{t-1} = \lambda(v-t+1)/(k-t+1)$).

Sometimes one can guarantee that an extension exists. Alltop (1975) shows that if t is even then any t-$(2k+1, k, \lambda)$ design extends to a $(t+1)$-$(2k+2, k+1, \lambda)$ design. (Proof: Invent a new point ∞ and add it to the existent blocks; add the complements of all blocks.) When t is odd, we have the following result due to Alltop (1975) and Dehon (1976): if a $(t+1)$-$(2k+1, k, \lambda')$ design is the union of two t-$(2k+1, k, \lambda)$ designs then a $(t+1)$-$(2k+2, k+1, \lambda)$ exists.

Cameron (1973) has determined what SBIBDs can have an extension. He finds that one necessarily has one of the following cases:

(i) $v = 4\lambda + 3$, $k = 2\lambda + 1$, a Hadamard design (cf. the appendix (2));
(ii) $v = (\lambda + 2)(\lambda^2 + 4\lambda + 2)$, $k = \lambda^2 + 3\lambda + 1$;
(iii) $v = 111$, $k = 11$, $\lambda = 1$;
(iv) $v = 495$, $k = 39$, $\lambda = 3$.

(Proof. The extension is a design with only two intersection numbers (namely 0 and $\lambda + 1$), i.e., is *quasi-symmetric*, and hence carries a strongly regular graph (cf. Goethals and Seidel 1970). Now apply the standard restrictions on parameters of strongly regular graphs.)

Concerning case (ii): one may show the following: If there exists a strongly regular graph Γ with parameters $(v, k, \lambda, \mu) = ((\lambda+1)^2(\lambda+4)^2, (\lambda+1)(\lambda^2 + 5\lambda + 5), 0, (\lambda+1)(\lambda+2))$ then it has $q_{11}^1 = 0$ so by Cameron et al. (1978) its subconstituents are again strongly regular, and we find a strongly regular graph Δ with parameters $(v, k, \lambda, \mu) = ((\lambda^2 + 4\lambda + 2)(\lambda^2 + 5\lambda + 5), (\lambda+1)^2(\lambda+3), 0, (\lambda+1)^2)$. By Haemers (1980) the μ-graphs in $\bar{\Gamma}$ are strongly regular and we find a strongly regular graph E with parameters $(v, k, \lambda, \mu) = ((\lambda+1)(\lambda+3)(\lambda^2 + 4\lambda + 2), (\lambda+1)(\lambda^2 + 3\lambda + 1), 0, \lambda(\lambda+1))$. Finally, on the set of neighbours of any vertex in Γ we have a 3-design 3-$(k_\Gamma, \mu_\Gamma, \lambda)$, the extension of a symmetric design with parameters as in (ii) above. Conversely, given a 3-design with these parameters we may construct a strongly regular graph with the parameters of Γ, so that extendible SBIBDs with parameters (ii) coexist with strongly regular graphs Γ. The only known examples are the cases $\lambda = 0$ (Clebsch graph; Petersen graph; $K_{2,2,2}$; degenerate design 3-$(5, 2, 0)$) and $\lambda = 1$ (Higman–Sims graph; 77-graph; Gewirtz graph; unique design $S(3, 6, 22)$). Probably there are no other examples; the case $\lambda = 2$ has been attacked by Bagchi (1988), corrected in Bagchi (1991), but unfortunately the amended proof still has a gap.

Concerning case (iii) (that of an extendible projective plane of order 10): Lam et al. (1983) have conducted an exhaustive computer search and found no partial geometry pg(6, 9, 4); this result implies that a projective plane of order 10 cannot have ovals, and a fortiori cannot be extendible.

In the case of an SBIBD $\mathcal{D}$ the standard terminology is slightly different: if $\mathcal{D}^*$ is the dual of $\mathcal{D}$ then one usually calls $\mathcal{D}_B = ((\mathcal{D}^*)_B)^*$ and $\mathcal{D}^B = \mathcal{D}^{*B*}$ the derived and residual design (at the block B). If $\mathcal{D}$ has parameters (b, v, r, k, λ) then $\mathcal{D}_B$ and $\mathcal{D}^B$ have parameters (in the same order) $(v-1, k, k-1, \lambda, \lambda-1)$ and

$(v-1, v-k, k, k-\lambda, \lambda)$, respectively. In particular, a residual design satisfies $r = k + \lambda$.

A BIBD is called *quasi-residual* if $r = k + \lambda$, or, equivalently, if $b + 1 = v + r$. Sometimes it is possible to prove that a quasi-residual design in fact must be a residual, i.e., must be embeddable in a SBIBD. It is very easy to see that an affine plane (in the finite case: a 2-$(n^2, n, 1)$ design) is embeddable in a projective plane (in the finite case: a 2-$(n^2 + n + 1, n + 1, 1)$ design).

Hall and Connor (1954) and Connor (1952) showed that also quasi-residual designs with $\lambda = 2$ are embeddable. (This can be used to prove nonexistence: for example, a 2-(15, 5, 2) BIBD cannot exist since it would be embeddable in a 2-(22, 7, 2) SBIBD, but by Bruck–Chowla–Ryser this latter design does not exist. The proof is by observing that such a design only has two possible intersection numbers (1 and 2) and hence is quasi-symmetric. The associated strongly regular graph has the parameters of the triangular graph and by Connor (1958), Shrikhande (1959a,b), Hoffman (1960), Chang (1959) must be triangular, except when $k = 6$. The latter case is settled by an ad-hoc argument (no design exists), and in the triangular case it is easily seen that the design is embeddable.)

For $\lambda \geqslant 3$ there exist examples of nonembeddable quasi-residual designs. E.g., there exist designs with parameters 2-(16, 6, 3) where some pair of blocks meet in 4 points (Bhattacharya 1944), and such designs obviously cannot be embedded in a 2-(25, 9, 3) design. Bose et al. 1976 (see also Neumaier 1982) showed for quasi-residual designs that given λ, if k is sufficiently large then the design is embeddable. Kelly (1982) considers the question in what cases an embedding must be unique (when it exists).

6. Existence and construction of *t*-designs with large *t*

Let us first look at the positive side: how does one construct t-designs with, say, $t > 3$?

The easiest way to obtain a t-(v, k, λ) design is by taking the orbit of a k-set under a group G acting t-transitively (or t-homogeneously) on a v-set; of course one obtains in this way precisely those designs that admit a block-transitive automorphism group. (See, e.g., Hughes 1965.) For $t \geqslant 2$ all such groups G (and corresponding actions) are known (see chapter 12), and for $t > 5$ there are no such groups other than the symmetric and alternating groups.

When this easy scheme does not work, one might try to combine several G-orbits, where G is a group that is less than t-transitive. This method has been used successfully by Doyen (1974), Kramer and Mesner (1976), Denniston (1976), Brouwer (1977a,b), Mills (1978), Kramer (1984), Leavitt and Magliveras (1984) and others. When the number of orbits to be combined gets larger than a few dozen then advanced techniques are necessary to keep the required computation time within realistic bounds; Kreher and Radziszowski (1986) used a variant of the L^3 (Lenstra–Lenstra–Lovász) lattice reduction technique (see chapter 19) to find a 6-(14, 7, 4) design, the smallest possible nontrivial 6-design. Today many

t-designs with $t \leqslant 5$ are known (infinite families of 4-designs have been constructed by Alltop (1969), Hubaut (1974) and Driessen (1978); Alltop (1972) constructs an infinite series of 5-designs). The first 6-designs (with parameters 6-(33, 8, 36) and 6-(20, 9, 112)) were constructed by Leavitt and Magliveras (1984), Kramer et al. (1985)

Another method that has produced some good designs was first given by Assmus and Mattson (1969) (see also Pless 1969, 1970, 1972 and MacWilliams and Sloane 1977, (chapter 16, section 8); if we have a linear code $\mathscr{C}$ with word length n and minimum distance d, and an integer t with $0 < t < d$ such that at most $d - t$ of the weights w of the dual code $\mathscr{C}^{\perp}$ are in the range $1 \leqslant w \leqslant n - t$, then the supports of the codewords of weight d in $\mathscr{C}$ form a t-design. In this way one finds several 5-designs with $v \in \{12, 24, 36, 48, 60\}$ such as, e.g., 5-(24, 9, 6), 5-(36, 12, 45) and 5-(48, 12, 8).

Finally one has recursive constructions: construct a design using smaller designs as building blocks. For $t = 2$ this approach is highly successful (cf. below, sections 7–10), and for $t = 3$ various constructions are known, but until recently there were only very few results for $t > 3$. For example, Tran van Trung (1984) shows that if there exists a t-(v, k, λ) design, and λ is not too large then there exists a t-$(v + 1, k, (v + 1 - t)\lambda)$ design (by putting $v + 1$ disjoint copies of the given design on a $v + 1$ set) – starting with Alltop's designs this yields a new infinite family of 5-designs.

A very ingenious recursive construction was developed by Teirlinck (1987) (see also Teirlinck 1989, 1992). Using the symmetric group in a highly nontrivial way he constructs t-designs without repeated blocks for all t.

When repeated blocks are allowed it is a simple exercise in linear algebra to show that t-designs with arbitrarily large t do exist. In fact Wilson (1973) shows that given t, k and v (with $t \leqslant k \leqslant v$) there is a constant $\bar{\lambda}$ such that a t-(v, k, λ) design (possibly with repeated blocks) exists for all $\lambda \geqslant \bar{\lambda}$ satisfying the obvious condition that each λ_i is integral.

Only finitely many Steiner systems $S(t, k, v)$ with $t \geqslant 4$ are known, namely $S(5, 6, 12)$, $S(5, 8, 24)$ (Witt 1938), $S(5, 6, 24)$, $S(5, 7, 28)$, $S(5, 6, 48)$, $S(5, 6, 84)$ (Denniston 1976), $S(5, 6, 72)$ (Mills 1978), $S(5, 6, 108)$ (Griggs) and the derived 4-designs. The other known Steiner systems are the following: with $t = 3$ we have the Möbius geometries $S(3, q + 1, q^n + 1)$; Hanani (1960) showed that an $S(3, 4, v)$ (called *Steiner quadruple system*) *exists iff* $v \equiv 2$ or 4 (mod 6). (For a survey on Steiner quadruple systems, see Lindner and Rosa 1978.) Hanani (1979) gave various recursive constructions for 3-designs. With $t = 2$ many constructions are known; the "geometric" ones are: affine geometries $S(2, q, q^n)$, projective geometries $S(2, q + 1, (q^{n+1} - 1)/(q - 1))$, unitals $S(2, q + 1, q^3 + 1)$ and arcs in a plane of even order (Denniston 1969) $S(2, 2^r, 2^{r+s} + 2^r - 2^s)$ $(r < s)$. Recursive constructions (see below) show that the necessary conditions $v \equiv 1 \pmod{k - 1}$ and $v(v - 1) \equiv 0 \pmod{k(k - 1)}$ for the existence of some $S(2, k, v)$ are sufficient if v is large enough, and in fact always suffice for $k \leqslant 5$ (Kirkman 1847, Hanani 1961, 1965). For a discussion of the situation for $k = 6$ see Mills (1984), Mullin et al. (1987) and Assaf (1988a). Infinite classes of cyclic Steiner 2-designs including a "unital"

$S(2, 7, 217)$ are constructed in Mathon (1987) and Bagchi and Bagchi (1989). Doyen and Rosa (1980) contains an extensive bibliography on Steiner systems.

Of course one conjectures that Steiner systems exist for arbitrarily large t, but the number of blocks quickly becomes too large for present-day computers to handle.

Concerning uniqueness, the following is known. The projective and affine planes of orders 2, 3, 4, 5, 7 and 8 are unique (MacInnes 1907, Hall 1953, Hall et al. 1956). (But one knows at least 4 projective planes and 7 affine planes of order 9, cf. Kamber 1976, Hurkens and Seidel 1985.) (In the meantime it has been shown by exhaustive computer search that there are precisely 4 projective planes of order 9, and no such planes of order 10.) It is possible that all projective planes of prime order are desarguesian; for all nonprime prime power orders larger than 8 also nondesarguesian planes are known (Dembowski 1968, p. 144). Also the affine space $AG(3, 2) = S(3, 4, 8)$ and the Möbius (or inversive) planes $MG(2, q) = S(3, q+1, q^2+1)$ with $q = 3$, 4, 5, 7 (Witt 1938, Chen 1972, Denniston 1973a,b) are unique. When q is an odd power of 2 (≥ 8), then one has the Suzuki inversive plane besides the classical one. No other Möbius planes are known. Finally, the Witt systems $S(4, 5, 11)$, $S(5, 6, 12)$, $S(4, 7, 23)$ and $S(5, 8, 24)$ (Witt 1938) are unique. In probably all others cases, when a Steiner system is known, then in fact at least two nonisomorphic systems are known.

On the negative side there are not too many results either. A first obvious restriction is that there cannot be a t-(v, k, λ) when no $(t-i)$-$(v-i, k-i, \lambda)$ exists. This observation (with $i = t-2$) together with Fisher's inequality yields $(k-t+1)(k-t+2) \leq \lambda(v-t+1)$ (Van Tilborg 1976). Some general restrictions on the sizes of codes and designs in association schemes have been given by Delsarte (1973) (the "linear programming bound") and some specific consequences are listed below. For Steiner systems we have the following additional results. Dembowski (1964) shows that inversive planes (i.e., designs $S(3, n+1, n^2+1)$) of even order n have order a power of two. Tits (1964) shows for Steiner systems $S(t, k, v)$ that $(k-t+1)(t+1) \leq v$ (which is satisfied with equality for $S(3, 4, 8)$, $S(5, 6, 12)$ and $S(5, 8, 24)$ but otherwise seems to be very weak – it rules out e.g., $S(10, 16, 72)$ but already nonexistence of $S(4, 10, 66)$ is known (Kantor 1974, Cameron 1977, Denniston 1978); see also section 3 of Cameron 1980). Other special nonexistence results for Steiner systems are: there is no $S(4, 6, 18)$ (Witt, 1938), and no $S(4, 5, 15)$ (Mendelsohn and Hung 1972). Concerning a hypothetical $S(4, 5, 17)$, Denniston (1980) has shown that any such system has trivial automorphism group. For arbitrary t-designs we have: there is no 3-$(11, 5, 2)$ (see Oberschelp 1972, Dehon 1976) and no 4-$(17, 8, 5)$, 6-$(19, 9, \lambda)$, $\lambda \leq 10$, 6-$(20, 10, \lambda)$, $\lambda = 7$, 14, 4-$(23, 8, 2)$, 4-$(23, 11, \lambda)$, $\lambda = 6, 12$ or 4-$(24, 12, 15)$ (Delsarte; Haemers and Weug 1974), or 4-$(17, 7, 2)$, 5-$(19, 9, 7)$, 12-$(29, 14, 4)$, 10-$(29, 13, 3)$, 22-$(53, 25, 5)$, 26-$(58, 28, 8)$ (Köhler 1985, 1988/89, see chapter 15).

For tables, see Brouwer (1977b) ($v \leq 18$), Driessen (1978), Gronau (1985), Chee et al. (1990) ($v \leq 30$), Hanani et al. (1983) ($t = 3, v \leq 32$) and Mathon and Rosa (1985) ($t = 2, r \leq 41$), with some additions in Abel (1994). Kramer and Mesner (1975) list admissible parameters for Steiner systems $S(t, k, v)$ with $v - t \leq 498$.

7. Mutually orthogonal Latin squares

As a preparation for the discussion of t-designs with $t = 2$ we must first consider an auxiliary structure that also has independent interest.

A *Latin square* (of order n) is an $n \times n$ matrix with entries in a set S of cardinality n such that each row and each column is a permutation of S. (A standard reference on Latin squares is Dénes and Keedwell 1976; a good survey can be found in Jungnickel 1984.)

Two Latin squares A and B with entries in S resp. T are called *orthogonal* if $S \times T = \{(a_{ij}, b_{ij}) \mid 1 \leq i, j \leq n\}$. In old (recreational) literature such a pair was often called a *Graeco–Latin* square, and one used $S = \{a, b, c, \ldots\}$ and $T = \{\alpha, \beta, \gamma, \ldots\}$. Example:

$$\begin{array}{cccc} a\alpha & b\delta & c\beta & d\gamma \\ b\gamma & a\beta & d\delta & c\alpha \\ c\delta & d\alpha & a\gamma & b\beta \\ d\beta & c\gamma & b\alpha & a\delta \end{array}$$

A lot of research has been devoted to the question: What is the maximum size $N(n)$ of a collection of mutually orthogonal Latin squares of order n?

The main results are the following:

- $N(0) = N(1) = +\infty$,
- $N(n) \leq n - 1$ *for* $n \geq 2$,
- $N(q) = q - 1$ *for prime powers* q,
- (MacNeish 1922) *If* $n = \Pi_i\, p_i^{e_i}$ *with* p_i *prime then* $N(n) \geq \max_i(p_i^{e_i} - 1)$,
- (Chowla, Erdős and Straus 1960) $\lim_{n\to\infty} N(n) = +\infty$.

In view of this last result we may define n_r as the largest integer for which no r mutually orthogonal Latin squares (MOLS) exist. (Thus, $N(n) \geq r$ for $n > n_r$.)

Our knowledge about n_r is summarized by $n_2 = 6$ (Tarry 1901, Bose et al. 1960), $n_3 \leq 10$ (Wang and Wilson 1978, Todorov 1985), $n_4 \leq 42$ (Abel 1991, Abel and Todorov 1993), $n_5 \leq 62$ (Hanani 1970), $n_6 \leq 76$ (Wilson 1974, Wojtas 1980), $n_7 \leq 780$, $n_8 \leq 2846$, $n_9 \leq 4030$, $n_{10} \leq 6148$, $n_{11} \leq 7222$, $n_{12} \leq 7286$, $n_{13} \leq 7288$, $n_{14} \leq 7874$, $n_{15} \leq 8360$, $n_{30} \leq 52502$ (Brouwer and van Rees 1982, Brouwer 1980), and for large r we have the estimate $n_r = O(r^{14.8})$ (Beth 1983); see also Chowla et al. 1960, Rogers 1964, Yuan 1966, Wilson 1974.

Usually one does not work with MOLS but translates the problem into the language of transversal designs – see the following two sections.

8. Group-divisible designs and transversal designs

A *parallel class* is a collection of blocks partitioning the point set of a design. A *group-divisible design* is a linear space in which a parallel class has been singled out. The blocks in this parallel class are renamed *groups* (and no longer counted as blocks). Note that it is meaningful to have groups consisting of a single point.

Remarks. (1) Various extensions of this concept exist (and are sometimes also called group-divisible); all are special cases of the concept of a partially balanced incomplete block design (PBIBD) where one asks for a system in which the point set carries an association scheme (cf. chapter 15) with n classes, and there are n indices λ_i such that two points in relation i are joined by precisely λ_i blocks. Our group-divisible designs have as underlying two class association scheme (strongly regular graph) the union of a number of disjoint cliques, and $\lambda_1 = 0$, $\lambda_2 = 1$. We shall use *group-divisible design of index* λ for the case where one has a partition into groups, and points from different groups are joined by λ blocks (i.e., $\lambda_1 = 0$, $\lambda_2 = \lambda$).

(2) Some authors write "part" or "level" or "groop" instead of "group". Many authors use "divisible" without qualifier.

There is a 1–1 correspondence between linear spaces on $v+1$ points with one singled out point and group-divisible designs on v points: Given the linear space, call the special point ∞ and throw it away; call the sets $G \setminus \{\infty\}$, where G was a line through ∞, groups. Conversely, given the group-divisible design, invent a new point ∞, add it to the groups and add the resulting sets to the collection of lines. This process is called "adding a point at infinity".

If a linear space has several mutually disjoint parallel classes then one may add a point at infinity for each of the parallel classes and finally join the points at infinity by a line at infinity. This process is often called "completion". Of course the obvious example is the construction of a projective plane starting from an affine plane.

A *transversal design* (of index λ) is a group-divisible design (of index λ) such that each block (i.e., nongroup) meets each group in precisely one point. It is easily seen that (if there are at least three groups then) all groups have the same size g, say, and all blocks have the same size k. Furthermore, we have $v = gk$, $b = \lambda g^2$, $r = \lambda g$.

There is a 1–1 correspondence between transversal designs (i.e., transversal designs of index unity) with parameters g, k and sets of $k-2$ mutually orthogonal Latin squares of order g: each of the g^2 blocks corresponds to a (the same) position in each of the Latin squares – given a block, its point on the first and second groups determine row and column, and its point on each of the remaining groups determines the entry in the corresponding Latin square.

One of the advantages of using transversal designs (instead of MOLS) is that one treats rows, columns and symbols uniformly. Another is that it brings some geometric flavour to the problems.

Let us look at some of the statements made about $N(n)$ in the previous section.

Given a block B and a point x outside, we see $k-1$ blocks (and one group) on x meeting B, so that $k-1 \leq g$, i.e., $N(g) \leq g-1$.

Starting with a projective plane of order q one finds (by throwing out a point ∞) a transversal design with $(k, g) = (q+1, q)$, which shows $N(q) = q-1$.

MacNeish's result follows by a simple direct product construction (formulated with equal ease in the language of Latin squares, transversal designs or quasigroups).

For the Chowla–Erdős–Straus theorem we need some recursive constructions considered in the next section.

9. PBD-closed sets

A set K of nonnegative integers is called *PBD-closed* if whenever we have a pairwise balanced design of index unity with all blocksizes in K then also $v \in K$. This concept was introduced by R.M. Wilson and proves to be extremely useful: one has strong information about PBD-closed sets and PBD-closed sets occur in many planes in a natural way.

Theorem (Wilson 1972). *Let K be PBD-closed. Then K is eventually periodic with period $\beta(K) := g.c.d.\{k(k-1) \mid k \in K\}$, i.e., if K intersects the residue class a (mod $\beta(K)$), then K contains almost all integers $k \equiv a$ (mod $\beta(K)$).*

Let us write $B(K, \lambda)$ for the set of all v for which a $B(K, \lambda; v)$ (that is, a PBD with v points, index λ and all blocksizes in K) exists. Write $B(K)$ for $B(K, 1)$ and $B(k, \lambda)$, $B(k)$ in case $K = \{k\}$. Now K is PBD-closed iff $B(K) = K$.

Some examples of PBD-closed sets are provided by the following.

Lemma (Breaking up blocks, Hanani 1961, (3.11)). *$B(B(K)) = B(K)$, and, more generally, $B(B(K, \lambda)) = B(K, \lambda)$ for all K.*

Hanani's lemma (Hanani 1961, (3.10)). *Let $k \in K$ and let $R_k = \{r \mid r(k-1)+1 \in B(K)\}$. Then $B(R_k) = R_k$.*

Theorem (Bose et al. 1960). *Let G_k be the set of integers g for which there exists a transversal design with blocks of size k, groups of size g and at least one parallel class of blocks. Then $B(G_k) = G_k$.*

(These three statements are easy exercises, best proved by a picture.)

Now we have enough material to sketch a proof of the Chowla–Erdős–Straus theorem: We wish to show that for all k the set G_k contains all sufficiently large integers. The easiest way to obtain transversal designs with a parallel class is to start with a transversal design with one more group and throw this group away; now the blocks passing through a discarded point will form a parallel class. Next, using MacNeish's theorem, we see that all numbers n with a factorization into prime powers all at least k are in G_k. Finally we need a few pairwise balanced designs in order to exploit the fact that G_k is PBD-closed. Given a transversal design with parameters g, k, discard $g-h$ points from one group to obtain a *truncated transversal design*: a group-divisible design with $k-1$ groups of size g and one group of size h, blocks of sizes k and $k-1$, and $v = (k-1)g + h$ points. This construction shows that if $N(g) \geqslant k-2$ and $g \geqslant h \geqslant 0$ and g, h, $k-1$, $k \in G_s$ then $(k-1)g + h \in G_s$. The rest is number theory, not combinatorics.

One may have wondered why we did not appeal to Wilson's theorem, but in fact Wilson's theorem is proved using Hanani's lemmas and the Chowla–Erdős–Straus theorem.

Wilson's theorem enables us to show that given k and λ a 2-(v, k, λ) exists whenever v satisfies the obvious divisibility restrictions and is large enough. More generally one has the following.

Theorem (Wilson 1972b). *$B(K, \lambda)$ contains all sufficiently large integers v with $\lambda(v-1) \equiv 0 \pmod{\alpha(K)}$ and $\lambda v(v-1) \equiv 0 \pmod{\beta(K)}$, where $\alpha(K) := g.c.d\{k-1 \mid k \in K\}$ and $\beta(K) := g.c.d.\{k(k-1) \mid k \in K\}$.*

Proof. By the foregoing it suffices to find at least one example in each residue class. Using difference sets in finite fields and an "unfolding" construction to make designs with $\lambda = 1$ and large v from designs with large λ and small v, enough examples can be found. See Wilson (1972a,b,c), Brouwer (1979), Beth (1985). □

10. Steiner triple systems

Let us try to work out an example and construct Steiner triple systems for all possible v. A *Steiner triple system* (STS(v)), is a Steiner system $S(2, 3, v)$, or equivalently, a linear space in which all lines have length 3. The condition that r and b should be integral forces $v \equiv 1, 3 \pmod 6$. To streamline our construction we shall also construct linear spaces with $v \equiv 5 \pmod 6$, this time requiring that all lines have size 3 except for one with size 5.

Extend the notation $B(K)$ by priming those block sizes that must not occur more than once. Then our claim is that $B(\{3, 5'\})$ is the set of all odd integers. Clearly all its members are odd. Let $R = \{r \mid 2r + 1 \in B(\{3, 5'\})\}$. Clearly R contains 0, 1, 2 – the corresponding linear space has at most one line. Also 3, 4, $5 \in R$ – the Fano plane PG(2, 2), the affine plane AG(2, 3) and the completion of a 1-factorization of the complete graph K_6 provide (the unique) examples. By a simple variation of Hanani's lemma we see that $B(\{3, 4, 5'\}) \subset R$. We show that $R = \mathbb{N}$. Given $r \in R$, if r is odd then by induction $r \in B(\{3, 5'\}) \subset R$ and we are done. If we can write $r = 3g + h$ with $0 \leqslant h \leqslant g$ where $g \equiv 0$ or $1 \pmod 3$ and either $g \neq 6$ or $h \leqslant 4$ then we find $r \in R$ using a truncated transversal design and induction. (Here we use the existence of a "truncated transversal design" on $3 \cdot 6 + 4$ points – i.e., a Latin square of order 6 with 4 pairwise disjoint transversals. This is guaranteed by a more general recursive construction due to Hanani (1961, (2.12)); let us give such a Latin square explicitly – the transversals are indicated with subscripts

A_1	B_3	C_2	D_4	E	F
B_2	A_4	D_1	C_3	F	E
C	D	F_3	E_1	A_2	B_4
D	C	E_4	F_2	B_1	A_3
E_3	F_1	A	B	C_4	D_2
F_4	E_2	B	A	D_3	C_1

.)

It remains to consider $r \in \{6, 8\}$, i.e., $v \in \{13, 17\}$. For $v = 13$ we may take the 26 blocks $\{1, 3, 9\}$, $\{2, 6, 5\}$ (mod 13) on the point set $\mathbb{Z}_{13}$. For $v = 17$ we complete the five parallel classes of the design on $\mathbb{Z}_{12}$ with blocks $\{0, 1, 4\}$, $\{0, 6\}$, $\{0, 5\}$, $\{0, 2\}$ (mod 12) and parallel classes $[\{0, 6\} + i]$, $[\{0, 5\} + 2i] + 0, 1$ and $[\{0, 2\}, \{1, 3\} + 4i] + 0, 2$ ($i \in \mathbb{Z}_{12}$).

This completes our construction and shows that there is a STS(v) for all $v \equiv 1, 3$ (mod 6), a result due to Kirkman (1847).

Many direct constructions of STSs exist, but the above provides one of the shorter existence proofs (assuming that $N(g) \geq 2$ for $g \neq 2, 6$ is already known) and shows the spirit of the constructions of other families of designs with small block sizes: a few small cases have to be handled explicitly, and the rest is done with the help of recursive constructions.

Let us, however, also mention some other types of construction.

A typical difference set construction in a finite field goes as follows: Let $q \equiv 1$ (mod 6) be a prime power, and let x be a primitive element of $\mathbb{F}_q$. Write $q = 6e + 1$. Then the set of blocks $\{\{x^i, x^{i+2e}, x^{i+4e}\} + y \mid 0 \leq i < e, y \in \mathbb{F}_q\}$ makes $\mathbb{F}_q$ into a STS(q). (Note that we used this construction above for the case $v = 13$.)

More generally one has (Wilson 1972a): If q is a prime power and $k|2\lambda$ or $(k-1)|2\lambda$ then the necessary condition $k(k-1) \mid \lambda(q-1)$ suffices for the existence of a 2-(q, k, λ) design. (The construction is similar.)

A completely different approach is that using Skolem sequences.

A *Skolem sequence* (of order n) is a partition of the numbers $1, \ldots, 2n$ into n disjoint pairs (a_i, b_i) such that $b_i - a_i = i$ ($i = 1, \ldots, n$). A Skolem sequence of order n exists iff $n \equiv 0$ or 1 (mod 4) (Skolem 1957). Given such a sequence, we find a partition of $1, \ldots, 3n$ into n pairwise disjoint triples $(i, a_i + n, b_i + n)$, and using these as base blocks (mod $6n + 1$) we find a STS($6n + 1$) (Skolem 1958). A *hooked Skolem sequence* (of order n) is a similar partition of the numbers $1, 2, \ldots, 2n - 1, 2n + 1$. Skolem (1958) conjectured and O'Keefe (1961) proved that such sequences exist iff $n \equiv 2$ or 3 (mod 4). Again we find Steiner triple systems – a hooked sequence provides a difference $3n + 1$ where an ordinary sequence gives $3n$, but mod $6n + 1$ these are the same. Hanani (1960) constructed similar partitions of $1, 2, \ldots, n, n + 2, \ldots, 2n + 1$ for $n \equiv 3$ (mod 4) and used these to construct cyclic STS($6n + 1$) for all n. But in fact Peltesohn (1938) had already twenty years earlier shown that a cyclic STS (i.e., one with $\mathbb{Z}_v$ in its group of automorphisms) exists for all $v \equiv 1$ or 3 (mod 6) except $v = 9$.

Many other results exist on STSs with specified automorphisms. Just to mention one example: Rosa (1972), Doyen (1972) and Teirlinck (1973) showed that STSs with an automorphism group containing an involution that fixes only one point ("reverse STSs") exist iff $v \equiv 1, 3, 9$ or 19 (mod 24).

Lindner and Rosa (1975) showed that an STS(v) with trivial automorphism group exists iff $v \equiv 1, 3$ (mod 6), $v \geq 15$, and Babai (1980), using Wilson's (1974b) estimate (see Aleksejev 1974) on the number $N(v)$ of nonisomorphic Steiner triple systems of order v (one has $\log N(v) \sim (n^2/6) \log n$) shows that almost all

STSs have a trivial group of automorphisms (i.e., if $A(v)$ is the number with trivial automorphism group then $A(v) \sim N(v)$).

Returning to the Skolem sequences, this has grown into a subject of its own, independent of STSs. Independent motivation is provided by the needs of radioastronomers who want a sequence of positionings of their arrays with antennas so as to do measurements on a wavelength and all its small integer multiples. See, e.g., Langford (1958), Davies (1959), Priday (1959), Baron (1970), Roselle (1972), Bermond et al. (1978a,b). There is also some relation with the problem of finding graceful numberings of certain graphs.

Instead of looking at automorphism groups one may ask for specific structural properties. Doyen and Wilson (1973) showed that if u, $v \equiv 1$ or 3 (mod 6) then there exists a STS(v) with a sub-STS(u) iff $v \geq 2u + 1$. Conversely it is sometimes useful to have systems without certain specified subconfigurations, see, e.g., Brouwer (1977c). One may introduce a dimension concept for linear spaces and study the possible values of the dimension function for STS, see, e.g., Teirlinck (1979). Probably the most important structural property studied is resolvability (see section 12). A resolvable STS is called a *Kirkman triple system*, and Ray-Chaudhuri and Wilson showed that a KTS(v) exists iff $v \equiv 3$ (mod 6), thus solving a one-century-old problem.

Finally one may consider systems of STSs. Teirlinck (1977) shows that given any two STSs of the same order, one may find isomorphic embeddings of them on the same point set so that they have no triple in common. Several people studied the maximum number of pairwise disjoint STSs on a given point set. Clearly this cannot be higher than $v - 2$, and it is now known that this upper bound is achieved for $v \neq 7$. Cayley (1850) showed that for $v = 7$ this maximum is 2. See Teirlinck (1973b, 1984, 1989), Schreiber (1973), Lu (1983, 1984).

11. 3-Designs

We are still very far from a satisfactory asymptotic theory for 3-designs, like Wilson's theory for 2-designs. Hanani (1963) constructs 3-designs with $k = 4$ for all admissible parameter sets. For some recursive constructions, see Hanani (1979). Some of the designs constructed there have been rediscovered by Assmus and Key (1986).

In the case of pairwise balanced designs one uses transversal designs in the recursive constructions, and the first thing to do here would be to prove the analogue of the Chowla et al. theorem for transversal 3-designs. (R.M. Wilson reports that J. Blanchard has recently done this, and more generally obtained the corresponding result for transversal t-designs.) Möbius planes provide 3-designs, and truncating them one obtains various triplewise balanced designs; Laguerre planes provide transversal 3-designs, and Minkowski planes provide nested transversal 3-designs. Using these it is possible to give recursive constructions for many 3-designs, but it seems that one does not yet have sufficient control to prove asymptotic existence results.

12. Resolvability

A *resolution* of a design is a partition of the set of blocks into parallel classes; a design is called *resolvable* when a resolution exists. (A parallel class is a 1-design with $r = 1$, a partition of the point set.) For example, affine planes, Hermitean unitals and the design of points and lines in projective 3-space are resolvable. The obvious necessary condition for resolvability of a design with constant block size k is that k divides v, and usually this condition is also sufficient. In particular, resolvable Steiner triple systems (known as *Kirkman triple systems* KTS(v)) have been constructed for all $v \equiv 3 \pmod 6$ by Ray-Chaudhuri and Wilson (1971) and resolvable Steiner systems $S(2, 4, v)$ for all $v \equiv 4 \pmod{12}$ by Hanani et al. (1972). For resolvable group-divisible designs with $k = 3$, see Rees and Stinton (1987). Asymptotic results are known.

The existence of such partitions implies strengthenings of Fisher's inequality. For example, for an (r, λ)-design, the point-block incidence matrix A satisfies $AA^{\mathrm{T}} = (r - \lambda)I + \lambda J$, and if $r > \lambda$ one concludes that $\mathrm{rk}\, A = v$. But in case the collection of blocks can be partitioned into c 1-designs, there are c essentially different ways of writing the all-one vector as a linear combination of rows of A, and it follows that $v = \mathrm{rk}\, A \leqslant b - c + 1$. The case of equality can be characterized, cf. Beutelspacher and Lamberty (1983).

An *affine resolvable BIBD* is a BIBD with a resolution such that for some constant $\mu > 0$ any two blocks from different parallel classes meet in precisely μ points. (And then $\mu = k^2/v$.) As we just saw, for a resolvable BIBD one has $b \geqslant v + r - 1$, and Bose (1942) shows that equality holds if and only if the design is affine resolvable. Examples of affine resolvable designs are the designs of points and hyperplanes in an affine space. Strong nonexistence results are known. For a survey, see Shrikhande (1976).

Much attention has been given to resolutions of the complete design $\binom{v}{k}$ of all k-subsets of a v-set. Indeed, Cameron (1976) is entirely devoted to them. First of all, such resolutions exist if and only if $k \mid v$ (Baranyai 1975, see also Brouwer 1976b, Baranyai and Brouwer 1977). (This is proved by a simple but ingenious repeated use of the Ford–Fulkerson max-flow min-cut and integral flow theorems.) In fact, Baranyai's theorem roughly speaking implies that all conceivable 1-design-like structures exist. For an application, see, e.g., Stinson (1983).

For $k = 2$ these resolutions are known as 1-factorizations of the complete graph, and for $v \leqslant 10$ all have been determined (Gelling 1973). Since there are many of them for larger v (for the number $f(v)$ of nonisomorphic 1-factorizations of K_v we have $\log f(v) \sim \frac{1}{2}n^2 \log n$), one might impose additional restrictions; a *perfect* 1-factorization is one in which the union of any two 1-factors (parallel classes) is a Hamilton circuit. It is conjectured that perfect 1-factorizations exist for all even v, and they are known for $v = p + 1$, $v = 2p$ (p an odd prime) and for all even $v < 40$. Cf. Seah and Stinson (1987). A survey of one-factorizations has been given by Mendelsohn and Rosa (1985), in fact the entire issue of the Journal of Graph Theory containing that paper is entirely devoted to factorizations of graphs.

An interesting one-factorization of the 4-subsets of a 24-set is provided by the Steiner system $S(5, 8, 24)$: given any 4-set ("tetrad") there is a unique partition of the 24-set containing this 4-set such that the union of any two elements of the partition is block of the Steiner system; see section 16.

One may also consider other partitions of the block set of a design. A design with constant block size k is called *separable* if its block set can be partitioned into 1-designs with $r \in \{1, k\}$. (This concept occurs in various results on the construction of sets of MOLS from a BIBD.)

A 3-design is called *doubly resolvable* if its set of blocks can be partitioned into resolvable 2-designs. Thus, Sylvester's school girl problem (solved by Denniston 1974) asked for a partition of the set of all triples on 15 points into resolvable Steiner triple systems $S(2, 3, 15)$. See also Denniston (1974a, 1979). Resolutions into more special designs can be found in Baker (1976, 1984), Beth (1974, 1979), Brouwer (1976a), Chouinard (1983), Denniston (1972, 1973c,d, 1983), Kramer and Mesner (1975b), Hartman (1987a,b). Partitions of the collection of all k-subsets of a v-set, such that no two k-sets in the same part have a $(k-1)$-set in common, play a role in the construction of good constant weight codes for $d = 4$, cf. Van Pul and Etzion (preprint).

Instead of asking for special resolutions, one may ask for several resolutions with specified mutual relationship. Two resolutions are called *orthogonal* when each parallel class of one resolution has at most one block in common with each parallel class of the other resolution. A pair of orthogonal resolutions of the complete design $\binom{v}{2}$ of all pairs in a v-set is called a *Room square*; see the first part of Wallis et al. (1972). For some variations, see Fuji-Hara (1986) and the references given there. Orthogonal factorizations of graphs are surveyed in Alspach et al. (1992); see also other papers in that collection.

13. Completing designs

Given a partial design of some kind that either has very few blocks, or is already almost a full design, it is often possible to prove that it can be completed to a design of the required type. Let us give some examples.

A *Latin rectangle* is a matrix with entries in some finite set S such that no element occurs twice in the same row or column. Thus, a Latin square of order n is an $n \times n$ Latin rectangle with $|S| = n$. Ryser (1951) gave necessary and sufficient conditions for the embeddability of Latin rectangles in Latin squares. Using this result, Evans (1960) proved that any partial Latin square of order n can be embedded in a Latin square of order m for each $m \geqslant 2n$. He conjectured that any partial Latin square of order n with at most $n - 1$ filled cells can be completed (to a Latin square of order n), and this was proved by Andersen and Hilton (1983) and Smetaniuk (1981). See also Damerell (1983), Andersen (1985) and Daykin and Häggkvist (1984).

Cruse (1974) gave necessary and sufficient conditions for the embeddability (1975a) of symmetric square Latin rectangles in symmetric Latin squares. Using his results, Lindner (1975a) showed that any partial Steiner triple system on m points can be completed to a Steiner triple system on $6m + 3$ points. (Finite embeddability had been shown already by Treash 1971. Lindner's bound has been improved later; many related results exist. See Lindner 1975b, Andersen et al. 1980.)

In a projective plane, any two lines have precisely one point in common. Many results say that if in a partial linear space any two lines meet, and there are enough lines, then it can be embedded in a projective plane. See, e.g., Dow (1983). A projective plane has equally many points and lines. Many results say that if the number of lines of a linear space is only slightly more than the number of points, then the design is either one of a few exceptions, or obtained by truncating a projective plane. See, e.g., De Witte (1976), Totten (1976, 1977). A projective plane has as many points per line as lines per point. Many results say that if the (maximum) number of lines per point is only slightly more than the (minimum) line size, then the design is obtained by truncating a projective plane. See, e.g., Shrikhande and Singhi (1985), Beutelspacher and Metsch (1986). The famous result by Bruck (1963) that a net of order n with roughly $n - (2n)^{1/4}$ parallel classes can be completed to an affine plane is another example of this phenomenon. Most of these results use Bose and Laskar's "claw and clique" technique in some way or another; let us very roughly sketch a version of it.

Given a linear space, look at the graph whose vertices are the lines and where two lines are adjacent when they are disjoint. Clearly, finding an embedding of the linear space into a projective plane is equivalent to finding a collection of cliques in this graph such that each edge is in precisely one of these cliques, any two cliques meet, and the lines in each given clique partition the point set of the linear space. Usually, in order to fulfil these requirements it is enough to find on each edge a sufficiently large clique. Now suppose that we can show that each j-claw in the graph can be extended to a $(j + 1)$-claw in at least M ways for $j < m$, but that no $(m + 1)$-claws exist. (This information might come from counting, but also eigenvalue arguments work in cases where the graph is sufficiently regular.) It follows that the M points that can be added to an $(m - 1)$-claw must be mutually adjacent, and in this way we find the required large cliques.

Somewhat stronger results have recently been obtained by Metsch.

14. Packing and covering

Corresponding to the problem of constructing any type of design, where there is "precisely λ" in the definition, one has problems of constructing maximal packings and minimal coverings, with "at most λ" and "at least λ" in the definitions. In particular, the packing and covering problems corresponding to the construction of Steiner systems $S(t,k,v)$ are the problems of determining the maximum size $D(t,k,v)$ of a collection of k-subsets of a v-set such that no t-subset is covered more than once, and the minimum size $C(t,k,v)$ of a collection of k-subsets of a v-set such that each t-set is covered at least once. Clearly, one has $C(t,k,v) \geq \binom{v}{t}/\binom{k}{t}$ and $D(t,k,v) \leq \binom{v}{t}/\binom{k}{t}$, in both cases with equality if and only if a Steiner system $S(t,k,v)$ exists. Rödl (1985) shows that for fixed t and k one has

$$\lim_{v\to\infty} C(t,k,v)\binom{k}{t}/\binom{v}{t} = \lim_{v\to\infty} D(t,k,v)\binom{k}{t}/\binom{v}{t} = 1$$

(see chapter 33). Slightly more precise bounds were given by Johnson (1962), Katona et al. (1964) and Schönheim (1964, 1966) who showed that

$$C_\lambda(t,k,v) \geq \left\lceil \frac{v}{k} \left\lceil \frac{v-1}{k-1} \left\lceil \cdots \left\lceil \frac{v-t+1}{k-t+1}\lambda \right\rceil \cdots \right\rceil\right\rceil\right\rceil$$

and

$$D_\lambda(t,k,v) \leq \left\lfloor \frac{v}{k} \left\lfloor \frac{v-1}{k-1} \left\lfloor \cdots \left\lfloor \frac{v-t+1}{k-t+1}\lambda \right\rfloor \cdots \right\rfloor\right\rfloor\right\rfloor .$$

These bounds are rather tight; for example, when $t=2$ and k, λ are fixed, then the difference between left-hand and right-hand sides remains bounded as v tends to infinity. For packings the above bound can be sharpened slightly by not only applying $D_\lambda(t,k,v) \geq (v/k)D_\lambda(t-1,k-1,v-1)$, but also the occasionally useful $D_\lambda(t,k,v) \geq (v/(v-k))D_\lambda(t,k,v-1)$. Another small improvement is obtained by observing that the collection of t-sets that are not covered precisely λ times is either empty or covers at least t points. For $t=2$ this yields: if $(k-1) \mid \lambda(v-1)$ and $\lambda v(v-1)/(k-1) \equiv -1 \pmod{k}$ then $C_\lambda(2,k,v) \geq \lambda v(v-1)/k(k-1)+1$, and if $(k-1) \mid \lambda(v-1)$ and $\lambda v(v-1)/(k-1) \equiv 1 \pmod{k}$ then $D_\lambda(2,k,v) \leq \lambda v(v-1)/k(k-1)-1$. The bounds resulting from this observations hold with equality for $(t,k)=(2,3)$ (Fort and Hedlund 1958, Schönheim 1966, Hanani, 1975) and for $(t,k)=(2,4)$, $v>19$ (Mills 1972, 1973, Brouwer et al. 1977, Brouwer 1979, Billington et al. 1984, Hartman 1986, Assaf 1986, 1988b); for $(t,k)=(3,4)$ equality holds for large v (Hartman et al. 1986). For $(t,k)=(2,5)$, see Mills (1983), Lamken et al. (1987), Mills and Mullin (1988), Assaf et al. (1993). For small v one has bounds like

$$D(t,k,v) \leq \frac{(k+1-t)v}{k^2-(t-1)v}$$

(provided the denominator is positive) for the packing numbers; Todorov (1985a) gives infinitely many cases with strict inequality for the covering numbers. Coverings using few sets are studied in Mills (1979), Todorov (1985b, 1986). See also Bate and Van Rees (1985), Todorov and Tonchev (1982). A recent survey is Mills and Mullin (1992).

These problems are also studied in various other guises. For example, finding

$D(t, k, v)$ is equivalent to finding the maximum size $A(n, d, w)$ of a binary code with word length n, constant weight w and minimum distance d, when $n = v$, $w = k$ and $d = 2(k - t + 1)$-tables for $n \leqslant 28$ are given in; finding $C(t, k, v)$ is equivalent to finding the *Turán number* $T(n, l, r)$, the minimum size of a collection of l-subsets of an n-set such that every r-subset containst at least one of these, when $n = v$, $l = v - k$, $r = v - t$.

A variation on this theme has recently gained popularity. One studies the function $g^{(k)}(t, \lambda; v)$, the minimum size of an exact t-covering, that is, of a t-wise balanced design, with largest block size k. For a survey, see Stanton (1988).

In this area one also encounters functions of interest to both mathematician and layman. The *lottery number* $L(n, l, r, t)$ is the minimum size of a collection of l-subsets of an n-set with the property that each r-set meets one of these l-subsets in at least t points. For suitable parameters n, l, r, t this is the minimum number of lotto forms one has to fill in to be assured of winning a prize. In Holland and Germany the values of $L(41, 6, 7, 4)$ and $L(49, 6, 6, 3)$ (respectively) are of special interest. See Hanani et al. (1964), Oberschelp (1972), Brouwer and Voorhoeve (1979). One also has the *football pool problem*, where one wants to find the size of the smallest code in Q^n with given covering radius δ. For $(n, \delta, |Q|) = (13, 1, 3)$ this can be interpreted (in Holland) as the minimum number of football toto forms one needs to complete to have a sure second prize. Cf. Kamps and Van Lint (1970). There is some recent progress here.

15. Codes and designs

There are close connections between coding theory and design theory. (See for example Cameron and Van Lint 1980, MacWilliams and Sloane 1977 (esp. chapter 6) and Bridges et al. 1981.) The most direct correspondences work as follows. Given a code $\mathscr{C}$ of word length n, take the collection of supports of the code words of weight w. This yields a design with block size w on n points. (We already mentioned the Assmus–Mattson theorem (in section 6), which guarantees that nice codes yield nice designs.) Conversely, given a design $(X, \mathscr{B})$ and a (usually finite) field F, consider the linear subspace of F^v spanned by the characteristic vectors of the elements of $\mathscr{B}$. This yields a (linear) code of word length v. Often it is easier to work with the code than with the design. Usually one tries to choose F in such a way that the resulting code becomes self-orthogonal, since there are powerful results on self-orthogonal codes. (To give a very trivial example: one can immediately conclude that $\dim \mathscr{C} \leqslant v/2$). For a design with block size k and intersection numbers $s_1, \ldots, s_l$ this means that one chooses the characteristic p such that $k \equiv s_j \equiv 0 \pmod p$ for all j, if possible. Examples of the use of this technique are Bagchi's attempted proof of the nonexistence of a design 3-$(57, 12, 2)$ (an extension of a biplane 2-$(56, 11, 2)$, cf. section 5) by studying the ternary code generated by such a design, the work of Calderbank on quasi-symmetric designs, and the characterizations by Tonchev of designs related to the Witt designs (cf. section 16).

In this context it is very useful to know dim $\mathscr{C}$, that is, to know the rank of the incidence matrix $\mathscr{A}$ of the design over the field F. For a BIBD, it follows from $AA^{\mathrm{T}} = (r-\lambda)I + \lambda J$ and $\det AA^{\mathrm{T}} = kr(r-\lambda)^{v-1}$ that A has p-rank v or $v-1$ unless $p \mid (r-\lambda)$, and p-rank v precisely when $p \nmid kr(r-\lambda)$. The p-rank of a projective plane of order n, with $p \mid n$, $p^2 \nmid n$ is $(n^2+n+2)/2$, see Sachar (1979), Klemm (1986). The most important (and most often rediscovered) result on p-ranks is that the p-rank of the incidence matrix of points and hyperplanes in PG(d, q), where $q = p^n$, equals $\binom{d+p-1}{d}^n + 1$ (Goethals and Delsarte 1968, Hamada 1968, MacWilliams and Mann 1967, Smith 1969) and that the corresponding p-rank for AG(d, q) is one less (Hamada 1973). Hamada (1968, 1973) also gives the ranks for the designs of points and t-flats in projective and affine spaces. He conjectured that these ranks are minimal for these geometries among the BIBDs with the same parameters, and later "Hamada's conjecture" was taken to mean that among several designs with the same parameters the "nicest" one would have the smallest p-rank. However, this is only a heuristic, and there are counterexamples for most interpretations of "nicest". For example, among the Steiner systems $S(2, 4, 28)$ the one with the largest automorphism group is the classical unital, which has 2-rank 21, while the Ree unital has 2-rank 19. For some positive results, see Hamada and Ohmori (1975); for some more counterexamples, see Tonchev (1986b). Lander (1983) gives more detailed results, using instead of p-rank the finer tool of elementary divisors.

16. Witt designs, Golay codes and Mathieu groups

By far the most important structure in design theory is the Steiner system $S(5, 8, 24)$. Together with the related designs $S(4, 7, 23)$, $S(5, 6, 12)$, $S(4, 5, 11)$ it was first constructed by Witt (1938) and these four Steiner systems are known as the *Witt designs*. We shall see that they are uniquely determined by their parameters. These designs are very closed related to the Golay codes, and we shall discuss these first.

16.1. The Golay codes

Let B be a vector space over $\mathbb{F}_q$ with fixed basis $e_1, \ldots, e_n$. A *code* $\mathscr{C}$ is a subset of V. A *linear code* is a subspace of V. The vector with all coordinates equal to zero (resp. one) will be denoted by $\mathbf{0}$ (resp. $\mathbf{1}$).

The *Hamming distance* $d_{\mathrm{H}}(u, v)$ between two vectors $u, v \in V$ is the number of coordinates where they differ: when $u = \sum u_i e_i$, $v = \sum v_i e_i$ then $d_{\mathrm{H}}(u, v) = |\{i \mid u_i \neq v_i\}|$. the *weight* of a vector u is its number of nonzero coordinates, i.e., $d_{\mathrm{H}}(u, \mathbf{0})$.

The *minimum distance* $d = d(\mathscr{C})$ of a code $\mathscr{C}$ is $\min\{d_{\mathrm{H}}(u, v) \mid u, v \in \mathscr{C}, u \neq v\}$. The *support* of a vector is the set of coordinate positions where it has a nonzero coordinate.

There exist codes, unique up to isomorphism, with the indicated values of n, q, $|\mathscr{C}|$ and d:

	n	q	$\lvert\mathscr{C}\rvert$	d	name of $\mathscr{C}$
(i)	23	2	4096	7	binary Golay code
(ii)	24	2	4096	8	extended binary Golay code
(iii)	11	3	729	5	ternary Golay code
(iv)	12	3	729	6	extended ternary Golay code

Codes with these parameters will be constructed in section 16.2. We shall show uniqueness in the binary case in section 16.4. For the ternary case, see Delsarte and Goethals (1975). Let us assume that the codes have been chosen such as to contain **0**. Then each of these codes is linear (the dimensions are 12, 12, 6, 6).

The codes (ii) and (iv) are *self dual*, i.e., with the standard inner product $(u, v) = \sum u_i v_i$ one has $\mathscr{C} = \mathscr{C}^\perp$ for these codes.

The codes (i) and (iii) are *perfect*, i.e., the spheres with radius $(d-1)/2$ around the code words partition the vector space. (Proof by counting: $\lvert\text{sphere}\rvert = 1 + \binom{23}{1} + \binom{23}{2} + \binom{23}{3} = 2048 = 2^{11}$ in case (i), and $\lvert\text{sphere}\rvert = 1 + 2\binom{11}{1} + 4\binom{11}{2} = 243 = 3^5$ in case (iii).)

Remark. Except for the repetition codes (with $\lvert\mathscr{C}\rvert = q$, $d = n$), there are no other perfect codes $\mathscr{C}$ with $d > 3$, cf. Van Lint (1971, 1974), Tietäväinen (1973, 1974), Tietäväinen and Perko (1971), Reuvers (1977), Best (1982), Hong (1984).

The *weight enumerators* $A(x) := \sum a_i x^i$, where a_i is the number of code words of weight i, are:

(i) $1 + 253x^7 + 506x^8 + 1288x^{11} + 1288x^{12} + 506x^{15} + 253x^{16} + x^{23}$,

(ii) $1 + 759x^8 + 2576x^{12} + 759x^{16} + x^{24}$,

(iii) $1 + 132x^5 + 132x^6 + 330x^8 + 110x^9 + 24x^{11}$,

(iv) $1 + 264x^6 + 440x^9 + 24x^{12}$.

(Proof: Cases (ii) and (iv) follow immediately from (i) and (iii) (cf. the first paragraph of the next section). In cases (i) and (iii) use the fact that the codes are perfect; e.g., in case (iii) the sphere around **0** covers the vectors of weight at most two. The $2^3\binom{11}{3}$ vectors of weight 3 must be covered by spheres around codewords of weight 5, so that $a_5 = 2^3 \cdot \binom{11}{3}/\binom{5}{3} = 132$; next $a_6 = (2^4 \cdot \binom{11}{4} - 132 \cdot \binom{5}{4} - 132 \cdot \binom{5}{3} \cdot 2)/\binom{6}{4} = 132$; etc.)

The supports of the code words of minimal nonzero weight form Steiner systems $S(4, 7, 23)$, $S(5, 8, 24)$, $S(4, 5, 11)$ and $S(5, 6, 12)$, respectively.

For those who know what a near polygon is: the partial linear space with as points the vectors of the extended ternary Golay code and as lines the cosets of 1-dimensional subspaces spanned by a vector of weight 12 is a near hexagon with $s + 1 = 3$ points/line and $t + 1 = 12$ lines/point and diagram (as distance transitive graph) as in fig. 16.1. It has quads (namely 3×3 grids GQ(2, 1)).

16.2. The Golay codes – Constructions

Given one of the extended codes one may *puncture* it by just deleting one coordinate position. This proces (i) and (iii) from (ii), (iv). Conversely, given (i)

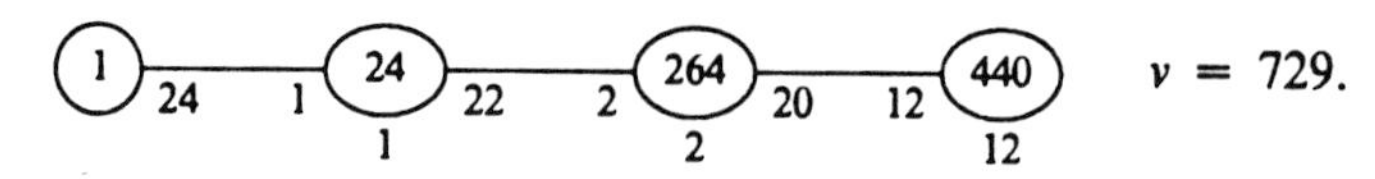

Figure 16.1.

one may construct (ii) by *extending* it, i.e., adding a *parity check* bit such as to make the weight of all code words even; and given (iii) (normalized by multiplying certain coordinate positions by -1 such that the normalized code contains the all-one vector) one may construct (iv) by adding a check trit such as to make the sum of all coordinates a multiple of three (but this latter fact is not easy to see; cf. Delsarte and Goethals 1975).

16.2.1. A construction of the extended binary Golay code

This code is the lexicographically first code with word length $n = 24$ and minimum distance 8: write down the numbers $0, 1, \ldots, 2^{24} - 1$ in binary and consider them as binary vectors of length 24. Cross out each vector that has distance less than 8 to a not previously crossed-out vector. The 4096 vectors not crossed out form the extended binary Golay code. (Proof. Just do it. Some work may be saved by observing (M.R. Best) that any lexicographically minimal binary code with a number of vectors that is a power of two is linear so that all one needs are the 12 base vectors. These turn out to be

0 0 0 0 0 0 0 0 0 0 0 0 0 0 0 0 1 1 1 1 1 1 1 1
0 0 0 0 0 0 0 0 0 0 0 0 1 1 1 1 0 0 0 0 1 1 1 1
0 0 0 0 0 0 0 0 0 0 1 1 0 0 1 1 0 0 1 1 0 0 1 1
0 0 0 0 0 0 0 0 0 1 0 1 0 1 0 1 0 1 0 1 0 1 0 1
0 0 0 0 0 0 0 0 1 0 0 1 0 1 1 0 0 1 1 0 1 0 0 1
0 0 0 0 0 0 1 1 0 0 0 0 0 0 1 1 0 1 0 1 0 1 1 0
0 0 0 0 0 1 0 1 0 0 0 0 0 1 0 1 0 1 1 0 0 0 1 1
0 0 0 0 1 0 0 1 0 0 0 0 1 1 0 0 0 1 1 1 0 1 0
0 0 0 1 0 0 0 1 0 0 0 1 0 0 0 1 0 1 1 1 1 0 0 0
0 0 1 0 0 0 0 1 0 0 0 1 0 0 1 0 0 0 0 1 1 1 0 1
0 1 0 0 0 0 0 1 0 0 0 1 0 1 0 0 0 1 0 0 1 1 1 0
1 0 0 0 0 0 0 1 0 0 0 1 0 1 1 1 0 0 1 0 0 1 0 0 .)

Remark. Deleting the columns with only one 1 and interchanging zeros and ones we find the incidence matrix of the unique symmetric group divisible design GD(5, 2, 2; 12) in Hanani's notation – see below.

16.2.2. Construction as quadratic residue codes

For $(n, q) = (11, 3)$ or $(23, 2)$ consider the linear code generated over $\mathbb{F}_q$ by the n vectors $c_i (1 \leq i \leq n)$ with coordinates

$$(c_i)_j = \begin{cases} 1 & \text{if } j - i \text{ is a nonzero square } (\text{mod } n)\,, \\ 0 & \text{otherwise}\,. \end{cases}$$

This yields the ternary and binary Golay codes. (Proof. The only nontrivial thing to check is the minimum distance. One easily sees that the extended code has all weights divisible by 3 resp. 4 so that all that remains is to prove that its minimum distance is not 3 resp. 4 and that is easy. For explicit details see Van Lint (1982, §6.9) (but notice that some of the statements there are valid only in the binary case).)

16.2.3. Constructions from the 2-(11, 5, 2) *biplane and the icosahedron*

Let B be the incidence matrix of a design with point set $\mathbb{Z}_{11}$ and blocks $\{1,3,4,5,9\}+i$ $(i \in \mathbb{Z}_{11})$ (i.e., the translates of the set of nonzero squares mod 11). (This design is a square block design 2-(11, 5, 2): any two points are on two blocks and dually.) Then the rows of the 12×24 matrix

$$\begin{pmatrix} I & \begin{matrix} 0 & \mathbf{1}^{\mathrm{T}} \\ \mathbf{1} & K \end{matrix} \end{pmatrix} \quad \text{where } K = J - B$$

generate the extended binary Golay code.

Let N be the adjacency matrix of the icosahedron (points: 12 vertices, adjacent: joined by an edge). Then the rows of the 12×24 matrix $(I \quad J-N)$ generate the extended binary Golay code. Conversely, given a generator matrix $(I \quad X)$ for the extended binary Golay code, either one of its rows has weight 12 and we are in the first situation, or all rows have weight 8 and X is the incidence matrix of the unique symmetric group divisible design GD(5, 2, 2; 12); by suitably ordering the rows and columns we may obtain $X = N$ and we are in the second situation.

16.2.4. A similar construction for the extended ternary Golay code

Let S be the 5×5 circulant matrix with first row $(0 \quad 1 \quad -1 \quad -1 \quad 1)$ (i.e., the quadratic residue character mod 5). Then the rows of the 6×12 matrix

$$\begin{pmatrix} I & \begin{matrix} 0 & \mathbf{1}^{\mathrm{T}} \\ -\mathbf{1} & S \end{matrix} \end{pmatrix}$$

generate the extended ternary Golay code (over $\mathbb{F}_3$). (Cf. Cameron and Van Lint 1980, chapter 13: Symmetry codes.) One checks easily that (up to permutating coordinate positions and multiplying columns by -1, i.e., up to monomial transformations) this is the only possibility for a generator matrix $(I \quad X)$.

16.3. Parameters of the Witt designs

For a t-(v, k, λ) design, the number of blocks containing a point set X and disjoint from a point set Y (where $X \cap Y = \emptyset$) can be expressed in the parameters t, v, k, λ, $|X|$, $|Y|$ when $|X \cup Y| \leq t$. Let us call these numbers $\mu(|X|, |Y|)$.

For $S(5, 8, 24)$ we have: $\lambda_5 = 1$, $\lambda_4 = 5$, $\lambda_3 = 21$, $\lambda_2 = 77$, $\lambda_1 = 253$, $\lambda_0 = 759$. The "intersection" triangle here gives the numbers $\mu(|X|, |Y|)$ with $|X \cup Y|$

constant in each row and $|X|$ increasing in each row, where $X \cup Y$ is contained in a block.

$$
\begin{array}{c}
759 \\
253 \quad 506 \\
77 \quad 176 \quad 330 \\
21 \quad 56 \quad 120 \quad 210 \\
5 \quad 16 \quad 40 \quad 80 \quad 130 \\
1 \quad 4 \quad 12 \quad 28 \quad 52 \quad 78 \\
1 \quad 0 \quad 4 \quad 8 \quad 20 \quad 32 \quad 46 \\
1 \quad 0 \quad 0 \quad 4 \quad 4 \quad 16 \quad 16 \quad 30 \\
1 \quad 0 \quad 0 \quad 0 \quad 4 \quad 0 \quad 16 \quad 0 \quad 30
\end{array}
$$

Given a block B_0 of $S(5,8,24)$, let n_i be the number of blocks B such that $|B_0 \cap B| = i$. Then $n_8 = 1$, $n_4 = 280$, $n_2 = 448$, $n_0 = 30$ and all other n_i are zero.

For those who know what a near polygon is: the partial linear space with as points the 759 blocks of the Steiner system $S(5,8,24)$ and as lines the partitions of the point set of the design in three pairwise disjoint blocks, is a near hexagon with $s+1=3$ points/line and $t+1=15$ lines/point and diagram (as distance transitive graph) as in fig. 16.2. It has quads (with 15 points and 15 lines: GQ(2,2), i.e., Sp(4,2) generalized quadrangles). A quad in the near polygon corresponds to a *sextet* in the design: a partition of the point set into six 4-sets such that the union of any two of them is a block. Distances 0, 1, 2, 3 in the near polygon corresponds to intersections of size 8, 0, 4, 2, respectively.

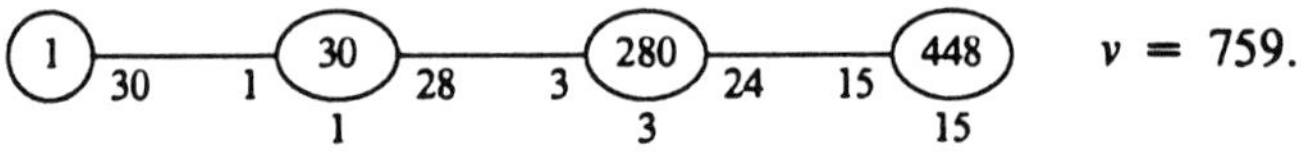

(1) 30 — 1 (30) 28 — 3 (280) 24 — 15 (448) $v = 759$.
1   3   15

Figure 16.2.

For $S(5,6,12)$ we have: $\lambda_5 = 1$, $\lambda_4 = 4$, $\lambda_3 = 12$, $\lambda_2 = 30$, $\lambda_1 = 66$, $\lambda_0 = 132$. Our intersection triangle becomes

$$
\begin{array}{c}
132 \\
66 \quad 66 \\
30 \quad 36 \quad 30 \\
12 \quad 18 \quad 18 \quad 12 \\
4 \quad 8 \quad 10 \quad 8 \quad 4 \\
1 \quad 3 \quad 5 \quad 5 \quad 3 \quad 1 \\
1 \quad 0 \quad 3 \quad 2 \quad 3 \quad 0 \quad 1
\end{array}
$$

Given a block B_0 of $S(5,6,12)$, let n_i be the number of blocks B such that $|B_0 \cap B| = i$. Then $n_6 = 1$, $n_4 = 45$, $n_3 = 40$, $n_2 = 45$, $n_0 = 1$ (and $n_5 = n_1 = 0$). In particular the complement of a block is again a block.

Note that all the above information follows directly from the parameters of these Steiner systems (and may thus be used in uniqueness proofs).

There exist unique designs $S(5,8,24)$, $S(4,7,23)$, $S(3,6,22)$, $S(2,5,21)$, $S(1,4,20)$, $S(5,6,12)$, $S(4,5,11)$, $S(3,4,10)$, $S(2,3,9)$, $S(1,2,8)$. ($S(2,5,21)$ is the projective plane of order 4, $S(2,3,9)$ the affine plane of order 3, $S(3,4,10)$ the Möbius plane of order 3.) (In view of the derivation $S(t,k,v)\to S(t-1,k-1,v-1)$ it suffices to construct $S(5,8,24)$ and $S(5,6,12)$, and we shall find these as the supports of the code words of minimal nonzero weight in the extended Golay codes; uniqueness will be shown as a corollary of the uniqueness of the Golay codes.)

16.4. The uniqueness of the extended binary Golay code $\mathscr{C}$

For a good account, of the uniqueness of the Golay codes and associated Steiner systems see MacWilliams and Sloane 1977, chapter 20). There the uniqueness of $S(5,8,24)$ is proven "by hand" – examining its structure in detail, and the uniqueness of $\mathscr{C}$ follows rather easily by use of the linear programming bound. Here we follow the opposite way, getting the uniqueness of $S(5,8,24)$ from that of $\mathscr{C}$, and proving the latter directly, without recourse to the theory of association schemes. Instead, the uniqueness of $\mathscr{C}$ will come as a consequence of the uniqueness of the 2-(11, 5, 2) biplane.

Theorem. *Let $\mathscr{C}$ be a binary code containing* $\mathbf{0}$, *with word length* 24, *minimum distance* 8 *and* $|\mathscr{C}|\geqslant 2^{12}$. *Then $\mathscr{C}$ is the extended binary Golay code.*

Proof. If we delete a coordinate position we find a code $\mathscr{C}_0$ with word length 23, minimum distance (at least) 7 and $|\mathscr{C}_0|\geqslant 2^{12}$. As we saw before, such a code must have $|\mathscr{C}_0|=2^{12}$ and weight enumerator coefficients $a_0=a_{23}=1$, $a_7=a_{16}=253$, $a_8=a_{15}=506$, $a_{11}=a_{12}=1288$ (by the sphere-packing argument it follows that $\mathscr{C}_0$ is perfect). Now if $\mathscr{C}$ contains a word of weight w not divisible by 4 then by suitably puncturing we would find a $\mathscr{C}_0$ containing a word of weight w or $w-1$ not 0 or -1 (mod 4), a contradiction. Hence $\mathscr{C}$ has weight enumerator coefficients $a_0=a_{24}=1$, $a_8=a_{16}=759$, $a_{12}=2576$. Giving an arbitrary vector in $\mathscr{C}$ the rôle of $\mathbf{0}$ we see that all distances between code words are divisible by 4. If $u,v\in\mathscr{C}$ then $d_{\mathrm{H}}(u,v)=\mathrm{wt}(u)+\mathrm{wt}(v)-2(u,v)$ so the inner product (u,v) is even and it follows that $\mathscr{C}$ is self-orthogonal. But $\mathscr{C}^{\perp}$ is a linear subspace of dimension $24-\dim\langle\mathscr{C}\rangle\leqslant 12$ so that $\mathscr{C}^{\perp}=\mathscr{C}$ and $\mathscr{C}$ is a linear code. Let u and $\bar{u}$ be two complementary weight 12 vectors in $\mathscr{C}$. The code $\mathscr{C}_u$ obtained from $\mathscr{C}$ by throwing away all coordinate positions where u has a 1, has word length 12 and dimension 11 and hence must be the even weight code (consisting of all vectors of even weight). This means that we can pick a basis for $\mathscr{C}$ consisting of $\bar{u}$ and 11 vectors v_j with $(u,v_j)=2$ so as to get a generator matrix of the form

$$\begin{pmatrix} 0 & \mathbf{0}^{\mathrm{T}} & \mathbf{1}^{\mathrm{T}} & 1 \\ \mathbf{1} & I & K & \mathbf{0} \end{pmatrix}$$

where I is an identity matrix of order 11. A little reflection shows that $J-K$ is the incidence matrix of a 2-(11, 5, 2) biplane. This shows uniqueness of $\mathscr{C}$ given the uniqueness of the 2-(11, 5, 2) biplane, and the latter is easily verified by hand. □

Theorem. *There is a unique Steiner system* $S(5, 8, 24)$.

Proof. (i) Existence: the words of weight 8 in $\mathscr{C}$ cover each 5-set at most once since $d = 8$, and exactly once since $\binom{24}{5} = 759 \cdot \binom{8}{5}$.

(ii) Uniqueness: Let $\mathscr{S}$ be such a system, and let $\mathscr{C}_1$ be the linear code over $\mathbb{F}_2$ spanned by its blocks. From the intersection numbers we know that $\mathscr{C}_1$ is self-orthogonal (i.e., $\mathscr{C}_1 \subset \mathscr{C}_1^\perp$) with all weights divisible by 4. In order to show that $|\mathscr{C}_1| \geq 2^{12}$ fix three independent coordinate positions, say 1, 2, 3 and look at the subcode $\mathscr{C}_2$ of $\mathscr{C}_1$ consisting of the vectors u with $u_1 = u_2 = u_3$. Then $\dim \mathscr{C}_1 = (\dim \mathscr{C}_2) + 2$. Thus, in order to prove $\dim \mathscr{C}_1 \geq 12$ it suffices to show that the code generated by the blocks of $S(5, 8\,24)$ containing 3 given points has dimension at least 10. In other words, we must show that the code generated by the lines of the projective plane PG(2, 4) (which is nothing but $S(2, 5, 21)$) has dimension at least 10, but that is the result of the next theorem.

The blocks of an $S(5, 8, 24)$ assume all possible 0–1 patterns on sets of cardinality at most 5 so that $\mathscr{C}_1^\perp$ has minimum weight at least 6. Since $\mathscr{C}_1$ has all weights divisible by 4 and $\mathscr{C}_1 \subset \mathscr{C}_1^\perp$ it follows that $d(\mathscr{C}_1) = 8$. Now apply the previous theorem to see that $\mathscr{C}_1$ is the extended binary Golay code, and $\mathscr{S}$ the set of its weight 8 vectors. □

Theorem. *The code over* $\mathbb{F}_2$ *spanned by the lines of the projective plane* PG(2, 4) *has dimension* 10.

Proof. Let *abcde* be a line in PG(2, 4). The set of ten lines consisting of all 5 lines on a, 3 more lines on b, and one more line on each of c, d, is linearly independent, so the dimension is at least 10. But the previous proof (or a simple direct argument showing that the extended code cannot be self-dual) shows that it is at most 10. □

16.5. Substructures of $S(5, 8, 24)$

Theorem. *Let* B_0 *be a fixed octad (block of* $S(5, 8, 24)$*). The* 30 *octads disjoint from* B_0 *form a self-complementary* 3-(16, 8, 3) *design, namely the design of the points and affine hyperplanes in* AG(4, 2), *the* 4*-dimensional affine space over* $\mathbb{F}_2$.

Proof. Let $\mathscr{B}$ be the collection of octads disjoint from B_0. We have seen already that $|\mathscr{B}| = 30$.

(i) The linear span of $\mathscr{B}$ is a code of dimension 5 and weight enumerator $1 + 30x^8 + x^{16}$. (Proof. Having zeros at the positions of B_0 gives 7 restrictions, so this span has codimension 7 in the extended binary Golay code $\mathscr{C}$.)

(ii) Each block $B \in \mathscr{B}$ is disjoint from a unique $B' \in \mathscr{B}$ and meets all other blocks in precisely 4 points. (Proof. Obvious from (i).)

(iii) $\mathcal{B}$ is a 3-(16, 8, 3) design. (Proof. Each triple is covered $30 \cdot \binom{8}{3}/\binom{16}{3} = 3$ times on average, but no triple is covered 4 times.)

(iv) We have AG(4, 2). (Proof. Invoke your favorite characterization of AG(4, 2) or PG(3, 2), say Dembowski–Wagner or Veblen and Young. An explicit construction of the vector space is also easy: choose a point $0 \notin B_0$ and regard it as origin. If x, y are nonzero points then the three blocks B_1, B_2, B_3 on 0, x, y have a fourth point z in common (for B_3 is the complement of $B_1 + B_2$ (i.e., $B_1 \triangle B_2$) hence contains $B_1 \cap B_2$) – now write $x + y = z$. If x, y, z are three arbitrary nonzero points and B_1, B_2, B_3 are the blocks containing 0, x, y then unless $z = x + y$ precisely one of the B_j, say B_1, also contains z. Now in order to check that $(x + y) + z = x + (y + z)$ we can do all computations within B_1 (using the induced 3-(8, 4, 1) design) – but clearly the 3-(8, 4, 1) design is unique (the extension of the Fano plane), i.e., is AG(3, 2) and addition is associative. This defines the vector space structure, and the blocks are the hyperplanes on 0 and their complements.) □

Theorem. *Let T_0 be a fixed tetrad (4-set). Then T_0 determines a unique sextet, i.e., partition of the 24-set into 6 tetrads T_i such that $T_i \cup T_j$ is a block for all i, j $(i \neq j)$.*

Proof. Since $\lambda_4 = 5$ there are 5 blocks B_i on T_0 $(i = 1, 2, 3, 4, 5)$ and with $T_i := B_i \backslash T_0$ we have $T_i \cup T_j = B_i + B_j$ $(0 \neq i \neq j \neq 0)$. Since $\lambda_5 = 1$ the 6 tetrads T_i are pairwise disjoint. □

Theorem. *Let B_0 be a fixed octad, $x \in B_0$, $y \notin B_0$, Z the complement of $B_0 \cup \{y\}$. Then there is a natural 1–1 correspondence between the $\binom{7}{3} = 35$ 3-sets in $B_0 \backslash \{x\}$ and the $(2^2 + 1)(2^2 + 2 + 1) = 35$ lines in the PG(3, 2) defined on Z. Triples meeting in a singleton correspond to intersecting lines.*

Proof. A line in the PG(3, 2) on Z is a $T \backslash \{y\}$ where T is a 4-set such that 3 of the blocks on it are disjoint from B_0. Of the remaining two blocks on T, precisely one contains the point x, and if B is this one then $B \cap B_0 \backslash \{x\}$ is the triple corresponding to the given line. □

Theorem. *Let D_0 be a fixed dodecad (support of a vector of weight 12 in $\mathscr{C}$). The 132 octads meeting D_0 in six points form the blocks of a Steiner system $S(5, 6, 12)$ on D_0.*

Proof. Each 5-set in D_0 is in a unique block of $S(5, 8, 24)$, and this block must meet D_0 in 6 points. □

Theorem. *Let D_0 be a fixed dodecad and $x \notin D_0$. The 22 octads meeting D_0 in six points and containing x form the blocks of a Hadamard 3-design 3-(12, 6, 2). There is a natural 1–1 correspondence between the $\frac{1}{2} \cdot 132 = 66$ pairs of disjoint blocks of the $S(5, 6, 12)$ on D_0 and the $\binom{12}{2} = 66$ pairs of points not in D_0.*

Proof. Given a pair of points x, y outside D_0, there are precisely two octads on $\{x, y\}$ meeting D_0 in six points, and these give disjoint blocks in the $S(5, 6, 12)$ (for: if these octads are B, B' then $B' = B + D_0$). Varying y we find 11 pairs of disjoint blocks, blocks from different pairs having precisely 3 points in common. □

16.6. The Mathieu group M_{24}

M_{24} is by definition the automorphism group of the extended binary Golay code $\mathscr{C}$ (or, what is the same, of the Witt design $S(5, 8, 24)$), i.e., the group of permutations of the 24 coordinate positions preserving the code. For a beautiful discussion of this and related groups, see Conway (1971).

Theorem. *M_{24} has order $24 \cdot 23 \cdot 22 \cdot 21 \cdot 20 \cdot 16 \cdot 3$ and acts 5-transitively on the 24 coordinate positions.*

Proof. Let N be the adjacency matrix of the icosahedron. Since $\operatorname{Aut} N$ is transitive on the 12 points, and N is nonsingular, so that if $(I \quad J - N)$ generates $\mathscr{C}$ then also $(J - N' \quad I)$ for some N' equivalent to N, it follows that M_{24} is transitive. (This immediately implies uniqueness of the binary Golay code – see next section.)

The representation as quadratic residue code gives an automorphism with cycle structure $1 + 23$, so M_{24} is 2-transitive. This same representation also gives $1 + 1 + 11 + 11$. The representation below exhibits an automorphism with cycle structure $1^3 7^3$ so that M_{24} is 3-transitive.

Let $\mathscr{H}$ be the extended binary Hamming code (with word length 8, dimension 4) consisting of the 8 rows of $\left(\begin{smallmatrix} 0 & \mathbf{0}^{\mathrm{T}} \\ \mathbf{1} & F \end{smallmatrix}\right)$ (where $F = \operatorname{circ}(0110100)$ is the incidence matrix of the Fano plane $\mathrm{PG}(2, 2)$) and their complements. Let $\mathscr{H}^*$ be the equivalent code obtained by replacing F by $F^* = \operatorname{circ}(0001011)$. Then $\mathscr{H} \cap \mathscr{H}^* = \{\mathbf{0}, \mathbf{1}\}$. Let $\mathscr{C} = \{(a + x, b + x, a + b + x) \mid a, b \in \mathscr{H}, x \in \mathscr{H}^*\}$. Then $\mathscr{C}$ has word length 24, dimension 12 and minimum distance 8 as one easily checks. Hence $\mathscr{C}$ is the extended binary Golay code.

From automorphisms with cycle structure $1^3 7^3$ and $1^4 5^5$ (the latter is easily seen in the icosahedral representation) we see that M_{24} is 4-transitive.

Both $\mathscr{H}$ and $\mathscr{H}^*$ have automorphism group $\mathrm{PSL}(2, 7)$ acting on the coordinates numbered ∞, 0, 1, 2, 3, 4, 5, 6. Elements in $\mathrm{PGL}(2, 7) \backslash \mathrm{PSL}(2, 7)$ interchange $\mathscr{H}$ and $\mathscr{H}^*$. Any automorphism of $\mathscr{H}$ of shape 4^2 (for definiteness, say $x \mapsto 2 - 1/(x + 2)$) yields an automorphism of $\mathscr{C}$ of shape 4^6.

From automorphisms of shape $1^4 5^5$ and 4^6 we see that M_{24} is transitive on 5-sets. The stabilizer of a 5-set contains permutations of shape $1^4 5^5$ and $1^8 2^8$ (the latter, e.g., by interchanging the first and second groups of 8 coordinates in the representation given above) inducing 5 and $1^3 2$ on the 5-set, but since (ABCDE) and (AB) generate the symmetric group Sym(5) on 5 symbols, this shows that M_{24} is 5-transitive.

Since M_{24} is transitive on 5-sets, and a 5-set determines a unique octad, M_{24} is transitive on octads.

Let us give yet another representation of $\mathscr{C}$ (due to Conway). Consider the set of 4×6 matrices with entries 0 or 1 satisfying the following two restraints:

(i) The six columns sums and the first row sum have the same parity.

(ii) If r_i denotes the ith row ($1 \leqslant i \leqslant 4$) and $\mathbb{F}_4 = \{0, 1, \omega, \bar{\omega}\}$ and $\mathscr{F}$ is the linear code (with word length 6, dimension 3 and minimum distance 4) over $\mathbb{F}_4$ generated by the rows of the matrix

$$\begin{pmatrix} 1 & 0 & 0 & 1 & \omega & \bar{\omega} \\ 0 & 1 & 0 & 1 & \bar{\omega} & \omega \\ 0 & 0 & 1 & 1 & 1 & 1 \end{pmatrix}$$

then $r_2 + \omega r_3 + \bar{\omega} r_4 \in \mathscr{F}$.

It is almost trivial to verify that these matrices form a linear code with word length 24, dimension 12 and minimum distance 8 over $\mathbb{F}_2$, i.e., we have the extended binary Golay code again.

Three octads are given in fig. 16.3.

Some easy automorphisms (with cycle structure):

- interchange rows r_2, r_3, r_4 cyclically $[3^6 1^6]$
- interchange the last two rows and the last two columns $[2^8 1^8]$
- interchange 1st and 2nd, 3rd and 4th, 5th and 6th column, and rows r_3 and r_4 $[2^{12}]$.

This representation shows that the pointwise stabilizer of a 5-set is transitive on the remaining 3 points of the octad containing it. Next observe that if π is a permutation fixing a certain octad pointwise and g fixes this octad setwise then $\pi^g := g^{-1}\pi g$ fixes the octad pointwise. It follows that if O is an orbit of the pointwise stabilizer of this octad, then gO is also an orbit. Since we can find g with shapes $4^2 + 4^4$ and $1^3 5 + 15^3$ and π with shape $1^5 3 + 13^5$ we see that the pointwise stabilizer of an octad is transitive on the remaining 16 points.

Let H be the subgroup of M_{24} fixing a block B (setwise) and a point $x \notin B$. By the above $|H| \geqslant 24 \cdot 23 \cdot 22 \cdot 21 \cdot 20 \cdot 16 \cdot 3/759 \cdot 16 = \frac{1}{2}8!$. The code words in $\mathscr{C}$ that are zero on the positions of $B \cup \{x\}$ form a subcode with codimension 8 in $\mathscr{C}$, i.e., with dimension 4. No nonidentity element of H can act trivially on this subcode (no two coordinate positions are dependent, e.g., because this is the code spanned by the complements of the hyperplanes in PG(3, 2) so $|H| \leqslant |\mathrm{PGL}(4,2)| = 15.14.12.8 = \frac{1}{2}8!$. Since equality must hold we have shown that $|M_{24}| = 24 \cdot 23 \cdot 22 \cdot 21 \cdot 16 \cdot 3$ and that $H \cong \mathrm{Alt}(8) \cong \mathrm{PGL}(4, 2)$. □

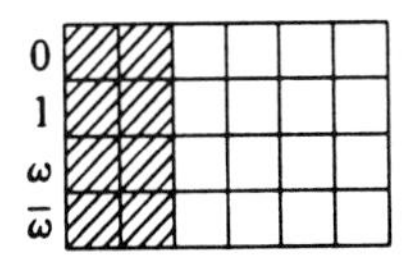

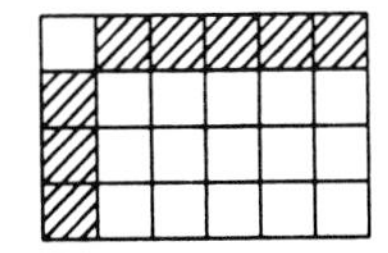
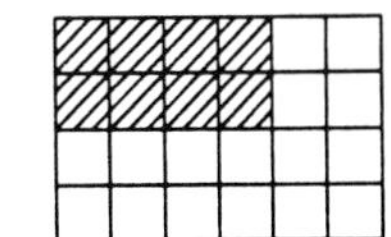

Figure 16.3.

Theorem. *M_{24} is transitive on trios (partitions of the point set into 3 octads), sextets and dodecads (vectors in $\mathscr{C}$ of weight 12).*

Proof. (i) PGL(4, 2) is transitive on the hyperplanes of PG(3, 2).

(ii) Any tetrad determines the sextet containing it, and M_{24} is 4-transitive.

(iii) Writing the dodecad as $B + B'$ where B and B' are octads with $|B \cap B'| = 2$ we see that it suffices to show that the pointwise stabilizer of B is transitive on the 16 blocks B' meeting B in a given pair. But this stabilizer is the elementary abelian group 2^4 and if some translation fixed B' then there would be an affine hyperplane meeting B' in 6 points – impossible. □

16.7. More uniqueness results

Theorem. *Let $\mathscr{C}_0$ be a binary code containing* **0** *with word length 23, minimum distance 7 and $|\mathscr{C}_0| \geqslant 2^{12}$. Then $\mathscr{C}_0$ is the (perfect) binary Golay code.*

Proof. Add a parity check to $\mathscr{C}_0$ to obtain $\mathscr{C}$. Thus $\mathscr{C}_0$ is obtained from $\mathscr{C}$ by suppressing some coordinate position, but all positions are equivalent since M_{24} is transitive. □

Theorem. *There is a unique Steiner system $S(4, 7, 23)$.*

Proof. The proof is very similar to that of the uniqueness of $S(5, 8, 24)$. Let $\mathscr{C}_0$ be the code spanned by the blocks and add a parity bit to obtain a self-orthogonal code $\mathscr{C}$ of word length 24. As before one identifies $\mathscr{C}$ as the extended binary Golay code, then $\mathscr{C}_0$ as the (perfect) binary Golay code, then the blocks of $S(4, 7, 23)$ as the words of weight 7 in this code. □

Theorem. *There is a unique Steiner system $S(3, 6, 22)$.*

Proof. Inspired by Lander (1983, esp. pp. 54 and 71), we first construct $\mathscr{D}$ as the extended linear code over $\mathbb{F}_2$ spanned by the lines of PG(2, 4). Then $\mathscr{D}$ has word length 22, and we have seen already that $\dim \mathscr{D} = 10$. $\mathscr{D}$ is self-orthogonal and hence there are three codes $\mathscr{D}_i$ of dimension 11 such that $\mathscr{D} \subset \mathscr{D}_i \subset \mathscr{D}^\perp$ $(i = 1, 2, 3)$. But $\mathscr{D}$ can be identified with the subcode of $\mathscr{C}$ defined by $u_1 = u_2 = u_3$, and the three codes $\mathscr{D}_i$ are found as subcodes defined by $u_2 = u_3$, $u_1 = u_3$ and $u_1 = u_2$, respectively. (More precisely, our codes are obtained from the subcodes of $\mathscr{C}$ just mentioned by dropping the first three coordinate positions and adding a parity bit; note that $\mathbf{1} \in \mathscr{D}$.) Now 3-transitivity of M_{24} tells us that the three codes $\mathscr{D}_i$ are equivalent; each has 77 words of weight 6. Given any Steiner system $S(3, 6, 22)$, its blocks must span one of the codes $\mathscr{D}_i$, and the blocks of the Steiner system are recovered as the supports of the code words of weight 6 in this code. □

Theorem. (a) *Let $\mathscr{C}^{(i)}$ be a binary code containing* **0** *with word length $24 - i$,* *minimum distance* 8, *and size at least* 2^{12-i}. *If* $0 \leqslant i \leqslant 3$ *then* $\mathscr{C}^{(i)}$ *is the i times shortened extended binary Golay code.*

(b) *Let* $\mathscr{C}_0^{(i)}$ *be a binary code containing* **0** *with word length* $23 - i$, *minimum distance* 7, *and size at least* 2^{12-i} *If* $0 \leqslant i \leqslant 3$ *then* $\mathscr{C}_0^{(i)}$ *is the i times shortened binary Golay code.*

The weight enumerators are (for $i > 0$*) given by*

i	n	dim	
1	23	11	$1 + 506x^8 + 1288x^{12} + 253x^{16}$
2	22	10	$1 + 330x^8 + 616x^{12} + 77x^{16}$
3	21	9	$1 + 210x^8 + 280x^{12} + 21x^{16}$
1	22	11	$1 + 176x^7 + 330x^8 + 672x^{11} + 616x^{12} + 176x^{15} + 77x^{16}$
2	21	10	$1 + 120x^7 + 210x^8 + 336x^{11} + 280x^{12} + 56x^{15} + 21x^{16}$
3	20	9	$1 + 80x^7 + 130x^8 + 160x^{11} + 120x^{12} + 16x^{15} + 5x^{16}$

Adding a parity check bit to $\mathscr{C}_0^{(i)}$ *we find* $\mathscr{C}^{(i)}$, *and for* $i > 0$ *the latter is the even weight subcode of* $\mathscr{C}_0^{(i)}$.

(c) *Let* $\mathscr{C}_{00}$ *be a binary self dual code with word length* 22 *and minimum distance* 6. *Then* $\mathscr{C}_{00}$ *is the once truncated binary Golay code.*

Proof (*Sketch*). Part (b) follows from part (a) since each of these codes has a group that acts transitively on the coordinate positions. For part (a), apply Delsarte's linear programming bound (enhanced by addition of a few obvious inequalities) to obtain uniqueness of all weight enumerators given. (Cf. Best et al. 1978.) The case $i = 0$ has been treated earlier. $\mathscr{C}^{(1)} \cup (\mathscr{C}^{(1)} + \mathbf{1})$ has minimum distance 7 hence is the binary Golay code. This settles the case $i = 1$. $\mathscr{C}^{(2)} \cup (\mathscr{C}^{(2)} + \mathbf{1})$ is self-orthogonal with minimum distance 6 and word length 22 with 2^{11} words hence is linear. But according to Pless and Sloane (1975) the unique such code is the once truncated binary Golay code. This settles the case $i = 2$ and part (c). If we extend $\mathscr{C}^{(3)} \cup (\mathscr{C}^{(3)} + \mathbf{1})$ with a parity check bit we obtain a self-orthogonal code $\mathscr{D}$ with minimum distance 6 and word length 22; $\mathscr{D}$ is contained in a self-dual code with $d = 6$ and $n = 22$, necessarily $\mathscr{C}_{00}$. Removing the parity bit again we find that $\mathscr{C}^{(3)}$ is contained in the twice truncated binary Golay code which has weight enumerator $(1 + x^{21}) + 21(x^5 + x^{16}) + 56(x^6 + x^{15}) + 120(x^7 + x^{14}) + 210(x^8 + x^{13}) + 280(x^9 + x^{12}) + 336(x^{10} + x^{11})$. Clearly $\mathscr{C}^{(3)}$ must consist of all vectors in this code with a weight divisible by 4, and hence is uniquely determined. □

(Dodunekov and Encheva 1988 show that linear codes $\mathscr{C}(4)$ and $\mathscr{C}(5)$ are unique, and that there are precisely two linear codes $\mathscr{C}(6)$. The weight enumerators are

i	n	dim	
4	20	8	$1 + 130x^8 + 120x^{12} + 5x^{16}$
5	19	7	$1 + 78x^8 + 48x^{12} + x^{16}$
6	18	6	$1 + 46x^8 + 16x^{12} + x^{16}$ or $1 + 45x^8 + 18x^{12}$.)

Starting from $S(5, 8, 24)$ and taking successive derived or residual designs we find designs with the following parameters:

5-(24,8,1)

4-(23,7,1) 4-(23,8,4)

3-(22,6,1) 3-(22,7,4) 3-(22,8,12)

2-(21,5,1) 2-(21,6,4) 2-(21,7,12) 2-(21,8,28)

Up to now we have seen uniqueness of the three largest Steiner systems (and used the uniqueness of $S(2, 5, 21) = PG(2, 4)$ – an easy exercise). Such strong results are not available for the remaining 6 designs. (In fact, observe that a 2-(21, 7, 3) design exists – e.g., the residual of an SBIBD 2-(31, 10, 3). Taking 4 copies of such a design, independently permuting the point sets in each case, produces large numbers of nonisomorphic designs with parameters 2-(21, 7, 12), so this structure is certainly not determined by its parameters alone.)

Tonchev (1986c) shows that there are unique quasi-symmetric designs 2-(22, 7, 16), 2-(21, 7, 12) and 2-(21, 6, 4). (A design is called *quasi-symmetric* if it has only two distinct block intersection numbers; here 0, 2 in case of 2-(21, 6, 4) and 1, 3 for the other two designs. Note that 3-(22, 7, 4) has $\lambda_2 = 16$.)

Tonchev (1986a) showed that there are unique designs 2-(23, 8, 56), 2-(22, 8, 40), 2-(21, 8, 28) with intersection numbers 0, 2, 4. (Note that 4-(23, 8, 4) has $\lambda_2 = 56$ and that 3-(22, 8, 12) has $\lambda_2 = 40$. Tonchev is not quite explicit about the middle case – he assumes 3-(22, 8, 12) – but his methods also work when only 2-(22, 8, 40) is given.)

The proofs are always by generating a self-orthogonal code and using the classification of binary self-dual codes with $n = 22$.

More information is contained in the following. Let $\mathscr{D}$ be a collection of k-subsets of an n-set such that any two k-subsets have distance at least 8. Then for each of the cases listed below we have $|\mathscr{D}| \leq b$, and when equality holds then the system is known to be unique, except in five cases. For $(n, k, b) = (19, 5, 12)$ there are precisely two nonisomorphic systems, corresponding to the two Latin squares of order 4. For $(n, k, b) = (18, 5, 9)$ there are precisely three nonisomorphic systems. For the three cases $(n, k, b) = (19, 6, 28)$, $(20, 7, 80)$, $(21, 8, 210)$ no information is available. In all cases other than these three the block intersection numbers are as shown in table 16.1.

Table 16.1

k	n							intersections
	18	19	20	21	22	23	24	
5	9	12	16	21				1
6		*28*	40	56	77			0,2
7			*80*	120	176	253		1,3
8				*210*	330	506	759	0,2,4

16.8. Related combinatorial structures

Many important combinatorial or group-theoretical structures are closely linked to the Witt designs or Golay codes. We find among other things two sporadic SBIBDs.

First of all, we have the Higman–Sims graph, the unique strongly regular graph with parameters $(v, k, \lambda, \mu) = (100, 22, 0, 6)$. It is constructed by taking a symbol ∞, the 22 points and 77 blocks of $S(3, 6, 22)$ as the 100 vertices of the graph, joining ∞ to the 22 points, each point to the blocks containing it, and two blocks when they are disjoint. The automorphism group of this graph is $\mathrm{HS} \cdot 2$, where HS denotes the simple Higman–Sims group. In this graph, the collection of points at distance 2 from a given point induces the unique strongly regular graph with parameters $(v, k, \lambda, \mu) = (77, 16, 0, 4)$, and in this latter graph again the points at distance 2 from a given point carry the unique strongly regular graph with parameters $(v, k, \lambda, \mu) = (56, 10, 0, 2)$. If we take the adjacency matrix of this last graph, and add in the diagonal, we obtain the incidence matrix of a *biplane* (SBIBD with $\lambda = 2$) with parameters 2-(56, 11, 2).

(Of course, more generally one obtains SBIBDs from strongly regular graphs with $\lambda = \mu$ or $\lambda = \mu - 2$. In fact one obtains precisely those SBIBDs that have a polarity with all points absolute, or without absolute points.)

Next, we have Higman's design. Let x and y be two points in the point set of $S(5, 8, 24)$. Construct a square design with as *points* the blocks containing x but not y and as *blocks* the blocks containing y but not x. A point B is incident with a block C when $|B \cap C| \in \{0, 4\}$. This produces an SBIBD with parameters 2-(176, 50, 14) and with HS as (2-transitive) automorphism group Higman (1969), Smith (1976).

Inside the Higman–Sims graph we can find the Hoffman–Singleton graph, and using $S(4, 7, 23)$ we can build the McLaughlin graph and the unique regular two-graph on 276 points, see Brouwer and Van Lint (1984). These structures have automorphism groups $P\Sigma U_3(5)$, $\mathrm{McL} \cdot 2$ and $\mathrm{Co} \cdot 3$, and conversely often provide the most convenient way to study these groups, just as one cannot study the Mathieu groups without knowledge of the Witt systems. Using $S(5, 8, 24)$ we can build the Leech lattice, see Leech (1967) and Conway and Sloane (1988). Finally, the Leech lattice is the first ingredient for building the Monster group, FG.

Appendix. The known SBIBDs

In the following, q will denote a prime power; $n := k - \lambda$.

(1) Hyperplanes in $\mathrm{PG}(d, q)$. $v = (q^{d+1} - 1)/(q - 1)$, $k = (q^d - 1)/(q - 1)$, $\lambda = (q^{d-1} - 1)/(q - 1)$, $n = q^{d-1}$. (Many other designs also have these parameters. Constructions with difference sets have been given by Singer 1938 and Gordon et al. 1962.)

(2) Hadamard designs. $v = 4n - 1$, $k = 2n - 1$, $\lambda = n - 1$. These exist iff there exists a Hadamard matrix of order $4n$. A *Hadamard matrix* is a matrix with ± 1 entries whose rows are mutually orthogonal. It is a long-standing conjecture that

Hadamard matrices of order $4n$ exist for all n. They are known to exist for many orders, e.g., for all $n \leqslant 166$, $n \neq 107$, and for $4n = 2^s m$ with $m^2 \leqslant 2^s$, see Wallis et al. (1972), Wallis (1976), Geramita and Seberry (1979), Sawade (1985) and Seberry and Yamada (1992). Constructions with difference sets have been given when v is a prime power by Paley (1933) and Todd (1933) (take the set of quadratic residues as a difference set), and by Hall (1956) (using cubic and sextic residues); and in case $v = q(q+2)$ where both q and $q+2$ are prime powers by Stanton and Sprott (1958). Many known construction techniques are discussed in the survey Hedayat and Wallis (1978).

(3) Regular Hadamard matrices with constant diagonal, $v = 4t^2$, $k = 2t^2 - t$, $\lambda = t^2 - t$, $n = t^2$, exist probably for all t; existence is known when $2t$ is the order of a Hadamard matrix, if both $2t-1$ and $2t+1$ are prime powers, and for $t = 7$; if existence is known for t_1 and t_2 then also for $t = 2t_1t_2$. (Cf. Menon 1962, Goethals and Seidel 1970, Wallis 1971, Wallis et al. 1972, section 5.3, Brouwer and Van Lint 1984, section 8D.)

(4) The adjacency matrix of the strongly regular graph with as vertices the elliptic or hyperbolic points off a quadric in $\mathrm{PG}(2m, 3)$ where two points are adjacent when they are joined by an elliptic line is the incidence matrix of a symmetric design with parameters $v = 3^m(3^m + \epsilon)/2$, $k = 3^{m-1}(3^m - \epsilon)/2$, $\lambda = 3^{m-1}(3^{m-1} - \epsilon)/2$, $n = 3^{2(m-1)}$ ($\epsilon = \pm 1$, $m \geqslant 2$).

(5) (Wallis 1971, McFarland 1973). $v = q^{d+1}((q^{d+1} - 1)/(q-1) + 1)$, $k = q^d(q^{d+1} - 1)/(q-1)$, $\lambda = q^d(q^d - 1)/(q-1)$, $n = q^{2d}$.

(6) (Brouwer 1983). $v = 2q(q^d - 1)/(q-1) + 1$, $k = q^d$, $\lambda = \frac{1}{2}q^{d-1}(q-1)$, $n = \frac{1}{2}q^{d-1}(q+1)$ (q an odd prime power, $d \geqslant 1$). These designs are for $d = 2$ equivalent to equidistant binary codes with word length v and distance $(v-1)/2$, Stinson and Van Rees (1984), Van Lint (1984). Denniston (1982) has determined all 78 such designs with $(d, q) = (2, 3)$.

(7) (Shrikhande and Singhi 1975, Van Leyenhorst 1980). $v = \lambda^3 + \lambda + 1$, $k = \lambda^2 + 1$, $n = \lambda^2 - \lambda + 1$ whenever both $\lambda - 1$ and $\lambda^2 - \lambda + 1$ are prime powers.

(8) (Mitchell 1979) $v = q^{h+1} - q + 1$, $k = q^h$, $\lambda = q^{h-1}$, $n = q^{h-1}(q-1)$ ($h \geqslant 2$, $q > 2$) whenever there exists an affine plane $\mathrm{AG}(2, q-1)$.

(9) Other designs with a difference set listed in Baumert (1971). (See also Hall 1974.) Examples are:

Biquadratic residues (Lehmer 1953, Hall 1956). $v = 4t^2 + 1$, $k = t^2$, $\lambda = (t^2 - 1)/4$, $n = (3t^2 + 1)/4$ (t odd, v a prime power). Difference sets with these same parameters have been constructed by Whiteman 1962 for t odd, $v = pq$ where both p and q are odd primes, $p \equiv 1 \pmod 4$ and $q = 3p + 2$.

Biquadratic residues with 0 (Lehmer 1953, Hall 1956). $v = 4t^2 + 9$, $k = t^2 + 3$, $\lambda = (t^2 + 3)/4$, $n = 3(t^2 + 3)/4$ (t odd, v a prime power).

Octic residues, $v = 8a^2 + 1 = 64b^2 + 9$ with v prime and a, b both odd, $k = a^2$, $\lambda = b^2$, $n = a^2 - b^2$.

Octic residues with 0, $v = 8a^2 + 49 = 64b^2 + 441$ with v prime and a odd, b even, $k = a^2 + 6$, $\lambda = b^2 + 7$, $n = a^2 - b^2 - 1$.

Many other cyclotomic constructions exist; e.g., for 16th powers see Lehmer (1954), Whiteman (1957). $v = 133$, $k = 33$, $\lambda = 8$, $n = 25$.

(10) Spence et al. (1993) constructed an SBIBD with $v = 160$, $k = 54$, $\lambda = 18$, $n = 36$. Spence (1993) generalized this to $v = (q+1)(q^{2t}-1)/(q-1)$, $k = q^{2t-1}(q+1)/2$, $\lambda = q^{2t-2}(q^2-1)/4$, $n = q^{2t-2}(q+1)^2/4$ when q is of the form $2^p - 1$.

Jungnickel and Pott further generalized this, and found $v = p^s(q^{2m}-1)/(q-1)$, $k = q^{2m-1}p^{s-1}$, $\lambda = p^{s-1}q^{2m-2}(p^{s-1}-1)/(p-1)$ where p is prime and $q = (p^s-1)/(p-1)$ is a prime power.

(11) Some special constructions.

$v = 56, k = 11, \lambda = 2, n = 9$	(Hall et al. 1970) ,
$v = 79, k = 13, \lambda = 2, n = 11$	(Aschbacher 1971) ,
$v = 49, k = 16, \lambda = 5, n = 11$	(Brouwer and Wilbrink 1984) ,
$v = 71, k = 15, \lambda = 3, n = 12$	(Haemers 1980, Beker and Haemers 1980) ,
$v = 71, k = 21, \lambda = 6, n = 15$	(Janko and Tran van Trung 1984) ,
$v = 66, k = 26, \lambda = 10, n = 16$	(Tran van Trung (1982), Bridges (1983),
$v = 70, k = 24, \lambda = 8, n = 16$	(Janko and Tran van Trung 1984),
$v = 78, k = 22, \lambda = 6, n = 16$	(Janko and Tran van Trung 1985, Tonchev 1987)
$v = 176, k = 50, \lambda = 14, n = 36$	(Higman 1969).

References

Abel, R.J.R.

[1991] Four mutually orthogonal Latin squares of orders 28 and 52, *J. Combin. Theory A* **58**, 306–309.

[1994] Forty-three balanced incomplete block designs, *J. Combin. Theory A* **65**, 252–267.

Abel, R.J.R., and D.T. Todorov

[1993] Four MOLS of orders 20, 30, 38 and 44, *J. Combin. Theory A* **64**, 144–148.

Aleksejev, V.O.

[1974] O čisle sistem trojek Štejnera, *Mat. Zametki* **15**, 767–774 [On the number of Steiner triple systems, *Math. Notes* **15**, 461–464] MR 50#12754; Zbl. 291.05006.

Alltop, W.O.

[1969] An infinite class of 4-designs, *J. Combin. Theory* **6**, 320–322.

[1972] An infinite class of 5-designs, *J. Combin. Theory A* **12**, 390–395.

[1975] Extending t-designs, *J. Combin. Theory A* **18**, 177–186. MR 51#10131; Zbl. 197.05028.

Alspach, B., K. Heinrich and G. Liu

[1992] Orthogonal factorizations of graphs, in: *Contemporary Design Theory,* eds. J.H. Dinitz and D.R. Stinson (Wiley, New York) pp. 13–40.

Andersen, L.D.

[1985] Completing partial Latin squares, *Mat. Fys. Medd. Dan. Vidensk. Selsk.* **41**, 23–69.

Andersen, L.D., and A.J.W. Hilton

[1983] Thank Evans!, *Proc. London Math. Soc. (3)* **47** 507–522. Zbl 557.05013.

Andersen, L.D., A.J.W. Hilton and E. Mendelsohn

[1980] Embedding partial Steiner triple systems, *Proc. London Math. Soc. (3)* **41**, 557–576. Zbl 461.05010.

Aschbacher, M.

[1971] On collineation groups of symmetric block designs, *J. Combin. Theory* **11**, 272–281.

Assaf, A.M.

[1986] On the covering of pairs by quadruples, *Discrete Math.* **61**, 119–132. MR 87h:05064.

[1988a] *Balanced incomplete Block Designs with $k = 6$ and $\lambda = 1$,* Preprint.

[1988b] *On the Packing of Pairs by Quadruples,* Preprint.

Assaf, A.M., N. Shalaby and L.P.S. Singh

[1993] Packing designs with block size 5 and index 2: the case v even, *J. Combin. Theory A* **63**, 43–54.

Assmus Jr, E.F., and J.D. Key

[1986] On an infinite class of systems with $t = 3$ and $k = 6$, *J. Combin. Theory A* **42**, 55–60. MR 87h:05043; Zbl 594.05016.

Assmus Jr, E.F., and H.F. Mattson Jr

[1969] New 5-designs, *J. Combin. Theory* **6**, 122–151. MR 42#7528; Zbl 179, p. 29.

Babai, L.

[1980] Almost all Steiner triple systems are asymmetric, in: *Topics on Steiner Systems,* eds. C.C. Lindner and A. Rosa, *Ann. Discrete Math.* **7**, 37–39. Zbl 462.05013.

Bagchi, Bh.

[1988] No extendable biplane of order nine, *J. Combin. Theory A* **49**, 1–12.

[1991] "Corrigendum", *J. Combin. Theory A* **57**, 162.

Bagchi, S., and Bh. Bagchi

[1989] Designs from pairs of finite fields, I. A cyclic unital $U(6)$ and other regular Steiner 2-designs, *J. Combin. Theory A* **52**, 51–61.

Baker, R.D.

[1976] Partitioning the planes of $AG_{2m}(2)$ into 2-designs, *Discrete Math.* **15**, 205–211. MR 54#4999; Zbl 326.05013.

[1984] Orthogonal line packings of $PG_{2m-1}(2)$, *J. Combin Theory A* **36**, 245–248. Zbl 531.05020.

Bannai, E.

[1977] On tight designs, *Quart. J. Math. (2)* **28**, 433–448. Zbl. 391.05005.

Baranyai, Zs.

[1975] On the factorization of the complete uniform hypergraph, in: *Infinite and Finite Sets, Proc. Coll. Keszthely, 1973,* eds. A. Hajnal, R. Rado and V.T. Sós, *Colloq. Math. Soc. János Bolyai* **10**, 91–108.

Baranyai, Zs., and A.E. Brouwer

[1977] *Extension of colourings of the edges of a complete (uniform hyper)graph,* Math. Centre Report ZW91 (Mathematisch Centrum. Amsterdam). Zbl. 362.05059.

Baron, G.

[1970] Über Verallgemeinerungen des Langford'schen Problems, in: *Combinatorial Theory & its Applications I, Proc. Conf. Balatonfüred, 1969, Colloq. Math. Soc. János Bolyai* **4**, 81–92.

Bate, J.A., and G.H.J. van Rees

[1985] Some results on N(4, 6, 10), N(4, 6, 11) and related coverings, in: *Proc. 16th Southeastern Int. Conf. on Combinatorics, Graph Theory and Computing, Boca Raton, 1985, Congress. Numerantium* **48**, 25–45.

Baumert, L.D.

[1971] *Cyclic Different Sets, Lecture Notes in Mathematics,* Vol. 182 (Springer, Berlin).

Beker, H.J., and W.H. Haemers

[1980] 2-Designs having an intersection number $k - n$, *J. Combin. Theory A* **28**, 64–81.

Bermond, J.-C., A.E. Brouwer and A. Germa

[1978a] Systèmes de triplets et différences associées, in: *Problèmes Combinatoires et Théorie des Graphes, Proc. Orsay 1976, Colloq. Int. du CNRS,* No. 260, eds. J.-C. Bermond, J.-C. Fournier, M. Las Vergnas and D. Sotteau (Gauthier-Villars, Paris) pp. 35–38.

Bermond, J.-C., A. Kotzig and J. Turgeon

[1978b] On a combinatorial problem of antennas in radioastronomy, in: *Combinatorics, Proc. 5th Hungarian Colloq., Keszthely, 1976,* eds. A. Hajnal and V.T. Sós, *Colloq. Math. Soc. János Bolyai* **18**, 135–149.

Best, M.R.

[1982] *A contribution to the nonexistence of perfect codes,* Ph.D. Thesis, University of Amsterdam (Mathematisch Centrum, Amsterdam, 1983).

Best, M.R., A.E. Brouwer, F.J. MacWilliams, A.M. Odlyzko and N.J.A. Sloane

[1978] Bounds for binary codes of length less than 25, *IEEE Trans. Inform. Theory* **IT-24**, 81–93. MR 57#19066; Zbl. 369.94011. Russian translation: 1980, *Kibern. Sb. Nov. Ser.* **17**, 28–59. Zbl. 456.94015.

Beth, T.

[1974] Algebraische Auflösungsalgorithmen für einige unendliche Familien von 3-Designs, *Le Matematiche* **29**, 105–135. MR 51#2948; Zbl 303.05016.

[1979] On resolutions of Steiner systems, *Mitt. Math. Sem. Gießen* **136**, 103 pp. MR 80f:05012; Zbl. 399.05020.

Beth, Th.

[1983] Eine Bemerkung zur Abschätzung der Anzahl orthogonaler lateinischer Quadrate mittels Siebverfahren, *Abh. Math. Sem. Univ. Hamburg* **53**, 284–288. MR 86f:05032.

Beth, Th., D. Jungnickel and H. Lenz

[1985] *Design Theory* (Bibliographisches Institut – Wissenschaftsverlag, Mannheim). MR 86j:05026.

Beutelspacher, A., and U. Lamberty

[1983] Strongly resolvable (r, λ)-designs, *Discrete Math.* **45**, 141–152.

Beutelspacher, A., and K. Metsch

[1986] Embedding finite linear spaces in projective planes, in: *Combinatorics 1984, Proc. Bari 1984, Ann. Discrete Math.* **30**, 39–56.

Bhattacharya, K.N.

[1944] A new balanced incomplete block design, *Sci. Culture* **9**, 508. MR 6, p. 163; Zbl 60, p. 313.

Billington, E.J., R.G. Stanton and D.R. Stinson

[1984] On λ-packings with block size 4, *Ars Combin.* **17A**, 73–84.

Block, R.E.

[1967] On the orbits of collineation groups, *Math. Z.* **96**, 33–49.

Bose, R.C.

[1939] On the construction of balanced incomplete block designs, *Ann. Eugenics* **9**, 353–399. MR 1, p. 199; Zbl 23, p. 1.

[1942] A note on the resolvability of incomplete block designs, *Sankhyā* **6**, 105–110.

Bose, R.C., S.S. Shrikhande and E.T. Parker

[1960] Further results on the construction of mutually orthogonal Latin squares and the falsity of Euler's conjecture, *Canad. J. Math.* **12**, 189–203. MR 23A#69; Zbl 93, p. 319.

Bose, R.C., S.S. Shrikhande and N.M. Singhi

[1976] Edge regular multigraphs and partial geometric designs with an application to the embedding of quasiresidual designs.

Brauer, R.

[1941] On the connection between the ordinary and the modular characters of groups of finite order, *Ann. of Math. (2)* **42**, 926–935. MR 3, p. 196.

Bremner, A.

[1979] A diophantine equation arising from tight 4-designs, *Osaka J. Math.* **16**, 353–356.

Bridges, W.G.

[1977] A characterization of Type-1 λ-designs, *J. Combin. Theory A* **22**, 361–367.

[1983] A (66, 26, 10)-design, *J. Combin. Theory A* **35**, 360–361. Zb. 517.05008.

Bridges, W.G., M. Hall Jr and J.L. Hayden

[1981] Codes and Designs, *J. Combin. Theory A* **31**, 155–174. Zbl 475.05012.

Brouwer, A.E.

[1976a] A note on the covering of all triples on 7 points with Steiner triple systems, Math. Centre Report ZN63 (Mathematisch Centrum, Amsterdam).

[1976b] A generalization of Baranyai's theorem, Math. Centre Report ZW81 (Mathematisch Centrum, Amsterdam). Zbl 341.05001.

[1977a] A new 5-design, Math. Centre Report ZW97 (Mathematisch Centrum, Amsterdam). MR 56#5314; Zbl 357.05016.

[1977b] The t-designs with $v < 18$, Math. Centre Report ZN76 (Mathematisch Centrum, Amsterdam).

[1977c] Steiner triple systems without forbidden subconfigurations, Math. Centre Report ZW104 (Mathematisch Centrum, Amsterdam). Zbl 367.05011.

[1979a] Optimal packings of K_4's into a K_n, *J. Combin. Theory A* **26**, 278–297. MR 80j:a05049;Zbl 412.05030.

[1979b] Wilson's theory, in: *Packing and Covering in Combinatorics, Mathematical Centre Tracts,* Vol. 106, ed. A. Schrijver (Mathematisch Centrum, Amsterdam) pp. 75–88. Zbl 438.05014.

[1980] On the existence of 30 mutually orthogonal Latin squares, II, Unpublished notes (København).

[1983] An infinite series of symmetric designs, Math. Centre Report ZW202 (Mathematisch Centrum, Amsterdam). MR 85h:05016; Zbl 534.05011.

Brouwer, A.E., and J.H. van Lint

[1984] Strongly regular graphs and partial geometries, in: *Enumeration and Design – Proc. Silver Jubilee Conf. on Combinatorics, Waterloo, 1982,* eds. D.M. Jackson and S.A. Vanstone (Academic Press, Toronto). MR 87c:05033; Zbl 555.05016.

Brouwer, A.E., and G.H.J. van Rees

[1982] More mutually orthogonal Latin squares, *Discrete Math.* **39**, 263–281. MR 84c:05019; Zbl 486.05015.

Brouwer, A.E., and M. Voorhoeve

[1979] Turán theory and the lotto problem, in: *Packing and Covering, Mathematical Centre Tracts,* Vol. 106, ed. A. Schrijver (Mathematisch Centrum, Amsterdam) pp. 99–105. Zbl 438.05025.

Brouwer, A.E., and H.A. Wilbrink

[1984] A symmetric design with parameters 2-(49, 16, 5), *J. Combin. Theory A* **37**, 193–194. MR 85i:05031;Zbl 541.05013.

Brouwer, A.E., H. Hanani and A. Schrijver

[1977] Group-divisible designs with block-size four, *Discrete Math.* **20**, 1–10. MR 57#5780;Zbl 371.62105.

Bruck, R.H.

[1963] Finite nets II. Uniqueness and embedding, *Pacific J. Math.* **13**, 421–457.

Bruck, R.H., and H.J. Ryser

[1949] The nonexistence of certain finite projective planes, *Canad. J. Math.* **1**, 88–93. MR 10, p. 319; Zbl 13, p. 375.

Cameron, P.J.

[1973] Extending symmetric designs, *J. Combin. Theory A* **14**, 215–220.

[1976] *Parallelisms of Complete Designs, London Mathematical Society Lecture Note Series,* Vol. 23 (Cambridge University Press, Cambridge). MR 54#7269; Zbl 333.05007.

[1977] Extensions of designs: variations on a theme, in: *Combinatorial Surveys, Proc. 6th British Combinatorial Conf.* (Academic Press, London) pp. 23–43.

[1980] Extremal results and configuration theorems for Steiner systems, in: *Topics on Steiner Systems,* eds. C.C. Lindner and A. Rosa, *Ann. Discrete Math.* **7**, 43–63. Zbl 441.51006.

Cameron, P.J., and J.H. van Lint

[1980] *Graphs, Codes and Designs, London Mathematical Society Lecture Note Series,* Vol. 43 (Cambridge University Press, Cambridge). Zbl 427.05001.

Cameron, P.J., J.-M. Goethals and J.J. Seidel

[1978] Strongly regular graphs having strongly regular subconstituents, *J. Algebra* **55**, 257–280. Zbl 444.05045.

Cayley, A.

[1850] On the triadic arrangements of seven and fifteen things, *London, Edinburgh and Dublin Philos. Mag. and J. Sci. (3)* **37**, 50–53. [*Collected Mathematical Papers,* Vol. I, pp. 481–484].

Chang, L.C.

[1959] The uniqueness and non-uniqueness of the triangular association scheme, *Sci. Record Peking Math. (New Ser.)* **3**, 604–613.

Chee, Y.M., C.J. Colbourn and D.L. Kreher

[1990] Simple t-designs with $v \leqslant 30$, *Ars Combinatoria* **29**, 193–258.

Chen, Yi

[1972] The Steiner system $S(3, 6, 26)$, *J. Geometry* **2**, 7–28. MR 46#3332; Zbl 227.50010; Zbl 231.50019.

Chouinard II, L.G.
[1983] Partitions of the 4-subsets of a 13-set into disjoint projective planes, *Discrete Math.* **45**, 297–300. MR 84h:05015; Zbl 509.05014.

Chowla, S., and H.J. Ryser
[1950] Combinatorial Problems, *Canad. J. Math.* **2**, 93–99.

Chowla, S., P. Erdős and E.G. Straus
[1960] On the maximal number of pairwise orthogonal Latin squares of a given order, *Canad. J. Math.* **12**, 204–208.

Connor Jr, W.S.
[1952] On the structure of balanced incomplete block designs, *Ann. Math. Statist.* **23**, 57–71. Correction: 1953, **24**, 135. MR 123, p. 617.
[1958] The uniqueness of the triangular association scheme, *Ann. Math. Statist.* **29**, 262–266. MR 20#3620.

Conway, J.H.
[1971] Three lectures on exceptional groups, in: *Finite Simple Groups*, eds. M.B. Powell and G. Higman (Academic Press, New York) pp. 215–247.

Conway, J.H., and N.J.A. Sloane
[1988] *Sphere Packings, Lattices and Groups* (Springer, New York).

Cruse, A.B.
[1974] On embedding incomplete symmetric Latin squares, *J. Combin. Theory A* **16**, 18–27. MR 48#8265.

Damerell, R.M.
[1983] On Smetaniuk's construction for Latin squares and the Andersen–Hilton theorem, *Proc. London Math. Soc. (3)* **47**, 523–526. Zbl 557.05014.

Davies, R.O.
[1959] On Langford's problem (II), *Math. Gaz.* **43**, 253–255.

Daykin, D.E., and R. Häggkvist
[1984] Completion of sparse partial Latin squares, in: *Graph Theory and Combinatorics, Proc. Cambridge, 1983*, pp. 127–132. Zbl 557.05016.

de Bruijn, N.G., and P. Erdős
[1948] On a combinatorial problem, *Indag. Math.* **10**, 421–423.

de Witte, P.
[1976] Finite linear spaces with two more lines than points, *J. Reine Angew. Math.* **288**, 66–73.

Dehon, M.
[1976a] Non-existence d'un 3-design de paramètres $\lambda = 2$, $k = 5$ et $v = 11$, *Discrete Math.* **15**, 23–25. MR 53#2705; Zbl 319.05011.
[1976b] Un théorème d'extension de t-designs, *J. Combin. Theory A* **21**, 93–99.

Delsarte, Ph.
[1973] An algebraic approach to the association schemes of coding theory, *Philips Res. Rep. Suppl.* **10**.

Delsarte, Ph., and J.-M. Goethals
[1975] Unrestricted codes with the Golay parameters are unique, *Discrete Math.* **12**, 211–224. Zbl 307.94013.

Dembowski, P.
[1958] Verallgemeinerungen von Transitivitätsklassen endlicher projektiver Ebenen, *Math. Z.* **69**, 59–89.
[1964] Möbiusebenen gerader Ordnung, *Math. Ann.* **157**, 179–205. MR 31#1607;Zbl 137, p. 401.
[1968] *Finite Geometries, Ergebnisse der Mathematik und ihre Grenzgebiete*, Vol. 44 (Springer, Berlin). MR 38#1597;Zbl 159, p. 500.

Dénes, J., and A.D. Keedwell
[1976] *Latin Squares and their Applications* (Academic Press, New York). MR 50#4338; Zbl 283.05014.

Denniston, R.H.F.
[1969] Some maximal arcs in finite projective planes, *J. Combin Theory* **6**, 317–319.
[1972] Some packing of projective spaces, *Rend. Accad. Naz. Lincei* **52**, 36–40.
[1973a] Cyclic pakcings of the projective space of order 8, *Accad. Naz. Lincei, Rend. Cl. Sci. Fis. Mat. Nat. (8)* **54**, 373–377.

[1973b] Packings of PG$(3,q)$, in: *Finite Structures and their Applications, C.I.M.E., II Ciclo, Bressanone, 1972* (Cremonese, Roma). MR 49#7160.
[1973c] Uniqueness of the inversive plane of order 5, *Man. Math.* **8**, 11–19. MR 48#1947; Zbl 251.05018.
[1973d] Uniquness of the inversive plane of order 7, *Man. Math.* **8**, 21–26. MR 48#1948; Zbl 251.05019.
[1974a] Double resolvability of some complete 3-designs, *Man. Math.* **12**, 105–112. MR 50#1910; Zbl 276.05022.
[1974b] Sylvester's problem of the 15 school-girls, *Discrete Math.* **9**, 229–233. MR 51#5322; Zbl 285.05002.
[1976] Some new 5-designs, *Bull. London Math. Soc.* **8**, 263–267. MR 58#276; Zbl 339.05019.
[1978] Non-existence of the Steiner system $S(4,10,66)$, *Utilitas Math.* **13**, 303–309. Zbl. 374.05011.
[1979] Further cases of double resolvability, *J. Combin. Theory A* **26**, 298–303. MR 80f:05010.
[1980] The problem of the higher values of *t*, in: *Topics on Steiner Systems*, eds. C.C. Lindner and A. Rosa, *Ann. Discrete Math.* **7**, 65–70.
[1982] Enumeration of symmetric designs $(25,9,3)$, in: *Algebraic and Geometric Combinatorics*, ed. E. Mendelsohn, *Ann. Discrete Math.* **15**, 111–127. MR 86d:05015, Zbl 488.05011.
[1983] A small 4-design, *Ann. Discrete Math.* **18**, 291–294.

Deza, M.
[1975] There exist only a finite number of tight *t*-designs $S_\lambda(t,k,v)$ for every block size *k*, *Utilitas Math.* **8**, 347–348. MR 56#8389.

Dodunekov, S.M., and S.B. Encheva
[1988] On the uniqueness of some linear subcodes of the binary extended Golay code, in: *Algebraic and Combinatorial Coding Theory, Proc. Varna 1988*, pp. 38–40.

Dow, S.
[1983] An improved bound for extending partial projective planes, *Discrete Math.* **45**, 199–207.

Doyen, J.
[1972] A note on reverse Steiner triple systems, *Discrete Math.* **1**, 315–319. MR 45#8542; Zbl 272.05013.
[1974] Construction methods in *t*-designs, in: *Proc. 5th Southeastern Conf. on Combinatorics, Graph Theory and Computing* (Utilitas Math., Winnipeg).

Doyen, J., and A. Rosa
[1980] An updated bibliography and survey of Steiner systems, in: *Topics on Steiner Systems*, eds. C.C. Lindner and A. Rosa, *Ann. Discrete Math.* **7**, 317–349 (A preprint dated February 1989 exists.) Zbl 422.51006.

Doyen, J., and R.M. Wilson
[1973] Embeddings of Steiner triple systems, *Discrete Math.* **5**, 229–239. MR 48#5881; Zbl 263.05017.

Driessen, L.H.M.E.
[1978] *t*-Designs, $t \geqslant 3$, Tech. Report (Dept. of Mathematics, Eindhoven University of Technology, Eindhoven).

Enomoto, H., N. Ito and R. Noda
[1979] Tight 4-designs, *Osaka J. Math.* **16**, 39–43.

Evans, T.
[1960] Embedding incomplete Latin squares, *Amer. Math. Monthly*, pp. 958–961.

Feit, W.
[1970] Automorphisms of symmetric balanced incomplete block designs, *Math. Z.* **118**, 40–49.

Fort Jr, M.K., and G.A. Hedlund
[1958] Minimal coverings of pairs by triples, *Pacific J. Math.* **8**, 709–719. MR 21#2595; Zbl 84, p. 14.

Fuji-Hara, R.
[1986] Mutually 2-orthogonal resolutions of finite projective space, *Ars. Combinatoria* **21**, 163–166.

Gelling, E.N.
[1973] *On 1-factorizations of the complete graph and the relationship to round-robin schedules*, M.Sc. Thesis (University of Victoria).

Geramita, A.V., and J. Seberry
[1979] *Orthogonal Designs, Lecture Notes in Pure and Applied Mathematics*, Vol. 45 (Marcel Dekker, New York).

Goethals, J.-M., and Ph. Delsarte

[1968] On a class of majority logic decodable cyclic codes, *IEEE Trans. Inform. Theory* **IT-14**, 182–188.

Goethals, J.-M., and J.J. Seidel

[1970a] Quasisymmetric block designs, in: *Combinatorial Structures and their Applications, Proc. Calgary, 1969* (Gordon and Breach, New York) pp. 111–116. MR 42#5818; Zbl 251.05009.

[1970b] Strongly regular graphs derived from combinatorial designs, *Canad. J. Math.* **22**, 597–614. MR 44#106; Zbl 198, p. 293.

Gordon, B., W.H. Mills and L.R. Welch

[1962] Some new difference sets, *Canad. J. Math.* **14**, 614–625.

Gronau, H.-D.O.F.

[1985] A survey of results on the number of t-(v, k, λ) designs, in: *Algorithms in Combinatorial Design Theory*, eds. C.J. Colbourn and M.J. Colbourn, *Ann. Discrete Math.* **26**, 209–219. Zbl 585.05001.

Guérin, R.

[1966] Existence er propriétes des carrés latins orthogonaux II, *Publ. Inst. Statist. Univ. Paris* **15**, 215–293. MR 35#4118.

Haemers, W.H.

[1980] Eigenvalue Techniques in Design and Graph Theory (Reidel, Dordrecht). Thesis (Technical University Eindhoven, 1979): *Mathematical Centre Tracts*, Vol. 121 (Mathematisch Centrum, Amsterdam, 1980).

Haemers, W.H., and C. Weug

[1974] *Linear Programming Bounds for Codes and Designs*, Technical Note N96 (M.B.L.E. Brussels).

Hall Jr, M.

[1953] Uniqueness of the projective plane with 57 points, *Proc. Amer. Math. Soc.* **4**, 912–916. MR 15, p. 460. Correction: 1954, *Proc. Amer. Math. Soc.* **5**, 994–997. Zbl 56, p. 385.

[1956] A survey of difference sets, *Proc. Amer. Math. Soc.* **7**, 975–986.

[1967] *Combinatorial Theory* (Blaisdell, Waltham, MA). MR 37#80; Zbl 196, p. 24.

[1974] Difference sets, in: *Combinatorics, Proc. 1974 Conf., Nijenrode Castle, Breukelen, Math. Centre Tracts*, Vol. 57, eds. M. Hall Jr and J.H. van Lint (Mathematisch Centrum, Amsterdam) pp. 1–26.

Hall Jr, M., and W.S. Connor

[1954] An embedding theorem for balanced incomplete block designs, *Canad. J. Math.* **6**, 35–41.

Hall Jr, M., and H.J. Ryser

[1951] Cyclic incidence matrices, *Canad. J. Math.* **3**, 495–502.

Hall Jr, M., J.D. Swift and R. Walker

[1956] Uniqueness of the projective plane of order eight, *Math. Tables Aids Comput.* **10**, 186–194. MR 18, p. 816; Zbl 73, p. 365.

Hall Jr, M., R. Lane and D. Wales

[1970] Designs derived from permutation groups, *J. Combin Theory* **8**, 12–22.

Hamada, N.

[1968] The rank of the incidence matrix of points and d-flats in finite geometries, *J. Sci. Hiroshima Univ. Ser. A-I* **32**, 381–396.

[1973] On the p-rank of the incidence matrix of a balanced or partially balanced incomplete block design and its application to error correction codes, *Hiroshima Math. J.* **3**, 153–226.

Hamada, N., and H. Ohmori

[1975] On the BIB design having the minimum p-rank, *J. Combin. Theory A* **18**, 131–140.

Hanani, H.

[1951] On the number of straight lines determined by n points, *Riveon Lematematika* **5**, 10–11. Hebrew, English summary. MR 13, p. 5.

[1954/1955] On the number of lines and planes determined by d points, *Technion Israel Inst. Tech. Sci. Publ.* **6**, 58–63. MR 17, p. 294.

[1960a] A note on Steiner triple systems, *Math. Scand.* **8**, 154–156. MR 23A#2330; Zbl 100, p. 18.

[1960b] On quadruple systems, *Canad. J. Math.* **12**, 145–157. MR 22#2558;Zbl 92, p. 12.

[1961] The existence and construction of balanced incomplete block designs, *Ann. Math. Statist.* **32**, 361–386. MR 29#4161; Zbl 107, p. 361.

[1963] On some tactical configurations, *Canad. J. Math.* **15**, 702–722.
[1965] A balanced incomplete block design, *Ann. Math. Statist.* **36**, 711. MR 30#4358; Zbl 131, p. 182.
[1970] On the number of orthogonal Latin squares, *J. Combin. Theory* **8**, 247–271.
[1975] Balanced incomplete block designs and related designs, *Discrete Math.* **11**, 255–369.
[1979] A class of three-designs, *J. Combin. Theory A* **26**, 1–19. MR 80b: 05010.

Hanani, H., D. Ornstein and V.T. Sós
[1964] On the lottery problem, *Magyar Tud. Akad. Mat. Kutató Int. Kőzl.* **9**, 155–158.

Hanani, H., D.K. Ray-Chaudhuri and R.M. Wilson
[1972] On resolve designs, *Discrete Math.* **3**, 343–357.

Hanani, H., A. Hartman and E.S. Kramer
[1983] On three-designs of small order, *Discrete Math.* **45**, 75–97.

Hartman, A.
[1986] On small packing and covering designs with block size 4, *Discrete Math.* **59**, 275–281. MR 87f:05048.
[1987a] Factorization of the complete uniform hypergraph, *Ars Combin.* **23B**, 239–250.
[1987b] Halving the complete design, *Ann. Discrete Math.* **34**, 207–224.

Hartman, A., W.H. Mills and R.C. Mullin
[1986] Covering triples by quadruples: an asymptotic solution, *J. Combin. Theory A* **41**, 117–138. MR 87e:05047.

Hedayat, A., and W.D. Wallis
[1978] Hadamard matrices and their applications, *Ann. Statist.* **6**, 1184–1238.

Higman, G.
[1969] On the simple group of D.G. Higman and C.C. Sims, *Illinois J. Math.* **13**, 74–80.

Hoffman, A.J.
[1960] On the uniqueness of the triangular association scheme, *Ann. Math. Statist.* **31**, 492–497.

Hong, Y.
[1984] On the nonexistence of unknown perfect 6- and 8-codes in Hamming schemes $H(n, q)$ with q arbitrary, *Osaka J. Math.* **21**, 687–700. MR 86b:94016;Zbl 551.94011.

Hubaut, X.
[1974] Two new families of 4-designs, *Discrete Math.* **9**, 247–249.

Hughes, D.R.
[1957] Collineations and generalized incidence matrices, *Trans. Amer. Math. Soc.* **86**, 284–296.
[1965] On t-designs and groups, *Amer. J. Math.* **87**, 761–778. MR 32#5727; Zbl 134, p. 30.

Hurkens, C.A.J., and J.J. Seidel
[1985] Conference matrices from projective planes of order 9, *European J. Combin.* **6**, 49–57.

Ito, N.
[1975] On tight 4-designs, *Osaka J. Math.* **12**, 493–522.
[1978] Corrections and Supplements to 'On tight 4-designs', *Osaka J. Math.* **15**, 693–697.

Janko, Z., and Tran van Trung
[1984a] Construction of two symmetric block designs for $(71, 21, 6)$, *Discrete Math.*, to appear.
[1984b] The existence of a symmetric block design for $(70, 24, 8)$, *Mitt. Math. Sem. Gießen* **165**, 17–18.
[1985] Construction of a new symmetric block design for $(78, 22, 6)$ with the help of tactical decompositions, *J. Combin. Theory A* **40**, 451–455. MR 87e:05026.

Johnson, S.M.
[1962] A new upper bound for error correcting codes, *IRE Trans. Inform. Theory* **IT-8**, 203–207.

Jungnickel, D.
[1984] Lateinische Quadrate, ihre Geometrien und ihre Gruppen, *Jber. Deutsch. Math.-Verein* **86**, 69–108. Zbl 581.05014.
[1992] Difference sets, in: *Contemporary Design Theory*, eds. J.H. Dinitz and D.R. Stinson (Wiley, New York) pp. 241–324.

Jungnickel, D., and A. Pott
[1994] A new class of symmetric (v, k, λ)-designs, *Designs, Codes and Cryptography* **4**, 319–325.

Jungnickel, D., and K. Vedder
[1984] On the geometry of planar difference sets, *European J. Combin.* **5**, 143–148.
Kamber, R.
[1976] *Untersuchungen über affine Ebenen der Ordnung 9,* Thesis (ETH Zürich).
Kamps, H.J.L., and J.H. van Lint
[1970] A covering problem, in: *Combinatorial Theory and its Applications., Colloq. Math. Soc. János Bolyai* **4**, pp. 679–685. Zbl 231.05030.
Kantor, W.M.
[1974] Dimension and embedding theorems for geometric lattices, *J. Combin. Theory A* **17**, 173–195.
Katona, G.O.H., T. Nemetz and M. Simonovits
[1964] On a problem of Turán in the theory of graphs, *Mat. Lapok* **15**, 228–238 (in Hungarian, English summary).
Kelly, G.
[1982] On the uniqueness of embedding a residual design, *Discrete Math.* **39**, 153–160.
Kibler, R.E.
[1978] A summary of non-cyclic difference sets, $k \leqslant 20$, *J. Combin. Theory A* **25**, 62–67.
Kirkman, T.P.
[1847] On a problem in combinations, *Cambridge and Dublin Math. J.* **2**, 191–204.
Klemm, M.
[1986] Über den p-Rang von Inzidenzmatrizen, *J. Combin. Theory A* **43**, 138–139. MR 87j:05030.
Köhler, E.
[1985] Über den allgemeinen Gray-Code und die Nichtexistenz einiger t-designs, in: *Graphen in Forschung und Unterricht, Festschrift K. Wagner,* ed. R. Bodendiek (Franzbecker, Salzdetfurth) pp. 102–111. MR 87g:05001.
[1988/1989] Allgemeine Schnittzahlen in t-Designs, *Discrete Math.* **73**, 133–142.
Kramer, E.S.
[1983] Some results on t-wise balanced designs, *Ars Combin.* **15**, 179–192.
[1984] The t-designs using $M_{21} \simeq \mathrm{PSL}_3(4)$ on 21 points, *Ars Combin.* **17**, 191–208. Zbl 548.05007.
Kramer, E.S., and D.M. Mesner
[1975a] Admissible parameters for Steiner systems $S(t,k,v)$ with a table for all $(v-t) \leqslant 498$, *Utilitas Math.* **7**, 211–222. MR 51#7899; Zbl 309.
[1975b] The possible (impossible) systems of 11 disjoint $S(2,3,13)$'s $(S(3,4,14)$'s) with automorphism of order 11, *Utilitas Math.* **7**, 55–58. MR 51#7898.
[1976] t-Designs on hypergraphs, *Discrete Math.* **15**, 263–296. MR 57#139.
Kramer, E.S., D.W. Leavitt and S.S. Magliveras
[1985] Construction procedures for t-designs and the existence of new simple 6-designs, in: *Algorithms in Combinatorial Design Theory,* eds. C.J. Colbourn and M.J. Colbourn, *Ann. Discrete Math.* **26**, 247-273. MR 87e:05037; Zbl 585.05002.
Kreher, D.L.
[1986] An incidence algebra for t-designs with automorphisms, *J. Combin. Theory A* **42**, 239–251. MR 87j:05031.
Kreher, D.L., and S.P. Radziszowski
[1986] The existence of simple 6-(14, 7, 4) designs, *J. Combin. Theory A* **41**, 237–243.
Lam, C.W.H., L. Thiel, S. Swiercz and J. McKay
[1983] The nonexistence of ovals in a projective plane of order 10, *Discrete Math.* **45**, 319–321. MR 84h:05028 (see also MR Annual index 1985, Author index, Errata and addenda).
Lamken, E.R., W.H. Mills, R.C. Mullin and S.A. Vanstone
[1987] Covering of pairs by quintuples, *J. Combin. Theory A* **44**, 49–68.
Lander, E.S.
[1983] *Symmetric Designs: An Algebraic Approach, London Mathematical Society Lecture Note Series,* Vol. 74 (Cambridge University Press, Cambridge).
Langford, C.D.
[1958] Problem, *Math. Gaz.* **42**, 28.

Leavitt, D.W., and S.S. Magliveras

[1984] Simple 6-(33, 8, 36)-designs from $P\Gamma L_2(32)$, in: *Computational Group Theory, Proc. Durham, 1982,* pp. 337–352. Zbl 548.05008.

Leech, J.

[1967] Notes on sphere packings, *Canad. J. Math.* **19**, 251–267.

Lehmer, E.

[1953] On residue difference sets, *Canad. J. Math.* **5**, 425–432.

[1954] On cyclotomic numbers of order sixteen, *Canad. J. Math.* **6**, 449–454.

Lindner, C.C.

[1975a] A partial Steiner triple system of order n can be embedded in a Steiner triple system of order $6n+3$, *J. Combin. Theory* **18**, 349–351. MR 52#129.

[1975b] A brief up-to-date survey of finite embedding theorems for partial quasigroups, in: *Proc. Conf. on Algebraic Aspects of Combinatorics, Sem. Toronto, 1975,* eds. D. Corneil and E. Mendelsohn, *Congress. Numerantium* 13, 53–78. MR 52#117.

Lindner, C.C., and A. Rosa

[1975] On the existence of automorphism free Steiner triple systems, *J. Algebra* **34**, 430–443. MR 53#2707; Zbl 308.05014.

[1978] Steiner quadruple systems, a survey, *Discrete Math.* **22**, 147–181. MR 80c:05055; Zbl 398.05015.

Lu, J.-X.

[1983] On large sets of disjoint Steiner triple systems, I–III, *J. Combin. Theory A* **34**, 140–146, 147–155, 156–182.

[1984] On large sets of disjoint Steiner triple systems, IV–VI, *J. Combin. Theory A* **37**, 136–163, 164–188, 189–192. Zbl 557.05012.

MacInnes, C.R.

[1907] Finite planes with less than eight points on a line, *Amer. Math. Monthly* **14**, 171–174.

MacNeish, H.F.

[1922] Euler Squares, *Ann. of Math.* **23**, 221–227.

MacWilliams, F.J., and N.J.A. Sloane

[1977] *The Theory of Error-Correcting Codes* (North-Holland, Amsterdam).

MacWilliams, J., and H.B. Mann

[1967] *On the p-rank of the design matrix of a difference set,* Math. Research Center Tech. Report 803 (University of Wisconsin, Madison, WI).

Majindar, K.N.

[1962] On the parameters and intersection of blocks of balanced incomplete block designs, *Ann. Math. Statist.* **33**, 1200–1205. MR 28#682.

Majumdar, K.N.

[1953] On some theorems in combinatorics relating to incomplete block designs, *Ann. Math. Statist.* **24**, 377–389. MR 15, p. 93.

Mann, H.B.

[1965] *Addition Theorems* (Wiley, New York).

[1969] A note on balanced incomplete block designs, *Ann. Math. Statist.* **40**, 679–680.

Mathon, R.

[1987] Constructions for cyclic Steiner 2-designs, *Ann. Discrete Math.* **34**, 353–362.

Mathon, R., and A. Rosa

[1985] Tables of parameters of BIBDs with $r \leqslant 41$ including existence, enumeration and resolvability results, in: *Algorithms in Combinatorial Design Theory,* eds. C.J. Colbourn and M.J. Colbourn, *Ann. Discrete Math.* **26**, 275–308.

McFarland, R.L.

[1973] A family of difference sets in non-cyclic groups, *J. Combin. Theory A* **15**, 1–10.

McFarland, R.L., and B.F. Rice

[1978] Translates and multipliers of abelian difference sets, *Proc. Amer. Math. Soc.* **68**, 375–379.

Mendelsohn, E., and A. Rosa

[1985] One-factorizations of the complete graph – a survey, *J. Graph Theory* **9**, 43–65.

Mendelsohn, N.S., and S.H.Y. Hung
[1972] On the Steiner systems $S(3,4,14)$ and $S(4,5,15)$, *Utilitas Math.* **1**, 5–95. MR 46#1618; Zbl 258.05017.

Menon, P.K.
[1962] On difference sets whose parameters satisfy a certain relation, *Proc. Amer. Math. Soc.* **13**, 739–745.

Mills, W.H.
[1972] On the covering of pairs by quadruples I, *J. Combin. Theory A* **13**, 55–78. MR 45#8534;Zbl 243.05024.
[1973] On the covering of pairs by quadruples II, *J. Combin. Theory A* **15**, 138–166. MR 47#8316;Zbl 261.05022.
[1978] A new 5-design, *Ars Combin.* **6**, 193–195. MR 81f:05023;Zbl 414.05013.
[1979] Covering designs I: Covering by a small number of subsets, *Ars Combin.* **8**, 199–315. Zbl 455.05041.
[1983] A covering of pairs by quintuples, *Ars Combin.*. **18**, 21–31.
[1984] Balanced incomplete block designs with $k=6$ and $\lambda=1$, in: *Enumeration and Design,* eds. D.M. Jackson and S.A. Vanstone (Academic Press, Toronto) pp. 239–244. MR 86f: 05024.

Mills, W.H., and R.C. Mullin
[1988] Covering pairs by quintuples: the case ν congruent to 3 (mod 4), *J. Combin. Theory A* **49**, 308–322.
[1992] Coverings and packings, in: *Contemporary Design Theory,* eds. J.H. Dinitz and D.R. Stinson (Wiley, New York) pp. 371–399.

Mitchell, C.J.
[1979] An infinite family of symmetric designs, *Discrete Math.* **26**, 247–250.

Motzkin, T.
[1951] The lines and planes connecting the points of a finite set, *Trans. Amer. Math. Soc.* **70**, 451–464. MR 12, p. 849.

Mullin, R.C., D.G. Hoffman and C.C. Lindner
[1987] A few more BIBDs with $k=6$ and $\lambda=1$, *Ann. Discrete Math.* **34**, 379–384.

Neumaier, A.
[1982] Quasi-residual 2-designs, $1^1/_2$-designs and strongly regular multigraphs, *Geom. Dedicata* **12**, 351–366.

Oberschelp, W.
[1972] Lotto-Garantiesysteme und Blockpläne, *Math.-Phys. Semesterber.* **19**, 55–67. MR 51#7901.

O'Keefe, E.S.
[1961] Verification of a conjecture of Th. Skolem, *Math. Scand.* **9**, 80–82. MR 23A#2331; Zbl 105, p. 250.

Paley, R.E.A.C.
[1933] On orthogonal matrices, *J. Math. Phys. Mass. Inst. Tech.* **12**, 311–320.

Parker, E.T.
[1957] On collineations of symmetric designs, *Proc. Amer. Math. Soc.* **8**, 350–351.

Peltesohn, R.
[1938] Eine Lösung der beiden Heffterschen Differenzenprobleme, *Compositio Math.* **6**, 251–257. Zbl 20, p. 39.

Peterson, Ch.
[1977] On tight 6-designs, *Osaka Math. J.* **14**, 417–435.

Petrenjuk, A.Ya.
[1968] On Fisher's inequality for tactical configurations, *Mat. Zametki* **4**, 417–425 (in Russian).

Pless, V.
[1969] On a new family of symmetry codes and related new five-designs, *Bull. Amer. Math. Soc.* **75**, 1339–1342.
[1970] The weight of the symmetry code for $p=29$ and the 5-designs contained therein, *Ann. New York Acad. Sci.* **175**(Article 1), 310–313.
[1972] Symmetry codes over GF(3) and new five-designs, *J. Combin. Theory* **12**, 119–142.

Pless, V., and N.J.A. Sloane
[1975] On the classification and enumeration of self-dual codes, *J. Combin. Theory A* **18**, 313–335.

Priday, C.J.
[1959] On Langford's problem (I), *Math. Gaz.* **43**, 250–253.
Pul, C.L.M., and T. Etzion
[preprint] New lower bounds for constant weight codes.
Ray-Chaudhuri, D.K., and R.M. Wilson
[1971] Solution of Kirkman's schoolgirl problem, in: *Combinatorics, Proc. Symp. on Pure Mathematics,* Vol. 19, ed. T.S. Motzkin (American Mathematical Society, Providence, RI) pp. 187–203.
[1975] On t-designs, *Osaka J. Math.* **12**, 737–744.
Rees, R., and D.R. Stinton
[1987] On resolvable group-divisible designs with block size 3, *Ars Combin.* **23**, 107–120.
Reuvers, H.F.H.
[1977] *Some non-existence theorems for perfect codes over arbitrary alphabets,* Ph.D. Thesis (Technical University Eindhoven, Eindhoven). Zbl. 352.94015.
Rödl, V.
[1985] On a packing and covering problem, *European J. Combin.* **6** (not 5) 69–78.
Rogers, K.
[1964] A note on orthogonal Latin squares, *Pacific J. Math.* **14**, 1395–1397.
Rosa, A.
[1972] On reverse Steiner triple systems, *Discrete Math.* **2**, 61–71. MR 45#4996; Zbl 242.05016.
Roselle, D.P.
[1972] Distributions of integers into s-tuples with given differences, in: *Proc. Manitoba Conf. on Numerical Mathematics, Winnipeg, 1971,* eds. R.S.D. Thomas and H.C. Williams, pp. 31–42. MR 49#211;Zbl 267.05018.
Ryser, H.J.
[1951] A combinatorial theorem with an application to Latin rectangles, *Proc. Amer. Math. Soc.* **2**, 550–552.
[1963] *Combinatorial Mathematics, Carus Mathemathical Monographs,* Vol. 14, (Wiley, New York). MR 27#51; Zbl 112, p. 248.
[1968] An extension of a theorem of de Bruijn and Erdős on combinatorial designs, *J. Algebra* **10**, 246–261.
Sachar, H.E.
[1979] The $\mathbb{F}_p$ span of a finite projective plane, *Geom. Dedicata* **8**, 407–415.
Sawada, K.
[1985] A Hadamard matrix of order 268, *Graphs Combin.* **1**, 185–187.
Schönheim, J.
[1964] On coverings, *Pacific J. Math.* **14**, 1405–1411. MR 30#1954; Zbl 128, p. 245.
[1966] On maximal systems of k-tuples, *Studia Sci. Math. Hungar.* **1**, 363–368. MR 34#2485; Zbl 146, p. 14.
Schreiber, S.
[1973] Covering all triples on n marks by disjoint Steiner systems, *J. Combin. Theory A* **15**, 347–350.
Schutzenberger, M.P.
[1949] A non-existence theorem for an infinite family of symmetrical block designs, *Ann. Eugenics* **14**, 286–287.
Seah, E., and D.R. Stinson
[1987] Some perfect one-factorizations of K_{14}, *Ann. Discrete Math.* **34**, 419–436.
Seberry, J., and M. Yamada
[1992] Hadamard matrices, sequences, and block designs, in: *Contemporary Design Theory,* eds. J.H. Dinitz and D.R. Stinson (Wiley, New York) pp. 431–560.
Seberry Wallis, J.
[1976] On the existence of Hadamard matrices, *J. Combin. Theory A* **21**, 188–195.
Seress, Á.
[1987] *A numerical characterization of type-I λ-designs,* Preprint.
Shrikhande, S.S.
[1950] The impossibility of certain symmetrical balanced incomplete block designs, *Ann. Math. Statist.* **21**, 106–111.

[1959a] On a characterization of the triangular association scheme, *Ann. Math. Statist.* **30**, 39–47.
[1959b] The uniqueness of the L_2 association scheme, *Ann. Math. Statist.* **30**, 781–798.
[1976] Affine resolvable balanced incomplete block designs: a survey, *Aequationes Math.* **14**, 251–269.

Shrikhande, S.S., and N.M. Singhi
[1975] Construction of geomatroids, *Utilitas Math.* **8**, 187–192.
[1976] On the λ-design conjecture, *Utilitas Math.* **9**, 301–318.
[1985] On a problem of Erdős and Larson, *Combinatorica* **5**, 351–358. MR 87j:05047.

Singer, J.
[1938] A theorem in finite projective geometry and some applications to number theory, *Trans. Amer. Math. Soc.* **43**, 377–385.

Skolem, Th.
[1957] On certain distributions of integers in pairs with given differences, *Math. Scand.* **5**, 57–68. MR 19, P. 1159; Zbl 84, p. 43.
[1958] Some remarks on the triple systems of Steiner, *Math. Scand.* **6**, 273–280. MR 21#5582; Zbl 105, p. 250.

Smetaniuk, B.
[1981] A new construction on Latin squares I: A proof of the Evans conjecture, *Ars Combin.* **11**, 155–172. Zbl 471.05013.

Smith, K.J.C.
[1969] On the p-rank of the incidence matrix of points and hyperplanes in a finite projective geometry, *J. Combin. Theory* **7**, 122–129.

Smith, M.S.
[1976] On the isomorphism of two simple groups of order 44 352 000, *J. Algebra* **41**, 172–174.

Spence, E.
[1993] A new family of symmetric (v, k, λ)-designs, *European J. Combin.* **14**, 131–136.

Spence, E., V.D. Tonchev and Tran van Trung
[1993] A symmetric 2-(160, 54, 18) design, *J. Combin. Designs* **1**, 65–68.

Stanton, R.G.
[1988] Of pairs and triples, *JCMCC* **3**, 29–39.

Stanton, R.G., and D.A. Sprott
[1958] A family of difference sets, *Canad. J. Math.* **10**, 73–77.

Stinson, D.R.
[1983] On scheduling perfect competitions, *Ars Combin.* **18**, 45–49.

Stinson, D.R., and G.H.J. van Rees
[1984] The equivalence of certain equidistant binary codes and symmetric BIBDs, *Combinatorica* **4**(4), 357–362.

Tarry, G.
[1900] Le problème des 36 officiers, *C.R. Assoc. Fr. Av. Sci. Nat.* **1**, 122–123.
[1901] Le problème des 36 officiers, part 2, *C.R. Assoc. Fr. Av. Sci. Nat.* **2**, 170–203.

Teirlinck, L.
[1973a] The existence of reverse Steiner triple systems, *Discrete Math.* **6**, 301–302. MR 48#8263; Zbl 266.05007.
[1973b] On the maximum number of disjoint Steiner triple systems, *Discrete Math.* **6**, 299–300.
[1977] On making two Steiner triple systems disjoint, *J. Combin. Theory A* **23**, 349–350. MR 56#8394; Zbl 401.05018.
[1979] On Steiner spaces, *J. Combin. Theory A* **26**, 103–114. MR 80j:51011;Zbl 411.05015.
[1984] On large sets of disjoint quadruple systems, *Ars Combin.* **17**, 173–176.
[1987] Nontrivial t-designs without repeated blocks exist for all t, *Discrete Math.* **65**, 301–311.
[1989] Locally trivial t-designs and t-designs without repeated blocks, *Discrete Math.* **77**, 345–356.
[1991] A completion of Lu's determination of the spectrum for large sets of disjoint Steiner triple systems, *J. Combin. Theory A* **57**, 302–305.
[1992] Large sets of disjoint designs and related structures, in: *Contemporary Design Theory*, eds. J.H. Dinitz and D.R. Stinson (Wiley, New York) pp. 561–592.

Tietäväinen, A.
[1973] On the nonexistence of perfect codes over finite fields, *SIAM J. Appl. Math.* **24**, 88–96. Zbl 251.94009 (233.94009).
[1974] A short proof for the nonexistence of unknown perfect codes over GF(q), $q > 2$, *Ann. Acad. Sci. Fenn. A (I. Mathematica)* **580**, 1–6. Zbl 289.94004.
Tietäväinen, A., and A. Perko
[1971] There are no unknown perfect binary codes, *Ann. Univ. Turku A I* **148**, 3–10.
Tits, J.
[1964] Sur les systèmes de Steiner associés aux trois 'grands' groupes de Mathieu, *Rend. Mat. e Appl. (5)* **23**, 166–184. MR 32#1262; Zbl 126, p. 263.
Todd, J.A.
[1933] A combinatorial problem, *J. Math. Phys. Mass. Inst. Technol.* **12**, 321–333.
Todorov, D.T.
[1985a] A class of covering numbers, in: *Mathematics and Mathematical Education* (in Russian), *Proc. 14th Spring Conf., Sl"nchev Bryag, 1985*, ed. G. Gerov (B"lgar. Akad. Nauk, Sofia). MR 87c:05043.
[1985b] On some covering designs, *J. Combin. Theory A* **39**, 83–101.
[1985c] Three mutually orthogonal Latin squares of order 14, *Ars Combin.* **20**, 45–47. MR 87d: 05045.
[1986] On the covering of triples by eight blocks, *Serdica* **12**, 20–29. MR 87h:05066.
Todorov, D.T., and V.D. Tonchev
[1982] On some coverings of triples, *C.R. Acad. Bulg. Sci.* **35**, 1209–1211.
Tonchev, V.D.
[1986a] A characterization of designs related to the Witt system $S(5,8,24)$, *Math. Z.* **191**, 225–230. MR 87g:05040.
[1986b] Quasisymmetric 2-(31, 7, 7) designs and a revision of Hamada's conjecture, *J. Combin. Theory A* **42**, 104–110. MR 87i:05043.
[1986c] Quasi-symmetric designs and self-dual codes, *European J. Combin.* **7**, 67–73.
[1987] Embedding of the Witt–Mathieu system $S(3,6,22)$ in a symmetric 2-$(78,22,6)$ design, *Geom. Dedicata* **22**, 49–75.
Totten, J.
[1976] Classification of restricted linear spaces, *Canad. J. Math.* **28**, 321–333.
[1977] Finite linear spaces with three more lines than points, *Simon Stevin* **51**, 35–47.
Tran van Trung
[1982] The existence of symmetric block designs with parameters $(41,16,6)$ and $(66,26,10)$, *J. Combin. Theory A* **33**, 201–204.
[1984] On the existence of an infinite family of simple 5-designs, *Math. Z.* **187**, 285–287.
Treash, C.
[1971] The completion of finite incomplete Steiner triple with systems with applications to loop theory, *J. Combin. Theory A* **10**, 259–265.
van Leyenhorst, D.C.
[1980] Manuscript.
van Lint, J.H.
[1971] Nonexistence theorems for perfect error-correcting codes, in: *Computers in Algebra and Number Theory, Symp. New York, 1970, SIAM-AMS Proc. IV*, eds. G. Birkhoff and M. Hall Jr (American Mathematical Society, Providence, RI) pp. 89–95.
[1974] Recent results on perfect codes and related topics, in: *Combinatorics (Part 1), Proc. Nijenrode Conf., Mathematical Centre Tracts*, Vol. 55, eds. M. Hall Jr and J.H. van Lint (Reidel, Dordrecht, 1975) pp. 158–178.
[1982] *Introduction to Coding Theory, Graduate Texts in Mathematics*, Vol. 86 (Springer, New York). Zbl 485.94015.
[1984] On equidistant binary codes of length $n = 4k + 1$ with distance $d = 2k$, *Combinatorica* **4**(4), 321–323.
van Tilborg, H.
[1976] A lower bound for v in a t-(v, k, λ) design, *Discrete Math.* **14**, 399–402.

Wallis, J.S.
[1976] On the existence of Hadamard matrices, *J. Combin. Theory A* **21**, 188–195.
Wallis, W.D.
[1971] Construction of strongly regular graphs using affine designs, *Bull. Aust. Math. Soc.* **4**, 41–49.
Wallis, W.D., A.P. Street and J.S. Wallis
[1972] *Combinatorics: Room Squares, Sum-Free Sets, Hadamard Matrices, Lecture Notes in Mathematics,* Vol. 292 (Springer, Berlin).
Wang, S.M.P., and R.M. Wilson
[1978] A few more squares II, in: *Proc. 9th South-East Conf. on Combinatorics, Graph Theory and Computing,* p. 688. Abstract.
Whiteman, A.L.
[1957] The cyclotomic numbers of order sixteen, *Trans. Amer. Math. Soc.* **86**, 401–413.
[1962] A family of difference sets, *Illinois J. Math.* **6**, 107–121.
Wilbrink, H.A.
[1985] A note on planar difference sets, *J. Combin. Theory A* **38**, 94–95.
Wilson, R.M.
[1972a] Cyclotomy and difference families in elementary abelan groups, *J. Number Theory* **4**, 17–47. MR 46#8860; Zbl 259.05011.
[1972b] An existence theory for pairwise balanced designs, I: Composition theorems and morphisms, *J. Combin. Theory A* **13**, 220–245. MR 46#3338; Zbl 263.05014.
[1972c] An existence theory for pairwise balanced designs, II: The structure of PBD-closed sets and the existence conjectures, *J. Combin. Theory A* **13**, 246–273. MR 46#3339; Zbl 263.05015.
[1973] The necessary conditions for t-designs are sufficient for something, *Utilitas Math.* **4**, 207–215.
[1974a] Concerning the number of mutually orthogonal Latin squares, *Discrete Math.* **9**, 181–198.
[1974b] Nonisomorphic Steiner triple systems, *Math. Z.* **135**, 303–313. MR 49#4803; Zbl 264.05009.
[1984] On the theory of t-designs, in: *Enumeration and Design, Proc. Waterloo, 1982,* eds. D.M. Jackson and S.A. Vanstone (Academic Press, Toronto) pp. 19–49. Zbl 557.05009.
Witt, E.
[1938a] Die 5-fach transitiven Gruppen von Mathieu, *Abh. Math. Sem. Univ. Hamburg* **12**, 256–264. Zbl 19, p. 251.
[1938b] Über Steinersche Systeme, *Abh. Math. Sem. Univ. Hamburg* **12**, 265–275. Zbl 19, p. 251.
Wojtas, M.
[1980] A note on mutually orthogonal Latin squares, *Analiza Dyskretna, Prace Nauk. Inst. Mat. Politech. Wroclaw* **19**, 11–14.
Woodall, D.R.
[1970] Square λ-linked designs, *Proc. London Math. Soc.* **20**, 669–687.
Yuan, W.
[1966] On the maximal number of pairwise orthogonal Latin square of order s; an application of the sieve method, *Chinese Math.* **8**, 422–432.

CHAPTER 15

Association Schemes

Andries E. BROUWER

Department of Mathematics, Eindhoven University of Technology, 5600 MB Eindhoven, The Netherlands

Willem H. HAEMERS

Department of Econometrics, Tilburg University, P.O. Box 90153, 5000 LE Tilburg, The Netherlands

Contents

HANDBOOK OF COMBINATORICS
Edited by R. Graham, M. Grötschel and L. Lovász

1. Introduction

Association schemes are perhaps the most important unifying concept in algebraic combinatorics. They provide a common view point for the treatment of problems in several fields, such as coding theory, design theory, algebraic graph theory and finite group theory.

Roughly, an association scheme is a very regular partition of the edge-set of a complete graph. Each of the partition classes defines a graph, and the adjacency matrices of these graphs (together with the identity matrix) form the basis of an algebra (known as the Bose–Mesner algebra of the scheme). But since this basis consists of 0–1 matrices, we see that the algebra is not only closed under matrix multiplication, but also under componentwise (Hadamard, Schur) multiplication. The interplay between these two algebra structures on the Bose–Mesner algebra yields strong information on the structure and parameters of an association scheme.

The relation of the theory of association schemes to the various fields varies. Finite group theory mainly serves as a source of examples – any generously transitive permutation representation of a group yields an association scheme – but also as a source of inspiration – many group-theoretic concepts and results can be mimicked in the (more general) setting of association schemes. On the other hand, in combinatorics association schemes form a tool. Using inequalities for the parameters of an association scheme, one finds upper bounds for the size of error-correcting codes, and lower bounds for the number of blocks of a t-design. Many uniqueness proofs for combinatorial structures depend crucially on the additional information one gets in case such inequalities hold with equality.

The graphs of the partition classes of an association scheme are very special. For instance distance-regular graphs and Kneser graphs are of this type. For these graphs the spectrum of the adjacency matrix can be computed from the parameters of the association scheme. Thus results about the eigenvalues of graphs (cf. chapter 31) can be applied.

An association scheme with just two classes is the same as a pair of complementary strongly regular graphs. Because strongly regular graphs are themselves important combinatorial objects and because their treatment forms a good introduction to the more general theory, we shall start with a section on strongly regular graphs. In the subsequent section we treat the elements of the general theory of association schemes: the adjacency matrices, the Bose–Mesner algebra, the eigenvalues, the Krein parameters, the absolute bound and Delsarte's inequality on subsets of an association scheme. The last section is devoted to some special association schemes and the significance to other fields of combinatorics, as indicated above.

Association schemes were introduced by Bose and Shimamoto (1952) and Bose and Mesner (1959) defined the algebra that bears their name. Delsarte (1973) found the linear programming bound, studied P- and Q-polynomial schemes, and applied the resulting theory to codes and designs. Higman (1974) introduced the more general concept of coherent configuration to study permutation representations of finite groups. There is much literature on association schemes and related topics. Bannai and Ito (1984) is the first text book on association schemes. On the more special topic of distance-regular graphs a lot of material can be found in Biggs (1974) and in Brouwer et al. (1989). Some introductory papers on association schemes are Delsarte (1974), Goethals (1979), chapter 21 of MacWilliams and Sloane (1977), Haemers (1979), chapter 17 of Cameron and Van Lint (1980) and Seidel (1983).

2. Strongly regular graphs

A simple graph of order v is *strongly regular* with parameters v, k, λ, μ whenever it is not complete or edgeless and

(i) each vertex is adjacent to k vertices,
(ii) for each pair of adjacent vertices there are λ vertices adjacent to both,
(iii) for each pair of non-adjacent vertices there are μ vertices adjacent to both.

For example, the pentagon is strongly regular with parameters $(v, k, \lambda, \mu) = (5, 2, 0, 1)$. One easily verifies that a graph G is strongly regular with parameters (v, k, λ, μ) if and only if its complement $\bar{G}$ is strongly regular with parameters $(v, v-k-1, v-2k+\mu-2, v-2k+\lambda)$. The line graph of the complete graph of order m, known as the *triangular graph* $T(m)$, is strongly regular with parameters $(\frac{1}{2}m(m-1), 2(m-2), m-2, 4)$. The complement of $T(5)$ has parameters $(10, 3, 0, 1)$. This is the Petersen graph.

A graph G satisfying condition (i) is called k-regular. It is well known and easily seen that the adjacency matrix of a k-regular graph has an eigenvalue k with eigenvector $\mathbf{1}$ (the all-one vector), and that every other eigenvalue ρ satisfies $|\rho| \leq k$ (see chapter 31 or Biggs 1974). For convenience we call an eigenvalue *restricted* if it has an eigenvector perpendicular to $\mathbf{1}$. We let I and J denote the identity and all-one matrices, respectively.

Theorem 2.1. *For a simple graph G of order v, not complete or empty, with adjacency matrix A, the following are equivalent*:

(i) *G is strongly regular with parameters (v, k, λ, μ) for certain integers k, λ, μ,*
(ii) *$A^2 - (\lambda - \mu)A + (k - \mu)J + \mu J$ for certain reals k, λ, μ,*
(iii) *A has precisely two distinct restricted eigenvalues.*

Proof. The equation in (ii) can be rewritten as

$$A^2 = kI + \lambda A + \mu(J - I - A)\,.$$

Now (i)$\Leftrightarrow$(ii) is obvious.

(ii)$\Rightarrow$(iii): Let ρ be a restricted eigenvalue, and u a corresponding eigenvector perpendicular to $\mathbf{1}$. Then $Ju = 0$. Multiplying the equation in (ii) on the right by u

yields $\rho^2 = (\lambda - \mu)\rho + (k - \mu)$. This quadratic equation in ρ has two distinct solutions. (Indeed, $(\lambda - \mu)^2 = 4(\mu - k)$ is impossible since $\mu \leqslant k$ and $\lambda \leqslant k - 1$.)

(iii)$\Rightarrow$(ii): Let r and s be the restricted eigenvalues. Then $(A - rI)(A - sI) = \alpha J$ for some real number α. So A^2 is a linear combination of A, I and J. □

As an application, we show that quasi-symmetric block designs give rise to strongly regular graphs. A quasisymmetric design is a 2-(v, k, λ) design such that any two blocks meet in either x or y points, for certain fixed x, y. (Cf. chapter 14.) Given this situation, we may define a graph G on the set of blocks, and call two blocks adjacent when they meet in x points. Then there exist coefficients $\alpha_1, \ldots, \alpha_7$ such that $NN^T = \alpha_1 I + \alpha_2 J$, $NJ = \alpha_3 J$, $JN = \alpha_4 J$, $A = \alpha_5 N^T N + \alpha_6 I + \alpha_7 J$, where A is the adjacency matrix of the graph G. (The α_i can be readily expressed in terms of v, k, λ, x, y.) Then G is strongly regular by (ii) of the previous theorem. (Indeed, from the equations just given it follows straightforwardly that A^2 can be expressed as a linear combination of A, I and J.) A large class of quasi-symmetric block designs is provided by the 2-(v, k, λ) designs with $\lambda = 1$ (also known as Steiner systems $S(2, k, v)$) – such designs have only two intersection numbers since no two blocks can meet in more than one point. This leads to a substantial family of strongly regular graphs, including the triangular graphs $T(m)$ (derived from the trivial design consisting of all pairs out of an m-set).

Another connection between strongly regular graphs and designs is found as follows: Let A be the adjacency matrix of a strongly regular graph with parameters (v, k, λ, λ) (i.e., with $\lambda = \mu$; such a graph is sometimes called a (v, k, λ) graph). Then by Theorem 2.1 (ii)

$$AA^T = A^2 = (k - \lambda)I + \lambda J ,$$

which reflects that A is the incidence matrix of a square ("symmetric") 2-(v, k, λ) design. (And in this way one obtains precisely all square 2-designs possessing a polarity without absolute points.) For instance, the triangular graph $T(6)$ provides a square 2-(15, 8, 4) design, the complementary design of the design of points and planes in the projective space PG(3, 2). Similarly, if A is the adjacency matrix of a strongly regular graph with parameters $(v, k, \lambda, \lambda + 2)$, then $A + I$ is the incidence matrix of a square 2-$(v, k, \lambda + 2)$ design (and in this way one obtains precisely all square 2-designs possessing a polarity with all points absolute).

Theorem 2.2. *Let G be a strongly regular graph and adjacency matrix A and parameters (v, k, λ, μ). Let r and s $(r > s)$ be the restricted eigenvalues of A and let f, g be their respective multiplicities. Then*

(i) $k(k - 1 - \lambda) = \mu(v - k - 1)$,

(ii) $rs = \mu - k$, $r + s = \lambda - \mu$,

(iii) $f, g = \frac{1}{2}(v - 1 \mp ((r + s)(v - 1) + 2k)/(r - s))$,

(iv) *r and s are integers, except perhaps when $f = g$, $(v, k, \lambda, \mu) = (4t + 1, 2t, t - 1, t)$ for some integer t.*

Proof. (i) Fix a vertex x of G. Let $\Gamma(x)$ and $\Delta(x)$ be the sets of vertices adjacent and non-adjacent to x, respectively. Counting in two ways the number of edges between $\Gamma(x)$ and $\Delta(x)$ yields (i). The equations (ii) are direct consequences of Theorem 2.1 (ii), as we saw in the proof. Formula (iii) follows from $f+g=\nu-1$ and $0=\operatorname{tr} A=k+fr+gs=k+\frac{1}{2}(r+s)(f+g)+\frac{1}{2}(r-s)(f-g)$. Finally, when $f\neq g$ then one can solve for r and s in (iii) (using (ii)) and find that r and s are rational, and hence integral. But $f=g$ implies $(\mu-\lambda)(\nu-1)=2k$, which is possible only for $\mu-\lambda=1$, $\nu=2k+1$. □

These relations imply restrictions for the possible values of the parameters. Clearly, the righthand sides of (iii) must be positive integers. These are the so-called *rationality conditions*. As an example of the application of the rationality conditions we can derive the following result due to Hoffman and Singleton (1960).

Theorem 2.3. *Suppose* $(v,k,0,1)$ *is the parameter set of a strongly regular graph. Then* $(v,k)=(5,2)$, $(10,3)$, $(50,7)$ *or* $(3250,57)$.

Proof. The rationality conditions imply that either $f=g$, which leads to $(\nu,k)=(5,2)$, or $r-s$ is an integer dividing $(r+s)(v-1)+2k$. By use of Theorem 2.2 (i)–(ii) we have

$$s=-r-1\,,\qquad k=r^2+r+1\,,\qquad v=r^4+2r^3+3r^2+2r+2\,,$$

and thus we obtain $r=1$, 2 or 7. □

The first three possibilities are uniquely realized by the pentagon, the Petersen graph and the Hoffman–Singleton graph. For the last case existence is unknown (but see Aschbacher 1971).

2.1. Conference matrices

Some important classes of strongly regular graphs are related to so-called conference matrices (the name stems from their application to networks for conference telephony). Since these matrices are not treated elsewhere in this Handbook, we shall elaborate a bit on them here. (See also Wallis et al. 1972).

A *conference matrix* of order n is an $n\times n$ matrix C with zero diagonal and 1 or -1 elsewhere, such that $CC^{\mathrm{T}}=(n-1)I$. Two conference matrices are *equivalent* if one can be obtained from the other by a permutation of rows and columns and multiplication of some rows and columns by -1. Delsarte et al. (1971) proved that every conference matrix C of order n is equivalent to one of the form

$$\begin{pmatrix} 0 & \mathbf{1}^{\mathrm{T}} \\ \mathbf{1} & S \end{pmatrix},$$

where S is skew-symmetric (i.e., $S=-S^{\mathrm{T}}$) in case $n=2$ or $n\equiv 0 \pmod 4$, and S is symmetric in case $n\equiv 2 \pmod 4$. We call S a *core* of C. If S is symmetric, then

$\frac{1}{2}(J - S - I)$ is the adjacency matrix of a strongly regular graph with parameters $(n-1, \frac{1}{2}(n-2), \frac{1}{4}(n-6), \frac{1}{4}(n-2))$. We call these graphs *conference graphs*; they are characterized by $f = g$, so they are the only cases in which non-integral eigenvalues can occur. Conversely, one can reconstruct a conference matrix from a conference graph. Belevitch (1950) (see also Van Lint and Seidel 1966) proved that for a symmetric conference matrix $n-1$ must be the sum of two squares. Thus we have a necessary condition for the existence of conference graphs, called the conference matrix condition.

If a conference matrix C is symmetric and has constant column sums, then $\frac{1}{2}(J - C - I)$ is the adjacency matrix of a strongly regular graph with parameters $(n = 4r^2 + 4r + 2, 2r^2 + 2, r^2 - 1, r^2)$, or the complement. (Here $n-1$ is a square, so the conference matrix condition is automatically fulfilled.) For many values of n conference matrices are known; see for instance Goethals and Seidel (1967) and Mathon (1978). Paley (1933) defined a matrix S whose rows and columns are indexed by the elements of a finite field $\mathbb{F}_q$ (q odd), by

$$S_{xy} = \chi(x - y)\,,$$

where χ is the quadratic residue character (i.e., $\chi(a) = 0$, 1 or -1 according to a being zero, a non-zero square, or a non-square respectively). It follows that S is the core of a conference matrix (S is symmetric if $q \equiv 1 \pmod 4$ and skew-symmetric otherwise). If q is a square, then C is equivalent to a conference matrix with constant row sum. Conference matrices are related to Hadamard matrices (see the appendix of chapter 14): if C is skew-symmetric, then $C + I$ is a Hadamard matrix, and if C is symmetric, then

$$\begin{pmatrix} -C + I & C + I \\ C + I & C - I \end{pmatrix}$$

is one.

2.2. Further restrictions

Except for the rationality conditions and the conference matrix condition, a few other restrictions on the parameters are known. We mention two of them. The Krein conditions, due to Scott (1973), can be stated as follows:

$$(r+1)(k + r + 2rs) \leqslant (k + r)(s + 1)^2\,,$$
$$(s+1)(k + s + 2rs) \leqslant (k + s)(r + 1)^2\,.$$

Seidel's absolute bound (see Delsarte et al. 1975, Koornwinder 1976, Seidel 1979) reads

$$v \leqslant \tfrac{1}{2} f(f + 3)\,,$$
$$v \leqslant \tfrac{1}{2} g(g + 3)\,.$$

The Krein conditions and the absolute bound are special cases of general inequalities for association schemes – we will meet them again in the next section;

the conference matrix condition is the analogue of the Bruck–Chowla–Ryser theorem for square 2-designs. In Brouwer and Van Lint (1984) one may find a list of all known restrictions; this paper gives a survey of the recent results on strongly regular graphs. It is a sequel to Hubaut's (1975) earlier survey of constructions. Seidel (1979) gives a good treatment of the theory. Some other references are Bose (1963), Cameron (1978) and Cameron and Van Lint (1980).

3. Association schemes

An *association scheme with d classes* is a finite set X together with $d+1$ relations R_i on X such that

(i) $\{R_0, R_1, \dots, R_d\}$ is a partition of $X \times X$;

(ii) $R_0 = \{(x, x) \mid x \in X\}$;

(iii) if $(x, y) \in R_i$, then also $(y, x) \in R_i$, for all $x, y \in X$ and $i \in \{0, \dots, d\}$;

(iv) for any $(x, y) \in R_k$ the number p_{ij}^k of $z \in X$ with $(x, z) \in R_i$ and $(z, y) \in R_j$ depends only on i, j and k.

The numbers p_{ij}^k are called the *intersection numbers* of the association scheme. The above definition is the original definition of Bose and Shimamoto (1952); it is what Delsarte (1973) calls a symmetric association scheme. In Delsarte's more general definition, (iii) is replaced by:

(iii′) for each $i \in \{0, \dots, d\}$ there exists a $j \in \{0, \dots, d\}$ such that $(x, y) \in R_i$ implies $(y, x) \in R_j$,

(iii″) $p_{ij}^k = p_{ji}^k$, for all $i, j, k \in \{0, \dots, d\}$.

Higman (1974, 1975, 1976) studied the even more general concept of a *coherent configuration*. He requires (i), (iii′), (iv) and

(ii′) $\{(x, x) \mid x \in X\}$ is a union of some R_j.

If (ii) holds, then the coherent configuration is called *homogeneous*. We shall not treat coherent configurations here, but content ourselves with the remark that a coherent configuration with at most five classes must be an association scheme (in the sense of Delsarte), see Higman (1975).

Define $n := |X|$, and $n_i := p_{ii}^0$. Clearly, for each $i \in \{1, \dots, d\}$, (X, R_i) is a simple graph which is regular of degree n_i.

Theorem 3.1. *The intersection numbers of an association scheme satisfy*

(i) $p_{0j}^k = \delta_{jk}$, $p_{ij}^0 = \delta_{ij} n_j$, $p_{ij}^k = p_{ji}^k$,

(ii) $\sum_i p_{ij}^k = n_j$, $\sum_j n_j = n$,

(iii) $p_{ij}^k n_k = p_{ik}^j n_j$,

(iv) $\sum_l p_{ij}^l p_{kl}^m = \sum_l p_{kj}^l p_{il}^m$.

Proof. (i)–(iii) are straightforward. The expressions at both sides of (iv) count quadruples (w, x, y, z) with $(w, x) \in R_i$, $(x, y) \in R_j$, $(y, z) \in R_k$, for a fixed pair $(w, z) \in R_m$. $\square$

It is convenient to write the intersection numbers as entries of the so-called *intersection matrices* $L_0, \ldots, L_d$:

$$(L_i)_{kj} = p_{ij}^k .$$

Note that $L_0 = I$. From the definition it is clear that an association scheme with two classes is the same as a pair of complementary strongly regular graphs. (With Delsarte's more general definition one also gets the skew-symmetric conference matrices.) If (X, R_1) is strongly regular with parameters (v, k, λ, μ), then the intersection matrices of the scheme are

$$L_1 = \begin{pmatrix} 0 & k & 0 \\ 1 & \lambda & k-\lambda-1 \\ 0 & \mu & k-\mu \end{pmatrix}, \qquad L_2 = \begin{pmatrix} 0 & 0 & v-k-1 \\ 0 & k-\lambda-1 & v-2k+\lambda \\ 1 & k-\mu & v-2k+\mu-2 \end{pmatrix}.$$

We see that (iii) generalizes Theorem 2.2(i).

3.1. *The Bose–Mesner algebra*

The relations R_i of an association scheme are described by their adjacency matrices A_i of order n defined by

$$(A_i)_{xy} = \begin{cases} 1 & \text{whenever } (x, y) \in R_i , \\ 0 & \text{otherwise} . \end{cases}$$

In other words, A_i is the adjacency matrix of the graph (X, R_i). In terms of the adjacency matrices, the axioms (i)–(iv) become

(i) $\sum_{i=0}^{d} A_i = J$,
(ii) $A_0 = I$,
(iii) $A_j = A_i^{\mathrm{T}}$, for all $i \in \{0, \ldots, d\}$,
(iv) $A_i A_j = \sum_k p_{ij}^k A_k$, for all $i, j, k \in \{0, \ldots, d\}$.

From (i) we see that the 0–1 matrices A_i are linearly independent, and by use of (ii)–(iv) we see that they generate a commutative $(d+1)$-dimensional algebra $\mathcal{A}$ of symmetric matrices with constant diagonal. This algebra was first studied by Bose and Mesner (1959) and is called the Bose–Mesner algebra of the association scheme.

Since the matrices A_i commute, they can be diagonalized simultaneously (see Marcus and Minc 1965), that is, there exists a matrix S such that for each $A \in \mathcal{A}$, $S^{-1}AS$ is a diagonal matrix. Therefore $\mathcal{A}$ is semi-simple and has a unique basis of minimal idempotents $E_0, \ldots, E_n$ (see Burrow 1965). These are matrices satisfying

$$E_i E_j = \delta_{ij} E_i ,$$

$$\sum_{i=0}^{d} E_i = I .$$

The matrix $(1/n)J$ is a minimal idempotent (idempotent is clear, and minimal follows since $\operatorname{rk} J = 1$). We shall take $E_0 = (1/n)J$. Let P and $(1/n)Q$ be the matrices relating our two bases for $\mathscr{A}$:

$$A_j = \sum_{i=0}^{d} P_{ij} E_i ,$$

$$E_j = \frac{1}{n} \sum_{i=0}^{d} Q_{ij} A_i .$$

Then clearly

$$PQ = QP = nI .$$

It also follows that

$$A_j E_i = P_{ij} E_i ,$$

which shows that the P_{ij} are the eigenvalues of A_j and the columns of E_i are the corresponding eigenvectors. Thus $\mu_i := \mathrm{rk}\, E_i$ is the multiplicity of the eigenvalue P_{ij} of A_j (provided that $P_{ij} \neq P_{kj}$ for $k \neq i$). We see that $\mu_0 = 1$, $\sum_{i=0}^{d} \mu_i = n$, and $\mu_i = \mathrm{tr}\, E_i = n(E_i)_{jj}$ (indeed, E_i has only eigenvalues 0 and 1, so $\mathrm{rk}\, E_k$ equals the sum of the eigenvalues).

Theorem 3.2. *The numbers P_{ij} and Q_{ij} satisfy*

(i) $P_{i0} = Q_{i0} = 1$, $P_{0i} = n_i$, $Q_{0i} = \mu_i$,

(ii) $P_{ij} P_{ik} = \sum_{l=0}^{d} p_{jk}^{l} P_{il}$,

(iii) $\mu_i P_{ij} = n_j Q_{ji}$, $\sum_i \mu_i P_{ij} P_{ik} = n n_j \delta_{jk}$, $\sum_i n_i Q_{ij} Q_{ik} = n \mu_j \delta_{jk}$,

(iv) $|P_{ij}| \leq n_j$, $|Q_{ij}| \leq \mu_j$.

Proof. Part (i) follows easily from $\sum_i E_i = I = A_0$, $\sum_i A_i = J = nE_0$, $A_i J = n_i J$, and $\mathrm{tr}\, E_i = \mu_i$.

Part (ii) follows from $A_j A_k = \sum_l p_{jk}^{l} A_l$.

The first equality in (iii) follows from $\sum_i n_j Q_{ji} P_{ik} = n n_j \delta_{jk} = \mathrm{tr}\, A_j A_k = \sum_i \mu_i P_{ij} P_{ik}$, since P is non-singular; this also proves the second equality, and the last one follows since $PQ = nI$.

The first inequality of (iv) holds because the P_{ij} are eigenvalues of the n_j-regular graphs (X, R_j). The second inequality then follows by use of (iii). □

Relations (iii) are often referred to as the *orthogonality relations*, since they state that the rows (and columns) of P (and Q) are orthogonal with respect to a suitable weight function.

If $d = 2$, and (X, R_1) is strongly regular with parameters (v, k, λ, μ), the matrices P and Q are

$$P = \begin{pmatrix} 1 & k & v-k-1 \\ 1 & r & -r-1 \\ 1 & s & -s-1 \end{pmatrix}, \qquad Q = \begin{pmatrix} 1 & f & g \\ 1 & f\dfrac{r}{k} & g\dfrac{s}{k} \\ 1 & -f\dfrac{r+1}{v-k-1} & -g\dfrac{s+1}{v-k-1} \end{pmatrix},$$

where r, s, f and g can be expressed in terms of v, k, λ by use of Theorem 2.2.

In general the matrices P and Q can be computed from the intersection numbers of the scheme, as follows from the following.

Theorem 3.3. *For* $i = 0, \ldots, d$, *the intersection matrix* L_j *has eigenvalues* P_{ij} $(0 \leq i \leq d)$.

Proof. Theorem 3.2(ii) yields

$$P_{ij} \sum_k P_{ik}(P^{-1})_{km} = \sum_{k,l} P_{il}(L_j)_{lk}(P^{-1})_{km} ,$$

hence $PL_jP^{-1} = \text{diag}(P_{0j}, \ldots, P_{dj})$. □

Thanks to this theorem, it is relatively easy to compute P, Q $(=(1/n)P^{-1})$ and μ_i $(=Q_{0i})$. It is also possible to express P and Q in terms of the (common) eigenvectors of the L_j. Indeed, $PL_jP^{-1} = \text{diag}(P_{0j}, \ldots, P_{dj})$ implies that the rows of P are left eigenvectors and the columns of Q are right eigenvectors. In particular, μ_i can be computed from the right eigenvector u_i and the left eigenvector v_i^{T}, normalized such that $(u_i)_0 = (v_i)_0 = 1$, by use of $\mu_i u_i^{\mathrm{T}} v_i = n$. Clearly, each μ_i must be an integer. These are the *rationality conditions* for an association scheme. As we saw in the case of a strongly regular graph, these conditions can be very powerful. Godsil (see chapter 31) puts the rationality conditions in a more general form, which is not restricted to association schemes.

3.2. The Krein parameters

The Bose–Mesner algebra $\mathscr{A}$ is not only closed under ordinary matrix multiplication, but also under componentwise (Hadamard, Schur) multiplication (denoted $\circ$). Clearly $\{A_0, \ldots, A_d\}$ is the basis of minimal idempotents with respect to this multiplication.

Write

$$E_i \circ E_j = \frac{1}{n} \sum_{k=0}^{d} q_{ij}^k E_k .$$

The numbers q_{ij}^k thus defined are called the *Krein parameters*. (Our q_{ij}^k are those of Delsarte, but differ from those of Seidel (1979) by a factor n.) As expected, we now have the analogue of Theorems 3.1 and 3.2.

Theorem 3.4. *The Krein parameters of an association scheme satisfy*

(i) $q_{0j}^k = \delta_{jk}$, $q_{ij}^0 = \delta_{ij}\mu_j$, $q_{ij}^k = q_{ji}^k$,

(ii) $\sum_i q_{ij}^k = \mu_j$, $\sum_j \mu_j = n$,

(iii) $q_{ij}^k \mu_k = q_{ik}^j \mu_j$,

(iv) $\sum_l q_{ij}^l q_{kl}^m = \sum_l q_{kj}^l q_{il}^m$,

(v) $Q_{ij}Q_{ik} = \sum_{l=0}^{d} q_{jk}^l Q_{il}$,

(vi) $n\mu_k q_{ij}^k = \sum_l n_l Q_{li} Q_{lj} Q_{lk}$.

Proof. Let $\sum(A)$ denote the sum of all entries of the matrix A. Then $JAJ = \sum(A)J$, $\sum(A \circ B) = \operatorname{tr} AB^{\mathrm{T}}$ and $\sum(E_i) = 0$ if $i \neq 0$; since then $E_i J = nE_iE_0 = 0$. Now (i) follows by use of $E_i \circ E_0 = (1/n)E_i$, $q_{ij}^0 = \sum(E_i \circ E_j) = \operatorname{tr} E_iE_j = \delta_{ij}\mu_j$, and $E_i \circ E_j = E_j \circ E_i$, respectively. Equation (iv) follows by evaluating $E_i \circ E_j \circ E_k$ in two ways, and (iii) follows from (iv) by taking $m = 0$. Equation (v) follows from evaluating $A_i \circ E_j \circ E_k$ in two ways, and (vi) follows from (v), using the orthogonality relation $\sum_l n_l Q_{lm} Q_{lk} = \delta_{mk}\mu_k n$. Finally, by use of (iii) we have

$$\mu_k \sum_j q_{ij}^k = \sum q_{ik}^j \mu_j = n \cdot \operatorname{tr}(E_i \circ E_k) = n \sum_l (E_i)_{ll}(E_k)_{ll} = \mu_i \mu_k ,$$

proving (ii). □

The above results illustrate a dual behaviour between ordinary multiplication, the numbers p_{ij}^k and the matrices A_i and P on the one hand, and Schur multiplication, the numbers q_{ij}^k and the matrices E_i and Q on the other hand. If two association schemes have the property that the intersection numbers of one are the Krein parameters of the other, then the converse is also true. Two such schemes are said to be (formally) dual to each other. One scheme may have several (formal) duals, or none at all (but when the scheme is invariant under a regular abelian group, there is a natural way to define a dual scheme, cf. Delsarte (1973). In fact usually the Krein parameters are not even integers. But they cannot be negative. These important restrictions, due to Scott (1973), are the so-called Krein conditions.

Theorem 3.5. *The Krein parameters of an association scheme satisfy* $q_{ij}^k \geq 0$ *for all* $i, j, k \in \{0, \ldots, d\}$.

Proof. The numbers $(1/n)q_{ij}^k$ $(0 \leq k \leq d)$ are the eigenvalues of $E_i \circ E_j$ (since $(E_i \circ E_j)E_k = (1/n)q_{ij}^k E_k$). On the other hand, the Kronecker product $E_i \otimes E_j$ is positive semidefinite, since each E_i is. But $E_i \circ E_j$ is a principal submatrix of $E_i \otimes E_j$, and therefore is positive semidefinite as well, i.e., has no negative eigenvalue. □

The Krein parameters can be computed by use of Theorem 3.4(vi). This equation also shows that the Krein condition is equivalent to

$$\sum_l n_l Q_{li} Q_{lj} Q_{lk} \geq 0 \quad \text{for all } i, j, k \in \{0, \ldots, d\} .$$

In case of a strongly regular graph we obtain

$$q_{11}^1 = \frac{f^2}{v}\left(1 + \frac{r^3}{k^2} - \frac{(r+1)^3}{(v-k-1)^2}\right) \geq 0 ,$$

$$q_{22}^2 = \frac{g^2}{v}\left(1 + \frac{s^3}{k^2} - \frac{(s+1)^3}{(v-k-1)^2}\right) \geq 0$$

(the other Krein conditions are trivially satisfied in this case), which is equivalent to the result mentioned in the previous section.

Neumaier (1981) generalized Seidel's absolute bound to association schemes, and obtained the following.

Theorem 3.6. *The multiplicities* μ_i $(0 \leqslant i \leqslant d)$ *of an association scheme with d classes satisfy*

$$\sum_{q_{ij}^k \neq 0} \mu_k \leqslant \begin{cases} \mu_i \mu_j & \text{if } i \neq j\,, \\ \frac{1}{2}\mu_i(\mu_i + 1) & \text{if } i = j\,. \end{cases}$$

Proof. The left-hand side equals $\mathrm{rk}(E_i \circ E_j)$. But $\mathrm{rk}(E_i \circ E_j) \leqslant \mathrm{rk}(E_i \otimes E_j) = \mathrm{rk}\, E_i \times \mathrm{rk}\, E_j = \mu_i \mu_j$. And if $i = j$, then $\mathrm{rk}(E_i \circ E_i) \leqslant \frac{1}{2}\mu_i(\mu_i + 1)$. Indeed, if the rows of E_i are linear combinations of μ_i rows, then the rows of $E_i \circ E_i$ are linear combinations of the $\mu_i + \frac{1}{2}\mu_i(\mu_i - 1)$ rows that are the elementwise products of any two of these μ_i rows. □

For strongly regular graphs with $q_{11}^1 = 0$ we obtain Seidel's bound: $v \leqslant \frac{1}{2}f(f + 3)$. But in case $q_{11}^1 > 0$, Neumaier's result states that the bound can be improved to $v \leqslant \frac{1}{2}f(f + 1)$.

3.3. Subsets of association schemes

The last subject of this section is a result of Delsarte (1973, Theorem 3.3, p. 26) on subsets in association schemes. For $Y \subseteq X$, $Y \neq \emptyset$, we define

$$a_i = \frac{1}{|Y|} \chi^{\mathrm{T}} A_i \chi\,,$$

where χ is the characteristic vector of Y. In other words, a_i is the average degree of the subgraph of (X, R_i) induced by Y. Clearly $a_0 = 1$, and $\sum_{i=0}^{d} a_i = |Y|$. The vector $a = (a_0, \ldots, a_d)$ is called the *inner distribution* of Y.

Theorem 3.7. *The inner distribution a of a non-empty subset of an association scheme satisfies* $aQ \geqslant 0$.

Proof. $|Y| \sum_{i=0}^{d} a_i Q_{ij} = \sum_{i=0}^{d} \chi^{\mathrm{T}} Q_{ij} A_i \chi = n\chi^{\mathrm{T}} E_j \chi \geqslant 0$, since E_j is positive semidefinite. □

This inequality leads to Delsarte's linear programming bound, as we shall see in the next section.

As an application we have the following result (Delsarte 1973, Theorem 3.9, p. 32).

Theorem 3.8. *Let* $\{\{0\}, I_1, I_2\}$ *be a partition of* $\{0, \ldots, d\}$, *and assume that y and Z are non-empty subsets of X such that the inner distribution b of Y satisfies* $b_i = 0$ *for* $i \in I_1$, *and the inner distribution c of Z satisfies* $c_i = 0$ *for* $i \in I_2$, *Then*

$|Y| \cdot |Z| \leqslant |X|$, *and equality holds if and only if for all* $i \neq 0$ *we have* $(bQ)_i = 0$ *or* $(cQ)_i = 0$.

Proof. Define $\beta_i = \mu_i^{-1}|Z|^{-1} \sum_j c_j Q_{ji}$. Then $\beta_0 = 1$, $\beta_i \geqslant 0$ for all i, and $\sum_i \beta_i Q_{ki} = n_k^{-1}|Z|^{-1} n c_k$. Now we have

$$|Y| = \sum_k b_k = (bQ)_0 \leqslant \sum_{i,k} b_k Q_{ki} \beta_i = \sum_i Q_{0i} \beta_i = \frac{n}{|Z|}. \qquad \square$$

Let us investigate a special case of this situation somewhat closer. Let, for $I \subseteq \{0, \ldots, d\}$, the *I-sphere* around the point $x \in X$ be the set $\{y \in X \mid (x, y) \in R_i$ for some $i \in I\}$. A non-empty subset Y of X is called *perfect* (more precisely, *I-perfect*) when the I-spheres around its points form a partition of X.

Theorem 3.9 ("Lloyd's theorem", cf. Lloyd 1957, Delsarte 1973, p. 63). *Let Y be I-perfect, with inner distribution a. Then* $\sum_{i \in I} P_{ji} = 0$ *for all* $j \neq 0$ *such that* $(aQ)_j \neq 0$.

Proof. Apply the previous theorem, with for Z an I-sphere. If c is the inner distribution of Z, then

$$|Z|\mu_j^{-1}(cQ)_j = \mu_j^{-1} \sum_i \sum_{g,h \in I} n_g p_{ih}^g Q_{ij} = \sum_i \sum_{g,h \in I} p_{gh}^i P_{ji} = \sum_{g,h \in I} P_{jg} P_{jh}$$
$$= \Big(\sum_{i \in I} P_{ji}\Big)^2. \qquad \square$$

4. Applications

In this section we discuss some special types of association scheme and their significance to other fields of combinatorics.

4.1. Distance regular graphs

Consider a connected simple graph with vertex set X of diameter d. Define $R_i \subset X^2$ by $(x, y) \in R_i$ whenever x and y have graph distance i. If this defines an association scheme, then the graph (X, R_1) is called *distance-regular*. The corresponding association scheme is called *metric*. By the triangle inequality, $p_{ij}^k = 0$ if $i + j < k$ or $|i - j| > k$. Moreover, $p_{ij}^{i+j} > 0$. Conversely, if the intersection numbers of an association scheme satisfy these conditions, then (X, R_1) is easily seen to be distance-regular.

Many of the association schemes that play a rôle in combinatorics are metric. In fact, all the examples treated in this chapter are metric. Strongly regular graphs are obviously metric. The line graph of the Petersen graph and the Hoffman–Singleton graph are easy examples of distance-regular graphs that are not strongly regular.

Any k-regular graph of diameter d has at most

$$1 + k + k(k-1) + \cdots + k(k-1)^{d-1}$$

vertices, as is easily seen. Graphs for which equality holds are called *Moore graphs*. Moore graphs are distance-regular, and those of diameter 2 were dealt with in Theorem 2.3. Using the rationality conditions Damerell (1977) and Bannai and Ito (1973) showed the following.

Theorem 4.1. *A Moore graph with diameter $d \geq 3$ is a $(2d+1)$-gon.*

A strong non-existence result of the same nature is the theorem of Feit and G. Higman (1964) about finite generalized polygons. A *generalized m-gon* is a point-line geometry such that the incidence graph is a connected, bipartite graph of diameter m and girth $2m$. It is called *regular* of order (s, t) for certain (finite or infinite) cardinal numbers s, t if each line is incident with $s+1$ points and each point is incident with $t+1$ lines. (It is not difficult to prove that if each point is on at least three lines, and each line has at least three points (and $m < \infty$), then the geometry is necessarily regular, and in fact $s = t$ in case m is odd.) From such a regular generalized m-gon of order (s, t), where s and t are finite and $m \geq 3$, we can construct a distance-regular graph with valency $s(t+1)$ and diameter $d = [\frac{1}{2}m]$ by taking the collinearity graph on the points.

Theorem 4.2. *A finite generalized m-gon of order (s, t) with $s > 1$ and $t > 1$ satisfies $m \in \{2, 3, 4, 6, 8\}$.*

Proofs of this theorem can be found in Feit and Higman (1964), Kilmoyer and Solomon (1973) and Roos (1980); again the rationality conditions do the job. The Krein conditions yield some additional information:

Theorem 4.3. *A finite regular generalized m-gon with $s > 1$ and $t > 1$ satisfies $s \leq t^2$ and $t \leq s^2$ if $m = 4$ or 8; it satisfies $s \leq t^3$ and $t \leq s^3$ if $m = 6$.*

This result is due to Higman (1971, 1974) and Haemers and Roos (1981). For each $m \in \{2, 3, 4, 6, 8\}$ infinitely many generalized m-gons exist. (For $m = 2$ we have trivial structures – the incidence graph is complete bipartite; for $m = 3$ we have (generalized) projective planes; an example of a generalized 4-gon of order (2, 2) with collinearity graph $T(6)$ can be described as follows: the points are the pairs from a 6-set, and the lines are the partitions of the 6-set into three pairs, with obvious incidence.)

Many association schemes have the important property that the eigenvalues P_{ij} can be expressed in terms of orthogonal polynomials. An association scheme is called *P-polynomial* if there exist polynomials f_k of degree k with real coefficients, and real numbers z_i such that $P_{ik} = f_k(z_i)$. Clearly we may always take $z_i = P_{i1}$.

By the orthogonality relation Theorem 3.2(iii) we have

$$\sum_i \mu_i f_j(z_i) f_k(z_i) = \sum_i \mu_i P_{ij} P_{ik} = n n_j \delta_{jk} ,$$

which shows that the f_k are orthogonal polynomials.

Theorem 4.4. *An association scheme is metric if and only if it is P-polynomial.*

Proof. Let the scheme be metric. Theorem 1.1 gives

$$A_1 A_i = p_{1i}^{i-1} A_{i-1} + p_{1i}^{i} A_i + p_{1i}^{i+1} A_{i+1} .$$

Since $p_{1i}^{i+1} \neq 0$, A_{i+1} can be expressed in terms of A_1, A_{i-1} and A_i. Hence for each j there exists a polynomial f_j of degree j such that

$$A_j = f_j(A_1) .$$

Using this we have $P_{ij} E_i = A_j E_i = f_j(A_1) E_i = f_j(A_1 E_i) E_i = f_j(P_{i1}) E_i$, hence $P_{ij} = f_j(P_{i1})$.

Now suppose that the scheme is P-polynomial. Then the f_j are orthogonal polynomials, and therefore they satisfy a 3-term recurrence relation (see Szegö 1959, p. 42)

$$\alpha_{j+1} f_{j+1}(z) = (\beta_j - z) f_j(z) + \gamma_{j-1} f_{j-1}(z) .$$

Hence

$$P_{i1} P_{ij} = -\alpha_{j+1} P_{ij+1} + \beta_j P_{ij} + \gamma_{j-1} P_{ij-1} \quad \text{for } i = 0, \ldots, d .$$

Since $P_{i1} P_{ij} = \sum_l p_{1j}^l P_{il}$ and P is non-singular, it follows that $p_{1j}^l = 0$ for $|l - j| > 1$. Now the full metric property easily follows by induction. □

This result is due to Delsarte (1973, Theorem 5.6, p. 61). There is also a result dual to this theorem, involving so-called Q-polynomial and cometric schemes. However, just as the intersection numbers p_{ij}^k have a combinatorial interpretation while the Krein parameters q_{ij}^k do not, the metric schemes have the combinatorial description of distance-regular graphs, while there is no combinatorial interpretation for the cometric property. For more information on P- and Q-polynomial association schemes, see Delsarte (1973) and Bannai and Ito (1984); for distance-regular graphs, see the book by Brouwer et al. (1989).

4.2. The Hamming scheme and the linear programming bound for error-correcting codes

Let $X = \mathcal{Q}^d$, the set of all vectors of length d with entries in $\mathcal{Q}$, where $\mathcal{Q}$ is some set of size q. Define $R_i \subset X^2$ by $(x, y) \in R_i$ if the Hamming distance between x and y (i.e., the number of coordinates in which x and y differ) equals i. This defines an association scheme, the *Hamming scheme* $H(d, q)$. The Hamming scheme is easily seen to be metric, and hence by Theorem 4.4 P-polynomial. The orthogonal polynomials involved are the Kravčuk polynomials $K_j(x)$.

Theorem. 4.5. *For the Hamming scheme $H(d, q)$ we have*

$$P_{ij} = Q_{ij} = K_j(i) = \sum_{k=0}^{j} (-1)^k (q-1)^{j-k} \binom{i}{k} \binom{n-i}{j-k} .$$

See Delsarte (1973, p. 38) or MacWilliams and Sloane (1977) for proofs. From Theorem 4.5 we see that $P = Q$, so the Hamming scheme is self-dual.

A subset $Y \subset X$ of $H(d, q)$, such that $Y^2 \cap R_i = \emptyset$ for $i = 1, \ldots, \delta - 1$ and $Y^2 \cap R_\delta \neq \emptyset$ (i.e., a subset Y such that the minimum Hamming distance between two vectors of Y equals δ), is nothing but an error-connecting code with parameters $(d, |Y|, \delta)$ over the alphabet $\mathcal{Q}$ (cf. chapter 16). Let $a = (a_0, \ldots, a_d)$ be the inner distribution of Y. Then $\sum_i a_i = |Y|$, $a_0 = 1$, $a_1 = \cdots = a_{\delta-1} = 0$, $a \geq 0$ (by definition), and $aQ \geq 0$ by Theorem 3.7.

Consider $a_\delta, \ldots, a_d$ as variables and define

$$a^* = 1 + \max \sum_{i=\delta}^{d} a_i$$

$$\text{subject to} \quad K_j(1) + \sum_{i=\delta}^{d} a_i K_j(i) \geq 0 , \quad j = 0, \ldots, d ,$$

$$a_i \geq 0 , \quad i = \delta, \ldots, d .$$

Then clearly $|Y| \leq a^*$. So a^* is an upper bound for the number of codewords with a given length and minimum distance. This bound, due to Delsarte (1973), is called the linear programming bound, since the value of a^* can be computed by linear programming. Of course, the above-mentioned inequalities are not the only ones satisfied by the a_i, and by adding extra inequalities to the system, one may obtain sharper bounds on $|Y|$. For details and more applications to coding theory, see Delsarte (1973), MacWilliams and Sloane (1977), Best et al. (1978), Best and Brouwer (1977), Brouwer (1993).

4.3. *The Johnson scheme and t-designs*

Let the set X consist of all subsets of size d of a set M, where $|M| = m \geq 2d$. Define relation $R_i \subset X^2$ by $(x, y) \in R_i$ if the Johnson distance between x and y (i.e., the cardinality of $x \setminus y$) equals i. This defines an association scheme, the *Johnson scheme $J(m, d)$*. Since the Johnson distance between x and y equals twice the Hamming distance between (the characteristic vectors of) x and y, it follows that also the Johnson distance satisfies the triangle inequality, so that the Johnson scheme is metric. Note that the graph (X, R_1) is complete for $d = 1$, and is the triangular graph $T(m)$ for $d = 2$. The following result (due to Ogasawara 1965 and Yamamoto et al. 1965) gives some parameters.

Theorem 4.6. *For the Johnson scheme $J(m, d)$ the following hold*:

$$P_{ij} = \frac{n_j}{\mu_i} Q_{ji} = E_j(i) = \sum_{k=0}^{j} (-1)^{j-k} \binom{d-k}{j-k} \binom{d-i}{k} \binom{m-d+k-i}{k}$$
$$= \sum_{k=0}^{j} (-1)^k \binom{i}{k} \binom{d-i}{j-k} \binom{m-d-i}{j-k},$$
$$\mu_i = \binom{m}{i} - \binom{m}{i-1}, \qquad n_j = \binom{d}{j} \binom{m-d}{j}.$$

Here $E_j(x)$ is a so-called Eberlein polynomial. It has degree $2j$ in the indeterminate x, and degree j in the indeterminate $x(m+1-x)$. Since $P_{i1} = d(m-1) - i(m+1-i)$, $E_j(i)$ indeed has degree j in P_{i1} as required by the definition of P-polynomial scheme.

The graph (X, R_d) of a Johnson scheme is called a Kneser graph. A subset Y of X such that any two elements of Y have non-empty intersection is a coclique (independent set) of the Kneser graph. By use of Theorem 3.7 and the above formulas, it can be deduced that

$$|Y| \leqslant \binom{m-1}{d-1},$$

the famous result of Erdős et al. (1961).

A t-(m, d, λ) design is a subset $Y \subset X$ of the Johnson scheme $J(m, d)$ such that each t-element subset of M is contained in precisely λ elements of Y. Delsarte (1973, Theorem 4.7, p. 51) proved the following.

Theorem 4.7. *A non-empty subset Y of the Johnson scheme $J(m, d)$ with inner distribution $a = (a_0, \ldots, a_d)$ is a t-(m, d, λ) design if and only if*

$$\sum_{i=0}^{d} a_i Q_{ij} = 0 \quad \text{for } j = 1, \ldots, t. \tag{$*$}$$

Just as we did in case of the Hamming scheme, we can define

$$a_* = 1 + \min \sum_{i=1}^{d} a_i$$

subject to $a_i \geqslant 0$ for $i = 1, \ldots, d$, and $\sum_{i=0}^{d} a_i Q_{ij} = 0$ for $j = 1, \ldots, t$, and

$$\sum_{i=0}^{d} a_i Q_{ij} \geqslant 0 \text{ for } j = t+1, \ldots, d,$$

where the Q_{ij} are found in Theorem 4.6. Now we have the linear programming (lower) bound for the number of blocks in a t-design; $|Y| \geqslant a_*$. Using the simplex algorithm, Haemers and Weug (1974) showed that this inequality implies the non-existence of the designs with parameters 4-(17, 8, 5), 4-(23, 11, λ), $\lambda = 6$, 12, 4-(24, 12, 15), 6-(19, 9, λ), $\lambda \leqslant 10$, 6-(20, 10, λ), $\lambda = 7$, 14. In certain other cases, such as 5-(19, 9, 7), Köhler (1985) ruled out the existence of t-designs by showing that no solution of the above system of inequalities corresponds to an actual design.

If we replace the restrictions $(aQ)_i = 0$ by restrictions $a_i = 0$ $(1 \leq i < \frac{1}{2}\delta)$ in this system of inequalities, and maximize $\Sigma\, a_i$, we obtain upper bounds for the cardinality of codes with minimum distance δ and constant weight d.

We might also require both $a_i = 0$ and $(aQ)_j = 0$ for suitable i, j, and obtain results for t-designs with restricted block intersections, such as quasi-symmetric block designs. By this method it follows for instance that a quasi-symmetric 2-(29, 7, 12) design with block intersections 1 and 3 cannot exist (Haemers 1975). For details and more results we refer to Delsarte (1973), MacWilliams and Sloane (1977), Best et al. (1978), Cameron and Van Lint (1980), Calderbank (1988). We also point out that several results of section 3 of chapter 31 can also be obtained by use of the framework of Johnson schemes.

Delsarte (1973) generalized the notion of t-designs to subsets of arbitrary association schemes satisfying (*). (Equivalently, a t-design is a subset Y of X such that its characteristic vector χ satisfies $E_j\chi = 0$ for $j = 1, \ldots, t$.) He shows that a t-design in the Hamming scheme is what is known as an orthogonal array of strength t (see chapter 14). Thus, we also have a linear programming bound for orthogonal arrays. More generally, one may give an interpretation of the classical concept of t-design in terms of ranked posets in the obvious way, and then prove for each of the eight known infinite families of P- and Q-polynomial association schemes that a subset is a classical t-design if and only if it is a Delsarte $f(t)$-design (where usually $f(t) = t$), see Delsarte (1976) and Stanton (1986).

4.4. Imprimitive schemes

In section 2 we saw how to associate to a quasi-symmetric 2-design an association scheme whose points are the blocks of the design. In the previous subsection we saw how t-designs give rise to interesting subsets of the Johnson scheme. Let us give one more example of a relation between designs and association schemes. Let N be the incidence matrix of a square 2-design. Then, defining $A_0 = 1$,

$$A_1 = \begin{pmatrix} 0 & N \\ N^{\mathrm{T}} & 0 \end{pmatrix}, \qquad A_2 = \begin{pmatrix} J & 0 \\ 0 & J \end{pmatrix}$$

and $A_3 = J - I - A_1 - A_2$, we obtain a 3-class association scheme. It is *imprimitive*, that is, the union of some of the R_i form a non-trivial equivalence relation (here R_2 is an equivalence relation). Another imprimitive association scheme we have seen in the line graph of the Petersen graph (where having maximal distance is an equivalence relation).

Given an imprimitive association scheme one may produce new association schemes; on the one hand, there is a natural way to give the set of equivalence classes the structure of an association scheme (the "quotient scheme"), and on the other hand, each equivalence class together with the restrictions of the original relations becomes an association scheme (a "subscheme" of the original scheme). See Brouwer et al. (1989, section 2.4) for details.

4.5. The group case

We very briefly discuss some relations between association schemes and finite permutation groups.

Let $\mathcal{G}$ be a permutation group acting on a set X. Then $\mathcal{G}$ has a natural action on X^2, the orbits of which are called orbitals. Suppose $\mathcal{G}$ acts generously transitive on X, that is, for any x, $y \in X$ there exists an element of $\mathcal{G}$ interchanging x and y. Then the orbitals form an association scheme.

(Without any requirements on $\mathcal{G}$, the orbitals form a coherent configuration (see section 3). The coherent configuration is homogeneous if $\mathcal{G}$ is transitive. We get an association scheme in the sense of Delsarte when the permutation character is multiplicity free.)

For any $x \in X$, the number of orbitals equals the number of orbits on X of $\mathcal{G}_x$ (the subgroup of $\mathcal{G}$ of permutations fixing x). This number is called the *rank* of $\mathcal{G}$. Thus, the number of classes in the association scheme is one less than the rank of $\mathcal{G}$. We can also transfer other permutation group theoretic terminology and results to the theory of association schemes. For instance, the Bose–Mesner algebra is in the group case known as the *centralizer algebra*, and all standard results on this centralizer algebra (cf., e.g., Wielandt 1964) have their direct analogue for the Bose–Mesner algebra.

The Hamming and Johnson schemes are derived from generously transitive permutation representations as discussed above; for instance, the Johnson scheme is derived from the representation of the symmetric group Sym (m) on the d-element subsets of an m-set.

If a metric association scheme belongs to the group case, then the corresponding distance-regular graph is called *distance-transitive*. In other words, a graph is distance-transitive when its group of automorphisms is transitive on pairs of vertices with a given distance. Distance-transitive strongly regular graphs are known as rank 3 graphs. A rank 3 permutation group is generously transitive if and only if it has even order; consequently every rank 3 permutation group of even order provides a strongly regular graph. All such strongly regular graphs have recently been classified, see Kantor and Liebler (1989), Liebeck (1985, 1987), and Liebeck and Saxl (1986).

4.6. Euclidean representations of graphs and association schemes

It is possible to represent a graph in a Euclidean space and obtain combinatorial information from facts in Euclidean geometry. For example, let G be a graph with adjacency matrix A with smallest eigenvalue s. If $m \geq -s$, then the matrix $A + mI$ is positive semidefinite, and hence we can write $A + mI = Z^{\mathrm{T}}Z$, for some real $f \times n$ matrix Z (where f is the rank of $A + mI$). This yields a representation $\bar{\ }: G \to \mathbb{R}^f$ sending the vertex x of G to column x of Z. For this representation we have for the inner products

$$(\bar{x}, \bar{y}) = (A + mI)_{xy} = \begin{cases} m & \text{if } y = x \\ 1 & \text{if } y \in \Gamma(x) , \\ 0 & \text{otherwise} . \end{cases}$$

The above shows that a lower bound on the smallest eigenvalue of G yields a Euclidean representation of G with known inner products. But conversely, suppose we have a Euclidean representation of G with inner products m, 1, 0 for equal, adjacent and non-adjacent vertices, respectively. Then the Gram matrix of this representation is $mI + A$, and since Gram matrices are positive semi-definite, it follows that A has smallest eigenvalue $\geqslant -m$.

As a very simple application, consider a graph G with smallest eigenvalue -1. We find a representation by vectors of norm 1 and it follows that having distance at most one is an equivalence relation on G, that is, G is a union of cliques.

As a second application, consider graphs G with smallest eigenvalue $\geqslant -2$. Our representation (for $m = 2$) satisfies

$$(\bar{x}, \bar{y}) = \begin{cases} 2 & \text{if } y = x , \\ 1 & \text{if } y \in \Gamma(x) , \\ 0 & \text{otherwise} \end{cases}$$

and hence the $\mathbb{Z}$-span of the image $\bar{G}$ is a root lattice. But all root lattices are known (they are direct sums of the lattices A_n, D_n, E_6, E_7, E_8) and this makes it possible to classify all such graphs G. See Cameron et al. (1976).

For an association scheme we have a natural supply of positive semi-definite matrices, namely the idempotents, and we can play the same game. Let $(X, \mathcal{R})$ be an association scheme, E an idempotent, and consider the map $\bar{\ }: X \to \mathbb{R}^n$ sending $x \in X$ to $E_{\cdot x}$, column x of E. (Then the dimension of the span of the image of X under this map is the rank of E.) For this representation we have for the inner products:

$$(\bar{x}, \bar{y}) = E_{x\cdot} E_{\cdot y} = e_{xy} .$$

In particular, if $E = E_j$, then $E = (1/n) \sum_i Q_{ij} A_i$ and we find that

$$(\bar{x}, \bar{y}) = Q_{ij} \quad \text{whenever } (x, y) \in R_i .$$

Suppose a graph G has a representation $\bar{\ }$ in $\mathbb{R}^f$ satisfying

$$(\bar{x}, \bar{y}) = \begin{cases} p & \text{if } y = x , \\ q & \text{if } d(x, y) = 1 , \\ r & \text{if } d(x, y) = 2 , \end{cases}$$

where $d(x, y)$ denotes the graph distance between the vertices x and y. Fix a vertex x of G, and define a representation $\hat{\ }$ of $\Gamma(x)$ by $\hat{y} = \bar{y} - (q/p)\bar{x}$. Then $(\hat{y}, \bar{x}) = 0$ for all $y \in \Gamma(x)$, and we have a representation $\hat{\ }$ of $\Gamma(x)$ in $\mathbb{R}^{f-1}$

satisfying

$$(\hat{x}, \hat{y}) = \begin{cases} p - q^2/p & \text{if } y = x , \\ q - q^2/p & \text{if } d(x, y) = 1 , \\ r - q^2/p & \text{otherwise} . \end{cases}$$

In a number of cases the above techniques suffice to determine all graphs with given parameters. For example, let G be a distance-regular graph with the same parameters as the Johnson graph $J(m, d)$ (with as vertices the d-subsets of a fixed m-sets, two d-subsets being adjacent when they meet in a $(d-1)$-set). From $Q_{ij} = P_{ji}\mu_j/k_i$ and the formulas given in Theorem 4.6 above, it follows that $Q_{i1} = m - 1 - im(m-1)/d(m-d)$, which is linear in i. It follows that every local graph $\Gamma(x)$ has smallest eigenvalue $\geqslant -2$, so that the theory developed in Cameron et al. (1976) applies. This enabled Neumaier (1985) and Terwilliger (1986) (independently) to completely classify such graphs G. (The result being that necessarily $G \cong J(m, d)$, except in the case $(m, d) = (8, 2)$, where there are three other graphs, known as the Chang graphs, with the same parameters as the triangular graph $T(8)$.)

As a second example, due to Ivanov and Shpectorov (1984), consider a distance-regular graph G with the same parameters as the dual polar graph for $U(2d, r)$ (that is, the graph with as vertices the maximal totally isotropic subspaces in $\mathrm{PG}(2d-1, q)$ provided with a non-degenerate hermitean form, where $q = r^2$; two vertices are adjacent when they meet in a subspace of codimension one in both). The graph G is regular of valency $k = rm$, its local graphs $\Gamma(x)$ are regular of valency $\lambda = r - 1$, and G has smallest eigenvalue $\theta_d = -m$ with multiplicity $f = k + 1 - m$, where $m = (q^d - 1)/(q - 1)$. Using the idempotent E_d we find a representation of G in $\mathbb{R}^f$, and then of $\Gamma(x)$ in $\mathbb{R}^{f-1}$, with inner products $(\bar{x}, \bar{y}) = Q_{id} = c((-r)^d - 1)/(-r - 1)$ (for some positive constant c) whenever $d(x, y) = i$. Now the Gram matrix M of the representation of $\Gamma(x)$ has eigenvalue 0 with multiplicity at least m, so that $\Gamma(x)$ has eigenvalue λ with multiplicity at least m, so that $\Gamma(x)$ has at least m connected components. It follows that $\Gamma(x)$ is a union of cliques, and that our graph G is the collinearity graph of a near polygon with classical parameters. If $d \geqslant 3$ then every such G is known, and we find that G must be the $U(2d, r)$ dual polar graph.

References

Aschbacher, M.

[1971] The non-existence of rank three permutation groups of degree 3250 and subdegree 57, *J. Algebra* **19**, 538–540.

Bannai, E., and T. Ito

[1973] On finite Moore graphs, *J. Fac. Sci. Univ. Tokyo, Sect IA* **20**, 191–208. Zbl 275.05121.

[1984] *Algebraic Combinatorics I: Association Schemes, Benjamin-Cummings Lecture Note Series,* Vol. 58 (Benjamin/Cummings, London). Zbl 555.05019.

Belevitch, V.

[1950] Theory of $2n$-terminal networks with application to conference telephony, *Electron. Comm.* **27**, 231–244.

Best, M.R., and A.E. Brouwer
[1977] The triply shortened binary Hamming code is optimal, *Discrete Math.* **17**, 235–245. MR 55#14400; Zbl 356.94009.

Best, M.R., A.E. Brouwer, F.J. MacWilliams, A.M. Odlyzko and N.J.A. Sloane
[1978] Bounds for binary codes of length less than 25, *IEEE Trans. Inform. Theory* **IT-24**, 81–93. MR 57#19066; Zbl 369.94011. Russian translation: 1980, *Kibern. Sb. Nov. Ser.* **17**, 28–59. Zbl 456.94015.

Biggs, N.L.
[1974] *Algebraic Graph Theory, Cambridge Tracts in Mathematics,* Vol. 67 (Cambridge University Press, Cambridge). Zbl 284.05101.

Bose, R.C.
[1963] Strongly regular graphs, partial geometries, and partially balanced designs, *Pacific J. Math.* **13**, 389–419.

Bose, R.C., and D.M. Mesner
[1959] On linear associative algebras corresponding to association schemes of partially balanced designs, *Ann. Math. Statist.* **30**, 21–38.

Bose, R.C., and T. Shimamoto
[1952] Classification and analysis of partially balanced incomplete block designs with two associate classes, *J. Am. Statist. Assoc.* **47**, 151–184.

Brouwer, A.E.
[1993] The linear programming bound for binary linear codes, *IEEE Trans. Inform. Theory* **IT-39**, 677–680.

Brouwer, A.E., and J.H. van Lint
[1984] Strongly regular graphs and partial geometries, in: *Enumeration and Design, Proc. Silver Jubilee Conf. on Combinatorics, Waterloo, 1982,* eds. D.M. Jackson and S.A. Vanstone (Academic Press, Toronto) pp. 85–122. MR 87c:05033; Zbl 555.05016. Russian translation: 1987, *Kibern. Sb. Nov. Ser.* **24**, 186–229. Zbl 636.05013.

Brouwer, A.E., A.M. Cohen and A. Neumaier
[1989] *Distance-Regular Graphs, Ergebnisse der Mathematik,* Vol. 3.18 (Springer, Heidelberg).

Burrow, M.
[1965] *Representation Theory of Finite Groups* (Academic Press, New York).

Calderbank, A.R.
[1988] Geometric invariants for quasi-symmetric designs, *J. Combin. Theory A* **47**, 101–110.

Cameron, P.J.
[1978] Strongly regular graphs, in: *Selected Topics in Graph Theory,* eds. L.W. Beineke and R.J. Wilson (Academic Press, New York) pp. 337–360. Zbl 468.05064.

Cameron, P.J., and J.H. van Lint
[1980] *Graphs, Codes and Designs, London Mathematical Society Lecture Note Series,* Vol. 43 (Cambridge University Press, Cambridge). Zbl 427.05001. Revised version: 1991, *Designs, Graphs, Codes and their Links, London Math. Soc. Student Texts,* Vol. 22 (Cambridge University Press, Cambridge).

Cameron, P.J., J.-M. Goethals, J.J. Seidel and E.E. Shult
[1976] Line graphs, root systems and elliptic geometry, *J. Algebra* **43**, 305–327. Zbl 337.05142.

Damerell, R.M.
[1973] On Moore graphs, *Proc. Cambridge Philos. Soc.* **74**, 227–236. MR 47#6553.

Delsarte, Ph.
[1973] An algebraic approach to the association schemes of coding theory, *Philips Res. Rep. Suppl.* **10**.
[1974] The association schemes of coding theory, in: *Combinatorics, Proc. 1974 Conf., Nijenrode Castle, Breukelen, Mathematical Centre Tracts,* Vol. 55, eds. M. Hall Jr and J.H. van Lint (Mathematisch Centrum, Amsterdam) pp. 143–161. Also published: 1975 (Reidel, Dordrecht).
[1976] Association schemes and t-designs in regular semilattices, *J. Combin. Theory A* **20**, 230–243.

Delsarte, Ph., and J.-M. Goethals
[1971] Orthogonal matrices with zero diagonal, II, *Canad. J. Math.* **23**, 816–832.

Delsarte, Ph., J.-M. Goethals and J.J. Seidel
[1975] Bounds for systems of lines and Jacobi polynomials, *Philips Res. Rep.* **30**, 91*–105*.

Erdős, P., C. Ko and R. Rado
[1961] Intersection theorems for systems of finite sets, *Quart. J. Math. Oxford (2)* **12**, 313–320.
Feit, W., and G. Higman
[1964] The non-existence of certain generalized polygons, *J. Algebra* **1**, 114–131.
Goethals, J.-M.
[1979] Association schemes, in: *Algebraic Coding Theory and Applications, CISM Courses and Lectures,* Vol. 258, ed. G. Longo (Springer, Wien) pp. 157–180.
Goethals, J.-M., and J.J. Seidel
[1967] Orthogonal matrices with zero diagonal, *Canad. J. Math.* **19**, 1001–1010. MR 36#5013.
Haemers, W.H.
[1975] *Sterke grafen en block designs,* MSc. Thesis (Eindhoven University of Technology, Eindhoven) (in Dutch).
[1979] Eigenvalue methods, *in: Packing and Covering in Combinatorics, Mathematical Centre Tracts,* Vol. 106, ed. A. Schrijver (Mathematisch Centrum, Amsterdam).
Haemers, W.H., and C. Roos
[1981] An inequality for generalized hexagons, *Geom. Dedicata* **10**, 219–222. Zbl 463.51012.
Haemers, W.H., and C. Weug
[1974] *Linear Programming Bounds for Codes and Designs,* Technical Note N96 (M.B.L.E. Brussels).
Higman, D.G.
[1971] Partial geometries, generalized quadrangles and strongly regular graphs, in: *Atti Convegno di Geometria, Combinatoria e sue Applicazioni, Perugia, 1970,* ed. A. Berlotti (Univ. degli Studi di Perugia, Perugia) pp. 263–293. MR 51#2945.
[1974] Invariant relations, coherent configurations and generalized polygons, in: *Combinatorics, Mathematical Centre Tracts,* Vol. 57, eds. M. Hall Jr and J.H. van Lint (Mathematisch Centrum, Amsterdam) pp. 27–43. Also published: 1975 (Reidel, Dordrecht) pp. 347–363.
[1975] Coherent configurations, Part I: Ordinary representation theory, *Geom. Dedicata* **4**, 1–32.
[1976] Coherent configurations, Part II: Weights, *Geom. Dedicata* **5**, 413–424. Zbl 353.05009.
Hoffman, A.J., and R.R. Singleton
[1960] On Moore graphs with diameters 2 and 3, *IBM J. Res. Develop.* **4**, 497–504.
Hubaut, X.
[1975] Strongly regular graphs, *Discrete Math.* **13**, 357–381.
Ivanov, A.A., and S.V. Shpectorov
[1989] The association schemes of dual polar spaces of type ${}^2A_{2d-1}(p^f)$ are characterized by their parameters if $d > 3$, *Lin. Algebra Appl.* **114**, 133–139.
Kantor, W.M., and R.A. Liebler
[1982] The rank three representations of the finite classical groups, *Trans. Amer. Math. Soc.* **271**, 1–71.
Kilmoyer, R., and L. Solomon
[1973] On the theorem of Feit–Higman, *J. Combin. Theory A* **15**, 310–322.
Köhler, E.
[1985] Über den allgemeinen Gray-Code und die Nichtexistenz einiger t-designs, in: *Graphen in Forschung und Unterricht, Festschrift K. Wagner,* ed. R. Bodendiek (Franzbecker, Salzdetfurth) pp. 102–111. MR 87g:05001.
Koornwinder, T.H.
[1976] A note on the absolute bound for systems of lines, *Proc. Kon. Nederl. Acad. Wet., Ser. A* **79** [= *Indag. Math.* **38**], 152–153.
Liebeck, M.W.
[1985] Permutation modules for rank 3 symplectic and orthogonal groups, *J. Algebra* **92**, 9–15. Zbl 559.20027.
[1987] The affine permutation groups of rank three, *Proc. London Math. Soc. (3)* **54**, 477–516. Zbl 621.20001.
Liebeck, M.W., and J. Saxl
[1986] The finite primitive permutation groups of rank three, *Bull. London Math. Soc.* **18**, 165–172. Zbl 586.20003.

Lloyd, S.P.
[1957] Binary block coding, *Bell System Tech. J.* **36**, 517–535.

MacWilliams, F.J., and N.J.A. Sloane
[1977] *The Theory of Error-Correcting Codes* (North-Holland, Amsterdam).

Marcus, M., and H. Minc
[1965] *Introduction to Linear Algebra* (MacMillan, New York).

Mathon, R.A.
[1978] Symmetric conference matrices of order pq^2+1, *Canad. J. Math.* **30**, 321–331.

Neumaier, A.
[1981] New inequalities for the parameters of an association scheme, in: *Combinatorics and Graph Theory, Proc. Calcutta, 1980, Lecture Notes in Mathematics,* Vol. 885, ed. S.B. Rao (Springer, Berlin) pp. 365–367. Zbl 479.05047.
[1985] Characterization of a class of distance regular graphs, *J. Reine Angew. Math.* **357**, 182–192. Zbl 552.05042.

Ogasawara, M.
[1965] *A Necessary Condition for the Existence of Regular and Symmetrical PBIB Designs of T_m Type, Mimeograph Series,* No. 418 (University of North Carolina, Institute of Statistics).

Paley, R.E.A.C.
[1933] On orthogonal matrices, *J. Math. Phys. Mass. Inst. Technol.* **12**, 311–320.

Roos, C.
[1980] An alternative proof on the Feit–Higman theorem on generalized polygons, *Delft Progr. Rep.* **5** 67–77. Zbl 452.05033.

Scott Jr, L.L.
[1973] A condition on Higman's parameters, *Notices Amer. Math. Soc.* **20**, A-97. (701–20–45).

Seidel, J.J.
[1979] Strongly regular graphs, in: *Surveys in Combinatorics, Proc. 7th British Combinatorial Conf., London Mathematical Society Lecture Note Series,* Vol. 38, ed. B. Bollobás (Cambridge University Press, Cambridge) pp. 157–180.
[1983] Delsarte's theory of association schemes, in: *Graphs and other Combinatorial Topics, Teubner Texte zur Mathematik,* Vol. 59, ed. M. Fiedler (Teubner, Leipzig) pp. 249–258.

Stanton, D.
[1986] t-Designs in classical association schemes, *Graphs Combin.* **2**, 283–286.

Szegö, G.
[1959] *Orthogonal Polynomials, Amer. Math. Soc. Colloq. Publ.* **23**. 1st Ed.: 1939; 4th Ed.: 1975.

Terwilliger, P.
[1986] The Johnson graph $J(d,r)$ is unique if $(d,r)\neq(2,8)$, *Discrete Math.* **58**, 175–189. Zbl 587.05038.

van Lint, J.H., and J.J. Seidel
[1966] Equilateral point sets in elliptic geometry, *Proc. Kon. Nederl. Acad. Wet., Ser. A* **69** [= *Indag. Math.* **28**], 335–348.

Wallis, W.D., A.P. Street and J.S. Wallis
[1972] *Combinatorics: Room Squares, Sum-Free Sets, Hadamard Matrices, Lecture Notes in Mathematics,* Vol. 292 (Springer, Berlin).

Wielandt, H.
[1964] *Finite Permutation Groups* (Academic Press, New York).

Yamamoto, S., Y. Fujii and N. Hamada
[1965] Composition of some series of association algebras, *J. Sci. Hiroshima Univ. Ser. A-I* **29**, 181–215.

CHAPTER 16

Codes

J.H. van LINT

Eindhoven University of Technology, Department of Mathematics and Computing Science, Eindhoven, The Netherlands

Contents

HANDBOOK OF COMBINATORICS
Edited by R. Graham, M. Grötschel and L. Lovász

1. Introduction

Error-correcting codes (sometimes called error-control codes) are used to protect digital data against errors that can occur when the data is transmitted over a so-called "noisy" communication channel. One of the modern applications is what is known as fault-tolerant computing. Here, a system of collaborating computers produces correct results, even though some of the intermediate results are communicated incorrectly from one computer to another. There are two other spectacularly successful applications of coding theory. The first are the pictures of several planets taken by satellites such as Voyager. The signals that the satellites send to earth are so distorted by noise that completely unintelligible pictures would result if one did not use error-correcting codes. The second application is the compact disc digital audio system that produces music of the finest quality, even if there are dust particles, scratches, etc. on the disc. Without Reed–Solomon codes (see section 5) this would not be the case. One of the reasons that the topic is increasing in importance is the fact that electronic circuits that realize the possibilities of the theory of error-control codes, have become sufficiently cheap to produce.

The subject of *coding theory* (and information theory) began in 1948 when Shannon (1948) published two papers on "A mathematical theory of communication" in the *Bell System Technical Journal*. As an introduction and in order to establish some terminology, we explain the idea of *block codes*. Consider a *source* that produces information in the form of a very long sequence of zeros and ones. We partition this sequence into blocks of length k and consider these blocks as elements of $\mathbb{F}_2^k$. The "*encoder*" is an injection from $\mathbb{F}_2^k$ to $\mathbb{F}_2^n$. The image of an information block is a block of length n, called a *codeword* $\boldsymbol{c}$, and the set of codewords is called the *code* C. A particularly simple situation occurs when the first k bits of $\boldsymbol{c}$ are the original information block. The remaining bits are the so-called *redundancy*. It is clear why the number k/n is called the *information rate* of C. (In general, if C is a binary code of length n, then $R := n^{-1} \log_2|C|$ is called the information rate of C.) Examples of channels are the glass fibre through which digitized telephone messages are communicated as light pulses, or the track on a compact disc. On this track the ones are stored as pits (produced by the write-laser). The zeros are so-called lands. (Actually, the situation is slightly more complicated but this suffices to understand the idea.) The reading is done with a weaker laser by measuring the reflection, thus distinguishing between pits and lands. Errors can be caused by dust, scratches, fingerprints, etc. These result in the interpretation of a 0 as a 1 or vice versa. In our theory, we consider the received word $\boldsymbol{r}$ as a sum $\boldsymbol{c} + \boldsymbol{e}$ in $\mathbb{F}_2^n$, where $\boldsymbol{e}$ is the so-called error-vector. The model we use for the channel is what is known as the *binary symmetric channel* (BSC). Here, each transmitted bit has a probability p of being received in error. The number of errors in a received word is equal to the number of ones in the vector $\boldsymbol{e}$, which is called the *weight* $w(\boldsymbol{e})$ of $\boldsymbol{e}$ (in general, the weight of a vector is the number of nonzero coordinates). A code C is called a t-error-correcting code if, for every $\boldsymbol{c} \in C$ and every $\boldsymbol{e}$ with $w(\boldsymbol{e}) \leq t$, the sum $\boldsymbol{c} + \boldsymbol{e} = \boldsymbol{r}$ uniquely determines $\boldsymbol{e}$ and $\boldsymbol{c}$. We use the following terminology. If $\boldsymbol{x}$ and $\boldsymbol{y}$ are vectors with n coordinates (in any vector space), then their *distance* $d(\boldsymbol{x}, \boldsymbol{y})$ is the number of places where they differ. Therefore, $d(\boldsymbol{x}, \boldsymbol{y}) = w(\boldsymbol{x} - \boldsymbol{y})$. In decoding a received message $\boldsymbol{r}$, one usually chooses a codeword $\boldsymbol{c}$ such that $d(\boldsymbol{r}, \boldsymbol{c})$ is minimal. If C is a code, then the minimal value of $d(\boldsymbol{x}, \boldsymbol{y})$ for all pairs of distinct codewords is called the *minimum distance* of C. If the distance is $2e + 1$, then the code is clearly e-error-correcting.

We mention without proof the following quite sensational theorem due to Shannon (1948). Assume that we are using a BSC with error probability p and a code C of length n for which every codeword has the same probability of occurring in a message. Decoding of received messages $\boldsymbol{r}$ is done by replacing $\boldsymbol{r}$ by a codeword $\boldsymbol{c}$ such that $w(\boldsymbol{r} - \boldsymbol{c})$ is minimal. Let P_C be the probability that a received codeword is interpreted incorrectly.

Theorem 1.1 (Shannon, 1948). *Let* $\varepsilon > 0$. *If* $0 < R < 1 + p \log p + (1-p) \log(1-p)$, *then there is an* n *and a code* C *of length* n *and rate* R, *such that* $P_C < \varepsilon$ (*logarithm to base* 2).

For a proof see Van Lint (1983). The proof is an existence proof and coding theorists are still trying to actually construct "good" codes that can be efficiently implemented in practice.

In this chapter we shall ignore large parts of coding theory such as encoding and decoding algorithms, convolutional codes, etc. After an initial period of development of the theory, coding theorists and combinatorialists became aware of each other's existence. Furthermore, it turned out that ideas from each of these areas could be used fruitfully in the other. After some introductory material, this chapter concentrates on those parts of coding theory that are strongly related to combinatorics (see also Assmus and Mattson 1974). As we saw above, the question of error-correction is essentially a question of *packing* spheres in a metric vector space. Connected to this there is the combinatorically interesting *covering* problem. If C is a code, then the *covering radius* of C is defined to be the smallest number ρ such that the spheres of radius ρ around the codewords cover the whole space. We shall consider this parameter in section 7.

In the References, we list several textbooks on coding theory. The major references for this chapter are Cameron and Van Lint (1991), Van Lint (1982), and MacWilliams and Sloane (1977).

2. Linear codes

We choose some field $\mathbb{F}_q$ as alphabet. Let $\boldsymbol{F}^{(n)}$ be the vector space $\mathbb{F}_q^n$. A subset C of $\boldsymbol{F}^{(n)}$ is called a k-dimensional *linear code* of length n or $[n, k]$ code if C is a k-dimensional linear subspace of $\boldsymbol{F}^{(n)}$. The rate of an $[n, k]$ code is k/n. If C has minimum distance d, then C is called an $[n, k, d]$ code. Since $d(\boldsymbol{x}, \boldsymbol{y}) = d(\boldsymbol{x} -$

$y, \mathbf{0}) = w(x - y)$, the minimum distance of a linear code is equal to the minimum weight of the code.

A *generator matrix* G for a linear code C is a k by n matrix for which the rows are a basis of C, i.e., $C = \{aG \mid a \in \mathbb{F}_q^k\}$. G is in *standard form* if $G = (I_k P)$, where I_k is the k by k identity matrix. In this case, the first k symbols of a codeword are called *information symbols* (these can be chosen arbitrarily). The remaining symbols, so-called "*check symbols*", are determined by the information symbols (and P).

An example is the [4, 2] ternary Hamming code with

$$G = \begin{pmatrix} 1 & 0 & 1 & 1 \\ 0 & 1 & 1 & 2 \end{pmatrix}.$$

In this example, all nonzero words have weight 3.

Two codes C_1 and C_2 (not necessarily linear) are called *equivalent* if C_2 is obtained from C_1 by a permutation of the positions in codewords (e.g., interchanging the first and second letters in all the codewords). Every linear code is equivalent to a code with a generator matrix in standard form. In general, a code C (not necessarily linear) is called *systematic* on k positions if $|C| = q^k$ and there is exactly one codeword for every possible choice of coordinates on these k positions. Since one can separate information symbols and redundant symbols, these codes are also called *separable*.

If a code C has length n and distance d, then deleting $d - 1$ symbols from every codeword yields a code of length $n - d + 1$ for which no two words are the same. Therefore, $|C| \leq q^{n-d+1}$. This is known as the *Singleton bound*. It can also be expressed as follows. If $|C| = q^k$, then $d \leq n - k + 1$. In the case of equality, the code C is called *maximum distance separable* (MDS). Such a code is systematic on any k-tuple of positions. Instead of writing a separate section on MDS codes, we have included the relevant information in appropriate places; we refer the reader to (i) the observation preceding Definition 2.2, and (ii) section 5.4 on RS codes.

If C is an $[n, k]$ code, we define the *dual code* $C^\perp$ by:

$$C^\perp := \{y \in F^{(n)} \mid \forall_{x \in C}[\langle x, y \rangle = 0]\},$$

where $\langle x, y \rangle := \sum_{i=1}^n x_i y_i$ is the usual inner product. $C^\perp$ is also a linear code, namely an $[n, n - k]$ code. If $C = C^\perp$, then C is called *self-dual*. If $C \subset C^\perp$, then C is called *self-orthogonal*. A generator matrix H of $C^\perp$ is called a *parity-check* matrix for the code C. We have

$$C = \{x \in F^{(n)} \mid xH^\top = \mathbf{0}\}.$$

If $G = (I_k P)$ is a generator matrix for C in standard form, then $H = (-P^\top I_{n-k})$ is a parity-check matrix for C.

Since the generator matrix G of the binary [4, 2] Hamming code, defined above, satisfies $GG^\top = 0$, the code is self-dual.

For any $x \in F^{(n)}$ we call the vector $xH^\top$ the *syndrome* of x. One of the standard methods of decoding linear codes is the so-called *syndrome decoding*. First, note that if $\boldsymbol{x} = \boldsymbol{c} + \boldsymbol{e}$, when $\boldsymbol{c}$ is a codeword and $\boldsymbol{e}$ the error-vector, then $\boldsymbol{x}H^{\top} = \boldsymbol{e}H^{\top}$. The space $\boldsymbol{F}^{(n)}$ is divided into cosets of the code C, and for each coset a word of minimum weight in the coset is chosen as "coset leader". Every word in a coset has the same syndrome as the coset leader. Decoding is done by subtracting the coset leader from the received word $\boldsymbol{x}$. So, the receiver only needs a list of syndromes with the corresponding coset leaders.

Let C be an $[n, k]$ code with parity-check matrix H. A codeword $\boldsymbol{x}$ corresponds to a linear combination of the columns of H equal to the $\boldsymbol{0}$ column. Therefore, the minimum distance of C equals the smallest number of columns of H that are linearly dependent. This observation leads to one of the best known classes of linear codes. We first observe that the columns of the l by n matrix H (where $l := n - k$) can be interpreted as points in $\mathrm{PG}(l-1, q)$. If these points are different, then the code $C^{\perp}$ is called *projective* and by the observation above, the code C has minimum distance at least 3. Now we fix l and then maximize n, keeping $C^{\perp}$ projective.

Definition 2.1. A linear code over $\mathbb{F}_q$ that has an l by n parity-check matrix [where $n = (q^l - 1)/(q-1)$], which has all the points of $\mathrm{PG}(l-1, q)$ as its columns, is called a *Hamming* code.

As we saw before, a Hamming code has minimum distance 3. For any $\boldsymbol{x} \in \boldsymbol{F}^{(n)}$ the set $\{\boldsymbol{y} \in \boldsymbol{F}^{(n)} \mid d(\boldsymbol{x}, \boldsymbol{y}) \leqslant e\}$ is called a *sphere* of radius e around $\boldsymbol{x}$. For a Hamming code, the spheres of radius 1 around codewords are clearly disjoint. Each sphere contains $1 + (q-1)n = q^l$ words and there are $q^k = q^{n-l}$ such spheres. Therefore, these spheres partition the space $\boldsymbol{F}^{(n)}$. Such a code is called a *perfect* code. In the case of an e-error-correcting perfect code, every word in $\boldsymbol{F}^{(n)}$ has distance $\leqslant e$ to exactly one codeword.

Now consider an $[n, k]$ code C that is MDS. Since C is systematic on any k positions, any k columns of the generator matrix of C are linearly independent. Therefore, $C^{\perp}$ has minimum distance $\geqslant k + 1 = n - (n-k) + 1$, i.e., the dual code is also MDS. Note that if we consider the columns of a generator matrix of an $[n, k, n-k+1]$ code as points in $\mathrm{PG}(k-1, q)$, then no k of these points lie in a hyperplane, i.e., we have an n-arc in $\mathrm{PG}(k-1, q)$.

Definition 2.2. If C is a code of length n over $\mathbb{F}_q$, we define the *extended* code $\bar{C}$ by:

$$\bar{C} = \left\{(c_1, c_2, \ldots, c_n, c_{n+1}) \mid (c_1, c_2, \ldots, c_n) \in C, \sum_{i=1}^{n+1} c_i = 0\right\}.$$

If C is linear with generator matrix G and parity-check matrix H, then $\bar{C}$ has a generator matrix $\bar{G}$ obtained from G by adding a column such that the sum of all columns is $\boldsymbol{0}$, and a parity-check matrix

$$\bar{H} = \begin{pmatrix} 1 & 1 & \cdots & 1 \\ & & & 0 \\ & H & & \vdots \\ & & & 0 \end{pmatrix}.$$

If C is binary code with odd minimum distance d, then $\bar{C}$ has minimum distance $d+1$. (In an extended binary code all weights and all distances are even.)

Let C be a code of length n over $\mathbb{F}_q$.

Definition 2.3. Let $A_i = |C|^{-1} \cdot |\{(\boldsymbol{x}, \boldsymbol{y}) \mid \boldsymbol{x} \in C, \boldsymbol{y} \in C, d(\boldsymbol{x}, \boldsymbol{y}) = i\}|$. Then $(A_0, A_1, \ldots, A_n)$ is called the *distance distribution* of C and $A(z) = \sum_{i=0}^{n} A_i z^i$ is called the *distance enumerator* of C.

If C is linear, then A_i equals the number of words of weight i in C and then $A(z)$ is usually called the *weight enumerator*. If we change the origin to another codeword (i.e., replace C by $C - \boldsymbol{c}$ for some codeword $\boldsymbol{c}$), the weight enumerator of a linear code does not change. Codes that have this property are said to be *distance invariant*.

One of the most useful theorems in the theory of linear codes is *MacWilliams's Theorem* (MacWilliams 1963). We shall see that it is crucial in the proof of Theorem 7.8; it can be used to show nonexistence of codes (nonintegral coefficients). It is used heavily in the nonexistence proof of the projective plane of order 10.

Theorem 2.4. *Let C be an $[n, k]$ code over $\mathbb{F}_q$ with weight enumerator $A(z)$ and let $B(z)$ be the weight enumerator of $C^{\perp}$. Then*

$$B(z) = q^{-k}(1 + (q-1)z)^n A\left(\frac{1-z}{1+(q-1)z}\right).$$

Proof. For a proof we refer to Van Lint (1982) or several of the textbooks mentioned in section 1. The essence of the proof is present in the proof for the binary case. We interpret words in $\mathbb{F}_2^n$ as elements of $\mathbb{Z}^n$. Then for any $\boldsymbol{u}$ we have

$$g(\boldsymbol{u}) := \sum_{\boldsymbol{v} \in \mathbb{F}_2^n} (-1)^{\langle \boldsymbol{u}, \boldsymbol{v} \rangle} z^{w(\boldsymbol{u})} = \sum_{(v_1, \ldots, v_n)} (-1)^{u_1 v_1 + \cdots + u_n v_n} z^{v_1 + \cdots + v_n}$$
$$= \prod_{i=1}^{n} [1 + (-1)^{u_i} z] = (1-z)^{w(\boldsymbol{u})}(1+z)^{n-w(\boldsymbol{u})}.$$

The result then follows from

$$\sum_{\boldsymbol{u} \in C} g(\boldsymbol{u}) = \sum_{\boldsymbol{v} \in \mathbb{F}_2^n} z^{w(\boldsymbol{v})} \sum_{\boldsymbol{u} \in C} (-1)^{\langle \boldsymbol{u}, \boldsymbol{v} \rangle}$$
$$= |C| \sum_{\boldsymbol{v} \in C^{\perp}} z^{w(\boldsymbol{v})} = |C| B(z),$$

because the inner term is 0 if $\boldsymbol{v} \notin C^{\perp}$ and it is $|C|$ for $\boldsymbol{v} \in C^{\perp}$. □

It is sometimes more convenient to replace $A(z)$ by a function of two variables.

For an arbitrary binary code C we shall also call

$$A(x, y) = \sum_{i=0}^{n} A_i x^{n-i} y^i \tag{2.5}$$

the distance enumerator of C and

$$A'(x, y) = \frac{1}{A(1,1)} A(x+y, x-y) \tag{2.6}$$

the *MacWilliams transform* of $A(x, y)$.

Note that for any binary code C, the number of codewords equals $A(1,1)$ and $A_0 = 1$. If C is linear, then $A'(x, y)$ is the distance enumerator of $C^{\perp}$. If $A(x, y)$ is any polynomial as in (2.5) with $A_0 = 1$, then $A''(x, y) = A(x, y)$.

There are several ways of obtaining new codes from old ones (not necessarily linear). We have already mentioned *extension*. The inverse of this procedure is called *puncturing*. If one replaces a code C by a subset of C, this is called *expurgating* the code. For example, the even-weight subcode of an $[n, k]$ code, which has words of odd weight, is the $[n, k-1]$ code obtained by deleting all odd-weight words. The inverse process is called *augmenting* a code (for example, by adding an extra row to the generator matrix of a linear code). If one takes all the codewords of C starting with a fixed symbol (say 0) and then deletes this symbol, the resulting code is called a *shortened* code. The inverse process, *lengthening*, is best illustrated by a linear code. Here one increases both the length and the dimension by 1.

We mention one more method that is often used: the so-called $(\boldsymbol{u}, \boldsymbol{u}+\boldsymbol{v})$-construction. Here one forms all words of the form $(\boldsymbol{u}, \boldsymbol{u}+\boldsymbol{v})$, where $\boldsymbol{u}$ is from a code C_1 of length n and distance d_1, and $\boldsymbol{v}$ is from a code C_2 of the same length and distance d_2. The new code has length $2n$, $|C_1| \cdot |C_2|$ words, and distance $\min\{2d_1, d_2\}$.

3. Bounds on codes

One of the central problems in combinatorial coding theory is the study of the numbers $A(n, d)$ defined below. We assume that a fixed alphabet of q symbols has been chosen. A code C is called an (n, M, d) code if it has M words of length n and minimum distance d.

Definition 3.1. $A(n, d) := \max\{M \mid \text{an } (n, M, d) \text{ code exists}\}$. A code C with $|C| = A(n, d)$ is called *optimal*.

In this section we shall treat some of the important theorems concerning $A(n, d)$ and some related problems. We restrict ourselves to $q = 2$. For the case of arbitrary q and for more details we refer to Van Lint (1982) and MacWilliams and Sloane (1977). If the proof of a theorem is no more than an exercise, we omit it. It is difficult to compare the bounds since the results depend heavily on the choice of n and d. Usually, bound 3.4 is better than 3.3 and nearly always, the best results are obtained from bound 3.16. Clearly, bound 3.10 is stronger than 3.4. All these bounds have an asymptotic version, obtained by taking $d = \delta n$ and then considering $\alpha(\delta) := \limsup_{n\to\infty} n^{-1}\log A(n, \delta n)$ (see Van Lint 1982).

Theorem 3.2.

$$A(n, 2l-1) = A(n+1, 2l)\,.$$

Theorem 3.3 (Singleton bound, see section 2).

$$A(n, d) \leqslant 2^{n-d+1}\,.$$

Theorem 3.4 (Hamming bound).

$$A(n, 2e+1) \leqslant 2^n \Big/ \sum_{i=0}^{e} \binom{n}{i}\,.$$

Theorem 3.5 (Gilbert–Varshamov bound). (*A lower bound!*)

$$A(n, d) \geqslant 2^n \Big/ \sum_{i=0}^{d-1} \binom{n}{i}\,.$$

Proof. If C is optimal, then every word $\boldsymbol{x} \in \boldsymbol{F}^{(n)}$ has distance less than d to some codeword. □

Remark. Suppose that C is linear and

$$|C| \cdot \sum_{i=0}^{d-1} \binom{n}{i} < 2^n\,.$$

Then there is a word $\boldsymbol{x} \notin C$ with distance $\geqslant d$ to every codeword. If we adjoin $\boldsymbol{x}$ to a basis of C, we find a new linear code of larger dimension that also has minimum distance $\geqslant d$. Therefore, the lower bound 3.5 also applies if we restrict ourselves to linear codes!

Theorem 3.6 (Plotkin bound). *If* $d > \frac{1}{2}n$, *then* $A(n, d) \leqslant 2d/(2d-n)$.

Proof. If C has M words, consider the M by n matrix that has the codewords of C as its rows. Let the column sums of this matrix be m_i $(1 \leqslant i \leqslant n)$. The sum of all distances of pairs of distinct codewords is at least $d\binom{M}{2}$. On the other hand this sum equals $\sum_{i=1}^{n} m_i(M - m_i)$, which is at most $\frac{1}{4}nM^2$. From this the inequality follows. □

We give one example to show how to use Theorem 3.6. We wish to find a bound for $A(13, 5)$. By Theorem 3.2 this equals $A(14, 6)$. If we shorten a $(14, M, 6)$ code three times, we obtain an $(11, M', 6)$ code with $M' \geqslant 2^{-3} \cdot M$. By Theorem 3.6 we have $M' \leqslant 12$. Therefore, $A(13, 5) \leqslant 96$. (Actually $A(13, 5) = 64$.)

Theorem 3.7 (Griesmer bound). *For a binary* $[n, k, d]$ *code we have* $n \geqslant \sum_{i=0}^{k-1} \lceil d/2^i \rceil$.

Proof. Let G be a generator matrix of the code. We may assume that the first row of G is the vector $\boldsymbol{c} = (11 \cdots 100 \cdots 0)$ of weight d. Every linear combination $\boldsymbol{x}$ of the remaining rows of G must have at least $\lceil d/2 \rceil$ ones in the last $n - d$ positions because either $\boldsymbol{x}$ or $\boldsymbol{x} + \boldsymbol{c}$ has at most $\lfloor d/2 \rfloor$ ones in the first d positions. The theorem follows by induction. □

Definition 3.8. $A(n, d, w)$ denotes the maximum number of codewords in a binary code of length n and minimum distance d for which all codewords have weight w.

Theorem 3.9.

$$A(n, 2l-1, w) = A(n, 2l, w) \leqslant \left[\frac{n}{w}\left[\frac{n-1}{w-1}\left[\cdots\left[\frac{n-w+l}{l}\right]\cdots\right]\right]\right].$$

Proof. Consider a shortened code and use induction. □

Theorem 3.10 (Johnson bound).

$$A(n, 2e+1) \leqslant 2^n \Bigg/ \left\{ \sum_{i=0}^{e} \binom{n}{i} + \frac{\binom{n}{e+1} - \binom{d}{e} A(n, d, d)}{\left[\frac{n}{e+1}\right]} \right\}.$$

Proof. If there are N_{e+1} words in $\boldsymbol{F}^{(n)}$ with distance $e + 1$ to the (n, M, d) code C, then

$$M \sum_{i=0}^{e} \binom{n}{i} + N_{e+1} \leqslant 2^n .$$

Consider an arbitrary codeword $\boldsymbol{c}$; w.l.o.g. $\boldsymbol{c} = \boldsymbol{0}$. C has at most $A(n, d, d)$ words of weight d, and each ot these has distance e to $\binom{d}{e}$ words of weight $e + 1$. Therefore, at least

$$\binom{n}{e+1} - \binom{d}{e} A(n, d, d)$$

words have distance $e + 1$ to $\boldsymbol{c}$ and to C. By varying $\boldsymbol{c}$, the same word at distance $e + 1$ to C can be counted at most $\lfloor n/(e+1) \rfloor$ times. The result follows. □

From Theorem 3.9 we have

$$\binom{d}{e} A(n, d, d) \leqslant \binom{n}{e} \left[\frac{n-e}{e+1}\right].$$

Combining this with Theorem 3.10 yields the original form of the Johnson bound

(Johnson 1962):

$$|C|\left\{\sum_{i=0}^{e}\binom{n}{i}+\frac{\binom{n}{e}}{\left[\frac{n}{e+1}\right]}\left(\frac{n-e}{e+1}-\left[\frac{n-e}{e+1}\right]\right)\right\}\leqslant 2^{n}. \tag{3.11}$$

Note that if $e+1$ divides $n+1$, then (3.11) becomes the Hamming bound. If $e=1$ and n is even, then (3.11) yields $A(n,3)\leqslant 2^{n}/(n+2)$. For example, $A(2^{l}-2,3)\leqslant 2^{2^{l}-l-2}$. Since equality holds for shortened Hamming codes, these codes are optimal.

Definition 3.12. A code C for which equality holds in (3.11) is called *nearly perfect.*

From the proof of Theorem 3.10 we see that if C is nearly perfect, then every word at distance $e+1$ to C has distance $e+1$ to exactly $\lfloor n/(e+1)\rfloor$ codewords. Similarly, a word at distance e to C has distance e or $e+1$ to the same number of codewords.

Many of the best known bounds for $A(n,d)$ known at present are based on a method that was developed by Delsarte (1972). It is known as the *linear programming bound.* Again, we restrict ourselves to $q=2$. We assume that n is fixed. (Also see chapter 15.)

Definition 3.13. The *Krawtchouk polynomial* $K_k(x)$ is defined by:

$$K_k(x)=\sum_{j=0}^{k}(-1)^{j}\binom{x}{j}\binom{n-x}{k-j},$$

where

$$\binom{x}{j}=\frac{x(x-1)\cdots(x-j+1)}{j!},\quad x\in\mathbb{R}.$$

Lemma 3.14. *If* $x\in\mathbb{F}_2^n$ *has weight* i, *then*

$$\sum_{\substack{y\in\mathbb{F}_2^n\\ \mathrm{wt}(y)=k}}(-1)^{\langle x,y\rangle}=K_k(i).$$

Proof. Trivial by counting. □

Lemma 3.15. *If* C *is a binary code of length* n *with distance distribution* $(A_0,A_1,\ldots,A_n)$, *then*

$$\sum_{i=0}^{n}A_iK_k(i)\geqslant 0,\quad k\in\{0,1,\ldots,n\}.$$

Proof. Let $|C| = M$. Then

$$M \sum_{i=0}^{n} A_i K_k(i) = \sum_{i=0}^{n} \sum_{\substack{(x,y)\in C^2 \\ d(x,y)=i}} \sum_{\substack{z\in \mathbb{F}_2^n \\ \mathrm{wt}(z)=k}} (-1)^{\langle z,x-y\rangle}$$

$$= \sum_{\substack{z\in \mathbb{F}_2^n \\ \mathrm{wt}(z)=k}} \Big(\sum_{x\in C} (-1)^{\langle z,x\rangle} \Big)^2 \geq 0 . \qquad \square$$

From Lemma 3.15 and the fact that $M = \sum_{i=0}^{n} A_i$, we see that we can bound $A(n, d)$ as follows.

Theorem 3.16.

$$A(n,d) \leq \max\Big\{ \sum_{i=0}^{n} A_i \,\Big|\, A_0 = 1, A_1 = A_2 = \cdots = A_{d-1} = 0, A_i \geq 0 \ (d \leq i \leq n), \sum_{i=0}^{n} A_i K_k(i) \geq 0 \textit{ for } k \in \{0, 1, \ldots, n\} \Big\} .$$

4. Cyclic codes

If C is a code of length n, then the subgroup of S_n consisting of those permutations of the n positions that map all codewords into codewords is called the *automorphism group* $\mathrm{Aut}(C)$ of the code. Many of the most interesting codes have the cyclic group of order n as subgroup of $\mathrm{Aut}(C)$.

Definition 4.1. A linear code C is called *cyclic* if a cyclic shift of a codeword is always again a codeword, i.e.,

$$\forall_{(c_0, c_1, \ldots, c_{n-1})\in C}[(c_{n-1}, c_0, \ldots, c_{n-2}) \in C] .$$

In the theory of cyclic codes, one usually makes the convention $(n, q) = 1$. In the following, this is assumed.

In order to introduce more algebraic tools, we use the following isomorphism between $\mathbb{F}_q^n$ and the additive group of the ring $\mathbb{F}_q[x]/(x^n - 1)$:

$$(a_0, a_1, \ldots, a_{n-1}) \rightleftarrows a_0 + a_1 x + \cdots + a_{n-1}x^{n-1} .$$

From Definition 4.1 it immediately follows that a cyclic code corresponds to an ideal in $\mathbb{F}_q[x]/(x^n - 1)$. Since this is a principal ideal ring, a cyclic code is a principal ideal generated by a polynomial $g(x)$ that divides $x^n - 1$. This is called the *generator* polynomial. (We observe that replacing $x^n - 1$ by $x^n + 1$ leads to a similar theory. The corresponding codes are called *negacyclic*.)

In the following, $x^n - 1 = f_1(x)f_2(x) \cdots f_r(x)$ will be the factorization of $x^n - 1$ into its irreducible factors [no repeated factors because $(n, q) = 1$]. So, there are 2^r possible cyclic codes of length n (where some may be equivalent). Clearly the dimension of the cyclic code C generated by $g(x)$ is n-degree $g(x)$. If $x^n - 1 = g(x)h(x)$, then $h(x)$ is called the *check* polynomial of C. If

$$g(x) = g_0 + g_1 x + \cdots + g_{n-k} x^{n-k}$$

and

$$h(x) = h_0 + h_1 x + \cdots + h_k x^k ,$$

then

$$G = \begin{pmatrix} g_0 & g_1 & \cdots & g_{n-k} & 0 & 0 & \cdots & 0 \\ 0 & g_0 & \cdots & & g_{n-k} & 0 & \cdots & 0 \\ \vdots & & \ddots & & & \ddots & & \vdots \\ 0 & & \cdots & g_0 & & \cdots & & g_{n-k} \end{pmatrix}$$

and

$$H = \begin{bmatrix} 0 & \cdots & & 0 & h_k & \cdots & h_1 & h_0 \\ 0 & \cdots & 0 & h_k & & \cdots & h_0 & 0 \\ \vdots & & & & & & & \vdots \\ & & \cdot^{\cdot^{\cdot}} & & & \cdot^{\cdot^{\cdot}} & & \\ 0 & & & & & & & \\ h_k & & \cdots & h_0 & 0 & & \cdots & 0 \end{bmatrix}$$

are a generator matrix, respectively parity-check matrix for C. Note that the code generated by $h(x)$ is equivalent to the dual of C (symbols in reversed order).

Consider the special case of a cyclic code generated by one of the irreducible factors $f_i(x)$. This code, denoted by M_i^+, is called a *maximal* cyclic code. The code generated by $(x^n - 1)/f_i(x)$, denoted by M_i^-, is a *minimal* cyclic code, also called *irreducible* cyclic code. Clearly an irreducible cyclic code has no zero divisors and is therefore a field (isomorphic to $\mathbb{F}_{q^k}$, where $k = \deg f_i(x) = \dim M_i^-$). Suppose that $n = 2^k - 1$, $q = 2$. Then M_i^- consists of $\mathbf{0}$ and the n cyclic shifts of the word $g(x) = (x^n - 1)/f_i(x)$. Therefore, all the words of M_i^- except $\mathbf{0}$ have the same weight. This weight must be 2^{k-1} (because the sum of all the weights is $n \cdot 2^{k-1}$). This is an example of a so-called *equidistant code* (any two distinct words have the same distance).

Note that any cyclic code C is the direct sum of the irreducible cyclic codes M_i^- that are subcodes of C. This is an example of orthogonal decomposition (of an algebra) since the product of a word from M_i^- and a word from M_j^- $(i \neq j)$ is zero.

It is often useful to have the following alternative description of an irreducible cyclic code.

Theorem 4.2. *Let k be the multiplicative order of p mod n, $q = p^k$, and let β be a primitive nth root of unity in $\mathbb{F}_q$. Then the set*

$$V = \{\mathbf{c}(\xi) = (\mathrm{Tr}(\xi), \mathrm{Tr}(\xi\beta), \ldots, \mathrm{Tr}(\xi\beta^{n-1})) \mid \xi \in \mathbb{F}_q\}$$

is an $[n, k]$ *irreducible cyclic code over* $\mathbb{F}_p$. *Here*

$$\mathrm{Tr}(\alpha) = \alpha + \alpha^p + \alpha^{p^2} + \cdots + \alpha^{p^{k-1}}$$

is the trace mapping from $\mathbb{F}_q$ *to* $\mathbb{F}_p$.

Proof. Since Tr is a linear mapping, V is linear. Since $\boldsymbol{c}(\xi\beta^{-1})$ is a cyclic shift of $\boldsymbol{c}(\xi)$ the code is cyclic. Since β is in no subfield of $\mathbb{F}_q$, we know that β is a zero of an irreducible cyclic polynomial $h(x) = h_0 + h_1 x + \cdots + h_k x^k$ of degree k. If $\boldsymbol{c}(\xi) = (c_0, c_1, \ldots, c_{n-1})$, then

$$\sum_{i=0}^{k} c_i h_i = \mathrm{Tr}(\xi h(\beta)) = \mathrm{Tr}(0) = 0 .$$

It follows that $x^k h(x^{-1})$ is the check polynomial for V, i.e., V is an irreducible cyclic code. □

Now, let β_i be a zero of $f_i(x)$ in some extension field of $\mathbb{F}_q$. The polynomial $f_i(x)$ is determined by one of its zeros. Let R be a set that contains at least one zero of each of the factors $f_i(x)$ of the generator of a cyclic code C. Then

$$C = \{c(x) \in \mathbb{F}_q[x]/(x^n - 1) \mid \forall_{\xi \in R}[c(\xi) = 0]\} .$$

We say that C is defined by the zero-set R. R is complete if it consists of all the zeros of $g(x)$. For the code M_i^+ we can take $R = \{\beta_i\}$.

Example 4.3. Let $n = (q^m - 1)/(q - 1)$ and let β be a primitive nth root of unity in $\mathbb{F}_{q^m}$. Furthermore, let $(m, q - 1) = 1$. The cyclic code

$$C = \{c(x) \mid c(\beta) = 0\}$$

is an $[n, n - m]$ Hamming code over $\mathbb{F}_q$.

In order to see this, consider a matrix H for which the columns are the representation of $1, \beta, \beta^2, \ldots, \beta^{n-1}$ as vectors in $\mathbb{F}_q^m$. By definition, every codeword $(c_0, c_1, \ldots, c_{n-1})$ has inner product 0 with every row of H. So H is a parity-check matrix for C. Since $(n, q - 1) = (m, q - 1) = 1$, the columns of H represent all the points of $\mathrm{PG}(m - 1, q)$. Therefore, C is a Hamming code.

Note that if R contains more than one element, the same idea can be used to construct a "parity-check" matrix for the cyclic code (but it may have more rows than necessary).

Every cyclic code has an *idempotent* element that generates the code. This is a consequence of the following theorem.

Theorem 4.4. *Let* C *be a cyclic code. Then there is a unique codeword that is an identity element for* C.

Proof. Let $x^n - 1 = g(x)h(x)$, where $g(x)$ generates C. Since $(g(x), h(x)) = 1$, there are polynomials $a(x)$ and $b(x)$ such that

$$a(x)g(x) + b(x)h(x) = 1 \quad \text{in } \mathbb{F}_q[x].$$

Let $c(x) = a(x)g(x) = 1 - b(x)h(x)$. Clearly, $c(x)$ is a codeword and since any codeword is a multiple of $g(x)$, we see that $c(x)$ is a unit for C (and therefore unique). □

Definition 4.5. The idempotent of an irreducible cyclic code M_i^- is called a *primitive* idempotent and denoted by $\theta_i(x)$.

Now, suppose $q = 2$. Let α be a primitive nth root of unity in some extension field of $\mathbb{F}_q$. If α^a is a zero of $f_i(x)$, then the zeros of $f_i(x)$ are the elements α^b, where b runs through the *cyclotomic coset* $\{a, 2a, 4a, \ldots, 2^{s-1}a\}$ and s is the minimal exponent such that $(2^s - 1)a \equiv 0 \pmod{n}$. If $c(x)$ is an idempotent of some cyclic code, then $c(\alpha^j) = 0$ or 1. If $c(x)$ is a primitive idempotent, then there is an irreducible factor $f_i(x)$ of $x^n - 1$ such that $c(\alpha^j) = 1$ if and only if $f_i(\alpha^j) = 0$. Therefore, the nonzeros of $c(x)$ are α^b, where b runs through some cyclotomic coset. This information along with the following theorem gives us an easy way to construct and describe all cyclic codes.

Theorem 4.6. *If C_1 and C_2 are binary cyclic codes with idempotents $c_1(x)$ and $c_2(x)$, then $C_1 \cap C_2$ has idempotent $c_1(x)c_2(x)$ and $C_1 + C_2$ has idempotent $c_1(x) + c_2(x) + c_1(x)c_2(x)$. For the primitive idempotents we have $\theta_i(x)\theta_j(x) = 0$ if $i \neq j$.*

The proof is an easy exercise.

We introduce one more tool that is often useful when treating cyclic codes. It is a discrete analogue of the Fourier transform. In coding theory it is usually referred to as the *Mattson–Solomon polynomial*. If $\mathbb{F}_q$ is our alphabet, let β be a primitive nth root of unity in the extension field $\boldsymbol{F}$ of $\mathbb{F}_q$. Let T be the set of polynomials over $\boldsymbol{F}$ of degree at most $n - 1$. We define a transform $\Phi : T \to T$ as follows. Let $a(x) \in T$. Then $A(X) = (\Phi a)(X)$ is defined by:

$$A(X) = \sum_{j=1}^{n} a(\beta^j)X^{n-j}. \tag{4.7}$$

If $\boldsymbol{a} = (a_0, a_1, \ldots, a_{n-1})$ is a codeword in a cyclic code, we identify it with $a(x) = a_0 + a_1x + \cdots + a_{n-1}x^{n-1}$ as usual, and call $A(X)$ the Mattson–Solomon polynomial of $\boldsymbol{a}$.

Theorem 4.8. *The inverse of Φ is given by*

$$a(x) = n^{-1}(\Phi A)(x^{-1}) \pmod{x^n - 1}.$$

Proof.

$$A(\beta^k) = \sum_{j=1}^{n} \sum_{i=0}^{n-1} a_i \beta^{ij} \cdot \beta^{-kj} = \sum_{i=0}^{n-1} a_i \sum_{j=1}^{n} \beta^{(i-k)j} = na_k . \qquad \square$$

5. Important classes of linear codes

We treat several classes of linear codes. The reader who wishes to compare these codes should consult table 1 at the end of this section.

5.1. Lexicodes

We define a *lexicographically least binary code* (lexicode) with distance d as follows. The word length depends on the number of codewords and is chosen such that there is no position where all codewords have a 0. Start with $\boldsymbol{c}_0 = \boldsymbol{0}$ and $\boldsymbol{c}_1 = (1, 1, \ldots, 1, 0, 0, \ldots, 0)$ of weight d. If $\boldsymbol{c}_0, \boldsymbol{c}_1, \ldots, \boldsymbol{c}_{l-1}$ have been chosen as codewords, then $\boldsymbol{c}_l$ is defined to be the first word in the lexicographic ordering with 1's as far to the left as possible, such that $d(\boldsymbol{c}_i, \boldsymbol{c}_l) \geq d$ for $0 \leq i \leq l-1$. (At each step the length may have to be adapted.) To show that one obtains linear codes in this way is somewhat tricky (cf. Van Lint 1982). The idea is to prove the following assertion by induction on k: "After 2^k codewords have been chosen, a lexicode is linear and during the process of choosing codewords the word length increased exactly after 2^i steps for $i < k$."

Consider the special case $d = 3$. Let C_k be the linear code that is obtained after 2^k codewords have been chosen and let n_k be the length. If C_k is not a perfect code, then there is a word $\boldsymbol{x}$ of length n_k that has distance 2 to C_k. Therefore, $(\boldsymbol{x}, 1)$ is a possible choice for $\boldsymbol{c}_{2^k}$. Therefore, we know that $n_{k+1} = n_k + 1$. If, on the other hand, C_k is perfect, we must have $\boldsymbol{c}_{2^k} = (1, 0, 0, \ldots, 0, 1, 1)$ and $n_{k+1} = n_k + 2$. It is now possible to prove by induction on a: $n_k = 2^a + i$ for $k = i + (2^a - a - 1)$ and $1 \leq i < 2^a$. This shows that the Hamming codes are lexicodes with $d = 3$ (take $i = 2^a - 1$).

For applications to combinatorial games see chapter 43.

5.2. Hamming codes

These codes, introduced in Definition 2.1, turned out to be cyclic (in many cases) in Example 4.3 and appeared as lexicodes above. Since the codes are perfect, they are also the optimal solutions to the following combinatorial *covering* problem. Find a set C of minimal cardinality in $\mathbb{F}_q^n$ such that every word $\boldsymbol{x}$ has distance $\leq e$ to some element of C. A well-known example is the so-called *football-pool problem*. Here, one has to forecast the outcome (win, lose, draw) of n football matches. The question is: how many forecasts of the n results are necessary to guarantee winning at least the second prize (one result may be incorrect). This is the covering problem mentioned above with $q = 3$ and $e = 1$. If $n = \frac{1}{2}(3^l - 1)$ a Hamming code provides an optimal solution. For more about football-pool problems see Blokhuis and Lam (1984), and Wille (1987).

5.3. BCH codes

One of the most widely studied classes of cyclic codes is the class of so-called BCH codes. The codes were introduced by Bose and Ray-Chaudhuri (1960), and Hocquenghem (1959). Their practical importance is due to the fact that there is an efficient decoding algorithm for these codes (when used on a BSC, see section 1).

Definition 5.1. A cyclic code of length n over $\mathbb{F}_q$ is called a *BCH code* of designed distance δ if, for a suitable choice of a primitive nth root of unity β, the complete zero-set of the code contains $\beta^l, \beta^{l+1}, \ldots, \beta^{l+\delta-2}$ for some l. If $l=1$, the code is called a *narrow-sense* BCH code. If $n = q^m - 1$ (i.e., β is a primitive element of $\mathbb{F}_{q^m}$) the BCH code is called *primitive*.

(Note that in Example 4.3 β^{q-1} is also primitive and the zero-set contains β and β^q, i.e., consecutive powers of β^{q-1}. So, the designed distance is at least 3. In fact 3 is the distance of the code.)

The expression "designed distance" is explained by the following theorem.

Theorem 5.2 (BCH bound). *The minimum distance of a BCH code with designed distance δ is at least δ.*

Proof. Let C be the code of Definition 5.1 and let $l=1$. For any codeword $\boldsymbol{a}$ the Mattson–Solomon polynomial (4.7) has degree $\leqslant n-\delta$, and hence by Theorem 4.8 $a_k = 0$ for a most $n-\delta$ values of k. Since the only zeros of $A(X)$ that play a role in this argument are nth roots of unity, working mod $X^n - 1$ does not alter the argument, i.e., taking $l=1$ was no restriction. □

For generalizations of this theorem, see Van Lint and Wilson (1986). By definition, a BCH code has the cyclic group of order n as subgroup of its automorphism group. We can show a lot more.

Theorem 5.3. *Every extended primitive BCH code of length $n+1 = q^m$ over $\mathbb{F}_q$ has* AGL$(1, q^m)$ *as a group of automorphisms.*

Proof. Let C be the BCH code and $(c_0, c_1, \ldots, c_{n-1})$ a codeword. We identify the position of c_i with the field element α^i in $\mathbb{F}_{q^m}$ and the position of c_n with the zero element of $\mathbb{F}_{q^m}$. Let $P_{u,v}$ be the element of AGL$(1, q^m)$ defined by:

$$P_{u,v}(X) = uX + v\ , \quad u \in \mathbb{F}_{q^m},\ v \in \mathbb{F}_{q^m},\ u \neq 0\ .$$

Suppose $P_{u,v}$ maps $(c_0, c_1, \ldots, c_n)$ into $(c_0', c_1', \ldots, c_n')$. Then for $0 \leqslant k \leqslant d-1$, we have

$$\sum_i c_i'(\alpha^k)^i = \sum_i c_i(u\alpha^i + v)^k = \sum_i c_i \sum_{l=0}^{k} \binom{k}{l} u^l \alpha^{il} v^{k-1}$$
$$= \sum_{l=0}^{k} \binom{k}{l} u^l v^{k-l} \sum_i c_i(\alpha^l)^i .$$

Here the inner sum is 0 for $0 \le l \le d-1$. If $l \ge d$, then $l > k$ and then $\binom{k}{l} = 0$. It follows that the permuted codeword is in $\bar{C}$. □

Corollary 5.4. *The minimum weight of a primitive binary BCH code is odd.*

Proof. Aut($\bar{C}$) is transitive on the positions. Therefore, there are codewords of minimum weight with a 0 in the check position. □

5.4. Reed–Solomon codes

The special case of Definition 5.1 in which $n = q - 1$ is called a Reed–Solomon code (RS code). Usually one takes $l = 1$. By Theorem 5.2 the minimum distance of the code is at least $\delta = n - k + 1$, where k is the dimension of the code. By the Singleton bound, the distance cannot be larger. Therefore, an RS code is MDS. Note that the dual of an RS code is again an RS code. If q is even, then the dual of the RS code with distance $\delta = \frac{1}{2}q$ has the same zeros as the code and furthermore it has 1 as a zero. Therefore, the extension of this code is self-dual. Another interesting property of an RS code is that its extension is again an MDS code as the following theorem states. These codes are widely used (e.g., in the compact disc audio system) because there is a very fast decoding algorithm.

Theorem 5.5. *Let C be an $[n, k, d]$ narrow-sense RS code over $\mathbb{F}_q$. Then $\bar{C}$ is an $[n+1, k, d+1]$ code.*

Proof. By definition, $n = q - 1$ and C is generated by $\prod_{i=1}^{d-1}(x - \alpha^i)$, where α is primitive in $\mathbb{F}_q$; $k = n - d + 1$. If $c(x)$ is a codeword, then the parity-check symbol that is added is $-c(1)$. If this is 0, then $c(x)$ is in fact a multiple of $\prod_{i=0}^{d-1}(x - \alpha^i)$ and therefore has weight at least $d + 1$ by Theorem 5.2. □

The same argument works if we use $-c(\alpha^d)$ as parity-check symbol, and by using both of them we find the following result.

Theorem 5.6. *An $[n, k, d]$ narrow-sense RS code can be extended twice to an $[n+2, k, d+2]$ code.*

Actually, it is possible to find cyclic codes with the same parameters (unless k is even and q is odd). In order to see this, observe that if α is primitive in $\mathbb{F}_{q^2}$, then $\beta = \alpha^{q-1}$ is a primitive $(q+1)$th root of unity and then β^i and $\beta^{q^i} = \beta^{-i}$ are the zeros of a polynomial of degree 2 in $\mathbb{F}_q[x]$. Therefore, it is possible to find a polynomial of degree $q-k+1$ in this way, which has consecutive powers of β as its zeros (unless k is even and q is odd).

Theorem 5.7. *If* $1 \leqslant k \leqslant q+1$, *then there exists a cyclic* $[q+1, k, q-k+2]$ *code* [*if* $q(k-1)$ *is even*].

If q is odd and k is even, then one can replace cyclic by negacyclic in Theorem 5.7.

We consider a special example, namely the $[8, 4, 5]$ extended RS code $\bar{C}$ over $\mathbb{F}_8$. As we saw above, $\bar{C}$ is self-dual. For $\mathbb{F}_8$ we take a primitive element α such that $\alpha^3 + \alpha^2 + 1 = 0$. Then the normal basis $(\beta_1, \beta_2, \beta_3) = (\alpha, \alpha^2, \alpha^4)$ has the property $\mathrm{Tr}(\beta_i \beta_j) = \delta_{ij}$. If we represent each element of $\mathbb{F}_8$ as a triple in $\mathbb{F}_2^3$ using this basis, then $\bar{C}$ is mapped into a $[24, 12]$ binary code G. It is easily seen that the special property of the basis ensures that G is also self-dual. By inspection of the generator of C, one also sees that all basis vectors of G have weight 8. Therefore, all codewords of G have weight divisible by 4 and since $\bar{C}$ has distance 5, it follows that G is a $[24, 12, 8]$ code. Therefore, G is the (unique) extended binary Golay code (see chapter 14). This construction is due to Pasquier (1980).

The original representation (and encoding) of RS codes was as follows. A sequence of k information symbols $a_0, a_1, \ldots, a_{k-1}$ is used to define the polynomial:

$$a(x) = a_0 + a_1 x + \cdots + a_{k-1} x^{k-1} .$$

The corresponding codeword $(c_0, c_1, \ldots, c_{n-1})$ has $c_i = a(\beta^i)$, where β is primitive in $\mathbb{F}_q$ and $n = q-1$. Since the polynomial $a(x)$ has at most $k-1$ zeros, $\boldsymbol{c}$ has weight at least $n-k+1$ (if $\boldsymbol{c} \neq \boldsymbol{0}$). To see that this defines the same code as above we set

$$c(x) = c_0 + c_1 x + \cdots + c_{n-1} x^{n-1}$$

and then the same proof as for Theorem 4.8 shows that $c(\beta^j) = 0$ for $1 \leqslant j < d = n-k+1$. The extended code has the extra symbol $a(0)$.

5.5. Reed–Muller codes

Consider the points of the affine geometry $\mathrm{AG}(m, 2)$ as column vectors in $\mathbb{F}_2^m$. We denote the standard basis by $\boldsymbol{u}_0, \boldsymbol{u}_1, \ldots, \boldsymbol{u}_{m-1}$. If $\sum_{i=0}^{m-1} \xi_{ij} 2^i$ $(0 \leqslant j < 2^m)$ is the binary representation of j, then we define $\boldsymbol{x}_j = \sum_{i=0}^{m-1} \xi_{ij} \boldsymbol{u}_i$. Let $n = 2^m$. Then the m by n matrix E with $\boldsymbol{x}_j$ as jth column represents the points of $\mathrm{AG}(m, 2)$. We define:

$$A_i = \{\boldsymbol{x}_j \in \mathrm{AG}(m, 2) \mid \xi_{ij} = 1\}, \quad \text{a hyperplane}, \tag{5.8}$$

$$\boldsymbol{v}_i = \text{the } i\text{th row of } E, \text{ i.e., the characteristic function of } A_i, \text{ considered as vector in } \mathbb{F}_2^n . \tag{5.9}$$

As usual we write **1** for the characteristic function of the space AG$(m, 2)$.

If $\boldsymbol{a} = (a_0, a_1, \ldots, a_{n-1})$ and $\boldsymbol{b} = (b_0, b_1, \ldots, b_{n-1})$ are words in $\mathbb{F}_2^n$, then we define:

$$\boldsymbol{ab} = (a_0 b_0, a_1 b_1, \ldots, a_{n-1} b_{n-1}) . \tag{5.10}$$

Note that if $i_1, i_2, \ldots, i_s$ are different, then $\boldsymbol{v}_{i_1} \boldsymbol{v}_{i_2} \cdots \boldsymbol{v}_{i_s}$ is the characteristic function of the $(m-s)$-flat $A_{i_1} \cap A_{i_2} \cap \cdots \cap A_{i_s}$. Therefore, the weight of this word is 2^{m-s}. The characteristic function of the set $\{\boldsymbol{x}_j\}$, i.e., the jth basis vector of $\boldsymbol{F}^{(n)}$, is

$$\boldsymbol{e}_j = \prod_{i=0}^{m-1} \{\boldsymbol{v}_i + (1 + \xi_{ij})\mathbf{1}\} ,$$

so the products $\boldsymbol{v}_{i_1} \boldsymbol{v}_{i_2} \cdots \boldsymbol{v}_{i_s}$ $(0 \leqslant s \leqslant m)$ form a basis of $\boldsymbol{F}^{(n)}$.

Definition 5.11. Let $0 \leqslant r < m$. The linear code of length $n = 2^m$ that has the products $\boldsymbol{v}_{i_1} \boldsymbol{v}_{i_2} \cdots \boldsymbol{v}_{i_s}$ with $s \leqslant r$ factors as basis is called the *rth order binary Reed–Muller code* [RM code; notation $\boldsymbol{R}(r, m)$].

The special case $\boldsymbol{R}(0, m)$ is the repetition code.

Theorem 5.12. *$\boldsymbol{R}(r, m)$ has minimum distance 2^{m-r}.*

Proof. From the definition one immediately sees that $\boldsymbol{R}(r+1, m+1)$ is obtained from $\boldsymbol{R}(r+1, m)$ and $\boldsymbol{R}(r, m)$ by the $(\boldsymbol{u}, \boldsymbol{u} + \boldsymbol{v})$ construction of section 2 (consider products without, respectively with, the factor $\boldsymbol{v}_{m-1}$). □

Theorem 5.13. *The dual of $\boldsymbol{R}(r, m)$ is $\boldsymbol{R}(m - r - 1, m)$.*

Proof. The dimensions of the two codes have sum 2^m. If $\boldsymbol{a}$ is a basis vector of $\boldsymbol{R}(r, m)$ and $\boldsymbol{b}$ a basis vector of $\boldsymbol{R}(m - r - 1, m)$, then by Definition 5.11 $\boldsymbol{ab}$ is the product of at most $m - 1$ factors $\boldsymbol{v}_i$ and hence the characteristic function of a flat of dimension at least 1, i.e., a set with an even number of points. Therefore, $\langle \boldsymbol{a}, \boldsymbol{b} \rangle = 0$. □

Corollary 5.14. *$\boldsymbol{R}(m-2, m)$ is the $[n, n-m-1]$ extended Hamming code.*

Our definition allows us to consider elements of AGL$(m, 2)$ as permutations of the positions of codewords in Reed–Muller codes.

Theorem 5.15. AGL$(m, 2) \subset$ Aut$(\boldsymbol{R}(r, m))$.

Proof. Let $\boldsymbol{a}$ be the characteristic function of an l-flat A in AG$(m, 2)$ and let $\boldsymbol{v}$ be a basis vector of the code $\boldsymbol{R}(l-1, m)$, i.e., $\boldsymbol{v}$ is the characteristic function of a flat V with dimension $\geqslant m - l + 1$. Therefore, $V \cap A$ is either empty or it has dimension at least 1. In both cases $|V \cap A|$ is even and therefore $\langle \boldsymbol{v}, \boldsymbol{a} \rangle = 0$. By Theorem 5.13 $\boldsymbol{a}$ is in $\boldsymbol{R}(m-l, m)$. So, a word is in $\boldsymbol{R}(r, m)$ if and only if it is a sum of characteristic functions of affine subspaces of dimension $\geqslant m-r$. Since $\mathrm{AGL}(m, 2)$ maps k-flats into k-flats (for every k) the proof is complete. $\square$

We now reorder the elements of $\mathrm{AG}(m, 2)$ as follows. Let α be a primitive element of $\mathbb{F}_{2^m}$. We form the matrix E^* with as columns the representations of $1, \alpha, \alpha^2, \ldots, \alpha^{n-1}$ as elements of $\mathbb{F}_2^m$. Finally, we add the $\mathbf{0}$ column to obtain a permutation of E. A cyclic shift of the columns of E^* corresponds to multiplication of elements of $\mathbb{F}_{2^m}$ by α. Since this is a linear mapping, our geometric description of Reed–Muller codes shows that, with this ordering of the coordinates, they are extensions of cyclic codes.

Theorem 5.16. *The codes $\boldsymbol{R}(r, m)$ are equivalent to extended cyclic codes.*

For more information about the connection between cyclic codes and finite geometries see Berlekamp (1984), MacWilliams and Sloane (1977), and Van Lint (1982).

Remark. Consider the Hamamard matrix H of order $n = 2^m$ obtained from $\begin{pmatrix} 1 & 1 \\ 1 & -1 \end{pmatrix}$ by repeatedly applying the Kronecker product (see chapter 14). In the rows of H and $-H$ we replace $+1$ by 0 and -1 by 1. We thus obtain the words of $\boldsymbol{R}(1, m)$.

5.6. Quadratic residue codes

In this subsection we restrict ourselves to binary codes. Much of the following is also valid over other alphabets (with slight modifications).

Let n be a prime $\equiv \pm 1 \pmod 8$. This ensures that 2 is a quadratic residue $\bmod\, n$. Let R_0 and R_1 denote the quadratic residues $\bmod\, n$ and the nonresidues, respectively. We define $\theta(x) = \sum_{r \in R_0} x^r$. If α is a primitive nth root of unity in an extension field of $\mathbb{F}_2$ and $i \in R_0$, then $\theta(\alpha^i)^2 = \theta(\alpha^i)$, i.e., $\theta(\alpha^i) = 0$ or 1. In the same way we see that $\theta(\alpha^j) + \theta(\alpha) = 1$ if $j \in R_1$. We choose α in such a way that $\theta(\alpha) = 0$. Since $2 \in R_0$ the polynomials $g_0(x) = \prod_{r \in R_0} (x - \alpha^r)$ and $g_1(x) = \prod_{r \in R_1} (x - \alpha^r)$ are in $\mathbb{F}_2[x]$ and $x^n - 1 = (x-1) g_0(x) g_1(x)$.

Definition 5.17. The binary cyclic codes of length n with generator $g_0(x)$, respectively $(x-1)g_0(x)$, are both called *quadratic residue codes* (QR codes). The extended QR code is obtained by adding a parity check to the code with generator $g_0(x)$.

Since $\theta(x)$ is an idempotent polynomial with the same zeros as $g_0(x)$ if $n \equiv -1 \pmod 8$, respectively as $(x-1)g_0(x)$ if $n \equiv 1 \pmod 8$, it is the idempotent of the corresponding code. We now form a matrix G as follows: take all cyclic shifts of the vector θ, add a column of zeros, respectively ones, if $n \equiv 1 \pmod 8$, respectively $n \equiv -1 \pmod 8$, and finally add a row of ones. From the facts above, it follows that the rows of G span the extended QR code of length $n+1$. Replacing 0 by -1 yields a Hadamard matrix of Paley type (see chapter 14) for which it is known that it is invariant under the permutations of PSL$(2, n)$. This proves the following theorem for which a direct proof can be found in MacWilliams and Sloane (1977) or Van Lint (1971).

Theorem 5.18. *The automorphism group of the extended binary QR code contains* PSL$(2, n)$.

Note that since PSL$(2, n)$ is transitive, the minimum weight of a binary QR code with generator $g_0(x)$ is odd.

Theorem 5.19. *The binary code with generator $g_0(x)$ has minimum distance d satisfying $d^2 > n$.*

Proof. Let $c(x)$ be a codeword of weight d. Replacing x by x^j with $j \in R_1$ yields a word $\hat{c}(x)$ of the same weight in the code with generator $g_1(x)$. Therefore, $c(x)\hat{c}(x)$ is a multiple of $g_0(x)g_1(x)$ but not of $(x-1)$ because d is odd. This implies $c(x)\hat{c}(x) = 1 + x + \cdots + x^{n-1}$. Since the product contains at most d^2 terms, the proof is complete. □

Remark. In the case $n \equiv -1 \pmod 8$ we can take $j = -1$ in this proof. Then in the product $c(x)\hat{c}(x)$ terms cancel four at a time and there are d terms equal to 1. So $n = d^2 - d + 1 - 4a$ for some a.

Example. Take $n = 23$. We find a $[23, 12]$ code with $d \geqslant 7$. By the Hamming bound, d cannot be larger than 7 so $d = 7$. This is the *binary Golay code* (see chapter 14).

5.7. *Symmetry codes over* $\mathbb{F}_3$

The following class of ternary codes was introduced by Pless (1972). Let C be a symmetric or skew conference matrix (cf. chapter 15) of order $m \equiv 0 \pmod 6$.

Definition 5.20. A *symmetry code* Sym$(2m)$ is a $[2m, m]$ ternary code with generator matrix $G = (I_m C)$.

(If $m-1$ is a power of a prime, then we take a Paley matrix for C; see chapter 14.) Since $CC^{\mathrm{T}} = -I_m$ (over $\mathbb{F}_3$) the code Sym$(2m)$ is self-dual.

Example. If we take $m = 6$ we find a ternary $[12, 6]$ code. Since it is self-dual it has minimum distance divisible by 3. A trivial analysis of G shows that $d = 3$ is impossible. So Sym(12) is a $[12, 6, 6]$ code. Puncturing yields a perfect code known as the *ternary Golay code* (see chapter 14). For small values of m it is

Table 1

Name	q	Length n	Dimension k	Distance d	Remarks
Hamming	q	$(q^l-1)/(q-1)$	$n-l$	3	Perfect
Primitive BCH	q	q^m-1	$\geqslant n-m\delta$	$\delta\leqslant d\leqslant 2\delta-1$	
Reed–Solomon	q	$q-1$	k	$n-k+1$	MDS
Reed–Muller	2	2^m	$\sum_{s=0}^{r}\binom{m}{s}$	2^{m-r}	
Quadratic residue	2	prime $\equiv\pm1$ (mod 8)	$\frac{1}{2}(n+1)$	$d^2>n$	Includes binary Golay
Symmetry	3	$2m\equiv 0$ (mod 12)	m	often large	Includes ternary Golay

relatively easy to find the minimum distance of Sym($2m$) by analysing G, e.g., $d=12$ for $m=18$.

Table 1 shows the information on the parameters of the codes treated in this section.

6. Some nonlinear codes

In this section we mention a few classes of nonlinear codes that are combinatorically interesting.

6.1. Hadamard codes

Let H_n be a Hadamard matrix of order n (cf. chapter 14). In H_n and in $-H_n$ we replace -1 by 0. In this way we find $2n$ rows which we consider as words in $\mathbb{F}_2^n$. Since any two rows differ in $\frac{1}{2}n$ positions or in all n positions, these words form an $(n, 2n, \frac{1}{2}n)$ binary code. From the Plotkin bound 3.6 one easily finds that $A(4m, 2m)\leqslant 8m$. Therefore, every Hadamard matrix of order $4m$ thus yields an optimal code of length $4m$. These codes are known as Hadamard codes. The same name is used for the $(n-1, 2n, \frac{1}{2}n-1)$ punctured codes and the $(n-1, n, \frac{1}{2}n)$ shortened codes.

6.2. Conference matrix codes

Let C be a symmetric conference matrix of order n ($n\equiv 2 \pmod 4$), cf. chapter 15). We assume C is normalized. By deleting the first row and column of C we obtain the matrix S (e.g., a Paley matrix). We now form a nonlinear code by taking as codewords the rows of $\frac{1}{2}(S+I+J)$ and $\frac{1}{2}(-S+I+J)$. In this way we obtain an $(n-1, 2n, \frac{1}{2}(n-2))$ code.

Example. Starting from the Paley matrix of order 9, we find a binary $(9, 20, 4)$ code. Note that an interesting modification of the method of Theorem 3.16 shows that this code is optimal. Using Theorem 3.16 for $n=8$, $d=3$, in combination with the obvious inequalities $A_8 \leq 1$, $A_6 + 4A_8 \leq 12$, yields $A(8,3) \leq 21$. But if M is odd, then in the proof of Lemma 3.15 the final inequality can be strengthened and then it follows that $M \leq 20$.

6.3. *Kerdock codes*

Consider the binary Reed–Muller codes of Definition 5.11 for $r = 1, 2$. If $L(\boldsymbol{x})$ is the form $\sum_{i=1}^{m} a_i x_i + \varepsilon$ (where $\varepsilon = 0$ or 1), then the codeword $\sum_{i=1}^{n} a_i \boldsymbol{v}_i + \varepsilon\mathbf{1}$ in $\boldsymbol{R}(1,m)$ is the characteristic function of the set $\{\boldsymbol{x} \in \mathrm{AG}(m,2) \mid L(\boldsymbol{x}) = 1\}$. If we replace $L(\boldsymbol{x})$ by $Q(\boldsymbol{x}) + L(\boldsymbol{x})$, where $Q(\boldsymbol{x}) = \sum_{1 \leq i < j \leq m} q_{ij} x_i x_j$ is a quadratic form, then we obtain a word in $\boldsymbol{R}(2,m)$. Clearly, $\boldsymbol{R}(2,m)$ is a union of cosets of $\boldsymbol{R}(1,m)$, where the coset $Q + \boldsymbol{R}(1,m)$ is obtained by fixing Q and letting L run over all forms of degree 1. If m is even and Q is nonsingular, then an affine transformation of coordinates in $\mathrm{AG}(m,2)$ puts Q into the standard form $x_1x_2 + x_3x_4 + \cdots + x_{m-1}x_m$. Then it is easy to show that the distance of a word in $\boldsymbol{R}(1,m)$ and a word in $Q + \boldsymbol{R}(1,m)$ is at least $2^{m-1} - 2^{m/2-1}$. To construct a large code that is a union of cosets of $\boldsymbol{R}(1,m)$ and that has this minimum distance, one needs a set of quadratic forms Q_i such that $Q_i - Q_j$ is nonsingular if $i \neq j$. Clearly, such a set cannot have more than 2^{m-1} elements [=the number of choices for $(q_{12}, q_{13}, \ldots, q_{1m})$].

Definition 6.1. A *Kerdock set* is a set of 2^{m-1} quadratic forms in $x_1, x_2, \ldots, x_m$ such that the difference of any two distinct elements is a nonsingular quadratic form.

Such sets were first constructed by Kerdock (1972). A construction of Kerdock sets depending on a strong connection between such sets, spreads on quadrics (orthogonal spreads) and translation planes (cf. chapter 13) was given by Kantor (1982). Also see Van Lint (1983) and MacWilliams and Sloane (1977).

Definition 6.2. A *Kerdock code* is a $(2^m, 2^{2m}, 2^{m-1} - 2^{m/2-1})$ nonlinear code that is the union of 2^{m-1} cosets $Q_i + \boldsymbol{R}(1,m)$ of the first-order Reed–Muller code, where Q_i runs through the quadratic forms of a Kerdock set.

6.4. *The Nordstrom–Robinson code*

Consider a representation of the [24, 12, 8] extended Golay code for which the word with eight ones followed by sixteen zeros is a codeword. There are 32 codewords with $x_1 = x_2 = \cdots = x_8 = 0$, and similarly there are 32 codewords with $x_i = x_8 = 1$, $x_j = 0$ for $1 \leq j \leq 7$ and $j \neq i$ $(1 \leq i \leq 7)$. Consider the union of these sets and subsequently delete the first eight positions. Clearly what remains is a $(16, 256, 6)$ code. This nonlinear code is known as the Nordstrom–Robinson code. It can be shown that this code is unique (Snover 1973). Therefore, this code is the unique example of Definition 6.2 with $m = 4$.

6.5. Preparata codes

Let m be odd ($m \geqslant 3$), $n = 2^m - 1$. We describe a code $\bar{\boldsymbol{P}}$ of length $2^{m+1} = 2n + 2$ as follows. Number the elements of $\mathbb{F}_{2^m}$ in some fixed order. If X and Y are subsets of $\mathbb{F}_{2^m}$, we use the notation (X, Y) for the word in $\boldsymbol{F}^{(2n+2)}$ that is the concatenation of the characteristic functions of the sets X and Y.

Definition 6.3. The extended *Preparata code* $\bar{\boldsymbol{P}}$ of length 2^{m+1} consists of all words (X, Y) for which

(i) $|X|$ is even, $|Y|$ is even,
(ii) $\sum_{x \in X} x = \sum_{y \in Y} y$,
(iii) $\sum_{x \in X} x^3 + (\sum_{x \in X} x)^3 = \sum_{y \in Y} y^3$.

Clearly, a set X satisfying (i) can be chosen in 2^n ways. Let M be a $2m$ by n $(0, 1)$-matrix in which the ith column is

$$\begin{pmatrix} a_i \\ a_i^3 \end{pmatrix},$$

where these elements are represented as column vectors in $(\mathbb{F}_2)^m$ and a_i runs through the nonzero elements of $\mathbb{F}_{2^m}$. If X is given, then $Y \backslash \{0\}$ can be found from (ii) and (iii) by solving a system of m linear equations of the form $M\,\boldsymbol{y} = \boldsymbol{c}$ [where $\boldsymbol{c}$ is found form the left-hand sides of (ii) and (iii)]. These have full rank. Therefore, there are 2^{n-2m} solutions for Y if X is given. Therefore, $\bar{\boldsymbol{P}}$ has 2^{2n-2m} codewords. If (X_1, Y_1) and (X_2, Y_2) are two codewords with $X_1 = X_2$, then their distance is at least 6 by the BCH bound 5.2. The assumption $|X_1 \Delta X_2| = |Y_1 \Delta Y_2| = 2$ easily leads to an equation of degree three that has no solutions. It follows that $\bar{\boldsymbol{P}}$ has distance 6. So, the Preparata code $\boldsymbol{P}$ is a $(2^{m+1} - 1, 2^k, 5)$ code with $k = 2^{m+1} - 2m - 2$.

Remark. The distance distribution of the code $\boldsymbol{P}$ of length 2^{m+1} is the MacWilliams transform [cf. (2.6)] of the distance distribution of the Kerdock code of the same length. Since the codes are not linear, this seems to be a coincidence.

An explanation of this fact was given in 1992 by A.R. Calderbank, R. Hammons, P. Vijay Kumar, N.J.A. Sloane, and P. Solé (see Calderbank et al. 1993, and Hammons et al. 1994).

They consider group codes over $\mathbb{Z}_4$. For these codes duality is defined in the usual way. Next, a so-called *symmetrized weight enumerator* is introduced which keeps track of the number of coordinates 0, 2, and 1 or 3 (the last two considered equivalent). They show that for a code C and its dual $C^{\perp}$ a generalization of Theorem 2.4 holds. Codes of length n over $\mathbb{Z}_4$ are mapped to binary codes of length $2n$ by the mapping ϕ defined by

$$\phi(0) = 00\,, \qquad \phi(1) = 01\,, \qquad \phi(2) = 11\,, \qquad \phi(3) = 10\,.$$

(Note that 1 and 3 lead to the same weight, namely 1).

Finally, certain extended cyclic codes and their duals (all over $\mathbb{Z}_4$) are introduced. It turns out that the mapping ϕ maps these codes into Kerdock codes and Preparata codes. A corresponding mapping sends the symmetrized weight enumerators of the codes over $\mathbb{Z}_4$ to the weight enumerators of the corresponding binary codes and one finds that indeed, the MacWilliams transform of the weight enumerator of a Kerdock code is the weight enumerator of the Preparata code of the same length.

7. Codes and designs

Many combinatorial designs (e.g., t-designs) have a strong connection to coding theory, in fact several of these designs were found by using appropriate codes. Quite an interesting theory of these connections has been developed, much of it due to Delsarte (1973b). We treat part of this theory in this section.

Let C be any binary code of length n with distance enumerator $A(x, y)$. If

$$A'(x, y) = \sum_{i=0}^{n} A'_i x^{n-i} y^i$$

is the MacWilliams transform of $A(x, y)$ [cf. (2.6)], then

$$A'_k = \frac{1}{|C|} \sum_{i=0}^{n} A_i \sum_{j=0}^{k} (-1)^j \binom{i}{j} \binom{n-i}{k-j} = \frac{1}{|C|} \sum_{i=0}^{n} A_i K_k(i) \tag{7.1}$$

(cf. Definition 3.13). So, by Lemma 3.15 the numbers A'_k are nonnegative. Clearly, $A'_0 = 1$.

Definition 7.2. The *dual distance* d' of a code C is defined by:

$$d' := \min\{i \geqslant 1 \mid A'_i \neq 0\}\ .$$

Note that if C is linear, then d' is the minimum distance of $C^{\perp}$. We can now show a relation between binary codes and *orthogonal arrays* (cf. chapter 14).

Theorem 7.3. *Let C be a binary code of length n with $|C| = M$. Let $[C]$ be the M by n array with the codewords of C as rows. Then if $r < d'$, any set of r columns of $[C]$ contains each r-tuple exactly $M/2^r$ times.*

Proof. Since $A'_k = 0$ for $1 \leqslant k < d'$, we have from (7.1)

$$\sum_{x \in C} (-1)^{\langle x,z \rangle} = 0 \quad \text{for every } z \in \mathbb{F}_2^n \text{ with } \mathrm{wt}(z) = k\ .$$

Taking $\mathrm{wt}(z) = 1$, we see that every column of $[C]$ must have $M/2$ ones and $M/2$ zeros. Then taking $\mathrm{wt}(z) = 2$, we conclude that every pair of columns must contain each of the four possible pairs exactly $M/4$ times. Proceeding by induction the result follows. □

Therefore, $[C]$ is an orthogonal array of strength $d' - 1$. From the distance distribution $(A_0, A_1, \ldots, A_n)$ of a code C and the MacWilliams transform $(A'_0, A'_1, \ldots, A'_n)$, we find the four so-called *fundamental parameters* of C (cf. Delsarte 1973a): the minimum distance d, the number s of distinct nonzero distances between codewords (i.e., besides A_0, there are s nonzero numbers A_i), the dual distance d', and the number s' of nonzero coefficients A'_i with $1 \leq i \leq n$. The number s' is called the *external distance* of C.

We wish to show some combinatorial significance of the external distance. To do this we introduce the set $N := \{i \mid A'_i \neq 0\}$ and the *characteristic polynomial*

$$F_C(x) := \frac{2^n}{|C|} \prod_{i \in N} \left(1 - \frac{x}{i}\right)$$

of the code. We call

$$F_C = \sum_{k=0}^{s'} \alpha_k K_k(x)$$

the Krawtchouk expansion of $F_C(x)$.

Theorem 7.4. *Let $V = \mathbb{F}_2^n$, $\boldsymbol{x} \in V$. Denote by $B(\boldsymbol{x}, k)$ the number of codewords at distance k from $\boldsymbol{x}$. Then*

$$\sum_{k=0}^{s'} \alpha_k B(\boldsymbol{x}, k) = 1 ,$$

and hence the covering radius of C is at most s'.

Proof. To prove the assertion, we manipulate in the group algebra $\boldsymbol{A} = \boldsymbol{C}V$ with $*$ as multiplication. Any subset S of V is identified with the element $\sum_{s \in S} s$ of $\boldsymbol{A}$. Let Y_k denote the set of words of weight k. The assertion of the theorem then can be stated as

$$C * \sum_{k=0}^{s'} \alpha_k Y_k = V .$$

To prove this assertion, we first observe that V is the only element in $\boldsymbol{A}$ with the property

$$\sum_{\boldsymbol{v} \in V} (-1)^{\langle \boldsymbol{w}, \boldsymbol{v} \rangle} = \begin{cases} 0 & \text{if } w(\boldsymbol{w}) \neq 0 , \\ 2^n & \text{if } \boldsymbol{w} = \boldsymbol{0} . \end{cases}$$

By Lemmas 3.14 and 3.15, and (7.1) we have

$$\sum_{\boldsymbol{c} \in C} \sum_{k=0}^{s'} \alpha_k \sum_{\boldsymbol{u}, w(\boldsymbol{u}) = k} (-1)^{\langle \boldsymbol{w}, \boldsymbol{u} + \boldsymbol{c} \rangle} = \sum_{\boldsymbol{c} \in C} (-1)^{\langle \boldsymbol{w}, \boldsymbol{c} \rangle} \sum_{k=0}^{s'} \alpha_k K_k(w) ,$$

where w is the weight of $\boldsymbol{w}$. The first factor is 0 if $w \notin N$, the second is 0 if $w \in N$

(unless $w=0$, in which case the expression is obviously 2^n). This completes the proof. □

In the following, $A_{i_1}, A_{i_2}, \ldots, A_{i_s}$ are there nonzero coefficients of $A(x, y)$.
An interesting situation arises when $s \leq d'$. In that case consider $A'(x, y)$ and substitute $x=1$. We find

$$1+\sum_{j=1}^{s} A_{i_j} y^{i_j}=\frac{|C|}{2^n} \sum_{i=0}^{n} A'_i(1+y)^{n-i}(1-y)^i . \tag{7.5}$$

Since $A'_i=0$ for $1<i<d'$, the jth derivative of the right-hand side of (7.5) in the point $y=1$ does not depend on the values of the nonzero A'_i for $0 \leq j \leq s-1$. By calculating these derivatives, we find s linearly independent equations for the numbers A_{i_j}. Therefore, these numbers depend only on the parameters n and $|C|$ and the numbers $i_1, i_2, \ldots, i_s$. Now, take any codeword as origin and consider the weight enumerator of C. Clearly, nonzero coefficients can only occur for indices 0 and i_j $(1 \leq j \leq s)$. Also, the MacWilliams transform of this weight enumerator must have zero coefficients for $1 \leq i \leq d'$. Therefore, the same equations that we found for the A_{i_j} apply. We have thus proved the following theorem.

Theorem 7.6. *If $s \leq d'$, then the code is distance invariant.*

By the same argument, it follows that if we know the values A_i for $0 \leq i \leq s'$, then $A(x, y)$ is determined by (7.5) because there are only s' unknowns on the right-hand side.

If C is distance invariant, then $A_n=0$ or $A_n=1$. Let us assume that we know A_n. We shall show that one can obtain t-designs by considering the words of some fixed weight as blocks. We take $t=d'-\bar{s}$, where $\bar{s}=s-A_n$. Consider some fixed t-tuple of positions and denote by λ_{i_j} the number of codewords of weight i_j that have ones in those t positions. Next, take $0 \leq j<\bar{s}$, $r=t+j$. We now count pairs $(\boldsymbol{c}, \boldsymbol{x})$, where $\boldsymbol{x}$ is a word of weight r with ones in the t fixed positions and $\boldsymbol{c}$ is a codeword with ones in all the positions of $\boldsymbol{x}$. From Theorem 7.3 we see that there are

$$\frac{M}{2^r}\binom{n-t}{j}$$

such pairs (where $M=|C|$). On the other hand, the number of pairs is

$$\sum_{j=1}^{s}\binom{i_j-t}{j} \lambda_{i_j} .$$

In this way we find $\bar{s}$ linearly independent equations for the $\bar{s}$ unknowns λ_{i_j}. It follows that these numbers are independent of the choice of the t-tuple. So, we have the following theorem.

Theorem 7.7. *If* $\bar{s} < d'$ *and* $i \geqslant d' - \bar{s}$, *then the words of weight i in the code form a t-design with* $t = d' - \bar{s}$.

An easy example is provided by $\boldsymbol{R}(1, m)$. This yields the expected 3-design. More interesting examples can be found from the Golay code and its derivatives (see chapter 14). From the Nordstrom–Robinson code we find three 3-designs (blocksize 6, 8 and 10).

There are several other theorems similar to Theorem 7.6. For these we refer to MacWilliams and Sloane (1977). We give one more theorem on designs obtained from codewords of fixed weight in a linear code – the celebrated *Assmus–Mattson theorem* (Assmus and Mattson 1969). We define the *support* of a codeword to be the set of positions where the nonzero coordinates occur. For a linear code over $\mathbb{F}_q$, the $q-1$ multiples of a fixed codeword have the same support. In the following, C is an $[n, k]$ code over $\mathbb{F}_q$ with minimum distance d, $C^{\perp}$ is the dual code, d' is the dual distance.

Theorem 7.8. *Let t be a positive integer less than d. If* $q > 2$ *let* v_0 *be the largest integer satisfying*

$$v_0 - \left[\frac{v_0 + q - 2}{q - 1}\right] < d$$

and let w_0 *be the largest integer satisfying*

$$w_0 - \left[\frac{w_0 + q - 2}{q - 1}\right] < d' ,$$

whereas $v_0 = w_0 = n$ *if* $q = 2$. *Suppose the number of nonzero weights of* $C^{\perp}$ *that are less than or equal to* $n - t$ *is* $\leqslant d - t$. *Then, for each weight v with* $d \leqslant v \leqslant v_0$, *the set of supports of codewords of weight v in C form a t-design. Furthermore, for each weight w with* $d' \leqslant w \leqslant \min\{n - t, w_0\}$, *the supports of codewords of weight w in* $C^{\perp}$ *form a t-design.*

Proof. By definition of v_0, two words of C with weight $\leqslant v_0$ that have the same support are scalar multiples of each other. Similarly for $C^{\perp}$. If we delete $d-1$ columns from a generator matrix of C, we still have a matrix of rank k. Consider a t-subset T of the positions. The code A is obtained from C by puncturing on T and B is obtained from $C^{\perp}$ by shortening on T. Then $B = A^{\perp}$. The conditions of the theorem imply that we can solve the MacWilliams equations of Theorem 2.4. This implies that the weight enumerator of B does not depend on the choice of T and the same holds for A. The assertion of the theorem now follows from straightforward counting arguments. □

Examples. (a) The symmetry code Sym($12l$) defined in Definition 5.20 is a ternary $[12l, 6l]$ self-dual code. Therefore, all weights are divisible by 3. If the minimum distance d is larger than $3l$, then we can take $t = 5$ and apply Theorem

7.8. This idea works for $l \leq 5$ and many of the resulting 5-designs were first constructed in this way (cf. Pless 1972).

(b) By the remark following Theorem 5.19, the extended QR code of length 48 has minimum distance 12 (14 is excluded by Theorem 3.4). Since the code is self-dual and the generator matrix has rows of weight 12, all weights are $\equiv 0 \pmod 4$. Again, we can apply Theorem 7.8 with $t = 5$. This yields four different 5-designs and their complements.

8. Perfect codes and uniformly packed codes

Perfect codes and nearly perfect codes were defined in section 2 and Definition 3.12. This concept was generalized by Semakov et al. (1971) and the theory was developed further by Goethals and Van Tilborg (1975) and Van Tilborg (1976). We shall give a brief survey. For details we refer to the papers and to Van Lint (1982), and MacWilliams and Sloane (1977). We restrict ourselves to codes in $\mathbb{F}_q^n$.

Definition 8.1. An e-error-correcting code C is called *uniformly packed* with parameters λ and μ if for any $\boldsymbol{x}$ with distance e to C, there are exactly λ codewords with distance $e+1$ to $\boldsymbol{x}$, and for any $\boldsymbol{x}$ with distance $\geq e+1$ to C there are exactly μ codewords with distance $e+1$ to $\boldsymbol{x}$.

By counting pairs $(\boldsymbol{x}, \boldsymbol{c})$ with $\boldsymbol{c} \in C$ and $d(\boldsymbol{x}, \boldsymbol{c}) = e+1$, we see that $\lambda \leq (n - e)(q-1)/(e+1)$, and equality implies $\mu = 0$ and C is perfect. Nearly perfect codes are binary uniformly packed codes with

$$\mu = \lambda + 1 = \left[\frac{n+1}{e+1}\right].$$

Note that if $e+1$ divides $n+1$, then the code is perfect.

The statement of Definition 8.1 can be translated into the group algebra terminology of the proof of Theorem 7.4. We find that C is uniformly packed with parameters λ and μ if and only if

$$C * \left\{\sum_{i=0}^{e-1} Y_i + \left(1 - \frac{\lambda}{\mu}\right) Y_e + \frac{1}{\mu} Y_{e+1}\right\} = V \tag{8.2}$$

in A. An argument analogous to the proof of Theorem 7.4 (see Van Tilborg 1976) then shows that (8.2) is equivalent to

$$F_C(x) = \sum_{k=0}^{e-1} K_k(x) + \left(1 - \frac{\lambda}{\mu}\right) K_e(x) + \frac{1}{\mu} K_{e+1}(x). \tag{8.3}$$

From (8.3) we see that for a uniformly packed code $s' = e+1$. Now assume that C has external distance $e+1$. Then from Theorem 7.4 we can calculate $B(\boldsymbol{x}, e+1)$ if $\boldsymbol{x}$ has distance e to C or distance $\geq e+1$ to C. Therefore, C is uniformly packed.

Theorem 8.4. *A code C is uniformly packed if and only if its external distance equals* $e+1$ *(where* $d=2e+1$ *or* $2e+2$*).*

A further consequence of this and section 7 is that for a uniformly packed code and any $\boldsymbol{x} \in V$, the numbers $B(\boldsymbol{x}, k)$ are completely determined by the distance from $\boldsymbol{x}$ to C.

If $e=1$ and the code is also linear, then a lot more is known. It can be shown (cf. Van Lint 1982) that a linear code with covering radius 2 and $d \geqslant 3$ is uniformly packed if and only if $C^{\perp}$ is a code which has $s=2$ (a two-weight code) and $i+j=n+1$, where $A_i \neq 0$, $A_j \neq 0$. For a survey of the connections of these codes to finite geometries see Calderbank and Kantor (1986).

Besides their interest as a packing problem in $\mathbb{F}_q^n$, the uniformly packed codes also produce designs. A q-ary $t-(n, k, \lambda)$-design is a collection S of vectors of weight k in $\mathbb{F}_q^n$ such that every vector of weight t in $\mathbb{F}_q^n$ is covered by exactly t elements of S.

Theorem 8.5. *If C is a uniformly packed code in* $\mathbb{F}_q^n$ *that contains* $\mathbf{0}$, *then the codewords of fixed weight in C form a q-ary e-design if* $d=2e+1$, *and a q-ary*$(e+1)$*-design if* $d=2e+2$.

Proof. For words of weight d this is a consequence of the definition. The proof is by induction on the weight using the fact that for any $\boldsymbol{x} \in V$ the numbers $B(\boldsymbol{x}, k)$ are completely determined by the distance from $\boldsymbol{x}$ to the code. □

We now give a few examples of uniformly packed codes.

8.1. A Hadamard code

Consider a $(12, 24, 6)$ Hadamard code (see section 6). Puncture to obtain an $(11, 24, 5)$ code C. If there were a word z with distance 2 or 3 to four codewords, then w.l.o.g. the original Hadamard matrix would have had four rows of the form

$$\begin{array}{cccccccccccc} + & + & + & + & + & + & + & + & + & + & + & + \\ + & + & + & + & + & + & - & - & - & - & - & - \\ + & + & + & - & - & - & + & + & + & - & - & - \\ - & - & - & + & + & + & + & + & + & - & - & - \end{array}$$

There is no row in $\{\pm 1\}^{12}$ that is orthogonal to these four. Furthermore, the words with distance >1 to C have an average of three codewords at distance 2 or 3 (easy counting). So this number is always three and C is uniformly packed.

8.2. Preparata codes

The Preparata codes $\boldsymbol{P}$ of Definition 6.3 have $n=2^{m+1}-1$ (m odd) and $e=2$. Therefore, they satisfy (3.11) with equality, i.e., they are nearly perfect.

8.3. *Two-weight codes*

Several two-weight codes can be obtained by starting with a first-order RM code (hyperplanes) and restricting to the positions of a nondegenerate elliptic quadric (since the hyperplanes have intersections with this quadric of only two different cardinalities). For details we refer to Van Tilborg (1976) or Calderbank and Kantor (1986).

8.4. *Nonexistence theorems* for perfect codes and uniformly packed codes

It was shown by Tietäväinen (1973) and Van Lint (1971) that the Golay codes are the only nontrivial e-error-correcting perfect codes with $e>1$ over any alphabet Q for which $|Q|$ is a prime power. For $e>2$, $e\neq 6$, the restriction on Q can be dropped, as was shown by Best (1982). The case $e=6$ was handled by Hong (1984). Van Tilborg (1976) showed that e-error-correcting uniformly packed codes with $e>3$ do not exist. For $e\leqslant 3$ much is known about nonexistence. The proofs for the binary case and for uniformly packed codes with $\lambda=\mu+1$ are easy (see Van Lint 1982). The problem of the existence of perfect codes that correct 1 or 2 errors for alphabets that are not fields looks hopeless at the moment. For a survey of perfect codes and several generalizations see Van Lint (1975).

9. Codes, finite geometries and block designs

The Reed–Muller codes of Definition 5.11 have of course a geometrical interpretation. Many ideas from the theory of affine spaces are used in treating, understanding and decoding these codes. Similarly, other finite geometries have led to codes with certain desirable properties. Many of these are in fact subcodes of generalizations of RM codes (in their cyclic representation). In this case the codes have not led to new insight into the combinatorics. For a treatment of geometric codes see Berlekamp (1984).

On the other hand, the methods of coding theory have led to quite a number of interesting theorems on block designs, including projective planes. Usually these results were nonexistence theorems. The idea is usually to study the binary code generated by the rows of the incidence matrix of a design that is not known but is assumed to exist. The properties of the alleged design lead to relations for the parameters of the corresponding binary code. Quite often, theorems of the type treated in this chapter show that such a code does not exist and hence the combinatorial design does not exist either. Below, we shall give a few examples of these methods. There have been recent applications by Bridges et al. (1981) and Tonchev (1986a,b). Also see Hall (1986, chapter 17).

Let A be the incidence matrix of a projective plane of order n. C is the code of length n^2+n+1 generated by the rows of C. If n is odd, the rows of A with a 1 in a fixed position have as sum a word with 0 in that position and 1's elsewhere. Therefore, C is the even-weight code. If n is even, we obtain more interesting results.

Theorem 9.1. *If* $n\equiv 2 \pmod 4$ *the code* C *has dimension* $\frac{1}{2}(n^2+n+2)$.

Proof. Clearly $\bar{C}$ is self-orthogonal. Hence $\dim C \leqslant \frac{1}{2}(n^2+n+2)$. Let $\dim C = r$, $\dim C^{\perp} = n^2+n+1-r = k$. Let H be a parity-check matrix for C in the form $H = (I_k P)$, possibly after a permutation of the places. Define

$$N := \begin{pmatrix} I_k & P \\ 0 & I_r \end{pmatrix}.$$

Over $\mathbb{Q}$ we have $\det AN^{\mathrm{T}} = \det A = (n+1)^{(n^2+n)/2}$. Since all entries in the first k columns of AN^{T} are even (by definition of H), we see that $\det A$ is divisible by 2^k. It follows that $k \leqslant \frac{1}{2}(n^2+n)$, and hence $r \geqslant \frac{1}{2}(n^2+n+2)$. $\square$

Note that Theorem 9.1 shows that $\bar{C}$ is self-dual.

Theorem 9.2. *C has minimum weight $n+1$ and furthermore*:
(i) *the vectors of weight $n+1$ in C are the lines of the plane*,
(ii) *the vectors of weight $n+2$ in C are the hyperovals of the plane.*

Proof. If $\boldsymbol{c}$ is a codeword of weight d, then count the intersections of $\boldsymbol{c}$ and the lines of the plane. The assertions follow immediately. If S is a hyperoval in the plane, consider the sum of the lines through pairs of points of S. This is clearly (the characteristic function of) S. $\square$

For an account of the attacks on the existence problem of the projective plane of order 10 (using similar methods) see chapter 13.

References

Textbooks on coding theory

Berlekamp, E.R.
[1984] *Algebraic Coding Theory* (Aegean Park Press, Laguna Hills, CA). Revised edition of one of the first major textbooks on the subject. Much about circuitry.

Blake, I.F., and R.C. Mullin
[1975] *The Mathematical Theory of Coding* (Academic Press, New York). Very much algebra; a lot on combinatorics.

Cameron, P.J., and J.H. van Lint
[1991] *Designs, Graphs, Codes and their Links, London Mathematical Society Student Texts,* Vol. 22 (Cambridge University Press, Cambridge). Written for combinatoralists; a major reference for this chapter.

Duske, J., and H. Jürgensen
[1977] *Codierungstheorie* (B.I. Wissenschaftsverlag, Mannheim). Very theoretical, no combinatorics, but a chapter on perfect codes.

Hamming, R.W.
[1980] *Coding and Information Theory* (Prentice Hall, Englewood Cliffs, NJ). Very little coding theory; elementary.

Heise, W., and P. Quattrocchi
[1983] *Informations- und Codierungstheorie* (Springer, Berlin). Written for third-year computer science students; much about channels.

Hill, R.
[1986] *A First Course in Coding Theory* (Oxford University Press, Oxford). For second-year mathematics students; no prerequisites; elementary exposition.

MacWilliams, F.J., and N.J.A. Sloane
[1977] *The Theory of Error-Correcting Codes* (North-Holland, Amsterdam). The encyclopedia of coding theory; nearly 1500 references, many tables.
McEliece, R.J.
[1977] *The Theory of Information and Coding* (Addison-Wesley, Reading, MA). For a large part information theory; many problems; no combinatorics.
Peterson, W.W., and E.J. Weldon
[1981] *Error-Correcting Codes* (MIT Press, Cambridge, MA). Revised edition of the first textbook on coding theory; many tables, no combinatorics.
Pless, V.
[1982] *Introduction to the Theory of Error-Correcting Codes* (Wiley, New York). For undergraduate mathematics and engineering students; elementary; chapter on designs.
van Lint, J.H.
[1973] *Coding Theory, Lecture Notes in Mathematics,* Vol. 201 (Springer, Berlin). Introduction; advanced exposition; slightly out of date.
[1982] *Introduction to Coding Theory, Graduate Texts in Mathematics,* Vol. 86 (Springer, Berlin). For mathematics graduates; terse exposition; very little about combinatorics.

Textbooks written mainly for communication engineers and computer scientists

Blahut, R.E.
[1983] *Theory and Practice of Error Control Codes* (Addison-Wesley, Reading, MA).
Clark, G.C., and J. Bibb Cain
[1982] *Error-Correcting Coding for Digital Communications* (Plenum Press, New York).
Cullman, G.
[1967] *Codes Détecteurs et Correcteurs d'Erreurs* (Dunod, Paris) (in French).
Furrer, F.J.
[1981] *Fehlerkorrigierende Block-Codierung für die Datenübertragung* (Birkhäuser, Basel) (in German).
Lin, S., and D.J. Costello
[1983] *Error Control Coding; Fundamentals and Applications* (Prentice Hall, Englewood Cliffs, NJ).

Articles referred to in this chapter

Assmus Jr, E.F., and H.F. Mattson Jr
[1969] New 5-designs, *J. Combin. Theory* **6**, 122–151.
[1974] Coding and combinatorics, *SIAM Rev.* **16**, 349–388.
Best, M.R.
[1982] *A contribution to the nonexistence of perfect codes,* Ph.D. Thesis (University of Amsterdam).
Blokhuis, A., and C.W.H. Lam
[1984] More coverings by rook domains, *J. Combin. Theory A* **36**, 240–244.
Bose, R.C., and D.K. Ray-Chaudhuri
[1960] On a class of error correcting binary group codes, *Inform. and Control* **3**, 68–79.
Bridges, W.G., M. Hall Jr and J.L. Hayden
[1981] Codes and designs, *J. Combin. Theory A* **31**, 155–174.
Calderbank, A.R., A.R. Hammons Jr, P.V. Kumar, N.J.A. Sloane and P. Solé
[1993] A linear construction for certain Kerdock and Preparata codes, *Bull. Amer. Math. Soc.* **29**, 218–222.
Calderbank, R., and W.M. Kantor
[1986] The geometry of two-weight codes, *Bull. London Math. Soc.* **18**, 97–122.
Delsarte, Ph.
[1972] Bounds for unrestricted codes, by linear programming, *Philips Res. Rep.* **27**, 272–289.
[1973a] Four fundamental parameters of a code and their combinatorial significance, *Inform. and Control* **23**, 407–438.
[1973b] An algebraic approach to the association schemes of coding theory, *Philips Res. Rep. Suppl.* **10**.

Goethals, J.-M., and H.C.A. van Tilborg
[1975] Uniformly packed codes, *Philips Res. Rep.* **30**, 9–36.
Hall Jr, M.
[1986] *Combinatorial Theory* (Wiley, New York) ch. 17.
Hammons Jr, A.R., P.V. Kumar, A.R. Calderbank, N.J.A. Sloane and P. Solé
[1994] The $\mathbb{Z}_4$-linearity of Kerdock, Preparata, Goethals, and related codes, *IEEE Trans. Inform. Theory* **IT-40**, 301–319.
Hocquengheim, A.
[1959] Codes correcteurs d'erreurs, *Chiffres (Paris)* **2**, 147–156.
Hong, Y.
[1984] On the nonexistence of unknown perfect 6- and 8-codes in Hamming schemes $H(n, q)$ with q arbitrary, *Osaka J. Math.* **21**, 687–700.
Johnson, S.M.
[1962] A new upper-bound for error-correcting codes, *IEEE Trans. Inform. Theory* **IT-8**, 203–207.
Kantor, W.M.
[1982] Spreads, translation planes and Kerdock sets I, II, *SIAM J. Algebra Discrete Methods* **3**, 151–165, 308–318.
Kerdock, A.M.
[1972] A class of low rate non-linear codes, *Inform. and Control* **20**, 182–187.
MacWilliams, F.J.
[1963] A theorem on the distribution of weights in a systematic code, *Bell. Syst. Tech. J.* **42**, 79–94.
Pasquier, G.
[1980] The Golay code obtained from an extended cyclic code over IF_8, *European J. Combin.* **1**, 369–370.
Pless, V.
[1972] Symmetry codes over GF(3) and new five-designs, *J. Combin. Theory* **12**, 119–142.
Semakov, N.V., V.A. Zinov'ev and G.V. Zaitsev
[1971] Uniformly packed codes, *Problems Inform. Transmission* **7**, 30–39.
Shannon, C.E.
[1948] A mathematical theory of communications, *Bell. Syst. Tech. J.* **27**, 379–423, 623–656.
Snover, S.L.
[1973] *The uniqueness of the Nordstrom–Robinson and the Golay binary codes,* Ph.D. Thesis (Michigan State University).
Tietäväinen, A.
[1973] On the nonexistence of perfect codes over finite fields, *SIAM J. Appl. Math.* **24**, 88–96.
Tonchev, V.D.
[1986a] A characterization of designs related to the Witt system $S(5, 8, 24)$, *Math. Z.* **191**, 225–230.
[1986b] Quasi-symmetric designs and self-dual codes, *European J. Combin.* **7**, 67–73.
van Lint, J.H.
[1971] Nonexistence theorems for perfect error-correcting codes, in: *Computers in Algebra and Number Theory,* Vol. IV (SIAM-AMS Proceedings).
[1975] A survey of perfect codes, *Rocky Mountain J. Math.* **5**, 189–224.
[1983] Kerdock codes and Preparata codes, *Congress. Numerantium* **39**, 25–41.
van Lint, J.H., and R.M. Wilson
[1986] On the minimum distance of cyclic codes, *IEEE Trans. Inform. Theory* **IT-32**, 23–40.
van Tilborg, H.C.A.
[1976] *Uniformly packed codes,* Ph.D. Thesis (Eindhoven University of Technology, Eindhoven).
Wille, L.T.
[1987] The football pool problem for 6 matches: a new upper bound optained by simulated annealing, *J. Combin. Theory A* **45**, 171–177.

CHAPTER 17

Extremal Problems in Combinatorial Geometry

Paul ERDŐS

Mathematical Institute of the Hungarian Academy of Sciences, Reáltanodo utca 13–15, Budapest 1053, Hungary

George PURDY

Department of Computer Science, 822 Old Chemistry, Mail Location 8, University of Cincinnati, Cincinnati, OH 45221-0008, USA

Contents

HANDBOOK OF COMBINATORICS
Edited by R. Graham, M. Grötschel and L. Lovász

1. Introduction

A conjecture both deep and profound
Is whether a circle is round.
In a paper of Erdős
Written in Kurdish
A counterexample is found.

Leo Moser

Although geometry has been studied for thousands of years, the term combinatorial geometry is of quite recent origin. [We do not know exactly when it first appeared, but Levi used it in the introduction to his 1929 book on configurations as if it were already an established term (Levi 1929).] Besides our present meaning of geometrical questions involving combinatorics, it is often used to cover the theory of convex bodies and the related problems of covering, packing, enumeration, and geometric inequalities. Although we do not specifically exclude these other topics, we are mainly interested in, and will mainly discuss, extremal problems involving finite sets of points in Euclidean space of a highly combinatorial nature. Within that context, we do not claim completeness and will naturally tend to discuss problems on which we ourselves have worked, and we will try to indicate the literature of related problems. We will give short sketches of proofs when we wish to illustrate the methods used to prove things. We refer the reader to the senior author's numerous survey papers on this subject (Erdős 1975, 1980, 1981, 1984, 1985a,b) and to Grünbaum (1972). See also Grünbaum (1967), Grünbaum and Shephard (1987), Gruber and Wills (1983), Croft (1970), and Edelsbrunner (1987).

We also want to point out the delightful book of Hadwiger and Debrunner which appeared in German and French and was later translated into English by Klee who added much original material and also brought the book up to date (Hadwiger et al. 1964). Klee has also informed us that his book with Stan Wagon, *Old and New Solved and Unsolved Problems from Plane Geometry and Number Theory* (Klee and Wagon 1991) has appeared.

Another source of problems and results is the 1986 edition of the collection "Research problems in discrete geometry" (Moser and Pach 1986). The nucleus of this collection was a list of problems (Moser 1966) compiled by W. O. J. Moser in memory of Leo Moser. Further expanded editions appear from time to time.

From 1954 until 1973, there was a section entitled "Ungelöste Probleme" (unsolved problems) edited by Hadwiger, in the Swiss journal *Elemente der Mathematik*. A total of 56 problems appeared. Stimulating geometrical problems also appear in the solved and unsolved problem sections in the *American Mathematical Monthly*.

Several journals have specialized in geometry: *Geometriae Dedicata*, published by Reidel, started in 1972; *Discrete and Computational Geometry*, published by Springer, started in 1986; and the *Journal of Geometry*, published by Birkhäuser, started in 1971.

In preparing this chapter we received help from many of the same people whose names appear in the references. We should particularly like to thank B. Grünbaum, L. M. Kelly, V. Klee, J. Pach, J.-P. Roudneff, and T. Zaslavsky. We also thank the editors for their assistance.

2. Sylvester–Gallai theorems

2.1. Sylvester's problem

Sylvester's problem derives its name from the fact that it was stated as a problem to be solved by the reader by J. J. Sylvester in a column of mathematical problems and solutions in the *Educational Times*, published in London, in 1893 (Sylvester 1893). Shortly thereafter an incorrect proof submitted by one of the readers appeared in the same column. Erdős thinks that Sylvester had a correct proof, but Coxeter (1948) seemed to assume that he did not.

The problem was to show that any configuration of n points in the plane, not all on a line, must determine a line containing exactly two of the points.

In posing this problem, Sylvester was probably motivated by the nine points of inflection of a general cubic curve. These nine (complex) points have the interesting property that a line through any two of them passes through a third – for more details see section 4.1.

Some 40 years later the problem was revived by Erdős and subsequently solved by the Hungarian mathematician T. Gallai, and also by several others. Gallai's proof may be found in de Bruijn and Erdős (1948). We shall call a two-point line a *Gallai line*, although the term *ordinary line*, introduced by Motzkin (1951), is also used.

Kelly's minimum-altitude proof is probably the nicest. It appeared in Coxeter (1948) along with Coxeter's own proof. Let the n points be given in the Euclidean plane, not all on a line, and form all of the connecting lines. Let the point P and the line L have the minimum positive distance that occurs among all point–line pairs. Then the claim is that L must be a Gallai line. To see this, let Q be the closest point on L and P. Suppose that L has at least three points on it. Then there will be two points P_1 and P_2 on the same closed half-line. Let us say that P_2 is between P_1 and Q, possibly coinciding with Q. Then it is elementary to see that the line PP_1 is closer to P_2 than P is to L, contrary to the original assumption, and so there is always a Gallai line.

By forming the polar reciprocal, interchanging relevant points and lines, one may arrive at the following dual form of the Sylvester–Gallai theorem: given n lines in $\mathbb{R}^2$, not all concurrent, and not all parallel, there is an intersection point determined by the lines that is incident with exactly two lines – a *simple* intersection point.

Alternatively, one may use the standard embedding of $\mathbb{R}^2$ into the real projective plane $\mathbb{R}^{2+}$ and form the projective dual of a system of n noncollinear points. That will be an arrangement of n lines not forming a pencil, and the dual of the Sylvester–Gallai theorem would say that the lines determine a simple crossing point.

The existence of a Gallai line leads to a simple induction proof of the fact that there are at least n lines: the result is true for $n = 3$; let $n > 3$ and assume the result is true for $n - 1$ points. Let L be a Gallai line, and let P be one of the two points on L. Removing P results in a configuration of $n - 1$ points. If they are all collinear, then the original configuration was a very special one called a "near-collinear set" with $n - 1$ points collinear, having n points and n lines. If the $n - 1$ points are not all collinear, then there are at least $n - 1$ lines by the induction hypothesis, and L is a new line. The result follows.

2.1.1. Conic sections

Wilson and Wiseman (1988) proved the following Sylvester theorem for conic sections: given n points in the plane, if no conic section contains them all, then there is a conic section containing exactly five of them. See section 5.10 for results on circles.

2.1.2. Compact convex sets

There have been a number of generalizations of Sylvester's theorem to versions in which the n points are replaced by n compact sets. In one of the most definitive of these, Herzog and Kelly (1960) proved the following:

The Herzog–Kelly theorem. *Let there be given n pairwise disjoint compact sets in $\mathbb{R}^d$, not all contained in a line. Suppose that at least one of the sets contains infinitely many points. Then there is a line intersecting exactly two of them.*

They remark that the result is not true if compactness of the sets is not assumed, as shown by the example of n parallel lines in the plane. If all of the n sets are finite, then the result is also false. For example, the nine points of the planar Pappus configuration may be arranged in three sets such that a line cutting any two intersects a third. See Herzog and Kelly (1960) for this and also for a proof of a weaker result in the finite case. Under the same hypothesis as the Herzog–Kelly theorem, Edelstein et al. (1963) proved that there is a *hyperplane* intersecting exactly two of the sets, which is a stronger result. We refer the reader to the references in Herzog and Kelly (1960) and Edelstein et al. (1963).

2.1.3. Infinite sets

Borwein (1984) proved Sylvester's theorem for compact countably infinite sets S in the plane. That is to say, if the line joining any two points of S intersects a third point of S, then S is contained in a line.

2.2. The Motzkin–Dirac conjecture on the number of Gallai lines

Erdős and de Bruijn asked whether $t_2(n)$, the minimum number of Gallai lines, where the minimum is taken over all configurations of n points, tends to infinity with n. Subsequent to this, Motzkin (1951) showed that $t_2(n)$ is at least the square root of n. Dirac (1951) also asked the same question and they both conjectured that $t_2(n) \geqslant [\frac{1}{2}n]$.

Kelly and Moser (1958) proved that $t_2(n) \geqslant \frac{3}{7}n$, with equality occurring for an example with 7 points.

The following example due to Motzkin (1975) shows that $t_2(n) \geqslant \frac{1}{2}n$ is best possible, if true, when n is even. If $n = 2m = 4k$, take the regular m-gon together with the m points on the line L at infinity where the lines joining pairs of the m-gon intersect L. The m Gallai lines arise when each vertex V_k of the m-gon is joined to the point at infinity corresponding to the direction determined by $V_{k-1}V_{k+1}$. If $n = 2n = 4k + 2$, the take a regular $2k$-gon and its center, together with $2k + 1$ points on the line at infinity.

Hansen (1981) "proves" that $t_2(n) \geqslant \frac{1}{2}n$ except when $n = 7$ and $n = 13$, where $t_2(7) = 3$ and $t_2(13) = 6$. Recently Csima and Sawyer (1993) corrected his proof so that it really shows $t_2(n) \geqslant \frac{6}{13}n$.

2.3. *Higher dimensions*

In $\mathbb{R}^d$ it must also be the case that n points not all collinear determine a Gallai line. This may be seen by projecting the points into the plane in one of the infinitely many directions which do not cause any image points to coincide or new collineations to occur, and then applying the result in the plane.

2.3.1. *The Motzkin conjectures*

What about planes? One might reasonably ask whether n points in Euclidean space, not all lying on a plane, always determine a plane with only three points – a so-called "elementary plane". Motzkin (1951) showed that unfortunately this is false by providing two counterexamples, but he saved the situation by proving the following (we use his numbering).

Theorem U_3. *If we define a* Gallai plane *to be one in which all but one of its points lie on a single line, then there is always a Gallai plane formed by n points in Euclidean three-space, which do not all lie on one plane.*

He observed that U_3 implies by an induction argument similar to the one for lines:

Theorem T_3. *Given n points in Euclidean space, not all lying on a plane, they determine at least n planes.*

He made the more general conjecture:

Conjecture U_d. If we define a *Gallai hyperplane* to be one in which all but one of its points lie in a $(d-2)$-dimensional flat, then n points in $\mathbb{R}^d$ always determine a Gallai hyperplane, unless they all lie in one hyperplane.

From this conjecture would follow:

Conjecture T_d. Given n points in $\mathbb{R}^d$, not all lying in one hyperplane, they determine at least n hyperplanes.

Motzkin was able to prove T_d for $d = 4, 5, \ldots$, but not the apparently harder U_d for $d > 3$. Hansen (1965) proved U_d for all d.

2.3.2. The Purdy conjectures

Let n points be given in $\mathbb{R}^3$, not all lying on a plane. The number of planes p is at least n, and the number of lines t, by projection into the plane, is at least n. Is it always the case that $p \geq t$? The example of two skew lines with k points on one line and $n - k$ on the other gives $p = n$ and $t = 2 + k(n - k)$, and the example of $n - 1$ points in a plane in general position and one point off the plane gives $p - t + n = 2$. We have shown (Erdős and Purdy 1975) that $p - t + n \geq 2$ if no three points are collinear and n is sufficiently large. Purdy (1986) has also shown that $p \geq ct$, where $c > 0$, provided that the points do not all lie on two skew lines. In fact, there is always one point with that many planes through it. Purdy conjectures that $p \geq r$ for n sufficiently large, except for the two infinite families of exceptions mentioned: the points lie on two skew lines, or all but one of the points lie on a plane. Grünbaum pointed out that the eight vertices of a cube form a counterexample, and also gave us an example with 15 points and another with 16 points, but thinks perhaps the conjecture is true for $n \geq 17$. See Grünbaum and Shephard (1984), where these examples are discussed in the dual setting.

In $\mathbb{R}^d$ Purdy extends the conjecture as follows: we call a configuration of points *irreducible* if points cannot be covered by a set of $r \geq 2$ nonempty flats $F_1, \ldots, F_r$, such that the sum of the ranks $\mathrm{rk}(F_1) + \cdots + \mathrm{rk}(F_r)$ is at most $d + 1 = \mathrm{rk}(\mathbb{R}^d)$. By "rank", we mean one more than the dimension (like the matroid rank): points have rank one, lines have rank two, etc. Thus a configuration of points on two skew lines in $\mathbb{R}^3$ is not irreducible, since the points are covered by two flats of rank two, and the configuration on $n - 1$ points on a plane and one point off the plane in $\mathbb{R}^3$ is not irreducible, being covered by a flat of rank three and a flat of rank one.

Purdy conjectures that if $n > N_d$ and the points form an irreducible configuration, then $W_d \geq W_{d-1}$, where W_k denotes the number of flats of rank k (the kth Whitney number). The conjecture implies by projection that $W_d \geq W_{d-1} \geq \cdots \geq W_2 \geq W_1$, which implies G. C. Rota's unimodality conjecture in this context. The conjecture is false in the more general context of geometric lattices (or equivalently, simple matroids), as may be seen by looking at d-dimensional projective space over the Galois field Z_p.

2.3.3. The number of Gallai planes

There are also results about the numbers of Gallai planes. Bonnice and Kelly (1971) proved that there are at least $\frac{3}{11}n$ Gallai planes in Euclidean space. In a later paper, Hansen (1980) improves this result to $\frac{2}{5}n$. It is not clear what the minimum number is. Probably more than cn Gallai planes exist with $c > \frac{2}{5}$. In the same paper, Hansen also gives further examples of configurations of points without elementary planes. As mentioned already, Motzkin (1951) observed that there are configurations in $\mathbb{R}^3$ without elementary planes: for example, two skew lines with at least three points on each. Further, he noted that the Desargues configuration in $\mathbb{R}^3$ with 10 points has none. Bonnice and Kelly (1971) furnish the following example: construct a vertical prism whose base is a regular m-gon for odd m. For the configuration take the $2m$ vertices of the prism, the m points on the plane at infinity where the (horizontal) sides of the parallel m-gon intersect it, and the single point on the plane at infinity where the p vertical edges of the prism intersect it. There are $3m + 1$ vertices. The configuration has no elementary planes. If $m = 3$ this is the Desargues configuration. Hansen's examples are somewhat similar: the prism is based on a regular $4m$-gon, and you can have a total of either $4m + 3$ or $10m + 2$ points.

Grünbaum and Shephard have studied these examples and many others in the dual setting where they give rise to simplicial arrangements of planes – all the cells are simplices. They also point out an error in one of Hansen's examples (Grünbaum and Shephard 1984).

3. Arrangements

3.1. Regions in the projective plane

Suppose that we have n lines in the real projective plane, not all passing through one point. Let t_k denote the number of points incident with exactly k of the lines for $k \geq 2$. The lines will divide the real projective plane into polygonal regions. Let p_k denote the number of regions having k sides.

There are two very useful formulas relating these quantities:

$$t_2 = 3 + t_4 + 2t_5 + \cdots + p_4 + 2p_5 + 3p_6 + \cdots$$

and

$$p_3 = 4 + 2(t_3 + 2t_4 + 3t_5 + \cdots) + p_5 + 2p_6 + 3p_7 + \cdots .$$

The first of these appeared in Melchior (1940). The second one appeared in Grünbaum (1972), and it implies the inequality

$$p_3 \geq 4 + p_5 + 2p_6 + 3p_7 + \cdots ,$$

with equality characterizing simple arrangements. See Grünbaum (1972) for sufficient conditions on the p_i that an arrangement exists. This relates to Eberhard's theorem – see Grünbaum (1967, p. 253).

We sketch a proof of the two identities. Let F denote the total number of regions. Let $V = t_2 + t_3 + \cdots$ be the total number of vertices. Then there is an underlying graph on V vertices which is embedded in the real projective plane, and let E denote the number of edges. The real projective plane may be derived from the sphere by identifying opposite points. The geodesic great circles on the sphere then correspond to lines in the projective plane. On the sphere everything will occur twice. There will be $V' = 2V$ vertices, $E' = 2E$ edges, and $F' = 2F$ regions. Euler's polyhedral formula gives $V' - E' + F' = 2$, so that in the projective plane $V - E + F = 1$. It is also easy to see that on the sphere, since the great circles are not all concurrent, there cannot be any digons (two-sided regions). Three nonconcurrent circles form only triangles, and the result follows by induction, since the addition of a circle to an arrangement having no digons will not introduce a digon.

Returning to the projective plane, we see that $2E = 3p_3 + 4p_4 + \cdots = 2(2t_2 + 3t_3 + 4t_4 + \cdots)$. Judicious substitution into the $V - E + F$ formula now gives us the two formulas.

Since the p_k are nonnegative, we also have Melchior's inequality

$$t_2 \geqslant 3 + t_4 + 2t_5 + 3t_6 + \cdots,$$

which holds, by duality, for n points in the real projective plane, not all collinear, where t_i is the number of lines incident with exactly i points. Obviously this also holds in the Euclidean plane. This implies that the number of Gallai lines is at least three, and so we have yet another solution to Sylvester's problem!

In section 4.1 we will discuss an inequality due to F. Hirzebruch which holds provided $0 = t_n = t_{n-1} = t_{n-2}$:

$$t_2 + \tfrac{3}{4}t_3 \geqslant n + t_5 + 3t_6 + 5t_7 + \cdots.$$

Euler's formula also gives us the number of regions $F = 1 - V + E = 1 - t_2 + 2t_3 + 3t_4 + \cdots$, so that the number of regions depends only on the t_i, and not on the particular embedding of the arrangement into the projective plane. In the Euclidean plane, n lines always determine $2n$ unbounded regions. When the line at infinity is removed, n regions are split in two to form them. (We are assuming that the example of n parallel lines has been excluded on the grounds that they are all concurrent on the line at infinity.) Thus in $\mathbb{R}^2$ the number of regions is $1 + n + t_2 + 2t_3 + 3t_4 + \cdots$.

The maximum number of regions that can be formed by an arrangement of n lines in the projective plane is $1 + \frac{1}{2}n(n-1)$. This may be seen by induction on n and the fact that three lines determine four regions. For n lines in the Euclidean plane the maximum number is n more than this, namely $1 + \frac{1}{2}n(n+1)$. In both cases the maximum occurs if the lines are in general position, no three lines are concurrent. Purdy (1980b) discusses the minimum number of regions f_2 formed if at most $n - k$ lines are concurrent. He shows that if $n \geqslant 4k^2 + k + 1$, then $f_2 \geqslant (k+1)(n-k)$.

3.2. Pseudolines

Many problems and results about arrangements of lines in the real projective plane make sense even if the lines are not perfectly straight. An *arrangement of pseudolines* is a family of n closed curves in the real projective plane such that any two meet at precisely one point, where they cross each other. There are arrangements of pseudolines with only nine points which cannot be drawn with straight lines. Grünbaum(1972) conjectured and Goodman and Pollack (1980a) proved that all arrangements of fewer than nine pseudolines can be drawn with straight lines.

Pseudolines were introduced by Levi in a 1926 paper on arrangements of lines (Levi 1926). He establishes the following "enlargement" lemma: given an arrangement of n pseudolines $L_1, \ldots, L_n$, and two points P and Q which do not both lie on any of them, there is an enlarged arrangement $L_1, \ldots, L_{n+1}$ in which L_{n+1} contains both P and Q. A proof is also contained in a monograph (Grünbaum 1972).

Melchior's identity and Grünbaum's identity above both hold for arrangements of pseudolines. Many results for points and lines, when changed to their dual formulations for lines and points, hold for arrangements of pseudolines as well. The result of Kelly and Moser (1958) mentioned in section 2.2 (that the number of Gallai lines t_2 is at least $\frac{3}{7}n$) is one such result. The corresponding statement for pseudolines is that an arrangement of n pseudolines always has at least $\frac{3}{7}n$ simple crossings. This was proved by Kelly and Rottenberg (1972). Their proof made use of Levi's enlargement lemma. Since then, many results have been extended to pseudolines, and we will mention these results when they arise naturally.

Kelly and Moser (1958) used Melchior's identity to show that if no line contains more than $n - k$ of the n points, where $27k^2 < 2n$, then the total number of lines t is at least $nk - \frac{3}{2}k^2$. Their proof works for pseudolines. Around that time, Erdős conjectured that there was a constant $c > 0$ such that $t > ckn$, no matter how large k was. Beck (1983), and simultaneously Szemerédi and Trotter (1983), proved this with very small c. Recently, c has improved (see section 3.6). Beck's proof also holds for pseudoline arrangements. The result is equivalent to a weakened conjecture of Dirac (see section 3.9).

Erdős asked the following question (Erdős 1983): if n lines are given in the projective plane, and $t_k = 0$ for $k > 3$, does there always exist a "Gallai" triangle, i.e., three lines of the arrangement whose three intersection points are all simple (Gallai) points? The example called B_n by Füredi and Palasti (see section 3.7) shows that the answer is no if n is at least four and not divisible by nine (Füredi and Palasti 1984). Another problem on simple points can be found in the above-mentioned article (Erdős 1983).

3.3. The Graham–Newman problem

Graham and Newman posed the following problem: given a finite set of points in the plane, each colored either red or blue and not all collinear, must there be a "monochromatic line"? That is, must there be a line containing only points of one color?

Motzkin (1967) and M. Rabin announced this result, but neither of them published a proof, and the first published proof is due to Chakerian. A shorter proof, which bears a resemblance to Kelly's elegant proof of Sylvester's theorem, appears in some class notes of Grünbaum (1974), where it is attributed to Sherman K. Stein. Since Chakerian's proof is nicely explained and readily available in Chakerian (1970), we present Stein's proof, previously unpublished. [Remark by Purdy: it seems that if you find a really short elegant proof, somebody else will write it up for you, because Coxeter wrote up Kelly's proof in Coxeter (1948) and Chakerian apparently wrote up Stein's proof for him.]

Theorem. *Given n lines in the real projective plane, colored by two colors, and not forming a pencil, there is an intersection point incident with only lines of one color.*

Proof. Let n lines be given, not forming a pencil, and colored by two colors. We wish to prove that there is an intersection point incident with only lines of one color, so let us suppose that there are no such monochromatic points. If the intersection points were all simple, i.e., no three lines concurrent, then any two lines of the same color would intersect in a monochromatic points. Hence the arrangement is not simple and there is an intersection point P incident with at least three of the lines.

Since the lines through P are not all of the same color, we may suppose that there are lines L, L', and L'' among the lines through P such that L and L' have color C, and L'' has color C''. If the lines not incident with P were all of color C, then any one of them would have to intersect L in a monochromatic point of that color. Hence there is a line L^* of color C'' not incident with P. If we ignore the other $n-4$ lines, then the two triangles $LL''L^*$ and $L''L'L^*$ share an edge which is part of the line L''. These two triangles would become merged into one larger triangle upon the removal of the line L''. The larger triangle we call a *characteristic triangle* of the arrangement. Any other triangle of this form, where L and L'' have one color and L'' and L^* have the other color, we also refer to as a *characteristic triangle* of the arrangement.

Since there are only finitely many characteristic triangles, there must be one which is minimal in the sense that there is not another one completely contained in it. Suppose that L, L', L'', and L^* as already described form a minimal characteristic triangle $LL'L^*$ which is cut into two triangles by the line L''. We claim that the intersection point Q of the lines L'' and L^* is monochromatic. To see this, first suppose not. Since L'' and L^* have color C'', there must be a line L^+ of color C passing through Q. The line L^+ will cut either triangle $LL''L^*$ or $L''L'L^*$. Without loss of generality we may suppose that L^+ cuts $LL''L^*$. But then the four lines L^+, L, L'', and L^* give rise to a characteristic triangle $LL''L^*$ which is contained in the supposedly smallest one $LL'L^*$, which is a contradiction. □

Remark. This proof and also Chakerian's proof work for pseudolines without any changes.

Can we insist that the monochromatic line always be red, say? Clearly not, since we can always place blue points so as to spoil the red lines.

Can we insist that the monochromatic line by a Gallai line (a 2-point line)? The example of $n-1$ red points on a line and one blue point off the line shows that we cannot.

Erdős asked the following question. Does there exist for every k a set of points in the plane so that if one colors the points by two colors in an arbitrary way, there always should be at least one line which contains at least k points, all of whose points have the same color? Graham and Selfridge gave an affirmative answer for $k=3$, but the problem is still open for $k>3$.

Is there a generalization of the Graham–Newman problem to three dimensions? The straightforward generalization is false: given n points in $\mathbb{R}^3$ colored by two colors, not all lying on one plane, there is *not* always a monochromatic plane. On a plane M put $r-1$ red points in general position, and place one red point off the plane. On every line formed by two or more red points in the plane M place a blue point. Clearly there are no monochromatic planes. Perhaps if neither color separately is entirely contained in a plane there must be a monochromatic plane. Perhaps if no three points are collinear.

Borwein has generalized the original result, replacing finite sets by countable compact sets (Borwein 1984).

3.4. *The two-coloring of regions*

If we are given n lines in the Euclidean (but not the real projective) plane, then an easy induction argument shows that the regions can be colored with two colors so that no two regions that have a side in common with the same color. If $n=1$, color the two sides of the line with opposite colors. Let $n>1$, and suppose that the result is true for $n-1$. Remove one line L, and use the induction hypothesis to color the resulting arrangement. Not put L back. Reverse the colors on one side of L and leave the colors alone on the other side. The result follows by induction.

The coloring is unique for each arrangement of lines, for once one region is colored, the colors of its neighbors are determined. Let us color the regions using r red regions and g green regions, where $g \leq r$. If you draw some examples, you will see that in general $g<r$. Fejes-Toth (1975) asked how large r can be compared to g. Palasti (1976) investigated this question for simple arrangements, applying a result from Palasti (1975) to show that $r/g<2$ for $n \leq 9$. A few years later, Grünbaum (1980), and simultaneously Simmons and Wetzel (1979), proved that $r \leq 2g-2$ for any arrangement. Examples due to Simmons, Palasti and other show that equality occurs for a few small values of n, but no infinite family is known. The example which Füredi and Palasti call A_n (see section 3.7) gives $r=2g-n-2$ if $n=3m$, which implies that $\limsup r/g=2$ as $n \to \infty$.

The regions formed by an arrangement of n pseudolines in the projective plane cannot in general be two-colored. But we may define *an arrangement of topological lines* in $\mathbb{R}^2$ in a manner similar to pseudolines. The above proofs can then be used with little modification to show that $r \leq 2g-2$ for arrangements of topological lines. Harborth (1985a) has shown that for such lines $r \leq \frac{1}{3}n(n+1)$ if n is odd, and $r \leq \frac{1}{6}n(2n+1)$ if n is even, with equality occurring infinitely often in both cases.

One can ask the same question in $\mathbb{R}^3$ and higher dimensions. In $\mathbb{R}^d$ Purdy and Wetzel (1980) showed that $\limsup r/g \leqslant \frac{1}{2}(d+1) + 1/d$ if the hyperplanes are in general position (no d having a line in common).

3.5. *Which sequences $\{p_i\}$ are realizable?*

In Levi's fundamental paper on pseudolines (Levi 1926) may be found a proof of the assertion that in an arrangement of pseudolines $p_3 \geqslant n$. Grünbaum (1972) conjectured that *for lines* equality only occurs for simple arrangements, and this was proved by Roudneff (1988). The result is false for pseudolines. There is a counterexample with $n=9$ due to Canham in Grünbaum (1972).

Grünbaum (1972) conjectured that $p_3 \leqslant \frac{1}{3}n(n-1)$ if n is sufficiently large. This was established by Canham for simple arrangements, and by Strommer (1977) for simplicial arrangements, i.e., arrangements in which every region is a triangle. Harborth and Roudneff have constructed simple arrangements of pseudolines achieving equality for infinitely many values of n (Harborth 1985a,b, Roudneff 1986). It is still not known whether this can be done with (straight) lines, a problem originally posed by Grünbaum. The example that Füredi and Palasti (1984) call B_n contains at least $4 + \frac{1}{3}n(n-3)$ triangles. This lower bound was also achieved by Strommer (1977) using a construction of Burr et al. (1974). If we insist on simple arrangements (no three lines concurrent), then Füredi and Palasti's A_n has at least $\frac{1}{3}n(n-1)$ triangles. See section 3.6 and Füredi and Palasti (1984).

Getting back to the conjecture itself, in 1980 Purdy established the general result

$$p_3 \leqslant \tfrac{7}{18}n(n-1) + \tfrac{1}{3} \quad \text{when } n \geqslant 6,$$

for all arrangements (Purdy 1980a). In 1987, Roudneff (1991) proved the full conjecture for $n \geqslant 10$, it being false for smaller n. One interesting feature of Roudneff's proof is that it perturbs the lines, changing them into pseudolines. This is the first case we know of where the proof cannot be carried out with (straight) lines alone. Unless another proof is found, pseudolines have now become indispensable.

If the sequence $p_3, p_5, p_6, p_7, \ldots$ satisfies the equation

$$p_3 = 4 + p_5 + 2p_6 + 3p_7 + \cdots,$$

then (Grünbaum 1972) there is a simple arrangement of (straight) lines realizing it, but it is not clear exactly what conditions p_4 must satisfy.

Grünbaum had conjectured that p_4 must be nonzero for simple arrangements if n is large enough. For pseudolines this was disproved by Harborth (1985a) and also by Roudneff (1987), but the question is still open for (straight) line arrangements.

The number p_4 of quadrilaterals satisfies the upper bound

$$p_4 \leqslant \tfrac{1}{2}n(n-3)$$

(Grünbaum 1972, Palasti 1975).

Any discussion of upper bounds on p_4 is complicated by minor disagreements in the literature. We will only say that Grünbaum (1972) and Palasti (1975) agree to the extent that $p_4 \leqslant \frac{1}{2}n(n-3)$ when $n \geqslant 8$, with equality achieved by simple arrangements when n is odd.

There are also mixed results. For example, Purdy (1980b) showed that if $n \geqslant 106$, then $\max\{p_3, p_4\} \geqslant 2n-8$. Strommer (1977) showed that $p_3 + p_4 > 0$ if the arrangement is simple and $n \geqslant 4$. Roudneff (1987) showed that $4p_4 + 5p_5 \geqslant 3n$ under the same conditions.

3.6. *The maximum of t_k*

In this section we discuss how large t_k can be as a function of n and k.

Croft and Erdős (Grünbaum 1972) constructed the following example, for which $t_k > cn^2/k^3$.

Example. For fixed k and m, take the m^2 points $P(x, y)$ with integer coordinates $1 \leqslant x, y \leqslant m$. The arrangement of lines will consist of all those lines that can be drawn through the points P in the k given directions. We shall choose k slopes so that as few lines as possible are needed to cover the m^2 lattice points. Only m horizontal lines are needed to cover the points, and only m vertical lines. More lines of slope 1 are needed, but $2m$ will suffice, a they will for lines of slope -1. More lines still will be needed for slope $\frac{1}{2}$, but $4m$ will suffice. Similarly $4m$ lines of slopes $-\frac{1}{2}$, 2, and -2 will cover the points. If i and j have no common factor, and $0 < i < j$, then $2jm$ lines of slopes i/j, $-i/j$, j/i, and $-j/i$ will cover the m^2 points. For fixed j, there are $\phi(j)$ choices for i. Let N be such that

$$k = 2 + 4\phi(2) + 4\phi(3) + \cdots + 4\phi(N-1) + r\,,$$

where $r < 4\phi(N)$. Then

$$n = 2m + 8\{2\phi(2) + 3\phi(3) + \cdots + (N-1)\phi(N-1) + rN\}m$$

lines can cover the m^2 points k times. The average value of Euler's $\phi(x)$ function is about 60% of x [actually $6x/\pi^2 + o(x)$], so asymptotically $k = 24N^2/\pi^2$ and $n = 16N^3m/\pi^2 = 2\pi k^{3/2}m/6^{3/2}$, $m^2 = 54n^2/(\pi^2k^3)$.

3.6.1. *Upper bounds*

What about upper bounds? Let $T_k = t_k + t_{k+1} + \cdots$. Simple counting arguments show that $t_k \leqslant T_k < cn^2/k^2$, and Erdős asked whether $T_k = O(n^2/k^3)$, i.e., whether the above Croft–Erdős construction is essentially optimal. Szemerédi and Trotter (1983) were able to show this if $k^2 < n$, with an enormous implied constant. Beck (1983) found the weaker result $T_k < cn^2/k^{2+d}$, where $d > 0$, but his proof has the virtue that it works for pseudolines as well. Both results are very impressive. Both papers independently deduce proofs of Erdős's conjecture that at least ckn lines are determined if no $n-k$ points are collinear (discussed in section 3.2), and then go on to deduce the weak form of Dirac's conjecture (discussed in section 3.9).

Szemerédi and Trotter deduced their upper bound on T_k from another inequality: let n points and t lines be given in the plane, with no particular relationship between them, and let $I(n, t)$ denote the total number of incidences, counting multiplicity, occurring between them. Then they showed

$$I(n, t) < cn^{2/3}t^{2/3},$$

with an enormous constant c. Recently, Clarkson et al. (1990) used completely different methods to show that

$$I(n, t) \leqslant 3(6)^{1/3}t^{2/3}n^{2/3} + 25n + 2t,$$

and their proof also works for pseudolines. The constant $3(6)^{1/3}$ in the leading terms is approximately 5.45. Their inequality can be used to show that

$$T_k \leqslant 16 \cdot 3^7 \frac{n^2}{k^3} \quad \text{if } k^2 \leqslant 8 \cdot 3^6 \frac{n}{25}.$$

3.7. *The orchard problem*

In a problem in the *Educational Times*, Sylvester (1867) considered the following question: how large can t_3 be, given that no four points are collinear. He also gave a construction showing that $t_3 > \frac{1}{8}n^2 + O(n)$. The following simplified version of Sylvester's construction, due to A. H. Stone, appears in Grünbaum (1972).

The cubic curve $y = x^3$ cannot have four collinear points on it. If $P(x)$ is the point (x, x^3), then it is not hard to see that three points $P(x_1)$, $P(x_2)$ and $P(x_3)$ are collinear if and only if $x_1 + x_2 + x_3 = 0$. If $n = 2m + 1$, then the points $P(-m)$, $P(-m + 1), \ldots, P(0), \ldots, P(m)$ provide at least $\frac{1}{2}m^2$ collinear triples. If $n = 2m$, then leave out $P(-m)$. This gives $t_3 \geqslant \frac{1}{8}n^2$.

Burr et al. (1979) improved this to $t_3 \geqslant \frac{1}{6}n^2 + O(n)$ using the Weierstrass elliptic function. (Sylvester claimed, without a published proof, this lower bound also.) They also given an account of the history of the "orchard problem", starting with J. Jackson in 1821, they discuss the connection with Kirkman–Steiner triples, and they have an extensive bibliography. Their constructions have applications to several related problems.

Subsequently, Füredi and Palasti (1984) obtained similar results by two purely geometrical constructions which they called A_n and B_n, and that is how we shall refer to them. A_n is a simple arrangement (no three lines concurrent). See Füredi and Palasti (1984) for the relationship of A_n and B_n to the constructions in Burr et al. (1979). We now give their constructions.

Let $P(\alpha)$ denote the point $(\cos\alpha, \sin\alpha)$ in the plane, and let $L(\alpha)$ denote the line through the two points $P(\alpha)$ and $P(\pi - 2\alpha)$. If $\alpha = \pi/3$, so that the two points coincide, then L is the tangent to the circle $x^2 + y^2 = 1$ at $P(\pi/3)$. Then A_n is the arrangement of lines $L(\pi/n)$, $L(3\pi/n), \ldots, L((2n-1)\pi/n)$, and B_n is the arrangement of lines $L(0)$, $L(2\pi/n), \ldots, L((n-1)\pi/n)$. The lines $L(\alpha)$, $L(\beta)$, and $L(\gamma)$ are concurrent if and only if $\alpha+\beta+\gamma$ is an even multiple of 2π. Füredi and Palasti showed $t_2(B_n)=n-3+\varepsilon$, where $\varepsilon=0, 2, 2$ if $n\equiv 0, 1, 2 \pmod 3$, and $t_k=0$ for $k>3$. The value of t_3 is then determined by counting pairs: $t_2+3t_3=\frac{1}{2}n(n-1)$. It is easy to see that A_n is a simple arrangement (no three lines concurrent), since $\alpha+\beta+\gamma$ cannot be an even multiple of 2π if α, β and γ are odd multiplies of 2π.

Erdős (1962) asked the more general question: how large can t_k be, given that $0=t_{k+1}=t_{k+2}=\cdots=t_{n-1}$. For $k>3$ there are no nontrivial upper bounds. For example, $t_4<\frac{1}{12}n^2$ is trivial, and not even $t_4<\frac{1}{12}(1-\varepsilon)n^2$ is known. For $k=[cn^{1/2}]$ one could ask whether $t_k>c'n^{1/2}/c$ is always possible, where c' is independent of both n and c. Erdős conjectured that if $t_5=0$, then $t_4=o(n^2)$. Perhaps even $t_4<cn^{3/2}$.

Karteszi (1963) constructed an example for all $k>3$ for which $t_k>c_kn\log n$ and $t_i=0$ for $i>k$. See section 5.6 for an explanation of Karteszi's method. Subsequently a better construction was found by Grünbaum (1976); this gives an arrangement having $t_k>c_kn^{(k-1)/(k-2)}$. When $k=4$, it gives $t_4>cn^{3/2}$, which Karteszi had also conjectured.

The same question can also be asked for arrangements of pseudolines, but the problem becomes much easier, as the following argument shows: take the dual of the Croft–Erdős construction, which will be an arrangement of n lines having at least cn^2/k^3 k-fold crossings, and now turn this into an arrangement of pseudolines such that $0=t_{k+1}=t_{k+2}=\cdots=t_{n-1}$, by the simple expedient of breaking up all the intersections of more than k lines. These actually occur on the line at infinity, due to the families of parallel lines.

3.8. Some conjectures

Let us return to n points in the plane, not all on a line, and as usual t_i denotes the number of lines incident with exactly i points. In Erdős and Purdy (1978) we find some new and perhaps unexpected properties of the family $\{t_i\}$. Erdős has conjectured that for $n>n_0$, there is always a k such that $t_k\geqslant n-1$. For example, $t_2\geqslant n-1$ in the near collinear set. Krier and Straus pointed out counterexamples for $n=6$, 9, and 13. The conjecture is true for $n\geqslant 25$. In fact, $\max\{t_2, t_3\}\geqslant n-1$ for $n\geqslant 25$.

We also made several conjectures in our paper. We conjectured that if $t_3\geqslant t_2$, then $t\geqslant cn^2$, where t is the total number of lines (unfortunately there is a misprint changing this to $t\geqslant cn$, which is trivial). It follows immediately from Melchior's inequality that $2t_2+t_3\geqslant t$. Our conjecture can therefore be stated in the stronger form: if $t_3\geqslant t_2$, then $t_3>cn^2$. We now have a proof of this conjecture (Erdős and Purdy 1990).

3.9. Dirac's problem

Dirac's problem: let n points be given in $\mathbb{R}^2$, not all lying on one line. Draw all the connecting lines, i.e., lines containing two or more of the points. Dirac (1951) asked whether there is always one of the n points which is incident with at least

$\frac{1}{2}n$ of the connecting lines. He offered the example of two intersecting lines with the points split as evenly as possible between them to show that $\frac{1}{2}n$ is best possible, if true. He also gave $n^{1/2}$ as a trivial lower bound. In Grünbaum (1972, p. 25) there is a list of six exceptions to this strict formulation of Dirac's question. If we define $r(n)$ to be the minimum taken over all arrangements of n points, then Grünbaum gives the upper bounds in table 1 (the third column gives the example(s) by their designation in a previous publication Grünbaum 1971).

Erdős (1961) asked whether the weaker lower bound $r(n) \geq cn$ for some $c > 0$ could be proved. (This question is in fact equivalent to Erdős's conjecture in section 3.2.) Beck (1983), and independently Szemerédi and Totter (1983), proved this with very small constants. A result of Clarkson et al. (1990) (see section 3.6) leads to a somewhat improved constant. The result also holds for pseudolines: given an arrangement of n pseudolines, one of the lines always has at least cn different crossing points on it.

What about higher dimensions? Given n points in $\mathbb{R}^d$, not all on a line, it follows by projection into the plane that there is a point incident with at least cn lines. If the points truly span $\mathbb{R}^3$, then Purdy (1981) showed that there is always a point incident with at least $c_1 n^{1-1/d}$ lines, and the proof works for all simple matroids. (See section 4.2 for a discussion of simple matroids.) Purdy also showed that if n points span $\mathbb{R}^3$, then one of them is always incident with at least $c_2 n$ planes. The constants c_1 and c_2 are close to 1.

3.10. *Directions*

Scott (1970), and independently also R. W. Shannon, asked the following question: given n points in the plane, not all on a line, what is the minimum number $D(n)$ of directions that the points can determine? If P, Q and R are collinear, then they determine one direction; if PQ and RS are parallel, then they determine one direction. Scott and Shannon were able to show

$$(2n)^{1/2} < D(n) \leq 2[\tfrac{1}{2}n] ,$$

the upper bound coming from the regular n-gon or the regular n-gon and its center, depending on the parity of n. Burton and Purdy (1979) achieved the first linear lower bound, showing that, if $n > 3$,

$$D(n) \geq \tfrac{1}{2}n .$$

Table 1

n	U_n	Example(s) $r(n) \leq U_n$
9	4	$A_1(9)$
15	6	$A_1(15)$
19	8	$A_1(19)$, $A_3(19)$
25	10	$A_5(25)$
31	12	$A_1(31)$
37	16	$A_2(37)$

Ungar (1982) showed that $D(n) = 2[\frac{1}{2}n]$. It does not happen very often that a problem in this subject gets a complete solution. His proof uses a method of representing arrangements of pseudolines called *allowable sequences*, introduced by Goodman and Pollack (see section 4.4).

Grünbaum has a related question about vectors: given n points in the plane, not all on a line, what is the minimum number $V(n)$ of different vectors that are determined by the points? Clearly $V(n) \geqslant D(n)$. Ungar's result and the near-collinear set ($n - 1$ points collinear) give therefore:

$$2[\tfrac{1}{2}n] \leqslant V(n) \leqslant 2n - 3 .$$

Define $V'(n)$ to be the same function under the additional assumption that no three points are collinear, so that $V'(n) \geqslant V(n)$. Danzer and Grünbaum (1967) showed that

$$V'(n) < cn^{\alpha} , \quad \text{where } \alpha = (\log 3)/(\log 2) = 1.58496\ldots ,$$

by means of the following example. For simplicity let $n = 2^k$ and take the n vertices of a cube in $\mathbb{R}^k$. There are only $2(3^k)$ different vectors, and this remains true if we now project the configuration into the plane in one of the infinitely many directions that will not cause three points to become collinear. J. Pach recently made the observation that the same construction shows that $h(n) < cn^{\alpha}$, where $h(n)$ denotes the minimum number of different distances among n points in general position in the plane (see section 5.4).

Jamison (1985) has written a survey of the directions problem, telling the history. Using Ungar's result, Jamison has also proved the curious result that n points in the plane not all collinear always determine a tree whose edges are line segments determined by the points having all different slopes, and he conjectures that there is always a path with $n - 1$ different slopes (Jamison 1987). It might be interesting to show that there was always a tree of small depth that covered most of the points. For example, the strong form of Dirac's conjecture (see section 3.9) would imply that there was always a tree of depth one (a star) which covered at least half of the points.

We have a related problem: let n red points be given in the plane, not all on a line, and form the t connecting lines. What is the minimum number of blue points that would be needed, so that each of the t lines has at least one blue point on it? Let the minimum number be denoted by $f(n)$. The weak Dirac conjecture, now a theorem, states that there is a red point incident with at least cn red lines, and these would all need different blue points. Hence $f(n) > cn$, but with a very poor constant. The near-collinear set, with $t = n$, shows that $f(n) \leqslant n$. Clearly $f(n) \leqslant D(n) = 2[\frac{1}{2}n]$. Dean Hickerson constructed an example showing that $f(n) \leqslant n - 2$.

3.11. Planes and hyperplanes

An *arrangement of hyperplanes* in $\mathbb{R}^d$ or P^d is a finite collection of hyperplanes not all having a point in common. The hyperplanes divide the space into

d-dimensional regions called d-cells, and they also divide each k-dimensional flat into k-dimensional regions called k-cells.

Let n points be given in $\mathbb{R}^d$ such that no $A_d n$ lie in a hyperplane. Then Beck (1983) showed that the points determine at least $c_d n^d$ hyperplanes.

Let n points and t lines be given in the plane, without any special relation to each other. As mentioned previously, Szemerédi and Trotter showed that the number of incidences $I(n, t)$ between them satisfies $I(n, t) < cn^{2/3}t^{2/3}$. Many results quoted in this chapter follow from their inequality.

It would be nice to find a three-dimensional version. If no three points are collinear, then it is not hard to show that the number of incidences between π planes and n points is less than $c(\log n)n^{3/4}\pi^{3/4}$. The factor $\log n$ can probably be eliminated. An example due to Noga Alon shows that the inequality is not true if we remove the restriction on collinear points.

Edelsbrunner et al. (1990) proved that the number of incidences is $O(n^{2/3}\pi \log \pi)$ if the n points are points of intersection formed by the π planes.

4. Other geometries

4.1. Complex space

Every nondegenerate homogeneous cubic equation with real coefficients can be put in the form

$$x^3 + y^3 + z^3 + cxyz = 0\,,$$

by a real homogeneous substitution. The nine points of inflection are then given by the scheme

$$\begin{array}{lll} (0, -1, 1)\,, & (0, -\omega^2, \omega)\,, & (0, -\omega, \omega^2)\,, \\ (1, 0, -1)\,, & (\omega, 0, -\omega^2)\,, & (\omega^2, 0, -\omega)\,, \\ (-1, 1, 0)\,, & (-\omega^2, \omega, 0)\,, & (-\omega, \omega^2, 0)\,, \end{array}$$

where ω is a primitive cubic root of unity.

Three points of inflection are collinear if (a) they lie in the same row of the scheme, (b) they lie in the same column, or (c) no two of the three points lie either in the same row or in the same column. For more details see, e.g., Hilton (1932, chapter 14).

The nine points of inflection of the general cubic have the interesting property that a line through any two of them contains a third. Presumably it was this example which prompted Sylvester to pose his now famous problem.

At this stage the reader might reasonably conjecture that if n is sufficiently large, then there is always a Gallai line. Such an expectation is contradicted by the following example in Motzkin (1951).

For any $n \geqslant 3$, let ω be a primitive complex nth root of unity. Let the $3n$ points

P_i, Q_j, and R_k have the following homogeneous coordinates:

$$P_i(\omega^i, -1, 0) \quad \text{for } 1 \leqslant i \leqslant n,$$
$$Q_j(0, \omega^j, -1) \quad \text{for } 1 \leqslant j \leqslant n,$$
$$R_k(-1, 0, \omega^k) \quad \text{for } 1 \leqslant k \leqslant n.$$

The three points P_i, Q_j, and R_k are collinear if and only if n divides $i+j+k$ exactly. This may be seen by observing that the three by three determinant formed from their coordinates, which is equal to $-1+\omega^{i+j+k}$, is zero precisely when $i+j+k$ is a multiple of n. Hence every line formed from a P_i and a Q_j will have an R_k on it, and so on. Of course $P_1, \ldots, P_n$ also lie on a line together, as do the Q's and the R's. Hence there are no Gallai lines, so long as n is at least 3.

4.1.1. *Some results from algebraic geometry*

Hirzebruch (1983) studied arrangements of lines in the complex projective plane by associating with each arrangement some algebraic surfaces, calculating their Chern numbers, and then applying the Miyaoka–Yau inequality (Miyaoka 1977, Yau 1977). Let t_i denote the number of points incident with exactly i of the lines. If n is the number of lines and $t_n = t_{n-1} = 0$, then the Miyaoka–Yau inequality implies that

$$t_2 + t_3 \geqslant n + t_5 + 2t_6 + 3t_7 + \cdots.$$

Hirzebruch (1984) obtains the following stronger result: if $t_n = t_{n-1} = t_{n-2} = 0$, then

$$t_2 + \tfrac{3}{4}t_3 \geqslant n + t_5 + 3t_6 + 5t_7 + \cdots.$$

If we are given n points in the complex plane $\mathbb{C}^2$, with at most $n-2$ of them on any line, and if t_i is the number of lines containing exactly i of the points, then by the usual duality argument applied in the complex projective plane the weaker Hirzebruch inequality must hold. If we make the stronger assumption that at most $n-3$ of the points are collinear, then the stronger inequality holds.

If the numbers t_j arise from real points $P_i(x_i, y_i)$, then the same t_j will result if the points P_i are situated in $\mathbb{C}^2$. This means that Hirzebruch's inequalities must also hold in $\mathbb{R}^2$. Erdős asks for an independent elementary proof of these inequalities in $\mathbb{R}^2$, since they do not seem to follow from Melchior's inequality. Purdy asks what Hirzebruch's methods would give if it were assumed that $t_n = t_{n-1} = \cdots = t_{n-r} = 0$.

4.1.2. *A problem of Serre*

No one has ever observed a finite set of points spanning $\mathbb{C}^3$ that do not determine a Gallai line, and Serre (1966) asked whether such a set can exist. Kelly (1986) answered this question by showing that indeed n points spanning $\mathbb{C}^3$ always determine a Gallai line. Here is a sketch of his proof, which uses Hirzebruch's weaker inequality: let S be a finite set of points spanning $\mathbb{R}^3$ without a Gallai line, and let $P \in S$. The lines through P and other points of S intersect the plane π at infinity in a set of points S^*. If the points of S^* formed a Gallai line, this would imply the existence of a Gallai line in the original configuration S. Hence $t_2 = 0$ for S^*, and $t_3 > 0$ even by Hirzebruch's weaker inequality. The 3-point line in the plane π corresponds to three lines through P lying in a plane π' and containing a smaller configuration of points also lacking a Gallai line. This very special configuration, consisting of a finite set of points confined to three lines in a plane, and the three lines passing through one of the points, and every two points collinear with a third point can be used to define a finite additive subgroup of the underlying field. This is absurd, since the field has characteristic zero.

4.2. *Sets of points as matroids*

4.2.1. *The Erdős–de Bruijn theorem*

Erdős and de Bruijn proved the following general theorem (de Bruijn and Erdős 1948): if $S_1, \ldots, S_m$ are $m > 1$ subsets of an n-element set S such that every pair of elements of S is in precisely one S_i, then $m \geqslant n$. Moreover, equality can only occur if $n - 1$ elements are in one S_i and two elements are in all the other S_i, or if $n = 1 + k(k-1)$ and $|S_1| = \cdots = |S_m| = k$, and each element occurs in exactly k of the sets. The second example is called a finite projective plane.

4.2.2. *Kelly's conjecture*

Consider the following situation: let $L_1, \ldots, L_n$ be n pairwise skew lines in $\mathbb{R}^4$ and let $H_1, \ldots, H_m$ be the hyperplanes spanned by them. If S_i is the set of lines L_j contained in H_i, then the Erdős–de Bruijn theorem applies and gives $m \geqslant n$. On the other hand, Kelly observed that the corresponding Sylvester–Gallai theorem is not true, i.e., there is not always a hyperplane incident with only two of the given lines. Kelly defines an S–G configuration (Sylvester–Gallai) to be a set of n points, for some n, not all collinear, such that the line joining any two contains at least one other (Kelly 1986), and he observed that the obvious one-to-one mapping between points of $\mathbb{C}^2$ and points of $\mathbb{R}^4$, together with duality in projective $\mathbb{C}^{2+}$ and $\mathbb{R}^{4+}$ provides a method of converting any S–G configuration in $\mathbb{C}^2$ into an analogous line arrangement in $\mathbb{R}^4$. Thus, if for some n there are n points in $\mathbb{C}^2$ not all on a line such that the line spanned by any two of them contains a third, then there are n pairwise skew lines in $\mathbb{R}^4$ such that the hyperplane spanned by any two of them contains a third. There are several S–G configurations discussed in section 4.1 and they all produce analogous examples of pairwise skew lines. Kelly (1986) also conjectures, with a misprint that is corrected in Boros et al. (1989), that there is always a hyperplane incident with either two or three of the lines, and points out that if this were proved without using the Hirzebruch inequalities, then it would give an independent proof of the Sylvester–Gallai theorem (see section 4.1) for n points spanning $\mathbb{C}^3$ – the reverse implications is not true. Unfortunately, Kelly's conjecture was disproved by Bokowski and Richter.

4.2.3. *Simple matroids*

For the rudiments of matroid theory we refer the reader to chapter 9 or the book Welsh (1976).

Given n points in $\mathbb{R}^d$ or $\mathbb{C}^d$ (or the corresponding projective spaces, or analogues over finite fields), we can define a simple matroid on these points in the following manner: if a subset S of these points lies in a k-dimensional flat, but not a $(k-1)$-dimensional flat, then S has rank k. In this matroid, the kth Whitney number W_k is the number of closed sets of rank k in the matroid, which will be the number of flats of dimension $k-1$ that are determined by the n points. It follows from the Erdős–de Bruijn theorem that for any simple matroid on n *elements* $W_2 \geqslant n$. It is also known that the number of hyperplanes W_{d-1} is at least n, see Welsh (1976). This is the generalization of Motzkin's result that the number of hyperplanes determined by n points spanning $\mathbb{R}^d$ is at least n (see section 2.3).

The Sylvester–Gallai theorem does not hold in the context of simple matroids. For example, the projective plane over Z_p, which forms a simple matroid of rank 3 on $n = p^2 + p + 1$ points, has exactly n lines, each containing $p+1$ points.

4.3. *The geometric lattice of flats*

Given a finite number of hyperplanes in P^d they form, together with their intersections, a geometric lattice ordered by reverse inclusion. The corresponding Möbius function becomes a useful tool for studying arrangements as the dimension increases. Hyperplanes in $\mathbb{R}^d$ only form a semilattice, since two hyperplanes might be parallel.

The flats of a simple matroid, when ordered by reverse inclusion, form a geometric lattice. All geometric lattices can be achieved in this way.

4.3.1. "*Facing up to arrangements*"

In his "Facing up to arrangements", Zaslavsky (1975) examined the geometric lattice of intersections of the flats ordered by reverse inclusion and derived formulas for the number of k-dimensional cells involving the standard lattice functions, such as the Möbius function $\mu(x, y)$.

For example, he shows that the number of regions determined by an arrangement A of hyperplanes in real projective space P^d is $c(A) = \frac{1}{2}(-1)^{r(A)}\chi(-1)$, where $\chi(y)$ is the characteristic polynomial

$$\chi(y) = \sum_t \mu(0, t) y^{r(A)-k(t)} ,$$

$r(A)$ is the rank of the lattice of intersections associated with A, and $k(t)$ is the length of the longest chain in the lattice below t. He also derives formulas for the number of bounded and unbounded regions in $\mathbb{R}^d$.

In addition to the application of his formulas that we will discuss below, they are used in the proof of a generalization of a theorem of Hadwiger due to Goodman and Pollack (1988), see section 6.5. The formulas also imply that the number of cells of an arrangement depends only on the lattice of intersections of the flats.

For a more geometrical treatment of formulas for the cell numbers and flat numbers, albeit limited to spaces of three of fewer dimensions, see Alexanderson and Wetzel (1981).

4.3.2. *Slimmest arrangements of hyperplanes*

In his book *Convex Polytopes*, Grünbaum (1967) conjectured that the smallest possible number of k-cells of a projective d-arrangement of n hyperplanes is that achieved by the *near pencil*, consisting of $n-d+1$ hyperplanes concurrent in a coline and $d-1$ more hyperplanes in general position to make the whole thing a d-arrangement. Shannon (1976) proved this conjecture, characterized all of the minimal arrangements, and showed that the same configuration also minimizes the number of k-flats. Independently of Shannon's solution, Zaslavsky had noticed that he could prove Grünbaum's conjecture by using the formulas in Zaslavsky (1975) and the following result of Dowling and Wilson (1974): the minimum value of $|w_k|$ over all geometric lattices with rank $r \geq 4$ and n atoms occurs when and only when the lattice is isomorphic to the direct product of a line and a free geometry. Here $|w_k|$ is the absolute value of w_k, which is the kth *Whitney number of the first kind*. The W_k are sometimes called Whitney numbers of the second kind instead of merely Whitney numbers. The minimum of W_k is also determined in Dowling and Wilson (1974), and this implies Shannon's result concerning the minimum number of k-flats.

Zaslavsky (1985) develops the necessary combinatorial machinery so that he can settle the analogue of Grünbaum's conjecture for arrangements of hyperplanes in $\mathbb{R}^d$. He shows that the minimum number of k-flats and k-cells is attained when the configuration is *a Euclidean near pencil*, which is more complicated then a projective near pencil.

4.3.3. *Back to P^d*

Makai and Martini have determined the minimum number of d-cells when the hyperplanes are assumed to be in general position. They showed (Makai and Martini 1989) the following. Let there be given an arrangement of n hyperplanes in real projective d-space, where no three of them have a $(d-2)$-dimensional intersection. Then the number of projective d-cells is smallest when the n hyperplanes are arranged as follows. Consider in a 2-flat an arrangement of $n-d+2$ lines $L_1, \ldots, L_{n-d+2}$, no three of them passing through one point. Say this 2-flat is spanned by the origin O and the first two unit vectors $\boldsymbol{e}_1$ and $\boldsymbol{e}_2$. Take now the third unit vector $\boldsymbol{e}_3$ and consider all 2-flats spanned by L's and $\boldsymbol{e}_3$, and the one spanned by O, $\boldsymbol{e}_1$, and $\boldsymbol{e}_2$. Then $n-d+3$ 2-flats are obtained. Consider the 3-flats spanned by them and the fourth unit vector $\boldsymbol{e}_4$, and the one spanned by O, $\boldsymbol{e}_1$, $\boldsymbol{e}_2$, and $\boldsymbol{e}_3$. Repeat this until n $(d-1)$-flats are constructed.

Perhaps more general results on geometric lattices could be proved which would imply the above result of Makai and Martini.

4.3.4. Matroids minimizing i_k

At the end of a paper containing more significant results, Elliott (1967) gave a simple proof that if n points are given in the plane, then the number of distinct noncollinear triples that they determine is smallest when $n-1$ of the points are collinear. Purdy (1982a) generalized this result to matroids, showing that the number i_k of independent sets of size k on a simple matroid of rank $d+1$ on n points is a minimum when $n-d+1$ points lie on one line and the other $d-1$ points are in general position, with a similar result for arbitrary matroids. This result was simultaneously proved by Björner (1980b), who applied some of his results in Björner (1980a) on shellable posets to prove the more general result that if a matroid has no circuits with fewer than c elements, then i_k is minimized if as many as possible of the points lie in a flat of rank c, and the rest are in general position (see chapter 34).

Björner's result for $c=3$ implies that for n points in $\mathbb{R}^3$, with no three points collinear, the number of triangles i_3 is minimized if all but one of the points are on a plane in general position. The same configuration minimizes i_4.

4.3.5. The unimodality of W_k and i_k

G. C. Rota conjectured that the Whitney numbers are unimodal. That is, $W_j \geqslant \min\{W_i, W_k\}$ if $i \leqslant j \leqslant k$. Welsh (1976) conjectures that the Whitney numbers are log concave, i.e., $W_i^2 \geqslant W_{i-1}W_{i+1}$. For simple matroids of rank four, this would say that $t^2 \geqslant n\pi$, where t is the number of lines, n is the number of points, and π is the number of planes. Mason (1972) has conjectured that $t^2 \geqslant 3n(n-1)\pi/2(n-2)$, with equality only when every plane contains exactly three points, and this has come to be known as the "points–lines–planes conjecture". Seymour (1982) proved it in the special case that no five points are collinear. His paper also contains a very simple proof of the weaker $t^2 \geqslant n\pi$ if no 4 points are collinear. Of course, both of these results apply to the simple matroid formed by n points spanning $\mathbb{R}^3$. In that special context Purdy (1986) showed that $t^2 \geqslant cn\pi$, without any other restrictions.

Mason (1972) also conjectured that

$$i_k^2 \geqslant i_{k-1}i_{k+1} \quad \text{for} \quad 0<k<n ,$$

and this was proved by Dowling (1980) for $k \leqslant 7$.

4.4. Allowable sequences

The *allowable n-sequences* of Goodman and Pollack are very closely related to arrangements of pseudolines in the real projective plane. We give here a brief account, and we refer the reader to the papers, e.g., the expository paper (Goodman and Pollack 1981), for more details.

Let $P_1, \ldots, P_n$ be n points in the plane, and let L be a line. If L is perpendicular to the line P_iP_j, then the orthogonal projections of P_i and P_j onto L will coincide. If L is not perpendicular to any of the lines P_iP_j, $1 \leqslant i < j \leqslant n$, then the projected points will be distinct and will occur in some order on L, defining a permutation of the indices 1, 2, . . . , n. If L rotates counterclockwise, the permutation changes every time L passes through a direction perpendicular to a line of the form P_iP_j. When L has turned through an angle π, the permutation will be n, $n-1, \ldots, 3, 2, 1$. The line L is to be imagined rotating forever, producing a periodic sequence of permutations $\Pi_1, \Pi_2, \ldots$.

An example of a sequence arising in this way is ..., 1(23)4(56), 13(246)5, (136)425, 63(14)25, 634(125), 6(345)21, (65)4(23)1, ... We have put parentheses around the blocks that are to be reversed in the next move.

Goodman and Pollack noticed that the infinite periodic sequence of permutations $\Pi_1, \Pi_2, \ldots$ arising in this way has the following properties:

(1) The move from Π_k to Π_{k+1} consists of reversing one or more disjoint blocks of indices.

(2) If (i, j) and (r, s) are two pairs of indices, then the pair (r, s) will reverse exactly once between every two reversals of (i, j), unless (r, s) and (i, j) are reversed together, in which case they will always be reversed together.

Any periodic sequence Σ of permutations of 1, 2, . . . , n satisfying (1) and (2) are called by Goodman and Pollack an *allowable sequence of permutations*, or an *allowable n-sequence*. One of their properties is that permutations Π and Π' a half-period apart are the reverses of each other.

Goodman and Pollack noticed that one could determine several geometric properties of the points $P_1, \ldots, P_n$ by applying rules to the allowable n-sequences:

(a) If P_k is on the line P_iP_j, then the indices i, j and k will be reversed together in the same block, and not otherwise.

(b) If the two lines P_rP_s and P_iP_j are parallel, then the pairs (r, s) and (i, j) will be reversed in different blocks, but at the same time, i.e., in the same move from Π_k to Π_{k+1}, and conversely.

They also noticed that two properties involving convexity could be determined from the allowable n-sequences. Let C be the convex closure of $P_1, \ldots, P_n$. Then

(c) P_i is an extreme point of C if and only if i is the first index in some term of Σ (and therefore the last in some other term).

(d) P_i is in the convex closure of $P_{i_1}, \ldots, P_{i_k}$ if and only if there is no term of Σ in which i precedes all of $i_1, \ldots, i_k$.

An allowable sequence can also be associated with an arrangement A of lines $\{L_1, \ldots, L_n\}$ in the real projective plane. Consider the intersection points to be in the finite part of the plane, and choose a direction to be horizontal which is not parallel to any of the lines. Let L be a horizontal line just above the highest intersection point, and write down the order (left to right) in which the lines L_i intersect L, and write down the different orders as L moves down parallel to itself. Whenever L passes an intersection point P the lines going into P will come out in the reverse order. When L has passed the bottom intersection point, one half-period of an allowable sequence Σ will have been traversed, and this is enough to define Σ.

It is easy to reverse the correspondence and assign an arrangement of pseudolines to every allowable sequence, getting a sort of "wiring diagram" with wires crossing whenever blocks are reversed. An example of an allowable sequence that is not realizable with straight lines (and therefore by duality not realizable by points either) is . . . , (12)345, 21(345), 214(35), 2(14)53, (24)(15)3, 4(25)(13), (45)(23)1, 54321, ...

Goodman and Pollack (1984b) show that one can determine whether two arrangements of pseudolines A and A' are isomorphic by looking at their associated allowable n-sequences. They also proved Helly's theorem, Radon's theorem, and Caratheodory's theorem, and some others in the context of allowable sequences (Goodman and Pollack 1980b,c). Peter Ungar completely solved the problem on the minimum number of directions determined by n points by using allowable sequences (see section 3.10).

5. Metric problems

5.1. Repeated distances

Let $d_k(n)$ be the maximum number of times that the distance 1 can occur among n points in $\mathbb{R}^k$. In 1946, Erdős showed that

$$n^{1+c/\log\log n} < d_2(n) < cn^{3/2}$$

and conjectured that the lower bound was the correct order of magnitude (Erdős 1946).

The lower bound is given by an m by m portion of the integer lattice, where m is approximately $n^{1/2}$, and the upper bound can be proved (although it was not in the original paper) by the following graph argument:

Let G be the *unit-distance graph*, i.e., the graph whose vertices are the n points, joined by an edge if and only if they are at distance 1. Then G cannot contain a complete bipartite Kuratowski $K_{2,3}$ subgraph, since this would require that there exist points X_1, X_2, Y_1, Y_2, Y_3 in the plane such that the six distances $d(X_i, Y_j) = 1$, which is impossible in the plane.

A theorem on extremal graph theory by Kővári et al. (1954) states that such a graph can have at most $cn^{3/2}$ edges. The general graph theorem is that a graph G not containing a Kuratowski $K_{r,s}$ subgraph with $r < s$ has at most $C_s n^{2-1/r}$ edges, where C_s is a positive constant depending only on s; see chapter 23.

Erdős also conjectured that if $X_1, \ldots, X_n$ are any n distinct points in the plane there is always an X_i that has fewer than $n^{c/\log\log n}$ points equidistant from it. The lattice points show that if true this is best possible. More generally, denote by $h(X_i)$ the largest number of points equidistant from X_i. Perhaps

$$h(X_1) + \cdots + h(X_n) < n^{1+c/\log\log n} .$$

In 1975, Jozsa and Szemerédi showed that $d_2(n) = o(n^{3/2})$. Beck and Spencer (1984) improved this to $d_2(n) < n^{13/9}$, and Spencer et al. (1984) improved this to $d_2(n) < cn^{4/3}$.

Erdős (1960) showed that in $\mathbb{R}^3$,

$$cn^{4/3} < d_3(n) < cn^{5/3} .$$

The upper bound can be proved by a similar graph argument, although it was proved differently in the original paper: the unit-distance graph G cannot have a complete bipartite Kuratowski $K_{3,3}$ subgraph, and so G has at most $cn^{5/3}$ edges by the same result of Kővári et al. (1954).

The lower bound comes as follows: take the points (x, y, z) such that $0 < x, y, z < n^{1/3}$. If d is a distance that does occur, then d^2 is an integer of the form $u^2 + v^2 + w^2$, where $0 \leqslant u, v, w < n^{1/3}$. Hence $d^2 \leqslant 3n^{2/3}$. Since every pair formed from the n points must have a distance, clearly some distance must occur at least $cn^{4/3}$ times. Erdős (1960) points out that the lower bound can be improved by a factor of $\log \log n$ using deep number-theoretic methods.

Beck (1983) improved the upper bound, so that

$$cn^{4/3} \log \log n < d_3(n) < n^{5/3-c} , \quad \text{where } c > 0 .$$

In the same paper, Beck mentioned that he can show that

$$d_3(n) < n^{13/8+\varepsilon} \quad \text{for } n > n_0(\varepsilon) \quad (13/8 = 1.625) .$$

Chung (1989) improved this to $d_3(n) < cn^{8/5}$. She achieved this by showing that the number of incidences between n points and t spheres is at most $c(n^{4/5}t^{4/5} + tn^{1/2} + n)$, provided no three of the spheres are collinear. She has generalized this result to higher-dimensional spheres. (See also section 5.10.)

In $\mathbb{R}^4$ things change drastically, thanks to a construction due to H. Lenz: let

$$P_i = (u_i, v_i, 0, 0) \quad \text{for } 1 \leqslant i \leqslant [\tfrac{1}{2}n]$$

and

$$Q_j = (0, 0, u_j, v_j) \quad \text{for } 1 \leqslant j \leqslant [\tfrac{1}{2}n] ,$$

where (u_i, v_i) are distinct solutions of the equation $u^2 + v^2 = \frac{1}{2}$. If necessary, the points P_i can be placed arbitrarily close together, as can the points Q_j. Every distance $d(P_i, Q_j)$ is unity. The construction shows that $d_4(n) \geqslant [\frac{1}{4}n^2]$.

For $k \geqslant 4$, Erdős (1960) showed that as n tends to infinity

$$d_k(n) = \tfrac{1}{2}n^2 - n^2\{2[\tfrac{1}{2}k]\}^{-1} + o(n^2) .$$

Here is a rough idea of the proof for even dimensions. The implied lower bound is obtained by a generalization of the Lenz example. To see the upper bound, let a configuration of n points in $\mathbb{R}^{2m}$ be given and construct the graph G with these points as vertices and joined by an edge if they are at distance 1. By elementary geometry you cannot have sets $S_1, \ldots, S_{m+1}$ of three points each such that the nine distances between the points of S_i and the points of S_j are unity for every i and j, since this would force the existence of $m+1$ mutually orthogonal planes, an impossibility in $\mathbb{R}^{2m}$. It is not even possible if S_1 has only two elements.

Erdős and Stone (1946) then have a result stating that such a graph G cannot have more than the requisite number of edges. Specifically they showed that every graph on n vertices having at least

$$\tfrac{1}{2}n^2\left(1-\frac{1}{d}\right)+cn^2$$

edges contains a complete $(d+1)$-partite graph with t vertices in each part for any fixed t. Subsequent work by Bollobás, Erdős, Simonovits, Szemerédi, and Chvátal has shown the result to be true even if t increases with n, provided $t \leqslant c' \log n$, where c' depends on c. We refer the reader to chapter 2.3.

Moser (1966) conjectured that among n points on the unit sphere in $\mathbb{R}^3$, no distance D can occur more than cn times. This was recently disproved in (Erdős et al. 1990a), where it is shown that for any distance D, $0<D<2$, there are n points on the unit sphere such that each one is at distance D from at least $c \log^* n$ others. Here $\log^* n$ denotes the minimum integer r such that, starting with n, if one iterates the logarithm function r times one obtains a value $\leqslant 1$. This result implies that the distance D occurs at least $\frac{1}{2}c(n-1)\log^* n$ times.

5.1.1. *Nearly equal distances*

Erdős et al. (1990b) prove that for a sufficiently large configuration of n points in the plane with minimum distance at least 1, the number of pairs whose distance is between t and $t+1$ cannot exceed $[\frac{1}{4}n^2]$ for any real t. Their method generalizes readily to higher dimensions.

5.2. *Maximum and minimum distances*

5.2.1. *Large distances*

Let $M_d(n)$ be the maximum number of times that the maximum distance can occur between n points in $\mathbb{R}^d$. Pannwitz and Hopf (1934) and Sutherland (1935) showed that $M_2(n)=n$. A. Vaszonyi conjectured that $M_3(n)=2n-2$, and this was proved independently by Grünbaum, Heppes, and Straszewicz. Lenz's construction done carefully shows that $M_4(n) \geqslant [\frac{1}{4}n^2]$. (The points on each circle must be kept close together.)

For which sets does equality occur? Let us call a set of n points in $\mathbb{R}^d$ (n,d)-best if it has diameter 1 and the distance 1 occurs $M_d(n)$ times. In $\mathbb{R}^2$ the $(n,2)$-best sets consist of points on the boundary of a Reuleaux m-gon. Neaderhouser and Purdy (1982) give a proof of this, and two infinite families of $(n,3)$-best sets. It seems to be a difficult problem to characterize the $(n,3)$-best sets.

Let n_k denote the maximum number of times that the kth largest distance occurs among n points in the plane. Then $n_1 = M_2(n)$. Vesztergombi (1985) has shown that $n_1 + 2n_2 \leqslant 3n$ and $n_2 \leqslant n + n_1$. In the same paper she showed that $n_k \leqslant kn$.

Let $d_1 > d_2 > d_3 > \cdots$ denote the different distances occurring among n points in the plane. Consider the graph G_k whose vertices are the n points, and two are joined by an edge if their distance is at least d_k. Erdős et al. (1989) show that the chromatic number $\chi(G_k)$ of G_k is at most 7 if $n \geqslant ck^2$, and 7 is best possible.

5.2.2. Farthest neighbors

Recently, Avis et al. (1988) have studied an interesting problem called the *farthest neighbor problem.* Given a set $S = \{X_1, \ldots, X_n\}$ of n points in $\mathbb{R}^k$, for each point X_i let $f(X_i)$ be the number of points of S at maximal distance from X_i, the farthest neighbors of X_i. Then define $F_k(n)$ to be the maximum of $f(X_1) + \cdots + f(X_n)$ over all possible sets S in $\mathbb{R}^k$. The problem is to estimate $F_k(n)$.

In 1984, Avis showed that $F_2(n) = 3n - 3$ if n is even (Avis 1984). Edelsbrunner and Skiena (1989) have shown that $F_2(n) = 3n - 4$ if n is odd. For $k \geqslant 4$ preliminary results suggest that the bounds for $F_k(n)$ are similar to those for $d_k(n)$, where the generalized Lenz construction provides the maximum.

However, $F_3(n)$ achieves its maximum value on an entirely different set: divide the n points evenly between a circle C and a line L through the center of C and orthogonal to it. They prove:

$$\tfrac{1}{4}n^2 + \tfrac{1}{2}3n < F_3(n) < \tfrac{1}{4}n^2 + \tfrac{3}{2}n + 255\,.$$

Chung (1989) showed that the number of farthest neighbors is at most $cn^{8/5}$ if no three points are collinear.

5.2.3. Minimum distance

Let $m_d(n)$ be the maximum number of times that the minimum distance can occur between n points in $\mathbb{R}^d$. That $m_2(n) < 3n$ follows from the observation that six equal circular discs may be placed around another disc of the same size so that the central one is touched by all others but no two overlap and that it is not possible to place seven discs in this way. Erdős showed that

$$3n - c_1 n^{1/2} < m_2(n) < 3n - c_2 n^{1/3}$$

and conjectured that

$$m_2(3n^2 + 3n + 1) = 9n^2 + 6n\,,$$

which comes from considering a large hexagonal chunk of a hexagonal packing of circles. Harborth (1974) proved this striking conjecture and showed that

$$m_2(n) = [3n - (12n - 3)^{1/2}] \quad \text{for all } n\,.$$

In an analogous three-dimensional situation around a given ball it is possible to place 12 balls of equal size, all touching the first one but not overlapping it or each other. It appears that with judicious positioning it might be possible to squeeze in a 13th sphere. This dilemma between 12 and 13 was a subject of a long-standing argument between Sir Isaac Newton and David Gregory starting in 1694. Newton believed 12 to be correct, but this was not proved until 180 years later. Among Gregory's unpublished papers H. W. Turnbull found notes dated 1694 of an actual conversation that took place between the two men. They had been discussing the distribution of stars of different magnitudes when the problem came up. It was R. Hoppe who eventually proved that the 52-year-old Newton had been right – see Bender (1874). There have been several proofs since then, of which the simplest is probably that due to Leech (1956).

Hoppe's result gives $m_3(n) < 6n$, and in fact $m_3(n) = 6n + O(n^{2/3})$. For $d > 3$ only two "touching numbers" r_d are known exactly ($r_8 = 240$ and $r_{24} = 196\,560$). For more details see Coxeter's (1963) survey. See also Fejes-Toth's excellent and more recent survey on packing and covering (Fejes-Toth 1983).

5.3. *Different distances*

A related but different question is to ask: given n points in $\mathbb{R}^k$, what is the minimum number of different distances that must occur. Let $D_k(n)$ denote the minimum taken over all configurations of n points in $\mathbb{R}^k$. The first result was due to Erdős:

$$cn^{1/2} < D_2(n) < cn/(\log n)^{1/2} .$$

The lower bound can be proved as follows. Let P be one of the n points. If there are at least $(n-1)^{1/2}$ distances from the point P, then the result follows. We may therefore assume that there are not. Then there must be a circle around P containing $m \geq (n-1)^{1/2}$ points. Let Q be one of those points, and let $Q_1, \ldots, Q_{m-1}$ be the others. There must be at least $\frac{1}{2}(m-1)$ different distances $d(Q, Q_i)$, and the result follows.

The upper bound is obtained from an m by m portion of the integer lattice, where $m^2 \leq n < (m+1)^2$. If d is a distance occurring between two of the points, then d^2 is an integer of the form $u^2 + v^2$, and d^2 does not extend $2n$. It is an old result of Landau (1909) that the number of such integers is less than $cn/(\log n)^{1/2}$, and the result follows.

Leo Moser improved the lower bound to $cn^{2/3}$ (Moser 1952), showing that in fact there is always one point from which there are at least that many distances (as the Erdős proof above also shows that $cn^{1/2}$ distances occur from one point). In 1982, Chung showed that $D_2(n) > cn^{5/7}$ (Chung 1984), but not necessarily from one point. Beck (1987) was able to show that there is always a point from which there are more than $n^{2/3+c}$ different distances. In fact, he showed that if $n > n_0(\varepsilon)$ then there is always a point from which there are at least $n^{58/81-\varepsilon}$ different distances. Since $58/81 > 5/7$, this is also an improvement on Chung's result. Chung, Szemerédi and Trotter have recently improved the result to $cn^{4/5}$.

Erdős (1957, 1970) asked the following question: what is the smallest number $f(k, d)$ so that $f(k, d)$ points in $\mathbb{R}^d$ always contain a set of k points whose distances are all different? He conjectured that $f(k, 1) = \{1 + o(1)\}k^2$, and pointed out that a result in Erdős and Turan (1941) showed that $f(k, 1) \geq \{1 + o(1)\}k^2$.

Erdős and Guy (1970) show that out of the n^2 points on the n by n integer lattice, a set of k points exists all of whose distances are distinct. Here $k > n^{2/3-e}$, for any $e > 0$, for all $n > n(e)$.

5.3.1. Two-distance sets

A set of n points in $\mathbb{R}^d$ is a *two-distance set* if there are two distinct distances α and β such that the distance between any pair of points in the set is either α or β and both distances are realized. Kelly (1947) posed the problem of determining $f(d)$, the largest possible cardinality of a two-distance set in $\mathbb{R}^d$, and he showed that $f(2)=5$, realized by the regular pentagon. Croft (1962) shows that $f(3)=6$, realized by the mid-points of the edges of a regular tetrahedron. The $\frac{1}{2}d(d+1)$ mid-points of the edges of a regular simplex show that $f(d) \geq \frac{1}{2}d(d+1)$. Larman et al. (1977) show that $f(d) \leq \frac{1}{2}(d+1)(d+4)$. They also show that $f(d) \leq 2d+3$ unless α^2/β^2 is the ratio of two consecutive integers. They use a result by Wilson (1975) on the existence of balanced incomplete block designs to show that $f(d) > \frac{1}{2}d(d+1)[1-4/d]$ if d is sufficiently large. Their construction gives $\alpha=1$ and $\beta=2^{1/2}$. Harborth and Piepmeyer (1990) show that if $\alpha=1$ and $\beta > \{2+3^{1/2}\}^{1/2}$, then the set has at most $d+1$ points. See also section 5.11. Blokhuis (1984) modified the proof in Larman et al. (1977) to show $f(d) \leq \frac{1}{2}(d+1)(d+2)$.

Let $f(s,d)$ be the largest possible cardinality of an s-distance set in $\mathbb{R}^d$. Bannai and Bannai (1981) showed that

$$f(s,d) \leq \binom{d+s}{s} + \binom{d+s-1}{s-1}.$$

Subsequently, the Bannais and Dennis Stanton (Bannai et al. 1983) eliminated the second term, showing:

$$f(s,d) \leq \binom{d+s}{s}.$$

5.4. General position

Many of the questions about the number of distances, etc. can be asked with the restriction of general position: no three points collinear, and sometimes also no four points on a circle.

Let $h(n)$ be the minimum number of distances determined by n points in the plane if not three points are collinear. Perhaps $h(n)/n$ tends to infinity as n tends to infinity, but we cannot even show that for $n > n_0$, $h(n) \geq n$.

Erdős had asked many times whether $h(d) = o(d^2)$, and recently J. Pach has proved this by showing that

$$h(n) < cn^{\alpha}, \quad \text{where } \alpha = (\log 2)/\log 3 = 1.58496\ldots.$$

The example he used to obtain this result is in fact the same one that Danzer and Grünbaum (1967) had used (see section 3.10) to obtain an upper bound on the number of different vectors $V'(n)$ determined by n points in general position. The Danzer–Grünbaum result implies Pach's result, since $2h(n) \leq V'(n) < cn^{\alpha}$, but we had not noticed this fact.

Let there be given n points in the plane which form a convex polygon. Erdős conjectured that these points determine at least $[\frac{1}{2}n]$ distinct distances. The regular polygon shows that if true this is best possible. Altman (1963) proved this.

See also Altman (1972). Erdős also conjectured that there is always one point so that the number of distinct distances from this point is also $[\frac{1}{2}n]$, and this is still open. Erdős also conjectured that there is one point that has no three points equidistant from it. L. Danzer constructed a convex 9-gon disproving this. The example is now published in Erdős (1985b). Erdős then conjectured that if $X_1, \ldots, X_n$ are the vertices of a convex polygon, then there is always an X_i that has no four vertices equidistant from it. This has neither been proved nor disproved.

Let $d_1 > d_2 > d_3 > \cdots$ denote the different distances occurring among the vertices of a convex n-gon. Consider the graph G_k whose vertices are the n points, and two are joined by an edge if their distance is at least d_k. Erdős et al. (1989) show that the chromatic number $\chi(G_k)$ of G_k is at most 3 if $n \geqslant ck^2$. Even if n is not assumed to be large, G_k has a vertex of degree at most $3k - 1$ and therefore $\chi(G_k) \leqslant 3k$. They conjecture that $\chi(G_k) \leqslant 2k + 1$. The regular $(2k+1)$-gon would show that this is best possible. It is possible that G_k always has a vertex of degree $\leqslant 2k$.

The conjecture that $\chi(G_k) \leqslant 2k + 1$ would imply Altman's result that there are at least $[\frac{1}{2}n]$ different distances as a special case. To see this, let k be the number of different distances, so that G_k is the complete graph and $\xi = \chi(G_k) \leqslant 2k + 1$. Then $k \geqslant \frac{1}{2}(n-1)$, which implies $k \geqslant [\frac{1}{2}n]$, since k is an integer.

Erdős once conjectured that if $X_1, \ldots, X_n$ are n points that determine only $o(n)$ distinct distances, then there always is a line which for $n > n_0(k)$ contains at least k points. Szemerédi proved this and then went on to conjecture that if $X_1, \ldots, X_n$ is such that no three of the points are on a line, then $X_1, \ldots, X_n$ determine at least $[\frac{1}{2}n]$ distinct distances, i.e., Altman's theorem has nothing to do with convexity (but only no three on a line). Unfortunately Szemerédi's proof only gives $[\frac{1}{3}n]$ instead of $[\frac{1}{2}n]$ and his conjecture remains open. For the proof, see Erdős (1975).

In Erdős and Purdy (1976) we used Karteszi's method to show the following. Let $f(n)$ denote the maximum number of times that unit distance can occur among n points in the plane if no three points lie on a line. Then $f(n) \geqslant cn \log n$, where c is approximately $2\{3 \log 3\}^{-1}$. See section 5.6 or Karteszi (1963) for an explanation of Karteszi's method.

5.5. Borsuk's problem

In 1932, Borsuk proved the following result: if the d-dimensional solid ball of diameter 1 is partitioned into d parts, then at least one of the parts has diameter 1. This is a consequence of the fact that if the sphere (the boundary of the solid ball) is covered by d closed sets, then at least one of the sets contains a pair of antipodal (i.e., diametrically opposite) points of the sphere.

The following year, Borsuk published his notorious problem: prove (or disprove) it is possible to partition every bounded set in $\mathbb{R}^d$ into $d + 1$ sets of smaller diameter. Clearly it is enough to consider closed sets of diameter 1. Since the convex hull of a set has the same diameter as the set itself, it would also be sufficient to solve the problem for closed convex sets.

In its generality, the problem has only been solved for $d=2$ and $d=3$. The most geometrically satisfying solution for the plane is probably the following. Every set of diameter 1 can be covered by a regular hexagon of edge length $1/\sqrt{3}$ and the hexagon can be split into three pieces of diameter $\frac{1}{2}\sqrt{3}=0.866\ldots$. There is a similar proof in three dimensions using a polyhedron obtained by a triple truncation of the regular octahedron. No one has found such a covering for sets in $\mathbb{R}^4$.

For $d \geqslant 4$ only special cases of Borsuk's problem have been solved. See Grünbaum's (1963) survey article for details and references. See also the survey article in the same volume by Danzer et al. (1963) on Helly's theorem and related questions.

The finite version of Borsuk's problem is particularly tantalizing: given n points in $\mathbb{R}^d$, can they always be partitioned into $d+1$ sets of smaller diameter? In other words, we assume that the given set in Borsuk's problem is finite. Whereas a negative answer to this question would of course imply a negative one for the original question, the reverse implication has never been established.

A simple induction argument shows that the finite Borsuk's problem has an affirmative answer in the plane. Let n points be given in $\mathbb{R}^2$. The result is true if $n=3$. Let $n>3$ and suppose the result is true for $n-1$. Since $M_2(n)=n$, it follows that the unit-distance graph G has at most n edges, so the average degree is $\leqslant 2$. Hence there is a vertex of degree $\leqslant 2$. That is, there is a point P which is not at distance 1 from more than two other vertices. With P removed, the remaining $n-1$ points can be divided into three sets of diameter less than 1, by the induction hypothesis. Put P in one of the three sets that does not contain any points at distance 1 from it. The result follows by induction.

A similar proof works for $d=3$. Since $M_3(n)=2n-2$, G has at most $2n-2$ edges and average degree less than 4. Hence there is a vertex of degree $\leqslant 3$ which will be a point having at most three points at distance 1 from it, and by induction the points can be divided into four sets of smaller diameter.

This approach fails at $d=4$, since $M_4(n) \geqslant [\frac{1}{4}n^2]$, and the problem is still open for $d \geqslant 4$.

5.6. The Hadwiger–Nelson problem

The problem of how many colors it takes to color the plane so that no two points of the same color are distance 1 apart has stimulated work on the chromatic number of unit-distance graphs and the related work on the dimension and faithful dimension of a graph.

More generally, E. Nelson and J. R. Isbell, and independently P. Erdős and H. Hadwiger, asked for the smallest number $h(d)$ such that the points of $\mathbb{R}^d$ can be colored by $h(d)$ colors so that no two points of the same color are distance 1 apart. Thus $h(d)=\chi(G(\mathbb{R}^d))$, i.e., the chromatic number of the unit-distance graph for the entire space. Clearly the number is the same if the distance 1 is replaced by any other distance. We could have just as well defined $h(d)$ to be the maximum of $\chi(G(S))$ over all finite subsets S of $\mathbb{R}^d$. This is a consequence of the following theorem.

Compactness theorem (de Bruijn and Erdős 1951). *Suppose that every finite subgraph of an infinite graph G has chromatic number* $\leqslant k$. *Then G itself has chromatic number at most k.*

The original proof used a theorem of Richard Rado on infinite matroids. All known proofs use the axiom of choice. The theorem is also a consequence of Gödel's compactness theorem.

5.6.1. In the plane

Hadwiger (1961) showed that $4 \leqslant h(2) \leqslant 7$. Erdős thinks that almost certainly $h(2) > 4$. The following seven-point example due to Leo Moser gives the simplest proof that $h(2) \geqslant 4$. Let O, P and Q be the vertices of an equilateral triangle of side 1. Let R be the reflection of O in the side PQ. Any 3-coloring must color O and R the same. Rotate the figure about O to obtain O, P', Q', and R' so that R and R' are at distance 1. Four colors are necessary to avoid a conflict between R and R'.

S. B. Stechkin has produced a decomposition of the plane into six sets, each of which fails to realize some distance, but unfortunately each distance is realized by some set, so it does not show that $h(2) < 7$ (see Larman and Rogers 1972).

The Moser graph has girth (smallest circuit size) 3. Wormald (1970) has a larger example with girth 5 that needs four colors.

5.6.2. In higher dimensions

Larman and Rogers (1972) used methods and results from the geometry of numbers to show that

$$cn^2 < h(d) < \{3 + o(1)\}^d .$$

After seeing their proof of the lower bound, Erdős made a conjecture about finite sets (there are at most $(2 - \varepsilon)^n$ subsets of an n-element set such that no two have an intersection of cardinality r, where $cn < r < c'n < \frac{1}{2}n$) which implied $h(d) > (1 + c)^d$. By proving a slightly different extremal-set-theory result, Frankl and Wilson (1981) were able to show that

$$h(d) \geqslant \{1 + 1/5 + o(1)\}^d .$$

Thus we know that $h(d)$ grows exponentially, but we do not know whether $h(d)^{1/d}$ tends to a limit as d tends to infinity. Frankl and Rödl (1987) did later prove Erdős's original set-theory conjecture. (See chapter 24.)

5.6.3. The dimension of a graph

Erdős et al. (1965) define the *dimension* of a graph G to be the smallest d such that the vertices of G can be represented as points in $\mathbb{R}^d$ so that points joining by an edge are distance 1 apart. Every graph has finite dimension, since the complete graph K_n on n vertices can be embedded in $\mathbb{R}^{d-1}$ as a regular simplex. They establish several facts:

(1) For any graph G, $\dim G \leqslant 2\chi(G)$.

(2) There exist graphs with arbitrarily high girth and chromatic number.

(3) If G has n vertices and girth $> c \log n$ for c sufficiently large, then $\chi(G) \leqslant 3$, and therefore $\dim G \leqslant 6$.

We make the remark that the above result of Larman and Rogers implies that $\dim G > c \log \chi(G)$ for some $c > 0$. Hence, (2) implies that there exist graphs of arbitrarily high dimension and girth.

Erdős and Simonovits (1980) show that $\dim G \leqslant \Delta(G) + 2$, where $\Delta(G)$ denotes the maximum degree of all the vertices of G. They also define the *faithful dimension* of a graph G to be the smallest d such that the vertices of G can be represented as points in $\mathbb{R}^d$ so that two points are joined by an edge if and only if they are at distance 1. They proved that the faithful dimension is bounded above by $2\Delta(G) + 1$. On the other hand, the faithful dimension is unbounded on the class of 2-chromatic graphs.

5.6.4. Regular graphs of faithful dimension 2

Karteszi's method has come to be a convenient term for the method used by Karteszi (1963) to construct a geometrical example, see section 3.7. We shall explain the method by applying it on the following problem of determining the smallest regular graphs of degree k that have faithful dimension 2.

That is, we define $f(k)$ to be the minimum number n so that there exists a set of n points in the plane so that the unit-distance graph G on these n points is regular of degree k, i.e., so that each point is at distance 1 from exactly k others.

Let S be a finite set of points such that the unit-distance graph $G(S)$ (in which points are joined by an edge when they are at distance 1) is regular of degree k. Let $S + \boldsymbol{a}$ denote the set obtained by translating S by the unit vector $\boldsymbol{a}$. Then every point $\boldsymbol{x}$ in S is at unit distance from its corresponding point $\boldsymbol{x} + \boldsymbol{a}$ in $S + \boldsymbol{a}$, so that one could almost say that the $2n$-element set T, obtained by combining the two sets S and $S + \boldsymbol{a}$, has an associated unit-distance graph $G(T)$ which is regular of degree $k + 1$. Unfortunately it might happen that a point $\boldsymbol{x}$ in S is at unit distance from a point $\boldsymbol{y} + \boldsymbol{a}$ in $S + \boldsymbol{a}$. Fortunately the simple procedure of rotating S can avoid this problem, since there are infinitely many possible rotated positions and only finitely many unit-distance occurrences to be avoided. This is essentially Karteszi's method, and it shows in this case that $f(k + 1) \leqslant 2f(k)$. Since $f(2) = 3$, this gives $f(3) \leqslant 6$, $f(4) \leqslant 12$, and more generally $f(k) \leqslant 3 \cdot 2^{k-2}$.

In 1974, Erdős asked whether $f(k) = 3 \cdot 2^{k-2}$, and Purdy, with the aid of a computer, produced an example G_4 showing that $f(4) \leqslant 9$. The regularity of G_4 as a line drawing makes it very beautiful, and it appears on the cover of at least one book (Bollobás 1986). After seeing G_4, which is the Cartesian product of two equilateral triangles, we realized that duplication can be replaced by triplication to show that $f(k + 2) \leqslant 3k$. More generally, by slightly perturbing the Cartesian product of a graph of degree r and one of degree k, we see that $f(k + r) \leqslant f(k)f(r)$, which shows the full power of Karteszi's method. Harborth (1985d) showed that the above constructions are optimal when $k \leqslant 5$. See also Harborth (1983).

5.6.5. Minimal forbidden unit-distance graphs

The computer search for small, regular unit-distance graphs leads naturally to the consideration of minimal graphs G that have dimension greater than 2. That is, if you remove an edge from G, it has dimension 2. The closely related problem of classifying the critical graphs of dimension 3 was raised in the paper Erdős et al. (1965).

Purdy and Purdy (1988), with the help of a CRAY computer, determined these minimal forbidden graphs for $n \leqslant 7$ and work on $n = 8$. They also prove (Purdy and Purdy 1988) that there are infinitely many minimal forbidden graphs. The work naturally generates interesting graphs, e.g., graphs which have dimension 2 but not faithful dimension 2, or forbidden graphs which do not contain the forbidden subgraphs K_4 or $K_{2,3}$.

5.7. Triangles of unit area

In 1967, A. Oppenheim asked us the following question: given n points in the plane, how many different triangles can have the same nonzero area A. Without loss of generality, of course, $A = 1$. Let $A_k(n)$ denote the maximum number of triangles of unit area, taken over all configurations of n points in $\mathbb{R}^k$. We were able to show (Erdős and Purdy 1971):

$$cn^2 \log\log n < A_2(n) < cn^{5/2} .$$

The lower bound is provided by points (i, j) with positive integer coordinates $i \leqslant a = [(\log n)^{1/2}]$ and $j \leqslant n/a$. It is easy to see that the area $\frac{1}{2}a!$ occurs cn^2 times from triangles with horizontal bases of lengths $1, 2, \ldots, a$. When the triangles with sloping bases are also counted, then estimates on the average behavior of Euler's totient function gives $cn^2 \log a$ triangles of that area. To see the upper bound, we fix a point P and show that at most $cn^{3/2}$ triangles PXX' can have unit area. We take the graph G whose vertices are the remaining $n-1$ points, with X and X' joined by an edge if triangle PXX' has unit area. Then G cannot have a $K_{2,3}$ subgraph, and so has at most $n^{3/2}$ edges.

Recently, Pach and Sharir (1990) have improved the exponent on the upper bound, showing that $A_2(n) = O(n^{7/3})$.

In $\mathbb{R}^3$ we were able to show $A_3(n) < cn^{8/3}$. In $\mathbb{R}^6$ a generalized Lenz construction gives us $3m$ points P_i, Q_j, etc. such that m^3 triangles are congruent and therefore have equal area. This left $\mathbb{R}^5$ in doubt. Purdy (1974) showed that $A_5(n) < cn^{3-d}$, where d was a small positive constant. Using algebraic geometry, he showed that there could not exist sets S, T, and U in $\mathbb{R}^5$ containing N points each, such that the N^3 triangles obtained by taking one vertex from each set all had unit area, where N was very large, but constant. The following extremal theorem on hypergraphs, due to Erdős (1964), then implies the result: given a collection of t triples of elements chosen from an n-element set, if $t > n^{3-f(k)}$, where $f(k) = k^{-2}$, then there are sets S, T, and U of size k so that all k^3 possible triples formed from them are in the original collection.

5.8. Triangles of different areas

One can also ask for the minimum number of different areas. Given n points in the plane, form all of the triangles. What is the minimum number of different areas that must occur if the points do not all lie on a straight line? (Degenerate triangles have zero area, of course). Let $T_2(n)$ denote the minimum number.

We showed

$$cn^{3/4} < T_2(n) < cn .$$

(Erdős and Purdy 1976): The lower bound was subsequently improved by Burton and Purdy (1979) to:

$$cn < T_2(n) .$$

Only the correct constant remains to be found.

5.9. Angles

5.9.1. Repeated angles

Given n points in $\mathbb{R}^k$, let $B_k(n, \alpha)$ be the maximum number of times that the angle α can occur among the ordered triples of points, and let $B_k(n)$ be the maximum of that for all α, $0 < \alpha < \pi$.

Conway et al. (1979) proved that

$$B_3(n) < cn^{3-1/3} .$$

Purdy (1988) showed that

$$B_4(n, \alpha) < cn^{3-1/25} \quad \text{except when } \alpha = \tfrac{1}{2}\pi .$$

Surprisingly, he showed that

$$B_4(3m, \tfrac{1}{2}\pi) \geqslant m^3$$

by means of the following example. Let u_i, v_i be m solutions of $u^2 + v^2 = 1$. Let

$$X_i = (u_i, v_i, 0, 0) \quad \text{for } 1 \leqslant i \leqslant m ,$$
$$Y_j = (0, 0, j, 0) \quad \text{for } 1 \leqslant j \leqslant m ,$$
$$Z_k = (-1, 0, 0, k) \quad \text{for } 1 \leqslant k \leqslant m .$$

Then $(Y_j - X_i) \cdot (Z_k - X_i) = u_i^2 - 1 + v_i^2 = 0$, so that the m^3 angles $Y_i X_j Z_k$ are all right angles.

A further extension of the Lenz example shows that $B_6(3m, \alpha) \geqslant m^3$ for all acute angles α, but the magnitude of $B_6(n, \alpha)$ is unclear for obtuse angles α. Another example shows that $B_7(3m, \alpha) \geqslant m^3$ for all angles α.

Purdy proved the upper bound on $B_4(n, \alpha)$, for α not a right angle, by showing that in $\mathbb{R}^4$ there cannot be 15 points X_i, Y_j, and Z_k such that the 125 angles $Y_j X_i Z_k$

all equal α. The hypergraph theorem mentioned in section 5.7 then implies the following.

Let H be the subset of the triples on a set S of n points in $\mathbb{R}^4$ defined by saying that XYZ is in H if and only if the angle XYZ equals α, and suppose that H contains more than $n^{3-1/25}$ triples. Then there are three 5-element subsets S_1, S_2, and S_3 of S, so that every triple $X_1X_2X_3$ with X_i in S_i is in H, and every angle $X_1X_2X_3$ equals α.

Recently, Pach and Sharir (1990) have shown that $cn^2 \log n < B_2(n) < c'n^2 \log n$. They also showed that for infinitely many angles α, there is a constant $c(\alpha)$ depending on α such that

$$B_2(n, \alpha) > c(\alpha)n^2 \log n .$$

5.9.2. Different angles

Corradi, Erdős and Hajnal asked: is it true that n points in the plane not all on a line determined at least $n-2$ different angles? This is trivial if not three points are on a line, but seems to present curious difficulties in general. The weak Dirac conjecture (see section 3.9), now a theorem, states that there is always a point through which cn different lines pass, with a very small constant c. There is therefore a fixed vertex with that many different angles involving it.

A related question is: is it true that the smallest (nonzero) angle determined by our points is $\leqslant \pi/n$? We have equality for the regular polygon. This is again trivial if no three of the points are on a line, but in the general case we only get c/n, by the weak Dirac conjecture.

5.9.3. Large angles

Erdős and Szekeres (1935) proved that given n points in the plane, if $n \geqslant 2^k$, then three of the points form an angle ϕ satisfying $\{1-1/k\}\pi < \phi \leqslant \pi$. They gave examples showing that this could not be improved.

Barany (1987) showed that under the right conditions there are many large angles: for any $\varepsilon > 0$ there is an $n(\varepsilon)$ such that every finite set S of points in the plane contains a subset T of cardinality at most $n(\varepsilon)$ such that every point in $S-T$ is the vertex of an angle θ formed with t_1 and t_2 in T such that $\pi - \varepsilon < \theta \leqslant \pi$. If $n(\varepsilon)$ is the smallest n that works, then

$$c_1/\varepsilon < \log_2 n(\varepsilon) < \{c_2/\varepsilon\} \log_2\{c_3/\varepsilon\} .$$

5.10. Problems involving circles

Motzkin (1951) solved Sylvester's problem for circles. If n points are given in the plane, not all on a line, and not all on a circle, he showed that there is either a circle or a line containing exactly three of the points. Just invert the plane around one of the points and apply the Sylvester–Gallai theorem. Erdős asked: what is the minimum number c of circles that are determined by the points. Elliott (1967) showed that, if n is sufficiently large, then $c \geqslant \frac{1}{2}(n-1)(n-2)$, i.e., the configuration of n points with $n-1$ collinear does indeed give the minimum. Serre had already pointed out that this was not the minimum for eight points – the eight vertices of two concentric squares determine fewer than 21 circles.

Let there be given n points in the plane such that no circle or line contains more than $n-k$ of them. Beck (1983) showed that the number of circles containing at least three and at most A of the points exceeds ckn^2. In fact, it is not hard to prove the same result with A replaced by 4. In the same paper he showed that the number of circles containing k or more points is at most cn^3/k^4, provided $k<n^{1/2}$.

Chung (1989) shows that the total number of incidences occurring between n points and t circles is at most $c(n^{3/4}t^{3/4}+n+t)$.

Erdős and Straus observed that if $n>\frac{1}{6}k^5$ and $X_1,\ldots,X_n$ are in general position (no three on a line, no four on a circle), then there are always k of them so that the $\frac{1}{6}k(k-1)(k-2)$ circles they determine all have different radii. Probably n can be much smaller.

5.10.1. Unit circles

A few years ago, Erdős asked how large is $f(n)$, where $f(n)$ is the largest number for which there are n points $X_1,\ldots,X_n$ in the plane so that there are $f(n)$ distinct circles of unit radius passing through pairs of them. Erdős observed

$$\tfrac{3}{2}<f(n)<n(n-1)\,.$$

The upper bound comes from observing that each pair of points determines at most two circles of unit radius.

Elekes (1984) used the following clever argument to show that

$$f(n)>cn^{3/2}\,.$$

Let $V_1,\ldots,V_m$ be unit vectors in general position in the sense that the $t=\frac{1}{6}m(m-1)(m-2)$ sums of the form $V_i+V_j+V_k$ are all distinct. It follows from this that the $n=\frac{1}{2}m(m-1)$ sums V_i+V_j are also all distinct. A unit circle with center $V_i+V_j+V_k$ will contain the three points V_i+V_j, V_i+V_k, and V_j+V_k. This gives us t unit circles containing at least three of the n points, so that $f(\frac{1}{2}m(m-1)\geq\frac{1}{6}m(m-1)(m-2)$, and the result follows.

Erdős conjectures that $f(n)/n^2$ tends to zero. It might be possible to find an asymptotic formula, or even an exact formula for $f(n)$. Harborth and Mengersen have determined exact values of $f(n)$ for $n\leq 8$, see table 2 (Harborth 1985c, Harborth and Mengersen 1986).

J. Beck proved the following in 1982. Let $C_1,\ldots,C_n$ be n circles of unit radius so that any two intersect. Then the number of distinct intersection points is at least cn^2. The best value of c is not known.

5.11. Integer distances

Anning and Erdős (1945) showed that for every n there are n points spanning the plane such that all of their distances are integers, but that there is no *infinite* set of such points, i.e., not all on a line, all integral distances.

Table 2

n	$f(n)$
3	1
4	4
5	4
6	8
7	12
8	16

Graham et al. (1974) asked whether there are $d+2$ points in $\mathbb{R}^d$ whose distances are all *odd* integers. They showed that there are such points if and only if d is of the form $16k-2$.

They point out that their examples all need three distances and ask whether there are sets of n points in $\mathbb{R}^d$ realizing only two distances, D and D', both integers. Their methods show that the answer is no if D and D' are both odd. Harborth and Piepmeyer (1991) show that this is only possible if $n \leqslant d+2$. They also show that if $D' > cD$, then $n \leqslant d+1$, where $c = \{2+3^{1/2}\}^{1/2} = 1.93\ldots$. See also section 5.3.1.

6. Helly-type theorems

6.0. Introduction

No discussion of combinatorial geometry would be complete without a discussion of Helly's theorem and its generalizations. It has been 30 years since the last two surveys were written, following a period of great activity. Xeroxed frequently, the encyclopedic survey article by Danzer et al. (1963) is one, and the delightful book of Hadwiger and Debrunner, translated and expanded by Klee (Hadwiger et al. 1964), now out of print, is the other. Our hope is that the present survey will shed some light on the relative "dark age" that followed these surveys and also make the reader aware of the recent "renascence". Nonetheless, one seven-year gap remains in the references, that from 1969 to 1975.

We have limited the discussion to finite collections of convex sets or to arbitrary collections of compact convex sets in $\mathbb{R}^d$. For a discussion of the intersection properties of infinite collections of arbitrary convex sets in $\mathbb{R}^d$ we refer the reader to the first section of Klee (1963).

For the results about *Helly families of sets* and their relation to Helly's theorem we refer the reader to Bollobás's (1986) book.

6.1. Helly's theorem

Helly's theorem (Helly 1923) says that if $n \geqslant d+1$ convex sets in E^d have the property that any $d+1$ of them have a nonempty intersection, then there is a point common to all of them.

On the line (i.e., $d = 1$) the theorem would say that if $n \geq 2$ line segments are given and any two of them intersect, then they all do. This is certainly true!

In the plane Helly's theorem says that if $n \geq 3$ convex sets are given such that any three intersect, then they all do. The example of n hyperplanes in general position in E^d shows that $d + 1$ cannot be replaced by d in Helly's theorem.

We shall deduce Helly's theorem from Radon's theorem, which asserts that $n \geq d + 2$ points in $\mathbb{R}^d$ can always be split into two sets A and B whose convex closures have a nonempty intersection.

The result is obvious if $n = d + 1$. We shall use induction on n. Suppose that $n > d + 1$, and consider a family $\{S_1, \ldots, S_n\}$ of convex sets in $\mathbb{R}^d$ each $d + 1$ of which intersect. The induction hypothesis may be applied to each of the smaller collections $\{S_1, \ldots, S_n\} - \{S_i\}$ to obtain the existence of a point P_i common to all of the sets except possibly S_i. We now apply Radon's theorem to $P_1, \ldots, P_n$. There is a partition of $\{P_1, \ldots, P_n\}$ into two sets A and B whose convex closures intersect at a point Q. Helly's theorem now follows by observing that Q belongs to all n sets of the family.

To prove Radon's theorem, let $\boldsymbol{x}_1, \ldots, \boldsymbol{x}_n$ be the vectors representing $n \geq d + 2$ points in $\mathbb{R}^d$. Since $n > d + 1$, there is a nontrivial solution $(y_1, \ldots, y_n)$ to the system of $d + 1$ homogeneous linear equations

$$
\begin{aligned}
&y_1\boldsymbol{x}_1 + \cdots + y_n\boldsymbol{x}_n = \boldsymbol{0}\,, \\
&y_1 + \cdots + y_n = 0\,.
\end{aligned}
$$

Some of the y's will be positive, some will be negative, and perhaps some will be zero. We may suppose that the points have been numbered so that $y_1, \ldots, y_k$ are nonnegative and $y_{k+1}, \ldots, y_n$ are negative, for some k, $1 < k < n$. Then $c = y_1 + \cdots + y_k > 0$. The point

$$Q = (y_1/c)\boldsymbol{x}_1 + \cdots + (y_k/c)\boldsymbol{x}_k$$

is in the convex closure of $A = \{\boldsymbol{x}_1, \ldots, \boldsymbol{x}_k\}$, and the same point

$$Q = (y_{k+1}/-c)\boldsymbol{x}_{k+1} + \cdots + (y_n/-c)\boldsymbol{x}_n$$

is in the convex closure of $B = \{\boldsymbol{x}_{k+1}, \ldots, \boldsymbol{x}_n\}$, and Radon's theorem is proved. We remark that if $n = d + 2$ points are in general position, then the partition $\{A, B\}$ is unique and is referred to as a *minimal Radon partition*. Tverberg (1966) extended Radon's theorem, showing that any set of $m(d + 1) - d$ or more points in $\mathbb{R}^d$ can be partitioned into m subsets whose convex closures all intersect.

6.1.1. Caratheodory's theorem

Helly's theorem may also be viewed as the dual of a theorem due to Caratheodory, which states the following: if S is a subset of $\mathbb{R}^d$ and $\boldsymbol{x}$ is in the convex closure of S, then there are points $\boldsymbol{x}_1, \ldots, \boldsymbol{x}_{d+1} \in S$ such that $\boldsymbol{x} = \lambda_1\boldsymbol{x}_1 + \cdots + \lambda_{d+1}\boldsymbol{x}_{d+1}$, where all $\lambda_i \geq 0$. See, e.g., Eggleston (1958).

6.1.2. Krasnoselskii's theorem

Krasnoselskii (alternative spelling Krasnosselsky) (Krasnosselsky 1946) proved the following: if S is a compact subset of $\mathbb{R}^d$ containing at least $d+1$ points such that for each $d+1$ points there is a point from which all $d+1$ are visible, then S is star-shaped. The point P is *visible* from Q if the segment PQ lies entirely in S. The set S is *star-shaped* if there is a point P from which every point of S is visible. Krasnoselskii's theorem has been regarded as a consequence of Helly's theorem for a long time, but Borwein (1977) showed that one can also derive Helly's theorem from Krasnoselskii's theorem.

In the plane the theorem may be described in the following picturesque manner: if a modern art galley is so shaped that for every three paintings there is a place where you can stand and see those three, then there is a place where you can stand and see all of the paintings.

Krasnoselskii's theorem is false for subsets of $\mathbb{R}^d$ that are not compact, and Peterson (1982) asked whether one could find a number f_d so that if any f_d+1 points are visible from some point, then the set must be *finitely starlike*, which he defined as follows.

Definition. A set is *finitely starlike* if for any finite number of points in the set there is a point in the set from which they are all visible.

Peterson's question has not been settled, but there are some partial results. Breen (1988) finds such numbers for sets S in the plane, with certain restrictions on the interior of S and the closure of S restricted to S. See also Breen (1985) for a survey of theorems of the Krasnoselskii type.

6.1.3. Some consequences of Helly's theorem

From Helly's theorem one can immediately deduce the following:

(a) If a convex set A can be translated so that it is contained in the intersection of any $d+1$ of $n \geqslant d+1$ given convex sets, then it can also be translated so that it is contained in all n of the convex sets. We shall deduce this from Helly's theorem to illustrate a technique: let $B_1, \ldots, B_n$ be the given convex sets. In any coordinate system let points be represented by vectors and for the set S let $S+\boldsymbol{a}$ denote the set of translates $\{\boldsymbol{x}+\boldsymbol{a} \mid \boldsymbol{x} \in S\}$. We define C_i to be the set of vectors $\boldsymbol{c}$ such that $A+\boldsymbol{c}$ is contained in B_i. The result follows by applying Helly's theorem to $C_1, \ldots, C_n$, which are also convex.

(b) If a convex set A can be translated so that it intersects the intersection of any $d+1$ of $n \geqslant d+1$ given convex sets, then it can also be translated so that it intersects the intersection of all n of the convex sets. To prove this, let $B_1, \ldots, B_n$ be the given sets, and let C_i be the set of vectors $\boldsymbol{c}$ such that $A+\boldsymbol{c}$ has a nonempty intersection with B_i. Apply Helly's theorem to the convex sets $C_1, \ldots, C_n$.

(c) If a convex set A can be translated so that it contains the intersection of any $d+1$ of $n \geqslant d+1$ given convex sets, then it can be translated so that it contains the intersection of all n of the convex sets. To prove this, define C_i to be the set of vectors $\boldsymbol{c}$ such that $A + \boldsymbol{c}$ contains B_i.

The above results were published in Vincensini (1939) and Klee (1953). Assertions (a), (b), and (c) all become false if A is allowed to be *rotated* as well as translated. To see this, let the family of convex sets of C_n of section 6.5, and let A be a line segment of length 2.

6.1.4. Compact sets

In the above arguments, we did not assume that the convex sets were compact, but sometimes it is convenient to do so. Helly's theorem for a finite family of compact convex sets implies the same result for a finite family of arbitrary convex sets by the following standard argument, which applies to essentially all the generalizations of Helly's theorem and will be taken as obvious from now on. Let $S_1, \ldots, S_n$ be arbitrary convex sets such that any $d+1$ of them have an intersection point $P = P(i_1, \ldots, i_{d+1})$ depending of course on the particular $d+1$ sets chosen. Let H be the polytope that is the convex closures of all such points P. For $i = 1, \ldots, n$ let T_i be the (compact, convex) intersection of H with the (topological) closure of S_i. Applying Helly's theorem for compact convex sets, we deduce that the intersection of the closures of all the S_i, which is the closure of their intersection, is nonempty. Helly's theorem then follows by observing that only a nonempty set can have a nonempty closure.

6.1.5. Infinite families

In Helly's theorem for compact sets there is no need to assume that the family of sets is finite. The same remark applies to virtually all (compact) generalizations of Helly's theorem. Suppose that we have a countably infinite family of compact convex sets in $\mathbb{R}^d$ such that any $d+1$ have a nonempty intersection. Let S_n be the intersection of the first n sets. Then by the finite version of Helly's theorem, S_n is nonempty. Let $\boldsymbol{x}_n$ be the point of S_n closest to the origin. The sequence $\{\boldsymbol{x}_n\}$ is contained in the compact set S_1 and therefore has a limit point $\boldsymbol{x}^*$, which must be in S^*, the intersection of the S_i. Hence $\boldsymbol{x}^*$ is in all of the sets of the original family. To obtain the result for uncountable families, we simply observe that in $\mathbb{R}^d$ the intersection of an infinite number of closed sets can be expressed as the intersection of some countable subcollection.

Helly's theorem is false for infinite families of bounded, but not closed convex sets, and for infinite families of closed but unbounded convex sets. For a discussion of the intersection properties of infinite families of noncompact convex sets in $\mathbb{R}^d$, see the first section of Klee (1963).

6.1.6. Rado's theorem

A theorem of Ramsey's (1930) asserts that if all of the r-membered subsets of an infinite set X are divided into m classes, then there is an infinite subset Y of X all of whose r-membered subsets belong to the same class. Using this theorem and Helly's theorem, Rado (1952) proved the following: every infinite family of convex sets in $\mathbb{R}^d$ contains an infinite subfamily with one of two properties: (a) any finite number of sets of the subfamily have a nonempty intersection; (b) for some $k \leqslant d$ every k members of the subfamily have nonempty intersection whereas every $k+1$ members have empty intersection.

6.1.7. *Multiplied Helly theorems*

The following *multiplied* version of Helly's theorem is attributed to Lovász in Barany (1982) (for simplicity we state it in $\mathbb{R}^2$).

Let r, b, $g \geqslant 3$, and suppose that we are given a red family of convex sets $R_1, \ldots, R_r$, a blue family $B_1, \ldots, B_b$, and a green family $G_1, \ldots, G_g$, such that R_i, B_j, and G_k have a nonempty intersection for any i, j, and k. Then for one of the colors it must be the case that all of the sets (of that color) intersect.

In the d-dimensional version there are families of $d+1$ different colors with at least $d+1$ in each family. Helly's theorem can be deduced from this by making the families all the same. Barany (1982) proves a dual version of this and remarks that both can be deduced from the following well-known fact: if the closed convex sets $S_1, \ldots, S_n$ do not have a common point, then there are closed halfspaces $H_1, \ldots, H_n$, with H_i containing S_i, for $i = 1, \ldots, n$, which do not have a common point. Many generalizations of Helly's theorem also have multiplied versions.

6.2. *Some generalizations*

De Santis (1957) extended Helly's theorem to show that if we are given $n \geqslant d + 1 - k$ convex sets in $\mathbb{R}^d$ such that any $d+1-k$ of them contain a common k-dimensional flat, then they all do. For $k=0$ this is Helly's theorem, and a 0-dimensional flat is simply a point.

If we refer to any translate of a subspace of codimension k as a *flat of deficiency* k, so that a hyperplane is a flat of deficiency 1, then De Santis's theorem may be more elegantly stated as follows: if we are given $n \geqslant k+1$ convex sets in $\mathbb{R}^d$ such that any $k+1$ of them contain a common flat of deficiency k, then they all do.

Grünbaum (1961) showed that if we are given $n \geqslant g(d,k)$ convex sets such that any $g(d,k)$ of them have a k-dimensional ball in common, then they all do. The function $g(d,k)$ is defined as follows:

$$g(d,0) = d+1\,, \quad g(d,1) = 2d\,,$$
$$g(d,k) = 2d-k \quad \text{for } 1<k<d\,,$$
$$g(d,d) = d+1\,.$$

For $k=0$ this is Helly's theorem, and a 0-dimensional ball is a point.

Horn (1949) and Klee (1951) showed that if $1 \leqslant k \leqslant d+1$, and you have $n \geqslant k$ compact convex sets in $\mathbb{R}^d$, then the following three assertions are equivalent:

(a) every k of the sets have a common intersection;

(b) every flat of deficiency $k-1$ in $\mathbb{R}^d$ admits a translate which intersects all the sets;

(c) every flat of deficiency k in $\mathbb{R}^d$ lies in a flat of deficiency $k-1$ that intersects all the sets.

Horn and Valentine (1949) added a fourth equivalent condition:

(d) for each set of k of the convex sets and for each flat F of deficiency $k-1$ there is a translate of F which intersects these k sets.

When $k=d+1$, the above reduces to Helly's theorem and assertion (c) is vacuous.

6.2.1. Parallelotopes and Helly dimension

If we are given n parallelotopes in $\mathbb{R}^d$ with edges parallel to the coordinate axes, with the property that any *two* of them intersect, then they all intersect. Thus the $d+1$ of Helly's theorem has been replaced by 2. In a sense these sets behave as if they were in one-dimensional space. To pursue this further we need some definitions.

Definition. A convex set K in $\mathbb{R}^d$ is a *convex body* if it is compact and contains a d-dimensional ball.

Definition. The set A is *homothetic* to B if $A=kB+\boldsymbol{c}$ for some vector $\boldsymbol{c}$ and scalar k. The set $kB := \{k\boldsymbol{x} \mid \boldsymbol{x} \in B\}$. (For example, if $k>1$, then kB is the result of expanding B about the origin of the coordinates.) Whether or not the sets A and B are homothetic is not affected by changing the origin.

Definition. The *Helly dimension* of a family of convex bodies F is the smallest number d such that the following statement is true: if A is a subfamily of F such that any $d+1$ members of A have a nonempty intersection, then all of the sets in A intersect. If F is the collection of all translates of a convex body K, then d is the *Helly (translate) dimension of* K. If F is the collection of homothetic images of a convex body K, then d is the *Helly (homothetic) dimension of* K. When the Helly (translate) dimension of K is the same as its Helly (homothetic) dimension, then we shall refer to their common value as the *Helly dimension of* K.

Clearly the collection of all parallelotopes in $\mathbb{R}^d$ with edges parallel to the coordinate axes has Helly dimension 1, and any single parallelotope has Helly dimension 1.

Definition. A *zonotope* Z is a convex set which can be expressed in the form $Z=S_1+S_2+\cdots+S_k$, where the S_i are (closed) line segments, and $A+B$ is the usual (Minkowski) sum given by $A+B=\{a+b \mid a \in A,\ b \in B\}$. For an exposition of zonotopes see McMullen (1971).

Zonotopes are in fact the projections of parallelotopes, and they have Helly dimension 1. It has been known for a long time that no other convex bodies have Helly (translate) dimension 1, see, e.g., Gritzmann (1988).

It would seem reasonable to conjecture that the convex bodies of Helly (translate) dimension $\leqslant d$ consist of the direct sums of compact convex sets that fit in $\mathbb{R}^d$. In the special case of centrally symmetric bodies, this has been proved by Boltyanskii (1976) when d is 2, and by Kincses (1987) when d is 3 or 4.

Barany and Kincses (1987) have characterized the Helly (homothetic) dimension of a convex body K as the dimension of any dense set of extreme points of the dual of K.

6.2.2. Quantitative Helly theorems

Barany et al. (1984) prove the following quantitative Helly-type theorem: given n convex sets in $\mathbb{R}^d$ such that the intersection of any $2d$ of them has volume at least unity, the intersection of all n of them must have volume at least $c(d)$, where $c(d)$ is a positive number depending only on d. Here $2d$ cannot be replaced by $2d-1$. To see this, take the family of the $2d$ halfspaces supporting a d-dimensional cube of side ε, where ε is small. The intersection of any $2d-1$ of them is unbounded, but the intersection of all $2d$ of them is only ε^d. Pach notes that similar quantitative Helly-type theorems can be proved with volume replaced by diameter, surface area, etc.

6.2.3. Eckhoff's conditions and fractional Helly theorems

What if most subcollections of size $d+1$ of a collection of convex sets in $\mathbb{R}^d$ have a nonempty intersection? Does it follow that most of the sets also intersect? Katchalski and Liu (1979) proved the following. Let $r>d$. For each ρ there is an α such that for any collection of n convex sets, if at least a proportion α of the r-tuples of sets have an intersection, then at least ρn of the sets have a common intersection.

Kalai (1984) made the above result quantitative by determining the smallest $\alpha=\alpha(\rho, d, r)$ that will work. For example, $\alpha(\rho, d, d+1)=1-(1-\rho)^{d+1}$.

Kalai deduced the above from Eckhoff's conditions, which we now explain. Let n convex sets be given in $\mathbb{R}^d$. We let f_i denote the number of subcollections of $i+1$ sets that have a nonempty intersection. Thus Helly's theorem says:

$$\text{if } f_d=\binom{n}{d+1}, \text{ then } f_{n-1}=1 .$$

Eckhoff conjectured that $\boldsymbol{f}=(f_0, \ldots, f_{n-1})$ would be an *f-vector* that actually occurred if and only if

$$f_{k-1} \leqslant \binom{r}{k}+(n-r)\binom{r}{k-1}+\cdots+\binom{n-r}{d}\binom{r}{k-d}$$

for all k, $1 \leqslant k \leqslant n$. Kalai proved the necessity of the inequalities in Kalai (1984) and their sufficiency in Kalai (1986). Eckhoff also had a proof of sufficiency, but did not publish it.

6.2.4. The Grünbaum and Motzkin conjecture

If one tries to generalize Helly's theorem to sets that are themselves unions of $\leqslant k$ disjoint convex sets, examples show that there is no $h(d, k)$ such that if any $h(d, k)$ members of a collection have a nonempty intersection then they all do.

Grünbaum and Motzkin (Grünbaum 1961) considered a modified situation in which not only are the sets expressible as the union of $\leq k$ disjoint convex sets, but the intersection of k or fewer sets can also be expressed as the union of $\leq k$ disjoint convex sets. They conjectured that for such sets there is a number $h(d, k)$, and that the smallest one that works is $h^{+}(d, k) = k(d+1)$. They proved their conjecture for $k = 2$, and seven years later, Larman (1968) proved it for $k = 3$. Larman also gives an example showing that $h^{+}(d, k) \geq k(d+1)$.

6.3. *Selection theorems*

Debrunner and Hadwiger (1957) introduced the following.

Definition. Let n convex sets be given. A *k-partition* of the sets is a partition of the sets into k classes so that the sets in each class have a common intersection.

Klee once asked (see Hadwiger et al. 1964), whether there exists a number h with the following property: given a collection of $n \geq h$ convex sets in $\mathbb{R}^2$, such that each h of them permit a 2-partition, the entire collection of n sets always permits a 2-partition. Unfortunately such an h does not exist, and the unitary rosette U_n of Hadwiger and Debrunner discussed in section 6.5 shows that. Each selection of $2n - 1$ sets from U_n permits a 2-partition, but the complete collection of $4n$ sets does not.

Let us ask whether there is any property at all of each selection of h sets which would imply that all of the sets permit a 2-partition. A necessary condition would be that given any three sets, two of them must lie in the same partition; therefore, two of them must have a nonempty intersection. Is this condition also sufficient?

Definition. A family of convex sets has *property* (p, q) if out of each selection of p sets some q have a nonempty intersection.

We can now restate our question: does a family of n convex sets with the $(3, 2)$ property necessarily permit a 2-partition? The answer is yes on the line [since $K(3, 2, 1) = 2$, see below], but no in the plane. In fact, in $\mathbb{R}^2$ there is no k for which a k-partition is guaranteed by the $(3, 2)$ property. To see this, take n lines in general position. Then any two intersect, but no three intersect. Hence they will only permit a k-partition if $n \leq 2k$.

Definition. For $p \geq q$, let $K(p, q, d)$ denote the smallest integer k for which a k-partition is admitted by each finite collection of convex sets in $\mathbb{R}^d$ having property (p, q).

It is not hard to see that $K(p, q, d) \geq p - q + 1$. Let us construct a family with a lopsided k-partition. One class has q sets in it, the other $p - q$ classes contain only one set. Let us suppose also that sets in different classes are indeed disjoint. Such a collection can be constructed already on the line. Clearly if any p sets are chosen, then at least q of them will have an intersection. Hence $K(p, q, d) \geq p - q + 1$. It is shown in Hadwiger et al. (1964) that $K(p, q, 1) = p - q + 1$. Hence $K(3, 2, 1) = 2$, whereas $K(3, 2, 2)$ is infinite.

It is not known whether $K(4, 3, 2)$ is finite. The above inequality gives $K(4, 3, 2) \geq 2$, but examples due to Danzer (in Hadwiger et al. 1964) and Grünbaum (1959) show that $K(4, 3, 2) \geq 3$. Grünbaum's example consists of nine translates of any centrally symmetric strictly convex body. It therefore also shows that $T(4, 3, 2) \geq 3$, where $T(p, q, d)$ is the analogue of $K(p, q, d)$ for families consisting of translates of a convex set.

The example of n hyperplanes in general position in $\mathbb{R}^d$ shows that $K(p, q, d)$ is infinite if $q \leq d$. Each d hyperplanes will have a nonempty intersection, but no $d + 1$ will, so that at least n/d partitions would be needed to cover all of the hyperplanes.

Helly's theorem asserts that $K(d + 1, d + 1, d) = 1$.

Debrunner and Hadwiger (1957) showed that $K(p, q, d) = p - q + 1$ whenever $p - (p - d - 1)/d \leq q \leq p$.

6.3.1. *Parallelotopes*

We can define a function $N(p, q, d)$ to be the analogue of $K(p, q, d)$ for families of mutually parallel parallelotopes in $\mathbb{R}^d$. Then

$$p - q + 1 \leq N(p, q, d) \leq \binom{p - q + d}{d}.$$

Moreover, if $1 + \frac{1}{2}p \leq q \leq p$, then $N(p, q, 2) = p - q + 1$.

For a discussion of these and related matters see Hadwiger et al. (1964).

6.4. *Common transversals*

An m-transversal of family of sets in $\mathbb{R}^d$ is an m-dimensional flat that intersects each set. We shall refer to a 1-transversal as simply a transversal.

One might ask whether there is a number s such that if any s convex sets in the plane have a transversal then they all do. This is false even if we restrict the sets to be compact and pairwise disjoint. For every n there is a family of $4n$ compact pairwise disjoint convex sets in $\mathbb{R}^2$ so that any $2n - 1$ have a common transversal, but there is no transversal common to all of them. This is the expanding rosette E_n of section 6.5.

Also in the plane, Vincensini defined a collection of convex sets to be *totally separable* if there exists a direction so that all the lines parallel to that direction intersect at most one of the sets. For example, the rosette E_n is not totally separable, even though the sets are pairwise disjoint.

Vincensini showed that there was indeed an s so that if each s of a totally separable family of compact convex sets in $\mathbb{R}^2$ admitted a common transversal, then they all did. Vincensini (1953) used $s = 4$, and Klee (1954) improved this to $s = 3$. We shall now sketch Klee's proof from Hadwiger et al. (1964).

Suppose that there is a line L so that all lines parallel to L intersect at most one set of the family. Thus any transversal of a pair of sets $\{A, B\}$ must intersect L at a nonzero angle ϕ. Indeed there is an interval of angles $I(A, B)$ corresponding to such transversal directions. With a little effort it follows from the hypothesis that any two intervals $I(A, B)$ and $I(C, D)$ intersect, so that Helly's theorem in $\mathbb{R}^1$ guarantees the existence of a direction common to all of the pairs. Projecting in that direction we see that any two projections A' of A and B' of B must intersect. One more application of Helly's theorem in $\mathbb{R}^1$ gives the result.

6.4.1. Hadwiger's theorem

Let A be an ordered family of convex sets in $\mathbb{R}^2$. A transversal L is *consistent with the ordering of* A if the directed line L encounters the sets of A in order (or reverse order).

Hadwiger (1957) proved that an ordered family of *disjoint* convex sets in $\mathbb{R}^2$ will have a transversal if any three sets have a transversal consistent with the ordering.

If a family of n compact convex sets is totally separable, then they are disjoint, and $n-1$ parallel lines can clearly be drawn separating the sets from each other. Any transversal of three sets must be consistent with the obvious order induced by the parallel line. Hadwiger's theorem provides the Klee–Vincensini theorem for finite families, and therefore for all families by a standard argument (see, e.g., section 6.1.5).

In a very pleasing development, Wenger (1990) proved Hadwiger's theorem without assuming that the sets of the family are disjoint. Wenger's proof is similar to that of the Klee–Vincensini theorem given above.

Wenger also makes the subtle observation that neither his theorem nor Hadwiger's asserts the existence of a transversal that is *consistent with the ordering*! He then goes on to prove that an ordered family of compact convex sets such that *every six* sets have a transversal consistent with the ordering will have a transversal *consistent with the ordering*. It follows immediately from this result that a finite ordered family of arbitrary convex sets in $\mathbb{R}^2$ will have a transversal consistent with the ordering if each six of the sets have a transversal consistent with the ordering (just use the argument in section 6.1.4). This result has the interesting corollary that an arrangement of labelled lines $L_1, \ldots, L_n$ will cross some line L in order if every six lines have this property.

6.4.2. Higher dimensions

Valentine (1963) proved the following theorem. Suppose that we are given a family of compact convex sets in $\mathbb{R}^3$ such that there exist three distinct planes Π_1, Π_2, and Π_3 containing a common line such that for each triple of convex sets M_1, M_2, M_3 in the family each pair of the triple is strictly separated from the remaining member of the triple by a translate of either Π_1 or Π_2 or Π_3, and this correspondence is cyclic. If every four members of the family have a common 2-transversal, then all the members of the family have a common 2-transversal. Valentine remarks that a corresponding result is true in $\mathbb{R}^d$.

In a recent development, Goodman and Pollack (1988) have proved a generalization of Hadwiger's theorem to $\mathbb{R}^d$, and their result implies Valentine's theorem. For the sake of simplicity we shall state it in $\mathbb{R}^3$ for finite families.

Given a finite family of compact convex sets in $\mathbb{R}^3$ such that no three of them have a common 1-transversal, then they all have a common 2-transversal if and only if there is a configuration of points $\{P_1, \ldots, P_n\}$ in general position in $\mathbb{R}^2$ such that for every four indices $i<j<k<l$ there is an oriented plane Π which meets B_i, B_j, B_k and B_l in points consistently with the order type of $\{P_1, \ldots, P_n\}$.

We need to define order type: two configurations of points $\{P_1, \ldots, P_n\}$ and $\{Q_1, \ldots, Q_n\}$ in the plane have the same *order type* if for any three indices i, j, and k the question "Does P_k lie on the left of the line from P_i to P_j?" always has the same answer as the question "Does Q_k lie on the left of the line from Q_i to Q_j?"

In $\mathbb{R}^d$ the statement of Goodman and Pollack's theorem is analogous. Their proof makes use of minimal Radon partitions and also a formula from Zaslavsky (1975) on the number of cells in a partition of $\mathbb{R}^d$ by hyperplanes (see section 4.3).

In $\mathbb{R}^d$ the above assumption that no three sets have a common 1-transversal generalizes to the assumption that no d sets have a $(d-2)$-transversal. When $d=2$, this is just Hadwiger's assumption that the sets are disjoint. Goodman and Pollack published their theorem before Wenger pointed out that this is superfluous in $\mathbb{R}^2$, so it is natural to ask whether it is possible to prove the d-dimensional result without that hypothesis. Pollack and Wenger (1990) have recently done just that. In fact, they proved the following even more general theorem: a finite family $\{S_1, \ldots, S_n\}$ of connected compact sets in $\mathbb{R}^d$ has a hyperplane transversal if and only if for some k there exists a set of points $\{P_1, \ldots, P_n\}$ that spans $\mathbb{R}^k$ and every $k+2$ sets of the family $\{S_1, \ldots, S_n\}$ are met by a k-flat consistent with the order type of $\{P_1, \ldots, P_n\}$.

6.4.3. *Parallelotopes*

Santalo (1940) proved that if we are given n parallelotopes in $\mathbb{R}^d$ with edges parallel to the coordinate axes, and if any $2^{d-1}(d+1)$ of them admit a $(d-1)$-transversal, then they all do.

6.4.4. *Translates*

Recently, Tverberg (1989) proved the following result: given any collection of pairwise disjoint translates of a convex set H such that any five permit a transversal, there is a transversal common to all the translates.

Danzer (1957) had proved this in the special case when H is a disc, and Grünbaum (1958) conjectured the general result. Danzer also gave the following example showing that five cannot be replaced by four: at the vertices of a regular pentagon of side 2 place the centers of discs of unit radius; any four of the discs have a common transversal, but the five discs do not have one.

The example C_n of section 6.5 shows that the translates must be assumed to be disjoint. Any $n-1$ of the discs in C_n admit a transversal, but there is no transversal for the entire family of n discs.

6.4.5. Flats

Lovász proved the following: given a family of r-dimensional subspaces of $\mathbb{R}^d$, such that any B_{rk} of them are intersected by the same k-dimensional subspace, then there is a k-dimensional subspace intersecting all of them, where

$$B_{rk} = \binom{r+k}{k} .$$

The result restated in terms of flats is the following: given a family of r-dimensional flats in $\mathbb{R}^d$ space, such that any F_{rk} of them have a k-transversal, there is always a k-transversal for the whole family, where

$$F_{rk} = \binom{r+k+2}{k+1} .$$

The families do not need to be finite or countable. The proof uses Grassmann products.

6.5. The rosette

An important example in the plane is the rosette, due to Hadwiger and Debrunner (Hadwiger et al. 1964).

For $i = 1, \ldots, 2n$ let S_i be the circular arc with polar coordinates (r, θ), where $r = r_i$, and

$$(i-n+1)\{\pi/(2n)\} \leq \theta \leq (i+n-1)\{\pi/(2n)\} .$$

Similarly, for $i = 1, \ldots, 2n$ let T_i be the circular arc with polar coordinates (r, θ), where $r = r_i$, and

$$(i+n-1)\{\pi/(2n)\} \leq \theta \leq (i+3n-1)\{\pi/(2n)\} .$$

Calculation shows that S_i and T_i are arcs of length $\pi - \pi/n$ radians, with S_i centered on $\theta = \pi i/(2n)$, and T_i centered on $\pi i/(2n) + \pi$.

Compact convex sets are obtained from the circular arcs by forming their convex closures, which are D-shaped regions, the petals of the rosette.

There are really two kinds of rosette:

(1) If r_i increases rapidly enough with i, then the petals of the rosette are pairwise disjoint, and no line intersects all $4n$ of them. Nonetheless, any $2n-1$ of them have a line in common. Let us call this example E_n, the expanding rosette.

(2) If $r_1 = r_2 = \cdots = r_{2n} = 1$, then we have $4n$ compact convex sets such that for each selection of $2n-1$ sets there is a pair of (antipodal) points P and P' so that each of the $2n-1$ selected sets contains at least one of P, P'. Nonetheless, there are not two points Q and Q' such that every one of the $4n$ sets contains at least one of Q, Q'. Let us call this rosette U_n, the unitary rosette.

Let the family C_n of n congruent discs be defined as follows for $n \geq 3$. Let the discs have radius R_n and center in polar coordinates

$$(r, \theta) = 1, 2i\pi/n) \quad \text{for } i = 1, \ldots, n .$$

Let

$$R_n = \cos^2(\pi/n) \quad \text{if } n \text{ is even} ,$$

$$R_n = \cos^2(\pi/n) + \cos^2(\pi/2n) - 1 \quad \text{if } n \text{ is odd} .$$

Any $n - 1$ of the discs have a transversal. In fact, it is possible to rotate and translate a line segment of length 2 so that it intersects any $n - 1$ of the discs. On the other hand, the n discs do not permit a transversal.

7. Some other problems

7.1. Euclidean Ramsey problems

A finite subset C of $\mathbb{R}^d$ is called r-Ramsey for $\mathbb{R}^d$ if for any partition of $\mathbb{R}^d$ into r sets $S_1, \ldots, S_r$, some S_i always contains a subset C' which is congruent to C. If C is r-Ramsey for every r for some $\mathbb{R}^d$, then it is said to be Ramsey.

The study of these problems was started by the six mathematicians R. L. Graham, P. Montgomery, B. Rothschild, J. Spencer, E. G. Straus, and P. Erdős. The six originators found two very general conditions for a set to be Ramsey – one is necessary and one is sufficient: if C is Ramsey, then C must lie on the surface of a sphere of some radius; a subset of the vertices of a d-dimensional parallelepiped (a "brick") is always Ramsey. It is conjectured by at least the first of the authors that the first condition actually characterizes Ramsey configurations. It has recently been shown by Křίž (1991) that the set of vertices of any regular n-gon is Ramsey.

Since every acute triangle is contained in a brick, it must be Ramsey. This raised the question whether obtuse triangles were Ramsey. Frankl and Rödl (1986) proved that all triangles are Ramsey. They mention that they are unable to prove the Ramseyness of any pentagon, and only of some quadrilaterals (although Křίž has now done this).

Many other challenging problems remain open. We shall present a few of them here.

What triangles are 2-Ramsey in $\mathbb{R}^2$? L. Shader proved that all right triangles are. On the other hand, equilateral triangles are not. Graham conjectures that equilateral triangles are the *only* triangles that are not 2-Ramsey in $\mathbb{R}^2$.

Color $\mathbb{R}^d$ with r different colors. Is it then true that at least one color contains the vertices of a k-dimensional simplex of unit volume? Graham (1980) proved and generalized this. His paper contains many of the relevant references on geometric Ramsey theory up to 1980. See also Graham's (1985, 1990) articles.

In a recent development, Frankl and Rödl prove that *all simplices* in arbitrary dimensions are exponentially Ramsey, which is a stronger property than just being Ramsey. A finite set A in $\mathbb{R}^d$ is *exponentially Ramsey* if there exists a positive real $\varepsilon(A)$ so that for every partition of $\mathbb{R}^n$ into r classes with $n \geq d$ and $r < (1+\varepsilon(A))^n$, one of the classes contains a congruent copy of A. We refer the reader to chapter 24.

Erdős posed the following question: if S is a subset of the plane such that no two points of S are at unit distance, is it then true that the complement of S contains the vertices of a unit square? Juhasz (1979) settled the question in the affirmative for the vertices of any quadrilateral, and she showed that the result was false for the vertices of a 12-gon.

7.2. Heilbronn's problem

An old problem of Heilbronn states the following. Let $X_1, \ldots, X_n$ be n points in the unit circle in the plane. Denote by $A_k(X_1, \ldots, X_n)$ the smallest area of all the k-gons formed from the n points, and let $g_k(n)$ denote the maximum of $A_k(X_1, \ldots, X_n)$ taken over all configurations of n points.

Heilbronn asked for the determination or estimation of $g_3(n)$. He of course observed that trivially $g_3(n) < c/n$ and suspected that the order of magnitude of $g_3(n)$ is $1/n^2$. Erdős observed that indeed $g_3(n) > c/n^2$. The first nontrivial result was due to K. F. Roth who proved that $g_3(n) < 1/n(\log\log n)^{1/2}$. This was improved by W. Schmidt to $1/n(\log n)^{1/2}$ and later by Roth to $1/n^{1+c}$, where $c = \frac{1}{10}$.

The competition between Schmidt and Roth was interrupted by Komlós et al. (1982) who improved c to $\frac{1}{7}$ and showed that

$$g_3(n) > (c \log n)/n^2 .$$

Their proof of the lower bound uses techniques for combinatorics and probability theory, and Erdős thinks that their method of proof will have further applications. Since Heilbronn had expressed the belief that $g_3(n)$ had order $1/n^2$, you could say that he had been proved wrong, but that would be a little harsh, and Szemerédi believes that it is quite possible that the new lower bound (with the $\log n$ factor) is best possible.

For $k > 3$, the first nontrivial results are due to Schmidt, who proved that

$$g_k(n) > c_k n^{-1-1/k}$$

and asked whether $g_4(n) = o(1/n)$. This problem is still open.

Erdős and Szemerédi posed the following problem. Denote by $D = D(X_1, \ldots, X_n)$ the smallest distance occurring between two of the X's, and by $A = A(X_1, \ldots, X_n)$ the smallest angle determined by any three of the points. If there are three collinear points, then A will be zero. Is it true that

$$DA = o(n^{-3/2})?$$

The answer to this question might throw some light on Heilbronn's problem. It is not too hard to show that $DA \leq cn^{-3/2}$. The regular polygon gives the largest known value for DA.

V. T. Sós, E. Straus, and Erdős have slightly modified Heilbronn's problem (see Erdős 1985a for details).

7.3. Diverse problems

Let $X_1, \ldots, X_n$ be n points in the plane not all on a line. Denote by $A = A(X_1, \ldots, X_n)$ the largest area and by $a = a(X_1, \ldots, X_n)$ the smallest positive area of the triangles $X_iX_jX_k$. Define $f(n)$ to be the minimum of A/a taken over all configurations of n points. Erdős et al. (1982) showed that $f(n) = [(n-1)/2]$ if $n > 37$. The corresponding function for d-dimensional simplices satisfies $f_d(n) \leqslant [(n-1)/d]$, and perhaps equality holds for $n > n_d$.

7.3.1. Convex polygons

Let $g(n)$ be the smallest integer such that any planar configuration of $g(n)$ points, no three on a line, must contain the vertices of a convex n-gon. Erdős and Szekeres (1935) proved that

$$2^{n-2} + 1 \leqslant g(n) \leqslant \binom{2n-4}{n-2}$$

and conjectured that $g(n)$ is equal to the lower bound. Their paper also contains an elegant proof of a much worse upper bound using Ramsey's theorem. Their conjecture has been verified for some small values. E. Klein proved that $g(4) = 5$, and E. Makai and P. Turán proved that $g(5) = 9$.

7.3.2. Empty polygons

Denote by $f(k)$ the smallest number (if it exists) for which every set of $f(n)$ points with no three on a line contains k which form a convex polygon that contains none of the other points in its interior. It is easy to see that $f(4) = 5$. Ehrenfeucht and Harborth proved that $f(5)$ was finite. Harborth (1979) showed that in fact $f(5) = 10$. It is not yet known if $f(6)$ exists. Horton (1983) showed that $f(7)$ does not exist.

7.3.3. The minimum number of empty triangles

Erdős asked the following question: given n points in $\mathbb{R}^2$, no three on a line, what is the minimum number $T(n)$ of empty triangles that they can form? A simple argument shows that $T(n) \geqslant \frac{1}{6}n(n-1)$, for any pair of points $\{P, Q\}$ determines a line L, and a closest point R to L. The triangle PQR must be empty, and triangles are counted at most three times in this way. That $T(n) < cn^2$ can be deduced from the example in Horton (1983) and was also proved independently by Purdy (1982b) and by Katchalski and Meir (1988), who also counted empty convex k-gons. Barany and Füredi (1987) looked at the corresponding problem in $\mathbb{R}^d$.

7.3.4. Mid-points of diagonal and convex n-gons

Erdős et al. (1990c) define $f(n)$ to be the minimum over all convex n-gons of the number of different mid-points of the diagonal. They show that

$$0.8\binom{n}{2} < f(n) < 0.9\binom{n}{2}.$$

7.4. Non-Archimedean valuations

Sometimes, p-adic valuations are useful for constructing examples. For any prime p, the p-adic norm $N_p(x)$ is defined for rational numbers x. A highly nonconstructive theory of homomorphisms requiring the axiom of choice then says that $N_p(x)$ extends to a non-Archimedean valuation $r(x)$ on the set of reals $\mathbb{R}$. We shall answer two geometric questions using $r(x)$.

(a) Is it possible to color the real projective plane P^2 with three colors so that there are three noncollinear points of each color and no line contains all three colors?

If we do not insist that each color has three noncollinear points, then the problem has a trivial solution. Let P be any point and L any line through P. The points of the plane are colored as follows: P is red, the rest of the line is green, and the rest of the plane is blue.

D. S. Carter and A. Vogt, and independently A. W. Hales and E. G. Straus, have shown that this is possible (with analogous results in higher dimensions). We give a sketch of the proof in Hales and Straus (1982), which is comparatively short, and refer the reader to that paper for further details and references.

Let $r(x)$ be any non-Archimedean valuation defined on the reals. Let (x, y, z) be the homogeneous coordinates of a point P. We color P as follows:

If $r(x) > \max\{r(y), r(z)\}$, then color the point C_1.

If $r(y) \geqslant r(x)$, and also $r(y) > r(z)$, then color C_2.

All remaining possibilities use color C_3.

Consider the points $P_j(x_i, y_i, z_i)$ for $i = 1, 2, 3$. Now look at the determinant

$$D = \begin{vmatrix} x_1 & y_1 & z_1 \\ x_2 & y_2 & z_2 \\ x_3 & y_3 & z_3 \end{vmatrix}.$$

This will be zero if and only if the three points are collinear. Suppose that point $P_i(x_i, y_i, z_i)$ is colored C_i for $i = 1, 2, 3$. Observe that

$$r(x_1) > \max\{r(y_1), r(z_1)\},$$
$$r(y_2) \geqslant r(x_2) \quad \text{and} \quad r(y_2) > r(z_2),$$
$$r(z_3) \geqslant \max\{r(x_3), r(y_3)\}.$$

This implies that $r(x_1y_2z_3) > r(x_iy_jz_k)$ for the $5 = 3! - 1$ other choices of distinct i, j, k. Hence $r(D) = r(x_1y_2z_3) \geqslant 1$. Hence the determinant D is nonzero, and the points are not collinear. Hence three points of different colors cannot be collinear.

The trivial valuation $t(x)$ is defined by saying that $t(0) = 0$, and $t(x) = 1$ if x is nonzero. It is not hard to show that if $r(x)$ is not the trivial valuation, then each color has three noncollinear points. If $r(x) = t(x)$, then we get the trivial coloring described above: C_1 will be points for which x is nonzero and $y = z = 0$. This will be the point on the line L at infinity where the x-axis intersects L. C_2 will be the points for which y is nonzero, $z = 0$, and x is anything. This will be the remainder of the line at infinity. C_3 will be the remainder of the projective plane.

Straus and Hale, and Carter and Vogt also showed that the implication goes the other way. That is, any 3-coloring of a projective plane over a field F having the above properties gives rise to a non-Archimedean valuation on F. Since the finite fields and their extensions do not possess such valuations, their projective planes do not have such colorings.

Thus, for example, if $P(2, p)$ is the projective plane formed from the field Z_p, where p is a prime number, and if the points of $P(2, p)$ are colored with three colors in such a way that there are three noncollinear points of each color, then there will always be a line with all three colors on it.

(b) Problem number 5479 in the *American Mathematical Monthly* posed by Fred Richman and John Thomas asked whether a square could be divided into an odd number of nonoverlapping triangles of equal area. The answer is no, as proved by Monsky (1970), using a non-Archimedean valuation $r(x)$ on the reals. John Thomas had already proved the result for the special case when the vertices of the triangles are rational, and he used a 2-adic valuation, but he did not extend it to the reals.

Mead (1979) extended Monsky's result to $\mathbb{R}^d$, showing that the unit hypercube can be divided into m simplices of equal volume if and only if m is a multiple of $d!$ Kasimatis (1989) proved that if $n \geq 5$, a regular n-gon can be dissected into m triangles of equal area if and only if m is a multiple of n. The result is false for $n = 3$. For $n = 4$ it follows from Monsky's result.

We now briefly sketch Monsky's proof, slightly modified.

Let $r(x)$ be a non-Archimedean valuation on the reals obtained by extending the 2-adic norm, so that $r(2) = \frac{1}{2}$. Color the points of the plane with three colors in the same manner as in (a) above. Let T be a triangle whose vertices are of three different colors. Then

$$r(\text{area } T) = r(\tfrac{1}{2}D) = r(\tfrac{1}{2}x_1 y_2 z_2) \geq r(\tfrac{1}{2}) = 2\,.$$

Suppose now that the unit square $0 \leq x,\ y \leq 1$ (all points have $z = 1$) is divided into m triangles of area $1/m$. Monsky proves that there must be a triangle with vertices of three different colors, which implies that $r(1/m) \geq 2$, and m is an even integer, as required. Incidentally, Monsky used the fact that no line had all three colors. He colored $\mathbb{R}^2$ rather than the projective plane, but it is the same coloring you obtain if you restrict that of Straus and Hales to $\mathbb{R}^2$ and only consider $p = 2$.

Stein (1989) conjectured that centrally symmetric $2n$-gons can never be dissected into an odd number of triangles of equal area, and proved this for $2n = 6$, 8. For $2n = 4$ this follows from Monsky's result, and for regular $2n$-gons it follows from Kasimatis's result. Monsky recently proved this conjecture by showing that in fact no symmetric subset of the plane can be dissected into an odd number of triangles of equal area.

7.5. Halving lines

Suppose we are given n points in the plane, no three on a line. It is easy to see that there is always at least one line L that is spanned by the points that also cuts the points in half. That is, so that as near as possible half of the remaining $n-2$ points are on each side of L. To find L, just rotate a line L' about a vertex of the convex closure of the n points, which will cause L' to cross the other $n-1$ points one at a time. This argument shows that there are at least three such *halving lines*.

Erdős et al. (1973) define $h_2(n)$ to be the *maximum* number of halving lines of a set of n points in the plane, no three collinear, and they give a construction showing that $h_2(n) \geqslant cn \log n$. They show that $h_2(n) = O(n^{3/2})$. Recently, Pach et al. (1992) improved this to $h_2(n) = O(n^{3/2}/\log^* n)$, where $\log^* n$ denotes the iterated logarithm function.

In Edelsbrunner's (1987) book *Algorithms in Combinatorial Geometry*, there is a discussion of $h_2(n)$ and the corresponding function $h_d(n)$ in $\mathbb{R}^d$, and a proof due to H. Edelsbrunner and R. Seidel that

$$h_d(n) \geqslant c_d n^{d-1} \log n \,.$$

Barany et al. (1990) show that

$$h_3(n) \leqslant c' n^{3-1/343} \,.$$

This was recently improved to

$$h_3(n) \leqslant cn^{8/3} \log^{5/3} n$$

by Aronov et al. (1991).

7.5.1. Edelsbrunner's problem

Edelsbrunner (1982) asked for bounds on the maximum number $g(k,n)$ of subsets of size *at most* k cut off by a line, for n points in the plane, no three on a line and $k \leqslant \frac{1}{2}n$. Goodman and Pollack (1984a) showed that $g(k,n) \leqslant 2nk - 2k^2 - k$. Alon and Györi (1986) determined that $g(k,n) = kn$. The problem may be stated more generally in terms of pseudolines, or allowable sequences, and Alon and Györi obtain the same value for the corresponding function analogous to $g(k,n)$.

References

Alexanderson, G.L., and J.E. Wetzel
[1981] Arrangements of planes in space, *Discrete Math.* **34**, 219–240.

Alon, N., and E. Györi
[1986] The number of small semispaces of a finite set of points in the plane, *J. Combin. Theory A* **41**, 154–157.

Altman, E.
[1963] On a problem of P. Erdős, *Amer. Math. Monthly* **70**, 148–157.
[1972] Some theorems on convex polygons, *Canad. Math. Bull.* **15**, 329–340.

Anning, N.H., and P. Erdős
[1945] Integral distances, *Bull. Amer. Math. Soc.* **51**, 598–600.

Aronov, B., B. Chazelle, H. Edelsbrunner, L. Guibas, M. Sharir and R. Wenger
[1991] *Discrete Comput. Geom.*, to appear.
Avis, D.
[1984] The number of farthest neighbor pairs of a finite planar set, *Amer. Math. Monthly* **91**, 417–420.
Avis, D., P. Erdős and J. Pack
[1988] Repeated distances in space, *Graphs and Combinatorics* **4**, 207–217.
Bannai, E., and E. Bannai
[1981] An upper bound for the cardinality of an s-distance subset in real Euclidean space, *Combinatorica* **1**, 99–102.
Bannai, E., E. Bannai and D. Stanton
[1983] An upper bound for the cardinality of an s-distance subset in real Euclidean space II, *Combinatorica* **3**, 147–152.
Barany, I.
[1982] A generalization of Caratheodory's theorem, *Discrete Math.* **40**, 141–152.
[1987] An extension of the Erdős–Szekeres theorem on large angles, *Combinatorica* **7**, 161–169.
Barany, I., and Z. Füredi
[1987] Empty simplices in Euclidean space, *Canad. Math. Bull.* **30**, 436–445.
Barany, I., and J. Kincses
[1987] A characterization of the Helly dimension of convex bodies, *Studia Sci. Math. Hungar.* **22**, 401–406.
Barany, I., M. Katchalski and J. Pach
[1984] Quantitative Helly-type theorems, *Amer. Math. Monthly* **91**, 362–365.
Barany, I., Z. Füredi and L. Lovász
[1990] On the number of halving planes, *Combinatorica* **10**, 175–183.
Beck, J.
[1983] On the lattice property of the plane and some problems of Dirac, Motzkin and Erdős in combinatorial geometry, *Combinatorica* **3**, 1 281–297.
[1987] *Different Distances,* Preprint.
Beck, J., and J. Spencer
[1984] Unit distances, *J. Combin. Theory A* **37**, 231–238.
Bender, C.
[1874] Bestemmung der grössten Anzahl gleich grosser Kugeln, welche sich auf eine Kugel von demselben Radius, wie die überigen, auflegen lassen, *Grunert Arch.* **56**, 302–313.
Björner, A.
[1980a] Shellable and Cohen–Macaulay partially ordered sets, *Trans. Amer. Math. Soc.* **260**, 159–183.
[1980b] Some matroid inequalities, *Discrete Math.* **31**, 101–103.
Blokhuis, A.
[1984] A new upper bound for the cardinality of 2-distance sets in euclidean space, in: *Convexity and Graph Theory, Jerusalem, 1981, North-Holland Math. Studies* **87**, 65–66.
Bollobás, B.
[1986] *Combinatorics* (Cambridge University Press, Cambridge).
Boltyanskii, V.G.
[1976] Generalization of a certain theorem of Szökefalvi-Nagy, *Dokl. Akad. Nauk SSSR* **228**, 265–268.
Bonnice, W.E., and L.M. Kelly
[1971] On the number of ordinary planes, *J. Combin. Theory A* **11**, 45–53.
Boros, E., Z. Füredi and L.M. Kelly
[1989] On representing Sylvester–Gallai designs, *Discrete Comput. Geom.* **4**, 345–348.
Borwein, J.
[1977] A proof of the equivalence of Helly's and Krasnosselsky's theorems, *Canad. Math. Bull.* **20**, 35–37.
Borwein, P.
[1984] Sylvester's problem and Motzkin's theorem for countable and compact sets, *Proc. AMS* **90**, 580–584.
Breen, M.
[1985] Krasnoselskii-type theorems, *Ann. New York Acad. Sci.* **440**, 142–146.
[1988] A weak Krasnoselskii theorem in $\mathbb{R}^d$, *Proc. Amer. Math. Soc.* **104**, 558–562.

Burr, S., B. Grünbaum and N. Sloane
[1979] The orchard problem, *Geom. Dedicata* **2**, 397–424.
Burton, G.R., and G. Purdy
[1979] The directions determined by n points in the plane, *J. London Math. Soc. (2)* **20**, 109–114.
Chakerian, G.D.
[1970] Sylvester's problem on collinear points and a relative, *Amer. Math. Monthly* **77**, 164–167.
Chung, F.R.K.
[1984] The number of different distances determined by n points in the plane, *J. Combin. Theory A* **36**, 342–354.
[1989] Sphere-and-point incidence relations in high dimensions with applications to unit distances and furthest neighbor pairs, *Discrete Comput. Geom.* **4**, 183–190.
Clarkson, K., H. Edelsbrunner, L. Guibas, M. Sharir and E. Welzl
[1990] Combinatorial complexity bounds for arrangements of curves and surfaces, *Discrete Comput. Geom.* **5**, 161–196.
Conway, J.H., H.T. Croft, M.J.T. Guy and P. Erdős
[1979] On the distribution of values of angles determined by coplanar points, *J. London Math. Soc. (2)* **19**, 137–143.
Coxeter, H.S.M.
[1948] A problem of collinear points, *Amer. Math. Monthly* **55**, 26–28.
[1963] An upper bound for the number of equal nonoverlapping spheres that can touch another of the same size, in: *Convexity, Proc. Symp. Pure Math.* **7**, pp. 53–71.
Croft, H.T.
[1962] 9-point and 7-point configurations in 3-space, *Proc. London Math. Soc. (3)* **12**, 400–424.
[1970] Some problems of combinatorial geometry, in: *Combinatorial Structures and their Applications, Proc. Calgary Int. Combin. Conf., Calgary, Alta., 1969* (Gordon and Breach, New York) pp. 47–52.
Csima, J., and E.T. Sawyer
[1993] There exist $6n/13$ ordinary points, *Discrete Comput. Geom.* **9**, 187–202.
Danzer, L.
[1957] Über ein Problem aus der kombinatorischen Geometrie, *Arch. Math.* **8**, 347–351.
Danzer, L., and B. Grünbaum
[1967] Problem 7, in: *Proc. Colloq. on Convexity, Copenhagen, 1965* (Kobenhavns Universitets Matematiske Institut).
Danzer, L., B. Grünbaum and V. Klee
[1963] Helly's theorem and its relatives, in: *Convexity, Proc. Symp. Pure Math.* **7**, 101–180.
de Bruijn, N.G., and P. Erdős
[1948] On a combinatorial problem, *Indag. Math.* **10**, 421–423.
[1951] A colour problem for infinite graphs and a problem in the theory of relations, *Indag. Math.* **13**, 369–373.
De Santis, R.
[1957] A generalization of Helly's theorem, *Proc. Amer. Math. Soc.* **8**, 336–340.
Debrunner, H., and H. Hadwiger
[1957] Über eine Variante zum Hellyschen Satz, *Arch. Math.* **8** 309–313.
Dirac, G.A.
[1951] Collinearity properties of sets of points, *Quart. J. Math. Oxford Ser. (2)* **2**, 221–227.
Dowling, T.A.
[1980] On the independent set numbers of a finite matroid, in: *Combinatorics 79, Proc. Colloq. Univ. Montreal, Montreal, Que., 1979,* Part I. *Ann. Discrete Math.* **8**, 21–28.
Dowling, T.A., and R.M. Wilson
[1974] The slimmest geometric lattices, *Trans. Amer. Math. Soc.* **196**, 203–216.
Edelsbrunner, H.
[1982] Letter to G. Purdy, January 29.
[1987] *Algorithms in Combinatorial Geometry, EATCS Monographs on Theoretical Computer Science,* Vol. 10 (Springer, Berlin).

Edelsbrunner, H., and S. Skiena
[1989] On the number of furthest neighbor pairs in a point set, *Amer. Math. Monthly* **96**, 614–618.
Edelsbrunner, H., L. Guibas and M. Sharir
[1990] The complexity of many cells in arrangements of planes and related problems, *Discrete Comput. Geom.* **5**, 197–216.
Edelstein, M., F. Herzog and L.M. Kelly
[1963] A Further theorem of the Sylvester type, *Proc. Amer. Math. Soc.* **14**, 359–363.
Eggleston, H.G.
[1958] *Convexity* (Cambridge University Press, Cambridge).
Elekes, G.
[1984] n points in the plane can determine $n^{3/2}$ unit circles, *Combinatorica* **4**, 131.
Elliott, P.D.T.A.
[1967] On the number of circles determined by n points, *Acta Math. Acad. Sci. Hungar.* **18**, 181–188.
Erdős, P.
[1946] On sets of distances of n points, *Amer. Math. Monthly* **53**, 248–250.
[1957] Néhany geometriai problémarol (in Hungarian), *Mat. Lapok* **8**, 86–92.
[1960] On sets of distances of n points in euclidean space, *Magyar Tud. Akad. Mat. Kut. Int. Kőzl* **5**, 165–168.
[1961] Some Unsolved problems, *Publ. Math. Inst. Hungar. Acad. Sci.* **6**, 221–254. MR 31#2106.
[1962] Néhany elemi geometriai problémarol, *Kőzépiskolai Matematikai Lapok* **24**, 193–201.
[1964] On extremal problems of graphs and generalized graphs, *Israel J. Math.* **2**, 183–190.
[1967] On some applications of graph theory to geometry, *Canad. J. Math.* **19**, 968–9f71.
[1970] On sets of distances of n points, in: *Research Problems,* ed. V. Klee, *Amer. Math. Monthly* **77**, 738–740.
[1975] On some problems of elementary and combinatorial geometry, *Ann. di Mat. (4)* **103**, 99–108.
[1980] Some combinatorial problems in geometry, in: *Proc. Conf. Haifa, Israel, 1979, Lecture Notes in Mathematics,* Vol 792 (Springer, Berlin) pp. 46–53.
[1981] Some applications of graph theory and combinatorial methods to number theory and geometry, in: *Algebraic Methods in Graph Theory, Colloq. Math. Soc. János Bolyai* **25**, 131–148.
[1983] Combinatorial problems in geometry, *Math. Chronicle* **12**, 35–54.
[1984] Some old and new problems in combinatorial geometry, *Ann. Discrete Math.* **20**, 129–136.
[1985a] Problems and results in combinatorial geometry, *Ann. New York Acad. Sci.* **440**, 1–11.
[1985b] Some Combinatorial and metric problems in geometry, in: *Intuitive Geometry, Siofok, Colloq. Math. Soc. János Bolyai* **48**, 167–177.
Erdős, P., and R. Guy
[1970] Distinct distances between lattice points, *Elemente Math.* **25**, 121–123.
Erdős, P., and G. Purdy
[1971] Some extremal problems in geometry, *J. Combin. Theory* **10**(3) 246–252.
[1975] Some extremal problems in geometry III, *Congress. Numerantium* **15**, 291–308.
[1976] Some extremal problems in geometry IV, *Congress. Numerantium* **17**, 307–322.
[1978] Some combinatorial problems in the plane, *J. Combin. Theory A* **25**, 205–210.
[1990] *Some More Combinatorial Problems in the Plane,* Preprint.
Erdős, P., and M. Simonovits
[1980] On the chromatic number of geometric graphs, *Ars Combin.* **9**, 229–246.
Erdős, P., and A.H. Stone
[1946] On the structure of linear graphs, *Bull. Amer. Math. Soc.* **52**, 1087–1091.
Erdős, P., and G. Szekeres
[1935] A combinatorial problem in geometry, *Comp. Math.* **2**, 463–470.
Erdős, P., and P. Turan
[1941] On a problem of Sidon, *J. London Math. Soc.* **16**, 292–296.
Erdős, P., F. Harary and W.T. Tutte
[1965] On the dimension of a graph, *Mathematika* **12**, 118–122.

Erdős, P., L. Lovász, G.J. Simmons and E.G. Straus
[1973] Dissection graphs of planar point sets, in: *A Survey of Combinatorial Theory*, eds. J.N. Shrivastava, F. Harary, C.R. Rao, G.-C. Rota and S.S. Shrikhande (North-Holland, Amsterdam) pp. 138–149.

Erdős, P., G. Purdy and E.G. Straus
[1982] On a problem in combinatorial geometry, *Discrete Math.* **40**, 45–52.

Erdős, P., L. Lovász and K. Vesztergombi
[1989] On the graph of large distances, *Discrete Comput. Geom.* **4**, 541–549.

Erdős, P., D. Hickerson and J. Pach
[1990a] A problem of L. Moser about repeated distances on the sphere, *Amer. Math. Monthly* **96**, 569–575.

Erdős, P., E. Makai, J. Pach and J. Spencer
[1990b] Gaps in difference sets, and the graph of nearly equal distances, in: *Applied Geometry and Discrete Mathematics, DIMACS Series in Discrete Mathematics and Theoretical Computer Science*, Vol. 4 (1991) pp. 265–273.

Erdős, P., P. Fishburn and Z. Füredi
[1990c] *Midpoint of Diagonals of Convex n-Gons*, Preprint.

Fejes-Toth, G.
[1983] New results in the theory of packing and covering, in: *Convexity and its Applications*, eds. P.M. Gruber and J.M. Wills (Birkhäuser, Basel) pp. 318–359.

Fejes-Toth, L.
[1975] A combinatorial problem concerning oriented lines in the plane, *Amer. Math. Monthly* **82**, 387–389.

Frankl, P., and V. Rödl
[1986] All triangles are Ramsey, *Trans. Amer. Math. Soc.* **297**, 777–779.
[1987] Forbidden intersections, *Trans. Amer. Math. Soc.* **300**, 259–286.
[1990] A partition property of simplices in euclidean space, *J. Amer. Math. Soc.* **3**, 1–7.

Frankl, P., and R.M. Wilson
[1981] Intersection theorems with geometric consequences, *Combinatorica* **1**, 357–368.

Füredi, Z., and I. Palasti
[1984] Arrangements of lines with a large number of triangles, *Proc. Amer. Math. Soc.* **92**, 561–566.

Goodman, J.E., and R. Pollack
[1980a] Proof of Grünbaum's conjecture on the stretchability of certain arrangements on pseudolines, *J. Combin. Theory A* **29**, 385–390.
[1980b] On the combinatorial classification of nondegenerate configurations in the plane, *J. Combin. Theory A* **29**, 220–235.
[1980c] Convexity theorems for generalized planar configurations, in: *Proc. Conf. on Convexity and Related Combinatorics, University of Oklahoma* (Marcel Dekker, New York).
[1981] A combinatorial perspective on some problems in geometry, *Congress. Numerantium* **32**, 383–394.
[1984a] On the number of k-subsets of a set of n points in the plane, *J. Combin. Theory A* **36**, 101–104.
[1984b] Semispaces of configurations, cell complexes of arrangements, *J. Combin. Theory A* **37**, 257–293.
[1988] Hadwiger's transversal theorem in higher dimensions, *J. Amer. Math. Soc.* **1**, 301–309.

Graham, R.L.
[1980] On partitions of E_n, *J. Combin. Theory A* **28**, 89–94.
[1985] Old and new euclidean Ramsey theorems, *Ann. New York Acad. Sci.* **440**, 20–30.
[1990] Topics in exclusive Ramsey theory, in: *Mathematics of Ramsey Theory*, eds. J. Nešetřil and V. Rödl (Springer, New York).

Graham, R.L., B.L. Rothschild and E.G. Straus
[1974] Are there $n+2$ points in E^n with odd integral distances? *Amer. Math. Monthly* **81**, 21–25.

Gritzmann, P.
[1988] MR 88f: 52, 010.

Gruber, P.M., and J.M. Wills
[1983] *Convexity and its Application* (Birkhäuser, Basel).

Grünbaum, B.
[1958] On common transversals, *Arch. Math.* **9**, 465–469.

[1959] On intersections of similar sets, *Portugal. Math.* **18**, 155–164.
[1961] The dimensions of intersections of convex sets, *Pacific J. Math.* **12**.
[1963] Borsuk's problem and related questions, in: *Convexity, Proc. Symp. Pure Math.* **VII**, 271–284.
[1967] *Convex Polytopes, Pure and Applied Mathematics,* Vol. 16 (Wiley/Interscience, New York).
[1971] Arrangements of hyperplanes, *Congress. Numerantium* **3**, 41–106.
[1972] *Arrangements and Spreads, CBMS Regional Conf. Series in Mathematics,* Vol. 10 (American Mathematical Society, Providence, RI).
[1974] *Arrangements of colored lines,* Class Notes.
[1976] New views on some old questions of combinatorial geometry, in: *Proc. Colloq. Int. sulle Teorie Combinatorie, Rome, 1973,* Vol. 1 (Accademia Nazionale dei Lincei) pp. 451–468.
[1980] Two-coloring the faces of arrangements, *Period. Math. Hungar.* **11**(3), 181–185.

Grünbaum, B., and G.C. Shephard
[1984] Simplicial arrangements in projective 3-space, *Mitt. Math. Semin. Univ., Giessen* **166**, 49–101.
[1987] *Tilings and Patterns* (Freeman, New York).

Hadwiger, H.
[1957] Über Eibereiche mit gemeinsamer Treffgeraden, *Portugal. Math.* **16**, 23–29.
[1961] Ungelöste Probleme N. 40, *Elemente Math.* **16**, 103–104.

Hadwiger, H., H. Debrunner and V. Klee
[1964] *Combinatorial Geometry in the Plane* (Holt, Rinehart & Winston, New York).

Hales, A.W., and E.G. Straus
[1982] Projective colorings, *Pacific J. Math.* **99**, 31–43.

Hansen, S.
[1965] A generalization of a theorem of Sylvester on lines determined by a finite set, *Math. Scand.* **16**, 175–180.
[1980] On configurations in 3-space without elementary planes and on the number of ordinary planes, *Math. Scand.* **47**, 181–194.
[1981] *Contributions to the Sylvester–Gallai-theory,* Dissertation for Habilitation (University of Copenhagen).

Harborth, H.
[1974] Solution to problem 664a, *Elemente Math.* **29**, 14–15.
[1979] Kovexe Fúnfecke in ebenen Punktmengen, *Elemente Math.* **33**, 116–118.
[1983] Aequidistante, regulare Punktmengen, in: *Collection: 2nd Colloq. for Geometry and Combinatorics, Part 1, 2, Karl-Marx-Stadt,* (Technische Hochschule Karl-Marx-Stadt) pp. 81–86. MR 87h:52019.
[1985a] Two-colorings of simple arrangements, *Colloq. Math. Soc. János Bolyai* **37**, 371–378.
[1985b] Some simple arrangements of pseudolines with a maximum number of triangles, *Ann. New York Acad. Sci.* **440**, 31–33.
[1985c] *Einheitskreise in ebenen Punktmengen, 3. Kolloq. über Diskrete Geometrie* (University of Salzburg) pp. 163–168.
[1985d] Regular point sets with unit distances, *Colloq. Math. Soc. János Bolyai* **48**, 239–253.

Harborth, H., and I. Mengersen
[1986] Point sets with many unit circles, *Discrete Math.* **60**, 193–197.

Harborth, H., and L. Piepmeyer
[1991] Point sets with small integral distances, in: *Applied Geometry and Discrete Mathematics, DIMACS Series in Discrete Mathematics and Theoretical Computer Science,* Vol. 4 (AMS, Providence, RI) pp. 319–324.

Helly, E.
[1923] Über Mengen konvexer Körper mit gemeinschaftlichen Punkten, *Jber. Deutsch. Math.-Verein.* **32**, 175–176.

Herzog, F., and L.M. Kelly
[1960] A generalization of the theorem of Sylvester, *Proc. Amer. Math. Soc.* **11**, 327–331.

Hilton, H.
[1932] *Plane Algebraic Curves,* 2nd Ed. (Oxford University Press, Oxford).

Hirzebruch, F.
[1983] Arrangements of lines and algebraic surfaces, in: *Arithmetic and Geometry, A Collection of Papers Dedicated to I.R. Shafarevich,* Vol. 2 (Birkhäuser, Basel) pp. 113–140.
[1984] *Singularities of Algebraic Surfaces and Characteristic Numbers* (Max Planck Inst. Math., Bonn).
Horn, A.
[1949] Some generalizations of Helly's theorem on convex sets, *Bull. Amer. Math. Soc.* **55**, 923–929.
Horn, A., and F.A. Valentine
[1949] Some properties of L-sets in the plane, *Duke Math. J.* **16**, 131–140.
Horton, J.D.
[1983] Sets with no empty convex 7-gons, *Canad. Math. Bull.* **26**, 482–484.
Jamison, R.
[1985] A survey of the slope problem, *Ann. New York Acad. Sci.* **440**, 34–51.
[1987] Direction trees, *Discrete Comput. Geom.* **2**, 249–254.
Juhasz, R.
[1979] Ramsey type theorems in the plane, *J. Combin. Theory A* **27**, 152–160.
Kalai, G.
[1984] Characterization of f-vectors of families of convex sets in $\mathbb{R}^d$, Part I: Necessity of Eckhoff's conditions, *Israel J. Math.* **48**, 175–195.
[1986] Characterization of f-vectors of families of convex sets in $\mathbb{R}^d$, Part II. Sufficiency of Eckhoff's conditions, *J. Combin. Theory A* **41**, 167–188.
Karteszi, F.
[1963] Sylvester egy tételéröl és Erdős egy sejtéséröl, *Közl. Mat. Lapok* **26**, 3–10.
Kasimatis, E.A.
[1989] Dissections of regular polygons into triangles of equal areas, *Discrete Comput. Geom.* **4**, 375–381.
Katchalski, M., and A. Liu
[1979] A problem of geometry in $\mathbb{R}^d$, *Proc. Amer. Math. Soc.* **75**, 284–288.
Katchalski, M., and A. Meir
[1988] On empty triangles determined by n points in the plane, *Acta Math. Hungar.* **51**, 323–328.
Kelly, L.M.
[1947] Elementary Problems and Solutions. Isosceles n-points, *Amer. Math. Monthly* **54**, 227–229.
[1986] On a problem of Serre on Sylvester–Gallai configurations, *Discrete Comput. Geom.* **1**, 101–104.
Kelly, L.M., and W.O.J. Moser
[1958] On the number of ordinary lines determined by n points, *Canad. J. Math.* **10**, 210–219.
Kelly, L.M., and R. Rottenberg
[1972] Simple points in pseudoline arrangements, *Pacific J. Math.* **40**, 617–622.
Kincses, J.
[1987] The classification of 3- and 4-Helly-dimensional convex bodies, *Geom. Dedicata* **22**, 283–301.
Klee, V.
[1951] On certain intersection properties of convex sets, *Canad. J. Math.* **3**, 272–275.
[1953] The critical set of a convex body, *Amer. J. Math.* **75**, 178–188.
[1954] Common secants for plane convex sets, *Proc. Amer. Math. Soc.* **5**, 639–641.
[1963] Infinite-dimensional intersection theorems, in: *Convexity, Proc. Symp. Pure Math.* **7**, 349–360.
Klee, V., and S. Wagon
[1991] *Old and New Solved and Unsolved Problems from Plane Geometry and Number Theory* (Mathematical Association of America).
Komlós, J., T. Pintz and E. Szemerédi
[1982] A lower bound for Heilbronn's problem, *J. London Math. Soc. (2)* **25**, 13–24.
Kővári, P., V.T. Sós and P. Turán
[1954] On a problem of K. Zarankiewicz, *Colloq. Math.* **3**, 50–57.
Krasnosselsky, M.A.
[1946] Sur un Critère pour qu'un domain soit étoilé (in Russian, with French summary), *Mat. Sb., N.S.* **19**, 309–310.

Kříž, I.

[1991] Permutation groups in Euclidean Ramsey theory, *Proc. Amer. Math. Soc.* **112**, 899–907.

Landau, E.

[1909] *Handbuch der Lehre von der Verteilung der Primzahlen,* Vol. 2 (Teubner, Leipzig). Also exists as Chelsea reprint: *Primazahlen,* two volumes in one.

Larman, D.G.

[1968] Helly type properties of unions of convex sets, *Mathematika* **15**, 53–59.

Larman, D.G., and C.A. Rogers

[1972] The realization of distances within sets in Euclidean space, *Mathematica* **19**, 1–24.

Larman, D.G., C.A. Rogers and J.J. Seidel

[1977] On two-distance sets in euclidean space, *Bull. London Math. Soc.* **9**, 261–2676.

Leech, J.

[1956] The problem of the thirteen spheres, *Math. Gaz.* **40**, 22–23.

Levi, F.

[1926] Die Teilung der projektiven Ebene durch Gerade oder Pseudogerade, *Ber. Math.-Phys. Kl. Sächs. Akad. Wiss. Leipzig* **78**, 256–267.

[1929] *Geometrische Konfigurationen* (Hirzel, Leipzig).

Makai, E., and H. Martini

[1989] A lower bound on the number of sharp shadow-boundaries of convex polytopes, *Period. Math. Hungar.* **20**, 249–260.

Mason, J.H.

[1972] Matroids: unimodal conjectures and Motzkin's theorem, in: *Combinatorics,* eds. D.J.A. Welsh and D.R. Woodall (Institute of Mathematics and Applications) pp. 207–221.

McMullen, P.

[1971] On zonotopes, *Trans. Amer. Math. Soc.* **159**, 91–109.

Mead, D.G.

[1979] Dissection of the hypercube into simplexes, *Proc. Amer. Math. Soc.* **76**, 302–304.

Melchior, E.

[1940] Über Vielseite der projektiven Ebene, *Deutsche Math.* **5**, 461–475.

Miyaoka, Y.

[1977] On the Chern numbers of surfaces of the general type, *Invent. Math.* **42**, 3225–3237.

Monsky, P.

[1970] On dividing a square into triangles, *Amer. Math. Monthly* **77**, 161–164.

Moser, L.

[1952] On the different distances determined by *n* points, *Amer. Math. Monthly* **59**, 85–91.

[1966] *Poorly Formulated Unsolved Problems of Combinatorial Geometry,* Mimeograph.

Moser, W.O.J., and J. Pach

[1986] *Research Problems in Discrete Geometry,* Manuscript (Department of Mathematics, McGill University, Montreal, Quebec).

Motzkin, T.

[1951] The lines and planes connecting the points of a finite set, *Trans. Amer. Math. Soc.* **70**, 451–464.

[1967] Nonmixed connecting lines, *Notices Amer. Math. Soc.* **14**, 837.

[1975] Sets for which no point lies on many connecting lines, *J. Combin. Theory A* **18**, 345–348.

Neaderhouser, C.C., and G.B. Purdy

[1982] On finite sets in E^k in which the diameter is frequently achieved, *Period. Math. Hungar.* **13**(3), 253–257.

Pach, J., and M. Sharir

[1990] *Repeated Angles in the Plane and Related Problems,* Manuscript.

Pach, J., W. Steiger and E. Szemerédi

[1992] An upper bound on the number of planar *k*-sets, *Discrete Comput. Geom.* **7**, 109–123.

Palasti, I.

[1975] The maximal number of quadrilaterals bounded by general straight lines in a plane, *Period. Math. Hungar.* **6**, 323–341.

[1976] The ratio of black and white polygons of a map generated by general straight lines, *Period. Math. Hungar.* **7**, 91–94.

Pannwitz, E., and H. Hopf
[1934] Aufgabe Nr. 167, *Jber. Deutsch. Math.-Verein.* **43**, 114.

Peterson, B.B.
[1982] Is there a Krasnoselskii theorem for finitely starlike sets?, in: *Proc. Conf. on Convexity and Related Combinatorial Geometry* (Marcel Dekker, New York).

Pollack, R., and R. Wenger
[1990] Necessary and sufficient conditions for hyperplane transversals, *Combinatorica* **10**, 307–311.

Purdy, C., and G. Purdy
[1988] Minimal forbidden distance one graphs, *Congress. Numerantium* **66**, 165–172.

Purdy, G.
[1974] Some extremal problems in geometry, *Discrete Math.* **7**, 305–315.
[1980a] Triangles in arrangements of lines II, *Proc. Amer. Math. Soc.* **79**, 77–81.
[1980b] On the number of regions determined by *n* lines in the projective plane, *Geom. Dedicata* **9**, 107–109.
[1981] A proof of a consequence of Dirac's conjecture, *Geom. Dedicata* **10**, 317–321.
[1982a] The independent sets of rank *k* of a matroid, *Discrete Math.* **38**, 87–91.
[1982b] The minimum number of empty triangles, *AMS Abstr.* **3**, 318.
[1986] Two results about points, lines and planes, *Discrete Math.* **60**, 215–218.
[1988] Repeated angles in E_4, *Discrete Comput. Geom.* **3**, 73–75.

Purdy, G., and J.E. Wetzel
[1980] Two-coloring inequalities for euclidean arrangements in general position, *Discrete Math.* **31**, 53–58.

Rado, R.
[1952] A theorem on sequences of convex sets, *Quart. J. Math. Oxford Ser. (2)* **3**, 183–186.

Ramsey, F.P.
[1930] On a problem in formal logic, *Proc. London Math. Soc. (2)* **30**, 264–286.

Roudneff, J.-P.
[1986] On the number of triangles in simple arrangements of pseudolines in the real projective plane, *Discrete Math.* **60**, 243–251.
[1987] Quadrilaterals and pentagons in arrangements of lines, *Geom. Dedicata* **23**, 221–227.
[1988] Arrangements of lines with a minimun number of triangles are simple, *Discrete Comput. Geom.* **3**, 97–102.
[1991] The maximum number of triangles in arrangements of (pseudo) lines, to appear. Alternative reference: *Matroides orientés et arrangements de pseudodroites,* Dissertation (Université Paris VI, June 1987).

Santaló, L.A.
[1940] Un teorema sobre conjuntos de paralelepipedos de aristas paralelas, *Publ. Inst. Mat. Univ. Nac. Litoral* **2**, 49–60.

Scott, P.R.
[1970] On the sets of directions determined by *n* points, *Amer. Math. Monthly* **70**, 502–505.

Serre, J.-P.
[1966] Problem, *Amer. Math. Monthly* **73**, 89.

Seymour, P.D.
[1982] On the points–lines–planes conjecture, *J. Combin. Theory B* **33**, 17–26.

Shannon, R.W.
[1976] A lower bound on the number of cells in an arrangement of hyperplanes, *J. Combin. Theory A* **20**, 327–335.f.

Simmons, G.J., and J.E. Wetzel
[1979] A two-coloring inequality for euclidean two-arrangements, *Proc. Amer. Math. Soc.* **77**, 124–127.

Spencer, J., E. Szemerédi and W.T. Trotter
[1984] Unit distances in the euclidean plane, in: *Graph Theory and Combinatorics: A Volume in Honour of P. Erdős,* ed. B. Bollobás (Academic Press, London) pp. 293–303.

Stein, S.
[1989] Equidissections of centrally symmetric octagons, *Aequationes Math.* **37**, 313–318.
Strommer, T.O.
[1977] Triangles in arrangements of lines, *J. Combin. Theory A* **23**, 314–320.
Sutherland, J.W.
[1935] *Jber. Deutsch. Math.-Verein.* **45**(2. Abt), 33.
Sylvester, J.J.
[1867] Problem 2473, *Mathematical Questions and Solutions from the Educational Times* **8**, 104–107.
[1893] Mathematical Question 11851, *Educational Times* **59**, 98–99.
Szemerédi, E., and W.T. Trotter
[1983] Extremal problems in discrete geometry, *Combinatorica* **3**, 381–392.
Tverberg, H.
[1966] A generalization of Radon's theorem, *J. London Math. Soc.* **41**, 123–128.
[1989] Proof of Grünbaum's conjecture on common transversals for translates, *Discrete Comput. Geom.* **4**, 191–203.
Ungar, P.
[1982] $2N$ noncollinear points determine at least $2N$ directions, *J. Combin. Theory A* **33**, 343–347.
Valentine, F.A.
[1963] The dual cone and Helly type theorems, in: *Convexity, Proc. Symp. Pure Math.* **7**, 473–493.
Vesztergombi, K.
[1985] On the distribution of distances in finite sets in the plane, *Discrete Math.* **57**, 129–145.
Vincensini, P.
[1939] Sur une extension d'un théorème de M.J. Radon sur les ensembles de corps convexes, *Bull. Soc. Math. France* **67**, 115–119.
[1953] Les ensembles d'arcs d'un même cercle dans leurs relations avec les ensembles de corps connexes du plan euclidien, *Atti IV. Congr. Un Mat. Ital.* **2**, 456–464.
Welsh, D.J.A.
[1976] *Matroid Theory, London Mathematical Society Monograph,* Vol. 8 (Academic Press, London).
Wenger, R.
[1990] A generalization of Hadwiger's transversal theorem to intersecting sets, *Discrete Comput. Geom.* **5**, 383–388.
Wilson, P.R., and J.A. Wiseman
[1988] A Sylvester theorem for conic sections, *Discrete Comput. Geom.* **3**, 295–305.
Wilson, R.M.
[1975] An existence theory for pairwise balanced designs, III: Proof of the existence conjectures, *J. Combin. Theory A* **18**, 71–79.
Wormald, N.
[1970] A 4-chromatic graph with a special plane drawing, *J. Aust. Math. Soc. A* **28**, 1–8.
Yau, S.-T.
[1977] Calabi's conjecture and some new results in algebraic geometry, *Proc. Nat. Acad. Sci. U.S.A.* **74**, 1798–1799.
Zaslavsky, T.
[1975] Facing up to arrangements: face-count formulas for partitions of space by hyperplanes, *Mem. Amer. Math. Soc.* **1**(1), No. 154.
[1985] Extremal arrangements of hyperplanes, *Ann. New York Acad. Sci.* **440**, 69–87.

CHAPTER 18

Convex Polytopes and Related Complexes

Victor KLEE*

Department of Mathematics, University of Washington, Seattle, WA 98195, USA

Peter KLEINSCHMIDT

Wirtschaftswissenschaftliche Fakultät, Universität Passau, 94030 Passau, Germany

Contents

*Supported in part by the National Science Foundation.

HANDBOOK OF COMBINATORICS
Edited by R. Graham, M. Grötschel and L. Lovász

1. Introduction and basic definitions

This survey is concerned with the combinatorial or facial structure of convex polytopes. The earliest result in the area is Euler's theorem asserting that if v, e, and f are the numbers of vertices, edges, and 2-faces of a 3-dimensional convex polytope, then $v - e + f = 2$. Attempts to understand and extend Euler's theorem were among the first stimuli for the developments on which we report. Since the early 1950s, the relationship to linear programming has been an important stimulus. It has been joined more recently by connections on the "pure" side with matroid theory, commutative algebra, and algebraic geometry, and on the "applied" side by problems from statistics, nonlinear optimization, computational geometry, and computational complexity. Thanks to all of these stimuli, as well as to the strong intuitive appeal of the subject, the combinatorial study of convex polytopes has advanced greatly in the past thirty years.

In studying the facial structure of polytopes, the natural setting is a finite-dimensional vector space E over an ordered field Φ. Of course, the real field $\mathbb{R}$ and its subfields are of special interest. Except where the contrary is stated, attention is confined here to the case in which $\Phi = \mathbb{R}$, but that is mainly for brevity as a large fraction of the facial or combinatorial theory can be developed for an arbitrary Φ.

As the terms are used here, a *polyhedron* is the intersection of a finite collection of closed halfspaces in E and a *polytope* is a bounded polyhedron. Equivalently, a polytope is the convex hull of a finite set. Since, for us, polyhedra and polytopes are all convex, the word *convex* is omitted in much of what follows.

A *face* of a polyhedron P is the empty set $\emptyset$, P itself, or the intersection of P with a supporting hyperplane. Prefixes indicate (affine) dimension, and the 0-, 1-, $(d-2)$- and $(d-1)$-faces of a d-polyhedron are respectively its *vertices, edges, ridges* and *facets.* A polyhedron is *pointed* if it has at least one vertex and hence has faces of all dimensions less than or equal to the dimension of the polyhedron itself.

With respect to the ordering given by set inclusion, the collection of all faces of a polyhedron P forms a lattice, the *face-lattice* of P. Two polyhedra P and Q are said to be *isomorphic*, or *combinatorially equivalent*, or *of the same combinatorial type*, if their face-lattices are isomorphic. When both are polytopes, this is equivalent to saying that there is a bijection between their vertex sets V and W such that a subset of V is the vertex set of a face of P if and only if its image in W is the vertex set of a face of Q.

A *complex*, as the term is used here, is a finite collection $\mathscr{C}$ of polytopes such that each face of a member of $\mathscr{C}$ is itself a member of $\mathscr{C}$, and the intersection of any two members of $\mathscr{C}$ is a face of each. The *dimension* of a complex $\mathscr{C}$ is the maximum of the dimensions of its members. The union of $\mathscr{C}$'s members, sometimes called *set* $\mathscr{C}$, is denoted by $\bigcup \mathscr{C}$. A d-complex $\mathscr{C}$ is *pure* if $\bigcup \mathscr{C}$ is the union of $\mathscr{C}$'s d-dimensional members. Note that when P is a d-polytope, the faces of P other than P itself form a pure $(d-1)$-complex $\mathscr{B}(P)$, the *boundary complex* of P.

A complex $\mathscr{C}$ in our sense is sometimes called a *cell complex* or a *polyhedral complex*, and its members are called *cells.* However, to emphasize our primary interest in the boundary complexes of polytopes, the members of $\mathscr{C}$ are here called *faces*, and when $\mathscr{C}$ is a pure d-dimensional complex its 0-, 1-, $(d-1)$-, and d-dimensional members are called *vertices*, *edges*, *ridges*, and *facets*.

Two complexes $\mathscr{C}$ and $\mathscr{D}$ are *isomorphic* if there is a bijection between their vertex sets V and W such that a subset S of V is the vertex set of a face of $\mathscr{C}$ if and only if the image of S in W is the vertex set of a face of $\mathscr{D}$. Thus the combinatorial equivalence of polytopes can be defined, according to taste, as isomorphism of their boundary complexes or isomorphism of their face-lattices. A complex $\mathscr{C}$ is *polytopal* if it is isomorphic to the boundary complex of a polytope. Polytopal complexes are the primary focus of this report. In order to convey the spirit of some of the methods that have been used to study such complexes, we include a few indications of proof. However, our primary concern is with definitions, statements of results, and references.

The literature contains detailed studies of the boundary complexes of a variety of specific polytopes, especially those that are regular or close to regular in various senses and those that arise from important types of optimization problems. That latter direction of research is often called *Polyhedral Combinatorics*, and is surveyed in chapter 30 in this Handbook. It has much in common with the subject matter of the present chapter, but its emphases are different. Here our primary concern is not with specific polytopes but with properties that hold for *all* polytopes or at least for very wide classes of them. Specific polytopes arise here mainly as the solutions of various extremum problems concerning combinatorial structure. Even though our treatment is restricted to the "general" (as opposed to the "specific") combinatorics of polytopes, we have been forced by limitations of space to ignore many important areas. However, by consulting the references listed here, the reader will gain some acquaintance with these areas as well as the ones discussed here.

Polytopes may be presented in a number of ways. In many cases, the initial description is not a matter of taste but is dictated by the nature of a problem that gives rise to a polytope. It is often desirable to be able to derive the full face-lattice or boundary complex from the initial presentation. This topic is discussed briefly in section 3.

Because of the many situations in which polytopes arise, and the frequency with which facial structure is of interest, polytopal complexes are important combinatorial objects. Many difficulties in studying them arise from the fact that they are defined in an extrinsic, geometric fashion, and it seems that a purely combinatorial characterization could be very useful. The problem of finding such a characterization is commonly called the *Steinitz problem*, because of a theorem in the book of Steinitz and Rademacher (1934) that solves the problem for 3-polytopes. The Steinitz problem is open for each dimension $d \geqslant 4$, but it has led to various 3-dimensional ramifications and higher-dimensional partial analogues. The most important of these are described in section 4.

Section 5 deals with the enumeration of combinatorial types. This has been a much-studied topic in the case of 3-polytopes, and for that case there are fairly complete results. In higher dimensions, on the other hand, the lack of a purely combinatorial characterization makes precise enumeration difficult even for relatively small numbers of vertices or facets. Nevertheless, there are a few precise enumerations as well as some rough asymptotic results, and they are described.

Among the combinatorial invariants associated with a d-polytope, the face-vector and the edge-graph are the ones that have been most thoroughly studied. They are the subjects of sections 6 and 7, respectively. The *face-vector* $f(P)$ is the sequence $(f_0, f_1, ..., f_{d-1})$ that lists for each i the number f_i of i-faces of P. In the edge-graph $\mathscr{E}(P)$, the nodes are P's vertices and the edges are pairs of vertices that share an edge of P. There are other natural ways of associating a graph with P, but when one speaks simply of *the graph* of P, the *edge-graph* $\mathscr{E}(P)$ is intended. A graph is *d-polytopal* if it is isomorphic to the graph of a d-polytope.

For results on polytopal complexes obtained before 1967, the most important reference is the book of Grünbaum (1967). Since we refer to this so frequently, it will be designated simply as Gr67. It, and a subsequent survey article (Grünbaum 1970) stimulated many later developments. There were later books by McMullen and Shephard (1971), Bartels (1973), Yemelichev et al. (1981), Brøndsted (1983), and Ziegler (1994), and survey articles by Ewald et al. (1979) and Bayer and Lee (1993).

We are indebted to M. Bayer, G. Kalai, L. Lovasz, B. Sturmfels, and G. Ziegler for helpful comments.

2. Additional terminology

A d-polyhedron is *simple* if it is pointed and each of its vertices is incident to precisely d edges or, equivalently, to precisely d facets. A polytope is *simplicial* if each of its facets is a *simplex* (the convex hull of an affinely independent set of points). (The simplices are the only polytopes that are both simple and simplicial.) The simple polytopes and the simplicial polytopes are of special importance because many problems from other areas lead naturally to one or the other sort of polytope, and because, for each d-polytope P in $\mathbb{R}^d$, there is an arbitrarily small perturbation of P's facets $\langle$vertices$\rangle$ that yields a nearby simple $\langle$simplicial$\rangle$ d-polytope.

When P is a d-polytope in $\mathbb{R}^d$ and the origin is interior to P, the *polar* P° of P is obtained by intersecting certain halfspaces corresponding to the points of P; specifically,

$$P^\circ = \{x \in \mathbb{R}^d : \langle x, y\rangle \leqslant 1 \text{ for all } y \in P\},$$

where $\langle \cdot, \cdot \rangle$ is the usual inner product. The same set P° is obtained when y ranges only over P's vertex-set, and hence P° is also a d-polytope with the origin in its interior. The boundary complex (or face-lattice) of P is *antiisomorphic* to that of P° – i.e., there is an inclusion-reversing bijection – and it is common also to say that the polytopes are *dual* to each other.

For an arbitrary d-polytope P, there is a translate Q such that the origin belongs to Q's *relative interior*, defined as the interior of Q relative to its affine hull A. It is then possible to form the polar Q° of Q in A, and this is a polytope whose face-lattice is antiisomorphic to that of P. Any polytope Q° formed in this way will be denoted by P^*. There are many different choices for P^*, but that will cause no confusion as they are all combinatorially equivalent.

Under polarity, simple polytopes are dual to simplicial polytopes, so combinatorial results on the facial structure of simplicial polytopes can be directly translated into results about their simple counterparts, and vice-versa. Thus much of the combinatorial study of polytopes can be approached from the "simple" viewpoint or, equivalently, from the "simplicial" viewpoint. Though formally equivalent, the two viewpoints lead to different emphases and different generalizations. From the viewpoint of linear programming and some other applications, the simple viewpoint is more natural. However, the simplicial viewpoint is emphasized here except in section 7, because it is more purely combinatorial in flavor and it enables us to use the well-developed language of simplicial complexes. Polytopes that are neither simple nor simplicial are also considered, but the initial focus for most topics is on the simplicial case.

A complex is *simplicial* if its members are all simplices. In particular, if P is a simplicial d-polytope, then the boundary complex $\mathcal{B}(P)$ is a pure simplicial $(d-1)$-complex. Simplicial complexes can be defined as *abstract complexes* in a purely combinatorial way, so that a pure simplicial $(d-1)$-complex, e.g., would consist of a finite collection of d-sets (sets of cardinality d) together with all subsets of those sets. However, here we stick to embedded complexes (i.e., to complexes as defined earlier), for they are more natural when considering the boundary complexes of nonsimplicial polytopes. When dealing with simplicial complexes, the reader may employ the geometric or the abstract definition, according to taste, for each abstract simplicial complex can be realized geometrically. In this Handbook, chapter 34 contains a good discussion of abstract simplicial complexes and the natural way of topologizing them.

Several other sorts of complexes arise naturally in the study of polytopal complexes. A complex $\mathcal{C}$ is called a *sphere* $\langle$*ball*$\rangle$ if $\bigcup \mathcal{C}$ is homeomorphic to a Euclidean sphere $\langle$ball$\rangle$. Note that if P is a d-polytope, then the boundary complex $\mathcal{B}(P)$ is a $(d-1)$-sphere and the complex $\mathcal{B}(P) \cup \{P\}$ is a d-ball. A complex $\mathcal{C}$ is called a *d-manifold* if the topological space $\bigcup \mathcal{C}$ is locally homeomorphic to $\mathbb{R}^d$. And $\mathcal{C}$ is a *d-pseudomanifold* $\langle$*closed d-pseudomanifold*$\rangle$ if $\mathcal{C}$ is a pure d-complex and each ridge of $\mathcal{C}$ is contained in at most $\langle$exactly$\rangle$ two facets of $\mathcal{C}$.

A *shelling* of a pure d-complex $\mathcal{C}$ is an ordering $F_1, \ldots, F_n$ of $\mathcal{C}$'s facets such that for $2 \leqslant j \leqslant n$ the complex $F_j \cap (\bigcup_{i=1}^{j-1} F_i)$ is a $(d-1)$-ball or $(d-1)$-sphere. A pure complex is *shellable* if it admits a shelling. This notion is discussed in detail in chapter 34.

For a complex $\mathscr{C}$ and a face F of $\mathscr{C}$, the *link* of F in $\mathscr{C}$ is the following subcomplex of $\mathscr{C}$:

$$\mathrm{link}(F,\mathscr{C}) := \{\, C \in \mathscr{C} : C \cap F = \emptyset \text{ and there exists } D \in \mathscr{C} \text{ for which } C \subset D \text{ and } F \subset D.\}$$

3. Presentations of polytopes

Polytopes and polyhedra can be presented in a variety of ways. In many applications, a polyhedron P is presented as an intersection of closed halfspaces, hence as the solution set of a system of linear inequalities $Ax \leqslant b$, where A is a real $m \times d$ matrix, $b \in \mathbb{R}^m$, and $x \in \mathbb{R}^d$. When P is d-dimensional and no row of the matrix $[A, b]$ is redundant in the representation, each row corresponds to a facet of P. See chapter 30 for more information on presentations of polyhedra.

A polytope P in $\mathbb{R}^d$ can also be represented as the set of all convex combinations of a finite set of points $\{x_1, \ldots, x_n\}$ in $\mathbb{R}^d$, hence as

$$P = \{\Sigma_{i=1}^{n} \lambda_i x_i \colon \lambda_i \geqslant 0, \Sigma_{i=1}^{n} \lambda_i = 1\}\,.$$

(By a theorem of Carathéodory, the same set is obtained when all but $d+1$ of the λ_i are required to be 0.) When the set $\{x_1, \ldots, x_n\}$ is minimal for this representation of P, it is P's vertex-set. Many authors have studied the problem of finding such a minimal representation, or, in other words, of finding the vertex set of a polytope that is presented as the convex hull of a given finite set of points. It seems that a clearly optimal algorithm is known only for the 2-dimensional case (Kirkpatrick and Seidel 1982). For higher-dimensional cases, usable algorithms and other references can be found in Swart (1985).

A common computational task is to pass from one sort of presentation of a polytope P to another. As is discussed in section 6, a d-polytope with n vertices $\langle$facets$\rangle$ may have as many as

$$\binom{n - \lfloor \frac{1}{2}(d+1) \rfloor}{n-d} + \binom{n - \lfloor \frac{1}{2}(d+2) \rfloor}{n-d}$$

facets $\langle$vertices$\rangle$, and hence one sort of presentation may be much longer than the other. Thus in measuring the efficiency of an algorithm for passing from one sort of presentation to another, it is reasonable to take into account the size of the output as well as the size of the input. The reader is referred to Dyer (1983), Swart (1985) and Seidel (1987) for studies of the computational complexity of some problems of this sort, for further references and for specific algorithms, and to Mattheiss and Rubin (1980) for computational experience with several algorithms.

Three other ways of presenting a d-polytope P are useful because they lead to lower-dimensional pictures of P's facial structure. Suppose that P is a d-polytope in $\mathbb{R}^d$, F is a facet of P, and x is a point of $\mathbb{R}^d \backslash P$ such that among the hyperplanes that are affine hulls of the facets of P, only that of F separates x from P (in the terminology of Gr67, x is *beyond* F but *beneath* all the other facets of P).

Let $\mathscr{C} = \mathscr{B}(P)\setminus\{F\}$. The projection of P onto F by rays issuing from x yields, when restricted to $\bigcup \mathscr{C}$, a homeomorphism of $\bigcup \mathscr{C}$ onto F, and this projection carries the members of $\mathscr{C}\setminus\{F\}$ onto the members of an isomorphic complex $\mathscr{C}_0$ with $\bigcup \mathscr{C}_0 = F$. The complex $\mathscr{C}$ is called a *Schlegel diagram* of P *based* on F.

Another useful dimension-reducing representation of a polytope is its *Gale diagram*, a notion developed by M. Perles (Gr67). We state the central result below and refer to McMullen (1979) for a comprehensive survey of Gale diagrams and their applications.

Suppose that P is a d-polytope in $\mathbb{R}^d$ and $X = \{x_1, \ldots, x_n\}$ is its vertex set. An *affine dependency* of X is a vector $(\alpha_1, \ldots, \alpha_n) \in \mathbb{R}^n$ with $\Sigma_{i=1}^n \alpha_i = 0$ and $\Sigma_{i=1}^n \alpha_i x_i = 0$. Suppose that the n-vectors $a_1, \ldots, a_{n-d-1}$ form a (linear) basis for the vector space of all affine dependencies of X. Let A be the $n \times (n-d-1)$ matrix whose columns are the a_i's, and let the $\overline{x}_i$'s, $1 \leqslant i \leqslant n$, be the row-vectors of A. The set $\overline{X} = \{\overline{x}_1, \ldots, \overline{x}_n\} \subset \mathbb{R}^{n-d-1}$ is called a *Gale transform* of P, and its normalized version $\tilde{X}$, obtained by replacing each nonzero $\overline{x}_i \in \overline{X}$ by the point $\tilde{x}_i = \overline{x}_i/\|\overline{x}_i\|$, is a *Gale diagram* of P. The following theorem from Gr67 shows that the entire boundary complex $\mathscr{B}(P)$ can be "read" from a Gale transform or Gale diagram. This is very useful when P's number n of vertices is not much larger than P's dimension d, for then the transform and diagram are low-dimensional objects.

Theorem 3.1. *A subset $\{x_{i_1}, \ldots, x_{i_k}\}$ of the set X of vertices of a d-polytope P is the vertex set of a face of P if and only if the origin belongs to the relative interior of the convex hull of the set $\overline{X}\setminus\{\overline{x}_{i_1}, \ldots, \overline{x}_{i_k}\}$ or, equivalently, to the relative interior of the convex hull of the corresponding subset of $\tilde{X}$.*

A more sophisticated relative of the Gale diagram is the affine Gale diagram of Sturmfels (1987a), which permits a further reduction in dimension. When P is a d-polytope P with n vertices, the Gale diagram is, in a sense, an $(n-d-1)$-dimensional object and the affine Gale diagram is an $(n-d-2)$-dimensional object. By means of appropriate rules of translation, P's combinatorial structure can be "read" from either diagram. The difference between $n-d-1$ and $n-d-2$ is unimportant when n is much larger than d, but it is important when, e.g., the larger number is 3 or 4.

4. Polytopal realizability of complexes

This section is divided into subsections as follows: Realization of 3-polytopes; The higher-dimensional Steinitz problem; Realization over incomplete ordered fields.

4.1. Realization of 3-polytopes

Steinitz's combinatorial characterization of 3-polytopes (Steinitz and Rademacher 1934) can be formulated in a number of ways. The following combines formulations of Grünbaum and Motzkin (1963) and Klee (1966). The important point is that condition (i), which arises in a geometric context, is equivalent to the purely graph-theoretic conditions (ii), (iii) and (iv). The latter three conditions and related properties of planar graphs are discussed in chapter 5 in this Handbook. (The reader should note that in the present chapter, graphs are always "simple" in the sense of not having any loops or multiple edges. However, we avoid the designation *simple graph* because of possible confusion with simple polytopes.)

Theorem 4.1. *For a graph G with v nodes and e edges, the following four conditions are equivalent:*

(i) *G is 3-polytopal;*

(ii) *G is planar and 3-connected;*

(iii) *if $\mathcal{K}$ is the set of all nonseparating circuits in G and f is the cardinality of $\mathcal{K}$, then*

(a) $v - e + f = 2$,

(b) *each edge of G belongs to exactly two members of $\mathcal{K}$, and*

(c) *whenever two members of $\mathcal{K}$ have more than a single node in common, their intersection consists of a single edge and its endpoints;*

(iv) *G contains a set $\mathcal{K}$ of circuits that satisfies* (a), (b) *and* (c).

Further, if G is 3-polytopal and $\mathcal{K}$ is as in (iv), *then $\mathcal{K}$ is the set of all nonseparating circuits in G, and in each realization of G by means of a 3-polytope P, the members of $\mathcal{K}$ correspond to the boundaries of the facets of P.*

For the interesting history and the proof of Theorem 4.1, see Gr67. The theorem, particularly the equivalence of (i) and (ii), has had a great impact on the study of 3-polytopes. Some of its uses are discussed in sections 5 and 7. It is very useful from an algorithmic as well as a mathematical viewpoint, because both the planarity and the 3-connectedness of a graph with n nodes can be tested in time $O(n)$ (Hopcroft and Tarjan 1973, 1974).

The following formulation is shown by Danaraj and Klee (1978b) to yield another linear-time algorithm for recognizing 3-polytopes.

Theorem 4.2. *A 2-complex is polytopal if and only if it is a shellable closed pseudomanifold.*

Another formulation, essentially equivalent to Theorem 4.2, is the following from Gr67.

Theorem 4.3. *A 2-complex is polytopal if and only if it is a sphere.*

For yet another equivalent formulation, we define a *d-diagram* as a complex $\mathcal{D}$ in $\mathbb{R}^d$ such that the union $D = \bigcup \mathcal{D}$ is a polytope, each face of D other than D itself is a member of $\mathcal{D}$, and $C \cap \partial D$ is a member of $\mathcal{D}$ for each $C \in \mathcal{D}$. Of course, each Schlegel diagram of a d-polytope is a $(d-1)$-diagram. The strict converse of this is false, because there are 2-diagrams that cannot be realized as Schlegel diagrams of 3-polytopes. (Chapter 36 characterizes plane graphs that are projections of polytopes.) However, as shown by Gr67, the converse is correct for $d = 3$ in the following combinatorial sense.

Theorem 4.4. *Every 2-diagram is isomorphic to a Schlegel diagram of a 3-polytope.*

The following theorem of Gr67 extends Steinitz's theorem to 3-polytopes with symmetries.

Theorem 4.5. *If G is a planar 3-connected graph, then:*

(i) *G is isomorphic to the graph of a centrally symmetric 3-polytope if and only if there exists an involution f of G such that each node v of G is separated from $f(v)$ by some circuit;*

(ii) *G is isomorphic to the graph of a 3-polytope with a plane of symmetry if and only if G admits an involution that reverses the orientations of its 2-faces.*

When a d-polytope P and a combinatorial automorphism f of the boundary complex $\mathscr{B}(P)$ are given, f is said to be *realizable by an isometry* if there exist in $\mathbb{R}^d$ a polytope P' combinatorially equivalent to P and an isometry of $\mathbb{R}^d$ such that $g(P') = P'$ and g induces the combinatorial automorphism f in $\mathscr{B}(P')$. The following is due to Mani (1971).

Theorem 4.6. *For each 3-polytope there is a combinatorially equivalent 3-polytope $P \subset \mathbb{R}^3$ such that every automorphism of P's graph is induced by an isometry of $\mathbb{R}^3$.*

The proofs of Theorems 4.1, 4.5 and 4.6 use several types of "reductions" of 3-connected planar graphs described by Gr67. These are inductive methods of constructing such graphs from ones that are in some sense simpler. The same is true of the proofs of the next four results.

For a facet F of a polytope P, we say that *the shape of F can be preassigned* if each polytope F' that is combinatorially equivalent to F is the image of F under some combinatorial isomorphism of P. The following is due to Barnette and Grünbaum (1970).

Theorem 4.7. *The shape of any facet of a 3-polytope can be preassigned.*

Another shape-assignment problem arises as follows. For a d-polytope Q in $\mathbb{R}^d$ and a point x of $\mathbb{R}^d \backslash Q$, let $C(x, Q)$ denote the union of all rays that issue from x and intersect the boundary but not the interior of Q. When each such ray intersects the boundary in a single point, the *profile* of Q from x is defined as the subcomplex of $\mathscr{B}(Q)$ consisting of all faces F of Q that are contained in $C(x, Q)$. This profile is always a $(d-2)$-sphere. A d-polytope P is said to have the *universal shadow-boundary property* if for each $(d-2)$-sphere $\mathscr{S}$ in $\mathscr{B}(P)$ there exists a d-polytope Q in $\mathbb{R}^d$, a point x of $\mathbb{R}^d \backslash Q$, and an isomorphism of $\mathscr{B}(P)$ onto $\mathscr{B}(Q)$ that carries $\mathscr{S}$ onto the profile of Q from x. The following is due to Barnette (1970).

Theorem 4.8. *Each 3-polytope has the universal shadow-boundary property.*

The next result appears in Gr67.

Theorem 4.9. *For each 3-polytope P, the boundary complex $\mathcal{B}(P)$ is isomorphic to the boundary complex of a polytope in the rational 3-space $\mathbb{Q}^3$.*

We end this subsection with a result that combines the discrete and the continuous aspects of polytopes. When V is the vertex-set of a d-polytope P in $\mathbb{R}^d$, a *realization of P in $\mathbb{R}^d$* is an injection $f : V \to \mathbb{R}^d$ such that $f(V)$ is the vertex-set of a polytope Q and f generates an isomorphism of $\mathcal{B}(P)$ onto $\mathcal{B}(Q)$. Two realizations f and g of P in $\mathbb{R}^d$ are *combinatorially isotopic* if there is a continuous one-parameter family of realizations that starts with f and ends with g. It is easily seen that each realization of P can be extended to a homeomorphism of $\mathbb{R}^d$ onto itself, and combinatorially isotopic realizations can be extended to isotopic homeomorphisms of $\mathbb{R}^d$. Hence each realization may be classified as orientation-preserving or orientation-reversing.

The *isotopy theorem for 3-polytopes*, due to Steinitz and Rademacher (1934), may be stated as follows.

Theorem 4.10. *If P is a 3-polytope in $\mathbb{R}^3$, then all orientation-preserving realizations of P in $\mathbb{R}^3$ are combinatorially isotopic.*

4.2. The higher-dimensional Steinitz problem

The above theorems give a fairly complete combinatorial picture of the face-lattices of 3-polytopes, and of several sorts of geometric realizability of these lattices. It is striking that *all* of them fail for higher-dimensional polytopes. The following result of Grünbaum and Sreedharan (1967) was the starting point for research on the higher-dimensional Steinitz problem.

Theorem 4.11. *There exists a simplicial 3-sphere with 8 vertices that is not isomorphic to the boundary complex of any 4-polytope.*

This result holds also for each $d > 4$, with 3, 8 and 4 replaced respectively by $d - 1$, $d + 4$, and d. In fact, it turns out that when d and $n - d$ are both large, the number of (combinatorial types of) d-polytopes with n vertices is greatly exceeded by the number of such types of nonpolytopal simplicial $(d - 1)$-spheres. (See section 5 for details.)

Theorem 4.11 was proved by showing that a certain simplicial 3-diagram, first presented by Brückner (1909) and assumed by him to arise as the Schlegel diagram of a 4-polytope, does not in fact arise in this way. When a d-diagram $\mathcal{D}$ is the Schlegel diagram of a polytope P, $\mathcal{D}$ has properties known as *invertibility* (arising from the fact that each facet of P may serve as the basis of a Schlegel diagram) and *dualizability* (arising from the existence of P's polar). Brückner's 3-diagram has neither of these properties, but it turns out also that neither is sufficient to guarantee that a simplicial 3-diagram is isomorphic to the Schlegel diagram of a 4-polytope (Schulz 1979, 1985, Barnette 1980).

Although Theorems 4.1–4.9 do not extend to general d-polytopes for $d \geqslant 4$, they do extend to those whose number of vertices is only slightly larger than d.

That is the import of Theorems 4.12–4.16 below, which should be compared with Theorems 4.2, 4.5(i), and 4.6–4.8.

Theorem 4.12. *Each piecewise linear $(d-1)$-sphere with at most $d+3$ vertices is polytopal.*

Theorem 4.13. *If a $(d-1)$-sphere with at most $2d$ vertices admits a combinatorial involution that has no fixed point then the sphere is isomorphic to the boundary complex of a centrally symmetric d-polytope.*

Theorem 4.14. *If a d-polytope has at most $d+3$ vertices, then the shape of each of its facets can be preassigned.*

Theorem 4.15. *If a d-polytope has at most $d+2$ vertices, then it has the universal shadow-boundary property.*

In each of Theorems 4.12–4.15, it is known that the bound on the number of vertices cannot be relaxed. It appears that Theorem 4.12 holds without the assumption of piecewise linearity, but that assumption is used in the proofs of Mani (1972) for the simplicial case and Kleinschmidt (1976b) for the general case. Theorems 4.13, 4.14, and 4.15 are due respectively to Kleinschmidt (1977), Kleinschmidt (1976a), and Shephard (1972).

The following is stated in Gr67.

Theorem 4.16. *If a d-polytope has at most $d+3$ vertices there is a combinatorially equivalent d-polytope $P \subset \mathbb{R}^d$ such that every automorphism of $\mathcal{B}(P)$ is induced by an isometry of $\mathbb{R}^d$.*

The strong isotopy theorem 4.10 does not extend beyond dimension 3. In fact, the following is only a small sample of the striking results of Mnëv (1988) and Vershik (1988) on the topological structure of spaces of realizations of polytopes.

Theorem 4.17. *For each $d \geqslant 4$ there exist a d-polytope $P \subset \mathbb{R}^d$ and an orientation-preserving realization f of P in $\mathbb{R}^d$ that is not combinatorially isotopic to the identity mapping on P's vertex-set.*

As far as the number of vertices is concerned, a smallest example for Theorem 4.17 is provided by a certain simplicial 4-polytope P with 10 vertices first constructed and studied by Bokowski et al. (1984). In this case, the f for Theorem 4.17 is provided by a bijection of P's vertex-set. This same P is such that although $\mathcal{B}(P)$ admits a fixed-point free involution and also admits several other combinatorial automorphisms, for each polytope Q combinatorially equivalent to P it is only the identity automorphism of $\mathcal{B}(Q)$ that comes from an affine mapping of Q onto itself. (Compare this with 4.6 and 4.16.)

Although 4.10 does not extend to higher dimensions, it has a consequence that may extend. Suppose that P and Q are combinatorially equivalent d-polytopes in $\mathbb{R}^d$, and Q' is Q's reflection in a hyperplane. Must there exist an isotopy h of $\mathbb{R}^d$

(a continuous one-parameter family h_t, $0 \leqslant t \leqslant 1$, of self-homeomorphisms of $\mathbb{R}^d$) such that h_0 is the identity mapping, $h_1(P)$ is Q or Q', and for each t, $h_t(P)$ is a polytope combinatorially equivalent to P? If so, can it be required in addition that when $t = 1$, h_t carries P's vertex-set onto that of $h_t(P)$? (When $d = 3$, the latter condition can be required for all t. However, that is not true for $d = 4$, because there is a 4-polytope $P \subset \mathbb{R}^4$ such that the automorphism group of $\mathscr{B}(P)$ is trivial and yet the isotopy theorem does not extend to P.)

Recalling 4.2, we turn now to the uses and limitations of shellability as a tool for recognizing d-polytopes when $d \geqslant 4$. The following is due to Bruggesser and Mani (1971).

Theorem 4.18. *Each polytope is shellable. That is, the facets of any polytope can be arranged in an order that constitutes a shelling.*

In fact, not only is the boundary complex of a polytope always shellable, but there exist shellings satisfying various restrictions on the order in which the facets appear. That was shown by Bruggesser and Mani (1971) and Danaraj and Klee (1974), and partly extended to oriented matroids by Mandel (1981) and Lawrence (1984). (It seems that although oriented matroid spheres are shellable, this is not known for their duals.) See chapter 9 for a detailed discussion of oriented matroids, and see also chapter 34 for the axiomatization of oriented matroids in terms of signed bases.

When $\mathscr{C}$ is a pure simplicial d-complex, shellings of $\mathscr{C}$ are easy to recognize, for a permutation $F_1, \ldots, F_n$ of $\mathscr{C}$'s facets is a shelling if and only if it is true whenever $1 \leqslant i < j \leqslant n$ that there exists $h < j$ for which $F_h \cap F_j$ is a $(d-1)$-simplex and $F_i \cap F_j \subseteq F_h \cap F_j$. This leads to a routine (though slow) algorithm for testing shellability and various strengthened versions of it. However, that does not provide a procedure for recognizing d-polytopes when $d \geqslant 4$, because all simplicial 3-spheres (polytopal or not) with at most 9 vertices turn out to be shellable (Danaraj and Klee 1978a).

Other necessary conditions for the polytopality of a complex $\mathscr{C}$ can be expressed in terms of $\mathscr{C}$'s f-vector and $\mathscr{C}$'s edge-graph (see sections 6 and 7), but these conditions are not sufficient. Indeed, it seems probable that there is no "nice" easily tested set of necessary and sufficient conditions for polytopality. A strong indication of this is the following theorem of Sturmfels (1988), which says that certain nonpolytopal spheres cannot be distinguished from polytopal spheres by any test of a "local" nature.

Theorem 4.19. *There exists a sequence $\{k_n\}_{n\in\mathbb{N}}$ of integers and a sequence $\{S_n\}_{n\in\mathbb{N}}$ of triangulated spheres such that the following two conditions are satisfied for each n:*

(i) *$\mathscr{S}_n$ is a nonpolytopal $(k_n - 1)$-sphere with $k_n + 4$ vertices;*

(ii) *each proper subcomplex $\mathscr{T}$ of $\mathscr{S}$ extends to a polytopal sphere with no additional vertices.*

Since there seems to be no hope of finding an easily understandable characterization of d-polytopality, it is natural to turn to the algorithmic approach. As noted in Gr67, the following is a consequence of the Tarski (1951) decision method for real-closed fields.

Theorem 4.20. *There exists an algorithm for deciding whether a given complex is polytopal.*

This is in contrast with the observation of S.P. Novikov (in Volodin et al. 1974) that there is no algorithm for deciding whether a given 5-manifold is a sphere. However, Tarski's method, though improved by Collins (1975) and Renegar (1992), seems to be still far from being applicable to test polytopality in significant cases. It seems likely that the problem of polytope recognition is NP-hard, and it does not seem to be known even whether the problem belongs to the class NP. (This class is discusssed in chapter 29.)

For the practical recognition of polytopal complexes, by far the most successful method has been the one initiated by Bokowski and developed further by Bokowski and Sturmfels (1987). It has been used to enumerate combinatorial types of 4-polytopes (see section 5), to prove part of Theorem 4.19, and in a number of other special problems. For example, it was used to establish polytopality of the Bokowski–Ewald–Kleinschmidt sphere mentioned after 4.17.

The Bokowski–Sturmfels approach is based on the theory of oriented matroids. They describe oriented matroids in terms of signed bases and call the resulting structure a *chirotope*. (By results of Lawrence 1982 and Dress 1986, this definition is equivalent to the definitions of oriented matroids formulated by Bland and Las Vergnas 1978 and Folkman and Lawrence 1978. However, the approach in terms of signed bases is better suited to computations that arise from the combinatorial study of polytopes.)

Before giving the formal definition of a chirotope, we note that if the vertices $y_1, \ldots, y_n$ of a k-polytope P in $\mathbb{R}^k$ are embedded in $\mathbb{R}^k \times \mathbb{R}$ as the points $y_i' = (y_i, 1)$, then many properties of P are reflected in algebraic relations among the y_i' (cf. 3.1). In particular, various combinatorial restrictions on $\mathcal{B}(P)$ lead to conditions that must be satisfied by the $(k+1) \times (k+1)$ subdeterminants of the matrix whose rows are the y_i'. Chirotopes provide an abstract framework for studying such conditions in terms of signs of determinants.

For $d \leqslant n$, let

$$\Lambda(n,d) = \{(\lambda_1, \ldots \lambda_d) \in \mathbb{N}^d : 1 \leqslant \lambda_1 < \cdots < \lambda_d \leqslant n\},$$

the set of all increasing d-tuples of integers between 1 and n. A mapping $\chi : \Lambda(n,d) \to \{-1, 0, 1\}$ is a *d-chirotope with n vertices* if for each $\mu \in \Lambda(n, d+2)$ there are vectors $x_{\mu_1}, \ldots, x_{\mu_{d+2}}$ in $\mathbb{R}^d$ such that

$$\chi(\lambda) = \operatorname{sign}(\det(x_{\lambda_1}, \ldots, x_{\lambda_d})) \tag{4.1}$$

for all $\lambda \in \Lambda(n,d)$ with $\lambda \subset \mu$. (Here χ is extended to $\{1, \ldots, n\}^d$ in the canonical alternating way.) The chirotope χ is *realizable* if there exist $x_1, \ldots, x_n \in \mathbb{R}^d$ such that 4.1 holds for *all* $\lambda \in \Lambda(n,d)$, and in this case χ is called *the chirotope of linear dependencies on* $(x_1, \ldots, x_n)$.

A vector $\gamma \in \{-1,0,+1\}^n$ is a *cocircuit* of the chirotope $\chi : \Lambda(n,d) \to \{-1,0,+1\}$ (or of the associated oriented matroid) if there exists $\mu \in \Lambda(n,d-1)$ such that for each i, the ith component of γ is equal to $\chi(\mu_1,\ldots,\mu_{d-1},i)$. A *facet* of χ is a set $F \subset \{1,\ldots,n\}$ that corresponds to a nonnegative cocircuit in the sense that for some $\mu \in \Lambda(n,d-1)$ it is true that

$$\chi(\mu_1,\ldots,\mu_{d-1},i) = \begin{cases} 0 & \text{for } i \in F, \\ 1 & \text{for } i \in \{1,\ldots,n\}\backslash F. \end{cases}$$

A set $G \subset \{1,\ldots,n\}$ is called a *face* of χ if G is the intersection of some facets of χ. The set of all such faces, ordered by inclusion, is denoted by $\mathrm{FL}(\chi)$.

For any convex k-polytope P in $\mathbb{R}^k$, with vertex set $y_1,\ldots,y_n$, P's face-lattice is isomorphic to $\mathrm{FL}(\chi)$ for the chirotope χ of linear dependencies on the set $\{(y_1,1),\ldots,(y_n,1)\} \subset \mathbb{R}^k \times \mathbb{R}$. That is almost obvious, for (interpreting χ as the determinant of $k+1$ points in $\mathbb{R}^{k+1}$) the definition of a facet of a chirotope generalizes the fact that the vertex set of a polytope's facet is the intersection of a supporting hyperplane with the polytope's vertex set.

The Bokowski–Sturmfels procedure accepts as input a $(d-2)$-sphere $\mathcal{S}$ with n vertices, and its aim is to decide whether $\mathcal{S}$ is polytopal. Success is not guaranteed, for the procedure may fail to terminate, but if it does terminate then it does so with a decision as to whether the sphere is polytopal. It is not feasible to include more than a very rough description of the method, but we do include that because of the special successes of the method.

The first step consists of computing the finite set $\mathcal{C}_{d,n}(\mathcal{S})$ of all d-chirotopes χ with n vertices such that $\mathrm{FL}(\chi)$ is isomorphic with $\mathcal{S}$. That is accomplished with the aid of the Plücker–Grassmann relations, algebraic relations among the subdeterminants of a matrix (Hodge and Pedoe 1968). The close connection between these and chirotopes is discussed by Bokowski and Sturmfels (1985). The second step is to test realizability: if there is a realizable $\chi \in \mathcal{C}_{d,n}(\mathcal{S})$, then $\mathcal{S}$ is polytopal, and otherwise $\mathcal{S}$ is nonpolytopal. This step is generally more difficult and time-consuming than the first one. However, as shown by Bokowski and Sturmfels (1986), the second step can also be based in part on the Plücker–Grassmann relations. For example, for each individual chirotope of the list produced in the first step, it may be possible to find a Plücker–Grassmann relation that is violated by the sign-patterns of the determinants. When this occurs, the sphere has been shown to be nonpolytopal. The computational difficulty consists of producing such a contradiction systematically.

4.3. *Realization over ordered fields other than $\mathbb{R}$*

Among ordered fields, only the real field $\mathbb{R}$ is complete and only the subfields of $\mathbb{R}$ are Archimedean. We turn now to the realization of polytopes in vector spaces over ordered fields other than $\mathbb{R}$. Of course, the subfields of $\mathbb{R}$ are of special interest. It is almost obvious (and also true!) that each *simplicial* d-polytope P in $\mathbb{R}^d$ is combinatorially equivalent to one in the rational d-space $\mathbb{Q}^d$, for the vertices of P can be moved, one by one, to nearby points of $\mathbb{Q}^d$ without disturbing P's combinatorial structure. Similarly, each *simple* polytope in $\mathbb{R}^d$ is combinatorially equivalent to one in $\mathbb{Q}^d$. By Theorem 4.9, it is true when $d \leqslant 3$ that *every* polytope in $\mathbb{R}^d$ is combinatorially equivalent to one in $\mathbb{Q}^d$. It is unknown whether this result is valid when $d = 4$. However, the following theorem shows that it fails for $d \geqslant 5$. Part (i) of Theorem 4.21 is shown by Richter–Gebert (1994) and part (ii) is shown by Perles in Gr67 (using the Gale diagrams described in section 3).

Theorem 4.21. (i) *For each $d, 5 \leqslant d \leqslant 7$, $\mathbb{R}^d$ contains a d-polytope with $d + 91$ vertices that is not realizable in $\mathbb{Q}^d$.*

(ii) *For each $d \geqslant 8$, $\mathbb{R}^d$ contains a d-polytope with $d + 4$ vertices that is not realizable in $\mathbb{Q}^d$.*

With $\mathbb{A}$ denoting the field of all real algebraic numbers, the following two results are due to Lindström (1971) and Mnëv (1983) respectively.

Theorem 4.22. *For an arbitrary ordered (even non-Archimedean) field Φ, each polytope in Φ^d can be combinatorially realized in $\mathbb{A}^d$.*

Theorem 4.23. *$\mathbb{A}$ is the smallest subfield of $\mathbb{R}$ such that every polytope in a real vector space is realizable in some vector space over $\mathbb{A}$.*

To prepare for the final result of this section, recall the fact (the negative solution of Hilbert's tenth problem by Matiyasevič in 1971) that there is no algorithm for deciding the integer solvability of Diophantine equations. As discussed by Mazur (1986) and Klee and Wagon (1991), it is still unknown whether this is true of the *rational* solvability of such equations. Sturmfels (1987c) proves the following, along with some related results that involve oriented matroids.

Theorem 4.24. *The following two statements are equivalent:*

(i) *there exists an algorithm that determines, for an arbitrary finite lattice L, whether L is isomorphic to the face-lattice of a polytope in a rational vector space;*

(ii) *there exists an algorithm that determines, for an arbitrary polynomial f with integer coefficients, whether f has zeros in the field of rational numbers.*

5. Enumeration of combinatorial classes

We use the term *combinatorial class* to mean an equivalence class of polytopes with respect to combinatorial equivalence or a more elaborate relationship that involves additional information. The problem is to count or estimate the number of such classes corresponding to certain specified parameters (e.g., dimension and number of vertices or edges). For the history of the subject, see Gr67.

This section is divided into three parts, as follows: Enumeration of 3-polytopes; Classes with few vertices; Higher-dimensional asymptotic results. Enumeration in the 3-dimensional case was first considered by Euler. With the aid of Steinitz's characterization of 3-polytopes, sharp asymptotic results have been obtained and several exact enumerations have been made. The second part deals with exact enumeration for $d \geqslant 4$, an area in which only a few results are known. The third part describes some higher-dimensional asymptotic estimates that focus on what happens as the dimension or number of vertices tends to infinity.

5.1. Enumeration of 3-polytopes

For 3-polytopes, Duivestijn and Federico (1981) provide a comprehensive survey of exact enumerations. We extract some of the most important results stated in their paper, but omit most further references as they can be found there.

All enumerations of classes of 3-polytopes count 3-connected planar graphs, the general method being to construct all members of the class (often by a graph operation that introduces a new edge splitting a 2-face), and then, using suitable numerical characteristics of the graphs obtained, avoid the multiple counting of isomorphic copies. Exact enumeration is most complete for $C_1(n)$, the number of isomorphism-types of 3-connected planar graphs with n edges, and for $R_1(n)$, the number of such rooted graphs. To *root* a graph that is embedded in the plane means to specify an edge as the root, to give the edge a direction, and to distinguish between the two sides of the edge.

If a planar graph G is 3-connected and has n edges, then (since G admits an essentially unique embedding in the plane) the number of rooted versions of G is $4n$. Thus the number of distinct rooted graphs corresponding to G is $4n/h$, where h is the order of G's automorphism group. If none of the graphs with n edges were symmetric, $C_1(n)$ would be equal to the quantity $A_1(n) := R_1(n)/4n$. As it is, $A_1(n)$ is a lower bound for $C_1(n)$ and is regarded as a good estimate of $C_1(n)$ for large n since the percentage of symmetric graphs with n edges is observed to fall drastically as n increases.

Table 1 gives some exact values for $C_1(n)$, $R_1(n)$, and $A_1(n)$. It is taken from Duivestijn and Federico (1981), and the actual graphs are available on tape from the first author of that article. The same article contains the values given in table 2 for $C_0(n)$, the number of combinatorial types of 3-polytopes with n vertices. In table 2, $S_0(n)$ is the number of types of *simplicial* 3-polytopes with n vertices; these values are taken from Grünbaum (1970). The rooted analogue of $S_0(n)$ is given precisely for all n by the following theorem of Tutte (1962), from which the values for $R_1(n)$ can be deduced with the aid of Euler's theorem.

Theorem 5.1. *The number of rooted simplicial* 3*-polytopes with* v *vertices is precisely*

$$\frac{2(4v-11)!}{(3v-7)!(v-2)!} \approx \frac{3}{16(6\pi v^5)^{1/2}} \left(\frac{256}{27}\right)^{v-2},$$

where $\approx$ *indicates that the ratio converges to* 1 *as* $v \to \infty$.

For unrooted 3-polytopes, sharp asymptotic formulas for the number of combinatorial types are established by Bender (1987).

Table 1

n	$C_1(n)$	$A_1(n)$	$R_1(n)$
6	1		1
8	1		4
9	2		6
10	2		24
11	4		66
12	12	12	214
13	22	13	676
14	58	40	2 209
15	158	122	7 296
16	448	333	24 460
17	1 342	1 220	82 926
18	4 199	3 946	284 068
19	13 384	12 920	981 882
20	43 708	42 767	3 421 318
21	144 810	142 948	12 007 554
22	485 704	482 002	42 416 488
23		1 638 248	150 718 770
24		5 608 558	538 421 590
25		19 328 566	1 932 856 590
26		67 017 765	6 969 847 484

Table 2

n	4	5	6	7	8	9	10	11	12
$C_0(n)$	1	2	7	34	257	2 606	32 300	$\geqslant$437 557	$\geqslant$6 363 115
$S_0(n)$	1	1	2	5	14	50	233	1 249	7 595

Theorem 5.2. *Let $C(i,j)$ denote the number of combinatorial types of 3-polytopes that have $i+1$ vertices and $j+1$ facets, and for $0 \leqslant i \leqslant 2$ let $C_i(n)$ denote the number of types with n i-faces. Let*

$$B(i,j) := \frac{1}{972\,ij(i+j)} \binom{2i}{j+3} \binom{2j}{i+3}.$$

Then

$$C(i,j) \approx B(i,j),$$

$$C_2(n) = C_0(n) \approx \left(\frac{\pi n(4+\sqrt{7})}{4\sqrt{7}} \right)^{1/2} B(n-1, \tfrac{1}{4}(n-1)(3+\sqrt{7})),$$

$$C_1(n) \approx \frac{\sqrt{\pi n}}{4} B(\frac{n}{2}, \frac{n}{2}),$$

where fractional factorials in binomial coefficients are to be approximated by Stirling's formula.

5.2. Classes with few vertices

The d-simplices are the only d-polytopes with $d+1$ vertices. Beyond that, the combinatorial types of d-polytopes have been enumerated for arbitrary d only when the number of vertices is $d+2$ or $d+3$. The results are based on Theorem 3.1 in conjunction with the low dimensionality of the relevant Gale diagrams.

Theorem 5.3. *Among d-polytopes, the number of combinatorial types with $d+2$ vertices is $\lfloor d^2/4 \rfloor$.*

There is also a formula for the number with $d+3$ vertices, but it is too complicated to state here, so we refer to Lloyd (1970).

Theorem 5.4. *Among simplicial d-polytopes, the number of combinatorial types with $d+2$ vertices is equal to $\lfloor d/2 \rfloor$ and the number with $d+3$ vertices is equal to*

$$2^{\lfloor d/2 \rfloor} - \left\lfloor \frac{d+4}{2} \right\rfloor + \frac{1}{4(d+3)} \sum_{\text{odd } h|(d+3)} \phi(h) 2^{(d+3)/h},$$

where ϕ is Euler's function.

The Gale diagram of a d-polytope with $d+2$ vertices is 1-dimensional. In the simplicial case, each vertex is mapped to -1 or 1 in the Gale diagram, and each of -1 and 1 receives at least two vertices. If $m_1 \langle m_{-1} \rangle$ denotes the number of vertices received by $1\langle -1 \rangle$, then two polytopes are isomorphic if and only if they yield the same unordered pair $\{m_{-1}, m_1\}$. That takes care of the first part of 5.4, and 5.3 is similar except that some vertices may be mapped to 0.

The second part of 5.4 is due to Perles (in Gr67). The results for $d+3$ vertices arise from complicated counting procedures for 2-dimensional Gale diagrams, using Polya's method.

It has been possible to go a bit farther in the 4-dimensional case. The following statement combines results of Grünbaum and Sreedharan (1967), Barnette (1973a), Altshuler et al. (1980) and Altshuler and Steinberg (1985).

Theorem 5.5. *Among 3-spheres, there are 1336 combinatorial types with 8 vertices, 1294 of which are polytopal. Among simplicial 3-spheres, there are 39 types with 8 vertices, 37 of which are polytopal, and 1296 types with 9 vertices, 1142 of which are polytopal.*

McMullen and Shephard (1970) and Ewald et al. (1976) count types of polytopes with few vertices and symmetry group G of given order g. A typical result states that for $k \geqslant 2$ and $d = (g-1)k+1$ there are exactly $\lfloor k^2/4 \rfloor$ types of d-polytopes for which the number of vertices is gk and the set of points fixed by G is 1-dimensional.

A *cyclic d-polytope* with n vertices is defined as the convex hull of n points of the *moment curve* M_d in $\mathbb{R}^d$, where $M_d = \{(t, t^2, \ldots, t^d) \colon t \in \mathbb{R}\}$. These d-polytopes are simplicial, and have the remarkable property of being *neighborly* in the sense that each set of $\lfloor d/2 \rfloor$ vertices is the vertex set of a face. (For even d, neighborliness implies simpliciality.) For a given dimension and number of vertices, all cyclic polytopes are combinatorially equivalent (Gale 1963), but other types of neighborly polytopes exist and there has been interest in counting them because of the central role played by neighborly polytopes in connection with the Upper-bound theorem (see 6.10–6.11). The formula in the following result was established by McMullen (1974), using Gale diagrams.

Theorem 5.6. *For even d, all neighborly d-polytopes with $d+2$ vertices are equivalent, as are all with $d+3$ vertices. For odd $d = 2k+1$, there are 2 combinatorial types with $d+2$ vertices, and the number of types with $d+3$ vertices is equal to*

$$\tfrac{1}{4}\left\{(5+(-1))^k \cdot 3^{\lfloor (k+1)/2 \rfloor} + 6\right\} + \frac{1}{4k+8}\left(\sum_{\text{odd } h|k+2} \phi(h)(3^{(k+2)/h} - 1)\right),$$

where ϕ is Euler's function.

Altshuler (1977) enumerated the neighborly 3-spheres with 10 vertices, and by checking these for polytopality, Bokowski and Sturmfels (1985) found that there are 431 types of neighborly 4-polytopes with 10 vertices. Bokowski and Shemer (1987) found that there are 37 types of neighborly 6-polytopes with 10 vertices.

5.3. *Asymptotic results for arbitrary dimension*

Theorem 5.7. *For each dimension d there are positive constants a, b and c (each depending on d) such that for all $n > d$:*

(i) *the number of combinatorial types of simplicial neighborly d-polytopes with n vertices is at least n^{an};*

(ii) *the number $c(n,d)$ of combinatorial types of d-polytopes with n vertices is at most $n^{d(d+1)n}$;*

(iii) *the number $s(n,d)$ of combinatorial types of triangulated $(d-1)$-spheres with n vertices is between $e^{bn^{\lfloor d/2 \rfloor}}$ and $n^{cn^{\lfloor d/2 \rfloor}}$.*

The lower bound (i) is due to Shemer (1982), who shows that for each d the constant a may be taken arbitrarily close to $\frac{1}{2}$ for sufficiently large n. The upper bound in (ii) actually applies to vertex-labeled polytopes. It is due to Goodman and Pollack (1986), whose original proof for the simplical case used a bound on the sum of the Betti numbers of a real algebraic variety. They (and Alon 1986) were later able to handle arbitrary polytopes, and to improve the bound. In particular, as is shown in chapter 32, it can be reduced to

$$\left(\frac{n}{d}\right)^{d^2 n \left(1+o\left(\frac{1}{\log(n/d)}\right)\right)} \quad \text{for } \frac{n}{d} \to \infty.$$

The upper bound in (iii) follows from the Upper-bound theorem for triangulated spheres (see 6.10), and the lower bound is proved by Kalai (1988a) by extending the construction of Billera and Lee (1981) of polytopes satisfying the McMullen conditions on f-vectors. (These tools are discussed in section 6.) The following consequences of (ii) and (iii) are noted by Kalai.

Theorem 5.8. *For each fixed $d \geqslant 4$ it is true that*

$$c(n,d)/s(n,d) \to 0 \quad \text{as } n \to \infty.$$

For each fixed $b \geqslant 4$, it is true that

$$c(d+b,d)/s(d+b,d) \to 0 \quad \text{as } d \to \infty.$$

6. Face-vectors

For a polytope or complex X, the *f-vector* (or *face-vector*) $f(X)$ is the sequence $(f_0(X), f_1(X), \ldots)$, where $f_i(X)$ is the number of i-dimensional faces of X. When X is a d-polytope $\langle d$-complex$\rangle$, the sequence may be terminated at $f_{d-1}(X)\langle f_d(X)\rangle$ without loss of information. Then the f-vector of a polytope is the f-vector of its boundary complex.

Since the time of Euler, the study of f-vectors has been a central topic in the combinatorial theory of polytopes. The notion itself is quite elementary, but the sharpest results concerning f-vectors have required deep applications of modern tools from commutative algebra and algebraic geometry. Grünbaum (1967, 1970) discusses the history of the subject up to 1970; Stanley (1985) and Björner (1986) provide more recent surveys.

This section is divided into the following subsections: General complexes and multicomplexes; Euler's relation and the Dehn–Sommerville equations; Face-vectors of simplicial polytopes; Face-vectors of general polytopes; Upper-bound and lower-bound theorems for polytopes; Upper-bound and lower-bound theorems for polytope-pairs.

6.1. General complexes and multicomplexes

The f-vectors of general complexes have been characterized, and the notions used there are used also in characterizing f-vectors of polytopes. For each pair of positive integers k and n, there is a unique representation of n in the form

$$n = \binom{a_k}{k} + \binom{a_{k-1}}{k-1} + \cdots + \binom{a_i}{i},$$

with $a_k > a_{k-1} > \cdots > a_i \geqslant i > 0$. This is called the *$k$-canonical representation* of n. In terms of it, we define

$$n^{(k)} = \binom{a_k}{k-1} + \binom{a_{k-1}}{k-2} + \cdots + \binom{a_i}{i-1},$$

$$n^{\langle k\rangle} = \binom{a_k+1}{k+1} + \binom{a_{k-1}+1}{k} + \cdots + \binom{a_i+1}{i+1},$$

and set

$$0^{\langle k \rangle} = 0^{(k)} = 0.$$

The following is proved by Kruskal (1963) and Katona (1968) for simplicial complexes, and extended by Wegner (1984) to general complexes. See chapter 24 for details.

Theorem 6.1. *For an ultimately vanishing sequence $f = (f_0, f_1, \ldots)$ of nonnegative integers, the following three conditions are equivalent:*

(i) *f is the f-vector of a simplicial complex;*

(ii) *f is the f-vector of a complex;*

(iii) *$f_k^{(k+1)} \leqslant f_{k-1}$ for each $k \geqslant 1$.*

A *multicomplex* is defined as a nonempty set $\mathcal{M}$ of monomials in a finite number of variables such that each divisor of a member of $\mathcal{M}$ is itself a member of $\mathcal{M}$. The *f-vector* of $\mathcal{M}$ is the sequence $(f_0(\mathcal{M}), f_1(\mathcal{M}), \ldots)$, where $f_i(\mathcal{M})$ is the number of monomials of degree i in $\mathcal{M}$. (Of course, $f_0(\mathcal{M}) = 1$.) When $\mathscr{C}$ is a simplicial complex with vertex set V, the corresponding multicomplex $\mathcal{M}_{\mathscr{C}}$ is formed by associating a variable x_v with each $v \in V$, and defining $\mathcal{M}_{\mathscr{C}}$ as the set of all monomials of the form $x_{v_0} x_{v_1} \cdots x_{v_k}$ where $\{v_0, v_1, \ldots, v_k\}$ is the vertex set of a face of $\mathscr{C}$. Then the f-vector of $\mathcal{M}_{\mathscr{C}}$ is just an index-shifted version of the f-vector of $\mathscr{C}$.

The following theorem is essentially due to Macaulay (1927), but the explicit formulation is that of Stanley (1978).

Theorem 6.2. *For an ultimately vanishing sequence $f = (f_0, f_1, \ldots)$ of nonnegative integers, the following three conditions are equivalent:*

(i) *f is the f-vector of a multicomplex;*

(ii) *there exists a field K and a finitely generated K-algebra $R = \bigoplus_{i \geqslant 0} R_i$ such that R_0 is isomorphic to K, R_1 generates R, and f lists the values of R's Hilbert function (i.e., $f_i = \dim R_i$ for each i);*

(iii) *$f_0 = 1$, and $f_{k+1} \leqslant f_k^{\langle k \rangle}$ for each $k \geqslant 1$.*

Sequences $f = (f_0, f_1, \ldots)$ satisfying the conditions of 6.2 are called *O-sequences*. They play a crucial role in characterizing the f-vectors of simplicial polytopes.

6.2. Euler's relation and the Dehn–Sommerville equations

Theorem 6.3. *For the f-vector of a d-polytope,*

$$\sum_{j=0}^{d-1} (-1)^j f_j = 1 + (-1)^{d-1}.$$

The *Euler relation* (Theorem 6.3) was established by L. Schläfli in 1852, assuming the shellability of polytopes (see 4.18). The first complete proof of 6.3 was that of Poincaré (1893), using topological methods and applying to more general situations. Theorem 6.3 is the only affine relation among the numbers $f_0, f_1, \ldots, f_{d-1}$

that holds for *all* d-polytopes, and hence the affine hull of the f-vectors of all d-polytopes is of dimension $d-1$ (Gr67). However, there are additional affine relations for *simplicial* polytopes. These are the *Dehn–Sommerville equations* formulated as follows in Gr67. (See Grünbaum 1967, 1970 for the history of these and the Euler relation.) (By convention, $f_{-1} = 1$.)

Theorem 6.4. *For the f-vector of a simplicial d-polytope, it is true that*

$$f_i = \sum_{j=i}^{d-1} (-1)^{d-1-j} \binom{j+1}{i+1} f_j, \quad -1 \leqslant i \leqslant d-2.$$

Here the Euler relation is obtained by setting $i = -1$. It turns out that if ρ is any affine relation for the sequence $(f_0, \ldots, f_{d-1})$ that holds for the f-vectors of all simplicial d-polytopes, then ρ is an affine combination of the relations in Theorem 6.4. Further, an appropriately chosen set of $\lfloor (d+1)/2 \rfloor$ of these relations forms a basis for all of them (Gr67), whence the affine space consisting of all ρ is of dimension $\lfloor d/2 \rfloor$.

Klee (1964a) establishes the Dehn–Sommerville equations for simplicial complexes much more general than the boundary complexes of simplicial polytopes. It is required only that the link of each face (including the empty face) has the same Euler characteristic as a sphere of the appropriate dimension. Complexes satisfying this condition are called *Eulerian manifolds*. See Stanley (1986) for a simple treatment in terms of Eulerian posets.

We digress to mention regular polytopes. If P is a 3-polytope, and j and k are positive integers such that each vertex of P is incident to j edges and each 2-face of P is incident to k edges, then $jf_0 = 2f_1 = kf_2$. It follows from Euler's theorem that

$$\frac{2}{j} + \frac{2}{k} - 1 = \frac{2}{f_1}.$$

There are only five solutions of this equation in integers greater than or equal to 3. These solutions correspond to the five regular 3-polytopes (Platonic solids), whose numbers of facets are 4, 6, 8, 12 and 20, respectively. For each $d \geqslant 3$, the simplices ($d+1$ facets), cross-polytopes (2^d facets) and cubes ($2d$ facets) satisfy the most stringent regularity conditions, and for $d \geqslant 5$ these are the only possibilities. However, for $d = 4$ there are also regular polytopes whose numbers of facets are 24, 120 and 600, respectively (see Coxeter 1963 for these). The various (equivalent) definitions of regularity of polytopes all involve metric notions, but McMullen (1967) defined a notion of combinatorial regularity of polytopes and proved that a polytope is combinatorially regular if and only if it is combinatorially equivalent to a regular polytope.

Gr67 discussed various homogeneity properties of polytopes that are related to but weaker then regularity. And several authors have defined purely combinatorial notions that are patterned after the theory of regular polytopes and provide suitable frameworks for the study of combinatorially regular structures. References to these appear in the paper of McMullen and Schulte (1990), which describes methods for constructing the regular incidence-polytopes introduced by Danzer and Schulte (1982).

6.3. *Face-vectors of simplicial polytopes*

If $\mathcal{S}_s^d$ denotes the collection of all simplicial d-polytopes, and $f(\mathcal{S}_s^d)$ the set of all f-vectors of such polytopes, then 6.4 determines the affine hull of $f(\mathcal{S}_s^d)$. It is a much deeper task to provide an intrinsic description (without reference to polytopes) of the set $f(\mathcal{S}_s^d)$ itself. Such a description was conjectured by McMullen (1971) and proved by Stanley (1980) and Billera and Lee (1980). The solution of this problem is a milestone in the combinatorial study of polytopes.

McMullen's conjecture is most conveniently stated in terms of h-vectors, where the *h-vector* $(h_0, \ldots, h_d)$ of a simplicial $(d-1)$-complex $\mathscr{C}$ is defined as follows:

$$h_i := \sum_{j=0}^{i} (-1)^{i-j} \binom{d-j}{d-i} f_{j-1}, \quad 0 \leqslant i \leqslant d.$$

Note that $h_0 = 1$ and $h_1 = f_0 - d$. The h_i have various interesting interpretations, to which we return later.

The f-vector can be recovered from the h-vector, for

$$f_j = \sum_{i=0}^{j+1} \binom{d-i}{d-j-1} h_i, \quad 0 \leqslant j < d.$$

Alternatively, the relationship between f-vectors and h-vectors may be summarized by means of the following equality in terms of generating polynomials:

$$\sum f_{i-1} x^i = \sum h_i (x+1)^i.$$

Here is McMullen's characterization.

Theorem 6.5. *An integer vector $(h_0, \ldots, h_d)$ is the h-vector of a simplicial d-polytope if and only if the following three conditions are satisfied:*

(i) $h_i = h_{d-i}, 0 \leqslant i \leqslant \lfloor d/2 \rfloor$;

(ii) $h_{i+1} \geqslant h_i, 0 \leqslant i < \lfloor d/2 \rfloor$;

(iii) $h_0 = 1$ *and* $h_{i+1} - h_i \leqslant (h_i - h_{i-1})^{\langle i \rangle}, 0 \leqslant i \leqslant \lfloor d/2 \rfloor$.

Condition (i) is equivalent to the Dehn–Sommerville equations for the f_i, and the equality $h_0 = h_d$ is the Euler relation. Note also that condition (iii) is equivalent to the assertion that

$$(h_0, h_1 - h_0, \ldots, h_{\lfloor d/2 \rfloor} - h_{\lfloor d/2 \rfloor - 1}, 0, 0, \ldots)$$

is an O-sequence as defined above. This fact is the crucial part of Stanley's proof of the "only if" part of Theorem 6.5. He shows that the quantities $h_i - h_{i-1} (i \leqslant \lfloor d/2 \rfloor)$ are the successive values of the Hilbert function of the rational cohomology ring (modulo a suitable ideal) of a certain algebraic (toric) variety associated with a simplicial polytope.

Using the polytope algebra, McMullen (1993) proved the "only if" part of 6.5 in a way that remains in the framework of convexity. His proof reveals interesting connections between the combinatorics of polytopes and the Brunn–Minkowski theory of convex bodies.

Starting from a sequence that satisfies condition (iii) of 6.5, Billera and Lee show that certain facets of a suitable cyclic $(d+1)$-polytope C form a shellable ball $\mathscr{B}$ that has the sequence as its h-vector. Then with an appropriate realization of C in $\mathbb{R}^{d+1}$, there is found a point z of $\mathbb{R}^{d+1}$ such that z is beyond the facets of C that form $\mathscr{B}$ and beneath C's other facets. The desired d-polytope is then obtained as the intersection of the convex hull of $\{z\} \cup C$ with a hyperplane that separates z from C.

There remain three important problems associated with 6.5:

(1) Because McMullen's and Stanley's constructions rely so heavily on the polytope's convexity, their proofs of "only if" do not extend directly to the case of the h-vectors of spheres. Nevertheless, it is believed that the characterization does extend.

(2) Which d-polytopes realize equality in (ii) for a given i? McMullen and Walkup (1971) conjecture that they are precisely the ones that are *i-stacked*, meaning that they can be triangulated without introducing any new j-faces for $j < d - i$. However, the conjecture has been proved only for $i = 0$ and $i = 1$. (A *stacked polytope* is one that is 1-stacked. Equivalently, a d-polytope P is stacked if and only if it appears in a sequence $P_0, P_1, \ldots$ of d-polytopes such that P_0 is a d-simplex and P_{i+1} is formed from P_i by adding a pyramidal cap over some facet of P_i.)

(3) Using Stanley's proof of the "only if" part of 6.5, Billera and Lee (1981) prove a similar result for the h-vector of what they call a simplicial polyhedral $(d-1)$-ball. They conjecture that their conditions are also sufficient for the existence of such a ball with the given h-vector, but this remains open.

6.4. Face-vectors of general polytopes

Theorem 6.6. *A triple of integers (f_0, f_1, f_2) is the f-vector of a 3-polytope if and only if it is true that*

$$f_0 - f_1 + f_2 = 2, \qquad 4 \leqslant f_0 \leqslant 2f_2 - 4 \quad \textit{and} \quad 4 \leqslant f_2 \leqslant 2f_0 - 4.$$

Theorem 6.6 is due to Steinitz (1906). For $d \geqslant 4$ there does not exist even a reasonable conjecture for a characterization of f-vectors of general d-polytopes. However, there have been two directions of progress, and we shall describe them.

Let $\mathscr{S}(d)$ denote the collection of all subsets of the set $\{0, 1, \ldots, d-1\}$. The *extended f-vector* of a d-polytope P is the function f defined on $\mathscr{S}(d)$ by setting $f(\emptyset) := 1$ and, when $S = \{i_1, i_2, \ldots, i_k\} \in \mathscr{S}(d)$ with $i_1 < i_2 < \cdots < i_k$, letting $f_S(P)$ denote the number of chains of faces of P of the form $F_1 \subset F_2 \subset \cdots \subset F_k$ with $\dim F_j = i_j$ for $1 \leqslant j \leqslant k$. Thus P's ordinary f-vector is just the restriction of the extended f-vector to the singletons in $\mathscr{S}(d)$. And when P is simplicial, the ordinary f-vector completely determines the extended one.

The following analogue of the Dehn–Sommerville equations is proved by Bayer and Billera (1985a) for Eulerian manifolds and hence for arbitrary polytopes.

Theorem 6.7. *Suppose that P is a d-polytope, $-1 \leqslant i < k \leqslant d$, $\{i,k\} \subseteq S \cup \{-1,d\}$ and the set $S \subseteq \mathscr{S}(d)$ does not include any j between i and k. Then*

$$\sum_{j=i+1}^{k-1} (-1)^{j-i-1} f_{S\cup\{j\}}(P) = f_S(P)\big[1-(-1)^{k-i-1}\big].$$

When $i=-1$, $k=d$, and $S=\emptyset$, the equality of 6.7 is just the Euler relation. Bayer and Billera find a basis for the affine span of the extended f-vectors of all d-polytopes, and show that the dimension of this span is $c_d - 1$, where c_d is the dth Fibonacci number ($c_0 = c_1 = 1$, $c_d = c_{d-1} + c_{d-2}$). (Another basis is found by Kalai 1988b.) Carrying the investigation further in the case of 4-polytopes, Bayer (1987) finds inequalities for extended f-vectors that lead to new conclusions concerning ordinary f-vectors. (Projections of f-vectors of 4-polytopes onto two coordinates are completely determined in Gr67 and papers of D. Barnette and J. Reay mentioned by Bayer.)

For an arbitrary d-polytope P, Stanley (1987a) introduces a new combinatorial invariant consisting of the polynomial $g_P(x)$ defined by the following recursion:

(i) $g_\emptyset(x) \equiv 1$;

(ii) $\deg g_P(x) \leqslant (\dim P)/2$;

(iii) $x^{d+1} g_P(1/x) - g_P(x) = \sum_{F\in\mathscr{B}(P)} g_F(x)\cdot(x-1)^{\dim P - \dim F}$.

It turns out that the coefficients of the polynomial $g_P(x)$ are linear functions of the numbers $f_S(P)$ defined earlier. And from the fact that $g_Q \equiv 1$ when Q is a simplex, it is deduced that in the case of simplicial polytopes, (iii) corresponds to an equation for a generating function defining the quantities $h_{i+1} - h_i$. Thus the coefficients of the polynomial g_P may be viewed as a generalization of the h-vector to nonsimplicial polytopes.

Theorem 6.8. *For each rational polytope P, all coefficients g_i of the polynomial g_P are nonnegative.*

This theorem, which may be regarded as a counterpart of the unimodality of the h_i for simplicial polytopes, is conjectured to apply to all polytopes. However, that remains open because the theory of toric varieties does not apply to nonrational polytopes (which do exist – see section 4) and Stanley's proof of 6.8 uses a computation (by Bernstein, Khovanskii and MacPherson) of the intersection homology groups of toric varieties.

It is known that $g_1 = f_0 - d - 1 \geqslant 0$, and g_2 is shown by Kalai (1987) to be the dimension of the space of stresses of a polytopal framework based on P. Hence the conjecture is correct at least for g_1 and g_2. Perhaps further results about the g_i will yield deeper insight into the behavior of the f-vectors of general polytopes.

6.5. Upper-bound and lower-bound theorems for polytopes

The vertices of a polytope P are of special interest with respect to the maximization of a convex function f on P, because f's maximum is attained at a vertex. There are good tests for recognizing local maxima, and when f is linear each local maximum is in fact a global maximum. (That is the basis of the simplex algorithm for linear programming.) However, if f is merely convex it may happen that each vertex provides a strict local maximum even though only one of them provides a global maximum. This leads to concern about how many vertices P may have when its dimension and number of facets are known, or, equivalently (by polarity), how many facets P may have when its dimension and number of vertices are known.

The proof of the following is due to Gale (1963).

Theorem 6.9. *For a cyclic d-polytope with n vertices, the total number of facets is*

$$\mu(n,d) = \binom{n - \lfloor\frac{1}{2}(d+1)\rfloor}{n-d} + \binom{n - \lfloor\frac{1}{2}(d+2)\rfloor}{n-d}.$$

The following was conjectured by T.S. Motzkin in 1957.

Theorem 6.10 (Upper-bound theorem). *Whenever $0 \leqslant i < d$, the cyclic d-polytopes with n vertices have the maximum number of i-faces among all d-polytopes with n vertices.*

It is easy to see, by "pulling" the vertices slightly, that the maxima in 6.10 are attained by simplicial d-polytopes. For these, the conclusion of 6.10 follows from 6.5's description of the f-vectors of simplicial polytopes. (In fact, as noted by Bayer and Billera 1985b, the cyclic polytopes maximize all of the numbers $f_S(P)$ mentioned above.) However, 6.10 is known in much greater generality than is 6.5. Working directly from the Dehn–Sommerville equations of 6.4, Klee (1964b) shows that for each d, 6.10 applies to all simplicial Eulerian manifolds in which the number of vertices is sufficiently large relative to the dimension (at least $d^2/4$ for the case $k = d-1$). Using the shellability of polytopes, Theorem 4.18, established by Bruggesser and Mani (1971), McMullen (1970) proves 6.10 for all polytopes. And with the aid of appropriate machinery from commutative algebra, Stanley (1975) extends 6.10 to all simplicial spheres.

In addition to shellability, McMullen's proof uses a nice interpretation of the quantities h_i defined just before 6.5. Let $F_1, \ldots, F_r$ be a shelling order of the facets of a simplicial d-polytope P, so that for $2 \leqslant k \leqslant r$ the intersection $F_k \cap (\bigcup_{i=1}^{k-1} F_i)$ is the union of certain facets of F_k. For $1 \leqslant k \leqslant r$, let G_k denote the intersection of all faces of F_k that are not faces of any F_i with $i < k$. Then h_i is just the number of G_k that have precisely i vertices. Using this interpretation of the h_i, it is seen that the Upper-bound theorem is equivalent to the system of inequalities

$$h_i \leqslant \binom{f_0 - d + i - 1}{i}, \quad 0 \leqslant i \leqslant \lfloor d/2 \rfloor,$$

which are direct consequences of 6.5 (iii).

Stanley's argument proceeds as follows for a simplicial complex $\mathscr{C}$ with vertices $v_1, \ldots, v_n$. Let K be an infinite field and let $A = K[x_1, \ldots, x_n]$, the polynomial ring in independent variables $x_1, \ldots, x_n$ over K. Let I be the ideal of A generated by the monomials $x_{i_1} \ldots x_{i_k}$ for which $\{v_{i_1}, \ldots, v_{i_k}\}$ is not the vertex set of any face of $\mathscr{C}$. Stanley shows that if the ring $A_{\mathscr{C}} := A/I$ (the *Stanley–Reisner* ring of $\mathscr{C}$) has the Cohen–Macaulay property then there is an ideal J in $A_{\mathscr{C}}$ such that $\mathscr{C}$'s h-vector is the Hilbert function of the graded K-algebra $A_{\mathscr{C}}/J$. It then follows from 6.2 that the h_i form an O-sequence, which implies the above inequalities on the h_i and hence the upper-bound theorem. The proof of the upper-bound theorem for simplicial spheres is completed by citing the theorem of Reisner (1976) that if $\mathscr{C}$ is a simplicial sphere then $A_{\mathscr{C}}$ does have the Cohen–Macaulay property.

Another important consequence of 6.5 is the following.

Theorem 6.11 (Lower-bound theorem). *Whenever $0 \leqslant i < d$ the stacked d-polytopes with n vertices have the minimum number of i-faces among all simplicial d-polytopes with n vertices.*

An equivalent formulation of 6.11 is that if $(f_0, f_1, \ldots)$ is the f-vector of a simplicial d-polytope, then

(i) $f_{d-1} \geqslant (d-1)f_0 - (d+1)(d-2)$,

(ii) $f_i \geqslant \binom{d}{i} f_0 - \binom{d+1}{i+1} i \quad (1 \leqslant i \leqslant d-2)$.

These inequalities can be derived directly from the following consequence of 6.5:

$$h_1 = f_0 - d \quad \text{and} \quad h_i \geqslant h_1 \text{ for } 1 \leqslant i \leqslant \lfloor d/2 \rfloor.$$

Like 6.10, Theorem 6.11 is a consequence of 6.5 but is known in greater generality than 6.5. Barnette's original proof (1971, 1973b) of 6.11 uses a reduction to polytopal graphs, and his (1982) paper extends the argument to homology manifolds. Klee (1975) proves (i) for closed simplicial pseudomanifolds. Using arguments from the rigidity theory of frameworks, Kalai (1987) establishes 6.11 for triangulated manifolds, and shows that equality holds in (i) or (ii) if and only if the manifold is equivalent to the boundary complex of a stacked polytope (a characterization proved for polytopes by Barnette (1973b) and Billera and Lee 1981). Kalai (1987) also proves (ii) for a certain class of closed pseudomanifolds, and shows that for general closed pseudomanifolds, (ii) is a consequence of an old unsettled conjecture on the rigidity of 2-manifolds.

The case of centrally symmetric polytopes is of special interest, because they arise in so many contexts – for example, as unit balls of finite-dimensional polyhedral Banach spaces. For such polytopes there is so far no reasonable "upper-bound conjecture". It is known that the degree of neighborliness of such polytopes is less than one might think. For example, when the number of vertices is large relative to the dimension there is always a pair of nonantipodal vertices that are not joined by an edge (Burton 1991).

A theorem of Stanley (1987b) about h-vectors of centrally symmetric polytopes yields the following lower-bound theorem. The stated bounds can be achieved by starting with the d-dimensional cross-polytope and applying $n-d$ successive pairs of stellar subdivisions of antipodal facets.

Theorem 6.12. *If* $(f_0, f_1, \ldots)$ *is the* f*-vector of a centrally symmetric simplicial* d*-polytope with* n *vertices, then*

(i) $f_{d-1} \geqslant 2^d + 2(n-d)(d-1)$,

(ii) $f_i \geqslant 2^{i+1}\binom{d}{i+1} + 2(n-d)\binom{d}{i} \quad (0 \leqslant i \leqslant d-2)$.

Not much is known in the nonsimplicial case, except for the following result of Figiel et al. (1977).

Theorem 6.13. *There exists a positive constant* c *(independent of* d*) such that*

$$\log f_0(P) \cdot \log f_{d-1}(P) \geqslant cd$$

for each centrally symmetric d*-polytope* P.

Inequality 6.12 (i) implies a result proved by Barany and Lovász (1982) with the aid of Borsuk's antipodal mapping theorem: If a d-polytope is simple and centrally symmetric then it has at least 2^d vertices.

Kalai (1989) conjectured that each centrally symmetric d-polytope has at least 3^d nonempty faces, and noted that this number is attained by cubes and a variety of other polytopes. Y. Kupitz conjectured that each d-polytope with no triangular face has at least 2^d vertices, and this was proved by Blind and Blind (1990).

6.6. Upper-bound and lower-bound theorems for polytope-pairs

For a simple d-polytope Q with a given number of facets, the dual equivalents of Theorems 6.10 and 6.11 determine the maximum and the minimum number of i-faces that Q can have $(0 \leqslant i \leqslant d-2)$. These extrema are also of interest when Q is an *unbounded* simple d-polyhedron, and then it is appropriate to take separate account of the number of bounded facets, the number of unbounded facets, and the dimension of Q's *recession cone* (the union of all rays in Q that issue from a fixed but arbitrary point of Q). In terms of the simplicial approach, this leads to the definitions and theorems below.

A pair (P, F) is called a *polytope-pair of type* (d, v, k, r) if P is a simplicial d-polytope with v vertices, F is a $(k-1)$-face of P, and $f_0(\text{link}(F, \mathcal{B}(P)) = r - k$. Under polarity, (P, F) corresponds to a pair (P^*, F^*) where P^* is a simple d-polytope with v facets and F^* is a $(d-k)$-face of P^* that has $r-k$ facets of its own. Let H be a supporting hyperplane of P^* with $H \cap P^* = F^*$, and let Q be the unbounded simple d-polyhedron obtained from P^* by applying a projective transformation that sends H into the hyperplane at infinity. If $k = 1$ then Q has $v-1$ facets, $r-1$ of which are unbounded. If $k > 1$ then Q has v facets, r of which are unbounded. In either case Q has a recession cone of dimension $d-k+1$. Every simple, pointed, convex d-polyhedron with $(d-k+1)$-dimensional recession cone can be obtained in this way.

As Q is unbounded, its face-lattice does not form a complex as we have used the term so far. However, our definition of a complex can be extended without any changes to allow cells to be unbounded pointed polyhedra. With this extension in mind, we can define the boundary complex $\mathcal{B}(Q)$, and the dual of this complex, just as in the bounded case. The dual of $\mathcal{B}(Q)$ is the simplicial ball

$$\Gamma = \mathcal{B}(P)\backslash F := \{G \in \mathcal{B}(P) : F \text{ is not contained in } G\}.$$

The triangulation of Γ's boundary is dual to the complex generated by the unbounded faces of Q.

We will present sharp upper and lower bounds for the number of faces of P and Γ, thus providing sharp upper and lower bounds for the number of faces of P^* and Q. The bounds are stated in terms of the h-vectors of P and Γ, but translating them into the context of P^* and Q is easy, using the facts that

$$f_j(P^*) = \sum_{i=0}^{d-j} \binom{d-i}{j} h_i(P), \quad 0 \leqslant j < d,$$

$$f_j(Q) = \sum_{i=0}^{d-j} \binom{d-i}{j} h_i(\Gamma), \quad 0 \leqslant j < d.$$

In particular, the second formula delivers the minimum and maximum possible numbers of faces of a simple d-polyhedron having a recession cone of dimension $d-k+1$ and v facets ($v-1$ if $k=1$), r of which are unbounded ($r-1$ if $k=1$).

Theorem 6.14 (Upper-bound theorem for polytope pairs). *Let $3 \leqslant d \leqslant r \leqslant v$ and $1 \leqslant k \leqslant d-2$. Put $n = \lfloor d/2 \rfloor$. Let (P, F) range over all polytope pairs of type (d, v, k, r). Set $\Gamma = \mathcal{B}(P)\backslash F$.*

If $1 \leqslant k \leqslant \lfloor (d-1)/2 \rfloor$ then

$$\max h_i(P) = \begin{cases} \binom{v-d+i-1}{i}, & 0 \leqslant i \leqslant k, \\ \binom{v-d+i-1}{i} - \binom{v-d+i-k-1}{i-k} + \binom{r-d+i-k-1}{i-k}, & k+1 \leqslant i \leqslant n, \\ \binom{v-i-1}{d-i} - \binom{v-i-k-1}{d-i-k} + \binom{r-i-k-1}{d-i-k}, & n+1 \leqslant i \leqslant d-k-1, \\ \binom{v-i-1}{d-i}, & d-k \leqslant i \leqslant d. \end{cases}$$

$$\max h_i(\Gamma) = \begin{cases} \binom{v-d+i-1}{i}, & 0 \leqslant i \leqslant k-1, \\ \binom{v-d+i-1}{i} - \binom{v-d+i-k-1}{i-k}, & k \leqslant i \leqslant n, \\ \binom{v-i-1}{d-i} - \binom{v-i-k-1}{d-i-k}, & n+1 \leqslant i \leqslant d-k-1, \\ \binom{v-i-1}{d-i} - r + d, & d-k \leqslant i \leqslant d-1, \\ 0, & i = d. \end{cases}$$

If $\lfloor (d-1)/2 \rfloor < k \leqslant d-2$ then

$$\max h_i(P) = \begin{cases} \binom{v-d+i-1}{i}, & 0 \leqslant i \leqslant n, \\ \binom{v-i-1}{d-1}, & n+1 \leqslant i \leqslant d; \end{cases}$$

$$\max h_i(\Gamma) = \begin{cases} \binom{v-d+i-1}{i}, & 0 \leqslant i \leqslant n, \\ \binom{v-i-1}{d-i}, & n+1 \leqslant i \leqslant k-1, \\ \binom{v-i-1}{d-i} - 1, & i = k, \\ \binom{v-i-1}{d-i} - r + d, & k+1 \leqslant i \leqslant d-1, \\ 0, & i = d. \end{cases}$$

Moreover, for each of the above equalities, the indicated maxima can be achieved simultaneously.

Portions of Theorem 6.14 appear in Klee (1974) and Lee (1984). It is proved completely (including the tightness of the bounds) in Barnette et al. (1986). Note that the cases $k = d - 1$ or d just correspond to the classical Upper-bound theorem. The following result is proved by Lee (1984), along with additional information about h-vectors of polytope-pairs.

Theorem 6.15 (Lower-bound-theorem for polytope pairs). *Let $4 \leqslant d \leqslant r \leqslant v$ and $2 \leqslant k \leqslant d-2$. Put $n = \lfloor d/2 \rfloor$ and $m = [(d-k)/2]$. Let (P, F) range over all polytope*
pairs of type (d, v, k, r). Set $\Gamma = \mathcal{B}(P) \backslash F$. Then

(i) $$\min h_i(P) = \begin{cases} 1, & i = 0, \\ v - d, & 1 \leqslant i \leqslant n; \end{cases}$$

(ii) $$\min h_i(\Gamma) = \begin{cases} 1, & i = 0, \\ v - d, & 1 \leqslant i \leqslant k-1, \\ v - d - 1, & i = k, \\ v - r, & k+1 \leqslant i \leqslant d-1, \\ 0, & i = d. \end{cases}$$

For $k = 1$ and $3 \leqslant d \leqslant r < v$, (i) *is the same but* (ii) *becomes*

$$\min h_i(\Gamma) = \begin{cases} 1, & i = 0, \\ v - d - 1, & i = 1, \\ v - r - 1, & 2 \leqslant i \leqslant d-1, \\ 0, & i = d. \end{cases}$$

All bounds in (i) *and* (ii) *can be achieved.*

7. Polytopal graphs

In the edge-graph $\mathscr{E}(\mathscr{C})$ of a complex $\mathscr{C}$, the nodes are $\mathscr{C}$'s vertices and the edges are pairs of vertices that share an edge of $\mathscr{C}$. Dually, in the *ridge-graph* $\mathscr{R}(\mathscr{C})$ (defined when $\mathscr{C}$ is pure), the nodes are facets of $\mathscr{C}$ and the edges are pairs of facets that share a ridge of $\mathscr{C}$. When P is a polytope, the graphs $\mathscr{R}(P) = \mathscr{R}(B(P))$ and $\mathscr{E}(P^*) = \mathscr{E}(\mathscr{B}(P^*))$ are isomorphic. Hence the d-polytopal graphs may be defined as those isomorphic to the edge-graph of a d-polytope or, equivalently, as those isomorphic to the ridge-graph of a d-polytope.

There is interest not only in polytopal graphs but also in polytopal digraphs, for these digraphs are closely associated with linear optimization, they provide a transparent approach to the shellability of simple polytopes (Danaraj and Klee 1974), and they are involved in a simple formulation of McMullen's proof of the Upper-bound theorem (Brøndsted 1983).

A linear functional φ is *admissible* for a d-polytope P if φ's domain contains P and φ does not attain the same value at any two vertices of P. The φ-digraph $\mathscr{D}(P, \varphi)$ of P is then formed by orienting each edge of $\mathscr{E}(P)$ in the direction of increasing φ. A digraph is *d-polytopal* if it is isomorphic to one formed in this way. Note that each polytopal digraph has a unique source (node with zero indegree) and a unique sink (node with zero outdegree).

The earlier sections have emphasized the "simplicial" approach to polytopes. However, in dealing with polytopal graphs we stress the "simple" approach because it leads naturally to a focus on edge-graphs, which seem more intuitive and more directly relevant to applications than ridge-graphs.

This section is divided into three subsections as follows: Results for arbitrary dimension; The d-step conjecture and its relatives; Special low-dimensional problems.

7.1. Results for arbitrary dimension

The following theorem of Balinski (1961) is the first significant one about arbitrary polytopal graphs.

Theorem 7.1. *The graph of a d-polytope is d-connected.*

There is an analogous result for polytopal digraphs.

Theorem 7.2. *In each d-polytopal digraph there are d independent dipaths from source to sink.*

To see that 7.2 implies 7.1, note that for any two nodes u and v of a d-polytope P there exist a d-polytope Q, an admissible φ for Q, and a projective bijection that carries P onto Q and u and v onto vertices of Q corresponding to the source and sink of $\mathscr{D}(Q, \varphi)$.

For each n and d, let $s(n, d)$ denote the maximum number of components into which a d-polytopal graph can be separated by removing n nodes. Klee (1964d) proves the following extension of 7.1, where $\mu(n, d)$ is the maximum number of vertices of d-polytopes with n facets (see 6.9–6.10).

Theorem 7.3. *For all d and n,*

$$s(n,d) = \begin{cases} 1, & \textit{if } n < d, \\ 2, & \textit{if } n = d, \\ \mu(n,d), & \textit{if } n > d. \end{cases}$$

Here is another property of d-polytopal graphs, due to Grünbaum and Motzkin (1963):

Theorem 7.4. *Each d-polytopal graph contains a subdivision of the complete graph on* $d+1$ *nodes.*

This theorem is extended as follows by Grünbaum (1965): *If* $\mathscr{C}$ *is a d-polytopal complex and* $0<k<d$ *then the k-skeleton of* $\mathscr{C}$ *contains a subdivision of the k-skeleton of a d-simplex.*

Various improvements and relatives of 7.4 provide other *necessary* conditions for d-polytopality of a graph (see Grünbaum 1970, 1975). However, it seems very unlikely that there exists an easily testable *characterization* of polytopal graphs. This pessimism seems justified by remarks in section 4 (4.19 ff.) in conjunction with the following fact, conjectured in 1970 by M. A. Perles and proved by Blind and Mani (1987) and by Kalai (1988c).

Theorem 7.5. *Each simple polytope is determined by its edge-graph. That is, two simple polytopes P and Q are isomorphic if and only if their edge-graphs are isomorphic.*

By the dual equivalent of 7.5, each simplicial polytope is determined by its ridge-graph. Of course, simplicial polytopes are not in general determined by their edge-graphs. (Consider the neighborly polytopes!) However, it is known that each simplicial d-polytope is determined by its $\lfloor d/2 \rfloor$-skeleton (Dancis 1984), while for an arbitrary d-polytope the best that can be said is that it is determined by its $(d-2)$-skeleton (Gr67).

Kalai (1988c) has a short proof of 7.5. Suppose that P is a simple d-polytope, and for each nonempty face F of P let $\mathscr{E}_F(P)$ denote the corresponding subgraph of $\mathscr{E}(P)$. Call an orientation $\ll$ of $\mathscr{E}(P)$ *good* if it is acyclic and with respect to it each $\mathscr{E}_F(P)$ has a unique sink. For each acyclic orientation $\ll$ of $\mathscr{E}(P)$, let $h_k^{\ll}$ denote the number of nodes of indegree k, and define the quantity

$$f^{\ll} = h_0^{\ll} + 2h_1^{\ll} + 4h_2^{\ll} + \cdots + 2^d h_d^{\ll}.$$

Note that if x is a node of $\mathscr{E}(P)$ of indegree k, then x is a sink in 2^k faces, because (by simplicity of P) each choice of i edges incident to x determines an i-face that includes them. It follows that if f is the number of nonempty faces of P and $\ll$ is any acyclic orientation, then $f^{\ll} \geqslant f$, with $f^{\ll} = f$ if and only if $\ll$ is good. Hence the good acyclic orientations of $\mathscr{E}(P)$ are those yielding the minimum value for $f^{\ll}$. To complete the proof, observe that the faces of P can be described in terms of the good acyclic orientations. Indeed, it is not hard to see that a subgraph H of $\mathscr{E}(P)$ corresponds to a k-face of P if and only if H is induced, k-regular, and connected, and its nodes precede all others with respect to some good acyclic orientation of $\mathscr{E}(P)$.

When P is a simple d-polytope, each facet of P corresponds to a subgraph of $\mathscr{E}(P)$ that is induced, connected, and $(d-1)$-regular. M. Perles has conjectured that, conversely, each such subgraph of $\mathscr{E}(P)$ arises from a facet. If this could be proved, even with "connected" replaced by "$(d-1)$-connected", it would provide a different proof of 7.5.

7.2. The d-step conjecture and its relatives

For a polytope P, let $\delta(P)$ denote the diameter of the graph $\mathscr{E}(P)$. When φ is admissible for P, let $\xi(P,\varphi)\langle$respectively $\eta(P,\varphi)\rangle$ denote the length of the shortest $\langle$respectively longest$\rangle$ dipath from source to sink in $\mathscr{D}(P,\varphi)$. For $n > d \geqslant 2$, let $\Delta(d,n)\langle$respectively $\Xi(d,n), H(d,n)\rangle$ denote the maximum of $\delta(P)$ $\langle$respectively $\xi(P,\phi)$, $\eta(P,\phi)\rangle$ over all P or (P,ϕ) for which P is a simple d-polytope with n facets. Obviously, $\Delta(d,n) \leqslant \Xi(d,n) \leqslant H(d,n)$. These functions are all of interest in connection with the behavior of edge-following algorithms for linear programming, and the case $n = 2d$ is of special interest. However, even though the diameter is one of the most intensively studied parameters of graph theory, little is known about these functions. They are difficult to study because the relevant graphs or digraphs are not defined in a purely combinatorial way but involve geometric notions as well.

The *d-step conjecture* asserts that $\Delta(d,2d) = d$ for all d, and the *Hirsch conjecture* asserts that $\Delta(d,n) \leqslant n-d$ for all d and n. Since a comprehensive survey of these conjectures is provided by Klee and Kleinschmidt (1987), we state only a few of the main results. The first part of 7.6 is due to Klee (1964b), its second part and 7.7 to Klee and Walkup (1967).

Theorem 7.6. $\Delta(3,n) = \lfloor 2n/3 \rfloor - 1$;

$$\Delta(d,d+k) = k \quad \textit{when } 1 \leqslant k \leqslant \min\{d,5\}.$$

Theorem 7.7. *The following three statements are equivalent:*

(i) *for all* $d, \Delta(d,2d) = d$;

(ii) *for all* d *and* $n, \Delta(d,n) \leqslant n-d$;

(iii) *any two vertices of a simple polytope can be joined by an edge-path that does not revisit any facet.*

Note that by 7.6, (ii) is correct for $d \leqslant 3$, (i) for $d \leqslant 5$. The dual equivalent of the latter fact is extended to simplicial spheres (and more general complexes) by Adler and Dantzig (1974). The dual equivalent of (iii) is that any two facets of a simplicial polytope are joined by a ridge-path that does not revisit any vertex. Mani and Walkup (1980) show that this fails for a certain nonpolytopal simplicial 3-sphere, and that leads to a counterexample to the sphere-analogue of the dual equivalent of (i). Specifically, there exists a nonpolytopal simplicial 11-sphere $\mathscr{S}$ with 24 vertices such that the diameter of the ridge-graph $\mathscr{R}(\mathscr{S})$ exceeds 12.

For large values of d and $n-d$, the following two results, due respectively to Larman (1970) and to Kalai and Kleitman (1992), provide upper bounds on $\Delta(d,n)$ that are the best known in certain directions.

Theorem 7.8. $\Delta(d,n) \leqslant 2^{d-3}n$; $\Delta(d,n) \leqslant d^{2+\log n}$.

Neither of these results rules out the possibility that as $d \to \infty$, $\Delta(d,2d) - d \to \infty$, but at least the second result provides a subexponential upper bound on $\Delta(d,2d)$. In any case, the following theorem of Lee (1984) shows that no drastic failure of the Hirsch conjecture (if there is such a failure) can be accounted for by strange behavior of f-vectors.

Theorem 7.9. *For each simple d-polytope P with n facets, there is a simple polytope Q such that P and Q have the same f-vector and the diameter of Q's edge-graph is at most $n - d + 1$.*

The polytopes that are dual to neighborly (or almost neighborly) polytopes are natural candidates for violation of the Hirsch conjecture. However, they can in any case not be "exponential violators", for Kalai (1991) has shown that their diameters are bounded by a polynomial in d and n.

Because of their relevance to many problems of combinatorial optimization, there is special interest in polytopes whose vertices have all coordinates equal to 0 or 1. A simple argument of Naddef (1989) shows that the Hirsch conjecture holds for such polytopes.

Even less is known about the functions H and Ξ. The first part of 7.10 is due to Klee (1965), its second part to Todd (1980). In 7.11, which is due to Klee and Minty (1972), the upper bound is a consequence of the Upper-bound theorem.

Theorem 7.10. $\Xi(3,n) = n - 3$; $\Xi(4,9) \geqslant 6$.

Theorem 7.11. *For each d there are positive constants α_d and β_d such that for all n, $\alpha_d n^{\lfloor d/2 \rfloor} < H(d,n) < \beta_d n^{\lfloor d/2 \rfloor}$. Also, $H(d,2d) \geqslant 2^d - 1$.*

For additional results related to diameters of polytopes, see Kalai (1991, 1992a,b).

7.3. Special low-dimensional problems

Our focus on high-dimensional results may have left the impression that most combinatorial questions about 3-polytopal graphs have been settled. That is far from being the case, and many interesting illustrations could be given. However, we limit ourselves to two old problems, both of which involve the *p-vector* of a 3-polytope. This is the ultimately zero sequence $(p_3, p_4, \ldots)$ in which p_i denotes the number of 2-faces with precisely i vertices. There are many open questions concerning p-vectors, and it seems unlikely that easily testable characterizations will ever be known. However, in the case of simple 3-polytopes the following result comes tantalizingly close to providing a complete characterization.

Theorem 7.12. *If $(p_3, p_4, \ldots)$ is the p-vector of a simple 3-polytope, then*

$$\sum_{k \geqslant 3} (6 - p_k) = 12.$$

For each sequence $(p_3, p_4, \ldots)$ *of nonnegative integers satisfying this equation, the value of* p_6 *can be adjusted so that the altered sequence is the p-vector of a simple 3-polytope.*

Thus the complete determination of the p-vectors of simple 3-polytopes reduces to the following: *For a sequence* $(p_3, p_4, p_5, p_7, \ldots)$ *such that* $\sum_{k \geqslant 3}(6 - p_k) = 12$, *which values can be filled in for* p_6 *so that the sequence* $(p_3, p_4, p_5, p_6, p_7, \ldots)$ *is the p-vector of a simple 3-polytope?*

The first part of 7.12 follows readily from Euler's theorem. The second part is an 1891 discovery of V. Eberhard, proved much more simply in Gr67 by use of the reductions of 3-connected planar graphs mentioned in section 4. Grünbaum (1968) shows that if $p_3 = p_4 = 0$, then p_6 may be required to have any preassigned value $\geqslant 8$. For other results on this, and on related problems concerning 3-polytopes that are not required to be simple, see Grünbaum (1967, 1968, 1970, 1975), Roudneff (1987), and their references.

In an early attack on the 4-color problem, P. G. Tait conjectured in 1880 that every simple 3-polytope is Hamiltonian (admits a circuit using all vertices). That is disproved by a 1946 example of Tutte, with 46 vertices, and there are later examples with only 38 vertices. (See Gr67 and Barnette 1983 for the history.) Garey et al. (1976) show that the problem of detecting Hamiltonicity in 3-polytopes is NP-complete, even when restricted to simple 3-polytopes for which $p_3 = p_4 = 0$. On the other hand, Hamiltonicity has been established for several broad classes of 3-polytopes. The successive classes in 7.13 are treated in the papers of Tutte (1956), Holton and McKay (1988), Holton et al. (1985), and Goodey (1975, 1977).

Theorem 7.13. *A* 3-*polytope P is Hamiltonian if P's graph is* 4-*connected, and also if P is simple and satisfies any of the following supplementary conditions:*

(i) *the number of vertices is at most* 36*;*

(ii) *the number of vertices is at most* 64 *and each* 2-*face has an even number of vertices;*

(iii) *the total number of* 2-*faces is equal to* $p_3 + p_4$ *or to* $p_3 + p_6$ *or to* $p_4 + p_6$.

Some of the above results are relevant to the conjecture of D. Barnette (in Grünbaum 1970) that a simple 3-polytope P is Hamiltonian if it has no odd 2-faces, or the conjecture of Goodey (1977) that P is Hamiltonian if no face has more than 6 vertices. Barnette conjectures also that each simple 4-polytope is Hamiltonian, which is discussed by Rosenfeld (1983).

References

Adler, I., and G.B. Dantzig

[1974] Maximum diameter of abstract polytopes, in: *Pivoting and Extensions*, ed. M. Balinski, *Math. Programming Study* **1**, 11–19.

Alon, N.

[1986] The number of polytopes, configurations and real matroids, *Mathematika* **33**, 62–71.

[1995] ch. 32, this volume.

Altshuler, A.
[1977] Neighborly 4-polytopes and neighborly combinatorial 3-manifolds with ten vertices, *Canad. J. Math.* **29**, 400–420.
Altshuler, A., and L. Steinberg
[1985] The complete enumeration of the 4-polytopes and 3-spheres with 8 vertices, *Pacific J. Math.* **117**, 1–16.
Altshuler, A., J. Bokowski and L. Steinberg
[1980] The classification of simplicial 3-spheres with nine vertices into polytopes and nonpolytopes, *Discrete Math.* **31**, 115–124.
Balinski, M.
[1961] On the graph structure of convex polyhedra in *n*-space, *Pacific J. Math.* **11**, 431–434.
Barany, I., and L. Lovász
[1982] Borsuk's theorem and the number of facets of centrally symmetric polytopes, *Acta Math. Acad. Sci. Hungar.* **40**, 323–329.
Barnette, D.
[1970] Projections of 3-polytopes, *Israel J. Math.* **8**, 304–308.
[1971] The minimum number of vertices of a simple polytope, *Israel J. Math.* **10**, 121–125.
[1973a] The triangulations of the 3-sphere with up to 8 vertices, *J. Combin. Theory* **14**, 37–52.
[1973b] A proof of the lower bound conjecture for convex polytopes, *Pacific J. Math.* **46**, 349–354.
[1980] An invertible non-polyhedral diagram, *Israel J. Math.* **36**, 89–96.
[1982] Decompositions of homology manifolds and their graphs, *Israel J. Math.* **41**, 203–212.
[1983] *Map Coloring, Polyhedra, and the Four-Color Problem, Dolciani Mathematical Expositions,* No. 8 (Mathematical Association of America, Washington, DC).
Barnette, D., and B. Grünbaum
[1970] Preassigning the shape of a face, *Pacific J. Math.* **32**, 299–306.
Barnette, D., P. Kleinschmidt and C.W. Lee
[1986] An upper bound theorem for polytope pairs, *Math. Oper. Res.* **11**, 451–464.
Bartels, H.
[1973] *A priori Informationen zur linearen Programmierung – Über Ecken und Hyperflachen auf Polyeder* (Verlag Anton Bain, Meisenheim-an-Glan).
Bayer, M.M.
[1987] The extended *f*-vectors of 4-polytopes, *J. Combin. Theory A* **44**, 141–151.
Bayer, M.M., and L. Billera
[1985a] Generalized Dehn–Sommerville relations for polytopes, spheres and Eulerian partially ordered sets, *Invent. Math.* **79**, 143–157.
[1985b] Counting faces and chains in polytopes and posets, in: *Combinatorics and Algebra, Contemporary Mathematics,* Vol. 34, ed. C. Greene (American Mathematical Society, Providence, RI) pp. 207–252.
Bayer, M.M., and C.W. Lee
[1993] Convex polytopes, in: *Handbook on Convex Geometry,* eds. P. Gruber and J.M. Wills (North-Holland, Amsterdam).
Bender, E.A.
[1987] The number of three-dimensional convex polyhedra, *Amer. Math. Monthly* **94**, 7–21.
Billera, L.J., and C.W. Lee
[1980] A proof of the sufficiency of McMullen's conditions for *f*-vectors of simplicial polytopes, *J. Combin. Theory A* **31**, 237–255.
[1981] The number of faces of polytope pairs and unbounded polyhedra, *European J. Combin.* **1**, 27–31.
Björner, A.
[1986] Face numbers of complexes and polytopes, in: *Proc. Int. Congr. of Mathematicians, Berkeley, 1986,* ed. A.M. Gleason (AMS, Providence, RI) pp. 1408–1416.
[1995] ch. 34, this volume.
Bland, R.G., and M. Las Vergnas
[1978] Orientability of matroids, *J. Combin. Theory B* **24**, 94–123.

Blind, G., and R. Blind
[1990] Convex polytopes without triangular faces, *Israel J. Math.* **71**, 129–134.
Blind, R., and P. Mani
[1987] On puzzles and polytope isomorphisms, *Aequationes Math.* **34**, 287–297.
Bokowski, J., and I. Shemer
[1987] Neighborly 6-polytopes with 10 vertices, *Israel J. Math.* **58**, 103–124.
Bokowski, J., and B. Sturmfels
[1985] *Problems of Geometrical Realizability – Oriented Matroids and Chirotopes*, Preprint Nr. 901 (Fachbereich Mathematik, Technische Hochschule Darmstadt).
[1986] On the coordinatization of oriented matroids, *Discrete Comput. Geom.* **1**, 296–306.
[1987] Polytopal and nonpolytopal spheres: an algorithmic approach, *Israel J. Math.* **57**, 257–272.
Bokowski, J., G. Ewald and P. Kleinschmidt
[1984] On combinatorial and affine automorphisms of polytopes, *Israel J. Math.* **47**, 123–130.
Brøndsted, A.
[1983] *An Introduction to Convex Polytopes* (Springer, New York).
Brückner, M.
[1909] Über die Ableitung der allgemeinen Polytope und die nach Isomorphismus verschiedenen Typen der allgemeinen Achtzelle, *Verhand. Kon. Akad. Wetenschap, Eerste Sectie* **10**(1).
Bruggesser, M., and P. Mani
[1971] Shellable decompositions of cells and spheres, *Math. Scand.* **29**, 197–205.
Burton, G.R.
[1991] The non-neighborliness of centrally symmetric convex polytopes having many vertices, *J. Combin. Theory A* **58**, 321–322.
Collins, G.
[1975] Quantifier elimination for real closed fields by cylindrical algebraic decomposition, in: *Proc. 2nd GI Conf. on Automata and Formal Languages, Lecture Notes in Computational Science*, Vol. 33 (Springer, Berlin) pp. 134–163.
Coxeter, H.S.M.
[1963] *Regular Polytopes*, 2nd Ed. (Macmillan, New York).
Danaraj, G., and V. Klee
[1974] Shellings of spheres and polytopes, *Duke Math. J.* **41**, 443–451.
[1978a] Which spheres are shellable?, in: *Algorithmic Aspects of Combinatorics*, ed. B. Alspach, *Ann. Discrete Math.* **2**, 33–52.
[1978b] A representation of two-dimensional pseudomanifolds and its use in the design of a linear-time shelling algorithm, in: *Algorithmic Aspects of Combinatorics*, ed. B. Alspach, *Ann. Discrete Math.* **2**, 53–63.
Dancis, J.
[1984] Triangulated manifolds are determined by their $[n/2]+1$-skeletons, *Topology Appl.* **18**, 17–26.
Danzer, L., and E. Schulte
[1982] Reguläre Inzidenzkomplexe I, *Geom. Dedicata* **13**, 295–308.
Dress, A.
[1986] Chirotopes and oriented matroids, *Bayreuth. Math. Schriften* **21**, 14–68.
Duivestijn, A.J.W., and P.D.J. Federico
[1981] The number of polyhedral (3-connected planar) graphs, *Math. Comput.* **37**, 523–532.
Dyer, M.E.
[1983] The complexity of vertex enumeration methods, *Math. Oper. Res.* **8**, 381–402.
Ewald, G., P. Kleinschmidt and C. Schulz
[1976] Kombinatorische Klassifikation symmetrischer Polytope, *Abh. Math. Seminar Univ. Hamburg* **45**, 191–206.
Ewald, G., P. Kleinschmidt, U. Pachner and C. Schulz
[1979] Neuere Entwicklungen in der kombinatorischen Konvexgeometrie, in: *Contributions to Geometry, Proc. Geometry Symp., Siegen, 1978*, eds. J. Tölke and J.M. Wills (Birkhäuser, Basel:) pp. 131–163.

Figiel, T., J. Lindenstrauss and V. Milman
[1977] The dimension of almost spherical sections of convex bodies, *Acta Math.* **139**, 53–94.
Folkman, J., and J. Lawrence
[1978] Oriented matroids, *J. Combin. Theory B* **25**, 199–236.
Frankl, P.
[1995] ch. 24, this volume.
Gale, D.
[1963] Neighborly and cyclic polytopes, in: *Convexity*, ed. V. Klee, *Amer. Math. Soc. Proc. Symp. Pure Math.* **7**, 225–232.
Garey, M.R., D.S. Johnson and R.E. Tarjan
[1976] The planar Hamiltonian circuit problem is NP-complete, *SIAM J. Comput.* **5**, 704–714.
Goodey, P.R.
[1975] Hamiltonian circuits in polytopes with even sided faces, *Israel J. Math.* **22**, 52–56.
[1977] A class of Hamiltonian polytopes, *J. Graph Theory* **1**, 181–185.
Goodman, J.E., and R. Pollack
[1986] Upper bounds for configurations and polytopes in $\mathbb{R}^d$, *Discrete Comput. Geom.* **1**, 219–227.
Grünbaum, B.
[1965] On the facial structure of convex polytopes, *Bull. Amer. Math. Soc.* **71**, 559–560.
[1967] *Convex Polytopes* (Wiley-Interscience, London).
[1968] Some analogues of Eberhard's theorem on convex polytopes, *Israel J. Math.* **6**, 398–411.
[1970] Polytopes, graphs and complexes, *Bull. Amer. Math. Soc.* **76**, 1131–1202.
[1975] Polytopal graphs, in: *Studies in Graph Theory, Part II, MAA Studies in Mathematics*, Vol. 12, ed. D.R. Fulkerson (Mathematical Association of America, Washington, DC) pp. 201–224.
Grünbaum, B., and V. Sreedharan
[1967] An enumeration of simplicial 4-polytopes with 8 vertices, *J. Combin. Theory* **2**, 437–465.
Grünbaum, G., and T.S. Motzkin
[1963] On polyhedral graphs, in: *Convexity*, ed. V. Klee, *Amer. Math. Soc. Proc. Symp. Pure Math.* **7**, 285–290.
Hodge, W.V.D., and D. Pedoe
[1968] *Methods of Algebraic Geometry* (Cambridge University Press, London).
Holton, D.A., and B.D. McKay
[1988] The smallest non-Hamiltonian 3-connected cubic planar graphs have 38 vertices, *J. Combin. Theory B* **45**, 305–319. Erratum: 1989, **47**, 248.
Holton, D.A., B. Manvel and B.D. McKay
[1985] Hamiltonian cycles in cubic 3-connected bipartite planar graphs, *J. Combin. Theory B* **38**, 279–297.
Hopcroft, J.E., and R.E. Tarjan
[1973] Dividing a graph into triconnected components, *SIAM J. Comput.* **2**, 135–158.
[1974] Efficient planarity testing, *J. Assoc. Comput. Mach.* **21**, 549–568.
Kalai, G.
[1987] Rigidity and the lower bound theorem I, *Invent. Math.* **88**, 125–151.
[1988a] Many triangulated spheres, *Discrete Comput. Geom.* **3**, 1–14.
[1988b] A new basis of polytopes, *J. Combin. Theory A* **49**, 191–209.
[1988c] A simple way to tell a simple polytope from its graph, *J. Combin. Theory A* **49**, 381–383.
[1989] The number of faces of centrally-symmetric polytopes, *Graphs and Combinatorics* **5**, 389–391.
[1990] On low-dimensional faces that high-dimensional polytopes must have, *Combinatorica* **10**, 271–280.
[1991] The diameter problem for convex polytopes and f-vector theory, in: *Applied Geometry and Discrete Mathematics: The Victor Klee Festschrift, DIMACS Series on Discrete Mathematics and Computer Science*, Vol. 4, eds. P. Gritzmann and B. Sturmfels (AMS/ACM, Providence, RI/New York) pp. 387–411.
[1992a] Upper bounds for the diameter and height of graphs of convex polyhedra, *Discrete Comput. Geom.* **8**, 363–372.
[1992b] A subexponential randomized simplex algorithm, in: *Proc. 24th Annu. Symp. on the Theory of Computing* (ACM Press, New York) pp. 475–482.

Kalai, G., and D. Kleitman
[1992] Quasi-polynomial bounds for the diameter of graphs of polyhedra, *Bull. Amer. Math. Soc.* **26**, 315–316.

Katona, G.O.H.
[1968] A theorem of finite sets, in: *Theory of Graphs, Proc. Tihany Conf., 1966,* eds. P. Erdős and G. Katona (Academic Press/Akadémia Kiadó, New York/Budapest) pp. 187–207.

Kirkpatrick, D.G., and R. Seidel
[1982] The ultimate planar convex hull algorithm? In: *Proc. 20th Annu. Allerton Conf. on Communication, Control and Computing,* pp. 35–42.

Klee, V.
[1964a] A combinatorial analogue of Poincaré's duality theorem, *Canad. J. Math.* **16**, 517–531.
[1964b] Diameters of polyhedral graphs, *Canad. J. Math.* **16**, 602–614.
[1964c] The number of vertices of a convex polytope, *Canad. J. Math.* **16**, 701–720.
[1964d] A property of d-polyhedral graphs, *J. Math. Mech.* **13**, 1039–1042.
[1965] Paths on polyhedra. I, *J. Soc. Ind. Appl. Math.* **13**, 946–956.
[1966] Convex polytopes and linear programming, in: *Proc. IBM Scientific Computing Symp. on Combinatorial Problems, White Plains, NY, 1984* (IBM Data Processing Division, White Plains, NY) pp. 123–158.
[1974] Polytope pairs and their relations to linear programming, *Acta Math.* **113**, 1–25.
[1975] A d-pseudomanifold with f_0 vertices has at least $df_0 - (d-1)(d+2)$ simplices, *Houston J. Math.* **1**, 81–86.

Klee, V., and P. Kleinschmidt
[1987] The d-step conjecture and its relatives, *Math. Oper. Res.* **12**, 718–755.

Klee, V., and G.J. Minty
[1972] How good is the simplex algorithm?, in: *Inequalities III,* ed. O. Shisha (Academic Press, New York) pp. 159–175.

Klee, V., and S. Wagon
[1991] *Old and New Unsolved Problems in Plane Geometry and Number Theory* (Mathematical Association of America, Washington, DC).

Klee, V., and D.W. Walkup
[1967] The d-step conjecture for polyhedra of dimension $d < 6$, *Acta Math.* **117**, 53–78.

Kleinschmidt, P.
[1976a] On facets with non-arbitrary shape, *Pacific J. Math.* **65**, 511–515.
[1976b] Sphären mit wenigen Ecken, *Geom. Dedicata* **5**, 307–320.
[1977] Konvexe Realisierbarkeit symmetrischer Sphären, *Arch. Math.* **28**, 433–435.

Kruskal, J.B.
[1963] The number of simplices in a complex, in: *Mathematical Optimization Techniques,* ed. R. Bellman (University of California Press, Los Angeles) pp. 251–278.

Larman, D.G.
[1970] Paths on polytopes, *Proc. London Math. Soc. (3)* **20**, 161–178.

Lawrence, J.
[1982] Oriented matroids and multiple ordered sets, *Linear Algebra Appl.* **48**, 1–12.
[1984] *Shellability of Oriented Matroid Complexes,* Unpublished Manuscript.

Lee, C.W.
[1984] Bounding the numbers of faces of polytope pairs and simple polyhedra, in: *Convexity and Graph Theory,* eds. M. Rosenfeld and J. Zaks, *Ann. Discrete Math.* **20**, 215–232.

Lindström, B.
[1971] On the realization of convex polytopes, Euler's formula, and Möbius functions, *Aequationes Math.* **6**, 235–240.

Lloyd, E.K.
[1970] The number of d-polytopes with few vertices, *Mathematika* **17**, 120–132.

Macaulay, F.S.
[1927] Some properties of enumeration in the theory of modular systems, *Proc. London Math. Soc.* **26**, 531–555.

Mandel, A.
[1981] *Topology of oriented matroids*, Ph.D. Dissertation (University of Waterloo, Ont.).

Mani, P.
[1971] Automorphismen von polyedrischen Graphen, *Math. Ann.* **192**, 279–303.
[1972] Spheres with few vertices, *J. Combin. Theory A* **13**, 346–352.

Mani, P., and D.W. Walkup
[1980] A 3-sphere counterexample to the W_v-path conjecture, *Math. Oper. Res.* **5**, 595–598.

Mattheiss, T.H., and D.S. Rubin
[1980] A survey and comparison of methods for finding all vertices of convex polyhedral sets, *Math. Oper. Res.* **5**, 167–185.

Mazur, B.
[1986] Arithmetic on curves, *Bull. Amer. Math. Soc.* **14**, 207–259.

McMullen, P.
[1967] Combinatorially regular polytopes, *Mathematika* **14**, 142–150.
[1970] The maximum number of faces of a convex polytope, *Mathematika* **17**, 179–184.
[1971] The number of faces of simplicial polytopes, *Israel J. Math.* **9**, 559–570.
[1974] The number of neighborly d-polytopes with $d+3$ vertices, *Mathematika* **21**, 26–31.
[1979] Representations and diagrams, in: *Contributions to Geometry, Proc. Geometry Symp., Siegen, 1978*, eds. J. Tölke and J. Wills (Birkhäuser, Basel) pp. 92–130.
[1993] On simple polytopes, *Inventiones Math.* **113**, 419–444.

McMullen, P., and E. Schulte
[1990] Constructions for regular polytopes, *J. Combin. Theory A* **53**, 1–28.

McMullen, P., and G.C. Shephard
[1970] Polytopes with an axis of symmetry, *Canad. J. Math.* **22**, 265–287.
[1971] *Convex Polytopes and the Upper Bound Conjecture, London Mathematical Society Lecture Note Series*, Vol. 3 (Cambridge University Press, London).

McMullen, P., and D.W. Walkup
[1971] A generalized lower bound conjecture for simplicial polytopes, *Mathematika* **18**, 264–273.

Mnëv, N.E.
[1983] On the realizability over fields of the combinatorial type of convex polytopes (in Russian), *Zap. Nauchn. Sem. Leningr. Otdel. Mat. Inst. Steklov* **123**, 203–207.
[1988] The universality theorems on the classification of configuration varieties, in: *Topology and Geometry – Rohlin Seminar, Lecture Notes in Mathematics*, Vol. 1346, eds. O.Ya. Viro and A.M. Vershik (Springer, Berlin) pp. 527–544.

Naddef, D.
[1989] The Hirsch conjecture is true for (0, 1)-polytopes, *Math. Programming B* **45**(1), 109–110.

Poincaré, H.
[1893] Sur la généralisation d'un théorème d'Euler relatif aux polyèdres, *C.R. Acad. Sci. Paris* **177**, 144–145.

Recski, A.
[1995] ch. 36, this volume.

Reisner, G.A.
[1976] Cohen–Macaulay quotients of polynomial rings, *Adv. in Math.* **21**, 30–49.

Renegar, J.
[1992] The computational complexity and geometry of the first order theory of the reals, *J. Symbolic Comput.* **13**, 255–352.

Richter-Gebert, J.
[1994] *Realization spaces of S-polytopes are universal*, Manuscript.

Rosenfeld, M.
[1983] Are all simple 4-polytopes Hamiltonian? *Israel J. Math.* **46**, 161–169.
Roudneff, J.-P.
[1987] An inequality for 3-polytopes, *J. Combin. Theory B* **42**, 156–166.
Schrijver, A.
[1995] ch. 30, this volume.
Schulz, C.
[1979] An invertible 3-diagram with 8 vertices, *Discrete Math.* **28**, 201–205.
[1985] Dual pairs of non-polytopal diagrams and spheres, *Discrete Math.* **55**, 65–72.
Seidel, R.
[1987] *Output-size sensitive algorithms for constructive problems in computational geometry*, Ph.D. Thesis (Department of Computer Science, Cornell University, Ithaca, NY).
Shemer, I.
[1982] Neighborly polytopes, *Israel J. Math.* **43**, 291–314.
Shephard, G.C.
[1972] Sections and projections of convex polytopes, *Mathematika* **19**, 144–162.
Stanley, R.P.
[1975] The upper bound conjecture and Cohen–Macaulay rings, *Studia Appl. Math.* **54**, 135–142.
[1978] Hilbert functions of graded algebras, *Adv. in Math.* **28**, 57–83.
[1980] The number of faces of simplicial convex polytopes, *Adv. in Math.* **35**, 236–238.
[1985] The number of faces of simplicial polytopes and spheres, in: *Discrete Geometry and Convexity*, ed. J.L. Goodman, *Ann. New York Acad. Sci.*, pp. 212–223.
[1986] *Enumerative Combinatorics*, Vol. 1 (Wadsworth & Brooks/Cole, Monterey, CA).
[1987a] Generalized h-vectors, intersection cohomology of toric varieties, and related results, in: *Commutative Algebra and Combinatorics, Kyoto, 1985, Advanced Studies in Pure Mathematics*, Vol. 11, eds. M. Nagata and H. Matsumura (Kinokuniya/North-Holland, Tokyo/Amsterdam) pp. 187–213.
[1987b] On the number of faces of centrally-symmetric simplicial polytopes, *Graphs and Combinatorics* **3**, 55–66.
Steinitz, E.
[1906] Über die Eulersche Polyederrelationen, *Arch. Math. Phys.* **11**, 86–88.
Steinitz, E., and H. Rademacher
[1934] *Vorlesungen über die Theorie der Polyeder* (Springer, Berlin).
Sturmfels, B.
[1987a] Boundary complexes of convex polytopes cannot be characterized locally, *J. London Math. Soc.* **35**, 314–326.
[1987b] *Oriented matroids and combinatorial convex geometry*, Dissertation (Fachbereich Mathematik, Technische Hochschule, Darmstadt).
[1987c] On the decidability of Diophantine problems in combinatorial geometry, *Bull. Amer. Math. Soc.* **17**, 121–124.
[1988] Some applications of at affine Gale diagrams to polytopes with few vertices, *SIAM J. Discrete Math.* 1, 121–133.
Swart, G.
[1985] Finding the convex hull facet by facet, *J. Algorithms* **6**, 17–48.
Tarski, A.
[1951] *A Decision Method for Elementary Algebra and Geometry* (University of California Press, Berkeley, CA). Second, revised edition: 1961.
Thomassen, C.
[1995] ch. 5, this volume.
Todd, M.
[1980] The monotonic bounded Hirsch conjecture is false for dimension at least 4, *Math. Oper. Res.* **5**, 599–601.
Tutte, W.T.
[1956] A theorem on planar graphs, *Trans. Amer. Math. Soc.* **82**, 99–116.
[1962] A census of planar triangulations, *Canad. J. Math.* **14**, 21–38.

Vershik, A.M.
[1988] Topology of the convex polytopes' manifolds, the manifold of projective configurations of a given combinatorial type and representations of lattices, in: *Topology and Geometry – Rohlin Seminar, Lecture Notes in Mathematics,* Vol. 1346, eds. O.Ya. Viro and A.M. Vershik (Springer, Berlin).

Volodin, I.A., V.E. Kuznetsov and A.T. Fomenko
[1974] The problem of discriminating algorithmically the standard three-dimensional sphere, *Uspekhi Mat. Nauk* **29**, 72–168 [*Russian Math. Surveys* **29**(5), 71–171].

Wegner, G.
[1984] Kruskal–Katona's theorem in generalized complexes, in: *Finite and Infinite Sets,* Vol. 2, *Colloq. Math. Soc. János Bolyai* **37**, 821–827.

Welsh, D.J.A.
[1995] ch. 9, this volume.

Yemelichev, V.A.M., M.M. Kovalev and M.K. Kravtsov
[1981] *Polytopes, Graphs and Optimisation* (in Russian) (Nauka, Moscow). English translation: G.H. Lawden, 1984 (Cambridge University Press, Cambridge).

Ziegler, G.M.
[1994] *Lectures on Polytopes* (Springer, Berlin).

CHAPTER 19

Point Lattices

Jeffrey C. LAGARIAS

AT&T Bell Laboratories, Room 2C-373, 600 Mountain Avenue, Murray Hill, NJ 07974, USA

Contents

HANDBOOK OF COMBINATORICS
Edited by R. Graham, M. Grötschel and L. Lovász

1. Introduction

A *point lattice* L (called simply a *lattice* in the following) is a set of points in $\mathbb{R}^n$ which form a discrete additive subgroup of $\mathbb{R}^n$. That is, L satisfies the two conditions:

(1) *Additivity*. If $\boldsymbol{v}_1, \boldsymbol{v}_2 \in L$, then $-\boldsymbol{v}_1 \in L$ and $\boldsymbol{v}_1 + \boldsymbol{v}_2 \in L$;

(2) *Discreteness*. There exists $c_0 > 0$ such that for any $\boldsymbol{v}_1, \boldsymbol{v}_2 \in L$ with $\boldsymbol{v}_1 \neq \boldsymbol{v}_2$, $|\boldsymbol{v}_1 - \boldsymbol{v}_2| \geqslant c_0$, where $|\cdot|$ is the Euclidean norm.

Elements in L are called *lattice points* or *lattice vectors*. The most familiar lattice is the integer lattice $\mathbb{Z}^n$. See fig. 1.1.

Figure 1.1. Part of the lattice $\mathbb{Z}^2$.

Equivalently, a lattice may be viewed as a discrete group G_L of translations acting on $\mathbb{R}^n$, in which case the orbit of $\mathbf{0}$ under G_L, i.e., $L = \{g(\mathbf{0}): g \in G_L\}$, is a point lattice in the sense above, and L determines G_L. The group G_L is a discrete subgroup of the group $T(\mathbb{R}^n)$ of all translations. $T(\mathbb{R}^n)$ is naturally identified with $\mathbb{R}^n$, and is a subgroup of the group $\mathbb{E}_n$ of all Euclidean motions or "isometries" of $\mathbb{R}^n$, which in turn is a subgroup $\mathbb{A}_n$ of all affine motions of $\mathbb{R}^n$.

There are several classes of combinatorial problems involving lattices. These problems are related in one way or another to convex bodies. The first class of problems concerns the packing or covering of space with translates of a given convex body. One may consider the special subclass of such problems in which the set of translates forms a lattice, which are called *lattice packings* and *lattice coverings*, respectively. The *geometry of numbers* is a branch of number theory concerned with finding the best lattice packings and coverings. The second class of problems is algebraic, and studies the possible symmetries of lattices, and more generally the problem of classifying discrete subgroups G of the group of isometries $\mathbb{E}_n$ for which the quotient space $\mathbb{E}_n/G$ is compact. This is the subject matter of *mathematical crystallography*. It is also closely related to the problem of classifying possible *tilings* of spaces with congruent polytopes. Polytopes of special shape admitting lattice tiling of space are called *parallelotopes*, and one problem is to classify the possible combinatorial types of parallelotopes. A third class of problems concerns the study of properties of convex polytopes all of whose vertices are in the lattice $\mathbb{Z}^n$, which are called *lattice polytopes*. There are many interesting relations between geometric invariants of such polytopes such as mixed volumes and combinatorial invariants such as the counting functions of the number of points in lattices $(1/d)\mathbb{Z}^n$ in such polytopes. These include *combinatorial reciprocity laws*. A fourth class of problems concerns the determination of when a polytope contains a lattice point of $\mathbb{Z}^n$, which is the *integer program-*

ming problem. Many combinatorial problems are naturally formulated as integer programs and such a formulation is often useful in their solution. For example, set covering problems, matching problems, existence of finite projective planes, existence of specific t-designs, and the problem of finding maximal independent sets in graphs are all naturally encodable as 0–1 integer programming problems. General methods from the geometry of numbers lead to integer programming algorithms, see section 7. These problem classes are the ones surveyed in this article.

Point lattices are one of the most pervasive structures in mathematics. In number theory they arise in studying the integral equivalence of positive definite integral quadratic forms (see section 2), in the geometry of numbers and Diophantine approximation, in algebraic number theory and class field theory. In algebraic geometry elliptic curves and more generally Abelian varieties have the structure $\mathbb{C}^n/L$, where L is a lattice. Modular functions, modular forms and their generalizations are invariants of isomorphism types of lattices. The theory of toric varieties and the problem of finding the number of solutions of systems of algebraic equations both use invariants attached to lattice polyhedra (section 6). The creation of group theory was partially motivated by the study of crystallography. Many of the sporadic simple groups have been constructed as (part of) the automorphism groups of particular lattices, e.g., the Monster group is the group of automorphisms of a certain 196884-dimensional lattice. In Riemannian geometry all compact flat manifolds arise as quotients $\mathbb{E}_n/G$ for certain crystallographic groups G. More generally, lattices viewed as groups of translations generalize to the study of discrete subgroups of arbitrary Lie groups and their quotient spaces. The lattice structure is evident in Fourier series, which exist on $\mathbb{R}^n/\mathbb{Z}^n$, and harmonic analysis is an important tool in the study of lattices. Lattices and their cosets play a fundamental role in algebraic coding theory. Finally, lattices arise in numerical analysis in the construction of multivariate splines and in numerical integration. These areas are for the most part not treated here; references to some of them are provided in section 8.

The lattice concept originally arose in two different areas, number theory and crystallography. In number theory it appears in the reduction theory of binary quadratic forms, which was created by Lagrange in 1773. This has a lattice interpretation due to Gauss in 1831. The reduction theory of n-dimensional positive definite quadratic forms, which is essentially reduction theory for lattices, was developed by Hermite in 1850 and Korkin and Zolotarev (1873, 1877). The full power of the lattice concept in number theory was indicated in Minkowski's development of the geometry of numbers in 1896. In crystallography Bravais (1850) began the mathematical development. In 1869, Jordan classified certain groups of movements and L. Sohnke pointed out its application to crystallography in 1879. The classification of all 3-dimensional space groups, the ones of direct relevance to crystallography, was accomplished by Fedorov (1889) and later, independently, by Schoenflies and Barlow. In 1900 Hilbert raised the problem of N-dimensional analogues of crystallography in the 18th problem on his famous problem list. Bieberbach (1910a,b, 1912) gave a solution.

General books and surveys for the individual sections are as follows. For geometry of numbers see Cassels (1971), Gruber (1979), Gruber and Lekkerkerker 1987). For packing and covering see Baranovskii (1971), Conway and Sloane (1988a), Fejes-Toth (1972), Rogers (1964), Ryshkov and Baranovskii (1979), Saaty and Alexander (1975). For mathematical crystallography see Charlap (1986), Delone et al. (1973), Hilton (1903), Schwartzenberger (1980). For tilings in low dimensions see Grünbaum and Shepard (1987). For integer polytopes see Ewald (1985) and Gritzmann and Wills (1992). For algorithmic geometry of numbers see Kannan (1987b) and Lovasz (1986). Lists of problems concerning lattice points appear in Erdős et al. (1989) and Hammer (1977). See section 8 for references to other areas of mathematics.

It is impossible to give more than an overview of this area in a short space. I have especially benefitted from and borrowed from other surveys, e.g., Ewald (1985), Milnor (1976) and Rogers (1964).

2. Lattices, convex bodies and reduction theory

A *(point) lattice* L is a discrete additive subgroup of $\mathbb{R}^n$. Let $\text{lin}(L)$ denote the vector space spanned by the elements of L. Let $\text{rank}(L)$ denote the dimension of $\text{lin}(L)$, i.e., the number of linearly independent vectors in L. A lattice is of full rank if rank $(L) = n$, i.e., if $\text{lin}(L) = \mathbb{R}^n$.

An *(ordered) basis* $B = [\boldsymbol{b}_1, \ldots, \boldsymbol{b}_r]$ of L is a set of vectors in L with

$$L = \{n_1\boldsymbol{b}_1 + \cdots + n_r\boldsymbol{b}_r : (n_1, \ldots, n_r) \in \mathbb{Z}^r\},$$

where $r = \text{rank}(L)$. We represent vectors in L as row vectors, and a basis

$$B = \begin{bmatrix} \boldsymbol{b}_1 \\ \cdot \\ \cdot \\ \cdot \\ \boldsymbol{b}_r \end{bmatrix}$$

as an $r \times n$ matrix. If B is a basis of a lattice L, so is $B' = UB$ for any $U \in \text{GL}(r, \mathbb{Z})$, and conversely all possible bases of L are obtained in this fashion. We associate with a basis B the *fundamental parallelepiped*

$$F(B) = \{\boldsymbol{w} = \alpha_1\boldsymbol{b}_1 + \cdots + \alpha_r\boldsymbol{b}_r : 0 \leq \alpha_i \leq 1\}.$$

The translates of $F(B)$ by elements of L give a perfect tiling of $\text{lin}(L)$.

Definition 2.1. The *determinant* $\det(L)$ of a lattice L is

$$\det(L) = \det(BB^{\mathrm{T}})^{1/2},$$

where B is any basis of L.

The determinant of L is independent of the choice of basis, and gives the

r-dimensional volume of the fundamental parallelepiped $F(B)$ regarded as embedded in $\mathbb{R}^n$.

A lattice L_1 is a *sublattice* of a lattice L_0 if $L_1 \subseteq L_0$ and $\operatorname{rank}(L_1) = \operatorname{rank}(L_0)$. The *index* of L_1 in L_0 is

$$[L_0 : L_1] = \frac{\det(L_1)}{\det(L_0)} .$$

A sublattice L_1 of a lattice L_0 is completely specified by an integer matrix M of determinant $\pm[L_0 : L_1]$ which takes a fixed basis B of L_0 to a basis $\tilde{B}$ of L_1. There are only a finite number of sublattices L_1 in a lattice L_0 having a fixed index $[L_0 : L_1] = k$, which are classified by the following result.

Theorem 2.2 (Hermite normal form). *For each integer matrix M of nonzero determinant, there exists a unique matrix $\tilde{M} = UM$ with matrix $U \in \mathrm{GL}(n, \mathbb{Z})$ such that $\tilde{M}$ is lower triangular with positive diagonal elements, i.e.,*

$$\tilde{M} = UM = \begin{bmatrix} a_{11} & & \mathbf{0} \\ a_{21} & a_{22} & \\ \vdots & & \\ a_{n1} & \cdots & a_{nn} \end{bmatrix}$$

with all $a_{ii} > 0$ and $0 \leqslant a_{ij} < a_{jj}$ for $1 \leqslant j < i \leqslant n$. The matrix $\tilde{M}$ is said to be in Hermite normal form.

In particular for such a lattice L_0 the complete set of sublattices of index n are given by the set of bases $\tilde{B} = \tilde{M}B$ where $\tilde{M}$ is in Hermite normal form with $\det(\tilde{M}) = n$. There are only a finite number of these, see Newman (1972).

The simplest example of a lattice is the *integer lattice* $\mathbb{Z}^n$ in $\mathbb{R}^n$, which has full rank and determinant $\det(\mathbb{Z}^n) = 1$. A sublattice of the integer lattice is called an *integral lattice*.

The fundamental invariants of a lattice describe how its points sit in space with respect to a norm, and more generally with respect to homothetic copies λK of a general body K.

A *body* K is a connected subset of $\mathbb{R}^n$ which is the closure of its interior int (K). It is *convex* if $\boldsymbol{x}, \boldsymbol{y} \in K$ implies that $\lambda\boldsymbol{x} + (1-\lambda)\boldsymbol{y} \in K$ for $0 \leqslant \lambda \leqslant 1$. It is *symmetric* (often called **0**-*symmetric* in the literature) if $\mathbf{0} \in \operatorname{int}(K)$ and $\boldsymbol{x} \in K$ implies that $-\boldsymbol{x} \in K$. It is a *star body* if $\mathbf{0} \in \operatorname{Int}(K)$ and $\boldsymbol{x} \in K$ implies that $\lambda\boldsymbol{x} \in K$ for $0 \leqslant \lambda \leqslant 1$. For $\lambda \in \mathbb{R}$ the *scaled body* λK is defined by

$$\lambda K = \{\lambda\boldsymbol{x} : \boldsymbol{x} \in K\} .$$

Definition 2.3. The *Minkowski sum* $K_1 + K_2$ of two bodies is

$$K_1 + K_2 = \{\boldsymbol{x}_1 + \boldsymbol{x}_2 : \boldsymbol{x}_1 \in K_1 \text{ and } \boldsymbol{x}_2 \in K_2\} .$$

In particular the *difference body* $DK = K - K$ is the Minkowski sum of K and $-K$. A difference body DK is always symmetric, and is convex if K is convex.

A *norm* (or *distance function*) $|\cdot|_K$ is a map $\mathbb{R}^n \to \mathbb{R}$ satisfying

(1) (Triangle inequality) $|x + y|_K \leq |x|_K + |y|_K$.

(2) $|\lambda x|_K = |\lambda|\,|x|_K$.

(3) If $x \neq \mathbf{0}$ then $|x|_K > 0$.

A norm is completely determined by its *unit ball*

$$K = K(|\cdot|) = \{x\colon |x|_K \leq 1\},$$

since one has

$$|x|_K = \inf\{\lambda\colon x \in \lambda K\}. \tag{2.1}$$

The unit ball of a norm is always a symmetric convex body K, and conversely any such body gives rise to a norm by (2.1). In fact (2.1) can be used to define $|\cdot|_K$ for any star body K, but conditions (1) and (3) above do not necessarily hold.

The symbol $|\cdot|$ will denote the Euclidean norm $|x|^2 = \langle x, x\rangle$, which is induced from the *Euclidean inner product* $\langle x, y\rangle = x^{\mathrm{T}} y = \sum_{i=1}^{n} x_i y_i$.

The *volume* of a convex body K in $\mathbb{R}^n$ is denoted $\mathrm{vol}(K)$ or $\mathrm{vol}_n(K)$. The *Euclidean unit ball* $B_n = \{x\colon |x|^2 \leq 1\}$ has volume

$$\mathrm{vol}(B_n) = \frac{2\pi^{n/2}}{n\Gamma(\frac{1}{2}n)},$$

where $\Gamma(\cdot)$ denotes the Gamma function.

Definition 2.4. Given a symmetric convex body K in $\mathbb{R}^n$ the ith *successive minimum* $\lambda_i(K, L)$ of a lattice L for $1 \leq i \leq \mathrm{rank}(L)$ *with respect to* K is the minimum value of λ such that λK contains i linearly independent lattice points of L (inside or on its boundary). The ith *successive minimum* $\lambda_i(L)$ refers to the case $K = B_n$, i.e., the Euclidean norm case.

The successive minima measure to what extent the lattice vectors "stretch" in various directions with respect to the convex body K. Here $\frac{1}{2}\lambda_1(K, L)$ gives the maximum size λK of a homothetic copy of K that can be packed in space without overlap with centers located at the points of L. The integer lattice $\mathbb{Z}^n$ has $\lambda_1(\mathbb{Z}^n) = \lambda_2(\mathbb{Z}^n) = \cdots = \lambda_n(\mathbb{Z}^n) = 1$.

Lattices have a duality theory related to the duality theory in Fourier analysis on $\mathbb{R}^n/\mathbb{Z}^n$. The Euclidean inner product plays an important role in this duality.

Definition 2.5. The *dual lattice* L^* (also called *reciprocal lattice* or *polar lattice*) to L is

$$L^* = \{y\colon y \in \mathrm{lin}(L) \text{ and } \langle x, y\rangle \in \mathbb{Z} \text{ for all } x \in L\}.$$

One has $\mathrm{rank}(L^*) = \mathrm{rank}(L)$ and $\det(L^*) = \det(L)^{-1}$.

An analogous duality notion can be defined for symmetric convex bodies.

Definition 2.6. The *dual body* K^* or (*polar body*) to a convex body K containing $\mathbf{0}$ in its interior is

$$K^* = \{y: \langle x, y\rangle \leq 1 \text{ for all } x \in K\} .$$

If K is symmetric then K^* is symmetric. The norm $|\cdot|_{K^*}$ is said to be *dual* to $|\cdot|_K$. Note that the L^p-norm is dual to the L^q-norm in this sense, when $1/p + 1/q = 1$, for $1 \leq p, q \leq \infty$.

Theorem 2.7. *There is a constant* $0 < c < 1$ *such that for all* n, *any symmetric convex body in* $\mathbb{R}^n$ *satisfies*

$$c^n \operatorname{vol}(B_n)^2 \leq \operatorname{vol}(K) \operatorname{vol}(K^*) \leq \operatorname{vol}(B_n)^2 .$$

The upper bound is due to Santaló (1949) and the lower bound to Bourgain and Milman (1987). Still unsolved is a conjecture of Mahler (1939) that the best possible lower bound is $4^n/n!$, which is attained by

$$K = \left\{x \in \mathbb{R}^n: \sum_{i=1}^{n} |x_i| \leq 1\right\} .$$

There are general relations between the successive minima of a lattice and its dual lattice, with respect to a given pair of dual norms.

Theorem 2.8. *The successive minima* $\lambda_i(L)$ *of a rank* n *lattice* L *and of its reciprocal lattice* $\lambda_i(L^*)$ *are related by*

$$1 \leq \lambda_i(L)\lambda_{n-i+1}(L^*) \leq n$$

for $1 \leq i \leq n$. *More generally, for any symmetric convex body* K,

$$1 \leq \lambda_i(K, L)\lambda_{n-i+1}(K^*, L^*) \leq n^{3/2}$$

for $1 \leq i \leq n$.

Here the well-known lower bounds go back to Mahler (1939), while the upper bounds are proved in Banaszczyk (1993). The lower bounds are sharp, and the upper bounds are best possible up to a constant, because a result of Conway and Thompson (given in Milnor and Husemoller 1973, chapter II, Theorem 9.5) gives lattices $L_n = L_n^*$ in $\mathbb{R}^n$ with

$$\lambda_1(L_n)\lambda_1(L_n^*) \geq \left(\frac{n}{2\pi e}\right)(1 + o(1)) \quad \text{as } n \to \infty .$$

Reduction theory deals with the choice of a "good" basis of a lattice L among the set of all possible bases of L. We describe reduction theory for the case of a full rank lattice L. The set of possible ordered bases of all full rank lattices L in

$\mathbb{R}^n$ make up the group $\mathrm{GL}(n, \mathbb{R})$, and the set of ordered bases of a single lattice L is a $\mathrm{GL}(n, \mathbb{Z})$-orbit in $\mathrm{GL}(n, \mathbb{R})$. A *reduction theory* consists of a choice of reduction domain $\mathcal{R}_n$ for the quotient space $\mathrm{GL}(n, \mathbb{Z})\backslash\mathrm{GL}(n, \mathbb{R})$. A *reduction domain* $\mathcal{R}_n$ is an open set $\mathcal{R}$ in $\mathrm{GL}(n, \mathbb{R})$ such that each lattice has at least one and at most $c(n)$ bases in $\bar{\mathcal{R}}_n$, where $c(n)$ is a finite bound depending on $\mathcal{R}_n$. An ordered basis lying in a reduction domain is called *reduced* (in the sense of a particular reduction theory).

Two classical reduction theories are due to Korkin and Zolotarev (1873, 1877) and to Minkowski (1905). They were originally developed as reduction theories for positive definite quadratic forms, and had the object of choosing reduced quadratic forms with small coefficients. A positive definite quadratic form $q(\boldsymbol{x}) = \boldsymbol{x}^{\mathrm{T}}M\boldsymbol{x}$, where M lies in the cone $\mathcal{P}_n^+$ of positive definite symmetric matrices. A reduction theory for positive definite quadratic forms consists of a choice of a reduction domain $\mathcal{F}_n$ for $\mathrm{GL}(n, \mathbb{Z})$ acting on $\mathcal{P}_n^+$, treating $q(\boldsymbol{x})$ and $q(U\boldsymbol{x})$ for $U \in \mathrm{GL}(n, \mathbb{Z})$ as the same quadratic form. A quadratic form reduction theory always arises from a lattice basis reduction theory using the map $\phi: GL(n, \mathbb{R}) \to \mathcal{P}_n^+$ which has $\phi(B) = BB^{\mathrm{T}}$. The region $\phi^{-1}(\mathcal{F})$ in $\mathrm{GL}(n, \mathbb{R})$ is a reduction domain for lattice bases which is invariant under the orthogonal group $O(n, \mathbb{R})$, i.e., if $M \in \phi^{-1}(\mathcal{F})$ then $MO \in \phi^{-1}(\mathcal{F})$ for all $O \in O(n, \mathbb{R})$, see Cassels (1978).

An ordered basis $[\boldsymbol{b}_1, \ldots, \boldsymbol{b}_n]$ is *Minkowski-reduced* if $\boldsymbol{b}_1$ is a shortest (nonzero) vector in L and is inductively defined by the requirement that $\boldsymbol{b}_i$ be a shortest vector in L subject to the condition that $[\boldsymbol{b}_1, \ldots, \boldsymbol{b}_i]$ can be extended to a basis of L. The domain $\mathcal{F}_M$ of Minkowski-reduced bases has a simple shape.

Theorem 2.9. *The domain $\mathcal{F}_M$ of Minkowski-reduced lattice bases has the property that*

$$Q_M = \{BB^{\mathrm{T}} : B \in \mathcal{F}_M\}$$

is a closed polyhedral cone having a finite number of facets contained in the cone $\mathcal{P}_n^+$ of positive definite symmetric $n \times n$ matrices.

In particular, one can write down a finite set of inequalities for a basis $B = [\boldsymbol{b}_1, \ldots, \boldsymbol{b}_n]$ of a lattice L of the form

$$|m_1\boldsymbol{b}_1 + \cdots + m_n\boldsymbol{b}_n|^2 \leq |m'_1\boldsymbol{b}_1 + \cdots + m'_n\boldsymbol{b}_n|^2$$

where $\boldsymbol{m} = (m_1, \ldots, m_n)$ and $\boldsymbol{m}' = (m'_1, \ldots, m'_n)$ are in $\mathbb{Z}^n$, which are necessary and sufficient conditions for a $B \in \mathcal{F}_M$. For $n \leq 4$ they are

$$|\boldsymbol{b}_i|^2 \leq |\boldsymbol{b}_j|^2 \qquad 1 \leq i < j < n ,$$

$$|\boldsymbol{b}_i|^2 \leq \left|\boldsymbol{b}_i + \sum_{j=0}^{i-1} m_j\boldsymbol{b}_j\right|^2 \qquad 2 \leq i \leq n ,$$

where m_i take all possible values from 1, 0, -1, see Cassels (1978, p. 257).

To define Korkin–Zolotarev reduction we first define projected lattices. A *lattice subspace* W of L is a subspace spanned by lattice points in L. Let P_V denote orthogonal projection onto the subspace V of $\mathbb{R}^n$ and let $W^{\perp} = \{y: \langle x, y\rangle = 0\}$ be the orthogonal complement of W in $\mathbb{R}^n$ (with respect to the Euclidean inner product $\langle\cdot,\cdot\rangle$). If W is a lattice subspace of L then the *projected lattice* $\pi_{W^{\perp}}(L)$ is a lattice in $\mathbb{R}^n$ contained in $W^{\perp}$, where $\pi_{W^{\perp}}(\cdot)$ denotes orthogonal projection onto $W^{\perp}$. (If W is not a lattice subspace then $\pi_{W^{\perp}}(L)$ need not be a lattice.) An ordered basis $[\boldsymbol{b}_1, \ldots, \boldsymbol{b}_n]$ of a lattice L is *Korkin–Zolotarev reduced* if

(1) $\boldsymbol{b}_1$ is a shortest nonzero vector in L.

(2) For $W_i = \mathbb{R}[\boldsymbol{b}_1, \ldots, \boldsymbol{b}_i]$, the vector $\pi_{W_i^{\perp}}(\boldsymbol{b}_{i+1})$ is a shortest vector in the projected lattice $\pi_{W_i^{\perp}}(L)$, for $1 \leq i \leq n-1$,

(3) *Size-reduction condition. For* $1 \leq i < j \leq n$,

$$|\langle \pi_{W_{i-1}^{\perp}}(\boldsymbol{b}_i), \pi_{W_{i-1}^{\perp}}(\boldsymbol{b}_j)\rangle| \leq \tfrac{1}{2}|\pi_{W_{i-1}^{\perp}}(\boldsymbol{b}_i)|^2 ,$$

where $W_0^{\perp} = \mathbb{R}^n$.

One can show that almost all lattices have a unique choice of Minkowski-reduced basis (resp. Korkin–Zolotarev reduced basis), up to the choice of sign $\pm\boldsymbol{b}_i$ for each basis vector. Furthermore, no rank n lattice has more than $c(n)$ such reduced bases, where $c(n)$ grows like $2^{O(n^2)}$.

The length of basis vectors in such reduced bases can be bounded in terms of the successive minima $\lambda_i(L)$. There are very good bounds for Korkin–Zolotarev reduced bases, cf. Lagarias et al. (1990).

Theorem 2.10. *If* $[\boldsymbol{b}_1, \ldots, \boldsymbol{b}_n]$ *is a Korkin–Zolotarev reduced basis of a lattice L then*

$$\frac{4}{i+3}\,\lambda_i(L)^2 \leq |\boldsymbol{b}_i| \leq \frac{i+3}{4}\,\lambda_i(L)^2 .$$

Weaker bounds of a similar type are known for Minkowski-reduced bases, with constants depending exponentially on i.

We end this section with an analytic invariant of a lattice.

Definition 2.11. The *zeta function* $\zeta_L(s)$ of a lattice L is the Dirichlet series

$$\zeta_L(s) = \sum_{\substack{x \in L \\ x \neq 0}} |\boldsymbol{x}|^{-s} .$$

Here $\zeta_L(s)$ coincides with the *Epstein zeta function* of the quadratic form

$$Q(y_1, \ldots, y_n) = \left|\sum_{i=1}^{n} y_i \boldsymbol{b}_i\right|^2$$

where $[\boldsymbol{b}_1, \ldots, \boldsymbol{b}_n]$ is a basis of L.

Theorem 2.12. *The Dirichlet series for* $\zeta_L(s)$ *converges in the half-plane* $\mathrm{Re}(s) > r$, *where* $r = \mathrm{rank}(L)$. $\zeta_L(s)$ *analytically continues to* $\mathbb{C}$ *except for a simple pole at* $s = r$ *which has residue* $\det(L)^{-1}$. *It satisfies the functional equation*

$$\xi_L(s) = \xi_{L^*}(r - s)\,,$$

where L^* *is the dual lattice and*

$$\xi_L(s) = \det(L)^{-s}\,\pi^{-rs}\Gamma\left(\frac{s}{2}\right)^r \zeta_L(s)\,.$$

Several authors have studied the problem of finding lattices L whose zeta functions $\zeta_L(s)$ have certain extremal properties, e.g., maximizing or minimizing

$$\lim_{s\to r}\,(s - r)^{-1}\det(L)\zeta_L(s)$$

over all rank r lattices, cf. Ryshkov and Baranovskii (1979) and Gruber (1979).

3. Geometry of numbers

The geometry of numbers studies relations between lattices and convex bodies (more generally star bodies), which frequently are expressed as inequalities.

Minkowski's fundamental theorem 3.1. *Let* K *be a symmetric convex body and* L *be a full rank lattice in* $\mathbb{R}^n$. *If*

$$\mathrm{vol}(K) \geq 2^n \det(L)\,,$$

then K *contains a nonzero lattice point of* L *(either inside it or on its boundary). That is, one always has*

$$\lambda_1(K, L)^n\,\mathrm{vol}(K) \leq 2^n \det(L)\,.$$

Minkowski provided this by observing that, if K contained no such lattice point, then the translates of $\mathrm{int}(\frac{1}{2}K)$ by all members of L are mutually disjoint. Suppose K is contained in a ball of radius c_0. Now the volume covered by such translates inside a large cube of size $T + c_0$ is 2^{-n} vol(K)(# lattice points in T) while

$$(\#\ \text{lattice points in}\ T) = \frac{T^n}{\det(L)} + \mathrm{O}(T^{n-1}) \quad \text{as}\ T \to \infty\,.$$

Since the cube of side $T + c_0$ has volume $T^n + \mathrm{O}(T^{n-1})$ one obtains

$$2^{-n}\,\mathrm{vol}(K)\frac{T^n}{\det(L)} + \mathrm{O}(T^{n-1}) \leq T^n + \mathrm{O}(T^{n-1})$$

and letting $T \to \infty$ proves the first part of the theorem. Since $\mathrm{vol}(\lambda K) = \lambda^n\,\mathrm{vol}(K)$ the second part of the theorem is equivalent to the first, by rescaling K so that $\lambda_1(K, L) = 1$.

Siegel (1935) observed, using Fourier analysis techniques (Parseval's theorem), that one obtains a sharpening of Minkowski's fundamental theorem.

Theorem 3.2. *Let K be a symmetric convex body and L a full rank lattice in $\mathbb{R}^n$, and suppose K contains no nonzero lattice point of L. Then*

$$\mathrm{vol}(K) + 4^n \,\mathrm{vol}(K)^{-1} \sum_{b \in L^* - \{0\}} \left| \int_{K/2} \exp(-2\pi \mathrm{i} \langle \boldsymbol{b}, \boldsymbol{x} \rangle)\, \mathrm{d}x \right|^2 = 2^n \det(L)\,,$$

where L^* is the dual lattice of L.

Fourier analysis provides powerful techniques for proving results in the geometry of numbers, and conversely geometry of numbers suggests general inequalities in harmonic analysis, e.g., Bourgain and Milman (1987) and Banaszczyk (1989, 1991, 1993).

Van der Corput (1935) gave a different extension of Minkowski's theorem.

Van der Corput's theorem 3.3. *Let K be a symmetric convex body and L a full rank lattice in $\mathbb{R}^n$. If*

$$\mathrm{vol}(K) \geqslant k 2^n \det(L)\,,$$

then K contains at least k pairs $\pm \boldsymbol{v}$ of nonzero lattice points in L.

The central result in the geometry of numbers is the following strengthening of Minkowski (1896) to Theorem 3.1.

Minkowski's second fundamental theorem 3.4. *Let K be a symmetric convex body and L a full rank lattice in $\mathbb{R}^n$. Then*

$$\frac{2^n}{n!} \det(L) \leqslant \lambda_1(K, L) \lambda_2(K, L) \cdots \lambda_n(K, L)\, \mathrm{vol}(K) \leqslant 2^n \det(L)\,.$$

The upper bound is attained for $L = \mathbb{Z}^n$ when K is a cube, the lower bound when $K = \{\boldsymbol{x} \in \mathbb{R}^n : \sum_{i=1}^n |x_i| \leqslant 1\}$. Proofs of the more difficult upper bound appear in Bambah et al. (1965), Cassels (1971, p. 208) and Gruber and Lekkerkerker (1987, p. 59). A generalization of this theorem to adele spaces appears in McFeat (1971) and in Bombieri and Vaaler (1983) and another generalization appears in Woods (1966).

An inhomogeneous extension of Theorem 3.1 is due to Blichfeldt (1914).

Blichfeldt's theorem 3.5. *Let K be any measurable subset of $\mathbb{R}^n$ and L a full rank lattice of $\mathbb{R}^n$. If*

$$\mathrm{vol}(K) \geqslant \det(L)$$

then K contains two different points $\boldsymbol{v}_1$, $\boldsymbol{v}_2$ such that $\boldsymbol{v}_1 - \boldsymbol{v}_2 \in L$.

Bombieri (1962) proved a Fourier-analytic generalization of Theorem 3.5 analogous to Siegel's theorem 3.2.

Blichfeldt also proved a kind of converse to this result.

Theorem 3.6. *If a convex body K in $\mathbb{R}^n$ contains j lattice points from a lattice L and these span $\mathbb{R}^n$, then*

$$\mathrm{vol}(K) \geqslant \frac{j-n}{n!}\det(L)\,.$$

Weaker versions of Minkowski's fundamental theorem 3.1 have been proved for nonsymmetric convex bodies. Define the *coefficient of asymmetry* $\sigma(K)$ of a convex body K containing $\mathbf{0}$ by

$$\sigma(K) = \sup_{|x|=1} \frac{\max\{\lambda\colon \lambda \boldsymbol{x} \in K\}}{\max\{\lambda\colon -\lambda \boldsymbol{x} \in K\}}\,.$$

Then $\sigma(K) \geqslant 1$ and equality holds if and only if K is symmetric.

Theorem 3.7 (Sawyer 1954). *A convex body K in $\mathbb{R}^n$ containing $\mathbf{0}$ contains a nonzero point of a lattice L if*

$$\mathrm{vol}(K) \geqslant (1+\sigma(K))^n\left[1-\left(1-\frac{1}{\sigma(K)}\right)^n\right]\det(L)\,.$$

Next we consider quantities which are associated to lattice packings of bodies. A full rank lattice L is called *admissible* for a set K in $\mathbb{R}^n$ if $\mathrm{Int}(K)$ contains no lattice points of L except $\mathbf{0}$. That is, L is admissible for K if $\{\frac{1}{2}K + \boldsymbol{v}\colon \boldsymbol{v} \in L\}$ is a (lattice) packing in $\mathbb{R}^n$ with copies of $\frac{1}{2}K$.

Definition 3.8. The *lattice constant* $\Delta(K)$ of a set K is

$$\Delta(K) = \inf\{\det(L)\colon L \text{ is an admissible lattice for } K\}\,.$$

A lattice L is called *critical* for K if it is admissible and $\det(L) = \Delta(K)$.

Mahler (1946) proved a convergence theorem that is often useful in showing that critical lattices exist. The set of lattices $L^{(i)}$ *converges* to a lattice L if there exist bases $\{\boldsymbol{b}_1^{(i)}, \ldots, \boldsymbol{b}_n^{(i)}\}$ of $L^{(i)}$ and $\{\boldsymbol{b}_1, \ldots, \boldsymbol{b}_n\}$ of L such that $\boldsymbol{b}_j^{(i)} \to \boldsymbol{b}_j$ as $i \to \infty$. This defines a topology on the set of all lattices L.

Mahler's selection theorem 3.9. *Let a sequence $L^{(i)}$ of lattices be given. If the sequence of determinants* $\det(L^{(i)})$ *is bounded and if there is a ball centered at* $\mathbf{0}$ *such that all $L^{(i)}$ are admissible for it, then the sequence contains a convergent subsequence.*

In particular any bounded symmetric convex body K has a critical lattice.

There are general results giving upper and lower bounds for lattice constants.

The following result was conjectured by Minkowski and proved by Hlawka (1944).

Minkowski–Hlawka theorem 3.10. *Let K be a Jordan measurable subset of $\mathbb{R}^n$. If* $\mathrm{vol}(K) < 1$ *then there exists a lattice L with* $\det(L) < 1$ *admissible for K. If K is star-shaped and symmetric around* $\mathbf{0}$ *and*

$$\mathrm{vol}(K) < 2\zeta(n)$$

there these exists such a lattice. Here $\zeta(n) = \sum_{k=1}^{\infty} k^{-n}$.

Siegel (1945) derived Theorem 3.10 from a stronger result. He defined a natural measure m on the set of all lattices of determinant 1 (which is the double coset space $\mathrm{GL}(n, \mathbb{Z}) \backslash \mathrm{GL}(n, \mathbb{R}) / \mathbb{R}^*$). Call a lattice point $\boldsymbol{x} \in L$ *primitive* if there are no lattice points on the segment $[\mathbf{0}, \boldsymbol{x}] = \{\lambda \boldsymbol{x}: 0 \leqslant \lambda \leqslant 1\}$ except $\mathbf{0}$ and $\boldsymbol{x}$.

Siegel's mean value theorem 3.11. *Let* $f: \mathbb{R}^n \to \mathbb{R}$ *be Riemann integrable and nonnegative. Then*

$$\int \Big(\sum_{\boldsymbol{a} \in L - \{\mathbf{0}\}} f(\boldsymbol{a}) \Big) \, \mathrm{d}m(L) = \int_{\mathbb{R}^n} f(x) \, \mathrm{d}x ,$$

where the integral on the left is taken over all lattices of determinant 1. *In addition*

$$\int \Bigg(\sum_{\substack{\boldsymbol{a} \in L - \{\mathbf{0}\} \\ \boldsymbol{a} \text{ primitive}}} f(\boldsymbol{a}) \Bigg) \, \mathrm{d}m(L) = \frac{1}{\zeta(n)} \int_{\mathbb{R}^n} f(x) \, \mathrm{d}x .$$

This result implies that the set of admissible lattices of determinant 1 for a general Jordan measurable set K has Siegel-measure $\geqslant 1 - \mathrm{vol}(K)$, giving the first part of Theorem 3.10.

The Minkowski–Hlawka theorem gives a criterion for admissible lattices to exist, and establishes the upper bound

$$\Delta(K) \leqslant \mathrm{vol}(K) .$$

Schmidt (1963) obtained the following result using Siegel-measure estimates.

Theorem 3.12. *For a general Jordan measurable set K in* $\mathbb{R}^2$,

$$\Delta(K) \leqslant \tfrac{15}{16} \, \mathrm{vol}(K) .$$

For a general Jordan measurable set in $\mathbb{R}^n$,

$$\Delta(K) \leqslant \frac{1}{\alpha n - \beta} \, \mathrm{vol}(K)$$

where $\alpha = 0.3465$ *and* β *are constants.*

In fact one can often exhibit admissible lattices in other cases, e.g., the star body $\{(x, y): |xy| \leq 1\}$ has infinite volume yet has admissible lattices, see Theorem 3.17 below.

Minkowski's fundamental theorem 3.1 gives a lower bound for the lattice constant

$$\Delta(K) \geq 2^{-n} \operatorname{vol}(K) .$$

A more general lower bound problem arises in connection with the following special case of Minkowski's theorem 3.1.

Minkowski's linear form theorem 3.13. *Let L be a lattice with* $\det(L) = 1$ *and let c_i be positive numbers with $\prod_{i=1}^{n} c_i = 1$. Then there is a nonzero lattice point of L in the box*

$$|x_i| \leq c_i , \quad \text{for } 1 \leq i \leq n .$$

This result includes Dirichlet's theorem, which asserts that given any reals $(\theta_1, \ldots, \theta_n)$ and a bound N there exists a denominator $1 \leq Q \leq N$ and numerators $(P_1, \ldots, P_n)$ with

$$\left|\theta_i - \frac{P_i}{Q}\right| \leq N^{-1-1/n} .$$

This is obtained by applying Theorem 3.13 with $L = \{(k_0, k_0\theta_1 - k_1, \ldots, k_0\theta_n - k_n): \boldsymbol{k} \in \mathbb{Z}^{n+1}\}$, on taking $c_0 = N$ and $c_i = N^{-1/n}$ for $1 \leq i \leq n$. Mordell conjectured the following result, later proved by Siegel, cf. Gruber and Lekkerkerker (1987, section 24).

Converse to Minkowski's linear form theorem 3.14. *There is a constant $\kappa(n)$ such that for any full rank lattice L in $\mathbb{R}^n$ with* $\det(L) = 1$ *there exists a box*

$$|x_i| \leq c_i \quad \text{for } 1 \leq i \leq n ,$$

with

$$\prod_{i=1}^{n} c_i \geq \kappa(n)$$

such that L is admissible for this box. One has $\kappa(n) \geq (2^2 3^3 \cdots (n-1)^{n-1} n!)^{-1}$.

There has been a great deal of work devoted to finding critical lattices for specific bodies. Concerning critical lattices, Swinnerton-Dyer (1953) proved the following result.

Theorem 3.15. *Any critical lattice of a symmetric convex body K contains at least $n(n+1)/2$ pairs $\pm\boldsymbol{x}$ of boundary points of K, which include n linearly independent points.*

Minkowski conjectured a form for the critical lattices for boxes and this was proved by Hajós (1942).

Theorem 3.16. *A lattice L with* $\det(L) = 1$ *is a critical lattice for the cube if and only if after a suitable permutation of the n coordinates it has a basis of the form*

$$\begin{aligned} \boldsymbol{b}_1 &= (1, 0, \ldots, 0)\,, \\ \boldsymbol{b}_2 &= (\beta_{21}, 1, 0, \ldots, 0)\,, \\ &\;\vdots \\ \boldsymbol{b}_n &= (\beta_{n1}, \ldots, \beta_{n,n-1}, 1)\,. \end{aligned}$$

In general it seems a very hard problem to determine the critical lattices of a convex body, even for very symmetrical bodies such as spheres.

Some unbounded star bodies have the property that a finite number of critical lattices exist and any admissible lattice with determinant sufficiently close to that of a critical lattice must be a scaled version λL_c of some critical lattice. In such a case the critical lattices are said to be *isolated*. The simplest example of this phenomenon is associated to the star body $K_0 = \{(x, y)\colon |xy| \leq 1\}$.

Theorem 3.17. *The unbounded star body* $K_0 = \{(x, y)\colon |xy| \leq 1\}$ *has lattice constant* $\Delta(K_0) = \sqrt{5}$, *with exactly two critical lattices*:

$$L_1 = \mathbb{Z}\left[(1, 1), \left(\frac{1-\sqrt{5}}{2}, \frac{1+\sqrt{5}}{2}\right)\right],$$

$$L_2 = \mathbb{Z}\left[(1, 1), \left(\frac{1+\sqrt{5}}{2}, \frac{1-\sqrt{5}}{2}\right)\right].$$

If L is an admissible lattice for K_0 with $\det(L) < \sqrt{8}$ *then $L = \lambda L_i$ for $i = 1$ or* 2.

See Gruber and Lekkerkerker (1987, p. 426) for a proof. A closely related problem is the determination of the Markoff spectrum in one-dimensional Diophantine approximation, see Cassels (1957, 1971, chapter II.4).

4. Packings and coverings

Packing and covering problems often lead to combinatorial structures of interest. Extremal packings in small dimensions tend to have very regular structures, with large groups of symmetries.

A collection of translates $\mathcal{K} = \{K + v_i\colon i \geq 1\}$ of a body K is a *packing* if

$$\operatorname{int}(K + v_i) \cap \operatorname{int}(K + v_j) = \emptyset \quad \text{if } i \neq j\,.$$

It is a *lattice packing* $\mathcal{K}(K, L)$ by a lattice L if in addition the translates are the set of vectors in L. To a given collection of sets $\mathcal{K}$ and a cube $C(r)$ of side r centered at $\mathbf{0}$ we set

$$\rho_+(\mathcal{K}, r) = \frac{1}{r^n} \sum_{K+v_i \cap C(r) \neq \emptyset} \operatorname{vol}(K + v_i),$$

$$\rho_-(\mathcal{K}, r) = \frac{1}{r^n} \sum_{K+v_i \subseteq C(r)} \operatorname{vol}(K + v_i)$$

and define the *upper* and *lower asymptotic densities* of $\mathcal{K}$ by

$$\rho_+(\mathcal{K}) = \limsup_{r\to\infty} \rho_+(\mathcal{K}, r),$$

$$\rho_-(\mathcal{K}) = \liminf_{r\to\infty} \rho_-(\mathcal{K}, r).$$

A packing $\mathcal{K}$ is said to have *density* $\rho_+(\mathcal{K})$. A lattice L packs K if $\lambda_1(K, L) \geq 2$. For lattice packings $\mathcal{K}(K, L)$ one has

$$\rho_+(\mathcal{K}(K, L)) = \rho_-(\mathcal{K}(K, L)) = \frac{\operatorname{vol}(K)}{\det(L)}.$$

Definition 4.1. The *packing constant* $\delta(K)$ for a body K is

$$\delta(K) = \sup\{\rho_+(\mathcal{K}) \colon \mathcal{K} \text{ a packing of } \mathcal{K}\}.$$

The *lattice packing constant* $\delta_L(\mathcal{K})$ for K is

$$\delta_L(K) = \sup\{\rho_+(\mathcal{K}(K, L)) \colon L \text{ a lattice packing for } K\}.$$

A symmetric convex body K in $\mathbb{R}^n$ has a lattice packing with a lattice L if and only if L is admissible for the body $2K$, whence

$$\delta_L(K) = \frac{\operatorname{vol}(K)}{\Delta(2K)} = 2^{-n} \frac{\operatorname{vol}(K)}{\Delta(K)}.$$

The difficulty of packing problems grows with the dimension, and much more is known in two dimensions than in higher dimensions.

Theorem 4.2. *For any convex body K in $\mathbb{R}^2$,*

$$\delta(K) = \delta_L(K).$$

If $H(K)$ denotes the volume of the smallest hexagon circumscribing K then

$$\delta(K) \leq \frac{\operatorname{vol}(K)}{H(K)}.$$

If K is also symmetric then

$$\delta(K) = \delta_L(K) = \frac{\operatorname{vol}(K)}{H(K)}.$$

The history of these results, which are due to Rogers, Fejes-Toth, and Reinhart, is detailed in Rogers (1964, chapter 1).

The least dense packing of any convex body in $\mathbb{R}^2$ is given by a triangle.

Theorem 4.3. *For any convex body K in $\mathbb{R}^2$*

$$\delta_L(K) = \delta(K) \geq \tfrac{2}{3} ,$$

with equality if and only if K is a triangle.

The problem of finding dense packings of spheres (i.e., of balls B_n) first arose in bounding the minimum of a positive definite quadratic form in terms of its determinant, and more recently in connection with coding theory, see Conway and Sloane (1988a). It appears as part of Hilbert's 18th problem. The densest sphere packing in $\mathbb{R}^2$ is the hexagonal packing, and by Theorem 4.2 its density is

$$\delta(B_2) = \frac{\pi}{2\sqrt{3}} = 0.9069 \ldots .$$

The densest sphere packing is not determined in any dimension ≥ 3. It remains a notorious unsolved problem (Kepler conjecture) to show that $\delta(B_3) = \delta_L(B_3)$, i.e., that the "cannonball" packing is the densest packing of 3-dimensional spheres. It is known that

$$0.7407 \ldots = \frac{\pi}{3\sqrt{2}} = \delta_L(B_3) \leq \delta(B_3) \leq 0.7730 \ldots .$$

For the latest word on this problem see Hales (1994). The relation of lattice packings of spheres to quadratic forms is incorporated in the following definition.

Definition 4.4. *Hermite's constant* γ_n is the smallest constant such that

$$\lambda_1(L)^2 \leq \gamma_n(\det(L))^{2/n}$$

holds for all lattices L of rank n.

That is, every positive definite quadratic form of determinant one takes on a nonzero value $\leq \gamma_n$ at an integer point. This shows that the lattice constant $\Delta(B_n) = \gamma_n^{-n/2}$ and that

$$\delta_L(B_n) = 2^{-n}\gamma_n^{n/2} \operatorname{vol}(B_n) .$$

The densest lattice packings of spheres and Hermite's constants γ_n are known in all dimensions ≤ 8, see table 4.1. Table 4.1 also lists critical lattices, using the notation of Conway and Sloane (1988a, chapter 5). This notation reflects the fact that these lattices are associated to root systems of certain Lie algebras; however, the association of critical lattices and Lie algebras breaks down in high dimensions.

Both lower and upper bounds are known for sphere-packing densities.

Table 4.1
Densest lattice packings with balls B_n

dim	Critical lattice	$(\gamma_n)^n$	$\delta_L(B_n)$
1	Z	1	1
2	A_2	$\frac{4}{3}$	$\frac{\pi}{2\sqrt{3}} = 0.90690$
3	A_3	2	$\frac{\pi}{3\sqrt{2}} = 0.74048$
4	D_4	4	$\frac{\pi^2}{16} = 0.61685$
5	D_5	8	$\frac{\pi^2}{15\sqrt{2}} = 0.46526$
6	D_6	$\frac{64}{3}$	$\frac{\pi^3}{48\sqrt{3}} - 0.37295$
7	E_7	64	$\frac{\pi^3}{105} = 0.29530$
8	E_8	128	$\frac{\pi^4}{384} = 0.25367$

Theorem 4.5. *There is a positive constant c_0 so that*

$$\delta_L(B_n) \geqslant \frac{c_0 n}{2^n},$$

so that one has

$$\gamma_n \geqslant \frac{n}{2\pi e}(1 + o(1)) \cong 0.05855n(1 + o(1)) \quad \text{as } n \to \infty .$$

Schmidt (1958) proved the same lower bound for $\delta_L(K)$ for any symmetric convex body K, for any $c_0 < \log 2$ and all sufficiently large n. His proof shows that a "random" lattice L drawn from the set of all lattices of determinant 1 has a suitably large value of $\lambda_1(L)$.

The best upper bound is due to Kabatyanskii and Levenshtein (1978), and arose from studying problems in coding theory.

Theorem 4.6. *As $n \to \infty$ one has*

$$\delta(B_n) \leqslant 2^{-(0.5990 + o(1))n},$$

so that

$$\gamma_n \leqslant 0.07731n(1+o(1)), \quad \text{as } n \to \infty .$$

The methods used to obtain these bounds were successfully applied by Levenshtein and Odlyzko and Sloane to solve the *kissing number problem* in dimensions 8 and 24, which is the problem of finding the maximum number of unit balls B_n that can touch one unit ball, see Conway and Sloane (1988a).

Theorem 4.7. *The maximum number of unit balls B_n that can touch a given unit ball in dimensions* 1, 2, 3, 8 *and* 24 *is* 2, 6, 12, 240, *and* 196560, *respectively.*

These solutions are unique up to rotations in dimensions 1, 2, 8 and 24, see Bannai and Sloane (1981). That the kissing number problem can be solved in dimensions 8 and 24 is fortuitous, and arises because particularly dense lattice sphere packings exist in these dimensions, which have large automorphism groups and attain the bound. These are the lattices E_8 and the Leech lattice Λ_{24}.

Next we consider packings of nonsymmetric bodies K. Minkowski observed that this case may in principle be reduced to that of symmetric bodies by using the *difference body* $DK = K - K$.

Lemma 4.8. *Any convex body K in $\mathbb{R}^n$ has*

$$\delta(K) = 2^n \frac{\text{vol}(K)}{\text{vol}(DK)} \delta(DK) ,$$

and

$$\delta_L(K) = 2^n \frac{\text{vol}(K)}{\text{vol}(DK)} \delta_L(DK) .$$

In applying this result it is useful to have bounds relating the volumes vol(K) and vol(DK), see Rogers (1964, Theorem 2.4).

Theorem 4.9. *Any convex body K in $\mathbb{R}^n$ has*

$$2^n \leqslant \frac{\text{vol}(DK)}{\text{vol}(K)} \leqslant \binom{2n}{n}$$

with equality on the right if and only if K is a simplex.

It seems possible that simplices are the most difficult convex bodies to pack. This is true for $\mathbb{R}^2$ by Theorem 4.3.

Theorem 4.10. *For a simplex T_n one has*

$$\frac{2(n!)^2}{(2n)!} \leqslant \delta_L(T_n) \leqslant \delta(T_n) \leqslant \frac{2^n(n!)^2}{(2n)!} .$$

This result is due to Rogers and Shephard, see Rogers (1964, chapter 6).

It is widely believed that in sufficiently large dimensions n there exist convex bodies K with $\delta_L(K) < \delta(K)$, but this has never been proved. Rogers (1964) conjectures that $\delta_L(B_n) < \delta(B_n)$ for sufficiently large n.

A set $\mathcal{K} = \{K + v_i : i \geq 1\}$ of translates of a body K is a *covering* if all points of $\mathbb{R}^n$ are in some translate. If in addition v_i are the vectors in a lattice L then it is a *lattice covering* $\mathcal{K}(K, L)$. A covering is said to have *density* $\rho_-(\mathcal{K})$.

Definition 4.11. The *covering radius* $\mu(K, L)$ of a body K by a lattice L is

$$\mu(\mathcal{K}, L) = \inf\left\{t : \mathbb{R}^n = \bigcup_{v \in L} (tK + \boldsymbol{v})\right\}.$$

The *covering radius* $\mu(L)$ of a lattice L is $\mu(B_n, L)$.

Alternatively, the covering radius $\mu(K, L)$ is the supremum of values t such that some translate $tK + v$ contains no lattice points.

A lattice L gives a lattice covering of a symmetric body K if $\mu(K, L) \leq 1$. For lattice coverings $\mathcal{K}(K, L)$ one has

$$\rho_-(\mathcal{K}(K, L)) = \mathrm{vol}(K) \min_L \left\{\frac{\mu(K, L)^n}{\det(L)}\right\}.$$

Definition 4.12. The *covering constant* $\vartheta(K)$ for a body K is

$$\vartheta(K) = \inf\{\rho_-(\mathcal{K}) : \mathcal{K} \text{ a covering with } K\}.$$

The *lattice covering constant* $\vartheta_L(K)$ is

$$\vartheta_L(K) = \inf\{\rho_-(\mathcal{K}(K, L)) : \mathcal{K} \text{ a lattice covering with } K\}.$$

The covering problem is most well understood in two dimensions. Results of Fary and Fejes-Toth give the following theorem, see Rogers (1964).

Theorem 4.13. *For any convex body in* $\mathbb{R}^2$

$$\frac{\mathrm{vol}(K)}{h_s(K)} \geq \vartheta_L(K) \geq \vartheta(K) \geq \frac{\mathrm{vol}(K)}{h(K)},$$

where $h(K)$ is the area of the largest hexagon inscribable in K, $h_s(K)$ is the area of the largest symmetric hexagon inscribable in K. Hence for symmetric bodies K,

$$\vartheta_L(K) = \vartheta(K) = \frac{\mathrm{vol}(K)}{h(K)}.$$

Furthermore, one has the following.

Theorem 4.14. *For any convex body in* $\mathbb{R}^2$,

$$\vartheta_L(K) \leq \tfrac{3}{2},$$

with equality if and only if K *is a triangle.*

Fejes-Toth also showed that the circle gives the most dense lattice covering among symmetrical convex bodies, i.e.,

$$\vartheta_L(K) \leq \frac{2\pi}{3\sqrt{3}}$$

if K is such a body.

The value of $\vartheta_L(B_n)$ giving the thinnest lattice covering with balls is known for $n \leq 5$. In each case the extremal lattice is a scaled version of A_n^*, the dual of the lattice A_n, which is the n-dimensional lattice in $\mathbb{R}^{n+1}$ defined by

$$A_n = \{x_0, x_1, \ldots, x_n) \in \mathbb{Z}^{n+1} : x_0 + x_1 + \cdots + x_n = 0\}$$

see Conway and Sloane (1988a, chapter 2).

Rogers obtained good upper bounds for covering densities and for lattice covering densities in arbitrary dimension.

Theorem 4.15. (Rogers 1957). *For any convex body* K *in* $\mathbb{R}^n$ *with* $n \geq 3$,

$$\vartheta(K) \leq n \log n + n \log\log n + 5n .$$

Rather weaker density bounds are known for lattice coverings.

Theorem 4.16 (Rogers 1959). *For any convex body* K *in* $\mathbb{R}^n$ *with* $n \geq 3$,

$$\vartheta_L(K) \leq n^{\log_2 \log n + c} .$$

for some constant c. *For the ball* B_n *one has*

$$\vartheta_L(B_n) \leq cn(\log n)^{[\log_2(\pi e)]/2}$$

for some constant c.

For the unit ball B_n, Coxeter et al. (1959) obtained a lower bound for the covering density.

Theorem 4.17. *One has*

$$\vartheta_L(B_n) \geq \vartheta(B_n) \geq \frac{n}{e\sqrt{e}}(1 + o(1))$$

as $n \to \infty$.

Transference theorems relate lattice packings of a symmetric body K by a lattice L to lattice coverings of the dual body K^* by the dual lattice L^*.

Theorem 4.18. *For the ball B_n one has*

$$\mu(B_n, L)^2 \lambda_1(B_n, L^*)^2 \leqslant \tfrac{1}{2}\Big(\sum_{i=1}^{n} (\gamma^*_{n-i})^2\Big) \leqslant \tfrac{1}{2} n^3 .$$

where $\gamma^*_i = \max(\gamma_1, \ldots, \gamma_i)$. For a general (not necessarily symmetric) body K in $\mathbb{R}^n$ one has

$$\mu(K, L)^2 \lambda_1(K^*, L)^2 \leqslant c_0 n^4 ,$$

for an absolute constant c_0.

The first bound appears in Lagarias et al. (1990), and the second follows from Kannan and Lovász (1988, Lemma 2.3 and Theorem 2.7).

5. N-dimensional crystallography and tilings

Crystalline structures provided one inspiration for formulating the concept of symmetry groups and for studying point lattices. The existence of crystals poses two distinct mathematical problems. The first is that of classifying the possible underlying arrangements in space of the atoms making up a crystal. This converts to the mathematical problem of classifying (up to isomorphism type) the possible discrete subgroups of the group of Euclidean motions $\mathbb{E}_3$ in $\mathbb{R}^3$ which have compact quotient space. The second problem is that of classifying the possible polyhedral shapes that crystals may take in nature. A third problem, not studied here, is the problem of inferring from physical measurements (e.g., angles of cleavage planes in crystals, X-ray diffraction patterns) a basis of the underlying lattice of atoms making up the crystal, which is of great interest to chemists.

The first question was initially studied by Bravais (1850), in the form of classifying the possible discrete subgroups of Euclidean motions $\mathbb{E}_3$ (space groups). Here the group $\mathbb{E}_n$ of Euclidean motions in $\mathbb{R}^n$ is the set of linear transformations

$$E(\boldsymbol{x}) = Q\boldsymbol{x} + \boldsymbol{n} \quad \boldsymbol{x} \in \mathbb{R}^n ,$$

where $Q \in O(n, \mathbb{R})$ is an orthogonal matrix and $\boldsymbol{n} \in \mathbb{R}^n$ is a translation. Early authors did not explicitly introduce the group structure but instead used the combined operations of rotations, reflections and translations. A complete classification of all possible 3-dimensional space groups was achieved by Fedorov (1885, 1889) in Russia and later independently by Schoenflies (1889, 1891) in Germany, and Barlow in England; they found 230 types of space groups. The methods used in this classification were special to three dimensions. Hilbert (1900) posed as the 18th of his celebrated problems that of showing the finiteness of the classification problem for the N-dimensional version of this problem. This was solved by Bieberbach (1910a,b, 1912), who proved three major theorems.

Bieberbach's first theorem 5.1. *Every discrete subgroup Γ of $\mathbb{E}_n$ having compact fundamental domain contains n linearly independent translations.*

Bieberbach's second theorem 5.2. *The subgroup T consisting of all translations $\boldsymbol{x} \mapsto \boldsymbol{x} + \boldsymbol{n}$ in Γ is a free abelian normal subgroup of Γ of finite index, and it is a maximal abelian subgroup of Γ.*

Hence the finite group $\Phi = \Gamma / T$ has a trivial center, and its action on $T \cong \mathbb{Z}^n$ by inner automorphisms defines a faithful representation of Φ in $\mathrm{GL}(n, \mathbb{Z})$. Bieberbach then used a result of Minkowski stating that there is an upper bound on the order of any finite subgroup of $\mathrm{GL}(n, \mathbb{Z})$ to prove the following result.

Finiteness theorem 5.3. *For each fixed n there are only finitely many isomorphism classes of groups Γ which can be obtained as an extension of $\mathbb{Z}^n$ by a finite subgroup of* $\mathrm{GL}(n, \mathbb{Z})$.

Such groups are now called *Bieberbach groups*. The embedding of such groups Γ into $\mathbb{E}_n$ is *not* uniquely determined up to conjugation $g\Gamma g^{-1}$ by an element $g \in \mathbb{E}_n$. However, Bieberbach (1912) proved the following result for the affine group $\mathbb{A}_n = \{h\colon h(\boldsymbol{x}) = M\boldsymbol{x} + \boldsymbol{n},\ M \in \mathrm{GL}(n, \mathbb{R}),\ \boldsymbol{n} \in \mathbb{R}^n\}$.

Bieberbach's third theorem 5.4. *Two discrete subgroups Γ_1 and Γ_2 of $\mathbb{E}_n$ having compact fundamental domain are isomorphic if and only if they are conjugate subgroups of $\mathbb{A}_n$, i.e., there exists $h \in \mathbb{A}_n$ such that $\Gamma_2 = h\Gamma_1 h^{-1}$.*

For proofs of Bieberbach's thoerems, see Buser (1985) or Charlap (1986). Zassenhaus (1948) pointed out much later that a converse to the Finiteness theorem 5.3 holds.

Theorem 5.5 (Zassenhaus). *Any group Γ which can be described as an extension of $\mathbb{Z}^n$ by a finite subgroup Φ of* $\mathrm{GL}\ (n, \mathbb{Z})$ *can actually be embedded as a discrete subgroup of the Euclidean group $\mathbb{E}_n$ so as to act discontinuously on $\mathbb{R}^n$ with compact fundamental domain.*

In the process of proving his third theorem, Bieberbach had to show that any finite group Φ has only finitely many embeddings into $\mathrm{GL}(n, \mathbb{Z})$ up to conjugating by an element of $\mathrm{GL}\ (n, \mathbb{Z})$, a result now called the Jordan–Zassenhaus theorem, after Zassenhaus (1938), although Bieberbach apparently first formulated it.

The number of Bieberbach groups are shown in table 5.1. This count of Bieberbach groups does not distinguish between two groups of motions that are mirror images of each other. If mirror images are counted as distinct, one adds 11

Table 5.1
Bieberbach groups

Dimension n	Number of distinct groups
1	2
2	17
3	219
4	4783

more groups to obtain the 230 space groups counted by crystallographers. The 4-dimensional result appears in Brown et al. (1978).

The two-dimensional Bieberbach groups are given in table 5.2, taken from Milnor (1976). The finite quotient Φ is either cyclic, consisting only of rotations, or dihedral, containing reflections also. (Of course the dihedral group of order 2 is isomorphic to the cyclic group of order 2, but these two groups are listed separately in this table, since they are not conjugate as subgroups of $GL(2, \mathbb{R})$.)

The 54 Bieberbach groups consisting only of orientation-preserving elements ($\det Q = 1$) are given in table 5.3. In addition there are 165 Bieberbach groups with orientation reversing elements. The quantities in parentheses give extra groups obtained if mirror image groups are counted as distinct.

An important open problem remaining is to prove appropriate analogues of Bieberbach's theorems that classify discrete subgroups of the affine group $\mathbb{A}_n$ for $n \geq 4$, see Charlap (1986). The case $n = 3$ is treated in Fried and Goldman (1983).

Now we turn to the second problem, that of classifying the possible polyhedral shapes that crystals may take. Crystalline structures fracture in preferential directions, called *cleavage planes*. It is experimentally observed that such cleavage planes satisfy a *law of rational indices*, which is as follows. Given three linearly independent cleavage planes L_1, L_2 and L_3, let ℓ_{12}, ℓ_{13} and ℓ_{23} denote lines through the origin $\mathbf{0}$ parallel to the lines of intersection $L_1 \cap L_2$, $L_1 \cap L_3$ and $L_2 \cap L_3$ respectively. Given any cleavage plane L let d_{12}, d_{13}, d_{23} denote the

Table 5.2
Bieberbach groups in the plane

Group Φ	Order	Number of embeddings in $GL(2, \mathbb{Z})$	Number of Bieberbach groups
Cyclic	1	1	1
Cyclic	2	1	1
Cyclic	3	1	1
Cyclic	4	1	1
Cyclic	6	1	1
Dihedral	2	2	3
Dihedral	4	2	4
Dihedral	6	2	2
Dihedral	8	1	2
Dihedral	12	1	1
Total		13	17

Table 5.3
Orientation-preserving Bieberbach groups in $\mathbb{R}^3$

Group Φ	Order	Number of embeddings in SL(3, $\mathbb{Z}$)	Number of Bieberbach groups
Cyclic	1	1	1
Cyclic	2	2	3
Cyclic	3	2	3 (+1)
Cyclic	4	2	5 (+1)
Cyclic	6	1	4 (+2)
Dihedral	4	4	9
Dihedral	6	3	5 (+2)
Dihedral	8	2	8 (+2)
Dihedral	12	1	4 (+2)
Tetrahedral	12	3	5
Octahedral	24	3	7 (+1)
Total		24	54 (+11)

distance from the origin $\mathbf{0}$ of the intersection $L \cap \ell_{ij}$ (which is ∞ if undefined.) Then the ratios d_{13}/d_{12} and d_{23}/d_{12} are rational numbers. The law of rational indices has a second formulation. Given a lattice Λ, we say an r-dimensional flat is *in a lattice direction* of Λ if it is parallel to an r-dimensional lattice flat. Then a polyhedron P satisfies the law of rational indices (in $\mathbb{R}^3$) if there is a lattice Λ such that each face and edge of P is in a lattice direction of Λ. This definition of the law of rational indices extends to n dimensions in an obvious way.

It is easy to see that any polytope in $\mathbb{R}^n$ whose vertices are lattice points of Λ, which we call a *lattice polytope*, satisfies the law of rational indices. There are an infinite number of combinatorially inequivalent types of such polytopes in all dimensions ≥ 2.

Cleavage planes of crystals in nature satisfy additional restrictions, apparently related to the requirement that the atoms in the underlying crystal lattice be closely spaced in the direction of the plane of cleavage; this leads the rational numbers that occur in the law of rational indices to have small denominators. A mathematically precise version of characterizing all polytopes using such conditions has apparently not been formulated. Instead, a related, narrower, mathematical problem was formulated by the crystallographer Fedorov, which is that of characterizing those convex polytopes that tile $\mathbb{R}^3$ by translations.

Definition 5.6. A closed convex polytope P is a *parallelotope* if it tiles $\mathbb{R}^n$ by translations (except for boundary points). It is *primitive* if each $(n-1)$-dimensional face coincides with the face of one translate $P + \boldsymbol{u}$ for $\boldsymbol{u} \in \Lambda$, and if each vertex belongs to $n+1$ edges of bodies $P + \boldsymbol{u}$.

The study of parallelotopes was initiated by Fedorov (1889, 1949). He showed that in two and three dimensions any parallelotope has a center about which it is centrally symmetric, hence its facets occur in pairs that are parallel. This motivates the choice of the name parallelotope. Two tilings by parallelotopes are *isomorphic* if there is an invertible affine transformation taking one to the other. Fedorov enumerated all parallelotopes in $\mathbb{R}^3$. One source of parallelotopes is as follows. Given a lattice Λ and an invertible $n \times n$ matrix A the *Voronoi region* or *honeycomb* associated to (A, Λ) is

$$V(A, \Lambda) = \{x \in \mathbb{R}^n : |Ax| \leq |A(x - u)| \text{ for all nonzero } u \in \Lambda\} .$$

Any Voronoi region is a parallelotope. Voronoi (1908) proved a partial converse.

Theorem 5.7 (Voronoi). *Every primitive parallelotope of* $\mathbb{R}^n$ *with center at* $\mathbf{0}$ *is the Voronoi region* $V(A, \Lambda)$ *for some lattice* Λ *and invertible matrix* A. *Every primitive parallelotope has exactly* $2(2^n - 1)$ *facets.*

This leads to the following famous problem.

Problem of Voronoi 5.8. Is every parallelotope in $\mathbb{R}^n$ a Voronoi region?

This was proved true for dimensions $n \leq 4$ by Delone (1926). A sharpening of this result is due to Zitomirskii (1929).

Theorem 5.9. *Every parallelotope in* $\mathbb{R}^n$ *whose* $(n-2)$*-dimensional faces each have* $(n-2)$*-dimensional intersection with and are a face of exactly* 2 *other tiles in the tiling is a Voronoi region.*

R. Penrose discovered finite sets of polygonal tiles in $\mathbb{R}^2$ that tile the plane, for which all such tilings are nonperiodic. De Bruijn (1981) developed an algebraic theory for such tiles and tilings, and shows that that can be constructed from using projections from lattices in higher dimensions, e.g., a lattice in $\mathbb{R}^5$ is used to construct Penrose tiles. The recent discovery of *quasicrystals* has a possible explanation using such nonperiodic tilings of $\mathbb{R}^3$, see De Bruijn (1986) and Gratias and Michel (1986) and Katz (1992).

6. Lattice polytopes

A *lattice polytope* or *integer polytope* is a convex polytope in $\mathbb{R}^n$ all of whose vertices are in the integer lattice $\mathbb{Z}^n$. In recent years many results have been proved relating the number of lattice points in such bodies K to geometric invariants such as its volume, $\mathrm{vol}_n(K)$.

Definition 6.1. For a lattice polytope P (not necessarily full-dimensional) the *lattice point enumerator* $L(P)$ counts the number of lattice points in P. Let $L^\circ(P)$

count the number of lattice points in the relative interior of P. Then

$$\dot{L}(P) = L(P) - L^{\circ}(P) = L(\partial P)$$

counts the number of lattice points on the boundary ∂P of P.

Definition 6.2. For a point $\boldsymbol{x}$ in an arbitrary polytope P of dimension d in $\mathbb{R}^n$ the *normalized internal angle* $\alpha(\boldsymbol{x}, P)$ of $\boldsymbol{x}$ is defined by

$$\alpha(\boldsymbol{x}, P) = \lim_{\varepsilon \to 0} \frac{\mathrm{vol}_d((\varepsilon B_d + \boldsymbol{x}) \cap P)}{\mathrm{vol}_d(\varepsilon B_d)} .$$

Then for a lattice polytope P the total *lattice internal angle function* $\hat{L}(P)$ is

$$\hat{L}(P) = \sum_{\boldsymbol{x} \in P \cap \mathbb{Z}^n} \alpha(\boldsymbol{x}, P) .$$

The geometric invariants considered are mostly generalizations of the notion of volume, which are special cases of mixed volumes.

Definition 6.3. Given convex bodies $K_1, \ldots, K_s$ in $\mathbb{R}^n$ the *mixed volumes* $V_n(K_{i_1}, \ldots, K_{i_s})$ are given by

$$\mathrm{vol}_n(t_1 K_1 + \cdots + t_s K_s) = \sum_{0 \leq i_1 \leq \cdots \leq i_s \leq n} \binom{n}{i_1, \ldots, i_s} V_n(K_{i_1}, \ldots, K_{i_s}) t_{i_1} t_{i_2} \cdots t_{i_s} ,$$

for all $t_1 \geq 0, \ldots, t_s \geq 0$.

A proof that mixed volumes are well-defined is given in Eggleston (1958). Note that $V_n(K, K, \ldots, K) = \mathrm{vol}_n(sK)$ is the ordinary n-dimensional volume of sK.

A special case of a mixed volume is the following.

Definition 6.4. The *quermassintegral* of a convex body K in $\mathbb{R}^n$ is

$$W_i(K) = V_n(K, K, \ldots, K, B_n, \ldots, B_n) ,$$

where B_n occurs i times and K $n - i$ times, i.e.,

$$\mathrm{vol}_n(K + tB_n) = \sum_{i=0}^{n} \binom{n}{i} W_i(K) t^i .$$

The *intrinsic r-volume* $\bar{V}_r(K)$ of K for $-0 \leq r \leq n$ is given by

$$\bar{V}_r(K) = (\mathrm{vol}_{n-r}(B_{n-r}))^{-1} \binom{n}{r} W_{n-r}(K) .$$

It can be shown that when K is an r-dimensional polytope in $\mathbb{R}^n$ then $\bar{V}_r(K)$ is the usual r-dimensional volume $\mathrm{vol}_r(K)$ of K, and $\bar{V}_{r-1}(K)$ is half the surface area of K.

Mixed volumes satisfy several important inequalities.

Theorem 6.5 (Brunn–Minkowski inequality). *Given two convex sets K_0 and K_1 in $\mathbb{R}^n$, set $K_\theta = (1-\theta)K_0 + \theta K_1$. Then $\mathrm{vol}(K_\theta)^{1/n}$ is a concave function of θ, i.e.,*

$$\mathrm{vol}(K_\theta)^{1/n} \geqslant (1-\theta)\,\mathrm{vol}(K_0)^{1/n} + \theta\,\mathrm{vol}(K_1)^{1/n}\,.$$

Now

$$\mathrm{vol}_n(K_\theta) = \sum_{i=1}^{n} \binom{n}{i} (1-\theta)^{n-i}\theta^i V(K_0, n-i, K_1, i)\,,$$

where $V(K_0, n-i, K_1, i) = V_n(K_0, \ldots, K_0, K_1, \ldots, K_1)$ with K_0 taken $(n-i)$ times and K_1 i times. One can deduce from Theorem 6.5 that

$$V(K_0, n-1, K_1, 1)^n \geqslant (\mathrm{vol}_n(K_0))^{n-1}\,\mathrm{vol}_n(K_1)\,,$$

which is called *Minkowski's inequality*.

Theorem 6.6 (Alexandrov–Fenchel inequality). *For n convex sets $K_1, \ldots, K_n$ in $\mathbb{R}^n$ one has*

$$V_n(K_1, \ldots, K_n)^2 \geqslant V_n(K_1, K_1, K_3, \ldots, K_n)V_n(K_2, K_2, K_3, \ldots, K_n)\,.$$

For a proof see Burago and Zalgaller (1988) or Schneider (1993). The Alexandrov–Fenchel inequalities have had a number of direct combinatorial applications, e.g., to show log-concavity of certain counting sequences (cf. Stanley 1981), and in the proof of the Van der Waerden conjecture for permanents by Falikman (1981) and Egorychev (1981) (see also Van Lint 1982). This asserts that the permanent of any doubly stochastic n by n non-negative matrix is always at least $n!/n^n$. In particular, this implies that any r-regular bipartite graph on $2n$ vertices has at least $n!\,(r/n)^n$ perfect matchings (cf. chapter 3).

A first set of results relates lattice point counting functions to the volume of an integer polytope. In 1899, G. Pick (according to Coxeter 1969) proved that the area of a lattice polygon is related to the number of lattice points it contains.

Pick's theorem 6.7. *The area of any simple polygon P in $\mathbb{R}^2$ (not necessarily convex) whose vertices are lattice points is*

$$\mathrm{area}(P) = L(P) + \tfrac{1}{2}\dot{L}(P) - 1\,.$$

This result is easily proved by showing that the right-hand side of this equation is additive, then splitting P into a union of lattice triangles containing no interior lattice points, and observing that any such lattice triangle has area $\frac{1}{2}$. Generalizations of Pick's theorem, and the problem of counting lattice points in lattice polytopes, are related to toric varieties, see Morelli (1993) and Pommersheim (1993).

J.E. Reeve found an extension to dimension 3; to do so he introduced the quantities

$$L(nP) = \#\{nP \cap \mathbb{Z}^n\} = \#\left\{P \cap \frac{1}{n}\mathbb{Z}^n\right\}$$

for $n \in \mathbb{Z}_+$, which count lattice points in a blown-up copy of P. A general result for computing the volume in $\mathbb{R}^n$ using these quantities was then found by MacDonald.

Theorem 6.8 (MacDonald 1963). *For a lattice polytope P in $\mathbb{R}^n$,*

$$\tfrac{1}{2}(n-1)n! \operatorname{vol}_n(P) = \sum_{j=0}^{d-1} (-1)^j \binom{d-1}{j} M(d-j-1, P),$$

where $M(j, P) = L(jP) - \frac{1}{2}L(j(\partial P))$ if $j \geq 1$ and $M(0, P) = 1$ or 0 according as d is even or odd.

This result applies more generally to any pure d-dimensional simplicial complex K composed of lattice simplices, except that one must take $M(0, K) = \chi(K) - \frac{1}{2}\chi(\partial K)$ where $\chi(K)$ is the Euler characteristic of K, see MacDonald (1971).

The lattice-point counting function $L(kP)$ has a very nice structure, found by Ehrhart (1967b), and extended by McMullen (1975a) and others.

Theorem 6.9. *For a lattice polytope P the generating function*

$$\sum_{k=0}^{\infty} L(kP)t^k = \frac{\sum_{j=0}^{n+1} c_j(P)t^j}{(1-t)^{n+1}},$$

where the quantities $c_j(P)$ are nonnegative integers. In particular $L(kP)$ is a polynomial of degree n in k, for all $k \geq 1$, i.e.,

$$L(kP) = \sum_{i=0}^{n} L_i(P)k^i,$$

for suitable coefficients $L_i(P)$.

The polynomial $L(kP)$ is called the *Ehrhart polynomial*. Similarly, it can be shown that one has polynomials

$$L^\circ(kP) = \sum_{i=0}^{n} L_i^\circ(P)k^i,$$

$$\dot{L}(kP) = \sum_{i=0}^{n} \dot{L}_i(P)k^i,$$

$$\hat{L}(kP) = \sum_{i=0}^{n} \hat{L}_i(P)k^i.$$

for $k \geq 1$. These coefficients have geometric interpretations, e.g.,

$$L_n(P) = L^{\circ}_{n-1}(P) = \hat{L}_n(P) = \mathrm{vol}_n(P) .$$

$$L_{n-1}(P) = L^{\circ}_{n-1}(P) = \tfrac{1}{2}\dot{L}_{n-1}(P) = \tfrac{1}{2}\sum_F \frac{\mathrm{vol}_{n-1}(F)}{\det(\mathbb{Z}^n \cap F)} ,$$

where F runs over all $(n-1)$-dimensional facets of P, and $\mathbb{Z}^n \cap F$ is the $(n-1)$-dimensional sublattice of $\mathbb{Z}^n$ generated by lattice points on the facet F. Also

$$L_0(P) = (-1)^n \dot{L}_0(P) = \chi(P) ,$$

where $\chi(P)$ is the Euler characteristic of P. Morelli (1993) obtains geometric formulae for all $L_i(P)$, see also Barvinok (1994).

These results extend further to Minkowski sums of several lattice polytopes (recall Definition 2.3). If P is an integer polytope in $\mathbb{R}^n$ let $L_i(P)$ denote the sum of the lattice points in $\mathbb{Z}^n$ in the union of i-faces of P.

Theorem 6.10 (Bernshtein 1976, McMullen 1975a). *Let $K_1, \ldots, K_q$ be lattice polytopes in $\mathbb{R}^n$. Then*

(a) *$L(k_1K_1 + \cdots + k_qK_q)$ is a polynomial in $k_1, \ldots, k_q$ for $k_1 \geq 0, \ldots, k_q \geq 0$.*

(b) *For any i, $0 \leq i \leq n-1$, $L_i(k_1K_1 + \cdots + k_qK_q)$ is a polynomial in $k_1, \ldots, k_q$, for $k_1 \geq 1, \ldots, k_q \geq 1$.*

The condition in (b) that all $k_i \geq 1$ is necessary because, e.g., for $q = 1$, choosing $k_1 = 0$ gives $L_i(0) = 1$, while the polynomial in (b) at $k_1 = 0$ gives the Euler characteristic of the i-skeleton of K, which need not be one. Bernshtein (1976) gave a combinatorial proof, and showed that the mixed volume of $K_1, \ldots, K_q$ can be expressed in terms of coefficients of these polynomials. There is a close connection between these results and counting roots of systems of multivariate polynomial equations in $(\mathbb{C}^*)^n$, see Bernshtein (1975). The most natural proof of Theorem 6.9 may be algebro-geometric. See Ewald (1985) for a survey, and a discussion of an algebro-geometric proof of the Alexandrov–Fenchel inequality 6.6 related to integer polytopes.

There are *reciprocity laws* relating the quantities $L(P)$, $L^{\circ}(P)$, and $\dot{L}(P)$ for P and its boundary ∂P, due to Ehrhart (1967a).

Theorem 6.11. *For a lattice polytope P in $\mathbb{R}^n$, for $0 \leq i \leq n$,*

$$L_i(P) = (-1)^{n-i} L^{\circ}_i(P) , \qquad \dot{L}_i(P) = (1 - (-1)^{n-i}) L_i(P) .$$

Stanley (1974) has given a "monster" combinatorial reciprocity law extending this kind of reciprocity law in many directions.

McMullen (1975b, 1986) related $L(P)$ to $\hat{L}(P)$ for lattice polytopes. Define the *internal angle* $\beta(F, P)$ of a face F of an arbitrary polytope P (of any dimension) to be $\alpha(\boldsymbol{x}, P)$ for any $\boldsymbol{x} \in F$ in the relative interior of F. Then for lattice polytopes

$$\hat{L}(P) = \sum_F (-1)^{\dim P - \dim F} \beta(F, P) L_i(F) .$$

He also obtains an *inversion law* for intrinsic r-volumes, as a special case of a much more general inversion law.

Theorem 6.12. *For a polytope P in $\mathbb{R}^n$,*

$$(-1)^r \delta_{r,\dim(P)} \bar{V}_r(P) = \sum_F (-1)^{\dim F} \alpha(F, P) \bar{V}_r(F) ,$$

where $\delta_{r,s}$ is the Kronecker delta function and the sum is over all faces of P.

Are there general inequalities relating the number of lattice points in a lattice polytope to its volume, and, more generally to its quermassintegrals? There are good lower bounds. For a convex body K in $\mathbb{R}^n$ with nonempty interior, Bokowski et al. (1972) showed that

$$L(K) \geq \bar{V}_n(K) - \bar{V}_{n-1}(K) ,$$

and observed that this is sharp. Hadwiger (1972) then obtained the following bound, also sharp.

Theorem 6.13. *For any convex body K in $\mathbb{R}^n$, with nonempty interior,*

$$L(K) \geq \sum_{i=0}^{n} (-1)^{n-i} \bar{V}_i(K) .$$

Schnell (1992) generalized the result of Bokowski et al. (1972) in another direction, to counting points in arbitrary lattices.

For upper bounds the situation is more complicated. An inequality of Davenport (1951) implies that

$$L(P) = \sum_{i=0}^{n} L_i(P) \leq \sum_{i=0}^{n} \binom{n}{i} \bar{V}_i(P) .$$

Wills (1973) conjectured that

$$L(P) \leq \sum_{i=0}^{n} \bar{V}_i(P) .$$

Overhagen (1974) proved Wills' conjecture for $n = 3$. Later McMullen (1975b) conjectured the stronger bounds

$$L_i(K) \leq \bar{V}_i(K) , \quad 0 \leq i \leq n .$$

and observed that the following holds.

Theorem 6.14. *If K is a convex body in $\mathbb{R}^n$ then*

$$L_i(K) \leq \bar{V}_i(K)$$

holds for $i = 0$, $n-1$ and n.

These inequalities need only be proved for lattice polytopes P, because if $P \subseteq K$ then $\bar{V}_i(P) \leq \bar{V}_i(K)$ by general properties of mixed volumes (Burago and Zalgaller 1988, p. 138). However, Hadwiger (1979) disproved Wills' conjecture, hence also McMullen's conjecture, showing that it fails for a regular n-simplex K_n of edge length $\sqrt{n}$ for $n \geq 441$. Gritzmann and Wills (1986) proved a weaker bound

$$L(K) \leq \sum_{i=0}^{n} \alpha_{i,n} \bar{V}_i(K)$$

for certain complicated constants $\alpha_{i,n}$. They conjectured that this inequality still holds with (smaller) values $\alpha_{i,n} = \sigma_i 2^i (\mathrm{vol}_i(B_i))^{-1}$, where σ_i is the fraction of the volume inside a regular simplex of side 2 covered by $d+1$ unit balls centered at its vertices. Wills (1990) conjectures that a similar inequality holds with $\alpha_{i,n} = \mathrm{vol}_i(B_i)\,\mathrm{vol}_{n-i}(B_{n-i})(\mathrm{vol}_n B_n)^{-1}$. Further inequalities on lattice-point enumerators appear in Betke and McMullen (1985), Wills (1991), and Gritzmann and Wills (1992).

In $\mathbb{R}^n$ for $n \geq 3$ there exist lattice simplices of arbitrarily large volume that contain no interior lattice points, e.g., the simplex in $\mathbb{R}^3$ with vertices $\{(0,0,1), (0,1,0), (1,0,0)$ and $(1,1,m)\}$, for $m \in \mathbb{Z}^+$. However, Hensley (1983) showed that if a lattice polytope P in $\mathbb{R}^n$ contains a fixed nonzero number k of lattice points, then its volume is bounded in terms of n and k. This result was sharpened by Lagarias and Ziegler (1991), as follows.

Theorem 6.15. *Any lattice polytope P in $\mathbb{R}^n$ that contains exactly $k \geq 1$ points in $d\mathbb{Z}^n$ in* int (P) *has*

$$\mathrm{vol}_n(P) \leq kd(7(kd+1))^{2^{n+1}}.$$

There are examples of lattice simplices S containing exactly one interior point in $d\mathbb{Z}^n$ having $\mathrm{vol}_n(S) \geq (d+1)^{2^{n-1}}/n!$. See Reznick (1986) for related results.

The simplest lattice polytope in $\mathbb{R}^n$ is a *minimal lattice simplex*, which is a lattice simplex of volume $1/n!$. Any lattice polytope in $\mathbb{R}^2$ can be cut up into minimal lattice simplices, but for $n \geq 3$ not all lattice polytopes can be cut up into minimal lattice simplices. However, one has the following result.

Theorem 6.16 (Kempf et al. 1973). *In any dimension n there exists an integer constant N_n such that for every lattice polytope P in $\mathbb{R}^n$ the homothetic image $(N_n!)P$ can be subdivided into minimal lattice simplices.*

Additional properties of lattice polytopes may be found in the book of Ehrhart

(1977) and surveys of Betke and Wills (1979) and McMullen and Schneider (1983).

7. Algorithmic geometry of numbers and integer programming

The fundamental theorems in the geometry of numbers are theorems asserting the existence of a lattice point (or set of lattice points) meeting certain conditions. The corresponding algorithmic problem is that of explicitly finding such lattice points or proving their nonexistence. The construction of such algorithms and the analysis of their computational complexity form the subject matter of the *algorithmic geometry of numbers*, cf. Kannan (1987b). These include algorithms for finding the Hermite normal form of an integer matrix, lattice basis reduction algorithms and integer programming algorithms.

Given a basis $B = [\boldsymbol{b}_1, \ldots, \boldsymbol{b}_n]$ of a lattice L_0, and a sublattice L_1 of finite index, there is a canonical choice of basis $\tilde{B} = [\tilde{\boldsymbol{b}}_1, \ldots, \tilde{\boldsymbol{b}}_n]$ for L_1 with respect to the basis B, for which $\tilde{B} = \tilde{M}B$ is in Hermite normal form, see Theorem 2.2. Frumkin (1977) and later Kannan and Bachem (1979) gave polynomial time algorithms for converting an integer matrix to Hermite normal form, or equivalently, given bases B, $\tilde{B}_0$ of L_0, L, respectively, finding $U \in \mathrm{GL}(n, \mathbb{Z})$ such that the basis $\tilde{B} = U\tilde{B}_0$ is in Hermite normal form with respect to B. Chou and Collins (1982) improved the algorithm of Kannan and Bachem, obtaining the following result.

Theorem 7.1. *There is an algorithm which when given an $n \times m$ integer matrix M of full row rank n will find the unique $U \in \mathrm{GL}(n, \mathbb{Z})$ such that $\tilde{M} = UM$ is in Hermite normal form. If B denotes the maximal absolute value of an element of M then the algorithm takes at most $\mathrm{O}(m^4(m + n \log nB)^2)$ bit operations using integers of length at most $\mathrm{O}(n \log(nB))$ bits.*

A *linear Diophantine equation* is a set of linear equations $A\boldsymbol{x} = \boldsymbol{b}$ in which A is an $m \times n$ integer matrix, $\boldsymbol{b}$ an $m \times 1$ integer vector, and only integer solutions $\boldsymbol{x} \in \mathbb{Z}^n$ are considered. If $L(A)$ denotes the lattice of integer solutions to the homogeneous linear Diophantine equation $A\boldsymbol{x} = \boldsymbol{0}$ then the general solution of $A\boldsymbol{x} = \boldsymbol{b}$ is $\{\boldsymbol{x}_0 + \boldsymbol{y}: \boldsymbol{y} \in L(A)\}$ where $\boldsymbol{x}_0$ is any solution. Frumkin (1976, 1977), and Von zur Gathen and Sieveking (1976), and later Chou and Collins (1982) construct polynomial time algorithms to decide if $A\boldsymbol{x} = \boldsymbol{b}$ has an integer solution $\boldsymbol{x}_0$, and if so to find it and a basis of the lattice $L(A)$.

The algorithmic version of Minkowski's fundamental theorem 3.1 is that of finding a shortest nonzero vector in a lattice L with respect to a given norm. An algorithm to answer this question easily allows one to decide if a given symmetric convex body K contains a nonzero point of L, and if so, to find one. However, this problem appears to be computationally difficult even for simple norms such as the supremum norm ($K = n$-cube) or Euclidean norm ($K = B_n$). Consider the following problems.

Sup-norm short vector problem 7.2.
Input: A basis $[\boldsymbol{b}_1, \ldots, \boldsymbol{b}_n]$ of a sublattice L of $\mathbb{Z}^n$ and an integer bound B.
Question: Does L contain a nonzero vector $\boldsymbol{v} = (v_1, \ldots, v_n)$ with all $|v_i| \leq B$?

Euclidean norm short vector problem 7.3.
Input: A basis $[\boldsymbol{b}_1, \ldots, \boldsymbol{b}_n]$ of a sublattice L of $\mathbb{Z}^n$ and an integer bound B.
Question: Does L contain a nonzero vector $\boldsymbol{v}$ with $|\boldsymbol{v}|^2 \leq B$.

The sup-norm short vector problem 7.2 is known to be NP-complete when the dimension n of the input may vary, see Van Emde Boas (1981). In particular, if $\mathrm{P} \neq \mathrm{NP}$ then no polynomial-time algorithm exists to find the shortest nonzero supremum norm vector in an integer lattice L, cf. Garey and Johnson (1979). Deciding the complexity of Euclidean norm short vector problem 7.3 is a fundamental open problem. All known algorithms for finding the shortest Euclidean norm vector in an integer lattice require exponential time.

It is natural to consider the weaker problem of finding a relatively short vector in a lattice. This problem can be shown to be nearly of the same computational difficulty as that of the apparently more difficult problem of finding a basis of short vectors of a lattice, see Lovász (1986, p. 24).

Theorem 7.4. *A polynomial-time algorithm for solving problem* (A) *below gives a polynomial-time algorithm for solving problem* (B). *Given as input a basis* $[\boldsymbol{b}_1, \ldots, \boldsymbol{b}_n]$ *of a sublattice* L *of* $\mathbb{Z}^n$:

(A) *Find a nonzero vector* $\tilde{\boldsymbol{b}}_1 \in L$ *with*

$$|\tilde{\boldsymbol{b}}_1| \leq a(n)\lambda_1(L)\,.$$

(B) *Find a basis* $[\tilde{\boldsymbol{b}}_1, \ldots, \tilde{\boldsymbol{b}}_n]$ *of* L *such that*

$$\prod_{i=1}^{n} |\tilde{\boldsymbol{b}}_i| \leq n^n a(n)^n \det(L)\,.$$

Lovász found a polynomial-time algorithm, now called Lovász' lattice basis reduction algorithm or L^3-algorithm, which finds a "good" basis, cf. Lenstra et al. (1982). This algorithm finds a basis of a lattice reduced in the following weak sense analogous to the Korkin–Zolotarev reduction conditions.

Definition 7.5. An ordered basis $[\boldsymbol{b}_1, \ldots, \boldsymbol{b}_n]$ is *Lovász-reduced* if its Gram–Schmidt orthogonalization

$$\boldsymbol{b}_i = \boldsymbol{b}_i^* + \sum_{1 \leq j < i} \mu_{ij} \boldsymbol{b}_j^*\,, \quad 1 \leq i \leq n$$

satisfies:

(i) *Size-reduction condition*: $|\mu_{ij}| \leq \frac{1}{2}$ for $j < i$.

(ii) For $1 \leq i \leq n-1$, the Gram–Schmidt vectors b_i^* satisfy

$$|b_i^*|^2 \leq \tfrac{4}{3}|b_{i+1}^* + \mu_{i+1,i} b_i^*|^2 .$$

It is relatively easy to prove the following bounds.

Theorem 7.6. *Any Lovász-reduced basis* $[b_1, \ldots, b_n]$ *of a lattice* L *satisfies*:
(i) *For* $1 \leq i \leq n$,

$$|b_i|^2 \leq 2^{n-2+i} \lambda_i(L)^2 .$$

(ii) $|b_1| \cdots |b_n| \leq 2^{n(n-1)/2} \det(L)$.

Lovász's algorithm proceeds to convert a given basis $[b_1, \ldots, b_n]$ to a Lovász-reduced basis as follows. In a *size-reduction step* one arranges that the size-reduced condition hold. Find the largest j having some i with $|\mu_{ij}| > \frac{1}{2}$. If m is the integer nearest μ_{ij}, form the new basis $[b_1, \ldots, b_{i-1}, b_i - mb_j, b_{i+1}, \ldots, b_n]$, which has the same associated Gram–Schmidt vectors and new $\mu'_{ik} = \mu_{ik}$ for $j < k \leq i$ and $\mu'_{ik} = \mu_{ik} - m\mu_{jk}$ for $1 \leq k \leq j$. In particular all $|\mu_{ik}|$ with $i \geq k > j$ are unaffected, while $|\mu'_{ij}| \leq \frac{1}{2}$. Continuing in this fashion one obtains a size-reduced basis in at most $\binom{n}{2}$ such steps. In an *exchange step* one looks for some b_i such that (ii) is violated (say the smallest such i) and exchanges b_i and b_{i+1} in the ordered basis. The algorithm alternates size-reduction steps and exchange steps until it halts. To bound the number of steps the algorithm takes, one defines the invariant

$$D(b_1, \ldots, b_n) = \prod_{i=1}^{n} |b_i^*|^{n-i+1} .$$

One can show that this invariant remains constant during a size-reduction step (since the b_i^* do not change) and decreases by a multiplicative factor of size at most $\frac{1}{3}\sqrt{2}$ at each exchange step. Since $D(b_1, \ldots, b_n) \geq \det(L)$, the number of exchange steps is finite, and the algorithm halts. If the input to the algorithm is a basis of an integer lattice L, then one obtains the following polynomial running time bound, see Lenstra et al. (1982).

Theorem 7.7. *Lovász' lattice basis reduction algorithm, when given an ordered basis* $[b_1, \ldots, b_n]$ *of a sublattice* L *of* $\mathbb{Z}^n$ *as input, produces a Lovász-reduced basis* $[\tilde{b}_1, \ldots, \tilde{b}_n]$ *of* L *in at most* $O(n^4 \log B)$ *arithmetic operations on integers requiring at most* $O(n \log B)$ *bits, where* $B = \max(|b_i|^2 : 1 \leq i \leq n)$.

This algorithm has found many applications. The motivating application was a polynomial-time algorithm to factor polynomials in $\mathbb{Q}[x]$, obtained by Lenstra et al. (1982). It has also been used to decide in polynomial time the solvability of the Galois group of the equation $f(x) = 0$ with $f(x) \in \mathbb{Q}[x]$ (Landau and Miller 1983), to find good simultaneous Diophantine approximations, (Lagarias 1985, Lovász 1986), to solve low-density subset sum problems (Lagarias and Odlyzko 1985), and to show that for many combinatorial optimization problems, polynomial-time solvability implies strongly polynomial-time solvability (Frank and Tardos 1987). It has also had various cryptographic applications, e.g., in cryptanalyzing knapsack public key cryptosystems and in predicting congruential pseudorandom number generators. Computer implementations of the algorithm have been used to disprove Merten's conjecture (Odlyzko and Te Riele 1985) and to find new t-designs (Kreher and Radziszowski 1987).

Schnorr (1987) found a hierarchy of successively stronger variants of this algorithm that find bases of lattices that are reduced in a stronger sense than Lovász-reduced, and which are guaranteed to find shorter vectors in the lattice than Theorem 7.6(i).

Lattice basis reduction ideas lead to a polynomial-time-computable version of a weakened form of Minkowski's convex body theorem. We consider convex bodies K defined by a *well-founded membership oracle*. Such an oracle gives a sphere of radius r contained in K, a sphere of radius R containing K, and for each $\boldsymbol{x} \in \mathbb{R}^n$ it correctly answers $\boldsymbol{x} \in K$ or $\boldsymbol{x} \notin K$. The input size of such an oracle for K in $\mathbb{R}^n$ is defincd to be $n + \lfloor |\log R| \rfloor + \lfloor |\log r| \rfloor$.

Theorem 7.8. *For variable dimension n there is an algorithm which when given a well-founded membership oracle for a convex body K in $\mathbb{R}^n$ symmetric around* $\mathbf{0}$ *with*

$$\operatorname{vol}(K) \geqslant 2^{n^2},$$

always finds a nonzero vector in $K \cap \mathbb{Z}^n$. The algorithm makes only a number of oracle calls polynomial in the input size of K, and uses at most a polynomial number of bit operations in this input size.

The algorithm uses a rounding procedure of Lovász (1986, Theorem 2.4.1) to find a ellipsoid E centered at $\mathbf{0}$ such that $E \subset K \subseteq (n+1)\sqrt{n}E$. This can be done so that the ellipsoid $E = \{\boldsymbol{x} : \boldsymbol{x}^{\mathrm{T}} B^{\mathrm{T}} B \boldsymbol{x} \leqslant 1\}$ has a rational matrix B with entries of polynomial length in the input size. Then one makes a linear change of variable of determinant 1 to take E to a sphere λB_n, taking $\mathbb{Z}^n$ to a rational lattice L. Now one uses Lovász' lattice basis reduction algorithm to find a short vector $\boldsymbol{x}$ in L. Since

$$\operatorname{vol}(E) \geqslant (n+1)^{-3n/2} \operatorname{vol}(K) \geqslant (n+1)^{-3n/2} 2^{n^2},$$

one has $\lambda \leqslant 2^{n - O(\log n)}$, and also $\lambda_1(L) \leqslant \gamma_n \det(L)^{1/n} \leqslant n$. Now the bound in Theorem 7.6 shows that $\boldsymbol{x} \in \lambda B_n$ for all $n \geqslant n_0$. Its inverse under the linear map is the desired vector in $\mathbb{Z}^n$. For $n < n_0$ one modifies this algorithm to find the shortest vector in L, using the method of Kannan (1987a).

Next we turn to the inhomogeneous problem, the closest vector problem.

Euclidean norm close vector problem 7.9.

Input: A basis $[b_1, \ldots, b_n]$ of a sublattice L of $\mathbb{Z}^n$, a vector $w \in \mathbb{Z}^n$ and an integer bound B.

Question: Does L contain a vector v such that $|v - w|^2 \leq B$?

This problem and its supremum norm variant were shown NP-complete by Van Emde Boas (1981). The Lovász basis reduction algorithm can easily be adapted to find a fairly close vector in polynomial time, cf. Babai (1986). If $\mu(v, L)$ denotes the distance of v to the nearest lattice point of L, then for an integer lattice L one can find a vector $b \in L$ with

$$|v - b|^2 \leq 2^n \mu(v, L)^2 ,$$

in polynomial time.

Integer programming studies the set of integer solutions $x = (x_1, \ldots, x_n) \in \mathbb{Z}^n$ in the polytope $P(A, b)$ determined by a set of linear inequality constraints

$$Ax \geq b .$$

Of particular interest is the 0–1 *integer programming problem* which arises when the extra condition that all $x_i \in \{0, 1\}$ is present, i.e., the $2n$ inequalities $x_i \geq 0$ and $-x_i \geq -1$ occur in $Ax \geq b$. Many basic combinatorial problems are naturally formulated as $\{0, 1\}$ integer programs, e.g., the existence of a finite projective plane of order n, the existence of specific t-designs, the existence of an independent set of given size of a graph, the existence of a coloring of a graph with k colors, and the existence of a perfect tiling of a planar region with polyominoes of given shapes. The existence question for integer programs is the following.

Integer program feasibility problem 7.10.

Input: (A, b) are $m \times n$ and $m \times 1$ matrices of rationals given in binary, with m arbitrary.

Question: Does $P = \{x: Ax \geq b\}$ contain a point $x \in \mathbb{Z}^n$?

This problem is well known to be NP-complete when the dimension n of the input is allowed to vary, see Garey and Johnson (1979). Indeed, extremely special subclasses of integer programs are NP-complete, including $\{0, 1\}$ integer programs and even those integer programs with at most two variables per constraint, cf. Lagarias (1985).

Lenstra (1983) used ideas from the geometry of numbers to obtain a polynomial time algorithm for solving the feasibility problem in a fixed dimension. Kannan (1987a) obtained the following improved bound.

Theorem 7.11. *For variable dimension n the Integer programming feasibility problem can be solved in* $O(n^{9n/2}L)$ *arithmetic operations on integers of* $O(n^{2n}L)$ *bits in size, where L is the number of bits in the input.*

Lenstra's basic idea is to either find a vector $\boldsymbol{v} \in \mathbb{Z}^n - \{\boldsymbol{0}\}$ such that

$$\max\{\langle \boldsymbol{v}, \boldsymbol{x}\rangle : \boldsymbol{x} \in P\} - \min\{\langle \boldsymbol{v}, \boldsymbol{x}\rangle : \boldsymbol{x} \in P\} \leqslant c^{n^2}, \tag{7.1}$$

or prove that no such $\boldsymbol{v}$ exists. This inequality says that the feasible solution polytope $P = P(A, \boldsymbol{b})$ is "narrow" in the direction perpendicular to $\boldsymbol{v}$. If no such $\boldsymbol{v}$ exists, then Lenstra finds a large cube inside the polytope P containing an integer feasible point. If such a $\boldsymbol{v}$ exists, then the quantity $\boldsymbol{v} \cdot \boldsymbol{x}$ can take at most c^{n^2} values. Now Lenstra can replace the original integer program by c^{n^2} integer programs in $(n-1)$-variables by adding an extra constraint

$$\langle \boldsymbol{v}, \boldsymbol{x}\rangle = m ,$$

where m runs over the c^{n^2} values allowed by the inequality (7.1). Lenstra reduces the problem of finding such a $\boldsymbol{v}$ to a short vector in a lattice problem, which can be solved in polynomial time. Kannan's improvements come from replacing c^{n^2} in (7.1) with $n^{5/2}$, at the cost of taking time $O(n^n)$ to decide if a solution to (7.1) exists, see also Kannan (1987b).

A related, more difficult problem, is to count the number of integer feasible solutions to an integer programming problem. This problem can be polynomial-time reduced to the problem of counting the lattice points in a lattice polytope, see Cook et al. (1992). Recently, Barvinok (1993b) showed that this latter problem is solvable in polynomial time when the dimension n is fixed.

As a final example, we consider the problem of computing $\mathrm{vol}(P)$ for a polytope P. Dyer and Frieze (1988) and Khachiyan (1989) showed that it is #P-complete to compute $\mathrm{vol}(P)$ for a lattice polytope P when given either a list of all its vertices or a list of all its facets as input. Some special cases where it can be computed quickly are given in Barvinok (1993a). Recently Dyer et al. (1989) showed there is a good random polynomial time approximation algorithm for the volume of a convex body. They assume that the body K is described by a well-founded membership oracle. When given $\varepsilon > 0$, the algorithm produces an estimate $\widehat{\mathrm{vol}}(K)$ such that

$$(1+\varepsilon)\,\mathrm{vol}(K) \geqslant \widehat{\mathrm{vol}}(K) \geqslant (1-\varepsilon)\,\mathrm{vol}(K)$$

with probability $\frac{2}{3}$, and the algorithm runs in time polynomial in $1/\varepsilon$ and the input size of K. This contrasts with a result of Barany and Furedi (1986) asserting that for fixed n, and for every polynomial-time deterministic algorithm give a convex body K in $\mathbb{R}^n$ specified by a well-founded membership oracle which produces both upper and lower bounds $\overline{\mathrm{vol}}(K)$ and $\underline{\mathrm{vol}}(K)$ for $\mathrm{vol}(K)$, there is some convex body K in $\mathbb{R}^n$ for which

$$\overline{\mathrm{vol}}(K) \geqslant \left(\frac{n}{\log n}\right)^n \underline{\mathrm{vol}}(K) .$$

Thus Dyer et al. (1989) provide an example where a probabilistic algorithm is theoretically superior to all deterministic ones.

8. Connections to other areas of mathematics

Lattices are ubiquitous. The following references give starting points for study of some of these other areas of mathematics.

(a) Quadratic forms: Cassels (1978), Conway and Sloane (1988a), Milnor and Husemoller (1973).

(b) Perfect and extreme forms: Coxeter (1951), Conway and Sloane (1988b).

(c) Diophantine approximation: Bombieri and Vaaler (1983), Cassels (1957), Schmidt (1980).

(d) Multi-dimensional continued fractions: Brentjes (1981), Hastad et al. (1989), Lagarias (1994).

(e) Transcendental number theory: Baker (1975), Wustholz (1987).

(f) Algebraic number theory and class field theory: Kubota (1987), Lang (1970).

(g) Abelian varieties, Jacobian varieties, elliptic curves: Lang (1973, 1978), Martens (1978), Mumford (1983, 1984, 1991), Swinnerton-Dyer (1974).

(h) Toric varieties and integral polyhedra: Danilov (1978), Fulton (1993), Kempf et al. (1973), Oda (1985).

(i) Lattice polyhedra and random walks: Handelman (1987).

(j) Classification of flat manifolds: Charlap (1986).

(k) Discrete subgroups of hyperbolic space: Beardon (1983), Mumford (1971), Patterson (1975).

(l) Discrete subgroups of Lie groups and symmetric spaces: Margulis (1991), Terras (1985, 1988), Zimmer (1984).

(m) Nonperiodic tilings, quasicrystals: De Bruijn (1981, 1986), Gratias and Michel (1986), Janot (1992), Katz (1992), Radin (1991).

(n) Coding theory: Calderbank (1991), Calderbank and Sloane (1987), Conway and Sloane (1988a), Forney (1988a,b).

(o) Numerical analysis: Dahmen and Micchelli (1988), Hua and Wang (1981), Neumeier and Seidel (1983).

(p) Optimization: Saaty and Alexander (1975), Schrijver (1986).

(q) Integral geometry: Santaló (1976, chapter 8).

References

Babai, L.

[1986] On Lovász' lattice reduction and the nearest lattice point problem, *Combinatorica* **6**, 1–14.

Baker, A.

[1975] *Transcendental Number Theory* (Cambridge University Press, Cambridge).

Bambah, R.P., R. Woods and H. Zassenhaus

[1965] Three proofs of Minkowski's second inequality in the geometry of numbers, *J. Aust. Math. Soc.* **5**, 453–462.

Banaszczyk, W.

[1989] Polar lattices from the point of view of nuclear spaces, *Rev. Mat. Univ. Comput. Madrid* **2**(Special Issue), 35–46.

[1991] *Additive Subgroups of Topological Vector Spaces, Lecture Notes in Mathematics,* Vol. 1466 (Springer, New York).
[1993] New bounds in some transference theorems in the geometry of numbers, *Math. Ann.* **296**, 625–636.

Bannai, E., and N.J.A. Sloane
[1981] Uniqueness of certain spherical codes, *Canad. J. Math.* **33**, 437–449.

Baranovskii, E.P.
[1971] Packings, coverings, partitionings and certain other distributions in spaces of constant curvature, in: *Algebra and Geometry, Progress in Mathematics,* Vol. 9, ed. R.V. Gamelkridze (Plenum Press, New York) pp. 209–250.

Barany, I., and Z. Furedi
[1986] Computing the volume is difficult, in: *Proc. 18th Annu. ACM Symp. on Theory of Computing* (ACM Press, New York) pp. 442–447.

Barvinok, A.
[1993a] Computing the volume, counting integral points, and exponential sums, *Discrete Comput. Geom.* **10**, 123–141.
[1993b] A polynomial time algorithm for counting integral points in polyhedra when the dimension is fixed, in: *Proc. 34th IEEE Conf. on Foundations of Computer Science* (IEEE Computer Society Press, Los Alamitos, CA) pp. 566–572. [*Math. Oper. Res.,* to appear.].
[1994] Computing the Ehrhart polynomial of a convex lattice polytope, *Discrete Comput. Geom.* **12**, 35–48.

Beardon, A.
[1983] *The Geometry of Discrete Groups* (Springer, New York).

Bernshtein, D.N.
[1975] The number of roots of a system of equations, *Funktsional'nyi Analiz i Ego Pril.* **9**, 1–4 (in Russian).
[1976] The number of integral points in integral polyhedra, *Funktsional'nyi Analiz i Ego Pril.* **10**, 72–73 (in Russian).

Betke, U., and P. McMullen
[1985] Lattice points in lattice polytopes, *Monatsh. Math.* **99**, 253–265.

Betke, U., and J.M. Wills
[1979] Stetige und diskrete Funktionale konvexer Körper, in: *Contributions to Geometry,* eds. J. Tölke and J.M. Wills (Birkhäuser, Basel) pp. 226–237.

Bieberbach, L.
[1910a] Über die Bewegungsgruppe der Euclidischen Raüme mit einem endlichen Fundamentalbereich, *Gött. Nachr.,* pp. 75–84.
[1910b] Über die Bewegungsgruppe der Euclidischen Raüme, *Math. Ann.* **70**, 297–336.
[1912] Über die Bewegungsgruppe der Euclidischen Raüme, *Math. Ann.* **72**, 400–412.

Blichfeldt, H.F.
[1914] A new principle in the geometry of numbers, with some applications, *Trans. Amer. Math. Soc.* **15**, 227–235.

Bokowski, J., H. Hadwiger and J.M. Wills
[1972] Eine Ungleichung zwischen Volumen, Oberfläche und Gitterpunktzahl konvexer Körper im n-dimensionalen Euklidischen Raum, *Mathematika* **2**, 127, 363–364.

Bombieri, E.
[1962] Sulla dimonstratione di C.L. Siegel del teorema fondamentale di Minkowski nella geometria die numeri, *Boll. Un. Mat. Ital.* **17**, 283–288.

Bombieri, E., and J. Vaaler
[1983] On Siegel's lemma, *Invent. Math.* **73**, 11–32. Addendum: **75**, 377.

Bourgain, J., and V.D. Milman
[1987] New volume ratio properties for convex symmetric bodies in $\mathbb{R}^n$, *Invent. Math.* **88**, 319–340.

Bravais, A.
[1850] Les systèmes formés par des points distribués regulièrement sur un plan un dans l'espace, *J. Ecole Polytech.* **19**, 1–128.

Brentjes, A.
[1981] *Multidimensional Continued Fraction Algorithms, Mathematical Centre Tract,* No. 145 (Mathematisch Centrum, Amsterdam).
Brown, H., R. Bülow, J. Neubüser, H. Wondratschek and H. Zassenhaus
[1978] *Crystallographic Groups in 4-dimensional Space* (Wiley, New York).
Burago, Yu.D., and V.A. Zalgaller
[1988] *Geometric Inequalities* (Springer, New York).
Buser, P.
[1985] A geometric proof of Bieberbach's theorems on crystallographic groups, *L'Enseignment Math.* **31**, 137–145.
Calderbank, A.R.
[1991] The mathematics of modems, *Math. Intelligencer* **13**, 56–65.
Calderbank, A.R., and N.J.A. Sloane
[1987] New trellis codes based on lattice and cosets, *IEEE Trans. Inform. Theory*. **IT-33**, 177–195.
Cassels, J.W.S.
[1957] *An Introduction to Diophantine Approximation* (Cambridge University Press, Cambridge).
[1971] *An Introduction to the Geometry of Numbers* (Springer, New York).
[1978] *Rational Quadratic Forms* (Academic Press, London).
Charlap, L.
[1986] *Bieberbach Groups and Flat Manifolds* (Springer, New York).
Chou, T.-W.J., and G.E. Collins
[1982] Algorithms for the solution of linear Diophantine equations, *SIAM J. Comput.* **11**, 687–708.
Conway, J.H., and N.J.A. Sloane
[1988a] *Sphere-packing, Lattices and Groups* (Springer, New York).
[1988b] Low dimensional lattices III. Perfect forms, *Proc. Roy. Soc. London A* **418**, 43–80.
Cook, W., M. Hartmann, R. Kannan and C. McDiarmid
[1992] On integer points in polyhedra, *Combinatorica* **12**, 27–37.
Coxeter, H.S.M.
[1951] Extreme forms, *Canad. J. Math.* **3**, 391–441.
[1969] *Introduction to Geometry,* 2nd Ed. (Wiley, New York).
Coxeter, H.S.M., L. Few and C.A. Rogers
[1959] Covering space with equal spheres, *Mathematika* **6**, 147–157.
Dahmen, W., and C.A. Micchelli
[1988] On the number of solutions to linear Diophantine equations and multivariate splines, *Trans. Amer. Math. Soc.* **308**, 509–532.
Danilov, V.I.
[1978] The geometry of toric varieties, *Russ. Math. Surveys* **33**, 97–154.
Davenport, H.
[1951] On a principle of Lipschitz, *J. London Math. Soc.* **26**, 79–83.
De Bruijn, N.G.
[1981] Algebraic theory of Penrose's non-periodic tilings of the plane I, II, *Kon. Nederl. Akad. Wetensch. Series A* **84**, 39–52, 53–66.
[1986] Quasicrystals and their Fourier transform, *Indag. Math.* **48**, 123–152.
Delone, B.N.
[1926] Sur les théories des paralleloèdres, *C.R. Acad. Sci. Paris* **183**, 464–467.
Delone, B.N., R.V. Galiulin and M.I. Shtogrin
[1973] On the Bravais type of lattices, *Itogi. Nauk. Tekh.* 120–252 [1975, *J. Soviet Math.* **4**, 79–156].
Dyer, M., and A. Frieze
[1988] On the complexity of computing the volume of a polyhedron, *SIAM J. Comput.* **17**, 967–974.
Dyer, M., A. Frieze and R. Kannan
[1989] A random polynomial time algorithm for approximating the volume of convex bodies, in: *Proc. 21st Annu. ACM Symp. on Theory of Computing* (ACM, New York) pp. 375–381.

Eggleston, H.G.
[1958] *Convexity* (Cambridge University Press, Cambridge).

Egorychev, G.P.
[1981] The affirmative answer to the van der Waerden problem for permanents, *Sibersk. Math. Z.* **22**, 65–71 (in Russian).

Ehrhart, E.
[1967a] Demonstration de loi de réciprocité du polyèdre rationnel, *C.R. Acad. Sci. Paris, Ser. A* **265**, 91–94.
[1967b] Sur une problème de géométrie diophantienne linéaire I, II, *J. Reine Angew. Math.* **226**, 1–29; **227**, 25–49.
[1977] *Polynômes Arithmetiques et Method des Polyèdres en Combinatoire* (Birkhäuser, Basel).

Erdős, P., P. Gruber and J. Hammer
[1989] *Lattice Points* (Longman Scientific and Technical, Essex, England).

Ewald, G.
[1985] Convex bodies and algebraic geometry, in: *Discrete Geometry and Convexity,* eds. J. Goodman et al., *Ann. New York Acad. Sci.* **440**, 196–204.

Falikman, D.I.
[1981] Proof of the van der Waerden conjecture on the permanent of a doubly stochastic matrix, *Math. Zametki* **29**, 931–938.

Fedorov, E.S.
[1885] The elements of the study of figures (in Russian), *Proc. St. Petersburg Mineralogical Soc.* **21**, 1–289.
[1889] Symmetry of finite figures (in: Russian), *Proc. St. Petersburg Mineralogical Soc.* **25**, 1–52.
[1949] *Structure of Crystals* (Akad. Nauk. SSSR, Moscow). Translation: 1971 (American Crystallographic Association).

Fejes-Toth, L.
[1972] *Lagerungen in der Ebene, auf der Kugel und im Raum,* 2nd Ed. (Springer, Berlin).

Forney, G.D.
[1988a] Coset codes, Part I: Introduction and geometrical classification, *IEEE Trans. Inform. Theory* **IT-34**, 1123–1151.
[1988b] Coset codes, Part II: Binary lattices and related codes, *IEEE Trans. Inform. Theory* **IT-34**, 1152–1187.

Frank, A., and E. Tardos
[1987] An application of simultaneous approximation in combinatorial optimization, *Combinatorica* **7**, 49–65.

Fried, D., and W.M. Goldman
[1983] Three-dimensional affine crystallographic groups, *Adv. in Math.* **47**, 1–49.

Frumkin, M.A.
[1976] An application of modular arithmetic to the construction of algorithms for solving systems of linear equations, *Dokl. Akad. Nauk SSSR* **229**, 1067–1070 [*Soviet Math. Dokl.* **17**, 1165–1168].
[1977] Polynomial time algorithms in the theory of linear Diophantine equations, in: *Fundamentals of Computation Theory, Lecture Notes in Computer Science,* Vol. 56 (Springer, New York) pp. 386–392.

Fulton, W.
[1993] *Introduction to Toric Varieties* (Princeton Univ. Press, Princeton, NJ).

Garey, M.R., and D.S. Johnson
[1979] *Computers and Intractability: A Guide to the Theory of NP-Completeness* (W.H. Freeman, San Francisco).

Gratias, D., and L. Michel
[1986] *International Workshop on Aperiodic Crystals, J. Phys. Paris Colloq.* **47**(7) Supp. C3.

Gritzmann, P., and J.M. Wills
[1986] An upper estimate for the lattice point enumerator, *Mathematika* **33**, 197–203.
[1992] Lattice points, in: *Handbook of Convex Geometry,* eds P. Gruber and J.M. Wills (North-Holland, Amsterdam) pp. 765–797.

Gruber, P., and J.M. Wills
[1992] eds., *Handbook of Convex Geometry,* 2 volumes (North-Holland, Amsterdam).
Gruber, P.M.
[1979] Geometry of numbers, in: *Contributions to Geometry, Proc. Geom. Symp., Siegen, 1978,* eds. J. Tölke and J.M. Wills (Birkhaüser, Basel) pp. 186–225.
Gruber, P.M., and C.G. Lekkerkerker
[1987] *Geometry of Numbers,* 2nd Ed. (North-Holland, Amsterdam).
Grünbaum, B., and G.C. Shephard
[1987] *Tilings and Patterns* (W.H. Freeman, New York).
Hadwiger, H.
[1972] Gitterperiodische Punktmengen und Isoperimetrie, *Monatsh. Math.* **76**, 410–418.
[1979] Gitterpunktzahl im Simplex und Wills'sche Vermutung, *Math. Ann.* **239**, 271–288.
Hajós, G.
[1942] Über einfache und mehrfache Bedeckung des n-dimensionalen Raumes mit einen Würfelgitter, *Math. Z.* **47**, 427–467.
Hales, T.
[1994] The status of the Kepler conjecture, *Math. Intelligencer* **6**(3), 47–58.
Hammer, J.
[1977] *Unsolved Problems concerning Lattice Points* (Pitman, London).
Handelman, D.E.
[1987] *Positive Polynomials, Convex Integral Polytopes, and a Random Walk Problem, Lecture Notes in Mathematics,* Vol. 1282 (Springer, New York).
Hastad, J., B. Just, J. Lagarias and C. Schnorr
[1989] Polynomial time algorithms for finding integer relations among real numbers, *SIAM J. Comput.* **18**, 859–881.
Hensley, D.
[1983] Lattice vertex polytopes with few interior lattice points, *Pacific J. Math.* **105**, 183–191.
Hilbert, D.
[1900] Mathematische Problem, *Göttinger Nachrichten,* pp. 253–297. Translation: 1902, *Bull. Amer. Math. Soc.* **8**, 437–479.
Hilton, H.
[1903] *Mathematical Crystallography.* Reprint: 1961 (Dover, New York).
Hlawka, E.
[1944] Zur geometrie der Zahlen, *Math. Z.* **49**, 285–312.
Hua, L.K., and Y. Wang
[1981] *Applications of Number Theory to Numerical Analysis* (Springer, New York).
Janot, C.
[1992] *Quasicrystals, A Primer* (Clarendon Press, Oxford).
Kabatyanskii, G.A., and V.I. Levenshtein
[1978] Bounds for packings in a sphere and in space, *Problemy Peredachii Informatsii* **14**, 3–25 (in Russian) [*Problems in Information Transmission* **14**, 1–17].
Kannan, R.
[1987a] Minkowski's convex body theorem and integer programming, *Math. Oper. Res.* **12**, 415–440.
[1987b] Algorithmic geometry of numbers, in: *Annu. Review of Computer Science* (Annual Reviews, Inc., New York) **3**, 231–267.
Kannan, R., and A. Bachem
[1979] Polynomial algorithms for computing the Smith and Hermite normal forms of an integer matrix, *SIAM J. Comput.* **8**, 499–507.
Kannan, R., and L. Lovász
[1988] Covering minima and lattice-point-free convex bodies, *Ann. of Math.* **128**, 577–602.
Katz, A.
[1992] A short introduction to quasicrystallography: in: *From Number Theory to Physics,* eds. M. Waldshmidt et al. (Springer, New York) pp. 496–537.

Kempf, G.F., D. Knudsen, D. Mumford and B. Saint-Donat
[1973] *Toroidal Embeddings, Lecture Notes in Mathematics,* Vol. 339 (Springer, Berlin).
Khachiyan, L.G.
[1989] The problem of calculating the volume of polytopes is #P-hard (in Russian), *Uspekhi Mat. Nauk* **44**(3), 179–180 [*Revision Math. Surveys* **44**(3), 199–200].
Korkin, A., and G. Zolotarev
[1873] Sur les formes quadratiques, *Math. Ann.* **6**, 366–389. [Names given as A. Korkine and G. Zolotareff].
[1877] Sur les formes quadratiques, *Math. Ann.* **11**, 242–292.
Kouchnirenko, A.G.
[1976] Polyèdres de Newton et nombres de Milnor, *Invent. Math.* **32**, 1–31.
Kreher, D.L., and S.P. Radziszowski
[1987] New t-designs found by basis reduction, *Congress. Numerantium* **59**, 155–164.
Kubota, T.
[1987] Geometry of numbers and class field theory, *Jpn. J. Math.* **13**, 235–275.
Lagarias, J.C.
[1985] The computational complexity of simultaneous Diophantine approximation problems, *SIAM J. Comput.* **14**, 196–209.
[1994] Geodesic multidimensional continued fractions, *Proc. London Math. Soc.*, to appear.
Lagarias, J.C., and A.M. Odlyzko
[1985] Solving low density subset sum problems, *J. Assoc. Comput. Mach.* **32**, 229–246.
Lagarias, J.C., and G. Ziegler
[1991] Bounds for lattice polytopes containing a fixed number of interior points in a sublattice, *Canad. J. Math.* **43**, 1022–1035.
Lagarias, J.C., H.W. Lenstra Jr and C.P. Schnorr
[1990] Korkin–Zolotarev bases and successive minima of a lattice and its dual lattice, *Combinatorica* **10**, 343–358.
Landau, S., and G. Miller
[1983] Solvability by radicals is in polynomial time, in: *Proc. 15th Annual ACM Conf. on Theory of Computing* (ACM, New York) pp. 140–151.
Lang, S.
[1970] *Algebraic Number Theory* (Addison-Wesley, Reading, MA).
[1973] *Elliptic Functions* (Addison-Wesley, Reading, MA).
[1978] *Elliptic Curves: Diophantine Analysis* (Springer, New York).
Lenstra, A.K., H.W. Lenstra Jr and L. Lovász
[1982] Factoring polynomials with rational coefficients, *Math. Ann.* **261**, 513–534.
Lenstra Jr, H.W.
[1983] Integer programming in a fixed number of variables, *Math. Oper. Res.* **8**, 538–548.
Lovász, L.
[1986] *An Algorithmic Theory of Numbers, Graphs and Convexity* (SIAM, Philadelphia, PA).
MacDonald, I.G.
[1963] The volume of a lattice polyhedron, *Proc. Cambridge Philos. Soc.* **59**, 719–726.
[1971] Polynomials associated with finite cell complexes, *J. London Math. Soc.* **4**, 181–192.
Mahler, K.
[1939] Ein Übertragungsprincip fur Konvexe Körper, *Casopis Pest. Mat. Fyz.* **68**, 93–102.
[1946] On lattice points in n-dimensional star domains I. Existence theorems, *Proc. R. Soc. London A* **187**, 151–187.
Margulis, G.A.
[1991] *Discrete Subgroups of Semisimple Lie Groups* (Springer, Berlin).
Martens, H.
[1978] Observations on morphisms of closed Riemann surfaces, *Bull. London Math. Soc.* **10**, 209–212.
McFeat, R.G.
[1971] Geometry of numbers in adele spaces, *Dissertationes Math. Rozprauny Mat.* **88**, 49pp.

McMullen, C., and R. Schneider

[1983] Valuations on convex bodies, in: *Convexity and its Applications,* eds. P.M. Gruber and J.M. Wills (Birkhäuser, Basel) pp. 170–247.

McMullen, P.

[1975a] Metrical and combinatorial properties of convex polytopes, in: *Proc. Int. Congr. of Mathematicians, Vancouver, B.C., 1974,* Vol. 1, pp. 491–495.

[1975b] Non-linear angle sum relations for polyhedral cones and polytopes, *Math. Proc. Cambridge Philos. Soc.* **78**, 247–261.

[1986] Angle-sum relations for polyhedral sets, *Mathematika* **33**, 173–188.

Milnor, J.

[1976] Hilbert's problem 18: On crystallographic groups, fundamental domains and on sphere packing, *Proc. Symp. Pure Math.* **28**, 491–506.

Milnor, J., and D. Husemoller

[1973] *Symmetric Bilinear Forms* (Springer, New York).

Minkowski, H.

[1896] *Geometrie der Zahlen* (Teubner, Leipzig).

[1905] Diskontinuitätsbereich für arithmetische Äquivalenz, *J. Reine Angew. Math.* **129**, 220–224.

Morelli, R.

[1993] Pick's Theorem and the Todd class of a toric variety, *Adv. in Math.* **100**, 183–231.

Mumford, D.

[1971] A remark on Mahler's compactness theorem, *Proc. Amer. Math. Soc.* **28**, 289–294.

[1983] *Tata lectures on theta I* (Birkhäuser, Boston).

[1984] *Tata lectures on theta II* (Birkhäuser, Boston).

[1991] *Tata Lectures on Theta III* (with M. Nori and P. Norman) (Birkhäuser, Boston).

Neumeier, A., and J.J. Seidel

[1983] Discrete hyperbolic geometry, *Combinatorica* **3**, 219–237.

Newman, M.

[1972] *Integral Matrices* (Academic Press, New York).

Oda, T.

[1985] *Convex Bodies and Algebraic Geometry* (Springer, New York).

Odlyzko, A.M., and H. te Riele

[1985] Disproof of the Mertens Conjecture, *J. Reine Angew. Math.* **357**, 138–160.

Overhagen, I.

[1974] Zur Gitterpunktzahl konvexer Körper im 3-dimensionalen Euklidischen Raum, *Math. Ann.* **208**, 221–232.

Patterson, S.J.

[1975] A lattice point problem for hyperbolic space, *Mathematika* **22**, 81–88.

Pommersheim, J.E.

[1993] Toric varieties, lattice points, and Dedekind sums, *Math. Ann.* **295**, 1–24.

Radin, C.

[1991] Global order from local sources, *Bull. Amer. Math. Soc.* **25**, 335–364.

Reznick, B.

[1986] Lattice point simplices, *Discrete Math.* **60**, 219–242.

Rogers, C.A.

[1957] A note on coverings, *Mathematika* **4**, 1–6.

[1959] Lattice coverings of space, *Mathematika* **6**, 33–39.

[1964] *Packing and Covering* (Cambridge University Press, Cambridge).

Ryshkov, S.S., and E.P. Baranovskii

[1979] Classical methods in the theory of lattice packings, *Russian Math Surveys* **34**, 2–68.

Saaty, T.L., and J.M. Alexander

[1975] Optimization and the geometry of numbers: packing and covering, *SIAM Review* **17**, 475–519.

Santaló, L.A.
[1949] Una invariante afin pasa os cuerpas convexos del espacio du n-dimensiones, *Portugal. Math.* **8**, 155–161.
[1976] *Integral Geometry and Geometric Probability* (Addison-Wesley, Reading, MA).
Sawyer, D.B.
[1954] The lattice determinants of asymmetric convex regions, *J. London Math. Soc.* **29**, 251–254.
Schmidt, W.
[1958] The measure of the set of admissible lattices, *Proc. Amer. Math. Soc.* **9**, 390–403.
[1963] On the Minkowski–Hlawka theorem, *Illinois J. Math.* **7**, 18–23. Correction: **7**, 714.
[1980] *Diophantine Approximation, Lecture Notes in Mathematics,* Vol. 785 (Springer, New York).
Schneider, R.
[1993] *Convex Bodies: The Brunn–Minkowski Theory* (Cambridge University Press, Cambridge).
Schnell, U.
[1992] Minimal determinants and lattice inequalities, *Bull. London Math. Soc.* **24**, 606–612.
Schnorr, C.P.
[1987] A hierarchy of polynomial time lattice basis reduction algorithms, *Theor. Comput. Sci.* **53**, 201–224.
Schoenflies, A.M.
[1889] Über Gruppen von Transformationen des Raumes in sich, *Math. Ann.* **34**, 172–203.
[1891] *Kristallsysteme und Kristallstructur* (Teubner, Leipzig).
Schrijver, A.
[1986] *Theory of Linear and Integer Programming* (Wiley, New York).
Schwartzenberger, R.L.E.
[1980] *N-dimensional Crystallography* (Pitman, London).
Siegel, C.L.
[1935] Über Gitterpunkte in konvexen Körpern und ein damit zusammenhängendes extremal Problem, *Acta Math.* **65**, 307–323.
[1945] A mean value theorem in the geometry of numbers, *Ann. of Math.* **46**, 340–347.
Stanley, R.P.
[1974] Combinatorial reciprocity theorems, *Adv. in Math.* **14**, 197–253.
[1981] Two combinatorial applications of the Alexandrov–Fenchel inequalities, *J. Combin. Theory A* **31**, 56–65.
Swinnerton-Dyer, H.P.F.
[1953] Extremal lattices of convex bodies, *Proc. Cambridge Philos. Soc.* **49**, 161–162.
[1974] *Analytic Theory of Abelian Varieties, London Mathematical Society Lecture Note Series,* Vol. 14 (Cambridge University Press, Cambridge).
Terras, A.
[1985] *Harmonic Analysis on Symmetric Spaces and Applications, I* (Springer, New York).
[1988] *Harmonic Analysis on Symmetric Spaces and Applications II* (Springer, New York).
Van der Corput, J.G.
[1935] Verallgemeinerung einer Mordellschen Beweismethode in der Geometrie der Zahlen, *Acta. Arithm.* **1**, 62–66.
van Emde Boas, P.
[1981] *Another NP-complete Partition Problem and the Complexity of Computing short Vectors in a Lattice,* Report 81–04 (Mathematisch Instituut, University of Amsterdam).
Van Lint, J.H.
[1982] The van der Waerden conjecture: two proofs in one year, *Math. Intelligencer* **4**, 72–77.
Von zur Gathen, J., and M. Sieveking
[1976] Weitere zum Erfullüngsproblem polynomial äequivalente kombinatorische Aufgaben, in: *Komplexität von Entscheidungsproblemen, Lecture Notes in Computer Science,* Vol. 43, eds. E. Specker and V. Strassen (Springer, New York) pp. 49–71.
Voronoi, G.
[1908] Nouvelles applications des paramètres continus à la theorie des formes quadratiques. Deuxième

mémoire. Recherches sur les paralleloèdres primitifs I, II, *J. Reine Angew. Math.* **134**, 198–287; **136**, 67–181.

Wills, J.M.

[1973] Zur Gitterpunktzahl konvexer Mengen, *Elem. Math.* **28**, 57–63.

[1990] Minkowski's successive minima and the zeros of a convexity function, *Monatsh. Math.* **109**, 157–164.

[1991] Bounds for the lattice point enumerator, *Geom. Dedicata* **40**, 237–244.

Woods, A.

[1966] A generalization of Minkowski's second inequality in the geometry of numbers, *J. Aust. Math. Soc.* **6**, 148–152.

Wüstholz, G.

[1987] A new approach to Baker's theorem on linear forms in logarithms I, II, in: *Diophantine Approximation and Transcendence Theory, Lecture Notes in Mathematics,* Vol. 1290, ed. G. Wüstholz (Springer, Berlin) pp. 189–202, 203–211.

Zassenhaus, H.

[1938] Neuer Beweis der Endlichkeit der Klassenzahl bei unimodular Äquivalenz endlichler ganzzähliger Substitutionsgruppen, *Abh. Math. Sem. Hamburg* **12**, 276–288.

[1948] Über einem Algorithmus zur Bestimmung der Raumgruppen, *Comment. Math. Helv.* **21**, 117–141.

Zimmer, R.J.

[1984] *Ergodic Theory and Semisimple Groups* (Birkhäuser, Boston).

Zitomirskii, O.K.

[1929] Verschärfung eines Satzes von Woronei, *Z. Leningrad Fiz. Math. Obxc* **2**, 131–151.

CHAPTER 20

Combinatorial Number Theory

Carl POMERANCE[1]

Department of Mathematics, University of Georgia, Athens, GA 30602, USA

András SÁRKÖZY[2]

Mathematical Institute of the Hungarian Academy of Sciences, Reáltanoda utca 13–15, Budapest 1364, Hungary

Contents

[1] Research partially supported by an NSF grant.
[2] Research partially supported by the Hungarian National Foundation for Scientific Research, Grant No. 1811.

HANDBOOK OF COMBINATORICS
Edited by R. Graham, M. Grötschel and L. Lovász

1. Introduction

What is the cardinality of the largest subset of $\{1, 2, \ldots, N\}$ that does not contain two relatively prime numbers? This is a typical problem in combinatorial number theory. That the problem is one in number theory, there is no doubt. But someone who leans towards combinatorics might prefer to think of it as a question of the largest complete subgraph of that graph on $\{1, 2, \ldots, N\}$ with edges that connect two numbers when they are not coprime.

The above question illustrates a common theme in combinatorial number theory. Namely, what arithmetic properties must a "dense" subset of the integers possess? One of the greatest theorems of this type, Szemerédi's theorem, is discussed in section 6. But combinatorial number theory also deals with other issues. For example, under what conditions is a subset of the natural numbers a basis, i.e., for some h, every number can be represented as a sum of h or fewer elements from the subset. Such issues are discussed in section 3. Combinatorial sieve methods, the subject of section 2, takes its starting point at the inclusion–exclusion principle. Its simpler aspects might be described as a device for controlling the "combinatorial explosion" in the number of terms involved in an inclusion–exclusion argument.

Combinatorial number theory can also deal with some classical problems of number theory when the methods used have a strong combinatorial flavor. In section 7 we present a proof of Wirsing's theorem on perfect numbers. This gem uses nothing but simple counting arguments from elementary combinatorics.

Combinatorial number theory is a relatively young field. In 1850, P. L. Chebyshev proved that

$$c_1 \frac{x}{\log x} < \pi(x) < c_2 \frac{x}{\log x} \quad \text{for } x \geq 2\,, \tag{1.1}$$

where $\pi(x)$ denotes the number of primes up to x, and c_1, c_2 are positive constants. This result constituted important progress towards the prime number theorem, $\pi(x) = (1 + o(1))x/\log x$, which was not proved until some 45 years later. Chebyshev's proof of (1.1) (which was later analyzed and simplified by Landau, Erdős and Diamond) had a certain combinatorial flavor.

In the period 1915–1924, V. Brun essentially single-handedly began the subject of combinatorial sieve methods. In 1927, B.L. van der Waerden proved his famous theorem that whenever the set of natural numbers is partitioned into two sets, then one set contains arbitrarily long arithmetic progressions. L.G. Schnirelmann, in 1930, used both Brun's results and his own ideas on the relationship between density and bases to prove that the set consisting of one and the primes is a basis. Inspired by these works, combinatorial number theory came into full flower in the 1930s and 1940s. Certainly the most significant force to shape and define the subject both then and now has been P. Erdős. We are much indebted to him for his generous help with this chapter.

In writing a chapter such as this, certain hard choices were necessarily forced upon us. The subject is very broad and does not have clearly delineated boundaries. It soon became clear that we had no chance of covering it all. Moreover, our philosophy for this chapter mandated the inclusion of representative proofs. Thus even the few areas that we do cover are not done encyclopaedically. Fortunately there are several excellent books that treat various aspects of combinatorial number theory with great thoroughness, books that we refer to frequently throughout for further problems, details, and references. These are *Sequences* by Halberstam and Roth (1983), *Old and New Problems and Results in Combinatorial Number Theory* by Erdős and Graham (1980), and *Unsolved Problems in Number Theory*, 2nd edition, by Guy (1994).

Some may consider the subject of integer partitions an important part of combinatorial number theory. Unfortunately, though, it is a subject we completely ignore. The interested reader is referred to the excellent monograph of Andrews (1976).

We now say a word about notation. If $n \in \mathbb{N}$ (the set of positive integers), then $\tau(n)$ is the number of positive divisors of n, $\nu(n)$ is the number of prime divisors of n, and $\Omega(n)$ is the number of prime and prime power divisors of n. For example, $\tau(12) = 6$, $\nu(12) = 2$, and $\Omega(12) = 3$. We say $n \in \mathbb{N}$ is *squarefree* if n has no square factor exceeding 1. We define the Möbius function $\mu(n)$ by $\mu(n) = (-1)^{\nu(n)}$ if n is squarefree, and $\mu(n) = 0$ if n is not squarefree. The sum of the positive divisors of n is denoted $\sigma(n)$. The number of integers in $\{1, 2, \ldots, n\}$ that are coprime to n is denoted $\varphi(n)$; this is Euler's function, of course.

The symbols $\mathcal{A}$, $\mathcal{B}, \ldots$ are reserved for sets of non-negative integers. If $\mathcal{A}$ is such a set, then $A(x)$ denotes the number of members of $\mathcal{A}$ not exceeding x. By $\mathcal{A} + \mathcal{B}$ we mean the set of numbers representable as $a + b$ with $a \in \mathcal{A}$, $b \in \mathcal{B}$. By $2\mathcal{A}$ we mean $\mathcal{A} + \mathcal{A}$, by $3\mathcal{A}$ we mean $2\mathcal{A} + \mathcal{A}$, etc. By $\mathcal{A} - \mathcal{A}$ we mean the set of numbers $a - a'$, where a, $a' \in \mathcal{A}$. By $|\mathcal{A}|$ we mean the cardinality of $\mathcal{A}$.

The letters p, q shall always denote primes. The function $\log x$ is the natural logarithm. When we say $f(x) \sim g(x)$ as $x \to \infty$, we mean $f(x) = (1 + o(1))g(x)$.

So as not to have too long a reference section, we often give only one or a few later references on a particular problem so that an interested reader may begin a literature search. We do not mean to imply that the articles for which we give bibliographic data are necessarily the most important ones. Sometimes we defer all references to the extensive listings in Erdős and Graham (1980) or Guy (1994).

Have you solved the problem at the start of the introduction? Suppose $N > 1$. If $\mathcal{A}$ is the set of even numbers in $\{1, 2, \ldots, N\}$, then $|\mathcal{A}| = \lfloor \frac{1}{2}N \rfloor$ and no two members of $\mathcal{A}$ are coprime. Moreover, if a set $\mathcal{B} \subset \{1, 2, \ldots, N\}$ has more than $\lfloor \frac{1}{2}N \rfloor$ members, then either $1 \in \mathcal{B}$ or $\mathcal{B}$ contains two consecutive numbers (which are clearly coprime). Thus the answer is $\lfloor \frac{1}{2}N \rfloor$ if $N > 1$; the case $N = 1$ is clearly degenerate. See section 7.4 for more on this problem.

2. Combinatorial sieve methods

Many number-theoretic problems can be reduced to a problem of the following type. A finite set $\mathcal{A} \subset \mathbb{N}$ and a finite set $\mathcal{P}$ of prime numbers are given. Estimate the number of of members of $\mathscr{A}$ that are not divisible by any primes belonging to $\mathscr{P}$. In other words, we "sift out" the multiples of the prime numbers belonging to $\mathscr{P}$ from $\mathscr{A}$ leaving the residual set whose cardinality $S(\mathscr{A}, \mathscr{P})$ we wish to estimate. As an illustration, we are going to consider the following three problems:

(i) Estimate $\pi(x)$;

(ii) Estimate the number of prime twins q, q' with $q' = q + 2$ and $q' \leqslant x$;

(iii) For each $x \in \mathbb{N}$, estimate the number of prime pairs q, r with $q + r = x$.

Problem (ii) is connected with the famous twin prime conjecture which asserts that there are infinitely many such pairs q, $q + 2$. Problem (iii) is connected with Goldbach's conjecture which asserts that if x is an even integer at least 4, then x is a sum of two primes. These are among the most famous unsolved problems in mathematics.

For any set $\mathscr{A} \subset \mathbb{N}$, let

$$\mathscr{A}(d) = \{n \in \mathscr{A} : d \mid n\} . \tag{2.1}$$

First we are going to study problem (i). Let $x \geqslant 1$, and let us write

$$\mathscr{A} = \{n \in \mathbb{N} : n \leqslant x\} , \tag{2.2}$$

so that

$$|\mathscr{A}(d)| = \lfloor x/d \rfloor . \tag{2.3}$$

Furthermore, let us write

$$\mathscr{P} = \{p \text{ prime: } p \leqslant \sqrt{x}\} . \tag{2.4}$$

Consider the following simple fact: an integer n with $\sqrt{x} < n \leqslant x$ is a prime if and only if there is no prime $p \in \mathscr{P}$ with $p \mid n$. Thus if we start out from the set $\mathscr{A}$ in (2.2) and we sift by the primes in the set $\mathscr{P}$ in (2.4), then the set left after the sifting procedure consists of the number 1 and the primes q with $\sqrt{x} < q \leqslant x$ [so that $S(\mathscr{A}, \mathscr{P}) = 1 + \pi(x) - \pi(\sqrt{x})$].

On the other hand, the number of integers left after the sifting procedure can be computed by the well-known inclusion–exclusion principle of elementary combinatorics (cf. chapter 21). In this way, we get the following formula:

$$\begin{aligned} &|\{1\} \cup \{q \text{ prime: } \sqrt{x} < q \leqslant x\}| \\ &\quad = |\mathscr{A}| + \sum_{k=1}^{\pi(\sqrt{x})} (-1)^k \sum_{p_1 < p_2 < \cdots < p_k \leqslant \sqrt{x}} |\mathscr{A}(p_1 p_2 \cdots p_k)| . \end{aligned} \tag{2.5}$$

In fact, to prove this identity, we have to show two facts:

(a) the contribution of 1 and of each prime q with $\sqrt{x} < q \leqslant x$ to the right-hand side of (2.5) is 1;

(b) if $n \leqslant x$ and it is divisible by at least one $p \in \mathscr{P}$, then its contribution to the right-hand side of (2.5) is 0.

[Note that only positive integers $n \leqslant x$ contribute to the right-hand side of (2.5).]

We have (a) immediately since in the first term, $|\mathscr{A}|$, every positive integer

$n \leqslant x$ is counted exactly once, while 1 and the primes q with $\sqrt{x} < q \leqslant x$ are not multiples of any prime $p \leqslant \sqrt{x}$, so are counted in none of the terms $(-1)^k |\mathcal{A}(p_1 p_2 \cdots p_k)|$.

To show (b), assume that $1 < n \leqslant x$ and $p_1', p_2', \ldots, p_l'$ are all the distinct prime divisors not exceeding $\sqrt{x}$ of n, where $l \geqslant 1$. Then the first term $|\mathcal{A}|$ on the right-hand side of (2.5) contributes with a weight 1. Any other term $(-1)^k |\mathcal{A}(p_1 p_2 \cdots p_k)|$ contributes with a weight $(-1)^k$ if and only if $p_1, p_2, \ldots, p_k$ are chosen from $p_1', p_2', \ldots, p_l'$; for a fixed k there are $\binom{l}{k}$ such terms with this property. Thus the total contribution of this n to the right-hand side of (2.5) is

$$1 + \sum_{k=1}^{l} (-1)^k \binom{l}{k} .$$

By the identity

$$\sum_{k=0}^{l} (-1)^k \binom{l}{k} = (1-1)^l = 0 , \tag{2.6}$$

this contribution is 0, which completes the proof of (2.5).

Writing

$$\prod_{p \leqslant z} p = P(z) , \tag{2.7}$$

we rewrite (2.5) in the following equivalent form:

$$1 + \pi(x) - \pi(\sqrt{x}) = \sum_{d \mid P(\sqrt{x})} \mu(d) |\mathcal{A}(d)| , \tag{2.8}$$

where μ is defined section 1. In view of (2.3), we obtain:

Theorem 2.9. *If* $x \geqslant 1$, *then*

$$1 + \pi(x) - \pi(\sqrt{x}) = \sum_{d \mid P(\sqrt{x})} \mu(d) \lfloor x/d \rfloor .$$

In fact, Legendre used this formula in his numerical studies of $\pi(x)$. The sieve method described above is called the *sieve of Eratosthenes*.

By choosing the set $\mathcal{A}$ in an appropriate way, problems (ii) and (iii) can be studied similarly. For example, in the case of problem (ii) we choose

$$\mathcal{A} = \{n(n+2) \colon n \in \mathbb{N}, n \leqslant x-2\} , \qquad \mathcal{P} = \{p \text{ prime} \colon p \leqslant \sqrt{x}\} . \tag{2.10}$$

Then by using the inclusion–exclusion principle, one may similarly derive the following formula analogous to (2.8), where $\mathcal{A}$ is now defined by (2.10) rather than (2.2):

$$|\{q \colon q, q+2 \text{ are primes}, \sqrt{x} < q \leqslant x-2\}| = \sum_{d \mid P(\sqrt{x})} \mu(d) |\mathcal{A}(d)| . \tag{2.11}$$

Here we have

$$|\mathcal{A}(d)| = |\{n(n+2): n \in \mathbb{N}, n \le x-2, d \mid n(n+2)\}| .$$

It is easy to see that $|\mathcal{A}(d)| \approx \omega(d)x/d$, where

$$\omega(d) = |\{n \in \mathbb{N}: 0 \le n < d, n(n+2) \equiv 0 \pmod d\}| . \tag{2.12}$$

Clearly, $\omega(p) = 1$ for $p = 2$, $\omega(p) = 2$ for any odd prime p, and by the Chinese Remainder Theorem, the function $\omega(n)$ is multiplicative [i.e., $\omega(mn) = \omega(m)\omega(n)$ when $(m, n) = 1$]. Thus for $\mu(d) \ne 0$, i.e., for d squarefree, we have

$$\big||\mathcal{A}(d)| - \omega(d)x/d\big| \le \omega(d) \le 2^{\nu(d)} , \tag{2.13}$$

where $\nu(d)$ is defined in section 1.

Finally, to attack problem (iii), one may choose

$$\mathcal{A} = \{n(x-n): n \in \mathbb{N}, n \le x\} \tag{2.14}$$

and $\mathcal{P}$ as in (2.4) and (2.10). We leave the further details to the reader.

The main problem with the sieve of Eratosthenes is that as in (2.8) and (2.12), it gives the number of integers left after the sifting in the form of a sum, and this sum has "too many" terms. For example, the sum on the right-hand side of (2.8) has $\tau(P(\sqrt{x})) = 2^{\pi(\sqrt{x})}$ terms and, in view of (1.1), this is much bigger than the number $\pi(x)$ which is being computed. As a consequence, one cannot use Theorem 2.9 for estimating $\pi(x)$ as $x \to \infty$. Indeed, the best we can do is use the approximation

$$\lfloor x/d \rfloor \approx x/d ,$$

whose error is less than 1 so that, in view of $|\mu(d)| \le 1$, the total error would be bounded by the number of terms, namely $2^{\pi(\sqrt{x})}$.

Ignoring the error, this approximation would lead to the estimate

$$\pi(x) \approx \sum_{d \mid P(\sqrt{x})} \mu(d)\frac{x}{d} = x \sum_{d \mid P(\sqrt{x})} \frac{\mu(d)}{d} = x \prod_{p \le \sqrt{x}} \left(1 - \frac{1}{p}\right) . \tag{2.15}$$

We now recall Mertens' theorem, an elementary result in prime number theory:

$$\prod_{p \le z} \left(1 - \frac{1}{p}\right) \sim \frac{1}{\mathrm{e}^{\gamma} \log z} \quad \text{as } z \to \infty , \tag{2.16}$$

where γ is Euler's constant. Thus, from (2.15) and (2.16), we get the approximation

$$\pi(x) \approx c \frac{x}{\log x} , \tag{2.17}$$

where $c = 2\mathrm{e}^{-\gamma} \approx 1.123$, but we get this approximation with an error term bounded only by $2^{\pi(\sqrt{x})}$ which is much greater than the approximating function. This error term is certainly not negligible since, by the prime number theorem, we have

$$\pi(x) = (1 + o(1)) \frac{x}{\log x},$$

while the constant c appearing in (2.17) is larger than 1. The heuristic (2.17) does give the correct order of magnitude for $\pi(x)$, a fact that we shall see can be expected (at least for upper bounds) from sieve methods.

The situation is even slightly worse in case of problems (ii) and (iii). For example, in the case of problem (ii), if we replace $|\mathcal{A}(d)|$ by $\omega(d)x/d$, our only bound for the error in each term of (2.11) is $2^{\nu(d)}$ as given by (2.13). This is worse than the error bound of 1 per term in our analysis of $\pi(x)$. [Note that the number of terms in (2.8) and in (2.11) are the same.] If we nevertheless make the approximation $|\mathcal{A}(d)| \approx \omega(d)x/d$ in (2.11) we get

$$|\{q\colon q, q+2 \text{ are primes}, q \leq x-2\}| \approx \sum_{d|P(\sqrt{x})} \mu(d)\omega(d)x/d\,. \tag{2.18}$$

Using the fact that $\mu(d)\omega(d)/d$ is a multiplicative function of d, we have for any z that

$$\sum_{d|P(z)} \mu(d)\omega(d)/d = \prod_{p\leq z} \left(1 - \frac{\omega(p)}{p}\right). \tag{2.19}$$

Then, using Mertens' theorem (2.16) and the fact that $\omega(2) = 1$, and $\omega(p) = 2$ for $p > 2$, we have by an easy calculation that

$$\prod_{p\leq z} \left(1 - \frac{\omega(p)}{p}\right) \sim e^{-2\gamma}\alpha/\log^2 z \quad \text{as } z \to \infty, \tag{2.20}$$

where

$$\alpha = 2 \prod_{p>2} [1 - (p-1)^{-2}] = 1.3202\ldots, \tag{2.21}$$

the so-called "twin prime constant". Putting (2.19) and (2.20) (with $z = \sqrt{x}$) into the heuristic approximation (2.18), we have the "conclusion" that the number of twin primes up to x is of order of magnitude $x/\log^2 x$ with the "suggestion" that the asymptotic constant is $4e^{-2\gamma}\alpha$. In fact, this is not far from the strong twin prime conjecture which asserts that the number of twin primes up to x is $(\alpha + o(1))x/\log^2 x$. That is, the above heuristic gives the conjectured order of magnitude, but not the conjectured constant.

In general, the above attempts at using a sieve lead us to the following thoughts. There is no hope of giving asymptotics for the number of integers left after the sifting process if we sieve by a "large" set of primes $\mathcal{P}$. On the other hand, one may hope to get good bounds for the number of these integers (even if $\mathcal{P}$ is "large") by some sort of refinement that reduces the number of terms in the sum $\sum_d \mu(d)|\mathcal{A}(d)|$.

V. Brun, working in the period 1915–1924 and at least in part basing his work on that of J. Merlin, was the first to succeed in modifying the sieve of Eratosthenes to prove highly non-trivial results with a sieve. First we are going to discuss a relatively simple version of Brun's method which is called *Brun's simple* (or *pure*) *sieve*. As we shall see it enables one to derive estimates very close to the conjectured best possible one in a quite cheap way.

Brun's first idea is to reduce the number of terms in the sum $\sum_d \mu(d)|\mathcal{A}(d)|$ by reducing the number of sifting primes. The simplest way to do this is to replace the condition $p \leq \sqrt{x}$ in the definitions (2.4) and (2.10) of $\mathcal{P}$ by $p \leq z$, where z is a parameter much smaller than $\sqrt{x}$ whose exact value should be fixed as some function of x depending on the problem being studied. This means that, for example, in the case of the twin prime problem, we sift out only those integers n for which $n(n+2)$ has a small prime factor (i.e., at most z). Since the remaining integers include all the twin primes larger than z, in this way we get an upper bound for the number of twin primes between z and x. On the other hand we do not get any lower bound for this number. We might only hope to get a lower bound for the number of twin "almost primes" up to x, where q in an "almost prime" if it has no prime factor up to z. (Usually in sieve methods the term "almost prime" is reserved for the case when $\log z/\log x$ is bounded away from 0, so that if $q \leq x$ is an almost prime, it has a bounded number of prime factors. We do not follow this convention here.)

So let us now study the general sieve problem: given $\mathcal{A} \subset \mathbb{N}$ finite and

$$\mathcal{P} = \{p \text{ prime: } p \leq z\}, \tag{2.22}$$

estimate

$$S(\mathcal{A}, \mathcal{P}) := |\{n \in \mathcal{A}: (n, P(z)) = 1\}|, \tag{2.23}$$

where $P(z)$ is defined by (2.7). Then by using the inclusion–exclusion principle, we get in the now familiar way that

$$S(\mathcal{A}, \mathcal{P}) = \sum_{d|P(z)} \mu(d)|\mathcal{A}(d)|,$$

where $\mathcal{A}(d)$ is given by (2.1).

The inclusion–exclusion formula is based on the identity (2.6). This identity is only a special case of the following more general identity:

$$\sum_{k=0}^{j} (-1)^k \binom{l}{k} = (-1)^j \binom{l-1}{j} \quad \text{for all } j, l \in \mathbb{N}, \tag{2.24}$$

which can be proved easily by induction on j. This identity implies that the sum on the left-hand side is ≥ 0 for even j and ≤ 0 for odd j.

The second idea of Brun is to utilize this alternation of sign for the sum in (2.24). We are going to show that for every $t \in \mathbb{N}$ we have both

$$S(\mathcal{A}, \mathcal{P}) \leq \sum_{\substack{d|P(z) \\ \nu(d) \leq 2t}} \mu(d)|\mathcal{A}(d)| \tag{2.25}$$

and

$$S(\mathcal{A}, \mathcal{P}) \geqslant \sum_{\substack{d|P(z) \\ \nu(d) \leqslant 2t-1}} \mu(d)|\mathcal{A}(d)| . \tag{2.26}$$

To prove (2.25) we will show that if $n \in \mathcal{A}$, then the contribution of n to the right-hand side of (2.25) is

(a) 1 for $(n, P(z)) = 1$,

(b) $\geqslant 0$ for $(n, P(z)) > 1$.

The assertion (a) is trivial since if $n \in \mathcal{A}$ and $(n, P(z)) = 1$, then n is counted only in the term $d = 1$ with weight $\mu(1) = 1$. To show (b), write $(n, P(z)) = p_1 p_2 \cdots p_l$, so that $p_1, p_2, \ldots, p_l$ are distinct primes up to z and $l \geqslant 1$. Then n is counted on the right-hand side of (2.25) only for d's of the form

$$d = p_1' p_2' \cdots p_k' | p_1 p_2 \cdots p_l , \quad k \leqslant 2t .$$

For a fixed $k \leqslant 2t$, the number of these d's is $\binom{l}{k}$, and they get counted with weight $(-1)^k$. Thus the total contribution of n to the right-hand side of (2.25) is

$$\sum_{k=0}^{2t} (-1)^k \binom{l}{k} .$$

By (2.24), this sum is non-negative, which completes the proof of (b) and (2.25). In a similar way we can prove (2.26).

As an application of Brun's simple sieve we adapt (2.25) and (2.26) to the twin prime problem, getting a relatively sharp bound in a relatively simple way.

Theorem 2.27. *The number N of integers $n \leqslant x - 2$ such that both n and $n + 2$ are free of prime factors up to z satisfies*

$$N \sim \mathrm{e}^{-2\gamma} \alpha x / \log^2 z ,$$

where α is the constant defined in (2.21), *and where the asymptotic relation holds as $x, z \to \infty$ in the region $z \leqslant x^{1/(20 \log \log x)}$.*

Thus Brun's simple sieve actually gives an asymptotic formula for the number of twin "almost primes" up to x. Moreover, by making the largest choice of z allowed, it gives a non-trivial upper estimate for the distribution of twin primes:

Corollary 2.28. *The number of primes $p \leqslant x$ with $p + 2$ also prime is $O(x(\log \log x)^2 / \log^2 x)$. In particular, the sum $\sum 1/p$ for primes p with $p + 2$ prime is either convergent or finite.*

The number $\sum 1/p$ described in the corollary is referred to as *Brun's constant.* Note that the O-estimate in the corollary is only off by a factor $(\log \log x)^2$ from the conjectured order of magnitude.

On the other hand, Theorem 2.27 does not give any lower bound for the distribution of twin primes. We still do not know if there are infinitely many; as mentioned above, this is one of the great unsolved problems in mathematics. We have witnessed a general pattern with sieve methods: they often give good upper bounds, but no or weak lower bounds, unless one is interested in "almost primes" of some kind.

Proof of Theorem 2.27. Define $\mathcal{A}$ by (2.10), $\mathcal{P}$ by (2.22), and $S(\mathcal{A}, \mathcal{P})$ by (2.23). Then the quantity N in the theorem is just $S(\mathcal{A}, \mathcal{P})$. In view of (2.19) and (2.20) it is thus sufficient to prove

$$S(\mathcal{A}, \mathcal{P}) = x \sum_{d|P(z)} \mu(d)\omega(d)/d + o(x/\log^2 z)\,. \tag{2.29}$$

Note that from the upper bound and lower bound for $S(\mathcal{A}, \mathcal{P})$ given by (2.25) and (2.26), we have

$$S(\mathcal{A}, \mathcal{P}) = \sum_{\substack{d|P(z)\\ \nu(d)\leq 2t}} \mu(d)|\mathcal{A}(d)| + O\Bigg(\sum_{\substack{d|P(z)\\ \nu(d)=2t}} |\mathcal{A}(d)|\Bigg)$$

for any choice of $t \in \mathbb{N}$. Thus by (2.13) we have

$$\begin{aligned}
S(\mathcal{A}, \mathcal{P}) &= x \sum_{\substack{d|P(z)\\ \nu(d)\leq 2t}} \mu(d)\frac{\omega(d)}{d} + O\Bigg(x \sum_{\substack{d|P(z)\\ \nu(d)=2t}} \frac{\omega(d)}{d}\Bigg) + O\Bigg(\sum_{\substack{d|P(z)\\ \nu(d)\leq 2t}} 2^{\nu(d)}\Bigg)\\
&= x \sum_{d|P(z)} \mu(d)\frac{\omega(d)}{d} + O\Bigg(x \sum_{\substack{d|P(z)\\ \nu(d)\geq 2t}} \frac{\omega(d)}{d}\Bigg) + O\Bigg(\sum_{\substack{d|P(z)\\ \nu(d)\leq 2t}} 2^{\nu(d)}\Bigg)\\
&= x \sum_{d|P(z)} \mu(d)\frac{\omega(d)}{d} + O(E_1) + O(E_2)\,,
\end{aligned} \tag{2.30}$$

say.

We shall show that E_1, E_2 are $O(x/\log^6 z)$, $O(x^{1/2})$, respectively, if we choose

$$t = \lfloor 5 \log\log z \rfloor$$

(and z large enough so that $t \geq 1$). Thus (2.30) will imply a strong version of (2.29) and thus prove the theorem.

To estimate E_1, we use a weak form of Mertens' theorem. In fact, by taking the logarithm of (2.16), we have

$$\sum_{p\leq z} \frac{1}{p} \leq \log\log z + c$$

for some constant c and all large z. (This inequality is actually a weak form of a much older theorem of Euler.) Thus using $\omega(d) \leq 2^{\nu(d)}$ and the multinomial theorem,

$$E_1 = x \sum_{\substack{d \mid P(z) \\ \nu(d) \geq 2t}} \frac{\omega(d)}{d} \leqslant x \sum_{l \geq 2t} 2^l \sum_{\substack{d \mid P(z) \\ \nu(d) = l}} \frac{1}{d}$$

$$\leqslant x \sum_{l \geq 2t} \frac{2^l}{l!} \Big(\sum_{p \leqslant z} \frac{1}{p} \Big)^l \leqslant x \sum_{l \geq 2t} \frac{1}{l!} (2 \log \log z + 2c)^l .$$

The terms in this last sum are decaying at least geometrically with a common ratio bounded below 1, so that the sum is majorized by its first term. Thus

$$E_1 = \mathrm{O}\Big(\frac{x}{[2t]!} [2 \log \log z + 2c]^{2t} \Big) = \mathrm{O}\Big(x \Big[\frac{2\mathrm{e} \log \log z + 2c\mathrm{e}}{2t} \Big]^{2t} \Big)$$

$$= \mathrm{O}(x(\tfrac{1}{5}\mathrm{e})^{10 \log \log z}) = \mathrm{O}(x/\log^6 z) .$$

The majorization of E_2 is easier. We have

$$E_2 = \sum_{\substack{d \mid P(z) \\ \nu(d) \leqslant 2t}} 2^{\nu(d)} \leqslant 2^{2t} \sum_{\substack{d \mid P(z) \\ \nu(d) \leqslant 2t}} 1 = 2^{2t} \sum_{l=0}^{2t} \binom{\pi(z)}{l} \leqslant 2^{2t} \pi(z)^{2t} .$$

For large z, we have $\pi(z) \leqslant \frac{1}{2} z$ (in fact, $z \geqslant 8$ will do), so that

$$E_2 \leqslant z^{2t} = \mathrm{O}(x^{1/2}) .$$

This estimate concludes our proof of Theorem 2.27. □

To eliminate the unwanted $(\log \log x)^2$ factor in Corollary 2.28, one needs *Brun's sieve* in its complete form. To explain the crucial idea of Brun's sieve we start out from the inequalities (2.25) and (2.26). These inequalities can be rewritten in the form

$$\sum_{d \mid P(z)} \chi_1(d) \mu(d) |\mathcal{A}(d)| \leqslant S(\mathcal{A}, \mathcal{P}) \leqslant \sum_{d \mid P(z)} \chi_2(d) \mu(d) |\mathcal{A}(d)| , \tag{2.31}$$

where

$$\chi_1(d) = \begin{cases} 1 & \text{for } \nu(d) \leqslant 2t - 1 \\ 0 & \text{for } \nu(d) > 2t - 1 , \end{cases} \qquad \chi_2(d) = \begin{cases} 1 & \text{for } \nu(d) \leqslant 2t , \\ 0 & \text{for } \nu(d) > 2t . \end{cases} \tag{2.32}$$

The idea is to replace these two functions by certain other functions $\chi_1(d)$, $\chi_2(d)$ with the following properties:

(a) $\chi_1(d)$, $\chi_2(d)$ satisfy (2.31),
(b) $\chi_1(1) = \chi_2(1) = 1$,
(c) $\chi_i(d) = 0$ or 1 if $d \mid P(z)$ for $i = 1, 2$.

The goal is to make the choice for χ_1, χ_2 so that we get better upper and lower bound estimates than that afforded by (2.32).

Brun succeeded in constructing functions $\chi_1(d)$, $\chi_2(d)$ with all these properties and giving very good estimates for $S(\mathcal{A}, \mathcal{P})$. The construction is too complicated to describe here; see, e.g., Halberstam and Richert (1974) for further details.

We now cite one general theorem that may be proved by Brun's sieve – it is a special case of Theorem 2.3 in Halberstam and Richert (1974).

Theorem 2.33. *Let $k \in \mathbb{N}$, let a_i, b_i be pairs of integers for $i = 1, \ldots, k$ such that each $(a_i, b_i) = 1$ and*

$$E = \left(\prod_{i=1}^{k} a_i\right)\left(\prod_{1 \leq r < s \leq k} (a_r b_s - a_s b_r)\right) \neq 0 .$$

Let $1 > \varepsilon > 0$, x, $y \in \mathbb{R}$ be arbitrary with $2 \leq y \leq x$, and let $z = y^{\varepsilon}$. Let

$$\mathcal{A} = \left\{\prod_{i=1}^{k} (a_i n + b_i) : n \in \mathbb{N}, x - y < n \leq x\right\} \tag{2.34}$$

and denote the number of solutions of $\prod_{i=1}^{k} (a_i n + b_i) \equiv 0 \pmod p$ by $\omega(p)$ for each prime p. Then if $\mathcal{P} = \{p \text{ prime}: p \leq z\}$,

$$S(\mathcal{A}, \mathcal{P}) \leq c \left(\prod_{\substack{p \mid E \\ p \leq y}} \left(1 - \frac{1}{p}\right)^{\omega(p) - k}\right) \frac{y}{\log^k y} ,$$

where the constant c depends only on k and ε.

Note that all of the sets $\mathcal{A}$ as in (2.2), (2.10), and (2.14) are of the form (2.34). For example, by choosing $k = 2$, $a_1 = 1$, $b_1 = 0$, $a_2 = 1$, $b_2 = 2$, $y = x$, and $z = \sqrt{x}$ we obtain:

Corollary 2.35. *There is a positive constant c such that if $x \geq 2$, then*

$$|\{q: q, q + 2 \text{ are primes}, q \leq x - 2\}| \leq c \frac{x}{\log^2 x} .$$

Similarly, if we let $k = 2$, $a_1 = 1$, $b_1 = 0$, $a_2 = -1$, $b_2 = n$, $x = y = n$, and $z = \sqrt{n}$, we get:

Corollary 2.36. *There is a positive constant c such that if $n \in \mathbb{N}$ and n is even, then*

$$|\{p: p \text{ and } n - p \text{ are primes}\}| \leq c \left(\prod_{p \mid n} \left(1 - \frac{1}{p}\right)^{-1}\right) \frac{n}{\log^2 n} .$$

Even for the simpler problem (i) where the prime number theorem gives us an asymptotic formula for $\pi(x)$, Theorem 2.33 can tell us something non-trivial when

y is small compared to x. By choosing $k=1$, $a_1=1$, and $b_1=0$, we have the following result originally due to Hardy and Littlewood.

Corollary 2.37. *If $2 \leq y \leq x$, there is an absolute constant c such that*

$$\pi(x) - \pi(x-y) \leq c\frac{y}{\log y}\,.$$

Brun's sieve has numerous applications and many of these are due to P. Erdős who, perhaps more than any other person, showed that sieve methods are indeed a powerful tool in number theory. For example, Erdős used Brun's sieve to estimate the differences between the consecutive primes. Let p_i denote the ith prime (so that $p_1=2$, $p_2=3$, $p_3=5$, etc.), write $d_n = p_n - p_{n-1}$ for $n>1$, and let $d_1 = 2$. It follows easily from the prime number theorem that

$$\liminf_{n\to\infty} d_n/\log n \leq 1\,.$$

We now prove the following result due to Erdős (1940).

Theorem 2.38. *There is a constant $c<1$ such that*

$$\liminf_{n\to\infty} d_n/\log n \leq c\,. \tag{2.39}$$

Proof. Let $\varepsilon>0$ be arbitrary, but fixed. Suppose that

$$\liminf_{n\to\infty} d_n/\log n > 1 - \tfrac{1}{2}\varepsilon\,. \tag{2.40}$$

Then there is some x_0 such that for $x \geq x_0$, if $n > \pi(x/\log x)$ we have $d_n > (1-\varepsilon)\log x$. Let $L = \pi(x) - \pi(x/\log x)$ and assume $\delta = \delta(x,\varepsilon)$ is such that there are exactly δL values of n with $\pi(x/\log x) < n \leq \pi(x)$ and d_n is between $(1-\varepsilon)\log x$ and $(1+\varepsilon)\log x$. Then for $x \geq x_0$, there are $(1-\delta)L$ values of n in this range with $d_n \geq (1+\varepsilon)\log x$. Thus

$$\delta L(1-\varepsilon)\log x + (1-\delta)L(1+\varepsilon)\log x \leq \sum_{n\leq\pi(x)} d_n \leq x\,,$$

so that

$$\varepsilon(1-2\delta) \leq \frac{x}{L\log x} - 1\,. \tag{2.41}$$

By the prime number theorem, $L\log x = (1+o(1))x$, so that (2.41) implies (since $\varepsilon>0$ is fixed) there is some $x_1(\varepsilon)$ such that if $x \geq x_1(\varepsilon)$ we have $\delta > \frac{1}{3}$.

We now use Theorem 2.33 to show that $\delta = O(\varepsilon)$ so that if ε is sufficiently small, (2.40) cannot hold. This will prove the theorem. For any $t \in \mathbb{N}$, let $D(t,x)$ denote the number of primes $p \leq x$ with $p+t$ prime. Thus

$$\delta L \leq \sum_{(1-\varepsilon)\log x < t < (1+\varepsilon)\log x} D(t,x)\,. \tag{2.42}$$

But by Theorem 2.33 with $k=2$, $a_1=1$, $b_1=0$, $a_2=1$, $b_2=t$, $y=x$, and $z=\sqrt{x}$, we have some absolute constant c' with

$$D(t,x) \leq c' \left(\prod_{\substack{p|t \\ p \leq x}} \left(1 - \frac{1}{p}\right)^{-1} \right) \frac{x}{\log^2 x}. \tag{2.43}$$

From elementary arguments it is not difficult to prove that

$$\sum_{t \leq u} \prod_{p|t} \left(1 - \frac{1}{p}\right)^{-1} \sim c''u \quad \text{as } u \to \infty$$

for some constant c''. Thus

$$\sum_{(1-\varepsilon)\log x < t < (1+\varepsilon)\log x} D(t,x) \leq c'(c'' + o(1)) 2\varepsilon \log x \frac{x}{\log^2 x}.$$

Since $L \sim x/\log x$, (2.42) thus implies for $x \geq x_2(\varepsilon)$

$$\delta \leq 3c'c''\varepsilon,$$

which is what we wanted to prove. □

Since 1942, the value of the constant c in (2.39) has been improved by several authors. The best estimate (derived by both combinatorial and analytic tools) has $c < \frac{1}{4}$ and is due to Maier (1988). Of course the twin prime conjecture implies that $c = 0$.

Theorem 2.33 can be proved also by another sieve method of less combinatorial nature which is due to A. Selberg. In some applications, Selberg's sieve is slightly superior to Brun's.

Returning to the three problems at the beginning of this section, note that in problem (i) we counted integers $n \not\equiv 0 \pmod p$ for primes $p \leq z$, in problem (ii) we counted integers satisfying $n \not\equiv 0 \pmod p$ and $n \not\equiv 2 \pmod p$, and in problem (iii) the excluded classes were $n \not\equiv 0 \pmod p$, $n \not\equiv x \pmod p$. In other words, there are 1, 2 and 2 "forbidden" residue classes, respectively. Brun's sieve and Selberg's sieve have a common feature: both methods can be used only in the case that the number of forbidden residue classes is bounded or it grows only very slowly in terms of p. If the number of forbidden residue classes grows rapidly (for example, more than cp residue classes for each p), then other sieve methods must be used. The most important sieve method of this type is the *large sieve* of Linnik and Rényi.

See Halberstam and Richert (1974) for detailed discussion of "small sieves" (Brun's and Selberg's sieves) and Montgomery (1971) for the large sieve. The former reference also contains proofs of J. Chen's remarkable theorems that (1) there are infinitely many primes p such that $p + 2$ is either prime or the product of two primes, and (2) every sufficiently large even number is the sum of a prime and another number which is either prime or the product of two primes.

Finally we mention Rosser's sieve, a general principle for a combinatorial small sieve. Iwaniec (1981) has done extensive work developing this general principle and has given details for some important special cases.

3. Bases and density theorems on addition of sets

As mentioned in the preceding section, it is conjectured that every even number exceeding 2 is a sum of two primes. This conjecture, which was stated by Goldbach in a letter to Euler in 1742, has the immediate corollary that *every* number exceeding 5 is a sum of three primes and that every number exceeding 1 is a sum of at most three primes.

In 1770, Waring stated without proof that for every n there is some number $g(n)$ such that every natural number is the sum of at most $g(n)$ positive nth powers. In that same year, Lagrange solved Waring's problem for $n = 2$, showing that every natural number is the sum of at most four squares. In 1909, Waring's conjecture was finally proved by Hilbert using a combinatorial argument.

Let $\mathbb{N}_0 = \mathbb{N} \cup \{0\}$ denote the set of non-negative integers. A set $\mathcal{A} \subseteq \mathbb{N}_0$ is said to be a *basis of order k* if every natural number can be represented as the sum of at most k elements of $\mathcal{A}$. If every sufficiently large integer can be represented as the sum of at most k elements of $\mathcal{A}$, then $\mathcal{A}$ is said to be an *asymptotic basis of order k*. Thus Goldbach's conjecture implies that the set of primes is an asymptotic basis of order 3 and that the set of primes together with 1 is a basis of order 3. Not only has Waring's problem been settled, the minimal choices for the numbers $g(n)$ are "known" for every n. If $G(n)$ is the least number such that the positive nth powers form an asymptotic basis of order $G(n)$, then *no* value of $G(n)$ is known except for $n = 1$, 2, and 4. See Vaughan (1981) and Balasubramanian et al. (1986) for more details.

If $\mathcal{A} \subseteq \mathbb{N}_0$, then the *lower asymptotic density* $\underline{d}(\mathcal{A})$, the *upper asymptotic density* $\overline{d}(\mathcal{A})$, and, if it exists, the *asymptotic density* $d(\mathcal{A})$ of $\mathcal{A}$ are defined by:

$$\underline{d}(\mathcal{A}) = \liminf_{n\to\infty} A(n)/n ,$$

$$\overline{d}(\mathcal{A}) = \limsup_{n\to\infty} A(n)/n ,$$

$$d(\mathcal{A}) = \lim_{n\to\infty} A(n)/n ,$$

respectively.

Assuming the Riemann hypothesis, Hardy and Littlewood proved in 1922 that every sufficiently large odd integer can be represented as the sum of three primes, which would imply, of course, that the set of primes is an asymptotic basis of order 4. The Hardy and Littlewood theorem was proved unconditionally in 1937 by Vinogradov. But the first to unconditionally prove that the set of primes is an asymptotic basis of some finite order was Schnirelmann in 1930.

This work of Schnirelmann opened up an important chapter in combinatorial number theory. His starting point was the following simple corollary of Brun's sieve and, in particular, Theorem 2.33.

Corollary 3.1. *The set of integers which can be represented as the sum of two primes has positive lower asymptotic density. That is, if $\mathcal{P}$ denotes the set of primes, then $\underline{d}(2\mathcal{P}) > 0$.*

Thus to prove that $\mathcal{P}$ is an asymptotic basis of finite order, it would be sufficient to show that any set of positive lower asymptotic density must necessarily be an asymptotic basis of finite order. Unfortunately, this is not so, as the set

$$\mathcal{A} = \{0, 2, 4, \ldots\} \tag{3.2}$$

of even non-negative integers show. This set has asymptotic density $\frac{1}{2}$, but no odd integer is a sum of members of $\mathcal{A}$.

However, Schnirelmann was able to save this idea with his concept of *Schnirelmann density*. If $\mathcal{A} \subseteq \mathbb{N}_0$ and we write $A^*(n)$ for the number of *positive* members of $\mathcal{A}$ up to n, then the Schnirelmann density $\sigma(\mathcal{A})$ of $\mathcal{A}$ is defined by:

$$\sigma(\mathcal{A}) = \inf_{n \in \mathbb{N}} A^*(n)/n \, .$$

Thus $\sigma(\mathcal{A}) > 0$ holds if and only if both $1 \in \mathcal{A}$ and $\underline{d}(\mathcal{A}) > 0$ hold. In addition, $\sigma(\mathcal{A}) = 1$ if and only if $\mathcal{A} = \mathbb{N}$ or $\mathbb{N}_0$. Schnirelmann proved the following theorems on the Schnirelmann density of sum sets.

Theorem 3.3. *If $\mathcal{A}$, $\mathcal{B} \subseteq \mathbb{N}_0$ with $0 \in \mathcal{A} \cap \mathcal{B}$, then*

$$\sigma(\mathcal{A} + \mathcal{B}) \geq \sigma(\mathcal{A}) + \sigma(\mathcal{B}) - \sigma(\mathcal{A})\sigma(\mathcal{B}) \, .$$

Proof. We may assume $\sigma(\mathcal{A}) > 0$. Let n be an arbitrary natural number and suppose

$$1 = a_1 < a_2 < \cdots < a_k \leq n$$

are the positive members of $\mathcal{A}$ that do not exceed n. Since $0 \in \mathcal{B}$, we have also $a_1, \ldots, a_k \in \mathcal{A} + \mathcal{B}$. What other members in $\mathcal{A} + \mathcal{B}$ do not exceed n? We now count those members of $\mathcal{A} + \mathcal{B}$ of the form $a_i + b$, where $i < k$, $b \in \mathcal{B}$, and $a_i < a_i + b < a_{i+1}$. This number is $B^*(a_{i+1} - a_i - 1) \geq (a_{i+1} - a_i - 1)\sigma(\mathcal{B})$. Similarly the number of $a_k + b$, where $b \in \mathcal{B}$ and $a_k < a_k + b \leq n$, is $B^*(n - a_k) \geq (n - a_k)\sigma(\mathcal{B})$. Thus the number of positive members of $\mathcal{A} + \mathcal{B}$ up to n is at least

$$\begin{aligned}
&A^*(n) + (n - a_k)\sigma(\mathcal{B}) + \sum_{i<k} (a_{i+1} - a_i - 1)\sigma(\mathcal{B}) \\
&\quad = A^*(n) + (n - a_1 - k + 1)\sigma(\mathcal{B}) \\
&\quad = A^*(n) + (n - A^*(n))\sigma(\mathcal{B}) \\
&\quad = (1 - \sigma(\mathcal{B}))A^*(n) + n\sigma(\mathcal{B}) \\
&\quad \geq n(1 - \sigma(\mathcal{B}))\sigma(\mathcal{A}) + n\sigma(\mathcal{B}) \, ,
\end{aligned}$$

which proves the theorem. □

Theorem 3.4. *If $\mathcal{A}$, $\mathcal{B} \subseteq \mathbb{N}_0$ with $0 \in \mathcal{A} \cap \mathcal{B}$ and $\sigma(\mathcal{A}) + \sigma(\mathcal{B}) \geq 1$, then $\sigma(\mathcal{A} + \mathcal{B}) = 1$. That is, every non-negative integer can be represented in the form $a + b$ with $a \in \mathcal{A}$, and $b \in \mathcal{B}$.*

Proof. If $n \in \mathcal{A} \cup \mathcal{B}$, then $n \in \mathcal{A} + \mathcal{B}$. So suppose $n \in \mathbb{N}_0$ and $n \notin \mathcal{A} \cup \mathcal{B}$. Then $n > 1$. We have

$$n \leq A^*(n) + B^*(n) = A^*(n-1) + B^*(n-1) .$$

Consider the positive integers $a \in \mathcal{A}$ for $a < n$, and $n - b$ for $b \in \mathcal{B}$, $0 < b < n$. There are at least n of these numbers and they all lie in $\{1, 2, \ldots, n-1\}$. Thus we have $a = n - b$ for some $a \in \mathcal{A}$, $b \in \mathcal{B}$, so that $n = a + b \in \mathcal{A} + \mathcal{B}$. □

Corollary 3.5. *If $\mathcal{A} \subseteq \mathbb{N}_0$ with $\sigma(\mathcal{A}) > 0$, then $\mathcal{A}$ is a basis of some finite order.*

Proof. Write $\mathcal{A}_0 = \mathcal{A} \cup \{0\}$. Then by Theorem 3.3 and induction, $\sigma(k\mathcal{A}_0) \geq 1 - (1 - \sigma(\mathcal{A}_0))^k$ for every $k \in \mathbb{N}$. Thus there is some k with $\sigma(k\mathcal{A}_0) \geq \frac{1}{2}$. Thus by Theorem 3.4, with $\mathcal{A} = \mathcal{B} = k\mathcal{A}_0$, we have $2k\mathcal{A}_0 = \mathbb{N} \cup \{0\}$. Thus $\mathcal{A}$ is a basis of order $2k$. □

Schnirelmann's theorem on the set of primes $\mathcal{P}$ follows easily from Corollaries 3.1 and 3.5 as we now see.

Theorem 3.6. *The set of primes is an asymptotic basis of finite order.*

Proof. Write $\mathcal{P}_1 = \mathcal{P} \cup \{0, 1\}$. Then by Corollary 3.1, $\sigma(2\mathcal{P}_1) > 0$, so that by Corollary 3.5, $2\mathcal{P}_1$ is a basis of some finite order. Thus $\mathcal{P}_1$ is a basis of some order k. Thus every $n \in \mathbb{N}$ may be represented in the form $s + v$, where s is a sum of at most k primes and $0 \leq v \leq k$.

We now show that every $n > 2$ may be represented as a sum of at most $\frac{3}{2}k + 1$ primes. Indeed, $n - 2 \in \mathbb{N}$, so we may write $n - 2 = s + v$, where s is a sum of at most k primes and $0 \leq v \leq k$. Then $n = s + (v + 2)$ and $2 \leq v + 2 \leq k + 2$. But it is trivial that every integer $m \geq 2$ can be represented as a sum of 2's and 3's with at most $\frac{1}{2}m$ summands. Thus $v + 2$ is a sum of at most $\frac{1}{2}(k + 2)$ primes, each prime being 2 or 3. Thus $n = s + (v + 2)$ is a sum of at most $\frac{3}{2}k + 1$ primes. □

In 1942, Mann improved on Schnirelmann's theorems 3.3 and 3.4 by proving the following result which had come to be known as the $\alpha + \beta$ conjecture.

Theorem 3.7. *If $\mathcal{A}$, $\mathcal{B} \subseteq \mathbb{N}_0$ with $0 \in \mathcal{A} \cap \mathcal{B}$, then $\sigma(\mathcal{A} + \mathcal{B}) \geq \min\{1, \sigma(\mathcal{A}) + \sigma(\mathcal{B})\}$.*

Thus in the case $\sigma(\mathcal{A}) > 0$, $\sigma(\mathcal{B}) > 0$, and $\sigma(\mathcal{A}) + \sigma(\mathcal{B}) < 1$, Theorem 3.7 gives a sharper result than Theorems 3.3 and 3.4. As we saw above, these latter theorems can be proved relatively easily. However, Mann's theorem 3.7 is much deeper.

A disadvantage of Schnirelmann's and Mann's theorems is that in both cases, the statement is formulated in terms of Schnirelmann density which is a fairly artificial concept. In particular, if we use these results for estimating the order of a basis, then we often get rather poor estimates. Thus in many applications it would be preferable to have an addition theorem involving asymptotic (lower) density. In 1953, Kneser proved the following (very deep) theorem.

Theorem 3.8. *If* $\mathscr{A}_0, \ldots, \mathscr{A}_k \subseteq \mathbb{N}_0$ *are infinite, then either*

$$\underline{d}(\mathscr{A}_0 + \cdots + \mathscr{A}_k) \geqslant \liminf_{n\to\infty} (A_0(n) + \cdots + A_k(n))/n$$

or there are natural numbers $g, a_0, \ldots, a_k$, *such that*

(i) *each* $\mathscr{A}_i$ *is contained in the union* $\mathscr{A}_i'$ *of* a_i *distinct congruence classes* (mod g),

(ii) *there are at most finitely many positive members of* $\mathscr{A}_0' + \cdots + \mathscr{A}_k'$ *not in* $\mathscr{A}_0 + \cdots + \mathscr{A}_k$,

(iii) $d(\mathscr{A}_0 + \cdots + A_k) \geqslant (a_0 + \cdots + a_k - k)/g$.

The following result of Nathanson and Sárközy (1989) gives a particularly simple application of Kneser's theorem 3.8.

Theorem 3.9. *If* $\mathscr{A} \subseteq \mathbb{N}$ *is an asymptotic basis of order* h *and if* $\underline{d}(\mathscr{A}) > 1/h$, *then for every* β *with* $1/h < \beta < \underline{d}(\mathscr{A})$, *there is a set* $\mathscr{B} \subset \mathscr{A}$ *with* $\underline{d}(\mathscr{B}) = \beta$ *and with* $\mathscr{B}$ *also an asymptotic basis of order* h.

Proof. Let $\mathscr{A}$ be as described and choose β with $1/h < \beta < \underline{d}(\mathscr{A})$. Let $\mathscr{C}$ be any subset of $\mathscr{A}$ with $\underline{d}(\mathscr{C}) = \beta$. Let $H = (h-1)/(h\beta - 1)$ and let $\mathscr{A}_0 \subset \mathscr{A}$ be a finite set such that for each $j \leqslant H$, $\mathscr{A}_0$ contains a representative of each residue class (mod j) that has at least one representative in $\mathscr{A}$. We claim that $\mathscr{B} = \mathscr{C} \cup \mathscr{A}_0$ fulfills the conditions of the theorem.

First, it is clear that $\underline{d}(\mathscr{B}) = \beta$, since $\mathscr{A}_0$ is finite. To show $\mathscr{B}$ is an asymptotic basis of order h, we apply Kneser's theorem 3.8 with h copies of $\mathscr{B}$. Since $\underline{d}(\mathscr{B}) > 1/h$, the first condition cannot hold. Thus there is some number g and a set $\mathscr{B}'$ of a residue classes (mod g) such that $\mathscr{B} \subseteq \mathscr{B}'$, all sufficiently large members of $h\mathscr{B}'$ are in $h\mathscr{B}$, and

$$1 \geqslant d(h\mathscr{B}) \geqslant \frac{ha - (h-1)}{g} \geqslant h\beta - \frac{h-1}{g}.$$

Thus $g \leqslant H$, so that $\mathscr{B}$ contains representatives of exactly the same residue classes (mod g) as does $\mathscr{A}$. Since $\mathscr{B} \subseteq \mathscr{B}'$ and $\mathscr{B}'$ is a union of complete residue classes (mod g), we have $\mathscr{A} \subseteq \mathscr{B}'$. Thus $\mathscr{B}'$ is an asymptotic basis of order h, as is $\mathscr{B}$. □

Kneser's theorem 3.8 serves a unifying role in additive number theory in that certain prior results with longer proofs can also be seen as fairly immediate corollaries of Theorem 3.8. For example, it is possible to prove via Schnirelmann's theorem 3.3 that if $\mathcal{A} \subseteq \mathbb{N}$, $\underline{d}(\mathcal{A}) > 0$, and $\mathcal{A}$ contains a finite subset that is relatively prime, then $\mathcal{A}$ is an asymptotic basis of some order – this is essentially what is done in the proof of Theorem 3.6. But assuming Kneser's theorem, the result is virtually immediate.

As one further example, we state the following corollary of Kneser's theorem, leaving the proof for the reader. [The proof is not completely trivial – for help, consult Halberstam and Roth (1983, pp. 54–55).] Parts of this result are due independently to Cauchy in 1813, Davenport in 1935, and I. Chowla in 1935. It is also possible to give a (relatively simple) direct proof, not using Kneser's theorem.

Theorem 3.10. (The Cauchy–Davenport–Chowla Theorem). *If $\mathcal{A}$, $\mathcal{B} \subseteq \mathbb{Z}/g$ with $|\mathcal{A}| = r$, $|\mathcal{B}| = s$, $0 \in \mathcal{B}$, and every other member of $\mathcal{B}$ is coprime to g, then $|\mathcal{A} + \mathcal{B}| \geq \min\{g, r + s - 1\}$.*

When we are studying a finite set $\mathcal{A} \subseteq \mathbb{N}$ and we need an addition theorem, then the only assumption that one might like to use is that $\mathcal{A} \subseteq \{1, 2, \ldots, N\}$ and $|\mathcal{A}|/N$ is large. Improving on a joint theorem with Nathanson, Sárközy has recently proved the following addition theorem of this type, see Sárközy (1989/1994).

Theorem 3.11. *Assume that $\mathcal{A} \subseteq \{1, 2, \ldots, N\}$ and that $|\mathcal{A}| > (N/k) + 1$, where $k \in \mathbb{N}$. Then there are d, $l \in \mathbb{N}$ with $d < k$, $l < 118/k$ such that $l\mathcal{A}$ contains an N-term arithmetic progression of multiples of d.*

Bourgain (1990) and Freiman et al. (1992) have recently shown that if $\mathcal{A}$ is "dense", then there are "long" arithmetic progressions in $2\mathcal{A}$ and considerably longer ones in $3\mathcal{A}$.

Using exponential sums and methods from the geometry of numbers, Freiman (1973) gave a deep analysis of the structure of sum sets of the form $k\mathcal{A}$ for finite sets $\mathcal{A}$ assuming that $|k\mathcal{A}|$ is not much greater than $|\mathcal{A}|$. Indeed, suppose $k_1, \ldots, k_d$ are integers at least 2, $u, v_1, \ldots, v_d$ are integers, and there are $k_1 \cdots k_d$ distinct integers of the form

$$n = u + \sum_{i=1}^{d} x_i v_i \quad \text{where } x_i \in \{1, \ldots, k_i\} \text{ for } i = 1, \ldots, d\,.$$

Then the set $\mathcal{P}$ of such numbers n is called a d-dimensional arithmetic progression of size $|\mathcal{P}| = k_1 \cdots k_d$. Freiman's most important result, the so-called "doubling theorem", says that if $|2\mathcal{A}|$ is not much greater than $|\mathcal{A}|$, then $|\mathcal{A}|$ can be well-covered by a generalized arithmetic progression:

Theorem 3.12. *For all $\alpha > 1$ there are constants $c_1 = c_1(\alpha)$, $c_2 = c_2(\alpha)$ such that if $|2\mathscr{A}| < \alpha|\mathscr{A}|$, then there is a generalized arithmetic progression $\mathscr{P}$ of dimension $d < c_1$ with $\mathscr{A} \subseteq \mathscr{P}$ and $|\mathscr{P}| < c_2|\mathscr{A}|$.*

Many further details, results, and problems on addition theorems and bases can be found in Halberstam and Roth (1983), Stöhr (1955), and Ostmann (1956).

A set $\mathscr{A} \subseteq \mathbb{N}_0$ is said to be a *minimal basis of order k* if $\mathscr{A}$ is a basis of order k, but no proper subset of $\mathscr{A}$ is a basis of order k. A set $\mathscr{A} \subseteq \mathbb{N}_0$ is said to be a *maximal nonbasis of order k* if $\mathscr{A}$ is not a basis of order k, but $A \cup \{a\}$ is a basis of order k for every $a \in \mathbb{N}_0 \setminus \mathscr{A}$. Stöhr, Härtter, Erdős, Nathanson and others have studied properties of minimal bases and maximal nonbases, see Stöhr (1955), Erdős and Nathanson (1987) and Nathanson (1989).

Similarly one can define the concept of a minimal asymptotic basis or maximal asymptotic nonbasis. Note that an immediate corollary of Theorem 3.9 is that if $\mathscr{A}$ is a minimal asymptotic basis of order h, then $\underline{d}(\mathscr{A}) \leqslant 1/h$. That this result is sharp is shown in Erdős and Nathanson (1988).

If $\mathscr{B} \subseteq \mathbb{N}_0$ is such that

$$\mathscr{A} \subseteq \mathbb{N}_0 \text{ and } 0 < \sigma(\mathscr{A}) < 1 \text{ imply } \sigma(\mathscr{A} + \mathscr{B}) > \sigma(\mathscr{A}),$$

then $\mathscr{B}$ is said to be an *essential component*. In 1933, Khintchin proved that the set of squares is an essential component. In 1936, Erdős generalized this theorem as follows by proving that every basis is an essential component.

Theorem 3.13. *If $\mathscr{B} \subseteq \mathbb{N}_0$ is a basis of order h and $\mathscr{A} \subseteq \mathbb{N}_0$ is arbitrary, then*

$$\sigma(\mathscr{A} + \mathscr{B}) \geqslant \sigma(\mathscr{A}) + \frac{1}{2h}(1 - \sigma(\mathscr{A}))\sigma(\mathscr{A}).$$

Landau slightly sharpened this result. Using a complicated graph-theoretic argument, Plünnecke improved the conclusion of Theorem 3.13 to the much sharper $\sigma(\mathscr{A} + \mathscr{B}) \geqslant \sigma(\mathscr{A})^{1-1/h}$. Ruzsa (1990/91) analyzed Plünnecke's method and gave further applications.

In 1942, Linnik proved the existence of a "thin" essential component $\mathscr{B}$ with

$$B(x) < \exp[(\log x)^{9/10+\varepsilon}].$$

This shows that an essential component need not be a basis. Wirsing improved on this estimate and Ruzsa (1987), using exponential sums, proved the following theorem (which settles the problem).

Theorem 3.14. *For every $\varepsilon > 0$ there is an essential component $\mathscr{B}$ with $B(x) = O((\log x)^{1+\varepsilon})$. Moreover, if $\mathscr{B}$ is any essential component, then there is some $c > 0$ such that $B(x) > (\log x)^{1+c}$ for all sufficiently large x.*

It is not known if $\mathscr{B} = \{2^i3^j\colon i, j \in \mathbb{N}_0\}$ is an essential component. Note that $B(x) \sim c \log^2 x$ for some $c > 0$.

4. Other additive problems

4.1. Sidon sets

As before, let $\mathbb{N}_0 = \mathbb{N} \cup \{0\}$. If $\mathcal{A} \subseteq \mathbb{N}_0$, let $s(\mathcal{A}, n)$ denote the number of solutions of

$$a + a' = n \quad \text{with } a, a' \in \mathcal{A}\ ,\ a \leq a'\ .$$

In 1931, S. Sidon posed the following two problems:

(i) How "dense" can $\mathcal{A}$ be if $s(\mathcal{A}, n) \leq 1$ for all n?

(ii) What is the slowest growing function $f(n)$ such that for some $\mathcal{A}$, $1 \leq s(\mathcal{A}, n) \leq f(n)$ holds for all $n \in \mathbb{N}_0$?

The first question motivates the following definition. If a set $\mathcal{A} \subseteq \mathbb{N}_0$ satisfies $s(\mathcal{A}, n) \leq 1$ for all n, then it is called a *Sidon set*. The first remarkable fact about Sidon sets is that the greedy algorithm provides a simple way to show the existence of relatively dense, finite Sidon sets.

Theorem 4.1. *If $N \in \mathbb{N}$, then there is a Sidon set $\mathcal{A} \subseteq \{1, 2, \ldots, N\}$ with $|\mathcal{A}| \geq \lfloor N^{1/3} \rfloor$.*

Proof. Clearly it suffices to show that if $\{a_1, a_2, \ldots, a_t\} \subseteq \{1, 2, \ldots, N\}$ is a Sidon set of cardinality t and

$$t \leq N^{1/3} - 1\ , \tag{4.2}$$

then there is an integer b such that

$$1 \leq b \leq N\ , \qquad b \not\in \{a_1, a_2, \ldots, a_t\}\ , \tag{4.3}$$

and

$$\{a_1, a_2, \ldots, a_t\} \cup \{b\} \text{ is a Sidon set}\ . \tag{4.4}$$

To show this, note that if an integer b satisfying (4.3) does not satisfy (4.4), then there are a_i, a_j, a_k with

$$a_i + b = a_j + a_k\ ,$$

or there are a_u, a_v with

$$b + b = a_u + a_v\ .$$

There are at most t^3 choices for triples a_i, a_j, a_k and at most t^2 choices for pairs a_u, a_v, so there are at most $t^3 + t^2$ "bad" b's. Thus in view of (4.2), the number of "good" b's satisfying (4.3) and (4.4) is at least

$$N - t - (t^3 + t^2) \geq N - (t + 1)^3 + 1 \geq 1\ ,$$

so there is at least one "good" b. □

This argument can be modified easily to give the existence of a dense *infinite* Sidon set.

Theorem 4.5. *There is a Sidon set* $\mathcal{A} \subseteq \mathbb{N}$ *such that* $A(n) \geq \lfloor n^{1/3} \rfloor$ *for all* $n \in \mathbb{N}$.

In the finite case, the estimate obtained in this way can be improved considerably. In fact, by using Singer's theorem on perfect difference sets, Erdős and Chowla independently proved in 1944 the following result.

Theorem 4.6. *For infinitely many* $N \in \mathbb{N}$, *there is a Sidon set* $\mathcal{A} \subseteq \{1, 2, \ldots, N\}$ *with* $|\mathcal{A}| > N^{1/2}$. *For all* $N \in \mathbb{N}$, *there is a Sidon set* $\mathcal{A} \subseteq \{1, 2, \ldots, N\}$ *with* $|\mathcal{A}| > N^{1/2} - cN^{5/16}$ (*for some absolute positive constant* c).

On the other hand, in 1941 Erdős and Turán had proved the following.

Theorem 4.7. *There is an absolute, positive constant* c *such that if* $N \in \mathbb{N}$ *and* $A \subseteq \{1, 2, \ldots, N\}$ *is a Sidon set, then* $|\mathcal{A}| < N^{1/2} + cN^{1/4}$.

Erdős conjectures that the expression $cN^{5/16}$ in Theorem 4.6 can be replaced with c and that the expression $cN^{1/4}$ in Theorem 4.7 can be replaced with $N^{o(1)}$.

Much less is known in the infinite case. The best-known lower bound, due to Ajtai et al. (1981) is annoyingly only slightly better than the near-trivial Theorem 4.5.

Theorem 4.8. *There is a Sidon set* $\mathcal{A} \subseteq \mathbb{N}$ *such that* $A(n) > 10^{-3}(n \log n)^{1/3}$ *for all sufficiently large* n.

We can do much better if we only want an Ω-result. Improving on a result of Erdős, Krückeberg showed the following in 1961.

Theorem 4.9. *There is a Sidon set* $\mathcal{A} \subseteq \mathbb{N}$ *such that*

$$\limsup_{n\to\infty} A(n)/\sqrt{n} \geq 1/\sqrt{2}\,.$$

Erdős conjectures that this lim sup is 1. Note that by Theorem 4.7, it is at most 1.

We do know the lim sup in Theorem 4.9 cannot be replaced by lim inf, as the following result of Erdős in 1955 shows.

Theorem 4.10. *There is an absolute constant* c *such that if* $\mathcal{A} \subseteq \mathbb{N}$ *is any infinite Sidon set, then*

$$\liminf_{n\to\infty} A(n)/\sqrt{n/\log n} \leq c\,.$$

The gap between Theorems 4.8 and 4.10 remains an important unsolved problem in the subject.

We can obtain interesting problems and results if we relax the condition $s(\mathscr{A}, n) \leq 1$ of a Sidon set to $s(\mathscr{A}, n) \leq g$. In 1960, using probability theory, Erdős and Rényi proved, among other interesting results, the following.

Theorem 4.11. *For every $\varepsilon > 0$, there is a number $g = g(\varepsilon)$ and an infinite set $\mathscr{A} \subseteq \mathbb{N}$ with $s(\mathscr{A}, n) \leq g$ for all $n \in \mathbb{N}$ and*

$$\lim_{n\to\infty} A(n)/n^{1/2-\varepsilon} = \infty .$$

In connection with problem (ii), in 1956 Erdős, using a probabilistic method, proved the following.

Theorem 4.12. *There are positive constants c_1, c_2 and a set $\mathscr{A} \subseteq \mathbb{N}$ such that*

$$c_1 \log n < s(\mathscr{A}, n) < c_2 \log n \quad \textit{for all } n \in \mathbb{N} .$$

It is not known if $s(\mathscr{A}, n) \sim c \log n$ for some positive constant c is possible. Another attractive problem is whether $s(\mathscr{A}, n) \geq 1$ for all $n \in \mathbb{N}$ implies $s(\mathscr{A}, n)$ is unbounded.

The following result due to Erdős and Fuchs in 1956 involves analytic methods. Let $r(\mathscr{A}, n)$ denote the total number of solutions of $a + a' = n$ with $a, a' \in \mathscr{A}$, where now, we do not require $a \leq a'$.

Theorem 4.13. *If $c > 0$ and $\mathscr{A} \subseteq \mathbb{N}$, then*

$$\sum_{n\leq N} r(\mathscr{A}, n) = cN + o(N^{1/4}(\log N)^{-1/2})$$

cannot hold.

As a corollary one can get an Ω-result for the error term in the circle problem; that is, for the quantity

$$\pi r^2 - \sum_{\substack{(i,j)\in\mathbb{Z}^2 \\ i^2+j^2\leq r^2}} 1 .$$

Recently, Montgomery and Vaughan (1990) have shown that the factor $(\log N)^{-1/2}$ in Theorem 4.13 may be dropped. See Halberstam and Roth (1983, including the footnote on p. 106) and Erdős (1956) for further information.

The results and problems discussed so far have been extended and generalized by many people. For further references, see Hayashi (1981) and Erdős et al. (1986).

4.2. The arithmetic structure of sum sets and difference sets

In 1934, Erdős and Turán proved the following theorem.

Theorem 4.14. *There is a positive constant c such that if* $\{a_1, a_2, \ldots, a_n\} \subset \mathbb{N}$, *then*

$$\nu\Big(\prod_{1 \leq i, j \leq n} (a_i + a_j)\Big) \geq c \log n .$$

Thus the set of integers with prime factors coming from a small set cannot contain a subset of the form $2\mathcal{A}$ with $|\mathcal{A}|$ large.

Since then many papers have been written on the arithmetic properties of sum sets $\mathcal{A} + \mathcal{B}$ and difference sets $\mathcal{A} - \mathcal{A}$. For example, in 1978–79, Fürstenberg and Sárközy (independently) studied the solvability of the equation $a - a' = n^2$ for a, $a' \in \mathcal{A}$. In these papers written on "hybrid" problems (i.e., problems involving both general sets and special sets of integers), combinatorial, analytic, and ergodic methods are used. See Sárközy (1989) for a survey of these results; see also Pintz et al. (1988) and Györy et al. (1988) for further recent efforts.

Recently it has been proved that if $\mathcal{A}$, $\mathcal{B}$ are "dense", then (i) the sum set $\mathcal{A} + \mathcal{B}$ contains an element $a + b$ all whose prime factors are "small" (Balog and Sárközy), (ii) there is a sum $a + b$ which is "almost prime" in the strong sense that it is the product of a prime and a bounded integer (Sárközy and Stewart), (iii) there is a sum $a + b$ with "many" distinct prime factors (Erdős et al. 1993), (iv) the members of $\mathcal{A} + \mathcal{B}$, weighted with respect to the number of their representations, behave like normal integers for the function $\nu(n)$ (Erdős et al. 1987). This last result has been sharpened and extended in various directions by Elliott and Sárközy and by Tenenbaum. Several authors have studied the structure of the difference set $\mathcal{A} - \mathcal{A}$ for "dense" sets $\mathcal{A}$; see Stewart and Tijdeman (1983) for references.

4.3. Complete sets and subset sums

For references, see Erdős and Graham (1980, pp. 53–60).

A set $\mathcal{A} \subseteq \mathbb{N}$ is said to be *complete* if every large integer can be written as the sum of the elements in some finite subset of $\mathcal{A}$. For example, the powers of 2 form a complete set. It is less well known that the squares form a complete set – indeed, every integer greater than 128 can be represented as a sum of distinct squares.

Must a dense enough set be complete? Improving on a result of Erdős, Folkman proved in 1966 the following result.

Theorem 4.15. *Suppose* $\mathcal{A} \subseteq \mathbb{N}$ *is such that* $A(x) > x^{1/2+\varepsilon}$ *for some* $\varepsilon > 0$ *and all large x. Suppose further that the set of subset sums from* $\mathcal{A}$ *contains a complete residue system* (mod m) *for every* $m \in \mathbb{N}$. *Then* $\mathcal{A}$ *is complete.*

A somewhat stronger statement, that is still open, was conjectured by Erdős in 1962. In some sense, though, Theorem 4.15 is best possible, for in 1960 Cassels showed that "1/2" cannot be replaced with any smaller number in the theorem. By using Theorem 3.11, Sárközy (1989/1994) proved a finite analog of Theorem 4.15 which has many applications. (Slightly later, Freiman independently proved nearly the same theorem.)

Let $s(\mathscr{A})$ be the largest number of subsets of $\mathscr{A}$ with the same sum. Improving on a result of Erdős and Moser, Sárközy and Szemerédi proved (by using Sperner's theorem) that if $\mathscr{A}$ is a set of positive reals with $|\mathscr{A}| = n$, then $s(\mathscr{A}) \leqslant c2^n n^{-3/2}$ for some absolute constant c. This result has since been improved by Nicolas, Beck, van Lint, and Stanley. In particular, Stanley showed that $s(\mathscr{A})$ is maximized over all sets $\mathscr{A}$ of n positive reals when $\mathscr{A}$ is an arithmetic progression and in this case the most popular subset sum is a subset sum closest to the average of the members of $\mathscr{A}$.

How dense can $\mathscr{A} \subseteq \{1, 2, \ldots, N\}$ be if the subset sums from $\mathscr{A}$ are distinct? One of Erdős's first conjectures is that

$$\max|\mathscr{A}| = \frac{\log N}{\log 2} + O(1)\,.$$

Towards this conjecture, Erdős and Moser proved that

$$\max|\mathscr{A}| \leqslant \frac{\log N}{\log 2} + \frac{\log\log N}{2\log 2} + O(1)\,.$$

In 1969, Conway and Guy gave the lower bound $(\log N)/(\log 2) + 2$ for $N = 2^k$ which is just 1 better than the trivial example of taking the powers of 2. However, no one has succeeded in giving *any* example that is 2 better than the trivial. Ryavec showed that

$$\sum_{a\in\mathscr{A}} \frac{1}{a} < 2$$

if $\mathscr{A}$ is a finite set of natural numbers with distinct subset sums.

In 1969, Erdős and Heilbronn showed that if $\mathscr{A} \subseteq \mathbb{Z}/p$, where p is prime and $|\mathscr{A}| > 3\sqrt{6p}$, then the subset sums of $\mathscr{A}$ cover all of $\mathbb{Z}/p$. Olson improved on the constant $3\sqrt{6}$ and Szemerédi (1970) extended the result to arbitrary finite Abelian groups. Sárközy (1989/1994) studied the case when the elements of $\mathscr{A}$ are not necessarily distinct.

5. Multiplicative problems

5.1. Primitive sets

For many references, see chapter V of Halberstam and Roth (1983) and Hall and Tenenbaum (1988).

Say $\mathscr{B} \subseteq \mathbb{N}$ has the property that whenever $b \in \mathscr{B}$, every positive multiple of b

is in $\mathscr{B}$. Examples of sets with this property include the set of even natural numbers, the set of composites, and the set of natural numbers with a divisor between 100 and 200. An example with historic interest is the set of abundant numbers, i.e., natural numbers n such that the sum of the positive divisors of n (other than n) exceeds n.

More generally, if $\mathscr{A} \subseteq \mathbb{N}$ is arbitrary, then the set $\mathscr{B}(\mathscr{A})$ of all positive multiples of members of $\mathscr{A}$ evidently has the property that if $b \in \mathscr{B}(\mathscr{A})$, then all positive multiples of b are in $\mathscr{B}(\mathscr{A})$. Is every set $\mathscr{B}$ with this property in the form $\mathscr{B}(\mathscr{A})$ for some $\mathscr{A}$? The answer is clearly yes. In fact, if $\mathscr{B}$ has this property and $\mathscr{A}$ is the set of primitive elements of $\mathscr{B}$, i.e., members of $\mathscr{B}$ not divisible by any other members of $\mathscr{B}$, then $\mathscr{B} = \mathscr{B}(\mathscr{A})$.

We say a set $\mathscr{A} \subseteq \mathbb{N}$ is *primitive* if no member of $\mathscr{A}$ divides another member of $\mathscr{A}$. Thus $\mathscr{A} \leftrightarrow \mathscr{B}(\mathscr{A})$ is a one-to-one correspondence between primitive sets and sets $\mathscr{B} \subseteq \mathbb{N}$ such that $kb \in \mathscr{B}$ whenever $b \in \mathscr{B}$.

The set of prime numbers is a primitive set. More generally, the set of $n \in \mathbb{N}$ with $\Omega(n) = k$ is primitive for any fixed $k \in \mathbb{N}$. The case $k = 2$ is the primitive set for the set of composites. If $N \in \mathbb{N}$, then

$$\mathscr{I}_N = \{n: \tfrac{1}{2}N < n \leq N\}$$

is primitive. Clearly any subset of a primitive set is primitive.

These considerations suggest several questions:

(i) Is $\mathscr{I}_N$ the most numerous primitive subset of $\{1, 2, \ldots, N\}$?

(ii) Must a primitive set have asymptotic density of 0?

(iii) Must a set $\mathscr{B}(\mathscr{A})$ have asymptotic density?

At least one of these questions is fairly easy as is seen in the following result.

Proposition 5.1. *$\mathscr{I}_N$ is the most numerous primitive subset of* $\{1, 2, \ldots, N\}$.

Proof. For each $n \in \mathbb{N}$, let n' denote the largest odd divisor of n. Clearly if $m' = n'$ and $m < n$, then $m \mid n$. Thus the mapping $n \to n'$ must be one-to-one on any primitive set $\mathscr{A}$. There are exactly $N - \lfloor \frac{1}{2}N \rfloor$ odd integers in $\{1, 2, \ldots, N\}$, so no primitive subset of $\{1, 2, \ldots, N\}$ can have more than $N - \lfloor \frac{1}{2}N \rfloor = |\mathscr{I}_N|$ members. □

The sets $\mathscr{I}_N$ can be essentially glued together to get a counter-example to question (ii). The key tool is the following, perhaps surprising, result of Erdős from 1935. We first note that, concerning question (iii), if $\mathscr{A} \subseteq \mathbb{N}$ is *finite*, then clearly $d(\mathscr{B}(\mathscr{A}))$ exists.

Theorem 5.2. *Let $\varepsilon_N = d(\mathscr{B}(\mathscr{I}_N))$; that is, ε_N is the asymptotic density of the integers with a divisor in $(\frac{1}{2}N, N]$. Then $\lim_{N\to\infty} \varepsilon_N = 0$.*

This result can be proved by consideration of the "normal" number of prime factors below N of a random integer. We still do not have an asymptotic formula for ε_N; the best results to date are due to Tenenbaum. Finally, it should be remarked that Erdős actually proved a stronger version of Theorem 5.2 where $(\frac{1}{2}N, N]$ is replaced by $(N^{1-\delta_N}, N]$ and $\delta_N \to 0$ arbitrarily slowly.

The following result is due to Besicovitch in 1934.

Theorem 5.3. *For each $\varepsilon > 0$, there is a primitive set $\mathscr{A}$ with $\overline{d}(\mathscr{A}) \geq \frac{1}{2} - \varepsilon$.*

Proof. We use Theorem 5.2; the original proof of Besicovitch used a weaker form of this result. Let $N_1 < N_2 < \cdots$ be a sequence of natural numbers with $\varepsilon_{N_i} \leq 2^{-i-1}\varepsilon$ and such that the number of integers up to N_i divisible by some member of $\mathscr{I}_{N_1} \cup \cdots \cup \mathscr{I}_{N_{i-1}}$ is at most

$$(2\varepsilon_{N_1} + \cdots + 2\varepsilon_{N_{i-1}})N_i < \varepsilon N_i\,. \tag{5.4}$$

Thus if we denote by $\mathscr{I}'_{N_i}$ the set of members of $\mathscr{I}_{N_i}$ not divisible by any member of $\mathscr{I}_{N_1} \cup \cdots \cup \mathscr{I}_{N_{i-1}}$, then $|\mathscr{I}'_{N_i}| > (\frac{1}{2} - \varepsilon)N_i$.

Further, if $\mathscr{A}$ is the union of all $\mathscr{I}'_{N_i}$, then $\mathscr{A}$ is evidently primitive and

$$A(N_i) \geq |\mathscr{I}'_{N_i}| \geq (\tfrac{1}{2} - \varepsilon)N_i$$

for each i. Thus $\overline{d}(\mathscr{A}) \geq \frac{1}{2} - \varepsilon$. $\square$

The set $\mathscr{A}$ just constructed (with $0 < \varepsilon < \frac{1}{6}$) also answers question (iii) in the negative. Indeed,

$$\overline{d}(\mathscr{B}(\mathscr{A})) \geq \overline{d}(\mathscr{A}) \geq \tfrac{1}{2} - \varepsilon > \tfrac{1}{3}\,,$$

and since the number of members of $\mathscr{B}(\mathscr{A})$ up to $\frac{1}{2}N_i$ is by (5.4) at most εN_i, we have

$$\underline{d}(\mathscr{B}(\mathscr{A})) \leq 2\varepsilon < \tfrac{1}{3}\,.$$

Thus $\mathscr{B}(\mathscr{A})$ does not possess asymptotic density.

It is a simple exercise to show the following.

Proposition 5.5. *If $\mathscr{A} \subseteq \mathbb{N}$ is such that $\sum_{a \in \mathscr{A}} 1/a < \infty$, then $d(\mathscr{B}(\mathscr{A}))$ exists.*

It is a slightly more difficult exercise to show that if furthermore $1 \notin \mathscr{A}$, then $d(\mathscr{B}(\mathscr{A})) < 1$; see Pomerance and Sárközy (1988).

As we have seen, a primitive set $\mathscr{A}$ need not have density 0. However, if one considers a weaker density than asymptotic density, namely logarithmic density, then every primitive set has density 0. The logarithmic density of a set $\mathscr{A} \subseteq \mathbb{N}$ is defined by

$$\delta(\mathscr{A}) = \lim_{N \to \infty} \frac{1}{\log N} \sum_{\substack{a \in \mathscr{A} \\ a \leq N}} \frac{1}{a}\,, \tag{5.6}$$

should this limit exist. It is not hard to show that if $d(\mathcal{A})$ exists, then so does $\delta(\mathcal{A})$ and it is equal to $d(\mathcal{A})$.

In 1935, Erdős proved the following result.

Theorem 5.7. *There is an absolute constant c such that*

$$\sum_{a \in \mathcal{A}} \frac{1}{a \log a} < c$$

for every primitive set $\mathcal{A} \subseteq \mathbb{N}$ *except* $\mathcal{A} = \{1\}$.

(Erdős conjectures that the maximal value of the sum in this theorem is attained when $\mathcal{A}$ is the set of primes.) It follows immediately that $\delta(\mathcal{A}) = 0$ for every primitive set $\mathcal{A}$. Since $\underline{d}(\mathcal{A}) \leqslant \delta(\mathcal{A})$ always (this is easy), another corollary is that the lower asymptotic density of any primitive set is 0.

In 1937, Davenport and Erdős proved the following result.

Theorem 5.8. (i) *If* $\mathcal{A} \subseteq \mathbb{N}$ *has positive upper logarithmic density, then* $\mathcal{A}$ *contains an infinite sequence* $a_1 < a_2 < \cdots$ *with* $a_i | a_{i+1}$ *for* $i = 1, 2, \ldots$.

(ii) *For any* $\mathcal{A} \subseteq \mathbb{N}$, $\delta(\mathcal{B}(\mathcal{A}))$ *exists.*

Of course, by "upper logarithmic density" we mean that the lim in (5.6) is replaced with lim sup. These results really underline the fact that logarithmic density is the "correct" measure when considering primitive sets and sets of multiples.

How large can $\sum_{\mathcal{A}} 1/a$ be for a primitive set $\mathcal{A} \subseteq \{1, 2, \ldots, N\}$? This question is partially answered by the following result of Behrend from 1935.

Theorem 5.9. *There is a positive constant* c_1 *such that if* $N \geqslant 3$, *then*

$$\sum_{a \in \mathcal{A}} \frac{1}{a} < c_1 (\log N)(\log \log N)^{-1/2}$$

for any primitive set $\mathcal{A} \subseteq \{1, 2, \ldots, N\}$.

Proof. Let $s(u)$ denote the cardinality of the largest primitive set made up of divisors of u. We begin our proof by using Sperner's theorem (see chapter 24) to compute $s(u)$ when u is squarefree. Indeed, if u is squarefree and $\nu(u) = k$, then each divisor of u corresponds to a subset of the k primes of u. Thus Sperner's theorem immediately gives

$$s(u) = \binom{k}{\lfloor \frac{1}{2} k \rfloor}. \tag{5.10}$$

We now show the connection of $s(u)$ to our problem. Let u' denote the largest squarefree divisor of u. If $\mathcal{A} \subseteq \{1, 2, \ldots, N\}$ is primitive and every member of $\mathcal{A}$

is squarefree, then

$$\sum_{u\leqslant N} s(u') \geqslant \sum_{u\leqslant N} \sum_{\substack{a|u\\ a\in\mathcal{A}}} 1 = \sum_{a\in\mathcal{A}} \sum_{\substack{u\leqslant N\\ a|u}} 1 = \sum_{a\in\mathcal{A}} \left[\frac{N}{a}\right]$$

$$> N\sum_{a\in\mathcal{A}} \frac{1}{a} - N\,.$$

Thus it will suffice to prove that

$$\sum_{u\leqslant N} s(u') \leqslant cN(\log N)(\log\log N)^{-1/2} \tag{5.11}$$

for some constant c and all $N\geqslant 3$.

To do this, we apply Stirling's formula to (5.10), getting

$$s(u) \leqslant c2^{\nu(u)}(\nu(u))^{-1/2}\,, \quad u \text{ squarefree}$$

for some constant c. Since $\nu(u)=\nu(u')$, we thus have (with $l=\lfloor\log\log N\rfloor$)

$$\begin{aligned}\sum_{u\leqslant N} s(u') &\leqslant c\sum_{u\leqslant N} 2^{\nu(u)}(\nu(u))^{-1/2}\\ &= c\sum_{\substack{u\leqslant N\\ \nu(u)\leqslant l}} 2^{\nu(u)}(\nu(u))^{-1/2} + c\sum_{\substack{u\leqslant N\\ \nu(u)>l}} 2^{\nu(u)}(\nu(u))^{-1/2}\\ &\leqslant c\cdot 2^l N + cl^{-1/2}\sum_{u\leqslant N} 2^{\nu(u)}\\ &\leqslant cN(\log N)^{\log 2} + cl^{-1/2}\sum_{u\leqslant N}\tau(u)\,,\end{aligned} \tag{5.12}$$

where $\tau(u)$ is defined in section 1. The final sum in (5.12) is easily majorized. We have

$$\sum_{u\leqslant N}\tau(u) = \sum_{u\leqslant N}\sum_{d|u} 1 = \sum_{d\leqslant N}\sum_{\substack{u\leqslant N\\ d|u}} 1 = \sum_{d\leqslant N}\lfloor N/d\rfloor$$

$$\leqslant N\sum_{d\leqslant N} 1/d \leqslant N(\log N+1)\,.$$

Putting this estimate in (5.12) gives (5.11) and thus the theorem for the case when every member of $\mathcal{A}$ is squarefree.

Now suppose $\mathcal{A}\subseteq\{1,2,\ldots,N\}$ is an arbitrary primitive set. For each $k\in\mathbb{N}$, let $\mathcal{A}_k$ denote the set of $a\in\mathcal{A}$ with largest square divisor being k^2. Thus $\{a/k^2\colon a\in\mathcal{A}_k\}$ is a primitive set of squarefree numbers not exceeding N/k^2. Thus

$$\sum_{a\in\mathcal{A}}\frac{1}{a} = \sum_{k=1}^{\infty}\sum_{a\in\mathcal{A}_k}\frac{1}{a} = \sum_{k=1}^{\infty}\frac{1}{k^2}\sum_{a\in\mathcal{A}_k}\frac{1}{a/k^2}$$

$$\leqslant c(\log N)(\log\log N)^{-1/2}\sum_{k=1}^{\infty}\frac{1}{k^2}\,,$$

by our theorem for primitive sets of squarefree numbers. Since $\sum 1/k^2$ is convergent, the general case of our theorem follows. □

That Theorem 5.9 is essentially best possible was shown by Pillai in 1939. Namely, Pillai showed that there is a positive constant c_2 such that for each large N there is a primitive set $\mathcal{A} \subseteq \{1, 2, \ldots, N\}$ with

$$\sum_{a \in \mathcal{A}} \frac{1}{a} > c_2(\log N)(\log \log N)^{-1/2} . \tag{5.13}$$

The gap between Theorem 5.9 and eq. (5.13) was eliminated by Erdős, Sárközy and Szemerédi in 1967. They show that if $L(N)$ is the maximum value of $\sum_{\mathcal{A}} 1/a$ for all primitive sets $\mathcal{A} \subseteq \{1, 2, \ldots, N\}$, then

$$L(N) = ((2\pi)^{-1/2} + o(1))(\log N)(\log \log N)^{-1/2} . \tag{5.14}$$

The lower bound in (5.14) had already been shown by Erdős in 1948 by taking $\mathcal{A}$ to be the set of $m \in \{1, 2, \ldots, N\}$ with $\Omega(m) = \lfloor \log \log N \rfloor$.

Interestingly, if $\mathcal{A}$ is an infinite primitive set, then we have

$$\sum_{\substack{a \in \mathcal{A} \\ a \leq N}} \frac{1}{a} = o((\log N)(\log \log N)^{-1/2}) \tag{5.15}$$

and no statement stronger than (5.15) is true, a result of Erdős, Szemerédi and Sárközy in 1967. For references, see Erdős et al. (1970).

5.2. Product sets and other multiplicative problems

If $\mathcal{A}, \mathcal{B} \subseteq \mathbb{N}$, we denote by $\mathcal{A}\mathcal{B}$ the set of products ab where $a \in \mathcal{A}$, $b \in \mathcal{B}$. Also we write $\mathcal{A}^2$ for $\mathcal{A}\mathcal{A}$, $\mathcal{A}^3$ for $\mathcal{A}^2\mathcal{A}$, etc.

In 1960, Erdős proved the following remarkable theorem.

Theorem 5.16. *If* $\mathcal{A} = \{1, 2, \ldots, N\}$, *then* $|\mathcal{A}^2| = N^2(\log N)^{-\alpha + o(1)}$, *where* $\alpha = 1 - (1 + \log \log 2)/\log 2$.

Thus there are only $o(N^2)$ distinct integers in the $N \times N$ multiplication table! This seeming paradox is explained by the fact that a "normal product" of integers $a_1, a_2 \leq N$ has about $2 \log \log N$ prime factors, which is quite *abnormal* for integers below N^2. The idea of looking at the normal number of prime factors of an integer is often fruitful; in fact this idea was mentioned above in connection with Theorem 5.2 whose proof is actually quite similar to the proof of Theorem 5.16.

What can one say about $|\mathcal{A}\mathcal{B}|$ if $\mathcal{A}$ and $\mathcal{B}$ are just "dense" subjects of $\{1, 2, \ldots, N\}$? This question is addressed in a recent paper of Pomerance and Sárközy (1990).

Theorem 5.17. *If* $\varepsilon > 0$ *is arbitrary and* $\mathcal{A}, \mathcal{B} \subseteq \{1, 2, \ldots, N\}$ *with* $|\mathcal{A}|, |\mathcal{B}| > \varepsilon N$, *then* $|\mathcal{A}\mathcal{B}| \geq N^2(\log N)^{1-2\log 2+o_\varepsilon(1)}$. *Moreover, there is a set* $\mathcal{A}_N \subseteq \{1, 2, \ldots, N\}$ *with* $|\mathcal{A}_N| \sim N$ *and* $|\mathcal{A}_N^2| = N^2(\log N)^{1-2\log 2+o(1)}$.

We say a set $\mathcal{A} \subseteq \mathbb{N}$ is a *multiplicative basis of order* k if $\mathcal{A}^k = \mathbb{N}$. See Wirsing (1957) for a study of density properties of multiplicative bases.

Following Theorem 4.12 we asked the Sidon problem: if $\mathcal{A}$ is a basis (additive) of order 2, must $s(n)$, the number of representations of n as $a_1 + a_2$ with a_1, $a_2 \in \mathcal{A}$, be unbounded? The multiplicative analog of this problem was solved by Erdős in 1965 (see Erdős and Graham 1980, p. 100).

Theorem 5.18. *If* $\mathcal{A}$ *is a multiplicative basis of order* 2, *then* $t(n)$, *the number of representations of* n *as* a_1a_2 *with* a_1, $a_2 \in \mathcal{A}$, *must be unbounded.*

How large a set $\mathcal{A} \subseteq \{1, 2, \ldots, N\}$ can we choose with all of the products a_1a_2 (with a_1, $a_2 \in \mathcal{A}$, $a_1 < a_2$) distinct? If $k(N)$ denotes the maximal cardinality of such a set $\mathcal{A}$, then Erdős has shown (with graph-theoretic tools) that

$$\pi(N) + c_1N^{3/4}/\log^{3/2}N < k(N) < \pi(N) + c_2N^{3/4}/\log^{3/2}N$$

for certain positive constants c_1, c_2 and all large N. Erdős and Posa have considered the analogous problem where all of the subset products from $\mathcal{A}$ are distinct. See Erdős and Graham (1980, p. 98) for references and a proof.

In 1976, Szemerédi proved the following attractive result (see Erdős and Graham 1980, pp. 98–99).

Theorem 5.19. *There is a constant* c *such that if* $\mathcal{A}$, $\mathcal{B} \subseteq \{1, 2, \ldots, N\}$ *and* $|\mathcal{A}\mathcal{B}| = |\mathcal{A}||\mathcal{B}|$, *then* $|\mathcal{A}\mathcal{B}| < cN^2/\log(N+1)$.

In contrast, it is easy to construct sets $\mathcal{A}$, $\mathcal{B} \subseteq \mathbb{N}$ such that each $n \in \mathbb{N}$ has a unique representation $n = ab$ with $a \in \mathcal{A}$, $b \in \mathcal{B}$. For example, we may choose $\mathcal{A}$ to be the powers of 2 and $\mathcal{B}$ to be the odd natural numbers, or more generally, $\mathcal{A}$ the natural numbers all of whose primes come from $\mathcal{P}_1$ and $\mathcal{B}$ the natural numbers all of whose primes come from $\mathcal{P}_2$, where $\mathcal{P}_1 \cup \mathcal{P}_2$ is an arbitrary partition of the set of primes. Erdős, Saffari, Vaughan, and Daboussi have studied this problem.

In 1975, Erdős and Selfridge (see Erdős and Graham 1980, p. 66) showed the following striking result.

Theorem 5.20. *The product of two or more consecutive positive integers is never a non-trivial power.*

There are many other problems and results concerning blocks of consecutive integers in Erdős and Graham (1980, section 8).

6. Van der Waerden's theorem and generalizations

For many references, see chapters 1 and 2 of Graham et al. (1980), and section 2 of Erdős and Graham (1980).

How much of the structure of the natural numbers must be preserved in a "dense" subset? In 1927, B. L. van der Waerden showed that if the natural numbers are partitioned into two subsets, then one subset contains arbitrarily long arithmetic progressions. In one sense, this theorem is best possible, since it is an easy exercise to partition the natural numbers into two subsets, neither of which contains an infinite arithmetic progression. But the theorem still leaves us wondering about our opening question, which we now repeat more specifically. How dense must a subset of the natural numbers be for it to contain arbitrarily long arithmetic progressions?

In particular, in 1936 Erdős and Turán conjectured that if $\mathscr{A} \subseteq \mathbb{N}$ has positive upper asymptotic density, then $\mathscr{A}$ contains arbitrarily long APs (we abbreviate "arithmetic progression" as AP). In 1952, Roth used the Hardy–Littlewood circle method from analytic number theory to prove the Erdős–Turán conjecture for three-term APs. In 1969, via a combinatorial argument, Szemerédi showed the conjecture for four-term APs, and in 1974, in what must be one of the most complex proofs in combinatorial number theory, he proved the complete conjecture. A few years later, Fürstenberg, using ergodic theory, gave another proof (also complicated) of what is now known as Szemerédi's theorem.

How much can Szemerédi's theorem be improved? An old conjecture of Erdös is that if $\mathscr{A} \subseteq \mathbb{N}$ satisfies only the weaker hypothesis

$$\sum_{a \in \mathscr{A}} \frac{1}{a} = \infty ,$$

rather than positive upper asymptotic density, then this is enough to force $\mathscr{A}$ to have arbitrarily long APs.

A corollary to this conjecture of Erdős is that the set of primes would contain arbitrarily long APs. It is unclear, though, that one should think of this prime number problem in terms of the Erdős conjecture. That is, Erdős is suggesting that the set of prime numbers contains arbitrarily long APs only because the prime numbers are fairly numerous and not because of any special properties of the prime numbers. This technique of generalizing a hard problem to put it in proper perspective is of course an often-successful trick in mathematics. However, the only progress we have had so far on showing the set of primes contains arbitrarily long APs is through intrinsic properties of the primes.

For example, Chudakov, Estermann, and van der Corput independently in 1937–38 used the circle method to show the following strengthening of Corollary 3.1: for any $A > 0$, the number of even integers up to x not the sum of two distinct primes is $O(x/\log^A x)$. From this we can prove the following result.

Corollary 6.1. *The set of primes contains infinitely many three-term APs.*

Proof. By the prime number theorem, the number of even integers up to x of the form $2p$ with p prime is $\sim x/(2\log x)$. Thus by the above-mentioned theorem, most of these numbers $2p$ can be represented as $q+r$, where $q<r$ are primes. But then q, p, r form a three-term AP of prime numbers. □

It is still unknown if there are infinitely many four-term APs of primes. The longest AP of primes ever found has length 22, a result of Pritchard et al. (1995).

We now look at several "equivalent" formulations of van der Waerden's theorem. The reason for the quotation marks is that logically, all theorems are equivalent. Here we mean it in the subjective sense that the proofs of interdependence are simple and fairly transparent.

Theorem 6.2. *The following statements are equivalent*:

(i) *If* $\mathbb{N}$ *is partitioned into two subsets, then one subset contains arbitrarily long APs.*

(ii) *For each* $k\in\mathbb{N}$, *there is a number* $W(k)$ *such that if* $\{1, 2, \ldots, W(k)\}$ *is partitioned into two subsets, then one subset contains a* k*-term AP.*

(iii) *For each* k, $r\in\mathbb{N}$, *there is a number* $W(k,r)$ *such that if* $\{1, 2, \ldots, W(k, r)\}$ *is partitioned into* r *subsets, then one subset contains a* k*-term AP.*

(iv) *For each* $r\in\mathbb{N}$, *if* $\mathbb{N}$ *is partitioned into* r *subsets, then one subset contains arbitrarily long APs.*

(v) *If* $\{a_n\}$ *is an infinite subsequence of* $\mathbb{N}$ *with* $\{a_{n+1}-a_n\}$ *bounded, then* $\{a_n\}$ *contains arbitrarily long APs.*

Proof. We first show that (i) $\Rightarrow$ (ii), which is probably the most difficult of the implications. It is an example of the "compactness principle" in Ramsey theory (cf. chapter 42). Suppose k is such that $W(k)$ does not exist. Thus for every N there is a subset $\mathcal{A}_N$ of $\{1, 2, \ldots, N\}$ such that neither $\mathcal{A}_N$ nor its complement contains a k-term AP. Obviously there is an infinite subsequence $N_{11}<N_{12}<\cdots$ of $\mathbb{N}$ such that

$$\mathcal{S}_1 := \mathcal{A}_{N_{11}} \cap \{1\} = \mathcal{A}_{N_{12}} \cap \{1\} = \cdots ;$$

that is, either 1 is in each $\mathcal{A}_{N_{1j}}$ or 1 is in no $\mathcal{A}_{N_{1j}}$. By passing to an infinite subsequence $N_{21}<N_{22}<\cdots$ of $\{N_{1j}\}$, we have

$$\mathcal{S}_2 := \mathcal{A}_{N_{21}} \cap \{1,2\} = \mathcal{A}_{N_{22}} \cap \{1,2\} = \cdots$$

and $\mathcal{S}_1 \subseteq \mathcal{S}_2$. Continuing in this fashion we find $\mathcal{S}_1 \subseteq \mathcal{S}_2 \subseteq \cdots \subseteq \mathbb{N}$ and an infinite subsequence $N_1<N_2<\cdots$ of $\mathbb{N}$ ($N_1=N_{11}$, $N_2=N_{21}$, etc.) with $\mathcal{S}_j=\mathcal{A}_{N_j}\cap\{1, 2, \ldots, j\}$ for each j. Thus if $\mathcal{S}=\bigcup \mathcal{S}_j$, then neither $\mathcal{S}$ nor $\mathbb{N}\backslash\mathcal{S}$ contains a k-term AP, contradicting (i).

Now we show (ii) $\Rightarrow$ (iii). We do this by induction on r, (ii) being the statement for $r=2$ (and the case $r=1$ being trivial). Suppose $W(k, r)$ exists for all k for some fixed $r\geq 2$. Say $\{1, 2, \ldots, N\}$ is partitioned into $r+1$ sets $\mathcal{A}_1, \ldots, \mathcal{A}_{r+1}$. If $N\geq W(l, r)$, then one of $\mathcal{A}_1\cup\mathcal{A}_2$, $\mathcal{A}_3, \ldots, \mathcal{A}_{r+1}$ contains an

l-term AP, call it $\mathscr{B}$. If $\mathscr{B} \subseteq \mathscr{A}_1 \cup \mathscr{A}_2$, then $\mathscr{A}_1 \cap \mathscr{B}$, $\mathscr{A}_2 \cap \mathscr{B}$ is a partition of $\mathscr{B}$ into two parts. Thus if $l = W(k, 2)$, then one part contains a k-term AP. Since clearly $W(k, 2) \geq k$, if $\mathscr{B}$ is contained in one of $\mathscr{A}_3, \ldots, \mathscr{A}_{r+1}$, then one of these sets contains a k-term AP. Thus, not only have we shown that $W(k, r+1)$ exists, but we have shown that the least choice for $W(k, r+1)$ is at most $W(W(k, 2), r)$.

It is obvious that (iii) $\Rightarrow$ (iv) $\Rightarrow$ (i).

It is also clear that (v) $\Rightarrow$ (i), since if $\mathbb{Z} = \mathscr{A}_1 \cup \mathscr{A}_2$, $\mathscr{A}_1 \cap \mathscr{A}_2 = \emptyset$, and $k \in \mathbb{N}$, then either $\mathscr{A}_1$ contains k consecutive integers or the maximal gap between consecutive members of $\mathscr{A}_2$ is at most k.

Finally we show (iv) $\Rightarrow$ (v), which will complete our proof. Suppose $\mathscr{A} \subseteq \mathbb{N}$ has maximal gap r between consecutive members. Let $\mathscr{A}_i = \mathscr{A} + \{i\}$ for $i = 0, 1, \ldots, r-1$. Then $\mathscr{A}_0 \cup \cdots \cup \mathscr{A}_{r-1} = \mathbb{N}$. Although these sets may not be disjoint, (iv) still implies that one of them contains a k-term AP. Then so does $\mathscr{A}_0 = \mathscr{A}$. □

Van der Waerden's theorem, as we originally stated it, is statement (i) of Theorem 6.2. We now give a proof of van der Waerden's theorem.

We begin with two definitions. If $x_1, x_2, \ldots, x_m$ and $x_1', x_2', \ldots, x_m'$ are two sequences where each term is in $\{0, 1, \ldots, l\}$, we say $\{x_i\}$ is *l-equivalent* to $\{x_i'\}$ if either l does not appear in either sequence or there is some k with $x_i = x_i'$ for $i \leq k$ and each $x_j, x_j' < l$ for $j > k$. That is, $\{x_i\}$ and $\{x_i'\}$ agree at least up to the last appearance of l.

Next, we define the statement $S(l, m)$ (where $l, m \in \mathbb{N}$) as the following assertion: for each $r \in \mathbb{N}$ there is a number $N(l, m, r)$ such that whenever $\{1, 2, \ldots, N(l, m, r)\}$ is partitioned into r parts $\mathscr{A}_1, \ldots, \mathscr{A}_r$, there exist a, $d_1, \ldots, d_m \in \mathbb{N}$ such that

$$a + l(d_1 + \cdots + d_m) \leq N(l, m, r) \tag{6.3}$$

and such that whenever $\{x_i\}$ and $\{x_i'\}$ are m-term sequences from $\{0, 1, \ldots, l\}$ that are l-equivalent, $a + x_1 d_1 + \cdots + x_m d_m$ and $a + x_1' d_1 + \cdots + x_m' d_m$ are in the same $\mathscr{A}_j$.

Note that the condition (6.3) guarantees that $a + \sum x_i d_i$ and $a + \sum x_i' d_i$ are in $\{1, 2, \ldots, N(l, m, r)\}$.

Note also that the assertion $S(l, 1)$ is essentially the same as statement (iii) of Theorem 6.2. Indeed, two integers $x, x' \in \{0, 1, \ldots, l\}$, considered as 1-term sequences, are l-equivalent if and only if both $x, x' < l$ or $x = x' = l$. The assertion $S(l, 1)$ says that there is some number $N(l, 1, r)$ such that if $\{1, 2, \ldots, N(l, 1, r)\}$ is partitioned into r subsets, then there are positive integers a, d with a, $a + d, \ldots, a + (l-1)d$ all in one of the parts. That is, one part contains an l-term AP. [Note that $S(l, 1)$ also carries the extra stipulation, not found in statement (iii) of Theorem 6.2, that $a + ld \leq N(l, 1, r)$.] Thus van der Waerden's theorem will follow from the following result.

Theorem 6.4. *For each l, $m \in \mathbb{N}$, the assertion $S(l, m)$ is a theorem.*

Proof. Our plan is as follows. First we show that if $S(l, k)$ is a theorem for $k = 1, \ldots, m$, then so too is $S(l, m+1)$. Next we show that if $S(l, m)$ is a theorem for all m, then so too is $S(l+1, 1)$. Thus our theorem will follow from this double induction and the fact that $S(1, 1)$ is trivially true.

Suppose $l, m \in \mathbb{N}$ and $S(l, k)$ is a theorem for $k = 1, \ldots, m$. Let $r \in \mathbb{N}$ be arbitrary, let $M = N(l, m, r)$, $M' = N(l, 1, r^M)$. We shall show that we may choose $N(l, m+1, r) = MM'$. Let $\mathcal{A}_1, \ldots, \mathcal{A}_r$ be a partition of $\{1, 2, \ldots, MM'\}$ and let C be the function that assigns to $i \in \{1, 2, \ldots, MM'\}$ the number $j \in \{1, 2, \ldots, r\}$ with $i \in \mathcal{A}_j$.

Consider now the matrix

$$A = \begin{bmatrix} C(1) & C(2) & \cdots & C(M) \\ C(M+1) & C(M+2) & & C(2M) \\ \vdots & & \ddots & \vdots \\ C((M'-1)M+1) & C((M'-1)M+2) & \cdots & C(M'M) \end{bmatrix}.$$

Each row of A is one of the r^M M-term sequences from $\{1, 2, \ldots, r\}$. We now partition $\{1, 2, \ldots, M'\}$ into r^M subsets where i, j are in the same subset if and only if the ith row and jth row of A are identical. Since $S(l, 1)$ is true and by our choice of M', one of these subsets contains an l-term arithmetic progression b, $b + d, \ldots, b + (l-1)d$, where b, d are positive integers with $b + ld \leqslant M'$. That is, rows $b + id$ for $i = 1, \ldots, l-1$ of A are identical.

We now apply $S(l, m)$ to $\{(b-1)M + 1, (b-1)M + 2, \ldots, bM\}$ (a translate of $\{1, 2, \ldots, M\}$) and the partition we already have of $\{1, 2, \ldots, MM'\}$ restricted to this subset. Thus there are natural numbers $a, d_1, \ldots, d_m$ such that

(1) $a \geqslant (b-1)M + 1$, $a + l(d_1 + \cdots + d_m) \leqslant bM$;

(2) whenever $\{x_i\}$, $\{x_i'\}$ are l-equivalent, m-term sequences from $\{0, 1, \ldots, l\}$, then $C(a + \sum x_i d_i) = C(a + \sum x_i' d_i)$.

Let $d_{m+1} = dM$. To prove assertion $S(l, m+1)$, we will show

(1′) $a + l(d_1 + \cdots + d_{m+1}) \leqslant MM'$;

(2′) whenever $\{x_i\}$, $\{x_i'\}$ are l-equivalent, $(m+1)$-term sequences from $\{0, 1, \ldots, l\}$, then $C(a + \sum x_i d_i) = C(a + \sum x_i' d_i)$.

For (1′), note that the left-hand side is

$$a + l(d_1 + \cdots + d_m) + ld_{m+1} \leqslant bM + ldM$$

by (1). But we noted above that $b + ld \leqslant M'$, so we have (1′).

For (2′) we may clearly assume that the l-equivalent sequences $\{x_i\}$ and $\{x_i'\}$ are not identical. Thus $x_{m+1}, x_{m+1}' < l$. Let

$$j = a - (b-1)M + \sum_1^m x_i d_i, \qquad j' = a - (b-1)M + \sum_1^m x_i' d_i. \tag{6.5}$$

Then $j, j' \in \{1, 2, \ldots, M\}$. We look now at columns j and j' of matrix A and how they intersect rows $b, b+d, \ldots, b + (l-1)d$. Of course, column j is constant on these rows, as is column j'. But from (6.5),

$$(b-1)M+j=a+\sum_{1}^{m} x_i d_i\,, \qquad (b-1)M+j'=a+\sum_{1}^{m} x_i' d_i\,,$$

and since $x_1, \ldots, x_m$ is l-equivalent to $x_1', \ldots, x_m'$, (2) implies $C((b-1)M+j) = C((b-1)M+j')$. Thus the constant value on the l special rows of column j is the same constant value as in column j'. Note that one of these rows is $b+x_{m+1}d$, whose jth entry is

$$C((b+x_{m+1}d-1)M+j)=C\left(a+\sum_{1}^{m+1} x_i d_i\right)$$

by (6.5). Another special row is $b+x_{m+1}'d$, whose j'th entry is

$$C((b+x_{m+1}'d-1)M+j')=C\left(a+\sum_{1}^{m+1} x_i' d_i\right).$$

Thus (2′) holds and we have proved $S(l, m+1)$.

For our second induction, assume $l \in \mathbb{N}$ and $S(l, m)$ is a theorem for all $m \in \mathbb{N}$. Choose $r \in \mathbb{N}$, $r \geqslant 2$ (since we clearly can take $N(l+1, 1, 1) = l+2$). Let $N = N(l\; r, r)$. We shall show we may take $N(l+1, 1, r) = 2N$. Indeed take any partition of $\{1, 2, \ldots, 2N\}$ into r subsets $\mathcal{A}_1, \ldots, \mathcal{A}_r$. Let $a, d_1, \ldots, d_r \in \mathbb{N}$ be such that $a + l(d_1 + \cdots + d_r) \leqslant N$ and such that whenever $\{x_i\}$, $\{x_i'\}$ are l-equivalent, m-term sequences from $\{0, 1, \ldots, l\}$, $a + \sum x_i d_i$ is in the same $\mathcal{A}_j$ as $a + \sum x_i' d_i$.

Since $r \geqslant 2$, we have $\binom{r+1}{2} > r$, so there are integers $u < v$ with $0 \leqslant u < v \leqslant r$ and such that

$$a+\sum_{i=1}^{u} ld_i\,, \qquad a+\sum_{i=1}^{v} ld_i$$

are in the same subset from $\mathcal{A}_1, \ldots, \mathcal{A}_r$, say $\mathcal{A}_j$. Let

$$a'=a+\sum_{i=1}^{u} ld_i\,, \qquad d'=\sum_{i=u+1}^{v} d_i\,.$$

We now claim that the $(l+1)$-term arithmetic progression $a', a'+d', \ldots, a'+ld'$ lies wholly in $\mathcal{A}_j$ and that $a'+(l+1)d' \leqslant 2N$. Indeed, we know already that a' and $a'+ld'$ both lie in $\mathcal{A}_j$ and if $x \in \{1, 2, \ldots, l-1\}$, then

$$\{\underbrace{l, \ldots, l}_{u}, \underbrace{0, \ldots, 0}_{r-u}\} \quad \text{and} \quad \{\underbrace{l, \ldots, l}_{u}, \underbrace{x, \ldots, x}_{v-u}, \underbrace{0, \ldots, 0}_{r-v}\}$$

are l-equivalent, so that a' and $a'+xd'$ both lie in $\mathcal{A}_j$. The assertion about $a'+(l+1)d'$ is trivial since $a'+ld' \leqslant a+l(d_1+\cdots+d_r) \leqslant N$. Thus $S(l+1, 1)$ is proved, as is the theorem. □

The above proof is a "fleshed-out" version of the short proof presented in

Graham et al. (1980, p. 32). It shows that the function $W(k, r)$ defined in statement (iii) of Theorem 6.2 is recursive, but it does not show it is primitive recursive. This has recently been done by Shelah (1988), thus solving a problem that had been outstanding for a long time.

For the record we now formally state Szemerédi's theorem.

Theorem 6.6. *If $\mathcal{A} \subseteq \mathbb{N}$ has positive upper asymptotic density, then $\mathcal{A}$ contains arbitrarily long APs.*

As van der Waerden's theorem contains several essentially equivalent forms, the following result which is superficially stronger than Szemerédi's theorem is easily proved to follow from it.

Corollary 6.7. *Let $k \in \mathbb{N}$ and $\varepsilon > 0$. For each sufficiently large N, depending on the choice of k and ε, if $\mathcal{A} \subseteq \{1, 2, \ldots, N\}$ with $|A| > \varepsilon N$, then $\mathcal{A}$ contains a k-term AP.*

Proof. We may assume $k \geqslant 3$. If the corollary is untrue, then there is an infinite sequence $N_1, N_2, \ldots$ and sets $\mathcal{A}_{N_i} \subseteq \{1, \ldots, N_i\}$ with $|\mathcal{A}_{N_i}| > \varepsilon N_i$ and $\mathcal{A}_{N_i}$ does not contain any k-term AP. Let $\mathcal{B}_{N_i} = \mathcal{A}_{N_i} + \{2N_i\}$, so that $|\mathcal{B}_{N_i}| > \varepsilon N_i$, $\mathcal{B}_{N_i} \subseteq \{2N_i + 1, \ldots, 3N_i\}$ and $\mathcal{B}_{N_i}$ does not contain any k-term AP. By passing to an infinite subsequence if necessary, we may assume $N_{i+1} \geqslant 5N_i$ for $i = 1, 2, \ldots$. Let

$$\mathcal{A} = \bigcup_{i=1}^{\infty} \mathcal{B}_{N_i} .$$

Then $\mathcal{A}$ has upper asymptotic density at least $\frac{1}{3}\varepsilon$. Furthermore, $\mathcal{A}$ contains no k-term AP, since no $\mathcal{B}_{N_i}$ does and since $\mathcal{A}$ does not contain any 3-term AP that is not wholly in some $\mathcal{B}_{N_i}$. However, Theorem 6.6 denies the existence of any such set $\mathcal{A}$, which proves the corollary. □

An interesting and still unsolved problem that is perhaps connected with these considerations is the following old problem of Erdős. Is there a sequence ε_n of 1's and -1's with

$$g(k, N) = \sum_{n=1}^{N} \varepsilon_{kn}$$

bounded for all $k, N \in \mathbb{N}$? Another form of this problem asks if $g(1, N)$ can be bounded for all N for some sequence ε_n of 1's and -1's that is also a multiplicative function.

The integers

$$4030, 4131, 4232, 4333, 4434, 4535, 4636, 4737, 4838, 4939 \qquad (6.8)$$

obviously form an AP of length 10. We generalize this idea as follows. Let C_t^N denote the set of N-term sequences from $\{0, 1, \ldots, t-1\}$. Then t distinct points

$P_1, P_2, \ldots, P_t$ in C_t^N are said to form a *line* if the sequence of jth terms P_1^j, $P_2^j, \ldots, P_t^j$ is either constant or is $0, 1, \ldots, t-1$. Thus the quadruplets formed by the digits of the integers in (6.8), viewed as members of C_{10}^4, form a line. The following result is due to Hales and Jewett in 1963.

Theorem 6.9. *For each $r, t \in \mathbb{N}$, there is a number $\mathrm{HJ}(r, t)$ such that if $N \geqslant \mathrm{HJ}(r, t)$ and C_t^N is partitioned into r subsets, then one subset contains a line.*

By writing the integers below t^N in base-t notation, we see that if $\{0, 1, \ldots, t^N - 1\}$ is partitioned into r subsets, where $N \geqslant \mathrm{HJ}(r, t)$, then one subset contains a t-term AP. That is, the Hales–Jewett theorem implies van der Waerden's theorem. In fact, Shelah's recent result mentioned above, that $W(k, r)$ is primitive recursive, was obtained by proving the stronger theorem that the function $\mathrm{HJ}(r, t)$ is primitive recursive.

In the 1930s, Gallai proved the following generalization of van der Waerden's theorem as a corollary of the latter. It is also possible to prove this result as a simple corollary of the Hales–Jewett theorem.

Theorem 6.10. *For all u, r, $k \in \mathbb{N}$, if $\mathbb{Z}^u$ is partitioned into r subsets, then one subset contains a set of the form $\mathcal{B}^u$ where $\mathcal{B} \subseteq \mathbb{Z}$ is an AP of length k.*

By $\mathcal{B}^u$ we mean of course all the points of $\mathbb{Z}^u$ whose coordinates lie in $\mathcal{B}$.

It is natural to ask if Theorems 6.9 and 6.10 can be generalized to "dense" sets. For Hales–Jewett, the following theorem was recently announced by Fürstenberg in the Abstracts of the 1990 International Congress of Mathematicians.

Theorem 6.11. *For each $t \in \mathbb{N}$, $\varepsilon > 0$, if N is sufficiently large and $\mathcal{A} \subseteq C_t^N$ with $|\mathcal{A}| > \varepsilon t^N$, then $\mathcal{A}$ contains a line.*

The proof is by ergodic methods. The following result is a Szemerédi-type analog of Gallai's theorem 6.10. It was first proved by Fürstenberg and Katznelson in 1978. However, it now may be viewed as a corollary of Theorem 6.11.

Theorem 6.12. *For each $k, u \in \mathbb{N}$ and each $\varepsilon > 0$, if N is sufficiently large and $\mathcal{A} \subseteq [-N, N]^u \cap \mathbb{Z}^u$ with $|\mathcal{A}| > \varepsilon N^u$, then $\mathcal{A}$ contains a subset $\mathcal{B}^u$ where $\mathcal{B} \subseteq \mathbb{Z}$ is an AP of length k.*

Another question one may attack is that of sequences of lattice points in $\mathbb{Z}^u$. If a sequence $\boldsymbol{v}_1, \boldsymbol{v}_2, \ldots$ in $\mathbb{Z}^u$ has "small gaps", what may be said? We should not expect to find long APs as can be seen from the following result of Dekking (1979).

Theorem 6.13. *There is an infinite sequence $\boldsymbol{v}_1, \boldsymbol{v}_2, \ldots$ in $\mathbb{Z}^2$ with each $\boldsymbol{v}_{i+1} - \boldsymbol{v}_i =$ (0, 1) or (1, 0) and such that no five of the $\boldsymbol{v}$'s form an AP of vectors.*

However, we may expect a large subset to be collinear, at least for $\mathbb{Z}^2$ (we mean "collinear" in the ordinary, geometric sense). The following result is due to Ramsey and Gerver (1979).

Theorem 6.14. *If $\boldsymbol{v}_1, \boldsymbol{v}_2, \ldots$ is an infinite sequence in $\mathbb{Z}^2$ with $|\boldsymbol{v}_{i+1} - \boldsymbol{v}_i|$ bounded, then for every k there are k of the $\boldsymbol{v}$'s which are collinear.*

As is also shown by Ramsey and Gerver, this result is not true for $\mathbb{Z}^3$.

Theorem 6.14 bears a similarity in appearance to statement (v) of Theorem 6.2 and thus may be thought of as an analog of van der Waerden's theorem. The following result of Pomerance (1980) would thus be a Szemerédi-analog.

Theorem 6.15. *If $\boldsymbol{v}_1, \boldsymbol{v}_2, \ldots$ is an infinite sequence in $\mathbb{Z}^2$ with*

$$\liminf_{N\to\infty} \frac{1}{N} \sum_{i=1}^{N} |\boldsymbol{v}_{i+1} - \boldsymbol{v}_i| < \infty ,$$

then for every k there are k of the $\boldsymbol{v}$'s that are collinear.

An old result of Schur says that if $\mathbb{N}$ is partitioned into finitely many classes, then one class must contain infinitely many triples x, y, z with $x + y = z$. This result has been starting point of many interesting Ramsey-type problems on the integers. One particularly interesting result is due to Hindman in 1974 who showed that if $\mathbb{N}$ is partitioned into finitely many classes, then one class contains an *infinite* set $\mathcal{A}$ such that all finite subset sums from $\mathcal{A}$ belong to the same class. For more on problems of this type, see chapter 3 of Graham et al. (1980).

7. Miscellaneous problems

As with combinatorics as a whole, combinatorial number theory is rich in attractive problems that defy precise classification. It is from the wealth of such problems in an area that we are sometimes able to discern patterns that become the broad outlines of a more mature branch of mathematics. In this section we take a very brief glimpse at a few of these problems.

7.1. Covering congruences

For references see section 3 of Erdős and Graham (1980) and section F13 of Guy (1994).

In 1934, Romanoff posed the following problem. Can every sufficiently large odd integer be written as a sum of a power of 2 and a prime? In 1950, Erdős and van der Corput independently answered this problem in the negative by showing that in fact there is an infinite arithmetic progression of positive odd numbers m not of the form $2^n + p$. Erdős's solution begins with the observation that *every* integer n is in at least one of the following residue classes:

$$\begin{aligned} &0 \pmod 2, \quad 0 \pmod 3, \quad 1 \pmod 4, \\ &3 \pmod 8, \quad 7 \pmod{12}, \quad 23 \pmod{24}. \end{aligned} \tag{7.1}$$

From this he was able to deduce that if n is a positive integer and m simultaneously satisfies

$$\begin{aligned} &m \equiv 1 \pmod{32}, \quad m \equiv 2^0 \pmod 3, \quad m \equiv 2^0 \pmod 7, \\ &m \equiv 2^1 \pmod 5, \\ &m \equiv 2^3 \pmod{17}, \quad m \equiv 2^7 \pmod{13}, \quad m \equiv 2^{23} \pmod{241}, \end{aligned} \tag{7.2}$$

then m is odd and not representable as $2^n + p$. Moreover, by the Chinese Remainder Theorem from elementary number theory, the set of integers m which satisfy all of the congruences in (7.2) form an infinite arithmetic progression with common difference $32 \cdot 3 \cdot 7 \cdot 5 \cdot 17 \cdot 14 \cdot 241$. (It is still not known if a positive odd integer m which is not representable as $2^n + p$ must belong to an infinite arithmetic progression of such integers.)

A finite system of residue classes, such as (7.1), which have *distinct* moduli and such that every integer belongs to at least one class, is called a *covering system of congruences*. Another example that is simpler than (7.1) (but that does not have relevance to the $2^n + p$ problem) is

$$0 \pmod 2, \quad 0 \pmod 3, \quad 3 \pmod 4, \quad 1 \pmod 6, \quad 5 \pmod{12}. \tag{7.3}$$

In both (7.1) and (7.3) the least modulus is 2. The following problem of Erdős from 1950 is the major unsolved problem in the area.

Problem 7.4. It is true that for every k there is a covering system of congruences with least modulus at least k?

An example due to Choi has least modulus 20. There are numerous other problems and some results concerning covering systems of congruences. We mention two more problems, the first due to Selfridge, the second to Erdős.

Problem 7.5. Is there a covering system of congruences with all moduli squarefree and greater than 2?

Problem 7.6. Is there a covering system of congruences with all moduli odd and greater than 1?

7.2. Graham's conjecture

If $\mathcal{S} \subseteq \mathbb{N}$, let $F(\mathcal{S}) = \{a/b\colon a, b \in \mathcal{S}\}$. Graham conjectured in 1970 that if $\mathcal{S}$ is finite, then

$$F(\mathcal{S}) \not\subseteq F(\{1, 2, \ldots, |\mathcal{S}| - 1\}).$$

That is, there are $a, b \in \mathcal{S}$ with $a/(a, b) \geq |\mathcal{S}|$. This problem has attracted much attention and there have been numerous partial results. One of the easier ones is due to Szemerédi who has proved Graham's conjecture in the case $|\mathcal{S}| = p$, a prime. Indeed, we may assume not every member of $\mathcal{S}$ is divisible by p (if not, replace each $a \in \mathcal{S}$ with a/p) so that there are two members $a, b \in \mathcal{S}$ with either $a \equiv b \not\equiv 0 \pmod p$ or $a \not\equiv b \equiv 0 \pmod p$. In either case, $a/b \not\in F(\{1, 2, \ldots, p - 1\})$.

Recently Graham's conjecture was independently proved by Szegedy (1986) and Zaharescu (1987) for all sufficiently large values of $|\mathcal{S}|$. They were even able to describe those sets $\mathcal{S}$ with $F(\mathcal{S}) = F(\{1, 2, \ldots, |\mathcal{S}|\})$.

Theorem 7.7. *There is some number n_0 such that if $n \geq n_0$, $\mathcal{S} \subseteq \mathbb{N}$, $|\mathcal{S}| = n$, then $F(\mathcal{S}) \not\subseteq F(\{1, 2, \ldots, n - 1\})$. If $F(\mathcal{S}) \subseteq F(\{1, 2, \ldots, n\})$, then there is some $k \in \mathbb{N}$ with either*

$$\mathcal{S} = \{k, 2k, \ldots, nk\} \quad \text{or} \quad \mathcal{S} = \left\{\frac{k}{1}, \frac{k}{2}, \ldots, \frac{k}{n}\right\}.$$

In Cheng and Pomerance (1994) it is shown that we may take $n_0 = 10^{50\,000}$ and in Balasubramanian and Soundararajan (1995) it is shown that we may take $n_0 = 5$. Note that the first claim in Theorem 7.7 is true for $n < 5$, but the second claim fails for $n = 4$, since $\mathcal{S} = \{2, 3, 4, 6\}$ has $F(\mathcal{S}) = F(\{1, 2, 3, 4\})$.

7.3. Perfect numbers – Wirsing's theorem

Let $\sigma(n)$ denote the sum of the positive divisors of n. If $\sigma(n) = 2n$, then n is said to be *perfect*. The first few examples are 6, 28, and 496. It has been known since Euclid that if $2^p - 1$ is prime, then $2^{p-1}(2^p - 1)$ is perfect. The three examples just cited fit this formula with $p = 2, 3, 5$. In fact, it has been shown by Euler that every *even* perfect number comes from Euclid's formula. The two big questions are: (1) are there infinitely many perfect numbers?, and (2) are there any odd perfect numbers?

From the Euclid–Euler results, the first question is equivalent to the existence of infinitely many Mersenne primes, that is, primes of the form $2^p - 1$. This is a very hard problem about which very little is known. We know 33 values of p for which $2^p - 1$ is prime, the largest being $p = 859\,433$.

We are still far from solving the second problem too. Numerous partial results are known, however. One of the more interesting theorems in the subject is due to Wirsing (1959), a paper which extends earlier joint work with Hornfeck.

Theorem 7.8. *There are absolute constants c_0, x_0 such that if $x \geq x_0$ and α is any rational number, then the number of $n \leq x$ with $\sigma(n) = \alpha n$ is at most $x^{c_0/\log\log x}$.*

Proof. We begin with the observation that $\sigma(n)$ is a multiplicative function $[\sigma(mn)=\sigma(m)\sigma(n)$ when $(m,n)=1]$. Suppose α is given; write α in reduced form u/v. Suppose x is large, $n\leqslant x$, and $\sigma(n)=\alpha n$. Write $n=ab$, where b is the largest divisor of n all of whose prime factors p satisfy $p\leqslant\log x$ or $p\mid v$. Then αb is an integer. The idea of the proof is to use the equation

$$\sigma(n)=\sigma(ab)=\sigma(a)\sigma(b)=a\cdot\alpha b\,. \tag{7.9}$$

The plan is to show that we must have $\sigma(b)\nmid\alpha b$, and use this to show that b just about determines a.

Suppose the prime factorization of a is $p_1^{\beta_1}p_2^{\beta_2}\cdots p_k^{\beta_k}$. Let l be the least integer $\geqslant\log x/\log\log x$. Since each $p_i>\log x$ and since $a\leqslant n\leqslant x$, we have

$$\beta_1+\beta_2+\cdots+\beta_k\leqslant l\,, \tag{7.10}$$

so that, in particular, $k\leqslant l$. Then

$$\begin{aligned}1\leqslant\frac{\sigma(a)}{a}&=\prod_{i=1}^{k}(1+p_i^{-1}+\cdots+p_i^{-\beta_i})<\prod_{i=1}^{k}\left(1+\frac{1}{p_i-1}\right)\\&<\exp\left(\sum_{i=1}^{k}\frac{1}{p_i-1}\right)<\exp\left(\frac{l}{(\log x)-1}\right)<2\end{aligned} \tag{7.11}$$

for $x\geqslant x_1$. Thus for $x\geqslant x_1$ we have $a\mid\sigma(a)$ if and only if $a=1$. Putting this into (7.9) we see that for $x\geqslant x_1$ we have either $a=1$ or $\sigma(b)\nmid\alpha b$. In fact we get even more. If $a'\mid a$, $(a',a/a')=1$, and $a'<a$, then applying (7.11) to a/a' we have $\sigma(a'b)\nmid a'\cdot\alpha b$.

Let us see how we can reconstruct the number a given only b and an ordered k-tuple (with $k\geqslant 0$) of positive integers $\beta_1,\beta_2,\ldots,\beta_k$ satisfying (7.10). First if $k=0$, then $a=1$, and we are done. So suppose $k>0$. Then $\sigma(b)\nmid\alpha b$ (if this fails then there can be no a at all), so let p_1 be the least prime that divides $\sigma(b)$ to a higher power than it divides αb. If $k=1$, then $a=p_1^{\beta_1}$, so suppose $k>1$. Let $b'=bp_1^{\beta_1}$. Then as before we may assume $\sigma(b')\nmid\alpha b'$; let p_2 be the least prime that divides $\sigma(b')$ to a higher power than it divides $\alpha b'$. If $k=2$, then $a=p_1^{\beta_1}p_2^{\beta_2}$. If $k>2$, we let $b''=bp_1^{\beta_1}p_2^{\beta_2}$ and continue as before. This procedure either terminates with an integer $a=p_1^{\beta_1}p_2^{\beta_2}\cdots p_k^{\beta_k}$ or proves no a can exist satisfying (7.9). If a is constructed, it may or may not satisfy (7.9). But if some a satisfying (7.9) does exist, this procedure will find it.

Thus for $x\geqslant x_1$ the number of $n\leqslant x$ satisfying $\sigma(n)=\alpha n$ is at most BC, where B is the number of $b\leqslant x$ such that $v\mid b$ and for every prime p in b we have $p\leqslant\log x$ or $p\mid v$ and C is the number of ordered tuples of natural numbers satisfying (7.10).

From elementary combinatorics we have $C=2^l$.

Note that we have $B\leqslant B_1B_2B_3$, where B_1 is the number of $b_1\leqslant x$ of the form $q_1^{\gamma_1}q_2^{\gamma_2}\cdots q_t^{\gamma_t}$ where $q_1,q_2,\ldots,q_t$ are all of the primes in v exceeding $\log x$ and $\gamma_1,\gamma_2,\ldots,\gamma_t$ are natural numbers, B_2 is the number of $b_2\leqslant x$ such that every prime in b_2 is in the interval $(\log^{3/4} x, \log x]$, and B_3 is the number of $b_3 \leq x$ divisible by no prime exceeding $\log^{3/4} x$.

An upper bound for B_1 is the number of sequences $\gamma_1, \gamma_2, \ldots, \gamma_t$ of natural numbers such that

$$\gamma_1 + \gamma_2 + \cdots + \gamma_t \leq l .$$

Thus $B_1 \leq \binom{l}{t-1} \leq 2^l$.

The total number of prime factors in a choice for b_2 is at most $(\log x)/\log(\log^{3/4} x) \leq 2l$. Say the primes in $(\log^{3/4} x, \log x]$ are $r_1, r_2, \ldots, r_m$. Then B_2 is at most the number of sequences $\delta_1, \delta_2, \ldots, \delta_m$ of non-negative integers with

$$\delta_1 + \delta_2 + \cdots + \delta_m \leq 2l .$$

Thus again using elementary combinatorics, we have

$$B_2 \leq \binom{m+2l}{m} \leq 2^{m+2l} .$$

However, $m = \pi(\log x) - \pi(\log^{3/4} x)$, so that from the prime number theorem we have $m \sim (\log x)/\log\log x$. Since we have yet to use such a "big gun", we could rely instead on the more elementary inequality (1.1). To be specific, we use $\pi(z) < 2z/\log z$, which holds for all $z > 1$. Thus we have $m < 2l$, so that $B_2 \leq 2^{4l}$. (In fact, the inequality $\pi(z) < 2z/\log z$ can be proved by a very easy argument involving binomial coefficients, but we suppress the details.)

If p is a prime and p^β divides some choice for b_3, then $p^\beta \leq x$ so that $\beta \leq (\log x)/\log 2$. Thus B_3 is at most the number of ordered $\pi(\log^{3/4} x)$-tuples with each coordinate a non-negative integer at most $(\log x)/\log 2$. Thus

$$B_3 \leq (1 + (\log x)/\log 2)^{\pi(\log^{3/4} x)} \leq (1 + (\log x)/\log 2)^{\log^{3/4} x} \leq 2^l$$

for $x \geq x_2$.

From the above, if $x \geq x_2$, then $B \leq B_1 B_2 B_3 \leq 2^{6l}$. Since $C = 2^l$, if we have $x \geq x_0 = \max\{x_1, x_2\}$, then the number of $n \leq x$ with $\sigma(n) = \alpha n$ is at most 2^{7l}, proving the theorem. □

While it is clear that a smaller value of c_0 may be found from a more careful proof, it would be more interesting to replace c_0 with some function tending to 0, perhaps only in the special case $\alpha = 2$ corresponding to perfect numbers. As for lower bounds, we know of no α for which we can prove $\sigma(n) = \alpha n$ has infinitely many solutions. In fact, we cannot even prove that the number of solutions is unbounded as α varies. It is known that if for some α and k there are infinitely many solutions to $\sigma(n) = \alpha n$ with $\nu(n) = k$, then there are infinitely many even perfect numbers, a result due to Kanold in 1956. Pomerance (1977a) proved the following effective form of this theorem.

Theorem 7.12. *For any α and k there is an effectively computable constant $N(\alpha, k)$*

such that if $n > N(\alpha, k)$, $\sigma(n) = \alpha n$, *and* $\nu(n) = k$, *then* $n = em$, *where e is an even perfect number* $(e, m) = 1$, *and* $m \leqslant N(\alpha, k)$.

The bound $N(\alpha, k)$ is not very friendly, although it is primitive recursive as a function of k. In the special case of odd solutions n, there is a somewhat more reasonable bound. For example, if n is an odd perfect number with $\nu(n) = k$, then Heath-Brown (1994), improving on a result in Pomerance (1977a) showed

$$n \leqslant 4^{4^k}.$$

That n is bounded by some function of k (for odd perfect numbers) was first shown by L. E. Dickson in 1913.

In 1932, D. H. Lehmer proposed the following problem that is similar in flavor to the odd perfect number problem. Lehmer asked if there are any composite natural numbers n with $\varphi(n) \mid n-1$, where φ is Euler's function from elementary number theory. This is still unsolved today. We do know that the number of composite integers $n \leqslant x$ with $\varphi(n) \mid n-1$, is $O(x^{1/2} \log^{3/4} x)$, and that if $\varphi(n)$ divides $n-1$, $\nu(n) = k$, and $k > 1$, then

$$n \leqslant k^{2^k},$$

see Pomerance (1977b). This can be improved to $n \leqslant 4^{2^k}$ using the method of Heath-Brown (1994).

Consider the function $s(n) = \sigma(n) - n$, the sum of the proper divisors of n. Thus a perfect number n satisfies $s(n) = n$. For any natural number n, one may consider the *aliquot sequence* for n: $n, s(n), s(s(n)), \ldots$. An old conjecture of Catalan and Dickson is that any such sequence either terminates at 0 (by hitting a prime and becoming 0 two steps later) or becomes periodic. This has been proved for all $n \leqslant 275$. Guy and Selfridge (1975) instead conjecture that the set of n whose aliquot sequence is unbounded has positive lower asymptotic density.

A cycle of length 2 for the function $s(n)$ is called an *amicable pair*. Namely, this is a pair of distinct integers a, b such that $s(a) = b$ and $s(b) = a$. Such numbers have been studied since Pythagoras who noted that 220 and 284 are an amicable pair. In 1955, Erdős showed that the numbers which belong to an amicable pair have asymptotic density 0. In Pomerance (1981) it is shown that the number of integers up to x which belong to an amicable pair is at most $x \cdot \exp(-(\log x)^{1/3})$ if x is sufficiently large.

7.4. *Graphs on the integers*

Consider the coprime graph on $\mathbb{Z}$. This is the graph whose vertex set is $\mathbb{Z}$ and two integers a, b are connected by an edge if $(a, b) = 1$.

The problem that opens this chapter can be reworded as follows. What is the largest set $\mathscr{A} \subseteq \{1, 2, \ldots, N\}$ such that the induced coprime graph on $\mathscr{A}$ contains no edges? This is the case $k = 2$ of the following famous problem of Erdős. Namely, what is the largest set $\mathscr{A} \subseteq \{1, 2, \ldots, N\}$ such that the induced coprime graph on $\mathscr{A}$ does not contain a complete graph on k vertices? Of course, the set of integers $n \leqslant N$ which have a prime factor among the first $k-1$ primes is such a set, and Erdős conjecture was that this set gives the maximum. This conjecture is fairly easily proved for $k=3$. The case $k=4$ was proved by Szabó and Tóth (1985). Finally this long-standing problem has been very recently settled completely by Ahlswede and Khachatrian (1995). First, they showed that there is a pair k, N for which the conjecture fails, and their example suggests that probably there are infinitely many integers k such that the conjecture fails for these k and certain small values of N. On the other hand, they proved that the following slightly weaker form of the conjecture is true: for every k there is a number $N_0 = N_0(k)$ such that for $N > N_0(k)$, up to N the set of multiples of the first $k-1$ primes gives the largest set with no k numbers pairwise coprime.

If $\mathscr{A} \subseteq \{1, 2, \ldots, N\}$ has $|\mathscr{A}| \geqslant \lfloor \frac{1}{2}N \rfloor + 1$, then we have seen that the coprime graph on $\mathscr{A}$ must contain an edge. Must it already contain many edges? The answer is yes, for as Erdős et al. (1980) show, there must be some $a \in \mathscr{A}$ with valence at least $cN/\log\log N$. Moreover, if $|\mathscr{A}| \geqslant (\frac{1}{2} + \varepsilon)N$, then the coprime graph on $\mathscr{A}$ contains at least $c(\varepsilon)N^2$ edges. They also show that if $|A| \geqslant (\frac{2}{3} + \varepsilon)N$, then the coprime graph on $\mathscr{A}$ contains at least $c(\varepsilon)N^3$ triangles, i.e. triplets a_1, a_2, a_3 with $(a_1, a_2) = (a_1, a_3) = (a_2, a_3) = 1$.

The coprime graph on $\mathbb{Z}$ has many edges so we might expect that if I_1 and I_2 are disjoint intervals of n consecutive integers, then the induced coprime graph on $I_1 \cup I_2$ contains a matching from I_1 to I_2. This is not the case, however. Suppose $I_1 = \{2, 3, 4\}$ and $I_2 = \{8, 9, 10\}$. Then any one-to-one correspondence between I_1 and I_2 must have at least one pair of even numbers in the correspondence. Another example: $I_1 = \{2, 3, 4, 5\}$, $I_2 = \{30, 31, 32, 33\}$. Here nothing can correspond with 30.

About 25 years ago, D. J. Newman conjectured that if $I_1 = \{1, 2, \ldots, n\}$ and I_2 is any interval of n consecutive integers, then there is a coprime matching from I_1 to I_2. (If $I_1 \cap I_2 \neq \emptyset$, we mean there is a one-to-one correspondence from I_1 to I_2 with corresponding numbers coprime. This can still be thought of as a matching in the coprime graph if we replace I_2 by $I_2 + \{n!\}$.) In Pomerance and Selfridge (1980), Newman's conjecture is proved by giving an algorithm for constructing a coprime matching and proving it works for every n. The proof involves effective estimates for the cardinality of the sets $S(x, u) = \{n \leqslant x\colon \varphi(n)/n \leqslant u\}$, where φ is Euler's function.

Consider now the divisor graph on $\mathbb{N}$. Here two distinct numbers a, b are connected by an edge if either $a \mid b$ or $b \mid a$. There are many attractive problems concerning the divisor graph; few of them are completely solved. The divisor graph is not as dense with edges as the coprime graph, so in general two n-element subsets of $\mathbb{N}$ should not be expected to contain a matching. Rather, we might consider the following. Let $f(n)$ be the least integer such that the divisor graph contains a matching from $\{1, 2, \ldots, n\}$ into $\{n+1, n+2, \ldots, f(n)\}$. The following result is due to Erdős and Pomerance (1980).

Theorem 7.13. *There are positive constants c_0, c_1 such that for all large n,*

$$c_0 n((\log n)/\log\log n)^{1/2} \leqslant f(n) \leqslant c_1 n(\log n)^{1/2}\,.$$

Proof. We present only the proof of the upper bound; the lower bound proof is much harder and not particularly combinatorial in flavor. We shall show that we may take c_1 as any number larger than 2.

Let $\varepsilon > 0$ be arbitrary. The divisor graph clearly contains a matching from the integers in $(n/\sqrt{\log n}, n]$ into the integers in $(n, n\lceil\sqrt{\log n}\rceil]$ – indeed, just multiply each number in the first interval by $\lceil\sqrt{\log n}\rceil$. It will thus be sufficient to show the divisor graph contains a matching from I to J, where

$$I = [1, n/\sqrt{\log n}] \cap \mathbb{N}\,, \qquad J = (n\lceil\sqrt{\log n}\rceil, (2+\varepsilon)n\sqrt{\log n}] \cap \mathbb{N}\,.$$

We consider in fact only the subgraph where $a \in I$, $b \in J$ are connected by an edge if b/a is prime.

If $a \in I$, the number of primes p such that $pa \in J$ is, by the prime number theorem,

$$\begin{aligned}
&\pi\left(\frac{1}{a}(2+\varepsilon)n\sqrt{\log n}\right) - \pi\left(\frac{1}{a}n\lceil\sqrt{\log n}\rceil\right) \\
&\quad > (1+\tfrac{1}{2}\varepsilon)\frac{n\sqrt{\log n}}{a\log(a^{-1}n\sqrt{\log n})} \\
&\quad \geqslant (1+\tfrac{1}{2}\varepsilon)\frac{\log n}{\log\log n}
\end{aligned}$$

if $n \geqslant n_0$, uniformly for al $a \in I$. On the other hand, if $b \in J$, the maximal number of $a \in I$ that can correspond to b is at most the number of primes p that divide b with $p \geqslant \log n$. Since $b \leqslant (2+\varepsilon)n\sqrt{\log n}$, this number evidently is at most

$$\frac{\log((2+\varepsilon)n\sqrt{\log n})}{\log\log n} < (1+\tfrac{1}{2}\varepsilon)\frac{\log n}{\log\log n}$$

for $n \geqslant n_1$. Thus for $n \geqslant \max\{n_0, n_1\}$, the König–Hall marriage lemma (see chapter 3) implies there is a matching from I into J. □

What is the length $H(n)$ of the longest simple path in the divisor graph on $\{1, 2, \ldots, n\}$? Hegyvári conjectured $H(n) = o(n)$ and this was proved by Pomerance (1983). It would be nice to get an asymptotic formula for $H(n)$. Recently, Saias and Tenenbaum have obtained fairly sharp estimates for $H(n)$. Other problems of a similar nature are considered in Erdős et al. (1983).

7.5. *Egyptian fractions*

The ancient Egyptians thought fractions $1/a$ where $a \in \mathbb{N}$ were especially nice. There is today a wide body of literature and many problems and results concerning Egyptian fractions – see section 4 in Erdős and Graham (1980) and section D11 of Guy (1994).

It has been known since Fibonacci that every positive rational r can be expressed as a finite sum of distinct Egyptian fractions. In fact, the greedy algorithm of choosing a_{n+1} minimal with $a_{n+1} > a_n$ and $1/a_1 + \cdots + 1/a_{n+1} \leqslant r$

always terminates. However, a representation of r as a sum of distinct Egyptian fractions is certainly not unique and this fact leads to many questions. For example, what is the fewest number of summands for r? Or, how many ways are there to write r as a sum of n distinct Egyptian fractions as $n \to \infty$? For the latter question, the case $r = 1$ has special interest.

It has been conjectured by Erdős and Straus that for every integer $n > 1$, $4/n$ can be written as a sum of three Egyptian fractions. This has been verified numerically for small values of n and has been shown true for all n but for a possible exceptional set of asymptotic density 0. More generally, Schinzel and Sierpiński have conjectured that every positive rational a/b can be expressed as a sum of three Egyptian fractions provided the denominator b is sufficiently large as a function of the numerator a. This is easily seen not to be true for a sum of two Egyptian fractions. For example, if p is a prime with $p \equiv 1 \pmod 3$, then $3/p = 1/x + 1/y$ is not solvable in integers.

7.6 Pseudoprimes

From Fermat's little theorem, if n is prime and $a \not\equiv 0 \pmod n$, then

$$a^{n-1} \equiv 1 \pmod n . \tag{7.14}$$

The congruence (7.14) is very useful for testing large numbers n for primality. Indeed, even if n is very large, it is a relatively simple matter to compute the least non-negative residue of $2^{n-1} \pmod n$; if this is not 1 (and $n > 2$), then n is composite. The residue may be found in $O(\log n)$ multiplications $\pmod n$ using the repeated squaring method. However, this method is not perfect – sometimes we come across composite numbers n that nevertheless satisfy (7.14) for some a. The least example with $a = 2$ is $n = 341$, and with $a = 3$ is $n = 91$. We say n is a *pseudoprime to the base a* if n is a composite natural number and (7.14) holds.

Let $P_a(x)$ denote the number of base a pseudoprimes up to x. Can we prove that for a fixed a we have $P_a(x) = o(\pi(x))$; that is, that base a pseudoprimes are rare compared with primes? Certainly not for $a = \pm 1$, since then (7.14) has many composite solutions. However, for $|a| > 1$, Erdős showed in 1956 that $P_a(x) = o(\pi(x))$ does in fact hold. The best result in this direction is due to Pomerance in 1981.

Theorem 7.15. *For each integer a with $|a| > 1$, there is a number $x_0(a)$ such that for $x \geq x_0(a)$ we have $P_a(x) \leq x^{1-\varepsilon(x)/2}$, where*

$$\varepsilon(x) = (\log\log\log x)/\log\log x . \tag{7.16}$$

In 1956, Erdős conjectured that $P_a(x) > x^{1-c\varepsilon(x)}$ for some $c > 0$ and x sufficiently large and where $\varepsilon(x)$ is defined in (7.16). This conjecture was refined by Pomerance to the following.

Conjecture 7.17. For each integer a with $|a| > 1$, we have

$$P_a(x) = x^{1-(1+o_a(1))\varepsilon(x)} .$$

One might wonder if a number n can be simultaneously a pseudoprime to the bases 2 and 3. This in fact can happen; the least example is $n = 1105$. It is not known if there are infinitely many such n. There are numbers n which are a pseudoprime to *every* base a with $(a, n) = 1$. Such numbers are called *Carmichael numbers*; the smallest example is $n = 561$. If $C(x)$ is the number of Carmichael numbers up to x, it is known that $C(x) \leqslant x^{1-\varepsilon(x)}$ for x sufficiently large and it is conjectured that $C(x) = x^{1-(1+o(1))\varepsilon(x)}$. Alford et al. (1994) recently proved there are infinitely many Carmichael numbers. In fact they showed the following.

Theorem 7.18. *For all sufficiently large values of* x, $C(x) > x^{2/7}$.

The proof, which roughly follows a heuristic argument given by Erdős in 1956, has some strong combinatorial elements.

We remark that any composite number n with $\varphi(n) \mid n - 1$ must also be a Carmichael number, but no such n are known to exist (see the earlier remarks on perfect numbers).

Acknowledgements

As mentioned in the Introduction, Paul Erdős played an important role in the writing of this chapter. In addition we gratefully acknowledge assistance from Adolf Hildebrand, Helmut Maier, and Melvyn Nathanson. We also thank the editors of this Handbook for their patience and encouragement.

References

Ahlswede, R., and L.H. Khachatrian
[1995] Maximal sets of numbers not containing $k+1$ pairwise coprime integers, *Acta Arithm.*, to appear.

Ajtai, M., J. Komlós and E. Szemerédi
[1981] A dense infinite Sidon sequence, *European J. Combin.* **2**, 1–11.

Alford, W.R., A. Granville and C. Pomerance
[1994] There are infinitely many Carmichael numbers, *Ann. of Math.* **140**, 703–722.

Andrews, G.E.
[1976] *The Theory of Partitions* (Addison-Wesley, Reading, MA).

Balasubramanian, R., and K. Soundararajan
[1995] On a conjecture of R.L. Graham, to appear.

Balasubramian, R., J.-M. Deshouillers and F. Dress
[1986] Probleme de Waring pour les bicarrés. I. Schéma de la solution, *C.R. Acad. Sci. Paris Sér. I Math.* **303**(4), 85–88.

Bourgain, J.
[1990] On arithmetic progressions in sums of sets of integers, in: *A Tribute to Paul Erdős,* eds. A. Baker, B. Bollobás and A. Hajnal (Cambridge University Press, Cambridge) pp. 105–109.

Cheng, F.Y., and C. Pomerance

[1994] On a conjecture of R.L. Graham, *Rocky Mountains J. Math.* **24**, 961–975.

Dekking, F.M.

[1979] Strongly non-repetitive sequences and progression-free sets, *J. Combin. Theory A* **27**, 181–185.

Erdős, P.

[1940] The difference of consecutive primes, *Duke Math. J.* **6**, 438–441.

[1956] Problems and results in additive number theory, in: *Colloque sur la Thórie des Nombres, Bruxelles, 1955* (Georges Thone/Masson and Cie, Liège/Paris) pp. 127–137.

Erdős, P., and R.L. Graham

[1980] *Old and New Problems and Results in Combinatorial Number Theory* (L'Enseignement Mathématique, Université de Genève, Monographie No. 28).

Erdős, P., and M.B. Nathanson

[1987] Problems and results on minimal bases in additive number theory, in: *Number Theory, New York 1984/1985, Lecture Notes in Mathematics,* Vol. 124, eds. D.V. Chudnovsky et al. (Springer, Berlin) pp. 87–96.

[1988] Minimal asymptotic bases with prescribed densities, *Ill. J. Math.* **32**, 562–574.

Erdős, P., and C. Pomerance

[1980] Matching the natural numbers up to n with distinct multiples in another interval, *Nederl. Akad. Wetensch. Proc. A* **83**, 147–161.

[1986] On the number of false witnesses for a composite number, *Math. Comp.* **46**, 259–279.

Erdős, P., A. Sárközy and E. Szemerédi

[1970] On divisibility properties of sequences of integers, in: *Number Theory, Colloq. Math. Soc. János Bolyai* **2**, 35–49.

[1980] On some extremal properties of sequences of integers. II, *Publ. Math. Debrecen* **27**, 117–125.

Erdős, P., R. Freud and N. Hegyvári

[1983] Arithmetical properties of permutations of integers, *Acta Math. Acad. Sci. Hungar.* **41**, 169–176.

Erdős, P., A. Sárközy and V.T. Sós

[1986] Problems and results on additive properties of general sequences, V, *Monatsh. Math.* **102**, 183–197.

Erdős, P., H. Maier and A. Sárközy

[1987] On the distribution of the number of prime factors of sums $a+b$, *Trans. Amer. Math. Soc.* **302**, 269–280.

Erdős, P., C. Pomerance, A. Sárközy and C.L. Stewart

[1993] On elements of sumsets with many prime factors, *J. Number Theory* **44**, 93–104.

Freiman, G.A.

[1973] *Foundations of a Structural Theory of Set Additions, Translations of Mathematical Monographs,* Vol. 37 (American Mathematical Society, Providence, RI).

Freiman, G.A., H. Halberstam and I.Z. Rusza

[1992] Integer sums containing long arithmetic progressions, *J. London Math. Soc.* **46**, 193–201.

Graham, R.L., B.L. Rothschild and J.H. Spencer

[1980] *Ramsey Theory* (Wiley, New York).

Guy, R.K.

[1994] *Unsolved Problems in Number Theory,* 2nd Ed. (Springer, New York).

Guy, R.K., and J.L. Selfridge

[1975] What drives an aliquot sequence? *Math. Comp.* **29**, 101–107. Corrigendum: 1980, **34**, 319–321.

Györy, K., C.L. Stewart and R. Tijdeman

[1988] On prime factors of sums of integers, III, *Acta Arithm.* **49**, 307–312.

Halberstam, H., and H.-E. Richert

[1974] *Sieve Methods* (Academic Press, London).

Halberstam, H., and K.F. Roth

[1983] *Sequences,* 2nd Ed. (Springer, New York).

Hall, R.R., and G. Tenenbaum

[1988] *Divisors* (Cambridge University Press, Cambridge).

Hayashi, E.K.
[1981] Omega theorems for the iterated additive convolution of a nonnegative arithmetic function, *J. Number Theory* **13**, 176–191.

Heath-Brown, D.R.
[1994] Odd perfect numbers, *Math. Proc. Cambridge Philos. Soc.* **115**, 191–196.

Iwaniec, H.
[1981] Rosser's sieve-bilinear forms of the remainder terms – some applications, in: *Recent Progress in Analytic Number Theory, Durham 1979,* Vol. 1, eds. H. Halberstam and C. Hooley (Academic Press, London) pp. 203–230.

Maier, H.
[1988] Small differences between prime numbers, *Michigan Math. J.* **35**, 323–344.

Montgomery, H.L.
[1971] *Topics in Multiplicative Number Theory, Lecture Notes in Mathematics,* Vol. 27 (Springer, New York).

Montgomery, H.L., and R.C. Vaughan
[1990] On the Erdős–Fuchs theorem, in: *A Tribute to Paul Erdős,* eds. A. Baker, B. Bollobás and A. Hajnal (Cambridge University Press, Cambridge) pp. 331–338.

Nathanson, M.B.
[1989] Additive problems in combinatorial number theory, in: *Number Theory, New York 1985/1988, Lecture Notes in Mathematics,* Vol. 1383, eds. D.V. Chudnovsky et al. (Springer, Berlin) pp. 123–139.

Nathanson, M.B., and A. Sárközy
[1989] On the maximum density of minimal asymptotic bases, *Proc. Amer. Math. Soc.* **105**, 31–33.

Ostmann, H.
[1956] *Additive Zahlentheorie,* Vols. I, II (Springer, Berlin).

Pintz, J., W.L. Steiger and E. Szemerédi
[1988] On sets of natural numbers whose difference set contains no squares, *J. London Math. Soc. (2)* **37**, 219–231.

Pomerance, C.
[1977a] Multiply perfect numbers, Mersenne primes and effective computability, *Math. Ann.* **226**, 195–206.
[1977b] On composite n for which $\varphi(n)\,|\,n-1$, II, *Pacific J. Math.* **69**, 177–186.
[1980] Collinear subsets of lattice point sequences – an analog of Szemerédi's theorem, *J. Combin. Theory A* **28**, 140–149.
[1981] On the distribution of amicable numbers, II, *J. Angew. Math.* **325**, 183–188.
[1983] On the longest simple path in the divisor graph, *Congress. Numerantium* **40**, 291–304.

Pomerance, C., and A. Sárközy
[1988] On homogeneous multiplicative hybrid problems in number theory, *Acta Arithm.* **49**, 291–302.
[1990] On products of sequences of integers, in: *Number Theory, Budapest 1987,* eds. K. György and G. Halász, Vol. 1, *Colloq. Math. Soc. János Bolyai* **51**, 447–467.

Pomerance, C., and J.L. Selfridge
[1980] Proof of D.J. Newman's coprime mapping conjecture, *Mathematika* **27**, 69–83.

Pritchard, P.A., A. Moran and A. Thyssen
[1995] Twenty-two primes in arithmetic progression, *Math. Comp.,* to appear.

Ramsey, L.T., and J.L. Gerver
[1979] On certain sequences of lattice points, *Pacific J. Math.* **83**, 357–363.

Ruzsa, I.Z.
[1987] Essential components, *Proc. London Math. Soc. (3)* **54**, 38–56.
[1990/91] An application of graph theory to additive number theory, *Scientia (Chile)* **4**, 93–94.

Sárközy, A.
[1989a] Hybrid problems in number theory, in: *Number Theory, New York 1985/1988, Lecture Notes in Mathematics,* Vol. 1383, eds. D.V. Chudnovsky et al. (Springer, Berlin) pp. 146–169.
[1989b] Finite addition theorems, I, *J. Number Theory* **32**, 114–130.
[1994] Finite addition theorems, II, *J. Number Theory* **48**, 197–218.

Shelah, S.
[1988] Primitive recursive bounds for van der Waerden numbers, *J. Amer. Math. Soc.* **1**, 683–697.
Stewart, C.L., and R. Tijdeman
[1983] On density-difference sets of integers, in: *Studies in Pure Mathematics – to the Memory of Paul Turán* (Birkhäuser/Hungarian Academy of Science, Basel/Budapest) pp. 701–710.
Stöhr, A.
[1955] Gelöste und ungelöste Fragen über Basen der natürlichen Zahlenreihe. I, II, *J. Reine Angew. Math.* **194**, 40–65; 111–140.
Szabó, C., and G. Tóth
[1985] Maximal sequences not containing 4 pairwise comprime integers (in Hungarian), *Mat. Lapok* **32**, 253–257.
Szegedy, M.
[1986] The solution of Graham's greatest common divisor problem, *Combinatorica* **6**, 67–71.
Szemerédi, E.
[1970] On a conjecture of Erdős and Heilbronn, *Acta Arithm.* **17**, 227–229.
Vaughan, R.C.
[1981] *The Hardy–Littlewood Method* (Cambridge University Press, Cambridge).
Wirsing, E.
[1957] Über die Dichte multiplikativer Basen, *Arch. Math.* **8**, 11–15.
[1959] Bemerkung zu der Arbeit über vollkommene Zahlen, *Math. Ann.* **137**, 316–318.
Zaharescu, A.
[1987] On a conjecture of Graham, *J. Number Theory* **27**, 33–40.

Author index

Subject index